28th International Colloquium on Group-Theoretical Methods in Physics 2010

GROUP 28: Physical and Mathematical Aspects of Symmetry

Journal of Physics: Conference Series Volume 284

Newcastle upon Tyne, UK
26-30 July 2010

Editors:

Maia Angelova
Wojciech Zakrzewski
Véronique Hussin
Bernard Piette

ISBN: 978-1-61782-665-8
ISSN: 1742-6588

Printed from e-media with permission by:

Curran Associates, Inc.
57 Morehouse Lane
Red Hook, NY 12571

Some format issues inherent in the e-media version may also appear in this print version.

Printed by Curran Associates, Inc. (2011)

For permission requests, please contact the Institute of Physics
at the address below.

Institute of Physics
Dirac House, Temple Back
Bristol BS1 6BE UK

Phone: 44 1 17 929 7481
Fax: 44 1 17 920 0979

techtracking@iop.org

Additional copies of this publication are available from:

Curran Associates, Inc.
57 Morehouse Lane
Red Hook, NY 12571 USA
Phone: 845-758-0400
Fax: 845-758-2634
Email: curran@proceedings.com
Web: www.proceedings.com

TABLE OF CONTENTS

PLENARY SESSIONS

PARALLEL SESSIONS

PREFACE

The XXVIIIth International Colloquium on Group-Theoretical Methods in Physics (ICGTMP), also known as the GROUP 28 conference, took place in Newcastle upon Tyne from 26–30 July 2010. There were 128 attendees from 22 countries – physicists, mathematicians, chemists and engineers who use mathematical and numerical methods based on geometry and symmetry in their work. The conference was organised jointly by Northumbria and Durham Universities. The conference website is http://group28.northumbria.ac.uk.

Symmetry plays an important role both in science and art, and group theory provides a mathematical framework for the study of symmetries. The series of ICGTMP meetings is the oldest conference series in the fields of geometry and physics. It started in 1972 in Marseilles, and so far 15 meetings have been held in Europe, 8 in North America, 2 in Asia, 1 in Australia and 2 in the Middle East. This was the second time that the conference took place in Britain, 28 years after the Xth Colloquium in Canterbury.

While the first meetings were centred on solid-state physics and crystallography, the main topics quickly became more diversified following the success of geometric and algebraic methods for modelling systems in particle physics, quantum mechanics, engineering and chemistry. The aim of the GROUP 28 meeting was to broaden and further diversify mathematical and numerical methods based on geometry and symmetry through their applications to biosciences, physical sciences, quantum information, nonlinearity and complexity. The conference included traditional and novel applications to mathematical and theoretical physics (including particle physics, conformal theory and cosmology), condensed matter, quantum optics and quantum information, complex and nonlinear systems, biosciences and other new exciting areas.

The scientific programme was delivered in the form of plenary talks, a public lecture, parallel sessions, a poster session and an open forum. An important feature of the meeting was that all plenary talks presented the state-of-the-art and were at the same time educational and exciting, promoting the multidisciplinary aspects of the research, and thus were inspirational for young scientists considering work in these fields. The plenary talks, each lasting 1 hour, were given by distinguished world experts and some young 'rising stars': Richard Ward, Ulf Leonhardt, Jens Eisert, Michael Berry, Shahn Majid, Arndt von Haeseler, Michio Jimbo, Katrin Wendland, Raymond Goldstein, Mark Trodden, Maria Vozmediano and Giulio Chiribella. The public lecture was given by Francesco Iachello and was open to participants and other people from the Newcastle-Durham region. In addition, talks of 30 minutes duration each and including more technical content, were given in four parallel sessions. Each parallel session had a designated time for informal interaction with the speakers, discussions of new directions of research and for forming new collaborations. The poster session, in a room where posters were exhibited for the duration of the conference, was easily accessible, and had a friendly and relaxed atmosphere, encouraging discussions of work and exchanges of new ideas. To secure a high quality scientific programme, all contributions were reviewed.

Another interesting feature of the conference was the Open Forum on the Friday afternoon, which was lively and well attended. It took the form of a question and answer session with a panel chaired by Allan Solomon, with the following members: Gerald Goldin, Jean-Pierre Gazeau, Mark Trodden and Giulio Chiribella. The focus was on new directions of research, novel applications and the further development of group theory, education, training and career opportunities for young researchers. The participants were asked to submit questions in advance; examples are: "What is the role of symmetries and conservation principles in deducing underlying physics from experimental data?" and "What are the most promising research areas where group theory and representations have applications?"

All the sessions were easily accessible and took place in the modern premises of the City Campus East of Northumbria University, conveniently located in the city centre of Newcastle.

In a separate ceremony on Tuesday evening in the Great Hall of Newcastle's Discovery Museum, the 2010 Wigner Medal was awarded to Michio Jimbo for his seminal work on quantum groups and for his study of affine Lie algebras, in connection with classical and quantum integrable systems. The Hermann Weyl prize was awarded to Giulio Chiribella, in recognition of his pioneering work on the application of group theoretical methods to the problem of quantum estimation, within the framework of quantum information theory. This was followed by a reception, which included a delightful, impromptu violin recital given by Gérard H E Duchamp, one of the conference participants.

The social programme included a variety of events. On the Monday evening, the Welcome Address by the Lord Mayor of Newcastle was followed by a Reception in Newcastle's Civic Centre. On the Wednesday afternoon the participants had the opportunity to visit beautiful Alnwick Castle. This was followed by a conference dinner in Newcastle's Assembly Rooms on the Thursday evening.

The conference was sponsored by the Engineering and Physical Sciences Research Council (grant EP/1009183/1), the London Mathematical Society (grant 1906), the Institute of Physics, Northumbria University, Durham University, Newcastle City Council and Newcastle's Discovery Museum. We thank our sponsors for their generous support.

It took two years to organise a conference of such importance. We express our gratitude to the International Advisory Committee for its help in selecting the plenary speakers and to the Standing Committee and its current and former Chairmen, Jean Pierre Gazeau and H D Doebner respectively.

We thank the Newcastle-Gateshead Convention Bureau for its support in organising the conference, Sarah Howells for her devoted secretarial, administrative and organisational work, Rosemary Zakrzewski for arranging the "Accompanying persons' programme" and to the research assistants and research students of the Intelligent Modelling research group at Northumbria University – Mu Niu, Sujan Rajbhandari, Helen Gibson, Ahmed Lawgali, Kushwanth Koya, Thuli Mazwi, Osemeke Mosindi and Sirichai Triamlumlerd – for their energetic assistance to the organisers. Finally, a big thank you to all the members of the Local Organising Committee for their devoted and tireless work in preparing and running the conference and our colleagues from both universities for their help and moral support.

Maia Angelova and Wojciech Zakrzewski
Chairs of the Local Organising Committee

INTRODUCTION

This volume contains contributions to the XXVIIIth International Colloquium on Group-Theoretical Methods in Physics, the GROUP 28 conference, which took place in Newcastle upon Tyne from 26–30 July 2010. All plenary and contributed papers have undergone an independent review; as a result of this review and the decisions of the Editorial Board most but not all of the contributions were accepted.

The volume is organised as follows: it starts with notes in memory of Marcos Moshinsky, followed by contributions related to the Wigner Medal and Hermann Weyl prize. Then the invited talks at the plenary sessions and the public lecture are published followed by contributions in the parallel and poster sessions in alphabetical order.

The Editors:
Maia Angelova, Wojciech Zakrzewski, Véronique Hussin and Bernard Piette

International Advisory Committee

Michael Baake University of Bielefeld, Germany
Gerald Dunne University of Connecticut, USA
J F (Frank) Gomes UNESP, Sao Paolo, Brazil
Peter Hanggi University of Augsburg, Germany
Jeffrey C Lagarias University of Michigan, USA
Michael Mackey McGill University, Canada
Nicholas Manton Cambridge University, UK
Alexei Morozov ITEP, Moscow, Russia
Valery Rubakov INR, Moscow, Russia
Barry Sanders University of Calgary, Canada
Allan Solomon Open University, Milton Keynes, UK
Christoph Schweigert University of Hamburg, Germany

Standing Committee

Twareque Ali Concordia University, Canada
Luis Boya Salamanca University, Spain
Enrico Celeghini Firenze University, Italy
Vladimir Dobrev Bulgarian Academy of Sciences, Bulgaria
Heinz-Dietrich Doebner Honorary Member, Clausthal University, Germany
Jean-Pierre Gazeau Chairman, Paris Diderot University, France
Mo-Lin Ge Nankai University. China

Gerald Goldin Rutgers University, USA
Francesco Iachello Yale University, USA
Joris Van der Jeugt Ghent University, Belgium
Richard Kerner Pierre et Marie Curie University, France
Piotr Kielanowski CINVESTAV, Mexico
Alan Kostelecky Indiana University, USA
Mariano del Olmo Valladolid University, Spain
George Pogosyan UNAM, Mexico, JINR, Dubna, Russia
Christoph Schweigert University of Hamburg, Germany
Reidun Twarock York University, UK
Luc Vinet Montréal University, Canada
Apostolos Vourdas Bradford University, UK
Kurt Wolf UNAM, Mexico

Local Organising Committee

Maia Angelova – Chair Northumbria University, Newcastle
Wojtek Zakrzewski – Chair Durham University, Durham
Sarah Howells – Secretary Northumbria University, Newcastle
Jeremy Ellman – Web Northumbria University, Newcastle
Véronique Hussin Northumbria, Durham and University of Montréal
Safwat Mansi Northumbria University, Newcastle
James McLaughlin Northumbria University, Newcastle
Bernard Piette Durham University, Durham
Ghanim Putrus Northumbria University, Newcastle
Sarah Rees Newcastle University, Newcastle
Petia Sice Northumbria University, Newcastle
Anne Taormina Durham University, Durham
Rosemary Zakrzewski Accompanying persons programme

Photograph by Bernard Piette: Souter Lighthouse, Marsden, Tyne and Wear, England

Journal of Physics: Conference Series **284** (2011) 011003 doi:10.1088/1742-6596/284/1/011003

Marcos Moshinsky (1921-2009)

Francesco Iachello

Center for Theoretical Physics, Sloane Physics Laboratory, Yale University, New Haven, CT 06520-8120, USA

E-mail: francesco.iachello@yale.edu

Abstract. This article contains a brief biography of Marcos Moshinsky together with a brief description of the legacy that he left behind.

1. Biography

Marcos Moshinsky was born in Kiev, Ukraine, on 20 April, 1921. His family emigrated to Mexico when he was six years old. He studied physics at the Universidad Autonoma de Mexico, where he entered as a research assistant in 1942 and where he obtained a B.Sc. Degree in 1944. At the end of World War 2 he went to the U.S. where he earned a M.S. in 1947 and a Ph.D. in 1949 at Princeton University under the supervision of Eugene Wigner with a thesis on "Relativistic interactions by means of boundary conditions".

From his return to Mexico in the 1950s up to the last few days of his life, Marcos Moshinsky made many important contributions to physics, in particular to two subjects:

- Symmetry in physics
- Harmonic oscillators

Some of Moshinsky's contributions are:

1. A seminal paper on "Transformation brackets for harmonic oscillator functions" later developed with T.A. Brody into the famous "Tables of transformation brackets" [1]. These brackets were used extensively in subsequent years and contributed considerably to the development of nuclear physics in the early 1960's.

2. Two fundamental papers with V. Bargmann on "Group theory of the harmonic oscillator" [2]. This was the beginning of the systematic use of harmonic oscillators to solve problems in physics.

3. Two seminal papers with Peter Kramer on "States of permutation symmetry" [3], a continuation of previous work, but now introducing states of permutation symmetry, thus allowing a description of many-body systems of identical particles.

4. Two papers with J.D. Louck and K. B. Wolf on "Group theory of the anisotropic oscillator" [4].

5. A series of papers on the group theory of collective motion in nuclei both within the framework of the liquid drop model of Bohr and Mottelson (with E. Chacón and R.T. Sharp) [5] and of the interacting boson model of Arima and Iachello (with O. Castaños, E. Chacón and A. Frank) [6]. These papers and others, written in subsequent years, clarified considerably the structure and nature of collective motion in nuclei.

Two other subjects where Moshinsky made important contributions are:

- Canonical transformations in classical and quantum mechanics
- Degeneracy algebras in quantum mechanics

Particularly important here are:

Journal of Physics: Conference Series **284** (2011) 011003 doi:10.1088/1742-6596/284/1/011003

1. A series of seminal papers with C. Quesne on "Canonical transformations and matrix elements" [7], and with J. Patera and P. Winternitz on "A unified approach to canonical transformations and accidental degeneracy" [8].

In addition to these papers, Marcos published more than 200 articles and 5 books the last one with Yu. F. Smirnov, summarizing 40 years of work on harmonic oscillators in modern physics.

For his work, Marcos received many national and international honors and awards, some of which are (i) the award Principe de Asturias in Spain in 1988, (ii) the Andrei Sakharov Medal for human rights in the USSR in 1991, and, most importantly, (iii) the Eugene Wigner Medal in 1998 presented to him at the XXII Group Theory Colloquium, in Hobart, Tasmania, Australia.

He was a member of several international academies including the Academia Pontificia at the Vatican in Rome. He was member and early President of the Academia Mexicana de Ciencias and a member of the Colegio Nacional de Mexico.

Marcos died on April 1, 2009, in Mexico City.

2. Legacy

Moshinsky was one of the initiators and founders of the Group Theory Colloquia, and as such he is remembered in this article.

He was a great scientist. He contributed greatly to the development of group theoretical methods in physics.

He was also a great teacher, the founder of a school of physics which has few equals in the world. In Mexico, he started in the 1950s a school of theoretical physics and brought up many students who later occupied important positions in Mexico City and elsewhere (Cuernavaca, Guadalajara, etc.). In Latin-America, he co-founded in 1959 the Escuela Latino-Americana de Fisica (ELAF). The "Moshinsky" school is not limited to Latin-America but it extends all over the World, including persons in Germany (Peter Kramer), in Belgium (Christianne Quesne), in Canada (Pavel Winternitz) to mention only a few. Marcos had a special personality that made him the ideal "teacher" and inspired many people.

He was a champion of human rights, especially for the country he was born in, part of the former USSR, and he received the Sakharov medal for it.

His "scientific" and "human" legacy will remain for many years to come in the memory of his students, his co-workers and all the persons, like the writer of this article, who had the fortune to know him and appreciate his warm and inspiring personality.

References

[1] Moshinsky M 1959 *Nucl. Phys.* **13** 104
Brody T A and Moshinsky M 1960 Tables of transformation brackets (Monografías del Instituto de Física, UNAM)
[2] Bargmann V and Moshinsky M 1960 *Nucl. Phys.* **18** 697
Bargmann V and Moshinsky M 1961 *Nucl. Phys.* **23** 177
[3] Kramer P and Moshinsky M 1966 *Nucl. Phys.* **82** 241
Kramer P and Moshinsky M 1968 *Nucl. Phys.* **A107** 481
[4] Louck J D, Moshinsky M and Wolf K B 1973 *J. Math. Phys.* **14** 692
Louck J D, Moshinsky M and Wolf K B 1973 *J. Math. Phys.* **14** 696
[5] Chacón E, Moshinsky M and Sharp R T 1976 *J. Math. Phys.* **17** 668
Chacón E and Moshinsky M 1977 *J. Math. Phys.* **18** 870
[6] Castaños O, Chacón E, Frank A and Moshinsky M 1979 *J. Math. Phys.* **20** 35
[7] Moshinsky M and Quesne C 1971 *J. Math. Phys.* **12** 1772
[8] Moshinsky M, Patera J and Winternitz P 1975 *J. Math. Phys.* **16** 82

Journal of Physics: Conference Series **284** (2011) 011003 doi:10.1088/1742-6596/284/1/011003

Figure 1. Marcos Moshinsky (1921-2009) at his age of 80 in a drawing by Renato Lemus Casillas.

Journal of Physics: Conference Series **284** (2011) 011004 doi:10.1088/1742-6596/284/1/011004

Wigner Medal 2010

Piotr Kielanowski

Departamento de Física, Centro de Investigación y de Estudios Avanzados del IPN
Mexico City, Mexico

The purpose of the Wigner Medal is to recognize outstanding contributions to the understanding of physics through group theory.

The Wigner Medal was established in 1977/78 and is administered by the Group Theory and Fundamental Physics Foundation, a publicly supported organization.

The founding members of the Group Theory and Fundamental Physics Foundation were

THE GROUP THEORY AND FUNDAMENTAL PHYSICS FOUNDATION

Chairman:	Arno Böhm, Physics Department The University of Texas at Austin
Vice-Chairman:	Morton Hamermesh, Physics Department The University of Minnesota
Treasurer:	John D. Dollard, Mathematics Department The University of Texas at Austin

PATRONS:

Hamad A. Alhendi
L. C. Biedenharn
Sharifa Böhm
H.B.G. Casimir
John D. Dollard
Lawrence Dresner
F. J. Dyson
Einstein Centennial Celebration Committee at Southern Illinois University
J. B. Garg
Marvin Goldberger
Morton Hamermesh
Richard Haymaker
Yehiel Ilamed
Erdal Inonu
Y. S. Kim
E. J. Konopinski
James D. Louck
W. M. MacDonald
Kazumi Maki
F. A. Matsen
Bernd T. Matthias
Esperanza de Moshinsky
Marcos Moshinsky
Tadao Nakano
Pedro L. Garcia Perez
T. O. Philips
Physical Chemistry Laboratory, ETH Zurich, Switzerland
Humitaka Sato
Marlan Scully
Frederick Seitz
The Solvay Institute, Brussels
SKM, Inc.
Edward Teller
Texas Nuclear
Tracor, Inc.
C. W. Ufford
Instituto de Fisica, Universidad Nacional Autonoma de Mexico
Janette Hegner Wheeler
John Archibald Wheeler

Journal of Physics: Conference Series **284** (2011) 011004 doi:10.1088/1742-6596/284/1/011004

SPONSORS:

R. K. Adair
Norman Austern
V. Bargmann
Joseph Birman
D. Allan Bromley
Harvey Brooks
Philip B. Burt
E.G.D. Cohen
B. L. Davies
R. H. Dicke
G. and S. Domokos
Leonard Eisenbud
Arnold M. Feingold
Stig Flodmark
Stephen A. Fulling
Rudolph Haag
Christopher J. Henrich
Emmett L. Hudspeth
J. H. Irving
John R. Klauder
Wolfgang Ludwig
Louis Michel
R. Nottrot
Robert F. O'Connell
S. Okubo
L. C. Papaloucas
Constantin Piron
Luigi A. Radicati
Frederick Reines
Fritz Rohrlich
Paul Roman
Frank Stern
Marshall H. Stone
Eric W. Vogt
Steven Weinberg
V. F. Weisskopf
A. S. Wightman
Pavel Winternitz
Chen N. Yang

CONTRIBUTORS:

Julius Ashkin
A. Bohm
Burt V. Bronk
Morrel H. Cohen
Volker Enss
U. Fano
William Faris
Hans Frauenfelder
R. H. Gluckstern
Robert Hofstadter
B. R. Judd
S. T. Kuroda
Francis J. Narcowich
Enrico Onofri
W. Opechowski
Herbert Piller
Arlan B. Ramsay
Martin G. Redlich
Peter Rejto
Yitzhak J. Sharon
L. C. Shepley
Eric Sheldon
A. Shimony
N. H. Zebouni

The Group Theory and Fundamental Physics Foundation is governed by the Board, whose members are at present:

A. Bohm (Chairman)
J.D. Dollard (Treasurer)
P. Kielanowski (Secretary)
F. Iachello
W. Schleich
L. Schulman
A. Wightman
B. Zumino

The Medal is awarded approximately every two years and the awarding ceremony is held at the colloquium on Group Theoretical Methods in Physics. The previous recipients of the Wigner Medal were:

1978 *E. P. Wigner*
1978 *V. Bargmann*
1980 *I.M. Gel'fand*
1982 *L. Michel*
1984 *Y. Ne'eman*
1986 *F. Gürsey*
1988 *I.M. Singer*

Journal of Physics: Conference Series **284** (2011) 011004 doi:10.1088/1742-6596/284/1/011004

1990 *F. Iachello*
1992 *J. Wess and B. Zumino*
1996 *V. Kac and R.V. Moody*
1998 *M. Moshinsky*
2000 *L. O'Raifeartaigh*
2002 *H.J. Lipkin*
2004 *E. Inönü*
2006 *S. Okubo*

The winner of the Wigner Medal is chosen by the Selection Committee which consists of 5 members. The chairman of the Group Theory and Fundamental Physics Foundation is the member ex officio.The remaining members are assigned for the 4-year term: 2 by the Standing Committee of the Group Theory Colloquia and 2 by the Board of the Group Theory and Fundamental Physics Foundation.

The 2010 International Selection Committee consisting of J. Patera (Canada) - Chairman, M. del Olmo (Spain), W. Schleich (Germany), J.-B. Zuber (France) and A. Bohm (USA) awarded the 2010 Wigner Medal to

Michio Jimbo
University of Tokyo

for his introduction of quantum groups and his study of affine Lie algebras in connection with classical and quantum integrable systems.

The ceremony for the Wigner Medal took place on July 27, 2010 in the Discovery Museum in Newcastle upon Tyne (UK) during the XXVIIIth Col-

Journal of Physics: Conference Series **284** (2011) 011004 doi:10.1088/1742-6596/284/1/011004

loquium on Group Theoretical Methods in Physics. The Laudatio for Professor Michio Jimbo was delivered by Professor Fedor Smirnov from CNRS (France). The full text of Laudatio is the following:

Laudatio for Professor Michio Jimbo

Fedor Smirnov
CNRS, France

It is a great honour and pleasure for me to deliver the laudatio for Professor Michio Jimbo who becomes this year recipient of the Wigner Medal. This is also an emotional task because Michio Jimbo is a friend and close collaborator for many years during which we shared our successes and our disappointments.

Before proceeding to chronological description of our laureate's contributions to mathematics and physics let me say a few words about the field of his scientific interests. The entire scientific carrier of Michio Jimbo is related with the study of classical and quantum integrable models. One may say that the integrable models present rather exotic class of physical systems existing mostly in two-dimensional space-time. However, this nihilistic point of view becomes rather marginal nowadays when the integrable models found various and unexpected applications. To give an example let us consider the Maldacena conjecture on the equivalence of super-symmetric Yang-Mills theory and the String Theory in Lobachevsky space. This equivalence relates the weak coupling regime of one theory to the strong coupling regime of another. Obviously, there is no hope to continue, for example, the perturbative series for the Yang-Mills theory to the domain of strong coupling. Only an exact solution may prove the Maldacena conjecture, and this solution was found to be given by an integrable model.

So, from the pragmatical point of view the integrable models provide a source of exact formulae which have been always valuable in physics. For instance, Jacobi's solution of the problems of two fixed centres always plays the role of the first approximation in celestial mechanics, and Onsager's exact solution of the Ising model is the cornerstone of the theory of the second order phase transitions. But there is one more feature of the integrable models due to which starting to work on the subject one finds himself trapped forever: the symmetry. Integrability always finds its explanation in the underlying symmetry. Uncovering this hidden symmetry is the most interesting and noble problem, to the solution of which Michio Jimbo made many fundamental

Journal of Physics: Conference Series **284** (2011) 011004 doi:10.1088/1742-6596/284/1/011004

contributions. Let me describe some of his results from this prospective. Certainly, I shall not be able to describe the content of all 150 publications by Michio, choosing the results which are the most valuable from my point of view. I apologize for this personal approach.

Michio Jimbo started his education at the University of Tokyo, and then moved to the University of Kyoto where he began to work with Professor Mikio Sato. I think meeting this great man influenced very much the scientific life of Michio Jimbo. At the same time Michio met Tetsuji Miwa who remained his collaborator for many years.

Together with Sato and Miwa they published a series of papers under the common title "Holonomic quantum fields". For me this papers provide a brilliant piece of mathematical physics. They originate from the famous work by Wu, McCoy, Tracy and Barouch in which the two-point correlation function for the scaling Ising model was related to the Painleve transcendent. Sato, Miwa and Jimbo put this result into much wider framework relating it to the isomonodromic deformation of differential equations. The theory developed in the papers "Holonomic quantum fields" became a classical part of mathematics.

The next series of papers by Michio Jimbo is written in collaboration with Date, Kashiwara and Miwa, it is entitled "Transformation groups for soliton equations". These papers are based on two ingenious observations: first, Hirota's way of rewriting the integrable equations in terms of τ*-functions, seconds, Sato's interpretation of* τ*-function through the infinite-dimensional Grassmanian. Starting from these two facts Date, Jimbo, Kashiwara, Miwa uncover the hidden symmetry of the soliton equations. According to their theory, for example, the famous Kadomtsev-Petviashvili equation possesses the symmetry described by the group* GL_∞*. The* τ*-function is nothing but the matrix element computed in the representations realized by fermions which roughly can be viewed as analogues of the fundamental representations of* GL_N*. Different solutions are parameterized by elements of* GL_∞*. Other integrable hierarchies are related to other infinite-dimensional groups. It is hard to overestimate these result which place the theory of soliton equations into the framework of the representation theory.*

The early eighties witnessed rapid development of the theory of the quantum integrable models due to works by Baxter, Faddeev, Sklyanin and others. Michio Jimbo with collaborators (Date, Miwa, Okado) started to work on the Interaction Round the Face (IRF) models. They produced the far-reaching generalizations of works by Baxter and collaborators. The development of physics and mathematics has its own logic, and often it so happens that the researches which at the first glance have nothing in common are in fact deeply related, and moreover take place together chronologically. In eighties the Conformal Field Theory (CFT) developed rapidly starting from the famous paper by Belavin, Polyakov and Zamolodchikov. Then it was realized that the IRF models found by Date, Jimbo, Miwa, Okado provide certain non-critical deformations of various complicated examples of CFT.

With rapid growth of the theory of quantum integrable models in eighties it was becoming more and more clear that some underlying algebraic structure was missing. The main ingredient of the hidden symmetry was found independently by Vladimir Drinfeld and Michio Jimbo, it is called the "quantum group". The discovery of quantum groups is a result of deep analysis the Yang-Baxter equations which lie in the foundation of quantum integrability. The quantum group is a deformation of usual Lie group which does not break the most important for physicist property: there is a rule called comultiplication allowing to compute the action of the quantum group on the tensor product of representation. This is similar to adding the angular momenta in the quantum mechanics. For this work Michio Jimbo received the prize of the Japan Academy.

The year 1991 became very important for the quantum integrable models community. In this year a special program Infinite Analysis was organized at RIMS (Kyoto University). I recall with a lot of nostalgia this time when all of us, physicist and mathematicians, had a wonderful opportunity to shear our insights. The most important result of this program was the paper by Jimbo, Miki, Miwa and Nakayashiki on the correlation functions for the six vertex model. All the ideas which Michio Jimbo developed during his life in science were combined in order to solve one of the most important problems in the theory of quantum integrable models. The solution is due to application of hidden symmetry under the infinite-dimensional quantum group. This paper, as well as the subsequent work on the critical six vertex model (Jimbo, Miwa, 1996), influenced very much further development of the field.

I am limited in time, so, I have to skip several further contributions of Michio Jimbo, and pass directly to 2004 when we started to work together. With H. Boos, T. Miwa and Y. Takeyama we continued the study of correlation functions. Our main result consists in finding a hidden fermionic structure of quantum integrable models. Surprisingly some of our new conclusions can be viewed as far reaching generalization of the subject of very first works by Michio Jimbo on the holonomic quantum fields This demonstrates one more time that the theory of quantum integrable models is a subject of great beauty in which very different methods and ideas happen to be interrelated.

I congratulate Michio Jimbo wishing him many new achievements and I hope everybody will join me in echoing this sentiment.

After the Wigner Medal there was a ceremony for the Weyl Prize and this was followed by the violin recital by a conference participant, Gerard H.E. Duchamp and a reception. In the name of the Board of the Group Theory and Fundamental Physics Foundation I would like to thank J.-P Gazeau for his commitment to the Wigner Medal concept during the selection process and to the organizers of the Colloquium, especially Professor Maia Angelova, for arranging such a memorable evening.

Journal of Physics: Conference Series **284** (2011) 011005 doi:10.1088/1742-6596/284/1/011005

The Hermann Weyl Prize

The Hermann Weyl Prize was created in 2000 by the Standing Committee of the International Group Theory Colloquium.

The purpose of the Weyl Prize is to provide recognition for young scientists (younger than 35) who have performed original work of significant scientific quality in the area of understanding physics through symmetries.

The Hermann Weyl prize consists of a certificate citing the accomplishments of the recipient, prize money of $500 and an allowance towards attendance of the bi-annual International Group Theory Colloquium at which the award is presented.

The previous winners of the award were: Edward Frenkel (2002), Nikita A. Nekrasov (2004), Boyko Bakalov (2006) and Mohammad M. Sheikh-Jabbari (2008).

The Selection Committee of the Weyl Prize 2010 consisted of S. T. Ali (Concordia University), E. Corrigan (Durham University), P. Kulish (St. Petersburg Mathematical Institute of the Russian Academy of Sciences), R. Mosseri (CNRS Paris) and M. A. del Olmo (University of Valladolid, chairman). This Committee has made the following announcement:

> *The Weyl Prize for the year 2010 was awarded to Dr Giulio Chiribella, in recognition of his pioneering work on the application of group theoretical methods in Quantum Information Theory. In particular, for providing a general solution to the problem of optimal estimation of symmetry transformations based on the notion of quantum entanglement between representation and multiplicity spaces, for the derivation of optimal protocols for the alignment of quantum reference frames, for the characterization of extreme quantum measurements in finite dimensions, for the proof of equivalence between asymptotic cloning and state estimation and for the proof of the optimality of measure-and-reprepare for quantum learning of unitary transformations.*

Journal of Physics: Conference Series **284** (2011) 011005 doi:10.1088/1742-6596/284/1/011005

Laudatio of Giulio Chiribella

M.A. del Olmo
Departamento de Física Teórica, Universidad de Valladolid, Valladolid, Spain.

Dear Colleagues,

It is an honour for me to deliver the Laudatio on the occasion of the award of the 2010 Hermann Weyl Prize to Giulio Chiribella in recognition of his pioneering work on the application of group theoretical methods in Quantum Information Theory.

Let me start by giving you some biographical notes about the awardee. Giulio Chiribela was born in 1980 in Mantova (Italy). In the period 1998–2002 he studied Physics at the Almo Collegio Borromeo in Pavia, and in June 2003 he received a Laurea degree *"cum laude"* in Physics with a thesis entitled *"Quantum State Estimation for Covariant Sets"* under the supervision of G. M. D'Ariano. During the years 2003–2006 he was a student at the Scuola Avanzata di Formazione Integrata in Pavia. During this period he studied for a Ph.D. at the University of Pavia under the supervision of G. M. D'Ariano with a grant of the Italian Ministry of Education. In January 2007 he was awarded this Ph.D. degree in Physics. His thesis was entitled *"Optimal estimation of quantum signals in the presence of symmetry"*. In the period 2007–2009 he was a postdoctoral fellow at the Quantum Information Theory Group of the University of Pavia.

At present he has a postdoctoral fellowship of the Quantum Foundations and Quantum Information groups at the Perimeter Institute for Theoretical Physics and he is also an affiliate member of the Institute for Quantum Computing, both in Waterloo (Canada).

The research work of G. Chiribella has focused on quantum information theory. More precisely, he has been involved in the resolution of problems of quantum estimation, measurement theory and quantum information processing, using the underlying group-theoretical and algebraic structures of the systems under consideration.

The main contributions made by Chiribella and his collaborators to quantum information theory are: the discovery of the role of equivalent group representations in quantum estimation, the derivation of the optimal protocols for the alignment of quantum reference frames, the characterization of extreme quantum measurements in finite dimensions, the proof of equivalence between asymptotic cloning and state estimation and the formulation of the theory of quantum combs and testers.

Quantum estimation theory deals with the problem of choosing the best estimation strategy, i.e., the experimental and/or theoretical procedure that optimizes the decoding of a signal state with respect to a given optimality criterion in order to recover the classical information encoded into it. A quantum system, prepared in some known state, undergoes an unknown transformation group, which has to be estimated with the maximum precision allowed by Quantum Mechanics. The estimation of physical parameters associated with transformation groups, such as time-translations, spatial-rotations and displacements in phase space, was treated in Chiribella's Ph.D. thesis. This work pointed out for the first time the role of multiple equivalent group representations in quantum estimation. In other words, quantum entanglement between representation spaces and multiplicity spaces associated with the irreducible representations of the group under consideration is shown to be the essential feature of optimal estimation. The general solution for a large class of problems of this kind was presented in different works covering the case of all finite and compact groups and, also, several examples of non-compact groups. As application of the theory some problems in quantum optics, where non-compact groups are involved, have been considered: the optimal estimation of squeezing ($\mathbb{R}$), the joint estimation of squeezing and displacement (affine group of the line), the study of quantum tomography ($SU(1,1)$), and the derivation of quantum benchmarks for the teleportation of squeezed states ($\mathbb{R}$).

The determination of the ultimate precision limits for the estimation of group parameters has important applications, in addition to its intrinsic interest, in high-sensitivity measurements, quantum metrology, uncertainty relations, and the study of quantum clocks and gyroscopes (*quantum reference frames*). The primitive systems that can be used for communication of reference frames are quantum spins, that can be considered as elementary quantum gyroscopes. Chiribella and his co-workers have shown that N spin-$1/2$ particles can be used as a quantum gyroscope to encode the directions of 3 cartesian axes with an error (variance) that vanishes as $1/N^2$, a scaling that would be impossible in classical statistics. They have recently shown how the $1/N^2$ scaling can be used in a cryptographic context, yielding new protocols that allow two distant parties to align their cartesian axes secretly and with minimum error. The scaling $1/N^2$ has been proven optimal in the most general class of possible quantum protocols, including protocols with many rounds of forward and backward communication, where the two communicating parties can exchange both quantum particles and classical messages.

Another contribution by Chiribella is the proof of the equivalence between asymptotic cloning and state estimation. In quantum information *asymptotic cloning* denotes a physical process that produces a large number of approximate copies of an unknown input state. The open problem consisted in proving that any such process can be simulated by the estimation of the unknown input state, followed by the subsequent repreparation of many identical copies of the estimate. Using permutation symmetry, Chiribella showed that the distance between a copy produced by cloning and a copy produced by estimation vanishes as $1/M$, where M is the number of copies. This result proves that the distribution of quantum information to a large number of indistinguishable users is essentially a classical task.

The formulation of the theory of quantum combs and testers has initiated a new line of research on *higher-order quantum information*. The idea is to generalize the theory of quantum measurements and transformations to the case where the input object is not a quantum state, but rather a quantum circuit (a physical device). This leads to the definition of *supermaps* (a general transformation that maps an input quantum operation into an output quantum operation). *Quantum combs* are a convenient representation of supermaps in terms of positive operators. A *quantum tester* is a special case of a comb that tests another comb, *i.e.*, it represents a device that tests a network of quantum devices. This provides an extension of the theory of quantum measurements to the case where the measured object is a whole circuit. The theory provided an advance in traditional quantum estimation problems, in particular in the estimation of an unknown transformation from a finite number of identical copies. Quantum combs and testers have been used in combination with group-theoretical and algebraical techniques to derive the optimal cloning of unitary transformations, the optimal tomography of quantum channels, the analysis of quantum protocols for the alignment of reference frames using an arbitrary number of rounds of forward and backward communication, and the characterization of multiround protocols that enjoy symmetry with respect to the action of a given group.

After addressing the merits of the laureate, I will take the liberty of encouraging the Group28 participants to promote the nomination of young scientists, who have performed original work of significant scientific quality in the area of understanding physics through symmetries, to the Herman Weyl Prize. It is necessary to continue in the path opened in 2002 by Edward Frenkel and that Giulio Chiribella now joins in such a brilliant way.

The future of this Colloquium depends on our ability to win over promising and enthusiastic young physicists and mathematicians to work in any of the multiple research topics where symmetry plays a relevant role. And this Weyl Prize is a way to boost their scientific careers.

Thank you very much for your attention!

PEER REVIEW STATEMENT

All papers published in this volume of *Journal of Physics: Conference Series* have been peer reviewed through processes administered by the proceedings Editors. Reviews were conducted by expert referees to the professional and scientific standards expected of a proceedings journal published by IOP Publishing.

Journal of Physics: Conference Series **284** (2011) 012001 doi:10.1088/1742-6596/284/1/012001

Group theoretic structures in the estimation of an unknown unitary transformation

Giulio Chiribella

Perimeter Institute for Theoretical Physics, 31 Caroline Street North, Waterloo, Ontario, N2L 2Y5, Canada

Abstract. This paper presents a series of general results about the optimal estimation of physical transformations in a given symmetry group. In particular, it is shown how the different symmetries of the problem determine different properties of the optimal estimation strategy. The paper also contains a discussion about the role of entanglement between the representation and multiplicity spaces and about the optimality of square-root measurements.

1. Introduction

The estimation of an unknown unitary transformation in a given symmetry group is a general problem related to many stimulating topics, such as high-precision measurements [1], coherent states and uncertainty relations [2], quantum clocks [3], quantum gyroscopes [4] and quantum reference frames [5]. The aim of this paper is to provide a synthetic account of the general theory of optimal estimation for symmetry groups, presenting new proofs and highlighting the underlying group theoretical and algebraic structures into play.

1.1. Prologue: dense coding

Let us start with a simple example. Suppose that we have at disposal a quantum system with two-dimensional Hilbert space $\mathcal{H} = \mathbb{C}^2$ and a black box that performs an unknown transformation $\mathcal{U}_{ij}, i,j \in \{0,1\}$, defined on the set $\mathsf{St}(\mathcal{H})$ of all density matrices on $\mathcal{H}$ as $\mathcal{U}_{ij}(\rho) := U_{ij}\rho U_{ij}^\dagger, \forall \rho \in \mathsf{St}(\mathcal{H})$, where $\{U_{ij}\}_{i,j=0}^1$ are the unitary matrices

$$U_{00} = \begin{pmatrix} 1 & 0 \\ 0 & 1 \end{pmatrix}, \quad U_{01} = \begin{pmatrix} 0 & 1 \\ 1 & 0 \end{pmatrix}, \quad U_{10} = \begin{pmatrix} 1 & 0 \\ 0 & -1 \end{pmatrix} \quad U_{11} = \begin{pmatrix} 0 & -1 \\ 1 & 0 \end{pmatrix}.$$

The value of the indices (i,j) is unknown to us and our goal is to find it out using the black box only once. Here the natural figure of merit is the minimization of the probability of error.

Clearly, if we apply the black box to a state $\rho \in \mathsf{St}(\mathcal{H})$ there will always be an error: the four states $\{\rho_{ij} := \mathcal{U}_{ij}(\rho)\}_{i,j=0}^1$ will never be perfectly distinguishable (in a two-dimensional Hilbert space there cannot be more than two perfectly distinguishable states!). However, if we introduce a second system $\mathcal{K} \simeq \mathcal{H}$ and apply the unknown transformation to the projector on the entangled vector $|\Phi\rangle = \frac{1}{\sqrt{2}}(|0\rangle|0\rangle + |1\rangle|1\rangle) \in \mathcal{H} \otimes \mathcal{K}$, where $\{|0\rangle, |1\rangle\}$ is the standard basis for $\mathbb{C}^2$, we obtain four states (the projectors on the vectors $|\Phi_{ij}\rangle := (U_{ij} \otimes I)|\Phi\rangle$) that are perfectly distinguishable: $\langle \Phi_{ij}|\Phi_{kl}\rangle = \mathrm{Tr}[U_{ij}^\dagger U_{kl}]/2 = \delta_{ik}\delta_{jl}$. This remarkable fact was observed for the first time by Bennett and Wiesner, who exploited it to construct a protocol known as *dense-coding* [6]. The moral of dense coding is that the use of entanglement with an ancillary system can improve dramatically the discrimination of unknown transformations. Despite the extreme simplicity of the mathematics, this curious example points out structures that are much deeper than it might seem to the first sight. The aim of the present paper is to illustrate these structures in the general context of the estimation of an unknown group transformation. Our analysis will include dense coding, where the group of interest is the Klein group $\mathbb{Z}_2 \times \mathbb{Z}_2$.

2. General problem: estimation of an unknown group transformation

2.1. The problem

Suppose that we have at disposal one use of a black box and that we want to identify the action of the black box. Suppose also that we have some prior knowledge about the black box: in particular, we know that

(i) the black box acts on a quantum system with finite dimensional Hilbert space $\mathcal{H} \simeq \mathbb{C}^d, d < \infty$.

(ii) it performs a deterministic transformation $\mathcal{U}_g$ belonging to a given representation of a given symmetry group G.

Mathematically, a *deterministic transformation* (also known as *quantum channel*) is described by a completely positive trace-preserving linear map $\mathcal{C}$ acting on the set $\mathsf{St}(\mathcal{H})$ of quantum states (non-negative matrices with unit trace) on the Hilbert space $\mathcal{H}$. The representation of the group G is then given a function $\mathcal{U} : g \mapsto \mathcal{U}_g$ from G to the set of quantum channels, with the usual requirements

$$\begin{aligned} \mathcal{U}_e &= \mathcal{I}_{\mathcal{H}} & \\ \mathcal{U}_g\mathcal{U}_{g^{-1}} &= \mathcal{I}_{\mathcal{H}} & \forall g \in \mathsf{G} \\ \mathcal{U}_g\mathcal{U}_h &= \mathcal{U}_{gh} & \forall g,h \in \mathsf{G} \end{aligned}$$

where $e \in \mathsf{G}$ is the identity element of the group and $\mathcal{I}_{\mathcal{H}}$ is the identity channel, given by $\mathcal{I}_{\mathcal{H}}(\rho) := \rho, \forall \rho \in \mathsf{St}(\mathcal{H})$. In this paper the group G will be always assumed to be either finite or compact.

2.2. Estimation strategies without ancillary systems

Since the group G and the representation $\{\mathcal{U}_g\}_{g\in\mathsf{G}}$ are both known, the problem here is to identify the group element $g \in \mathsf{G}$ that specifies the action of the black box. How can we accomplish this task? A first idea is to prepare the system in a suitable input state $\rho \in \mathsf{St}(\mathcal{H})$ and to apply the unknown transformation $\mathcal{U}_g$ to it, thus obtaining the output state $\rho_g := \mathcal{U}_g(\rho)$. The procedure can be represented diagrammatically as

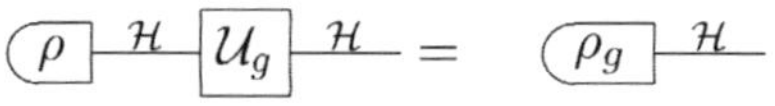

In this way, the output state ρ_g will carry some information about g and we can try to extract this information with a quantum measurement and to produce an estimate $\hat{g} \in \mathsf{G}$. The combination of the quantum measurement with our classical data processing can be described by a single mathematical object, namely a *positive operator-valued measure* (*POVM*, for short).

Let us denote by $\mathsf{Lin}(\mathcal{H})$ the set of linear operators on $\mathcal{H}$ and by $\mathsf{Lin}_+(\mathcal{H}) \subset \mathsf{Lin}(\mathcal{H})$ the set of non-negative operators. If G is a finite group, a POVM with outcomes in G is just a function $P : \mathsf{G} \to \mathsf{Lin}_+(\mathcal{H})$ sending the element $\hat{g} \in \mathsf{G}$ to the non-negative operator $P_{\hat{g}} \in \mathsf{Lin}_+(\mathcal{H})$ and satisfying the requirement $\sum_{\hat{g}\in\mathsf{G}} P_{\hat{g}} = I_{\mathcal{H}}$, where $I_{\mathcal{H}}$ is the identity on $\mathcal{H}$. The conditional probability of inferring $\hat{g}$ when the true value is g is then given by the Born rule $p(\hat{g}|g) = \mathrm{Tr}[P_{\hat{g}}\rho_g]$.

Journal of Physics: Conference Series **284** (2011) 012001 doi:10.1088/1742-6596/284/1/012001

In general, if G is a compact group and $\sigma(\mathsf{G})$ is the collection of its measurable subsets [7], a POVM with outcomes in G is an operator-valued measure $P : \sigma(\mathsf{G}) \to \mathsf{Lin}_+(\mathcal{H})$ sending the measurable subset $B \in \sigma(\mathsf{G})$ to the non-negative operator $P_B \in \mathsf{Lin}_+(\mathcal{H})$ and satisfying the requirements

- $P_{\mathsf{G}} = I_{\mathcal{H}}$
- if $B = \bigcup_{i=1}^{\infty} B_i$ and $\{B_i\}_{i=1}^{\infty}$ are disjoint, then $P_B = \sum_{i=1}^{\infty} P_{B_i}$.

The conditional probability that the estimate $\hat{g}$ lies in the subset $B \in \sigma(\mathsf{G})$ is given by the Born rule $p(B|g) = \mathrm{Tr}[P_B \rho_g]$.

In the following I will frequently use the notation $P(\mathrm{d}\hat{g})$ to indicate the POVM P. Accordingly, I will also write $p(\mathrm{d}\hat{g}|g) = \mathrm{Tr}[P(\mathrm{d}\hat{g})\rho_g]$ and $P_B = \int_B P(\mathrm{d}\hat{g})$. In the case of finite groups, the notation $P(\mathrm{d}\hat{g})$ will be understood as synonym of $P_{\hat{g}}$ and the integral over a subset B will be understood as a finite sum. The pair (ρ, P) of an input state and a POVM will be referred to as an *estimation strategy.*

2.3. *Estimation strategies with ancillary systems*

In the previous paragraph we discussed strategies where the unknown transformation was applied to a suitable state $\rho \in \mathsf{St}(\mathcal{H})$. However, these strategies are not most general ones: we can introduce an ancillary system with Hilbert space $\mathcal{K}$, prepare a *bipartite* input state $\rho \in \mathsf{St}(\mathcal{H} \otimes \mathcal{K})$, and then apply the black box on system $\mathcal{H}$, thus obtaining the output state $\rho_g := (\mathcal{U}_g \otimes \mathcal{I}_{\mathcal{K}})(\sigma)$. The schematic of this procedure is

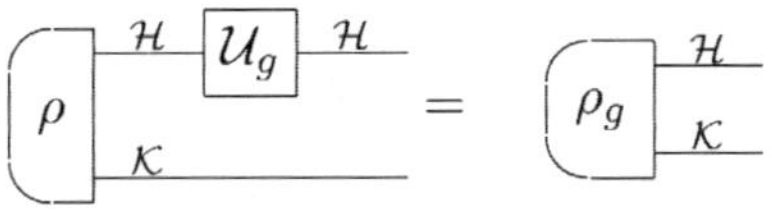

In this case, to obtain an estimate $\hat{g}$ of the unknown group element g we will use a POVM P on the tensor product Hilbert space $\mathcal{H} \otimes \mathcal{K}$.

Classically, we would not expect any improvement from the use of an ancillary system because this system will not carry any information about the unknown transformation. However, here the state ρ can be an entangled state, that is, a state that cannot be written as a convex combination of tensor product states. As the example of dense coding teaches, the use of entanglement can improve the estimation dramatically. Note, however, that mathematically every estimation strategy with ancillary system can be reduced to the form of an estimation strategy without ancillary system by choosing $\mathcal{H}' := \mathcal{H} \otimes \mathcal{K}$ and $\mathcal{U}'_g := \mathcal{U}_g \otimes \mathcal{I}_{\mathcal{K}}$.

2.4. *Cost of an estimation strategy*

To quantify how far is the guess $\hat{g}$ from the true value one typically introduces a *cost function* $c : \mathsf{G} \times \mathsf{G} \to \mathbb{R}, (\hat{g}, g) \mapsto c(\hat{g}, g)$ which takes its minimum value when $\hat{g} = g$. If the true value is g, the expected cost of the estimation strategy (ρ, P) is

$$c(\rho, P|g) := \int_{\mathsf{G}} c(\hat{g}, g) p(\mathrm{d}\hat{g}|g) = \int_{\mathsf{G}} c(\hat{g}, g)\, \mathrm{Tr}[P(\mathrm{d}\hat{g})\rho_g].$$

Ideally, we would like to minimize the cost simultaneously for every g. However, in general this is not possible because the number of states in the orbit $\{\rho_g\}_{g \in \mathsf{G}}$ is larger than the dimension of the Hilbert space and therefore there is no way in which the states $\{\rho_g\}_{g \in \mathsf{G}}$ can be perfectly distinguishable.

Since the true value $g \in \mathsf{G}$ is unknown, one approach to optimization is to minimize the *worst-case cost* of the strategy, defined as

$$c_{wc}(\rho, P) = \max_{g \in \mathsf{G}} c(\rho, P|g). \qquad (1)$$

An alternative (and less conservative) approach consists in assuming a prior distribution $\pi(dg)$ for the true value $g \in \mathsf{G}$ and in minimizing the average cost. Given that g is completely unknown, a natural choice of prior is the Haar measure dg, normalized as $\int_{\mathsf{G}} dg = 1$. With this choice, the average cost is given by

$$c_{ave}(\rho, P) := \int_{\mathsf{G}} dg \ c(\rho, P|g). \tag{2}$$

For finite groups we can understand the integral as a finite sum and dg as synonymous of $1/|\mathsf{G}|$, where $|\mathsf{G}|$ is the cardinality of the group G.

In general, the worst-case and the average approach are very different: for example, the functional $c_{ave}(\rho, P)$ is linear in both arguments while the functional $c_{wc}(\rho, P)$ is not. The two approaches may lead to very different optimal strategies: for example, since $c_{ave}(\rho, P)$ is linear in ρ the optimal input state can be searched without loss of generality in the set of pure states $\rho = |\varphi\rangle\langle\varphi|$, while this reduction is generally not possible for the minimization of $c_{wc}(\rho, P)$. However, when the cost function enjoys suitable group symmetries it is possible to show that there is a strategy (ρ, P) that is optimal for both the worst-case and the average approach. This point will be illustrated in Section 4.

3. Basic notions and notations

3.1. Bipartite states

Let $\{|\phi_i\rangle\}_{i=1}^{d_{\mathcal{H}}}$ (resp. $\{|\psi_j\rangle\}_{j=1}^{d_{\mathcal{K}}}$) be a fixed orthonormal bases for the Hilbert space $\mathcal{H}$ (resp. $\mathcal{K}$). The choice of orthonormal bases induces a bijective correspondence between linear operators from $\mathcal{K}$ to $\mathcal{H}$ and bipartite vectors in $\mathcal{H} \otimes \mathcal{K}$. Following the "double-ket notation" of Ref. [8], if $A \in \mathsf{Lin}(\mathcal{K}, \mathcal{H})$ is a linear operator from $\mathcal{K}$ to $\mathcal{H}$ we define the bipartite vector $|A\rangle\rangle \in \mathcal{H} \otimes \mathcal{K}$ as

$$|A\rangle\rangle := \sum_{i=1}^{d_{\mathcal{H}}} \sum_{j=1}^{d_{\mathcal{K}}} \langle\phi_i|A|\psi_j\rangle \ |\phi_i\rangle|\psi_j\rangle. \tag{3}$$

Two useful properties of this correspondence are given by the equations

$$\langle\langle A|B\rangle\rangle = \mathrm{Tr}[A^\dagger B] \qquad \forall A, B \in \mathsf{Lin}(\mathcal{K}, \mathcal{H}) \tag{4}$$

$$\begin{aligned} |A\rangle\rangle &= (A \otimes I_{\mathcal{K}})|I_{\mathcal{K}}\rangle\rangle \\ &= (I_{\mathcal{H}} \otimes A^T)|I_{\mathcal{H}}\rangle\rangle \qquad \forall A \in \mathsf{Lin}(\mathcal{K}, \mathcal{H}), \end{aligned} \tag{5}$$

where A^T denotes the transpose of A with respect to the fixed bases. Using the singular value decomposition of A, it is easy to see that the state $|A\rangle\rangle$ is of the product form $|A\rangle\rangle = |\phi\rangle|\psi\rangle$ if and only if A is rank-one. Moreover, if the dimension of $\mathcal{K}$ is larger than $d_{\mathcal{H}}$, then for every state $|A\rangle\rangle$ there exists a $d_{\mathcal{H}}$-dimensional subspace $\mathcal{K}_A$ such that $|A\rangle\rangle \in \mathcal{H} \otimes \mathcal{K}_A$. This is clear from Eq. (5): the dimension of the image of A^T cannot exceed the dimension of its domain $\mathcal{H}$.

3.2. Group representations

The unknown transformation in our estimation problem belongs to a group representation $\{\mathcal{U}_g\}_{g \in \mathsf{G}}$, where each map $\mathcal{U}_g$ is a completely positive trace preserving map sending operators in $\mathsf{Lin}(\mathcal{H})$ to operators in $\mathsf{Lin}(\mathcal{H})$. Now, the condition $\mathcal{U}_{g^{-1}}\mathcal{U}_g = \mathcal{I}_{\mathcal{H}}$ implies that the map $\mathcal{U}_g$ must have the form $\mathcal{U}_g(\rho) = U_g \rho U_g^\dagger$, where $U_g \in \mathsf{Lin}(\mathcal{H})$ is a unitary matrix. In particular, for $g = e$ we can choose $U_e = I_{\mathcal{H}}$. Moreover, the condition $\mathcal{U}_g\mathcal{U}_h = \mathcal{U}_{gh}, \forall g, h \in \mathsf{G}$ implies that the unitaries $\{U_g\}_{g \in \mathsf{G}}$ define a projective unitary representation, that is, a function $U : g \mapsto U_g$ sending elements of the group to unitary operators and satisfying the relation

Journal of Physics: Conference Series **284** (2011) 012001 doi:10.1088/1742-6596/284/1/012001

$U_g U_h = \omega(g,h) U_{gh}, \forall g,h \in \mathsf{G}$, where $\omega(g,h) \in \mathbb{C}$ is a multiplier, satisfying the properties

$$\begin{array}{ll} |\omega(g,h)| = 1 & \forall g,h \in \mathsf{G} \\ \omega(g,h)\omega(gh,k) = \omega(g,hk)\omega(h,k) & \forall g,h,k \in \mathsf{G} \\ \omega(g,e) = \omega(e,g) = 1 & \forall g \in \mathsf{G}. \end{array}$$

Since the group G is compact and the space $\mathcal{H}$ is finite dimensional, with a suitable choice of basis the Hilbert space can be decomposed as

$$\mathcal{H} = \bigoplus_{\mu \in \mathsf{Irr}(U)} \mathcal{H}_\mu \otimes \mathbb{C}^{m_\mu}, \tag{6}$$

where the sum runs over the set $\mathsf{Irr}(U)$ of all irreducible representations contained in the isotypic decomposition of U, $\mathcal{H}_\mu$ is a *representation space* of dimension d_μ, carrying the irreducible representation U^μ, and $\mathbb{C}^{m_\mu}$ is a *multiplicity space*, where m_μ is the multiplicity of the irreducible representation U^μ in the decomposition of U. Accordingly, the representation U can be written in the block diagonal form

$$U = \bigoplus_{\mu \in \mathsf{Irr}(U)} U^\mu \otimes I_{m_\mu}. \tag{7}$$

Note that all irreducible representations U^μ must be projective unitary representations and must have the same multiplier ω.

In the optimization of the estimation strategy (ρ, P) we can take the multiplicities m_μ as large as we want. Indeed, introducing an ancillary system $\mathcal{K}$ means replacing the Hilbert space $\mathcal{H}$ with the tensor product $\mathcal{H}' := \mathcal{H} \otimes \mathcal{K}$ and replacing the unitary U_g with the unitary $U'_g := U_g \otimes I_{\mathcal{K}}$ for every $g \in \mathsf{G}$. In Eqs. (6) and (7) this means replacing the multiplicity space $\mathbb{C}^{m_\mu}$ with $\mathbb{C}^{m_\mu} \otimes \mathcal{K}$ and the multiplicity m_μ with $m'_\mu := m_\mu d_{\mathcal{K}}$. In particular, we are free to choose the ancillary system $\mathcal{K}$ so that the condition $m_\mu \geq d_\mu$ is met for every $\mu \in \mathsf{Irr}(U)$.

3.3. Bipartite states and group representations

Here we combine the facts observed in the paragraphs 3.1 and 3.2. Let us consider a vector $|\Psi\rangle \in \mathcal{H}$ in a Hilbert space where the representation U acts. Using the isotypic decomposition of Eq. (6) and the correspondence of Eq. (3) for each tensor product $\mathcal{H}_\mu \otimes \mathbb{C}^{m_\mu}$, we can write

$$|\Psi\rangle = \bigoplus_{\mu \in \mathsf{Irr}(U)} |\Psi_\mu\rangle\!\rangle \in \bigoplus_{\mu \in \mathsf{Irr}(U)} \mathcal{H}_\mu \otimes \mathbb{C}^{m_\mu},$$

where Ψ_μ is a linear operator in $\mathsf{Lin}(\mathbb{C}^{m_\mu}, \mathcal{H}_\mu)$. In other words, the vector $|\Psi\rangle$ can be written as a linear superposition of bipartite vectors. As we will see in the next sections, the fact that we can have entanglement between each representation space $\mathcal{H}_\mu$ and the corresponding multiplicity space $\mathbb{C}^{m_\mu}$ is the key ingredient to the construction of the optimal estimation strategies.

Suppose that U has been chosen so that $m_\mu \geq d_\mu$ for every $\mu \in \mathsf{Irr}(U)$. From the last observation of paragraph 3.1 we know that each vector $|\Psi_\mu\rangle\!\rangle$ will be contained in a subspace $\mathcal{H}_\mu \otimes \mathcal{K}_\mu$, where $\mathcal{K}_\mu \simeq \mathbb{C}^{d_\mu} \simeq \mathcal{H}_\mu$ is a suitable subspace of $\mathbb{C}^{m_\mu}$. Therefore, every vector $|\Psi\rangle$ belongs to a suitable subspace $\widetilde{\mathcal{H}} \simeq \bigoplus_{\mu \in \mathsf{Irr}(U)} \mathcal{H}_\mu \otimes \mathcal{H}_\mu$. Note that the whole orbit $\{|\Psi_g\rangle := U_g |\Psi\rangle\}_{g \in \mathsf{G}}$ belongs to the subspace $\widetilde{\mathcal{H}}$. In conclusion, as long as we are interested in *pure* input states we can effectively replace $\mathcal{H}$ with the space $\widetilde{\mathcal{H}} := \bigoplus_{\mu \in \mathsf{Irr}(U)} \mathcal{H}_\mu \otimes \mathcal{H}_\mu$ where $m_\mu = d_\mu$ for every $\mu \in \mathsf{Irr}(U)$ and we can replace the representation U with the representation $\widetilde{U}$ defined by $\widetilde{U}_g := \bigoplus_{\mu \in \mathsf{Irr}(U)} U^\mu_g \otimes I_\mu$, where I_μ denotes the identity on $\mathcal{H}_\mu$.

3.4. The class states

An important family of states in $\widetilde{\mathcal{H}} = \bigoplus_{\mu\in\mathsf{Irr}(U)} \mathcal{H}_\mu \otimes \mathcal{H}_\mu$ is the family of states of the form

$$|\Phi\rangle = \bigoplus_{\mu\in\mathsf{Irr}(U)} \frac{c_\mu}{\sqrt{d_\mu}} |I_\mu\rangle\rangle, \qquad c_\mu \in \mathbb{C}, \ \sum_{\mu\in\mathsf{Irr}(U)} |c_\mu|^2 = 1 \tag{8}$$

These states are linear superpositions of maximally entangled states in the sectors $\mathcal{H}_\mu \otimes \mathcal{H}_\mu$. In the following I will call these states *class states*. The reason for the name comes from the fact that these states correspond to *class functions* (see definition below) through the Fourier-Plancherel theory (see e.g. [9]). The remaining part of this paragraph is aimed at making this correspondence clear.

Let us start from Fourier-Plancherel theory. Denote by $\mathsf{Irr}(\mathsf{G},\omega)$ the set of all irreducible representations of G with multiplier ω and denote by $u^\mu_{ij}(g) := \langle \mu, i|U^\mu_g|\mu, j\rangle$ the matrix element of the operator U^μ_g with respect to a fixed orthonormal basis $\{|\mu,i\rangle\}_{i=1}^{d_\mu}$ for the representation space $\mathcal{H}_\mu$. Recall that the functions $\{\tilde{u}^{\mu^*}_{ij}(g) := \sqrt{d_\mu} u^{\mu*}_{ij}(g)\}_{\mu\in\mathsf{Irr}(\mathsf{G},\omega); i,j=1,\dots,d_\mu}$ are an orthonormal basis for the Hilbert space $L_2(\mathsf{G}, \mathrm{d}g)$ endowed with the scalar product $\langle f_1, f_2\rangle := \int_\mathsf{G} \mathrm{d}g \ f_1^*(g) f_2(g)$. Using this fact, we can expand every function $f \in L_2(\mathsf{G}, \mathrm{d}g)$ as

$$|f\rangle = \bigoplus_{\mu\in\mathsf{Irr}(\mathsf{G},\omega)} \sum_{i,j=1}^{d_\mu} f^\mu_{ij} |\tilde{u}^{\mu*}_{ij}\rangle. \tag{9}$$

On the Hilbert space $L_2(\mathsf{G}, \mathrm{d}g)$ we consider the action of G given by the left-regular representation T with multiplier ω, defined by $(T_g f)(h) := \omega(g, g^{-1}h) f(g^{-1}h), \quad \forall f \in L_2(\mathsf{G}, \mathrm{d}g), \forall g, h \in \mathsf{G}$. [10]. In particular, the transformation of the functions $\tilde{u}^{\mu*}_{ij}(g)$ is given by

$$T_g |\tilde{u}^{\mu*}_{ij}\rangle = \sum_{k=1}^{d_\mu} u^\mu_{ki}(g) |\tilde{u}^{\mu*}_{kj}\rangle, \tag{10}$$

that is, for every fixed value of j the vectors $\{|\tilde{u}^{\mu*}_{ij}\rangle\}_{i=1}^{d_\mu}$ span an invariant subspace carrying the irreducible representation U^μ. It is then easy to see that the unitary operator $V : L_2(\mathsf{G}, \mathrm{d}g) \to \bigoplus_{\mu\in\mathsf{Irr}(\mathsf{G},\omega)} \mathcal{H}_\mu \otimes \mathcal{H}_\mu$ defined by

$$V|u^{\mu^*}_{ij}\rangle := |\mu, i\rangle|\mu, j\rangle \qquad \forall \mu \in \mathsf{Irr}(\mathsf{G},\omega), \forall i, j = 1, \dots, d_\mu \tag{11}$$

intertwines the two representations T_g and $W_g = \bigoplus_{\mu\in\mathsf{Irr}(\mathsf{G},\omega)} U^\mu_g \otimes I_\mu$: indeed, using Eqs. (10) and (11) one has $VT_g|u^{\mu*}_{ij}\rangle = U^\mu_g|\mu, i\rangle|\mu, j\rangle = (U^\mu_g \otimes I_\mu)V|u^{\mu*}_{ij}\rangle$ for every $\mu \in \mathsf{Irr}(\mathsf{G},\omega), \forall i, j = 1, \dots, d_\mu$, and therefore $VT_g = W_g V$. Applying the unitary V to the expansion of Eq. (9) we then obtain

$$V|f\rangle = \bigoplus_{\mu\in\mathsf{Irr}(\mathsf{G},\omega)} \sum_{i,j=1}^{d_\mu} f^\mu_{ij} |\mu, i\rangle|\mu, j\rangle.$$

Let us come now to class functions. A class function is a function that is constant on conjugacy classes, that is, $f(hgh^{-1}) = f(g)$ for every $g, h \in \mathsf{G}$. Any class function can be written as a linear combination of irreducible characters, namely $f(g) = \sum_{\mu\in\mathsf{Irr}(\mathsf{G})} f_\mu \chi^*_\mu(g)$, where $\chi^*_\mu(g) = \mathrm{Tr}[U^{\mu*}_g]$. Exploiting the correspondence given by the unitary V we obtain

$$V|f\rangle = \bigoplus_{\mu\in\mathsf{Irr}(\mathsf{G},\omega)} \frac{f_\mu}{\sqrt{d_\mu}} |I_\mu\rangle\rangle.$$

It is now clear that the class states of Eq. (8) are nothing but the projection of the class functions to the finite dimensional subspace $\widetilde{\mathcal{H}} \subset \bigoplus_{\mu \in \mathsf{Irr}(\mathsf{G},\omega)} \mathcal{H}_\mu \otimes \mathcal{H}_\mu$.

Besides providing a justification for the choice of the name *class states*, the Fourier-Plancherel theory also provides a hint of the fact that the class states are optimal for estimation. Indeed, if we take the scalar product of two states in the orbit $\{|\Phi_g\rangle := \widetilde{U}_g|\Phi\rangle\}$ we obtain $\langle\Phi_g|\Phi_h\rangle = \sum_{\mu \in \mathsf{Irr}(U)} \frac{|c_\mu|^2}{d_\mu} \chi_\mu(g^{-1}h)$. If c_μ is chosen so that $c_\mu = \lambda d_\mu$ for some constant $\lambda \in \mathbb{C}$ we then obtain $\langle\Phi_g|\Phi_h\rangle \propto \sum_{\mu \in \mathsf{Irr}(U)} d_\mu \chi_\mu(g^{-1}h)$, which is nothing but the truncated version of the Dirac delta $\delta(g,h) = \sum_{\mu \in \mathsf{Irr}(\mathsf{G},\omega)} d_\mu \chi_\mu(g^{-1}h)$. This provides an heuristic argument for the optimality of the class states, although the choice of the coefficients $\{c_\mu\}$ to provide the "best" finite dimensional approximation of the Dirac delta will depend on the choice of the cost function $c(\hat{g}, g)$.

3.5. Notations

If $V \in \mathsf{Lin}(\mathcal{H})$ is a unitary matrix, the symbol $\mathcal{V}$ will always denote the completely positive map $\mathcal{V}(\rho) = V\rho V^\dagger$. Let $\mathcal{C} : \mathsf{Lin}(\mathcal{H}) \to \mathsf{Lin}(\mathcal{K})$ be a completely positive map and let $\mathcal{C}(\rho) = \sum_{i=1}^r C_i \rho C_i^\dagger$, $C_i \in \mathsf{Lin}(\mathcal{H},\mathcal{K})$ be a Kraus form for $\mathcal{C}$. The notation $\mathcal{C}^\dagger, \mathcal{C}^*, \mathcal{C}^T$ will be used for the following maps

$$\mathcal{C}^\dagger(\rho) := \sum_{i=1}^r C_i^\dagger \rho C_i$$

$$\mathcal{C}^*(\rho) := \sum_{i=1}^r C_i^* \rho C_i^T$$

$$\mathcal{C}^T(\rho) := \sum_{i=1}^r C_i^T \rho C_i^* = (\mathcal{C}^*)^\dagger(\rho),$$

where the complex conjugation $*$ and the transpose T are defined with respect to the fixed bases for $\mathcal{H}$ and $\mathcal{K}$. Note that the map $\mathcal{C}^\dagger$ is the adjoint of $\mathcal{C}$ with respect to the Hilbert-Schmidt scalar product $\langle\langle A|B\rangle\rangle = \mathrm{Tr}[A^\dagger B], A, B, \in \mathsf{Lin}(\mathcal{H})$: indeed, we have $\langle\langle A|\mathcal{C}(B)\rangle\rangle = \langle\langle \mathcal{C}^\dagger(A)|B\rangle\rangle$.) Note also that the definition of the maps $\mathcal{C}^\dagger, \mathcal{C}^*, \mathcal{C}^T$ does not depend on the choice of a particular Kraus form.

Consider the Hilbert space $\widetilde{\mathcal{H}} = \bigoplus_{\mu \in \mathsf{Irr}(U)} \mathcal{H}_\mu \otimes \mathcal{H}_\mu$, the representation $\widetilde{U} = \bigoplus_{\mu \in \mathsf{Irr}(U)} U^\mu \otimes I_\mu$, and let

$$\widetilde{U}' = \left\{ \bigoplus_{\mu \in \mathsf{Irr}(U)} I_\mu \otimes B_\mu \mid B_\mu \in \mathsf{Lin}(\mathcal{H}_\mu) \right\}$$

$$\widetilde{U}'' = \left\{ \bigoplus_{\mu \in \mathsf{Irr}(U)} A_\mu \otimes I_\mu \mid A_\mu \in \mathsf{Lin}(\mathcal{H}_\mu) \right\}$$

be the commutant and the bicommutant of $\widetilde{U}$, respectively. Every class state in Eq. (8) then defines a modular conjugation: for every operator $A = \bigoplus_{\mu \in \mathsf{Irr}(U)} A_\mu \otimes I_\mu \in \widetilde{U}''$ I will denote by $A^R \in \widetilde{U}'$ its modular conjugate, given by

$$A^R := \bigoplus_{\mu \in \mathsf{Irr}(U)} I_\mu \otimes A_\mu^*. \tag{12}$$

Every class state $|\Phi\rangle \in \widetilde{\mathcal{H}}$ then enjoys the symmetry

$$\widetilde{U}_g \widetilde{U}_g^R |\Phi\rangle = |\Phi\rangle \qquad \forall g \in \mathsf{G}, \tag{13}$$

as it can be easily verified using Eq. (5) for every $\mu \in \mathsf{Irr}(U)$. The modular conjugate representation will be denoted by $\widetilde{U}^R$. For a completely positive map $\mathcal{C}(\rho) = \sum_{i=1}^r C_i \rho C_i^\dagger$ with $C_i \in \widetilde{U}''$ for every $i = 1, \dots, r$ we will denote by $\mathcal{C}^R$ the completely positive map $\mathcal{C}^R(\rho) := \sum_{i=1}^r C_i^R \rho C_i^{R\dagger}$. Note that in general we have

$$\mathcal{C}(|\Phi\rangle\langle\Phi|) = \mathcal{C}^{R\dagger}(|\Phi\rangle\langle\Phi|). \tag{14}$$

Introducing the *swap map* $\mathcal{S}$, defined by $\mathcal{S}\left(\bigoplus_{\mu \in \mathsf{Irr}(U)} A_\mu \otimes B_\mu\right) := \bigoplus_{\mu \in \mathsf{Irr}(U)} B_\mu \otimes A_\mu$, we obtain the relations

$$A = \mathcal{S}\left(A^{R*}\right) \qquad \forall A \in \widetilde{U}'' \tag{15}$$

$$\mathcal{C} = \mathcal{S}\mathcal{C}^{R*}\mathcal{S} \qquad \forall \mathcal{C} : \mathcal{C}(\rho) = \sum_{i=1}^r C_i \rho C_i^\dagger, \quad C_i \in \widetilde{U}'' \ \forall i = 1, \dots, r. \tag{16}$$

4. Optimal estimation

4.1. Left-invariant cost functions: optimality of covariant measurements

In this paragraph I will shortly review two classic results about the structure of covariant measurements and about their optimality. The proofs of these results can be found in the monographs [11, 12, 13].

Let us start from the definition:

Definition 1 *A POVM $P : \sigma(\mathsf{G}) \to \mathsf{Lin}_+(\mathcal{H})$ is* covariant *with respect to the representation $\{U_g\}_{g \in \mathsf{G}}$ if $P_{gB} = \mathcal{U}_g(P_B) \quad \forall g \in \mathsf{G}, \forall B \in \sigma(\mathsf{G})$.*

Covariant POVMs have a very simple structure:

Theorem 1 (Structure of covariant POVMs) *Let G be a compact group, U be a projective unitary representation of G on the Hilbert space $\mathcal{H}$. A POVM $P : \sigma(\mathsf{G}) \to \mathsf{Lin}_+(\mathcal{H})$ is covariant with respect to U if and only if it has the form*

$$P(\mathrm{d}\hat{g}) = \mathcal{U}_{\hat{g}}(\xi)\ \mathrm{d}\hat{g} \tag{17}$$

where $\mathrm{d}\hat{g}$ is the normalized Haar measure and $\xi \in \mathsf{Lin}_+(\mathcal{H})$ is a suitable operator, called the seed *of the covariant POVM.*

The special form of covariant POVMs provides a great simplification in optimization problems, because it reduces the optimization of the whole POVM to the optimization of a single non-negative operator $\xi \in \mathsf{Lin}_+(\mathcal{H})$. Luckily, this simplification can be done without loss of generality in most situations. Precisely, covariant measurement are optimal whenever the cost function $c(\hat{g}, g)$ is *left-invariant*, that is $c(h\hat{g}, hg) = c(\hat{g}, g) \forall g, h \in \mathsf{G}$.

Theorem 2 (Optimality of covariant POVMs) *Let G be a compact group, $\{U_g\}_{g \in \mathsf{G}}$ be a unitary projective representation of G on the Hilbert space $\mathcal{H}$ and $c(\hat{g}, g)$ be a left-invariant cost function. Then, for every input state $\rho \in \mathsf{St}(\mathcal{H})$ the optimal POVM P for both the worst-case and uniform average approach can be assumed without loss of generality to be covariant. With this choice one has $c_{ave}(\rho, P) = c_{wc}(\rho, P) = c(\rho, P|g) \forall g \in \mathsf{G}$.*

4.2. Right-invariant cost functions: optimality of class states

In this paragraph I illustrate with a new proof a recent result on the optimality of class states [14]. This result is dual to the classic result on the optimality of covariant POVMs showed in the previous paragraph.

A central feature of quantum theory is the validity of the *purification principle* [15]: for every mixed state $\rho \in \mathsf{St}(\mathcal{H})$ there exists a Hilbert space $\mathcal{K}$ and a pure state $|\Psi\rangle\!\rangle \in \mathcal{H}\otimes\mathcal{K}$ such that $\rho = \mathrm{Tr}_{\mathcal{K}}[|\Psi\rangle\!\rangle\langle\!\langle\Psi|]$ The pure state Ψ is referred to as a *purification* of ρ. From Eq. (5) we have $\rho = \Psi\Psi^\dagger$ and $\mathsf{Supp}(\rho) = \mathsf{Rng}(\Psi)$, where Supp and Rng denote the support and the range, respectively. The simplest example of purification of a state $\rho \in \mathsf{St}(\mathcal{H})$ is the *square-root purification* $|\rho^{\frac{1}{2}}\rangle\!\rangle \in \mathcal{H}\otimes\mathcal{H}'$, where $\mathcal{H}'\simeq\mathcal{H}$.

A large number of quantum features, such as teleportation and no cloning, are simple consequences of the purification principle [15]. The structure of the optimal states for the estimation of group parameters is no exception to that. In the following I will give a simple proof of the optimality of the states in Eq. (8) based on the purification principle. To this purpose I will use a remarkable consequence of purification, namely the fact, first noticed by Schrödinger [16], that every ensemble decomposition of a state can be induced from the purification via a quantum measurement on the purifying system:

Lemma 1 *Let $(X,\sigma(X))$ be a measurable space and let $\rho(\mathrm{d}x) : \sigma(X) \to \mathsf{Lin}_+(\mathcal{H}), B \mapsto \rho_B$ be an ensemble, that is, an operator-valued measure such that $\rho_X \in \mathsf{St}(\mathcal{H})$. If $|\Psi\rangle\!\rangle \in \mathcal{H}\otimes\mathcal{K}$ is a purification of ρ_X then there exists a POVM $P : \sigma(X) \to \mathsf{Lin}_+(\mathcal{K}), B\in\sigma(X) \mapsto P_B$ such that*

$$\rho_B = \mathrm{Tr}_{\mathcal{K}}[(I_{\mathcal{H}}\otimes P_B)|\Psi\rangle\!\rangle\langle\!\langle\Psi|] \qquad \forall B\in\sigma(X). \tag{18}$$

Proof. Let Ψ^{-1} be the inverse of Ψ on its support and define the POVM P via $P_B := \left[\Psi^{-1}\rho_B(\Psi^\dagger)^{-1}\right]^T$. The POVM P provides a resolution of the projector on the range of Ψ^T and can be easily extended to a resolution of the identity on $\mathcal{K}$. Using the identity $\Psi\Psi^{-1} = P_{\mathsf{Supp}(\rho_X)}$ where $\Pi_{\mathsf{Supp}(\rho_X)}$ is the projector on the support of ρ_X we have

$$\begin{aligned}\mathrm{Tr}_{\mathcal{K}}[(I_{\mathcal{H}}\otimes P_B)|\Psi\rangle\!\rangle\langle\!\langle\Psi|] &= \mathrm{Tr}_{\mathcal{K}}[(I_{\mathcal{H}}\otimes(\Psi^{-1}\rho_B)^T|\Psi\rangle\!\rangle\langle\!\langle\Psi|(I_{\mathcal{H}}\otimes(\Psi^*)^{-1})] \\ &= \mathrm{Tr}_{\mathcal{H}'}[(|\Psi\Psi^{-1}\rho_B\rangle\!\rangle\langle\!\langle\Psi\Psi^{-1}|] \\ &= \mathrm{Tr}_{\mathcal{H}'}[|\Pi_{\mathsf{Supp}(\rho_X)}\rho_B\rangle\!\rangle\langle\!\langle\Pi_{\mathsf{Supp}(\rho_X)}] \\ &= \rho_B.\end{aligned}$$

■

Suppose that the cost function $c(\hat{g},g)$ is *right-invariant*, that is $c(\hat{g}h,gh) = c(\hat{g},g), \forall\hat{g},g,h\in\mathsf{G}$. Then we have the following

Theorem 3 (Optimality of the purification of invariant states) *Let G be a compact group, U be a projective unitary representation of G on the Hilbert space $\mathcal{H}$ and $c(\hat{g},g)$ be a right-invariant cost function. Then, the optimal input state for both the worst-case and uniform average approach can be assumed without loss of generality to be the purification of an invariant state $\rho\in U'$. Denoting by $|\Psi\rangle\in\mathcal{H}\otimes\mathcal{L}$ such a purification, there exists an optimal POVM P on $\mathcal{H}\otimes\mathcal{L}$ such that $c_{ave}(\Psi,P) = c_{wc}(\Psi,P) = c(\Psi,P|g), \forall g\in\mathsf{G}$.*

Proof. Suppose that $\sigma\in\mathsf{St}(\mathcal{H}\otimes\mathcal{K})$ is an optimal input state and that $Q(\mathrm{d}\hat{g})$ is an optimal POVM for the average approach (resp. for the worst case approach). We now prove that there is a Hilbert space $\mathcal{L}$ and another state $|\Psi\rangle\in\mathcal{H}\otimes\mathcal{L}$ that is the purification of an invariant state $\rho\in U'$ and is optimal as well. Consider the orbit $\{\sigma_h := (\mathcal{U}_h\otimes\mathcal{I}_{\mathcal{K}})(\sigma)\}_{h\in\mathsf{G}}$ and the operator-valued measure $\sigma(\mathrm{d}h) = \sigma_h\mathrm{d}h$. Clearly the state $\sigma_{\mathsf{G}} = \int_{\mathsf{G}}\mathrm{d}h\ \sigma_h$ is invariant under $U\otimes I_{\mathcal{K}}$. Let $|\Psi\rangle\in\mathcal{H}\otimes\mathcal{K}\otimes\mathcal{H}'\otimes\mathcal{K}'$, with $\mathcal{H}'\simeq\mathcal{H},\mathcal{K}'\simeq\mathcal{K}$, be the square-root purification of σ_{G} and define $\mathcal{L} := \mathcal{K} \otimes \mathcal{H}' \otimes \mathcal{K}'$. Note that $|\Psi\rangle \in \mathcal{H} \otimes \mathcal{L}$ is also a purification of the invariant state $\rho := \mathrm{Tr}_{\mathcal{K}}[\sigma_{\mathsf{G}}] \in U'$. We now have to show that $|\Psi\rangle$ is optimal for the uniform average (resp. worst-case) approach. By lemma 1, we know that there is a POVM $M_h dh$ on $\mathcal{H}' \otimes \mathcal{K}'$ such that $\rho_h = \mathrm{Tr}_{\mathcal{H}'\otimes\mathcal{K}'}[(I_{\mathcal{H}} \otimes I_{\mathcal{K}} \otimes M_h)|\Psi\rangle\langle\Psi|] \quad \forall h \in \mathsf{G}$. Define now the POVM $P(\mathrm{d}\hat{g}) := \int_{\mathsf{G}} \mathrm{d}h\ Q(\mathrm{d}(\hat{g}h)) \otimes M_h$. It is easy to see that the average cost of the strategy (Ψ, P) is equal to the average cost of the strategy (σ, Q):

$$\begin{aligned}
c_{ave}(\Psi, P) &= \int_{\mathsf{G}} \mathrm{d}g \int_{\mathsf{G}} c(g,\hat{g})\ \langle\langle \Psi_g | P(\mathrm{d}\hat{g}) | \Psi_g \rangle\rangle \\
&= \int_{\mathsf{G}} \mathrm{d}g \int_{\mathsf{G}} \mathrm{d}h \int_{\mathsf{G}} c(g,\hat{g})\ \langle\langle \Psi | (U_g^\dagger Q(\mathrm{d}(\hat{g}h)) U_g \otimes M_h) | \Psi \rangle\rangle \\
&= \int_{\mathsf{G}} \mathrm{d}h \int_{\mathsf{G}} \mathrm{d}g \int_{\mathsf{G}} c(g,\hat{g})\ \mathrm{Tr}[U_g^\dagger Q(\mathrm{d}(\hat{g}h)) U_g \sigma_h] \\
&= \int_{\mathsf{G}} \mathrm{d}h \int_{\mathsf{G}} \mathrm{d}(gh) \int_{\mathsf{G}} c(gh,\hat{g}h)\ \mathrm{Tr}[Q(\mathrm{d}(\hat{g}h)) \sigma_{gh}] \\
&= c_{ave}(\sigma, Q),
\end{aligned}$$

having used the right-invariance of the Haar measure and of the cost function. On the other hand, the worst-case cost of the strategy (Ψ, P) cannot be larger than the worst-case cost of the strategy (σ, Q):

$$\begin{aligned}
c_{wc}(\Psi, P) &= \max_{g\in\mathsf{G}} \int_{\mathsf{G}} c(\hat{g},g)\ \langle\ \Psi_g | P(\mathrm{d}\hat{g}) | \Psi_g \rangle \\
&= \max_{g\in\mathsf{G}} \int_{\mathsf{G}} \mathrm{d}h \int_{\mathsf{G}} c(\hat{g},g)\ \langle\ \Psi | (U_g^\dagger Q(\mathrm{d}(\hat{g}h)) U_g \otimes M_h) | \Psi' \rangle \\
&= \max_{g\in\mathsf{G}} \int_{\mathsf{G}} \mathrm{d}h \int_{\mathsf{G}} c(g,\hat{g})\ \mathrm{Tr}[U_g^\dagger Q(\mathrm{d}(\hat{g}h)) U_g \sigma_h] \\
&= \max_{g\in\mathsf{G}} \int_{\mathsf{G}} \mathrm{d}h \int_{\mathsf{G}} c(\hat{g}h, gh)\ \mathrm{Tr}[Q(\mathrm{d}(\hat{g}h)) \sigma_{gh}] \\
&= \max_{g\in\mathsf{G}} \int_h \mathrm{d}h\ c(\sigma, Q|gh) \\
&\leq \int_h \mathrm{d}h \max_{g\in\mathsf{G}}\ c(\sigma, Q|gh) \\
&= c_{wc}(\sigma, Q).
\end{aligned}$$

Hence, we proved that, both for the average and for the worst case approach, whichever the optimal strategy (σ, Q) is, the strategy (Ψ, P) cannot be worse. Finally, we observe that for the strategy (Ψ, P) we have $c(\Psi, P|g) = \int_{\mathsf{G}} \mathrm{d}h \int_{\mathsf{G}} c(\hat{g}h, gh)\ \mathrm{Tr}[Q(\mathrm{d}(\hat{g}h))\sigma_{gh}] = \int_h \mathrm{d}h\ c(\sigma, Q|h)$. This proves that $c(\Psi, P|g)$ is independent of $g \in \mathsf{G}$, whence $c(\Psi, P|g) = c_{ave}(\Psi, P) = c_{wc}(\Psi, P)$. ■

Theorem 4 (Optimality of the class states) *Let G be a compact group, U be a unitary projective representation of G on the Hilbert space $\mathcal{H}$ satisfying the property $d_\mu \leq m_\mu \forall \mu \in \mathsf{Irr}(U)$ and $c(\hat{g}, g)$ be a right-invariant cost function. Then, the optimal input state for both the worst-case and uniform average approach can be assumed without loss of generality to be a class state of the form*

$$|\Phi\rangle = \bigoplus_{\mu\in\mathsf{Irr}(U)} \sqrt{\frac{p_\mu}{d_\mu}} |I_\mu\rangle\rangle \in \widetilde{\mathcal{H}} = \bigoplus_{\mu\in\mathsf{Irr}(U)} \mathcal{H}_\mu \otimes \mathcal{H}_\mu \subseteq \mathcal{H}, \tag{19}$$

where $p_\mu \geq 0$ and $\sum_\mu p_\mu = 1$.

Journal of Physics: Conference Series **284** (2011) 012001 doi:10.1088/1742-6596/284/1/012001

Proof. By theorem 3, we know that there exists an optimal state that is the purification of an invariant state $\rho \in U'$. Now the form of an invariant state is $\rho = \bigoplus_{\mu\in\mathsf{Irr}(U)} \frac{p_\mu}{d_\mu} I_\mu \otimes \rho_\mu$, with $\rho_\mu \in \mathsf{St}(\mathbb{C}^{m_\mu})$. The square-root purification of ρ is then given by

$$|\Psi\rangle = \bigoplus_{\mu\in\mathsf{Irr}(U)} \sqrt{\frac{p_\mu}{d_\mu}} |I_\mu\rangle\!\rangle |\rho_\mu^{\frac{1}{2}}\rangle\!\rangle \in \bigoplus_{\mu\in\mathsf{Irr}(U)} \mathcal{H}_\mu \otimes \mathcal{H}_\mu \otimes \mathbb{C}^{m_\mu} \otimes \mathbb{C}^{m_\mu}.$$

Here the first copy of $\mathcal{H}_\mu$ is the representation space, while the remaining spaces contribute to the multiplicity. The action of the representation $\{U_g \otimes I_{\mathcal{H}}\}_{g\in\mathsf{G}}$ then generates the orbit $|\Psi_g\rangle := \bigoplus_{\mu\in\mathsf{Irr}(U)} \sqrt{\frac{p_\mu}{d_\mu}}\ |U_g^\mu\rangle\!\rangle |\rho_\mu^{\frac{1}{2}}\rangle\!\rangle$. This orbit can be obtained from the orbit $|\Phi_g\rangle := \bigoplus_{\mu\in\mathsf{Irr}(U)} \sqrt{\frac{p_\mu}{d_\mu}}\ |U_g^\mu\rangle\!\rangle \in \widetilde{\mathcal{H}}$ by applying the isometry $V : \bigoplus_{\mu\in\mathsf{Irr}(U_g)} \mathcal{H}_\mu \otimes \mathcal{H}_\mu \to \bigoplus_{\mu\in\mathsf{Irr}(U)} \mathcal{H}_\mu \otimes \mathcal{H}_\mu \otimes \mathbb{C}^{m_\mu} \otimes \mathbb{C}^{m_\mu}$ defined by $V(\bigoplus_{\mu\in\mathsf{Irr}(U)} |v_\mu\rangle\!\rangle) = \bigoplus_{\mu\in\mathsf{Irr}(U)} |v_\mu\rangle |\rho_\mu^{\frac{1}{2}}\rangle\!\rangle$. This proves that the state $|\Phi\rangle = \bigoplus_{\mu\in\mathsf{Irr}(U)} \sqrt{\frac{p_\mu}{d_\mu}} |I_\mu\rangle\!\rangle$ is an optimal input state. ■

4.3. Invariant cost functions and isotropic states: optimality of isotropic seeds

Suppose that the cost function $c(\hat{g}, g)$ is *invariant*, that is, both left- and right-invariant. Equivalently, this means that $c(\hat{g}, g) = f(\hat{g}^{-1}g)$ where f is a class function. From the discussion of the previous paragraphs we already know that *i)* the optimal input state will be class state and *ii)* the optimal POVM will be covariant. In fact, in this case the optimal POVM has an additional symmetry that reflects the symmetries of class states (see Eq. (13)). Furthermore, this additional property appears not only for class states but also for a more general family of input states that we call *isotropic*.

Definition 2 (Isotropic operators) *An operator $A \in \mathsf{Lin}(\widetilde{\mathcal{H}})$ with $\widetilde{\mathcal{H}} = \bigoplus_{\mu\in\mathsf{Irr}(U)} \mathcal{H}_\mu \otimes \mathcal{H}_\mu$ is* isotropic *if*

$$\mathcal{U}_g\mathcal{U}_g^R(A) = A \qquad \forall g \in \mathsf{G}.$$

The set of isotropic states (i.e. non-negative isotropic operators with unit trace) is convex. Clearly, every class state is isotropic. As a consequence, every mixture of class states is an isotropic state.

Theorem 5 (Optimality of isotropic seeds) *Let G be a compact group, $\widetilde{U}$ be a projective unitary representation of G on the Hilbert space $\widetilde{\mathcal{H}} = \bigoplus_{\mu\in\mathsf{Irr}(U)} \mathcal{H}_\mu \otimes \mathcal{H}_\mu$, $c(\hat{g}, g)$ be an invariant cost function and $\rho \in \mathsf{St}(\mathcal{H})$ be an isotropic state. Then, the seed $\xi \in \mathsf{Lin}_+(\widetilde{\mathcal{H}})$ of the optimal covariant measurement for the input state ρ is isotropic.*

Proof. Since the cost function is left-invariant, the optimal POVM can be assumed to be covariant (theorem 2) and there is no difference between the average and worst case approaches. Moreover, for any covariant POVM $P(\mathrm{d}g) = \widetilde{\mathcal{U}}_g(\xi)\mathrm{d}g$ we can define another covariant POVM $P'(\mathrm{d}g) = \widetilde{\mathcal{U}}_g(\xi')\mathrm{d}g$ with ξ where ξ' is isotropic. Indeed, it is enough to choose $\xi' := \int_{\mathsf{G}} \mathrm{d}h\, \widetilde{\mathcal{U}}_h\widetilde{\mathcal{U}}_h^R(\xi)$. Note that $P'(\mathrm{d}g)$ is still a resolution of the identity. The uniform average cost of the strategy (ρ, P') is the same of the strategy (ρ, P):

$$\begin{aligned}
c_{ave}(\rho, P') &= \int_{\mathsf{G}} \mathrm{d}g \int_{\mathsf{G}} \mathrm{d}\hat{g}\ c(\hat{g}, g)\,\mathrm{Tr}[\widetilde{\mathcal{U}}_{\hat{g}}(\xi')\widetilde{\mathcal{U}}_g(\rho)] \\
&= \int_{\mathsf{G}} \mathrm{d}g \int_{\mathsf{G}} \mathrm{d}\hat{g} \int_{\mathsf{G}} \mathrm{d}h\ c(\hat{g}, g)\,\mathrm{Tr}[\widetilde{\mathcal{U}}_{\hat{g}h}(\xi)\widetilde{\mathcal{U}}_g\widetilde{\mathcal{U}}_h^{R\dagger}(\rho)] \\
&= \int_{\mathsf{G}} \mathrm{d}h \int_{\mathsf{G}} \mathrm{d}g \int_{\mathsf{G}} \mathrm{d}\hat{g}\ c(\hat{g}h, gh)\,\mathrm{Tr}[\widetilde{\mathcal{U}}_{\hat{g}h}(\xi)\widetilde{\mathcal{U}}_{gh}(\rho)] \\
&= c_{ave}(\rho, P),
\end{aligned}$$

having used the fact that ρ is isotropic (and hence $\widetilde{\mathcal{U}}_h^{R\dagger}(\rho) = \widetilde{\mathcal{U}}_h(\rho)$) and the right-invariance of the cost function. ■

An example of isotropic seed is given by the *class seed* $\xi = |\eta\rangle\langle\eta|$, where $|\eta\rangle \in \mathcal{H}$ is the class vector

$$|\eta\rangle = \bigoplus_{\mu \in \mathsf{Irr}(U)} \sqrt{d_\mu} |I_\mu\rangle\!\rangle. \tag{20}$$

We will now see that this seed plays a special role: any isotropic seed can be obtained from it by applying a suitable physical transformation.

Theorem 6 *$\xi \in \mathsf{Lin}_+(\widetilde{\mathcal{H}})$ is an isotropic seed if and only if $\xi = \mathcal{C}(|\eta\rangle\langle\eta|)$ for some completely positive map $\mathcal{C} : \mathsf{Lin}(\widetilde{\mathcal{H}}) \to \mathsf{Lin}(\widetilde{\mathcal{H}})$ that is bistochastic (i.e. trace-preserving and unit-preserving) and leaves all elements in the algebra U' invariant.*

Proof. Let us write $\xi = \sum_{i=1}^r |\eta_i\rangle\langle\eta_i|$ and $|\eta_i\rangle = \bigoplus_{\mu \in \mathsf{Irr}(U)} \sqrt{d_\mu} |\eta_{i\mu}\rangle\!\rangle$. Defining the operators $C_i := \bigoplus_{\mu \in \mathsf{Irr}(U)} \eta_{i\mu} \otimes I_\mu \in U''$ and defining the completely positive map $\mathcal{C}$ as $\mathcal{C}(\rho) := \sum_{i=1}^r C_i \rho C_i^\dagger$ we then obtain $\xi = \mathcal{C}(|\eta\rangle\langle\eta|)$. Clearly, one has $\mathcal{C}(A) = A$ for every $A \in U'$. The requirement that $\mathcal{C}$ be bistochastic is equivalent to the requirement that ξ be an isotropic seed. First, we can prove that $\mathcal{C}$ is unital:

$$\begin{aligned}\mathcal{C}(I) &= \int_{\mathsf{G}} \mathrm{d}g\ \mathcal{C}\widetilde{\mathcal{U}}_g(|\eta\rangle\langle\eta|) = \int_{\mathsf{G}} \mathrm{d}g\ \mathcal{C}\widetilde{\mathcal{U}}_g^{R\dagger}(|\eta\rangle\langle\eta|)\\ &= \int_{\mathsf{G}} \mathrm{d}g\ \widetilde{\mathcal{U}}_g^{R\dagger}\mathcal{C}(|\eta\rangle\langle\eta|) = \int_{\mathsf{G}} \mathrm{d}g\ \widetilde{\mathcal{U}}_g^{R\dagger}(\xi) = \int_{\mathsf{G}} \mathrm{d}g\ \widetilde{\mathcal{U}}_g(\xi) = I\end{aligned}$$

To prove that $\mathcal{C}$ is trace-preserving, we equivalently prove that $\mathcal{C}^\dagger$ is unital. Using Eq. (14) it is immediate to obtain $\mathcal{C}^{R\dagger}(I) = I$: indeed, we have

$$\begin{aligned}\mathcal{C}^{R\dagger}(I) &= \int_{\mathsf{G}} \mathrm{d}g\ \mathcal{C}^{R\dagger}\widetilde{\mathcal{U}}_g(|\eta\rangle\langle\eta|) = \int_{\mathsf{G}} \mathrm{d}g\ \widetilde{\mathcal{U}}_g\mathcal{C}^{R\dagger}(|\eta\rangle\langle\eta|)\\ &= \int_{\mathsf{G}} \mathrm{d}g\ \widetilde{\mathcal{U}}_g\mathcal{C}(|\eta\rangle\langle\eta|) = \int_{\mathsf{G}} \mathrm{d}g\ \widetilde{\mathcal{U}}_g(\xi) = I\end{aligned}$$

Finally, it is enough to observe that $\mathcal{C}^\dagger = \mathcal{S}\left(\mathcal{C}^{R\dagger}\right)^* \mathcal{S}$ (see Eq. (15)) and, therefore, $\mathcal{C}^\dagger(I) = I$.■

Corollary 1 *Let $\xi \in \mathsf{Lin}(\mathcal{H})$ be an isotropic seed, written as $\xi = \sum_{i=1}^r |\eta_i\rangle\langle\eta_i|$, $|\eta_i\rangle = \bigoplus_{\mu \in \mathsf{Irr}(U)} \sqrt{d_\mu}|\eta_{i\mu}\rangle\!\rangle$. Then we have $\sum_{i=1}^r \eta_{i\mu}\eta_{i\mu}^\dagger = \sum_{i=1}^r \eta_{i\mu}^T\eta_{i\mu}^* = I_\mu$.*

In the following the covariant POVM $E(\mathrm{d}\hat{g}) = \widetilde{\mathcal{U}}_{\hat{g}}(|\eta\rangle\langle\eta|)\mathrm{d}\hat{g}$ generated by the class vector $|\eta\rangle$ in Eq. (20) will be called *class POVM*.

Corollary 2 *Any covariant POVM $P(\mathrm{d}\hat{g}) = \widetilde{\mathcal{U}}_g(\xi)$ with isotropic seed can be obtained from the class POVM $E(\mathrm{d}\hat{g}) = \widetilde{\mathcal{U}}_{\hat{g}}(|\eta\rangle\langle\eta|)\mathrm{d}\hat{g}$ by application of a bistochastic map $\mathcal{C}^{R\dagger}$.*

Note that the bistochastic map $\mathcal{C}^{R\dagger}$ is *covariant*, that is $\widetilde{\mathcal{U}}_g\mathcal{C}^{R\dagger} = \mathcal{C}^{R\dagger}\widetilde{\mathcal{U}}_g \forall g \in \mathsf{G}$.

Corollary 3 *Let $\rho \in \mathsf{St}(\widetilde{\mathcal{H}})$ be an isotropic state and $P(\mathrm{d}\hat{g}) = \widetilde{\mathcal{U}}_g(\xi)$ be a covariant POVM with isotropic seed. Then, the estimation strategy (ρ, P) has the same cost of the estimation strategy $(\mathcal{C}^{R\dagger}(\rho), E)$, where E is the class POVM $E(\mathrm{d}\hat{g}) = \widetilde{\mathcal{U}}_{\hat{g}}(|\eta\rangle\langle\eta|)\mathrm{d}\hat{g}$.*

The question now is: why should we perform a physical transformation on the input state ρ before performing the measurement? Intuitively, one might expect that the application of the transformation $\mathcal{C}^{R\dagger}$ can only degrade the information contained in the signal states $\rho_g := \widetilde{\mathcal{U}}_g(\rho)$. In the next paragraph I will show that the intuition is correct in the case where ρ is a class state and the cost function $c(\hat{g}, g)$ has negative Fourier transform.

4.4. Invariant cost functions with negative Fourier transform: optimality of square-root measurements

Here we will confine our attention to a special family of invariant cost functions, that is, to cost functions of the form

$$c(\hat{g},g) = \sum_{\sigma\in\mathsf{UIrr(G)}} a_\sigma \chi^*_\sigma(\hat{g}^{-1}g) \qquad a_\sigma \le 0 \quad \forall\sigma\in\mathsf{UIrr(G)}. \tag{21}$$

where $\mathsf{UIrr(G)}$ denotes the set of all unitary irreducible representations of G (the complex conjugate of the character has been introduced just for later convenience). Optimization problems with cost functions in this family are the maximization of the *entanglement fidelity*, corresponding to the cost function given by $c(\hat{g},g) = -|\operatorname{Tr}[V_{\hat{g}^{-1}g}]/n|^2$, where $V:\mathsf{G}\to\mathsf{Lin}(\mathbb{C}^n)$ is a unitary representation of G, and the *maximum likelihood* [17], corresponding to the cost function $c(\hat{g},g) = -\delta(\hat{g}^{-1}g)$.

Incidentally, we observe that each cost function of the form in Eq. (21) defines a positive semidefinite product on the vector space of class functions, given by

$$\begin{aligned}\langle f_1,f_2\rangle_c &:= -\int_{\mathsf{G}} \mathrm{d}\hat{g}\int_{\mathsf{G}}\mathrm{d}g\; c(\hat{g},g) f_1^*(\hat{g}) f_2(g)\\ &= -\sum_{\sigma\in\mathsf{UIrr(G)}} a_\sigma f^*_{1,\sigma} f_{2,\sigma}.\end{aligned}$$

The next theorem summarizes the structures presented so far:

Theorem 7 Optimal estimation strategy for invariant functions with negative Fourier transform. *Let G be a compact group, $\widetilde{U}$ be a unitary projective representation of G on the Hilbert space $\widetilde{\mathcal{H}} = \bigoplus_{\mu\in\mathsf{Irr}(U)} \mathcal{H}_\mu\otimes\mathcal{H}_\mu$ and $c(\hat{g},g)$ be an invariant cost function of the form of Eq. (21). Then, the class POVM $E(\mathrm{d}\hat{g}) = \mathcal{U}_g(|\eta\rangle\langle\eta|)\mathrm{d}\hat{g}$ is optimal for every class state $|\Phi\rangle$ of the form $|\Phi\rangle = \bigoplus_{\mu\in\mathsf{Irr}(U)}\sqrt{p_\mu/d_\mu}|I_\mu\rangle\rangle$. The cost is given by $c_{ave} = \sum_{\sigma\in\mathsf{UIrr(G)}} a_\sigma m_\sigma^{\mu\nu^*}\sqrt{p_\mu p_\nu}$, where $m_\sigma^{\mu\nu^*}$ is the multiplicity of the irreducible representation U^σ in the tensor product $U^\mu\otimes U^{\nu *}$.*

Proof. By theorem 5 we can assume without loss of generality that the optimal POVM is covariant, i.e. $P(\mathrm{d}\hat{g}) = \mathcal{U}_g(\xi)\mathrm{d}\hat{g}$, and that the seed ξ is isotropic. Writing the seed as $\xi = \sum_{i=1}^r |\eta\rangle\langle\eta|$ with $|\eta\rangle = \bigoplus_{\mu\in\mathsf{Irr}(U)}\sqrt{d_\mu}|\eta_{i\mu}\rangle\rangle$ we obtain for the cost

$$\begin{aligned}
c(P,\Phi|e) &= \int_{\mathsf{G}}\mathrm{d}\hat{g}\; c(\hat{g},e)\langle\Phi|\widetilde{\mathcal{U}}_{\hat{g}}(\xi)|\Phi\rangle\\
&= \sum_{\sigma\in\mathsf{UIrr(G)}}\sum_{\mu,\nu\in\mathsf{Irr}(U)} a_\sigma\sqrt{\frac{p_\mu p_\nu}{d_\mu d_\nu}}\int_{\mathsf{G}}\mathrm{d}\hat{g}\;\chi^*_\sigma(\hat{g})\;\langle\langle I_\mu|\widetilde{\mathcal{U}}_{\hat{g}}(\xi)|I_\nu\rangle\rangle\\
&= \sum_{\sigma\in\mathsf{UIrr(G)}}\sum_{\mu,\nu\in\mathsf{Irr}(U)} a_\sigma\sqrt{p_\mu p_\nu}\sum_{i=1}^r\int_{\mathsf{G}}\mathrm{d}\hat{g}\;\chi^*_\sigma(\hat{g})\;\langle\langle I_\mu|(U^\mu_{\hat{g}}\otimes I_\mu)|\eta_{i\mu}\rangle\rangle\langle\langle\eta_{i\nu}|(U^{\nu\dagger}_{\hat{g}}\otimes I_\nu)|I_\nu\rangle\rangle\\
&= \sum_{\sigma\in\mathsf{UIrr(G)}}\sum_{\mu,\nu\in\mathsf{Irr}(U)} a_\sigma c\sqrt{p_\mu p_\nu}\sum_{i=1}^r\int_{\mathsf{G}}\mathrm{d}\hat{g}\;\chi^*_\sigma(\hat{g})\;\langle\langle I_\mu|U^\mu_{\hat{g}}\eta_{i\mu}\rangle\rangle\langle\langle U^\nu_{\hat{g}}\eta_{i\nu}|I_\nu\rangle\rangle\\
&= \sum_{\sigma\in\mathsf{UIrr(G)}}\sum_{\mu,\nu\in\mathsf{Irr}(U)} a_\sigma\sqrt{p_\mu p_\nu}\sum_{i=1}^r\int_{\mathsf{G}}\mathrm{d}\hat{g}\;\chi^*_\sigma(\hat{g})\;\operatorname{Tr}[U^\mu_{\hat{g}}\eta_{i\mu}\otimes U^{\nu *}_{\hat{g}}\eta^*_{i\nu}]\\
&= \sum_{\sigma\in\mathsf{UIrr(G)}}\sum_{\mu,\nu\in\mathsf{Irr}(U)} \frac{a_\sigma\sqrt{p_\mu p_\nu}}{d_\sigma}\sum_{i=1}^r\operatorname{Tr}[\Pi_\sigma^{\mu\nu^*}(\eta_{i\mu}\otimes\eta^*_{i\nu})],
\end{aligned}\tag{22}$$

where $P_\sigma^{\mu\nu^*}$ is the projector on the direct sum of all irreducible subspaces of $\mathcal{H}_\mu \otimes \mathcal{H}_\nu$ that carry the irreducible representation U^σ as a subrepresentation of the tensor product $U^\mu \otimes U^{\nu *}$. Since $a_\sigma \leq 0$ for every σ, we have the bound

$$\begin{aligned} c(P,\Phi|e) &\leq \sum_{\sigma\in\mathsf{UIrr(G)}} \sum_{\mu,\nu\in\mathsf{Irr}(U)} \frac{a_\sigma\sqrt{p_\mu p_\nu}}{d_\sigma} \left|\sum_{i=1}^r \mathrm{Tr}[\Pi_\sigma^{\mu\nu^*}(\eta_{i\mu}\otimes\eta_{i\nu}^*)]\right|, \\ &= \sum_{\sigma\in\mathsf{UIrr(G)}} \sum_{\mu,\nu\in\mathsf{Irr}(U)} \frac{a_\sigma\sqrt{p_\mu p_\nu}}{d_\sigma} \left|\sum_{i=1}^r \mathrm{Tr}[\Pi_\sigma^{\mu\nu^*}(\eta_{i\mu}\otimes I_\nu)(I_\mu\otimes\eta_{i\nu}^*)\Pi_\sigma^{\mu\nu^*}]\right|, \\ &\leq \sum_{\sigma\in\mathsf{UIrr(G)}} \sum_{\mu,\nu\in\mathsf{Irr}(U)} \frac{a_\sigma\sqrt{p_\mu p_\nu}}{d_\sigma} \sqrt{\sum_{i=1}^r \mathrm{Tr}[\Pi_\sigma^{\mu\nu^*}(\eta_{i\mu}\eta_{i\mu}^\dagger\otimes I_\nu)]}\sqrt{\sum_{j=1}^r \mathrm{Tr}[(I_\nu\otimes\eta_{i\nu}^T\eta_{i\nu}^*)\Pi_\sigma^{\mu\nu^*}]}, \end{aligned}$$

having used the Cauchy-Schwarz inequality $|\mathrm{Tr}[A^\dagger B]| \leq \sqrt{\mathrm{Tr}[A^\dagger A]\,\mathrm{Tr}[B^\dagger B]}$ for block matrices $A = \bigoplus_{i=1}^r A_i$ and $B = \bigoplus_{i=1}^r B_i$. Finally, using the normalization conditions of corollary 1 we obtain

$$c(P,\Phi,e) \leq \sum_{\sigma\in\mathsf{UIrr(G)}} \sum_{\mu,\nu\in\mathsf{Irr}(U)} \frac{a_\sigma\sqrt{p_\mu p_\nu}}{d_\sigma}\mathrm{Tr}[\Pi_\sigma^{\mu\nu^*}] \leq \sum_{\sigma\in\mathsf{UIrr(G)}} \sum_{\mu,\nu\in\mathsf{Irr}(U)} a_\sigma m_\sigma^{\mu\nu^*}\sqrt{p_\mu p_\nu},$$

where $m_\sigma^{\mu\nu^*}$ is the multiplicity of U^σ in the tensor product of $U^\mu \otimes U^{\nu *}$. On the other hand, by substituting in Eq. (22) it is immediate to see that the class POVM $E(\mathrm{d}\hat g) = \mathcal{U}_{\hat g}(|\eta\rangle\langle\eta|)$ achieves the bound and hence is the best possible POVM. ■

The optimality of the class POVM can be rephrased as a result about the optimality of *square-root measurements* [18] (see also [13] for a surprising *ante litteram* example). The general definition of square-root measurement that we will consider here is the following:

Definition 3 *Let $(X,\sigma(X))$ be a measurable space and let $\rho(\mathrm{d}x) : \sigma(X) \to \mathsf{Lin}_+(\mathcal{H}), B \mapsto \rho_B$ be an ensemble. The square-root measurement associated to ρ is the POVM $P : \sigma(B) \mapsto \mathsf{Lin}_+(\mathsf{Supp}(\rho_X))$ given by $P_B := \rho_X^{-\frac{1}{2}}\rho_B\rho_X^{-\frac{1}{2}}$, where $\rho_X^{-\frac{1}{2}}$ is the inverse of $\rho_X^{\frac{1}{2}}$ on its support.*

It is easy to see that the class POVM $E(\mathrm{d}\hat g) = \widetilde{\mathcal{U}}_{\hat g}(|\eta\rangle\langle\eta|)\mathrm{d}\hat g$ is the square-root measurement associated to the ensemble $\rho(\mathrm{d}g) := |\Phi_g\rangle\langle\Phi_g|\mathrm{d}g$ where $|\Phi_g\rangle = \widetilde{U}_g|\Phi\rangle$ for some class state $|\Phi\rangle = \bigoplus_{\mu\in\mathsf{Irr}(U)}\sqrt{\frac{p_\mu}{d_\mu}}|I_\mu\rangle\rangle$. Indeed, we have $\rho_{\mathsf{G}} = \int_{\mathsf{G}}\mathrm{d}g\ |\Phi_g\rangle\langle\Phi_g| = \bigoplus_{\mu\in\mathsf{Irr}(U)}\frac{p_\mu}{d_\mu^2}\ I_\mu\otimes I_\mu$, and therefore $\rho_{\mathsf{G}}^{-\frac{1}{2}}|\Phi_{\hat g}\rangle = \widetilde{U}_{\hat g}\rho_{\mathsf{G}}^{-\frac{1}{2}}|\Phi\rangle = \widetilde{U}_{\hat g}\left(\sum_{\mu\in\mathsf{Irr}(U)}\sqrt{d_\mu}|I_\mu\rangle\rangle\right) = \widetilde{U}_{\hat g}|\eta\rangle$, which means $E(\mathrm{d}\hat g) = \rho_{\mathsf{G}}^{-\frac{1}{2}}|\Phi_{\hat g}\rangle\langle\Phi_{\hat g}|\rho_{\mathsf{G}}^{-\frac{1}{2}}\mathrm{d}\hat g$.

A slight generalization of Theorem 7 is then given by the following:

Corollary 4 *Let G be a compact group, $\widetilde{U}$ be a projective unitary representation of G on the Hilbert space $\widetilde{\mathcal{H}} = \bigoplus_{\mu\in\mathsf{Irr}(U)}\mathcal{H}_\mu\otimes\mathcal{H}_\mu$ and $c(\hat g, g)$ be an invariant cost function of the form of Eq. (21). Then, the square-root measurement is optimal for every state $|\Psi\rangle \in \widetilde{\mathcal{H}}$ of the form $|\Psi\rangle = V|\Phi\rangle$ where $|\Phi\rangle \in \widetilde{\mathcal{H}}$ is a class state as in Eq. (19) and $V \in \mathsf{Lin}(\widetilde{\mathcal{H}})$ is a unitary matrix in U'.*

Proof. The square-root measurement associated to the ensemble $|\Psi_g\rangle\langle\Psi_g|\mathrm{d}g$ is $P(\mathrm{d}\hat g) = \mathcal{U}_g\mathcal{V}(|\eta\rangle\langle\eta|)$. It is easy to see that the cost of the strategy (Ψ,P) equal the cost of the strategy

(Φ, E): indeed, the conditional probabilities are the same:

$$\begin{aligned} p'(\mathrm{d}\hat{g}|g) &:= \mathrm{Tr}[P(\mathrm{d}\hat{g})\mathcal{U}_g(|\Psi\rangle\langle\Psi|) \\ &= \mathrm{Tr}[\mathcal{U}_{\hat{g}}\mathcal{V}(|\eta\rangle\langle\eta|)\mathcal{U}_g\mathcal{V}(|\Phi\rangle\langle\Phi\rangle]\mathrm{d}\hat{g} \\ &= \mathrm{Tr}[\mathcal{U}_{\hat{g}}(|\eta\rangle\langle\eta|)\mathcal{U}(|\Phi\rangle\langle\Phi|)]\mathrm{d}\hat{g} := p(\mathrm{d}\hat{g}|g) \end{aligned}$$

This means that the square-root measurement $P(\mathrm{d}\hat{g})$ must be optimal for $|\Psi\rangle$, otherwise E would not be optimal for $|\Phi\rangle$. ■

5. Conclusions and open problems

This paper provided a synthesis of the main group theoretical structures appearing in the estimation of unitary transformations on finite dimensional Hilbert spaces. I deliberately avoided any mention to infinite dimensions, in order to keep the exposition at the simplest possible level and to highlight the structures that underly optimal estimation in the neatest possible way. The development of a general theory of optimal estimation for infinite dimensional systems is mostly an open problem, especially in the case of non-compact groups, for which there are many particular results (especially for abelian groups) but still no unifying framework. The development of a general theory of estimation for non-abelian non-compact groups rests as a stimulating challenge for future research.

6. Acknowledgments

This paper contains an advanced elaboration of some ideas present in my PhD thesis "Optimal estimation of quantum signals in the presence of symmetry" (Pavia, October 2006). I would like to take this opportunity to express my gratitude to my PhD advisor and friend G. M. D'Ariano and to my extraordinary collaborators P. Perinotti and M. F. Sacchi. Research at Perimeter Institute for Theoretical Physics is supported in part by the Government of Canada through NSERC and by the Province of Ontario through MRI.

[1] D. Leibfried *et al*, Nature **438**, 639 (2005). T. Nagata, R. Okamoto, J. L. OBrien, K. Sasaki, and S. Takeuchi, Science **316**, 726 (2007). B. L. Higgins, D. W. Berry, S. D. Bartlett, H. M. Wiseman, G. J. Pryde, Nature **450**, 393 (2007).
[2] S. T. Ali, J. P. Antoine and J. P. Gazeau, *Coherent States, Wavelets and Their Generalizations*, Springer-Verlag, 2000.
[3] V. Bužek, R. Derka, and S. Massar, Phys. Rev. Lett. **82**, 2207, (1999).
[4] G. Chiribella, G. M. D'Ariano, P. Perinotti, and M. F. Sacchi, Phys. Rev. Lett. **93**, 180503 (2004).
[5] S. D. Bartlett, T. Rudolph, R. W. Spekkens Rev. Mod. Phys. **79**, 555 (2007).
[6] C. H. Bennett and S. J. Wiesner, Phys. Rev. Lett. **69**, 2881 (1992).
[7] Precisely, $\sigma(\mathsf{G})$ denotes the Borel σ-algebra generated by the topology of G.
[8] G. M. D'Ariano, P. Lo Presti, and M. F. Sacchi, Phys. Lett. A **272**, 32 (2000).
[9] G. B. Folland, editor. *A course in abstract harmonic analysis*. Studies in Advanced Mathematics. CRC Press, Boca Raton, FL, 1995.
[10] The definition is more intuitive if we use formally the Dirac vectors $\{|h\rangle\}_{h\in\mathsf{G}}$, with $\langle h_1|h_2\rangle = \delta(h_1^{-1}h_2)$ and write $|f\rangle = \int_{\mathsf{G}} \mathrm{d}h\, f(h)|h\rangle$. With this notation we have $T_g|h\rangle = \omega(g,h)|gh\rangle$, which shows clearly that T is a projective representation with multiplier ω.
[11] C. W. Helstrom, *Quantum detection and estimation theory* (Academic Press, New York, 1976).
[12] A. S. Holevo, *Probabilistic and Statistical Aspects of Quantum Theory* (North Holland, Amsterdam, 1982).
[13] V. P. Belavkin and V. P. Maslov, *Design of Optimal Dynamic Analyzers: Mathematical Aspects of Wave Pattern Recognition*, Mathematical Aspects Of Computer Engineering Advances in Science and Technology in the USSR, Mir Publisher (1988), http://arxiv.org/abs/quant-ph/041203.
[14] G. Chiribella, G. M. D'Ariano, and M. F. Sacchi, Phys. Rev. A **72**, 042338 (2005).
[15] G. Chiribella, G. M. D'Ariano, and P. Perinotti, Phys. Rev. A **81**, 062348 (2010).
[16] E. Schrödinger, Proc. Camb. Phil. Soc. **31**, 555 (1935).
[17] G. Chiribella, G. M. D'Ariano, P. Perinotti, M. F. Sacchi, Int. J. Quantum Inf. **4**, 453 (2006).
[18] P. Hausladen and W. K. Wootters, J. Mod. Opt. **41**, 2385 (1994).

Journal of Physics: Conference Series **284** (2011) 012002 doi:10.1088/1742-6596/284/1/012002

Symmetry: the search for order in Nature

F Iachello

Center for Theoretical Physics, Sloane Physics Laboratory, Yale University, New Haven, CT 06520-8120, USA

E-mail: francesco.iachello@yale.edu

Abstract. After an historical introduction of the concept of symmetry, the many ways in which symmetry is used today in physics are briefly reviewed. A new concept, super-symmetry, introduced in the 1970's is also briefly reviewed, and the only experimental example of supersymmetry in physics presented. The future of symmetry in physics is briefly discussed.

1. The notion of symmetry

Symmetry, from the Greek σύμμετρος (well-ordered, well-proportioned) was originally introduced to describe certain properties of artifacts (Polykleitos, Περι βελοποιϊκών, IV, 2). All ancient civilizations used this concept. Two examples are shown in figures 1 and 2.

Figure 1. Decorative motif (Sumerian, circa 2000 B.C.): translation symmetry.
(From F. Iachello, in *Symmetry in Science II*, Edited by B. Gruber and R. Lenczewcki, Plenum Press, New York, 1986)

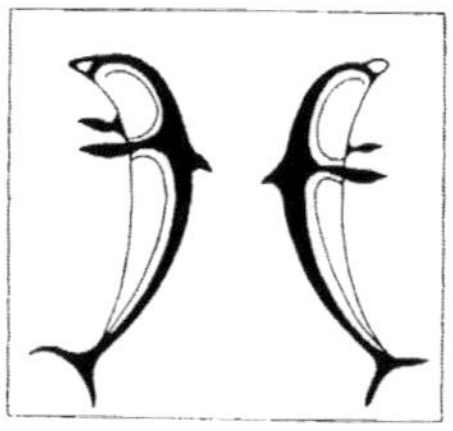

Figure 2. Tile found at the Megaron in Tyrins, Greece (Late Helladic, circa 1200 B.C.): reflection symmetry.
(From F. Iachello, *loc. cit.*, 1986)

The language of symmetry is mathematics. The Greeks developed mathematics (geometry) in the 5th Century B.C., introduced the five regular polyhedra, the tetrahedron, the octahedron, the cube, the icosahedron and the dodecahedron, and associated them with the constituents of the Universe: fire (tetra-), air (octa-), earth (cube), water (icosa-) and the Universe itself (penta-dodeca-hedron). (Plato, *Timaeus*, 55C).

The study of symmetry took another step forward during the Italian Renaissance. Regular polyhedra were complemented by more complex structures, the Archimedean polyhedra, one of which is shown in figure 3. The symmetries of the regular bodies were described in detail (Piero della Francesca, *De Quinque corporibus regularis*, 1482). A mathematical description was introduced

Journal of Physics: Conference Series **284** (2011) 012002 doi:10.1088/1742-6596/284/1/012002

(projective transformation), which, translated into modern mathematical language, is the theory of group transformations, or simply group theory, figure 4.

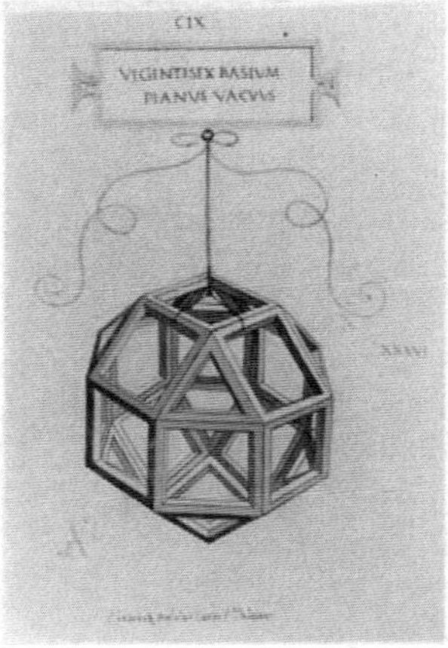

Figure 3. An Archimedean polyhedron with 26 bases (after Leonardo da Vinci). (From Luca Pacioli, *Divina Proportione*, Venice, 1509). (Facsimile of the original figure CIX.)

Figure 4. Luca Pacioli instructing Guidobaldo da Montefeltro in mathematics (geometry). (Oil painting by Jacopo de' Barbari, 1494). (From a print at the *Museo e Gallerie Nazionali di Capodimonte*, Napoli, Italy). An Archimedean polyhedron is shown in the top left, and the projection method is shown in the tablet to which Pacioli is pointing.

Symmetry became so important that in 1595 Kepler stated: the planetary system (known at the time), Saturn, Jupiter, Mars, Earth, Venus and Mercury can be reduced to the motion of regular bodies, figure 5. Kepler concluded his book with the sentence: *Credo spatioso numen in orbe*, that is, I believe in a geometric order of the Universe.

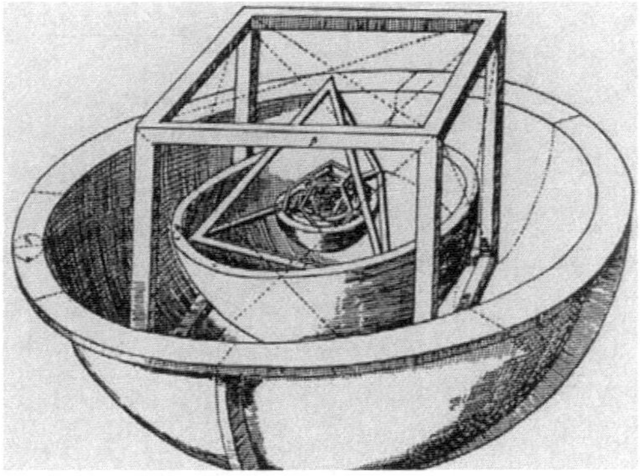

Figure 5. The celestial spheres. (From Kepler, *Mysterium Cosmographicum*, Tübingen, 1595.) (As reproduced in F. Iachello, in *Symmetries in Science VII,* Edited by B. Gruber and T. Otsuka, Plenum Press, New York, 1994)

2. Symmetry in Physics

Symmetry was originally introduced in physics to describe certain properties of the constituents of matter: crystals, molecules (figure 6), etc., much in the same way as in art, especially when physics changed from a description of the macroscopic to that of the microscopic world.

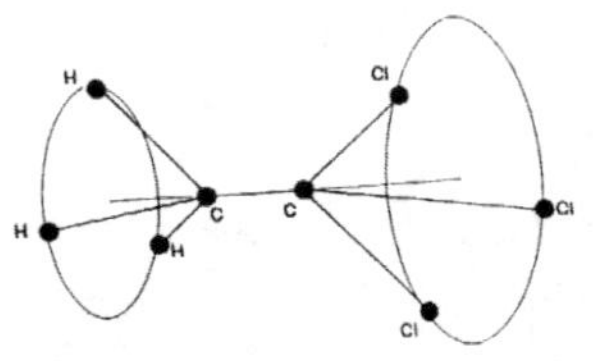

Figure 6. The molecule $H_3C_2Cl_3$ with C_3 symmetry: rotations of 120° around the C-C axis. (From F. Iachello, in *Symmetry in Science VII*, Edited by B. Gruber and T. Otsuka, Plenum Press, New York, 1994).

Symmetry acquired importance when it became apparent that the laws of Nature appear to possess symmetry properties (end of 19th Century). For example, Maxwell equations are invariant under Lorentz transformations

$$x' = \frac{x - vt}{\sqrt{1 - \frac{v^2}{c^2}}}; t' = \frac{t - \left(\frac{v}{c^2}\right)x}{\sqrt{1 - \frac{v^2}{c^2}}} \tag{1}$$

This was the beginning of the symmetry approach to physics. Symmetry became the guiding principle in constructing theories of the Universe, the most notable one being that of special relativity (Einstein, 1905). At about the same time, the mathematical language needed to describe symmetries was further developed with the introduction of Lie algebras and groups (Lie, 1880's) and their classification (Cartan, 1890's). This was the beginning of the group theory approach to physics. Group theory became one of the major tools for studying physics problems and their solutions.

3. The many ways of symmetry in physics

Symmetry and its language group theory are used today in a variety of ways.

3.1. Geometric symmetry

Geometric symmetry describes the arrangement of constituent particles into a structure, for example atoms in a molecule. The mathematical framework to describe these symmetries is point groups. A recent example of a molecule where symmetry plays a crucial role is fullerene discovered in 1985 (figure 7).

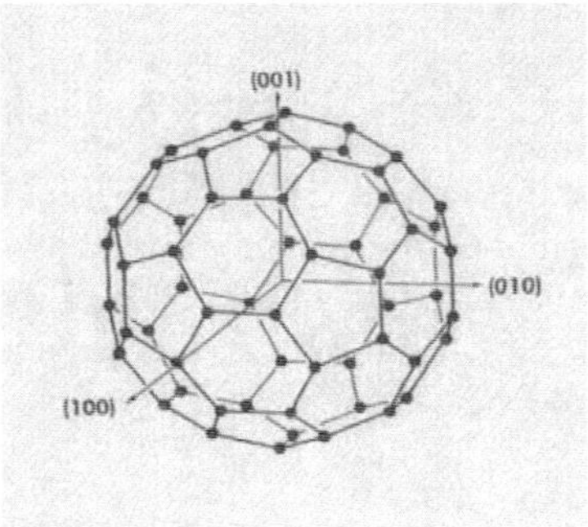

Figure 7. The molecule C_{60} with icosahedral I_h symmetry. (Discovered by Curl, Kroto and Smalley, 1985). (From F. Iachello, in *Group Theoretical Methods in Physics*, V.V. Dodonov and V.I. Man'ko, eds., Lecture Notes in Physics Vol. 382, Springer, Berlin, 1991).

3.2. Permutation symmetry

This symmetry describes properties of systems of identical particles. Its mathematical framework is the permutation group, S_n. Permutation symmetry became particularly important with the development of quantum mechanics (1920's). The wave function of two identical particles is either symmetric or antisymmetric under interchange of the two particles,

$$\begin{aligned} \psi(1,2) &= +\psi(2,1) \\ \psi(1,2) &= -\psi(2,1) \end{aligned} \quad . \tag{2}$$

The first equation applies to bosons and the second to fermions. A visualization of permutation symmetry is given by the Escher drawing of figure 8.

Figure 8. M.C. Escher, *Study of the regular division of the plane with horsemen*, 1946. (© M.C. Escher Heirs, c/o Cordon Art, Baarn, Holland)

3.3. Space-time (or fundamental) symmetry

This symmetry fixes the form of the equations of motion. Its mathematical framework is continuous Lie groups. For example the free Dirac equation

$$\left(i\gamma^{\mu}\partial_{\mu} - m\right)\psi(x) = 0 \tag{3}$$

is invariant under the group of Lorentz transformations, *SO(3,1)*, or, in general, under the Poincare' group, *ISO(3,1)*. All laws of Nature appear to be invariant under Lorentz transformations.

3.4. Gauge symmetry

This symmetry fixes the form of the interaction between particles and external fields, and the form of the equations satisfied by the fields. Its mathematical framework is continuous Lie groups. For example, the laws of electrodynamics, Maxwell equations, are invariant under *U(1)* gauge transformations

$$A_{\mu} \rightarrow A_{\mu} + \partial_{\mu}\Lambda \tag{4}$$

where Λ is an arbitrary function of space and time. The external electromagnetic field, A_{μ}, appears in the coupling of charged particles with the external field as in the Dirac equation

$$\left[\gamma_{\mu}\left(i\partial_{\mu} - eA_{\mu}\right) - m\right]\psi(x) = 0 \quad . \tag{5}$$

A major discovery of the 2nd part of the 20th Century has been that strong, weak and electromagnetic interactions all appear to be governed by gauge symmetries, the overall gauge group being

$$SU_c(3) \otimes SU_w(2) \otimes U(1) \quad . \tag{6}$$

3.5. Dynamic symmetry

This symmetry fixes the interaction between constituent particles. Its main role is to determine the spectral properties of quantum systems (patterns of energy levels). Its mathematical framework is continuous Lie groups. This type of symmetry was introduced implicitly by Pauli [1] for the hydrogen atom. The Hamiltonian with Coulomb interactions is invariant under a set of transformations, *G*, larger than rotations (Runge-Lenz transformations, *SO(4)*). It can be written in terms of Casimir operators of *G*,

$$H = \frac{\vec{p}^2}{2m} - \frac{e^2}{r} = -\frac{A}{C_2\left(SO(4)+1\right)} \quad , \tag{7}$$

resulting in an explicit expression for the eigenvalues in terms of quantum numbers (figure 9)

$$E(n,\ell,m) = -\frac{A}{n^2} \ . \tag{8}$$

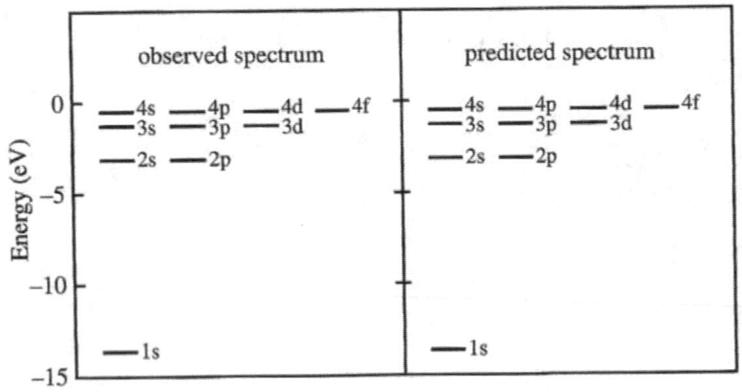

Figure 9. The spectrum of the hydrogen atom is shown as an example of SO(4) dynamic symmetry in atoms. States are labeled by their spectroscopic notation, 1s, 2s, 2p, etc.

This type of symmetry assumed an important role in physics with the introduction of flavor symmetry, $SU_f(3)$, by Gell'Mann and Ne'eman in 1962 [2]. The mass spectrum of particles known at that time appeared to be well described by the Gell'Mann-Okubo mass formula

$$M = a + b\left[C_1(U(1))\right] + c\left[C_2(SU(2)) - \frac{1}{4}C_1^2(U(1))\right] \tag{9}$$

producing an explicit expression of the masses in terms of quantum numbers (figure 10)

$$M(Y,I,I_3) = a + bY + c\left[I(I+1) - \frac{1}{4}Y^2\right]. \tag{10}$$

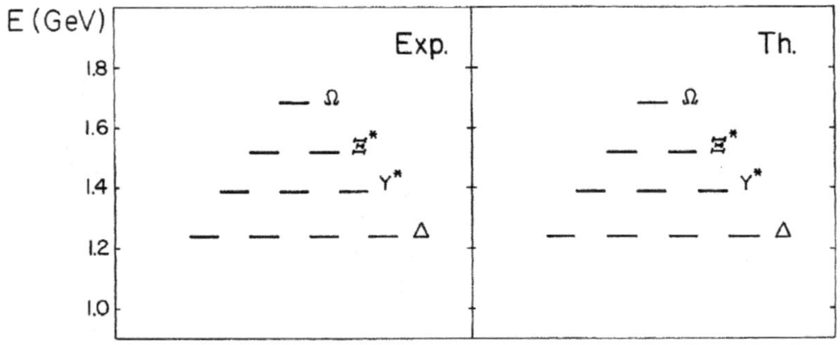

Figure 10. The spectrum of the baryon decuplet, $^4 10$, is shown as an example of $SU_f(3)$ dynamic symmetry in hadrons.

A major discovery of the 2nd part of the 20th Century has been that dynamic symmetries are pervasive in physics. An example is atomic nuclei. The constituent of nuclei, neutrons and protons appear to bind in pairs with angular momentum 0 and 2 (s and d pairs). Pairs behave as bosons, hence the name

Interacting Boson Model given to the model that describes the systems (Iachello, 1974; Arima and Iachello, 1976) [3]. The algebraic structure of the model is u(6) and its dynamic symmetries, obtained by breaking u(6) into its subalgebras, are

$$\begin{aligned} &u(6) \supset u(5) \supset so(5) \supset so(3) \supset so(2) \\ &u(6) \supset su(3) \supset so(3) \supset so(2) \\ &u(6) \supset so(6) \supset so(5) \supset so(3) \supset so(2) \end{aligned} \tag{11}$$

When a dynamic symmetry occurs, all properties of the system can be calculated in explicit analytic form. In particular, the energies of the states are given in terms of quantum numbers. For the three symmetries of Eq.(11), the explicit expressions are

$$\begin{aligned} &E^{(I)}\left(N,n_d,v,n_\Delta,L,M_L\right) = E_0 + \varepsilon n_d + \alpha n_d(n_d+1) + \beta v(v+3) + \gamma L(L+1) \\ &E^{(II)}\left(N,\lambda,\mu,K,L,M_L\right) = E_0 + \kappa\left(\lambda^2+\mu^2+\lambda\mu+3\lambda+3\mu\right) + \kappa' L(L+1) \\ &E^{(III)}\left(N,\sigma,\tau,v_\Delta,L,M_L\right) = E_0 + A\sigma(\sigma+4) + B\tau(\tau+3) + CL(L+1) \end{aligned} \tag{12}$$

The various terms in these expressions are the eigenvalues of the Casimir operators in the appropriate irreducible representations. In the last 30 years, many examples of dynamic symmetries in nuclei have been found. One of them is shown in figure 11.

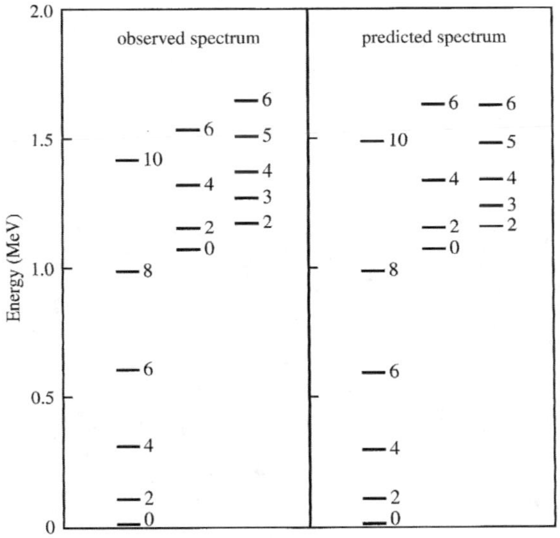

Figure 11. The spectrum of ^{156}Gd is shown as an example of SU(3) dynamic symmetry in nuclei.

Another example is molecules. Spectra of molecules can be built from elementary excitations, called vibrons. The corresponding model is called vibron model and, in the case of diatomic molecules, has algebraic structure u(4) (Iachello, 1980) [4]. Dynamic symmetries in this model are obtained by breaking u(4) into its subalgebras

Journal of Physics: Conference Series **284** (2011) 012002 doi:10.1088/1742-6596/284/1/012002

$$u(4) \supset u(3) \supset so(3) \supset so(2)$$
$$u(4) \supset so(4) \supset so(3) \supset so(2) \tag{13}$$

Several examples of dynamic symmetries in molecules have been found an example of which is shown in figure 12.

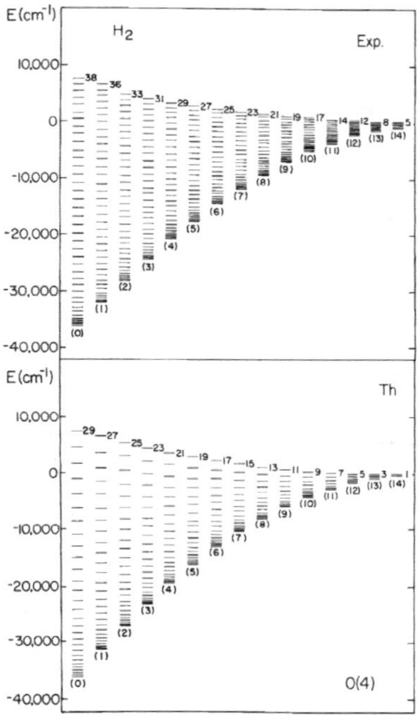

Figure 12. The spectrum of the H_2 molecule is shown as an example of SO(4) dynamic symmetry in molecules.

4. Supersymmetry in physics

In the 1970's, in an attempt to further unify the laws of physics, a new concept was introduced: supersymmetry [5]. In Sect. 3.2 when discussing permutational symmetry, bosons and fermions were introduced. The symmetry operations described in the previous section pertain to systems of bosons *or* of fermions. The symmetry operations change bosons into bosons *or* fermions into fermions. In supersymmetry, in addition to those symmetry operations there are others which change bosons into fermions and vice-versa. This very strange type of symmetry is appropriate for mixed systems of bosons *and* fermions. It is difficult to visualize supersymmetry. The drawing of Escher in figure 13 conveys the spirit. Supersymmetry implies that the figure be invariant under interchange of pair of fish with individual fish.

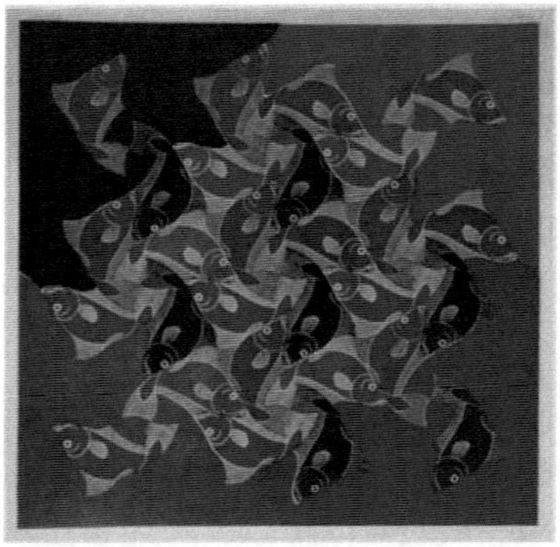

Figure 13. M.E. Escher, *Fish*, circa 1942. (© M.C. Escher Heirs, c/o Cordon Art, Baarn, Holland)

Also in the 1970's the mathematical language needed to describe supersymmetry, Graded Lie algebras and groups was developed and a classification of these algebraic structures given [6]. Supersymmetry and its language, Graded Lie algebras and groups, is used today in a variety of ways. One of them is the use of the algebra of supersymmetry as a tool to solve problems in quantum mechanics, called Supersymmetric Quantum Mechanics (Witten, 1970's) [7]. Most importantly, supersymmetry has become a guiding principle in constructing theories of Nature.

5. Some of the ways of supersymmetry

5.1. Space-time (fundamental) supersymmetry

This type of supersymmetry is a generalization of Lorentz-Poincare' invariance. In addition to space-time coordinates x,t , which are bosonic, there are super space-time coordinates, θ, which are fermionic, i.e. Grassmann varaiables. The supersymmetry transformations mix x,t and θ. The mathematical framework to describe these supesrsymmetries is the SuperPoincare' group. A consequence of supersymmetry is that to each particle there corresponds a superparticle (quarks-squarks, etc.)

5.2. Gauge supersymmetry

This type of supersymmetry fixes the form of the equations satisfied by the fields. An example is the Wess-Zumino Lagrangean [5]

$$L = L_B + L_F + L_{BF} \tag{14}$$

with

$$\begin{aligned} L_B &= -\frac{1}{2}\left(\partial_\mu A(x)\right)^2 - \frac{1}{2}\left(\partial_\mu B(x)\right)^2 - \frac{1}{2}m^2A^2(x) - \frac{1}{2}m^2B^2(x) \\ &-gmA(x)\left[A^2(x)+B^2(x)\right] - \frac{1}{2}g^2\left[A^2(x)+B^2(x)\right] \\ L_F &= -\frac{1}{2}i\overline{\psi}(x)\gamma^\mu\partial_\mu\psi(x) - \frac{1}{2}im\overline{\psi}(x)\psi(x) \\ L_{BF} &= -ig\overline{\psi}(x)\left[A(x)-\gamma_5B(x)\right]\psi(x) \end{aligned} \tag{15}$$

In this Lagrangean, written in terms of two scalar fields $A(x)$ and $B(x)$ and a spinor field $\psi(x)$, all couplings are given in terms of a single coupling constant g. To each bosonic field there corresponds a fermionic field (gluons and gluinos, etc.). Fundamental space-time and gauge supersymmetries are one of the most active areas of research in particle physics at the present time. The Large Hadron Collider (LHC) has been built to search, in part, for supersymmetric partners of the known particles. LHC is

Journal of Physics: Conference Series **284** (2011) 012002 doi:10.1088/1742-6596/284/1/012002

now (2010) operative and we will know (hopefully soon) if fundamental and gauge supersymmetries exist.

5.3. Dynamic supersymmetry

This supersymmetry fixes the boson-boson, fermion-fermion and boson-fermion interaction in a mixed system of bosons and fermions. It determines the spectral properties of mixed systems of bosons and fermions. The corresponding Hamiltonian

$$H = H_B + H_F + V_{BF} \tag{16}$$

is invariant under Bose-Fermi transformations. An example of this type of supersymmetry is provided by atomic nuclei (Iachello, 1980; Balantekin, Bars, Iachello, 1980) [8]. In nuclei some of the constituents bind in pairs (s and d pairs, bosons) while others remain unpaired (fermions). The corresponding model, called Interacting Boson-Fermion Model [9] has a graded algebraic structure, u(6/Ω). A consequence of dynamic supersymmetry is that all properties of mixed systems of bosons and fermions can be calculated in explicit analytic form, and all states can be classified in a given irreducible representation of a supergroup. Dynamic supersymmetries in the Interacting Boson-Fermion Model are obtained by breaking u(6/Ω) into its subalgebras (graded or not). An example is shown in figure 14.

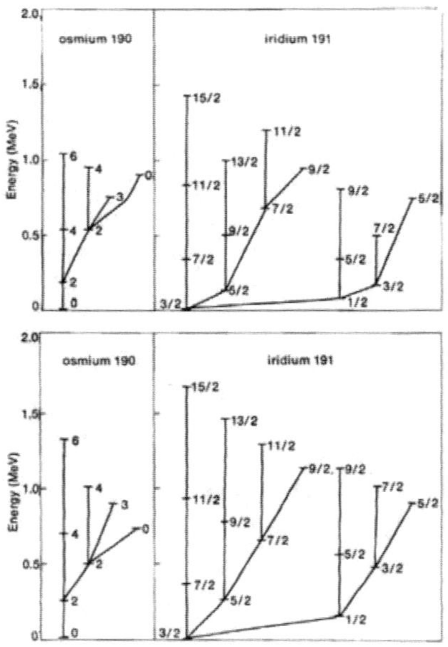

Figure 14. Spectra of ^{190}Os and ^{191}Ir nuclei are shown as an example of U(6/4) supersymmetry in nuclei.

Dynamic supersymmetries in nuclei, discovered in 1980, have been confirmed recently (2000) in a series of experiments involving several laboratories worldwide, especially the Ludwig Maximilians Universität in München, Germany [10]. This is the only experimentally confirmed example of supersymmetry in physics.

6. Recent developments

New mathematical frameworks have been developed in recent years to enlarge the application of the concept of symmetry in physics.

6.1. Infinite dimensional Lie algebras (Kac-Moody)

These have been used to construct fundamental theories in elementary particle physics (fundamental symmetries). The use of this mathematical construction to study dynamic symmetries remains to be done.

6.2. Deformations of Lie algebras
Also called "quantum groups", these have been used to construct theories of elementary particles and as a tool to study dynamic symmetries.

7. The future of symmetry in physics

7.1. Fundamental symmetries: experimental verification of supersymmetry in particle physics
Supersymmetry has been observed in nuclei. Important questions that need to be answered are: Is this observation an accidental fact? Does supersymmetry, even if badly broken, exists in particle physics? Does supersymmetry play a fundamental role in Nature?

7.2. Geometric symmetry: unravelling further the role of symmetry in complex materials
As new materials are discovered, more and more the role of geometric symmetries becomes apparent. An example is shown in figure 15.

Figure 15. Crystal structure of a molybdenum oxide nanowheel. (H.N. Miras *et al.*, Science, Vol. 327, 72 (2010)).
(From the cover of Science, 1 Jan 2010. Image: Leroy Cronin, Ryo Tsunashima, Haralampos Miras, University of Glasgow).

7.3. Dynamic symmetry: application to complex systems. Simplicity in complexity program
Most applications so far of dynamic symmetry have been in molecules, atoms, nuclei (figure 16) and hadrons. Are there other applications? Interesting systems for possible application of dynamic symmetry methods are: macromolecules, polymers, atomic (Bose and Bose-Fermi) condensates, biological molecules.

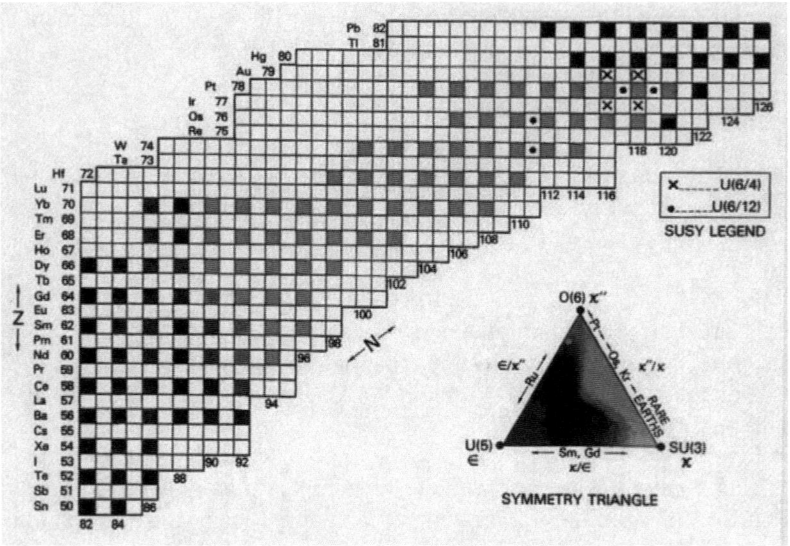

Figure 16. Classification of spectra of nuclei according to dynamic symmetry groups. (Figure from R.F. Casten and D.H. Feng, Physics Today 37, 26 (1984)).
(As adapted by F. Iachello, Rivista Nuovo Cimento 19, 1 (1996)).

At the macroscopic level, many forms of Nature, even the most complex, are often ordered (Figure 17). Is there order in the bio-world at the microscopic level?

Figure 17. E. Haeckel, *Kunstformen der Natur*, Leipzig, 1899. Living forms of the family of Discoidea.
(From F. Iachello, in *Symmetries in Science VII*, Edited by B. Gruber and T. Otsuka, Plenum Press, New York, 1994).

8. Conclusions

Symmetry in its various forms has become a guiding principle in the description of Nature. The 20th Century has seen the development of space-time and gauge symmetries as a tool in determining the fundamental laws of physics. It has also seen the emergence of dynamic symmetry (and supersymmetry) as a way to classify the structure of physical systems. The 20th Century has also seen the development of new mathematical tools needed to describe symmetries (and supersymmetries). As Galileo Galilei wrote: *The book of Nature is written in the language of mathematics.*

Nature and physics appear to display order at all levels: the fundamental laws of Nature are dictated by symmetry principles (space-time and gauge symmetries); spectra of quantum systems are often ordered as seen in molecules, atoms, nuclei, hadrons (dynamic symmetries); constituent particles often

Journal of Physics: Conference Series **284** (2011) 012002 doi:10.1088/1742-6596/284/1/012002

aggregate in ordered structures (geometric symmetries). As Herman Weyl wrote: *Nature loves symmetry*.

In the 21th Century, as the complexity of the phenomena that we are studying increases, symmetry may play an equally important role. In fact, one of the lessons we have learned is that the more complex the structure, the more useful is the concept of symmetry. Figure 18 shows how starting from the regular polyhedra introduced by the Greeks in the 5th Century B.C., we can construct more complex structures.

Figure 18. Five regular tetrahedra whose 20 vertices are those of a regular dodecahedron (From Table IX, no.11, *Vielecke und Vielfläche*, Dr. Max Brückner, Leipzig, B.G. Teubner, 1900, as copied by M.C. Escher, 1950).

Symmetry helps unravelling this structure even if it of such a complexity that its properties are not self-evident, as in the case shown in figure 19.

Figure 19. M.E. Escher, *Circle Limit III*, 1959. Tessellation of the hyperbolic Poincare' plane. The symmetry group here is the tessellation group [6,4].
(© M.C. Escher Heirs, c/o Cordon Art, Baarn, Holland)

The future of symmetry in physics is thus still open.

References

[1] Pauli W 1926 *Z. Physik* **36** 336
[2] Gell-Mann M 1962 *Phys. Rev.* **125** 1067
Ne'eman Y 1962 *Nucl. Phys.* **26** 222
[3] Iachello F 1974 *Proc. Int. Conf. on Nuclear Physics (Amsterdam)* ed H P Blok and A E L Dieperink (Amsterdam: Scholar's Press) p 163
Arima A and Iachello F 1975 *Phys. Rev. Lett.* **35** 1069
Iachello F and Arima A 1987 The interacting boson model (Cambridge, Cambridge University Press)
[4] Iachello F 1981 *Chem. Phys. Lett.* **78** 581

Iachello F and Levine R D 1995 Algebraic theory of molecules (Oxford, Oxford University Press)
[5] Ramond P 1971 *Phys. Rev.* D **3** 2415
Neveu A and Schwartz 1971 *Nucl. Phys.* B **31** 86
Volkov D V and Akulov V P 1974 *Phys. Lett.* B **46** 109
Wess J and Zumino B 1974 *Nucl. Phys.* B **70** 39
[6] Kac V C 1975 *Funct. Anal. Appl.* **9** 91
Kac V C 1977 *Comm. Math. Phys.* **53** 31
[7] Witten E 1981 *Nucl. Phys.* B **188** 513
[8] Iachello F 1980 *Phys. Rev. Lett.* **44** 772
Balantekin B, Bars I and Iachello F 1981 *Phys. Rev. Lett.* **47** 19
[9] Iachello F and Van Isacker P 1991 The interacting boson-fermion model (Cambridge, Cambridge University Press)
[10] Metz A, Jolie J, Graw G, Hertenberger R, Gröger J, Günther C, Warr N, Eisermann Y 1999 *Phys. Rev.Lett.* **83** 1542
Metz A, Eiserman Y, Gollwitzer A, Hertenberger R, Vaniion B D, Graw G, Jolie J 2000 *Phys. Rev.* C **61** 064313

Journal of Physics: Conference Series **284** (2011) 012003 doi:10.1088/1742-6596/284/1/012003

Scaling Limit of the Noncommutative Black Hole

Shahn Majid[1]
Queen Mary University of London
School of Mathematics, Mile End Rd, London E1 4NS, UK
E-mail: s.majid@qmul.ac.uk

Abstract. We show that the 'quantum' black hole wave operator in the κ-Minkowski or bicrossproduct model quantum spacetime introduced in [1] has a natural scaling limit $\lambda_p \to 0$ at the event horizon. Here λ_p is the Planck time and the geometry at the event horizon in Planck length is maintained at the same time as the limit is taken, resulting in a classical theory with quantum gravity remnants. Among the features is a frequency-dependent 'skin' of some $\frac{\omega}{\nu}$ Planck lengths just inside the event horizon for $\omega > 0$ and just outside for $\omega < 0$, where ν is the frequency associated to the Schwarzschild radius. We use bessel and hypergeometric functions to analyse propagation through the event horizon and skin in both directions. The analysis confirms a finite redshift at the horizon for positive frequency modes in the exterior.

1. Introduction

First applications of quantum groups and noncommutative geometry to model quantum gravity effects have been around for some 2 decades now[2]. The work [3] proposed how noncommutative spacetime might tame infinities in physics while in 1994 we introduced the particular bicrossproduct or κ-Minkowski spacetime [4] which has since attracted a fair amount of attention, mainly because it 'almost commutative' and hence easier to work with. In this model the spatial generators x_i mutually commute but the time variable t does not. Rather,

$$[x_i, t] = \imath\lambda_p x_i \tag{1}$$

where λ_p is a real parameter. If the noncommutative geometry arises as a model of quantum gravity corrections to classical geometry then presumably $\lambda_p \sim 10^{-44}s$ as Planck time. The model has a bicrossproduct form of quantum Poincaré group and Section 2 provided a little more background to its construction.

In spite of its success, which includes the proposal of its testing by time of flight data being collected for gamma-ray bursts by FERMI-GLAST currently in orbit, the main criticism of the bicrossproduct model remains that it is rather special to flat spacetime, relying as it does on (quantum group) Fourier transform and plane waves[5]. This does not sit too well as a quantum gravity effect as it effectively applies in some weak field limit. In [1] we have proposed an answer to this criticism; a general construction of a noncommutative spacetime $C(M) \rtimes_\tau \mathbb{R}$ in which space is now any classical Riemannian 3-manifold $(M, \bar{g})$ equipped with a conformal Killing vector field τ and a static metric of classical form

$$\beta^{-1}\bar{\mathrm{d}}t\bar{\otimes}\bar{\mathrm{d}}t + \bar{g} \tag{2}$$

[1] The author was supported by a Leverhulme Senior Research Fellowship

where $\beta \in C(M)$ is a function on M. In our new theory the noncommutative differential geometry replaces the quantum Poincaré group as the guiding force, though the latter would be expected in the fullness of time to reemerge as the fiber of an associated affine quantum group frame bundle using the formalism of [6]. In fact the noncommutative geometry does not contain a full development of the noncommuative metric and noncommutative Riemannian geometry, although again this should emerge in the fullness of time. Rather, it takes a shortcut straight to the physically important noncommutative spacetime wave operator.

This theory constructs the noncommutative wave-operator on $C(M) \rtimes_\tau \mathbb{R}$ viewed as quantizing any static spacetime. This includes the Schwarzschild black hole. However, computations are quite formidable and hence in practice we work with a simplified version in which we use the flat bicrossproduct model spacetime (1) but 'bolt on' the nontrivial Riemannian 3-geometry by a process of minimal coupling. This allows us to treat the black hole using the flat spacetime bicrossproduct model as the ambient noncommutative space. The derivation in [1] is outlined briefly in Section 3 to the point of the noncommutative wave equation (6).

The present paper studies the black hole noncommutative wave operator in a certain scaling limit where distance ρ in Planck lengths from the event horizon is maintained while at the same time sending $\lambda_p \to 0$. This leads to a classical and much more amenable wave equation

$$\left(-2\left(\frac{\imath}{\nu}\frac{\partial}{\partial t} + \rho\ln(1 - \frac{\imath}{\rho\nu}\frac{\partial}{\partial t})\right) + \frac{\partial}{\partial\rho} + \rho\frac{\partial^2}{\partial\rho^2}\right)\psi = 0$$

where the parameter ν is the Schwarzschild frequency. This still retains quantum gravity effects and has the same qualitative behaviour around the horizon as the original theory in [1]. Our findings are summarized at the end of the paper, see Figure 3. The physical significance of the novel 'noncommutative skin' layer at the event horizon remains more fully to be understood. It is perfectly possible to read the paper starting at the scaling limit wave equation (7).

2. Background to the bicrossproduct model quantum spacetime

We first explain some of the background to the model. Indeed, quantum groups arose out of physics in two different classes. The first and most well known are the Drinfeld-Jimbo quantum groups $U_q(\mathfrak{g})$ associated to complex semisimple Lie algebras $\mathfrak{g}$ and coming out of integrable systems. Professor Jimbo at this conference rightly obtained the Weyl prize in part for that. The other less well-known class are the 'bicrossproduct' quantum groups introduced by the author as associated to Lie group factorisations and coming out of ideas for quantum gravity and Planck scale physics[2]. At the time Planck scale physics was considered unreachable by experiment but in recent years this view has changed and these bicrossproduct quantum groups have gained in interest particularly as deformations of Poincaré and other inhomogenenous non-simple groups.

The general construction here has semidirect product form $C(M) \blacktriangleright\!\!\triangleleft U(\mathfrak{g})$ where M is a Lie group on which a Lie algebra $\mathfrak{g}$ acts and where M acts back on g and on the set of its associated group G so as to form a 'matched pair'. The latter 'back-reaction' induces a semdirect coproduct or, equivalently, a dual quantum group $U(\mathfrak{m}) \triangleright\!\!\blacktriangleleft C(G)$, where $\mathfrak{m}$ is the Lie algebra of M, making everything symmetric between the two factors. There is also a canonical action of the first quantum group on $U(\mathfrak{m})$ and its of dual on $U(\mathfrak{g})$, which are therefore naturally arising noncommative geometries with these quantum groups as symmetries. Such data arise from the local factorisation of any group X into GM where neither subgroup need be normal, a very common occurrence in mathematics and physics.

As well as the general construction, including specifics for the Iwasawa factorisation, my 1988 PhD thesis explicitly covered[2, 7, 8] the case $SO(1,3) \approx SO(3)\mathcal{H}_3$ where $\mathcal{H}_3 = \mathbb{R}^2 \rtimes \mathbb{R}$ is a curved (nonAbelian) version of $\mathbb{R}^3$. Its Lie algebra h_3 has relations (1) for $i = 1, 2$, $t = x_3$ providing

Journal of Physics: Conference Series **284** (2011) 012003 doi:10.1088/1742-6596/284/1/012003

a noncommutative space on which the bicrossproduct quantum group $C(\mathcal{H}_3) \blacktriangleright\!\!\triangleleft U(so_3)$ acts. The 4D case studied a bit later in [4] is similarly based on $SO(2,3) \approx SO(1,3)\mathcal{H}_{1,3}$ where the Lie algebra $h_{1,3}$ again has the relations (1) but with $i = 1, 2, 3$. The associated bicrossproduct quantum Poincaré group $C(\mathcal{H}_{1,3}) \blacktriangleright\!\!\triangleleft U(so_{1,3})$ was shown to be isomorphic as an abstract Hopf algebra to a 'κ-Poincaré quantum group' obtained by contraction of $U_q(so(2,3))$ in [9] but had a different and inequivalent interpretation of the generators and therefore should not be confused with that. Moreover, the bicrossproduct model not only provided the correct interpretation of the Hopf algebra but provided the quantum spacetime (1) on which it acts as the enveloping algebra $U(h_{1,3})$. This is the bicrossproduct spacetime or 'κ-Minkowski' model introduced by the author and H. Ruegg[4].

Since then, the bicrossproduct model quantum spacetime has attracted a great deal of attention, not least due to its testable variable speed of light prediction[5]. That there might be such a prediction had been hoped for on the basis that the deformed relations of the bicrossproduct Poincaré algebra involve a deformed form of the Casimir, but this of itself has no meaning as one can just rename the generators; merely calling a generator P_0 or P_i does not make it energy or momentum. However, such speculations were turned into an actual prediction in [5] by analysing the associated wave operator on noncommutative spacetime and its action on noncommutative plane waves $e^{\imath \vec{k}\cdot x}e^{\imath \omega t}$. We made the reasonable assumption that physical plane waves would, in this model of quantum gravity effects, be identified with these normal ordered mathematical plane waves.

Another feature of the bicrossproduct family is that the classical data of a matched pair of actions is entirely classical, consisting in the Poincaré case of a certain non-linear action of the Lorentz group on the nonAbelian momentum group $\mathcal{H}_{1,3}$ and vice versa. As a result the orbits of the Lorentz group in momentum space are classical but 'squashed' into regions with limiting surfaces or accumulation boundaries. This phenomenon is identical to that first discovered in the $iso(3)$ case [2, 7, 8], where the spherical orbits were deformed into non-concentrically nested spheres squashed into a region of $\mathbb{R}^3$; in the $iso(1,3)$ case the mass hyperbolae are similarly squashed into a cylinder $|\vec{k}| < \hbar/(c\lambda_p)$. In recent years some authors have sought to rebrand the bicrossproduct model as 'doubly special' because of this remarkable feature of squashed orbits and the feature of being able to refer everything to classical nonlinear equations, but without actually producing any genuinely new model with quantum group symmetry. Both features are in fact typical of bicrossproduct models with noncompact factorising groups if one wants other examples, and I refer to [10] for a recent review. Of perhaps more interest is how the bicrossproduct model emerges as a limit of a q-deformed geometry and how this in turn emerges from quantum gravity with cosmological constant, matters which are explored in the 2+1 case in [11].

3. Background to the noncommutative black hole wave operator

Given a manifold M, a metric $\bar{g}$ with Levi-Civita connection $\bar{\nabla}$ and a conformal Killing vector field τ, we define in [1] a noncommutative spacetime coordinate algebra $C(M) \rtimes_\tau \mathbb{R}$ consisting of elements of the form $\psi = \sum \psi_n t^n$ where $\psi_n \in C(M)$ by which we mean, eg, smooth functions on M, and product defined via commutation relations

$$[f, t] = \imath\lambda_p\tau(f), \quad \forall f \in C(M).$$

One can think of this as a cross product of $C(M)$ by the group $\mathbb{R}$ with action defined by τ and at least when M is compact make everything precise as a cross product C^*-algebra. Next, given any pointwise invertible function $\beta \in C(M)$ we define a differential calculus as, by definition, a space of 1-forms $\Omega^1(C(M) \rtimes_\tau \mathbb{R})$ spanned by usual 1-forms from M along with two extra 1-forms

$\mathrm{d}t, \theta'$, with the commutation relations between functions and differentials[1]

$$\begin{array}{rcl} [f,g]=0, & & [f,t]=\imath\lambda_p\tau(f), \quad [\mathrm{d}f,g]=\imath\lambda_p\bar{g}^{-1}(\bar{\mathrm{d}}f,\bar{\mathrm{d}}g)\theta' \\ {[\theta',f]=0,\quad [\theta',t]} & = & \alpha\imath\lambda_p\theta', \quad [\mathrm{d}f,t]=\imath\lambda_p(\mathrm{d}\tau(f)-\mathrm{d}f) \\ {[f,\mathrm{d}t]} & = & \imath\lambda_p\mathrm{d}f, \quad [\mathrm{d}t,t]=\imath\lambda_p(\beta\theta'-\mathrm{d}t). \end{array} \quad (3)$$

for all $g, f \in C(M)$, where $\alpha = \frac{2}{n}\mathrm{div}(\tau) - 1$ if M has dimension n. One of the Jacobi identities will quickly lead you to see that you need τ to be a conformal Killing vector. We do rather more in [1], building Ω^1 directly on the space of classical 1-forms $\bar{\Omega}^1(M)$ extended by two extra dimensions $\theta', \mathrm{d}t$. The general principle of defining a differential structure via a bimodule of 1-forms (or an entire differential graded algebra) is common to most approaches to noncommutative geometry[13].

We also consider 2-forms $\Omega^2(C(M) \rtimes \mathbb{R})$ and show in [1] that the calculus is locally inner, i.e. in the neighbourhood of any point one can find a function $h \in C(M)$ such that

$$\theta = \mathrm{d}t - h\theta', \quad [\psi, \theta] = \imath\lambda_p\mathrm{d}\psi, \quad \{\omega, \theta\} = -\imath\lambda_p\mathrm{d}\omega$$

for all ψ in the noncommutative coordinate algebra and all noncommutative 1-forms ω. Finally, we show that for all normal ordered $\psi \in C(M) \rtimes_\tau \mathbb{R}$ the exterior derivative has the form

$$\mathrm{d}\psi = \bar{\mathrm{d}}\psi + \partial_0\psi\mathrm{d}t + \imath\frac{\lambda_p}{2}\theta'(\bar{\Delta}_{LB} - \frac{1}{2}\beta^{-1}\bar{g}^{-1}(\bar{\mathrm{d}}\beta))\psi + \imath\lambda_p\Delta_0\psi\theta'$$

where $\bar{\mathrm{d}}$ is the classical spatial exterior derivative, ∂_0 is the finite difference operator

$$\partial_0\psi(t) = \frac{\psi(t) - \psi(t - \imath\lambda_p)}{\imath\lambda_p}$$

and $\bar{\Delta}_{LB}$ is the classical Laplace-Betrami operator on M. We use the inverse metric to view the classical 1-form $\bar{\mathrm{d}}\beta$ as a vector field, which we regard as a differential operator. Finally $2\Delta_0$ is a deformation of the Laplacian or double-derivative in the t direction. We define the spacetime wave operator by

$$\mathrm{d}\psi = \bar{\mathrm{d}}\psi + \partial_0\psi\mathrm{d}t + \imath\frac{\lambda_p}{2}\Box\psi\theta'$$

in keeping with a philosophy that the extra dimension θ' plays the role of the conjugate to the wave operator regarded as a 'noncommutative vector field'.

One can apply this to the spherically symmetric 3-geometry $(M, \bar{g})$ underlying the Schwarzschild black hole. Here $\bar{g} = h(r)^2\bar{\mathrm{d}}r\bar{\otimes}\bar{\mathrm{d}}r + r^2\bar{\mathrm{d}}^2\Omega$ where

$$h(r)^2 = \frac{1}{1 - \frac{\gamma}{r}}, \quad \gamma = \frac{2GM}{c^2}$$

for a black hole of mass M centred at the origin. Then the classical black hole has the form of our static metric with

$$\beta = -\frac{1}{c^2(1 - \frac{\gamma}{r})}. \quad (4)$$

Moreover, the 3-geometry has a conformal killing vector field $\tau = \frac{r}{h}\frac{\partial}{\partial r}$. Hence by the above we have a quantum spacetime and a wave operator on it deforming that of a classical black hole.

As this wave operator and noncommutative geometry is a little hard to compute, we also propose[1] and in practice work with a more down to earth version of the theory where we let $M = \mathbb{R}^3$ with its conformal Killing vector field $\tau = r\frac{\partial}{\partial r}$. Then the spacetime is the usual flat spacetime bicrossproduct model with relations (1). But for the calculus we take β as in (4) which is no longer the 5D flat space calculus[12] for this model. Its wave operator is then not the flat space one used in [5] but nor is it quite the black hole as we took a flat 3-geometry. To fix this we use the correct $\bar\Delta_{LB} - \frac{1}{2}\beta^{-1}\bar g^{-1}(\bar d\beta)$ in place of the one computed from the flat metric. We call this the 'minimally coupled black hole' as we basically adapted a flat space computation by replacing derivatives by covariant ones. Explicitly,

$$\Box_{BH}\psi(t) = 2\Delta_0\psi(t) + \left((\frac{2}{r}-\frac{\gamma}{r^2})\frac{\partial}{\partial r} + (1-\frac{\gamma}{r})\frac{\partial^2}{\partial r^2} + e_i e_i\right)\psi(t+\imath\lambda_p) \tag{5}$$

is our 'minimally coupled' noncommutative black hole wave operator. Here $e_i = \frac{\partial}{\partial x_i} - \frac{x_i}{r}\frac{\partial}{\partial r}$ is the 2-sphere covariant derivative such that $e_i e_i = -\frac{l(l+1)}{r^2}$ on the spherical harmonics Y^l_m. The nontrivial part is to compute Δ_0. Let $L\psi(t) = \psi(t+\imath\lambda_p)$. Then[1]

$$\Delta_0 = \frac{L}{c^2\lambda_p^2}\left(-X + \frac{X^2}{2} + (1-\frac{\gamma}{r})(X - \frac{\gamma}{r}\ln(1-\frac{X}{1-\frac{\gamma}{r}}))\right), \quad X = \imath\lambda_p\partial_0.$$

If we denote the expression in large parentheses as $\mathcal{D}(X)$ then the noncommutative wave equation becomes[1]

$$\left(\frac{2}{c^2\lambda_p^2}\mathcal{D}(\imath\lambda_p\partial_0) + (\frac{2}{r}-\frac{\gamma}{r^2})\frac{\partial}{\partial r} + (1-\frac{\gamma}{r})\frac{\partial^2}{\partial r^2} + e_i e_i\right)\psi = 0 \tag{6}$$

on normal-ordered ψ with t-dependence to the right. We shall study this wave equation under the physical hypothesis that observed fields are to be identified in our model with normal ordered fields as a measure of quantum gravity corrections.

Note that as $X\to 0$ we have $\mathcal{D}(X) = \frac{X^2}{2(1-\frac{\gamma}{r})} + O(X^3)$ so we obtain $-\frac{1}{c^2(1-\frac{\gamma}{r})}\partial_0^2$ in this limit. This has the correct classical limit for the black hole wave equation. Also in the limit $r\to\infty$ we have $\mathcal{D}(X) = -\frac{X^2}{2} + O(\frac{\gamma}{r})$ so we obtain the standard bicrossproduct flat space wave operator used in [5], which replaces time derivatives by the finite difference ones ∂_0.

However, the function $\mathcal{D}(X)$ also has a nice limit $\mathcal{D}(X) = -X + \frac{X^2}{2} + O(r-\gamma)$ as $r\to\gamma$, i.e. close to the event horizon. So the noncommutative geometry as visible in the wave operator smooths out the coordinate singularity normally occurring there; in some sense it makes it one dimension better as a log in place of a $1/(r-\gamma)$. It similarly improves matters at the other boundary $r = \gamma e^{-\omega\lambda_p}$ which is the other side if an interregnum or 'skin' region in which the logarithm has a negative argument. We will take a close look at both boundaries by means of a scaling limit. The geometrical picture that we refer to is shown in Figure 1 for positive and negative ω.

4. Solutions of the wave equation near the classical event horizon

To study this limit $r\to\gamma$ further, we set

$$r - \gamma = l_p\rho$$

where $l_p = c\lambda_p$ is the Planck length and ρ is the distance from the horizon in Planck units. Then

$$\frac{X}{1-\frac{\gamma}{r}} = \imath(\frac{\gamma + l_p\rho}{c\rho})\partial_0.$$

Hence rl_p times the wave equation (6) becomes

$$\left(\frac{2(\gamma+l_p\rho)}{l_p}\mathcal{D}(\imath(\frac{\gamma+l_p\rho}{c\rho})\partial_0) + (1+\frac{l_p\rho}{\gamma+l_p\rho})\frac{\partial}{\partial\rho} + \rho\frac{\partial^2}{\partial\rho^2} + l_p(\gamma+\rho l_p)e_i e_i\right)\psi = 0.$$

Journal of Physics: Conference Series **284** (2011) 012003 doi:10.1088/1742-6596/284/1/012003

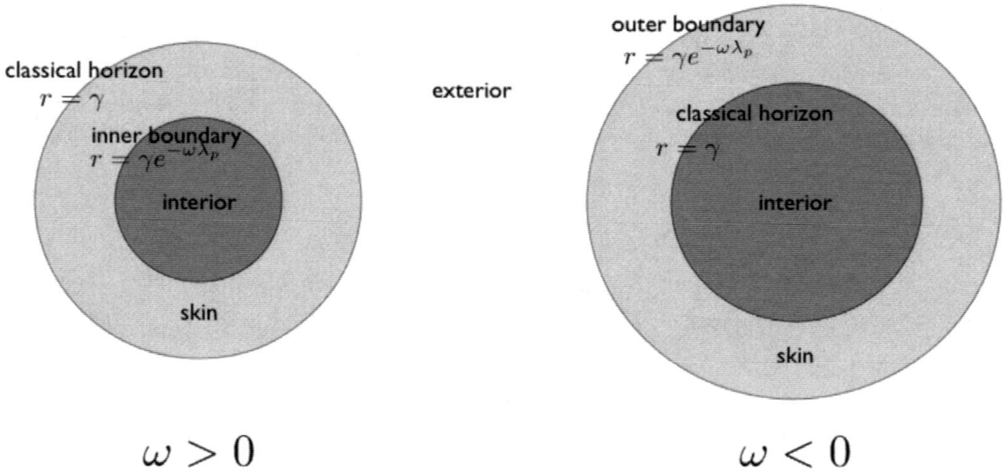

Figure 1. Picture of the noncommutative black hole. The coordinate singularity at the event horizon is resolved into two boundaries with a frequency dependent 'interregnum' or skin between them of depth $\gamma|e^{-\omega\lambda_p} - 1|$ where γ is the Schwarzschild radius. The picture is different for positive and negative frequency ω modes.

Although l_p is what it is (about $10^{-33}cm$) we can simplify this equation from a mathematical point of view by setting $l_p \to 0$. This should be viewed as no more than a certain scaling limit approximation in which some quantum gravity effects are ignored but a certain remnant remains. In this limit the wave equation in these coordinates becomes

$$\left(-2\left(\frac{\imath}{\nu}\frac{\partial}{\partial t} + \rho\ln(1 - \frac{\imath}{\rho\nu}\frac{\partial}{\partial t})\right) + \frac{\partial}{\partial\rho} + \rho\frac{\partial^2}{\partial\rho^2}\right)\psi = 0 \tag{7}$$

if we assume bounded angular momentum so that $l_p e_i e_i \to 0$. Here

$$\nu = \frac{c}{\gamma}$$

is, roughly speaking, the frequency of a wave of wavelength the size of the black hole, as γ is the Schwarzschild radius.

We look at ψ with time-dependence $e^{\imath\omega t}$ (separating variables). Then the equation (7) becomes for each mode of frequency ω,

$$\left(-2\left(-\frac{\omega}{\nu} + \rho\ln(1 + \frac{\omega}{\rho\nu})\right) + \frac{\partial}{\partial\rho} + \rho\frac{\partial^2}{\partial\rho^2}\right)\psi = 0 \tag{8}$$

where ψ is now a function just of ρ. Remembering the original physical interpretation of ρ, we see that for a given sign of ω there is a strip of width $|\omega|/\nu$ Planck lengths where the argument of the logarithm is negative. Specifically, this happens if

$$\omega > 0: \quad -\frac{|\omega|}{\nu} < \rho < 0, \quad \omega < 0: \quad 0 < \rho < \frac{|\omega|}{\nu}.$$

Thus we have qualitatively the same kind of 'quantum skin' or 'interregnum' region in our scaling limit as for the original theory in Figure 1, but now with width $|\omega|/\nu$. This lies just inside the

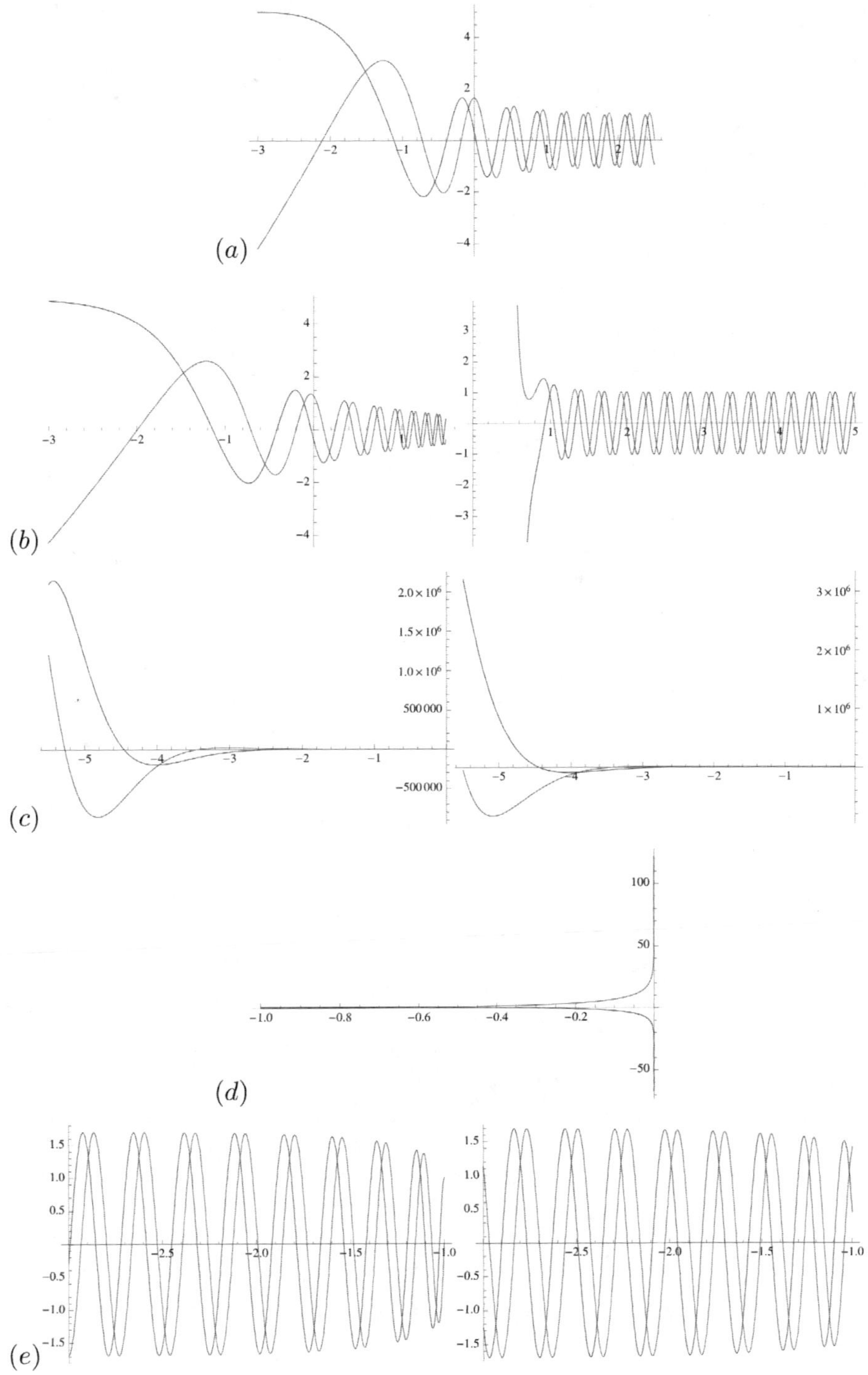

Figure 2. Scaling limit solutions against distance ρ Planck lengths from event horizon. (a) Numerical solution for $\frac{\omega}{\nu} = 10$ with differing boundary conditions from $\rho = 10^5$ to $\rho = 10^{-3}$ on log scale (b) Comparison with real and imaginary parts bessel solution for $\rho >> \frac{\omega}{\nu} = 10^1$ (right) and $10^{-3} < \rho << \frac{\omega}{\nu}$ (left). (c) Real and imaginary parts numerical solution (left) and comparison with hypergeometric solution (right) for $\frac{\omega}{\nu} = 5.5$ in 'skin' $-\frac{\omega}{\nu} < \rho < 0$ driven by real boundary condition at the horizon, on linear scale. (d) Ditto numerical solution but for $\frac{\omega}{\nu} = 1$ driven by boundary condition at the interior boundary. (e) Numerical solution (left) for $\frac{\omega}{\nu} = 10$ with differing boundary conditions from $\rho = -10^1$ to $\rho = -10^3$, and comparison (right) with $|\rho| >> \frac{\omega}{\nu}$ bessel solution, on log scale.

event horizon for positive frequencies and just outside it for negative frequencies. As long as we stay out of these regions, we have real solutions. We use MATHEMATICA to compute numerical solutions, but our goal is to gain an analytic understanding of them in different regimes. We focus on $\omega > 0$ but return to the negative frequency case at the end.

(i) $\rho >> \frac{\omega}{\nu}$ *Far outside the event horizon* If we assume that

$$\frac{|\omega|}{\nu} << \rho \tag{9}$$

then we can expand the logarithm. At lowest nontrivial order we obtain the conventional wave equation in unusual coordinates

$$\left(-\frac{1}{\rho\nu^2}\frac{\partial^2}{\partial t^2} + \frac{\partial}{\partial\rho} + \rho\frac{\partial^2}{\partial\rho^2}\right)\psi = 0.$$

Thus, if we write $\rho = e^{\frac{x}{\gamma}}$ then the wave equation has solutions

$$\psi_{\pm}(t,\rho) = e^{\imath\omega(t\pm\frac{x}{c})} = e^{\imath\omega t}\rho^{\pm\imath\frac{\omega}{\nu}}.$$

If we do a little better and expand to the next cubic order, the wave equation takes the approximate form

$$\left(-\frac{1}{\rho\nu^2}\frac{\partial^2}{\partial t^2} - 2\frac{\imath}{3\rho^2\nu^3}\frac{\partial^3}{\partial t^3} + \frac{\partial}{\partial\rho} + \rho\frac{\partial^2}{\partial\rho^2}\right)\psi = 0. \tag{10}$$

This has solutions

$$\psi_{\pm}(t,\rho) = e^{\imath\omega t} I_{\pm 2\imath\frac{\omega}{\nu}}\left(\sqrt{\frac{8\omega^3}{3\rho\nu^3}}\right)\Gamma\left(1 \pm 2\imath\frac{\omega}{\nu}\right)$$

where $I_n(z)$ is a bessel I function but extended to complex n and the Γ function is a normalisation. In another context one might consider taking $\imath\omega = \omega'$ real in which case note the poles in the normalisation factor of $\psi_{\pm}$ at $\mp$ positive integer $\frac{2\omega'}{\nu}$. Figure 2(b) on the right confirms, however, that the case of real ω is wave-like and indeed asymptotes to the plane waves just found when ρ is large. Note that the plot is against $\log\rho$ with $\frac{\omega}{\nu} = 10$. The approximation (9) means that we can only approach the horizon to ω/ν Planck lengths before quantum gravity effects render the weak field (i.e. small γ or large ν) expansion invalid. Thus, the right side of Figure 2(b) is only valid for $\rho >> 10^1$ and this is confirmed on comparison with the numerical solution in Figure 2(a).

(ii)$0 < \rho << \frac{\omega}{\nu}$ *Just outside the event horizon* The numerical solution in Figure 2(a) continues down to $\rho = 10^{-3}$ or so on the log plot. At some point we enter the regime

$$|\rho| << \frac{\omega}{\nu} \tag{11}$$

where the logarithm is suppressed and the wave equation becomes

$$2\frac{\imath}{\nu}\frac{\partial}{\partial t}\psi = \left(\frac{\partial}{\partial\rho} + \rho\frac{\partial^2}{\partial\rho^2}\right)\psi. \tag{12}$$

As a result the wave stops being so exponentially stretched in the direction of increasing ρ (it begins to be stretched in the logarithmic plot) and starts to behave as a linear combination of

$$\psi \sim J_0\left(\sqrt{\frac{8\rho\omega}{\nu}}\right),\quad Y_0\left(\sqrt{\frac{8\rho\omega}{\nu}}\right)$$

where J_0, Y_0 are bessel J and Y functions. These have respectively a finite value and zero slope at the event horizon $\rho = 0$ or an infinite value and finite slope at $\rho = 0$ in the logarithmic plot. Thus we conclude that the equation (7) has a full set of solutions outside the event horizon becoming exponentially stretched plane waves for large ρ and tending to either a constant or a constant $\rho\frac{\partial}{\partial\rho}\psi$ (the slope in the logarithmic plot) at the event horizon $\rho = 0$. A generic boundary condition set at large ρ will stimulate the latter Y_0 mode and result in a log divergence at the horizon, so this is presumably the fate of an infalling wave. However, a wave created at the event horizon (which would be relevant to Beckenstein-Hawking radiation) could take the form of the regular J_0 mode.

(iii)$-\frac{\omega}{\nu} < \rho < 0$ *In the skin layer* We now look inside the event horizon. Initially, for small negative ρ we are in the same regime as in (ii), except that there is now a term $-2\pi\imath\rho$ term in the wave operator if we take the standard choice of branch of the logarithm. This system can be solved using hypergeometric functions as

$$\psi \sim e^{\rho\sqrt{2\pi\imath}}{}_1F_1\left(\frac{1}{2}+\frac{\omega}{\sqrt{2\pi\imath}},1,-2\rho\sqrt{2\pi\imath}\right), e^{\rho\sqrt{2\pi\imath}}F_U\left(\frac{1}{2}+\frac{\omega}{\sqrt{2\pi\imath}},1,-2\rho\sqrt{2\pi\imath}\right)$$

where F_U is the confluent hypergeometric function. We focus for the moment on ${}_1F_1$ which has a finite value and zero slope at the event horizon and can therefore match the J_0 (bessel J) on the other side. Its continuation further into the 'skin' region acquires an imaginary part and both parts grow rapidly as we go deeper into the skin (more negative ρ). The values at $\rho = -\frac{\omega}{\nu}$ are (according to MATHEMATICA) given by a Laguerre L function and appear to grow very rapidly with $\frac{\omega}{\nu}$. In Figure 2(c) we show this analytic approximate solution on the right and for comparison the full numerical solution, which is similar. The boundary condition is a real value 1 and zero slope at or close to $\rho = 0$ and we have reduced to $\frac{\omega}{\nu} = 5.5$ in order to have a visible comparison at a point where significant 'amplification' starts to appear.

For solutions driven by real conditions at (or rather, just inside) the interior boundary $\rho = -\frac{\omega}{\nu}$ the numerical solution is shown in Figure 2(d) for a modest $\frac{\omega}{\nu} = 1$. In general a significant visible amplification in the numerical solution (say a factor of some 10^6) sets in around $\frac{\omega}{\nu} = 4$ and a comparable imaginary part is similarly acquired, i.e. much as we had going the other way. However, close inspection of the numerical solution appears to indicate that even for the example shown, there is in fact a log divergence at $\rho = 0$. (This is more visible in the middle scenario in our later Figure 3 where we have used a log scale so that a log divergence is a straight line and where we have suppressed a very large imaginary component in the plot allowing us to see a bit more detail than in Figure 2(d).) This brings us back to the hypergeometric mode F_U. This indeed has a log divergence in its real and imaginary part and a similar form in the region $-\frac{\omega}{\nu} << \rho < 0$. Even though the approximation used breaks down as we approach $\rho = -\frac{\omega}{\nu}$, as confirmed by the detail in Figure 3 mentioned, a generic boundary condition there can be expected to propagate into the region where the approximation (11) is valid and thereafter drive this mode to produce a log divergence. The divergence can be matched in real slope at $\rho = 0$ on the log plot to that of a Y_0 (bessel Y) mode outside, although the role of the similarly divergent imaginary part is not entirely clear. It is presumably related to the logarithmic asymptotic form. Overall, we see that there is a significant asymmetry in the modes crossing in the 'skin layer' from the black hole interior, which appear to diverge as F_U just inside the horizon, and J_0 modes crossing from outside the event horizon which are regular at the horizon and match to ${}_1F_1$ modes inside.

(iv)$\rho << -\frac{\omega}{\nu}$ *Black hole interior* As we continue further with negative ρ we return again to the regime $|\rho| >> \frac{\omega}{\nu}$ and we are back to (10), this time with $\rho < 0$. If we change variables to $-\rho$ we have the same equation but with $-\omega$. Hence the cubic correction enters with the opposite sign. We have the same solutions Bessel functions I_n with imaginary n as before but now an $\imath$ in the argument. As Figure 2(e) on the right confirms, we still have wave-like

Journal of Physics: Conference Series **284** (2011) 012003 doi:10.1088/1742-6596/284/1/012003

behaviour and these tend for large negative ρ to plane waves in log plot. The difference is that whereas the bessel function in regime (i) showed a slight stretching of the waves in the log plot compared to large ρ, now we see a slight compression in the log plot compared to large negative ρ. We again see a good fit with the numerical solutions, shown for real boundary conditions of finite amplitude/zero derivative, or vice-versa. We have returned to $\frac{\omega}{\nu} = 10$ and have matched normalisation at large negative ρ.

(v)*Negative frequencies* We have focussed on $\omega > 0$. If we suppose instead that $\omega < 0$ then note in (8) that we can change back to $\omega > 0$ if we also change ρ to $-\rho$. Hence the story for $\omega < 0$ is exactly the same as above but with a left-right reflection in all of Figure 2. This time a mode entering from far outside the event horizon is slightly (but finitely) compressed compared to the overall frequency-independent compression expressed in the log plot, and diverges just above the classical event horizon. Meanwhile, a mode entering from the black hole 'interior' is slightly (but finitely) stretched in the log plot, crosses the classical event horizon without coordinate singularity and is amplified as it passes out the other side.

All of this behaviour is remarkably similar to that of the full theory in [1], suggesting that our simplified scaling limit wave equation (7) shares many qualitative features with the full wave equation. The only thing missing is that we cannot investigate 'standing waves' in the black hole interior. This is because our scaling limit covers only the geometry near the horizon and a fixed number of Planck lengths from it (this includes the skin region). In effect, on entering the black hole interior the wave solutions in Figure 2(e) continue on towards the origin but this is considered as infinitely far away for a macroscopic black hole. On the other hand, one could consider a different scaling limit in which $\gamma = nl_p$ and n is fixed as $l_p \to 0$ to analyse Planck size black holes in the model. This will be looked at elsewhere.

Meanwhile, the propagation in the two directions across the event horizon is summarised in Figure 3. Because the orbital angular momentum was washed out in our scaling limit and because we are far from the centre of the black hole, the system is more like propagation in a line. More precisely, two lines laid end to end. Thus, one could change variables to $x = \gamma \log \rho$ for $\rho > 0$ and $y = -\gamma \log(-\rho)$ for $\rho < 0$ and arrive at a more conventional $\frac{\partial^2}{\partial x^2}$ or $\frac{\partial^2}{\partial y^2}$ for the spatial part. We have refrained from this only in order to keep the original context. In the log plot in Figure 3 the interior and exterior propagations each span a factor of 10^6 in ρ and are set at $\frac{\omega}{\nu} = 10$. However, this would result in extremely high 'amplification' factors in the skin region and hence for purposes of illustration only, we have set $\frac{\omega}{\nu} = 1.6$ for solution in the skin region. The boundary conditions are generally mixed ones setting amplitude and gradient, with the arrows indicating the location of the boundary condition and solving the equations from there, rather than indicating 'direction of travel', which we have not particularly analysed. Although it does not appear to be too singular the equation at $\rho = -\frac{\omega}{\nu}$ is numerically ill-conditioned and as a result the boundary conditions are set just inside the skin region rather than right on the boundary.

We should also say that the physical interpretation of all of these features remains problematic. We have not discussed here issues such as unitarity (this is likely to be a problem) but it should be remembered that the entire noncommutative model is intended as an effective description of quantum gravity effects rather than a self-contained closed system. However, at least at first sight it would appear that waves from infinity are likely to log diverge at the classical event horizon $\rho = 0$. We have seen that one can also create modes at the event horizon which are regular there and extend to waves outside. Inside the skin they grow rapidly and acquire an imaginary component, and the (now complex) solution appears to be matchable to waves continuing into the interior. This is shown in the lower part of Figure 3. Meanwhile, in the middle of Figure 3 we show waves in the interior which appear to be matchable to a solution inside which becomes complex and has a log divergence at $\rho = 0$, this time on the inside of the

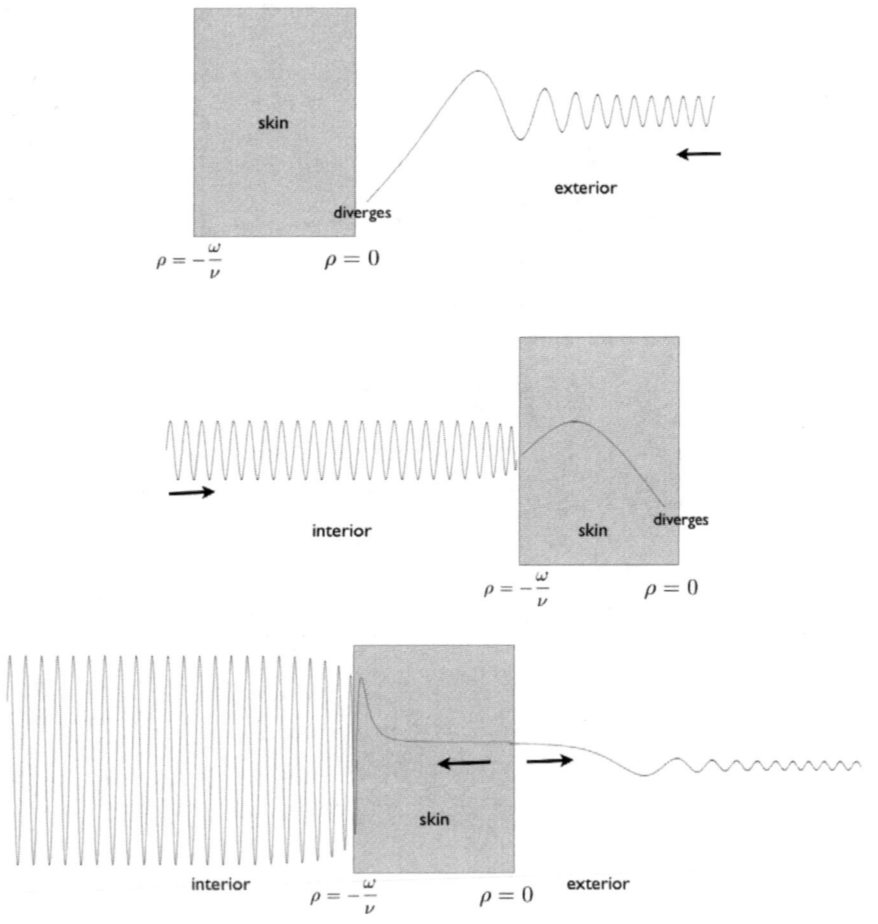

Figure 3. Pastice of black hole wave solutions in scaling limit at the classical event horizon $\rho = 0$ for $\omega > 0$ on a log plot. The system is driven by the location of the boundary conditions as indicated by the arrow. In the interior and exterior regions we have set $\frac{\omega}{\nu} = 10$ but have artificially reduced this in the skin regions to $\frac{\omega}{\nu} = 1.6$ in order to have a viable diagram with visible detail, and also note that solutions in the skin region acquire a comparable imaginary part which we omit for clarity. At top, we show a typical mode set far in the exterior, resulting in a log divergence but finite redshift just outside the horizon $\rho = 0$. In middle we show a mode set in the 'far' interior, resulting in a log divergence just inside the horizon. At bottom we show solutions set at the horizon. The pictures for $\omega < 0$ are the same with 'exterior' and 'interior' interchanged, but note that the boundary condition illustrated would still be driving the skin region from the classical event horizon which is now just below the skin of the black hole.

classical event horizon. All of this is for positive frequency with a reversed picture for negative frequency. This represents a significant time-asymmetry, which is to be expected. Whether the log divergences at $\rho = 0$ are important is not clear as they could be artefacts of the scaling limit; in the actual quantum gravity model presumably other physical effects would take over.

References

[1] Majid S 2010, Almost commutative Riemannian geometry, I: wave operators, arXiv:1009.2201 (math.QA), 50pp.
[2] Majid S 1988, Hopf algebras for physics at the Planck scale, J. Class. Quant. Gravity 5, 1587-1607.
[3] Majid S 1990, On q-regularization, Int. J. Modern Physics A. 5, 4689-4696
[4] Majid S& Ruegg H 1994, Bicrossproduct structure of the κ-Poincare group and non-commutative geometry, Phys. Lett. B. 334, 348-354
[5] Amelino-Camelia G & Majid S 2000, Waves on noncommutative spacetime and gamma-ray bursts, Int. J. Mod. Phys. A15, 4301-4323.
[6] Majid S 1999, Quantum and braided group Riemannian geometry, J. Geom. Phys. 30, 113-146
[7] Majid S 1990, Matched pairs of Lie groups associated to solutions of the Yang-Baxter equations, Pacific J. Math 141, 311-332.
[8] Majid S 19991, Hopf-von Neumann algebra bicrossproducts, Kac algebra bicrossproducts, and classical Yang-Baxter equations, J. Functional Analysis 95, 291-319.
[9] Lukierski J, Nowicki A, Ruegg H, & Tolstoy VN 1991, q-Deformation of Poincaré algebra. Phys. Lett. B, 268, 331-338.
[10] Majid S 2009, Algebraic approach to quantum gravity II: noncommutative spacetimes, in Approaches to Quantum Gravity, ed. D. Oriti. C.U.P. 466-492
[11] Majid S & Schroers B 2009, q-Deformation and semidualisation in 3D quantum gravity, J. Phys A 42, 425402 (40pp)
[12] Sitarz A 1995, Noncommutative differential calculus on the κ-Minkowski space, Phys.Lett. B 349, 42-48.
[13] Connes A 1994, *Noncommutative Geometry*, Academic Press.

Journal of Physics: Conference Series **284** (2011) 012004 doi:10.1088/1742-6596/284/1/012004

Cosmic acceleration and the challenge of modifying gravity

Mark Trodden

Center for Particle Cosmology, Department of Physics and Astronomy, University of Pennsylvania, Philadelphia, Pennsylvania 19104, USA

E-mail: trodden@physics.upenn.edu

Abstract. I briefly discuss the challenges presented by attempting to modify general relativity to obtain an explanation for the observed accelerated expansion of the universe. Foremost among these are the questions of theoretical consistency - the avoidance of ghosts in particular - and the constraints imposed by precision local tests of gravity within the solar system. For those models that clear these highly constraining hurdles, modern observational cosmology offers its own suite of tests, improving with upcoming datasets, that offer the possibility of ruling out modified gravity approaches or providing an intriguing hint of new infrared physics. In the second half of the talk, I discuss a recent approach to extracting cosmology from higher-dimensional induced gravity models.

1. Introduction

The discovery of late-time cosmic acceleration has led cosmologists to carefully examine the possible contributions to the mass-energy of the universe that might source this behavior within general relativity (GR). Perhaps more provocatively, a second possibility has also been considered, namely that GR itself may not provide the correct set of rules with which to understand how the known matter and radiation content affects the universe on the largest scales. It may be that curvatures and length scales in the observable universe are only now reaching values at which an infrared modification of gravity can make itself apparent by driving self-acceleration (for reviews see [1, 2, 3, 4]).

General relativity is very well tested in the solar system, in measurements of the period of the binary pulsar, and in the early universe, via primordial nucleosynthesis. None of these tests, however, probes the ultra-large length scales and low curvatures characteristic of the Hubble radius today. It is therefore *a priori* conceivable that gravity is modified in the very far infrared, in such a way that the universe begins to accelerate at late times.

In practice, however, as I will describe below, it is difficult to construct a simple model that embodies this hope. A straightforward possibility is to modify the usual Einstein-Hilbert action by adding new covariant terms constructed from the scalar invariants of the theory. Such theories can lead to late-time acceleration, but unfortunately typically lead to one of two problems. Either they are in conflict with tests of GR in the solar system, due to the existence of additional dynamical degrees of freedom, or they contain ghost-like degrees of freedom that seem difficult to reconcile with fundamental theories. Nevertheless, a restricted class of such theories remain viable, and should be further constrained by upcoming cosmological missions.

A more dramatic strategy is to imagine that we live on a brane embedded in a large extra dimension. Although such theories can lead to perfectly conventional gravity on large scales, it is also possible to choose the dynamics in such a way that new effects show up exclusively in the far infrared. Such theories can naturally lead to late-time acceleration, but may have strong-coupling or ghost issues. Nevertheless, these models hold out the possibility of having interesting and testable predictions that distinguish them from models of dynamical dark energy.

In this talk, delivered at the 28th International Colloquium on Group Theoretical Methods in Physics (ICGTMP), I attempted to provide an overview of theoretical approaches to this problem, and to describe some of the challenges, both theoretical and observational, faced by attempts to address cosmic acceleration in this way. The presentation was intended to explain the key ideas and to highlight some very recent work of my own. For this reason, and because this writeup is necessarily brief, I have chosen to reference only selected review articles, those papers to which I directly referred in the talk itself, and several that came out directly afterwards..

2. Modifying Gravity

Although, within the context of General Relativity (GR), one doesn't think about it too often, the metric tensor contains, in principle, more degrees of freedom than the usual spin-2 *graviton*. The reason why one doesn't hear of these degrees of freedom in GR is that the Einstein-Hilbert action is a very special choice, resulting in second-order equations of motion, which constrain away the scalars and the vectors, so that they are non-propagating. However, this is not the case if one departs from the Einstein-Hilbert form for the action. When using any modified action (and the usual variational principle) one inevitably frees up some of the additional degrees of freedom. In fact, this can be a good thing, in that the dynamics of these new degrees of freedom may be precisely what one needs to drive the accelerated expansion of the universe. However, there is often a price to pay.

The problems may be of several different kinds. First, there is the possibility that along with the desired deviations from GR on cosmological scales, one may also find similar deviations on solar system scales, at which GR is rather well-tested. Second is the possibility that the newly-activated degrees of freedom may be badly behaved in one way or another; either having the wrong sign kinetic terms (ghosts), and hence being unstable, or leading to superluminal propagation, which may lead to other problems.

These constraints are surprisingly restrictive when one tries to create viable modified gravity models yielding cosmic acceleration. In the next few sections I will describe several ways in which one might modify the action, and in each case demonstrate how cosmic acceleration emerges. However, I will also point out how the constraints I have mentioned rule out these simple examples, and mention how one must complicate the models to recover viable models.

2.1. A Simple Model: $f(R)$ Gravity

The simplest way one could think to modify GR to obtain cosmic acceleration is to replace the Einstein-Hilbert Lagrangian density by a general function $f(R)$ of the Ricci scalar R [5, 6].

$$S = \frac{M_{\rm Pl}^2}{2}\int d^4x\sqrt{-g}\,[R + f(R)] + \int d^4x\sqrt{-g}\,\mathcal{L}_{\rm m}[\chi_i, g_{\mu\nu}]\ , \tag{1}$$

where $M_{\rm Pl} \equiv (8\pi G)^{-1/2}$ is the (reduced) Planck mass and $\mathcal{L}_{\rm m}$ is the Lagrangian density for the matter fields χ_i.

Here, I have written the matter Lagrangian as $\mathcal{L}_{\rm m}[\chi_i, g_{\mu\nu}]$ to make explicit that in this frame - the *Jordan* frame - matter falls along geodesics of the metric $g_{\mu\nu}$.

The equation of motion obtained by varying the action (1) is

$$(1 + f_R)\,R_{\mu\nu} - \frac{1}{2}g_{\mu\nu}\,(R + f) + \left(g_{\mu\nu}\nabla^2 - \nabla_\mu\nabla_\nu\right) f_R = \frac{T_{\mu\nu}}{M_{\rm Pl}^2}\ , \tag{2}$$

where I have defined $f_R \equiv \partial f/\partial R$.

Further, if the matter content is described as a perfect fluid, with energy-momentum tensor,

$$T^m_{\mu\nu} = (\rho_m + p_m)U_\mu U_\nu + p_m g_{\mu\nu} \ , \tag{3}$$

where U^μ is the fluid rest-frame four-velocity, ρ_m is the energy density and p_m is the pressure, then the fluid equation of motion is the usual continuity equation.

When considering the background cosmological evolution of such models, the metric can be taken as the flat Robertson-Walker form, $ds^2 = -dt^2 + a^2(t)d\mathbf{x}^2$. In this case, the usual Friedmann equation of GR is modified to become

$$3H^2 - 3f_R(\dot{H} + H^2) + \frac{1}{2}f + 18f_{RR}H(\ddot{H} + 4H\dot{H}) = \frac{\rho_m}{M_{\rm Pl}^2} \tag{4}$$

and the continuity equation is

$$\dot{\rho}_m + 3H(\rho_m + p_m) = 0 \ . \tag{5}$$

When supplied with an equation of state parameter w, the above equations are sufficient to solve for the background cosmological behavior of the space-time and it's matter contents. For appropriate choices of the function $f(R)$ it is possible to obtain late-time cosmic acceleration without the need for dark energy, although evading bounds from precision solar-system tests of gravity turns out to be a much trickier matter, as we shall see.

While one can go ahead and analyze this theory in the Jordan frame, it is more convenient to perform a carefully-chosen conformal transformation on the metric, in order to render the gravitational action in the usual Einstein Hilbert form of GR. Consider writing

$$\tilde{g}_{\mu\nu} = \Omega(x^\alpha)g_{\mu\nu} \ , \tag{6}$$

and construct the function $r(\Omega)$ that satisfies

$$1 + f_R[r(\Omega)] = \Omega \ . \tag{7}$$

Defining a rescaled scalar field by $\Omega \equiv e^{\beta\phi}$, with $\beta M_{\rm Pl} \equiv \sqrt{2/3}$, the resulting action becomes

$$\tilde{S} = \frac{M_{\rm Pl}}{2}\int d^4x\sqrt{-\tilde{g}}\,\tilde{R} \ + \ \int d^4x\sqrt{-\tilde{g}}\left[-\frac{1}{2}\tilde{g}^{\mu\nu}(\partial_\mu\phi)\partial_\nu\phi - V(\phi)\right] + \int d^4x\sqrt{-\tilde{g}}\,e^{-2\beta\phi}\mathcal{L}_{\rm m}[\chi_i, e^{-\beta\phi}\tilde{g}_{\mu\nu}] \ , \tag{8}$$

where the potential $V(\phi)$ is determined entirely by the original form (1) of the action and is given by

$$V(\phi) = \frac{e^{-2\beta\phi}}{2}\left\{e^{\beta\phi}r[\Omega(\phi)] - f(r[\Omega(\phi)])\right\} \ . \tag{9}$$

The equations of motion in the Einstein frame are much more familiar than those in the Jordan frame, although there are some crucial subtleties. In particular, note that in general, test particles of the matter content χ_i do not freely fall along geodesics of the metric $\tilde{g}_{\mu\nu}$.

The equations of motion in this frame are those obtained by varying the action with respect to the metric $\tilde{g}_{\mu\nu}$

$$\tilde{G}_{\mu\nu} = \frac{1}{M_{\rm Pl}^2}\left(\tilde{T}_{\mu\nu} + T^{(\phi)}_{\mu\nu}\right) \ , \tag{10}$$

with respect to the scalar field ϕ

$$\tilde{\nabla}^2\phi = -\frac{dV}{d\phi}(\phi) \ , \tag{11}$$

and with respect to the matter fields χ_i, described as a perfect fluid.

Once again, I will specialize to consider background cosmological evolution in this frame. The Einstein-frame line element can be written in familiar FRW form as

$$ds^2 = -d\tilde{t}^2 + \tilde{a}^2(\tilde{t})d\mathbf{x}^2 \ , \tag{12}$$

where $d\tilde{t} \equiv \sqrt{\Omega}\, dt$ and $\tilde{a}(t) \equiv \sqrt{\Omega}\, a(t)$. The Einstein-frame matter energy-momentum tensor is then given by

$$\tilde{T}^m_{\mu\nu} = (\tilde{\rho}_m + \tilde{p}_m)\tilde{U}_\mu \tilde{U}_\nu + \tilde{p}_m \tilde{g}_{\mu\nu} \ , \tag{13}$$

where $\tilde{U}_\mu \equiv \sqrt{\Omega}\, U_\mu$, $\tilde{\rho}_m \equiv \rho_m/\Omega^2$ and $\tilde{p}_m \equiv p_m/\Omega^2$.

Now, as I mentioned in the introduction, any modification of the Einstein-Hilbert action must, of course, be consistent with the classic solar system tests of gravity theory, as well as numerous other astrophysical dynamical tests. We have chosen the coupling constant μ to be very small, but we have also introduced a new light degree of freedom. As shown by Chiba [7], the simple model above is equivalent to a Brans-Dicke theory with $\omega = 0$ in the approximation where the potential was neglected, and would therefore be inconsistent with solar system measurements [8].

To construct a realistic $f(R)$ model requires a more complicated function, with more than one adjustable parameter in order to fit the cosmological data [9] and satisfy solar system bounds.

2.2. *Extensions: Higher-Order Curvature Invariants*

It is natural to consider generalizing the action of [5] to include other curvature invariants [10]. There are, of course, any number of terms that one could consider, but for simplicity, focus on those invariants of lowest mass dimension that are also parity-conserving $P \equiv R_{\mu\nu}\, R^{\mu\nu}$ and $Q \equiv R_{\alpha\beta\gamma\delta}\, R^{\alpha\beta\gamma\delta}$.

The action then takes the form

$$S = \int d^4x \sqrt{-g}\,[R + f(R, P, Q)] + \int d^4x\, \sqrt{-g}\, \mathcal{L}_M \ , \tag{14}$$

where $f(R, P, Q)$ is a general function describing deviations from general relativity.

Actions of the form (14) generically admit a maximally-symmetric solution that is often unstable to another accelerating power-law attractor. It has been shown that solar system constraints, of the type I have described for $f(R)$ models, can be evaded by these more general models when, for example, the Q terms are relevant on those scales. However, these theories generically contain ghosts and/or superluminally propagating modes [11, 12, 13, 14]. I therefore will not discuss them further here.

2.3. *Induced Gravity Models*

In the Dvali-Gabadadze-Porrati (DGP) model [15], our observed $4D$ universe is embedded in an infinite empty fifth dimension. Despite the fact that the extra dimension is infinite in extent, the inverse-square law is nevertheless recovered at short distances on the brane due to an intrinsic, four-dimensional Einstein-Hilbert term in the action

$$S_{\rm DGP} = \int_{\rm bulk} {\rm d}^5x \sqrt{-g_5}\frac{M_5^3}{2} R_5 + \int_{\rm brane} {\rm d}^4x \sqrt{-g_4}\left(\frac{M_4^2}{2} R_4 + \mathcal{L}_{\rm matter}\right) . \tag{15}$$

The Newtonian potential on the brane scales as $1/r$ at short distances, as in $4D$ gravity, and asymptotes to $1/r^2$ at large distances, characteristic of $5D$ gravity. The cross-over scale m_5^{-1} between these two behaviors is set by the bulk and brane Planck masses (M_5 and M_4 respectively) via $m_5 = \frac{M_5^3}{M_4^2}$.

In this picture, the higher-dimensional nature of gravity affects the $4D$ brane through deviations from general relativity on horizon scales, that may give rise to the observed accelerated expansion. This model faces its own challenges however. The branch of solutions that include self-acceleration suffers from ghost-like instabilities, and on the observational front, DGP cosmology is statistically disfavored in comparison to ΛCDM and is significantly discordant with constraints on the curvature of the universe.

3. Cascading Cosmology

These facts, among others, have led to the idea of *cascading gravity* [16, 17, 18] — a higher-dimensional generalization of the DGP idea, which is free of divergent propagators and ghost instabilities. In this model one embeds a succession of higher-codimension branes into each other, with energy-momentum confined to the $4D$ brane and gravity living in higher-dimensional space. An important first test, which I've been working on recently [19], is whether such models can reproduce a successful cosmological evolution.

3.1. A Proxy Theory for Cascading Gravity

The main idea is to embed a 3-brane in a succession of higher-dimensional DGP branes, each with their own Einstein-Hilbert term. Denote coordinates in the full six dimensional spacetime by $x^0, x^1, x^2, x^3, x^5, x^6$. Indices $M, N, ...$ run over 0,1,2,3,5 (i.e. the $4+1D$ coordinates), indices $\mu, \nu, ...$ run over 0,1,2,3 (i.e. the $3+1D$ coordinates), and indices $i, j, ...$ run over $1, 2, 3$ (i.e. the $3D$ spatial coordinates). Further denote the fifth and sixth dimensional coordinates by $y = x^5$ and $z = x^6$, where convenient. The action is then

$$\begin{aligned} S_{\rm cascade} &= \int_{\rm bulk} {\rm d}^6x\sqrt{-g_6}\frac{M_6^4}{2}R_6 + \int_{4-{\rm brane}} {\rm d}^5x\sqrt{-g_5}\frac{M_5^3}{2}R_5 \\ &+ \int_{3-{\rm brane}} {\rm d}^4x\sqrt{-g_4}\left(\frac{M_4^2}{2}R_4 + \mathcal{L}_{\rm matter}\right) . \end{aligned} \tag{16}$$

As a result, the force law on the 3-brane "cascades" from $1/r^2$ to $1/r^3$ to $1/r^4$ as one moves increasingly far from a source, with the $4D \to 5D$ and $5D \to 6D$ cross-over scales given respectively by m_5^{-1} and m_6^{-1}, with $m_6 = \frac{M_6^4}{M_5^3}$

The next question is, of course, whether the resulting cosmology is consistent with current observations, and whether it offers distinguishing signatures from ΛCDM cosmology. Unfortunately, finding analytical solutions is a significant challenge, even in the simplest $6D$ case, as the bulk metric is generally expected to depend on all extra-dimensional coordinates plus time [20].

To proceed, consider, in analogy with a useful decoupling limit for the DGP model, the limit $M_5, M_6 \to \infty$, with the strong-coupling scale $\Lambda_6 = (m_6^4 M_5^3)^{1/7}$ kept fixed. In this limit, the action (16) may be expanded around flat space, and reduces to a local theory on the 4-brane, describing $5D$ weak-field metric perturbations h_{MN} and an interacting scalar field π. The resulting action is [16]

$$\begin{aligned} S_{\rm decouple} &= \frac{M_5^3}{2}\int_{\rm bulk} {\rm d}^5x\left[-\frac{1}{2}h^{MN}(\mathcal{E}h)_{MN} + \pi\eta^{MN}(\mathcal{E}h)_{MN} - \frac{27}{16m_6^2}(\partial\pi)^2\Box_5\pi\right] \\ &+ \int_{\rm brane} {\rm d}^4x\left[-\frac{M_4^2}{4}h^{\mu\nu}(\mathcal{E}h)_{\mu\nu} + \frac{1}{2}h^{\mu\nu}T_{\mu\nu}\right] , \end{aligned} \tag{17}$$

where $(\mathcal{E}h)_{MN}$ is the linearized Einstein tensor in $5D$, and $(\mathcal{E}h)_{\mu\nu}$ that in $4D$. Nearly all of the interesting features of DGP gravity are due to the helicity-0 mode π and can be understood at the level of the decoupling theory.

Of course (17) is restricted to weak-field gravity and therefore cannot be used to find cosmological solutions. As a "proxy" brane-world scenario, we proposed to complete (17) into a covariant, non-linear theory of gravity in $5D$ coupled to a 3-brane, using

$$\begin{aligned} S &= \frac{M_5^3}{2}\int_{\rm bulk} {\rm d}^5x\sqrt{-g_5}\left[e^{-3\pi/2}R_5 - \frac{27}{16m_6^2}(\partial\pi)^2\Box_5\pi\right] \\ &+ \int_{\rm brane} {\rm d}^4x\sqrt{-g_4}\left[\frac{M_4^2}{2}R_4 + \mathcal{L}_{\rm matter}\right]. \end{aligned} \tag{18}$$

By construction this theory reduces to (17) in the weak-field limit, and therefore agrees with cascading gravity to leading order in $1/M_5$. The proposed $5D$ completion is by no means unique, but the hope is that the salient features of cascading cosmology are captured by the $5D$ effective theory, and that the resulting predictions are at least qualitatively robust to generalizations of (18).

3.2. Covariant Equations of Motion On and Off the Brane

The bulk Einstein equations are

$$\begin{aligned} e^{-3\pi/2}G_{MN} &= -\frac{27}{16m_6^2}\left[\partial_{(M}(\partial\pi)^2\partial_{N)}\pi - \frac{1}{2}g_{MN}\partial^K(\partial\pi)^2\partial_K\pi - \partial_M\pi\partial_N\pi\Box_5\pi\right] \\ &- \left(g_{MN}\Box_5 - \nabla_M\nabla_N\right)e^{-3\pi/2}\,, \end{aligned} \tag{19}$$

where G_{MN} is the $5D$ Einstein tensor, and the π equation of motion is

$$(\Box_5\pi)^2 - (\nabla_M\partial_N\pi)^2 - R^{MN}\partial_M\pi\partial_N\pi = \frac{4}{9}m_6^2e^{-3\pi/2}R_5, \tag{20}$$

where R_{MN} is the $5D$ Ricci tensor and R_5 is the Ricci scalar. Remarkably, even though the cubic π interaction has four derivatives, all higher-derivative terms cancel in the variation, yielding a second-order equation of motion for π. This is typical of a larger class of "galileon" theories [21]. I won't have time to speak in detail about these here, but would like to highlight that we are close to completing an extension of these theories to the multi-field case, which will allow the exploration of the low-energy limit of braneworld models in co-dimension greater than one[1].

We also require the Israel junction condition

$$\begin{aligned} 2M_5^3e^{-3\pi/2}\left(Kq_{\mu\nu} - K_{\mu\nu} - \frac{3}{2}q_{\mu\nu}\mathcal{L}_n\pi\right) &= \frac{27}{8}\frac{M_5^3}{m_6^2}\left(\partial_\mu\pi\partial_\nu\pi\mathcal{L}_n\pi + \frac{1}{3}q_{\mu\nu}\left(\mathcal{L}_n\pi\right)^3\right) \\ &+ T_{\mu\nu}^{(4)} - M_4^2G_{\mu\nu}^{(4)}\,, \end{aligned} \tag{21}$$

where

$$T_{\mu\nu}^{(4)} \equiv -\frac{2}{\sqrt{-q}}\frac{\delta(\sqrt{-q}\mathcal{L}_{\rm matter})}{\delta q^{\mu\nu}} \tag{22}$$

is the matter stress-energy tensor on the brane, and $G_{\mu\nu}^{(4)}$ is the Einstein tensor derived from the induced metric $q_{\mu\nu}$. Similarly, the boundary condition for π on the brane is

$$e^{-3\pi/2}K + \frac{9}{8m_6^2}\left(K_{\mu\nu}\partial^\mu\pi\partial^\nu\pi + 2\mathcal{L}_n\pi\Box_4\pi + K(\mathcal{L}_n\pi)^2\right) = 0\,. \tag{23}$$

(Note that equations (21) and (23) are not independent, of course; the divergence of (21) can be shown to be proportional to (23) after using the bulk momentum constraint equation.)

[1] This work [22] (see also [23, 24, 25, 26]), and some initial explorations [27, 28], were recently completed.

3.3. The Cosmological Evolution on the Brane

The study of brane-world cosmology requires using the equations of motion to obtain a Friedmann equation on the brane, assuming homogeneity and isotropy along the 3+1 world-volume dimensions. The junction conditions (21) and (23) do not form a closed system of equations for $q_{\mu\nu}$, hence deriving an induced Friedmann equation requires knowledge of the bulk geometry [20].

Because of the bulk scalar field, there is no Birkhoff theorem to ensure that the bulk solutions are necessarily static under the assumption of homogeneity and isotropy on the brane — the most general bulk geometry depends on both the extra-dimensional coordinate *and* time. For concreteness, however, focus here on a static warped geometry with Poincaré-invariant slices, representing a tractable first step for which, as we will see, the resulting phenomenology is already surprisingly rich.

Writing

$$ds^2_{\rm bulk} = a^2(y)(-d\tau^2 + d\vec{x}^2) + dy^2 \,, \tag{24}$$

the brane motion is governed by two functions, $y(t)$ and $\tau(t)$, describing the embedding, where t is proper time on the brane. The induced metric is of the Friedmann-Robertson-Walker (FRW) form, with spatially-flat ($k = 0$) constant-time hypersurfaces $ds^2_{\rm brane} = -dt^2 + a^2(y)d\vec{x}^2$, where, by virtue of t being the proper time,

$$\left(\frac{dt}{d\tau}\right)^2 = a^2 - \left(\frac{dy}{d\tau}\right)^2 . \tag{25}$$

Given a solution $a(y)$ to the bulk equations (19)–(20), the covariant junction conditions (21) and (23) allow us to solve for the embedding $(y(t), \tau(t))$, and hence the cosmology induced by brane motion through the warped bulk.

For the stress energy on the brane, we assume a collection of (non-interacting) perfect fluids with energy densities $\rho^{(i)}_{\rm m}$ and pressures $P^{(i)}_{\rm m}$, obeying the standard continuity equations. To derive the Friedmann equation on the brane, we use the $(0,0)$ component of (21). Since $\partial_0\pi = \pi' dy/dt$ and $dy/dt = aH/a'$, we can write the resulting equation as the standard Friedmann equation with an additional effective energy density ρ_π resulting from the π field,

$$3H^2M_4^2 = \sum_i \rho^{(i)}_{\rm m} + \rho_\pi \,, \tag{26}$$

where

$$\rho_\pi \equiv M_5^3\sqrt{a'^2 + a^2H^2}\left\{\frac{9}{8m_6^2}\left(2\left(\frac{aH}{a'}\right)^2 - 1\right)\frac{\pi'^3}{a'} - 6e^{-3\pi/2}\left(\frac{\pi'}{2a'} - \frac{1}{a}\right)\right\} , \tag{27}$$

encoding all the complexity and new physics of the cascading cosmology model. Given a solution $a(y)$, $\pi(y)$ to the bulk equations, this relation may be inverted to obtain $y(a)$, and used to express all y-dependent terms in ρ_π as functions of a. Equation (26), together with the continuity equations for the brane matter, then forms a closed system for the brane scale factor $a(t)$.

3.4. Numerical Solutions

To get a feel for the type of cosmological solutions that this model may exhibit requires numerically evolving the full bulk and brane equations in the presence of matter on the brane. To be explicit, assume zero spatial curvature on the brane, include relativistic and pressureless components consistent with the standard cosmological model: $\Omega_{\rm m} = 0.3$, $\Omega_{\rm r} = 8.5 \times 10^{-4}$, and fix the scale factor today to be $a_0 = 1$.

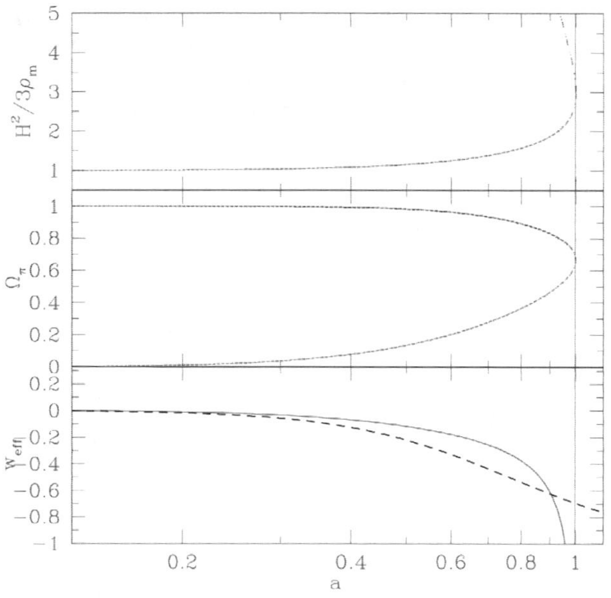

Figure 1. See text for explanation

In the figure, example evolution histories in which no cosmological constant is present to drive cosmic acceleration are plotted. The top panel shows the deviation of the expansion history from that derived from standard matter (for which $3H^2/\rho_m = 1$). The curves each show consistent solutions to the modified Friedmann equation (26): one solution (red, thick line) recovers the standard expansion history at early times and then undergoes accelerated expansion at late times; the other solution (blue, dotted line) has an expansion history entirely inconsistent with that of standard ΛCDM, with the π field dominating the expansion at all eras, and undergoing heavily decelerated expansion at late times.

The center panel shows the evolution of the effective fractional energy density, $\Omega_\pi = 8\pi G\rho_\pi/3H^2$, for these two solutions. For the accelerating solution, the phantom-like behavior in the matter era allows the π field to dominate and drive cosmic acceleration at late times. The model is not physical, however, since as $\Omega_\pi \to 2/3$ one finds $\dot{H} \to \infty$ and a singularity occurs. This singularity is of an unusual nature - the bulk geometry is smooth and it is the brane embedding that is singular. It is possible therefore that this singularity could be circumvented by the use of a more general metric ansatz, but this has yet to be demonstrated.

Finally, the bottom panel shows a comparison of the effective equation of state for the expansion, $w_{\rm eff} = -1 - (2/3){\rm d}\ln H/{\rm d}\ln a$, for the accelerating π (red, full line) and fiducial ΛCDM (black, dashed line) scenarios. The π driven expansion histories assume the numerical values $H_0 = 2.33\times10^{-4}$ Mpc^{-1}, $m_6 = 3.5\times10^{-18}$ Mpc^{-1} and $m_5 = 4.4\times10^{-31}$Mpc^{-1} for which the maximum singularity occurs just after $a = 1$.

4. Conclusion

Among the possible explanations for the observed accelerated expansion of the universe, the possibility that general relativity may become modified on the largest scales is a particularly intriguing one. In this talk I have outlined a number of modern approaches to this problem, focusing, as expected, on those that I have been involved with in one way or another. I have also described how the combined constraints of theoretical consistency, solar system measurements, and cosmological observations tightly bound the possible viable models. I have also discussed very recent work on cascading cosmology, and have shown that cosmic acceleration can arise within this model without the need for a cosmological constant. However, for the limited solutions explored thus far, one hits a singularity in the expansion history so that the universe cannot smoothly transition towards $\Omega_\pi \rightarrow 1$

As I mentioned in the talk, out of the higher dimensional constructions, such as the DGP model, an interesting set of four dimensional effective field theories - the *galileons* - arises, encapsulating the effects of modifying gravity. The work which was underway, generalizing that work to multi-galileon theories, is now completed, and we have been hard at work on constraining the resulting Lagrangians. Now as then, much remains to be done to understand the full range of acceptable models.

Acknowledgments

I would like to thank the organizers of the ICGTMP-2010 conference, and in particular Wojciech Zakrzewski, for inviting me to deliver this talk. I would also like to thank my collaborators, from whose joint work with me I have borrowed liberally in putting together this summary. This work is supported in part by NASA ATP grant NNX08AH27G, NSF grant PHY-0930521, Department of Energy grant DE-FG05-95ER40893-A020, and by the Fay R. and Eugene L. Langberg chair.

[1] E. J. Copeland, M. Sami and S. Tsujikawa, Int. J. Mod. Phys. D **15**, 1753 (2006) [arXiv:hep-th/0603057].
[2] J. Frieman, M. Turner and D. Huterer, Ann. Rev. Astron. Astrophys. **46**, 385 (2008) [arXiv:0803.0982 [astro-ph]].
[3] A. Silvestri and M. Trodden, Rept. Prog. Phys. **72**, 096901 (2009) [arXiv:0904.0024 [astro-ph.CO]].
[4] R. R. Caldwell and M. Kamionkowski, Ann. Rev. Nucl. Part. Sci. **59**, 397 (2009) [arXiv:0903.0866 [astro-ph.CO]].
[5] S. M. Carroll, V. Duvvuri, M. Trodden and M. S. Turner, Phys. Rev. D **70**, 043528 (2004) [arXiv:astro-ph/0306438].
[6] S. Capozziello, S. Carloni and A. Troisi, Recent Res. Dev. Astron. Astrophys. **1**, 625 (2003) [arXiv:astro-ph/0303041].
[7] T. Chiba, Phys. Lett. B **575**, 1 (2003) [arXiv:astro-ph/0307338].
[8] B. Bertotti, L. Iess and P. Tortora, Nature **425**, 374 (2003).
[9] R. Bean, D. Bernat, L. Pogosian, A. Silvestri and M. Trodden, Phys. Rev. D **75**, 064020 (2007) [arXiv:astro-ph/0611321].
[10] S. M. Carroll, A. De Felice, V. Duvvuri, D. A. Easson, M. Trodden and M. S. Turner, Phys. Rev. D **71**, 063513 (2005) [arXiv:astro-ph/0410031].
[11] T. Chiba, JCAP **0503**, 008 (2005) [arXiv:gr-qc/0502070].
[12] I. Navarro and K. Van Acoleyen, "Consistent long distance modification of gravity from inverse powers of the JCAP **0603**, 008 (2006) [arXiv:gr-qc/0511045].
[13] A. De Felice, M. Hindmarsh and M. Trodden, JCAP **0608**, 005 (2006) [arXiv:astro-ph/0604154].
[14] G. Calcagni, B. de Carlos and A. De Felice, Nucl. Phys. B **752**, 404 (2006) [arXiv:hep-th/0604201].
[15] G. R. Dvali, G. Gabadadze and M. Porrati, Phys. Lett. B **485**, 208 (2000) [arXiv:hep-th/0005016].
[16] C. de Rham, G. Dvali, S. Hofmann, J. Khoury, O. Pujolas, M. Redi and A. J. Tolley, "Cascading gravity: Extending the Dvali-Gabadadze-Porrati model to higher Phys. Rev. Lett. **100**, 251603 (2008) [arXiv:0711.2072 [hep-th]].
[17] C. de Rham, S. Hofmann, J. Khoury and A. J. Tolley, JCAP **0802**, 011 (2008) [arXiv:0712.2821 [hep-th]].
[18] N. Kaloper and D. Kiley, JHEP **0705**, 045 (2007) [arXiv:hep-th/0703190].
[19] N. Agarwal, R. Bean, J. Khoury and M. Trodden, Phys. Rev. D **81**, 084020 (2010) [arXiv:0912.3798 [hep-th]].
[20] N. Chatillon, C. Macesanu and M. Trodden, Phys. Rev. D **74**, 124004 (2006) [arXiv:gr-qc/0609093].
[21] A. Nicolis, R. Rattazzi and E. Trincherini, Phys. Rev. D **79**, 064036 (2009) [arXiv:0811.2197 [hep-th]].
[22] K. Hinterbichler, M. Trodden and D. Wesley, arXiv:1008.1305 [hep-th].
[23] C. Deffayet, S. Deser and G. Esposito-Farese, Phys. Rev. D **80**, 064015 (2009) [arXiv:0906.1967 [gr-qc]].
[24] C. Deffayet, S. Deser and G. Esposito-Farese, Phys. Rev. D **82**, 061501 (2010) [arXiv:1007.5278 [gr-qc]].
[25] A. Padilla, P. M. Saffin and S. Y. Zhou, arXiv:1007.5424 [hep-th].
[26] A. Padilla, P. M. Saffin and S. Y. Zhou, arXiv:1008.3312 [hep-th].
[27] M. Andrews, K. Hinterbichler, J. Khoury and M. Trodden, arXiv:1008.4128 [hep-th].

[28] G. L. Goon, K. Hinterbichler and M. Trodden, arXiv:1008.4580 [hep-th].

Journal of Physics: Conference Series **284** (2011) 012005 doi:10.1088/1742-6596/284/1/012005

Skyrmions and monopoles — isolated and arrayed

R S Ward

Department of Mathematical Sciences, Durham University, Durham DH1 3LE, UK

E-mail: richard.ward@durham.ac.uk

Abstract. There are two basic types of three-dimensional topological soliton, namely skyrmions and monopoles. This article reviews some of the features of such solitons, concentrating on static multi-skyrmions and multi-monopoles with structure group SU(2), and focusing in particular on periodic arrays. An isolated multi-skyrmion typically has a regular, for example polyhedral, structure; and there are multi-monopoles which look very similar. However, when it comes to infinite arrays of solitons — chains, walls or crystals — then the skyrmion and monopole systems behave rather differently.

1. Introduction: 3-dimensional topological solitons

Topological solitons come in two basic types: skyrmions, which wind around all of space; and monopoles, which wind around spatial infinity. The general background to both may be found in the monograph [1]. Skyrmions were introduced fifty years ago, as a model of nucleons [2]. Magnetic monopoles were introduced even longer (eighty years) ago, by Dirac [3]. Monopoles are predicted in many cosmological models, but as yet there has been no confirmed sighting, and much of the interest in them over the last thirty years has been mathematical: the partial differential equations for static BPS monopoles are completely-integrable, and are associated with some fascinating geometric structure.

Both types of object (skyrmions and monopoles) also occur in condensed-matter contexts. But these applications are more commonly 2-dimensional rather than 3-dimensional, and will not be pursued here.

Topological solitons are stable classical objects, described by fields $\phi_\alpha(x^j)$. Here $x^j = (x^1, x^2, x^3)$ are the standard coordinates on $\mathbb{R}^3$; we shall concentrate here on the the static case, so the fields are functions only of these spatial coordinates (but there will be a brief mention of time-dependence later on).

Static solitons correspond to local minima of an energy functional $E[\phi_\alpha] = \int \mathcal{E}(\phi_\alpha, \partial_j\phi_\alpha)\, d^3x$, where the energy density $\mathcal{E}$ is localized in space. In the best-known cases, one has a high degree of internal symmetry: in particular, the system is invariant under the action of a Lie group G (for example a gauge group). Soliton stability comes partly from non-trivial topology of the fields ϕ_α; but topology is scale-free, and so some additional mechanism is needed to fix the soliton size. In our case, this is achieved by the balance of terms in $\mathcal{E}$: typically between a term which provides an expanding force and one which provides a contracting force. For each of the two systems — skyrmions and monopoles — we will review isolated solitons in $\mathbb{R}^3$ with finite total energy, before discussing arrays in $\mathbb{R}^3$: singly-periodic chains, doubly-periodic walls, and triply-periodic crystals.

2. Skyrmions in $\mathbb{R}^3$

In the Skyrme system, we have a field $U(x^j)$ which takes values in a Lie group G; from now on, take the simplest nontrivial case $G = \mathrm{SU}(2)$. The boundary condition is $U(x^j) \to \mathbf{1}$ as $r \to \infty$. The energy density is defined to be

$$\mathcal{E} = -\tfrac{1}{2}\,\mathrm{tr}(\xi_j \xi_j) - \tfrac{1}{16}\,\mathrm{tr}([\xi_j, \xi_k][\xi_j, \xi_k]),$$

where $\xi_j = U^{-1}\partial_j U$ (note that ξ_j is Lie-algebra-valued). Topology arises because U has a winding number $N \in \mathbb{Z}$; in the nuclear-physics application, N is the baryon number. The energy satisfies a topological lower bound $E_N = E/(12\pi^2 N) \geq 1$.

For $N \leq 4$ one gets the following minimal-energy solutions, the energy densities of which are plotted in figure 1:

- $N = 1$: $E_N = 1.232$, sphere;
- $N = 2$: $E_N = 1.179$, torus;
- $N = 3$: $E_N = 1.146$, tetrahedron;
- $N = 4$: $E_N = 1.120$, cube.

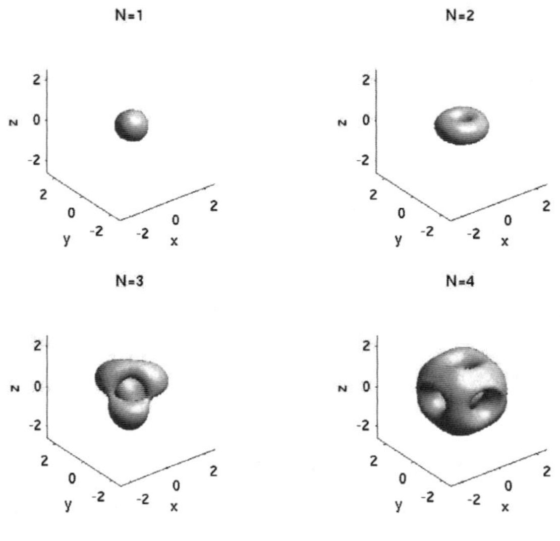

Figure 1. Skyrmions for $1 \leq N \leq 4$: energy densities.

In general, for $N \gg 1$, skyrmions attract and coalesce. There are many local minima, which are only accessible numerically. Typically, the solution resembles a hollow shell ('buckyball'), or a solid crystal chunk. For some shell-like examples with $N = 32, 37, 47, 67, 97$, see [4, 5]. In the case $N = 32$, the solid crystal chunk has an energy which is slightly higher than that of the shell [5] ($E_N = 1.072$ for the shell, versus $E_N = 1.076$ for the crystal); but the belief is that for very large N, the lowest-energy solution should be a solid crystal chunk.

3. Arrays of skyrmions

The lowest possible of E_N, namely $E_N = 1.038$, is attained for a triply-periodic configuration, in fact a cubic lattice of half-skyrmions. In what follows, this will be referred to as the Skyrme crystal; its energy density is plotted in figure 2, where one sees eight half-skyrmions in a

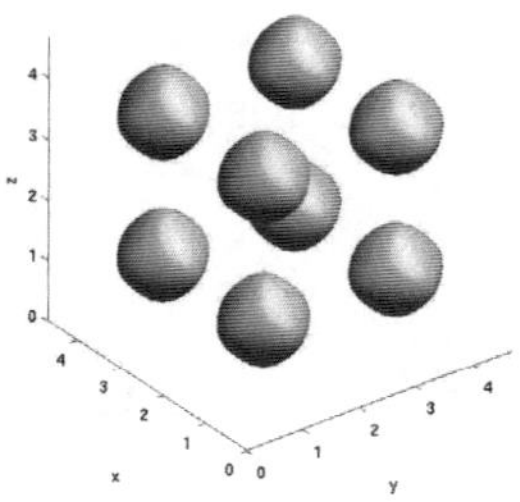

Figure 2. The energy density of the Skyrme crystal.

fundamental cell of the crystal. Skyrme crystals (triply-periodic) are in fact closely related [6] to doubly-periodic Skyrme walls, and also to singly-periodic Skyrme chains, each of which will now be described.

If the shell-like solutions of the previous section are visualized, in the language of carbon chemistry, as buckyballs (spherical fullerenes), then a Skyrme wall corresponds to graphene. For a single wall, the field U is doubly-periodic in x and y, and has the asymptotic behaviour $U \to \pm\mathbf{1}$ as $z \to \pm\infty$. The corresponding 2-dimensional array of half-skyrmions can be either square or hexagonal, with the latter having an energy-per-charge which is slightly lower than the former: $E_N = 1.062$ for the hexagonal wall, and $E_N = 1.068$ for the square wall [1]. The energy densities of these two solutions are plotted in figure 3. Parallel walls attract one another, so there are also doubly-periodic solutions representing p parallel multi-walls [7]. As one might expect, the p-square-wall solution tends to the Skyrme crystal as $p \to \infty$.

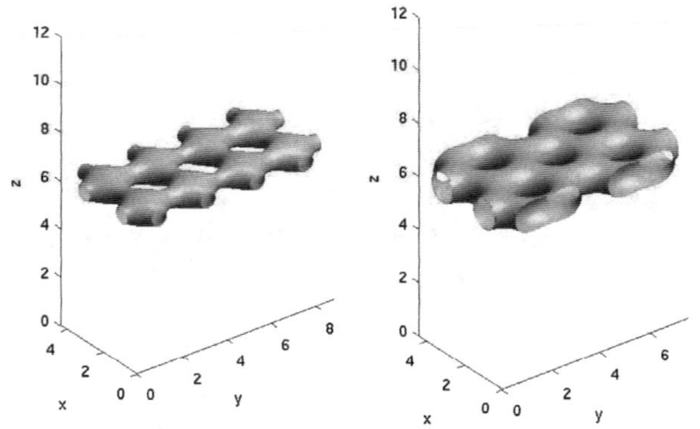

Figure 3. The energy densities of the square wall and the hexagonal wall.

Finally, one can have Skyrme chains: in this case, $U(x, y, z)$ is periodic in z, and $U \to \pm\mathbf{1}$ as $x^2 + y^2 \to \pm\infty$. The simplest chain may be viewed as a chain of half-skyrmion pairs, and also as a bound vortex-antivortex pair [8], and its energy-per-charge is $E_N = 1.143$. More generally, there exist stable chains in which each link consists of $2p$ half-skyrmions; and (as before) these solutions tend to the Skyrme crystal as $p \to \infty$. The simplest case, a $p = 1$ chain along the z-axis, is illustrated in figure 4. This plots the energy density of the solution over two periods — four links of the chain, with each link being antiperiodic.

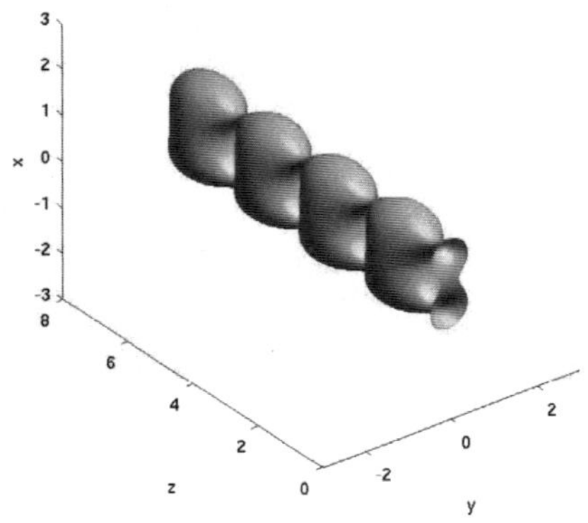

Figure 4. The energy density of the simplest Skyrme chain.

To summarize: in principle, all of nuclear physics is described by quantum chromodynamics (QCD); but direct calculations of nuclear properties in QCD is still rather difficult, and it is worth investigating a simpler approximations such as the Skyrme model. However, fitting to actual nuclear data involves modifications to the basic system, for example, adding quantum corrections and a pion-mass term. The Skyrme system admits arrays of half-skyrmions of various sorts — one may 'add up' infinitely many skyrmions without encountering divergences — and such arrays may be relevant to nuclei and exotic nuclear states. In this context, one would need to study the effect of modifications such as those mentioned above. Another interesting question, on which little work has been done, is to see what happens for larger groups $G > \mathrm{SU}(2)$.

4. Magnetic monopoles in $\mathbb{R}^3$

The setting here is gauge theory, and more specifically that of a Yang-Mills-Higgs system. The first step is to specify the gauge group G, a Lie group; in what follows, this will be either U(1) (in which case the equations are linear) or SU(2) (in which case they are nonlinear). The gauge potential is a 1-form $A_j(x^k)$ on $\mathbb{R}^3$, taking values in the Lie algebra $\mathfrak{g}$ of G. The corresponding magnetic field is $B_j = \varepsilon_{jkl}(\partial_k A_l + A_k A_l)$, and the energy density of the magnetic field is $\mathcal{E}_B = |B_j|^2$.

The case $\mathfrak{g} = \mathfrak{u}(1)$ corresponds to electromagnetism, and then B_j is the usual magnetic field vector. Stationary points of the magnetic-field energy are solutions of $\partial_j B_j = 0$. The simplest solution incorporating a point source at $x^j = 0$ is

$$B_j = \tfrac{1}{2} N \, r^{-3} \, x_j \, , \tag{1}$$

where N, for topological reasons, has to be an integer. This is the Dirac monopole [3]. Note that it is singular at $r = 0$, and it has infinite energy.

The generalization of this to non-Abelian gauge groups allows smooth finite-energy solutions. The simplest version is that of Bogomolny-Prasad-Sommerfield (BPS) monopoles, with gauge algebra $\mathfrak{g} = \mathfrak{su}(2)$. In addition to the gauge potential $A_j(x^k)$, we now also have a scalar (Higgs) field $\Phi(x^k)$ taking values in $\mathfrak{g}$. The covariant derivative of Φ is defined to be $D_j\Phi = \partial_j\Phi + [A_j, \Phi]$, and the system satisfies the Bogomolny equations

$$D_j\Phi = B_j. \tag{2}$$

This system of nonlinear partial differential equations is completely-integrable, and consequently has many remarkable properties.

The energy density is defined to be $\mathcal{E} = -\frac{1}{2}\operatorname{tr}[B_j B_j + (D_j\Phi)(D_j\Phi)]$; and the boundary condition is $-\operatorname{tr}(\Phi^2) \to 1$, $\mathcal{E} \to 0$ as $r \to \infty$. The asymptotic Higgs field $\Phi|_{r=\infty}$ is, in effect, a map from S^2 to S^2, and so it has a winding number $N \in \mathbb{Z}$: in other words, configurations are classified topologically by the integer N. The energy satisfies the topological lower bound $E \geq 2\pi N$; and $E = 2\pi N$ if and only if the equations (2) hold. For each $N \in \mathbb{Z}$ there exists a $4N$-parameter family of smooth solutions of (2). These $4N$ moduli correspond to the position in space for each of the N monopoles (there are no forces between static monopoles, so each of them can have any location), plus a phase for each one.

One consequence of the integrability of (2) is the existence of the Nahm Transform for BPS monopoles. This is a generalized integral transform $(A_j, \Phi) \longleftrightarrow (\widehat{A}_\alpha, \widehat{\Phi})$ between Yang-Mills-Higgs systems. In summary, one first solves a linear partial differential equation $\Delta\psi = 0$ in which the coefficients of the operator Δ are constructed from (A_j, Φ); and then one does an integral involving ψ to get the transformed field $(\widehat{A}_\alpha, \widehat{\Phi})$. For monopoles on $\mathbb{R}^3$, the Nahm transform of an N-monopole is a set of Nahm data consisting of three $N \times N$ matrices $\Phi_j(s)$ on $(-1, 1)$, satisfying the ordinary differential equations

$$\Phi_j' = \varepsilon_{jkl}\Phi_k\Phi_l\,. \tag{3}$$

Since it is generally easier to solve the ODEs (3) than to solve the PDEs (2), this transform is a highly-effective way of understanding and constructing monopole solutions. For example, there are explicit solutions (involving trigonometric and elliptic functions) of (3) corresponding to spherical, toroidal, tetrahedral and cubic monopoles for $N = 1, 2, 3, 4$ respectively; and there are many similar explicit examples for $N > 4$. Each $4N$-dimensional moduli space $\mathcal{M}$ has a natural hyper-Kähler metric (for example, the Atiyah-Hitchin metric in the case $N = 2$), and slow-motion dynamics of monopole systems corresponds to geodesic motion on $\mathcal{M}$. More details on all this may be found in [1].

5. Arrays of monopoles

In this section, we consider what happens when one tries to form periodic arrays of monopoles. The first issue is that of convergence. Suppose that we have an infinite array of Dirac monopoles of the form (1). The monopole at $\mathbf{x}_n$ has $\Phi_n = -1/(2|\mathbf{x} - \mathbf{x}_n|)$, and $\sum_n \Phi_n$ diverges for a regular array $\{\mathbf{x}_n\}$. So we need to regularize this infinite series, by (in effect) subtracting an infinite constant. For example, if the poles are equally-spaced along the z-axis chain — say $\mathbf{x}_n = \mathbf{x} - n\,\mathbf{e}$, where $\mathbf{e} = (0, 0, 2\pi)$ — then the series

$$\Phi = -\frac{1}{2r} - \frac{1}{2}\sum_{n\neq 0}\left[\frac{1}{|\mathbf{x} - \mathbf{x}_n|} - \frac{1}{2\pi n}\right] \tag{4}$$

converges (except at the poles), and $\Phi \sim \frac{1}{4\pi}\log(x^2 + y^2)$ as $x, y \to \infty$. So (4) represents a chain of Dirac monopoles. And it motivates the boundary condition that one uses in the non-abelian case [9]: again Φ behaves logarithmically as $x, y \to \infty$.

The Nahm transform of an SU(2) N-monopole chain is a solution of the U(N) Hitchin equations on the cylinder $\mathbb{R} \times S^1$ [9]. The latter consists of a U(N) gauge potential $(\widehat{A}_1, \widehat{A}_2)$, plus a Higgs field $\widehat{\Phi}$ taking values in the complexification of SU(2), satisfying the Hitchin equations

$$D_{\bar{\zeta}}\widehat{\Phi} = 0, \quad \widehat{B} = \tfrac{1}{2}\mathrm{i}\,[\widehat{\Phi}, \widehat{\Phi}*]. \tag{5}$$

Here ζ is a complex coordinate on $\mathbb{R} \times S^1$, $\widehat{\Phi}*$ denotes the Hermitian conjugate of $\widehat{\Phi}$, and $\widehat{B}$ is the magnetic field. In addition, $(\widehat{A}_\alpha, \widehat{\Phi})$ are required to satisfy various boundary conditions and constraints, of which one is

$$\det \widehat{\Phi} = (-1)^{N+1} C \cosh(2\pi\zeta) + K, \tag{6}$$

where C and K are complex constants. In fact, $|C|$ corresponds to the ratio *monopole size/chain period*; so $|C| \gg 1$ corresponds to a monopole chain in which the individual monopoles spread out and overlap (see below).

The simplest case is $N = 1$, corresponding to a chain of 1-monopoles, and in that case (5,6) gives us an explicit solution of the Hitchin equations, namely

$$\widehat{\Phi} = C \cosh(2\pi\zeta) + K, \quad \widehat{A}_\alpha = 0.$$

The (inverse) Nahm transform of this can be implemented numerically, and we get the 1-monopole chain illustrated in figure 5. This plots the quantity $|\Phi|^2 = -\frac{1}{2}\operatorname{tr}(\Phi^2)$ over one period in the z-direction, for various values of the parameter C. If $C = 1$, the individual monopoles are localized, but for $C \gg 1$ they spread out to form a strip in the z-direction. More details of this may be found in [10].

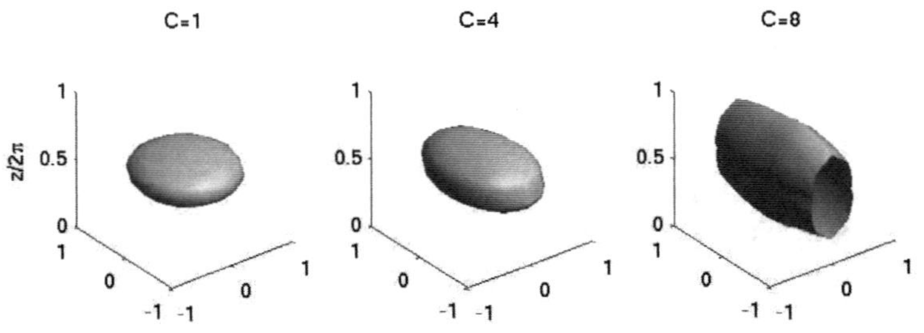

Figure 5. $|\Phi|^2$ over one period, for $C = 1, 4, 8$.

For $N \geq 2$, things are not quite as explicit, but one can still use the Nahm transform to obtain useful information. For example, a special case of $N = 2$ describes the slow-motion dynamics of interacting parallel monopole chains, corresponding to geodesics on the $N = 2$ moduli space [11]. In general, the procedure is to:

- identify a geodesic Γ in the moduli space;
- for each $t \in \Gamma$, compute the corresponding solution $(\widehat{A}_\alpha, \widehat{\Phi})$ of the SU(2) Hitchin equations;
- Nahm-transform to compute the monopole field (A_j, Φ);
- display these as a function of t.

The simplest example is obtained by fixing the centre-of-mass of the 2-monopole system, and imposing a D_2 symmetry corresponding to invariance under rotations by π about each of the x-, y- and z-axes: this selects a particular geodesic in the moduli space. The Hitchin equations then reduce to an elliptic sinh-Gordon equation on the cylinder, which can be solved numerically. Finally, the Nahm transform can be implemented numerically, and $|\Phi|^2 = -\operatorname{tr}\Phi^2$ plotted for a sequence of t-values, to visualize the dynamics. This was the process used to produce figure 6, which shows snapshots of the scattering of two monopole chains (each plotted over three z-periods). One sees the right-angle scattering which is typical of topological-soliton interactions.

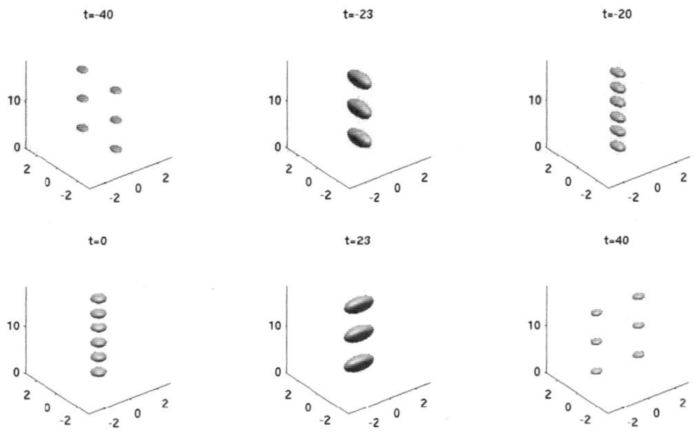

Figure 6. Scattering of two monopole chains.

Our final topic is that of doubly-periodic monopole solutions, or monopole walls. (It is worth pointing out that smooth triply-periodic monopoles, or monopole crystals, cannot exist — the divergences become uncontrollable. So double periodicity is as far as one can go.) So we impose periodicity in the x- and y-directions. There are now two topological charges N_+ and N_-, corresponding to the winding number of the Higgs field Φ as $z \to \pm\infty$ respectively. The appropriate boundary condition in z turns out to be

$$\Phi \sim 2\pi N_\pm \, z \text{ as } z \to \pm\infty,$$

with the energy density behaving like $\mathcal{E} \to 8\pi^2 (N_\pm)^2$. The interpretation is that one has a wall of monopoles separating two domains, in each of which the energy density is constant. In this case, the Nahm transform maps walls to walls, with (in general) different gauge group G and topological charges $N_\pm$. For example, the simplest non-trivial case of a Nahm correspondence is

$$\{G = \mathrm{U}(1), \quad N_\pm = \pm 1\} \leftrightarrow \{G = \mathrm{SU}(2), \quad N_- = 0, \quad N_+ = 1\}.$$

The energy density $\mathcal{E}$ of the SU(2) monopole wall in this case is illustrated in figure 7: the

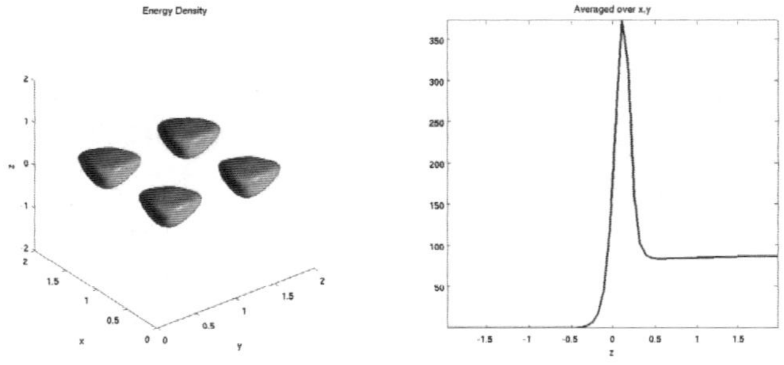

Figure 7. $\{G = \mathrm{SU}(2), N_- = 0, N_+ = 1\}$: an example solution

left-hand plot is a 3-dimensional plot of $\mathcal{E}$ over two periods in x and y, so one sees four of the monopoles in the wall; while the right-hand plot shows $\mathcal{E}$ integrated over x and y, as a function of z, revealing how the wall separates two domains, one with zero energy and one with constant positive energy density.

Monopole arrays have many other nice mathematical features which follow from the complete-integrability of the equations (2). For example, associated with any solution is a set of "spectral data" which are usually much simpler than the field itself, are invariant under the Nahm transform, and give one a lot of information about the solution. For example, in the monopole-wall case, the holonomy of $D_x + \mathrm{i}\Phi$ over one period in the x-direction is holomorphic in $z - \mathrm{i}y$ (as a consequence of (2)), and it has a very simple explicit form. All this structure ensures that BPS monopoles will remain of great mathematical interest, even if their physical importance remains unclear.

References

[1] Manton N S and Sutcliffe P M 2004 *Topological Solitons* (Cambridge: Cambridge University Press)
[2] Skyrme T H R 1961 A non-linear field theory *Proc. Roy. Soc. Lond.* A **260** 127–38
[3] Dirac P A M 1931 Quantized singularities in the electromagnetic field *Proc. Roy. Soc. Lond.* A **133** 60–72
[4] Battye R A, Houghton C and Sutcliffe P M 2003 Icosahedral skyrmions *J. Math. Phys.* **44** 3543–54
[5] Battye R A and Sutcliffe P M 2005 Skyrmions and the pion mass *Nucl. Phys.* B **705** 384–400
[6] Silva Lobo J 2010 Deformed Skyrme Crystals *J. High Energy Phys.* JHEP10(2010)029
[7] Silva Lobo J and Ward R S 2009 Skyrmion multi-walls *J. Phys.* A **42** 482001
[8] Harland D and Ward R S 2008 Chains of skyrmions *J. High Energy Phys.* JHEP12(2008)093
[9] Cherkis S and Kapustin A (2001) Nahm transform for periodic monopoles and $N = 2$ super Yang-Mills theory *Commun. Math. Phys.* **218** 333–71
[10] Ward R S 2005 Periodic monopoles *Phys. Lett.* B **619** 177–83
[11] Harland D and Ward R S 2009 Dynamics of periodic monopoles *Phys. Lett.* B **675** 262–6

Journal of Physics: Conference Series **284** (2011) 012006 doi:10.1088/1742-6596/284/1/012006

A BPS Skyrme model

C Adam[1], J Sanchez-Guillen[1] and A Wereszczynski[2]

[1] Departamento de Física de Partículas, Universidad de Santiago, and Instituto Galego de Física de Altas Enerxías (IGFAE) E-15782 Santiago de Compostela, Spain

[2]Institute of Physics, Jagiellonian University, Reymonta 4, Kraków, Poland

E-mail: adam@fpaxp1.usc.es, joaquin@fpaxp1.usc.es, wereszczynski@th.if.uj.edu.pl

Abstract. Within the set of generalized Skyrme models, we identify a submodel which has both infinitely many symmetries and a Bogomolny bound which is saturated by infinitely many exact soliton solutions. Concretely, the submodel consists of the square of the baryon current and a potential term only. Further, already on the classical level, this BPS Skyrme model reproduces some features of the liquid drop model of nuclei. Here, we review the properties of the model and we discuss the semiclassical quantization of the simplest Skyrmion (the nucleon).

1. Introduction

The derivation of the correct low energy hadron dynamics from the underlying fundamental theory (QCD) is, without doubt, one of the most prominent challenges of strong interaction physics, due to the non-perturbative nature of the quark and gluon interactions in the low energy limit. In the large N_c limit it is known that QCD becomes equivalent to an effective theory of mesons [1], [2], where baryons (hadrons as well as atomic nuclei) appear as solitonic excitations, where the topological charge is identified with the baryon number [3]. One of the most popular realizations of this idea is the Skyrme model [4]-[6], i.e., a version of a phenomenological chiral Lagrangian where the primary fields are mesons. The original Lagrangian has the following form (here U is an SU(2) valued field which is related to the meson fields)

$$L = L_2 + L_4 + L_0 \tag{1}$$

(the subindices refer to the number of derivatives in each term), where the sigma model part is

$$L_2 = -\frac{f_\pi^2}{4} \, \mathrm{Tr} \, (U^\dagger \partial_\mu U \, U^\dagger \partial^\mu U) \tag{2}$$

and a quartic term, referred to as the Skyrme term, has to be added, to circumvent the standard Derrick argument against the existence of static solutions

$$L_4 = -\frac{1}{32e^2} \, \mathrm{Tr} \, ([U^\dagger \partial_\mu U, U^\dagger \partial_\nu U]^2). \tag{3}$$

The last term is a potential

$$L_0 = -\mu^2 V(U, U^\dagger), \tag{4}$$

which explicitly breaks the the chiral symmetry. Its particular form is usually adjusted to a concrete physical situation. Finally, static properties of baryons as well as nuclei are derived with the help of the

Journal of Physics: Conference Series **284** (2011) 012006 doi:10.1088/1742-6596/284/1/012006

semiclassical quantization of the solitonic zero modes [7], [8], [9].
Within this framework, it has been established that topological solitons (Skyrmions) possess qualitative properties of baryons [7], [8]. Quantitatively, for the original Skyrme model nucleon properties are reproduced within a typical precision of about 30%. For higher nuclei, one of the principal problems of the Skyrme model is that the binding energies of the nuclei result too large already at the classical level. This problem may be traced back to the fact that the solitons in the Skyrme model are not BPS, i.e., they do not saturate the Bogomolny bound of the model. Other related problems are too strong internuclear forces at intermediate distances and the cristal like behaviour of large nuclei in the standard Skyrme model. The situation seems to be better for the excitation spectra of nuclei. Indeed, by identifying the symmetries of higher Skyrmions and by the use of the semiclassical quantization method the (iso-) rotational excitations of several light nuclei have been described rather successfully in [9].
In any case, as an effective field theory the Skyrme model should be viewed as a derivative expansion where higher derivative terms have been neglected. Further, baryons, i.e., extended (solitonic) solutions, have regions where derivatives are not small, so it may not be justified to omit these higher terms. Indeed, many *generalized* Skyrme models, i.e., extensions of the original Skyrme Lagrangian with higher derivative terms have been investigated [10]-[16].
Among them, the simplest and most natural generalization of the Skyrme model is provided by the addition of the following sextic derivative term

$$L_6 = \frac{\lambda^2}{24^2} \left[\mathrm{Tr}\, (\epsilon^{\mu\nu\rho\sigma} U^\dagger \partial_\mu U \; U^\dagger \partial_\nu U \; U^\dagger \partial_\rho U) \right]^2 . \tag{5}$$

Phenomenologically, this term effectively appears if one considers a massive vector field coupled to the chiral field via the baryon density [10], [16]. It turns out that such a modified Skyrme model gives slightly better predictions [10], [11], [13], [15]. Besides, this sextic term is at most quadratic in time derivatives, so it allows for a standard hamiltonian interpretation. It is, in fact, nothing but the topological (or baryon) current density squared,

$$L_6 = \lambda^2 \pi^4 \mathbb{B}_\mu \mathbb{B}^\mu \tag{6}$$

where

$$\mathbb{B}^\mu = \frac{1}{24\pi^2} \mathrm{Tr}\, (\epsilon^{\mu\nu\rho\sigma} U^\dagger \partial_\nu U \; U^\dagger \partial_\rho U \; U^\dagger \partial_\sigma U). \tag{7}$$

Within the set of generalized Skyrme models, the standard Skyrme model may be viewed as an approximation which is mode tractable and sufficient for the description of certain aspect of nuclei. The *BPS Skyrme model* is another approximation proposed in [17], [18], which consists of the potential and the sextic term only. In this contribution, we want to discuss its main properties as well as its viability for the description of nuclei. The lagrangian is

$$L_{06} = \frac{\lambda^2}{24^2} \left[\mathrm{Tr}\, (\epsilon^{\mu\nu\rho\sigma} U^\dagger \partial_\mu U \; U^\dagger \partial_\nu U \; U^\dagger \partial_\rho U) \right]^2 - \mu^2 V(U, U^\dagger), \tag{8}$$

where the subindex 06 refers to the fact that only a potential term without derivatives and a term sextic in derivatives appear in this lagrangian. The model is, by construction, more topological in nature than the standard Skyrme model. As in the standard Skyrme model, soliton solutions are stabilized by a higher derivative term. The model is an example of a BPS theory where the topological soliton solutions saturate its Bogomolny bound and, therefore, binding energies are zero at the classical level. Further, this BPS model has an infinite number of symmetries, among which there are the volume preserving diffeomorphisms on base space, which endowes the BPS Skyrme model with the symmetries of an incompressible liquid. Moreover, the solitons of this theory are of the compact type which results (classically) in a finite range of their interactions (contact-like interactions). As a consequence, the BPS Skyrme model apparently is a good guess for the description of certain properties of nuclei, as we

discuss below. For a more detailed discussion we refer to [17], [18].
We want to remark that there exists another BPS generalization of the Skyrme model [19], which is based on a dimensional reduction from a (4+1) dimensional Yang-Mills theory, where the $SU(2)$ Skyrme field is accompanied by an infinite tower of vector and tensor mesons.

2. The BPS Skyrme model

2.1. Field equations

We use the standard parametrization of U by means of a real scalar ξ and a three component unit vector $\vec{n}$ field ($\vec{\tau}$ are the Pauli matrices), $U = \exp[i\xi\vec{n}\cdot\vec{\tau}]$. Further, the vector field may be related to a complex scalar u by the stereographic projection

$$\vec{n} = \frac{1}{1+|u|^2}\left(u+\bar{u}, -i(u-\bar{u}), |u|^2-1\right).$$

For the topological current the resulting expression is

$$\mathbb{B}^\mu = \frac{-i}{\pi^2}\frac{\sin^4\xi}{(1+|u|^2)^4}\,\epsilon^{\mu\nu\rho\sigma}\xi_\nu u_\rho \bar{u}_\sigma,$$

and, assuming for the potential $V = V(\mathrm{tr}(U+U^\dagger)) = V(\xi)$ (which we assume for the rest of the paper) we get for the lagrangian

$$L_{06} = -\frac{\lambda^2\sin^4\xi}{(1+|u|^2)^4}\left(\epsilon^{\mu\nu\rho\sigma}\xi_\nu u_\rho\bar{u}_\sigma\right)^2 - \mu^2 V(\xi). \tag{9}$$

The Euler–Lagrange equations read

$$\frac{\lambda^2\sin^2\xi}{(1+|u|^2)^4}\partial_\mu(\sin^2\xi\, H^\mu) + \mu^2 V'_\xi = 0\,, \qquad \partial_\mu\left(\frac{K^\mu}{(1+|u|^2)^2}\right) = 0\,,$$

where

$$H_\mu = \frac{\partial(\epsilon^{\alpha\nu\rho\sigma}\xi_\nu u_\rho\bar{u}_\sigma)^2}{\partial\xi^\mu}, \quad K_\mu = \frac{\partial(\epsilon^{\alpha\nu\rho\sigma}\xi_\nu u_\rho\bar{u}_\sigma)^2}{\partial\bar{u}^\mu}.$$

These objects obey the useful formulae

$$H_\mu u^\mu = H_\mu\bar{u}^\mu = 0,\ \ K_\mu\xi^\mu = K_\mu u^\mu = 0,\ \ H_\mu\xi^\mu = K_\mu\bar{u}^\mu = 2(\epsilon^{\alpha\nu\rho\sigma}\xi_\nu u_\rho\bar{u}_\sigma)^2.$$

2.2. Symmetries

Apart from the standard Poincare symmetries, the model has an infinite number of target space symmetries. The sextic term alone is the square of the pullback of the volume form on the target space S^3, where this target space volume form reads explicitly

$$dV = -i\frac{\sin^2\xi}{(1+|u|^2)^2}d\xi du d\bar{u} \tag{10}$$

and the exterior (wedge) product of the differentials is understood. Therefore, the sextic term alone is invariant under all target space diffeomorphisms which leave this volume form invariant (the volume-preserving diffeomorphisms on the target S^3). The potential term in general does not respect all these symmetries, but depending on the specific choice, it may respect a certain subgroup. Concretely, for $V = V(\xi)$, the potential is invariant under those volume-preserving target space diffeomorphisms which do not change ξ,

$$\xi\to\xi\,,\quad u\to\tilde{u}(u,\bar{u},\xi)\,,\quad (1+|\tilde{u}|^2)^{-2}d\xi d\tilde{u}d\bar{\tilde{u}} = (1+|u|^2)^{-2}d\xi u d\bar{u}.$$

Since u spans a two-sphere in target space, these transformations form a one-parameter family of the groups of the area-preserving diffeomorphisms on the corresponding target space S^2 (one-parameter family because the transformations may still depend on ξ, although they act nontrivially only on $u, \bar{u}$). Both the Poincare transformations and this family of area-preserving target space diffeomorphisms are symmetries of the full action, so they are Noether symmetries with the corresponding conserved currents. The energy functional for static fields has the additional infinite-dimensional group of volume-preserving diffeomorphisms on base space as symmetries, as we want to discuss now. These symmetries are not symmetries of the full action, so they are not of the Noether type, but nevertheless they are very interesting from a physical point of view, as we will see below. The energy functional for static fields reads

$$E = \int d^3x \left(\frac{\lambda^2 \sin^4 \xi}{(1+|u|^2)^4} (\epsilon^{mnl} i \xi_m u_n \bar{u}_l)^2 + \mu^2 V \right) \tag{11}$$

and we observe that both d^3x and $\epsilon^{mnl} i \xi_m u_n \bar{u}_l$ are invariant under coordinate transformations of the base space coordinates x_j which leave the volume form d^3x invariant.

2.3. Bogomolny bound

In the BPS Skyrme model, there exists a Bogomolny bound for static finite energy solutions [20], [17], [18], [21]. The most concise version of the proof has been given by M. Speight [21], so here we use this version of the proof. In a first step, the energy functional has to be rewritten as a complete square plus a remainder,

$$\begin{aligned} E &= \int d^3x \left(\frac{\lambda^2 \sin^4 \xi}{(1+|u|^2)^4} (\epsilon^{mnl} i \xi_m u_n \bar{u}_l)^2 + \mu^2 V \right) \\ &= \int d^3x \left(\frac{\lambda \sin^2 \xi}{(1+|u|^2)^2} \epsilon^{mnl} i \xi_m u_n \bar{u}_l \pm \mu \sqrt{V} \right)^2 \\ &\mp \int d^3x \frac{2\mu\lambda \sin^2 \xi \sqrt{V}}{(1+|u|^2)^2} \epsilon^{mnl} i \xi_m u_n \bar{u}_l \\ &\geq \pm (2\lambda\mu\pi^2) \left[\frac{-i}{\pi^2} \int d^3x \frac{\sin^2 \xi \sqrt{V}}{(1+|u|^2)^2} \epsilon^{mnl} \xi_m u_n \bar{u}_l \right] \equiv E_{\rm BPS}, \end{aligned} \tag{12}$$

and in a second step it remains to prove that the remainder $E_{\rm BPS}$ is topological and does not depend on the field configuration. For this we note that the integrand in the above expression is not only a three-form on the base space but, in addition, it is the pullback under the map U of a three-form on target space. Therefore, the base space integral of the base space three-form is equal to the target space integral of the target space three-form times the number of times the target space is covered while the base space is covered once (i.e., times the winding number $|B|$),

$$\begin{aligned} E_{\rm BPS} &= 2\lambda\mu\pi^2 |B| \frac{-i}{\pi^2} \int_{S^3} \frac{d\xi du d\bar{u} \sin^2 \xi}{(1+|u|^2)^2} \sqrt{V(\xi)} \\ &\equiv 2\lambda\mu\pi^2 < \sqrt{V} >_{S^3} |B| \end{aligned} \tag{13}$$

where $< \sqrt{V} >_{S^3}$ is the average value of $\sqrt{V}$ on the target space S^3,

$$< \sqrt{V} >_{S^3} = \frac{\int_{S^3} d\Omega \sqrt{V}}{\int_{S^3} d\Omega}$$

and $d\Omega$ is the volume form on S^3 (and $\int_{S^3} d\Omega = \pi^2$). The corresponding Bogomolny (first order) equation is

$$\frac{\lambda \sin^2 \xi}{(1+|u|^2)^2} \epsilon^{mnl} i \xi_m u_n \bar{u}_l = \mp \mu \sqrt{V} \tag{14}$$

and is satisfied by all soliton solutions which we will encounter below.

2.4. Exact solutions

We are interested in static topologically nontrivial solutions. Thus u must cover the whole complex plane ($\vec{n}$ covers at least once S^2) and $\xi \in [0, \pi]$. The natural axially symmetric ansatz is

$$\xi = \xi(r), \quad u(\theta, \phi) = g(\theta) e^{in\phi}. \tag{15}$$

Then, the field equation for u reads

$$\frac{1}{\sin\theta} \partial_\theta \left(\frac{g^2 g_\theta}{(1+g^2)^2 \sin\theta} \right) - \frac{g g_\theta^2}{(1+g^2)^2 \sin^2\theta} = 0,$$

and the solution with the right boundary condition is

$$g(\theta) = \tan\frac{\theta}{2}. \tag{16}$$

Observe that this solution holds for all values of n. The equation for the real scalar field reads

$$\frac{n^2 \lambda^2 \sin^2\xi}{2r^2} \partial_r \left(\frac{\sin^2\xi \, \xi_r}{r^2} \right) - \mu^2 V_\xi = 0$$

or, after introducing the new variable $z = (\sqrt{2}\mu r^3)/(3|n|\lambda)$,

$$\sin^2\xi \, \partial_z \left(\sin^2\xi \, \xi_z \right) - V_\xi = 0. \tag{17}$$

It may be integrated to

$$\frac{1}{2} \sin^4\xi \, \xi_z^2 = V, \tag{18}$$

where we chose a vanishing integration constant to get finite energy solutions. This first integration of the field equation is just the dimensionally reduced Bogomolny equation.

For a further evaluation we have to specify a potential. Here we shall restrict ourselves to the case of the standard Skyrme potential

$$V = \frac{1}{2}\text{Tr}(1 - U) \quad \rightarrow \quad V(\xi) = 1 - \cos\xi. \tag{19}$$

Imposing the boundary conditions for topologically non-trivial solutions we get

$$\xi = \begin{cases} 2\arccos\sqrt[3]{\frac{3z}{4}} & z \in [0, \frac{4}{3}] \\ 0 & z \geq \frac{4}{3}. \end{cases} \tag{20}$$

The solution is of the compacton type, i.e., it has a finite support (see [17], [18] for details; similar compact solutions in the baby Skyrme model have been found in [22], [23], [24]).

We remark that the compact nature of the solutions is a consequence of the specific potential we chose (more concretely, its quadratic behaviour near the vacuum). Other potentials (with a different behaviour near the vacuum) lead to non-compact solutions.

The corresponding energy is

$$E = 8\sqrt{2}\pi\mu\lambda|n| \int_0^{4/3} \left(1 - \left(\frac{3z}{4} \right)^{\frac{2}{3}} \right) dz = \frac{64\sqrt{2}\pi}{15}\mu\lambda|n| \tag{21}$$

and is linear in the topological charge $|B| = |n|$. The energy density and topological charge density per unit volume are

$$\begin{aligned} \mathcal{E} &= 8\sqrt{2}\mu\lambda(1-|n|^{-\frac{2}{3}}\tilde{r}^2) \quad \text{for} \quad 0 \leq \tilde{r} \leq |n|^{\frac{1}{3}} \\ &= 0 \quad \text{for} \quad \tilde{r} > |n|^{\frac{1}{3}}. \end{aligned} \tag{22}$$

$$\begin{aligned} \mathcal{B} &= \text{sign}(n)\frac{4}{\pi^2}(1-|n|^{-\frac{2}{3}}\tilde{r}^2)^{\frac{1}{2}} \quad \text{for} \quad 0 \leq \tilde{r} \leq |n|^{\frac{1}{3}} \\ &= 0 \quad \text{for} \quad \tilde{r} > |n|^{\frac{1}{3}} \end{aligned} \tag{23}$$

where

$$\tilde{r} = \left(\frac{\sqrt{2}\mu}{4\lambda}\right)^{\frac{1}{3}} r \equiv \frac{r}{R_0} \tag{24}$$

(here R_0 is the compacton radius for the soliton with charge one), see Figure 1.

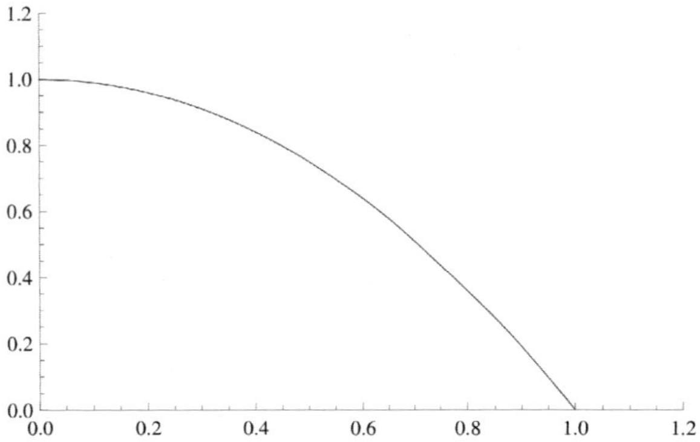

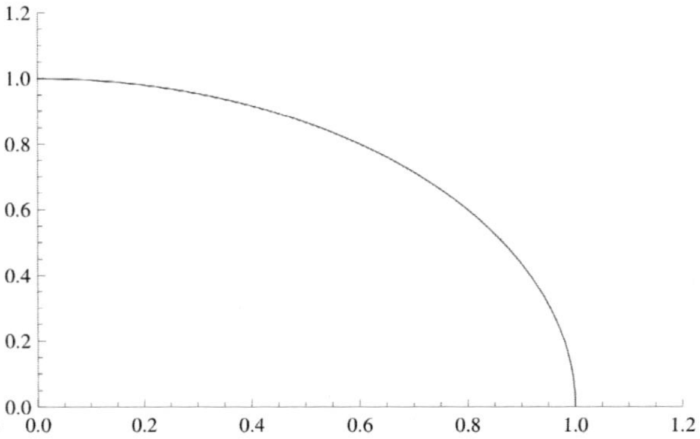

Figure 1. Normalized energy density (left figure) and topological charge density (right figure) as a function of the rescaled radius $\tilde{r}$, for topological charge n=1. For $|n| > 1$, the height of the densities remains the same, whereas their radius grows like $|n|^{\frac{1}{3}}$

3. Some phenomenology of nuclei

For simplicity, we choose the standard Skyrme potential $V = 1 - \cos\xi$ throughout this section.

3.1. Classical aspects

Already the classical solutions of the BPS Skyrme model describe quite well some static properties of nuclei [17], [18]. Concretely, we find

- A linear mass - baryon number relation: $E = 64\sqrt{2}\mu\lambda|B|/15 \equiv E_0|B|$, which is a well-known relation for nuclei
- No binding energies due to the BPS property of soliton solutions. Binding energies for physical nuclei are small ($\leq 1\%$), while in the standard Skyrme model they are significantly bigger.
- No forces between solitons: due to the BPS and compact nature of solutions there are no long range forces. Physical nuclei have a very short range interaction.

- Radii of nuclei: the compact solutions have a definite radius $R = (2\sqrt{2}\lambda|B|/\mu)^{(1/3)} \equiv R_0 \sqrt[3]{|B|}$ which, again, is a well-known relation for physical nuclei.
- Incompressible fluid: the static energy functional has the volume-preserving diffeos (VPDs) as symmetries, which are the symmetries of an incompressible ideal fluid. Physical nuclei do not have this symmetry; they have definite shape, and deformations cost energy. But volume-preserving deformations cost much less energy, as a consequence of the liquid drop model of nuclei.

To summarize, the classical model already reproduces some features of the nuclear liquid drop model (mass and volume proportional to baryon charge, strictly finite size, VPDs). Clearly, this picture is still too naive (there are no pions and, therefore, no long range interactions and no pion cloud; there are no quantum corrections, etc.), and both the quantization and the inclusion of further terms in the lagrangian will be necessary for a more precise description of nuclear properties.

3.2. Quantization

Here we perform the first steps towards a semiclassical (collective coordinates) quantization of solitons, as is done for the standard Skyrme model. Concretely, the standard procedure consists in quantizing the (iso-)rotational degrees of freedom of each soliton. In principle, one can quantize all solitons in this way and obtain information about the binding energies and rotational spectra of nuclei, but here we shall restrict to the simplest $B = 1$ case (the nucleon). Following the standard treatment, we introduce the collective coordinates of the isospin by including a time-dependent iso-rotation of the classical soliton configuration

$$U(x) = A(t) U_0(x) A^\dagger(t), \tag{25}$$

where $A(t) = a_0 + i a_i \tau_i \in SU(2)$ and $a_0^2 + \vec{a}^2 = 1$. The $B = 1$ hedgehog is invariant under a combined rotation in base and isotopic space, therefore it is enough to introduce the collective coordinates of one of the two. The standard procedure consists in inserting these iso-rotations into the lagrangian, introducing the conjugate momenta to the SU(2) variables a_n and applying the standard canonical quantization to the latter. This results in an energy expression which is the sum of the static soliton energy E_0 and an isospin (or equivalently spin) contribution

$$H = E_0 + \frac{\hbar^2 I^2}{2\mathcal{I}},$$

where I^2 is the isospin operator (the spherical laplacian on S^3) [18], [15]. Further, $\mathcal{I}$ is the moment of inertia

$$\mathcal{I} = \frac{4\pi}{3} \int dr \sin^4 \xi \, \xi_r^2 = \frac{2^8 \sqrt{2}\pi}{15 \cdot 7} \lambda\mu \left(\frac{\lambda}{\mu}\right)^{\frac{2}{3}}$$

The soliton with baryon number one is quantized as a fermion. Concretely, the nucleon has spin and isospin 1/2, whereas the Δ resonance has spin and isospin 3/2, so we find for their masses

$$M_N = E_0 + \frac{3\hbar^2}{8\mathcal{I}}, \quad M_\Delta = E_0 + \frac{15\hbar^2}{8\mathcal{I}} \quad \Rightarrow \quad M_\Delta - M_N = \frac{3\hbar^2}{2\mathcal{I}}, \tag{26}$$

like in the standard Skyrme model. These expressions may now be fitted to the physical masses of the nucleon ($M_N = 938.9$ MeV) and the Δ resonance ($M_\Delta = 1232$ MeV), which determines the coupling constants

$$\lambda\mu = 45.70\text{MeV}\,, \qquad \frac{\lambda}{\mu} = 0.2556\text{fm}^3$$

These may now be used to "predict" further physical quantities like, e.g. the charge radii of the nucleons. For the isoscalar (baryon) charge density per unit r we find

$$\rho_0 = 4\pi r^2 B^0 = -\frac{2}{\pi} \sin^2 \xi \xi_r' \tag{27}$$

radius	BPS Skyrme	massive Skyrme	exp.
compacton	0.90	-	-
electric isoscalar $r_{e,0}$	0.64	0.68	0.72
electric isovector $r_{e,1}$	0.67	1.04	0.88
magnetic isoscalar $r_{m,0}$	0.71	0.95	0.81
$r_{e,1}/r_{e,0}$	1.05	1.53	1.22
$r_{m,0}/r_{e,0}$	1.12	1.40	1.13
$r_{e,1}/r_{m,0}$	0.94	1.10	1.09

Table 1. Compacton radius and some charge radii and their ratios for the nucleon. The numbers for the massive Skyrme model are from Ref. [8]. All radii are in fm.

and for the isovector charge density per unit r

$$\rho_1 = \frac{1}{\mathcal{I}}\frac{4\pi}{3}\lambda^2 \sin^4 \xi \xi_r'^2 . \tag{28}$$

The corresponding isoscalar and isovector mean square electric radii are

$$< r^2 >_{e,0} = \int dr r^2 \rho_0 = \left(\frac{\lambda}{\mu}\right)^{2/3} , \tag{29}$$

$$< r^2 >_{e,1} = \int dr r^2 \rho_1 = \frac{10}{9}\left(\frac{\lambda}{\mu}\right)^{2/3} . \tag{30}$$

Further, the isoscalar magnetic radius is defined as the ratio

$$< r^2 >_{m,0} = \frac{\int dr r^4 \rho_0}{\int dr r^2 \rho_0} = \frac{5}{4}\left(\frac{\lambda}{\mu}\right)^{2/3} . \tag{31}$$

The numerical values are displayed in Table 1.
In the BPS Skyrme model, all radii are bound by the compacton radius $R_0 = \sqrt{2}(\lambda/\mu)^{(1/3)}$. Further, all radii in the BPS Skyrme model are significantly smaller than their physical values, as well as significantly smaller than the values predicted in the standard massive Skyrme model. This has to be expected, because we know already that the pion cloud is absent in the BPS Skyrme model. Besides, we see that the error caused by the absence of the pion cloud in the model partly cancels in the ratios, as one would expect.

4. Conclusions

In this contribution we reviewed the properties of a certain submodel within the set of generalized Skyrme models, the *BPS Skyrme model*. The model is quite interesting on its own, because of its infinitely many symmetries and conserved charges, its solvability for any form of the potential, and because all solutions are of the BPS type. They obey a first order differential equation and saturate a Bogomolny bound.
Further, we found some evidence that the model may be a good starting point for an effective field theory description of nuclei. For a further development of this application of the model, however, the following problems have to be resolved or further investigated.

- Symmetry breaking: the infinitely many symmetries of the model are not shared by physical nuclei. In addition, it is not clear how to select the correct soliton from the infinitely many ones related by symmetries or how to quantize these infinitely many symmetries. Therefore, a realistic phenomenological application will require the breaking of these symmetries. The challenge will be to identify a breaking mechanism which effectively breaks the unwanted symmetries without perturbing too much the good properties of the model (like weak binding energies, weak internuclear forces, etc.).

- Quantization of higher nuclei: the semiclassical quantization of higher solitons should be performed and applied to higher nuclei, along the lines of what was done for the standard Skyrme model, e.g., in [9]. Recently in [25], the authors used our BPS Skyrme model with a different potential for this purpose. Concretely, they calculated the exact static soliton solutions plus the (iso-) rotational energies in the rigid rotor quantization for general baryon charge $B = n$ for the axially symmetric ansatz. Then they allowed for small perturbations of the total energies by the quadratic and quartic Skyrme terms and fitted the resulting binding energies to the experimental binding energies of the most abundant isotopes of higher nuclei, assuming, as is usually done, that these correspond to the states with the lowest possible value of the isospin. The resulting agreement between calculated and experimentally determined masses and binding energies is impressive, lending further support to the viability of the BPS Skyrme model as the leading contribution to an effective theory for the properties of nuclear matter. Their specific choice of the potential was motivated by the two requirements to avoid the compact nature of the solitons in order to simplify the analytical calculations, on the one hand, and to have the standard pion tail in the full model with the quadratic Skyrme term included, on the other hand. One can see, however, that these two requirements are mutually incompatible (but this issue is completely irrelevant for the remaining results of that paper). Their potential has, in fact, two vacua and behaves quadratically about one vacuum, but sextic around the other. Further, their solitons approach the sextic vacuum at large distances. This implies that the pure BPS model solutions are not exactly compact but approach the (sextic) vacuum by a steep exponential tail. On the other hand, in the full model there is no pion tail (the pion tail would be induced by the quadratic vacuum which is, however, attained at the center of the soliton, and not in the asymptotic large distance region). In any case, a further investigation of the semiclassical quantization with the inclusion of additional physical effects like higher excitations of nuclei or electrostatic interactions, among others, is the next necessary step.
- Motivation from QCD: it would be very interesting to see whether the rather good phenomenological properties of the model can be justified in a more rigorous manner from the fundamental theory of strong interactions, i.e., QDC. In this context we want to emphasize that the two term of the BPS Skyrme model are rather specific. The sextic term is essentially topological in nature and is, therefore, naturally related to collective excitations of the underlying microscopic degrees of freedom. The potential, on the other hand, provides the chiral symmetry breaking and might therefore be related to collective degrees of freedom of the quarks, like the quark condensate. A related issue is the behaviour for a large number of colors N_c. Indeed, the good phenomenological properties of the BPS Skyrme model seem to further improve in the limit of large N_c, but a deeper understanding of this fact is still missing. For a more detailed discussion we refer to [18]. In any case, these issues currently are under investigation.

To summarize, we have demonstrated that BPS solutions can be obtained in a Skyrme type theory. We think that we have found and solved an interesting submodel in the space of generalized Skyrme models, which is singled out both by its capacity to reproduce qualitative properties of the liquid drop model of nuclei and by its unique mathematical structure. The model directly relates the nuclear mass to the topological charge, and it naturally provides both a finite size for the nuclei and the liquid drop behaviour, which probably is not easy to get from an effective field theory. Finally, our exact solutions might provide a calibration for the demanding numerical computations in physical applications of more general Skyrme models.

Acknowledgments
C.A. and J.S.-G. thank the Ministry of Science and Investigation, Spain (grant FPA2008-01177), and the Xunta de Galicia (grant INCITE09.296.035PR and Conselleria de Educacion) for financial support. A.W. acknowledges support from the Ministry of Science and Higher Education of Poland grant N N202 126735 (2008-2010).

References
[1] t'Hooft G 1974 *Nucl. Phys.* B **72** 461
[2] Witten E. 1979 *Nucl. Phys.* B **160** 57
[3] Witten E 1983 *Nucl. Phys.* B **223** 433
[4] Skyrme THR 1961 *Proc. Roy. Soc. Lon.* **260** 127
[5] Skyrme THR 1962 *Nucl. Phys.* **31** 556
[6] Skyrme THR 1971 *J. Math. Phys.* **12** 1735
[7] Adkins GS, C. R. Nappi CR and Witten E 1983 *Nucl. Phys.* B **228** 552
[8] Adkins GS and Nappi CR 1984 *Nucl. Phys.* B**233** 109
[9] Battye RA, Manton NS, Sutcliffe PM and Wood SW 2009 *Phys. Rev.* C **80** 034323
[10] Adkins GS and Nappi CR 1984 *Phys. Lett.* B **137** 251
[11] Jackson A, Jackson AD, Goldhaber AS, Brown GE and Castillejo LC 1985 *Phys. Lett.* B **154** 101
[12] Ishii N, Aoki S and Hatsuda T 2007 *Phys. Rev. Lett.* 99 022001
[13] Floratos I and Piette B 2001 *Phys. Rev.* D **64** 045009
[14] Floratos I and Piette B 2001 *J. Math. Phys.* **42** 5580
[15] Ding GL and Yan ML 2007 *Phys. Rev.* C **75** 034004
[16] Sutcliffe P 2009 *Phys. Rev.* D **79** 085014
[17] Adam C, Sanchez-Guillen J and Wereszczynski A 2010 *Phys. Lett.* B **691** 105 (*Preprint* arXiv: 1001.4544)
[18] Adam C, Sanchez-Guillen J and Wereszczynski A 2010 *Phys. Rev.* D **82** 085015 (*Preprint* arXiv:1007.1567)
[19] Sutcliffe P 2010 *J. High Energy Phys.* JHEP08(2010)019 (*Preprint* arxive:1003.0023)
[20] de Innocentis M and Ward RS 2001 *Nonlinearity* **14** 663
[21] Speight JM 2010 *J. Phys.* A **43** 405201 (*Preprint* arXiv:1006.3754)
[22] Gisiger T and Paranjape MB 1997 *Phys. Rev.* D **55** 7731
[23] Adam C, Klimas P, Sanchez-Guillen J and Wereszczynski A 2009 *Phys. Rev.* D **80** 105013
[24] Adam C, Romanczukiewicz T, Sanchez-Guillen J and Wereszczynski A 2010 *Phys. Rev.* D **81** 085007
[25] Bonenfant E and Marleau L 2010 *Phys. Rev.* D **82** 054023 (*Preprint* arXiv:1007.1396)

Journal of Physics: Conference Series **284** (2011) 012007 doi:10.1088/1742-6596/284/1/012007

Lowest weight representations of super Schrödinger algebras in low dimensional spacetime

N Aizawa

Department of Mathematics and Information Sciences, Graduate School of Science, Osaka Prefecture University, Nakamozu Campus, Sakai, Osaka 599-8531, Japan

E-mail: aizawa@mi.s.osakafu-u.ac.jp

Abstract. We investigate the lowest weight representations of the super Schrödinger algebras introduced by Duval and Horváthy. This is done by the same procedure as the semisimple Lie algebras. Namely, all singular vectors within the Verma modules are constructed explicitly then irreducibility of the associated quotient modules is studied again by the use of singular vectors. We present the classification of irreducible Verma modules for the super Schrödinger algebras in $(1+1)$ and $(2+1)$ dimensional spacetime with $\mathcal{N} = 1, 2$ extensions.

1. Introduction

Throughout the last four decades nonrelativistic conformal symmetries have been observed in wide area of physics ranging from condensed matter to high energy (see for example [1–18] and references therein). In many cases the symmetries are described by the groups extending the Galilei symmetry group of nonrelativistic systems [19, 20]. Those extended Galilei groups are non-semisimple as the original Galilei group. This may be an obstacle to develop systematic study of representations for such groups.

The simplest conformal extension of the Galilei group may be the Schrödinger groups [1, 2] introduced in the studies of symmetry for free Schrödinger equation. Geometrical interpretation of the Schrödinger group revealed that the AdS/CFT correspondence has its nonrelativistic analogue [14, 15] (see also [8]). Supersymmetric extensions of the Schrödinger group and its Lie algebra have also been discussed in connection with various physical systems such as fermionic oscillator [21, 22], spinning particles [23], nonrelativistic Chern-Simons matter [24–26], Dirac monopole and magnetic vortex [25], many-body quantum systems [27] and so on. Some super Schrödinger algebras were constructed from the viewpoint of infinite dimensional Lie super algebra [28] or by embedding them to conformal superalgebras [29, 30].

Despite the numerous publications on the Schrödinger algebra and its supersymmetric extensions, the representation theories of them are not studied well. Especially there are a few works based on representation theoretic viewpoint. We mention the followings: Projective representations of the Schrödinger group in 3 spatial dimension were constructed in [31]. Irreducible representations of the Schrödinger algebras up to 3 spatial dimension were investigated in [32–34]. The author of [35] studied the highest weight representations of the $\mathcal{N} = 2$ super Schrödinger algebra in 2 spatial dimension ("exotic" algebra in the terminology of [25]). To apply the (super) Schrödinger algebras for physical problems detailed study of irreducible representations of them will be helpful.

In the present work, we provide a classification of the Verma modules (a lowest weight representations) over the super Schrödinger algebras for simple cases, namely, one and two spatial dimensions with $\mathcal{N} = 1, 2$ supersymmetric extensions. The Schrödinger algebra discussed in this paper is centrally extended one. The central term corresponds to the mass of a system. In general, the Schrödinger algebra of fixed spacetime dimension has some distinct supersymmetric extensions even for fixed value of $\mathcal{N}$. A systematic method to extend the Schrödinger algebra in $(n+1)$ dimensional spacetime to arbitrary $\mathcal{N}$ has developed in [25]. We restrict our study of the representations to the super Schrödinger algebras introduced in [25]. We denote the centrally extended Schrödinger algebra in $(n+1)$ dimensional spacetime by $\mathfrak{s}(n)$. The $\mathcal{N} = 1, 2$ extension of $\mathfrak{s}(1)$ are denoted by $\mathfrak{s}(1/1)$ and $\mathfrak{s}(1/2)$, respectively. The $\mathcal{N} = 1$ extension of $\mathfrak{s}(2)$ is also denoted by $\mathfrak{s}(2/1)$. On the other hand there exist two $\mathcal{N} = 2$ extensions for $\mathfrak{s}(2)$. We denote one of them so-called standard by $\mathfrak{s}(2/2)$ and the another one so-called exotic by $\hat{\mathfrak{s}}(2/2)$. The superalgebra $\mathfrak{s}(n/2)$ corresponds to the $N_+ = 1$, $N_- = 0$ extended algebra in the notation of [25].

The plan of this paper is as follows. We present the definition and structure of the super Schrödinger algebras in the next section. After giving the Verma modules in section 3 we give all the singular vectors and list of irreducible lowest weight modules without detailed calculation. The detailed proof is found in [36, 37]. The employed procedure in the references is the same one as [32, 33]. This shows that the standard method of representation theory for semisimple Lie algebras is able to apply the super Schrödinger algebras, too.

2. Definition and structure of super Schrödinger algebras

The bosonic Schrödinger algebra in $(n + 1)$ dimensional spacetime $\mathfrak{s}(n)$ is generated by the following transformations: time translation (H), space translations (P_a), Galilei boosts (G_a), spatial rotations (J_{ab}), dilatation (D), conformal transformation (K) and the central element (M). Contrast to the relativistic conformal algebra, there exists only one conformal generator. In this section we give the definition of the super Schrödinger algebras studied in this paper and their triangular decomposition which will be the key to study representations.

2.1. Algebras in $(1 + 1)$ *dimensional spacetime*

The bosonic sector has six generators. Their nonvanishing commutators are given by

$$\begin{aligned} &[H, D] = 2H, \qquad [H, K] = D, \qquad [D, K] = 2K, \qquad [P, G] = M, \\ &[H, G] = P, \qquad [D, G] = G, \qquad [P, D] = P, \qquad [P, K] = G. \end{aligned} \tag{1}$$

The $\mathcal{N} = 1$ extension $\mathfrak{s}(1/1)$ has three fermionic generators: $\mathscr{Q}, \mathscr{S}, \mathscr{X}$. They satisfy the anti-commutation relations

$$\begin{aligned} &\{\mathscr{Q}, \mathscr{Q}\} = -2H, \qquad \{\mathscr{S}, \mathscr{S}\} = -2K, \qquad \{\mathscr{X}, \mathscr{X}\} = -M, \\ &\{\mathscr{Q}, \mathscr{X}\} = -P, \qquad \{\mathscr{S}, \mathscr{X}\} = -G, \qquad \{\mathscr{Q}, \mathscr{S}\} = -D. \end{aligned} \tag{2}$$

As seen from the relations none of them are nilpotent. Nonvanishing bosonic-fermionic relations are listed below:

$$\begin{aligned} &[\mathscr{Q}, D] = \mathscr{Q}, \qquad [\mathscr{Q}, K] = \mathscr{S}, \qquad [D, \mathscr{S}] = \mathscr{S}, \qquad [H, \mathscr{S}] = \mathscr{Q}, \\ &[\mathscr{Q}, G] = \mathscr{X}, \qquad [P, \mathscr{S}] = \mathscr{X}. \end{aligned} \tag{3}$$

The $\mathcal{N} = 2$ extension $\mathfrak{s}(1/2)$ has six fermionic generators $\mathscr{Q}_a, \mathscr{S}_a, \mathscr{X}_a$ $(a = \pm)$ and one additional bosonic generator R corresponding to the R-parity. The generator R commutes with all bosonic elements of $\mathfrak{s}(1/2)$. The nilpotent fermionic generators satisfy the anti-commutation relations

$$\begin{aligned} &\{\mathscr{Q}_\pm, \mathscr{Q}_\mp\} = -2H, && \{\mathscr{S}_\pm, \mathscr{S}_\mp\} = -2K, && \{\mathscr{X}_\pm, \mathscr{X}_\mp\} = -M, \\ &\{\mathscr{Q}_\pm, \mathscr{X}_\mp\} = -P, && \{\mathscr{S}_\pm, \mathscr{X}_\mp\} = -G, && \{\mathscr{Q}_\pm, \mathscr{S}_\mp\} = -D \mp R. \end{aligned} \tag{4}$$

Bosonic-fermionic commutators are given by

$$\begin{aligned} &[\mathscr{Q}_\pm, D] = \mathscr{Q}_\pm, && [\mathscr{Q}_\pm, K] = \mathscr{S}_\pm, && [D, \mathscr{S}_\pm] = \mathscr{S}_\pm, && [H, \mathscr{S}_\pm] = \mathscr{Q}_\pm \\ &[\mathscr{Q}_\pm, G] = \mathscr{X}_\pm, && [P, \mathscr{S}_\pm] = \mathscr{X}_\pm, && [R, \mathscr{A}_\pm] = \pm\mathscr{A}_\pm, \end{aligned} \tag{5}$$

where $\mathscr{A}_a = \mathscr{Q}_a, \mathscr{S}_a, \mathscr{X}_a$.

2.2. Algebras in $(2+1)$ dimensional spacetime

In this spacetime there exists one spatial rotation J. The bosonic algebra $\mathfrak{s}(2)$ has nine generators. Their nonvanishing commutators are given by

$$\begin{aligned} &[H, D] = 2H, && [H, K] = D, && [D, K] = 2K, && [P_\pm, G_\mp] = 2M, \\ &[H, G_\pm] = P_\pm, && [D, G_\pm] = G_\pm, && [P_\pm, D] = P_\pm, && [P_\pm, K] = G_\pm, \\ &[J, G_\pm] = \pm G_\pm, && [J, P_\pm] = \pm P_\pm. \end{aligned} \tag{6}$$

The $\mathscr{N} = 1$ extension $\mathfrak{s}(2/1)$ has four fermionic elements $\mathscr{Q}, \mathscr{S}, \mathscr{X}_\pm$ subject to the relations:

$$\begin{aligned} &\{\mathscr{Q}, \mathscr{Q}\} = -2H, && \{\mathscr{S}, \mathscr{S}\} = -2K, && \{\mathscr{X}_\pm, \mathscr{X}_\mp\} = -2M, \\ &\{\mathscr{Q}, \mathscr{X}_\pm\} = -P_\pm, && \{\mathscr{S}, \mathscr{X}_\pm\} = -G_\pm, && \{\mathscr{Q}, \mathscr{S}\} = -D. \end{aligned} \tag{7}$$

The generators $\mathscr{X}_\pm$ are nilpotent, however, $\mathscr{Q}, \mathscr{S}$ are not. Nonvanishing bosonic-fermionic commutators are given by

$$\begin{aligned} &[\mathscr{Q}, D] = \mathscr{Q}, && [\mathscr{Q}, K] = \mathscr{S}, && [D, \mathscr{S}] = \mathscr{S}, && [H, \mathscr{S}] = \mathscr{Q}, \\ &[\mathscr{Q}, G_\pm] = \mathscr{X}_\pm, && [P_\pm, \mathscr{S}] = \mathscr{X}_\pm, && [J, \mathscr{X}_\pm] = \pm\mathscr{X}_\pm. \end{aligned} \tag{8}$$

For $\mathfrak{s}(2)$ there exist two distinct $\mathscr{N} = 2$ extensions, called *standard* and *exotic*. The standard extension $\mathfrak{s}(2/2)$ has eight fermionic generators $\mathscr{Q}_\pm, \mathscr{S}_\pm, \mathscr{X}_{\pm\pm}, \mathscr{X}_{\pm\mp}$ and additional bosonic R-parity operator. The fermionic sector is defined by the relations:

$$\begin{aligned} &\{\mathscr{Q}_\pm, \mathscr{Q}_\mp\} = -2H, && \{\mathscr{S}_\pm, \mathscr{S}_\mp\} = -2K, && \{\mathscr{X}_{++}, \mathscr{X}_{--}\} = -2M,, \\ &\{\mathscr{X}_{+-}, \mathscr{X}_{-+}\} = -2M, && \{Q_\pm, \mathscr{X}_{\sigma\mp}\} = -P_\sigma, && \{\mathscr{S}_\pm, \mathscr{X}_{\sigma\mp}\} = -G_\sigma, \\ &\{\mathscr{Q}_\pm, \mathscr{S}_\mp\} = -D \mp R, \end{aligned} \tag{9}$$

where $\sigma = +, -$. Nonvanishing bosonic-fermionic relations are listed below:

$$\begin{aligned} &[\mathscr{Q}_\pm, D] = \mathscr{Q}_\pm, && [\mathscr{Q}_\pm, K] = \mathscr{S}_\pm, && [D, \mathscr{S}_\pm] = \mathscr{S}_\pm, && [H, \mathscr{S}_\pm] = \mathscr{Q}_\pm \\ &[G_\pm, \mathscr{Q}_\pm] = -\mathscr{X}_{\pm\pm}, && [G_\pm, \mathscr{Q}_\mp] = -\mathscr{X}_{\pm\mp}, && [P_\pm, \mathscr{S}_\pm] = \mathscr{X}_{\pm\pm}, && [P_\pm, \mathscr{S}_\mp] = \mathscr{X}_{\pm\mp}, \\ &[R, \mathscr{A}_\pm] = \pm\mathscr{A}_\pm, && [R, \mathscr{X}_{\sigma\pm}] = \pm\mathscr{X}_{\sigma\pm}, && [J, \mathscr{X}_{\sigma\pm}] = \sigma\mathscr{X}_{\sigma\pm}, \end{aligned} \tag{10}$$

where $\mathscr{A} = \mathscr{Q}, \mathscr{S}$.

The exotic extension $\hat{\mathfrak{s}}(2/2)$ has six fermionic generators $\mathscr{Q}_\pm, \mathscr{S}_\pm, \mathscr{X}_\pm$ and additional bosonic R-parity operator. The fermionic generators are subject to the relations:

$$\begin{aligned} &\{\mathscr{Q}_\pm, \mathscr{Q}_\mp\} = -2H, && \{\mathscr{S}_\pm, \mathscr{S}_\mp\} = -2K, && \{\mathscr{X}_\pm, \mathscr{X}_\mp\} = -M, \\ &\{\mathscr{Q}_\pm, \mathscr{X}_\mp\} = -P_\mp, && \{\mathscr{S}_\pm, \mathscr{X}_\mp\} = -G_\mp, && \{\mathscr{Q}_\pm, \mathscr{S}_\mp\} = -D \mp (J+R), \end{aligned} \tag{11}$$

and those for the bosonic-fermionic sector are

$$\begin{array}{llll} [\mathscr{Q}_\pm, D] = \mathscr{Q}_\pm, & [\mathscr{Q}_\pm, K] = \mathscr{S}_\pm, & [D, \mathscr{S}_\pm] = \mathscr{S}_\pm, & [H, \mathscr{S}_\pm] = \mathscr{Q}_\pm \\ [\mathscr{Q}_\pm, G_\pm] = 2\mathscr{X}_\pm, & [P_\pm, \mathscr{S}_\pm] = 2\mathscr{X}_\pm, & [J, \mathscr{Q}_\pm] = \mp\mathscr{Q}_\pm, & [J, \mathscr{S}_\pm] = \mp\mathscr{S}_\pm, \\ [R, \mathscr{A}_\pm] = \pm 2\mathscr{A}_\pm, & & & \end{array} \tag{12}$$

where $\mathscr{A} = \mathscr{Q}, \mathscr{S}, \mathscr{X}$. We remark that the R-parity operator for $\mathcal{N} = 2$ extensions also commute with all bosonic generators.

2.3. *Triangular decomposition*

The grading and triangular decomposition for $\mathfrak{s}(n)$ introduced in [32] are easily extended to the super Schrödinger algebras discussed in this paper. We define the following degree for $\mathfrak{s}(1/1)$:

$$\begin{array}{lll} \deg K = 2, & \deg G = \deg \mathscr{S} = 1, & \deg D = \deg M = \deg \mathscr{X} = 0, \\ \deg P = \deg \mathscr{Q} = -1, & \deg H = -2. & \end{array} \tag{13}$$

It is immediate to see that $\mathfrak{s}(1/1)$ is $\mathbb{Z}$-graded by this degree. According to the sign of the degree one may define the triangular decomposition of $\mathfrak{s}(1/1)$ as follow:

$$\begin{aligned} \mathfrak{s}(1/1) &= \mathfrak{s}(1/1)^+ \oplus \mathfrak{s}(1/1)^0 \oplus \mathfrak{s}(1/1)^- \\ &= \{\, K,\ G \,\} \oplus \{\, D,\ M,\ \mathscr{X} \,\} \oplus \{\, P,\ \mathscr{Q} \,\}. \end{aligned} \tag{14}$$

Appropriate definitions of degree for other algebras also enable us to define the triangular decomposition. Below we give the decomposition. The algebras $\mathfrak{s}(1/2)$ and $\mathfrak{s}(2/1)$ are $\mathbb{Z}^2$-graded:

decomposition	generator (degree)
$\mathfrak{s}(1/2)^+$	$K(2,0),\ G(1,0),\ \mathscr{S}_+(1,1),\ \mathscr{S}_-(1,-1),\ \mathscr{X}_+(0,1)$
$\mathfrak{s}(1/2)^0$	$D(0,0),\ R(0,0),\ M(0,0)$
$\mathfrak{s}(1/2)^-$	$H(-2,0),\ P(-1,0),\ \mathscr{Q}_+(-1,1),\ \mathscr{Q}_-(-1,-1),\ \mathscr{X}_-(0,-1)$
$\mathfrak{s}(2/1)^+$	$K(2,0),\ G_+(1,1),\ G_-(1,-1),\ \mathscr{S}(1,0),\ \mathscr{X}_+(0,1)$
$\mathfrak{s}(2/1)^0$	$D(0,0),\ J(0,0),\ M(0,0)$
$\mathfrak{s}(2/1)^-$	$H(-2,0),\ P_+(-1,1),\ P_-(-1,-1),\ \mathscr{Q}(-1,0),\ \mathscr{X}_-(0,-1)$

While the algebras $\mathfrak{s}(2/2)$ and $\hat{\mathfrak{s}}(2/2)$ are $\mathbb{Z}^3$-graded:

decomposition	generator (degree)
$\mathfrak{s}(2/2)^+$	$K(2,0,0),\ G_+(1,1,0),\ G_-(1,-1,0),\ \mathscr{S}_+(1,0,1),$
	$\mathscr{S}_-(1,0,-1),\ \mathscr{X}_{++}(0,1,1),\ \mathscr{X}_{+-}(0,1,-1)$
$\mathfrak{s}(1/2)^0$	$D(0,0,0),\ J(0,0,0),\ M(0,0,0),\ R(0,0,0)$
$\mathfrak{s}(1/2)^-$	$H(-2,0,0),\ P_+(-1,1,0),\ P_-(-1,-1,0),\ \mathscr{Q}_+(-1,0,1)$
	$\mathscr{Q}_-(-1,0,-1),\ \mathscr{X}_{-+}(0,-1,1),\ \mathscr{X}_{--}(0,-1,-1)$
$\hat{\mathfrak{s}}(2/2)^+$	$K(2,0,0),\ G_+(1,1,0),\ G_-(1,-1,0),$
	$\mathscr{S}_+(1,-1,2),\ \mathscr{S}_-(1,1,-2),\ \mathscr{X}_+(0,0,2)$
$\hat{\mathfrak{s}}(1/2)^0$	$D(0,0,0),\ J(0,0,0),\ M(0,0,0),\ R(0,0,0)$
$\hat{\mathfrak{s}}(1/2)^-$	$H(-2,0,0),\ P_+(-1,1,0),\ P_-(-1,-1,0),$
	$\mathscr{Q}_+(-1,-1,2),\ \mathscr{Q}_-(-1,1,-2),\ \mathscr{X}_-(0,0,-2)$

One can introduce an algebra anti-automorphism for the super Schrödinger algebras. It is defined for the bosonic generators of $\mathfrak{s}(1/\mathcal{N})$ by

$$\omega(P) = G, \quad \omega(H) = K, \quad \omega(D) = D, \quad \omega(M) = M, \quad \omega(R) = R, \tag{15}$$

and those for the algebras in $(2+1)$ dimension by

$$\omega(P_a) = G_{-a}, \quad \omega(H) = K, \quad \omega(D) = D, \quad \omega(M) = M, \quad \omega(J) = J, \quad \omega(R) = R. \tag{16}$$

The mapping of the fermionic elements of $\mathfrak{s}(1/1)$ is given by

$$\omega(\mathscr{Q}) = \mathscr{S}, \qquad \omega(\mathscr{X}) = \mathscr{X}. \tag{17}$$

All the fermionic elements of the other algebras are obey the same transformation rule together with all the signature of the suffices being changed, e.g., $\omega(\mathscr{X}_{+-}) = \mathscr{X}_{-+}$. The mapping is involutive: $\omega^2 = id$. Let $\mathfrak{g}$ be one of the superalgebras considered in this paper and let us denote its triangular decomposition by $\mathfrak{g} = \mathfrak{g}^+ \oplus \mathfrak{g}^0 \oplus \mathfrak{g}^-$. Then one see that $\omega(\mathfrak{g}^+) = \mathfrak{g}^-$, $\omega(\mathfrak{g}^0) = \mathfrak{g}^0$.

3. Verma modules and their reducibility

The algebras under consideration admit the triangular decomposition. This allows us to define the lowest weight modules for the algebras. The lowest weight vector v_0 is defined by

$$X v_0 = 0 \quad {}^\forall X \in \mathfrak{g}^-, \qquad X v_0 = \Lambda(X) v_0 \quad {}^\forall X \in \mathfrak{g}^0, \tag{18}$$

where $\Lambda(X)$ is an eigenvalue of X. The Verma module over $\mathfrak{g}$ is defined by

$$V^\Lambda = U(\mathfrak{g}^+) v_0, \tag{19}$$

where $U(\mathfrak{g}^+)$ is the enveloping algebra of $\mathfrak{g}^+$. Reducibility of the Verma modules may be investigated by singular vectors as in the cases of semisimple Lie algebras and the bosonic Schrödinger algebras [32, 33]. A singular vector v_s is defined as a homogeneous element of V^Λ such that $v_s \neq \mathbb{C} v_0$ and $\mathfrak{g}^- v_s = 0$. In this section we provide explicit formulae of the singular vectors and reducibility of the Verma modules for *massive* case, that is, the generator M has nonvanishing eigenvalues. Here we give only the outline of the calculation. Detailed proof for $(1+1)$ dimensional algebras is found in [36] and for $(2+1)$ dimension will be published separately [37].

3.1. $(1+1)$ *Dimensional algebras*

We start with $\mathfrak{s}(1/1)$. The lowest weight condition (18) for $\mathfrak{s}(1/1)$ yields

$$\begin{aligned} &\mathscr{Q} v_0 = P v_0 = 0, \\ &D v_0 = -d v_0, \quad M v_0 = m v_0, \quad \mathscr{X} v_0 = \chi v_0, \end{aligned} \tag{20}$$

where $d \in \mathbb{R}$ is the conformal weight and the minus sign is for later convenience. The variable χ is of odd parity relating to the mass eigenvalue by the relation $m = 2\chi^2$. The Verma module (19) for $\mathfrak{s}(1/1)$ is given by

$$V^d = \{\, G^k K^\ell v_0,\ G^k K^\ell \mathscr{S} v_0 \mid k, \ell \in \mathbb{Z}_{\geq 0} \,\}. \tag{21}$$

It is easy to see that the generator D is diagonal on the basis given in (21). It follows that V^d is decomposed into a direct sum of the subspaces V_n^d spanned by the vectors v satisfying $Dv = nv$. The singular vector v_s belongs to V_n^d. One may set

$$v_s = \sum_k a_k v_k^{(n)}, \quad v_k^{(n)} \in V_n^d, \tag{22}$$

and impose the conditions

$$\mathscr{Q} v_s = P v_s = 0. \tag{23}$$

This yields some relations for a_k then by solving the relations one may determine the singular vectors up to overall constant.

Proposition 1 *The Verma module V^d over $\mathfrak{s}(1/1)$ has precisely one singular vector iff $d+1/2 \in \mathbb{Z}_{\geq 0}$ and it is given by*

$$v_s = (G^2 - 2mK)^{d+1/2}(G - 2\chi\mathscr{S})v_0. \tag{24}$$

When the Verma module has singular vectors one find invariant subspaces constructed on the singular vector $\mathcal{I} = U(\mathfrak{g}^+)v_s$. Thus the module is reducible. To find irreducible modules one may consider the quotient module $V^d/\mathcal{I}$ then seeks singular vectors in the quotient module. In the present case it turns out that there is no singular vector in the quotient module. Therefore $V^d/\mathcal{I}$ is irreducible.

Proposition 2 *All irreducible lowest weight modules over $\mathfrak{s}(1/1)$ are listed as follows:*

(i) The Verma module V^d for $d + 1/2 \notin \mathbb{Z}_{\geq 0}$

(ii) The quotient module $V^d/\mathcal{I}$ for $d + 1/2 \in \mathbb{Z}_{\geq 0}$

All irreducible modules given are infinite dimensional.

We now turn to the superalgebra $\mathfrak{s}(1/2)$. The lowest weight vector v_0 is defined by

$$\begin{aligned} &\mathscr{Q}_{\pm} v_0 = P v_0 = \mathscr{X}_- v_0 = 0, \\ &D v_0 = -d v_0, \quad M v_0 = m v_0, \quad R v_0 = r v_0. \end{aligned} \tag{25}$$

The Verma modules over $\mathfrak{s}(1/2)$ are given by

$$V^{d,r} = \{\ G^k K^\ell \mathscr{S}_+^a \mathscr{S}_-^b \mathscr{X}_+^c v_0 \mid k, \ell \in \mathbb{Z}_{\geq 0},\ a, b, c \in \{0, 1\}\ \}. \tag{26}$$

The same procedure as $\mathfrak{s}(1/1)$ leads us to the following propositions:

Proposition 3 *The Verma module $V^{d,r}$ over $\mathfrak{s}(1/2)$ has precisely one singular vector iff $d - 1/2 \in \mathbb{Z}_+$ and it is given by*

$$\begin{aligned} &v_s^p = (G^2 - 2mK)^{d-1/2} u_0, \\ &u_0 = (G\mathscr{S}_- \mathscr{X}_+ + m\mathscr{S}_+ \mathscr{S}_- + 2mK)v_0 + \frac{d + r + 1}{2d + 1}(G^2 - 2mK)v_0. \end{aligned} \tag{27}$$

Proposition 4 *All irreducible lowest weight modules over $\mathfrak{s}(1/2)$ are listed as follows:*

(i) The Verma module $V^{d,r}$ for $d - 1/2 \notin \mathbb{Z}_+$

(ii) The quotient module $V^{d,r}/\mathcal{I}$ for $d - 1/2 \in \mathbb{Z}_+$

where $\mathcal{I} \subset V^{d,r}$ is the invariant submodule constructed on the singular vector (27): $\mathcal{I} = U(\mathfrak{s}(1/2)^+)v_s$. All irreducible modules given are infinite dimensional.

3.2. (2 + 1) *Dimensional algebras*

More involved in this case, however, one may use the same procedure as (1 + 1) dimensional algebras to study the reducibility of the Verma modules.

We start with the superalgebra $\mathfrak{s}(2/1)$. The lowest weight for $\mathfrak{s}(2/1)$ is defined by

$$\begin{aligned} &\mathscr{Q}v_0 = P_{\pm}v_0 = \mathscr{X}_- v_0 = 0, \\ &Dv_0 = -dv_0, \quad Jv_0 = -jv_0, \quad Mv_0 = mv_0. \end{aligned} \tag{28}$$

The Verma modules are constructed on the lowest weight vector:

$$V^{d,j} = \{\ G_+^k G_-^h K^\ell \mathscr{S}^a \mathscr{X}_+^b v_0 \mid k, h, \ell \in \mathbb{Z}_{\geq 0},\ a, b \in \{0, 1\}\ \}. \tag{29}$$

The Verma modules contain a singular vector for discrete values of d :

Proposition 5 *The Verma module $V^{d,j}$ over $\mathfrak{s}(2/1)$ has precisely one singular vector iff $d + 1 \in \mathbb{Z}_+$ and it is given by*

$$v_s = (G_+G_- - 2mK)^{d+1}(G_-\mathscr{X}_+ - 2m\mathscr{S})v_0. \tag{30}$$

It turns out that the factor modules by the invariant submodule constructed on v_s do not contain any singular vectors. We thus have a classification of irreducible modules over $\mathfrak{s}(2/1)$.

Proposition 6 *All irreducible lowest weight modules over $\mathfrak{s}(2/1)$ are listed as follows:*

(i) The Verma module $V^{d,j}$ for $d + 1 \notin \mathbb{Z}_+$

(ii) The quotient module $V^{d,j}/\mathcal{I}$ for $d + 1 \in \mathbb{Z}_+$

where $\mathcal{I} \subset V^{d,r}$ is the invariant submodule constructed on the singular vector (30): $\mathcal{I} = U(\mathfrak{s}(2/1)^+)v_s$. All irreducible modules given are infinite dimensional.

Next we investigate the standard $\mathcal{N} = 2$ extension $\mathfrak{s}(2/2)$. The lowest weight vector for $\mathfrak{s}(2/2)$ is defined by

$$\begin{aligned} &\mathscr{Q}_{\pm}v_0 = P_{\pm}v_0 = \mathscr{X}_{-+}v_0 = \mathscr{X}_{--}v_0 = 0, \\ &Dv_0 = -dv_0, \quad Jv_0 = -jv_0, \quad Rv_0 = rv_0, \quad Mv_0 = mv_0, \end{aligned} \tag{31}$$

and application of a element of $U(\mathfrak{s}(2/2)^+)$ on v_0 generates a basis of a Verma module over $\mathfrak{s}(2/1)$:

$$V^{d,j,r} = \{\ G_+^k G_-^h K^\ell \mathscr{S}_+^{a_1} \mathscr{S}_-^{a_2} \mathscr{X}_{++}^{b_1} \mathscr{X}_{+-}^{b_2} v_0 \mid k, h, \ell \in \mathbb{Z}_{\geq 0},\ a_i, b_i \in \{0, 1\}\ \}. \tag{32}$$

Reducibility of the Verma modules is similar to the other cases. We can show the following proposition on the existence of singular vectors.

Proposition 7 *The Verma module $V^{d,j,r}$ over $\mathfrak{s}(2/2)$ has precisely one singular vector iff $d \in \mathbb{Z}_+$ and it is given by*

$$\begin{aligned} &v_s = (G_+G_- - 2mK)^d(G_-\mathscr{X}_+ - 2m\mathscr{S})u_0, \\ &u_0 = \{G_-^2\mathscr{X}_{++}\mathscr{X}_{+-} - 2mG_-(\mathscr{S}_+\mathscr{X}_{+-} - \mathscr{S}_-\mathscr{X}_{++}) + 4m^2(\mathscr{S}_+\mathscr{S}_- + K)\}v_0. \end{aligned} \tag{33}$$

Further search of singular vectors in the quotient modules conclude the following classification of irreducible modules:

Proposition 8 *All irreducible lowest weight modules over $\mathfrak{s}(2/2)$ are listed as follows:*

(i) The Verma module $V^{d,j,r}$ for $d \notin \mathbb{Z}_+$

Journal of Physics: Conference Series **284** (2011) 012007 doi:10.1088/1742-6596/284/1/012007

(ii) The quotient module $V^{d,j,r}/\mathcal{I}$ for $d \in \mathbb{Z}_+$

where $\mathcal{I} \subset V^{d,r}$ is the invariant submodule constructed on the singular vector (33): $\mathcal{I} = U(\mathfrak{s}(2/2)^+)v_s$. All irreducible modules given are infinite dimensional.

We now turn to investigation of the exotic $\mathcal{N} = 2$ extension $\hat{\mathfrak{s}}(2/2)$. The singular vectors and irreducible modules have been studied in [35]. Here we give a closed form expression of the singular vectors and more precise classification of irreducible modules. The lowest weight vector v_0 and the Verma modules $V^{d,j,r}$ are defined as usual:

$$\begin{aligned} &\mathscr{Q}_\pm v_0 = P_\pm v_0 = \mathscr{X}_- v_0 = 0, \\ &Dv_0 = -dv_0, \quad Jv_0 = -jv_0, \quad Rv_0 = 2rv_0, \quad Mv_0 = mv_0. \end{aligned} \tag{34}$$

$$V^{d,j,r} = \{\ G_+^k G_-^h K^\ell \mathscr{S}_+^a \mathscr{S}_-^b \mathscr{X}_+^c v_0, \ \mid k,h,\ell \in \mathbb{Z}_{\geq 0},\ a,b,c \in \{0,1\}\ \}. \tag{35}$$

The classification of irreducible modules is summarized in the following two propositions.

Proposition 9 *The Verma module $V^{d,j,r}$ over $\hat{\mathfrak{s}}(2/2)$ has precisely one singular vector iff $d \in \mathbb{Z}_+$ and $r = -(d-j+2)/2$. It is given by*

$$\begin{aligned} v_s &= (G_+G_- - 2mK)^d u_0, \\ u_0 &= (G_-\mathscr{S}_-\mathscr{X}_+ + M\mathscr{S}_+\mathscr{S}_- + 2mK)v_0. \end{aligned} \tag{36}$$

Proposition 10 *All irreducible lowest weight modules over $\hat{\mathfrak{s}}(2/2)$ are listed as follows:*

(i) The Verma module $V^{d,j,r}$ if $d \notin \mathbb{Z}_+$ nor $r \neq -(d-j+2)/2$

(ii) The quotient module $V^{d,j,r}/\mathcal{I}$ if $d \in \mathbb{Z}_+$ and $r = -(d-j+2)/2$

where $\mathcal{I} \subset V^{d,r}$ is the invariant submodule constructed on the singular vector (36): $\mathcal{I} = U(\hat{\mathfrak{s}}(2/2)^+)v_s$. All irreducible modules given are infinite dimensional.

4. Concluding remarks

We investigate the lowest weight representations of the super Schrödinger algebras in $(1+1)$ and $(2+1)$ dimensional spacetime with $\mathcal{N} = 1, 2$ extensions. Our main result is a classification of irreducible lowest weight modules for nonvanishing mass. This was done by the use of singular vectors. All irreducible modules are infinite dimensional. Finite dimensional irreducible modules appear when the mass is set equal to zero [36]. One may introduce a bilinear form analogous to the Shapovalov form [38] of the semisimple Lie algebra to the Verma modules. Let $\mathfrak{g}$ be a super Schrödinger algebra discussed in this paper and V be a Verma module over it. We define the bilinear form $(\ ,\ ) : V \otimes V \to \mathbb{C}$ by the relations:

$$(Xv_0, Yv_0) = (v_0, \omega(X)Yv_0), \qquad (v_0, v_0) = 1, \quad X, Y \in U(\mathfrak{g}) \tag{37}$$

By the bilinear form one may define unitary representations of the super Schrödinger algebras. It is also easy to see that if $v_m, v_n \in V$ have different weight of the generator D, then they are orthogonal with respect to this form:

$$(v_m, v_n) = 0. \tag{38}$$

It follows that a singular vector of $\mathfrak{g}$ is orthogonal to any other vectors in V. This is the same property as semisimple case. Thus one may analyze the reducibility of the Verma modules also via the bilinear form.

One may use the singular vectors to obtain invariant partial differential equations. This is an analogy to the cases of semisimple Lie algebras [39, 40] and the bosonic Schrödinger algebras [32, 33, 41, 42]. In the latter case the obtained partial differential equations are a hierarchy of free Schrödinger equations. To obtain the invariant equations we need a vector field representation of the algebra. Vector field representations for $\mathfrak{s}(1/1)$ and $\mathfrak{s}(1/2)$ are found in [36]. Another link of the present work to physical problem is a a supersymmetric extension of the group theoretical approach to nonrelativistic holography discussed in [43]. It may also require vector field representations of the super Schrödinger algebras.

We restrict ourselves to the super Schrödinger algebras in this work. However there exist other algebras which may be regarded as nonrelativistic analogue of conformal algebras [19, 20] (see also [6]) and their supersymmetric extensions [44]. Physical importance of such algebras has been recognized widely [6, 7, 10–12, 18], however, representation theory of such algebras are far from completion. It is an important issue to develop representation theory for the algebras along the line of the present work. We announce that it has been done for one of the so-called conformal Galilei algebras in [45].

References

[1] Niederer U 1972 *Helv. Phys. Acta.* **45** 802
[2] Hagen C R 1972 *Phys. Rev.* D**5** 377
[3] Henkel M 1994 *J. Stat. Phys.* **75** 1023 hep-th/9310081
[4] Dunne G 1995 *Self-dual Chern-Simons theories* (Springer-Verlag)
[5] Sachdev S 1999 *Quantum Phase Transitions* (Cambridge University Press)
[6] Duval C and Horváthy P A 2009 *J. Phys.* **A**:*Math. Theor.* **42** 465206 arXiv:0904.0531 [math-ph]
[7] Horváthy P A, Martina L and Stichel P C 2010 *SIGMA* **6** 060 arXiv:1002.4772 [hep-th]
[8] Duval C, Gibbons G W and Horváthy P A 1991 *Phys. Rev.* **D43** 3907 hep-th/0512188
[9] Mehen T, Stewart I W and Wise M B 2000 *Phys. Lett.* **B474** 145 hep-th/9910025
[10] Stichel P C and Zakrzewski W J 2004 *Annals Phys.* **310** 158 arXiv:hep-th/0309038
[11] Lukierski J, Stichel P C and Zakrewski W J 2006 *Phys. Lett.* **A357** 1 hep-th/0511259
[12] Lukierski J, Stichel P C and Zakrewski W J 2007 *Phys. Lett.* **B650** 203 arXiv:hep-th/0702179
[13] Nishida Y and Son D T 2007 *Phys. Rev.* D**76** 086004 arXiv:0706.3746 [hep-th]
[14] Son D T 2008 *Phys. Rev.* D**78** 046003 arXiv:0804.3972 [hep-th]
[15] Balasubramanian K and McGreevy J 2008 *Phys. Rev. Lett.* **101** 061601 arXiv:0804.4053 [hep-th]
[16] Bagchi A and Gopakumar R 2009 *J. High Energy Phys.* JHEP07(2009)037 arXiv:0902.1385 [hep-th]
[17] Alishahiha M, Davody A and Vahedi A 2009 *J. High Energy Phys.* JHEP08(2009)022 arXiv:0903.3953 [hep-th]
[18] Martelli D and Tachikawa Y 2010 *J. High Energy Phys.* JHEP05(2010)091 arXiv:0903.5184 [hep-th]
[19] Havas P and Plebański J 1978 *J. Math. Phys.* **19** 482
[20] Negro J, del Olmo M A and Rodriguez-Marco A 1997 *J. Math. Phys.* **38** 3786, *ibid*, 3810
[21] Beckers J and Hussin V 1986 *Phys. Lett.* **A118** 319
[22] Beckers J, Dehin D and Hussin V 1987 *J. Phys.* **A**:*Math. Gen.* **20** 1137
[23] Gauntlett J P, Gomis J and Townsend P K 1990 *Phys. Lett.* **B248** 288
[24] Leblanc M, Lozano G and Min H 1992 *Ann. Phys. (N.Y.)* **219** 328 hep-th/9206039
[25] Duval C and Horváthy P A 1994 *J. Math. Phys.* **35** 2516 hep-th/0508079
[26] Nakayama Yu , Sakaguchi M and Yoshida K 2009 *J. High Energy Phys.* JHEP04(2009)096 arXiv:0902.2204 [hep-th]
[27] Galajinsky A 2009 *Phys. Lett.* **B675** 116 arXiv:0902.2910 [hep-th]
[28] Henkel M and Unterberger J 2006 *Nucl. Phys.* **B746**[FS] 155 arXiv:math-ph/0512024
[29] Sakaguchi M and Yoshida K 2008 *J. Math. Phys.* **49** 102302 arXiv:0805.2661 [hep-th]
[30] Sakaguchi M and Yoshida K 2008 *J. High Energy Phys.* JHEP08(2008)049 arXiv:0806.3612 [hep-th]
[31] Perroud M 1977 *Helv. Phys. Acta* **50** 233
[32] Dobrev V K, Doebner H D and Mrugalla Ch 1997 *Rep. Math. Phys.* **39** 201
[33] Mrugalla Ch 1997 PhD Thesis, Technical University Clausthal
[34] Feinsilver P, Kocik J and Schott R 2004 *Fort. Physik* **52** 343 math-ph/0008035.
[35] Nakayama Yu 2008 *J. High Energy Phys.* JHEP10(2008)083 arXiv:0807.3344 [hep-th]
[36] Aizawa N 2010 Lowest weight representations of super Schrödinger algebras in one dimensional space *Preprint* arXiv:1009.0085 [math-ph]
[37] Aizawa N in preparation
[38] Shapovalov N N 1972 *Funct. Anal. Appl.* **6** 307

[39] Dobrev V K *Rep. Math. Phys.* **25** 159
[40] Kostant B 1975 *Lecture Notes in Mathematics* **466** (Springer-Verlag: Berlin) p 101
[41] Aizawa N, Dobrev V K and Doebner H D 2002 *Proc. 2nd Int. Symp Quantum Theory and Symmetries* ed E Kapuścik and A Horzela (World Scientific) p 222
[42] Aizawa N, Dobrev V K, Doebner H D and Stoimenov S 2008 *Proc. VII Int. Workshop Lie Theory and Its Applications in Physics* ed H D Doebner and V K Dobrev (Heron Press) p 372
[43] Aizawa N and Dobrev V K 2010 *Nucl. Phys.* **B828**[PM] 581 arXiv:0906.0257 [hep-th]
[44] Lukierski J 2010 $D = 4$ Extended Galilei superalgebras with central charges *Preprint* arXiv:1009.0182 [hep-th]
[45] Aizawa N and Isaac P S 2010 On irreducible representations of the exotic conformal Galilei algebra *Preprint* arXiv:1010.4075 [math-ph] to appear in *J. Phys. A:Math. Theor.*

Group Approach to the Quantization of Non-Abelian Stueckelberg Models

V. Aldaya[1], M. Calixto[2,1] and F.F. López-Ruiz[1]

[1] Instituto de Astrofísica de Andalucía (IAA-CSIC), Apartado Postal 3004, 18080 Granada, Spain

[2] Departamento de Matemática Aplicada y Estadística, Universidad Politécnica de Cartagena, Paseo Alfonso XIII 56, 30203 Cartagena, Spain

E-mail: valdaya@iaa.es, Manuel.Calixto@upct.es, flopez@iaa.es

Abstract. The quantum field theory of Non-Linear Sigma Models on coadjoint orbits of a semi-simple group G are formulated in the framework of a Group Approach to Quantization. In this scheme, partial-trace Lagrangians are recovered from two-cocycles defined on the infinite-dimensional group of sections of the jet-gauge group $J^1(G)$. This construction is extended to the entire physical system coupled to Yang-Mills fields, thus constituting an algebraic formulation of the Non-Abelian Stueckelgerg formalism devoid of the unitarity/renormalizability obstruction that this theory finds in the standard Lagrangian formalism under canonical quantization.

1. Introduction

Non-Linear σ-fields have attracted a renewed attention as an essential ingredient in generalized, non-Abelian Stueckelberg models [1] formulated in an attempt to replace the Higgs mechanism in non-abelian gauge theories [2, 3, 4, 5]. Unfortunately, those models can accomodate either unitarity or renormalizability but not both [6]. In this paper we propose a non-perturbative description of Non-Linear Sigma Fields minimaly coupled to Yang-Mills vector potentials on the framework of a Group Approach to Quantization (GAQ) [7, 8, 9].

Group quantization of non-Abelian gauge groups had been only achieved consistently in $1+1$ dimensions by representing the corresponding Kac-Moody group [10]. In fact, the non-trivial cohomology of such groups provided a co-cycle [11] which is easily translated to a WZW Lagrangian. In $3+1$ dimensions, however, the Mickelsson two-cocycle is absent and a bit more involved construction is required. One of the new ingredients will be the consideration of the jet-gauge group $J^1(G)$ of the original rigid symmetry group G as the tarjet space for the physical fields, so that those fields are mapps from the space-time manifold M onto $J^1(G)$, providing the Goldston-like scalar fields $\varphi^a(x)$ along with the corresponding vector potentials $A^a_\mu(x)$ [12, 13]. These fields constitute de new local group $G^1(M)$. The other ingredient refers to the use of a class of two-cocycles that, even though they are coboundaries, and, therefore trivial from some mathematical points of view, they define a central extension endowed with a canonical (left- or right-) invariant form [14] which gives a physical Lagrangian for fields living on a coadjoint orbit of $G^1(M)$. The corresponding Lagrangian can then be seen as a (covariant) partial trace of the standard σ-model full-trace (chiral) Lagrangian $Tr(U^{-1}\partial_\mu U U^{-1}\partial^\mu U)$, $U \in G$ coupled to the vector potentials according to a Minimal Coupling prescription addressed by one of the two-cocycles involved in the central extension of the local group.

In this paper we review very briefly, in Sec.2, the standard classical approach to the gauge theory of internal interactions, and revisite this classical theory under the equivalent scheme of jet-gauge groups. Then, in Sec.3, we propose the new framework, based entirely on symmetries, to account for a non-canonical group quantization of massive gauge theories without Higgs.

2. Internal gauge interactions: the standard Utiyama theory revisited

In the standard formulation of gauge theories, the well-known Minimal Coupling Principle for internal gauge symmetries establishes that if the action of some matter fields ψ^α, $\alpha = 1, ..., N$

$$S = \int \mathcal{L}_{\text{matt}}(\psi^\alpha, \partial_\mu \psi^\alpha) d^4x,$$

is invariant under a rigid internal Lie group G, then the modified action

$$\widehat{S} = \int [\mathcal{L}_{\text{matt}}(\psi^\alpha, D_\mu \psi^\alpha) + \mathcal{L}_0(F^{(a)}_{\mu\nu})] d^4x$$

is invariant under the local group $G(M)$, where (M is the Minkowski space-time)

$$D_\mu \psi^\alpha \equiv \partial_\mu \psi^\alpha - g A^{(a)}_\mu X^\alpha_{(a)\beta} \psi^\beta$$

is usually known as covariant derivative, and

$$F^{(a)}_{\mu\nu} \equiv \partial_\mu A^{(a)}_\nu - \partial_\nu A^{(a)}_\mu + \frac{g}{2} C^a_{bc} (A^{(b)}_\mu A^{(c)}_\nu - A^{(b)}_\nu A^{(c)}_\mu)$$

is known as curvature of the connection $A^{(a)}_\mu$ (the gauge potentials or Yang-Mills fields).

The constant matrix $X^\alpha_{(a)\beta}$ is the representative of the group generator in the direction, $(a) = 1, ..., \dim G$, acting on the field components, so that, the infinitesimal variation of the matter fields is

$$\delta_{(a)} \psi^\alpha = X^\alpha_{(a)\beta} \psi^\beta$$

Note that in terms of this infinitesimal variation the covariant derivative is written as

$$D_\mu \psi^\alpha \equiv \partial_\mu \psi^\alpha - g A^{(a)}_\mu \delta_{(a)} \psi^\alpha$$

2.1. Including the group parameters in Gauge Theory

We now present a conceptual and structural revision of the standard formulation of gauge theories.

Let G be a Lie group of internal symmetry, the rigid group, and $G(M)$ the gauge group (also named current group), whose elements are the mappings from Minkowski space-time M into G:

$$G(M) \equiv \{M \to G\}.$$

Let us introduce the group of the 1-jets of the gauge group $G(M)$ [12],

$$J^1(G(M)).$$

The formal definition of $J^1(G(M))$ is fully analogous to that of the $(q^i, \dot{q}^j)$-space in Lagrangian Mechanics, or $(\psi^\alpha, \psi^\beta_\mu)$-space in Lagrangian Field Theory, when one desires to vary independently coordinates and velocities (momenta) according to the modified Hamilton principle. That is:

$$J^1(G(M)) \equiv G(M) \times M / \sim^1$$

where the equivalence relation $\sim^1$ is defined as follows:

$$(\varphi, m) \sim^1 (\varphi', m')$$

for all (φ, m), (φ', m') belonging to $G(M) \times M$, if and only if

$$m = m'$$

$$\varphi(m) = \varphi'(m)$$

$$\partial_\mu \varphi(m) = \partial_\mu \varphi'(m).$$

This definition may easily be extended to the order rth. A natural coordinate system for $J^1(G(M))$ would be

$$\{x^\mu, \varphi^a, \varphi^a_\mu\}$$

where x^μ parametrize locally M, and φ^a and φ^a_μ correspond to $g^a(\varphi(m))$ and $\partial_\mu(g^a(\varphi(m))$, respectively, under the quotient by the equivalence relation $\sim^1$, for a local coordinate system in G, $\{g^a\}$.

Thus, the relevant group in Gauge Theory, extending the standard gauge group $G(M)$, will be that constituted by the mapping from M to $J^1(G(M))$,

$$G^1(M) \equiv \Gamma(J^1(G(M))) \equiv \{M \to J^1(G(M))\},$$

parametrized by fields $\varphi^a(x)$, $\varphi^a_\mu(x)$, where φ^a_μ is not necessarily the derivative of φ^a except for *jet-extensions*, that is:

$$j^1 : G(M) \to G^1(M)$$

$$\varphi \mapsto j^1(\varphi) \,/\, \varphi^a_\mu = \partial_\mu \varphi^a \,.$$

In this formalism φ^a_μ are essentially the gauge potentials $A^{(a)}_\mu$:

$$A^{(a)}_\mu \equiv \theta^{(a)}_b \varphi^b_\mu \,,$$

which are (not) pure gauge if φ^a_μ are (not) jet-extensions, i.e. are (not) of the form $\partial_\mu \varphi^a$, where, $\theta^{(a)}_b$ is the (non-constant) invertible matrix defining the (right-) invariant canonical 1-form on the group

$$\theta^{R(a)} = \theta^{(a)}_b d\varphi^b \quad (dUU^{-1} \text{ for linear groups})$$

dual to the (right-invariant) generators $X^R_{(a)} = X^b_{(a)} \frac{\partial}{\partial \varphi^b}$ $(X^b_{(a)} \equiv \delta_{(a)} \varphi^b)$, that is:

$$\theta^{(a)}_b X^b_{(c)} = \delta^a_c \,.$$

Notice that for pure gauge we have $A^{(a)}_\mu = \theta^{(a)}_b \partial_\mu \varphi^b \equiv \theta^{(a)}_\mu$ and, hence, the curvature $F(A_\mu) = F(\theta_\mu) = 0$, that is, θ_μ can be seen as a flat connection.

2.2. Lagrangian formalism on jet-gauge groups: gauge parameters $\varphi^a \in G(M)$ as "matter" fields

Standard variational calculus varies Lagrangians depending on the "field configuration" space E, with generic field coordinates $\{x^\mu, \psi^\alpha\}$, and the corresponding field derivatives $\{\partial_\mu \psi^\alpha\}$. Our configuration space will be now $J^1(G(M))$, with coordinates $\{x^\mu, \varphi^a, A^{(a)}_\mu\}$,. Lagrangians are accordingly functions

$$\mathcal{L} = \mathcal{L}(x^\mu, \varphi^a, A^{(a)}_\mu; \partial_\nu \varphi^a, \partial_\nu A^{(a)}_\mu).$$

We can proceed to formulate some sort of Utiyama's Theorem on $J^1(G(M))$:

If the action of some "matter" fields φ^a, $a = 1, ..., \dim(G)$

$$S = \int \mathcal{L}_{\text{"matt"}}(\varphi^a, \partial_\mu \varphi^b) d^4x,$$

is invariant under a rigid internal Lie group G, then the modified action

$$\widehat{S} = \int [\mathcal{L}_{\text{"matt"}}(\varphi^a, D_\mu \varphi^b) + \mathcal{L}_0(F^{(a)}_{\mu\nu})] d^4x$$

is invariant under the local group $G(M) \in G^1(M)$, where

$$D_\mu \varphi^a \equiv \partial_\mu \varphi^a - g A^{(b)}_\mu X^a_{(b)} ,$$

the new "covariant" derivative, $X^a_{(b)}$ are the components of the generators of the group G acting on itself and

$$F^{(a)}_{\mu\nu} \equiv \partial_\mu A^{(a)}_\nu - \partial_\nu A^{(a)}_\mu + \frac{g}{2} C^a_{bc} (A^{(b)}_\mu A^{(c)}_\nu - A^{(b)}_\nu A^{(c)}_\mu)$$

is the curvature of the connection $A^{(a)}_\mu \equiv \theta^{(a)}_b \varphi^b_\mu$ (the gauge potentials or the Yang-Mills fields).

We should remark that the Minimal Coupling now occurs in an affine manner. In fact, in the expression

$$\partial_\mu \varphi^a - g A^{(b)}_\mu X^a_{(b)}$$

the matrix $X^a_{(b)}$ is invertible and this way, the minimal coupling above is proportional to

$$\theta^{(a)}_b [\partial_\mu \varphi^b - g X^b_{(c)} A^{(c)}_\mu] \;=\; \theta^{(a)}_\mu - g A^{(a)}_\mu$$

where $\theta^{(a)}$ are, for concreteness, the components of the right-invariant canonical 1-form on G, θ^R, dual to the right-invariant vector fields $X^R_{(a)}$ ($X^{R\ b}_{(a)} = [\theta^{R(b)}_{\ \ a}]^{-1}$). We shall omit the R script if not required.

2.3. Non-linear σ-Lagrangians for "matter" fields: The non-Abelian Stueckelberg formalism

The new minimal coupling, when written in the form $\theta - A$, strongly suggests the introduction of "matter" of the σ-model type:

$$\mathcal{L}^G_\sigma = \frac{1}{2} \text{Tr}[\theta_\mu \theta^\mu], \;\; \theta_\mu = \partial_\mu U U^{-1}.$$

In fact, this Lagrangian is G-invariant (left- and right-invariant, indeed,; that is, chiral) and the new minimal coupling gives rise to

$$\hat{\mathcal{L}}^G_\sigma = \frac{1}{2} \text{Tr}[(\theta_\mu - g A_\mu)(\theta^\mu - g A^\mu)]$$

which is gauge-invariant even though it contains mass terms for the A_μ fields, with total Lagrangian

$$\mathcal{L}^G_{\rm MYM} = -\frac{1}{4}F^{\mu\nu}F_{\mu\nu} + m^2\hat{\mathcal{L}}^G_\sigma$$

This Lagrangian $\mathcal{L}^G_{\rm MYM}$ actually corresponds to the non-Abelian Stueckelberg formalism for massive Yang-Mills fields although, to be precise, the scalar fields φ^a are here the parameters of the local group $G(M)$ themselves, whereas in the standard (non-Abelian) Stueckelberg formalism the corresponding scalar fields are external fields behaving under G in exactly the same way as our φ's.

Unfortunately the σ-sector Lagrangian

$$\mathcal{L}^G_\sigma = \frac{1}{2}{\rm Tr}(\theta^R_\mu\theta^{R\mu}) = \frac{1}{2}{\rm Tr}(\theta^L_\mu\theta^{L\mu}) = \frac{1}{2}{\rm Tr}(\partial_\mu U\partial^\mu U^\dagger) \equiv \frac{1}{2}g_{ab}(\varphi)\partial_\mu\varphi^a\partial^\mu\varphi^b$$

$[U(x) = e^{i\varphi^a(x)T_a} \in G(M), \quad T_a, a = 1,\dots,\dim(G)$ are the Lie-algebra generators of G with commutation relations $[T_a,T_b] = C^c_{ab}T_c,\ Tr(T_aT_b) = \delta_{ab}]$ is not properly quantizable due essentially to its non-polynomic character; in fact, the "metric" function $g_{ab}(\varphi)$ may depend on any power of φ so that, standard perturbation theory would require, roughly speaking, an infinite number of bare natural constants; that is as many constants as coefficients in the power expansion of $g_{ab}(\varphi)$ [15]. In more accurate terms, we might conclude that physical field theories involving a non-linear sigma sector of scalars can be either renormalizable or unitary but not both [6].

3. A "partial-trace" alternative to the non-Abelian Stueckelberg formalism

A way out to the troubles found in facing the canonical quantization of Lagrangians containing those non-quadratic kinetic terms associated with non-Abelian sigma fields consists in rearranging the theory in such a manner that we shall be able to find an infinite number of symmetries providing as many non-trivial Noether invariants as parameters there are in the solution manifold of the physical problem. With this extended symmetry at hand one may try to formulate enough (generalized) Ward-like identities to reduce the infinities appearing in standard canonical perturbation theory or, as we shall make here, to resort to a non-standard, symmetry based, Group Approach to Quantization [7, 8, 9].

3.1. The partial-trace Lagrangian

We restrict the whole trace on G to a partial trace on a quotient manifold G/H, where H is the isotropy subgroup of a direction $\lambda = \lambda^a T_a$ in the Lie algebra of G under the adjoint action

$$\lambda \to U\lambda U^\dagger$$

where λ^a are real numbers subjected to ${\rm Tr}(\lambda^2) = 1$.

Defining $\Lambda \equiv U\lambda U^\dagger$ the claimed $G/H - \sigma$ Lagrangian has the expression:

$$\mathcal{L}^{G/H}_\sigma = \frac{1}{2}{\rm Tr}([-iU^\dagger\partial_\mu U,\lambda]^2) \equiv \frac{1}{2}{\rm Tr}([\theta^L_\mu,\lambda]^2) = \frac{1}{2}{\rm Tr}([\theta_\mu,\Lambda]^2) = \frac{1}{2}{\rm Tr}((\partial_\mu\Lambda)^2)$$

and the minimally coupled version:

$$\hat{\mathcal{L}}^{G/H}_\sigma = \frac{1}{2}{\rm Tr}([-iU^\dagger D_\mu U,\lambda]^2) = \frac{1}{2}{\rm Tr}([\theta_\mu - A_\mu,\Lambda]^2)$$

is again invariant under the gauge transformations

$$U \to U'U, A_\mu \to U'A_\mu U'^\dagger - i\partial_\mu U'U'^\dagger$$

The total Lagrangian for the Massive Yang-Mills theory is then

$$\mathcal{L}^G_{\text{MYM}} = \mathcal{L}^G_{\text{YM}} + m^2 \hat{\mathcal{L}}^G_\sigma$$

with ordinary $\mathcal{L}^G_{\text{YM}} \equiv -\frac{1}{4}\text{Tr}(F^{\mu\nu}F_{\mu\nu})$.

Let us remark that under the change of variables

$$\tilde{A}_\mu = U^\dagger(A_\mu - \theta_\mu)U = U^\dagger A_\mu U + iU^\dagger\partial_\mu U$$

(along with analogous change $\phi = U^\dagger\psi$ for eventual fermions) the Lagrangian above may be written as

$$\mathcal{L}^{G/H}_{\text{MYM}} = -\frac{1}{4}Tr(F^{\mu\nu}(\tilde{A})^2) + \frac{1}{2}m^2 Tr([\tilde{A}_\mu, \lambda]^2)$$

It should be noted that for the case $G = SU(2)$ with standard spherical basis $T_\pm, T_0$ and $\lambda = \lambda^0 T_0$, that is $H = U(1)$, the mass term would be written

$$\frac{1}{2}m^2 Tr([\tilde{A}_\mu, \lambda]^2) = m^2_W \tilde{W}^+_\mu \tilde{W}^{\mu-}$$

where, $\tilde{W}^\pm_\mu = \tilde{A}^1_\mu \pm i\tilde{A}^2_\mu$ and $m_W = m\lambda^0$ remaining $\tilde{W}^0_\mu = \tilde{A}^3_\mu$ massless (see [16] for more details).

3.2. Quantum symmetries: an attempt to Group Quantization

For linear systems, the canonical quantization rules realize the Lie algebra of the Heisenberg-Weyl group. This means postulating equal-time commutation relations

$$[\phi(x), \pi(y)] = i\delta(x - y).$$

For non-linear systems the Heisenberg-Weyl group must be replaced with a (more involved) symmetry group of the solution manifold, keeping the general idea of considering as basic conjugate operators those group generators giving central terms under commutation.

We should be able to identify such symmetry group by analyzing the symplectic potential (or Liouville 1-form) in the solution manifold (rather than the Lagrangian) [17], which generalizes $p_i dq^i$ from particle mechanics . The symplectic potential can be obtained by integrating the Lagrangian Poincaré-Cartan [18] form on a Cauchy hypersurface Σ

The symplectic potential for the σ-sector is given by (see Ref. [16])

$$\Theta^{G/H}_\sigma = \int_\Sigma \text{Tr}([\theta_\mu, \Lambda][-i\delta U U^\dagger, \Lambda])d\sigma^\mu$$

which is (semi-)invariant under the local "Euclidean group"

$$\begin{aligned} U &\rightarrow U'U \\ \theta_\mu &\rightarrow U'\theta_\mu U'^\dagger + \theta'_\mu \end{aligned}$$

Even more, the total symplectic potential

$$\Theta^{G/H}_{MYM} = \int_\Sigma \text{Tr}(F^{\mu\nu}(\tilde{A})\delta\tilde{A}_\nu + m^2\tilde{A}^\mu[-i\delta U U^\dagger, \lambda])d\sigma_\mu$$

is invariant under the local "Massive Yang-Mills group" $G(M)^{G/H}_{MYM}$ with group law given by (n^μ characterizes the Cauchy surface):

$$\begin{aligned}
U''(x) &= U'(x)U(x), \\
\theta''_\mu(x)n^\mu &= U'(x)\theta_\mu(x)n^\mu U'^\dagger(x) + \theta'_\mu(x)n^\mu, \\
A''_\mu(x) &= U'(x)A_\mu(x)U'^\dagger(x) + A'_\mu(x), \\
F''_{\mu\nu}(x) &= U'(x)F_{\mu\nu}(x)U'^\dagger(x) + F'_{\mu\nu}(x), \\
\zeta'' &= \zeta'\zeta \exp\left(i\int_\Sigma d\sigma^\mu(x)J_\mu(U',A',F';U,A,F)\right), \\
J_\mu &= J^{\rm YM}_\mu + J^\sigma_\mu, \\
J^{\rm YM}_\mu &= \frac{1}{2}{\rm Tr}\left((A'^\nu - \theta'^\nu)U'F_{\mu\nu}U'^\dagger - F'_{\mu\nu}U'(A^\nu - \theta^\nu)U'^\dagger\right), \\
J^\sigma_\mu &= m^2{\rm Tr}\left(\lambda(U'(A_\mu - \theta_\mu)U'^\dagger - (A_\mu - \theta_\mu))\right).
\end{aligned} \tag{1}$$

In fact, $\Theta^{G/H}_{MYM}$ is nothing other than the left-invariant canonical 1-form $\theta^{L(\zeta)}$ of this $U(1)$-centrally extended group.

Denoting $\hat{E}^\mu_a(x)$ the generator of translations in $A^a_\mu(x)$ and $\hat{A}^a_i(x)$ the generator of translations in $F^{0i}_a(x) \equiv E^i_a(x)$ (in much the same way $\hat{p}$ is the generator of translations in q in standard Quantum Mechanics) the Lie algebra of $G(M)^{G/H}_{MYM}$ can be adopted as equal-time ($d\sigma_\mu \to d\sigma_0$) basic "canonically conjugate" commutators:

$$\begin{aligned}
\left[\hat{G}_a(x), \hat{G}_b(y)\right] &= iC^c_{ab}\hat{G}_c(x)\delta(x-y), \\
\left[\hat{A}^a_j(x), \hat{E}^k_b(y)\right] &= i\delta^k_j\delta^a_b\delta(x-y), \\
\left[\hat{G}_a(x), \hat{A}^b_j(y)\right] &= iC^b_{ac}\hat{A}^c_j(x)\delta(x-y) - i\delta^b_a\partial^x_j\delta(x-y), \\
\left[\hat{G}_a(x), \hat{E}^\mu_b(y)\right] &= iC^c_{ab}\hat{E}^\mu_c(x)\delta(x-y) - im^2\delta^\mu_0 C^c_{ab}\lambda_c\delta(x-y)
\end{aligned} \tag{2}$$

The symmetry above requires some comments concerning the structure of the central extension by $U(1)$, which difines, after all, the structure of our proposal for the quantization of the Massive Yang-Mills Theory. On the one hand, the dynamical content of the fields A^a_i, $i = 1,2,3$ and their canonical momenta E^j_b, reflected by the second line in (2), is a consequence of the (current) cocycle J^{YM}_μ, and reproduces the ordinary duality coordinate-momentum for Yang-Mills fields; on the other, the dynamical character of the gauge group parameters themselves, $\varphi^a(x)$, as well as the associated momenta $E^0_a(x)$ is dictated by the cocycle J^σ_μ, responsible for the central term in the last line of (2). This cocycle is trivial from the strictic mathematical point of view but highly non-trivial from the physical point of view as it contributes with a non-trivial piece to the symplectic potential and/or Lagrangian at the classical level and to the corresponding quantization form $\Theta^{G/H}_{MYM} + \frac{d\zeta}{i\zeta}$. The interest of this sort of coboundaries with non-trivial physical content was first remarked by Saletan [19] and widely used in GAQ and in the representations of semisimple Lie groups as the Virasoro group and Kac-Moody groups (see [14] and references therein).

A unitary, irreducible representation of this (infinite-dimensional) Lie algebra on wave functionals $\Psi(E^\mu)$ in the "electric-field representation" E^μ_a ($E^0_a \equiv m^2 Tr(T_a\Lambda)$) can be achieved as:

$$\begin{aligned}\hat{E}^{\mu}_{a}\Psi(E) &= (E^{\mu}_{a} - m^{2}\delta^{\mu}_{0}\lambda_{a})\Psi(E),\\ \hat{A}^{a}_{j}\Psi(E) &= i\frac{\delta}{\delta E^{j}_{a}}\Psi(E),\\ \hat{G}_{a}\Psi(E) &= \left(\vec{\nabla}\cdot\vec{E}_{a} + iC^{c}_{ab}(E^{\mu}_{c} - m^{2}\delta^{\mu}_{0}\lambda_{c})\frac{\delta}{\delta E^{\mu}_{b}}\right)\Psi(E)\end{aligned}$$

where the last line accounts for the non-Abelian "Gauss law" when the constraint condition $\hat{G}_{a}\Psi(E) = 0$ is required. A more detailed study of the quantization process will be presented elsewhere [20].

Acknowledgements
Work partially supported by the Spanish MICINN (FIS 2008-06078-C03-01), Fundación Séneca (08814/PI/08) and Junta de Andalucía (FQM219, FQM1951). M. Calixto thanks the "Universidad Politécnica de Cartagena" for the award "Intensificación de la Actividad Investigadora 2009-2010" and the Fundación Séneca for financial support (PCTRM 2007-2010: INFO and FEDER).

References

[1] H. Ruegg and M. Ruiz-Altaba, Int. J. Mod. Phys. **A19**, 3265-3348 (2004).
[2] M. Bando, T. Kugo and K. Yamawaki, Nucl. Phys. **B259**, 493 (1985); Phys. Rep. **164**, 217 (1988).
[3] A.C. Longhitano, Nucl. Phys. **B188**, 118 (1981).
[4] W.A. Bardeen and K. Shizuya, Phys. Rev. **D18**, 1969 (1978).
[5] T. Kunimasa and T. Goto, Prog. Theor. Phys. **37**, 452-464 (1967).
[6] R. Delbourgo and G. Thompson, Phys. Rev. Lett. **57**, 2610-2612 (1986); J. Kubo, Phys. Rev. Lett. **58**, 2000 (1987); P. Kosinski and L. Szymanowski, Phys. Rev. Lett. **58**, 2001 (1987); R. Delbourgo, S. Twisk and G. Thompson, Int. J. Mod. Phys. **A3**, 435-449 (1988).
[7] V. Aldaya and J.A. de Azcárraga, J. Math. Phys. **23**, 1297 (1982)
[8] V. Aldaya, M. Calixto and J. Guerrero, Commun. Math. Phys. **178**, 399 (1996)
[9] V. Aldaya, J.L. Jaramillo and J. Guerrero, J. Math. Phys. **44**, 5166 (2003)
[10] J. Mickelsson *Current Algebras and Groups*, Plenum Press (1989).
[11] J. Mickelsson, Phys. Rev. Lett. **55**, 2099 (1985)
[12] V. Aldaya, E. Sanchez-Sastre and M. Calixto, Rep. Math. Phys. **59**, 83 (2007)
[13] V. Aldaya and E. Sanchez-Sastre, Int. J. Geom. Meth. Mod. Phys. **5**, 197 (2008)
[14] J. Guerrero, J.L. Jaramillo and V. Aldaya, J. Math. Phys. **45**, 2051 (2004)
[15] S.V. Ketov, *Quantum Non-Linear Sigma-Models*, Texts and Monographs in Physics, Springer (2000)
[16] V. Aldaya, M. Calixto and F.F. Lopez-Ruiz, Mod. Phys. Lett. **A24**, 2731 (2009)
[17] V.P. Nair, *Quantum Field Theory*, Graduate Texts in Contemporary Physics, Springer (2005)
[18] V. Aldaya and J. Azcarraga, J. Math. Phys. *19*, 1869 (1978)
[19] E.J. Saletan, J. Math. Phys. **2**, 1 (1961)
[20] V. Aldaya, M. Calixto and F.F. Lopez-Ruiz, *Symmetry Group of Massive Yang-Mills Theories without Higgs and their Quantization*, in preparation

Journal of Physics: Conference Series **284** (2011) 012009 doi:10.1088/1742-6596/284/1/012009

One dimensional systems with singular perturbations.

J.J. Alvarez[1], M. Gadella[2], L.M. Glasser[3], L.P. Lara[4], L.M. Nieto[2].

[1]Departamento de Informática, E.U. de Informática, Universidad de Valladolid, 40005 Segovia, Spain

[2]Departamento de FTAO, University of Valladolid, 47071 Valladolid, Spain

[3]Department of Physics, Clarkson University, Potsdam, NY 13699-5820, USA

[4]Departamento de Sistemas, FRRO, Zevallos 1345, Rosario, Argentina.

E-mail: jjalvarez@infor.uva.es, manuelgadella1@gmail.com, laryg@clarkson.edu, lplara@fceia.unr.edu.ar, luismi@metodos.fam.cie.uva.es

Abstract. This paper discusses some one dimensional quantum models with singular perturbations. Eventually, a mass discontinuity is added at the points that support the singular perturbations. The simplest model includes an attractive singular potential with a mass jump both located at the origin. We study the form of the only bound state. Another model exhibits a hard core at the origin plus one or more repulsive deltas with mass jumps at the points supporting these deltas. We study the location and the multiplicity of these resonances for the case of one or two deltas and settle the basis for a generalization. Finally, we consider the harmonic oscillator and the infinite square well plus a singular potential at the origin. We see how the energy of bound states is affected by the singular perturbation.

1. Introduction

In this brief review, we shall introduce some one dimensional models showing singular interactions type Dirac delta and its derivative plus mass jumps at the points supporting these singularities. The motivation for this study is in one side to construct solvable models in basic quantum mechanics and on the other to use these simple models as building blocks for more complicated but still solvable models showing interesting features like resonances. Then, it seems quite natural that the first situation under our consideration is the simplest that includes the above mentioned ingredients: one singularity at the origin, in this case a linear combination of a delta and its first derivative, plus a mass jump at the same point.

The singular potential is introduced by making use of the formalism of self adjoint extensions of symmetric operators, which has been extensively studied in the literature [1, 2, 3]. In physics models, singular potentials have been studied by several authors [4]. Hamiltonians with position dependent mass have been object of study in numerous publications [6, 7]. A particular case of particular interest appears when the dependence of the mass with the position is constant except for a jump at one or several points. This is a situation of physical interest. For instance in one dimensional models in which a particle moves along a wire formed by two distinct materials that join abruptly at some point [8].

In the simplest problem under our consideration mentioned above, we use an attractive delta in order to study the resulting bound state. With no surprise we obtain that its energy depends on the chosen self adjoint extension.

Now, assume a one dimensional free particle being allowed to move on the positive semiaxis only. This implies the existence of an impenetrable barrier at the origin (also called a hard core). If we add a potential formed by two repulsive Dirac deltas located at points $a, b > 0$, then the system shows resonances, which can be double (double poles of the S matrix) in the case of two deltas [5]. Then, it is quite interesting to study the influence of the mass jumps in the behavior of the resonances and on their multiplicity.

In the last section of the present review, we include a brief discussion on the influence of a singular interaction on two well know exactly solvable models: the harmonic oscillator and the infinite square well, both in one dimension. Now, the singularity can be either an attractive or a repulsive delta plus its derivative, although no discontinuity in the mass is included in these latter cases.

2. Delta potential with a mass jump

To start with, let us consider the most general form of a one dimensional free Hamiltonian with non constant mass. It is given by

$$K = \frac{1}{2}\, m^\alpha(x) p m^\beta(x) p m^\alpha(x)\,, \tag{1}$$

with $2\alpha + \beta = -1$ and $m(x)$ is the function that gives the mass in terms of the position. In this paper, we are going to assume that the mass is constant save for discontinuities at some given points (mass jumps). A mass jump at the origin gives the following function $m(x)$:

$$m(x) = m_1 H(-x) + m_2 H(x) = \left\{ \begin{array}{lll} m_1 & \text{if} & x < 0\,, \\ m_2 & \text{if} & x > 0 \end{array} \right. , \tag{2}$$

where $H(x)$ is the Heaviside step function.

In this first simple example the total Hamiltonian has the form $K + W$, where W is the singular perturbation given by

$$W(x) := -a\delta(x) + b\delta'(x)\,, \tag{3}$$

with $a > 0$, $b \in \mathbf{R}$. This gives the following formal Hamiltonian:

$$H = K + W = \frac{1}{2}\, m^\alpha(x) p m^\beta(x) p m^\alpha(x) - a\delta(x) + b\delta'(x)\,, \tag{4}$$

where the singular part has to be defined by using the formalism of self adjoint extensions of symmetric operators [13, 2, 3]. These self adjoint extensions are characterized by their domains. In the present case, the domain is specified by choosing, for the functions in this domain, convenient matching conditions at the origin. Then, functions in the domain $\mathcal{D}(H)$ of H as in (4) are expected to have a discontinuity at the origin. Any wave function $\psi(x) \in \mathcal{D}(H)$ should satisfy the Schrödinger equation for H, which is

$$K\psi(x) - a\delta(x)\psi(x) + b\delta'(x)\psi(x) = E\psi(x)\,. \tag{5}$$

Since $\psi(x)$ is in general discontinuous at the origin, we need a prescription to determine expressions like $\delta(x)\psi(x)$ and $\delta'(x)\psi(x)$. The simplest uses the following averages [3]:

$$\psi(x)\delta(x) = \frac{\psi(0+) + \psi(0-)}{2}\, \delta(x)\,,$$

$$\psi(x)\delta'(x) = \frac{\psi(0+) + \psi(0-)}{2}\, \delta'(x) - \frac{\psi'(0+) + \psi'(0-)}{2}\, \delta(x)\,, \tag{6}$$

where $\varphi(0+)$ and $\varphi(0-)$ denote right and left limits of the function $\varphi(x)$ discontinuous at the origin, respectively. Note that (5) is an equation in distributions.

Although this problem can be solved for arbitrary real values of α and β with $2\alpha + \beta = -1$ [9], here we shall discuss the simpler case in which the kinetic term K is given by

$$K = \begin{cases} -\frac{\hbar^2}{2m_1}\frac{d^2}{dx^2} & \text{if} \quad x < 0 \\ -\frac{\hbar^2}{2m_2}\frac{d^2}{dx^2} & \text{if} \quad x > 0 \end{cases}, \tag{7}$$

which corresponds to the choice $\alpha = 0$, $\beta = -1$. Then, our goal is to solve the Schrödinger equation (5) with the kinetic term given in (7). To do it, we just write in (5) $\psi(x) = f_1(x)H(-x) + f_2(x)H(x)$ and and use (6) and (7). Then, we obtain a linear combination of $H(-x)$, $H(x)$, $\delta(x)$ and $\delta'(x)$, which is equal to zero. This is only possible if the coefficients of this linear combination are all equal to zero. The resulting expressions give the matching conditions for $\psi(x)$ at the origin

$$\begin{pmatrix} \psi(0+) \\ \psi'(0+) \end{pmatrix} = \begin{pmatrix} \frac{\hbar^2+m_1 b}{\hbar^2-m_2 b} & 0 \\ \frac{-a\hbar^2(m_1+m_2)}{\hbar^4-m_2^2 b^2} & \frac{\hbar^2-m_1 b}{\hbar^2+m_2 b} \end{pmatrix} \begin{pmatrix} \psi(0-) \\ \psi'(0-) \end{pmatrix}, \tag{8}$$

the energy of the bound state,

$$E = -\frac{\hbar^2}{2}\frac{(m_1+m_2)^2 a^2 \hbar^4}{[(\hbar^2 - m_1 b)(\hbar^2 - m_2 b)\sqrt{m_1} + (\hbar^2 + m_1 b)(\hbar^2 + m_2 b)\sqrt{m_2}]^2} \tag{9}$$

and the wave function of the bound state

$$\psi(x) = A\,[(\hbar^2 - m_2 b)\, e^{k_1 x}\, H(-x) + (\hbar^2 + m_1 b)\, e^{-k_2 x}\, H(x)]\,, \tag{10}$$

with $k_i = \sqrt{-2Em_i/\hbar^2}$, $i = 1, 2$ and A is a normalization constant [9].

Note that (9) is not symmetric with respect to the interchange of m_1 and m_2, but is symmetric with respect to the interchange of the notions of left and right for which we have to interchange the value of the masses as well as the sign of b. In the equal masses limit, $m_1 = m_2 = m$, the value of the energy of the unique bound state is the same which has been obtained by a direct calculation in [10]. This is:

$$E = -\frac{1}{2}\frac{m a^2 \hbar^6}{(\hbar^4 + m^2 b^2)^2}\,. \tag{11}$$

In the next section, we shall study the conditions under which the Hamiltonian $H = K + V$ is self adjoint. One unexpected consequence of these conditions is that b in (3,4) cannot be arbitrary, although for $m_1 \neq m_2$ it must have the form [9]:

$$b^2 = \frac{\hbar^4}{m_1^2 + m_2^2 + m_1 m_2}\,. \tag{12}$$

For $m_1 = m_2$ this restriction makes no sense [9]. This conditions also imply that if $b = 0$, then $m_1 = m_2$. The presence of the mass jump plus the delta at the origin requires the presence of the term $b\delta'(x)$. Without this term with b given by (11), the proposed model has no solution [9].

3. Self adjointness of the Hamiltonian

In order to define the Hamiltonian H given in the previous section, we start with a domain for the kinetic term K in which K be symmetric with equal deficiency indices. One such domain could be

$$\mathcal{D} := \{\psi(x) \in W_2^2(\mathbf{R}) \ / \ \psi(0) = \psi'(0) = 0\}\,, \tag{13}$$

where $W_2^2(\mathbf{R})$ is the linear space of functions $\psi(x) : \mathbf{R} \longrightarrow \mathbf{C}$, $\mathbf{C}$ being the field of complex numbers, such that i.) $\psi(x)$ admits a first continuous derivative; ii.) the second derivative of $\psi(x)$ exists almost everywhere and iii.) both $\psi(x)$ and its second derivative $\psi''(x)$ are square integrable, i.e.,

$$\int_{-\infty}^{\infty} \{|\psi(x)|^2 + |\psi''(x)|^2\}\, dx < \infty\,. \tag{14}$$

The adjoint of K, $K^\dagger$, has domain $\mathcal{D}^* \equiv W_2^2(\mathbf{R}/\{0\})$. The functions in $W_2^2(\mathbf{R}/\{0\})$ satisfy the same properties of the functions in $W_2^2(\mathbf{R})$ except that they and their derivatives may admit a finite jump at the origin. It is straightforward that K has deficiency indices (2, 2) and therefore it has an infinite number of self-adjoint extensions. These self-adjoint extensions are restrictions of $K^\dagger$ to certain subdomains, which are determined by the matching conditions at the origin [13, 2, 3]. For any two $\varphi, \psi \in \mathcal{D}^*$, we have:

$$\langle\psi|K^\dagger\varphi\rangle = \mathcal{G} + \langle K^\dagger\psi|\varphi\rangle\,, \tag{15}$$

where

$$\mathcal{G} = \frac{\hbar^2}{2m_2}\,[\psi^*(0+)\varphi'(0+) - \psi'^*(0+)\varphi(0+)] - \frac{\hbar^2}{2m_1}\,[\psi^*(0-)\varphi'(0-) - \psi'^*(0-)\varphi(0-)]\,. \tag{16}$$

If we define T as

$$\begin{pmatrix} \psi(0+) \\ \psi'(0+) \end{pmatrix} = T \begin{pmatrix} \psi(0-) \\ \psi'(0-) \end{pmatrix}, \tag{17}$$

we have

$$\mathcal{G} = (\psi^*(0-), \psi'^*(0-))[-M_1 + T^\dagger M_2 T] \begin{pmatrix} \varphi(0-) \\ \varphi'(0-) \end{pmatrix}, \tag{18}$$

with

$$M_i = \frac{\hbar^2}{2m_i} \begin{pmatrix} 0 & 1 \\ -1 & 0 \end{pmatrix}, \quad i = 1, 2\,. \tag{19}$$

From (15), one sees that the domains of the self adjoint extensions of K are those that make $\mathcal{G} \equiv 0$, nontrivially, i.e., allowing non zero limit values of the wave function and its first derivative at the origin. Therefore, according to (18) and taking into account on the arbitrariness of the functions $\psi(x)$ and $\varphi(x)$ the domains of the self adjoint extensions of K are those characterized by the following property of T:

$$M_1 = T^\dagger M_2 T\,. \tag{20}$$

Note that because of (17), once we have defined T satisfying (20), we have defined a domain and viceversa. Therefore, (20) gives a necessary and sufficient condition for the domain characterized by T be a domain of a self adjoint extension of K.

In the case of the Hamiltonian $H = K + W$ studied in the previous section, matrix T was given in equation (8). It is straightforward to show that it satisfies (20) and that consequently H as in (4) is well defined as a legitimate self adjoint extension of K. Furthermore, equation (20) restricts the possible values of b to one (and the same with the opposite sign), which in the case of K given by (7) is precisely (12).

4. A one dimensional model of resonances with mass jumps

The next example of one dimensional potential with a singular perturbation is given by

$$H = K + V_0 + \gamma\delta(x-a)\,, \qquad a > 0\,, \tag{21}$$

where K and V_0 have the following form:

$$K = \begin{cases} -\frac{\hbar^2}{2m_1}\frac{d^2}{dx^2} & \text{if} \quad 0 < x < a \\ -\frac{\hbar^2}{2m_2}\frac{d^2}{dx^2} & \text{if} \quad x > a \end{cases}\,, \qquad V_0(x) = \begin{cases} \infty & \text{if} \quad x \le 0 \\ 0 & \text{if} \quad x > 0 \end{cases}\,. \tag{22}$$

Thus, H defines an impenetrable barrier supported on the negative real semiaxis (hard core) and has a repulsive delta and a mass jump at the point $x = a$. The simplest and most natural matching conditions at the point $x = a$ that gives account for our situation would be the following:

$$\psi(a-) = \psi(a+) = \psi(a)\,, \qquad \frac{1}{m_2}\,\psi'(a+) - \frac{1}{m_1}\,\psi'(a-) = \frac{2\gamma}{\hbar^2}\,\psi(a)\,, \tag{23}$$

since this is the minimal generalization of the matching conditions defining a repulsive delta interaction at $x = a$ that includes the mass jump. Formulas (23) can be written in matrix form as in (17) with 0 replaced by a. In this case, matrix T is quite simple

$$T = \begin{pmatrix} 1 & 0 \\ 2m_2\gamma/\hbar^2 & m_2/m_1 \end{pmatrix}. \tag{24}$$

This T in (24) satisfies (20) and therefore, defines a self adjoint Hamiltonian H.

In the case of a repulsive delta, the model has no bound states, but due to the presence of the hard core, it has resonances. The situation is similar to the constant mass case. If we denote by I and II the regions $0 < x < a$ and $a < x$ respectively, the solution of the Schrödinger equation corresponding to the Hamiltonian in (21) with boundary conditions $\psi_I(0) = 0$ and $\psi'_I(0) = 1$ is

$$\psi_I(x) = \frac{1}{k_1}\,\sin(k_1 x)\,, \qquad \psi_{II}(x) = C\,e^{ik_2x} + D\,e^{-ik_2x}\,, \tag{25}$$

with $k_i = \sqrt{2m_iE/\hbar^2}$, $i = 1, 2$. Purely outgoing boundary condition will give resonances. These conditions imply the existence of outgoing wave only, which means that $D(k) = 0$ with $k = k_1 = k_2\sqrt{m_1/m_2}$. Resonances come in pairs of poles. In fact, if k_0 is a solution of $D(k) = 0$, then $-k_0^*$ is also a solution [11]. If we introduce the new variables z and ξ, defined as $z = ka$, $m_2 = m_1\xi^2$, $\xi > 0$, and if we choose units in such a way that $\hbar^2 = 2\gamma a m_1$, equation $D = 0$ gives [11]:

$$(z + i\xi)\sin z + i\xi z\cos z = 0\,. \tag{26}$$

Solutions in z of this equation are the resonances of the model. In Figs. 1 and 2, we plot the zeroes of the real part of (26) (solid curves) and the zeroes of its imaginary part (dashed curves) for two different values of $\xi = \sqrt{m_2/m_1}$. Thus, intersections of solid and dashed lines represent resonances.

Search for degenerate resonances.- A resonance of order n is given by a pole of n-th order of the S-matrix or, equivalently, by a zero of n-th order of $D(k)$. Resonances of order $n \ge 2$ are called multiple pole resonances or degenerate resonances [5]. Note that a necessary condition for the existence of a degenerate resonance is the existence of a simultaneous solution k_0 of the transcendent equations $D(k) = D'(k) = 0$. Thus, if there is no such a solution, no degenerate resonances may exist.

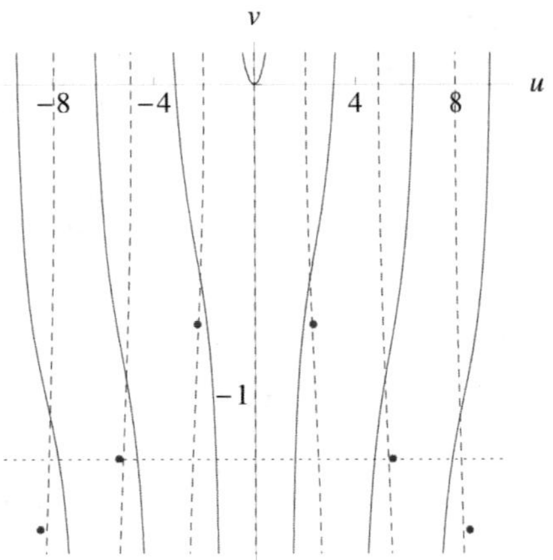

Figure 1. Resonances for $\xi = \sqrt{m_2/m_1} = 1.2$. Dots represent resonances in the limit of equal masses ($\xi = 1$). Here $k = \mu + i\nu$.

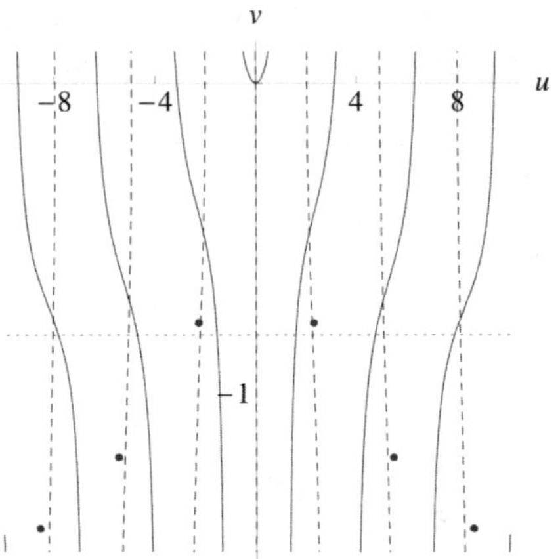

Figure 2. Resonances for $\xi = \sqrt{m_2/m_1} = 1.5$. Dots represent resonances in the limit of equal masses ($\xi = 1$). Here $k = \mu + i\nu$.

Then, to search for degenerate resonances, we first need to derive the left hand side of (26) with respect to k and then equalize to zero the resulting expression. It can be shown [11] that both equations do not have simultaneous solutions and that therefore, no degenerate resonances may exist in this model. Consequently, a jump of mass along a delta barrier never has the effect of a double delta barrier [5] in the sense that it does not produce degenerate resonances.

We would like to remark that, in opposition with the situation described in Section 2, derivatives of the delta are not necessary in the present case. In the model described in Section 2, the derivative of the delta was necessary because it naturally appeared when we obtained the second derivative of the wave function with a discontinuity at the origin. In this other case, matching conditions (23) impose a jump in the derivative of the wave function and not on the wave function itself. Thus, the derivative of the delta cancels out when we calculate the explicit form of the Schrödinger equation in this case. In any case and for further consideration, one may also include a perturbation of the type $a\delta(x-a) + b\delta'(x-a)$ along with a mass discontinuity at $x = a$.

4.1. More than one singular point

Once we have solved the problem of finding the resonances of the previous model, we may investigate the situation in which more than one delta is present in the potential. In addition, we consider respective mass jumps at the points supporting the deltas. Thus, if the singular potential is given by $\gamma_1\delta(x-a_1) + \gamma_2\delta(x-a_2) + \ldots + \gamma_n\delta(x-a_n)$, with $0 < a_1 < a_2 < \ldots < a_n$, the function of the mass $m(x)$ should be given by

$$m(x) := \begin{cases} m_1 & \text{if} \quad 0 < x < a_1 \\ m_2 & \text{if} \quad a_1 < x < a_2 \\ \ldots & \ldots \quad \ldots\ldots \\ m_{n+1} & \text{if} \quad a_n < x \end{cases} . \tag{27}$$

The generalization of the kinetic term K as in (5) to this situation is obvious. Now, in order to obtain the self adjoint Hamiltonian with the proposed singular potential, we need to determine matching conditions at all points in the sequence $\{a_1, a_2, \ldots, a_n\}$. This matching conditions will be given by matrix relations, so that if $\psi(x)$ is a wave function on the domain of self adjointness of H, it should fulfill at each a_i, $i = 1, 2, \ldots, n$, the following relation:

$$\begin{pmatrix} \psi_{i+1}(a_i) \\ \psi'_{i+1}(a_i) \end{pmatrix} = T_i(a_i) \begin{pmatrix} \psi_i(a_i) \\ \psi'_i(a_i) \end{pmatrix}, \tag{28}$$

were we have denoted by $\psi_i(x)$ the solution of the Schrödinger equation between a_{i-1} and a_i, $a_0 = 0$. Note that $\psi_i(x) = C_i\, e^{k_i x} + D_i\, e^{-ik_i x}$, where $k_i = \sqrt{2m_i E/\hbar^2}$. Then,

$$\begin{pmatrix} \psi_i(a_i) \\ \psi'_i(a_i) \end{pmatrix} = \begin{pmatrix} e^{ik_i a_i} & e^{-ik_i a_i} \\ ik_i\, e^{ik_i a_i} & -ik_i\, e^{-ik_i a_i} \end{pmatrix} \begin{pmatrix} C_i \\ D_i \end{pmatrix} = \mathcal{M}_i(a_i) \begin{pmatrix} C_i \\ D_i \end{pmatrix}. \tag{29}$$

Matrices $\mathcal{M}_i(a_i)$ are defined in (29). With the help of (29) and (28), we can derive the following recurrence relation:

$$\begin{pmatrix} C_{i+1} \\ D_{i+1} \end{pmatrix} = \mathcal{M}_i^{-1}(a_i)\, T_i(a_i)\, \mathcal{M}_i(a_i) \begin{pmatrix} C_i \\ D_i \end{pmatrix}, \tag{30}$$

The matrices, $T_i(a_i)$ have the following form:

$$T_i(a_i) := \begin{pmatrix} 1 & 0 \\ \gamma_i m_{i+1} & \frac{m_{i+1}}{m_i} \end{pmatrix}. \tag{31}$$

The simplest case with $n \neq 1$ is $n = 2$. Then, we have two deltas and their respective mass jumps. The half line $x > 0$ is divided into three regions. The wave function in each region is given by $\psi_i(x) = C_i\, e^{ik_i x} + D_i\, e^{-ik_i x}$ with $i = 1, 2, 3$. The relation among the coefficients C_1, D_1 and C_3, D_3 is

$$\begin{pmatrix} C_3 \\ D_3 \end{pmatrix} = \left(\mathcal{M}_3^{-1}(a_2)\, T_2(a_2)\, \mathcal{M}_2(a_2)\right) \left(\mathcal{M}_2^{-1}(a_1)\, T_1(a_1)\, \mathcal{M}_1(a_1)\right) \begin{pmatrix} C_1 \\ D_1 \end{pmatrix} \tag{32}$$

where,

$$\begin{pmatrix} C_1 \\ D_1 \end{pmatrix} = \begin{pmatrix} \frac{1}{2ik} \\ -\frac{1}{2ik} \end{pmatrix} \tag{33}$$

are the initial conditions. Purely outgoing boundary conditions determine resonances, which are now given by solutions of the transcendental equation $D_3(k) = 0$. In the constant mass case was shown that doubly degenerate resonances should exist [5], which are solutions of the equations $D(k_0) = D'(k_0) = 0$ with $D''(k_0) \neq 0$. Therefore, this kind of resonances should be expected when to the two deltas we add the mass jumps. Indeed this is the case and numerical estimations have shown the presence of doubly degenerate resonances at the time that deny the existence of resonances of order higher than two [12]. Also, preliminary results have indicated the existence of resonances of order at most three in the case of three deltas with their respective mass jumps at the same points.

5. Exactly solvable models with singular interactions

In this last Section, we shall analyze both the one dimensional harmonic oscillator and the infinite square well with a singular perturbation of the type $-a\delta(x)+b\delta'(x)$. The coefficient a could be either positive or negative, and results will be slightly different in each case, and b is real. Thus, let us take the following Hamiltonian

$$H=-\frac{d^2}{dx^2}+V_0(x)+W(x)\,, \qquad W(x)=-a\delta(x)+b\delta'(x)\,, \qquad a>0\,, \tag{34}$$

and either

$$V_0(x)=\frac{\kappa}{2}\,x^2\,, \qquad \text{or} \qquad V_0(x)=\left\{\begin{array}{ll} 0 & \text{if} \quad |x|<c \\ \infty & \text{if} \quad |x|\geq c \end{array}\right. . \tag{35}$$

In the first case we have an harmonic oscillator and in the second an infinite square well. The singular potential is supported at the origin. In both models no resonances can be expected, so that we shall focus our attention on the behavior of bound states.

This study can be performed in two equivalent manners. Either we may solve directly the Schrödinger equation for each case or we apply the Green function method. Let us briefly outline the second method.

Let us consider a one-dimensional quantum system whose dynamical evolution is governed by two different Hamiltonian operators

$$H_0=-\frac{d^2}{dx^2}+V_0(x), \qquad H=H_0+W(x)\,, \tag{36}$$

the first one is the regular free Hamiltonian H_0 and the second one the perturbed Hamiltonian H (units are usually chosen such that $\hbar=2m=1$). Let us assume that H_0 admits a Green function given by $G_0(x,x',E)$ (the dimension of x and x' is here irrelevant). Then, $\psi(x)$ represents a bound state of the total Hamiltonian H with energy E if and only if the Lippman-Schwinger bound state equation

$$\psi(x)=\int G_0(x,x',E)W(x')\psi(x')\,dx'\,. \tag{37}$$

is satisfied. In consequence, bound states of the total hamiltonian H should satisfy the equation given by

$$\psi(x)=\int G_0(x,x',E)[-a\delta(x')+b\delta'(x')]\,\psi(x')\,dx'\,. \tag{38}$$

This integral equation is easily solvable in our case. In fact if we use the short notation:

$$G_0^{ij}(x,x'):=\frac{\partial^{i+j}}{\partial x^i\partial x'^j}\,G_0(x,x',E)\,, \quad i,j=0,1 \tag{39}$$

and

$$\begin{array}{ll} A:=G_0(0+,0)=G_0(0-,0)\,, & B:=G_0^{01}(0+,0)=G_0^{10}(0-,0)\,, \\ C:=G_0^{01}(0-,0)=G_0^{10}(0+,0)\,, & D:=G_0^{11}(0+,0)=G_0^{11}(0-,0)\,, \end{array} \tag{40}$$

we can insert (39) and (40) into (38). In order to obtain the product of the delta and its derivative times the wave function $\psi(x')$, we use equations (6). Thus, we obtain an homogeneous system of equations in the unknowns $\psi(0\pm)$ and $\psi'(0\pm)$, where $\varphi(0+)$ and $\varphi(0-)$ are the right and left limits of the function $\varphi(x)$ at the origin. This systems admits a solution if and only if the determinant of the coefficients vanishes. This is a determinant of a 4×4 matrix, which can be simplified to

$$\det \begin{pmatrix} 2 & -1+\frac{b}{2}(B-C) & 0 \\ -1 & 1+\frac{a}{2}A+\frac{b}{2}C & \frac{b}{2}A \\ 0 & \frac{a}{2}(B+C)+bD & 1+\frac{b}{2}(B+C) \end{pmatrix} = 0\,. \tag{41}$$

This equation (41) can give the energy of bound states for the cases under our study, described by either Hamiltonian in (34) and (35).

The Green function of the harmonic oscillator is given by

$$G_0(y,y',E) = \frac{D_\nu(y_>)\,D_\nu(y_<)}{2D_\nu(0)\,D'_\nu(0)}\,, \tag{42}$$

where $D_\nu(y)$ is the parabolic cylinder equation with index $\nu = (2E/\omega - 1)/2$ and we use the standard notation $y_>, y_< = \max, \min(y, y')$. This function, which admits an integral representation, is described in [13]. Using the identities (40), we arrive at

$$A = -\frac{D_\nu(0)}{2\,D'_\nu(0)}\,, \qquad B = -C = -\frac{1}{2}\,, \qquad D = \frac{D'_\nu(0)}{2\,D_\nu(0)} = -\frac{1}{4A}\,. \tag{43}$$

For odd oscillator levels, the wave functions of the free oscillator are odd and therefore, they vanish at the origin. In this case, $D_\nu(0) = 0 \Longleftrightarrow A = 0$. Here, equation $A = 0$ replaces to (41) and therefore, it should give the bound states for the odd oscillator levels plus the singular potential. Since $A = 0$ is equivalent to $D_\nu(0) = 0$, which happens if and only if $E = \omega[(2n+1)1/2]$ $(\hbar = 1)$, it results that the energy of odd levels does not change under the action of the singular perturbation. For even levels, let us go back to (41). Since now, $B = -C$ this gives $1 + \alpha A - \beta^2 AD = 0$, which together with (43) this yields to

$$-\frac{D_\nu(0)}{D'_\nu(0)} = \frac{4+\beta^2}{2\alpha}\,. \tag{44}$$

Then, using the properties of the parabolic cylinder functions [13], we obtain:

$$\frac{\Gamma[(1-2E/\omega)/4]}{\sqrt{2}\,\Gamma[(3-2E/\omega)/4]} = \frac{4+\beta^2}{2\alpha} = \frac{4\omega^{1/2} + b^2\omega^{-1/2}}{2a}\,. \tag{45}$$

Here, we obtain the following result: the odd oscillator levels remain invariant while the even levels are lowered in energy if $a > 0$ and raised if $a < 0$. It is interesting to remark that the higher the level the smaller is the effect of the perturbation.

For the infinite square well, the Green function is given by [14]

$$G_0(x,x',k^2) = \frac{\cos[2kc - k|x-x'|] - \cos[k(x+x')]}{2k\sin(2kc)}\,, \tag{46}$$

where $E = k^2$. Now, a similar analysis that we did for the harmonic oscillator produces the following results [14]: the energy levels with odd wave function remain unaltered while the energy levels with even wave function are either lowered $(a > 0)$ or raised $(a < 0)$.

Acknowledgments

Partial financial support is acknowledged to the Spanish Junta de Castilla y León (Project GR224) and the Ministry of Science and Innovation (Project MTM2009-10751).

References

[1] Albeverio S, Gesztesy F, Hoegh-Krohn R and Holden H 1988 *Solvable Models in Quantum Mechanics* (Springer, Berlin)
[2] Albeverio S and Kurasov P 2000 *Singular perturbations of differential operators. Solvable Schrödinger type operators* (London Mathematical Society Lecture Notes 271, Cambridge UP, Cambridge, UK)
[3] Kurasov P 1996 Jour. Math. Anal. Appl. **201** 297
[4] Seba P 1986 Rep. Math. Phys. **24** 111
Coutinho FAB, Nogami Y and Fernando Pérez J 1997 J. Phys. A: Math. Gen. **30** 3937
Christiansen P L, Arnbak H C, Zolotaryuk A V, Ermakov V N and Gaididei Y B 2003 J. Phys. A: Math. Gen. **36**, 7589
Toyama F M and Nogami Y, J. 2007 Phys. A: Math. Theor. **40** F685
Albeverio S, Dabrowski L and Kurasov P 1998 Lett. Math. Phys. **45** 33
Fülöp T and Tsutsui I 2000 Phys. Lett. A **264** 366
Hejcik P and Cheon T 2006 Phys. Lett. A **356** 290
Fernández C, Palma G and Prado H (2005) J. Phys A: Math. Gen., **38**, 7509
[5] Hernández E, Jáuregui A and Mondragón A 2000 Journal of Physics A: Mathematical and General, **33** 4507
Antoniou I E et al. 2001 Chaos, Solitons and Fractals, **12** 2719
Hernández E, Jáuregui A and Mondragón A, Physical Review A 2003 **67** 022721
Hernández E, Jáuregui A and Mondragón A 2005 Physical Review E, **72** 026221
[6] Moiseyev N and Lefebvre R 2001 Phys. Rev. A **64** 052711
Cruz S C Y, Negro J and Nieto L M Phys. Lett. A 2007 **369** 400
Ganguly A and Nieto L M 2007 J. Phys. A: Math. Theor, **40** 7265
Ganguly A, Kuru S Negro J and Nieto L M 2006 Phys. Lett. A, **360** 228
Ganguly A, Ioffe M V and Nieto L M 2006 J. Phys. A: Math. Gen., **39** 14659
Gadella M, Kuru S and Negro J 2007 Physics Letters A, **362**, 265
[7] Lévy-Leblond J M 1995 Phys. Rev. A **52** 1845
[8] Morrow R A and Brownstein K R 1984 Phys. Rev. B **30** 678
Einevoll G T and Hemmer P C 1988 J. Phys. C **21** L1193
Thomsen J, Einevoll G T and Hemmer P C 1989 Phys. Rev. B **39** 12783
Einevoll G T, Hemmer P C and Thomsen J 1990 Phys. Rev. B **42** 3485
[9] Gadella M, Heras F J H, Negro J and Nieto L M 2009 J. Phys. A: Math. Theor. **42** 465207.
[10] Gadella M, Negro J and Nieto L M 2009 Phys. Lett. A **373** 1310
[11] Alvarez J J, Gadella M, Heras F J H, Nieto L M 2009 Phys. Lett. A **373** 4022
[12] Alvarez J J, Gadella M, Nieto L M 2010 A study of resonances in a one-dimensional model with singular Hamiltonian and mass jumps *Submitted*
[13] Abramovich M and Stegun I A 1970 *Handbook of Mathematical Functions* (Dover, New York)
[14] Gadella M, Glasser M L and Nieto L M 2010 The infinite square well with a singular perturbation *Submitted*

Super Vector Bundles

L Balduzzi 1, C Carmeli 1 and G Cassinelli 2

1 Department of Physics, University of Genova, via Dodecaneso 33 16146 Genova , IT
2 Department of Physics and I.N.F.N., Sezione di Genova, University of Genova, via Dodecaneso 33 16146 Genova , IT

E-mail: gianni.cassinelli@ge.infn.it

Abstract. We review the basic aspects of the theory of smooth super vector bundles. In particular we give the precise link existing between the geometrical and the sheaf theoretical approaches.

1. Introduction

In [1, § 3.2] it is stated very concisely that, if M is a (smooth) supermanifold, a (smooth) super vector bundle of rank $m|n$ over M can be defined in two equivalent ways as:

(i) a super fiber bundle E over M, with typical fiber $\mathbb{R}^{m|n}$ and structural group $\mathrm{GL}(m|n)$ (*geometrical approach*);

(ii) a sheaf of $\mathcal{O}_M$-modules $\mathcal{V}$, locally free of dimension $m|n$ (*algebraic approach*).

In this paper we are going to examine how this objects can be defined, how the two approaches are linked and how it is possible to recover the ideas of fiber and section. The main problem is that a super vector bundle cannot be described as a family of super vector spaces parameterized by the points of $\widetilde{M}$. Indeed, as usual, working pointwise, one loses information about the nilpotent part of the sheaf. So we always have to consider neighbourhoods of the points.

2. Supermanifolds

A (smooth) *supermanifold* M of dimension $p|q$ is a locally compact, second countable, Hausdorff topological space $|M|$ endowed with a sheaf $\mathcal{O}_M$ of superalgebras, locally isomorphic to $\mathcal{C}^\infty(\mathbb{R}^p)\otimes\Lambda(\vartheta_1,\dots,\vartheta_q)$. A morphism $\psi\colon M\to N$ between supermanifolds is a pair of morphisms $(|\psi|,\psi^*)$ where $|\psi|\colon|M|\to|N|$ is a continuous map and $\psi^*\colon\mathcal{O}_N\to\mathcal{O}_M$ is a sheaf morphism above $|\psi|$.

Remark 2.1. We will consider only smooth supermanifolds. It can be proved that in this category a morphism of supermanifolds is determined once we know the corresponding morphism on the superalgebra of global sections (see, for example, [2] and [3]). In other words, a morphism $\psi\colon M\to N$ can be identified with a superalgebra map $\psi^*\colon\mathcal{O}_N(|N|)\to\mathcal{O}_M(|M|)$. We will tacitly use this fact several times. Moreover, in the following, we will denote with $\mathcal{O}(M)$ the superalgebra of global sections $\mathcal{O}_M(|M|)$.

Suppose now U is an open subset of $|M|$ and let $\mathcal{J}_M(U)$ be the ideal of the nilpotent elements of $\mathcal{O}_M(U)$. It is possible to prove that $\mathcal{O}_M/\mathcal{J}_M$ defines a sheaf of purely even algebras over $|M|$

locally isomorphic to $\mathcal{C}^\infty(\mathbb{R}^p)$. Therefore $\widetilde{M} := (|M|, \mathcal{O}_M/\mathcal{J}_M)$ defines a classical manifold, called the *reduced manifold* associated to M. The projection $s \mapsto \widetilde{s} := s + \mathcal{J}_M(U)$, with $s \in \mathcal{O}_M(U)$, is the pullback of the embedding $j : \widetilde{M} \to M$. In the following we denote with $\mathrm{ev}_x(s) := \widetilde{s}(x)$ the *evaluation* of s at $x \in U$. It is also possible to check that, given a morphism $\psi \colon M \to N$, $|\psi|^*(\widetilde{s}) = \widetilde{\psi^*(s)}$, so that the map $|\psi|$ is automatically smooth. As a consequence, it is possible to associate a *reduced map* $\widetilde{\psi} = |\psi| \colon \widetilde{M} \to \widetilde{N}$ to each supermanifold morphism ψ.

An important and very used tool in working with supermanifolds is the functor of points. Given a supermanifold M one can construct the functor

$$M(\,\cdot\,) \colon \mathsf{SMan}^{\mathrm{op}} \longrightarrow \mathsf{Set}$$

from the opposite of the category of supermanifolds to the category of sets defined by $S \mapsto M(S) := \mathrm{Hom}(S, M)$ and called the *functor of points* of M. In particular, for example, $M(\mathbb{R}^{0|0}) \cong |M|$ as sets. Each supermanifold morphism $\psi \colon M \to N$ defines the natural transformation $\psi(\,\cdot\,) \colon M(\,\cdot\,) \to N(\,\cdot\,)$ given by $[\psi(S)](x) := \psi \circ x$. Due to Yoneda's lemma, each natural transformation between $M(\,\cdot\,)$ and $N(\,\cdot\,)$ arises from a unique morphism of supermanifolds in the way just described. The category of supermanifolds can thus be embedded onto a full subcategory of the category $[\mathsf{SMan}^{\mathrm{op}}, \mathsf{Set}]$ of functors from the opposite of the category of supermanifolds to the category of sets. Let

$$\begin{aligned} \mathcal{Y} \colon \mathsf{SMan} &\longrightarrow [\mathsf{SMan}^{\mathrm{op}}, \mathsf{Set}] \\ M &\longmapsto M(\,\cdot\,) \end{aligned}$$

denote such embedding. It is a fact that the image of **SMan** under $\mathcal{Y}$ is strictly smaller than $[\mathsf{SMan}^{\mathrm{op}}, \mathsf{Set}]$. The elements of $[\mathsf{SMan}^{\mathrm{op}}, \mathsf{Set}]$ isomorphic to elements in the image of $\mathcal{Y}$ are called *representable*. Supermanifolds can thus be thought as the representable functors in $[\mathsf{SMan}^{\mathrm{op}}, \mathsf{Set}]$. For all the details we refer to [1, 2, 4, 5, 6].

Example 2.2. $\mathbb{R}^{p|q}$ is the super manifold whose reduced manifold is $\mathbb{R}^p$, and with the sheaf of sections given by the restriction of $\mathcal{C}^\infty(\mathbb{R}^p) \otimes \Lambda(\mathbb{R}^q)$. Notice that $\mathbb{R}^{p|q}$ denotes also the super vector space $\mathbb{R}^{p|q} = \mathbb{R}^p \oplus \mathbb{R}^q$. Using the functor of points one can prove that the two concepts can be identified (in the sheaf–theoretical approach, the linear structure of $\mathbb{R}^{p|q}$ is encoded by the linear sections $(\mathbb{R}^{p|q})^*$).

3. Constant rank morphisms and Subsupermanifolds

The definition of immersion and submersion does not present particular problems.

Definition 3.1. Let $\psi \colon M \to N$ be a supermanifold morphism and $x \in \widetilde{M}$. ψ is an *immersion* (resp. a *submersion*) *at* x if $(d\psi)_x$ is injective (resp. surjective). ψ is an *immersion* (resp. a *submersion*), if this happens at each point. ψ is an *injective immersion* (resp. a *surjective submersion*), if, in addition, $\widetilde{\psi}$ is injective (resp. surjective).

Remark 3.2. If $\psi \colon M \to N$ is a submersion and U is an open subsupermanifold of N, then $\psi^{-1}(U)$ is defined as the open subsupermanifold of M on $\widetilde{\psi}^{-1}(\widetilde{U})$.

In the super context the concept of constant rank is more subtle (see, in particular, [4, § 2.3]).

We recall that, if M is a supermanifold, $\mathrm{Aut}_{\mathcal{O}(M)}(\mathcal{O}(M)^{p|q})$ denotes the group of the even invertible $(p|q) \times (p|q)$ matrices $\begin{pmatrix} s_{i_1,j_1} & t_{k_1,l_1} \\ t_{k_2,l_2} & s_{i_2,j_2} \end{pmatrix}$ with entries $s_{i_r,j_r} \in \mathcal{O}(M)_0$ and $t_{k_r,l_r} \in \mathcal{O}(M)_1$.

Definition 3.3. Let U be a superdomain. If $J = \left(\begin{smallmatrix} A & B \\ C & D \end{smallmatrix}\right)$ is an even $(p|q) \times (m|n)$ matrix with entries in $\mathcal{O}(U)$, then we say that J has *constant rank* $r|s$ if there exist $G_1 \in \mathrm{Aut}_{\mathcal{O}(U)}(\mathcal{O}(U)^{p|q})$ and $G_2 \in \mathrm{Aut}_{\mathcal{O}(U)}(\mathcal{O}(U)^{m|n})$ such that $G_1 J G_2$ has the form $\left(\begin{smallmatrix} A' & 0 \\ 0 & D' \end{smallmatrix}\right)$ with $A' = \left(\begin{smallmatrix} \mathbb{1}_{\mathbb{K}^r} & 0 \\ 0 & 0 \end{smallmatrix}\right)$ and $D' = \left(\begin{smallmatrix} \mathbb{1}_{\mathbb{K}^s} & 0 \\ 0 & 0 \end{smallmatrix}\right)$. If $\psi\colon M \to N$ is a morphism between supermanifolds, we say that ψ has *constant rank* $r|s$ *at* $x \in \widetilde{M}$, if there exists a coordinate neighborhood of x such that the super Jacobian matrix J_ψ has constant rank $r|s$. ψ has *constant rank* if it has constant rank at every point.

Proposition 3.4. *Let $\psi\colon M \to N$ be a supermanifold morphism and $x \in \widetilde{N}$, such that ψ has constant rank $r|s$ at each point of $\widetilde{\psi}^{-1}(x)$ (see definition 3.3). If*

$$\mathcal{J}_x := \{\, s \in \mathcal{O}(N) \mid \widetilde{s}(x) = 0 \,\}$$

and

$$\mathcal{J}_x^\psi := \mathcal{O}(M)\psi^*(\mathcal{J}_x)$$

is the ideal in $\mathcal{O}(M)$ generated by $\psi^*(\mathcal{J}_x)$, then there exists a unique closed subsupermanifold S of M such that $\mathcal{J}_x^\psi = \mathcal{J}_S$. S has dimension $\dim M - r|s$.

Definition 3.5. $\psi^{-1}(x)$ denotes the closed subsupermanifold S in proposition 3.4.

4. Super Vector Bundles

4.1. Geometrical approach

In this paragraph we discuss the geometrical formulation of the concept of super vector bundle. We begin with the definition of super fiber bundle.

Before we proceed, let us fix some notation. We recall that, if M is a supermanifold, $\widetilde{M}$ denotes the associated reduced manifold defined in the previous section. In order to avoid possible confusions, we distinguish between a open subset $\widetilde{U}$ of $\widetilde{M}$ and the corresponding open subsupermanifold $U := (\widetilde{U}, \mathcal{O}_{M|\widetilde{U}})$ of M (see also section 3).

As usual, $!_M$ (or simply !) denotes the unique map from some supermanifold M to the terminal object $\mathbb{R}^{0|0}$. If $x \in \widetilde{M}$, $\hat{x}$ denotes the constant map from some supermanifold to M whose image is x or, in other words, $\hat{x}$ is the morphism whose pullback is the evaluation ev_x.

Definition 4.1. Let E, M, and N be supermanifolds and let

$$\pi\colon E \longrightarrow M$$

be a surjective submersion. E is called *super fiber bundle* with *base* M and *typical fiber* N if, for all $x \in \widetilde{M}$, there is an open neighborhood $\widetilde{U}$ of x and an isomorphism

$$t\colon \pi^{-1}(U) \longrightarrow U \times N$$

such that $\mathrm{pr}_U \circ t = \pi$ (here and often in the following the necessary restrictions are understood). The morphism π is called *projection* and a pair (U, t) *local trivialization.*

Notice that, according to remark 3.2,

$$\pi^{-1}(U) = \left(\widetilde{\pi}^{-1}(\widetilde{U}), \mathcal{O}_{E|\widetilde{\pi}^{-1}(\widetilde{U})}\right)$$

Since π is a submersion, it is a constant rank morphism. The following definition is hence well posed.

Definition 4.2. Let $x \in \widetilde{M}$. The subsupermanifold $\pi^{-1}(x)$ of E specified by definition 3.5 is called *fiber* of E at x. We denote the embedding $\pi^{-1}(x) \to E$ by i_x.

The next proposition shows that the above definition is coherent.

Proposition 4.3. *Let $\pi^{-1}(x)$ be as in the foregoing definition. Then*

(i) $\begin{array}{ccc} \pi^{-1}(x) & \xrightarrow{i_x} & E \\ \downarrow{!} & & \\ \mathbb{R}^{0|0} & & \end{array}$ *is the pullback of* $\begin{array}{ccc} & & E \\ & & \downarrow{\pi} \\ \mathbb{R}^{0|0} & \xrightarrow{\hat{x}} & M \end{array}$;

(ii) $\pi^{-1}(x)$ *is (not canonically) isomorphic to typical fiber* N*.*

Example 4.4. $E := M \times N$ with $\pi := \mathrm{pr}_M$ is a super fiber bundle (*trivial super fiber bundle*) with typical fiber N.

Definition 4.5. We say that (E, π, M, V) is a *super vector bundle* if

(i) $V = V_0 \oplus V_1$ is a finite-dimensional super vector space;

(ii) (E, π, M, V) is a fiber bundle with typical fiber V;

(iii) there exists a family of local trivializations $\{ (U_i, t_i) \}$ such that

 (a) $\bigcup_i \widetilde{U}_i = \widetilde{M}$;

 (b) if $\widetilde{U}_{i,j} := \widetilde{U}_i \cap \widetilde{U}_j$, there exists a map (called *transition function*)

$$\varphi_{i,j} \colon U_{i,j} \longrightarrow \mathrm{GL}(V)$$

 such that the isomorphism $t_i \circ t_j^{-1}$ is of the form

$$\begin{array}{ccc} U_{i,j} \times V & \xrightarrow{t_i \circ t_j^{-1}} & U_{i,j} \times V \\ & {\scriptstyle \langle \mathbb{1}_{U_{i,j}}, \varphi_{i,j} \rangle \times \mathbb{1}_V} \searrow \quad \nearrow {\scriptstyle \mathbb{1} \times \alpha} & \\ & U_{i,j} \times \mathrm{GL}(V) \times V & \end{array} \tag{1}$$

 where α the action of $\mathrm{GL}(V)$ on V.

Remark 4.6. The condition expressed by diagram (1) is stronger than the request that changes of trivialization are fiberwise linear. Example 4.15 will further clarify this aspect.

Proposition 4.7. *If (E, π, M, V) is a super vector bundle, then, for all $x \in \widetilde{M}$, $\pi^{-1}(x) \cong V$ (in not canonical way) as super vector spaces.*

Proof. From proposition 4.3, $\pi^{-1}(x) \cong V$ as supermanifolds and so $\pi^{-1}(x)$ inherits a super vector space structure from V. The choice of another trivialization involves an even linear isomorphism that does not alter this structure. Indeed, if (U, t) and (U, t') are local trivializations in a neighborhood of x, $\varphi \colon U \to \mathrm{GL}(V)$ is the transition function between these, and $u, u' \colon \pi^{-1}(x) \to V$ are the isomorphisms induced by t and t', we have

$$u \circ u'^{-1} = \alpha \circ (\varphi \times \mathbb{1}_V) \circ \langle \hat{x}, \mathbb{1}_V \rangle \colon V \longrightarrow V$$

that is an even linear isomorphism since $\widetilde{\varphi}(x) \in \widetilde{\mathrm{GL}}(V) = \mathrm{GL}(V_0) \times \mathrm{GL}(V_1)$. $\square$

Let now M be a supermanifold, V a finite-dimensional super vector space, and $\{ \widetilde{U}_i \}$ an open covering of $\widetilde{M}$.

Definition 4.8. A family $\{\varphi_{i,j}\}$ of morphisms

$$\varphi_{i,j}\colon U_{j|\widetilde{U}_i\cap\widetilde{U}_j} \longrightarrow \mathrm{GL}(V)$$

is said to satisfy the *cocycle condition* if, for each i,j,k, $\varphi_{i,j}\cdot\varphi_{j,k}=\varphi_{i,k}$ in $\mathrm{GL}(V)(U_{i,j,k})$, where $U_{i,j,k}$ denotes the open subsupermanifold on $\widetilde{U}_i\cap\widetilde{U}_j\cap\widetilde{U}_k$.

Proposition 4.9. *Transition functions of a super vector bundle (E,π,M,V) satisfy the cocycle conditions. Vice-versa if M, V, and $\{\widetilde{U}_i\}$ are as above, and a family of functions $\{\varphi_{i,j}\}$ is given satisfying the cocycle conditions, then there exists a unique super vector bundle (E,π,M,V) (up to isomorphism) with typical fiber V and $\{\varphi_{i,j}\}$ as transition functions.*

Proof. First statement is immediate using functor of points. Indeed if $\{\varphi_{i,j}\}$ is the family of transition functions of E, we have[1]

$$\begin{array}{ccccc}
 & & \overset{t_i\circ t_k^{-1}}{\curvearrowright} & & \\
(U_{i,j,k}\times V)(S) & \xrightarrow{t_j\circ t_k^{-1}} & (U_{i,j,k}\times V)(S) & \xrightarrow{t_i\circ t_j^{-1}} & (U_{i,j,k}\times V)(S)\\
(x,v) & \longmapsto & \Big(x,\alpha\big(\varphi_{j,k}(x),v\big)\Big) & \longmapsto & \begin{array}{l}\big(x,\alpha\big(\varphi_{i,j}(x),\alpha(\varphi_{j,k}(x),v)\big)\big)=\\ =\big(x,\alpha\big(\varphi_{i,j}(x)\cdot\varphi_{j,k}(x),v\big)\big)\end{array}
\end{array}$$

for each supermanifold T, $x\colon T\to U_{i,j,k}$, and $v\colon T\to V$; taking $x=\mathbb{1}_{U_{i,j,k}}$ and comparing with eq. (1), we have $\varphi_{i,j}\cdot\varphi_{j,k}=\varphi_{i,k}$.

Vice-versa, define $E_i:=U_i\times V$ and

$$\chi_{i,j}\colon E_{j|\widetilde{U}_{i,j}\times\widetilde{V}} \longrightarrow E_{i|\widetilde{U}_{i,j}\times\widetilde{V}}$$

as $\chi_{i,j}:=(\mathbb{1}_{U_{i,j}}\times\alpha)\circ(\langle\mathbb{1}_{U_{i,j}},\varphi_{i,j}\rangle\times\mathbb{1}_V)$. Now we can glue supermanifolds $\{E_i\}$ into a super vector bundle E. Indeed, by a similar classical result (see [7, prop. 2.2.5]), reduced manifolds $\{\widetilde{E}_i\}$ can be glued into a classical vector bundle $\widetilde{E}$ on $\widetilde{M}$ with $\{\widetilde{\varphi}_{i,j}\}$ as transition functions. Sheaves on them can also be glued by $\{\chi_{i,j}\}$ into a supermanifold sheaf on $\widetilde{E}$ (using, as above, the functor of points language, it is easy to check that the $\{\chi_{i,j}\}$ satisfy cocycle condition stated in the proposition); finally we define $\pi\colon E\to M$ as the map such that $\pi_{|E_i}=\mathrm{pr}_{U_i}$ for all U_i (this makes sense since $\mathrm{pr}_{U_i}\circ\chi_{i,j}=\mathrm{pr}_{U_j}$). For how (E,π,M,V) has been constructed, it is the required super vector bundle. □

4.2. Morphisms

Definition 4.10. Let (E,π_E,M) and (F,π_F,N) be two fiber bundles. A morphism $\psi\colon E\to F$ is a *morphism of super fiber bundles* if there exists $\psi_\flat\colon M\to N$ such that

$$\begin{array}{ccc}
E & \xrightarrow{\psi} & F\\
{\scriptstyle\pi_E}\downarrow & & \downarrow{\scriptstyle\pi_F}\\
M & \xrightarrow{\psi_\flat} & N
\end{array}$$

commutes.

[1] Here we write formally $\varphi(x)$ for $\varphi\circ x$ (and so on).

Journal of Physics: Conference Series **284** (2011) 012010 doi:10.1088/1742-6596/284/1/012010

Since π_E is an epimorphism, if $\psi_\flat$ exists, it is unique.

If $x \in \widetilde{M}$ and $y = \widetilde{\psi_\flat}(x)$, ψ induces a map

$$\psi_x \colon \pi_E^{-1}(x) \longrightarrow \pi_F^{-1}(y)$$

that is defined as the unique morphism that makes the diagram

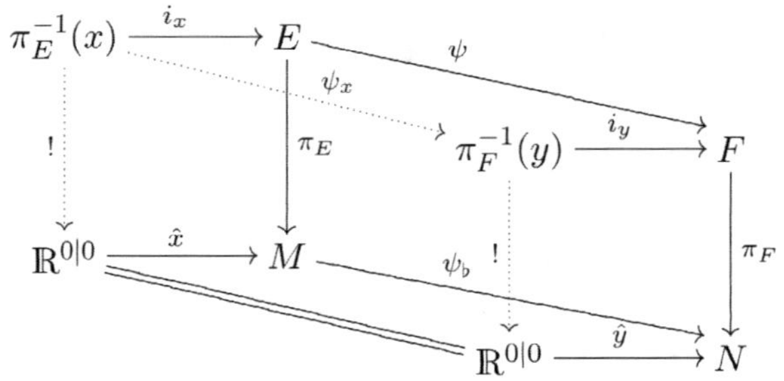

commuting. Such morphism exists and it is unique since $\pi_F^{-1}(y)$ is a pullback and

$$\pi_F \circ \psi \circ i_x = \psi_\flat \circ \pi_E \circ i_x = \psi_\flat \circ \hat{x} \circ ! = \hat{y} \circ !$$

Remark 4.11. In example 2.2 we saw that a super vector space V can be thought as a supermanifold whose functor of points is

$$S \longmapsto \big(\mathcal{O}(S) \otimes V\big)_0$$

If V and W are super vector spaces, and $\mathbf{Hom}(V, W)$ is the super vector space of internal homomorphisms between them, we can define

$$\beta \colon \mathbf{Hom}(V, W) \times V \longrightarrow W$$

through the natural transformation

$$\begin{aligned} \big(\mathcal{O}(S) \otimes \mathbf{Hom}(V, W)\big)_0 \times \big(\mathcal{O}(S) \otimes V\big)_0 &\longrightarrow \big(\mathcal{O}(S) \otimes W\big)_0 \\ (s \otimes K, t \otimes v) &\longmapsto (-1)^{|K||t|} st \otimes K(v) \end{aligned}$$

where S is a generic supermanifold.

Definition 4.12. Let E and F be super vector bundles with typical fiber V and W respectively, and $\beta \colon \mathbf{Hom}(V, W) \times V \to W$ as in remark 4.11. A *morphism of super vector bundles* from E to F is a morphism ψ of super fiber bundles such that for all $x \in \widetilde{M}$ there exists an open neighborhood $\widetilde{U}$ with the following properties:

(i) there are two local trivializations (U, t_E) on E and (U', t_F) on F with $\widetilde{\psi_\flat}(\widetilde{U}) \subseteq \widetilde{U}'$;

(ii) there is a map $\xi \colon U \to \mathbf{Hom}(V, W)$ such that

$$\begin{array}{ccccc} \pi_E^{-1}(U) & & \xrightarrow{\quad \psi_{|\pi_E^{-1}(U)} \quad} & & \pi_F^{-1}(U') \\ \Big\downarrow t_E & & & & \Big\downarrow t_F \\ U \times V & \xrightarrow{\langle \mathbb{1}_U, \xi\rangle \times \mathbb{1}_V} & U \times \mathbf{Hom}(V, W) \times V & \xrightarrow{\psi_\flat \times \beta} & U' \times W \end{array}$$

commutes.

It is easy to see that the above definition does not depend on the chosen trivializations.
From definition 4.12, it follows immediately:

Proposition 4.13. *If $\psi\colon E \to F$ is a morphism of super vector bundles, the map*

$$\psi_x \colon \pi_E^{-1}(x) \longrightarrow \pi_F^{-1}(\widetilde{\psi_\flat}(x))$$

is a morphism of super vector spaces.

Remark 4.14. Also in this case, saying that a morphism of super fiber bundles $\psi\colon E \to F$ is a morphism of super vector bundles it is not sufficient that ψ_x is linear for all x (see the next example).

Example 4.15. Let M be a supermanifold, $E := M \times \mathbb{R}^{1|1}$, and $\psi\colon E \to E$ a morphism of super vector bundles with $\psi_\flat = \mathbb{1}_M$. Then there is $\xi\colon M \to \mathbf{Hom}(\mathbb{R}^{1|1}, \mathbb{R}^{1|1})$ such that

$$\psi = (\mathbb{1}_M \times \beta) \circ (\langle \mathbb{1}_M, \xi \rangle \times \mathbb{1}_{\mathbb{R}^{1|1}})$$

As usual ξ can be identified with a matrix $\left(\begin{smallmatrix} f_1 & g_1 \\ g_2 & f_2 \end{smallmatrix}\right)$ with $f_i \in \mathcal{O}(M)_0$ and $g_j \in \mathcal{O}(M)_1$, which is the matrix of pullbacks of canonical coordinates of $\mathbf{Hom}(\mathbb{R}^{1|1}, \mathbb{R}^{1|1})$ under ξ^*. So, if $\{ x_i, \vartheta_j \}$ are coordinates on M and $\{ y, \eta \}$ on $\mathbb{R}^{1|1}$, ψ is defined by

$$\begin{aligned} \psi^* \colon \mathcal{O}(M \times \mathbb{R}^{1|1}) &\longrightarrow \mathcal{O}(M \times \mathbb{R}^{1|1}) \\ x_i, \vartheta_j &\longmapsto x_i, \vartheta_j \\ y &\longmapsto f_1 y + g_1 \eta \\ \eta &\longmapsto g_2 y + f_2 \eta \end{aligned}$$

If $x \in \widetilde{M}$, we have

$$\begin{aligned} \psi_x^* \colon \mathcal{O}(\mathbb{R}^{1|1}) &\longrightarrow \mathcal{O}(\mathbb{R}^{1|1}) \\ y &\longmapsto \widetilde{f_1}(x) y \\ \eta &\longmapsto \widetilde{f_2}(x) \eta \end{aligned}$$

Another morphism of super fiber bundles $\psi'\colon E \to E$ with, for example, $\psi'^*(\eta) = g_2 y^2 + f_2 \eta$ gives the same map on fibers, but it is not a morphism of super vector bundles.

Proposition 4.16. *Let (F, π, N, W) be a super fiber bundle, M a supermanifold and $\zeta\colon M \to N$ a morphism. Then the pullback*

$$\begin{array}{ccc} \zeta^*(F) & \xrightarrow{\ \bar{\zeta}\ } & F \\ {\scriptstyle \bar{\pi}} \big\downarrow & & \big\downarrow {\scriptstyle \pi} \\ M & \xrightarrow{\ \zeta\ } & N \end{array}$$

exists and $\zeta^*(F)$ is a super fiber bundle on M with projection $\bar{\pi}$ and typical fiber W. If F is a super vector bundle with transition functions $\{ \varphi_{i,j} \}$, then $\zeta^*(F)$ is a super vector bundle with transition functions $\{ \varphi_{i,j} \circ \zeta \}$ and $(\bar{\zeta}, \zeta)$ is a morphism of super vector bundles.

Proof. Let (U, t) be a local trivialization of F and $\widetilde{U}' = \widetilde{\zeta}^{-1}(\widetilde{U})$. The diagram below is a pullback diagram:

$$\begin{array}{ccc} U' \times W & \xrightarrow{\ \zeta \times \mathbb{1}\ } & U \times W \\ {\scriptstyle \mathrm{pr}_{U'}} \big\downarrow & & \big\downarrow {\scriptstyle \mathrm{pr}_U} \\ U' & \xrightarrow{\ \zeta\ } & U \end{array}$$

This fact can be checked directly in the category of sets and then be extended to each category with products by means of Yoneda's lemma. From the uniqueness of pullbacks, it follows that $\zeta^*(F)$ is locally isomorphic to $U' \times W$; each local trivialization $t\colon \pi^{-1}(U) \to U \times W$ induces a local trivialization $t'\colon \bar{\pi}^{-1}(U') \to U' \times W$.

If F is a super vector bundle, t_1 and t_2 are two local trivializations of U, and $\varphi_{1,2}$ is the relative transition function, we have

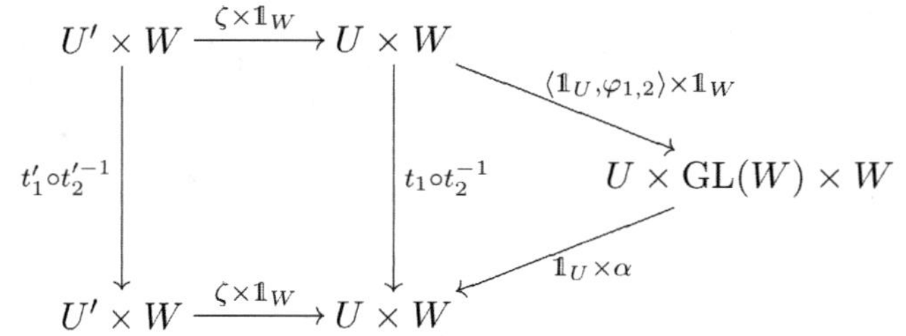

where t_1' and t_2' are the local trivializations of $\zeta^*(F)$ induced by t_1 and t_2; using functor of points one can see that $\varphi_{1,2} \circ \zeta$ is the relative transition function. The fact that $\bar{\zeta}$ is locally of the form $\zeta \times \mathbb{1}$ proves the last statement. □

Corollary 4.17. *Every morphism* $(\psi, \psi_\flat)\colon (E, \pi_E, M) \to (F, \pi_F, N)$ *splits into the two super fiber bundle morphisms*

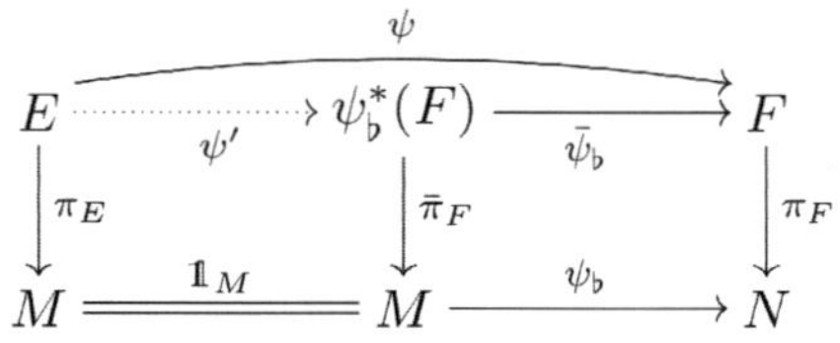

one on the identity and the other fixed by $\psi_\flat$.

4.3. Sections

Let (E, π, M, V) be a super vector bundle with $\dim V = m|n$. It is natural to define sections, in analogy with the classical case, as morphisms

$$\sigma\colon M \longrightarrow E$$

such that $\pi \circ \sigma = \mathbb{1}_M$. We denote the set of such maps by $\Gamma_E(M)$ and we call them *geometrical sections* on M. Clearly it is possible to define local geometrical sections

$$\Gamma_E(U) := \left\{ \sigma\colon U \longrightarrow \pi^{-1}(U) \;\middle|\; \pi \circ \sigma = \mathbb{1}_U \right\}$$

for each open subsupermanifold U of M.

It is easy to check that Γ_E is a locally free sheaf of $\mathcal{O}_{M,0}$-modules on $\widetilde{M}$. The problem is that $\Gamma_E(M)$ is an ungraded object, and it cannot encode the information contained in the super vector bundle E. We want to construct a super vector space $\mathbf{\Gamma}_E(M)$ with $\mathbf{\Gamma}_E(M)_0 \cong \Gamma_E(M)$ that we will call the *full sections* of E.

As usual, let $\widetilde{\pi}_*\mathcal{O}_E$ be the push-forward sheaf given by

$$\widetilde{\pi}_*\mathcal{O}_E\colon W \longmapsto \mathcal{O}_E\big(\widetilde{\pi}^{-1}(W)\big)$$

for each open subset W of $\widetilde{M}$. We want to define a subsheaf of $\widetilde{\pi}_*\mathcal{O}_E$ that could be called the sheaf of sections that are linear on the fibers. If $E := M \times V$ is a trivial super vector bundle, the natural candidate is $\mathcal{O}(M) \otimes V^*$. In general, we can give the following definition.

Definition 4.18. If W is an open subset of $\widetilde{M}$, let $\mathcal{O}_E^{\text{lin}}(W)$ be the set of $s \in \mathcal{O}_E\big(\widetilde{\pi}^{-1}(W)\big)$ such that for all $x \in W$ there exists a local trivialization (U, t) with $x \in \widetilde{U} \subseteq W$ and

$$(t^{-1})^*\big(s_{|\widetilde{\pi}^{-1}(\widetilde{U})}\big) \in \mathcal{O}(U) \otimes V^* \tag{2}$$

It is easy to check that $\mathcal{O}_E^{\text{lin}}$ is a subsheaf of $\widetilde{\pi}_*\mathcal{O}_E$ (notice that $\mathcal{O}_E^{\text{lin}}$ is a sheaf on $\widetilde{M}$), and that the above definition makes sense.

The sheaf $\widetilde{\pi}_*\mathcal{O}_E$ is a sheaf of $\mathcal{O}_M$-modules on $\widetilde{M}$. For each $W \subseteq \widetilde{M}$, $f \in \mathcal{O}_M(W)$, and $s \in \mathcal{O}_E\big(\widetilde{\pi}^{-1}(W)\big)$, we define

$$f \cdot s := \pi^*(f)s$$

$\mathcal{O}_E^{\text{lin}}$ is a locally free sheaf of $\mathcal{O}_M$-modules of dimension $\dim V = m|n$.

So we can define the sheaf of full sections of E by duality.

Definition 4.19. For each open subset W of $\widetilde{M}$ let

$$\mathbf{\Gamma}_E(W) := \mathbf{Hom}_{\mathcal{O}_M(W)}\big(\mathcal{O}_E^{\text{lin}}(W), \mathcal{O}_M(W)\big)$$

i. e. $\mathbf{\Gamma}_E(W)$ is the set of all $\mathcal{O}_M(W)$-linear maps $\mathcal{O}_E^{\text{lin}}(W) \to \mathcal{O}_M(W)$. We call $\mathbf{\Gamma}_E(W)$ the space of *full sections* of E on W.

Since we are in the smooth category and sheaf morphisms are determined by maps of global sections, it is not hard to define restriction maps for $\mathbf{\Gamma}_E$ and to check that it defines a sheaf on $\widetilde{M}$.

For each W, $\mathbf{\Gamma}_E(W)$ is a graded $\mathcal{O}_M(W)$-module and, if (U, t) is a local trivialization,

$$\mathbf{\Gamma}_E(U) \cong \mathbf{Hom}_{\mathcal{O}(U)}\big(\mathcal{O}(U) \otimes V^*, \mathcal{O}(U)\big) \cong \mathcal{O}(U) \otimes V$$

Then also $\mathbf{\Gamma}_E$ is a locally free sheaf of $\mathcal{O}_M$-modules of dimension $m|n$.

In the classical context, $\mathcal{O}_E^{\text{lin}}$ is the sheaf of sections of the dual vector bundle. Using proposition 4.9, one could prove a similar result in the supergeometrical setting.

Proposition 4.20. *$\mathcal{O}_E^{\text{lin}}$ is the sheaf of full sections of a super vector bundle with typical fiber V^* and transition functions $(\varphi_{i,j}^*)^{-1}$ that we can call* dual vector bundle *of E.*

Moreover it is possible to prove the following (non surprising) result.

Proposition 4.21. $\Gamma_M(M) \cong \mathbf{\Gamma}_E(M)_0$.

4.4. Algebraic approach

If $\mathcal{V}$ and $\mathcal{W}$ are two (algebraic) super vector bundles on M, a *morphism of super vector bundles*

$$\rho\colon \mathcal{V} \longrightarrow \mathcal{W}$$

is a $\mathcal{O}_M$-module morphism.

We end this section examining the relation between the geometrical definition of super vector bundle and the algebraic one (for the classical case see [8, 7]).

Proposition 4.22. *The category of geometrical super vector bundles on a supermanifold M whose morphisms are of the form $(\psi, \mathbb{1}_M)$ (see definition 4.12) is equivalent to the above described category of algebraic super vector bundles on M.*

Sketch of Proof. If (E, π, M, V) is a geometrical super vector bundle, $\mathbf{\Gamma}_E$ is a locally free sheaf of $\mathcal{O}_M$-modules of dimension $\dim V$. $E \mapsto \mathbf{\Gamma}_E$ gives the equivalence functor on the objects.

Vice-versa, let $\mathcal{V}$ an algebraic super vector bundle on M of dimension $m|n$. Since $\mathcal{V}$ is locally free, for all $x \in \widetilde{M}$ there exists an open neighborhood $\widetilde{U}$ of x such that $\mathcal{V}(\widetilde{U}) \cong \mathcal{O}_M(\widetilde{U})^{m|n} = \mathcal{O}_M(\widetilde{U}) \otimes \mathbb{R}^{m|n}$. Let $\{ \widetilde{U}_i \}$ be an open covering of $\widetilde{M}$ of such neighborhoods and let $t_i \colon \mathcal{V}(\widetilde{U}_i) \to \mathcal{O}_M(\widetilde{U}_i)^{m|n}$ be the corresponding isomorphisms. If $\widetilde{U}_{i,j} := \widetilde{U}_i \cap \widetilde{U}_j$, we have an automorphism $\varphi_{i,j} = t_i \circ t_j^{-1}$ of $\mathcal{O}_M(\widetilde{U}_{i,j})^{m|n}$ that is an $(m+n) \times (m+n)$ invertible matrix $\begin{pmatrix} f_{k,l} & g_{r,s} \\ g_{r',s'} & f_{k',l'} \end{pmatrix}$ with $f_{k,l}, f_{k',l'} \in \mathcal{O}_M(\widetilde{U}_{i,j})_0$ and $g_{r,s}, g_{r',s'} \in \mathcal{O}_M(\widetilde{U}_{i,j})_1$. As usual, this matrix identifies a morphism $U_{i,j} \to \mathrm{GL}(m|n)$, which we denote equally by $\varphi_{i,j}$. Restricting to $\widetilde{U}_{i,j,k} := \widetilde{U}_i \cap \widetilde{U}_j \cap \widetilde{U}_k$, the composition of such module automorphisms is the usual matrix product. So one can see that the morphisms $\varphi_{i,j} \colon U_{i,j} \to \mathrm{GL}(m|n)$ satisfy cocycle condition (definition 4.8) and, according to proposition 4.9, a geometrical super vector bundle can be constructed with typical fiber $\mathbb{R}^{m|n}$ and $\{ \varphi_{i,j} \}$ as transition functions.

Now we have to construct the equivalence functor on morphisms. Let $(\psi, \mathbb{1}_M) \colon (E, \pi_E, M, V) \to (F, \pi_F, M, W)$ be a morphism of geometrical super vector bundles, and let $\widetilde{U} \subseteq \widetilde{M}$ and $\xi \colon U \to \mathbf{Hom}(V, W)$ be as in definition 4.12. We know that ξ is identified with an element

$$\sum_i s_i \otimes K_i \in \left(\mathcal{O}_M(\widetilde{U}) \otimes \mathbf{Hom}(V, W)\right)_0$$

(compare example 2.2 and remark 4.11) and then it determines an even map of $\mathcal{O}_M(\widetilde{U})$-modules

$$\begin{aligned} \mathcal{O}_M(\widetilde{U}) \otimes V &\longrightarrow \mathcal{O}_M(\widetilde{U}) \otimes W \\ t \otimes v &\longmapsto \sum_i (-1)^{|K_i||t|} s_i t \otimes K_i(v) \end{aligned}$$

(notice that, since $|s_i| = |K_i|$, $(s_i \otimes K_i)(t \otimes v) = (t \otimes 1)(s_i \otimes K_i)(1 \otimes v)$).

Vice-versa an even map

$$\mathcal{O}_M(\widetilde{U}) \otimes V \longrightarrow \mathcal{O}_M(\widetilde{U}) \otimes W$$

of $\mathcal{O}_M(\widetilde{U})$-modules is an even matrix with entries in $\mathcal{O}_M(\widetilde{U})$ and it identifies a morphism $U \to \mathbf{Hom}(V, W)$. Therefore, proceeding locally, we can construct a geometrical morphism from an algebraic one. □

References

[1] Deligne P and Morgan J W 1999 *Quantum fields and strings: a course for mathematicians, Vol. 1, 2 (Princeton, NJ, 1996/1997)* (Providence, RI: Amer. Math. Soc.) pp 41–97
[2] Kostant B 1977 *Differential geometrical methods in mathematical physics (Proc. Sympos., Univ. Bonn, Bonn, 1975)* (Berlin: Springer) pp 177–306. Lecture Notes in Math., Vol. 570
[3] Bartocci C, Bruzzo U and Hernández Ruipérez D 1991 *The geometry of supermanifolds* (*Mathematics and its Applications* vol 71) (Dordrecht: Kluwer Academic Publishers Group) ISBN 0-7923-1440-9
[4] Leĭtes D A 1980 *Uspekhi Mat. Nauk* **35** 3–57, 255 ISSN 0042-1316
[5] Manin Y I 1988 *Gauge field theory and complex geometry* (*Grundlehren der Mathematischen Wissenschaften [Fundamental Principles of Mathematical Sciences]* vol 289) (Berlin: Springer-Verlag) ISBN 3-540-18275-6 translated from the Russian by N. Koblitz and J. R. King
[6] Varadarajan V S 2004 *Supersymmetry for mathematicians: an introduction* (*Courant Lecture Notes in Mathematics* vol 11) (New York: New York University Courant Institute of Mathematical Sciences) ISBN 0-8218-3574-2
[7] Osborn H 1982 *Vector bundles. Vol. 1* (*Pure and Applied Mathematics* vol 101) (New York: Academic Press Inc. [Harcourt Brace Jovanovich Publishers]) ISBN 0-12-529301-1 foundations and Stiefel–Whitney classes
[8] Swan R G 1962 *Trans. Amer. Math. Soc.* **105** 264–277 ISSN 0002-9947

Journal of Physics: Conference Series **284** (2011) 012011 doi:10.1088/1742-6596/284/1/012011

New superintegrable models with position-dependent mass from Bertrand's Theorem on curved spaces

A Ballesteros[1], A Enciso[2], F J Herranz[1], O Ragnisco[3] and D Riglioni[3]

[1] Departamento de Física, Universidad de Burgos, E-09001 Burgos, Spain
[2] Departamento de Física Teórica II, Universidad Complutense, E-28040 Madrid
[3] Dipartimento di Fisica, Università di Roma Tre and Instituto Nazionale di Fisica Nucleare sezione di Roma Tre, Via Vasca Navale 84, I-00146 Roma, Italy

E-mail: angelb@ubu.es, aenciso@fis.ucm.es, fjherranz@ubu.es, ragnisco@fis.uniroma3.it, riglioni@fis.uniroma3.it

Abstract. A generalized version of Bertrand's theorem on spherically symmetric curved spaces is presented. This result is based on the classification of $(3+1)$-dimensional (Lorentzian) Bertrand spacetimes, that gives rise to two families of Hamiltonian systems defined on certain 3-dimensional (Riemannian) spaces. These two systems are shown to be either the Kepler or the oscillator potentials on the corresponding Bertrand spaces, and both of them are maximally superintegrable. Afterwards, the relationship between such Bertrand Hamiltonians and position-dependent mass systems is explicitly established. These results are illustrated through the example of a superintegrable (nonlinear) oscillator on a Bertrand-Darboux space, whose quantization and physical features are also briefly addressed.

1. Introduction

Bertrand's theorem, which dates back to the XIX century [1], is a landmark result in classical mechanics characterizing the Kepler and harmonic oscillator potentials in terms of their qualitative dynamics. More precisely [2], Bertrand's theorem asserts that any spherically symmetric natural Hamiltonian system $H = \frac{1}{2}|\mathbf{p}|^2 + V(|\mathbf{q}|)$ in (a subset of) $\mathbb{R}^3$ that has a stable circular trajectory passing through each point in its configuration space and all whose bounded trajectories are closed is either a harmonic oscillator ($V(r) = A\,r^2 + B$) or a Kepler system ($V(r) = A/r + B$).

Surprisingly, the classical theorem of Bertrand found a natural extension in and application to general relativity some fifteen years ago thanks to a remarkable paper of Perlick [3]. Indeed, the author undertook the classification of all *Bertrand spacetimes*, which, roughly speaking, are spherically symmetric and static spacetimes whose timelike geodesics satisfy properties analogous to those of the trajectories of the harmonic oscillator or Kepler systems. The connection between timelike geodesics in spacetime and the trajectories of a classical Hamiltonian system is that, if one writes the Lorentzian metric as

$$\eta = g_{ij}(q)\,\mathrm{d}q^i\,\mathrm{d}q^j - V^{-1}(q)\,\mathrm{d}t^2,$$

where g is a Riemannian metric on a 3-manifold, the timelike geodesics in spacetime are naturally related to the trajectories of the Hamiltonian system

$$H = g^{ij}(q)p_ip_j + V(q).$$

Journal of Physics: Conference Series **284** (2011) 012011 doi:10.1088/1742-6596/284/1/012011

Perlick's classification of Bertrand spacetimes consisted of two multi-parametric families. When one additionally imposes that the spacial part of the metric be Euclidean, Perlick's result becomes tantamount to the classical Bertrand's theorem. Moreover, the general case includes a number of other systems that have received considerable attention, particularly in connection with integrable monopole motion and the existence of generalized Runge–Lenz vectors. For instance, Perlick's classification includes spacetimes constructed over the 3-sphere, the hyperbolic 3-space and the Iwai–Katayama spaces [4, 5], which generalize the Taub–NUT spacetime.

In this paper we review several recent results related to this problem and we also establish a natural connection between curved Bertrand systems and position-dependent mass (PDM) Hamiltonians [6]–[20]. We begin by showing, in the next section, that the rather complicated families of Bertrand spacetimes admit a strikingly simple physical interpretation [21]: they correspond to either an intrinsic oscillator or an intrinsic Kepler system. Next we address in section 3 the superintegrability of the associated Hamiltonian system and recall the construction of additional integrals of motion of Runge–Lenz type, which settled in a satisfactory way a problem with a large body of related literature (cf. [22, 23] and references therein). This allows us to state an optimal version of Bertrand's theorem on Riemannian manifolds. Furthermore, we present in section 4 the application of the above results to PDM Hamiltonians by rewriting the previous Bertrand Hamiltonians in terms of a variable mass function. All of these results are explicitly illustrated by discussing in section 5 the N-dimensional (ND) version of one of the most interesting Bertrand Hamiltonians: the Darboux III Hamiltonian [24]. The associated quantum mechanical problem together with some possible physical applications of these new PDM integrable models are also considered in the last section.

2. Bertrand spacetimes

To begin with, let us recall Perlick's definition of a Bertrand spacetime [3]. We will consider a spherically symmetric, static spacetime $(\mathcal{M}\times\mathbb{R},\eta)$, where $\mathcal{M}$ is a 3-manifold. This ensures that the Lorentzian metric η can be written as

$$\eta = h(r)^2\,\mathrm{d}r^2 + r^2\big(\mathrm{d}\theta^2+\sin^2\theta\,\mathrm{d}\varphi^2\big) - \frac{\mathrm{d}t^2}{V(r)}\,,$$

where V is a smooth scalar function and

$$g = h(r)^2\,\mathrm{d}r^2 + r^2(\mathrm{d}\theta^2+\sin^2\theta\,\mathrm{d}\varphi^2) \tag{1}$$

defines a Riemannian metric on $\mathcal{M}$. Following Perlick, by a *trajectory* in spacetime we mean the projection of an inextendible timelike geodesic to a (fixed but otherwise arbitrary) constant time leaf $\mathcal{M}\{t_0\}$. This terminology is motivated by the fact that a trajectory in spacetime actually corresponds to a trajectory (in configuration space) of the Hamiltonian

$$H = \frac{1}{2}g^{ij}p_ip_j + V = \frac{1}{2}\left(\frac{p_r^2}{h(r)^2}+\frac{p_\theta^2}{r^2}+\frac{p_\varphi^2}{r^2\sin^2\theta}\right)+V(r) \tag{2}$$

in $\mathcal{M}$. As customary, g^{ij} is the inverse matrix to g_{ij}, $p\in T^*\mathcal{M}$ is the momentum and (p_r,p_θ,p_φ) are the conjugate momenta of the coordinates (r,θ,φ).

Definition 1. The Lorentzian 4-manifold $(\mathcal{M}\times\mathbb{R},\eta)$ is a *Bertrand spacetime* if:

(i) There is a circular ($r=$ const.) trajectory passing through each point of $\mathcal{M}$.

(ii) The above circular trajectories are stable, that is, any initial condition sufficiently close to that of a circular trajectory gives a periodic trajectory.

Perlick's classification of all Bertrand spacetimes [3] can then be stated as follows:

Theorem 2 (Perlick). *The metric of a Bertrand spacetime can be expressed in exactly one of the following forms:*

(i) Type I:

$$g = \frac{m^2 \mathrm{d}r^2}{n^2\,(1+Kr^2)} + r^2(\mathrm{d}\theta^2 + \sin^2\theta\,\mathrm{d}\varphi^2)\,, \qquad V = \sqrt{r^{-2}+K} + G\,.$$

(ii) Type II:

$$g = \frac{2m^2\left(1 - Dr^2 \pm \sqrt{(1-Dr^2)^2 - Kr^4}\right)}{n^2\left((1-Dr^2)^2 - Kr^4\right)}\,\mathrm{d}r^2 + r^2(\mathrm{d}\theta^2 + \sin^2\theta\,\mathrm{d}\varphi^2)\,,$$
$$V = G \mp r^2\left(1 - Dr^2 \pm \sqrt{(1-Dr^2)^2 - Kr^4}\right)^{-1}.$$

Here m, n are coprime nonnegative integers and D, G, K are real constants.

Notice that in [3, 21] this result was expressed through the rational number $\beta = n/m$.

To provide an interpretation of the above (rather involved) formulas for the Bertrand spacetimes, it is convenient to define the concept of harmonic oscillator and Kepler potential in any spherically symmetric 3-manifold. To this end, let us use the notation (1) for a spherically symmetric metric on $\mathcal{M}$ and consider its associated Laplacian, which we denote by Δ_g. It is standard that if $u(r)$ is function on $\mathcal{M}$ that only depends on the radial coordinate, then its Laplacian is also radial and given by

$$\Delta_g u(r) = \frac{1}{r^2 h(r)}\frac{\mathrm{d}}{\mathrm{d}r}\left(\frac{r^2}{h(r)}\frac{\mathrm{d}u}{\mathrm{d}r}\right).$$

Then the symmetric Green function $u(r)$ is obtained as the solution of the equation $\Delta_g u(r) = 0$ on $\mathcal{M}\backslash\{\mathbf{0}\}$, namely

$$u(r) = \int^r \frac{h(r')}{r'^2}\,\mathrm{d}r'. \tag{3}$$

As the Kepler potential in 3D Euclidean space is simply the radial Green function $u(r)$ of the Laplacian and the harmonic oscillator is its inverse square, it is natural to make the following

Definition 3. The *Kepler* and the *harmonic oscillator potentials* in $(\mathcal{M}, g)$ are respectively given by the radial functions

$$V_{\mathrm{K}}(r) = A_1\left(\int_a^r r'^{-2}h(r')\,\mathrm{d}r' + B_1\right), \qquad V_{\mathrm{O}}(r) = A_2\left(\int_a^r r'^{-2}h(r')\,\mathrm{d}r' + B_2\right)^{-2},$$

where A_j and B_j $(j = 1, 2)$ are constants.

This definition is obviously valid in higher dimensions as well.

Example 4. Let $(\mathcal{M}, g)$ be the simply connected 3D space of constant sectional curvature κ. In this case the metric has the form (1) with $h(r)^2 = 1/(1-\kappa r^2)$, so that the corresponding Kepler and harmonic oscillator potentials are

$$V_{\mathrm{K}} = \sqrt{r^{-2} - \kappa}\,, \qquad V_{\mathrm{O}} = (r^{-2} - \kappa)^{-1},$$

up to additive and multiplicative constants. In terms of the radial geodesic distance ρ_κ to the origin point $r = 0$, that is, $r = \sin(\sqrt{\kappa}\,\rho_\kappa)/\sqrt{\kappa}$ [21], these can be rewritten as

$$V_{\mathrm{K}} = \sqrt{\kappa}\cot\left(\sqrt{\kappa}\,\rho_\kappa\right), \qquad V_{\mathrm{O}} = \tan^2(\sqrt{\kappa}\,\rho_\kappa)/\kappa,$$

thus reproducing the known prescriptions for the sphere and the hyperbolic space [25, 26].

This readily gives the following interpretation of the Bertrand spacetimes [21]:

Theorem 5. *In a type I (resp. type II) Bertrand spacetime, V is the intrinsic Kepler (resp. harmonic oscillator) potential associated with g.*

3. Bertrand's theorem in 3D curved spaces

Let us now consider the maximal superintegrability (MS) of the Bertrand Hamiltonians (2). It is well known that, the Bertrand Hamiltonians being spherically symmetric, to establish their MS it suffices to obtain a functionally independent additional integral. As it turns out, the most convenient way to obtain this additional integral is as a 'generalized Runge–Lenz tensor'.

Indeed, Bertrand Hamiltonians are somehow similar to the usual harmonic oscillator and Kepler systems in Euclidean space. It is classical that the MS of the Kepler system can be readily proved using that the Runge–Lenz vector is conserved [27]. In the case of the harmonic oscillator, there is no natural way of defining a conserved Runge–Lenz *vector*, but the elements of the symmetric *Fradkin matrix* $C_{ij} = 2\omega^2 q_i q_j + p_i p_j$ are constants of the motion and encode the main algebraic properties of the model [28]. A straightforward computation shows that the standard Kepler system is obtained from the type I Bertrand Hamiltonian by setting $K = 0$ and $n = m = 1$, while the harmonic oscillator is the type II Hamiltonian with parameters $K = D = 0$, $n = 2$ and $m = 1$. The statement of the following key theorem, which proves the MS of Bertrand Hamiltonians [22], is therefore not surprising:

Theorem 6. *There exists a (nontrivial) rank-n symmetric tensor field invariant under the flow of the Bertrand Hamiltonian, where n is the parameter introduced in theorem 2.*

The proof of this theorem has been fully discussed in [22]. Here we want to stress that the parameter n plays a crucial role in the construction of the first integrals. The basic observation, which goes back to Fradkin [29], is that any spherically symmetric Hamiltonian $H_0 = \frac{1}{2}|\mathbf{p}|^2 + U(|\mathbf{q}|)$ preserves the unit vector field

$$\mathbf{a} = \frac{\cos\varphi}{r}\mathbf{q} + \frac{\sin\varphi}{rJ}\mathbf{q}\times(\mathbf{q}\times\mathbf{p}), \qquad J = p_\varphi = r^2\dot{\varphi}.$$

Of course, this only shows the existence of a *local* additional integral, which is trivial in view of the flow-box theorem, but the point is that this provides a bona fide global first integral provided one can express $\cos\varphi$ and $\sin\varphi/J$ in terms of $\mathbf{p}$ and $\mathbf{q}$. In the case of the Kepler problem, this readily yields a conserved vector field which is essentially the Runge–Lenz vector divided by its norm. In the case of the harmonic oscillator, the above quantities are not well defined functions of $\mathbf{p}$ and $\mathbf{q}$, but they do define an analogous bi-valued conserved vector field. By taking the two-fold symmetrized tensor product of this vector field one can remove this indeterminacy, and this procedure yields an invariant 2-tensor essentially analogous to the conserved matrix C_{ij} mentioned above. The proof of theorem 6 follows this line of thought a bit further. Ultimately, the role n plays in the proof merely reflects the properties of the trajectories of the Hamiltonian (studied in [3, 21]), which depend crucially on the numbers n and m. It should be noticed that the dependence of the constants of motion on the momenta could be extremely complicated, which is the reason why only a few among the Bertrand Hamiltonians had previously been identified as MS systems.

To summarize, let us state the complete version of the optimal extension of Bertrand's theorem to Riemannian manifolds [22]:

Theorem 7. *Let $H = \frac{1}{2}g^{ij}p_i p_j + V$ be a Bertrand Hamiltonian, i.e., a spherically symmetric, natural Hamiltonian system on a Riemannian 3-manifold $(\mathcal{M}, g)$ that has a stable circular trajectory passing through each point in its configuration space and whose bounded trajectories are all closed. Then the following statements hold:*

(i) The metric g and the potential V are of the form given in theorem 2 for some coprime positive integers n, m.

(ii) The potential is the intrinsic Kepler or the harmonic oscillator potential of $(\mathcal{M}, g)$.

(iii) H is superintegrable. More precisely, there exists a nontrivial rank n tensor field which is invariant and plays the role of the Runge–Lenz vector.

4. Curved Bertrand systems as classical PDM Hamiltonians

So far, by starting from Perlick's classification of (3+1)D Bertrand spacetimes, we have obtained two families of MS Hamiltonians on 3D Riemannian manifolds, which are either of Kepler or oscillator type. These results can also be translated into the language of PDM systems [6]–[20]. For this purpose, we remark that Bertrand Hamiltonians are initially expressed in the form (2), that is, $H = \frac{1}{2}g^{ij}p_i p_j + V(r)$, with the potential V being determined by theorem 2. Hence it is necessary to rewrite the Hamiltonian in terms of a variable mass $M(|\mathbf{q}|)$ in the form

$$H = \frac{\mathbf{p}^2}{2M(|\mathbf{q}|)} + V(|\mathbf{q}|). \tag{4}$$

At this point we stress that the 'radial' Bertrand coordinate r is by no means $|\mathbf{q}|$. Therefore, the translation has to be achieved by defining the appropriate change of coordinates $r \leftrightarrow |\mathbf{q}|$ in the underlying metric (1) of the Bertrand Hamiltonians (as in example 4), thus giving rise to a conformally flat metric:

$$g = h(r)^2\,\mathrm{d}r^2 + r^2(\mathrm{d}\theta^2 + \sin^2\theta\,\mathrm{d}\varphi^2) = f(|\mathbf{q}|)^2\,\mathrm{d}\mathbf{q}^2. \tag{5}$$

This yields the relations

$$r = |\mathbf{q}| f(|\mathbf{q}|), \qquad f(|\mathbf{q}|)\mathrm{d}|\mathbf{q}| = h(r)\mathrm{d}r, \qquad \frac{1}{|\mathbf{q}|}\,\mathrm{d}|\mathbf{q}| = \frac{h(r)}{r}\,\mathrm{d}r. \tag{6}$$

Consequently, the variable mass can thus be related to the conformal factor by setting

$$M(|\mathbf{q}|) = m_0 f(|\mathbf{q}|)^2 = \frac{m_0 r^2}{\mathbf{q}^2}, \tag{7}$$

where m_0 is a positive real constant that hereafter we shall fix to 1. In the following we present this relationship for each of the two types of Bertrand Hamiltonians.

4.1. Type I: Bertrand-Kepler Hamiltonians

In this case, by taking into account that

$$h(r) = \frac{m}{n\sqrt{1+Kr^2}}$$

and by applying relations (6) we obtain that

$$|\mathbf{q}| = \left(\frac{r}{1+\sqrt{1+Kr^2}}\right)^{m/n}, \qquad r = \frac{2}{|\mathbf{q}|^{-(n/m)} - K|\mathbf{q}|^{(n/m)}}.$$

Therefore the PDM function is given by

$$M(|\mathbf{q}|) = \frac{4}{\left(|\mathbf{q}|^{-(n/m)} - K|\mathbf{q}|^{(n/m)}\right)^2 \mathbf{q}^2}, \tag{8}$$

and the resulting PDM Hamiltonian provided by theorem 2 turns out to be

$$H = \frac{1}{8}\left(|\mathbf{q}|^{-(n/m)} - K|\mathbf{q}|^{(n/m)}\right)^2 \mathbf{q}^2\,\mathbf{p}^2 + A_1\left\{K + \frac{1}{4}\left(|\mathbf{q}|^{-(n/m)} - K|\mathbf{q}|^{(n/m)}\right)^2\right\}^{1/2}. \tag{9}$$

Example 8. Riemannian spaces of constant sectional curvature κ arise in the Bertrand Hamiltonians of type I when $n = m = 1$ and $K = -\kappa$ [21]. Hence the variable mass (8) and the Hamiltonian (9) reduce to

$$M(|\mathbf{q}|) = \frac{4}{(1+\kappa \mathbf{q}^2)^2}, \qquad H_\kappa = \frac{1}{8}\left(1+\kappa \mathbf{q}^2\right)^2 \mathbf{p}^2 + A_1 \frac{1-\kappa \mathbf{q}^2}{2|\mathbf{q}|},$$

which is the known Kepler system written in Poincaré coordinates [25, 30] on the spherical ($\kappa > 0$), hyperbolic ($\kappa < 0$) and Euclidean ($\kappa = 0$) spaces. Clearly, we can scale the Hamiltonian as $4H_\kappa$. The corresponding Runge–Lenz vector in these coordinates can be found in [25, 30].

4.2. Type II: Bertrand-oscillator Hamiltonians
Now the function $h(r)$ is given by

$$h(r) = \frac{m\sqrt{2}}{n}\left(\frac{1 - Dr^2 \pm \sqrt{(1-Dr^2)^2 - Kr^4}}{(1-Dr^2)^2 - Kr^4}\right)^{1/2}, \tag{10}$$

and the relation between the variables $|\mathbf{q}|$ and r defined through (6) can be obtained as

$$|\mathbf{q}| = \exp\left\{ r\,u(r) - \int^r u(r')\mathrm{d}r' \right\}, \tag{11}$$

(up to an additive constant coming from the integral) where $u(r)$ is the Green function (3). For these type II systems $u(r)$ reads as

$$u(r) = \mp \frac{m\sqrt{2}}{nr}\left(1 - Dr^2 \pm \sqrt{(1-Dr^2)^2 - Kr^4}\right)^{1/2}.$$

Therefore, the explicit general result for $M(|\mathbf{q}|) = r^2/\mathbf{q}^2$ is quite cumbersome and we omit it here. However, for some particular cases it adopts a simple form, as it is shown in the sequel.

Example 9. The three Riemannian spaces of constant curvature κ now appear by setting $n = 2$, $m = 1$, $K = 0$, $D = \kappa$ and by taking the positive sign within the two posibilities '$\pm$' in (10) [21]. In this way we find that

$$h(r) = \frac{1}{\sqrt{1-\kappa r^2}}, \qquad u(r) = -\frac{\sqrt{1-\kappa r^2}}{r},$$

so that

$$r = \frac{2|\mathbf{q}|}{1+\kappa \mathbf{q}^2}, \qquad |\mathbf{q}| = \frac{r}{1+\sqrt{1-\kappa r^2}},$$

provided that we have dropped an additive constant $\ln 2$ in the integral (11). Then, as expected, we obtain the same variable mass as in the previous example and the corresponding Bertrand oscillator in Poincaré variables is given by

$$H_\kappa = \frac{1}{8}\left(1+\kappa \mathbf{q}^2\right)^2 \mathbf{p}^2 + A_2 \frac{2\mathbf{q}^2}{(1-\kappa \mathbf{q}^2)^2}.$$

The integrals $C_{ii} \equiv I_i$ in the diagonal of the conserved matrix C_{ij} can be found in [25].

5. The Darboux III oscillator

In what follows we focus our attention on a particular system of the family of Bertrand Hamiltonians of type II, the so called Darboux III oscillator. The underlying Bertrand space is the 3D version of the Darboux surface of type III [31, 32], for which an ND spherically symmetric generalization was constructed in [33, 34]. Such a 3D Darboux-Bertrand space corresponds to choose the + sign in (10) and to set $n = 2$, $m = 1$, $K = D^2$ and $D = -2\lambda$, where λ is a real parameter. This yields

$$h(r) = \frac{1+\sqrt{1+4\lambda r^2}}{2\sqrt{1+4\lambda r^2}}, \qquad u(r) = -\left(\frac{1+\sqrt{1+4\lambda r^2}}{2r}\right).$$

Therefore the transformations between the radial variables r and $|\mathbf{q}|$ turn out to be

$$r = |\mathbf{q}|\sqrt{1+\lambda\mathbf{q}^2}, \qquad |\mathbf{q}| = \left(\frac{\sqrt{1+4\lambda r^2}-1}{2\lambda}\right)^{1/2}.$$

Then the variable mass function reads $M(|\mathbf{q}|) = 1+\lambda\mathbf{q}^2$ and the resulting Darboux Hamiltonian is given by

$$H_\lambda = \frac{\mathbf{p}^2}{2(1+\lambda\mathbf{q}^2)} + \frac{\omega^2\mathbf{q}^2}{2(1+\lambda\mathbf{q}^2)}, \tag{12}$$

where we have written $A_2 = \omega^2/2$. According to section 3, the 3D Hamiltonian H_λ is a MS system, since it is endowed with a conserved matrix C_{ij} (a curved Fradkin tensor). In fact, this result can directly be extended to arbitrary dimension N, as it has been proven in [24, 35]:

Theorem 10. *(i) The Hamiltonian H_λ (12), for any dimension N and for any real value of λ, is endowed with the following constants of motion.*
• $(2N-3)$ *angular momentum integrals:*

$$C^{(m)} = \sum_{1\le i<j\le m} (q_ip_j - q_jp_i)^2, \qquad C_{(m)} = \sum_{N-m<i<j\le N} (q_ip_j - q_jp_i)^2, \tag{13}$$

where $m = 2,\dots,N$ and $C^{(N)} = C_{(N)}$.
• N^2 *integrals given by the components of the ND curved Fradkin tensor:*

$$C_{ij} = p_ip_j - \big(2\lambda H_\lambda(\mathbf{q},\mathbf{p}) - \omega^2\big)q_iq_j, \tag{14}$$

where $i,j = 1,\dots,N$ and such that $H_\lambda = \frac{1}{2}\sum_{i=1}^N C_{ii}$.
(ii) Each of the three sets $\{H_\lambda, C^{(m)}\}$, $\{H_\lambda, C_{(m)}\}$ $(m = 2,\dots,N)$ and $\{C_{ii}\}$ $(i = 1,\dots,N)$ is formed by N functionally independent functions in involution.
(iii) The set $\{H_\lambda, C^{(m)}, C_{(m)}, C_{ii}\}$ for $m = 2,\dots,N$ with a fixed index i is constituted by $2N-1$ functionally independent functions.

We remark that the constants of motion (13) and (14) can also be obtained [35] from the *free Euclidean motion* by means of a Stäckel transform or coupling constant metamorphosis (see [36, 37] and references therein).

It is also worth stressing that although the above statement holds for any real value of λ, the specific resulting system does depend on such a value, in such a manner that H_λ comprises, in fact, *three* different nonlinear physical systems [35]:

- *Nonlinear hyperbolic oscillator.* For $\lambda > 0$ the Darboux space is the complete Riemannian manifold $\mathcal{M}^N = (\mathbb{R}^N, g)$ with metric $g_{ij} = (1 + \lambda \mathbf{q}^2)\,\delta_{ij}$. The scalar curvature $R(|\mathbf{q}|)$ has a minimum at the origin $R(0) = -2\lambda N(N-1)$, which coincides with the scalar curvature of the ND *hyperbolic space* with negative constant sectional curvature $\kappa = -2\lambda$.
- *Nonlinear spherical oscillator.* For $\lambda < 0$ we firstly consider the interior Darboux space defined by $\mathcal{M}^N = (B_{|\mathbf{q}|_c}, g)$, where $g_{ij} = (1 - |\lambda|\mathbf{q}^2)\,\delta_{ij}$ and $B_{|\mathbf{q}|_c}$ denotes the ball centered at 0 of radius $|\mathbf{q}|_c = 1/\sqrt{|\lambda|}$ (the critical value for the metric and for H_λ). Now $R(0) = 2|\lambda|N(N-1)$ is exactly the scalar curvature of the ND *spherical space* with positive constant sectional curvature $\kappa = +2|\lambda|$.
- *Nonlinear exterior potential.* For $\lambda < 0$ we can also consider the exterior Darboux space defined by $\mathcal{M}^N = (\mathbb{R}^N \backslash \overline{B_{|\mathbf{q}|_c}}, g)$; this implies to reverse the sign of the metric and, therefore, of the Hamiltonian itself, namely, $g_{ij} = (|\lambda|\mathbf{q}^2 - 1)\,\delta_{ij}$. This system can naturally be interpreted as an infinite barrier potential rather than an oscillator one.

6. Quantization of the Darboux III oscillator

The quantization problem for Bertrand Hamiltonians arises as a challenging research program, whose first steps in the case of the Darboux III oscillator are summarized as follows.

Let us consider the quantum Cartesian coordinates and momenta, $\hat{\mathbf{q}}$, $\hat{\mathbf{p}}$, with Lie brackets and differential representation given by

$$[\hat{q}_i, \hat{p}_j] = \mathrm{i}\hbar\delta_{ij}, \qquad \hat{q}_i = q_i, \qquad \hat{p}_i = -\mathrm{i}\hbar\partial_i = -\mathrm{i}\hbar\frac{\partial}{\partial q_i}, \qquad \Delta = \frac{\partial^2}{\partial^2 q_1} + \cdots + \frac{\partial^2}{\partial^2 q_N}.$$

Our aim now is to construct the quantum mechanical counterpart of the ND classical Hamiltonian (12): $H_\lambda(\mathbf{q}, \mathbf{p}) \to \hat{H}_\lambda(\hat{\mathbf{q}}, \hat{\mathbf{p}})$. As it is well known, the crucial point is to obtain the quantum analogue of the kinetic term, since there is an order ambiguity in its quantization. This task can be faced by applying three different quantization procedures [38]: (i) the 'Schrödinger quantization'; (ii) the Laplace–Beltrami quantization (which makes use of the Laplace operator on curved spaces); and (iii) a PDM quantization.

We stress that if we impose that the quantum Hamiltonian $\hat{H}_\lambda$ keeps the MS property (that is, the existence of $2N-2$ algebraically independent operators that commute with $\hat{H}_\lambda$), then only the Schrödinger quantization yields, in a direct way, to fulfill this condition. Nevertheless, the Laplace–Beltrami and PDM quantizations also lead to MS quantum Hamiltonians once an additional 'pure' quantum potential term is added to the initial quantum Hamiltonian, and such potential terms are related through gauge transformations to the Schrödinger quantization. The resulting MS Schrödinger quantization of H_λ (12) is characterized as follows [39] (this result is worth to be compared with theorem 10).

Theorem 11. *Let $\hat{H}_\lambda$ be the ND quantum Bertrand-Darboux Hamiltonian given by*

$$\hat{H}_\lambda = \frac{1}{2(1+\lambda\hat{\mathbf{q}}^2)}\,\hat{\mathbf{p}}^2 + \frac{\omega^2\hat{\mathbf{q}}^2}{2(1+\lambda\hat{\mathbf{q}}^2)} = \frac{1}{2(1+\lambda\mathbf{q}^2)}\big(-\hbar^2\Delta + \omega^2\mathbf{q}^2\big). \tag{15}$$

For any real value of λ the following statements hold:
(i) $\hat{H}_\lambda$ commutes with the following observables:
• $(2N-3)$ *quantum angular momentum operators,*

$$\hat{C}^{(m)} = \sum_{1\le i<j\le m} (\hat{q}_i\hat{p}_j - \hat{q}_j\hat{p}_i)^2, \qquad \hat{C}_{(m)} = \sum_{N-m<i<j\le N} (\hat{q}_i\hat{p}_j - \hat{q}_j\hat{p}_i)^2, \tag{16}$$

where $m = 2.\ldots, N$ and $\hat{C}^{(N)} = \hat{C}_{(N)}$.

• N^2 *operators which form an ND quantum Fradkin tensor,*

$$\hat{C}_{ij} = \hat{p}_i\hat{p}_j - 2\lambda\hat{q}_i\hat{q}_j\hat{H}_\lambda(\hat{\mathbf{q}},\hat{\mathbf{p}}) + \omega^2\hat{q}_i\hat{q}_j, \tag{17}$$

where $i, j = 1, \dots, N$ *and such that* $\hat{H}_\lambda = \frac{1}{2}\sum_{i=1}^N \hat{C}_{ii}$.
(ii) Each of the three sets $\{\hat{H}_\lambda, \hat{C}^{(m)}\}$, $\{\hat{H}_\lambda, \hat{C}_{(m)}\}$ $(m = 2, \dots, N)$ *and* $\{\hat{C}_{ii}\}$ $(i = 1, \dots, N)$ *is formed by* N *algebraically independent commuting observables.*
(iii) The set $\{\hat{H}_\lambda, \hat{C}^{(m)}, \hat{C}_{(m)}, \hat{C}_{ii}\}$ *for* $m = 2, \dots, N$ *with a fixed index* i *is formed by* $2N - 1$ *algebraically independent observables.*
(iv) $\hat{H}_\lambda$ *is formally self-adjoint on the Hilbert space* $L^2(\mathcal{M}^N)$, *endowed with the scalar product*

$$\langle\Psi|\Phi\rangle_\lambda = \int_{\mathcal{M}^N} \overline{\Psi(\mathbf{q})}\Phi(\mathbf{q})(1 + \lambda\mathbf{q}^2)\mathrm{d}\mathbf{q}.$$

Clearly, the results of theorem 11 should be adapted to each of the three different systems described in the previous section. In particular, we consider here the quantum hyperbolic-type Hamiltonian, that has been fully solved in [39].

6.1. The nonlinear hyperbolic oscillator: quantum case

The quantum Hamiltonian $\hat{H}_\lambda$ with $\lambda > 0$ has recently been shown to give rise to a new exactly solvable quantum model in N dimensions [39] which has both a discrete and a continuous spectrum. The discrete spectrum depends on a single principal quantum number $n = 0, 1, 2 \dots$ and its eigenvalues are

$$\begin{aligned} E_n &= -\hbar^2\lambda\left(n + \frac{N}{2}\right)^2 + \hbar\left(n + \frac{N}{2}\right)\sqrt{\hbar^2\lambda^2\left(n + \frac{N}{2}\right)^2 + \omega^2} \\ &= -\hbar^2\lambda\left(n + \frac{N}{2}\right)^2\left\{1 - \sqrt{1 + \frac{\omega^2}{\hbar^2\lambda^2\left(n + \frac{N}{2}\right)^2}}\right\}. \end{aligned} \tag{18}$$

Therefore, the degeneracy of this model is exactly the same as in the ND isotropic oscillator (whose spectrum is recovered under the limit $\lambda \to 0$), as it should be expected from the beginning due to its MS property. Notice also that the bound states of this system are hydrogen-like since

$$\lim_{n\to\infty} E_n = \frac{\omega^2}{2\lambda}, \qquad \lim_{n\to\infty}(E_{n+1} - E_n) = 0.$$

Therefore $\hat{H}_\lambda$ has an infinite number of eigenvalues contained in $(0, \frac{\omega^2}{2\lambda})$ and their only accumulation point is $\frac{\omega^2}{2\lambda}$ which is, in turn, the bottom of the continuous spectrum given by $[\frac{\omega^2}{2\lambda}, \infty)$. The corresponding wave functions can be explicitly found in [39].

Finally, a short disgression on possible physical applications of this kind of exactly solvable quantum PDM Hamiltonians is in order. Firstly, we recall that the MIC-Kepler and oscillator potentials on the 3D sphere have been shown in [10] to be suitable as effective models for strong and weak confinement regimes in spherical quantum dots [40]. Secondly, it is worth stressing that a parabolic mass function has been proposed in [13, 14] in order to describe a 1D quantum well formed by a GaAs/Al_xGa_{1-x}As heterostructure. In fact, if the concentration x grows in terms of a given spatial coordinate q_1 as $x(q_1) = \alpha\, q_1^2$ and in this type of material $m^*(x) = m_0(a + b\, x)$ [13], then an effective mass function of the type $m^*(q_1) = m_0(a + b\,\alpha\, q_1^2)$ arises. Thus, we have obtained a realistic quantum exactly solvable model coming from the Bertrand-oscillator potential on a hyperbolic space with non-constant curvature.

Acknowledgments

This work was partially supported by the Spanish MICINN under grants MTM2010-18556 and FIS2008-00209, by the Junta de Castilla y León (project GR224), by the Banco Santander–UCM (grant GR58/08-910556) and by the Italian–Spanish INFN–MICINN (project ACI2009-1083).

References

[1] Bertrand J 1873 *C. R. Math. Acad. Sci. Paris* **77** 849
[2] Féjoz J and Kaczmarek L 2004 *Ergod. Th. & Dynam. Sys.* **24** 1583
[3] Perlick V 1992 *Class. Quant. Grav.* **9** 1009
[4] Iwai T and Katayama N 1994 *J. Math. Phys.* **35** 2914
[5] Iwai T and Katayama N 1995 *J. Math. Phys.* **36** 1790
[6] von Roos O 1983 *Phys. Rev. B* **27** 7547
[7] Lévy-Leblond J M 1995 *Phys. Rev. A* **52** 1845
[8] Chetouani L, Dekar L and Hammann T F 1995 *Phys. Rev. A* **52** 82
[9] Plastino A R, Rigo A, Casas M, Gracias F and Plastino A 1999 *Phys. Rev. A* **60** 4318
[10] Gritsev V V and Kurochkin Yu A 2001 *Phys. Rev. B* **64** 035308
[11] Quesne C and Tkachuk V M 2004 *J. Phys. A: Math. Gen.* **37** 4267
[12] Bagchi B, Banerjee A, Quesne C and Tkachuk V M 2005 *J. Phys. A: Math. Gen.* **38** 2929
[13] Koc R, Koca M and Sahinoglu G 2005 *Eur. Phys. J. B* **48** 583
[14] Schmidt A G M 2006 *Phys. Lett. A* **353** 459
[15] Mustafa O and Mazharimousavi S H 2006 *Phys. Lett. A* **358** 259
[16] Quesne C 2006 Ann. Phys. **321** 1221
[17] Cruz y Cruz S, Negro J and Nieto L M 2007 *Phys. Lett. A* **369** 400
[18] Schmidt A G M, Azeredo A D and Gusso A 2008 *Phys. Lett. A* **372** 2774
[19] Midya B and Roy B 2009 *Phys. Lett. A* **373** 4117
[20] Lévai G and Özer O 2010 *J. Math. Phys.* **51** 092103
[21] Ballesteros A, Enciso A, Herranz F J and Ragnisco O 2008 *Class. Quant. Grav.* **25** 165005
[22] Ballesteros A, Enciso A, Herranz F J and Ragnisco O 2009 *Comm. Math. Phys.* **290** 1033
[23] Ngome J P 2009 *J. Math. Phys.* **50** 122901
[24] Ballesteros A, Enciso A, Herranz F J and Ragnisco O 2008 *Physica D* **237** (2008) 505
[25] Ballesteros A and Herranz F J 2007 *J. Phys. A: Math. Theor.* **40** F51
[26] Shchepetilov A V 2005 *J. Math. Phys.* **46** 114101
[27] Guillemin V and Sternberg S 1990 *Variations on a theme by Kepler* (Providence: AMS)
[28] Fradkin D M 1965 *Amer. J. Phys.* **33** 207
[29] Fradkin D M 1967 *Prog. Theor. Phys.* **37** 798
[30] Ballesteros A and Herranz F J 2009 *J. Phys. A: Math. Theor.* **42** 245203
[31] Koenigs G 1972 *Leçons sur la théorie générale des surfaces* vol 4, ed G Darboux (New York: Chelsea) p 368
[32] Kalnins E G, Kress J M, Miller W Jr and Winternitz P 2003 *J. Math. Phys.* **44** 5811
[33] Ballesteros A, Enciso A, Herranz F J and Ragnisco O 2007 *Phys. Lett. B* **652** 376
[34] Ballesteros A, Enciso A, Herranz F J and Ragnisco O 2009 *Ann. Phys.* **324** 1219
[35] Ballesteros A, Enciso A, Herranz F J, Ragnisco O and Riglioni D 2010 arXiv:1010.3358
[36] Hietarinta J, Grammaticos B, Dorizzi B and Ramani A 1984 *Phys. Rev. Lett.* **53** 1707
[37] Kalnins E G, Miller W Jr and Post S 2010 *J. Phys. A: Math. Theor.* **43** 035202
[38] Ballesteros A, Enciso A, Herranz F J, Ragnisco O and Riglioni D 2010 submitted to *Ann. Phys.*
[39] Ballesteros A, Enciso A, Herranz F J, Ragnisco O and Riglioni D 2010 arXiv:1007.1335
[40] Harrison P 2009 *Quantum wells, wires and dots* (New York: Wiley)

Poisson Deformed Super-Algebras and Moyal Quantization

A. Ballesteros[1], **E. Celeghini**[2] **and M.A. del Olmo**[3]

[1] Departamento de Física, Universidad de Burgos, E-09006, Burgos, Spain
[2] Dipartimento di Fisica, Università di Firenze and INFN–Sezione di Firenze, I50019 Sesto Fiorentino, Firenze, Italy
[3] Departamento de Física Teórica, Universidad de Valladolid, E-47005, Valladolid, Spain.

E-mail: angelb@ubu.es, celeghini@fi.infn.it, olmo@fta.uva.es

Abstract.
The procedure to obtain the semi-classical limit of Lie-Hopf and quantum algebras introduced in preceding papers has been extended to Lie and quantum super-algebras. With a different definition of the limit the approach is the same: an $\hbar$-depending family of Hopf super-algebras is constructed starting from the super-bialgebra; then the singular limit $\hbar \to 0$ allows to move from quantum to semi-classical framework in the spirit of Moyal quantization. The Poisson undeformed $osp^P(1|2)$ and deformed $osp_q^P(1|2)$ super-algebras are exhibited, both as a paradigmatic example and because of their interest in applications.

1. Introduction

Group theory provides the natural mathematical language to formulate symmetry principles and to derive their consequences in Mathematics and Physics. Among groups, Lie ones are the most considered, because structures particularly manageable owing to their very high level of symmetry and of their holomorphic nature [1]. It is sufficient to consider the connection between quantum mechanics and the rotation group to see that quantum theories are almost inseparable from the underlying group structure.

This so advantageous approach has in itself a paradox: we assume that two objects (two elementary particles, two reference frames, etc..) are absolutely equal and, for this reason, indistinguishable. But this means that we are not able to identify them as separate objects and, thus, to state any relation between them. The solution of the paradox is, of course, the idea that a symmetry breaking always exists but that it is irrelevant for the problem we are considering: the electron at our left has the same properties of the electron at our right because the fact to be at our left is unimportant at the considered degree of accuracy. Also from this epistemological argument -that could find room in a presocratic fragment- an interest is born in describing a symmetry with a breaking that can be adiabatically reduced to zero.

For this reason, from the beginning of development of quantum groups we worked to find in terms of quantum groups a mathematical description of broken symmetries in quantum physics, the field where symmetry has played the largest role up to now. However, as symmetry is relevant also in classical physics, we have recently considered the idea of extending the quantum group approach to Hamiltonian mechanics [2, 3]. A broken symmetry has been thus described in a consistent way also at this level.

Journal of Physics: Conference Series **284** (2011) 012012 doi:10.1088/1742-6596/284/1/012012

Technically the idea is to include the deformation in the Moyal quantization [4] to relate Poisson brackets and algebraic structure of quantum mechanics. Starting from a quantum algebra (where a deformation parameter z is already present) another parameter -the Planck constant $\hbar$- has been thus included in the scheme in such a way that it is possible to consider the limit where the Planck constant goes to zero in order to introduce in the semi-classical description an analytically broken symmetry.

However in the Hilbert space framework the full product of operators is defined. Thus, when we fix an algebra (and not also its representation) we have to choose between commutators and anticommutators. We are used to taking into account the commutators because a null commutator of two observables implies that the two observables can be simultaneously diagonalized but this argument does not play any role for not null cases. Indeed, super-algebras, where both commutators and anticommutators appear -i.e. where products are partially antisymmetrical and partially symmetrical- should be taken into account on the same footing. For this reason we focus here on the semi-classical limit of super-algebras and their deformation.

In sect. 2 a brief resumé of the results for the Moyal quantization in algebras is presented to introduce the notations and the basic ideas of the approach. For a better understanding, the method is illustrated in sect. 3 by a brief description of the Poisson-Hopf standard quantum deformations $su_q^P(2)$ and $su_q^P(3)$. Sect. 4 introduces the peculiarities of grading in connection with the relations between quantum and classical schemes. In sect. 5 a particular attention has been dedicated to the connection between the relevant cases of $osp(1|2)$ and $osp_q(1|2)$, and their semi-classical limits $osp^P(1|2)$ and $osp_q^P(1|2)$ respectively (that look interesting in describing the configuration and momentum spaces of quantum mechanics). Harmonic oscillator and its deformation are explicitly described from this new point of view. Finally sect. 6 illustrates the general philosophy of the approach.

2. Quantum (super)-algebras and Poisson-Hopf limit

In Moyal quantization [4] the non commutativity of the algebra of functions of the Poisson-Lie groups G is written as

$$[f_1, f_2] := f_1 \star f_2 - f_2 \star f_1 = \hbar \{f_1, f_2\} + o(\hbar^2),$$

where $\star$ denotes the non-commutative star product and $\{f_1, f_2\}$ is the Poisson bracket in Fun(G), related to the commutator by the Poisson limit

$$\{f_1, f_2\} = \lim_{\hbar \to 0} \frac{1}{\hbar} [f_1, f_2] .$$

Let us remember that quantization tries to associate to a commutative algebra of physical observables -Fun(M)- of a classical mechanical system (with phase space M) a non-commutative algebra of linear operators -$Op(\mathcal{H})$- on the suitable complex Hilbert space, $\mathcal{H}$, of the quantum system under consideration. More precisely, the idea of quantization is to find a linear bijective map $\mathbb{Q}$: Fun$(M) \mapsto Op(\mathcal{H})$ such that for all $f \in$ Fun(M)

$$\mathbb{Q}(\{f_1, f_2\}) = \frac{\mathbf{i}}{\hbar} [\mathbb{Q}(f_1), \mathbb{Q}(f_2)] . \qquad (1)$$

This condition comes from the fact that if f_2 is the classical Hamiltonian of the system and $\mathbb{Q}(f_2)$ its quantum counterpart then $\{f_1, f_2\}$ and $[\mathbb{Q}(f_1), \mathbb{Q}(f_2)]$ give the time evolution of the physical observable f_1 in the classical and in the quantum case, respectively. Unfortunately such a map $\mathbb{Q}$ is not known in general, even for very simple cases.

When a Lie algebra g is involved at the level of quantum operators a solution of (1) can be found. Indeed when there is a set of quantum observables X_i closing a Lie algebra

$$[X_i, X_j] \;=\; f_{ij}^{k}\, X_k$$

in the (classical) Poisson-Hopf manifold a corresponding algebra can be defined by

$$\{\mathbb{Q}^{-1}(X_i), \mathbb{Q}^{-1}(X_j)\} \;:=\; \lim{}_{\hbar\to 0}\; \frac{\mathbf{i}}{\hbar}\; \mathbb{Q}^{-1}([X_i, X_j])\,,$$

$$\Delta(\mathbb{Q}^{-1}(X_i)) \;:=\; \lim{}_{\hbar\to 0}\; \mathbb{Q}^{-1}(\Delta(X_i))\,.$$

The dual space of g -g^*- can thus be equipped with a Poisson-Hopf structure whose fundamental brackets are

$$\{x^i, x^j\} \;=\; f_k^{ij}\, x^k, \tag{2}$$

where the x^i can be considered as local coordinates on g^*. So for any two functions on Fun(g^*) their Poisson bracket will be then given by

$$\{f_1, f_2\} \;=\; \frac{\partial f_1}{\partial x^i}\,\frac{\partial f_2}{\partial x^j}\; f_{ij}^{k}\, x^k\,.$$

Taking in mind all these ideas we see that we can quantize the Kostant-Kirillov-Souriau structure [5, 6] as follows

$$\{x^i, x^j\} = f_k^{ij} x^k \qquad \Longrightarrow \qquad [X_i, X_j] = \hbar\, f_{ij}^{k}\, X_k\,,$$

obtaining a "new" Lie algebra $g_\hbar$ isomorphic to g^* (and also to g as $f_{ij}^{k} = f_k^{ij}$) . Considering $\hbar$ as a free parameter we have thus a family of equivalent Lie algebras $g_\hbar$ determined by

$$[X_i, X_j] = \hbar\, f_{ij}^{k}\, X_k \tag{3}$$

that coincide with the original algebra when $\hbar = 1$.

This approach can be iterated for deformed algebras. Indeed, quantum groups were initially introduced as "quantizations" and the deformation parameter q was supposed to be related to the Planck constant $\hbar$ (for a detailed discussion, see [7, 8] and references therein). However, soon after, q became a parameter whose geometric and/or physical meaning has to be found for each particular case. In fact, quantum groups were soon seen as "abstract" Hopf algebras in order to explore whether they can be considered as symmetries of some physically relevant systems. The keystone of this idea was the discovery of the $su_q(2)$ invariance of the Heisenberg spin XXZ chain [9, 10], that was followed by a number of results exploiting quantum algebra symmetries in two-dimensional models [11]. Indeed, in the XXZ chain the "quantum" deformation parameter $z := \log(q)$ is identified with the anisotropy of the chain, which is completely independent of the (truly quantum) Planck constant. The same independence with respect to $\hbar$ can be traced in many other physical applications of quantum algebras and groups, like for instance lattice systems (where z is related to the lattice length) [12, 13], deformations of kinematical symmetries (in which the deformation parameter is a fundamental scale, see [14, 15] and references therein), effective nuclear models [16, 17], etc.

All these results suggest that $\hbar$ and q can be introduced separately in physics. It is exactly starting from this idea that a mathematical framework in which both the deformation parameter $z := \log q$ and the Planck constant $\hbar$ are independently and simultaneously considered has been introduced in quantum algebras [2, 3]. From an arbitrary quantum algebra $U_z(g)$ a one-parameter family of equivalent quantizations $U_{z,\hbar}(g)$, that depends explicitly on both $\hbar$ and z, have been constructed. Then, a q-Poisson-Hopf algebra was obtained as the $\hbar \to 0$ limit, getting -as in the undeformed case- a proper classical-mechanical limit of quantum-mechanical structure of symmetry.

From a technical point of view, to a one-parameter family of algebras $g_\hbar$, eq. (3), a one-parameter family of Lie bialgebras $(g_\hbar, \delta)$ is associated. By using the method of the analytical deformation, described in [18, 19], a two-parameter quantum algebra $(U_{z,\hbar}(g), \Delta_{z,\hbar})$, depending explicitly on $\hbar$ (as well as on z), is thus obtained. Poisson-Hopf super-algebra is then derived in the limit $\hbar \to 0$. Here we limit ourselves to summarize the procedure (see [2, 3] for details).

The starting point is the bialgebra (g, δ) [20], related to the quantum algebra, where g is a Lie algebra

$$[X_i, X_j] = f_{ij}^k \, X_k \tag{4}$$

and δ is the antisymmetric compatible cocommutator map $\delta : g \longrightarrow g \otimes g$

$$\delta(X_i) \; := \; z \, c_i^{jk} \, X_j \otimes X_k \, . \tag{5}$$

From the general properties of Hopf algebras and considering expression (4) and (5) as first orders in z of the quantum algebra, i.e.

$$[X_i, X_j] \; = \; f_{ij}^k \, X_k + o(z^2)$$

$$\Delta_z(X) = (X \otimes 1 + 1 \otimes X) + z \, \delta(X) + o(z^2),$$

the analytical form of the quantum algebra $(U_z(g), \Delta_z)$ was derived in [18, 19] as a first step.

Now, as a second step, eq.(3) has been substituted to (4) without varying (5), or -equivalently- the mapping

$$X_i \mapsto X_i/\hbar \, , \quad z \mapsto \hbar z \, . \tag{6}$$

has been realized inside (4) and (5). As the above described procedure does not change, a two-parameter quantum algebra $(U_{z,\hbar}(g), \Delta_{z,\hbar})$ was thus obtained.

As a third and last step, from $(U_{z,\hbar}(g), \Delta_{z,\hbar})$, the Poisson-Hopf algebra $(\mathrm{Fun}(g_z), \Delta_z^P)$ has been derived in the Poisson limit (i.e., $\hbar \to 0$). $(\mathrm{Fun}(g_z), \Delta_z^P)$ is just a Poisson-Lie structure on the quantum algebra $(U_z(g), \Delta_z)$ whose Lie bialgebra limit is characterized by the structure tensor f_i^{jk} of eq. (2). Concretely the Poisson bracket on $\mathrm{Fun}(g_z)$ is obtained by

$$\{\mathcal{X}_i, \mathcal{X}_j\} \; := \; \lim_{\hbar \to 0} \frac{[X_i, X_j]}{\hbar} \, , \tag{7}$$

and the coproduct map by

$$\Delta_z^P(\mathcal{X}_i) \; := \; \lim_{\hbar \to 0} \Delta_{z,\hbar}(X_i) \, , \tag{8}$$

where $\mathcal{X}_i (:= \mathbb{Q}^{-1}(X_i))$ can be seen as the local coordinates on the manifold ${g_z}^*$, so the Poisson brackets (7) are the so-called fundamental brackets and the coproduct (8) is a Poisson algebra homomorphism between $\mathrm{Fun}(g_z)$ and $\mathrm{Fun}(g_z) \otimes \mathrm{Fun}(g_z)$. Of course in the limit $z \to 0$ (i.e. starting from undeformed Lie algebra), coproducts are primitive and commutation rules linear: in this case, the Poisson limits (7) and (8) lead to the same formal structure where commutation rules have been just replaced by Poisson brackets. However, in quantum algebras nonlinear functions of the generators appear both in commutation rules and in coproducts: the q-Poisson structure given by the limit $\hbar \to 0$ can be formally different from the original quantum algebra structure and this is exactly what happens because, in the Poisson limit, contributions that arise from reordering disappear.

From a mathematical point of view, the dependence from $\hbar$ is regular for $\hbar \neq 0$ but the limit $\hbar \to 0$ is singular and does not commute with the procedure defined in eqs.(7,8), we can formally resume as

$$(\mathrm{Fun}(g_z), \Delta_z^P) := \lim_{\hbar\to 0} (U_{z,\hbar}, \Delta_{z,\hbar}),$$

that is just the Poisson-Lie structure on g_z whose algebra is determined by the dual δ^* of the Lie bialgebra map (5), (see [20, 8]).

The approach above described is extended in this paper to super-algebras introducing an adequate rescaling of odd generators (see eq.(13)). In conclusion, we have the following commutative diagram where g is an arbitrary algebra or super-algebra:

$$\begin{array}{ccccc}
(U(g), \Delta_0) & \xleftarrow{\hbar=1} & (U_\hbar(g), \Delta_0) & \xrightarrow{\hbar\to 0} & (\mathrm{Fun}(g), \Delta_0^P) \\
\uparrow{\scriptstyle z\to 0} & & \uparrow{\scriptstyle z\to 0} & & \uparrow{\scriptstyle z\to 0} \\
(U_z(g), \Delta_z) & \xleftarrow{\hbar=1} & (U_{z,\hbar}(g), \Delta_{z,\hbar}) & \xrightarrow{\hbar\to 0} & (\mathrm{Fun}(g_z), \Delta_z^P)\,.
\end{array}$$

Our interest is focused on the lower right corner, $(\mathrm{Fun}(g_z), \Delta_z^P)$, that defines in a constructive way from any (super)-bialgebra a related Poisson-Hopf deformed (super)-algebra.

3. Standard q-Poisson algebras $su_q^P(2)$ and $su_q^P(3)$

The general preceding discussion is now illustrated by the $su_q^P(2)$ and $su_q^P(3)$ examples. The $\hbar$-dependent family of Lie bialgebras $(su(2)_\hbar, \delta)$ related to the standard bialgebra is

$$[J_+, J_-] = 2\,\hbar\, J_3, \qquad [J_3, J_\pm] = \pm \hbar\, J_\pm\,, \tag{9}$$

$$\delta(J_3) = 0, \qquad \delta(J_\pm) = z\; J_3 \wedge J_\pm\,. \tag{10}$$

Since eqs.(10) are the same of the standard bialgebra and we have not reordering in the quantization, coproducts are not related to $\hbar$,

$$\Delta_z(J_3) = \Delta_0(J_3) = J_3 \otimes 1 + 1 \otimes J_3, \qquad \Delta_z(J_\pm) = e^{z\,J_3} \otimes J_\pm + J_\pm \otimes e^{-z\,J_3}. \tag{11}$$

However, starting from (9,11) we get

$$[J_+, J_-] = \hbar\,\frac{\sinh(2\,z\,J_3)}{z}, \qquad [J_3, J_\pm] = \pm\,\hbar\,J_\pm$$

and the $(\hbar, z)$-deformed Casimir operator

$$C_{z,\hbar} = \frac{\sinh^2(zJ_3)}{z^2}\cosh(z\hbar) + \frac{1}{2}\,[J_+\,, J_-]_+\,. \tag{12}$$

As the limit $\hbar \to 0$ is singular, the parameter $\hbar$ becomes essential to obtain the Poisson algebras. Introducing $\mathcal{J}_i := \mathbb{Q}^-1(J_i)$, the limit $\hbar \to 0$ gives us the q-Poisson-Hopf algebra $su_q^P(2) \equiv (\mathrm{Fun}(g_z), \Delta_z^P)$ as the commutative algebra of functions on g_z endowed with the Poisson bracket and coproduct map obtained by performing, respectively, the limits (7) and (8):

$$\{\mathcal{J}_+, \mathcal{J}_-\} = \frac{\sinh(2\,z\,\mathcal{J}_3)}{z}, \qquad \{\mathcal{J}_3, \mathcal{J}_\pm\} = \pm\mathcal{J}_\pm\,;$$

$$\Delta_z^P(\mathcal{J}_3) = \Delta_0^P(\mathcal{J}_3) = \mathcal{J}_3 \otimes 1 + 1 \otimes \mathcal{J}_3\,, \qquad \Delta_z^P(\mathcal{J}_\pm) = e^{z\,\mathcal{J}_3} \otimes \mathcal{J}_\pm + \mathcal{J}_\pm \otimes e^{-z\,\mathcal{J}_3}$$

and the q-deformed Casimir function obtained from (12) is:

$$C_z^P = \lim_{\hbar\to 0} C_{q,\hbar} = \frac{\sinh^2(z\,\mathcal{J}_3)}{z^2} + \mathcal{J}_+\,\mathcal{J}_- \,.$$

For the deformed $su_q(3)$ the procedure is the same. We start from the standard $\hbar$-depending bialgebra for $su(3)$ $(i,j,k=1,2,3)$:

$$[H_i, H_j] = 0, \qquad [H_i, F_{jk}] = \hbar\,(\delta_{ij} - \delta_{ik})F_{jk},$$
$$[F_{ij}, F_{kl}] = \hbar\,(\delta_{jk}F_{il} - \delta_{il}F_{kj}) + \hbar\,\delta_{jk}\delta_{il}(H_i - H_j)$$

$$\delta(H_i) = 0,$$
$$\delta(F_{ij}) = \tfrac{1}{2}\,(H_i - H_j)\wedge F_{ij} + \textstyle\sum_{k=i+1}^{j-1} F_{ik}\wedge F_{kj} \qquad (i<j),$$
$$\delta(F_{ij}) = \tfrac{1}{2}\,(H_j - H_i)\wedge F_{ij} - \textstyle\sum_{k=j+1}^{i-1} F_{ik}\wedge F_{kj} \qquad (i>j).$$

The analytical approach of [18, 19] allows to obtain the quantum algebra $su_{z,\hbar}(3)$ on which the related $\hbar\to 0$ limit is defined. All the details can be found in [2, 3]. As an example of the fact that commutators and not only the Universal Enveloping Algebra are, also at formal level, different in semi-classical limit we exhibit the quantum commutator

$$[F_{12}, F_{13}] = \frac{4}{\hbar}\,(\sinh\frac{z\hbar}{2})^2\, F_{12}\,F_{23}\,F_{12}$$

that, because the factor $\frac{4}{\hbar}\,(\sinh\frac{z\hbar}{2})^2$ is related to the reordering, reduces at semi-classical level to the Poisson bracket

$$\{\mathcal{F}_{12}, \mathcal{F}_{13}\} = z^2\,\mathcal{F}_{12}^2\,\mathcal{F}_{23}\,.$$

4. (Quantum) Super-algebras and (Quantum) Super-Poisson

In quantum mechanics what is defined is the product of operators i.e. both commutator and anticommutator, but the commutator has been considered the relevant one because if two operators commute they are simultaneously diagonalizable, disregarding the fact that if they do not commute the commutator is an operator like all the others including anticommutator. Other reasons of our interest in commutators are that the elements of a Lie group are related to the commutators of the generators and that we are used to consider as the most relevant super-algebra in physics the super-algebra of the fermions $h(1|1)$.

However super-algebras are objects that can play a large role in the description of symmetry as it is well known and also shown in the following section. The preceding approach has thus been extended to super-algebras, both Lie and quantum, introducing a generalization of eq. (6) to describe the relation between Poisson anti-brackets and anticommutators. The $\hbar$ dependence is indeed different for even (we denote by X_i) and odd (we denote by v_l) generators. The super-algebra $\{X_i; v_l\}$ is defined by

$$[X_i, X_j] = f_{ij}^k\,X_k \qquad [X_i, v_l] = g_{il}^m\,v_m \qquad [v_l, v_m]_+ = h_{lm}^k\,X_k$$

where $[\bullet,\bullet]_+$ is the anticommutator, $f_{ij}^k = -f_{ji}^k$ and $h_{lm}^k = +h_{ml}^k$ but in the classical limit, while commutators go to zero, anticommutators reduce to the simple product of non ordered objects. I.e.

$$[\bullet,\bullet] \propto \hbar^1 + o(\hbar^2) \qquad [\bullet,\bullet]_+ \propto \hbar^0 + o(\hbar^1).$$

So eq. (6) is still valid for the even elements of the bialgebra but, to obtain the correct $\hbar \to 0$ limit, it must be improved by an appropriate relation for the odd part:

$$X_i \;\mapsto\; X_i/\hbar\,, \qquad v_l \;\mapsto\; v_l/\sqrt{\hbar}, \qquad z \;\mapsto\; \hbar\, z. \tag{13}$$

As shown in the following examples, this recipe is all we need to extend the limit to Lie and quantum super-algebras.

5. The Poisson-Hopf super-algebras $osp^P(1|2)$ and its quantization $osp_q^P(1|2)$

The smallest super-algebra is $h(1|1)$, the algebra of fermions. The transition to semi-classical world is trivial and the only possible deformation has not found application up to now. We thus consider $osp(1|2)$, the second super-algebra in order of "complication". It has three even generators $\{H, X_\pm\}$ and two odd ones $\{v_\pm\}$. $X_\pm$ can be considered as elements of the Universal Enveloping Algebra as, in the appropriate real form,

$$X_\pm \;=\; \frac{1}{2}[v_\pm, v_\pm]_+ \;=\; {v_\pm}^2.$$

The full algebra can be thus described in terms of $\{H, v_\pm\}$ by

$$[H, v_\pm] \;=\; \pm\, v_\pm \qquad [v_+, v_-]_+ \;=\; 2\,H.$$

The best known $osp(1|2)$ representation is $\{|n\rangle\,;\; n \in I_+\}$ where

$$H|n\rangle = |n\rangle(n+1/2) \qquad v_+|n\rangle = |n+1\rangle\sqrt{n+1} \qquad v_-|n\rangle = |n-1\rangle\sqrt{n}\,. \tag{14}$$

We find thus an alternative description of the harmonic oscillator in this representation of $osp(1|2)$. All other representations -characterized by a positive highest weight $k \neq 1/2$- have not, up to now, found applications in physics.

Using the recipe (13), the semi-classical form $osp^P(1|2)$ is easily obtained and it is formally equal to $osp(1|2)$. For this reason, we describe here only its quantum deformation $osp_q(1|2)$ and its semi-classical limit. $osp_q(1|2)$ is well known [21] and we simply apply the preceding procedure. The $\hbar$-depending super-bialgebra is

$$[H, v_\pm] \;=\; \pm\,\hbar\, v_\pm, \qquad [v_+, v_-]_+ \;=\; 2\,H$$
$$\delta(H) = 0\,, \qquad \delta(v_\pm) \;=\; z\; H \wedge v_\pm\;.$$

The two-parameters family of analytical quantum super-deformation is described by

$$\begin{aligned} &[H, v_\pm] \;=\; \pm\,\hbar\, v_\pm, && [v_+, v_-]_+ \;=\; \tfrac{1}{z}\sinh(2\,z\,H) \\ &\Delta_z(H) = \Delta_0(H) = H \otimes 1 + 1 \otimes H, && \Delta_z(v_\pm) = v_\pm \otimes e^{-zH} + e^{zH} \otimes v_\pm, \end{aligned} \tag{15}$$

and the deformed representation with $k = 1/2$ is $\{|n\rangle\,;\; n \in I_+\}$

$$\frac{H}{\hbar}|n\rangle = |n\rangle(n{+}1/2) \quad \frac{v_+}{\sqrt{\hbar}}|n\rangle = |n{+}1\rangle\; f(n{+}1) \quad \frac{v_-}{\sqrt{\hbar}}|n\rangle = |n{-}1\rangle\; f(n) \qquad f(n) = \sqrt{\frac{\sinh(2zn)}{2\,z\cosh(z)}}$$

Now in the limit $\hbar \to 0$ we get the q-Poisson-Hopf super-algebra $osp_q^P(1|2)$

$$\begin{aligned} &\{\mathcal{H}, \mathbf{v}_\pm\} \;=\; \pm\mathbf{v}_\pm && \{\mathbf{v}_+, \mathbf{v}_-\}_+ = \tfrac{1}{z}\,\sinh(2\,z\,\mathcal{H}), \\ &\Delta_z^P(\mathcal{H}) = \Delta_0^P(\mathcal{H}) = \mathcal{H} \otimes 1 + 1 \otimes \mathcal{H}\,, && \Delta_z^P(\mathbf{v}_\pm) = \mathbf{v}_\pm \otimes e^{-z\mathcal{H}} + e^{+z\mathcal{H}} \otimes \mathbf{v}_\pm \end{aligned}$$

formally identical to (15).

6. General Philosophy

The algebraic structures here presented illustrate the full "hierarchy of complexity" that Hopf (super)-algebras provide. Quantum (super)-algebras $U_{z,\hbar}(g)$ (as well as $U_z(g) \equiv U_{z,\hbar=1}(g)$) would be the richest and most complex structures (both non commutative since $\hbar \neq 0$ and non cocommutative since $z \neq 0$), the q-Poisson (super)-algebras would be somehow intermediate (being commutative and non cocommutative) and the Lie (super)-algebras would be the "simplest" ones (non-commutative and cocommutative).

In this context, the procedure to obtain a q-Poisson analogue of a given quantum (super) algebra is canonically defined. Firstly, one goes from $U_z(g)$ to $U_{z,\hbar}(g)$ taking into account explicitly from the beginning both parameters $(\hbar, z)$ in the analytical construction of the deformation. At first sight, this two-parameter deformation seems to be irrelevant since for any finite value of $\hbar$, by using the Lie algebra automorphism $\{ X \mapsto \hbar X\ ,\ \ v \mapsto \sqrt{\hbar}\, v \}$ and by transforming the deformation parameter as $z \mapsto z/\hbar$, (inverse of (13)), the quantum algebra $(U_{z,\hbar}(g), \Delta_{z,\hbar})$ can be converted into $(U_z(g), \Delta_z)$. However, in order to obtain the Poisson analogue, one has to perform the $\hbar \to 0$ limit that *does not commute* with this automorphism and gives rise to the right result. The extension to counity and antipode has not been discussed because it is trivial. Note that a similar singular limit, that time in the deformation parameter z, allows to construct the contraction of quantum algebras [22].

With respect to the Poisson limit (7) and (8) we would like finally to stress that, because of the $\hbar$-dependence, it is essential to write the quantum algebra in terms of products with well defined symmetry. The so called q-commutators, without precise symmetry, cannot be used since their Poisson limit is not well-defined. This fact did not allow in the past the construction of q-Poisson analogues for algebras of rank greater than one. However, the analytical bases approach presented in [18, 19] provides a quantization framework based on pure commutators and anticommutators and lead thus to a well-defined Poisson limit for any quantum (super)-algebras. This is another argument to consider analytical bases as the right generalization for quantum algebras of Cartan bases.

Acknowledgments

This work was partially supported by the Ministerio de Educación y Ciencia of Spain (Projects FIS2009-09002 and MTM2007-67389, all of them with EU-FEDER support), by the Junta de Castilla y León and by INFN-MICINN (Italy-Spain).

[1] Weyl H 1939 *The Classical Groups* (Priceton Univ. Press, Princeton).
[2] Ballesteros A, Celeghini E and del Olmo M A 2009 *J. Phys. A: Math. Theor.* **42** 275202.
[3] Ballesteros A, Celeghini E and del Olmo M A 2010 *J. Phys. A: Conference Series* **237** 012003.
[4] Moyal J E 1949 *Proc. Phyl. Cambridge Soc. Inst.* **45** 99.
[5] Souriau J M 1970 *Structures des systemes dynamiques* (Dunod, Paris)
[6] Kirillov A A 1976 *Elements of the Theory of Representations* (Springer, Berlin)
[7] Kulish P P and Reshetikhin N Yu 1981 *Proc. Steklov Math. Inst.* **101** 101; Engl. Transl. 1983 *J. Soviet Math.* **23** 4.
[8] Majid S 1995 *Foundations of Quantum Group Theory* (Cambridge Univ. Press, Cambridge)
[9] Pasquier V and Saleur H 1990 *Nucl. Phys.* **B330** 523
[10] Kulish P P and Sklyanin E K 1991 *J. Phys. A: Math. Gen.* **24** L435
[11] Gómez C, Ruiz-Altaba M and Sierra G 1995 *Quantum Groups in Two-Dimensional Physics* (Cambridge Univ. Press, Cambridge)
[12] Bonechi F, Celeghini E, Giachetti R, Sorace E and Tarlini M 1992 *Phys. Rev. Lett.* **68** 3718
[13] Bonechi F, Celeghini E, Giachetti R, Sorace E and Tarlini M 1992 *Phys. Rev.* **B46** 5727
[14] Lukierski J, Nowicky A and Ruegg H 1992 *Phys. Lett.* **B293**, 344
[15] Ballesteros A, Herranz F J , del Olmo M A and Santander M 1994 *J. Math. Phys.* **35**, 4928
[16] Celeghini E, Giachetti R, Sorace E and Tarlini M 1992 *Phys. Lett.* **B280** 180
[17] Bonatsos D and Daskaloyannis C 1999 *Prog. Part. Nucl. Phys.* **43** 537
[18] Celeghini E, Ballesteros A and del Olmo M A 2008 *J. Phys. A: Math. Theor.* **41** 304038
[19] Ballesteros A, Celeghini E and del Olmo M A 2008 *J. Phys. C. S.* **128** 012049
[20] Chari V and Pressley A 1994 *A Guide to Quantum Groups* (Cambridge Univ. Press, Cambridge)
[21] Kulish P P and Reshetikhin N Yu 1989 *Lett. Math. Phys.* **18** 143.
[22] Celeghini E, Giachetti R, Sorace E and Tarlini M 1992 *Lecture Notes in Mathematics* **1510** (Springer, Berlin)

Journal of Physics: Conference Series **284** (2011) 012013 doi:10.1088/1742-6596/284/1/012013

Correlations between transfer reactions in nuclear supersymmetry

Roelof Bijker

Instituto de Ciencias Nucleares, Universidad Nacional Autónoma de México, A.P. 70-543, 04510 México, D.F., México

E-mail: bijker@nucleares.unam.mx

Abstract. Nuclear supersymmetry establishes precise correlations among the spectroscopic properties of different nuclei. It is shown that these correlations can be derived in terms of $SU(3)$ isoscalar factors by developing the concept of generalized F-spin. An application to one-neutron transfer reactions shows that the mixed symmetry $L = 2$ state is predicted to be excited more strongly than the first excited $L = 2$ state.

1. Introduction

Symmetries have played an important role in nuclear physics. Examples are isospin symmetry, the Wigner supermultiplet theory, special solutions to the Bohr Hamiltonian, the Elliott model, pseudo-spin symmetries and, more recently, dynamical symmetries and supersymmetries of the interacting boson model and its extensions.

Nuclear supersymmetry was proposed more than twenty five years ago [1] in the context of the interacting boson model (IBM) and the interacting boson-fermion model (IBFM) which have proved remarkably successful in providing a unified framework of even-even [2] and odd-even nuclei [3], respectively. One of its most attractive features is that it gives rise to a simple algebraic description, in which dynamical symmetries and supersymmetries play a central role, both as a way to improve our basic understanding of the importance of (super)symmetry in nuclear dynamics, and as a starting point for more precise calculations. Nuclear supersymmetry provides a theoretical framework in which different nuclei are treated as members of the same supermultiplet and whose spectroscopic properties are described by a single Hamiltonian and a single set of transition and transfer operators.

Aside from their esthetic appeal, (super)symmetries provide energy formula, selection rules and closed expressions for electromagnetic transition rates and transfer strengths which can be used as benchmarks to study and interpret the experimental data, even if these symmetries may be valid only approximately. In contrast to the situation in particle physics in which experimental evidence for the existence of supersymmetric particles is yet to be found, nuclear supersymmetry has been verified experimentally.

The purpose of this contribution is to study correlations between different one- and two-nucleon transfer reactions. It is shown that these correlations can be obtained in an explicit and elegant way in the framework of generalized F-spin. The ideas are illustrated by an application to one-neutron transfer reaction intensities.

Table 1. Overview of algebraic models

Model	Generators	Invariant	Symmetry
IBM	$b_i^\dagger b_j$	N	$U(6)$
IBFM	$b_i^\dagger b_j\,,\ a_i^\dagger a_j$	$N,\ M$	$U(6)\otimes U(\Omega)$
SUSY	$b_i^\dagger b_j\,,\ a_i^\dagger a_j\,,\ b_i^\dagger a_j\,,\ a_i^\dagger b_j$	$\mathcal{N}$	$U(6/\Omega)$

2. Nuclear supersymmetry

Dynamical supersymmetries were introduced in nuclear physics in the context of the Interacting Boson Model (IBM) and its extensions [1]. The IBM describes collective excitations in even-even nuclei in terms of a system of interacting monopole ($s^\dagger$) and quadrupole ($d^\dagger$) bosons [2]. The bosons are associated with the number of correlated proton and neutron pairs, and hence the number of bosons N is half the number of valence nucleons. For odd-mass nuclei the IBM was extended to include single-particle degrees of freedom [4]. The ensuing Interacting Boson-Fermion Model (IBFM) has as its building blocks N bosons ($b_l^\dagger$) with $l = 0, 2$ and $M = 1$ fermion ($a_j^\dagger$) with $j = j_1, j_2, \ldots$ [3]. The IBM and IBFM can be unified into a supersymmetry (SUSY)

$$U(6/\Omega) \supset U(6) \otimes U(\Omega) \ , \tag{1}$$

where $\Omega = \sum_j (2j+1)$ is the dimension of the fermion space [1, 5, 6, 7] . In this framework, even-even and odd-even nuclei form the members of a supermultiplet which is characterized by $\mathcal{N} = N + M$, *i.e.* the total number of bosons and fermions. Supersymmetry distinguishes itself from other symmetries in that it includes, in addition to transformations among fermions and among bosons, also transformations that change a boson into a fermion and *vice versa* (see Table 1).

The concept of nuclear SUSY was extended in 1985 to include the neutron-proton degree of freedom [8]. In this case, a supermultiplet consists of an even-even, an odd-proton, an odd-neutron and an odd-odd nucleus. The best experimental evidence of a supersymmetric quartet is provided by the ^{194,195}Pt and ^{195,196}Au nuclei as an example of the $U_\nu(6/12) \otimes U_\pi(6/4)$ supersymmetry [9, 10, 11, 12, 13, 14], in which the odd neutron is allowed to occupy the $3p_{1/2}$, $3p_{3/2}$ and $2f_{5/2}$ orbits of the 82-126 shell, and the odd proton the $2d_{3/2}$ orbit of the 50-82 shell. This supermultiplet is characterized by $\mathcal{N}_\nu = 5$ and $\mathcal{N}_\pi = 2$. The interpretation of these four nuclei as members of a supersymmetric quartet made it possible to predict [8] the properties of ^{196}Au almost 15 years before they were measured experimentally [9, 11].

In its most general form, the Hamiltonian is written in terms of the generators of the graded Lie algebra $G = U_\nu(6/12) \otimes U_\pi(6/4)$ which can be reduced to the rotation group as

$$\begin{aligned}
U_\nu(6/12) \otimes U_\pi(6/4) \ &\supset \ U^{B_\nu}(6) \otimes U^{F_\nu}(12) \otimes U^{B_\pi}(6) \otimes U^{F_\pi}(4) \\
&\supset \ U^{B}(6) \otimes U^{F_\nu}(6) \otimes U^{F_\nu}(2) \otimes U^{F_\pi}(4) \\
&\supset \ U^{BF_\nu}(6) \otimes U^{F_\nu}(2) \otimes U^{F_\pi}(4) \\
&\supset \ SO^{BF_\nu}(6) \otimes SU^{F_\pi}(4) \otimes U^{F_\nu}(2) \\
&\supset \ Spin(6) \otimes U^{F_\nu}(2) \\
&\supset \ Spin(5) \otimes U^{F_\nu}(2) \\
&\supset \ Spin(3) \otimes SU^{F_\nu}(2) \\
&\supset \ SU(2) \ .
\end{aligned} \tag{2}$$

This group chain corresponds to the decomposition of the neutron orbits with $j = 1/2,\ 3/2,\ 5/2$ into an pseudo-orbital part with $l = 0,\ 2$ and a pseudo-spin part $s = 1/2$. Since the pseudo-orbital part of the odd neutron has the same values of angular momentum as the bosons, the boson and fermion chains can be combined at the level of $U(6)$.

If the Hamiltonian is expressed in terms of Casimir invariants of the subgroups appearing in the reduction shown in Eq. (2) a dynamical supersymmetry arises. For example, the Hamiltonian

$$H = A\,C_{2U^{BF_\nu}(6)} + B\,C_{2SO^{BF_\nu}(6)} + B'\,C_{2Spin(6)} + C\,C_{2Spin(5)} + E\,C_{2Spin(3)} + F\,C_{2SU(2)} \,, \tag{3}$$

describes the excitation spectra of a quartet of nuclei characterized by $\mathcal{N}_\nu$ and $\mathcal{N}_\pi$ [8]. The quartet consists of an even-even nucleus with $N_\nu = \mathcal{N}_\nu$, $M_\nu = 0$ and $N_\pi = \mathcal{N}_\pi$, $M_\pi = 0$, an odd-proton nucleus with $N_\nu = \mathcal{N}_\nu$, $M_\nu = 0$ and $N_\pi = \mathcal{N}_\pi - 1$, $M_\pi = 1$, an odd-neutron nucleus with $N_\nu = \mathcal{N}_\nu - 1$, $M_\nu = 1$ and $N_\pi = \mathcal{N}_\pi$, $M_\pi = 0$, and an odd-odd nucleus with $N_\nu = \mathcal{N}_\nu - 1$, $M_\nu = 1$ and $N_\pi = \mathcal{N}_\pi - 1$, $M_\pi = 1$. The energy spectra of the four nuclei belonging to the supersymmetric quartet are described by a single energy formula in terms of the eigenvalues of the Casimir operators

$$\begin{aligned} E \;=\; & A\left[N_1(N_1+5) + N_2(N_2+3) + N_1(N_1+1)\right] \\ & + B\left[\Sigma_1(\Sigma_1+4) + \Sigma_2(\Sigma_2+2) + \Sigma_3^2\right] + B'\left[\sigma_1(\sigma_1+4) + \sigma_2(\sigma_2+2) + \sigma_3^2\right] \\ & + C\left[\tau_1(\tau_1+3) + \tau_2(\tau_2+1)\right] + D\,L(L+1) + E\,J(J+1) \,. \end{aligned} \tag{4}$$

The coefficients A, B, B', C, D, and E can be determined from the experimental excitation energies.

Recently, the structure of the odd-odd nucleus ^{194}Ir was investigated by a series of transfer and neutron capture reactions [15]. In particular, the new data from the polarized $(\vec{d},\alpha)$ transfer reaction provided crucial new information about and insight into the structure of the spectrum of ^{194}Ir which led to significant changes in the assignment of levels as compared to previous work [16]. Fig. 1 shows the negative parity levels of ^{194}Ir in comparison with the theoretical spectrum in which it is assumed that these levels originate from the $3p_{1/2}$, $3p_{3/2}$, $2f_{5/2} \otimes 2d_{3/2}$ configuration. The theoretical energy spectrum is calculated using the energy formula of Eq. (4) with $A = 26.3$, $B = 8.7$, $B' = -33.6$, $C = 35.1$, $D = 6.3$, and $E = 4.5$ (all in keV). Given the complex nature of the spectrum of heavy odd-odd nuclei, the agreement is remarkable. There is an almost one-to-one correlation between the experimental and theoretical level schemes [15].

The successful description of the odd-odd nucleus ^{194}Ir opens the possibility of identifying a second quartet of nuclei in the $A \sim 190$ mass region with $U_\nu(6/12) \otimes U_\pi(6/4)$ supersymmetry. The new quartet consists of the nuclei ^{192,193}Os and ^{193,194}Ir and is characterized by $\mathcal{N}_\nu = 5$ and $\mathcal{N}_\pi = 3$ [17].

3. Correlations

The nuclei belonging to a supersymmetric quartet are described by a single Hamiltonian, and hence their wave functions are strongly related. Moreover, inspection of the wave functions with $U_\nu(6/12) \otimes U_\pi(6/4)$ supersymmetry shows that there is a large similarity between the quantum numbers of the even-even and the odd-neutron nuclei

$$\begin{aligned} |\mathrm{ee}\rangle &= \left|[\mathcal{N}_\nu],[\mathcal{N}_\pi];[\mathcal{N}-f,f],\alpha,L\right\rangle \,, \\ |\mathrm{on}\rangle &= \left|[\mathcal{N}_\nu-1],[\mathcal{N}_\pi];[\mathcal{N}-1-f,f],[1];[\mathcal{N}-f-g,f+g-h,h],\alpha,L,\tfrac{1}{2};J\right\rangle \,. \end{aligned} \tag{5}$$

with $\mathcal{N} = \mathcal{N}_\nu + \mathcal{N}_\pi$ and $0 \le h \le g \le 1$. The square brackets denote $U(6)$ representations and the coefficients α represent all labels in the reduction from $U^{BF_\nu}(6)$ to the rotation group. In the case of the odd-neutron nucleus, the angular momentum L is coupled with the pseudo-spin $s = 1/2$ of the neutron orbits to the total angular momentum J. We note, that for $h = 0$

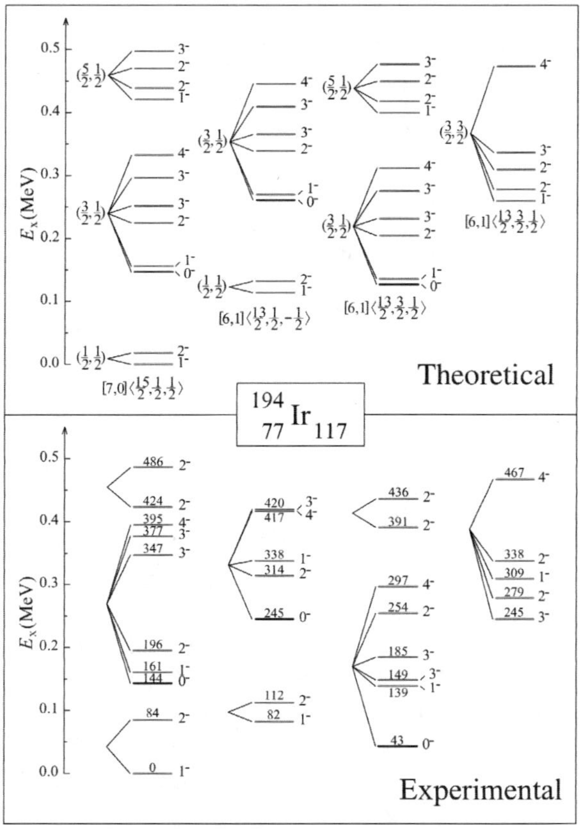

Figure 1. Comparison between the theoretical and experimental spectrum of ^{194}Ir.

the wave functions share the quantum numbers $[\mathcal{N}-f,f],\alpha,L$ (for the odd-neutron nucleus the label f is replaced by $f+g$). The main difference is the way in which the different $U(6)$ representations are coupled. Similar observations hold for the wave functions of the odd-proton and odd-odd nuclei

$$
\begin{aligned}
|\mathrm{op}\rangle &= |[\mathcal{N}_\nu],[\mathcal{N}_\pi-1];[\mathcal{N}-1-f,f],\alpha,L\rangle\ , \qquad (6)\\
|\mathrm{oo}\rangle &= \left|[\mathcal{N}_\nu-1],[\mathcal{N}_\pi-1];[\mathcal{N}-2-f,f],[1];[\mathcal{N}-1-f-g,f+g-h,h],\alpha,L,\frac{1}{2};J\right\rangle\ .
\end{aligned}
$$

The correspondence between the quantum numbers of the wave functions of nuclei belonging to a supersymmetric quartet suggests that the correlations between transition rates and transfer intensities can be obtained in an explict manner. In this section, I show how this can be achieved in the framework of the concept of generalized F-spin and discuss an application to one-neutron transfer reactions.

3.1. Generalized F-spin

The correlations between different transfer reactions can be derived in an elegant and explicit way by a generalization of the concept of F-spin which was introduced in the neutron-proton IBM (or IBM-2) [18] in order to distinguish between proton and neutron bosons. The eigenstates of the $U(6/12)_\nu \otimes U(6/4)_\pi$ supersymmetry are characterized by the irreducible representations $[N_1,N_2,N_3]$ of $U^{BF_\nu}(6)$ which arise from the coupling of three different $U(6)$ representations,

Table 2. $SU(3)$ quantum numbers of the creation and annihilation operators

	(λ,μ)	F	F_z	Y
$b_\pi^\dagger$	$(1,0)$	$\frac{1}{2}$	$\frac{1}{2}$	$\frac{1}{3}$
$b_\nu^\dagger$	$(1,0)$	$\frac{1}{2}$	$-\frac{1}{2}$	$\frac{1}{3}$
$a_\nu^\dagger$	$(1,0)$	0	0	$-\frac{2}{3}$
b_ν	$(0,1)$	$\frac{1}{2}$	$\frac{1}{2}$	$-\frac{1}{3}$
$-b_\pi$	$(0,1)$	$\frac{1}{2}$	$-\frac{1}{2}$	$-\frac{1}{3}$
a_ν	$(0,1)$	0	0	$\frac{2}{3}$

$[N_\nu]$ for the neutron bosons, $[N_\pi]$ for the proton bosons and $[N_\rho]$ for the pseudo-orbital angular momentum of the odd neutron ($N_\rho = 0$ for the even-even and odd-proton nuclei of the quartet, and $N_\rho = 1$ for the odd-neutron and the odd-odd nuclei). In analogy with the three quark flavors in the quark model (u, d and s), also in this case there are three different types of identical objects ($b_\pi^\dagger$, $b_\nu^\dagger$ and $a_\nu^\dagger$), which can be distinguished by F-spin and hypercharge Y. The two kinds of bosons form an F-spin doublet, $F = 1/2$, with charge states $F_z = 1/2$ for protons ($b_\pi^\dagger$) and $F_z = -1/2$ for neutrons ($b_\nu^\dagger$) [18]. In addition, the bosons carry hypercharge $Y = 1/3$. The pseudo-orbital part of the angular momentum of the odd neutron ($a_\nu^\dagger$) has $F = F_z = 0$ and $Y = -2/3$ [19, 20]. The three creation operators span the fundamental triplet representation of $SU(3)$ with $(\lambda,\mu) = (1,0)$, and the corresponding annihilation operators the antitriplet with $(\lambda,\mu) = (0,1)$ (see Table 2).

Group theoretically, the generalized F-spin is defined by the reduction

$$\begin{array}{ccccc} U(18) & \supset & U(6) & \otimes & U(3) \\ \downarrow & & \downarrow & & \downarrow \\ [N] & & [N_1,N_2,N_3] & & [N_1,N_2,N_3] \end{array} \tag{7}$$

with $N = N_1+N_2+N_3$. Here $U(6)$ is to be identified with the $U^{BF_\nu}(6)$ of the group reduction of Eq.(2) which is the result of first coupling the bosons at the level of $U(6)$, followed by coupling the orbital part

$$|[N_\nu],[N_\pi];[N_\nu+N_\pi-i,i],[N_\rho];[N_1,N_2,N_3]\rangle \ . \tag{8}$$

This sequence of $U(6)$ couplings can be described in a completely equivalent way by the three-dimensional index group $U(3)$ of Eq. (7) which can be reduced to

$$\begin{array}{ccccccccc} U(3) & \supset & SU(3) & \supset & [SU(2) & \supset & SO(2)] & \otimes & U(1) \\ \downarrow & & \downarrow & & \downarrow & & \downarrow & & \downarrow \\ |\,[N_1,N_2,N_3] & , & (\lambda,\mu) & , & F & , & F_z & , & Y\,\rangle \end{array} \tag{9}$$

The relation between the two sets of quantum numbers is given by

$$\begin{aligned} (\lambda,\mu) &= (N_1-N_2,N_2-N_3) \ , \\ F &= \frac{1}{2}(N_\pi+N_\nu-2i) \ , \\ F_z &= \frac{1}{2}(N_\pi-N_\nu) \ , \\ Y &= \frac{1}{3}(N_\pi+N_\nu-2N_\rho) \ . \end{aligned} \tag{10}$$

The one-to-one correspondence between the $U(6)$ couplings and the generalized F-spin basis shown in Eq. (10) makes it possible to reexpress the wave functions of Eqs. (5) and (7) in the $SU(3)$ basis of Eq. (9). As a result, one finds

$$
\begin{aligned}
|\mathrm{ee}\rangle &= \left|(\mathcal{N}-2f,f),\frac{\mathcal{N}-2f}{2},\frac{\mathcal{N}_\pi-\mathcal{N}_\nu}{2},\frac{\mathcal{N}}{3};\alpha,L\right\rangle , \qquad (11)\\
|\mathrm{on}\rangle &= \left|(\mathcal{N}-2f-2g+h,f+g-2h),\frac{\mathcal{N}-1-2f}{2},\frac{\mathcal{N}_\pi-\mathcal{N}_\nu+1}{2},\frac{\mathcal{N}-3}{3};\alpha,L,\frac{1}{2};J\right\rangle ,
\end{aligned}
$$

for the even-even and the odd-neutron nuclei, and

$$
\begin{aligned}
|\mathrm{op}\rangle &= \left|(\mathcal{N}-1-2f,f),\frac{\mathcal{N}-1-2f}{2},\frac{\mathcal{N}_\pi-\mathcal{N}_\nu-1}{2},\frac{\mathcal{N}-1}{3};\alpha,L\right\rangle , \qquad (12)\\
|\mathrm{oo}\rangle &= \left|(\mathcal{N}-1-2f-2g+h,f+g-2h),\frac{\mathcal{N}-2-2f}{2},\frac{\mathcal{N}_\pi-\mathcal{N}_\nu}{2},\frac{\mathcal{N}-4}{3};\alpha,L,\frac{1}{2};J\right\rangle ,
\end{aligned}
$$

for the odd-proton and odd-odd nuclei.

Matrix elements of $SU(3)$ tensor operators can now be obtained by applying the Wigner-Eckart theorem in generalized F-spin space, and correlations between different matrix elements can be expressed in terms of ratios of $SU(3)$ isoscalar factors.

3.2. One-neutron transfer

As an example of correlations in nuclear supersymmetry, let us consider one-neutron transfer reactions between the even-even and odd-neutron nuclei of the same supermultiplet, which are relevant for the reactions ^{194}Pt $\leftrightarrow$ ^{195}Pt and ^{192}Os $\leftrightarrow$ ^{193}Os. These transfer reactions are described by the fermionic generators of the graded Lie algebra. In a study of the ^{194}Pt $\rightarrow$ ^{195}Pt stripping reaction, it was found [21] that one-neutron $j = 3/2,\ 5/2$ transfer reactions can be described by the operator

$$
\tilde{P}_j = \frac{\alpha_j}{\sqrt{2}}\left[\left(\tilde{s}_\nu \times a^\dagger_{\nu,j}\right)^{(j)} - \left(\tilde{d}_\nu \times a^\dagger_{\nu,\frac{1}{2}}\right)^{(j)}\right] = \alpha_j\, T^{(1,1),\frac{1}{2},\frac{1}{2},-1}_{(2,0,0),(1,0),2,\frac{1}{2},j} . \qquad (13)
$$

The right-hand side of the equation shows the tensorial character under generalized F-spin (λ,μ), F, F_z, Y (upper indices) and under the groups appearing in Eq. (2) in the reduction of $U^{BF_\nu}(6)$ to the rotation group $(\sigma_1,\sigma_2,\sigma_3)$, (τ_1,τ_2), L, S, J (lower indices). Since the quantum numbers of the ground state in the even-even nucleus are given by

$$
\left|(\mathcal{N},0),\frac{\mathcal{N}}{2},\frac{\mathcal{N}_\pi-\mathcal{N}_\nu}{2},\frac{\mathcal{N}}{3};(\mathcal{N},0,0),(0,0),0\right\rangle , \qquad (14)
$$

the only states in the odd-neutron nucleus that can be populated by the transfer operator of Eq. (13) are the ones with $f = h = 0$, $g = 0, 1$, $(\tau_1,\tau_2) = (1,0)$, $L = 2$ and $J = j$. The corresponding matrix elements can be derived by first decoupling the pseudo-spin part of the neutron orbitals and then applying the $SU(3)$ Wigner-Eckart theorem which gives rise to an expression in terms of a $SU(3)$ isoscalar factor and a $SU(3)$ reduced matrix element

$$
\begin{aligned}
ME_1 &= \Big\langle (\mathcal{N}-2g,g),\frac{\mathcal{N}-1}{2},\frac{\mathcal{N}_\pi-\mathcal{N}_\nu+1}{2},\frac{\mathcal{N}}{3}-1;(\sigma_1,\sigma_2,\sigma_3),(1,0),2,\frac{1}{2};j \;\| \\
&\qquad T^{(1,1),\frac{1}{2},\frac{1}{2},-1}_{(2,0,0),(1,0),2,\frac{1}{2},j} \;\|\; (\mathcal{N},0),\frac{\mathcal{N}}{2},\frac{\mathcal{N}_\pi-\mathcal{N}_\nu}{2},\frac{\mathcal{N}}{3};(\mathcal{N},0,0),(0,0),0\Big\rangle \\
&= -\sqrt{\frac{2j+1}{5}}\left\langle \begin{array}{cc} (\mathcal{N},0) & (1,1) \\ \frac{\mathcal{N}}{2},\frac{\mathcal{N}_\pi-\mathcal{N}_\nu}{2},\frac{\mathcal{N}}{3} & \frac{1}{2},\frac{1}{2},-1 \end{array}\right| \left.\begin{array}{c} (\mathcal{N}-2g,g) \\ \frac{\mathcal{N}-1}{2},\frac{\mathcal{N}_\pi-\mathcal{N}_\nu+1}{2},\frac{\mathcal{N}}{3}-1 \end{array}\right\rangle \qquad (15)\\
&\qquad \Big\langle (\mathcal{N}-2g,g);(\sigma_1,\sigma_2,\sigma_3),(1,0),2 \;|||\; T^{(1,1)}_{(2,0,0),(1,0),2} \;|||\; (\mathcal{N},0);(\mathcal{N},0,0),(0,0),0\Big\rangle .
\end{aligned}
$$

The inverse reaction is described by the transfer operator

$$P_j^\dagger = \frac{\alpha_j}{\sqrt{2}}\left[\left(s_\nu^\dagger \times \tilde{a}_{\nu,j}\right)^{(j)} - \left(d_\nu^\dagger \times \tilde{a}_{\nu,\frac{1}{2}}\right)^{(j)}\right] = \alpha_j\, T^{(1,1),\frac{1}{2},-\frac{1}{2},1}_{(2,0,0),(1,0),2,\frac{1}{2},j} \,. \tag{16}$$

The matrix elements of the operator of Eq. (16) for one-neutron transfer reactions from the ground state of the odd-neutron nucleus to the final even-even nucleus are given by

$$\begin{aligned} ME_2 &= \Big\langle (\mathcal{N}-2g,g), \frac{\mathcal{N}-2g}{2}, \frac{\mathcal{N}_\pi-\mathcal{N}_\nu}{2}, \frac{\mathcal{N}}{3}; (\sigma_1,\sigma_2,\sigma_3), (1,0), 2 \parallel T^{(1,1),\frac{1}{2},-\frac{1}{2},1}_{(2,0,0),(1,0),2,\frac{1}{2},j} \\ &\qquad \parallel (\mathcal{N},0), \frac{\mathcal{N}-1}{2}, \frac{\mathcal{N}_\pi-\mathcal{N}_\nu+1}{2}, \frac{\mathcal{N}}{3}-1; (\mathcal{N},0,0), (0,0), 0, \frac{1}{2}; \frac{1}{2}\Big\rangle \\ &= \sqrt{\frac{2j+1}{5}} \left\langle \begin{array}{cc|c} (\mathcal{N},0) & (1,1) & (\mathcal{N}-2g,g) \\ \frac{\mathcal{N}-1}{2}, \frac{\mathcal{N}_\pi-\mathcal{N}_\nu+1}{2}, \frac{\mathcal{N}}{3}-1 & \frac{1}{2},-\frac{1}{2},1 & \frac{\mathcal{N}-2g}{2}, \frac{\mathcal{N}_\pi-\mathcal{N}_\nu}{2}, \frac{\mathcal{N}}{3} \end{array} \right\rangle \\ &\qquad \Big\langle (\mathcal{N}-2g,g); (\sigma_1,\sigma_2,\sigma_3), (1,0), 2 \parallel\!| T^{(1,1)}_{(2,0,0),(1,0),2} \parallel\!| (\mathcal{N},0); (\mathcal{N},0,0), (0,0), 0 \Big\rangle \,. \end{aligned} \tag{17}$$

Since the reduced matrix elements appearing in Eqs. (16) and (18) are the same, the matrix elements for one-neutron stripping and pick-up reactions are strongly correlated. Their ratio is proportional to that of two $SU(3)$ isoscalar factors

$$\frac{ME_2}{ME_1} = -\frac{\left\langle \begin{array}{cc|c} (\mathcal{N},0) & (1,1) & (\mathcal{N}-2g,g) \\ \frac{\mathcal{N}-1}{2}, \frac{\mathcal{N}_\pi-\mathcal{N}_\nu+1}{2}, \frac{\mathcal{N}}{3}-1 & \frac{1}{2},-\frac{1}{2},1 & \frac{\mathcal{N}-2g}{2}, \frac{\mathcal{N}_\pi-\mathcal{N}_\nu}{2}, \frac{\mathcal{N}}{3} \end{array} \right\rangle}{\left\langle \begin{array}{cc|c} (\mathcal{N},0) & (1,1) & (\mathcal{N}-2g,g) \\ \frac{\mathcal{N}}{2}, \frac{\mathcal{N}_\pi-\mathcal{N}_\nu}{2}, \frac{\mathcal{N}}{3} & \frac{1}{2},\frac{1}{2},-1 & \frac{\mathcal{N}-1}{2}, \frac{\mathcal{N}_\pi-\mathcal{N}_\nu+1}{2}, \frac{\mathcal{N}}{3}-1 \end{array} \right\rangle} \,. \tag{18}$$

The isoscalar factors can be derived in the standard way from the recursion relations for the $SU(3)$ step operators combined with the orthonormality conditions, De Swart's phase convention [22] and Racah's factorization lemma

$$\begin{aligned} &\left\langle \begin{array}{cc|c} (\lambda_1,\mu_1) & (\lambda_2,\mu_2) & (\lambda,\mu) \\ F_1, F_{1z}, Y_1 & F_2, F_{2z}, Y_2 & F, F_z, Y \end{array} \right\rangle \\ &\qquad = \left\langle \begin{array}{cc|c} (\lambda_1,\mu_2) & (\lambda_2,\mu_2) & (\lambda,\mu) \\ F_1, Y_1 & F_2, Y_2 & F, Y \end{array} \right\rangle \langle F_1, F_{1z}, F_2, F_{2z} | F, F_z \rangle \,. \end{aligned} \tag{19}$$

Table 3 shows the results for some relevant isoscalar factors. A proof by induction made it possible to obtain closed expressions as a function of k which are valid for all allowed values of F-spin and hypercharge Y [23].

The correlations between one-neutron stripping and pick-up reactions can be obtained in an explicit form by applying Racah's factorization lemma to Eq. (18) and inserting the appropriate values of the $SU(3)$ isoscalar factors from Table 3 and the $SU(2)$ Clebsch-Gordan coefficients

$$\frac{ME_2}{ME_1} = \begin{cases} -1 & g=0 \\ \sqrt{\frac{\mathcal{N}_\pi}{\mathcal{N}_\nu(\mathcal{N}-1)}} & g=1 \end{cases} \tag{20}$$

It is important to emphasize, that Eqs. (18) and (20) are parameter-independent predictions which are a direct consequence of nuclear supersymmetry and which can be tested experimentally. Tables 4 and 5 show the intensities of one-neutron transfer reactions between even-even and odd-neutron nuclei and *vice versa*.

Journal of Physics: Conference Series **284** (2011) 012013 doi:10.1088/1742-6596/284/1/012013

Table 3. $SU(3)$ isoscalar factors

F_1	Y_1	F_2	Y_2	(λ,μ)	F	Y	$\left\langle \begin{matrix} (N,0) & (1,1) \\ F_1,Y_1 & F_2,Y_2 \end{matrix} \middle\| \begin{matrix} (\lambda,\mu) \\ F,Y \end{matrix} \right\rangle$
$\frac{N-k-1}{2}$	$\frac{N}{3}-k-1$	$\frac{1}{2}$	1	$(N,0)$	$\frac{N-k}{2}$	$\frac{N}{3}-k$	$-\sqrt{\frac{3(k+1)(N-k)}{2N(N+3)}}$
$\frac{N-k+1}{2}$	$\frac{N}{3}-k+1$	$\frac{1}{2}$	-1	$(N,0)$	$\frac{N-k}{2}$	$\frac{N}{3}-k$	$\sqrt{\frac{3k(N-k+2)}{2N(N+3)}}$
$\frac{N-k+1}{2}$	$\frac{N}{3}-k+1$	$\frac{1}{2}$	-1	$(N-2,1)$	$\frac{N-k}{2}$	$\frac{N}{3}-k$	$\sqrt{\frac{(N-k)(N-k+2)}{N(N+2)}}$
$\frac{N-k}{2}$	$\frac{N}{3}-k$	$\frac{1}{2}$	1	$(N-2,1)$	$\frac{N-k-1}{2}$	$\frac{N}{3}-k+1$	$\sqrt{\frac{k}{N+2}}$

Table 4. Intensities of one-neutron transfer reactions between even-even and odd-neutron nuclei $I_n(\mathrm{ee}\to\mathrm{on}) = |\langle \mathrm{on} \parallel \tilde{P}_j \parallel \mathrm{ee}\rangle|^2$.

n	(λ,μ)	$(\sigma_1,\sigma_2,\sigma_3)$	$I_n(\mathrm{ee}\to\mathrm{on})$
1	$(\mathcal{N},0)$	$(\mathcal{N},0,0)$	$\frac{2(2j+1)(\mathcal{N}+4)\mathcal{N}_\nu}{5\mathcal{N}(\mathcal{N}+1)^2}\,\alpha_j^2$
2	$(\mathcal{N},0)$	$(\mathcal{N}-2,0,0)$	$\frac{(2j+1)(\mathcal{N}-2)(\mathcal{N}-1)(\mathcal{N}+3)\mathcal{N}_\nu}{10\mathcal{N}^2(\mathcal{N}+1)^2}\,\alpha_j^2$
3	$(\mathcal{N}-2,1)$	$(\mathcal{N}-1,1,0)$	$\frac{(2j+1)(\mathcal{N}-1)(\mathcal{N}+4)^2\mathcal{N}_\nu}{5\mathcal{N}(\mathcal{N}+1)(\mathcal{N}+2)}\,\alpha_j^2$
4	$(\mathcal{N}-2,1)$	$(\mathcal{N}-2,0,0)$	$\frac{(2j+1)(\mathcal{N}-2)^2(\mathcal{N}-1)(\mathcal{N}+3)\mathcal{N}_\nu}{10\mathcal{N}^2(\mathcal{N}+1)(\mathcal{N}+2)}\,\alpha_j^2$

It is convenient to take ratios of intensities, since they do not depend on the value of the coefficients α_j and hence provide a direct test of the wave functions. The transfer I_1 corresponds to the excitation of the ground state band of the final nucleus, whereas the other transfers involve the excited bands. The largest intensity is that of the final state with $(\sigma_1,\sigma_2,\sigma_3)=(\mathcal{N}-1,1,0)$ with relative intensity [17, 21]

$$\frac{I_3(\mathrm{ee}\to\mathrm{on})}{I_1(\mathrm{ee}\to\mathrm{on})} = \frac{(\mathcal{N}-1)(\mathcal{N}+1)(\mathcal{N}+4)}{2(\mathcal{N}+2)}\,, \tag{21}$$

which gives 29.3 for $^{194}\mathrm{Pt}\to{}^{195}\mathrm{Pt}$ ($\mathcal{N}_\pi=2$ and $\mathcal{N}_\nu=5$) to be compared to the experimental value of 19.0 for $j=5/2$, and 37.8 for $^{192}\mathrm{Os}\to{}^{193}\mathrm{Os}$ ($\mathcal{N}_\pi=3$ and $\mathcal{N}_\nu=5$). The equivalent ratio for the inverse reactions is given by

$$\frac{I_3(\mathrm{on}\to\mathrm{ee})}{I_1(\mathrm{on}\to\mathrm{ee})} = \frac{I_3(\mathrm{ee}\to\mathrm{on})}{I_1(\mathrm{ee}\to\mathrm{on})}\,\frac{\mathcal{N}_\pi}{\mathcal{N}_\nu(\mathcal{N}-1)}\,, \tag{22}$$

which gives 1.96 for $^{195}\mathrm{Pt}\to{}^{194}\mathrm{Pt}$ and 3.24 for $^{193}\mathrm{Os}\to{}^{192}\mathrm{Os}$. It is important to note that, whereas the final state with $(\sigma_1,\sigma_2,\sigma_3)=(\mathcal{N}-1,1,0)$ in the odd-neutron nucleus $^{195}\mathrm{Pt}$ represents the first excited doublet of states at an excitation energy of approximately 100 keV, in the even-even nucleus $^{194}\mathrm{Pt}$ this configuration corresponds to a highly excited $L=2$ state with mixed symmetry at an excitation energy of approximately 2.5 - 3.0 MeV [24]. Despite the reduction in relative intensity due to the correlation factor $\mathcal{N}_\pi/\mathcal{N}_\nu(\mathcal{N}-1)$, the mixed symmetry state is nevertheless predicted to be excited more strongly than the first excited $L=2$ state. This prediction could be tested in future transfer experiments.

Table 5. Intensities of one-neutron transfer reactions between odd-neutron and even-even nuclei $I_n(\text{on} \rightarrow \text{ee}) = |\langle \text{ee} \parallel P_j^\dagger \parallel \text{on}\rangle|^2$.

n	(λ, μ)	$(\sigma_1, \sigma_2, \sigma_3)$	$I_n(\text{on} \rightarrow \text{ee})$
1	$(\mathcal{N}, 0)$	$(\mathcal{N}, 0, 0)$	$I_1(\text{ee} \rightarrow \text{on})$
2	$(\mathcal{N}, 0)$	$(\mathcal{N}-2, 0, 0)$	$I_2(\text{ee} \rightarrow \text{on})$
3	$(\mathcal{N}-2, 1)$	$(\mathcal{N}-1, 1, 0)$	$I_3(\text{ee} \rightarrow \text{on}) \frac{\mathcal{N}_\pi}{\mathcal{N}_\nu(\mathcal{N}-1)}$
4	$(\mathcal{N}-2, 1)$	$(\mathcal{N}-2, 0, 0)$	$I_4(\text{ee} \rightarrow \text{on}) \frac{\mathcal{N}_\pi}{\mathcal{N}_\nu(\mathcal{N}-1)}$

4. Summary and conclusions

In this contribution, I discussed the concept of supersymmetry as used in nuclear physics. Nuclear supersymmetry establishes precise links among the spectroscopic properties of different nuclei belonging to a supermultiplet. Since the wave functions of the members of a supermultiplet are connected by symmetry, there exists a high degree of correlations between different one- and two-nucleon transfer reaction intensities. In order to discuss these correlations in an explicit manner, the concept of generalized F-spin was developed. As a consequence, the correlations can be expressed in an elegant way in terms of $SU(3)$ isoscalar factors. The relevant $SU(3)$ isoscalar factors were derived in a concise way by combining the usual methods based on recurrence relations with a proof by induction.

As an application, I discussed the correlations between one-neutron stripping and pick-up reactions. Nuclear supersymmetry predicts that the $L = 2$ mixed symmetry states in the even-even nuclei ^{194}Pt and ^{192}Os are excited much stronger (two to three times as strong) than the first excited $L = 2$ state. These predictions can be tested experimentally by combining $(\vec{d}, p)$ stripping and (p, d) pick-up reactions.

Finally, it is important to note, that the technique of the generalized F-spin is valid for any system whose wave functions are described in terms of couplings of three different symmetric representations of $U(6)$, as for example in nuclear supersymmetry (this contribution), in an algebraic description of the neutron skin in neutron-rich nuclei [25] and in an isospin-invariant extension of the IBM for light nuclei called IBM-3 [26].

Acknowledgments

It is a pleasure to thank José Barea, Alejandro Frank and Franco Iachello and Ruslán Magaña for interesting discussions on supersymmetric nuclei. This work was supported in part by PAPIIT-UNAM (grant IN113808).

References

[1] Iachello F 1980 *Phys. Rev. Lett.* **44** 772
[2] Iachello F and Arima A 1987 *The interacting boson model* (Cambridge: Cambridge University Press)
[3] Iachello F and Van Isacker P 1991 *The interacting boson-fermion model* (Cambridge: Cambridge University Press)
[4] F. Iachello and O. Scholten, *Phys. Rev. Lett.* **43**, 679 (1979).
[5] Balantekin A B, Bars I and Iachello F 1981 *Phys. Rev. Lett.* **47** 19
[6] Balantekin A B, Bars I, Bijker R and Iachello F 1983 *Phys. Rev.* C **27** 1761
[7] Sun H Z, Frank A and Van Isacker P 1983 *Phys. Rev.* C **27** 2430
[8] Van Isacker P, Jolie J, Heyde K and Frank A 1985 *Phys. Rev. Lett.* **54** 653
[9] Metz A, Jolie J, Graw G, Hertenberger R, Gröger J, Günther C, Warr N and Eisermann Y 1999 *Phys. Rev. Lett.* **83** 1542

[10] Metz A, Eisermann Y, Gollwitzer A, Hertenberger R, Valnion B D, Graw G and Jolie J 2000 *Phys. Rev.* C **61** 064313
[11] Gröger J, Jolie J, Krücken R, Beausang C W, Caprio M, Casten R F, Cederkall J, Cooper J R, Corminboeuf F, Genilloud L, Graw G, Günther C, De Huu M, Levon A I, Metz A, Novak J R, Warr N and Wendel T 2000 *Phys. Rev.* C **62** 064304
[12] Wirth H F, Graw G, Christen S, Eisermann Y, Gollwitzer A, Hertenberger R, Jolie J, Metz A, Möller O, Tonev D and Valnion B D 2004 *Phys. Rev.* C **70** 014610
[13] Barea J, Bijker R and Frank A 2004 *J. Phys. A: Math. Gen.* **37** 10251
[14] Barea J, Bijker R and Frank A 2005 *Phys. Rev. Lett.* **94** 152501
[15] Balodis M, Wirth H F, Graw G, Hertenberger R, Bērziņš J, Krāmere N, Jolie J, Christen S, Möller O, Tonev D, Barea J, Bijker R, Frank A and Von Egidy T 2008 *Phys. Rev.* C **77** 064602
[16] Jolie J and Garrett P 1996 *Nucl. Phys.* A **596** 234
[17] Barea J, Bijker R, Frank A, Graw G, Hertenberger R, Wirth H F, Christen S, Jolie J, Tonev D, Balodis M, Bērziņš J, Krāmere N and Von Egidy T 2009 *Phys. Rev.* C **79** 031304(R)
[18] Arima A, Otsuka T, Iachello F and Talmi I 1977 *Phys. Lett.* B **66** 205
[19] Bijker R, Barea J and Frank A 2009 *Rev. Mex. Fís.* S **55** (2) 30 [arXiv:0902.4863]
[20] Bijker R 2010 *J. Phys.: Conf. Ser.* **237** 012005
[21] Bijker R and Iachello F 1985 *Ann. Phys.* **161** 360
[22] De Swart J J 1963 *Rev. Mod. Phys.* **35** 916
[23] R. Magaña and R. Bijker, work in progress.
[24] Bijker R, Dieperink A E L, Scholten O and Spanhoff R 1980 *Nucl. Phys. A* **344** 207;
Von Brentano P *et al* 1996 *Phys. Rev. Lett.* **76** 2029
[25] Warner D D and Van Isacker P 1997 *Phys. Lett.* B **395** 145
[26] García-Ramos J E and Van Isacker P 1999 *Ann. Phys.* **274** 45

GROUP 28: Physical and Mathematical Aspects of Symmetry IOP Publishing
Journal of Physics: Conference Series **284** (2011) 012014 doi:10.1088/1742-6596/284/1/012014

Group-Theoretical Revision of the Unruh Effect

M. Calixto[1,2], E. Pérez-Romero[2] and V. Aldaya[2]

[1] Departamento de Matemática Aplicada y Estadística, Universidad Politécnica de Cartagena, Paseo Alfonso XIII 56, 30203 Cartagena, Spain

[2] Instituto de Astrofísica de Andalucía (IAA-CSIC), Apartado Postal 3004, 18080 Granada, Spain

E-mail: Manuel.Calixto@upct.es

Abstract. We revise the Unruh effect (vacuum radiation in uniformly relativistic accelerated frames) in a group-theoretical setting by constructing a conformal SO(4,2)-invariant quantum field theory and its spontaneous breakdown when selecting Poincaré invariant degenerated vacua (namely, coherent states of conformal zero modes). Special conformal transformations (accelerations) destabilize the Poincaré vacuum and make it to radiate.

1. Introduction

The Fulling-Davies-Unruh effect [1, 2, 3] has to do with *vacuum radiation* in a non-inertial reference frame and shares some features with the (black-hole) Hawking [4] effect. In simple words, whereas the Poincaré invariant vacuum $|0\rangle$ in QFT looks the same to any inertial observer (i.e., it is stable under Poincaré transformations), it converts into a thermal bath of radiation with temperature

$$T = \frac{\hbar a}{2\pi v k_B} \tag{1}$$

in passing to a uniformly accelerated frame (a denotes the acceleration, v the speed of light and k_B the Boltzmann constant).

This situation is always present when quantizing field theories in curved space as well as in flat space, whenever some kind of global mutilation of the space is involved (viz, existence of horizons). This is the case of the natural quantization in Rindler coordinates [2, 5], which leads to a quantization inequivalent to the normal Minkowski quantization (see next Section), or that of a quantum field in a box, where a dilatation produces a rearrangement of the vacuum [1].

In the reference [6], it was showed that the reason for the Planckian radiation of the Poincaré invariant vacuum under uniform accelerations (that is, the Unruh effect) is more profound and related to the spontaneous breakdown of the conformal symmetry in *quantum* field theory. From this point of view, a Poincaré invariant vacuum will be regarded as a coherent state of *conformal zero modes*, which are undetectable ("dark") by inertial observers but unstable under special conformal transformations

$$x^\mu \to x'^\mu = \frac{x^\mu + c^\mu x^2}{1 + 2cx + c^2x^2}, \tag{2}$$

which can be interpreted as transitions to systems of relativistic, uniformly accelerated observers with acceleration $a = 2c$ (see e.g. Ref. [7, 8, 9] and later on Eq. (14)). In the reference [6] a quite involved "second quantization formalism on a group G" was developed, which was applied to the conformal group in (1+1) dimensions, $SO(2,2) \simeq SO(2,1) \times SO(2,1)$, which consists of two copies of the pseudo-orthogonal group $SO(2,1)$ (left- and right-moving modes, respectively). Here we shall use more conventional methods of quantization and we shall work in realistic (3+1) dimensions, using the (more involved) conformal group $SO(4,2) \simeq SU(2,2)/\mathbb{Z}_4$.

The point of view exposed in this paper is consistent with the idea that quantum vacua are not really empty to every observer. Actually, the quantum vacuum is filled with zero-point fluctuations of quantum fields. The situation is similar to quantum many-body condensed mater systems describing, for example, superfluidity and superconductivity, where the ground state mimics the quantum vacuum in many respects and quasi-particles (particle-like excitations above the ground state) play the role of matter. Moreover, we know that zero-point energy, like other non-zero vacuum expectation values, leads to observable consequences as, for instance, the Casimir effect, and influences the behavior of the Universe at cosmological scales, where the vacuum (dark) energy is expected to contribute to the cosmological constant, which affects the expansion of the universe (see e.g. [10] for a nice review). Indeed, dark energy is the most popular way to explain recent observations that the universe appears to be expanding at an accelerating rate.

The organization of the paper is as follows. In Section 2 we briefly review the standard explanation for the Unruh effect, which has to do with space-time mutilation and Bogolyubov transformations. In Section 3 we construct a conformal-invariant quantum theory in 3+1 dimensions and in Section 4 we discuss its (spontaneous) breakdown to a Poincaré-invariant quantum theory by selecting Poincaré-invariant pseudo-vacua which are coherent states of conformal zero modes. We compute the mean energy of the accelerated Poincaré-invariant pseudo-vacua. This is part of a work in preparation [11], where the reader will find more details and additional results.

2. Vacuum radiation as a consequence of space-time mutilation

The existence of event horizons in passing to accelerated frames of reference leads to unitarily inequivalent representations of the quantum field canonical commutation relations and to a (ill-)definition of particles depending on the state of motion of the observer.

2.1. Field decompositions and vacua

To use an explicit example, let us consider a real scalar massless field $\phi(x)$, satisfying the Klein-Gordon equation

$$\eta^{\mu\nu}\partial_\mu\partial_\nu\phi(x) = 0\,. \tag{3}$$

Let us denote by a_k, a_k^* the Fourier coefficients of the decomposition of ϕ into positive and negative frequency modes:

$$\phi(x) = \int dk(a_k f_k(x) + a_k^* f_k^*(x)). \tag{4}$$

The Fourier coefficients a_k, a_k^* are promoted to annihilation and creation operators $\hat{a}_k, \hat{a}_k^*$ of particles in the quantum field theory. The Minkowski vacuum $|0\rangle_M$ is defined as the state nullified by all annihilation operators

$$\hat{a}_k|0\rangle_M = 0,\ \forall k. \tag{5}$$

2.2. Rindler coordinate transformations

Let us consider now the Rindler coordinate transformation (see e.g. [5]):

$$t = a^{-1}e^{az'}\sinh(at'),\quad z = a^{-1}e^{az'}\cosh(at'). \tag{6}$$

Journal of Physics: Conference Series **284** (2011) 012014 doi:10.1088/1742-6596/284/1/012014

The worldline $z' = 0$ has constant acceleration a (in natural unities). This transformation entails a mutilation of Minkowski spacetime into patches or charts with event horizons.

The new coordinate system provides a new decomposition of ϕ into Rindler positive and negative frequency modes:

$$\phi(x') = \int dq(a'_q f'_q(x') + a'^*_q f'^*_q(x')). \tag{7}$$

The Rindler vacuum $|0\rangle_R$ is defined as the state nullified by all Rindler annihilation operators:

$$\hat{a}'_q|0\rangle_R = 0\,\forall q. \tag{8}$$

Let us see that the Minkowki vacuum $|0\rangle_M$ and the Rindler vacuum $|0\rangle_R$ are not identical. In fact, the Minkowski vacuum $|0\rangle_M$ has a nontrivial content of Rindler particles.

2.3. Bogolyubov transformations

The Fourier components a'_q, a'^*_q of the field ϕ in the new (accelerated) reference frame are expressed in terms of both a_k, a^*_k through a Bogolyubov transformation:

$$\begin{aligned} a'_q &= \int dk\,(\alpha_{qk} a_k + \beta_{qk} a^*_k)\,, \\ \alpha_{qk} &= \langle f'_q | f_k \rangle, \;\; \beta_{qk} = \langle f'_q | f^*_k \rangle. \end{aligned} \tag{9}$$

The vacuum states $|0\rangle_M$ and $|0\rangle_R$, defined by the conditions (5) and (8), are not identical if the coefficients β_{qk} in (9) are not zero. In this case the Minkowski vacuum has a non-zero average number of Rindler particles given by:

$$N_R = \langle 0|\hat{N}_R|0\rangle_M = \langle 0| \int dq \hat{a}'^\dagger_q \hat{a}'_q |0\rangle_M = \int dk dq |\beta_{qk}|^2 \tag{10}$$

That is, in the second quantized theory, the vacuum states $|0\rangle_M$ and $|0\rangle_R$ are not identical if the coefficients β_{qk} are not zero. Both quantizations are inequivalent.

3. Vacuum radiation as a spontaneous breakdown of de conformal symmetry

In this section we shall offer an alternative explanation for the Unruh effect based on symmetry grounds. Actually, in Quantum Field Theory, the vacuum state is expected to be stable under some underlying group of symmetry transformations G (namely, the Poincaré group). Then the action of some spontaneously broken symmetry transformations can destabilize/excitate the vacuum and make it to *radiate*. We shall see that this is precisely the case of the Planckian radiation of the Poincaré invariant vacuum under uniform accelerations. Here, the Poincaré invariant vacuum looks the same to every inertial observer but converts itself into a thermal bath of radiation with temperature (1) in passing to a uniformly accelerated frame. In fact, in the reference [6], it was shown that the reason for this radiation is related to the spontaneous breakdown of the conformal symmetry in *quantum* field theory.

3.1. The conformal group and the compactified Minkowski space

The conformal group in 3+1 dimensions, $SO(4,2)$, is composed by Poincaré (spacetime translations $b^\mu \in \mathbb{R}^4$ and Lorentz $\Lambda^\mu_\nu (\in SO(3,1))$ transformations augmented by dilations ($e^\tau \in \mathbb{R}_+$) and relativistic uniform accelerations (special conformal transformations, $c^\mu \in \mathbb{R}^4$) which, in Minkowski spacetime, have the following realization:

$$\begin{aligned} x'^\mu &= x^\mu + b^\mu, \;\; x'^\mu = \Lambda^\mu_\nu(\omega) x^\nu, & (11) \\ x'^\mu &= e^\tau x^\mu, \;\; x'^\mu = \frac{x^\mu + c^\mu x^2}{1 + 2cx + c^2 x^2}, & (12) \end{aligned}$$

respectively. The interpretation of special conformal transformations

$$x'^{\mu} = \frac{x^{\mu} + c^{\mu}x^2}{1 + 2cx + c^2x^2} \tag{13}$$

as transitions from inertial reference frames to systems of relativistic, uniformly accelerated observers was identified many years ago by [7, 8, 9], although alternative meanings have also been proposed [12, 13]. For simplicity, let use take an acceleration along the "z" axis: $c^{\mu} = (0, 0, 0, \alpha)$, and the temporal path $x^{\mu} = (t, 0, 0, 0)$. Then the transformation (13) reads:

$$t' = \frac{t}{1 - \alpha^2 t^2}, \qquad z' = \frac{\alpha t^2}{1 - \alpha^2 t^2}. \tag{14}$$

Writing z' in terms of t' gives the usual formula for the relativistic uniform accelerated (hyperbolic) motion:

$$z' = \frac{1}{a}(\sqrt{1 + a^2 t'^2} - 1)$$

with $a = 2\alpha$.

The infinitesimal generators (vector fields) of the transformations (12) are easily deduced:

$$\begin{array}{rcl} P_{\mu} &=& \frac{\partial}{\partial x^{\mu}}, \quad M_{\mu\nu} = x_{\mu}\frac{\partial}{\partial x^{\nu}} - x_{\nu}\frac{\partial}{\partial x^{\mu}}, \\ D &=& x^{\mu}\frac{\partial}{\partial x^{\mu}}, \quad K_{\mu} = -2x_{\mu}x^{\nu}\frac{\partial}{\partial x^{\nu}} + x^2\frac{\partial}{\partial x^{\mu}} \end{array} \tag{15}$$

and they close into the conformal Lie algebra

$$\begin{array}{rcl} [M_{\mu\nu}, M_{\rho\sigma}] &=& \eta_{\nu\rho}M_{\mu\sigma} + \eta_{\mu\sigma}M_{\nu\rho} - \eta_{\mu\rho}M_{\nu\sigma} - \eta_{\nu\sigma}M_{\mu\rho}, \\ [P_{\mu}, M_{\rho\sigma}] &=& \eta_{\mu\rho}P_{\sigma} - \eta_{\mu\sigma}P_{\rho}, \quad [P_{\mu}, P_{\nu}] = 0, \\ [K_{\mu}, M_{\rho\sigma}] &=& \eta_{\mu\rho}K_{\sigma} - \eta_{\mu\sigma}K_{\rho}, \quad [K_{\mu}, K_{\nu}] = 0, \\ [D, P_{\mu}] &=& -P_{\mu}, \quad [D, K_{\mu}] = K_{\mu}, \quad [D, M_{\mu\nu}] = 0, \\ [K_{\mu}, P_{\nu}] &=& 2(\eta_{\mu\nu}D + M_{\mu\nu}). \end{array} \tag{16}$$

The quadratic Casimir operator has the following expression:

$$C_2 = -2D^2 + M_{\mu\nu}M^{\mu\nu} - P_{\mu}K^{\mu} - K_{\mu}P^{\mu} = -2D^2 + M_{\mu\nu}M^{\mu\nu} - 2P_{\mu}K^{\mu} - 8D. \tag{17}$$

Any group element $g \in SO(4, 2)$ (near the identity element 1) could be written as the exponential map

$$g = \exp(u), \; u = \tau D + b^{\mu}P_{\mu} + a^{\mu}K_{\mu} + \omega^{\mu\nu}M_{\mu\nu}, \tag{18}$$

of the Lie-algebra element u.

One would be tempted to blame the singular character of the special conformal transformation (13,14) to be responsible for the radiation effect, in much the same way the (singular) Rindler transformations (6) are supposedly responsible for the Unruh effect (i.e., existence of horizons). However, one could always work with the compactified Minkowski space $\mathbb{M} = \mathbb{S}^3 \times_{\mathbb{Z}_2} \mathbb{S}^1 \simeq U(2)$, which can be obtained as a coset $\mathbb{M} = SO(4, 2)/\mathbb{W}$, where $\mathbb{W}$ denotes the Weyl subgroup generated by $K_{\mu}, M_{\mu\nu}$ and D (i.e., a Poincaré subgroup $\mathbb{P} = SO(3, 1)$Ⓢ$\mathbb{R}^4$ augmented by the dilations $\mathbb{R}^+$). The Weyl group $\mathbb{W}$ is the stability subgroup (the little group in physical usage) of $x^{\mu} = 0$. Now the conformal group acts transitively on $\mathbb{M}$ and free from singularities. So, where would the radiation come from now?. Let us see that the reason for vacuum radiation in relativistic uniformly accelerated frames is more profound an related to the spontaneous breakdown of the conformal symmetry.

3.2. SU(2,2) as the four-cover of SO(4,2)
Instead of $SO(4,2)$, we shall work by convenience with its four cover:

$$SU(2,2) = \left\{ g = \begin{pmatrix} A & B \\ C & D \end{pmatrix} \in \mathrm{Mat}_{4\times4}(\mathbb{C}) : g^\dagger \Gamma g = \Gamma, \det(g) = 1 \right\}, \tag{19}$$

where Γ denotes a hermitian form of signature $(++,--)$.

The conformal Lie algebra (16) can also be realized in terms of gamma matrices in, for instance, the Weyl basis

$$\gamma^\mu = \begin{pmatrix} 0 & \sigma^\mu \\ \check{\sigma}^\mu & 0 \end{pmatrix}, \quad \gamma^5 = i\gamma^0\gamma^1\gamma^2\gamma^3 = \begin{pmatrix} -\sigma^0 & 0 \\ 0 & \sigma^0 \end{pmatrix}, \tag{20}$$

where $\check{\sigma}^\mu \equiv \sigma_\mu$ (we are using the convention $\eta = \mathrm{diag}(1,-1,-1,-1)$) and σ^μ are the standard Pauli matrices

$$\sigma^0 = I = \begin{pmatrix} 1 & 0 \\ 0 & 1 \end{pmatrix}, \sigma^1 = \begin{pmatrix} 0 & 1 \\ 1 & 0 \end{pmatrix}, \sigma^2 = \begin{pmatrix} 0 & -i \\ i & 0 \end{pmatrix}, \sigma^3 = \begin{pmatrix} 1 & 0 \\ 0 & -1 \end{pmatrix}. \tag{21}$$

Indeed, the choice

$$\begin{aligned} D &= \frac{\gamma^5}{2}, \; M^{\mu\nu} = \frac{[\gamma^\mu, \gamma^\nu]}{4} = \frac{1}{4}\begin{pmatrix} \sigma^\mu\check{\sigma}^\nu - \sigma^\nu\check{\sigma}^\mu & 0 \\ 0 & \check{\sigma}^\mu\sigma^\nu - \check{\sigma}^\nu\sigma^\mu \end{pmatrix}, \\ P^\mu &= \gamma^\mu\frac{1+\gamma^5}{2} = \begin{pmatrix} 0 & \sigma^\mu \\ 0 & 0 \end{pmatrix}, \; K^\mu = \gamma^\mu\frac{1-\gamma^5}{2} = \begin{pmatrix} 0 & 0 \\ \check{\sigma}^\mu & 0 \end{pmatrix} \end{aligned} \tag{22}$$

fulfils the commutation relations (16). These are the Lie algebra generators of the fundamental representation of $SU(2,2)$.

The group $SU(2,2)$ acts transitively on the compactified Minkowski space $\mathbb{M}_4 = U(2)$, with (matrix) coordinates X, as

$$X \to X' = (AX+B)(CX+D)^{-1}. \tag{23}$$

Setting $X = x_\mu\sigma^\mu$ (with σ^μ Pauli matrices) the transformations (12) can be recovered from (23) as follows:

i) Standard Lorentz transformations, $x'^\mu = \Lambda^\mu_\nu(\omega)x^\nu$, correspond to $B = C = 0$ and $A = D^{-1\dagger} \in SL(2,\mathbb{C})$.
ii) Dilations correspond to $B = C = 0$ and $A = D^{-1} = a^{1/2}\sigma^0$
iii) Spacetime translations equal $A = D = \sigma^0$, $C = 0$ and $B = b_\mu\sigma^\mu$.
iv) Special conformal transformations correspond to $A = D = \sigma^0$ and $C = c_\mu\sigma^\mu, B = 0$ by noting that $\det(CX+I) = 1 + 2cx + c^2x^2$:

$$X' = X(CX+I)^{-1} \leftrightarrow x'^\mu = \frac{x^\mu + c^\mu x^2}{1 + 2cx + c^2x^2}$$

3.3. Unirreps of de conformal group: discrete series
We shall consider the complex extension of $\mathbb{M}_4 = U(2)$ to the 8-dimensional conformal (phase) space:

$$\mathbb{D}_4 = U(2,2)/U(2)^2 = \{Z \in \mathrm{Mat}_{2\times2}(\mathbb{C}) : I - ZZ^\dagger > 0\}$$

and the Unirrep

$$[U_\lambda(g)\phi](Z) = |CZ+D|^{-\lambda}\phi(Z'),\ Z' = (AZ+B)(CZ+D)^{-1} \tag{24}$$

on the space $\mathcal{H}_\lambda(\mathbb{D}_4)$ of square-integrable holomorphic functions ϕ with invariant integration measure

$$d\mu_\lambda(Z,Z^\dagger) = \pi^{-4}(\lambda-1)(\lambda-2)^2(\lambda-3)\det(I-ZZ^\dagger)^{\lambda-4}|dZ|,$$

where the label $\lambda \in \mathbb{Z}, \lambda \geq 4$ is the conformal, scale or mass dimension ($|dZ|$ denotes the Lebesgue measure in $\mathbb{C}^4$).

3.4. *The Hilbert space of our conformal particle*

It has been proved in [14] that the infinite set of homogeneous polynomials

$$\varphi^{j,m}_{q_1,q_2}(Z) = \sqrt{\frac{2j+1}{\lambda-1}\binom{m+\lambda-2}{\lambda-2}\binom{m+2j+\lambda-1}{\lambda-2}}\det(Z)^m\mathcal{D}^j_{q_1,q_2}(Z), \tag{25}$$

with

$$\mathcal{D}^j_{q_1,q_2}(Z) = \sqrt{\frac{(j+q_1)!(j-q_1)!}{(j+q_2)!(j-q_2)!}}\sum_{p=\max(0,q_1+q_2)}^{\min(j+q_1,j+q_2)}\binom{j+q_2}{p}\binom{j-q_2}{p-q_1-q_2}z_{11}^p z_{12}^{j+q_1-p}z_{21}^{j+q_2-p}z_{22}^{p-q_1-q_2} \tag{26}$$

the standard Wigner's $\mathcal{D}$-matrices ($j \in \mathbb{N}/2$), verifies the following closure relation (the reproducing Bergman kernel):

$$\sum_{j\in\mathbb{N}/2}\sum_{m=0}^{\infty}\sum_{q_1,q_2=-j}^{j}\overline{\varphi^{j,m}_{q_1,q_2}(Z)}\varphi^{j,m}_{q_1,q_2}(Z') = \frac{1}{\det(I-Z^\dagger Z')^\lambda} \tag{27}$$

and constitutes an orthonormal basis of $\mathcal{H}_\lambda(\mathbb{D}_4)$, (the sum on j accounts for all non-negative half-integer numbers)

3.5. *Hamiltonian and energy spectrum*

In [15] we have argued that the dilation operator D plays the role of the Hamiltonian of our quantum theory. Actually, the replacement of time translations by dilations as dynamical equations of motion has already been considered in [16] and in [17], when quantizing field theories on space-like Lorentz-invariant hypersurfaces $x^2 = x^\mu x_\mu = \tau^2$ =constant. In other words, if one wishes to proceed from one surface at $x^2 = \tau_1^2$ to another at $x^2 = \tau_2^2$, this is done by scale transformations; that is, D is the evolution operator in a proper time τ.

From the general expression (24), we can compute the finite left-action of dilations $g = e^{\tau D}$ ($B = 0 = C$ and $A = e^{-\tau/2}\sigma^0 = D^{-1}$) on wave functions,

$$[U_\lambda(g)\phi](Z) = e^{\lambda\tau}\phi(e^\tau Z). \tag{28}$$

The infinitesimal generator of this transformation is the Hamiltonian operator:

$$H = \lambda + \sum_{i,j=1}^{2} Z_{ij}\frac{\partial}{\partial Z_{ij}} = \lambda + z_\mu\frac{\partial}{\partial z_\mu}. \tag{29}$$

The set of functions (25) constitutes a basis of Hamiltonian eigenfunctions (homogeneous polynomials) with energy eigenvalues E_n^λ (the homogeneity degree) given by:

$$H\varphi^{j,m}_{q_1,q_2} = E_n^\lambda\varphi^{j,m}_{q_1,q_2},\ E_n^\lambda = \lambda+n,\ n = 2j+2m. \tag{30}$$

Actually, each energy level E_n^λ is $(n+1)(n+2)(n+3)/6$ times degenerated. This degeneracy coincides with the number of linearly independent polynomials $\prod_{i,j=1}^2 Z_{ij}^{n_{ij}}$ of fixed degree of homogeneity $n = \sum_{i,j=1}^2 n_{ij}$. This proves that the set of polynomials (25) is a basis for analytic functions $\phi \in \mathcal{H}_\lambda(\mathbb{M})$. The spectrum is equi-spaced and bounded from below, with ground state $\varphi_{0,0}^{0,0} = 1$ and zero-point energy $E_0^\lambda = \lambda$.

3.6. The ground state is Poincaré-stable and polarized by accelerations

Under a general conformal transformation, the excited ground state is:

$$\tilde{\varphi}_{0,0}^{0,0}(Z) = [U_\lambda(g)\varphi_{0,0}^{0,0}](Z) = \det(CZ+D)^{-\lambda}. \tag{31}$$

Therefore, for Poincaré transformations we have $C = 0$ and $\det(D) = 1$, which means that the ground state $\varphi_{0,0}^{0,0}$ looks the same to every inertial observer (it is stable for zero acceleration $C = 0$, i.e., it is Poincaré invariant). We shall restrict from now on to pure accelerations: $D = A = I, B = 0$ and $C = c_\mu \sigma^\mu$. For arbitrary accelerations, $C \neq 0$, we can decompose the accelerated ground state $\tilde{\varphi}_{0,0}^{0,0}$ using the Bergman kernel expansion (27) as:

$$\tilde{\varphi}_{0,0}^{0,0}(Z) = \sum_{j\in\mathbb{N}/2}\sum_{m=0}^{\infty}\sum_{q_1,q_2=-j}^{j} \varphi_{q_2,q_1}^{j,m}(-C)\varphi_{q_1,q_2}^{j,m}(Z). \tag{32}$$

From here, we interpret the coefficient $\varphi_{q_2,q_1}^{j,m}(-C)$ as the probability amplitude of finding the accelerated ground state in the excited level $\varphi_{q_1,q_2}^{j,m}$ of energy $E_n^\lambda = \lambda + 2j + 2m = \lambda + n$

3.7. Mean Energy of the accelerated ground state

The energy expectation value in the accelerated ground state (32) can be calculated as:

$$\mathcal{E}(C) = \frac{\sum_{j=0}^{\infty}\sum_{n=0}^{\infty}\sum_{q_1,q_2=-j}^{j} |\varphi_{q_1,q_2}^{j,n}(C)|^2(\lambda+2j+2n)}{\sum_{j=0}^{\infty}\sum_{n=0}^{\infty}\sum_{q_1,q_2=-j}^{j} |\varphi_{q_1,q_2}^{j,n}(C)|^2}. \tag{33}$$

Using (27) and its derivatives in (30), with the Hamiltonian operator given by (29), we obtain:

$$\begin{aligned}\sum_{j=0}^{\infty}\sum_{n=0}^{\infty}\sum_{q_1,q_2=-j}^{j} |\varphi_{q_1,q_2}^{j,n}(C)|^2 &= \frac{1}{\det(I-C^\dagger C)^\lambda}\\ \sum_{j=0}^{\infty}\sum_{n=0}^{\infty}\sum_{q_1,q_2=-j}^{j} |\varphi_{q_1,q_2}^{j,n}(C)|^2(\lambda+2j+2n) &= \lambda\frac{1-\det(C^\dagger C)}{\det(I-C^\dagger C)^{\lambda+1}},\end{aligned} \tag{34}$$

so that the mean energy is:

$$\mathcal{E}(C) = \lambda\frac{1-\det(C^\dagger C)}{\det(I-C^\dagger C)}. \tag{35}$$

For the particular case of an acceleration along the "z" axis, $C = c_\mu\sigma^\mu = c\sigma^3$, we have:

$$\mathcal{E}(c) = \lambda\frac{1-c^4}{1-2c^2+c^4} = \lambda\frac{1+c^2}{1-c^2} = \lambda + 2\lambda\frac{c^2}{1-c^2}. \tag{36}$$

The mean energy (36) is of Planckian type for

$$c^2(T) \equiv e^{-\frac{h\nu}{k_B T}}, \tag{37}$$

where we have introduced dimensions, with $h\nu$ the quantum of energy for our harmonic oscillator. The last (*ad hoc*) assignment (37) can be in fact obtained from first thermodynamical principles (see [11] for the actual computation of the partition function, entropy and temperature). Here we shall just point out that the formula (37) gives

$$T = -T_E/\ln(c^2),\ T_E \equiv h\nu/k_B, \tag{38}$$

which differs from the Unruh formula (1). It also implies the existence of a *maximal acceleration* $a_{\max}$, so that the dimension-full acceleration is $a = a_{\max}c$ (see [18] for a discussion on the physical consequences of a maximal acceleration and [15] and [11] for more details).

4. Second-Quantized Theory, Conformal Zero Modes and θ-Vacua

We have discussed the effect of relativistic accelerations in first (one particle) quantization. However, the proper setting to analyze radiation effects is in the second quantized theory. Let us denote (for space-saving notation) by $n = \{j, m, q_1, q_2\}$ the multi-index of the basis functions (25). The Fourier coefficients a_n (and a_n^*) of the expansion in energy modes of a state

$$\phi = \sum_n a_n \varphi_n, \tag{39}$$

are promoted to annihilation $\hat{a}_n$ (and creation $\hat{a}_n^\dagger$) operators in second quantization. The fact that the ground state of first quantization, φ_0, is invariant under Poincaré transformations implies that $\hat{a}_0$ commutes with all Poincaré generators. It also commutes with all annihilation operators and creation operators

$$[\hat{a}_0, \hat{a}_n^\dagger] = 0,\ n > 0 \tag{40}$$

of particles with non-zero energy. Therefore, by Schur's Lemma, $\hat{a}_0$ must behave as a multiple of the identity in the broken theory, which means that Poincaré "θ-vacua" fulfil

$$\hat{a}_0|\theta\rangle = \theta|\theta\rangle \Rightarrow |\theta\rangle = e^{\theta\hat{a}_0 - \bar{\theta}\hat{a}_0^\dagger}|0\rangle. \tag{41}$$

That is, Poincaré "θ-vacua" are *coherent states of conformal zero modes* (see [19] and [20] for a thorough exposition on coherent states).

The second-quantized version of (32) for an acceleration $C = c\sigma^3$ along the third axis is:

$$\hat{a}'_0 = \sum_{n=0}^{\infty} \varphi_n(c)\hat{a}_n, \tag{42}$$

so that accelerated Poincaré θ-vacua become:

$$|\theta'\rangle = e^{\theta\hat{a}'_0 - \bar{\theta}\hat{a}'^\dagger_0}|0\rangle. \tag{43}$$

The average number of particles with energy E_n in the accelerated vacuum (43) is then given by

$$N(c) = \langle\theta'|\hat{a}_n^\dagger\hat{a}_n|\theta'\rangle = |\theta|^2|\varphi_n(c)|^2, \tag{44}$$

where $|\theta|^2$ is the total average number of particles in $|\theta\rangle$, and $|\varphi_n(c)|^2$ is the probability of finding a particle in the energy state E_n of the accelerated vacuum $|\theta'\rangle$. One can also compute the mean energy per mode and see that it reproduces the Black-Body spectrum for (37). See more details in the coming paper [11].

Acknowledgements

Work partially supported by the Fundación Séneca (08814/PI/08), Spanish MICINN (FIS2008-06078-C03-01) and Junta de Andalucía (FQM219, FQM1951). M. Calixto thanks the "Universidad Politécnica de Cartagena" and C.A.R.M. for the award "Intensificación de la Actividad Investigadora 2009-2010". "Este trabajo es resultado de la ayuda concedida por la Fundación Séneca, en el marco del PCTRM 2007-2010, con financiación del INFO y del FEDER de hasta un 80%".

References

[1] S.A. Fulling, Nonuniqueness of Canonical Field Quantization in Riemannian Space-Time, Phys. Rev. **D7** (1973) 2850-2862
[2] P.C.W. Davies, Scalar production in Schwarzschild and Rindler metrics, J. Phys. **A8** (1975) 609-616
[3] W. G. Unruh, Notes on black-hole evaporation, Phys. Rev. **D14** (1976) 870
[4] S.W. Hawking, Black-hole explosions?, Nature **248** (1974) 30
[5] N. D. Birrell and P. C. W. Davies, Quantum Fields in Curved Space, Cambridge Monographs on Mathematical Physics (1994).
[6] V. Aldaya, M. Calixto and J.M. Cerveró, Commun. Math. Phys. **200** (1999) 325-354
[7] E. L. Hill, On accelerated coordinate systems in classical and relativistic mechanics, Phys. Rev. **67** (1945) 358-363.
[8] L. J. Boya and J. M. Cerveró, Contact Transformations and Conformal Group. I. Relativistic Theory, International Journal of Theoretical Physics **12** (1975) 47-54
[9] T. Fulton, F. Rohrlich and L. Witten, Physical consequences to a coordinate transformation to a uniformly accelerationg frame, Nuovo Cimento **26** (1962) 652-671
[10] G.E. Volovik: Vacuum energy: myths and reality, Int. J. Mod. Phys. **D15** (2006) 1987-2010
[11] M. Calixto, E. Pérez-Romero and V. Aldaya, The Unruh Effect and the Spontaneous Breakdown of the Conformal Symmetry, in preparation.
[12] H. Weyl, "Space, Time and Matter". Dover. NY. First Edition 1922.
[13] H.A. Kastrup, Gauge properties of the Minkowski space, Phys. Rev. **150**, 1183 (1966).
[14] M. Calixto and E. Pérez-Romero, Extended MacMahon-Schwinger's Master Theorem and Conformal Wavelets in Complex Minkowski Space, Applied and Computational Harmonic Analysis, (to appear), arXiv:1002.3498
[15] M. Calixto and E. Pérez-Romero, Conformal Spinning Quantum Particles in Complex Minkowski Space as Constrained Nonlinear Sigma Models in U(2,2) and Born's Reciprocity , arXiv:1006.5958, 33pp.
[16] V. Aldaya and J.A. de Azcárraga: Group manifold analysis of the structure of relativistic quantum dynamics, Annals of Physics **165** (1985) 484-504
[17] S. Fubini, A.J. Hanson and R. Jackiw: New approach to field theory, Phys. Rev. **D7** (1973) 1732-1760
[18] E.R. Caianiello, Is there a maximal aceleration?, Lettere al Nuovo Cimento **32** (1981) 65-70
[19] A. Perelomov: Generalized Coherent States and Their Apllications, Springer-verlag (1986)
[20] J.R. Klauder and Bo-Sture Skagerstam: Coherent States: Applications in Physics and Mathematical Physics, World Scientific (1985)

Journal of Physics: Conference Series **284** (2011) 012015 doi:10.1088/1742-6596/284/1/012015

Hidden quartic symmetry in $N = 2$ supersymmetry

Rutwig Campoamor-Stursberg†
† I.M.I-U.C.M, Plaza de Ciencias 3, E-28040 Madrid, Spain
E-mail: `rutwig@pdi.ucm.es`

Michel Rausch de Traubenberg‡
‡IPHC-DRS, UdS, CNRS, IN2P3, 23 rue du Loess, F-67037 Strasbourg Cedex, France
E-mail: `Michel.Rausch@IReS.in2p3.fr`

Abstract. It is shown that for $N = 2$ supersymmetry a hidden symmetry arises from the hybrid structure of a quartic algebra. The implications for invariant Lagrangians and multiplets are explored.

1. Introduction

The consideration of quadratic algebras – Lie (super-)algebras – largely dominates the algebraic formalism underlying theoretical physics, although non-quadratic structures emerge, in more or less natural form, in various descriptions of physical phenomena. For instance, generalized algebraic approaches are given by the n-linear algebras in Quantum Mechanics [1], ternary structures in the description of multiple M_2-branes [2, 3] or higher order extensions of the Poincaré algebra [4].

The notion of Lie algebras of order $F > 2$ was first considered and later intensively studied in [4, 5, 6, 7], motivated by the fact that these structures, beyond their formal mathematical properties, constitute a cornerstone for the construction of higher order extensions of the Poincaré algebra. In this context, a specific cubic extension in arbitrary space-time dimension was shown to be of interest in the frame of Quantum Field Theory [8, 9, 10]. Finally, it was realised that it was possible to associate a group [6] and an adapted superspace associated to these structures [11].

Probably, the main technical difficulty related to Lie algebras of order F is their hybrid structure: the "algebra" is partially quadratic and partially of order F. In this work, we would like to show that to graded Lie superalgebras of certain form, one can naturally associate a quartic algebra. It is shown that along the lines of this construction, one can associate to $N = 2$ supersymmetry a quartic extension of the Poincaré algebra. This construction indicates that some kind of hidden quartic symmetry appears in usual supersymmetry, which further means that invariant Lagrangians constructed so far are also invariant under the induced quartic structure.

We illustrate the fact that on the top of the representations of supersymmetry, a hierarchy of representations can be constructed. The work presented in this note was obtained in [12] in more detail.

2. Lie algebras of order four – quartic extensions of the Poincaré algebra

There are various types of extensions of Lie algebras that can be considered. The case under inspection here, enabling us to construct non-trivial extensions of the Poincaré algebra, are related to the quartic case, for which we recall the main properties.

The vector space $\mathfrak{g} = \mathfrak{g}_0 \oplus \mathfrak{g}_1$ with basis $\{X_i, i = 1, \cdots, \dim \mathfrak{g}_0\}$, $\{Y_a, a = 1, \cdots, \dim \mathfrak{g}_1\}$ is called an elementary Lie algebra of order four if it satisfies the following brackets [4]

$$\begin{aligned} \big[X_i, X_j\big] = f_{ij}{}^k X_k, \qquad & \big[X_i, Y_a\big] = R_{ia}{}^b Y_b, \\ \big\{Y_{a_i}, Y_{a_2}, Y_{a_3}, Y_{a_4}\big\} \ = \ & \sum_{\sigma \in S_4} Y_{\sigma(a_1)} Y_{\sigma(a_2)} Y_{\sigma(a_3)} Y_{\sigma(a_4)} = Q_{a_1 a_2 a_3 a_4}{}^i X_i, \end{aligned} \tag{1}$$

S_4 being the permutation group with four elements. In addition, we have also the following generalised Jacobi identities:

$$\begin{aligned} & \big[Y_{a_1}, \big\{Y_{a_2}, Y_{a_3}, Y_{a_4}, Y_{a_5}\big\}\big] + \big[Y_{a_2}, \big\{Y_{a_3}, Y_{a_4}, Y_{a_5}, Y_{a_1}\big\}\big] + \big[Y_{a_3}, \big\{Y_{a_4}, Y_{a_5}, Y_{a_1}, Y_{a_2}\big\}\big] + \\ & \qquad \big[Y_{a_4}, \big\{Y_{a_5}, Y_{a_1}, Y_{a_2}, Y_{a_3}\big\}\big] + \big[Y_{a_5}, \big\{Y_{a_1}, Y_{a_2}, Y_{a_3}, Y_{a_4}\big\}\big] = 0. \end{aligned} \tag{2}$$

Let us note that the structure defined by equations (1) and (2) is neither an algebra nor a 4−algebra in the usual sense, but a kind of hybrid structure. Some of the brackets will be quadratic $[\mathfrak{g}_0, \mathfrak{g}_0] \subseteq \mathfrak{g}_0, [\mathfrak{g}_0, \mathfrak{g}_1] \subseteq \mathfrak{g}_1$, while some others will be quartic $\{\mathfrak{g}_1, \mathfrak{g}_1, \mathfrak{g}_1, \mathfrak{g}_1\} \subseteq \mathfrak{g}_0$. This feature obviously generates the question whether from this hybrid structure we can extract some additional properties that cannot be codified either by the binary or quartic structure alone.

In the preceding context, the quartic extensions of the Poincaré algebra in $D = 4$ dimensions are realised by means of two Majorana spinors. Using the $\mathfrak{sl}(2, \mathbb{C}) \cong \mathfrak{so}(1, 3)$ notations of dotted and undotted indices, a left-handed spinor is given by $\psi_L{}^\alpha$ and a right-handed spinor by $\bar{\psi}_{R\dot{\alpha}}$. The spinor conventions to raise/lower indices are the following: $\psi_{L\alpha} = \varepsilon_{\alpha\beta}\psi_L{}^\beta$, $\psi_L{}^\alpha = \varepsilon^{\alpha\beta}\psi_{L\beta}$, $\bar{\psi}_{R\dot{\alpha}} = \varepsilon_{\dot{\alpha}\dot{\beta}}\bar{\psi}_R{}^{\dot{\beta}}$, $\bar{\psi}_R{}^{\dot{\alpha}} = \varepsilon^{\dot{\alpha}\dot{\beta}}\bar{\psi}_{R\dot{\beta}}$ with $(\psi_\alpha)^\star = \bar{\psi}_{\dot{\alpha}}$, $\varepsilon_{12} = \varepsilon_{\dot{1}\dot{2}} = 1$, $\varepsilon^{12} = \varepsilon^{\dot{1}\dot{2}} = -1$. The $4D$ Dirac matrices, in the Weyl representation, are

$$\Gamma^\mu = \begin{pmatrix} 0 & \sigma^\mu \\ \bar{\sigma}^\mu & 0 \end{pmatrix}, \tag{3}$$

with $\sigma^\mu{}_{\alpha\dot{\alpha}} = (1, \sigma^i)$, $\bar{\sigma}^{\mu\dot{\alpha}\alpha} = (1, -\sigma^i)$, σ^i $(i = 1, 2, 3)$ being the Pauli matrices. With these notations, we introduce two series of Majorana spinors $Q^I{}_\alpha, \bar{Q}_{I\dot{\alpha}}$ satisfying the relation $(Q^I{}_\alpha)^\dagger = \bar{Q}_{I\dot{\alpha}}$. The Lie algebra of order four with $\mathfrak{g}_0 = I\mathfrak{so}(1, 3)$ (the Poincaré algebra) and $\mathfrak{g}_1 = \langle Q^I{}_\alpha, \bar{Q}_{I\dot{\alpha}}\rangle$ define the following quartic extension of the Poincaré algebra (we only give the quartic brackets explicitly)

$$\begin{aligned} \big\{Q^{I_1}{}_{\alpha_1}, Q^{I_2}{}_{\alpha_2}, Q^{I_3}{}_{\alpha_3}, Q^{I_4}{}_{\alpha_4}\big\} &= 0, \\ \big\{Q^{I_1}{}_{\alpha_1}, Q^{I_2}{}_{\alpha_2}, Q^{I_3}{}_{\alpha_3}, \bar{Q}_{I_4\dot{\alpha}_4}\big\} &= 2i\Big(\delta^{I_1}{}_{I_4}\varepsilon^{I_2I_3}\varepsilon_{\alpha_2\alpha_3}\sigma^\mu{}_{\alpha_1\dot{\alpha}_4} + \delta^{I_2}{}_{I_4}\varepsilon^{I_1I_3}\varepsilon_{\alpha_1\alpha_3}\sigma^\mu{}_{\alpha_2\dot{\alpha}_4} \\ &+ \ \delta^{I_3}{}_{I_4}\varepsilon^{I_1I_2}\varepsilon_{\alpha_1\alpha_2}\sigma^\mu{}_{\alpha_3\dot{\alpha}_4}\Big)P_\mu, \\ \big\{Q^{I_1}{}_{\alpha_1}, Q^{I_2}{}_{\alpha_2}, \bar{Q}_{I_3\dot{\alpha}_3}, \bar{Q}_{I_4\dot{\alpha}_4}\big\} &= 0, \end{aligned} \tag{4}$$

the remaining brackets involving three $\bar{Q}$ and one Q or four $\bar{Q}$ being obtained immediately (the tensor ε^{IJ} is defined by $-\varepsilon^{12} = \varepsilon^{21} = \varepsilon_{12} = -\varepsilon_{21} = 1$).

As noted previously, the quartic extension of the Poincaré algebra obtained is neither an algebra nor a four-algebra. This feature represents one of the difficulties to handle with these algebraic structures. Consequently one natural question we should address concerns the possibility to associate appropriate quadratic structures to Lie algebras of order four. Its has to be mentioned that a similar analysis has been performed for Lie algebras of order three, were it has been shown that, no possibility to construct an associated quartic structure exists.

3. Quartic structures associated to Lie superalgebras

As mentioned earlier, higher order extensions are no fully satisfactory, in spite of various interesting results derived for them [8, 9, 10]). Thus one may wonder whether or not some quadratic structure should be related to the algebra (4). This question is partially motivated by the fact that some ternary algebras [14] of the Filippov type considered in the Bagger-Lambert-Gustavsson model are equivalent to certain Lie (super-)algebras [2, 13, 14].
We consider the $\mathbb{Z}_2 \times \mathbb{Z}_2$−graded Lie superalgebra

$$\mathfrak{g} = \big(\mathfrak{g}_{(0,0)} \oplus \mathfrak{g}_{(1,1)}\big) \oplus \big(\mathfrak{g}_{(1,0)} \oplus \mathfrak{g}_{(0,1)}\big), \tag{5}$$

where $(a,b) \in \mathbb{Z}_2 \times \mathbb{Z}_2$ and $\mathfrak{g}_{(a,b)}$ is even (resp. odd) when $a+b = 0 \bmod 2$ (resp. $a+b = 1 \bmod 2$). Introduce the corresponding bases for the grading blocks:

$$\begin{array}{ll} \mathfrak{g}_{(1,1)} = \langle B_i, i = 1, \cdots, \dim \mathfrak{g}_{(0,0)} \rangle, & \mathfrak{g}_{(0,0)} = \langle Z \rangle, \\ \mathfrak{g}_{(1,0)} = \langle F_a^+, a = 1, \cdots, \dim \mathfrak{g}_{(1,0)} \rangle, & \mathfrak{g}_{(0,1)} = \langle F_a^-, a = 1, \cdots, \dim \mathfrak{g}_{(0,1)} \rangle, \end{array} \tag{6}$$

the corresponding commutation relations are

$$\begin{array}{ll} \big[B_i, B_j\big] = f_{ij}{}^k B_k, & [B_i, Z] = 0, \\ \big[B_i, F_a^\pm\big] = R^\pm_i{}_a{}^b F_b^\pm, & \big[Z, F_a^\pm\big] = 0, \\ \big\{F_i^+, F_j^-\big\} = Q_{ij}{}^a B_a, & \big\{F_i^\pm, F_j^\pm\big\} = g^\pm_{ij} Z. \end{array} \tag{7}$$

We mention that, the superalgebra defined by (7) satisfies also appropriate Jacobi identities (that we do not recall here since they will not be relevant for our purpose). It is important to notice that $\mathfrak{g}_{(0,0)}$ commutes with all remaining factors, in other words, that Z acts like a central charge.

We now show that to the algebra (7) one can naturally and simply associate a quartic structure which share some similarities with the algebra (1). Indeed, using the obvious relation,

$$\{A_1, A_2, A_3, A_4\} = \{\{A_1, A_2\}, \{A_3, A_4\}\} + \{\{A_1, A_3\}, \{A_2, A_4\}\} + \{\{A_1, A_4\}, \{A_2, A_3\}\},$$

the relations

$$\begin{array}{rcl} \{F_{a_1}^+, F_{a_2}^+, F_{a_3}^+, F_{a_4}^+\} &=& \Big(g^+_{a_1a_2} g^+_{a_3a_4} + g^+_{a_1a_3} g^+_{a_2a_4} + g^+_{a_1a_4} g^+_{a_2a_3}\Big) Z^2 \\ \{F_{a_1}^+, F_{a_2}^+, F_{a_3}^+, F_{a_4}^-\} &=& 2Z(g^+_{a_1a_2} Q_{a_3a_4}{}^i + g^+_{a_1a_3} Q_{a_2a_4}{}^i + g^+_{a_3a_3} Q_{a_1a_4}{}^i) B_i \\ \{F_{a_1}^+, F_{a_2}^+, F_{a_3}^-, F_{a_4}^-\} &=& \Big(Q_{a_1a_3}{}^i Q_{a_2a_4}{}^j + Q_{a_1a_4}{}^i Q_{a_2a_3}{}^j \{B_i, B_j\} + 2g^+_{i_1i_2} g^-_{i_3i_4} Z^2 , \end{array} \tag{8}$$

(plus similar relations involving either three F^- and one F^+ or four F^-) follow at once.

Since we are constructing an analogue of the four-Lie algebra (1), we also assume that the algebra is partially quadratic and partially quartic. This means that in addition to the brackets (8), we have also to define the quadratic brackets $[\mathfrak{g}_0, \mathfrak{g}_0] \subseteq \mathfrak{g}_0, [\mathfrak{g}_0, \mathfrak{g}_1] \subseteq \mathfrak{g}_1$. We simply assume that these brackets are the same of the corresponding brackets of the Lie superalgebra. As the quartic brackets are concerned, we observe that $\{\mathfrak{g}_1, \mathfrak{g}_1, \mathfrak{g}_1, \mathfrak{g}_1\}$ close quadratically in $\mathfrak{g}_0$. The next step in the construction is to impose the Jacobi identities (2). This is an extra condition. Indeed, one can show that the Jacobi identity of the Lie superalgebras do not reproduce the generalised Jacobi identity (2). This will not lead to any contradiction since it happens that if we have a finite dimensional representation of (8), the identities (2) are trivially satisfied. Moreover, for the case under inspection in this work this will not be a constraint, since the generalised Jacobi identity will be trivially satisfied as well. This happens because the four-brackets $\{\mathfrak{g}_1, \mathfrak{g}_1, \mathfrak{g}_1, \mathfrak{g}_1\}$ close upon P_μ or Z (see below) thus we automatically have $[\{\mathfrak{g}_1, \mathfrak{g}_1, \mathfrak{g}_1, \mathfrak{g}_1\}, \mathfrak{g}_1] = 0$.

Finally since *by construction* the relations (8) are just a consequence of (7), this means that the Lie algebras of the form (7) present some hidden quartic symmetry. In some sense we could say that the algebra (8) is the "square" of the algebra (6). Furthermore, the fact that quadratic relations imply quartic relations means that any representation of the Lie superalgebra (6) will also be a (non-faithful) representation of the quartic algebra (8). Of course, the converse is not necessarily true.

We now focus on the relevant case of the $N = 2$ supersymmetric extension of the Poincaré algebra with central charge, and show that it is of the form (7). Indeed, for the even part, we define $\mathfrak{g}_{(1,1)} = I\mathfrak{so}(1,3) = \langle L_{\mu\nu}, P_\mu \rangle$ to be the Poincaré algebra in four-dimensions and $\mathfrak{g}_{(0,0)} = \langle Z \rangle$ the central charge. Although, for the odd part we introduce two series of Majorana spinors $Q^I{}_\alpha, \bar{Q}_{I\dot{\alpha}}, I = 1,2$ $((Q^I{}_\alpha)^\dagger = \bar{Q}_{I\dot{\alpha}})$ such that $\mathfrak{g}_{(1,0)} = \langle Q^I{}_\alpha \rangle$ and $\mathfrak{g}_{(0,1)} = \langle \bar{Q}_{I\dot{\alpha}} \rangle$. Since the $N = 2$ supersymmetric extension of the Poincaré algebra takes the form

$$\begin{aligned} \left\{Q^I{}_\alpha, \bar{Q}_{J\dot{\alpha}}\right\} &= -2i\delta^I{}_J \sigma^\mu{}_{\alpha\dot{\alpha}} P_\mu \\ \left\{Q^I{}_\alpha, Q^I{}_\beta\right\} &= 2Z\varepsilon^{IJ}\varepsilon_{\alpha\beta} \\ \left\{\bar{Q}_{I\dot{\alpha}}, \bar{Q}_{J\dot{\beta}}\right\} &= -2Z\varepsilon_{IJ}\varepsilon_{\dot{\alpha}\dot{\beta}}, \end{aligned} \tag{9}$$

which is analogous to (7), the results of the previous section give rise to the four order quartic extension of the Poincaré algebra

$$\begin{aligned} \{Q^{I_1}{}_{\alpha_1}, Q^{I_2}{}_{\alpha_2}, Q^{I_3}{}_{\alpha_3}, Q_{I_4\alpha_4}\} &= 2Z^2\Big(\varepsilon_{\alpha_1\alpha_2}\varepsilon_{\alpha_3\alpha_4}\varepsilon^{I_1I_2}\varepsilon^{I_3I_4} \\ &\quad +\varepsilon_{\alpha_1\alpha_3}\varepsilon_{\alpha_2\alpha_4}\varepsilon^{I_1I_3}\varepsilon^{I_2I_4} + \varepsilon_{\alpha_1\alpha_4}\varepsilon_{\alpha_2\alpha_3}\varepsilon^{I_1I_4}\varepsilon^{I_2I_3}\Big), \\ \{Q^{I_1}{}_{\alpha_1}, Q^{I_2}{}_{\alpha_2}, Q^{I_3}{}_{\alpha_3}, \bar{Q}_{I_4\dot{\alpha}_4}\} &= -2iZ\Big(\delta^{I_1}{}_{I_4}\varepsilon^{I_2I_3}\varepsilon_{\alpha_2\alpha_3}\sigma^\mu{}_{\alpha_1\dot{\alpha}_4} + \delta^{I_2}{}_{I_4}\varepsilon^{I_1I_3}\varepsilon_{\alpha_1\alpha_3}\sigma^\mu{}_{\alpha_2\dot{\alpha}_4} \\ &\quad + \delta^{I_3}{}_{I_4}\varepsilon^{I_1I_2}\varepsilon_{\alpha_1\alpha_2}\sigma^\mu{}_{\alpha_3\dot{\alpha}_4}\Big)P_\mu, \\ \{Q^{I_1}{}_{\alpha_1}, Q^{I_2}{}_{\alpha_2}, \bar{Q}_{I_3\dot{\alpha}_3}, \bar{Q}_{I_4\dot{\alpha}_4}\} &= 2\Big(\delta^{I_1}{}_{I_3}\delta^{I_2}{}_{I_4}\sigma^\mu{}_{\alpha_1\dot{\alpha}_3}\sigma^\nu{}_{\alpha_2\dot{\alpha}_4} + \delta^{I_1}{}_{I_4}\delta^{I_2}{}_{I_3}\sigma^\mu{}_{\alpha_1\dot{\alpha}_4}\sigma^\nu{}_{\alpha_2\dot{\alpha}_3}\Big)P_\mu P_\nu \\ &\quad + 2Z^2\varepsilon_{\alpha_1\alpha_2}\varepsilon_{\dot{\alpha}_3\dot{\alpha}_4}\varepsilon^{I_1I_2}\varepsilon_{I_3I_4}. \end{aligned}$$

A representation of the super-Poincaré algebra will automatically be a representation of the induced quartic algebra, as shown in the general case. This fact provides us with an interesting consequence, namely, that the invariant $N = 2$ Lagrangians constructed so far are moreover invariant with respect to the transformations induced by the quartic algebra. Thus, the corresponding $N = 2$ supermultiplets and their associated transformations laws will automatically be invariant multiplets of the corresponding quartic structure with the same transformation properties. This construction can thus be interpreted, in some sense, as a possibility to circumvent the constraints of the Haag-Lopuszanski-Sohnius theorem [16]. This analogy should however not be pushed too far: the construction of quartic algebras executed in this work depends essentially on the supersymmetric algebra formalism and the associated constraints.

Journal of Physics: Conference Series **284** (2011) 012015 doi:10.1088/1742-6596/284/1/012015

4. Representation of quartic extensions of the Poincaré algebra

As we have seen, the ansatz linking algebras of order four to Lie superalgebras has remarkable consequences concerning their respective representation theories, in the sense that superalgebra representations automatically induce representations of the order four structures. We point out that the converse of this statement is not true. Consider for instance massive representations. The little algebra is generated by $P^0 = -im$ and $Q^I{}_\alpha, \bar{Q}_{I\dot{\alpha}}$ and the four-brackets take the form

$$\begin{aligned}
\{Q_{\alpha_1}{}^{I_1}, Q_{\alpha_2}{}^{I_2}, Q_{\alpha_3}{}^{I_3}, Q_{\alpha_4}{}^{I_4}\} &= 2Z^2\Big(\varepsilon_{\alpha_1\alpha_2}\varepsilon_{\alpha_3\alpha_4}\varepsilon^{I_1I_2}\varepsilon^{I_3I_4} \\
&\quad +\varepsilon_{\alpha_1\alpha_3}\varepsilon_{\alpha_2\alpha_4}\varepsilon^{I_1I_3}\varepsilon^{I_2I_4} + \varepsilon_{\alpha_1\alpha_4}\varepsilon_{\alpha_2\alpha_3}\varepsilon^{I_1I_4}\varepsilon^{I_2I_3}\Big), \\
\{Q^{I_1}{}_{\alpha_1}, Q^{I_2}{}_{\alpha_2}, Q^{I_3}{}_{\alpha_3}, \bar{Q}_{I_4\dot{\alpha}_4}\} &= 2mZ\Big(\delta^{I_1}{}_{I_4}\varepsilon^{I_2I_3}\varepsilon_{\alpha_2\alpha_3}\sigma^0{}_{\alpha_1\dot{\alpha}_4} + \delta^{I_2}{}_{I_4}\varepsilon^{I_1I_3}\varepsilon_{\alpha_1\alpha_3}\sigma^0{}_{\alpha_2\dot{\alpha}_4} \\
&\quad + \delta^{I_3}{}_{I_4}\varepsilon^{I_1I_2}\varepsilon_{\alpha_1\alpha_2}\sigma^0{}_{\alpha_3\dot{\alpha}_4}\Big), \\
\{Q^{I_1}{}_{\alpha_1}, Q^{I_2}{}_{\alpha_2}, \bar{Q}_{I_3\dot{\alpha}_3}, \bar{Q}_{I_4\dot{\alpha}_4}\} &= 2m^2\Big(\delta^{I_1}{}_{I_3}\delta^{I_2}{}_{I_4}\sigma^0{}_{\alpha_1\dot{\alpha}_3}\sigma^0{}_{\alpha_2\dot{\alpha}_4} + \delta^{I_1}{}_{I_4}\delta^{I_2}{}_{I_3}\sigma^0{}_{\alpha_1\dot{\alpha}_4}\sigma^0{}_{\alpha_2\dot{\alpha}_3}\Big) \\
&\quad + 2Z^2\varepsilon_{\alpha_1\alpha_2}\varepsilon_{\dot{\alpha}_3\dot{\alpha}_4}\varepsilon^{I_1I_2}\varepsilon_{I_3I_4}.
\end{aligned}$$

If we now make the following substitutions (analogous to the corresponding substitution for the $N = 2$ supersymmetric extension with central charge):

$$\begin{aligned}
a^1 &= Q^1{}_1 - \bar{Q}_{2\dot{2}}\,, \quad a^3 = Q^1{}_1 + \bar{Q}_{2\dot{2}}\,, \\
a^2 &= Q^1{}_2 + \bar{Q}_{2\dot{1}}\,, \quad a^4 = Q^1{}_2 - \bar{Q}_{2\dot{1}}\,,
\end{aligned} \tag{10}$$

one observes that $a^1, \cdots, a^4, a^\dagger_1, \cdots, a^\dagger_4$ generate the Clifford algebra of the polynomial

$$\begin{aligned}
P^2(x_1, \cdots, x_4, y^1, \cdots, y^4) &= \Big(2(2m+Z)x_1y^1 + 2(2m+Z)x_2y^2 + 2(2m-Z)x_3y^3 \\
&\quad +2(2m-Z)x_4y^4\Big)^2
\end{aligned}$$

in the sense that

$$\Big(x_Ia^I + y^Ia^\dagger_I\Big)^4 = P^2(x_1, \cdots, x_4, y^1, \cdots, y^4). \tag{11}$$

The representations of the $N = 2$ supersymmetric algebra in four dimensions are obtained from the study of representations of the Clifford algebra, *i.e.* when the a's satisfy the quadratic relation,

$$\Big(x_Ia^I + y^Ia^\dagger_I\Big)^2 = P(x_1, \cdots, x_4, y^1, \cdots, y^4), \tag{12}$$

which is obviously compatible with (11). On the contrary, one can construct representations of (11) such that the condition (12) is not satisfied.[1] It can be shown that to any polynomial f, a Clifford algebra $\mathcal{C}_f$ can be associated to it, and that a matrix representation can be obtained [18]. Uniqueness is however lost for degree higher than two, which complicates considerably the analysis of representations (see, for instance, [19]). Being still an unsolved problem, some structural results have been already obtained in the general frame [20].

In the context that occupies us, this procedure may give new representations corresponding to interesting quartic extensions of the Poincaré algebra. The hierarchy of representations on the top of the standard representations obtained in supersymmetric theories might share some similarities with the parafermionic extension of the Poincaré algebra considered in [21].

[1] The algebra (11) is called the Clifford algebra of the polynomial P^2 [17].

5. Concluding remarks

We have pointed out the existence of a formal way to associate a quartic algebra (which closes with fully symmetric quartic brackets) to a graded Lie superalgebra of a certain type, their fundamental interest being their application to standard supersymmetric theories. This specifically alludes to the fact that any representation of $N = 2$ supersymmetric algebras shares a hidden symmetry arising from the quartic structure. Further, for massive representations, it turns out that the role of central charge is essential to the argument.

Clearly these results can be generalised to other space-time dimensions [12]. For example, it turns out that that the quartic extensions in ten space-time dimensions is exceptional and related to type IIA supersymmetry. Although we have focused on the case where $\mathfrak{g}_{(0,0)}$ is one dimensional, there is no reason for restricting to only one central charge. The straightforward generalisation to a higher number of charges is however subjected to finding the appropriate algebraic structures having a physical significance.

References

[1] Nambu Y 1973 *Phys. Rev. D* **7** 2405
Curtright T and Zachos C K 2003 *Phys. Rev. D* **68** 085001 [arXiv:hep-th/0212267].
[2] Bagger J and Lambert N 2008 *Phys. Rev. D* **7**, 065008 [arXiv:0711.0955 [hep-th]].
Bagger J and Lambert N 2009 *Phys. Rev. D* **79** 025002 [arXiv:0807.0163 [hep-th]].
Gustavsson A 2009 *Nucl. Phys. B* **811** 66 [arXiv:0709.1260 [hep-th]].
[3] Azcárraga J A de and Izquierdo J M 2010 *J. Phys. A: Math. Theor.* **43** 293001 [arXiv:math-ph/1005.1028v1]
[4] Rausch de Traubenberg M and Slupinski M J 2000 *J. Math. Phys.* **41** 4556 [arXiv:hep-th/9904126].
Rausch de Traubenberg M and Slupinski M J 2002 *J. Math. Phys.* **43** 5145 [arXiv:hep-th/0205113].
[5] Goze M, Rausch de Traubenberg M and Tanasa A 2007 *J. Math. Phys.* **48** 093507 [arXiv:math-ph/0603008].
[6] Rausch de Traubenberg M 2008 *J. Phys. Conf. Ser.* **128** 012060 [arXiv:0710.5368 [math-ph]];
Goze M and Rausch de Traubenberg M 2009 *J. Math. Phys.* **50** 063508 [arXiv:0809.4212 [math-ph]].
[7] Campoamor-Stursberg R and Rausch de Traubenberg M 2009 *J. Gen. Lie Theory Appl.* **3** 113 [arXiv:0811.3076 [math-ph]].
[8] Mohammedi N, Moultaka G and Rausch de Traubenberg M 2004 *Int. J. Mod. Phys.* **A19** 5585 [arXiv:hep-th/0305172].
[9] Moultaka G, Rausch de Traubenberg M and Tanasa A 2005 *Int. J. Mod. Phys.* **A20** 5779 [arXiv:hep-th/0411198].
[10] Moultaka G, Rausch de Traubenberg M and Tanasa A 2004, in *Proceedings of the XIth International Conference Symmetry Methods in Physics, Prague 21-24 June 2004* [arXiv:hep-th/0407168].
[11] Campoamor-Stursberg R and Rausch de Traubenberg M 2009, *J. Phys. A; Math. Theor.* **42** 495202 [arXiv:0907.2149 [hep-th]].
[12] Campoamor-Stursberg R and Rausch de Traubenberg M 2010, to appear in *J. Phys. A; Math. Theor* [arXiv:1005.4994 [hep-th]].
[13] Filippov V 1968 *Sibirsk. Mat. Zh.* **39** 660
[14] Figueroa-O'Farrill J 2009 *J. Phys. A: Math. Theor.* **42**, 445206 [arXiv:0905.4900 [hep-th]].
[15] Coleman S and Mandula J 1967 *Phys. Rev.* **159** 1251
[16] Haag R, Lopuszanski J T and Sohnius M F 1975 *Nucl. Phys.* **B88** 257
[17] Roby N 1969 *C. R. Acad. Sc. Paris* **268** 484
Roby N 1970 *Bull. Sc. Math.* **94** 49
Revoy Ph 1977 *C. R. Acad. Sc. Paris* **284** 985
[18] Fleury N and Rausch de Traubenberg M 1992 *J. Math. Phys.* **33** 3356
Fleury N and Rausch de Traubenberg M 1994 *Adv. Appl. Cliff. Alg.* **4** 123
[19] Revoy Ph 1993 *Adv. Appl. Cliff. Alg.* **3** 39
[20] Haile D H and Tesser S 1988 *J. Alg.* **116** 372
[21] Jarvis P D 1978 *Aust. J. of Phys.* **31** 461

Coherent states on the circle

G Chadzitaskos, P Luft and J Tolar

Department of Physics
Faculty of Nuclear Sciences and Physical Engineering
Czech Technical University in Prague
Břehová 7, CZ - 115 19 Prague, Czech Republic

E-mail: jiri.tolar@fjfi.cvut.cz, goce.chadzitaskos@fjfi.cvut.cz

Abstract.
We present a possible construction of coherent states on the unit circle as configuration space. In our approach the phase space is the product $\mathbb{Z}\times S^1$. Because of the duality of canonical coordinates and momenta, i.e. the angular variable and the integers, this formulation can also be interpreted as coherent states over an infinite periodic chain. For the construction we use the analogy with our quantization over a finite periodic chain where the phase space was $\mathbb{Z}_M \times \mathbb{Z}_M$. Properties of the coherent states constructed in this way are studied and the coherent states are shown to satisfy the resolution of unity.

1. Introduction

Coherent states belong to the most important tools in numerous applications of quantum theory. The problem of coherent states on the circle was investigated by S. de Bièvre [1], and also by J. A. González and M. A. del Olmo [2], based on canonical coherent states [3]. The Weil-Brezin-Zak transform of canonical coherent states on the real line was then the essential ingredient leading to a set of coherent states on the circle labeled by the variables of the cylinder $\mathbb{R} \times \mathbf{S^1}$.

A different approach was employed by C. J. Isham and J. R. Klauder [4]. They constructed coherent states on the circle by using the Euclidean group $E(2)$, which is the semi-direct product of groups $\mathbb{R}^2$ and $SO(2)$ (see also [5]) However, they observed that there does not exist an irreducible representation of $E(2)$ such that the resolution of unity holds. Therefore they considered only reducible representations and extended the method to the case of the n-dimensional sphere.

It should be noted that a very detailed study of deformation quantization on the cylinder as classical phase space was recently given in [6]. Their arguments confirm that in quantum theory the 'quantum' phase space $\mathbb{Z} \times S^1$ is involved, which is our starting point in this investigation.

2. Position and momentum operators on the circle

Let the configuration space be the unit circle. The position variable can be identified with the angle, i.e. it takes real values modulo 2π. Let x take a value from the interval $\langle -\pi, \pi)$. In the Dirac formalism the *position operator* is defined

$$\hat{Q} = \int_{-\pi}^{\pi} x|x\rangle\langle x|dx, \text{ with } \langle x|y\rangle = \delta(x-y).$$

The position operator $\hat{Q}$ has continuous spectrum $x \in \langle -\pi, \pi)$ with the corresponding eigenvectors $\{|x\rangle\}$. An arbitrary quantum state $|\psi\rangle$ can be expressed in the form

$$|\psi\rangle = \int_{-\pi}^{\pi} \psi(x)|x\rangle, \quad \text{where} \quad \psi(x) = \langle x|\psi\rangle.$$

It is useful to expand $\psi(x)$ in the Fourier series

$$\psi(x) = \sum_{n=-\infty}^{\infty} a_n e^{inx}.$$

The function $\psi(x)$ is the wavefunction and the position operator acts on it $\hat{Q}\psi(x) = x\psi(x)$. The momentum eigenvectors are then defined via Fourier transform

$$|p\rangle = \int_{-\pi}^{\pi} e^{ipx}|x\rangle dx \tag{1}$$

and its inverse

$$|x\rangle = \frac{1}{2\pi} \sum_{k=-\infty}^{\infty} e^{ikx}|k\rangle. \tag{2}$$

In the Hilbert space formulation, where

$$\mathcal{H} = L^2(\mathbf{S}^1, d\varphi),$$

we shall follow our approach in [7]. The momentum operator will be defined using the unitary representation $\hat{V}(\alpha)$ of the group of rotations $U(1)$ of the unit circle,

$$[\mathbf{V}(\alpha)\psi](\beta) = \psi(\beta - \alpha), \quad \psi \in L^2(\mathbf{S}^1, d\varphi), \quad \alpha, \beta \in U(1). \tag{3}$$

Unitary operators $\mathbf{V}(\alpha)$ shift the argument of functions in $L^2(\mathbf{S}^1, d\varphi)$. The momentum operator is then by Stone's theorem

$$\widehat{P} = -i\frac{d}{d\varphi}. \tag{4}$$

The position operator

$$(\widehat{\mathbf{Q}}\psi)(\varphi) = \varphi\psi(\varphi), \tag{5}$$

formally satisfies the well known commutation relation

$$[\widehat{Q}, \widehat{P}] = i\mathbb{I}, \tag{6}$$

but it is not well-defined on $\mathcal{H}$.

3. Construction of coherent states

In order to define coherent states directly by Perelomov's method [3], we should first construct a system of unitary operators labeled by elements of the group $\mathbb{Z} \times U(1)$; second, it is necessary to determine the 'vacuum' vector $|0, 0\rangle$. The system of unitary operators will be defined

$$\widehat{W}(m, \alpha) := e^{im\widehat{Q}} e^{-i\alpha\widehat{P}} = e^{im\widehat{Q}}\mathbf{V}(\alpha), \quad \alpha \in [\pi, \pi), \ m \in \mathbb{Z}. \tag{7}$$

The factors do not commute

$$e^{im\widehat{\mathbf{Q}}}e^{-i\alpha\widehat{\mathbf{P}}} = e^{im\alpha}e^{-i\alpha\widehat{\mathbf{P}}}e^{im\widehat{\mathbf{Q}}}, \quad \alpha \in [\pi,\pi),\ m \in \mathbb{Z}, \tag{8}$$

but the operator $e^{im\widehat{Q}}$ is now well defined on $\mathcal{H}$,

$$e^{im\widehat{Q}}\psi(\varphi) = e^{im\varphi}\psi(\varphi). \tag{9}$$

Here, the system of operators $\widehat{W}(m,\alpha)$ does not create a representation of group $\mathbb{Z} \times U(1)$ but thanks to the Weyl form of commutation relations we obtain a projective representation of $\mathbb{Z} \times U(1)$.

The vacuum vector $|0,0\rangle$ will be determined in analogy with canonical coherent states on $L^2(\mathbb{R})$. The requirement that the vacuum state be an eigenvector of annihilation operator with eigenvalue 0 is written here in exponential form

$$e^{\widehat{Q}+i\widehat{P}}|0,0\rangle = |0,0\rangle. \tag{10}$$

Using the Baker-Campbell-Hausdorff formula the operator $e^{\widehat{Q}+i\widehat{P}}$ could be separated in the product of two operators $e^{\widehat{Q}}$ and $e^{i\widehat{P}}$ in arbitrary order. Such change will not influence the final vacuum state $|0,0\rangle$.

Condition (10) leads to the Gauss exponential function

$$|0,0\rangle = \mathcal{A}e^{-\frac{\varphi^2}{2}}, \quad \varphi \in [-\pi,\pi). \tag{11}$$

Hence the vacuum state is an element of our Hilbert space, $|0,0\rangle \in L^2(\mathbf{S}^1, d\varphi)$. At $\varphi = \pm\pi$ it is continuous but its derivative has a small discontinuity ($\approx e^{-5}$). The normalization constant $\mathcal{A}$ is given by

$$\mathcal{A} = \frac{1}{\sqrt{\int_{-\pi}^{\pi} exp(-\varphi^2)d\varphi}} \doteq 0.751128. \tag{12}$$

The system of coherent states on $L^2(\mathbf{S}^1, d\varphi)$ is now obtained by the action of the system of operators $\widehat{W}(m,\alpha)$ on the vacuum state $|0,0\rangle$:

$$|m,\alpha\rangle := \widehat{W}(m,\alpha)|0,0\rangle. \tag{13}$$

The functional form of our coherent states is given by

$$|m,\alpha\rangle = \mathcal{A}e^{im\varphi}e^{-\frac{(\varphi-\alpha)^2}{2}}, \quad \varphi \in [-\pi,\pi), \tag{14}$$

i.e. for $\alpha \neq 0$ they are displaced and phased versions of (11) with a discontinuity at $\varphi = \pm\pi$.

4. Properties of our coherent states on $L^2(\mathbf{S}^1, d\varphi)$

In this section we shall check several properties of our coherent states which are known to hold for canonical coherent states on $L^2(\mathbb{R})$ [3, 8]. First of all, we shall look at the resolution of unity, i.e. we shall prove the following equality for our coherent states:

$$\sum_{k\in\mathbb{Z}} \int_{\mathbf{S}^1} |k,\alpha\rangle\langle k,\alpha| d\alpha = c\widehat{I}, \tag{15}$$

Journal of Physics: Conference Series **284** (2011) 012016 doi:10.1088/1742-6596/284/1/012016

where c is a constant. Thus for an arbitrary normalized vector η from our Hilbert space $L^2(\mathbf{S}^1, d\varphi)$ the identity

$$\sum_{k\in\mathbb{Z}} \int_{\mathbf{S}^1} \langle\eta|k,\alpha\rangle\langle k,\alpha|\eta\rangle d\alpha = c\widehat{I}, \tag{16}$$

should hold. Since the inner product of $|\eta\rangle$ with some coherent state $|k,\alpha\rangle$ has the integral form

$$\langle k,\alpha|\eta\rangle = \mathcal{A}\int_{\mathbf{S}^1} e^{-ik\varphi}e^{-\frac{(\varphi-\alpha)^2}{2}}\eta(\varphi)d\varphi, \tag{17}$$

one can calculate the constant c with the result

$$c = 2\pi, \tag{18}$$

so that the resolution of unity for our set of coherent states takes the form

$$\sum_{k\in\mathbb{Z}} \int_{\mathbf{S}^1} |k,\alpha\rangle\langle k,\alpha| d\alpha = 2\pi\widehat{I}. \tag{19}$$

The next properties to be checked concern the overlaps of our (normalized) coherent states given by inner products in $L^2(\mathbf{S}^1, d\varphi)$. Here it is necessary to correctly realize the way how the operator $\mathbf{V}(\alpha) = exp(-i\alpha\widehat{P})$ acts on the Hilbert space $L^2(\mathbf{S}^1, d\varphi)$ when the circle $\mathbf{S^1}$ — our configuration space — is identified with the interval $[-\pi,\pi)$. Then the action of operator $e^{-i\alpha\widehat{P}}$ on function $\psi(\varphi) \in L^2(\mathbf{S}^1, d\varphi)$ for $\alpha\in[0,\pi]$ has the form :

$$e^{-i\alpha\widehat{P}}\psi(\varphi) = \begin{cases} \psi(\varphi-\alpha) & \varphi\in[-\pi+\alpha,\pi], \\ \psi(\varphi-\alpha+2\pi) & \varphi\in[-\pi,-\pi+\alpha]. \end{cases} \tag{20}$$

For $\alpha\in[-\pi,0]$ we have

$$e^{-i\alpha\widehat{P}}\psi(\varphi) = \begin{cases} \psi(\varphi-\alpha) & \varphi\in[-\pi,\pi+\alpha], \\ \psi(\varphi-\alpha-2\pi) & \varphi\in[\pi+\alpha,\pi]. \end{cases} \tag{21}$$

Note that one has to consider addition modulo 2π in the argument of function ψ. This is the reason why the inner product *cannot* be calculated simply according to

$$\langle m,\alpha|n,\beta\rangle = \mathcal{A}^2\int_{-\pi}^{\pi} e^{-i\varphi(n-m)}e^{-\frac{(\varphi-\alpha)^2}{2}}e^{-\frac{(\varphi-\beta)^2}{2}}d\varphi. \tag{22}$$

From now on we shall restrict ourselves only to the cases when α and β are non-negative:

$$\alpha\in[0,\pi], \quad \beta\in[0,\pi]. \tag{23}$$

Without loss of generality we may also assume

$$\beta\geq\alpha. \tag{24}$$

Taking into account (20) and (21), we have to split the inner product of two coherent states into two terms

$$\langle m,\alpha|n,\beta\rangle = \mathcal{A}I_1(\alpha,\beta,n-m) + \mathcal{A}I_2(\alpha,\beta,n-m), \tag{25}$$

where

$$I_1(\alpha,\beta,n-m) := \int_{\alpha-\pi}^{\beta-\pi} e^{i\varphi(n-m)} e^{-\frac{(\varphi-\alpha)^2}{2}} e^{-\frac{(\varphi-\beta+2\pi)^2}{2}} d\varphi \tag{26}$$

and

$$I_2(\alpha,\beta,n-m) := \int_{\beta-\pi}^{\pi+\alpha} e^{i\varphi(n-m)} e^{-\frac{(\varphi-\alpha)^2}{2}} e^{-\frac{(\varphi-\beta)^2}{2}} d\varphi. \tag{27}$$

In order to evaluate the integrals $I_1(\alpha,\beta,n-m)$ and $I_2(\alpha,\beta,n-m)$, we use the error function of a complex variable z,

$$\operatorname{erf}(z) := \frac{2}{\sqrt{\pi}} \int_{\Gamma(z)} e^{-\psi^2} d\psi; \tag{28}$$

here $\Gamma(z)$ denotes an arbitrary continuous path of finite length which connects the origin $0 \in \mathbb{C}$ with a complex number $z \in \mathbb{C}$.

The integral $I_1(\alpha,\beta,n-m)$, after the substitution

$$\omega = \varphi + \pi - \frac{\alpha+\beta}{2}, \tag{29}$$

takes the form

$$\begin{aligned} I_1(\alpha,\beta,n-m) &= e^{-(\frac{\beta-\alpha}{2})^2-\pi} e^{(\frac{\alpha+\beta}{2}-\pi)(m-n)} \times \\ &\times \int_{\frac{\alpha-\beta}{2}}^{\frac{\beta-\alpha}{2}} e^{i\omega(n-m)} e^{-\omega^2} d\omega \end{aligned} \tag{30}$$

which leads to the formula

$$\begin{aligned} I_1(\alpha,\beta,n-m) &= (-\frac{\sqrt{\pi}}{2}) e^{-(\frac{\beta-\alpha}{2})^2-\pi} e^{i(\frac{\alpha+\beta}{2}-\pi)(m-n)} e^{-\frac{(n-m)^2}{4}} \times \\ &\times [\operatorname{erf}(\frac{\alpha-\beta}{2} + \frac{i(n-m)}{2}) \\ &+ \operatorname{erf}(\frac{\alpha-\beta}{2} - \frac{i(n-m)}{2})]. \end{aligned} \tag{31}$$

The other integral $I_2(\alpha,\beta,n-m)$, after the substitution $\omega = \varphi - \frac{\alpha+\beta}{2}$, yields

$$\begin{aligned} I_2(\alpha,\beta,n-m) &= (-\frac{\sqrt{\pi}}{2}) e^{-(\frac{\beta-\alpha}{2})^2} e^{i(\frac{\alpha+\beta}{2})(m-n))} e^{-\frac{(n-m)^2}{4}} \times \\ &\times [\operatorname{erf}(\frac{\alpha-\beta}{2} - \pi + \frac{i(n-m)}{2}) \\ &+ \operatorname{erf}(\frac{\alpha-\beta}{2} - \pi - \frac{i(n-m)}{2})]. \end{aligned} \tag{32}$$

We have to admit that, unfortunately, we do not see any way how to further simplify the above analytic expressions of the integrals $I_1(\alpha,\beta,n-m)$ and $I_2(\alpha,\beta,n-m)$ to see whether the coherent states are mutually non-orthogonal. However, we have numerically computed absolute values of the inner products for many pairs of coherent states and have drawn the graphs for several different values of $n-m$ fixed in each graph. It was apparent — for all plotted cases — that the overlap never vanishes. Because of restriction on the length of the contribution we intend to publish a detailed account of the matter including a number of plots confirming this property.

The next checked quantities are the expectation values of the position operator in the coherent states $|m,\alpha\rangle$ for $\alpha \geq 0$. Explicitly we have

$$\langle m,\alpha|\widehat{Q}|m,\alpha\rangle = \mathcal{A}^2 \int_{-\pi}^{-\pi+\alpha} \varphi e^{-(\varphi-\alpha+2\pi)^2} d\varphi + \mathcal{A}^2 \int_{-\pi+\alpha}^{\pi} \varphi e^{-(\varphi-\alpha)^2} d\varphi, \tag{33}$$

or, using the error function,

$$\langle m,\alpha|\widehat{Q}|m,\alpha\rangle = \alpha - \mathcal{A}^2\sqrt{\pi^3}(\mathrm{erf}(\pi) - \mathrm{erf}(\pi-\alpha)). \tag{34}$$

One observes that for α positive the expectation values do not depend on m (the same result is obtained also for negative values of α). It is interesting that the expectation value of position is nonlinear in α for $\alpha \neq 0$. It is clear that for $\alpha = 0$ the quasi-Gaussian (11) is symmetric around $\varphi = 0$, so the expectation value is an integral of an odd function, evidently equal to zero. However, if $\alpha \in (0,\pi]$, the displaced quasi-Gaussian is no more symmetric around $\varphi = 0$ and the integration leads to a deviation depending on a difference of error functions (34); maximal deviation is attained for $\alpha = \pi$.

Important among the checked quantities are the expectation values of the momentum operator in coherent states $|m,\alpha\rangle$. The explicit form

$$\begin{aligned}\langle m,\alpha|\widehat{P}|m,\alpha\rangle &= \mathcal{A}^2 \int_{-\pi}^{-\pi+\alpha} (m + i(\varphi-\alpha+2\pi))e^{-(\varphi-\alpha+2\pi)^2} d\varphi \\ &+ \mathcal{A}^2 \int_{-\pi+\alpha}^{\pi} (m + i(\varphi-\alpha))e^{-(\varphi-\alpha)^2} d\varphi\end{aligned}$$

can be simplified into the formula

$$\langle m,\alpha|\widehat{P}|m,\alpha\rangle = m, \tag{35}$$

which could be anticipated by analogy with the canonical coherent states on $L^2(\mathbb{R})$.

Finally, the expectation values of the square of momentum operator in coherent states $|m,\alpha\rangle$ are determined by computing the integrals

$$\begin{aligned}\langle m,\alpha|\widehat{P}^2|m,\alpha\rangle &= \mathcal{A}^2 \int_{-\pi}^{-\pi+\alpha} [1 + (m + i(\varphi-\alpha+2\pi))^2]e^{-(\varphi-\alpha+2\pi)^2} d\varphi \\ &+ \mathcal{A}^2 \int_{-\pi+\alpha}^{\pi} [1 + (m + i(\varphi-\alpha))^2]e^{-(\varphi-\alpha)^2} d\varphi\end{aligned} \tag{36}$$

with the result

$$\langle m,\alpha|\widehat{P}^2|m,\alpha\rangle = m^2 + \frac{1}{2} + \mathcal{A}^2\pi e^{-\pi^2} \tag{37}$$

which does not depend explicitly on α.

5. Conclusion

For our system of coherent states their overlaps and matrix elements were expressed analytically, then some calculated numerically or evaluated with the help of MATHEMATICA. The property of resolution of unity was shown to hold.

Acknowledgements

The support by the Ministry of Education of Czech Republic (projects MSM6840770039 and LC06002) is acknowledged. The authors are grateful to the referee for several constructive remarks which helped to improve the presentation.

Journal of Physics: Conference Series **284** (2011) 012016 doi:10.1088/1742-6596/284/1/012016

References

[1] de Bièvre S 1989 Coherent states over symplectic homogenous spaces, *J Math Phys* **30** 1401–1407
[2] González J A and del Olmo M A 1998 Coherent states on the circle and quantization *J Phys A: Math Gen* **31** 884
[3] Perelomov M A 1986 *Generalized Coherent States and Their Applications* (Berlin: Springer)
[4] Isham C J and Klauder J R 1991 Coherent states for n-dimensional Euclidean groups $E(n)$ and their application *J Math Phys* **32**(3) 607–620
[5] Nieto L M, Atakishiyev N M, Chumakov S M and Wolf K B 1998 Wigner distribution function for Euclidean systems *J Phys A: Math Gen* **31** 3875–3895
[6] González J A, del Olmo M A and Tosiek J 2003 Quantum mechanics on the cylinder *J Opt B: Quantum Semiclass Opt* **5** S306–S315
[7] Tolar J and Chadzitaskos G 1997 Quantization on $\mathbf{Z}_M$ and coherent states over $\mathbf{Z}_M \times \mathbf{Z}_M$ *J Phys A: Math Gen* **30** 2509–2517
[8] Klauder J R and Skagerstam B S 1985 *Coherent States: Applications in Physics and Mathematical Physics* (Singapore: World Scientific)

Spectral conditions for positive maps and entanglement witnesses

Dariusz Chruściński and Andrzej Kossakowski
Institute of Physics, Nicolaus Copernicus University,
Grudziądzka 5/7, 87–100 Toruń, Poland
E-mail: darch@fizyka.umk.pl

Abstract.
We provide partial classification of entanglement witnesses and positive maps which is based on a family of spectral conditions. Interestingly many well known examples of positive maps (e.g. transposition, reduction map, the family of Choi maps) belong to the class constructed *via* spectral conditions.

1. Introduction

One of the most important problems of Quantum Information Theory [1] is the characterization of mixed states of composed quantum systems. In particular it is of primary importance to test whether a given quantum state exhibits quantum correlation, i.e. whether it is separable or entangled. For low-dimensional systems there exists simple necessary and sufficient condition for separability. The celebrated Peres-Horodecki criterion [2, 3] states that a state of a bipartite system living in $\mathbb{C}^2 \otimes \mathbb{C}^2$ or $\mathbb{C}^2 \otimes \mathbb{C}^3$ is separable iff its partial transpose is positive. Unfortunately, for higher-dimensional systems there is no single *universal* separability condition.

It turns out that the above problem may be reformulated in terms of positive linear maps in operator algebras: a state ρ in $\mathcal{H}_1 \otimes \mathcal{H}_2$ is separable iff $(\mathrm{id} \otimes \Phi)\rho$ is positive for any positive map Φ which sends positive operators on $\mathcal{H}_2$ into positive operators on $\mathcal{H}_1$. Therefore, a classification of positive linear maps between operator algebras $\mathcal{B}(\mathcal{H}_1)$ and $\mathcal{B}(\mathcal{H}_2)$ is of primary importance. Unfortunately, in spite of the considerable effort, the structure of positive maps is rather poorly understood [4]-[18] (for the recent review paper see [16]). Positive maps play an important role both in physics and mathematics providing generalization of $*$-homomorphisms, Jordan homomorphisms and conditional expectations. Normalized positive maps define affine mappings between sets of states of $\mathbb{C}^*$-algebras.

In the present paper we perform partial classification of positive linear maps which is based on spectral conditions. Actually, the presented method enables one to construct maps with a desired degree of positivity. These are so called k-positive maps where an integer k runs from 1 (simply positive maps) up to $d = \min\{\dim \mathcal{H}_1, \dim \mathcal{H}_2\}$ (so called completely positive (CP) maps). It is well known [19] that CP maps are fully classified due to Stinespring's theorem. It turns out that any positive map which is not CP can be written as $\Phi = \Phi_+ - \Phi_-$, with $\Phi_\pm$ being CP maps. However, there is no general method to recognize the positivity of Φ from $\Phi_+ - \Phi_-$. We show that suitable spectral conditions satisfied by the pair (Φ_+, Φ_-) guarantee the k-positivity of $\Phi_+ - \Phi_-$. Equivalently, we provide partial classification of entanglement witnesses which define basic tool in entanglement theory [20, 21].

Interestingly many well known examples of positive maps (e.g. transposition, reduction map, the family of Choi maps) belong to the class constructed *via* spectral conditions. It should be stressed, however, that there are positive maps which do not belong to our class. Therefore, in order to provide full characterization of positive maps further analysis is needed.

2. Positive maps vs. entanglement witnesses

Let $M_d(\mathbb{C})$ denotes a $\mathbb{C}^*$-algebra of $d \times d$ complex matrices and let $M_d^+(\mathbb{C})$ denotes a subset of positive elements in $M_d(\mathbb{C})$. It is clear that $M_d^+(\mathbb{C})$ defines a convex cone in $M_d(\mathbb{C})$. Now, a linear map

$$\Phi \ : \ M_{d_1}(\mathbb{C}) \ \longrightarrow \ M_{d_2}(\mathbb{C}) \ , \tag{1}$$

is said to be positive if $\Phi(M_{d_1}^+(\mathbb{C})) \subset M_{d_2}^+(\mathbb{C})$. Φ is called unital if

$$\Phi(\mathbb{I}_{d_1}) = \mathbb{I}_{d_2} \ , \tag{2}$$

where $\mathbb{I}_d$ is a unit matrix in $M_d(\mathbb{C})$. It is called trace preserving if

$$\mathrm{Tr}\, \Phi(X) = \mathrm{Tr}\, X \ , \tag{3}$$

for all $X \in M_{d_1}(\mathbb{C})$. Note that trace preserving positive maps map quantum states of d_1–level system into quantum states of d_2–level system. It is clear that a map (1) is positive if it satisfies

$$\langle \psi | \Phi(|\phi\rangle\langle\phi|) |\psi\rangle \geq 0 \ , \tag{4}$$

for each $\phi \in \mathbb{C}^{d_1}$ and each $\psi \in \mathbb{C}^{d_2}$. Unfortunately, the above condition is very difficult to check. Consider now an extended map

$$\mathrm{id}_k \otimes \Phi \ : \ M_k(\mathbb{C}) \otimes M_{d_1}(\mathbb{C}) \ \longrightarrow \ M_k(\mathbb{C}) \otimes M_{d_2}(\mathbb{C}) \ , \tag{5}$$

where id_k is an identity map in $M_k(\mathbb{C})$. A map Φ is called k-positive if $\mathrm{id}_k \otimes \Phi$ is positive. Finally, Φ is CP if it is d-positive form $d = \min\{d_1, d_2\}$. Trace preserving CP map is called quantum channel.

Consider now a composite system living in $\mathbb{C}^{d_1} \otimes \mathbb{C}^{d_2}$ and denote by $\mathcal{P}$ the space of positive operators in $\mathcal{B}(\mathbb{C}^{d_1} \otimes \mathbb{C}^{d_2})$. Let us recall [23] that for any normalized positive operator σ on $\mathcal{H}_1 \otimes \mathcal{H}_2$ one may define its Schmidt number

$$\mathrm{SN}(\sigma) = \min_{p_k, \psi_k} \left\{ \max_k \ \mathrm{SR}(\psi_k) \right\} \ , \tag{6}$$

where the minimum is taken over all possible pure states decompositions

$$\sigma = \sum_k p_k \, |\psi_k\rangle\langle\psi_k| \ , \tag{7}$$

with $p_k \geq 0$, $\sum_k p_k = 1$ and ψ_k are normalized vectors in $\mathcal{H}_1 \otimes \mathcal{H}_2$. The Schmidt rank $\mathrm{SR}(\psi)$ denotes the number of non-vanishing Schmidt coefficients in the Schmidt decomposition of ψ. This number characterizes the minimum Schmidt rank of the pure states that are needed to construct such density matrix. It is evident that $1 \leq \mathrm{SN}(\rho) \leq d$, with $d = \min\{d_1, d_2\}$, and $d_k = \dim \mathcal{H}_k$. Moreover, ρ is separable iff $\mathrm{SN}(\rho) = 1$. It was proved [23] that the Schmidt number is non-increasing under local operations and classical communication.

This notion enables one to introduce the following family of positive cones:

$$\mathrm{V}_r = \{\, \rho \in \mathcal{P} \mid \mathrm{SN}(\rho) \leq r \,\} \ . \tag{8}$$

One has the following chain of inclusions

$$\mathrm{V}_1 \subset \ldots \subset \mathrm{V}_d \equiv \mathcal{P} \ , \tag{9}$$

where $d = \min\{d_1, d_2\}$. Clearly, V_1 is a cone of separable (unnormalized) states and $V_d \setminus V_1$ stands for a set of entangled states. Note, that a partial transposition $(\mathrm{id} \otimes \tau)$ gives rise to another family of cones:

$$\mathrm{V}^l = (\mathrm{id} \otimes \tau)\mathrm{V}_l \ , \tag{10}$$

such that $\mathrm{V}^1 \subset \ldots \subset \mathrm{V}^d$. One has $\mathrm{V}_1 = \mathrm{V}^1$, together with the following hierarchy of inclusions:

$$\mathrm{V}_1 = \mathrm{V}_1 \cap \mathrm{V}^1 \subset \mathrm{V}_2 \cap \mathrm{V}^2 \subset \ldots \subset \mathrm{V}_d \cap \mathrm{V}^d \ . \tag{11}$$

Note, that $\mathrm{V}_d \cap \mathrm{V}^d$ is a convex set of PPT (unnormalized) states (recall, that a state ρ is PPT (Positive Partial Transpose) if $(\mathrm{id} \otimes \tau)\rho \geq 0$). Finally, $\mathrm{V}_r \cap \mathrm{V}^s$ is a convex subset of PPT states ρ such that $\mathrm{SN}(\rho) \leq r$ and $\mathrm{SN}[(\mathbb{1} \otimes \tau)\rho] \leq s$.

Let us observe that a linear map Φ from $M_{d_1}(\mathbb{C})$ into $M_{d_2}(\mathbb{C})$ is k-positive iff

$$(\mathrm{id} \otimes \Phi)(\mathrm{V}_k) \subset \mathcal{P} \ . \tag{12}$$

and it is k-copositive[1] iff

$$(\mathrm{id} \otimes \Phi)(\mathrm{V}^k) \subset \mathcal{P} \ . \tag{13}$$

Denoting by P_k (P^k) a convex cone of k-positive (k-copositive) maps one has the following chains of inclusions

$$\mathrm{P}_d \subset \mathrm{P}_{d-1} \subset \ldots \subset \mathrm{P}_2 \subset \mathrm{P}_1 \ , \tag{14}$$

and

$$\mathrm{P}^d \subset \mathrm{P}^{d-1} \subset \ldots \subset \mathrm{P}^2 \subset \mathrm{P}^1 \ , \tag{15}$$

where P_d (P^d) stands for a set of completely positive (copositive) maps.

A positive map $\Phi : M_{d_1}(\mathbb{C}) \longrightarrow M_{d_2}(\mathbb{C})$ is *decomposable* iff $\Phi \in P_d + P^d$, that is, Φ can be written as $\Phi = \Phi_1 + \Phi_2$, with $\Phi_1 \in P_d$ and $\Phi_2 \in P^d$. Otherwise Φ is *indecomposable.* Indecomposable maps can detect entangled state from $V_d \cap V^d \equiv \mathrm{PPT}$, that is, bound entangled states.

Using well known Choi-Jamiołkowski isomorphism we may assign to any linear map $\Phi : M_{d_1}(\mathbb{C}) \longrightarrow M_{d_2}(\mathbb{C})$ the following Hermitian operator $W \in M_{d_1}(\mathbb{C}) \otimes M_{d_2}(\mathbb{C})$:

$$W = (\mathrm{id} \otimes \Phi)P^+ \in M_{d_1}(\mathbb{C}) \otimes M_{d_2}(\mathbb{C}) \ , \tag{16}$$

where P^+ stands for (unnormalized) maximally entangled state in $\mathbb{C}^{d_1} \otimes \mathbb{C}^{d_1}$. If e_i $(i = 1, \ldots, d_1)$ is an orthonormal basis in $\mathbb{C}^{d_1}$, then

$$W = \sum_{i,j=1}^{d_1} e_{ij} \otimes \Phi(e_{ij}) \ , \tag{17}$$

where $e_{ij} = |e_i\rangle\langle e_j|$. Any operator W constructed this way is called an entanglement witness. One has the following

[1] A map Φ is k-copositive if $\tau \circ \Phi$ is k-positive.

Proposition 1 *A state represented by a density operator ρ in $\mathcal{H}_1 \otimes \mathcal{H}_2$ is entangled if and only if there exists an entanglement witness W such that*

$$\mathrm{Tr}(\rho W) < 0 . \tag{18}$$

One says that W detects ρ.

Let us observe that W is an entanglement witness if and only if

$$\langle \psi | W | \psi \rangle \geq 0 , \tag{19}$$

for any $\psi \in \mathbb{C}^{d_1} \otimes \mathbb{C}^{d_2}$ such that $\mathrm{SR}(\psi) \leq 1$, that is, $\psi = \psi_1 \otimes \psi_2$. Similarly, if Φ is k-positive that the corresponding entanglement witness W satisfies $\langle \psi | W | \psi \rangle \geq 0$ for any $\psi \in \mathbb{C}^{d_1} \otimes \mathbb{C}^{d_2}$ such that $\mathrm{SR}(\psi) \leq k$. Note, that if Φ is CP then $\langle \psi | W | \psi \rangle \geq 0$ for any ψ. It implies that Φ is CP iff the corresponding W is a positive operator.

3. Spectral family of positive maps

Consider the space $\mathcal{L}(\mathcal{H}_1, \mathcal{H}_2)$ of linear operators $a : \mathcal{H}_1 \longrightarrow \mathcal{H}_2$, or equivalently the space of $d_1 \times d_2$ matrices, where $d_i = \dim \mathcal{H}_i < \infty$. Let us recall that $\mathcal{L}(\mathcal{H}_1, \mathcal{H}_2)$ is equipped with a family of Ky Fan k-norms [22]: for any $a \in \mathcal{L}(\mathcal{H}_1, \mathcal{H}_2)$ one defines

$$||\, a \,||_k^2 := \sum_{i=1}^{k} s_i(a)^2 , \tag{20}$$

where $s_1(a) \geq \ldots \geq s_d(a)$ ($d = \min\{d_1, d_2\}$) are the singular values of a. Clearly, for $k = 1$ one recovers the operator norm $||\, a \,||_1 = ||\, a \,||$ and if $d_1 = d_2 = d$, then for $k = d$ one reproduces the trace norm $||\, a \,||_d = ||\, a \,||_{\mathrm{tr}}$. The family of k-norms satisfies:

(i) $||\, a \,||_k \leq ||\, a \,||_{k+1}$,
(ii) $||\, a \,||_k = ||\, a \,||_{k+1}$ if and only if $\mathrm{rank}\, a = k$,
(iii) if $\mathrm{rank}\, a \geq k + 1$, then $||\, a \,||_k < ||\, a \,||_{k+1}$.

Consider now the Hilbert space being the tensor product $\mathcal{H}_1 \otimes \mathcal{H}_2$. It is clear that $\mathcal{H}_1 \otimes \mathcal{H}_2$ is isomorphic to the space of linear operators from $\mathcal{L}(\mathcal{H}_1, \mathcal{H}_2)$, i.e. each $\psi \in \mathcal{H}_1 \otimes \mathcal{H}_2$ corresponds to some linear operator $F : \mathcal{H}_1 \longrightarrow \mathcal{H}_2$. Indeed, taking orthonormal basis $\{e_i\}$ ($i = 1, \ldots, d_1$) in $\mathcal{H}_1$ and $\{f_\alpha\}$ ($\alpha = 1, \ldots, d_2$) in $\mathcal{H}_2$, one has

$$\psi = \sum_{i=1}^{d_1} \sum_{\alpha=1}^{d_2} \psi_{i\alpha}\, e_i \otimes f_\alpha , \tag{21}$$

where $\psi_{i\alpha}$ are complex coefficients. Defining

$$F e_i := \sum_{\alpha=1}^{d_2} \psi_{i\alpha}\, f_\alpha ,$$

one finds

$$\psi = \sum_{i=1}^{d_1} e_i \otimes F e_i , \tag{22}$$

which establishes the above-mentioned isomorphism. Let us note that the normalization condition

$$\langle\psi|\psi\rangle = \sum_{i=1}^{d_1}\sum_{\alpha=1}^{d_2} |\psi_{i\alpha}|^2 = 1 \ , \tag{23}$$

gives rise to

$$\mathrm{tr}\, FF^* = 1 \ , \tag{24}$$

for the corresponding operator F. It is therefore clear that the rank-1 projector $P := |\psi\rangle\langle\psi|$ may be represented in the following way

$$P = \sum_{i,j=1}^{d_1} e_{ij} \otimes F e_{ij} F^* \ . \tag{25}$$

Now, it is easy to see that

$$\mathrm{SR}(\psi) = \mathrm{rank}\, F \ , \tag{26}$$

where $\mathrm{SR}(\psi)$ denotes the Schmidt rank of ψ ($1 \leq \mathrm{SR}(\psi) \leq d$), i.e. the number of non-vanishing Schmidt coefficients in the Schmidt decomposition of ψ.

It is well known that any CP map may be represented in the so-called Kraus form [24]

$$\Phi_{\mathrm{CP}}(a) = \sum_\alpha K_\alpha a K_\alpha^* \ , \tag{27}$$

where (Kraus operators) $K_\alpha \in \mathcal{L}(\mathcal{H}_1, \mathcal{H}_2)$. Any positive map is a difference of two CP maps $\Phi = \Phi_+ - \Phi_-$. However, there is no general method to recognize the positivity of Φ from $\Phi_+ - \Phi_-$. Consider now an orthonormal basis F_α in $\mathcal{L}(\mathcal{H}_1, \mathcal{H}_2)$, that is

$$(F_\alpha, F_\beta)_{\mathrm{HS}} = \delta_{\alpha\beta} \ . \tag{28}$$

Moreover, let us take $\lambda_1, \ldots, \lambda_D \geq 0$, with $D = d_1 d_2$, and construct two CP maps:

$$\Phi_-(a) := \sum_{\alpha=1}^{L} \lambda_\alpha F_\alpha a F_\alpha^* \ , \tag{29}$$

and

$$\Phi_+(a) := \sum_{\alpha=L+1}^{D} \lambda_\alpha F_\alpha a F_\alpha^* \ . \tag{30}$$

We assume that $\lambda_1, \ldots, \lambda_L > 0$, where $1 < L < D$. Finally, let us consider

$$\Phi(a) = \sum_{\alpha=L+1}^{D} \lambda_\alpha F_\alpha a F_\alpha^* - \sum_{\alpha=1}^{L} \lambda_\alpha F_\alpha a F_\alpha^* \ . \tag{31}$$

One proves [17] the following

Theorem 1 *Suppose that* $\sum_{\alpha=1}^{L} ||F_\alpha||_k^2 < 1$, *and define*

$$\mu_\ell = \frac{\sum_{\alpha=1}^{L} \lambda_\alpha \, ||F_\alpha||_\ell^2}{1 - \sum_{\alpha=1}^{L} ||F_\alpha||_\ell^2} \ , \tag{32}$$

for $\ell = 1, 2, \ldots, k$*. If*

$$\lambda_\alpha \geq \mu_k \ , \qquad \alpha = L+1, \ldots, D \ , \tag{33}$$

then the map (31) is k-positive. Moreover, if $\sum_{\alpha=1}^{L} ||F_\alpha||_{k+1}^2 < 1$*, and*

$$\lambda_\alpha < \mu_{k+1} \ , \qquad \alpha = L+1, \ldots, D \ , \tag{34}$$

then the map (31) is not $(k+1)$*-positive.*

This theorem provides simple method for constructing positive maps (using only spectral conditions. It is clear that Theorem 1 provides the following construction of entanglement witnesses: let us introduce two positive operators W_- and W_+

$$W_- = \sum_{\alpha=1}^{L} \lambda_\alpha P_\alpha \ , \tag{35}$$

and

$$W_+ = \sum_{\alpha=L+1}^{D} \lambda_\alpha P_\alpha \ , \tag{36}$$

where the rank-1 projectors P_α are defined as follows

$$P_\alpha = \sum_{i,j=1}^{d_1} e_{ij} \otimes F_\alpha e_{ij} F_\alpha^* \ . \tag{37}$$

Finally, define W by

$$W = W_+ - W_- \ . \tag{38}$$

One has the following result: if the condition (33) is satisfied then

$$\langle \psi | W | \psi \rangle \geq 0 \ , \tag{39}$$

for any vector ψ satisfying $\mathrm{SR}(\psi) \leq k$.

4. Examples

Interestingly many well known examples of positive maps/entanglement witnesses belong to the class constructed *via* Theorem 1.

Example 1 Transposition in $M_d(\mathbb{C})$:

$$\tau(X) = X^{\mathrm{T}} \ . \tag{40}$$

The map is trivially positive since transposition does not change the spectrum of the Hermitian operator. Moreover, it is trace preserving and unital. For example for $d = 2$ the corresponding entanglement witness reads as follows

$$W = \left(\begin{array}{cc|cc} 1 & \cdot & \cdot & \cdot \\ \cdot & \cdot & 1 & \cdot \\ \hline \cdot & 1 & \cdot & \cdot \\ \cdot & \cdot & \cdot & 1 \end{array} \right) \ , \tag{41}$$

where dots represent zeros. Its spectrum is given by $\{-1, 1, 1, 1\}$ and hence τ is not CP. It is well known that transposition is not CP for any d (actually, it is even not 2-positive). For $d = 2$ one finds

$$\Phi_-(X) = \lambda_1 F_1 X F_1^* \ , \tag{42}$$

and

$$\Phi_+(X) = \sum_{\alpha=2}^{4} \lambda_\alpha F_\alpha X F_\alpha^* \ , \tag{43}$$

where $\lambda_1 = \ldots = \lambda_4 = 1$, and

$$F_1 = \frac{1}{\sqrt{2}} \begin{pmatrix} 0 & 1 \\ -1 & 0 \end{pmatrix} , \ F_2 = \begin{pmatrix} 1 & 0 \\ 0 & 0 \end{pmatrix} , \ F_3 = \begin{pmatrix} 0 & 0 \\ 0 & 1 \end{pmatrix} , \ F_4 = \frac{1}{\sqrt{2}} \begin{pmatrix} 0 & 1 \\ 1 & 0 \end{pmatrix} . \tag{44}$$

It is easy to check that

$$(F_\alpha, F_\beta)_{\rm HS} = \delta_{\alpha\beta} \ . \tag{45}$$

Moreover

$$||F_1||_1^2 = \frac{1}{2} \ , \qquad ||F_1||_2^2 = 1 \ . \tag{46}$$

One easily finds $\mu_1 = 1$ and hence condition (33) for $k = 1$ is trivially satisfied $\lambda_\alpha \geq \mu_1$ for $\alpha = 2, 3, 4$.

Example 2 A map in $M_d(\mathbb{C})$ known as the reduction map

$$R_d(X) = \mathbb{I}_d {\rm Tr}\, X - X \ . \tag{47}$$

It is easy to check for positivity: indeed for any rank-1 projector $P = |\psi\rangle\langle\psi|$ one gets

$$R_d(P) = \mathbb{I}_d - P = P^\perp \ , \tag{48}$$

where $P^\perp$ defines a projector onto a hyperplane orthogonal to ψ. Hence, $R_d(P)$ being a projector is semi-positive definite. Again, for $d = 2$ it is easy to find the corresponding entanglement witness W

$$W = \left(\begin{array}{cc|cc} \cdot & \cdot & \cdot & -1 \\ \cdot & 1 & \cdot & \cdot \\ \hline \cdot & \cdot & 1 & \cdot \\ -1 & \cdot & \cdot & \cdot \end{array} \right) , \tag{49}$$

with the spectrum given by $\{-1, 1, 1, 1\}$. It shows that reduction map is positive but not CP. Interestingly, one may show that composing reduction with transposition one obtains a CP map. One finds

$$\Phi_-(X) = \lambda_1 F_1 X F_1^* \ , \tag{50}$$

and

$$\Phi_+(X) = \sum_{\alpha=2}^{D} \lambda_\alpha F_\alpha X F_\alpha^* \ , \tag{51}$$

where $\lambda_1 = d - 1$ and $\lambda_2 = \ldots = \lambda_D = 1$. Moreover

$$F_1 = \frac{1}{\sqrt{d}} \mathbb{I}_d \ , \tag{52}$$

and the remaining F_α are arbitrary $d \times d$ matrices such that

$$(F_\alpha, F_\beta)_{\rm HS} = \delta_{\alpha\beta} \ , \tag{53}$$

and $\mathrm{Tr}\, F_\alpha = 0$ (for $\alpha = 2,\ldots,D$). Again, one finds $\mu_1 = 1$ and hence condition (33) is trivially satisfied $\lambda_\alpha \geq \mu_1$ for $\alpha = 2,\ldots,D = d^2$. Now, since F_1 corresponds to the maximally entangled state one has $1 - ||F_1||_2^2 = (d-2)/d < 1$. Hence, condition (34)

$$\mu_2 = 2\,\frac{d-2}{d-1} > \lambda_\alpha\ , \quad \alpha = 2,\ldots,D\ , \tag{54}$$

implies that Φ is not a 2-positive.

Example 3 A family of Choi maps in $M_d(\mathbb{C})$

$$\Phi_\mu(X) = \mathbb{I}_d \mathrm{Tr}\, X - \mu\, X\ , \quad \mu \leq 1\ . \tag{55}$$

Note, that for $\mu = 1$ it reproduces the reduction map (47). Now, conditions (33) and (34) imply that if

$$\frac{1}{k+1} < \mu \leq \frac{1}{k}\ , \tag{56}$$

then Φ_μ is k-positive but not $(k+1)$-positive. It reproduces well known result [23]. Hence, if $\mu \leq 1/d$, then Φ_μ is CP. Again, for $d = 2$ it is easy to find the corresponding W_μ matrix

$$W_\mu = \left(\begin{array}{cc|cc} 1-\mu & \cdot & \cdot & -\mu \\ \cdot & 1 & \cdot & \cdot \\ \hline \cdot & \cdot & 1 & \cdot \\ -\mu & \cdot & \cdot & 1-\mu \end{array}\right)\ , \tag{57}$$

with the spectrum given by $\{1-2\mu, 1, 1, 1\}$. Clearly, for $\mu \leq 1/2$ the map is CP.

Example 4 A family of entanglement witnesses in $\mathbb{C}^3 \otimes \mathbb{C}^3$ defined by [25]

$$W[a,b,c] = \left(\begin{array}{ccc|ccc|ccc} a & \cdot & \cdot & \cdot & -1 & \cdot & \cdot & \cdot & -1 \\ \cdot & b & \cdot & \cdot & \cdot & \cdot & \cdot & \cdot & \cdot \\ \cdot & \cdot & c & \cdot & \cdot & \cdot & \cdot & \cdot & \cdot \\ \hline \cdot & \cdot & \cdot & c & \cdot & \cdot & \cdot & \cdot & \cdot \\ -1 & \cdot & \cdot & \cdot & a & \cdot & \cdot & \cdot & -1 \\ \cdot & \cdot & \cdot & \cdot & \cdot & b & \cdot & \cdot & \cdot \\ \hline \cdot & \cdot & \cdot & \cdot & \cdot & \cdot & b & \cdot & \cdot \\ \cdot & \cdot & \cdot & \cdot & \cdot & \cdot & \cdot & c & \cdot \\ -1 & \cdot & \cdot & \cdot & -1 & \cdot & \cdot & \cdot & a \end{array}\right)\ , \tag{58}$$

with $a,b,c \geq 0$. Necessary and sufficient conditions for $W[a,b,c]$ to be an entanglement witness are

(i) $0 \leq a < 2$,
(ii) $a + b + c \geq 2$,
(iii) if $a \leq 1$, then $bc \geq (1-a)^2$.

A family $W[a,b,c]$ generalizes celebrated Choi indecomposable witness corresponding to $a = b = 1$ and $c = 0$. It was shown [25] that $W[a,b,c]$ is decomposable if and only if $a \geq 0$ and

$$bc \geq \frac{(2-a)^2}{4}\ . \tag{59}$$

Hence $W[a,b,c]$ from our spectral class is always decomposable. In particular $W[0,1,1]$ reproduces the entanglement witness corresponding to the reduction map in $d = 3$. Note, that there are entanglement witnesses $W[a,b,c]$ which are decomposable, i.e. satisfy (59), but do not belong to or spectral class. Similarly one can check when $W[a,b,c]$ defines an entanglement witness corresponding to 2-positive map. One finds $\mu_2 = 2(2-a)$ and hence condition (33) implies

(i) $1 \leq a < 2$,

(ii) $b, c \geq 2(2-a)$,

which shows that condition (59) immediately follows.

5. Conclusions

We provide partial classification of positive linear maps based on spectral conditions. From the physical point of view our scheme provides a simple method for constructing entanglement witnesses. It turns out [17] that this scheme may be easily generalized to the multipartite setting. Interestingly many well known examples of positive maps (e.g. transposition, reduction map, the family of Choi maps) belong to the class constructed *via* spectral conditions. Therefore, in order to provide full characterization of positive maps further analysis is needed.

Acknowledgement

This work was partially supported by the Polish Ministry of Science and Higher Education Grant No 3004/B/H03/2007/33.

References

[1] M. A. Nielsen and I. L. Chuang, *Quantum computation and quantum information*, Cambridge University Press, Cambridge, 2000.
[2] A. Peres, Phys. Rev. Lett. **77**, 1413 (1996).
[3] P. Horodecki, Phys. Lett. A **232**, 333 (1997).
[4] E. Størmer, Acta Math. **110**, 233 (1963).
[5] W. Arveson, Acta Math. **123**, 141 (1969).
[6] M.-D. Choi, Lin. Alg. Appl. **10**, 285 (1975); *ibid* **12**, 95 (1975).
[7] S.L. Woronowicz, Rep. Math. Phys. **10**, 165 (1976).
[8] S.L. Woronowicz, Comm. Math. Phys. **51**, 243 (1976).
[9] A.G. Robertson, Quart. J. Math. Oxford (2), **34**, 87 (1983).
[10] A.G. Robertson, Proc. Roy. Soc. Edinburh Sect. A, **94**, 71 (1983). **34**, 87 (1983).
[11] W.-S. Tang, Lin. Alg. Appl. **79**, 33 (1986).
[12] H. Osaka, Lin. Alg. Appl. **153**, 73 (1991); *ibid* **186**, 45 (1993).
[13] K.-C. Ha, Publ. RIMS, Kyoto Univ., **34**, 591 (1998).
[14] D. Chruściński and A. Kossakowski, J. Phys. A: Math. Theor. **41**, 215201 (2008); *ibid* **41**, 145301 (2008).
[15] E. Størmer, J. Funct. Anal. **254**, 2303 (2008).
[16] D. Chruściński and A. Kossakowski, Open Systems and Inf. Dynamics, **14**, 275 (2007).
[17] D. Chruściński and A. Kossakowski, Comm. Math. Phys. **290**, 1051 (2009).
[18] Chruściński and A. Kossakowski, Phys. Lett. A **373**, 2301 (2009).
[19] V. Paulsen, *Completely Bounded Maps and Operator Algebras*, Cambridge University Press, 2003.
[20] B.M. Terhal, Phys. Lett. A **271**, 319 (2000); Linear Algebr. Appl. **323**, 61 (2000).
[21] M. Lewenstein, B. Kraus, J. I. Cirac, and P. Horodecki, Phys. Rev. A **62**, 052310 (2000).
[22] R.A. Horn and C.R. Johnson, *Topics in Matrix Analysis*, (Cambridge University Press, New York, 1991).
[23] B. Terhal and P. Horodecki, Phys. Rev. A **61**, 040301 (2000).
[24] K. Kraus, *States, Effects and Operations: Fundamental Notions of Quantum Theory*, Springer Verlag, 1983.
[25] S. J. Cho, S.-H. Kye, and S.G. Lee, Lin. Alg. Appl. **171**, 213 (1992).

Journal of Physics: Conference Series **284** (2011) 012018 doi:10.1088/1742-6596/284/1/012018

How to embed 'open' Lie superalgebras into affine Kac-Moody superalgebras

Claudia Daboul [1], **Jamil Daboul** [2], **Marc de Montigny** [3]

[1] IBP Ltd, 17 Perryn Road, W3 7LR London, United Kingdom,
[2] Physics Department, Ben Gurion University of the Negev, 84105 Beer Sheva, Israel,
[3] Campus Saint-Jean and Theoretical Physics Institute University of Alberta, 8406 - 91 Street Edmonton, Alberta, Canada T6C 4G9

E-mail: claudia@ibp.de, jdaboul@gmail.com, montigny@phys.ualberta.ca

Abstract. We discuss the concepts of *open* Lie superalgebras and their *closure*, and describe the procedure of embedding the closure into affine twisted and untwisted Kac-Moody *super*algebras. We apply the formalism to an open algebra of a relativistic model obtained by Cotăescu and Visinescu [J. Phys. A **40** (2007) 11987]. This yields an intuitive explanation of why the *hydrogen algebras*, studied by Daboul and Slodowy, must be 'twisted-like' KM subalgebras.

1. Introduction

Recently, we examined the concepts of *open* Lie superalgebras and their *closure*, and some applications [1]. In order to introduce these concepts, consider the dynamical symmetry algebra of the 3-dimensional non-relativistic hydrogen atom, which is spanned by the conserved angular momentum $\vec{L}$ and the Runge-Lenz vector $\vec{A}$. The components of $\vec{L}$ and $\vec{A}$ satisfy the following commutation relations (or Poisson brackets, in the classical treatment) [2, 3, 4]:

$$[L_i, L_j] = \mathrm{i}\,\varepsilon_{ijk} L_k, \qquad i,j,k = 1,2,3\,, \tag{1a}$$
$$[L_i, A_j] = \mathrm{i}\,\varepsilon_{ijk} A_k, \tag{1b}$$
$$[A_i, A_j] = \mathrm{i}\,\varepsilon_{ijk}\, h\, L_k, \qquad \text{where} \quad h := -2mH\,, \tag{1c}$$
$$[h, L_i] \;= [h, A_i] = 0\,, \tag{1d}$$

where m is the reduced mass, and H is the Hamiltonian. It is the appearance of H which leads to an open algebra and its closure through embedding into an infinite-dimensional algebra.

The appearance of the factor H in (1c), which can be either an operator (or a function), but not a constant, together with the commutation relations (1), makes the following set

$$\breve{\mathbb{H}}_3 := \{\vec{L}, \vec{A}; h\} \equiv \{L_i, A_i \;;\; h \mid i = 1,2,3\}\,, \tag{2}$$

a Lie algebra which is not closed. For this reason, we shall call such algebras *open Lie (super) algebras*. The usual way to deal with this dilemma is to replace the Hamiltonian H by its energy eigenvalues E, and then to rescale the Runge-Lenz vector. This procedure yields three closed (real) Lie algebras, $\mathfrak{so}(4), \mathfrak{so}(3,1)$ and $\mathfrak{e}(3)$, depending on the sign of the energy: $E < 0, E > 0$ and $E = 0$, respectively (see references in [1]-[4]).

Instead, if we iterate the commutation relations of $\{\vec{L}, \vec{A}; h\}$, we generate the following closed infinite-dimensional Lie algebra, which we call the *closure* of $\{\vec{L}, \vec{A}; h\}$:

$$\mathbb{H}_3^* := \overline{\{\vec{L}, \vec{A}; h\}} = \langle h^n L_i, h^n A_i \mid n \geq 0 \rangle_{\mathbb{R}}. \tag{3}$$

Daboul and Slodowy *identified* $\mathbb{H}_3^*$ as the positive part of a twisted Kac-Moody algebra (KMA) $D_2^{(2)+}$ [4]. Later Daboul and Daboul noted that $D_2^{(2)+}$ is isomorphic to $A_1^{(1)+}$ [5, 1]. In other words,

$$\mathbb{H}_3^* \simeq \mathbb{H}_3 := \langle \hat{L}_i^{2n}, \hat{A}_i^{2n+1} \mid n \geq 0 \rangle_{\mathbb{R}} \simeq D_2^{(2)+} \simeq A_1^{(1)+}, \tag{4}$$

where the notation $\mathfrak{s}^{(m)}$ will be defined in (11), and

$$\hat{L}_i^{2n} := \chi(h^n L_i) \quad \text{and} \quad \hat{A}_i^{2n+1} := \chi(h^n A_i),$$

where χ is a homomorphism, which we shall explain in (22) below.

The dynamical algebras for the N-dimensional hydrogen atoms were later identified in [5]: they were denoted as $\mathbb{H}_N$ and called *hydrogen algebras.* We review this identification in Sec. 4.3 below.

The purpose of this paper is to show how the above procedure can be generalized to embeddings of open *super*algebras into loop *super*algebras. The proofs of the relevant theorems can be found in our article [1].

2. Affine Lie superalgebras

In this section, we briefly review the concepts associated with affine Lie superalgebras that pertain to our discussion.

2.1. *Lie superalgebras*

A *Lie superalgebra* (over a field K either $\mathbb{R}$ or $\mathbb{C}$), $\mathfrak{s} = \mathfrak{s}_{\bar{0}} \oplus \mathfrak{s}_{\bar{1}}$, is a superalgebra if its product $[\![\cdot,\cdot]\!]$, called *supercommutator*, satisfies the following three conditions among its *homogeneous* elements:

(1) *algebraic grading:*

$$[\![\mathfrak{s}_{\bar{i}}, \mathfrak{s}_{\bar{j}}]\!] \subset \mathfrak{s}_{\overline{i+j}}, \quad (\bar{i}, \bar{j} \in \mathbb{Z}_2),$$

(2) *super skew-symmetry:*

$$[\![x, y]\!] = -(-1)^{|x||y|}[\![y, x]\!],$$

(3) *super Jacobi-identity:*

$$(-1)^{|z||x|}[\![x, [\![y, z]\!]\,]\!] + (-1)^{|x||y|}[\![y, [\![z, x]\!]\,]\!] + (-1)^{|y||z|}[\![z, [\![x, y]\!]\,]\!] = 0 .$$

where to an element $x \in \mathfrak{s}_{\bar{i}}$, one assigns a *parity*, by $|x| = i \in \mathbb{Z}_2$ [6, §1.5]. The elements with $|x| = 0$ and $|x| = 1$ are called *even* and *odd*, respectively.

Example: An interesting example of simple finite-dimensional superalgebras, which also illustrates the even-odd grading, is given by the $(m+n)$ square matrices over $\mathbb{C}$:

$$\mathfrak{gl}(m|n) = \mathfrak{gl}(m|n)_{\bar{0}} \oplus \mathfrak{gl}(m|n)_{\bar{1}} := \left\{\begin{pmatrix} a & 0 \\ 0 & d \end{pmatrix}\right\} \oplus \left\{\begin{pmatrix} 0 & b \\ c & 0 \end{pmatrix}\right\}, \tag{5}$$

where the matrices a and d are m and n square matrices, respectively.

Journal of Physics: Conference Series **284** (2011) 012018 doi:10.1088/1742-6596/284/1/012018

The matrices in (5) become a Lie superalgebra, called the *general linear Lie superalgebra*, if we define its supercommutator by

$$[\![X,Y]\!] := XY - (-1)^{|X||Y|}YX\,, \qquad \text{for} \quad X \in \mathfrak{gl}(m|n)_{|X|}\,.$$

Hence, the supercommutator is related to the ordinary commutator, $[X,Y] := XY - YX$, as follows:

$$[\![X_1,X_2]\!] = [X_1,X_2] + 2\begin{pmatrix} b_2c_1 & 0 \\ 0 & c_2b_1 \end{pmatrix}, \quad \text{where} \quad X_i = \begin{pmatrix} a_i & b_i \\ c_i & d_i \end{pmatrix}. \tag{6}$$

In particular, Eq. (6) shows that supercommutators and commutators become equal, if both X_1 and X_2 have diagonal form, *i.e.* either $b_1 = b_2 = 0$ or $c_1 = c_2 = 0$.

2.2. *Affine untwisted Kac-Moody superalgebras*

A *loop superalgebra* $\mathcal{L}(\mathfrak{s})$ can be defined from *any* finite-dimensional Lie superalgebra $\mathfrak{s}$ as follows:

$$\mathcal{L}(\mathfrak{s}) = \mathsf{K}[t,t^{-1}] \otimes_{\mathsf{K}} \mathfrak{s} \simeq \langle x^n \mid x \in \mathfrak{s}, n \in \mathbb{Z}\rangle_{\mathsf{K}}\,, \tag{7}$$

where $\mathsf{K}[t,t^{-1}]$ denotes the algebra of Laurent polynomials in the variable t, and $x^n := t^n x$. The generators of $\mathcal{L}(\mathfrak{s})$ satisfy the *supercommutation relations*

$$[\![x^m, y^n]\!] = [\![x,y]\!]^{m+n}\,. \tag{8}$$

Note that, hereafter, we only deal with loop superalgebras, not with genuine affine Kac-Moody algebras (which have central terms). Nevertheless, we shall use the terminology and notations of Kac-Moody-algebras to describe these algebras. In particular, we will use the notation $s^{(1)}$ for $\mathcal{L}(\mathfrak{s})$.

2.3. *Affine twisted Kac-Moody superalgebras*

Definition 1. Let $\mathfrak{s}$ be a finite-dimensional Lie superalgebra with a $\mathbb{Z}_m$-grading that commutes with the parity grading on $\mathfrak{s}$,

$$\mathfrak{s} = \bigoplus_{k=0}^{m-1} \mathfrak{s}_k, \qquad m \geq 2\,, \tag{9}$$

such that

$$[\![\mathfrak{s}_i, \mathfrak{s}_j]\!] \subset \mathfrak{s}_{i+j}, \quad i,j,i+j \ (\mathrm{mod}\ m).$$

and such that

$$\mathfrak{s}_k = (\mathfrak{s}_k)_{\bar{0}} \oplus (\mathfrak{s}_k)_{\bar{1}}, \quad (\mathfrak{s}_k)_{\bar{p}} = \mathfrak{s}_k \cap \mathfrak{s}_{\bar{p}}, \quad k \in \mathbb{Z}_m, \bar{p} \in \mathbb{Z}_2. \tag{10}$$

Then a *Lie super subalgebra* $\mathfrak{s}^{(m)}$ of $\mathfrak{s}^{(1)}$, is defined as follows:

$$\mathfrak{s}^{(m)} := \bigoplus_{n\in\mathbb{Z},k=0}^{m-1} t^{nm+k} \otimes \mathfrak{s}_k = \mathfrak{s}^{(m)}_{\bar{0}} \oplus \mathfrak{s}^{(m)}_{\bar{1}}\,, \tag{11}$$

where

$$\mathfrak{s}^{(m)}_{\bar{p}} = \bigoplus_{n\in\mathbb{Z},k=0}^{m-1} t^{nm+k} \otimes (\mathfrak{s}_k)_{\bar{p}}\,.$$

Note that $\mathfrak{s}^{(m)}_{\bar{0}}$ is a normal (not super) subalgebra of $\mathfrak{s}^{(m)}$.

Any $\mathfrak{s}^{(m)}$ of the form (11) will be called *twisted-like KMA*, since it may still be isomorphic to an untwisted KMA. Only when this is not the case, we will refer to it as a *twisted KMA* or "genuinely twisted KMA". We shall also call such $\mathfrak{s}^{(m)}$ *twisted KM* sub*algebra*, if its connection to $\mathfrak{s}^{(1)}$ is relevant.

2.4. Level-grading and positive subalgebras of KM superalgebras

Eq. (8) yields the *natural grading* for the (twisted or untwisted) loop superalgebras $\mathfrak{s}^{(m)}$

$$[\![\mathfrak{s}^{[m]}, \mathfrak{s}^{[n]}]\!] \subseteq \mathfrak{s}^{[m+n]}, \tag{12}$$

where the homogeneous subspaces

$$\mathfrak{s}^{[n]} := \mathfrak{s}^{(m)} \cap \langle x^n | x \in \mathfrak{s} \rangle_{\mathsf{K}} \tag{13}$$

will be called *levels.* To their elements, we assign a grade, which we call *level-grade* (or *l-grade*), and denote it by

$$\lfloor x^n \rfloor = \lfloor t^n x \rfloor = n\,. \tag{14}$$

We define the *positive subalgebra* $\mathfrak{s}^{(m)+}$ as the subalgebra of a (twisted or untwisted) KMA $\mathfrak{s}^{(m)}$ which is the sum of all its non-negative levels,

$$\mathfrak{s}^{(m)+} := \langle x^n \in \mathfrak{s}^{(m)} \,|\, n \geq 0 \rangle_{\mathsf{K}} = \bigoplus_{n=0}^{\infty} \mathfrak{s}^{[n]}\,. \tag{15}$$

3. Open algebras and their closures

Equipped with the previous definitions, we now turn to the concepts of *open algebras*, their *closures* and their *embedding.*

We shall also show in this section that by simply solving the system of linear equations (23) (which depends only on the integer coefficients r_{ij}^k in the definition (18) of open algebras), we can determine in general whether an open algebra leads to a twisted-like or to an untwisted KMA. We apply this formalism first to the hydrogen algebras, $\mathbb{H}_N$, which were mentioned in the introduction, and then to the dynamical symmetry algebra of the Dirac-Taub-NUT (DTN) model [7]. It turns out that the hydrogen algebras lead to twisted-like KM algebras while the DTN model leads to an untwisted KM algebra. Furthermore we will see how both systems are closely connected.

3.1. Definitions and notation

Consider a finite-dimensional Lie superalgebra,

$$\mathfrak{s} := \langle x_1, \ldots, x_N \rangle_{\mathsf{K}}\,, \qquad \text{with} \quad [\![x_i, x_j]\!] = \sum_{k=1}^{N} c_{ij}^k\, x_k, \quad i,j = 1, \ldots, N\,, \tag{16}$$

over the field K, with x_i homogeneous with respect to the inherent $\mathbb{Z}_2$-grading of the superalgebra.

Definition 2. Let

$$\breve{\mathfrak{s}} = \{X_1, \ldots, X_N; v\} \tag{17}$$

be a set of generators X_i and v a variable or an *operator of infinite order* (analogous to h in Eq. (2)) that commutes with all the X_i, and such that

$$[\![X_i, X_j]\!] = \sum_{k=1}^{N} c_{ij}^k\, v^{r_{ij}^k} X_k, \quad \text{for } i,j = 1, \ldots, N\,, \tag{18}$$

where the powers $r_{ij}^k \in \mathbb{N}_0$ are such that the supercommutators in (18) satisfy the super-grading, the super-skew-symmetry and the super-Jacobi identity, and with the parity of the generators defined by

$$|v^n X_i| := |x_i|. \tag{19}$$

We call $\breve{\mathfrak{s}}$ an *open algebra* if at least one of the r^k_{ij} in (18) is greater than zero, and the finite-dimensional Lie superalgebra $\mathfrak{s}$ in (16) is its *underlying algebra.*

By commuting the generators on the r.h.s. of Eq. (18) with each other and with the original X_i, we generate higher powers of v, such as

$$[\![v^{r^k_{ij}} X_k, v^{r^l_{ij}} X_l]\!] = \sum_p c^p_{kl} v^{r^k_{ij}+r^l_{ij}+r^p_{kl}} X_p \,.$$

We refer to this procedure as *repeated iterations.*

Definition 3. The *closure* of $\breve{\mathfrak{s}}$ is defined as *the smallest Lie superalgebra which can be generated from* $\breve{\mathfrak{s}}$ *by repeated iterations of its commutation relations.* It is denoted by $\overline{\breve{\mathfrak{s}}}$ or, more explicitly, by $\overline{\{\breve{\mathfrak{s}}; v\}}$.

Note that we do not include the operator v itself as a generator. Therefore the closure $\overline{\{\breve{\mathfrak{s}}; v\}}$ of an open algebra does not have to include *all* the powers $v^n X_i$ with $n \geq 0, i = 1, \ldots, N$. Note that if the underlying finite-dimensional algebra $\mathfrak{s}$ is nilpotent, then $\overline{\{\breve{\mathfrak{s}}; v\}}$ is finite dimensional. See another counter-example in Eq.(31). Thus,

$$\overline{\{\breve{\mathfrak{s}}; v\}} \subseteq \mathcal{L}(\breve{\mathfrak{s}}) := \langle v^n X_i, n \in \mathbb{Z}, i = 1, \ldots, N \rangle \,. \tag{20}$$

Lemma 1. The following mapping

$$\varphi(vX_i) = \varphi(X_i) = x_i \,, \quad \forall\ X_i \in \breve{\mathfrak{s}}, \tag{21}$$

of $\overline{\{\breve{\mathfrak{s}}; v\}}$ onto the superalgebra $\mathfrak{s}$ in (16), is a homomorphism [1, Lemma 2].

3.2. Embedding and identification of open algebras into affine Kac-Moody algebras

An *embedding* is an injective and structure-preserving map, denoted by $f : A \hookrightarrow B$. This map defines an isomorphism between a Lie algebra A and its image $f(A) \subseteq B$. Thus, we shall say that A can be *identified* with B if $f(A) \simeq B$.

In this section, we study the embedding of the closure of an open algebra $\overline{\breve{s}}$ into the loop algebra $\mathfrak{s}^{(1)} \equiv \mathcal{L}(\mathfrak{s})$ in (7), which is generated by the $\mathfrak{s}$ of (16).

Theorem 1. The closure $\overline{\breve{\mathfrak{s}}}$ of an open algebra $\breve{\mathfrak{s}}$ of the form (18) can be embedded into the affine KM algebra $\mathfrak{s}^{(1)}$, $\overline{\breve{\mathfrak{s}}} \hookrightarrow \mathfrak{s}^{(1)}$, by the following homomorphism:

$$\chi(v^p X_i) = t^{p\lfloor v \rfloor + \lfloor X_i \rfloor} x_i =: x_i^{p\lfloor v \rfloor + \lfloor X_i \rfloor} , \quad i = 1, \ldots, N, \quad X_i \in \breve{\mathfrak{s}}, \tag{22}$$

if and only if the homogeneous system of linear equations

$$\lfloor X_i \rfloor + \lfloor X_j \rfloor - \lfloor X_k \rfloor - r^k_{ij} \lfloor v \rfloor = 0 \,, \qquad \forall\ c^k_{ij} \neq 0 \quad i, j, k = 1, \ldots, N \,, \tag{23}$$

has a solution, which can always be chosen such that $\lfloor X_i \rfloor \in \mathbb{Z}$, $\lfloor v \rfloor \in \mathbb{N}$ and such that the greatest common divisor (GCD) of the $\lfloor X_k \rfloor$ and $\lfloor v \rfloor$ is 1.

As in (14) we call the integers $\lfloor X_i \rfloor$ and $\lfloor v \rfloor$ level grades or l-grades of the X_i and v respectively.

Theorem 2. Let $\mathfrak{g}$ be an ordinary (*i.e.* not super) Lie algebra, with all its structure constants totally antisymmetric, $c^k_{ij} = c_{ijk}$. Then all the generators $X_i \in \breve{\mathfrak{g}}$ must have non-negative l-grades, *i.e.* $\lfloor X_i \rfloor \geq 0$.

An example: Let us apply the above theorem to the open algebra of the 3-dimensional hydrogen atom, discussed in the introduction. From Eq. (1a), we obtain $\lfloor L\rfloor + \lfloor L\rfloor = \lfloor L\rfloor$, so that $\lfloor L\rfloor = 0$. Eq. (1b) leads to $\lfloor L\rfloor + \lfloor A\rfloor = \lfloor A\rfloor$, which yields no conditions. From Eq. (1c), we find $2\lfloor A\rfloor = \lfloor h\rfloor + \lfloor L\rfloor$. Under the restrictions of the theorem this has the unique solution

$$\lfloor L\rfloor = 0, \lfloor A\rfloor = 1, \lfloor h\rfloor = 2. \tag{24}$$

3.3. Twisted and untwisted open Lie algebras

Our motivation to undertake the work presented in Ref. [1] was the observation that, despite the claim in Ref. [7] that the open superalgebra obtained therein was of twisted type, that superalgebra is actually untwisted (More in Section 4).

Hereafter, we derive more results which will enable us to distinguish between the two types of open algebras.

Lemma 2. If an open Lie algebra $\breve{\mathfrak{s}}$ has a solution with $\lfloor v\rfloor = m$, then the underlying finite-dimensional Lie algebra

$$\mathfrak{s} = \langle x_1, \ldots, x_N\rangle_{\mathsf{K}} = \varphi(\breve{\mathfrak{s}})$$

admits a $\mathbb{Z}_m$-grading, defined by the subspaces

$$\mathfrak{s}_k = \langle x_i = \varphi(X_i) \mid \lfloor X_i\rfloor \equiv k\,(m)\,\rangle_{\mathsf{K}}\,. \quad \text{where} \quad 0 \le k \le m-1\;, \tag{25}$$

where the map φ is defined in Eq. (21) [1, Lemma 5].

This result is important, because it enables us to define a twisted-like KMSA $\mathfrak{s}^{(\lfloor v\rfloor)}$, according to the definition 1. This in turn allows us to embed $\bar{\breve{\mathfrak{s}}}$ into twisted-like KMSA:

Theorem 3. The embedding χ in (22) maps the closure $\bar{\breve{\mathfrak{s}}}$ into the KMA $\mathfrak{s}^{(\lfloor v\rfloor)}$.

Corollary 1 If $\mathfrak{g}$ satisfies the conditions of theorem 4, then $\bar{\breve{\mathfrak{g}}}$ can be embedded into a *positive* twisted-like subalgebras, as defined in (15):

$$\bar{\breve{\mathfrak{g}}} \hookrightarrow \mathfrak{g}^{(\lfloor v\rfloor)+}$$

Since $\mathfrak{s}^{(\lfloor v\rfloor)}$ for $v \geq 2$ is a subalgebra of the untwisted KMA $\mathfrak{s}^{(1)}$, the embedding χ can be extended naturally to an embedding of *the loop algebra* $\mathcal{L}(\breve{\mathfrak{s}})$ *of the open algebra* $\breve{\mathfrak{s}}$

$$\mathcal{L}(\breve{\mathfrak{s}}) := \langle v^n X_i, n \in \mathbb{Z}, i = 1, \ldots, N\rangle$$

into $\mathfrak{s}^{(1)}$ by

$$\chi(v^n X_i) := t^{n\lfloor v\rfloor}\chi(X_i). \tag{26}$$

Corollary 2: The loop algebra $\mathcal{L}(\breve{\mathfrak{s}})$ is isomorphic to $\mathfrak{s}^{(\lfloor v\rfloor)}$ by the following natural extension of the embedding χ in (26):

$$\chi(\mathcal{L}(\breve{\mathfrak{s}})) = \mathfrak{s}^{(\lfloor v\rfloor)}.$$

Journal of Physics: Conference Series **284** (2011) 012018 doi:10.1088/1742-6596/284/1/012018

3.4. Identification of $\mathbb{H}_N$

As in (24) for $\mathbb{H}_3$, the linear system of equations (23) yields *unique* solutions with $\lfloor h \rfloor = 2$ also for the hydrogen algebras $\mathbb{H}_N$. Thus, according to the above corollary, the closure $\mathbb{H}_N^*$ of $\breve{\mathbb{H}}_N$, which is the dynamical algebra of the N-dimensional hydrogen atom, has a characteristic embedding which extends to an isomorphism of its loop algebra with the twisted-like subalgebra $\mathfrak{so}(N+1)^{(2)}$ of $\mathfrak{so}(N+1)^{(1)}$. This enabled us to identify $\mathbb{H}_N^*$ with the twisted-like $\mathfrak{so}(N+1)^{(2)}$ [5]. These are real KMA, which are isomorphic to the following real affine KMA:

$$\begin{aligned} \mathbb{H}_{2l-1}^* &\simeq \mathbb{H}_{2l-1} \simeq D_l^{(2)+}, & l \geq 2, \\ \mathbb{H}_{2l}^* &\simeq \mathbb{H}_{2l} \simeq B_l^{(2)+} \subseteq B_l^{(1)+}, & l \geq 2, \end{aligned} \tag{27}$$

where

$$\mathbb{H}_N^* := \overline{\{L_{ij}, A_i | 1 \leq i,j \leq N\}} = \langle h^n L_{ij}, h^n A_i | n \in \mathbb{N}_0 \rangle_{\mathbb{R}},$$

$$\mathbb{H}_N := \langle \hat{L}_{ij}^{2n}, \hat{A}_i^{2n+1} | 1 \leq i,j \leq N; n \in \mathbb{N}_0 \rangle_{\mathbb{R}}.$$

For a more details see eqn. (36) in [5].

Note also that $D_l^{(2)}$ for $l \geq 3$ are *genuinely twisted.* In contrast, $B_l^{(2)}$ are not genuine; they can be untwisted, *i.e.* $B_l^{(2)} \simeq B_l^{(1)}$.

An interesting exception is the algebra $D_2^{(2)}$, since $D_2^{(2)} \simeq A_1^{(1)}$, as we proved in the appendix of [1]. This is because D_2 is not a simple Lie algebra.

4. Open superalgebra of DTN model

In Ref. [7], Cotăescu and Visinescu examined a relativistic Dirac theory on manifolds carrying the Gross-Perry-Sorkin monopole defined on the Euclidean Taub-NUT space. This 'Dirac-Taub-NUT' (DTN) model has the following dynamical algebra,

$$\mathcal{S}_0 = \{J_i, K_i, \tilde{I}, M, Q, Q_i; b\}, \qquad i = 1,2,3, \tag{28}$$

where, *by definition*, all the generators in (28) commute with the relativistic operator b, where b^2 is defined by

$$b^2 := {P_4}^2\, \tilde{I} - H_D^2, \tag{29}$$

where H_D in Eq. (29) is a Dirac Hamiltonian and $P_4 := -\mathrm{i}\,\frac{\partial}{\partial x_4}$ [7]. Moreover, the operators $\tilde{I}$ and M in Eq. (28) are Casimir-type bosonic operators, which commute with all the operators in $\mathcal{S}_0$.

The generators in (28) satisfy the following supercommutation relations:

$$[J_i, J_j] = \mathrm{i}\,\varepsilon_{ijk} J_k, \tag{30a}$$
$$[J_i, K_j] = \mathrm{i}\,\varepsilon_{ijk} K_k, \tag{30b}$$
$$[K_i, K_j] = \mathrm{i}\,\varepsilon_{ijk}\, b^2\, J_k, \tag{30c}$$
$$[J_i, Q_j] = \mathrm{i}\,\varepsilon_{ijk} Q_k, \tag{30d}$$
$$[K_i, Q_j] = \mathrm{i}\,\varepsilon_{ijk} b Q_k, \tag{30e}$$
$$\{Q_i, Q_j\} = 2\delta_{ij}\tilde{I}, \tag{30f}$$
$$[J_i, Q] = [J_i, Q] = 0, \tag{30g}$$
$$\{Q, Q_j\} = 2(bJ_i + K_i), \tag{30h}$$
$$\{Q, Q\} = 2b^2 M, \tag{30i}$$

so that according to our definition the $\mathcal{S}_0$ is an open superalgebra.

According to our definition in Sec. 3.2, the closure $\overline{\mathcal{S}_0}$ is the subalgebra of $\mathcal{L}(\mathcal{S}_0)$ generated by $\mathcal{S}_0$. Hence,

$$\overline{\mathcal{S}_0} = \langle b^n J_i, b^n K_i, b^n \tilde{I}, b^n Q_i \mid n \geq 0 \rangle_{\mathsf{K}} + \langle Q, M, b^2 M \rangle_{\mathsf{K}}\,, \tag{31}$$

is a genuine subalgebra of $\mathcal{L}(\mathcal{S}_0)^+$, since instead of $\langle b^n Q, b^n M | n \geq 0\rangle_{\mathsf{K}}$ we can only generate the three elements Q, M and $b^2 M$.

4.1. Embedding of $\overline{\mathcal{S}_0}$

The linear system (23) in Theorem 1 has a unique solution with integer values, positive level grade of b and $\mathrm{GCD}(\lfloor X_i \rfloor, \lfloor b \rfloor) = 1$:

$$\lfloor K_i \rfloor = \lfloor Q \rfloor = \lfloor b \rfloor = 1\,, \qquad i = 1,2,3, \qquad \text{and}$$
$$\lfloor J_i \rfloor = \lfloor Q_i \rfloor = \lfloor \tilde{I} \rfloor = \lfloor M \rfloor = 0\,. \tag{32}$$

As in [7] let $\mathcal{W}_0 := \varphi(\mathcal{S}_0)$ denote the underlying finite dimensional super algebra of the open algebra $\mathcal{S}_0$. Since $\lfloor b \rfloor = 1$ in (32), the associated embedding χ maps the closure $\overline{\mathcal{S}_0}$ into the affine *untwisted* KM algebra $\mathcal{W}_0^{(1)}$, but not into a twisted-like subalgebra $\mathcal{W}_0^{(m)}$, $m > 1$.

5. Summary and insight

We showed that $\mathcal{S}_0$ *cannot* be embedded into a *twisted* KMSA, *in contradiction to the conclusion of CV* in Ref [7].

The relativistic DTN provides us with an intuitive explanation of why $\lfloor h \rfloor = 2$, and consequently explains why $\mathbb{H}_N$ must be twisted-like. We argue as follows: Eqs. (30a)-(30c) are similar to Eqs. (1a)-(1c). Hence, the operator b^2 in (30c) corresponds to the non-relativistic Hamiltonian operator h in Eq. (1c). But since the variable b also appears linearly (in Eqs. (30e) and (30h)), a $\lfloor h \rfloor = \lfloor b^2 \rfloor = 1$ would yield $\lfloor b \rfloor = 1/2$, which contradicts the condition $\lfloor b \rfloor \in \mathbb{N}$.

We can summarize the embedding procedure by the following algorithm:

(i) Map the open algebra $\{\check{\mathfrak{s}}, v\}$ onto $\mathfrak{s}$ by the map φ in (21).

(ii) Find $\lfloor v \rfloor$ by solving the system of linear equation in (23).

(iii) Determine the associated $\mathbb{Z}_{\lfloor v \rfloor}$-grading of $\mathfrak{s}$.

(iv) Construct the twisted-like KMSA $\mathfrak{s}^{(\lfloor v \rfloor)}$.

(v) Finally, map $\overline{\{\check{\mathfrak{s}}, v\}}$ into $\mathfrak{s}^{(\lfloor v \rfloor)}$ by using the map χ in (26).

The above algorithm is illustrated in the following diagram:

$$\begin{array}{ccc}
\chi(v^n X_i) & := & (t^{\lfloor v \rfloor})^n \cdot \left(t^{\lfloor X_i \rfloor}\, x_i\right) = t^{n\lfloor v \rfloor + \lfloor X_i \rfloor}\, x_i \\
\Cap & & \Cap \\
\bar{\check{\mathfrak{s}}} = \langle v^n X_1, \ldots, v^n X_N \rangle_{\mathsf{K}} & \overset{\chi}{\hookrightarrow} & \mathfrak{s}^{(\lfloor v \rfloor)} \subseteq \mathfrak{s}^{(1)} = \langle t^n x_1, \ldots, t^n x_N \rangle_{\mathsf{K}} \\
\Uparrow \textbf{(iteration)} & & \Uparrow \textbf{(looping)} \\
\check{\mathfrak{s}} = \{X_1, \ldots, X_N; v\} & \overset{\varphi}{\longmapsto} & \mathfrak{s} := \langle x_1, \ldots, x_N \rangle_{\mathsf{K}}
\end{array}$$

References

[1] Daboul C, Daboul J and de Montigny M 2010 Embedding of "open" superalgebras into affine Kac-Moody superalgebras *J. Math. Phys.* **51** 053521(15pp)

[2] Schiff L I 1968 *Quantum Mechanics, 3rd edition* (New York: McGraw-Hill); Sudbery A 1986 *Quantum Mechanics and the Particles of Nature* (Cambridge: Cambridge University Press)

[3] Goldstein H 1980 *Classical Mechanics* (Reading: Addison-Wesley)

[4] Daboul J, Slodowy P and Daboul C 1993 The hydrogen algebra as centerless twisted Kac-Moody algebra *Phys. Lett. B* **317** 321-328;
Daboul C, Daboul J and Slodowy P 1995 The dynamical algebra of the hydrogen atom as a twisted loop algebra, in *Group Theoretical Methods in Physics, Proceedings of the 20th International Colloquium*, Eds. A. Arima, T. Eguchi, N. Nakanishi (Singapore: World Scientific) e-print ArXiv: `hep-th/9408080`

[5] Daboul C and Daboul J 1998 From hydrogen atom to generalized Dynkin diagrams *Phys. Lett. B* **425** 135-144

[6] M. Wakimoto, *Infinite-Dimensional Lie Algebras, (Translations of Mathematical Monographs, Vol. 195)* (American Mathematical Society, New York, 2001)

[7] Cotăescu I I and Visinescu M 2007 Infinite loop superalgebras of the Dirac theory on the Euclidian Taub-NUT space *J. Phys. A: Math. Theor.* **40** 11987-99

Journal of Physics: Conference Series **284** (2011) 012019 doi:10.1088/1742-6596/284/1/012019

Topics on n-ary algebras

J.A. de Azcárraga

Dept. of Theoretical Physics and IFIC (CSIC-UVEG), University of Valencia, 46100-Burjassot (Valencia), Spain

E-mail: j.a.de.azcarraga@ific.uv.es

J. M. Izquierdo

Dept. of Theoretical Physics, University of Valladolid, 47011-Valladolid, Spain

E-mail: izquierd@fta.uva.es

Abstract. We describe the basic properties of two n-ary algebras, the Generalized Lie Algebras (GLAs) and, particularly, the Filippov ($\equiv n$-Lie) algebras (FAs), and comment on their n-ary Poisson counterparts, the Generalized Poisson (GP) and Nambu-Poisson (N-P) structures. We describe the Filippov algebra cohomology relevant for the central extensions and infinitesimal deformations of FAs. It is seen that semisimple FAs do not admit central extensions and, moreover, that they are rigid. This extends the familiar Whitehead's lemma to all $n \geq 2$ FAs, $n = 2$ being the standard Lie algebra case. When the n-bracket of the FAs is no longer required to be fully skewsymmetric one is lead to the n-Leibniz (or Loday's) algebra structure. Using that FAs are a particular case of n-Leibniz algebras, those with an anticommutative n-bracket, we study the class of n-Leibniz deformations of simple FAs that retain the skewsymmetry for the first $n-1$ entires of the n-Leibniz bracket.

1. Introduction

The Jacobi identity (JI) for Lie algebras $\mathfrak{g}$, $[X,[Y,Z]] + [Y,[Z,X]] + [Z,[X,Y]] = 0$, may be looked at in two ways. First, it may be seen as a consequence of the associativity of the composition of the generators in the Lie bracket. Secondly, it may be viewed as the statement that the adjoint map is a derivation of the Lie algebra, $ad_X\,[Y,Z] = [ad_X\,Y, Z] + [Y, ad_X\,Z]$.

A natural generalization is to consider n-ary algebras $\mathfrak{H}$. In its more general form, the problem goes back to the multioperator linear algebras of Kurosh (see [1,2]). In our Lie algebra context, we have to look for the possible characteristic identities that a fully antisymmetric n-ary bracket,

$$(X_1, \ldots, X_n) \in \mathfrak{H} \times \cdots \times \mathfrak{H} \mapsto [X_1, \ldots, X_n] \in \mathfrak{H} \,, \tag{1.1}$$

may satisfy (the n-Leibniz case is discussed in Sec. 7). When $n > 2$ the above two aspects of the JI are no longer equivalent and two n-ary generalizations of the Lie algebra structure immediately suggest themselves. These depend on which aspect of the $n = 2$ JI is retained to define their corresponding *characteristic identity*. This leads to[1]:

[1] It is also possible to consider intermediate possibilities between the two below: see [3,4].

(a) ***Higher order Lie algebras or generalized Lie algebras (GLAs)*** $\mathcal{G}$.
They were proposed independently in [5–7] and [8–11]. Their bracket is defined by the full antisymmetrization

$$[X_{i_1}, \ldots, X_{i_n}] := \sum_{\sigma \in S_n} (-1)^{\pi(\sigma)} X_{i_{\sigma(1)}} \ldots X_{i_{\sigma(n)}} \, . \tag{1.2}$$

For n even, this definition implies the *generalized Jacobi identity* (GJI)

$$\sum_{\sigma \in S_{2n-1}} (-1)^{\pi(\sigma)} \left[[X_{i_{\sigma(1)}}, \ldots, X_{i_{\sigma(n)}}], X_{i_{\sigma(n+1)}} \ldots, X_{i_{\sigma(2n-1)}} \right] = 0 \, , \quad i = 1, \ldots, \dim \mathcal{G} \, , \tag{1.3}$$

which follows from the associtivity of the products in (1.2) (for n odd, the *r.h.s* is $n!(n-1)!\,[X_{i_1}, \ldots, X_{i_{2n-1}}]$ rather than zero giving rise to a *mixed GJI*[2]). Chosen a basis of $\mathcal{G}$, the bracket may be written as $[X_{i_1}, \ldots, X_{i_{2p}}] = \Omega_{i_1 \ldots i_{2p}}{}^j X_j$, where the $\Omega_{i_1 \ldots i_{2p}}{}^j$ are the *GLA structure constants.* Thus, for n even, a GLA is defined by an n-linear antisymmetric bracket (1.2) closed in $\mathcal{G}$ that satisfies the GJI (1.3).

(b) ***n-Lie or Filippov algebras (FAs)*** $\mathfrak{G}$.
The characteristic identity that generalizes the $n = 2$ JI is the *Filippov identity* (FI) [13]

$$[X_1, \ldots, X_{n-1}, [Y_1, \ldots Y_n]] = \sum_{a=1}^{n} [Y_1, \ldots Y_{a-1}, [X_1, \ldots, X_{n-1}, Y_a], Y_{a+1}, \ldots Y_n] \, . \tag{1.4}$$

Let $\mathscr{X} = (X_1, \ldots, X_{n-1})$ be antisymmetric in its $(n-1)$ entries, $\mathscr{X} \in \wedge^{n-1}\mathfrak{G}$. The $\mathscr{X}$s will be called [14] *fundamental objects*, and act on $\mathfrak{G}$ by

$$\mathscr{X} \cdot Z \equiv ad_{\mathscr{X}} Z := [X_1, \ldots, X_{n-1}, Z] \quad \forall Z \in \mathfrak{G} \, . \tag{1.5}$$

Thus, the FI just expresses that (note the dot) $\mathscr{X} \cdot \equiv ad_{\mathscr{X}}$ is a derivation of the FA bracket,

$$ad_{\mathscr{X}} [Y_1, \ldots, Y_n] = \sum_{a=1}^{n} [Y_1, \ldots, ad_{\mathscr{X}} Y_a, \ldots, Y_n] \, . \tag{1.6}$$

Chosen a basis, $\mathfrak{G}$ may be defined by the *FA structure constants*,

$$[X_{a_1} \ldots X_{a_n}] = f_{a_1 \ldots a_n}{}^d X_d \, , \quad a, d = 1, \ldots \dim \mathfrak{G} \, , \tag{1.7}$$

in terms of which the FI is written as

$$f_{b_1 \ldots b_n}{}^l f_{a_1 \ldots a_{n-1} l}{}^s = \sum_{k=1}^{n} f_{a_1 \ldots a_{n-1} b_k}{}^l f_{b_1 \ldots b_{k-1} l b_{k+1} \ldots b_n}{}^s \quad . \tag{1.8}$$

Note. There is a considerable confusion in the literature concerning the names of the above two n-ary algebras and those of the characteristic identities they satisfy; we refer to Sec. 1 in [15] for a justification of the terminology we advocate.

[2] When more than two nested brackets are used, other identities follow from associativity; see [12].

2. Some definitions and properties of FAs

The definitions of ideals, solvable ideals and semisimple Lie algebras can be extended to the $n > 2$ case [13, 16–18] following the pattern of the Lie algebra one (for a review of FAs and their applications with further references, see [15]). For instance, a subalgebra $I \subset \mathfrak{G}$ is an ideal of $\mathfrak{G}$ if $[X_1, \dots, X_{n-1}, Z] \subset I \quad \forall X \in \mathfrak{G}, \forall Z \in I$. An ideal I is (n-)solvable if the series

$$I^{(0)} := I \ , \ I^{(1)} := [I^{(0)}, \dots, I^{(0)}] \ , \dots , \ I^{(s)} := [I^{(s-1)}, \dots, I^{(s-1)}] \ , \dots \tag{2.9}$$

terminates. A FA $\mathfrak{G}$ is then semisimple if it does not have solvable ideals, and simple if $[\mathfrak{G}, \dots, \mathfrak{G}] \neq \{0\}$ and does not contain non-trivial ideals. There is also a Cartan-like criterion for semisimplicity [18]: a FA is semisimple iff

$$k(\mathscr{X}, \mathscr{Y}) = k(X_1, \dots, X_{n-1}, Y_1, \dots, Y_{n-1}) := Tr(ad_{\mathscr{X}} ad_{\mathscr{Y}}) \tag{2.10}$$

is non-degenerate in the sense that

$$k(Z, \mathfrak{G}, \overset{n-2}{\dots}, \mathfrak{G}, \mathfrak{G}, \overset{n-1}{\dots}, \mathfrak{G}) = 0 \ \Rightarrow \ Z = 0 \ . \tag{2.11}$$

A semisimple FA is the sum of simple ideals $\mathfrak{G} = \mathfrak{G}_{(1)} \oplus \dots \oplus \mathfrak{G}_{(k)}$.

The derivations of a FA $\mathfrak{G}$ generate a Lie algebra. To see it, introduce first the composition of fundamental objects [19],

$$\mathscr{X} \cdot \mathscr{Y} := \sum_{a=1}^{n-1} (Y_1, \dots, Y_{a-1}, [X_1, \dots, X_{n-1}, Y_a], Y_{a+1}, \dots, Y_{n-1}) \ , \tag{2.12}$$

which reflects that $\mathscr{X}$ acts as a derivation. It is then seen that the FI implies that

$$\mathscr{X} \cdot (\mathscr{Y} \cdot \mathscr{Z}) - \mathscr{Y} \cdot (\mathscr{X} \cdot \mathscr{Z}) = (\mathscr{X} \cdot \mathscr{Y}) \cdot \mathscr{Z} \qquad , \quad \forall \mathscr{X}, \mathscr{Y}, \mathscr{Z} \in \wedge^{n-1} \mathfrak{G} \ , \tag{2.13}$$

$$ad_{\mathscr{X}} ad_{\mathscr{Y}} Z - ad_{\mathscr{Y}} ad_{\mathscr{X}} Z = ad_{\mathscr{X} \cdot \mathscr{Y}} Z \qquad , \quad \forall \mathscr{X}, \mathscr{Y} \in \wedge^{n-1} \mathfrak{G}, \forall Z \in \mathfrak{G} \ , \tag{2.14}$$

which means that $ad_{\mathscr{X}} \in \mathrm{End}\, \mathfrak{G}$ satisfies $ad_{\mathscr{X} \cdot \mathscr{Y}} = -ad_{\mathscr{Y} \cdot \mathscr{X}}$. These two identities show that the inner derivations $ad_{\mathscr{X}}$ associated with the fundamental objects $\mathscr{X}$ generate (the ad map is not necessarily injective) an ordinary Lie algebra, the Lie algebra $\mathrm{Lie}\, \mathfrak{G}$ associated with the FA $\mathfrak{G}$.

An important type of FAs is the class of metric Filippov algebras. These are relevant in physical applications (where a scalar product is needed), as in the Bagger-Lambert-Gustavsson model [20–22] in M-theory. These FAs are endowed with a metric $< \ , \ >$, $\langle Y \ , \ Z \rangle = g_{ab} Y^a Z^b$, $\forall \ Y, Z \in \mathfrak{G}$ which is invariant *i.e.*,

$$\begin{aligned} \mathscr{X} \cdot \langle Y \ , \ Z \rangle &= \langle \mathscr{X} \cdot Y \ , \ Z \rangle + \langle Y \ , \ \mathscr{X} \cdot Z \rangle \\ &= \langle [X_1, \dots, X_{n-1}, Y] \ , \ Z \rangle + \langle Y \ , \ [X_1, \dots, X_{n-1}, Z] \rangle = 0 \ . \end{aligned} \tag{2.15}$$

As a result, the structure constants with all indices down $f_{a_1 \dots a_{n-1} bc}$ are completely antisymmetric since the invariance of g above implies $f_{a_1 \dots a_{n-1} b}{}^l g_{lc} + f_{a_1 \dots a_{n-1} c}{}^l g_{bl} = 0$. The $f_{a_1 \dots a_{n+1}}$ define a skewsymmetric invariant tensor f under the action of $\mathscr{X}$, since the FI implies

$$\sum_{i=1}^{n+1} f_{a_1 \dots a_{n-1} b_i}{}^l f_{b_1 \dots b_{i-1} l b_{i+1} \dots b_{n+1}} = 0 \qquad \text{or} \qquad L_{\mathcal{X}} . f = 0 \ . \tag{2.16}$$

3. Examples of n-ary algebras

3.1. Examples of GLAs

Let n be *even*, $n = 2p$. We look for structure constants $\Omega_{i_1 \ldots i_{2p}}{}^j$ satisfying the GJI (1.3) *i.e.*, such that

$$\Omega_{[j_1 \ldots j_{2p}}{}^l \Omega_{j_{2p+1} \ldots j_{4p-1}]l}{}^s = 0 \quad , \quad i = 1, \ldots, \dim \mathcal{G}. \tag{3.17}$$

It turns out [7, 6] that given a simple compact Lie algebra, the coordinates of the (odd) cocyles for the corresponding Lie algebra cohomology satisfy the GJI identity (3.17). Thus, these provide the structure constants of an infinity of GLAs with brackets with $n = 2(m_i - 1)$ entries, where $i = 1, \ldots, \ell$, ℓ is the rank of the algebra and the m_i are the ranks of the ℓ Casimir-Racah primitive symmetric invariants associated with the corresponding $(2m_i - 1)$-cocycles; see further [15, 23].

3.2. Examples of FAs

A very important class of finite Filippov algebras is provided by the real simple n-Lie algebras defined on $(n+1)$-dimensional vector spaces [13]. Chosen a basis $\{e_a\}$ $(a = 1, \ldots, n+1)$, their n-brackets are given by

$$[e_1 \ldots \hat{e}_a \ldots e_{n+1}] = (-1)^{i+1} \varepsilon_a e_a \quad \text{or} \quad [e_{a_1} \cdots \, e_{a_n}] = (-1)^n \sum_{a=1}^{n+1} \varepsilon_a \epsilon_{a_1 \ldots a_n}{}^a e_a \;, \tag{3.18}$$

where, using Filippov's notation, the $\varepsilon_a = \pm 1$ are sign factors. In particular, the Euclidean $(\varepsilon_a = +1)$ simple FAs A_{n+1} are constructed on Euclidean $(n+1)$-dimensional vector spaces. Thus, in contrast with the $n = 2$ (Lie) algebra case, simple n-Lie algebras have a very rigid structure for $n \geq 3$: they reduce to the Euclidean (A_{n+1}) and Lorentzian $(A_{s,t}\ ,\ s+t = n+1)$ generalizations of the $n = 2$ $so(3)$ and $so(1, 2)$ Lie algebras, $[e_i, e_j] = \sum_k \varepsilon_k \epsilon_{ijk} e_k$, $i, j, k = 1, 2, 3$.

There are also infinite-dimensional FAs that generalize the ordinary Poisson algebra by means of the bracket of n functions $f_i = f_i(x_1, x_2, \ldots, x_n)$ defined by

$$[f_1, f_2, \ldots, f_n] := \epsilon^{i_1 \ldots i_n}_{1 \ldots n}\, \partial_{i_1} f^1 \ldots \partial_{i_n} f^n = \left| \frac{\partial(f_1, f_2, \cdots, f_n)}{\partial(x^1, x^2, \ldots, x^n)} \right| \quad . \tag{3.19}$$

This bracket was considered by Nambu [24] (who discussed it specially for the $n = 3$ case) and by Filippov [13]. The above Jacobian n-bracket satisfies the FI, which can be checked *e.g.* by using the 'Schouten identities' trick; we denote the resulting FA by $\mathfrak{N}$. These FAs are also metric FAs. For the simple infinite-dimensional FAs see further [25] and references therein.

For $n = 2$, GLAs, FAs and Lie algebras coincide.

4. n-ary Poisson structures

Both GLAs and FAs have their n-ary Poisson structure counterparts. These satisfy the associated GJI and FI characteristic identities, to which Leibniz's rule is added.

(a) *Generalized Poisson structures (GPS)*

The generalized Poisson structures [5, 6] (GPS) are naturally introduced for $n = 2s$ even (see [26] for n odd and [27] for the Z_2-graded case). They are defined by brackets $\{f_1, \ldots, f_n\}$ where the f_i, $i = 1. \ldots, n = 2s$, are functions on a manifold. They are fully antisymmetric

$$\{f_1, \ldots, f_i, \ldots, f_j, \ldots, f_n\} = -\{f_1, \ldots, f_j, \ldots, f_i, \ldots, f_n\} \;, \tag{4.20}$$

satisfy Leibniz's rule ,

$$\{f_1, \ldots, f_{n-1}, gh\} = g\{f_1, \ldots, f_{n-1}, h\} + \{f_1, \ldots, f_{n-1}, g\}h \;, \tag{4.21}$$

and the characteristic identity of the GLAs, the GJI (1.3), which now reads

$$\sum_{\sigma\in S_{4s-1}} (-1)^{\pi(\sigma)}\{f_{\sigma(1)},\dots,f_{\sigma(2s-1)},\{f_{\sigma(2s)},\dots,f_{\sigma(4s-1)}\}\} = 0 \ . \tag{4.22}$$

As with ordinary Poisson structures, there are linear GPS given *e.g.* by the coordinates of the primitive, odd cocyles of the compact simple $\mathfrak{g}$. *Linear GPS* are defined by *linear GPS tensors i.e.*, by multivectors of the form

$$\Lambda = \frac{1}{(2m-2)!}\Omega_{i_1\dots i_{2m-2}}{}^{\sigma} x_\sigma \partial^{i_1}\wedge\dots\wedge\partial^{i_{2m-2}} \tag{4.23}$$

which have zero Schouten-Nijenhuis bracket with themselves [6,7]. Indeed, as it may be checked, $[\Lambda,\Lambda]_{SN}=0$ expresses the GJI (eq. (3.17)); this is satisfied when the $\Omega_{i_1\dots i_{2m-2}}{}^{\sigma}$ are the $(2m-1)$-cocycle coordinates [6,7]. In fact, all the $(2m_i-2)$-GLAs associated with the simple Lie algebras cohomology $(2m_i-1)$-cocycles define linear GPS.

(b) ***Nambu-Poisson structures (N-P)***
These are defined by relations (4.20) and (4.21), but now the characteristic identity is the FI,

$$\begin{aligned}&\{f_1,\dots,f_{n-1},\{g_1,\dots,g_n\}\} = \{\{f_1,\dots,f_{n-1},g_1\},g_2,\dots,g_n\} + \\ &\{g_1,\{f_1,\dots,f_{n-1},g_2\},g_3,\dots,g_n\}+\dots+\{g_1\dots,g_{n-1},\{f_1,\dots,f_{n-1},g_n\}\} \ .\end{aligned} \tag{4.24}$$

N-P structures were studied in general in [28].

For $n=2$, the two n-ary Poisson structures above reproduce the standard Poisson one.

The Filippov identity for the jacobian of n functions was first written by Filippov [13], and later by Sahoo and Valsakumar [29] and by Takhtajan [28] (who called it fundamental identity) in the context of Nambu mechanics [24]. Physically, the FI is a consistency condition for the time evolution [29,28], which is given in terms of $(n-1)$ 'hamiltonian' functions that determine an $ad_{\mathscr{X}}$ derivation of the Nambu FA $\mathfrak{N}$. Every even N-P structure is also a GPS, but the converse does not hold (see [15]).

As it is the case of the finite-dimensional FAs, the $n>2$ N-P Poisson structures are extremely rigid; the N-P tensors defining them have the property of being decomposable *i.e.*, they may be given locally by $\partial_{x_1}\wedge\partial_{x_2}\wedge\dots\wedge\partial_{x_n}$ [30,19] so that the 'canonical form' of the N-P bracket has the form (3.19) (see [15] for more references on this point).

The question of the quantization of N-P mechanics has been the subject of a vast amount of literature. It is fair to say that (for arbitrary $n>2$) it remains a problem in general, aggravated by the fact that there are not so many physical examples of N-P mechanical systems waiting to be quantized when $n\neq 2$. We just refer here to [26,31,32] and to [15] for further discussion and references.

5. Central extensions and deformations of FAs

It is well known that the Whitehead lemma for semisimple Lie algebras states the vanishing of the second cohomology groups, $H_0^2(\mathfrak{g})=0$, $H_\rho^2(\mathfrak{g},\mathfrak{g})=0$, where ρ is a representation of $\mathfrak{g}$ (in particular, ad or trivial, $\rho=0$). Hence, semisimple Lie algebras do not admit non-trivial central extensions and are moreover rigid (non-deformable) since their central extensions and infinitesimal deformations are governed, respectively, by $H_0^2(\mathfrak{g})$ and $H_{ad}^2(\mathfrak{g},\mathfrak{g})$. Let us now turn to the $n>2$ FA case [14].

5.1. Central extensions of a FA

Given a Filippov algebra $\mathfrak{G}$ with n-bracket (1.7), a central extension $\widetilde{\mathfrak{G}}$ of $\mathfrak{G}$ is a FA of the form

$$\begin{aligned}&[\widetilde{X}_{a_1},\dots,\widetilde{X}_{a_n}] := f_{a_1\dots a_n}{}^{d}\,\widetilde{X}_d + \alpha^1(X_1,\dots,X_n)\Xi \ , \\ &[\widetilde{X}_1,\dots,\widetilde{X}_{n-1},\Xi]=0 \quad , \quad \widetilde{X}\in\widetilde{\mathfrak{G}} \ , \ \alpha^1\in\wedge^{n-1}\mathfrak{G}^*\wedge\mathfrak{G}^* \ ,\end{aligned} \tag{5.25}$$

where $\mathfrak{G}^*$ is the dual of the $\mathfrak{G}$ vector space. If we now introduce p-cochains as maps

$$\alpha^p \in \wedge^{n-1}\mathfrak{G}^* \otimes \cdots \otimes \wedge^{n-1}\mathfrak{G}^* \wedge \mathfrak{G}^* \ , \ \alpha^p : (\mathscr{X}_1, \ldots, \mathscr{X}_p, Z) \mapsto \alpha^p(\mathscr{X}_1, \ldots, \mathscr{X}_p, Z) \ , \tag{5.26}$$

the above $\alpha^1(X_1, \ldots, X_n) = \alpha^1(\mathscr{X}, X_n)$ is a one-cochain. Note that the order of the p-cochains α^p for the cohomology of FAs $\mathfrak{G}$ ($n \geq 3$) is naturally defined as the number p of fundamental objects among the arguments of the cochain (for a Lie algebra $\mathfrak{g}$, $\mathscr{X} = X$ and p counts the number of algebra elements so that the α above would be a two- rather than a one-cocyle on $\mathfrak{g}$).

Since the centrally extended $\widetilde{\mathfrak{G}}$ is a FA, the FI for the n-bracket in $\widetilde{\mathfrak{G}}$ implies that the one-cochain $\alpha^1(\mathscr{X}, Z)$ in (5.25) (with $X_n = Z$) has to satisfy the condition

$$\alpha^1(\mathscr{X}, \mathscr{Y} \cdot Z) - \alpha^1(\mathscr{X} \cdot \mathscr{Y}, Z) - \alpha^1(\mathscr{Y}, \mathscr{X} \cdot Z) \equiv (\delta\alpha^1)(\mathscr{X}, \mathscr{Y}, Z) = 0 \ . \tag{5.27}$$

A central extension is actually trivial if it is possible to find new generators $\widetilde{X}' = \widetilde{X} - \beta(X)\Xi$ (where β is a zero-cochain, $\beta \in \mathfrak{G}^*$) such that

$$[\widetilde{X}'_{a_1}, \ldots, \widetilde{X}'_{a_n}] = f_{a_1 \ldots a_n}{}^d \widetilde{X}'_d = f_{a_1 \ldots a_n}{}^d \widetilde{X}_d - \beta([X_{a_1}, \ldots, X_{a_n}])\Xi$$

i.e., $\alpha^1(X_1, \ldots, X_{n-1}, Z) = -\beta([X_1, \ldots, X_{n-1}, Z])$, again with $X_{a_n} = Z$. This may be rewritten in the form

$$\alpha^1(\mathscr{X}, Z) = -\beta([X_1, \ldots, X_{n-1}, Z]) \equiv (\delta\beta)(X_1, \ldots, X_{n-1}, Z) \equiv (\delta\beta)(\mathscr{X}, Z) \ , \tag{5.28}$$

where β is the zero-cochain generating the trivial one-cocycle, $\alpha^1 = \delta\beta$. Therefore, central extensions of FAs are characterized by one-cocycles modulo one-coboundaries.

The above suffices to infer the form of the full FA cohomology complex suitable for central extensions. Let α^p be a generic p-cochain. Then, $(C_0^\bullet(\mathfrak{G}), \delta)$ is defined by (see [26])

$$\begin{aligned}(\delta\alpha)(\mathscr{X}_1, \ldots, \mathscr{X}_{p+1}, Z) &= \sum_{1 \leq i < j}^{p+1} (-1)^i \alpha(\mathscr{X}_1, \ldots, \hat{\mathscr{X}_i}, \ldots, \mathscr{X}_i \cdot \mathscr{X}_j, \ldots, \mathscr{X}_{p+1}, Z) \\ &+ \sum_{i=1}^{p+1} (-1)^i \alpha(\mathscr{X}_1, \ldots, \hat{\mathscr{X}_i}, \ldots, \mathscr{X}_{p+1}, \mathscr{X}_i \cdot Z) \ ,\end{aligned} \tag{5.29}$$

which, for $n = 2$, reproduces the Lie algebra cohomology complex for the trivial action. Defining p-cocycles and p-coboundaries as usual, the p-th FA cohomology group (for the trivial action) is $H_0^p(\mathfrak{G})$ $=Z_0^p(\mathfrak{G})/B_0^p(\mathfrak{G})$. Therefore, a FA $\mathfrak{G}$ admits non-trivial central extensions when $H_0^1(\mathfrak{G}) \neq 0$.

5.2. *Infinitesimal deformations of FAs*

A similar approach may be used for deformations. An infinitesimal deformation in Gerstenhaber's sense [33] of a FA is obtained by modifying the n-bracket as

$$[X_1, \ldots, X_n]_t = [X_1, \ldots, X_n] + t\alpha^1(X_1, \ldots, X_n) \ , \tag{5.30}$$

where t is the deformation parameter and $\alpha^1 : \wedge^{n-1}\mathfrak{G} \wedge \mathfrak{G} \to \mathfrak{G}$ is now $\mathfrak{G}$-valued, so that $\mathfrak{G}$ will act on it. Again, the FI for the deformed FA n-bracket $[X_1, \ldots, X_n]_t$ constrains α^1. The FI is

$$[X_1, \ldots, X_{n-1}, [Y_1, \ldots, Y_n]_t]_t = \sum_{a=1}^{n} [Y_1, \ldots, Y_{a-1}, [X_1, \ldots, X_{n-1}, Y_a]_t, Y_{a+1}, \ldots, Y_n]_t \ ; \tag{5.31}$$

with $Y_n = Z$, it may we rewritten as

$$[\mathscr{X}, (\mathscr{Y} \cdot Z)_t]_t = [(\mathscr{X} \cdot \mathscr{Y})_t, Z]_t + [\mathscr{Y}, (\mathscr{X} \cdot Z)_t]_t \ . \tag{5.32}$$

At first order in t, the FI gives the following condition on the one-cochain α^1:

$$\begin{aligned}
&[X_1, \ldots, X_{n-1}, \alpha^1(Y_1, \ldots, Y_n)] + \alpha^1(X_1, \ldots, X_{n-1}, [Y_1, \ldots, Y_n]) \\
&= \sum_{a=1}^{n} [Y_1, \ldots, Y_{a-1}, \alpha^1(X_1, \ldots, X_{n-1}, Y_a), Y_{a+1}, \ldots, Y_n] \\
&+ \sum_{a=1}^{n} \alpha^1(Y_1, \ldots, Y_{a-1}, [X_1, \ldots, X_{n-1}, Y_a], Y_{a+1}, \ldots, Y_n) \ .
\end{aligned} \tag{5.33}$$

In terms of fundamental objects and with $Y_n = Z$, this may be read as a one-cocycle conditon for α^1,

$$\begin{aligned}
(\delta\alpha^1)(\mathscr{X}, \mathscr{Y}, Z) = ad_{\mathscr{X}}\alpha^1(\mathscr{Y}, Z) - ad_{\mathscr{Y}}\alpha^1(\mathscr{X}, Z) - (\alpha^1(\mathscr{X}, \quad) \cdot \mathscr{Y}) \cdot Z \\
-\alpha^1(\mathscr{X} \cdot \mathscr{Y}, Z) - \alpha^1(\mathscr{Y}, \mathscr{X} \cdot Z) + \alpha^1(\mathscr{X}, \mathscr{Y} \cdot Z) = 0 \quad ,
\end{aligned} \tag{5.34}$$

where, for instance for n=3,

$$\begin{aligned}
\alpha^1(\mathscr{X}, \quad) \cdot \mathscr{Y} &:= (\alpha^1(\mathscr{X}, \quad) \cdot Y_1, Y_2) + (Y_1, \alpha^1(\mathscr{X}, \quad) \cdot Y_2) \\
&= (\alpha^1(\mathscr{X}, Y_1), Y_2) + (Y_1, \alpha^1(\mathscr{X}, Y_2)) \ .
\end{aligned} \tag{5.35}$$

To see whether the $\mathfrak{G}$-valued cocycle α^1 is a one-coboundary, we look for the possible triviality of the infinitesimal deformation. It will be trivial if new generators can been found in terms of a $\beta : \mathfrak{G} \to \mathfrak{G}$, $X_i' = X_i - t\beta(X_i)$, such that

$$[X'_1, \ldots, X'_n]_t = [X_1, \ldots, X_n]' \equiv [X_1, \ldots, X_n] - t\beta([X_1, \ldots, X_n]) \ . \tag{5.36}$$

At first order in t this implies

$$\begin{aligned}
[X'_1, \ldots, X'_n]_t &= [X_1, \ldots, X_n]_t - t\sum_{a=1}^{n} [X_1, \ldots, X_{a-1}, \beta(X_a), X_{a+1}, \ldots, X_n]_t \\
&= [X_1, \ldots, X_n] + t\alpha^1(X_1, \ldots, X_n) - t\sum_{a=1}^{n} [X_1, \ldots, X_{a-1}, \beta(X_a), X_{a+1}, \ldots, X_n] \ .
\end{aligned} \tag{5.37}$$

Therefore, a deformation is trivial if

$$(\alpha^1)(X_1, \ldots, X_n) := -\beta([X_1, \ldots, X_n]) + \sum_{a=1}^{n} [X_1, \ldots, X_{a-1}, \beta(X_a), X_{a+1}, \ldots, X_n] \equiv (\delta\beta)(\mathscr{X}, X_n) \tag{5.38}$$

i.e., when the one-cocycle α^1 is the one-coboundary $\alpha^1 = \delta\beta$,

$$\alpha^1(\mathscr{X}, Z) = (\delta\beta)(\mathscr{X}, Z) = -\beta(\mathscr{X} \cdot Z) + (\beta(\quad) \cdot \mathscr{X}) \cdot Z + \mathscr{X} \cdot \beta(Z) \ . \tag{5.39}$$

If all one-cocycles are trivial, the FA is *stable* or *rigid*.

The above may be used to write the full complex $(C^{\bullet}_{ad}(\mathfrak{G}, \mathfrak{G}), \delta)$ adapted to the deformations of FA problem (see [14, 15] for details), introduced by Gautheron [19] in the context of Nambu-Poisson cohomology and also considered by Rotkiweicz [34], but it will not be needed here. We shall just mention that general p-cochains are now $\mathfrak{G}$-valued maps $\alpha^p : \wedge^{(n-1)}\mathfrak{G} \otimes \overset{p}{\cdots} \otimes \wedge^{(n-1)} \mathfrak{G} \wedge \mathfrak{G} \to \mathfrak{G}$.

Journal of Physics: Conference Series **284** (2011) 012019 doi:10.1088/1742-6596/284/1/012019

6. Whitehead lemma for FAs

It follows from the above discussion that an analogue of the Whitehead lemma for FAs would require having $H_0^1(\mathfrak{G}) = 0$ and $H_{ad}^1(\mathfrak{G}, \mathfrak{G}) = 0$ for $\mathfrak{G}$ semisimple. That this is indeed the case was proven in [14], taking advantage of the fact that all simple FAs have the same general structure [17,13] given in eq. (3.18).

Characterizing the real-valued $Z_0^1(\mathfrak{G})$ and the $\mathfrak{G}$-valued $Z_{ad}^1(\mathfrak{G}, \mathfrak{G})$ one-cocycles for central extensions and deformations of a FA by their coordinates,

$$\alpha^1_{a_1 \dots a_n} = \alpha^1(e_{a_1}, \dots, e_{a_n}) \quad , \quad \alpha^1_{a_1 \dots a_n}{}^d = \alpha^1(e_{a_1}, \dots, e_{a_n})^d \, , \quad a, d = 1, \dots, (n+1) \tag{6.40}$$

and using the explicit form of the n-brackets of the simple FAs, it is possible to show [14] that the above one-cocycles are necessarily one-coboundaries generated, respectively, by zero-cochains of coordinates β_a , β_a^d .

Therefore, $H_0^1(\mathfrak{G}) = 0$, $H_{ad}^1(\mathfrak{G}, \mathfrak{G}) = 0$ for simple FAs, which therefore do not admit non-trivial central extensions nor deformations. Using now that a semisimple FA is the sum of simple ideals the following lemma is proved in [14]:

Lemma (*Whitehead lemma for semisimple n-Lie algebras*)
Semisimple Filippov algebras, $n \geq 2$, do not admit non-trivial central extensions and are, moreover, rigid.

7. Relaxing anticommutativity: n-Leibniz algebras and cohomology

Leibniz (Loday's) algebras [35] $\mathscr{L}$ are a non-commutative version of Lie algebras: their bracket need not be anticommutative ($[X, Y] \neq -[Y, X]$) but still satisfies the (left, say) 'Leibniz' identity

$$[X, [Y, Z]] = [[X, Y], Z] + [Y, [[X, Z]] \quad ; \tag{7.41}$$

right Leibniz algebras are defined in an analogous form.

Similarly, (left, say) n-Leibniz algebras $\mathfrak{L}$ [36,37] are defined by removing the anticommutatitivity requirement for the n-Leibniz bracket while keeping the (left) FI. Introducing also fundamental objects for n-Leibniz algebras $\mathfrak{L}$, the identity reads

$$\mathscr{X} \cdot (\mathscr{Y} \cdot \mathscr{Z}) = (\mathscr{X} \cdot \mathscr{Y}) \cdot \mathscr{Z} + \mathscr{Y} \cdot (\mathscr{X} \cdot \mathscr{Z}) \qquad \forall \mathscr{X}, \mathscr{Y}, \mathscr{Z} \in \otimes^{n-1} \mathfrak{L} \, . \tag{7.42}$$

Note that now $\mathscr{X} \in \otimes^{n-1}\mathfrak{L}$ since, in contrast with FAs, the anticommutativity of the $(n-1)$ arguments in the fundamental object $\mathscr{X}$ is no longer assumed since the n-bracket in (1.5) is no longer antisymmetric for $\mathfrak{L}$. Nevertheless, the above is still the (left) FI (1.4) previously defining FAs. As a result, the characteristic FI

$$\mathscr{X} \cdot (\mathscr{Y} \cdot \mathscr{Z}) - \mathscr{Y} \cdot (\mathscr{X} \cdot \mathscr{Z}) = (\mathscr{X} \cdot \mathscr{Y}) \cdot \mathscr{Z} \qquad \forall \, \mathscr{X}, \mathscr{Y}, \mathscr{Z} \in \otimes^{n-1} \mathfrak{L} \quad , \tag{7.43}$$

which already guaranteed the nilpotency of the coboundary operator δ for the different FA cohomology complexes (as the JI does in the ordinary Lie algebra cohomology), will also do the job for the various n-Leibniz cohomologies. Therefore, and with the proper definition of p-cochains on $\mathfrak{L}$, the n-Leibniz [38,37] and the FA cohomological complexes have the same structure (see [15] for details): n-Leibniz cohomology underlies n-Lie cohomology. This is why the N-P cohomology may be studied from the point of view of n-Leibniz cohomology, as pointed out and discussed by Daletskii and Takhtajan [36].

Our proof of the Whitehead Lemma above for FAs [14], however, relied on the antisymmetry of the FA n-bracket, and thus it does not hold when the anticommutativity is relaxed. On the other hand, n-Lie algebras $\mathfrak{G}$ may be considered as a particular case of n-Leibniz ones $\mathfrak{L}$: FAs are n-Leibniz algebras with a fully skewsymmetric n-bracket. Thus, we may look for n-Leibniz central extensions and deformations of FAs considering these as n-Leibniz ones and expect, in general, to find a richer structure. This has been observed explicitly for the $n = 2$ case [39] by looking at Leibniz deformations of the Heisenberg Lie algebra; also, for $n = 3$, a specific 3-Leibniz deformation of the simple Euclidean 3-Lie algebra A_4 has been given in [40]. Thus, a natural extension of our work above is to look for n-Leibniz deformations of simple n-Lie algebras to see whether this opens more possibilities.

It is natural to relax the skewsymmetry of the FA n-bracket in such a way that we remain within the class of n-Leibniz algebras that have fully skewsymmetric fundamental objects; this corresponds (see eq. (1.5)) to having n-Leibniz brackets that are antisymmetric in their first $n-1$ arguments. For $n = 3$, this type of real Leibniz algebras have in fact appeared in the study of multiple M2-branes [41]. Other examples of weakening the skewsymmetry have been considered in the same M-theory context, as the complex 'hermitean (right) three-algebras' introduced by Bagger and Lambert [42] that are behind the Aharony, Bergman, Jafferis, and Maldacena theory [43]; see further [44].

Our results on the class of real n-Leibniz deformations and central extensions of simple FAs which retain the skewsymmetry of the FA fundamental objects may be summarized by the following two theorems, both proven in [45]:

Theorem 1 (*A class of n-Leibniz deformations of simple FAs*)
The n-Leibniz algebra deformations of the $(n + 1)$-dimensional simple FA's that preserve the skewsymmetry of the $(n - 1)$ first elements in the n-Leibniz bracket (or that of the fundamental objects) are all trivial for $n > 3$. For $n = 3$, there is a non-trivial one-cocycle with coordinates

$$\alpha^1_{a_1 a_2 c d} \propto \epsilon_{a_1 a_2}{}^{eg} \varepsilon_c \epsilon_{egcd} = 2\varepsilon_c(\delta_{a_1 c}\delta_{a_2 d} - \delta_{a_1 d}\delta_{a_2 c}) \ .$$

Further, all $n = 2$ semisimple Filippov (*i.e.*, Lie) algebras are rigid as Leibniz algebras.

For the $n = 3$ Euclidean simple FA A_4, the above is the deformation given in [40].

Theorem 2 (*A class of n-Leibniz central extensions of simple FAs*)
The n-Leibniz algebra central extensions of simple FA's that preserve the skewsymmetry of the $(n-1)$ first entries of the n-bracket (or of the fundamental objects) are all trivial for any $n > 2$.

For $n = 2$ the fundamental objects have only one algebra element and therefore there are no skewsymmetry restrictions. Our proof of the $n > 2$ theorem also extends to the $n = 2$ simple algebras A_3 ($so(3)$) and $A_{1,2}$ ($so(1, 2)$); the case of arbitrary simple Lie algebras is covered in [46] and [47] (Prop. 3.2 and Cor. 3.7).

8. Final comments

We have outlined some properties of n-ary algebras and, in particular, of Filippov algebras. Although these structures are mathematically interesting in themselves, they have also appeared in physics as n-ary Poisson structures and, recently, in the mentioned Bagger-Lambert-Gustavsson model in the case of FAs.

Our theorems 1 and 2 above apply to a (natural) class of n-Leibniz deformations of FAs. Other possibilities will arise if the deformations are not restricted to n-Leibniz algebras with antisymmetric fundamental objects but, obviously, each type will require separate study.

Contractions of FAs have recently been introduced in [48].

Acknowledgments

This work has been partially supported by the research grants FIS2008-01980 and FIS2009-09002 from the Spanish MICINN and by VA013C05 from the Junta de Castilla y León (Spain).

References

[1] A. G. Kurosh, *A cycle of papers on multioperator rings and algebras: multioperator rings and algebras*, Russian Math. Surveys **24** (1969), 1-13.
[2] T. M. Baranovič and M. S. Burgin, *Linear Ω-algebras*, Uspekhi Mat. Nauk **30** (1975), 61-106 [Russian Math. Surveys **30** (1975) 65-113].
[3] P. Gautheron, *Simple facts concerning Nambu algebras*, Commun. Math. Phys. **195** (1998) 417-434.
[4] A. M. Vinogradov and M. M. Vinogradov, *On multiple generalizations of Lie algebras and Poisson manifols*, Contemp. Math. **219** (1998) 273-287; *Graded multiple analogs of Lie algebras*, Acta Applicandae Math. **72** (2002) 183-197.
[5] J.A. de Azcárraga, A. Perelomov and J.C. and Pérez Bueno, *New generalized Poisson structures*, J. Phys. **A29** (1996) L151-L157 [arXiv:q-alg/9601007].
[6] J. A. de Azcárraga, A. M. Perelomov and J. C. Pérez Bueno, *The Schouten-Nijenhuis bracket, cohomology and generalized Poisson structures*, J. Phys. **A29** (1996) 7993-8010 [arXiv:hep-th/9605067].
[7] J. A. de Azcárraga and J. C. Pérez-Bueno, *Higher-order simple Lie algebras*, Commun. Math. Phys. **184**, (1997) 669-681 [arXiv:hep-th/9605213].
[8] P. Hanlon and H. Wachs, *On Lie k-algebras*, Adv. in Math. **113** (1995) 206-236 .
[9] V. Gnedbaye, *Les algèbres k-aires el leurs opérads*, C. R. Acad. Sci. Paris, Série I, **321** (1995) 147-152.
[10] J.-L. Loday, *La renaissance des opérades*, Sem. Bourbaki **792** (1994-95) 47-54.
[11] P. W. Michor and A. M. Vinogradov, *n-ary Lie and associative algebras*, Rend. Sem. Mat. Univ. Pol. Torino **53** (1996) 373-392.
[12] M. Bremner, *Varieties of anticommutative n-ary algebras*, J. Algebra **191** (1997) 76-88; *Identities for the ternary commutator*, J. Algebra **206** (1998) 615-613.
[13] V. Filippov, *n-Lie algebras*, Sibirsk. Mat. Zh. **26** (1985) 126-140 [Siberian Math. J. **26** (1985) 879-891].
[14] J.A. de Azcárraga and J.M. Izquierdo, *Cohomology of Filippov algebras and an analogue of Whitehead's lemma*, J. Phys. Conf. Ser. **175** (2009) 012001 [arXiv:0905.3083[math-ph]].
[15] J.A. de Azcárraga and J.M. Izquierdo, *n-ary algebras: a review with applications*, J. Phys. **A43** (2010) 293001-1-117 [arXiv:1005.1028 [math-ph]].
[16] S. M. Kasymov, *Theory of n-Lie algebras*, Algebra i Logika **26** (1987) 277-297 (Algebra and Logic **26** (1988) 155-166].
[17] W. X. Ling, *On the structure of n-Lie algebras*. PhD thesis, Universität-Gesamthochshule-Siegen, Siegen, 1993.
[18] S. M. Kasymov, *On analogues of Cartan criteria for n-Lie algebras*, Algebra i Logika **34** (1995), 274-287 [Algebra and Logic **34** (1995) 147-154].
[19] P. Gautheron, *Some remarks concerning Nambu mechanics*, Lett. Math. Phys. **37** (1996) 103-116.
[20] J. Bagger and N. Lambert, *Modelling multiple M2's*, Phys. Rev. **D75** (2007) 045020 [arXiv:hep-th/0611108].
[21] J. Bagger and N. Lambert, *Comments On Multiple M2-branes*, JHEP **02** (2008) 105 [arXiv:0712.3738 [hep-th]].
[22] A. Gustavsson, *One-loop corrections to Bagger-Lambert theory*, Nucl. Phys. **B807** (2009) 315-333 [arXiv:0805.4443 [hep-th]].
[23] G. Pinczon and R. Ushirobira, *New applications of graded Lie algebras to Lie algebras, generalized Lie algebras, and cohomology*, J. Lie Theory **17** (2007) 633-667 [arXiv:math.RT/0507387].
[24] Y. Nambu, *Generalized Hamiltonian dynamics*, Phys. Rev. **D7** (1973) 2405-2414.
[25] N. Cantarini and V. G. Kac, *Classification of simple linearly compact n-Lie superalgebras*, Commun. Math. Phys. **298** (2010) 833-853 [arXiv:0909.3284 [math.QA]].
[26] J. A. de de Azcárraga, J. M. Izquierdo and J. C. Pérez Bueno, *On the higher-order generalizations of Poisson structures*, J. Phys. **A30** (1997) L607-L616 [arXiv:hep-th/9703019].
[27] J. A. de Azcárraga, J. M. Izquierdo, A. M. Perelomov, and J. C. Pérez Bueno, *The Z_2-graded Schouten-Nijenhuis bracket and generalized super-Poisson structures*, J. Math. Phys. **38** (1997) 3735-3749 [arXiv:hep-th/9612186].
[28] L. Takhtajan, *On foundation of the generalized Nambu mechanics*, Commun. Math. Phys.**160** (1994) 295-316 [hep-th/9301111].
[29] D. Sahoo and M. C. Valsakumar, *Nambu mechanics and its quantization*, Phys. Rev. **A46** (1992) 4410-4412.
[30] D. Alekseevsky and P. Guha, *On decomposability of Nambu-Poisson tensor*, Acta Math. Univ. Comenianae **LXV** (1996) 1-9.
[31] H. Awata, M. Li, D. Minic and T. Yoneya, *On the quantization of Nambu brackets*, JHEP **02** (2001) 013 [arXiv:hep-th/9906248].
[32] T. Curtright and C. Zachos, *Classical and quantum Nambu mechanics*, Phys. Rev. **D68** (2003) 085001 [arXiv:hep-th/0212267].
[33] M. Gerstenhaber, *On the Deformation of Rings and Algebras*, Annals Math. **79** (1964) 59-103.
[34] M. Rotkiewicz, *Cohomology ring of n-Lie algebras*, Extracta Math. **20** (2005) 219-232.
[35] J.-L. Loday, *Une version non-commutative des algèbres de Lie*, L'Ens. Math. **39** (1993) 269-293.
[36] Y. L. Daletskii and L. Takhtajan, *Leibniz and Lie algebra structures for Nambu algebra*, Lett. Math. Phys. **39** (1997) 127-141.
[37] Casas, J.M., Loday, J.-L. and Pirashvili, T., *Leibniz n-algebras*, Forum Math. **14** (2002) 189-207.

[38] J.-L. Loday and T. Pirashvili, *Universal enveloping algebras of Leibniz algebras and (co)homology*, Mat. Annalen **296**, (1993) 139-158.

[39] A. Fialowski and A. Mandal, *Leibniz algebra deformations of a Lie algebra*, J. Math. Phys. **49** (2008) 093511 [arXiv:0802.1263 [math.KT]].

[40] J. M. Figueroa-O'Farrill, *Three lectures on 3-algebras*, arXiv:0812.2865 [hep-th].

[41] S. Cherkis, Vladimir Dotsenko, and C. Sämann, *On superspace actions for multiple M2-branes, metric 3-algebras and their classification*, Phys. Rev. **D79** (2009) 086002 [arXiv:0812.3127 [hep-th]].

[42] J. Bagger and N. Lambert, *Three-algebras and N=6 Chern-Simons gauge theories*, Phys. Rev. **D79** (2009) 025002 [arXiv:0807.0163 [hep-th]].

[43] O. Aharony, O. Bergman, D. L. Jafferis, and J. Maldacena, *N=6 superconformal Chern-Simons-matter theories, M2-branes and their gravity duals*, JHEP **10** (2008) 091 [arXiv:0806.1218 [hep-th]].

[44] A. Gustavsson and S.-J. Rey, *Enhanced N=8 supersymmetry of ABJM theory on R(8) and R(8)/Z(2)*, arXiv:0906.3568 [hep-th].

[45] J. A. de Azcárraga and J. M. Izquierdo, *On a class of n-Leibniz deformations and rigidity of the simple n-Lie algebras*, to appear in J. Math. Phys. [arXiv:1009.2709 [math-ph]].

[46] J.-L. Loday and T. Pirashvili, *Leibniz representations of Lie algebras*, J. Alg. **181** (1996) 414-425.

[47] Naihong Hu, Yufeng Pei and Dong Liu, *A cohomological characterization of Leibniz central extensions of Lie algebras*, Proc. Amer. Math. Soc. **136** (2008) 437-447 (see further arXiv:math/0605399 [math.QA]).

[48] J. A. de Azcárraga, J. M. Izquierdo, M. Picon, *Contractions of Filippov algebras*, to appear in J. Math. Phys [arXiv:1009.0372 [math-ph]].

GROUP 28: Physical and Mathematical Aspects of Symmetry IOP Publishing
Journal of Physics: Conference Series **284** (2011) 012020 doi:10.1088/1742-6596/284/1/012020

The Tamm-Dancoff Approximation as the boson limit of the Richardson-Gaudin equations for pairing

Stijn De Baerdemacker

Ghent University, Department of Physics and Astronomy, Proeftuinstraat 86, 9000 Gent, Belgium

Department of Physics, University of Toronto, Toronto, Ontario M5S 1A7, Canada

Department of Physics, University of Notre Dame, Notre Dame, IN 46556-5670, USA

E-mail: stijn.debaerdemacker@ugent.be

Abstract. A connection is made between the exact eigen states of the BCS Hamiltonian and the predictions made by the Tamm-Dancoff Approximation. This connection is made by means of a parametrised algebra, which gives the exact quasi-spin algebra in one limit of the parameter and the Heisenberg-Weyl algebra in the other. Using this algebra to construct the Bethe Ansatz solution of the BCS Hamiltonian, we obtain parametrised Richardson-Gaudin equations, leading to the secular equation of the Tamm-Dancoff Approximation in the bosonic limit. An example is discussed in depth.

1. Introduction

The description of superconductive properties in many-particle systems such as metals [1] or atomic nuclei [2] involves the process of pairing. Particles in time-reversed single-particle states will lower their energy by forming a pair, constituting a condensate in the ground state of the system. This ground state is very-well approximated by the Bardeen-Cooper-Schrieffer (BCS) wavefunction [1], which is a variational solution of the BCS Hamiltonian in the grand canonical ensemble. While this solution is essentially sufficient for the thermodynamic limit, $N \to \infty$, it is not fit for the description of finite-size systems. This is due to the increasing relevance of the pair fluctuations with decreasing number of particles, which cannot be taken into account in the BCS approximation [3]. This is not of paramount importance for modestly small systems, such as *e.g.* single-species nucleon pairing within a single major nuclear shell [4], because these can be diagonalised numerically on a modern computer with relative ease. However, for mesoscopic systems, such as metallic nanograins [5], or nucleons in multiple open shells, the dimension of the Hilbert space is huge, such that straightforward numerical diagonalisation becomes practically intractable.

Fortunately, alternative methods to extract the spectroscopy of the model (BCS) Hamiltonian exist. Richardson and Sherman [6, 7] have shown that the reduced BCS Hamiltonian, with a level-independent interaction, is exactly solvable by means of a Bethe Ansatz product wave function, provided the variables in the Ansatz wavefunction provide a solution of the set of non-linear Richardson-Gaudin (RG) equations. As a result, the problem of finding the eigenvalues and eigenstates of the BCS Hamiltonian is reduced from diagonalising a matrix that scales *factorial* with the size of the system to solving a set of algebraic equations that scales *linear* with the system size. However, the RG equations are highly non-linear in the variables, and proved to be very challenging to solve [8, 9], due to the occurrence of singularities at the critical interaction values. It was only recently that this obstacle has been sufficiently removed [10] by means of a change of variables method, canceling the singularities at the critical points. Alongside the numerical usefulness, the method also contributed to a better understanding of the critical points [11], which finds its place among other recent studies of the structure of the eigenstates of the exactly solvable and integrable models [12, 13]. For an extensive review on the subject, I would like to refer to the recent *colloquium* [14].

The purpose of the present manuscript is to contribute to the understanding of the Bethe Ansatz eigenstates of the BCS Hamiltonian and the associated RG equations by making a connection with the Tamm-Dancoff Approximation (TDA) [15]. In the next section, the BCS model and its algebraic properties will be introduced. In section 3, the TDA is introduced and its relation with RG is made. Conclusions are presented in section 4

2. The reduced BCS model, an exactly solvable model

Consider a fermionic system with n single-particle energy levels ε_i $(1 \le i \le n)$ with degeneracy Ω_i. For condensed matter physics the levels are two-fold degenerate ($\Omega_i = 2$), corresponding to the two projections of the electron spin. In the context of nuclear structure physics, we can assign a half-integer total angular momentum j_i with degeneracy $\Omega_i = 2j_i + 1$ to every level [4]. Naturally, and in order to make the discussion as general as possible, we will treat the case of condensed matter as $j_i = \frac{1}{2}$ systems. The reduced BCS Hamiltonian for such a general system is given by

$$\hat{H} = \sum_{i=1}^{n} \varepsilon_i \hat{n}_i + g \sum_{i,k=1}^{n} \hat{S}_i^\dagger \hat{S}_k, \tag{1}$$

with the operators in the Hamiltonian given by

$$\hat{n}_i = \sum_{m_i=-j_i}^{j_i} \hat{a}^\dagger_{j_i m_i} \hat{a}_{j_i m_i}, \quad \hat{S}_i^\dagger = \tfrac{1}{2} \sum_{m_i=-j_i}^{j_i} (-)^{j_i+m_i} \hat{a}^\dagger_{j_i m_i} \hat{a}^\dagger_{j_i -m_i}, \quad \hat{S}_i = (\hat{S}_i^\dagger)^\dagger, \tag{2}$$

which are known respectively as the particle counting, pair creation and pair annihilation operator. The underlying algebraic structure of these operators is the $su(2)$ quasi-spin algebra. This algebra is spanned by the pair creation/annihilation operators, completed with the operator $\hat{S}_i^0 = \frac{1}{2}\hat{n}_i - \frac{1}{4}\Omega_i$. We obtain

$$[\hat{S}_i^0, \hat{S}_k^\dagger] = \delta_{ik}\hat{S}_i^\dagger, \qquad [\hat{S}_i^0, \hat{S}_k] = -\delta_{ik}\hat{S}_i, \qquad [\hat{S}_i^\dagger, \hat{S}_k] = 2\delta_{ik}\hat{S}_i^0. \tag{3}$$

So, the spectrum generating algebra of the pairing problem is $\bigoplus_{i=1}^n su(2)_i$, with $su(2)_i$ the quasi-spin algebra for the level i. The associated Hilbert space is constructed from the direct product of the physically allowed $su(2)_i$ representations $|s_i, m_{s_i}\rangle$, with the quantum numbers[1] defined by the eigenvalues of the quadratic Casimir operator $\hat{C}_i^2$ of $SU(2)_i$ and its $SO(2)_i$ projection $\hat{S}_i^0$

$$\hat{C}_i^2 |s_i, m_{s_i}\rangle = s_i(s_i+1)|s_i, m_{s_i}\rangle \tag{4}$$

$$\hat{S}_i^0 |s_i, m_{s_i}\rangle = m_{s_i}|s_i, m_{s_i}\rangle. \tag{5}$$

The physically allowed representations are those for which the following relations hold

$$m_{s_i} = \tfrac{1}{2}n_i - \tfrac{1}{4}\Omega_i, \quad s_i = \tfrac{1}{4}\Omega_i - \tfrac{1}{2}v_i, \tag{6}$$

[1] I will denote the quasi-spin projection quantum number m of the quasi-spin s_i explicitly by m_{s_i} to rule out any confusion with the total angular momentum projection m_i of j_i.

with n_i the number of particles in a level i and v_i the *seniority* of the level i, which is the number of particles in a level i not pairwise coupled [16]. Obviously, for systems with a definite total number of particles N, we require the extra condition $\sum_i n_i = N$. For simplicity, we will assume that there are no unpaired particles present in the system ($v_i = 0, \forall i$), so the total number of pairs is given by $N/2$.

The key ingredient of the Richardson-Gaudin treatment of the pairing model is that the Hamiltonian (1) can be diagonalised by means of a Bethe Ansatz wavefunction [7, 8]

$$|\psi\rangle = \prod_{\alpha=1}^{N/2} \left(\sum_{i=1}^{n} \frac{S_i^\dagger}{2\varepsilon_i - E_\alpha} \right) |\theta\rangle, \tag{7}$$

provided the Richardson variables, given as the free parameters E_α, are a solution of the RG equations

$$1 + 2g \sum_{i=1}^{n} \frac{s_i}{2\varepsilon_i - E_\alpha} - 2g \sum_{\beta\neq\alpha}^{N/2} \frac{1}{E_\beta - E_\alpha} = 0, \qquad (1 \leq \alpha \leq N/2). \tag{8}$$

The energy of the eigen state is then given by $E = \sum_{\alpha=1}^{N/2} E_\alpha$. In the next section, it will be discussed how the Richardson-Gaudin treatment can be related to the Tamm-Dancoff Approximation, which is a well-studied method in many-body physics.

3. Richardson-Gaudin and the Tamm-Dancoff Approximation

3.1. The Tamm-Dancoff Approximation

Most commonly, the Tamm-Dancoff Approximation is used in the context of particle-hole excitations across a closed Hartree-Fock vacuum [15, 4], but we can interprete it here somewhat larger as the description of elementary excitations which are also related to the Random Phase Approximation or Equations of Motion methods [17]. The basic idea behind the approximation is that we can treat all the excited states of the Hamiltonian as harmonic excitations or multiphonon states of the elementary eigen modes or phonons of the system. In the case of the BCS Hamiltonian (1), the elementary excitations can be found by solving the eigen value equation for the 1-pair excitations $\hat{b}_k^\dagger = \sum_i Y_{ki} \hat{S}_i^\dagger$

$$\hat{H} \hat{b}_k^\dagger |\theta\rangle = \hbar\omega_k \hat{b}_k^\dagger |\theta\rangle. \tag{9}$$

Once this is solved, the excited states are approximated by the following multiphonon states

$$|\psi\rangle \approx |\phi_{[\nu_k]}\rangle = \prod_{k=1}^{n} (\hat{b}_k^\dagger)^{\nu_k} |\theta\rangle, \tag{10}$$

with $\sum_k \nu_k = N/2$ and the notation $[\nu_k]$ is short-hand for the array $[\nu_1, \nu_2, \ldots, \nu_n]$. Similarly, the energy of this state is equal to

$$E_{[\nu_k]}^{\mathrm{TDA}} = \sum_{k=1}^{n} \hbar\omega_k \nu_k. \tag{11}$$

Obviously, this is a gross simplification compared to the exact Bethe Ansatz wave function (7). However, it is worthwhile to scrutinise the differences. For a 1-pair system, the RG and TDA approach are identical because the action of the elementary phonons $\hat{b}_k^\dagger$ on the vacuum is equivalent to the Bethe Ansatz wavefunction (7) for a single pair ($N/2 = 1$). Moreover, it is

Table 1. The parameters of a simple 3-level pairing system that serves as an example throughout the text.

i	j_i	Ω_i	ε_i
1	$\frac{5}{2}$	6	0.0000
2	$\frac{3}{2}$	4	0.7796
3	$\frac{9}{2}$	10	2.1024

well-known that the TDA eigen modes ($\hbar\omega_k$) for the reduced BCS Hamiltonian can be identified as the solutions of the secular TDA equation [4, 15].

$$1 + 2g \sum_{i=1}^{n} \frac{s_i}{2\varepsilon_{j_i} - \hbar\omega} = 0, \tag{12}$$

which is identical to the RG equation for a single pair (8). Naturally, the situation is different for multi-pair excitations. It is instructive to continue with an example. Consider a system of 4 pairs in 3 levels with the parameters given in Table 1. The left-hand side of the secular TDA equation (12) for the parameter set of Table 1 is plotted in Figure 1, revealing quite a bit of information at a glance. It is readily understood that there are as many eigen modes as there are levels and one eigen mode is structurally different from the others in the sense that it is not bound in a domain between two single-particle poles. This solution is unbound below and is generally referred to as the *collective* solution.

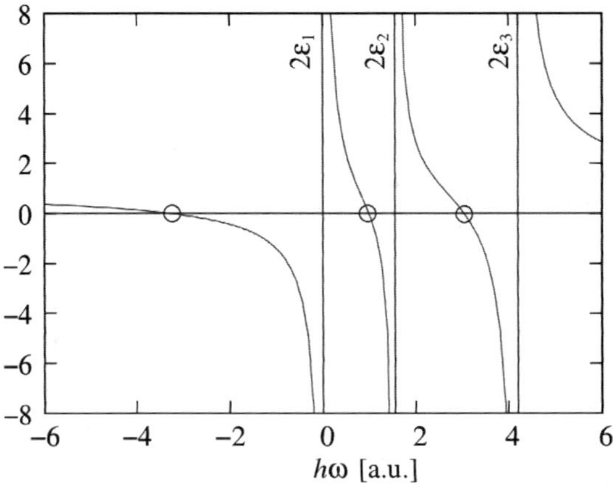

Figure 1. The TDA secular equation (eq. (12)) for the parameters given in Table 1 and an interaction strength $g = -0.5$. The poles at the single-particle energies are accentuated by vertical lines and the 3 solutions of the equation are marked graphically by open circles.

So, we gather that there are 3 eigen modes for the example under discussion. Since we are accommodating 4 pairs in the system, we can construct a total of 15 distinct TDA multiphonon states $[\nu_1, \nu_2, \nu_3]$ with $\nu_1 + \nu_2 + \nu_3 = 4$. Comparing this with the real dimension of the Hilbert space, which is 11, we immediately realise that TDA overestimates the total number of physically allowed states. The question arises which ones are physical and which are not. To investigate this question, it is instructive to plot the exact excitation spectrum of the system against the

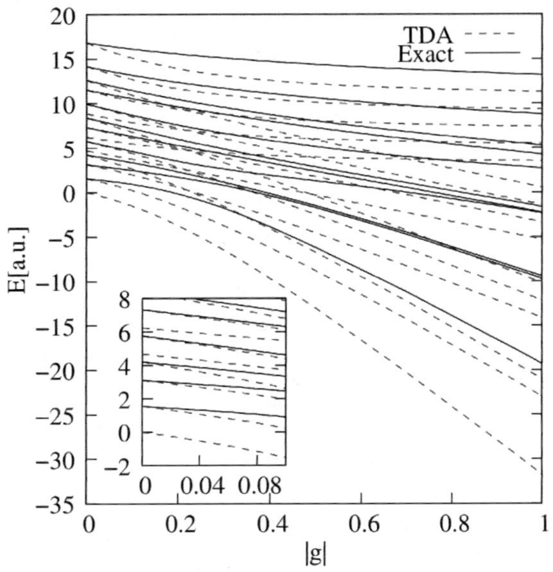

Figure 2. The exact spectrum of the BCS Hamiltonian (1) in (black) solid lines and the predictions of the TDA in (blue) dashed lines are plotted for a range of interaction strength g. In the inset, we find a zoom of the plot for small values of g.

predictions made by TDA, which is done in Figure 2. We can roughly divide the Figure in 2 different domains, *i.e.* the small-interaction domain $|g| \sim 0$ and the large-interaction domain $|g| \gg 0$. The former domain is enlarged and depicted in the inset of the Figure. From this inset, it is clear which TDA states are unphysical for the small-interaction domain. We see that for $|g| \to 0$, the exact ground-state energy is given by approximately $6\varepsilon_1 + 2\varepsilon_2 = 1.5592$, whereas the TDA predicts the ground state at $8\varepsilon_1 = 0.0000$. This is because the TDA can put all 4 elementary excitations in the lowest level, $[\nu_1, \nu_2, \nu_2] = [4, 0, 0]$, whereas this is fundamentally forbidden in the Hilbert space because of the Pauli principle. The degeneracy of the lowest level $\Omega_1 = 6$ only allows for 3 pairs, so one pair needs to be promoted to the 2nd level. As a result, we can identify $[4, 0, 0]$ as physically forbidden in this domain. Further investigation reveals that $[1, 3, 0]$, $[0, 4, 0]$ and $[0, 3, 1]$ are also forbidden on similar grounds. The situation is less clear for the large-interaction domain because we cannot rely on pure combinatorial arguments. However, we can draw qualitative conclusions by noticing that the excitation spectrum tends to reorganise in small clusters around $|g| = 1$. We count clusters of 1, 2 and 3 states from the bottom up and 5 more states that have no cluster characterisation. Inspecting the predictions by TDA, we notice a similar pattern where every cluster can be associated with a distinct number of collective phonons ($[4, 0, 0]$ for the ground state, $[3, 1, 0]$ and $[3, 0, 1]$ for the next cluster, etc.). Nevertheless, we also notice that TDA systematically underestimates the excitation energy of the corresponding states. Again, this is a manifestation of the Pauli exclusion principle, however in a different way. The correct clustering for the lowest states reveals that TDA gives a correct idea about the collective structure of the lowest states, but TDA allows Pauli-forbidden configurations to contribute to the energy of the eigen state. Anyhow, these qualitative arguments do not answer the question which TDA states are related to which exact eigenstates and which TDA states are not physical. In the next subsection, I will present a construction for this particular purpose.

3.2. An algebraic approach to go from Tamm-Dancoff to Richardson-Gaudin

We may conclude from the previous subsection that the difference between TDA and the exact solution are due to the Pauli principle. In algebraic phrasing, the fundamental algebra of the exact pairing model is quasi-spin $su(2)$ whereas TDA is characterised by the bosonic Heisenberg-Weyl algebra $hw(1)$, given by [3].

$$[\hat{b}_k, \hat{b}_l^\dagger] = \delta_{kl}\hat{1}, \quad [\hat{b}_k, \hat{1}] = [\hat{b}_k^\dagger, \hat{1}] = 0. \tag{13}$$

From a physical point of view, one can conclude that the fermion pairs $\hat{S}_i^\dagger$ (eq. (2)) have been replaced by genuine bosons $\hat{b}_k$ in the TDA. This substitution is rather abrupt, so it would be interesting to have a more adiabatic means to *bosonise* the fermion pairs. This can be realised by introducing the following parametrised algebra

$$[\hat{S}_i^0, \hat{S}_k^\dagger] = \delta_{ik}\hat{S}_i^\dagger, \qquad [\hat{S}_i^0, \hat{S}_k] = -\delta_{ik}\hat{S}_i, \qquad [\hat{S}_i^\dagger, \hat{S}_k] = \delta_{ik}(\xi 2\hat{S}_i^0 + (\xi - 1)\tfrac{1}{2}\Omega_i\hat{1}), \tag{14}$$

with ξ a real-valued parameter between 0 and 1. On the one hand, we retain the quasi-spin algebra for $\xi = 1$, and on the other hand, we get the Heisenberg-Weyl algebra in the $\xi = 0$ limit, if we scale the boson creation- and annihilation operators with a factor $\sqrt{2/\Omega_i}$, and interprete the operator $\hat{S}_i^0$ as a boson counting operator. In the intermediate region, we can use the parameter ξ as a continuous switch to transform the fermion pairs into bosons and the other way around. This construction is closely related to the contraction of $u(2)$ to $hw(1)$, as defined by Arecchi *et. al.* [18, 19]. The question is now whether this functionality is also transferred to the pairing system. Remarkably, if we use the parametrised algebra (14) instead of the original quasi-spin algebra (3), we observe that the system remains exactly solvable by means of a Bethe Ansatz wavefunction, provided the free variables in the wavefunction are a solution of the following parametrised RG equations

$$1 + g\sum_{i=1}^{n} \frac{\frac{1}{2}\Omega_i - \xi v_i}{2\varepsilon_i - E_\alpha} - 2g\xi \sum_{\beta\neq\alpha}^{N/2} \frac{1}{E_\beta - E_\alpha} = 0, \qquad (1 \leq \alpha \leq N/2). \tag{15}$$

It is readily verified that this set of equations reduces to the correct corresponding expressions in the limits for ξ. For $\xi = 1$, we retain the set of RG equations (eq. (8)), and for $\xi = 0$, we obtain $N/2$ identical copies of a (seniority zero ($v_i = 0, \forall i$)) secular TDA equation. This means in the latter case, that every individual pair in the Bethe Ansatz wavefunction needs to satisfy the TDA secular equation, independent from the other pairs because the coupling term between the variables E_α in the eqs. (15) has disappeared at $\xi = 0$. More details on the technical aspects of these results, as well as on the nature of the algebra (14) will be given in a forthcoming paper.

Once we have the ξ-parametrised RG equations, we can investigate how the exact eigen states of the pairing problem are related to the solutions of the TDA. For this purpose, we can start from a known solution of the pairing problem [10] and see which TDA solution corresponds to this solution, or *vice versa*, we can start from a particular multiphonon solution of TDA and reconstruct the corresponding exact solution, if it exists. This can be done in both cases by adiabatically switching off/on the ξ parameter in the eqs. (15) and solving numerically along the trajectory. One particular result for the present example (with parameters from Table 1 and $g = -0.5$) can be found in Figure 3. We start from the $[4, 0, 0]$ multiphonon TDA solution of the system (see also Figure 1 for the TDA eigen modes) and, by adiabatically switching on the parameter ξ, obtain the solution of the RG equations. In the Figure, the ξ-trajectory of the solution in the complex plane is depicted.

Finally, we can use this method to make a one-to-one connection between the exact eigen states and the predicted states by TDA, plotted in Figure 2. This is done extensively in Figure 4

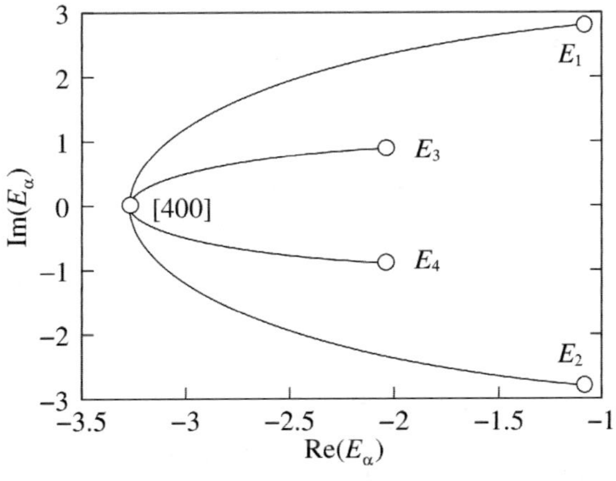

Figure 3. The ξ-trajectory of a solution of the parametrised RG equations (15) in the complex plane. We start from the $[4, 0, 0]$ TDA solution and adiabatically go to the corresponding solution of the RG equations (8) for $\xi = 1$ (denoted by E_α, $\alpha = 1 \ldots 4$). The parameters for the model are given in Table 1 and $g = -0.5$.

and 5. Each row of panels corresponds to a particular TDA state, denoted by $[\nu_1, \nu_2, \nu_3]$. In the first panel, the energy of the elementary TDA eigen modes (blue dashed line) and the real part of the corresponding RG variables (black full line) are plotted as a function of the interaction strength $|g|$. The second panel shows the imaginary part of the RG variables, and the third panel shows the total energy of the exact eigenstate $\sum_{\alpha=1}^{4} E_\alpha$ (black full line) as well as the TDA prediction of the total energy (eq. (11)) (blue dashed line) as a function of $|g|$. For every TDA state $[\nu_1, \nu_2, \nu_3]$, the interaction strengths g is scanned and the corresponding RG variables are given if they exist. In some cases, such as *e.g.* for the low-interaction domain of $[4, 0, 0]$ (first row of Figure 4), the TDA state did not produce an exact solution and it is only for values larger than $|g| \sim 0.190$ that it leads to an exact state (the ground state). It is readily seen from the second row that the ground state below this value $|g| \sim 0.190$ is connected to the $[3, 1, 0]$ TDA state. Dotted lines are added to the figures to guide the eye at those transitional points where the TDA origin of the exact states changes from one state to another. In addition, a label $|i)$ is added in the third panel of every row to denote the corresponding exact state. We can draw multiple conclusions from these figures. It can be observed that the TDA states $[4, 0, 0]$, $[1, 3, 0]$, $[0, 3, 1]$ (not depicted because it does not lead to any exact solution for $|g| \in [0, 1]$) and $[0, 4, 0]$ do not correspond to an exact solution in the small-interaction domain, which is consistent with what we derived earlier. Some states are clearly connected to a given state for the whole domain of $|g| \in [0, 1]$ (such as *e.g.* $[3, 0, 1]$ with the 3rd excited state), whereas other TDA states have no corresponding exact solution over the domain (such as *e.g.* $[0, 3, 1]$). It can also occur that one TDA state can connect to different exact states, however for different domains of $|g|$ (such as *e.g.* $[3, 1, 0]$, which gives rise to the ground state for small $|g|$ as well as to the 2nd excited state for larger $|g|$). In addition, we can verify that the observed clustering around $|g| \sim 1$ is indeed related to the number of collective TDA eigen modes ν_1, consistent with the earlier observations.

4. Conclusions

In conclusion, an algebraic technique to connect the exact eigen states of the BCS Hamiltonian with the predictions of Tamm-Dancoff Approximation is presented. The connection is made via a parametrised algebra, giving rise to the quasi-spin $su(2)$ algebra and the Heisenberg-Weyl

Journal of Physics: Conference Series **284** (2011) 012020 doi:10.1088/1742-6596/284/1/012020

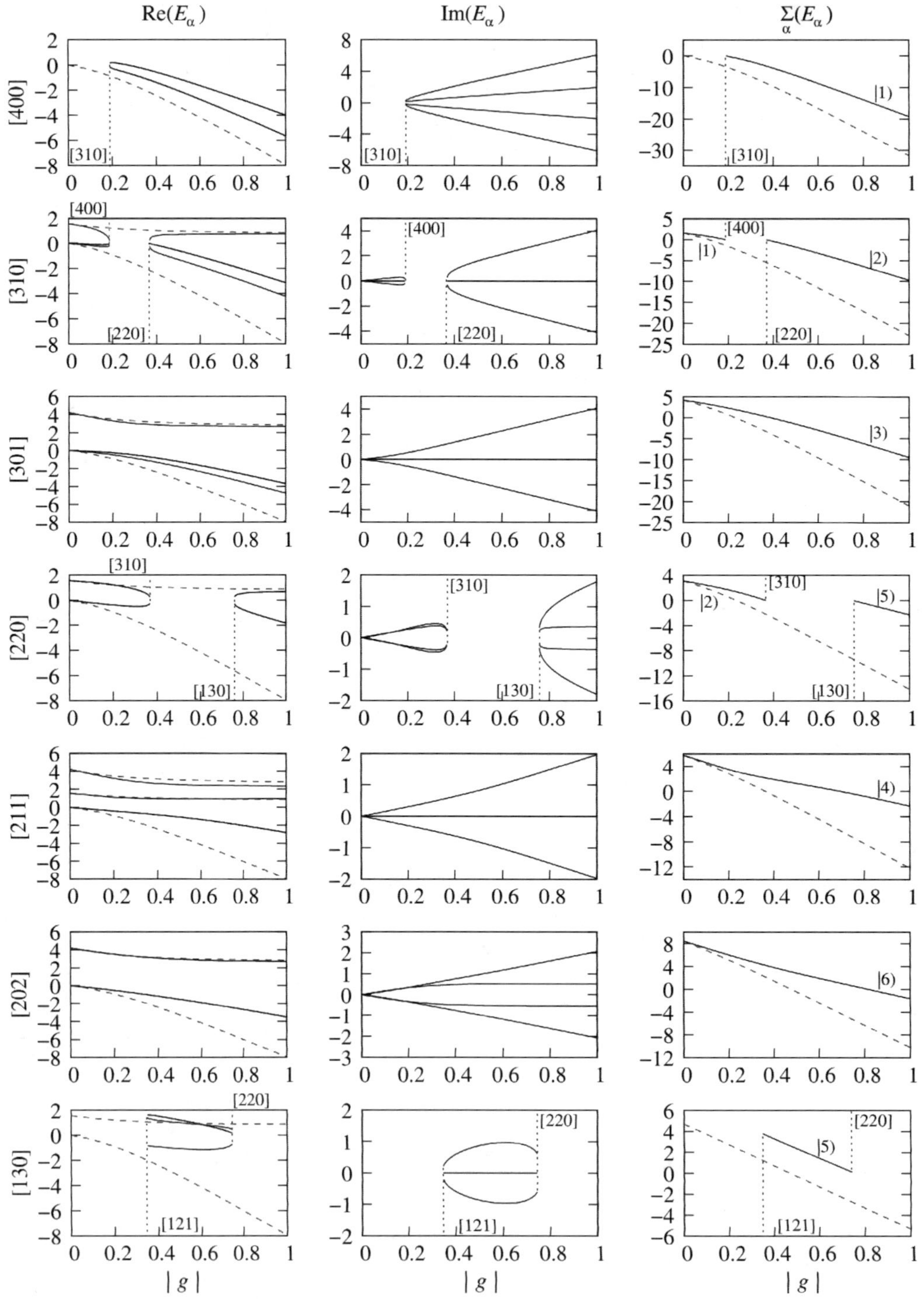

Figure 4. The TDA eigen modes and corresponding RG variables for given TDA multiphonon states as a function of the interaction $|g|$. More information is discussed in the text.

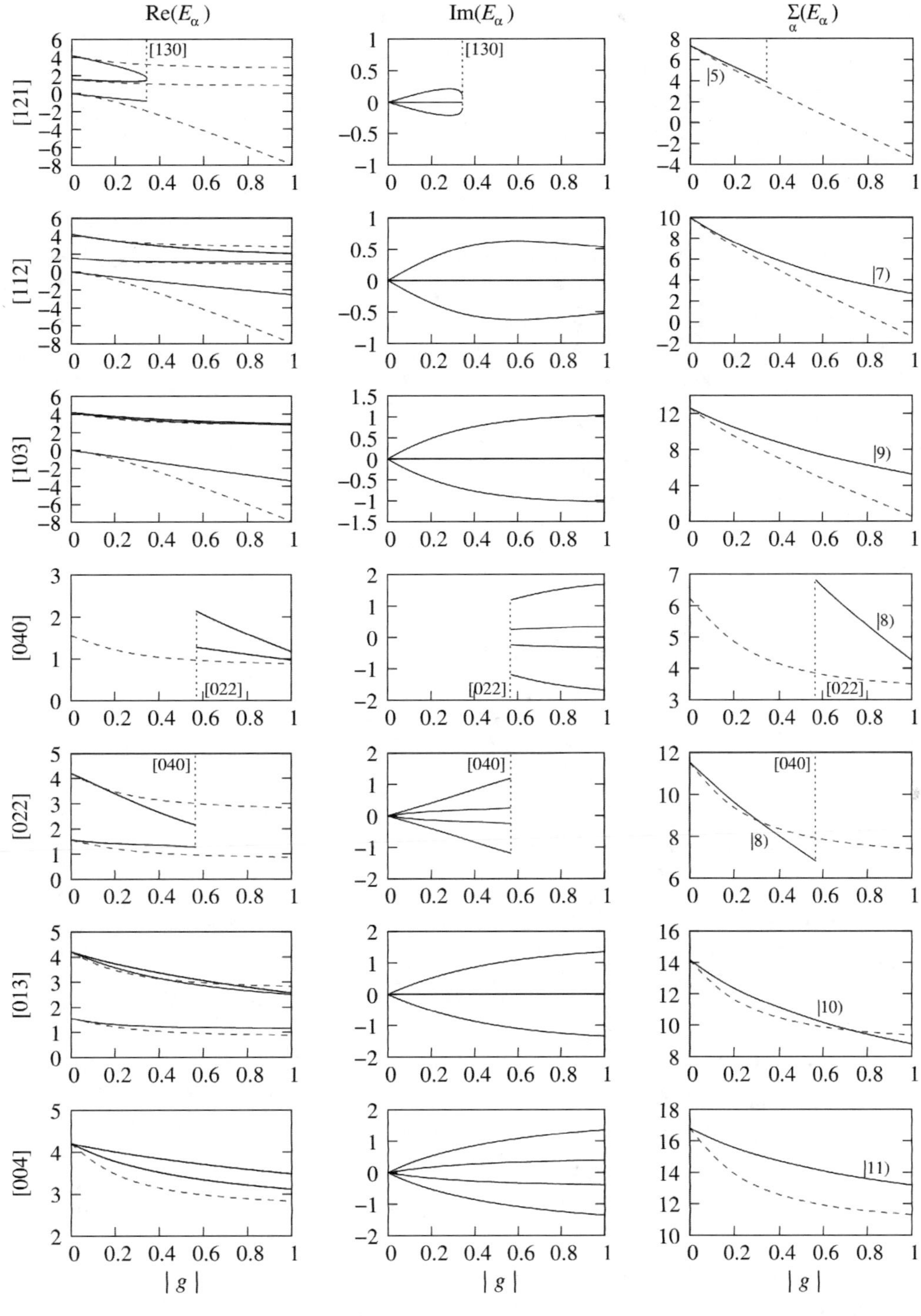

Figure 5. The TDA eigen modes and corresponding RG variables for given TDA multiphonon states as a function of the interaction $|g|$. More information is discussed in the text.

$hw(1)$ in the two limits of the parameter. Using this parametrised algebra for the Bethe Ansatz wavefunction, we obtain the corresponding parametrised Richardson-Gaudin equations, which reduce to the exact Richardson-Gaudin equations in the quasi-spin limit and the Tamm-Dancoff secular equation in the Heisenberg-Weyl limit. These parametrised RG equations allow us to make a clear-cut connection between the exact states of the BCS Hamiltonian and the TDA predictions.

Acknowledgments
It is a pleasure to acknowledge P. Van Isacker, S. Rombouts, D. J. Rowe and D. Van Neck for many stimulating and interesting discussion sessions. I would also like to thank V. Hellemans, K. Heyde and B. Verstichel for discussions and interesting suggestions. Financial support comes from the "FWO-Vlaanderen" as well as 2 scholarships for a "Long stay abroad" to support my stay at the University of Toronto and the University of Notre Dame.

References

[1] Bardeen J, Cooper L N and Schrieffer J 1957 *Phys. Rev.* **108** 1175
[2] Bohr A, Mottelson B R and Pines D 1958 *Phys. Rev.* **110** 936
[3] Rowe D J and Wood J 2010 *Fundamentals of Nuclear Models* (Singapore: World Scientific)
[4] Heyde K 1994 *The nuclear shell model* (Berlin: Springer-Verlag)
[5] von Delft J and Ralph D C 2001 *Phys. Rep.* **345** 61
[6] Richardson R W 1963 *Phys. Lett.* **3** 277
[7] Richardson R W and Sherman N 1964 *Nucl. Phys.* **52** 221
[8] Richardson R W and Sherman N 1964 *Nucl. Phys.* **52** 253
[9] Richardson R W 1966 *Phys. Rev.* **141** 949
[10] Rombouts S, Van Neck D and Dukelsky J 2004 *Phys. Rev. C* **69** 061303(R)
[11] Domínguez F, Esebbag C and Dukelsky J 2006 *J. Phys. A: Math. Gen.* **39** 11349
[12] Román J M, Sierra G and Dukelsky J 2003 *Phys. Rev. B* **67** 064510
[13] Sambataro M 2007 *Phys. Rev. C* **75** 054314
[14] Dukelsky J, Pittel S and Sierra G 2004 *Rev. Mod. Phys.* **76** 643
[15] Rowe D J 1970 *Nuclear Collective Motion* (London: Methuen)
[16] Talmi I 1993 *Simple models of complex nuclei* (Chur: Harwood academic publishers)
[17] Rowe D J 1968 *Rev. Mod. Phys.* **40** 153
[18] Arecchi F T, Courtens E, Gilmore R and Harry T 1972 **6** 2211
[19] Gilmore R 2008 *Lie Groups, Physics and Geometry, An Introduction for Physicists, Engineers and Chemists* (Cambridge: Cambridge University Press)

Creating subgroups of $U(2^w)$ for quantum-minus computers

Alexis De Vos and Michiel Boes

Imec v.z.w. and Vakgroep elektronika en informatiesystemen
Universiteit Gent
Sint Pietersnieuwstraat 41
B - 9000 Gent, Belgium

E-mail: alex@elis.UGent.be

Abstract. Classical reversible computers on w bits are isomorphic to the (finite) symmetric group $\mathbf{S}_{2^w}$; quantum computers on w qubits are isomorphic to the (Lie) unitary group $U(2^w)$. We investigate and classify groups $\mathbf{X}$ which represent computers intermediate between classical reversible computers and quantum computers. Such intermediate groups $\mathbf{X}$ may exist in three flavours:

- finite groups of order larger than $(2^w)!$,
- infinite but discrete groups, and
- Lie groups of dimension smaller than $(2^w)^2$.

The larger the group, the more powerful the computer may be, but the smaller the group, the easier it can be to build the computer hardware. In the present paper, we investigate the first two flavours only.

For our purpose, we start from 1-qubit transformations, represented by 2×2 unitary matrices. We call this group the creator. Its members are called gates and act on one qubit. Controlled gates are quantum circuits acting on w qubits, such that the 1-qubit transformation (applied to a particular qubit) depends on the state of the $w-1$ other qubits. The controlled gates generate the group $\mathbf{X}$ of $2^w \times 2^w$ matrices, called the creation.

We discuss all creators of order up to 8. Additionally a creator of order 16 and one of order 192 are discussed.

1. Introduction

Reversible logic circuits, acting on w bits, form a group, isomorphic to the symmetric group $\mathbf{S}_m$ of degree m and order $m!$, where m is equal to 2^w. For $w = 1$, the group $\mathbf{S}_2$ consists of two 2×2 matrices:

$$e = \begin{pmatrix} 1 & 0 \\ 0 & 1 \end{pmatrix} \text{ and } n = \begin{pmatrix} 0 & 1 \\ 1 & 0 \end{pmatrix} ,$$

where e is called the `IDENTITY` gate and n is called the `NOT` gate. The group may be generated by a single generator, i.e. by n. We note that $n^2 = e$.

We now enlarge this matrix group, either by replacing the generator n by another generator or by adding extra generator(s). In the former case, we make sure that the `NOT` gate n is member of the group. The resulting new group of 2×2 unitary matrices, we denote by $\mathbf{X}_2$. We have

$$\mathbf{S}_2 \subset \mathbf{X}_2 \subset U(2) . \quad (1)$$

As the two matrices e and n form a subgroup of $\mathbf{X}_2$, Lagrange's theorem tells us that the order of $\mathbf{X}_2$ is even. We denote this number by p.

Logic circuits of w qubits are described by $2^w \times 2^w$ unitary matrices, which form a group isomorphic to the Lie group $\mathrm{U}(2^w)$. Within this group we define a particular set, represented by $2^w \times 2^w$ matrices with all matrix entries equal to 0, except for 2^{w-1} submatrices of size 2×2, each member of U(2). We impose that those submatrices are either equal to e or equal to some given unitary matrix v, element of $\mathbf{X}_2$:

$$v = \begin{pmatrix} v_{11} & v_{12} \\ v_{21} & v_{22} \end{pmatrix} .$$

Such logic gates are called controlled `V`s. One qubit k is called the controlled qubit; the others are called the controlling qubits. The matrices with equal k form a group isomorphic to the direct product $\mathbf{X}_2 \times \mathbf{X}_2 \times ... \times \mathbf{X}_2 = \mathbf{X}_2^{2^{w-1}}$, of order $p^{2^{w-1}}$. The figure shows an arbitrary example of a group element: a controlled `V` with $w = 3$ and $k = 2$. The 1-qubit transformation v is applied to qubit A_2, iff a given Boolean function $f(A_1, A_3)$ equals 1. We call this function f the control function of the gate. If e.g. f is the `OR` function A_1 `OR` A_3, then the circuit is represented by the transformation matrix

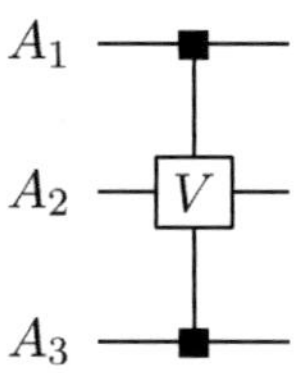

$$= \begin{pmatrix} \mathbf{1} & 0 & \mathbf{0} & 0 & 0 & 0 & 0 & 0 \\ 0 & \mathbf{v_{11}} & 0 & \mathbf{v_{12}} & 0 & 0 & 0 & 0 \\ \mathbf{0} & 0 & \mathbf{1} & 0 & 0 & 0 & 0 & 0 \\ 0 & \mathbf{v_{21}} & 0 & \mathbf{v_{22}} & 0 & 0 & 0 & 0 \\ 0 & 0 & 0 & 0 & \mathbf{v_{11}} & 0 & \mathbf{v_{12}} & 0 \\ 0 & 0 & 0 & 0 & 0 & \mathbf{v_{11}} & 0 & \mathbf{v_{12}} \\ 0 & 0 & 0 & 0 & \mathbf{v_{21}} & 0 & \mathbf{v_{22}} & 0 \\ 0 & 0 & 0 & 0 & 0 & \mathbf{v_{21}} & 0 & \mathbf{v_{22}} \end{pmatrix} ,$$

with the four submatrices bold-faced (one equal to e, the three others to v).

The four generators of the group with arbitrary $f(A_1, A_3)$ are

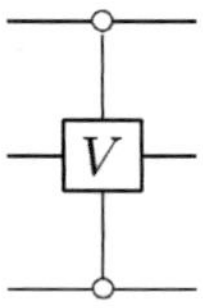

,
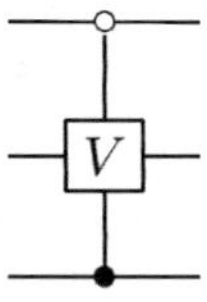

,
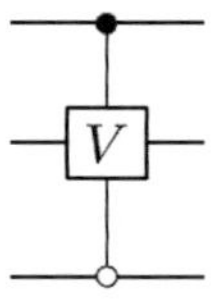

, and
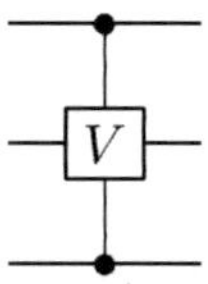

.

These gates have control functions f equal to $\overline{A_1}\ \overline{A_3}$, $\overline{A_1}\, A_3$, $A_1\, \overline{A_3}$, and $A_1\, A_3$, respectively. Here $\overline{A}$ is a short-hand notation for `NOT` A. In order to decompose an arbitrary control gate into generators, it suffices to decompose its control function f into minterms.

Each of the w choices of the number k gives rise to $g2^{w-1}$ generators, where g is the number of 2×2 matrices necessary to generate $\mathbf{X}_2$. The complete set of $gw2^{w-1}$ generators generates the group $\mathbf{X}$. If $\mathbf{X}_2$ is $\mathbf{S}_2$, then $\mathbf{X}$ is nothing else but $\mathbf{S}_{2^w}$. If $\mathbf{X}_2$ is not $\mathbf{S}_2$, then $\mathbf{X}$ is not necessarily $\mathbf{S}_{2^w}$. We call the group $\mathbf{X}$ the creation of $\mathbf{X}_2$; we call the group $\mathbf{X}_2$ the creator of $\mathbf{X}$. Above, we have seen that $\mathbf{S}_2$ is the creator of $\mathbf{S}_{2^w}$. Below, we investigate which groups are created by groups $\mathbf{X}_2$. Because of (1), we automatically have

$$\mathbf{S}_{2^w} \subset \mathbf{X} \subset \mathrm{U}(2^w) .$$

2. Results for small creators

Here follows a full classification of creations $\mathbf{X}$, created by small creators $\mathbf{X}_2$, i.e. creators of order p smaller than 10.

Table 1. The orders of small creator groups ($w = 1$) and the orders of the corresponding creation groups (where m stands for 2^w with $w > 1$).

$p =$ order of creator	2	4	4	6	6	6	8	8	8	8
isomorphism of creator	$\mathbf{Z}_2$	$\mathbf{Z}_2\times \mathbf{Z}_2$	$\mathbf{Z}_4$	$\mathbf{S}_3$	$\mathbf{Z}_6$	$\mathbf{Z}_6$ $\mathbf{S}_3$	$\mathbf{D}_4$	$\mathbf{D}_4$	$\mathbf{Z}_2\times \mathbf{Z}_4$	$\mathbf{Z}_8$ $\mathbf{Z}_2\times \mathbf{Z}_4$ $\mathbf{D}_4$
order of creation	$m!$	$\frac{m!2^m}{2}$	$\aleph_0$	$\frac{m!3^m}{3}$	$m!3^m$	$\aleph_0$	$m!2^m$	$\frac{m!4^m}{4}$	$\frac{m!4^m}{2}$	$\aleph_0$

2.1. Creator of order 2
The case $p = 2$ is trivial. The creator $\mathbf{X}_2$ is generated by a single generator, i.e. n, and consists of two 2×2 matrices: e and n. The creations $\mathbf{X}$ are the groups of $2^w \times 2^w$ permutation matrices and thus are isomorphic to the symmetric groups $\mathbf{S}_{2^w}$, of order $(2^w)!$. This constitutes the case of reversible classical computers.

2.2. Creators of order 4
There exist three groups of 2×2 unitary matrices, containing the `NOT` matrix n and being of order 4.

The first creator of order 4 is generated by two generators: $-e$ and n. It consists of the matrices e, $-e$, n, and $-n$ and is isomorphic to $\mathbf{Z}_2\times \mathbf{Z}_2$. Its creations consist of $2^w \times 2^w$ matrices with all entries from $\{0, 1, -1\}$, such that the number of -1s is even. The order of the creation is $(2^w)!2^{2^w-1}$.

The second one is generated by a single generator, i.e.

$$s = \frac{1}{2}\begin{pmatrix} 1+i & 1-i \\ 1-i & 1+i \end{pmatrix} .$$

We call this matrix the 'square root of `NOT`' [1] [2] [3]. It generates a group isomorphic to the cyclic group $\mathbf{Z}_4$. The members of the group are s, $s^2 = n$, $s^3 = \overline{s}$, and $s^4 = e$. This creator gives rise to creations of a countably infinite order [4].

The third order-4 group is generated by $-s$ and consists of the four matrices $-s$, n, $-\overline{s}$, and e. This creator behaves similar to the previous one.

2.3. Creators of order 6
There exist four abelian groups of 2×2 unitary matrices, containing the `NOT` matrix n and being of order 6. All are isomorphic to the cyclic group $\mathbf{Z}_6$. They are generated by a single generator, respectively

$$\frac{1}{4}\begin{pmatrix} 3+\lambda & 1-\lambda \\ 1-\lambda & 3+\lambda \end{pmatrix}, \frac{1}{4}\begin{pmatrix} -3+\lambda & 1+\lambda \\ 1+\lambda & -3+\lambda \end{pmatrix}, \frac{1}{2}\begin{pmatrix} \lambda & -1 \\ -1 & \lambda \end{pmatrix}, \text{ and } \frac{1}{2}\begin{pmatrix} 0 & -1+\lambda \\ -1+\lambda & 0 \end{pmatrix},$$

where λ is a short-hand notation for $i\sqrt{3}$. The fourth creator creates groups of order $(2^w)!3^{2^w}$ (the formula however not holding in the trivial case $w = 1$), consisting of $2^w \times 2^w$ matrices with all entries from $\{0, 1, (-1+\lambda)/2, (-1-\lambda)/2\}$. The three other creators create infinite creations.

Besides this finite number of abelian creators, there exist a (non-countable) infinity of non-abelian creators of order 6. They are isomorphic to the symmetric group $\mathbf{S}_3$. They are generated by two generators: n and

$$\begin{pmatrix} \cos(\theta) & \sin(\theta)\exp(i\alpha) \\ \sin(\theta)\exp(-i\alpha) & -\cos(\theta) \end{pmatrix},$$

where α is an arbitrary angle (different from $\pm\pi/2$), but θ obeys

$$\sin(\theta) = -\frac{1}{2\cos(\alpha)} .$$

These order-6 creators yield infinite creations, except if $\alpha = \pm\pi/3$, in which case the creations have merely order $(2^w)!3^{2^w-1}$.

2.4. Creators of order 8

There exist eight different creator groups isomorphic to $\mathbf{Z}_8$, each one generated by a matrix of the form

$$\frac{1}{4}\begin{pmatrix} \lambda_1+\lambda_2 & \lambda_1-\lambda_2 \\ \lambda_1-\lambda_2 & \lambda_1+\lambda_2 \end{pmatrix},$$

where λ_1 is $2i^k$ with an integer k and λ_2 is $\sqrt{2} \pm i\sqrt{2}$. All eight creators yield infinite creations.

There exist three different creators isomorphic to $\mathbf{Z}_2\times\ \mathbf{Z}_4$: one is generated by the two matrices $-e$ and s, one by the generators n and is, and one by the two generators ie and n. The last creator yields groups of order $(2^w)!4^{2^w-1/2}$ (consisting of matrices with entries from $\{0, 1, -1, i, -i\}$, with the restriction that the number of entries equal to $\pm i$ is even), the other two yielding creations of infinite order.

Finally, there exist ∞^1 matrix groups isomorphic to the dihedral group $\mathbf{D}_4$, generated by the two generators n and

$$\begin{pmatrix} \cos(\theta) & -i\sin(\theta) \\ i\sin(\theta) & -\cos(\theta) \end{pmatrix} .$$

All create infinite creations, except in the cases $\theta = 0$ (in which case the order is $(2^w)!2^{2^w}$) and $\theta = \pi/2$ (in which case the order is $(2^w)!4^{2^w-1}$).

3. Results for a few larger creators

Below, some creators of order p higher than 9 are discussed, because they are well-known in quantum computer theory.

3.1. A creator with $p = 16$

The notorious Pauli group [5] [6] consists of the sixteen matrices $\pm e$, $\pm n$, $\pm\sigma_2$, $\pm\sigma_3$, $\pm ie$, $\pm in$, $\pm i\sigma_2$, and $\pm i\sigma_3$, where σ_2 and σ_3 stand for the Pauli matrices[1]

$$\begin{pmatrix} 0 & -i \\ i & 0 \end{pmatrix} \text{ and } \begin{pmatrix} 1 & 0 \\ 0 & -1 \end{pmatrix},$$

respectively. The group is generated by the three generators n, σ_2, and σ_3. It creates finite creations (of order $(2^w)!4^{2^w-1/2}$), consisting of the same $2^w \times 2^w$ matrices with entries from $\{0, 1, -1, i, -i\}$, created by the order-8 creator generated by ie and n (see Section 2.4).

We note that the creations are much larger than the w-qubit Pauli group (of order 4×4^w), consisting of the tensor products of the one-qubit Pauli matrices.

[1] Sometimes n, i.e. the `NOT` matrix, is also called a Pauli matrix and then denoted σ_1.

3.2. A creator with $p = 192$

The notorious Clifford group [6] [7] [8] consists of the 192 matrices generated by the two generators

$$\frac{1}{\sqrt{2}} \begin{pmatrix} 1 & 1 \\ 1 & -1 \end{pmatrix} \text{ and } \begin{pmatrix} 1 & 0 \\ 0 & i \end{pmatrix} ,$$

known as the `HADAMARD` gate and the `PHASE` gate, respectively. The group creates infinite creations. Again, the creations are much larger than the w-qubit Clifford groups (of order $2^{w^2+2w+3} \prod_{j=1}^{w} 4^j - 1)$, consisting of the tensor products of the one-qubit Clifford matrices.

4. Conclusion

In the above examples, the finite creators lead to creations, which can be subdivided into two different classes:

- either finite creations
- or (countably) infinite creations.

We shortly discuss both categories.

4.1. Finite creations

Whenever all generators (and thus automatically all elements) of the creator are unitary 2×2 matrices with two zero entries, then the creations are groups of $2^w \times 2^w$ matrices with all entries equal zero, except one entry in each row and one entry in each column, with a modulus equal to 1 (and a phase angle equal to some rational multiple of π). Such matrices are monomial and thus strongly resemble permutation matrices. Creation group orders turn out to be of the form $(2^w)!\, c^{2^w}/d$, with both c and d appropriate integer constants.

Such matrices lead to computational power hardly exceeding the power of classical computers. Indeed, any such matrix transforms a classical input pattern into a classical output pattern, up to a phase factor, as illustrated by a $w = 2$ example:

$$\begin{pmatrix} 0 & \exp(i\varphi_1) & 0 & 0 \\ 0 & 0 & 0 & \exp(i\varphi_2) \\ \exp(i\varphi_3) & 0 & 0 & 0 \\ 0 & 0 & \exp(i\varphi_4) & 0 \end{pmatrix} \begin{pmatrix} 0 \\ 0 \\ 1 \\ 0 \end{pmatrix} = \exp(i\varphi_4) \begin{pmatrix} 0 \\ 0 \\ 0 \\ 1 \end{pmatrix} .$$

In passing, we mention that also infinite groups of such matrices (displaying φ angles which are not rational multiples of π) have such low computational power.

4.2. Infinite creations

In all above examples, as soon as one generator of the creator is a 2×2 unitary matrix u with all four entries different from zero, the creations are of infinite order. We conjecture that this result is generally valid. There exists a proof [9] for the case where u is of the form

$$\begin{pmatrix} \cos(\theta) & \sin(\theta)\exp(i\alpha) \\ \sin(\theta)\exp(-i\alpha) & -\cos(\theta) \end{pmatrix} ,$$

with both non-zero $\cos(\theta)$ and non-zero $\sin(\theta)$. There also exists a proof [4] if u is the matrix s, i.e. the $\sqrt{\texttt{NOT}}$ gate. This very same proof remains valid for all u of the form

$$\frac{1}{2} \begin{pmatrix} (1+i)\exp(i\eta) & 1-i \\ 1-i & (1+i)\exp(-i\eta) \end{pmatrix} .$$

The infinite (but discrete) creations are computationally more powerful than classical reversible computers. Their extra power is e.g. illustrated by the fact that the 3-bit Toffoli gate is not decomposable into simpler parts within the framework of classical computing, but is decomposable into 2-qubit gates if the creator is the order-4 group with the square root of `NOT` [10] [11] [12]. Nevertheless, they do not have the full computational power of quantum computing. Therefore, any creation of order $\aleph_0$, we call quantum-minus computers.

References

[1] Deutsch D 1992 *Physics World* **5** 57–61
[2] Deutsch D, Ekert A and Lupacchini A 2000 *The Bulletin of Symbolic Logic* **3** 265–83
[3] Galindo A and Martín-Delgado M 2002 *Review of Modern Physics* **74** 347–423
[4] De Vos A, De Beule J and Storme L 2009 *Serdica Journal of Computing* **3** 359–70
[5] Nielsen M and Chuang I 2000 "Quantum computation and quantum information", Cambridge University Press, pp 65, 454, and 611
[6] Planat M 2010 *Invertis Journal of Science and Technology* **3** 1–4
[7] Planat M and Solé P 2009 *Journal of Physics A: Mathematical and Theoretical* **42** 042003
[8] Sloane N `http://public.research.att.com/~njas/sequences/?id%3A003956`
[9] Boes M, De Vos A and De Beule J 2010 *Proc. 9 th Int. Workshop on Boolean Problems*, Freiberg (September 2010), pp 51–6
[10] Barenco A, Bennett C, Cleve R, DiVincenzo D, Margolus N, Shor S, Sleator T, Smolin J and Weinfurter H 1995 *Physical Review* C **52** 3457–67
[11] Wille R, Saeedi M and Drechsler R 2009 *Proc. 18 th Int. Workshop on Logic and Synthesis*, Berkeley (July 2009) pp 43–9
[12] Miller M and Sasanian Z 2010 *Proc. 9 th Int. Workshop on Boolean Problems*, Freiberg (September 2010), pp 37–41

Journal of Physics: Conference Series **284** (2011) 012022 doi:10.1088/1742-6596/284/1/012022

Indecomposable representations of the Euclidean algebra $\mathfrak{e}(3)$ from irreducible representations of the symplectic algebra $\mathfrak{sp}(4,\mathbb{C})$.

Andrew Douglas[1], **Joe Repka**[2]

[1] Department of Mathematics, New York City College of Technology, City University of New York, 300 Jay Street, Brooklyn, New York, USA 11201.

[2] Department of Mathematics, University of Toronto, 40 St. George Street, Toronto, ON, Canada, M5S 2E4.

E-mail: adouglas@citytech.cuny.edu

Abstract. The Euclidean group $E(3)$ is the noncompact, semidirect product group $E(3) \cong SO(3) \ltimes \mathbb{R}^3$. It is the Lie group of orientation-preserving isometries of 3-dimensional Euclidean space. The Euclidean algebra $\mathfrak{e}(3)$ is the complexification of the Lie algebra of $E(3)$. We embed $\mathfrak{e}(3)$ into the 10-dimensional symplectic algebra $\mathfrak{sp}(4,\mathbb{C})$, the simple Lie algebra of type C_2. We show that, up to conjugation by an element of $Sp(4,\mathbb{C})$, there is only one embedding of $\mathfrak{e}(3)$ into $\mathfrak{sp}(4,\mathbb{C})$, and then prove that the irreducible representations of $\mathfrak{sp}(4,\mathbb{C})$ remain indecomposable upon restriction to $\mathfrak{e}(3)$, thus creating a new class of indecomposable $\mathfrak{e}(3)$-representations.

1. Introduction

The Euclidean group $E(3)$ is the noncompact, semidirect product group $E(3) \cong SO(3) \ltimes \mathbb{R}^3$. It is the Lie group of orientation-preserving isometries of 3-dimensional Euclidean space. The Euclidean algebra $\mathfrak{e}(3)$ is the complexification of the Lie algebra of $E(3)$. Both $E(3)$ and $\mathfrak{e}(3)$ have found significant applications in physics, examples of which are described in [4].

The finite-dimensional irreducible representations of $\mathfrak{e}(3)$ are not very interesting, but classifying its indecomposable representations remains a significant challenge. We remind the reader that a representation is *irreducible* if it has no proper subrepresentations. It is *indecomposable* if it is not isomorphic to a direct sum of two nonzero subrepresentations. A representation may be indecomposable without being irreducible.

Although a full classification of $\mathfrak{e}(3)$-indecomposable representations remains elusive, constructing large classes of indecomposable representations that may be classified is a viable option. One interesting possibility is to embed $\mathfrak{e}(3)$ into an algebra for which the irreps are known, and then consider their restrictions to $\mathfrak{e}(3)$, regarded as a subalgebra. In some cases, it may be possible to show that the resulting representations of $\mathfrak{e}(3)$ are indecomposable.

This direction of research has been pursued, for instance, by Douglas and Premat [3], who show that irreducible $\mathfrak{sl}(3,\mathbb{C})$-modules remain $\mathfrak{e}(2)$-indecomposable, and later in a tour de force by Casati et al. [2] who established that irreducible $\mathfrak{sl}(3,\mathbb{C})$ and $\mathfrak{sp}(4,\mathbb{C})$-modules remain indecomposable modules of the Diamond Lie algebra under appropriate embeddings. The Diamond Lie algebra is a central extension of the Poincaré Lie algebra in two dimensions.

In the current article we begin by embedding the Euclidean algebra $\mathfrak{e}(3)$ into the 10-dimensional symplectic algebra $\mathfrak{sp}(4,\mathbb{C})$, the simple Lie algebra of type C_2. It is the smallest simple Lie algebra into which $\mathfrak{e}(3)$ embeds. In fact, we are able to show that, up to conjugation by an element of $Sp(4,\mathbb{C})$, there is only one such embedding. We then show that the finite-dimensional irreducible representations of $\mathfrak{sp}(4,\mathbb{C})$ remain indecomposable upon restriction to $\mathfrak{e}(3)$ under this embedding, thus creating a new class of indecomposable $\mathfrak{e}(3)$-representations.

The article is organized as follows. In section 2 we describe the basis and commutation relations of $\mathfrak{e}(3)$. Section 3 records information about the simple Lie algebra $\mathfrak{sp}(4,\mathbb{C})$ and its irreducible representations that will be employed in the following section. In section 4 we prove that, up to conjugation by an element in $Sp(4,\mathbb{C})$, there is only one embedding of $\mathfrak{e}(3)$ into $\mathfrak{sp}(4,\mathbb{C})$. In section 5 we prove that irreducible representations of $\mathfrak{sp}(4,\mathbb{C})$ remain $\mathfrak{e}(3)$-indecomposable under the (essentially) unique embedding of $\mathfrak{e}(3)$ into $\mathfrak{sp}(4,\mathbb{C})$.

2. The Euclidean algebra $\mathfrak{e}(3)$

The *Euclidean algebra* $\mathfrak{e}(3)$ is the complexification of the Lie algebra of the Euclidean Lie group $E(3)$. For a more detailed discussion of $E(3)$ and the calculation of its Lie algebra we refer the reader to Hall [6]. The Euclidean algebra $\mathfrak{e}(3)$ has basis $E, H, F, P_0, P_\pm$, and nonzero commutation relations

$$\begin{array}{llllll} [H,E] & = 2E, & [H,F] & = -2F, & [E,F] & = H, \\ [H,P_\pm] & = \pm 2P_\pm, & [E,P_0] & = -P_+, & [F,P_0] & = -P_-, \\ [F,P_+] & = -2P_0, & [E,P_-] & = -2P_0. & & \end{array} \tag{1}$$

One can easily see that $\langle E,H,F\rangle \cong \mathfrak{sl}(2,\mathbb{C}) \cong \mathfrak{so}(3)_\mathbb{C}$, and that $\langle P_0, P_\pm\rangle$ is an abelian ideal of $\mathfrak{e}(3)$.

Moreover, under the adjoint action, the action of the $\mathfrak{so}(3)_\mathbb{C}$ subalgebra $\langle E,H,F\rangle$ decomposes as the direct sum of two three-dimensional irreps, $\langle E,H,F\rangle$ and $\langle P_0,P_\pm\rangle$.

3. The symplectic algebra $\mathfrak{sp}(4,\mathbb{C})$ and its irreducible representations

The symplectic algebra $\mathfrak{sp}(4,\mathbb{C})$ is the Lie algebra of 4×4 complex matrices X satisfying $JX^{\mathrm{tr}}J = X$, where J is the 4×4 matrix

$$J = \begin{pmatrix} 0 & I \\ -I & 0 \end{pmatrix}. \tag{2}$$

It is the 10-dimensional simple Lie algebra of type C_2 (equivalent to the simple Lie algebra of type B_2). Let

$$\{x_i, y_i, h_j : 1 \le i \le 4, 1 \le j \le 2\} \tag{3}$$

be the Chevalley basis of $\mathfrak{sp}(4,\mathbb{C})$ with nontrivial commutation relations

$$\begin{array}{llllllll} [h_i,h_j] & = 0, & [h_i,x_j] & = C_{ji}x_j, & [h_i,y_j] & = -C_{ji}y_j, & [x_i,y_j] & = \delta_{ij}h_i, \\ [h_1,x_4] & = 2x_4, & [h_1,y_4] & = -2y_4, & [h_2,x_3] & = x_3, & [h_2,y_3] & = -y_3, \\ [x_1,x_2] & = -x_3, & [x_1,x_3] & = -2x_4, & [x_1,y_3] & = 2y_2, & [x_1,y_4] & = y_3, \\ [x_2,y_3] & = -y_1, & [x_3,y_1] & = 2x_2, & [x_3,y_2] & = -x_1, & [x_3,y_3] & = h_1+2h_2, \\ [x_3,y_4] & = -y_1, & [x_4,y_1] & = x_3, & [x_4,y_3] & = -x_1, & [x_4,y_4] & = h_1+h_2, \\ [y_1,y_2] & = y_3, & [y_1,y_3] & = 2y_4, & & & & \end{array} \tag{4}$$

for $1 \le i,j \le 2$ and where C is the Cartan matrix of type C_2

$$C = \begin{pmatrix} 2 & -1 \\ -2 & 2 \end{pmatrix}. \tag{5}$$

For $i = 1$ or 2, define $\Lambda_i \in \mathfrak{h}*$ by $\Lambda_i(h_j) = \delta_{ij}$. For each $\lambda = m_1\Lambda_1 + m_2\Lambda_2 \in \mathfrak{h}*$ with nonnegative integers m_1, m_2, there exists a finite dimensional irreducible $\mathfrak{sp}(4,\mathbb{C})$-module $V(m_1, m_2)$ which can be realized as the quotient of the universal enveloping algebra $\mathcal{U}(\mathfrak{sp}(4,\mathbb{C}))$ by the left ideal J_λ generated by x_i, $h_i - \lambda(h_i)$, $y_i^{1+\lambda(h_i)}$, $1 \leq i \leq 2$ (here the action of $\mathcal{U}(\mathfrak{sp}(4,\mathbb{C}))$ on itself and on $V(m_1, m_2)$ is given by left multiplication). We will denote the element $1 + J_\lambda$ of $V(m_1, m_2)$ by v_λ. The dimension of $V(m_1, m_2)$ follows in a straightforward manner from Weyl's character formula [7]:

$$\dim(V(m_1, m_2)) = \frac{1}{3!}(m_1 + 1)(m_2 + 1)(m_1 + m_2 + 2)(m_1 + 2m_2 + 3). \tag{6}$$

Bliem [1] showed that $V(m_1, m_2)$ has basis

$$\mathcal{B} = \{y_1^{a_1} y_2^{a_2} y_1^{a_3} y_2^{a_4} v_\lambda\}, \tag{7}$$

subject to the relations

$$\begin{aligned} &0 \leq a_4 \leq m_2, \\ &2a_4 \leq a_3 \leq m_1 + 2a_4, \\ &\frac{1}{2}a_3 \leq a_2 \leq m_2 + a_3 - 2a_4, \\ &0 \leq a_1 \leq m_1 + 2a_2 - 2a_3 + 2a_4. \end{aligned} \tag{8}$$

Bliem's basis, however, is not convenient for the present purposes. We use the spanning set of $V(m_1, m_2)$ described in the following Lemma.

Lemma 3.1 *The $\mathfrak{sp}(4,\mathbb{C})$-module $V(m_1, m_2)$ is spanned by the (not necessarily linearly independent) set*

$$\mathcal{S}_\lambda = \{y_1^a\ y_2^b\ y_3^c\ y_4^d\ v_\lambda\} \tag{9}$$

subject to the relations

$$b \leq m_2, \qquad c \leq m_1 + 2m_2, \qquad d \leq m_1 + m_2, \qquad a + c + 2d \leq 2m_1 + 2m_2. \tag{10}$$

Proof: The set $\{y_1^a y_2^b y_3^c y_4^d v_\lambda\}$ with nonnegative integers a, b, c, d spans $V(m_1, m_2)$ [7]. Since y_2, y_3 and y_4 commute, a defining a relation of $V(m_1, m_2)$ implies $b \leq m_2$.

Note that $(h_1 + 2h_2) \cdot y_1^a y_2^b y_3^c y_4^d v_\lambda = (m_1 + 2m_2 - 2(b + c + d))v_\lambda$ and $(h_1 + h_2) \cdot y_1^a y_2^b y_3^c y_4^d v_\lambda = (m_1 + m_2 - (a + c + 2d))v_\lambda$. Hence, since $\langle x_3, h_1 + 2h_2, y_3\rangle$ and $\langle x_4, h_1 + h_2, y_4\rangle$ are each isomorphic to $\mathfrak{sl}(2,\mathbb{C})$, the representation theory of $\mathfrak{sl}(2,\mathbb{C})$ implies the remaining conditions given in (10). □

4. Classification of Embeddings of $\mathfrak{e}(3)$ into $\mathfrak{sp}(4,\mathbb{C})$

4.1. Restriction to $\mathfrak{so}(3)_\mathbb{C}$

Suppose $T : \mathfrak{e}(3) \hookrightarrow \mathfrak{sp}(4,\mathbb{C})$ is an embedding of Lie algebras. The restriction of T to the $\mathfrak{so}(3)_\mathbb{C}$ subalgebra is a four-dimensional representation of $\mathfrak{so}(3)_\mathbb{C} \cong \mathfrak{su}(2)_\mathbb{C}$, so it must be one of the following: $4V^0$, $V^{1/2} \oplus 2V^0$, $V^1 \oplus V^0$, $2V^{1/2}$, or $V^{3/2}$, where V^j is the $(2j+1)$-dimensional representation of $\mathfrak{so}(3)_\mathbb{C} \cong \mathfrak{su}(2)_\mathbb{C}$. Of these, the first is not an embedding.

From (1), we note that, when H acts on V^1, its highest eigenvalue is 2. Accordingly, its highest eigenvalue on V^j is $2j$. If $T|_{\mathfrak{so}(3)_\mathbb{C}} \cong V^{3/2}$, then $T(H)$ will have eigenvalues $3, 1, -1, -3$. Under its action on $\mathfrak{sl}(4,\mathbb{C})$ under the $\mathfrak{sl}(4,\mathbb{C})$-adjoint action, $T(H)$ will have highest eigenvalue $3 - (-3) = 6$, which means that the action of $T(\mathfrak{so}(3)_\mathbb{C})$ on $\mathfrak{sl}(4,\mathbb{C})$ will include a copy of V^3. Since V^3 has dimension 7, it cannot be complementary to $\mathfrak{sp}(4,\mathbb{C})$, which has dimension 10 inside the 15-dimensional space $\mathfrak{sl}(4,\mathbb{C})$. This implies that the action of $T(\mathfrak{so}(3)_\mathbb{C})$ on $\mathfrak{sp}(4,\mathbb{C})$ contains a copy of V^3. But this is impossible, since we know that the $\mathfrak{so}(3)_\mathbb{C}$ action on $\mathfrak{e}(3)$ and hence the $T(\mathfrak{so}(3)_{\mathbb{C}})$ action on $T(\mathfrak{e}(3)) \subset \mathfrak{sp}(4,\mathbb{C})$ decomposes as $V^1 \oplus V^1$, which has dimension 6.

If $T|_{\mathfrak{so}(3)_{\mathbb{C}}} \cong V^{1/2} \oplus 2V^0$, then $T(H)$ has eigenvalues $1,-1,0,0$. Under the $\mathfrak{sl}(4,\mathbb{C})$-adjoint action, $T(H)$ will have highest eigenvalue $1-(-1)=2$, but only with multiplicity 1. This means that the action of $T(\mathfrak{so}(3)_{\mathbb{C}})$ on $T(\mathfrak{e}(3)) \subset \mathfrak{sl}(4,\mathbb{C})$ cannot contain $V^1 \oplus V^1$, a contradiction.

If $T|_{\mathfrak{so}(3)_{\mathbb{C}}} \cong V^1 \oplus V^0$, then $T(H)$ has eigenvalues $2,-2,0,0$. It must be conjugate by an element of $Sp(4,\mathbb{C})$ to the diagonal matrix $\mathrm{diag}(2,0,-2,0)$. Under the $\mathfrak{sp}(4,\mathbb{C})$ adjoint action, this matrix has highest eigenvalue $2-(-2)=4$, implying that the action of $T(\mathfrak{so}(3)_{\mathbb{C}})$ on $\mathfrak{sp}(4,\mathbb{C})$ contains a copy of V^2. Since $\dim(V^2)=5$, $\mathfrak{sp}(4,\mathbb{C})$ cannot also contain $V^1 \oplus V^1$, so this case is also impossible.

The only remaining possibility is that $T|_{\mathfrak{so}(3)_{\mathbb{C}}}$ must decompose as $V^{1/2} \oplus V^{1/2}$.

4.2. Embeddings of $\mathfrak{so}(3)_{\mathbb{C}}$ *into* $\mathfrak{sp}(4,\mathbb{C})$

Without loss of generality, we can assume that T takes the element $H \in \mathfrak{so}(3)_{\mathbb{C}} \subset \mathfrak{e}(3)$ into the diagonal Cartan subalgebra of $\mathfrak{sp}(4,\mathbb{C})$. The result of the preceding subsection shows that we can assume that $T(H) = \mathrm{diag}(1,-1,-1,1)$. The element $T(H)$ has eigenspaces under the adjoint action in $\mathfrak{sp}(4,\mathbb{C})$ as follows:

The 2-eigenspace is spanned by

$$E_0 = \begin{pmatrix} 0 & 1 & 0 & 0 \\ 0 & 0 & 0 & 0 \\ 0 & 0 & 0 & 0 \\ 0 & 0 & -1 & 0 \end{pmatrix}, \quad E' = \begin{pmatrix} 0 & 0 & -1 & 0 \\ 0 & 0 & 0 & 0 \\ 0 & 0 & 0 & 0 \\ 0 & 0 & 0 & 0 \end{pmatrix}, \quad E'' = \begin{pmatrix} 0 & 0 & 0 & 0 \\ 0 & 0 & 0 & 0 \\ 0 & 0 & 0 & 0 \\ 0 & 1 & 0 & 0 \end{pmatrix}; \tag{11}$$

The -2-eigenspace is spanned by

$$F_0 = \begin{pmatrix} 0 & 0 & 0 & 0 \\ 1 & 0 & 0 & 0 \\ 0 & 0 & 0 & -1 \\ 0 & 0 & 0 & 0 \end{pmatrix}, \quad F' = \begin{pmatrix} 0 & 0 & 0 & 0 \\ 0 & 0 & 0 & 1 \\ 0 & 0 & 0 & 0 \\ 0 & 0 & 0 & 0 \end{pmatrix}, \quad F'' = \begin{pmatrix} 0 & 0 & 0 & 0 \\ 0 & 0 & 0 & 0 \\ -1 & 0 & 0 & 0 \\ 0 & 0 & 0 & 0 \end{pmatrix}; \tag{12}$$

The 0-eigenspace is spanned by $Z = \begin{pmatrix} 1 & 0 & 0 & 0 \\ 0 & 1 & 0 & 0 \\ 0 & 0 & -1 & 0 \\ 0 & 0 & 0 & -1 \end{pmatrix}$ and

$$H_0 = \begin{pmatrix} 1 & 0 & 0 & 0 \\ 0 & -1 & 0 & 0 \\ 0 & 0 & -1 & 0 \\ 0 & 0 & 0 & 1 \end{pmatrix}, \quad H' = \begin{pmatrix} 0 & 0 & 0 & 1 \\ 0 & 0 & 1 & 0 \\ 0 & 0 & 0 & 0 \\ 0 & 0 & 0 & 0 \end{pmatrix}, \quad H'' = \begin{pmatrix} 0 & 0 & 0 & 0 \\ 0 & 0 & 0 & 0 \\ 0 & 1 & 0 & 0 \\ 1 & 0 & 0 & 0 \end{pmatrix}. \tag{13}$$

One obvious embedding of $\mathfrak{so}(3)_{\mathbb{C}}$ in $\mathfrak{sp}(4,\mathbb{C})$ is spanned by H_0, E_0, F_0. In fact, up to conjugacy by elements of $Sp(4,\mathbb{C})$, it is the only one that can arise as the restriction of an embedding of $\mathfrak{e}(3)$.

Proposition 4.1 *Any embedding T of $\mathfrak{so}(3)_{\mathbb{C}}$ into $\mathfrak{sp}(4,\mathbb{C})$ which decomposes as $V^{1/2} \oplus V^{1/2}$ is conjugate by an element of $Sp(4,\mathbb{C})$ to the embedding given by $T_0(H) = H_0$, $T_0(E) = E_0$, $T_0(F) = F_0$.*

Proof: As above, we can assume that $T(H) = H_0$. Then $T(E)$, an element of the 2-eigenspace for $T(H)$, must be a linear combination of E_0, E', E'', say $T(E) = aE_0 + bE' + cE''$, with $a, b, c \in \mathbb{C}$. Likewise, $T(F) = rF_0 + sF' + tF''$, where $r, s, t \in \mathbb{C}$.

We calculate that

$$[T(E), T(F)] = \begin{pmatrix} ar+bt & 0 & 0 & as+br \\ 0 & -cs-ar & as+br & 0 \\ 0 & cr+at & -ar-bt & 0 \\ cr+at & 0 & 0 & cs+ar \end{pmatrix}. \tag{14}$$

For this to equal $T(H) = H_0$, we must have $as+br = 0$, $cr+at = 0$, $ar+bt = 1$, and $cs+ar = 1$. The last two conditions imply $bt = cs$. From this we find that the vector (r, s, t) is a complex scalar multiple of $(a, -b, -c)$. If the constant is λ, it all comes down to $\lambda(a^2 - bc) = 1$.

In particular, this cannot happen if $a^2 - bc = 0$, and otherwise it determines the value of λ, namely $\lambda = \frac{1}{a^2-bc}$. Note that $a^2 - bc \neq 0$ is equivalent to requiring that $T(E)$ have rank 2, so it is a necessary condition for a representation of the form $V^{1/2} \oplus V^{1/2}$. We conclude that $(r, s, t) = \left(\frac{a}{a^2-bc}, \frac{-b}{a^2-bc}, \frac{-c}{a^2-bc}\right)$.

The question arises whether there is a $g \in Sp(4, \mathbb{C})$ such that $T(X) = gT_0(X)g^{-1}$, for all $X \in \mathfrak{so}(3)_{\mathbb{C}}$. In particular, such a g would have to satisfy $gH_0g^{-1} = H_0$. Combined with the condition $g \in Sp(4, \mathbb{C})$, this forces g to be of the form

$$g = \begin{pmatrix} \alpha & 0 & 0 & \beta \\ 0 & u\alpha & u\beta & 0 \\ 0 & \gamma & \delta & 0 \\ \frac{\gamma}{u} & 0 & 0 & \frac{\delta}{u} \end{pmatrix}, \tag{15}$$

where $\begin{pmatrix} \alpha & \beta \\ \gamma & \delta \end{pmatrix} \in SL(2, \mathbb{C})$ and $u \in \mathbb{C}^\times$. A simple calculation then shows that

$$gE_0g^{-1} = \begin{pmatrix} 0 & \frac{\alpha\delta+\beta\gamma}{u} & -2\alpha\beta & 0 \\ 0 & 0 & 0 & 0 \\ 0 & 0 & 0 & 0 \\ 0 & \frac{2\gamma\delta}{u^2} & -\frac{\alpha\delta+\beta\gamma}{u} & 0 \end{pmatrix}. \tag{16}$$

The question is whether this can be made equal to

$$T(E) = aE_0 + bE' + cE'' = \begin{pmatrix} 0 & a & -b & 0 \\ 0 & 0 & 0 & 0 \\ 0 & 0 & 0 & 0 \\ 0 & c & -a & 0 \end{pmatrix}, \tag{17}$$

subject to the condition $a^2 - bc \neq 0$.

Assuming that $b \neq 0$, we can solve and find that if α is an arbitrary nonzero complex number, u is a square root of $\frac{1}{a^2-bc}$, $\beta = \frac{b}{2\alpha}$, $\gamma = \frac{\alpha c}{a/u+a^2-bc}$, and $\delta = \frac{1+au}{2\alpha}$, then the displayed matrices in (16) and (17) are equal. If $b = 0$, then it works with $\alpha = 1$, $\beta = 0$, $u = 1$, $\gamma = \frac{c}{2a}$, and $\delta = a$.

In any event, provided $a^2 - bc \neq 0$, it is possible to find $g \in Sp(4, \mathbb{C})$ so that $gT_0(X)g^{-1} = T(X)$, for $X = H, E$. Another straightforward calculation shows that the same relation then also holds for $X = F$, and hence for all $X \in \mathfrak{so}(3)_{\mathbb{C}}$. □

4.3. Embeddings of $\mathfrak{e}(3)$ *into* $\mathfrak{sp}(4,\mathbb{C})$

We may embed $\mathfrak{e}(3)$ into $\mathfrak{sp}(4,\mathbb{C})$ as follows:

$$\begin{aligned} \psi(H) &= H_0, \quad \psi(E) = E_0, \quad \psi(F) = F_0 \\ \psi(P_0) &= H', \quad \psi(P_+) = 2E', \quad \psi(P_-) = -2F'. \end{aligned} \tag{18}$$

In the following theorem we establish that ψ is (essentially) the unique embedding of $\mathfrak{e}(3)$ into $\mathfrak{sp}(4,\mathbb{C})$.

Theorem 4.2 *Up to conjugacy by elements of* $Sp(4,\mathbb{C})$*,* ψ *is the only embedding of* $\mathfrak{e}(3)$ *into* $\mathfrak{sp}(4,\mathbb{C})$*.*

Proof: Let $T : \mathfrak{e}(3) \hookrightarrow \mathfrak{sp}(4,\mathbb{C})$ be an embedding of $\mathfrak{e}(3)$ into $\mathfrak{sp}(4,\mathbb{C})$. Then, T restricts to an embedding of $\mathfrak{so}(3)_\mathbb{C}$ which decomposes as $V^{1/2} \oplus V^{1/2}$. One copy of $V^{1/2}$ is the adjoint action of $\mathfrak{so}(3)_\mathbb{C}$ on its image $T(\mathfrak{so}(3)_\mathbb{C})$. The other is the action of $\mathfrak{so}(3)_\mathbb{C}$ on the image $T(\mathfrak{p})$ of the abelian ideal $\mathfrak{p} = \langle P_0, P_+, P_- \rangle$.

As has been shown in Proposition 4.1, we can assume that $T(\mathfrak{so}(3)_\mathbb{C})$ is the span of H_0, E_0, F_0. The element $T(P_+)$ must be a linear combination of the elements E_0, E', E'' in the 2-eigenspace, say $T(P_+) = xE_0 + yE' + zE''$, for some $x, y, z \in \mathbb{C}$.

In this case, we must have

$$\begin{aligned} T(P_0) &= -\frac{1}{2}[F_0, T(P_+)] = -\frac{1}{2}(x[F_0,E_0] + y[F_0,E'] + z[F_0,E'']) = \frac{x}{2}H_0 + \frac{y}{2}H' + \frac{z}{2}H'', \\ T(P_-) &= -[F_0, T(P_0)] = -\frac{1}{2}(x[F_0,H_0] + y[F_0,H'] + z[F_0,H'']) = -xF_0 - yF' - zF''. \end{aligned}$$

Since $\mathfrak{p}$ is an abelian ideal, its image under T must be abelian. We find that

$$[T(P_+), T(P_-)] = \begin{pmatrix} -x^2 - yz & 0 & 0 & -2xy \\ 0 & x^2 + yz & -2xy & 0 \\ 0 & -2xz & x^2 + yz & 0 \\ -2xz & 0 & 0 & -x^2 - yz \end{pmatrix}, \tag{19}$$

so the requirement is

$$x^2 + yz = 0, \tag{20}$$

$$xy = 0, \tag{21}$$

$$xz = 0. \tag{22}$$

It is easy to check that these conditions imply that $T(\mathfrak{p})$ is abelian, i.e., that $T(P_0)$, $T(P_+)$, and $T(P_-)$ all commute.

If $x \neq 0$, then (21) and (22) force $y = z = 0$, in which case (20) fails. So we must have $x = 0$. From (20), it then follows that $y = 0$ or $z = 0$.

So any embedding of $\mathfrak{e}(3)$ into $\mathfrak{sp}(4,\mathbb{C})$ is $Sp(4,\mathbb{C})$-conjugate either to the embedding T (which is precisely ψ) or to T':

$$\begin{aligned} T(H) &= H_0 & T'(H) &= H_0 \\ T(E) &= E_0 & T'(E) &= E_0 \\ T(F) &= F_0 & T'(F) &= F_0 \\ T(P_0) &= H' & T'(P_0) &= H'' \\ T(P_+) &= 2E' & T'(P_+) &= 2E'' \\ T(P_-) &= -2F' & T'(P_-) &= -2F'' \end{aligned}$$

But it is easy to check that these two embeddings are conjugate by the following element of $Sp(4, \mathbb{C})$:

$$w = \begin{pmatrix} 0 & 0 & 0 & 1 \\ 0 & 0 & -1 & 0 \\ 0 & -1 & 0 & 0 \\ 1 & 0 & 0 & 0 \end{pmatrix}, \tag{23}$$

i.e., for all $X \in \mathfrak{e}(3)$, $wT(X)w^{-1} = T'(X)$. We have just established the result. □

5. Representations of $\mathfrak{e}(3)$ from representations of $\mathfrak{sp}(4, \mathbb{C})$

With respect to the Chevalley basis of $\mathfrak{sp}(4, \mathbb{C})$ given in Eq. (4), we may explicitly embed $\mathfrak{e}(3)$ into $\mathfrak{sp}(4, \mathbb{C})$ as follows:

$$\begin{array}{lcr} \phi : \mathfrak{e}(3) & \hookrightarrow & \mathfrak{sp}(4, \mathbb{C}) \\ E & \mapsto & x_3 \\ H & \mapsto & h_1 + 2h_2 \\ F & \mapsto & y_3 \\ P_+ & \mapsto & -2x_2 \\ P_0 & \mapsto & y_1 \\ P_- & \mapsto & 2y_4. \end{array} \tag{24}$$

Hence, any embedding of $\mathfrak{e}(3)$ into $\mathfrak{sp}(4, \mathbb{C})$ is equivalent to ϕ under conjugation by an element in $Sp(4, \mathbb{C})$. In particular, ϕ is equivalent to ψ defined in Eq. (18). The following Lemma will be used to prove the indecomposability of $V(m_1, m_2)$ in Theorem 5.2 below. Its proof is omitted since it may be established in a straightforward manner by direct computation.

Lemma 5.1 *Let $y_1^a\, y_2^b\, y_3^c\, y_4^d\, v_\lambda$ be a nonzero element of $\mathcal{S}_\lambda$ ($\mathcal{S}_\lambda$ defined in Lemma 3.1). Then*

$$H \cdot y_1^a\, y_2^b\, y_3^c\, y_4^d\, v_\lambda = (m_1 + 2m_2 - 2(b + c + d))y_1^a\, y_2^b\, y_3^c\, y_4^d\, v_\lambda, \tag{25}$$

$$\begin{array}{l} P_+ \cdot y_1^a\, y_2^b\, y_3^c\, y_4^d\, v_\lambda = -2b(m_2 - b + 1)y_1^a y_2^{b-1} y_3^c y_4^d v_\lambda \\ -2c(c-1)y_1^a y_2^b y_3^{c-2} y_4^{d+1} v_\lambda + 2c y_1^{a+1} y_2^b y_3^{c-1} y_4^d v_\lambda, \end{array} \tag{26}$$

The above equation is interpreted in the sense that if $b = 0$, the first term on the right side is zero, if $c = 0$, the second and third terms on the right are zero, and if $c = 1$, the second term on the right is zero.

Let $b + c + d = m_2$, then

$$\begin{array}{l} (P_+)^{m_2} \cdot y_1^a\, y_2^b\, y_3^c\, y_4^d\, v_\lambda \\ = \begin{cases} (-2)^{m_2}(\Pi_{i=1}^{m_2} i(m_2 - i + 1))\; y_1^a v_\lambda & \text{for } a \le m_1,\ c = d = 0, \\ 0 & \text{for } (a + c > m_1,\ d = 0) \text{ or } d > 0, \\ \alpha\; y_1^{a+c} v_\lambda & \text{for } d = 0,\ a + c \le m_1, \end{cases} \end{array} \tag{27}$$

where α is a (possibly zero) scalar.

Theorem 5.2 *The irreducible $\mathfrak{sp}(4, \mathbb{C})$-module $V(m_1, m_2)$ is $\mathfrak{e}(3)$-indecomposable.*

Proof: Suppose $V(m_1, m_2) \cong M \oplus M'$. Upon restriction to the subalgebra $\langle E, H, F\rangle \cong \mathfrak{sl}(2, \mathbb{C})$, the representations $V(m_1, m_2)$, M and M' decompose into H-weight spaces. One of M or M' must contain an element

$$y_2^{m_2} v_\lambda + \sum_{a,b,c,d} \alpha(a, b, c, d)\; y_1^a y_2^b y_3^c y_4^d v_\lambda, \tag{28}$$

for $y_1^a y_2^b y_3^c y_4^d v_\lambda \in \mathcal{S}_\lambda$, scalars $\alpha(a,b,c,d)$, and the sum over indices (a,b,c,d) such that $b+c+d=m_2$, $(a,b,c,d) \neq (0,m_2,0,0)$. This element has H-weight equal to m_1. Without loss of generality, let this element belong to M. Considering Eqs. (27) and (28),

$$\begin{aligned}(P_+)^{m_2} \cdot (y_2^{m_2} v_\lambda + \sum_{a,b,c,d} \alpha(a,b,c,d)\ y_1^a y_2^b y_3^c y_4^d v_\lambda) \\ = (-2)^{m_2} (\Pi_{i=1}^{m_2} i(m_2-i+1))\ v_\lambda + \sum_{a,c} \beta(a,c)\ y_1^{a+c} v_\lambda \in M,\end{aligned} \tag{29}$$

where $1 \leq a+c \leq m_1$ and $\beta(a,c)$ are scalars. One can easily show that $v_\lambda, y_1 v_\lambda, ..., y_1^{m_1} v_\lambda$ is a linearly independent set. For $m_2 > 0$, $(-2)^{m_2}(\Pi_{i=1}^{m_2} i(m_2-i+1)) \neq 0$, so that the coefficient of v_λ after the application of $(P_+)^{m_2}$ to Eq. (28) is nonzero. Recalling that $\phi(P_0) = y_1$, a straightforward application of P_0 to the right hand side of Eq. (29) then yields

$$y_1^a v_\lambda \in M, \text{ for } 0 \leq a \leq m_1. \tag{30}$$

We will now proceed by induction on the H-weight of the H-weight spaces of $V(m_1,m_2)$ to show that each H-weight space belongs to M.

The H-weight space $V(m_1,m_2)_{m_1+2m_2}$ has basis v_λ, $y_1 v_\lambda$, $y_1^2 v_\lambda$,...,$y_1^{m_1} v_\lambda$. Eq. (25) implies that $V(m_1,m_2)_{m_1+2m_2}$ is the H-weight space of highest H-weight. Eq. (30) implies that $V(m_1,m_2)_{m_1+2m_2} \subseteq M$.

Let w, such that $-(m_1+2m_2) \leq w < m_1+2m_2$, be a fixed H-weight of $V(m_1,m_2)$, and suppose $V(m_1,m_2)_{w+2i} \subseteq M$ for all $i>0$ such that $w+2i \leq m_1+2m_2$. Let the nonzero element $y_1^a y_2^b y_3^c y_4^d v_\lambda \in \mathcal{S}_\lambda$ be such that $m_1+2m_2-2(b+c+d)=w$. We proceed in cases to show that $y_1^a y_2^b y_3^c y_4^d v_\lambda \in M$:

Case 1. $d>0$: The H-weight of $y_1^a y_2^b y_3^c y_4^{d-1} v_\lambda$ is $w+2$ and the vector is thus contained in M. Noting that $\phi(P_-) = 2y_4$ and $[y_4, y_i] = 0$, for $i=1,2,3$, we have $P_- \cdot y_1^a y_2^b y_3^c y_4^{d-1} v_\lambda = 2y_1^a y_2^b y_3^c y_4^d v_\lambda \in M$.

Case 2. $d=0$ and $c>0$: The H-weight of $y_2^b y_3^{c-1} v_\lambda$ is $w+2$ and hence the vector is contained in M. Then, noting that $\phi(F) = y_3$, $[y_3, y_2] = 0$ and $\phi(P_0) = y_1$, $(P_0^a F) \cdot y_2^b y_3^{c-1} v_\lambda = y_1^a y_2^b y_3^c v_\lambda \in M$.

Case 3. $d=0$, $c=0$ and $b>0$: In this case $w = m_1+2m_2-2b$. As mentioned after Eq. (28), we have

$$y_2^{m_2} v_\lambda + \sum_{a',b',c',d'} \alpha'(a',b',c',d')\ y_1^{a'} y_2^{b'} y_3^{c'} y_4^{d'} v_\lambda \in M, \tag{31}$$

where the sum is over a', b', c', d' such that $b'+c'+d' = m_2$ and $(a',b',c',d') \neq (0,m_2,0,0)$. From Eq. (26) we then have

$$\begin{aligned}(P_+)^{m_2-b} \cdot (y_2^{m_2} v_\lambda + \sum_{a',b',c',d'} \alpha'(a',b',c',d')\ y_1^{a'} y_2^{b'} y_3^{c'} y_4^{d'} v_\lambda) \\ = (-2)^{m_2-b} (\Pi_{i=1}^{m_2-b} i(m_2-i+1))\ y_2^b v_\lambda \\ + \sum_{a'',b'',c'',d''} \alpha''(a'',b'',c'',d'')\ y_1^{a''} y_2^{b''} y_3^{c''} y_4^{d''} v_\lambda \in M,\end{aligned} \tag{32}$$

for scalars $\alpha''(a'',b'',c'',d'')$ where $b''+c''+d'' = b$ and $(a'',b'',c'',d'') \neq (0,b,0,0)$. For $m_2-b>0$, $(-2)^{m_2-b}(\Pi_{i=1}^{m_2-b} i(m_2-i+1)) \neq 0$, hence the coefficient of $y_2^b v_\lambda$ after the application of $(P_+)^{m_2-b}$

to Eq. (31) is nonzero. Each nonzero element $y_1^{a''}y_2^{b''}y_3^{c''}y_4^{d''}v_\lambda$ in the sum above is contained in M by the above two cases, except for terms where $(a'', b'', c'', d'') = (a'', b, 0, 0)$, $a'' > 0$. Hence,

$$y_2^b v_\lambda + \sum_{a''>0} \beta(a'')\, y_1^{a''} y_2^b v_\lambda \in M, \tag{33}$$

for scalars $\beta(a'')$. Since $V(m_1, m_2)$ is finite-dimensional and $y_2^b v_\lambda \neq 0$, there exists $N \geq 0$ such that $y_1^N y_2^b v_\lambda \neq 0$, but $y_1^{N+1} y_2^b v_\lambda = 0$. It is easily shown that the set $\{y_2^b v_\lambda, y_1 y_2^b v_\lambda, ..., y_1^N y_2^b v_\lambda\}$ is linearly independent.

Recalling that $\phi(P_0) = y_1$, a straightforward application of P_0 to Eq. (33) yields

$$y_2^b v_\lambda \in M. \tag{34}$$

Finally we have

$$P_0^a \cdot y_2^b v_\lambda = y_1^a y_2^b v_\lambda \in M. \tag{35}$$

Hence, each H-weight space of $V(m_1, m_2)$ is contained in M and thus $V(m_1, m_2) \subseteq M$. Hence $V(m_1, m_2) = M$ and $M' = 0$. Thus $V(m_1, m_2)$ is indecomposable. □

An interesting consequence of the proof of Theorem 5.2 is that $V(m_1, m_2)$ is generated by the single element $y_2^{m_2} v_\lambda$ as an $\mathfrak{e}(3)$-module.

6. Conclusions

We have shown that the irreducible $\mathfrak{sp}(4, \mathbb{C})$-modules $V(m_1, m_2)$ remain indecomposable as $\mathfrak{e}(3)$-modules under the embedding described in Eq. (24), which we have shown is essentially the only embedding of $\mathfrak{e}(3)$ into $\mathfrak{sp}(4, \mathbb{C})$. Thus, we have created a new class of indecomposable $\mathfrak{e}(3)$-modules. An important and interesting consequence of this construction is that we can decompose the tensor product of representations within this class: The tensor product decomposition of indecomposable $\mathfrak{e}(3)$-modules in this class follows from the well-known tensor product decomposition of irreducible $\mathfrak{sp}(4, \mathbb{C})$-modules. One illustrative example is the following 400-dimensional tensor product

$$V(2,0) \otimes V(1,2) \cong_{\mathfrak{e}(3)} V(3,2) \oplus V(1,3) \oplus V(3,1) \oplus 2V(1,2) \oplus V(3,0) \oplus V(1,1). \tag{36}$$

The above decomposition was calculated with the assistance of the computer algebra system GAP [5].

The embedding of $\mathfrak{e}(3)$ into $\mathfrak{sp}(4, \mathbb{C})$ does not correspond to an embedding of the group $E(3)$ into $Sp(4, \mathbb{C})$. In fact, it corresponds to an embedding into $Sp(4, \mathbb{C})$ of the two-fold cover $\mathrm{SU}(2) \ltimes \mathbb{R}^3$ of $E(3) = \mathrm{SO}(3) \ltimes \mathbb{R}^3$. In the terminology of Douglas and Repka [4], it is a "spinor" embedding.

It has recently been pointed out to the authors that the techniques of this paper can be improved and generalized, an idea that will be pursued in future work.

Acknowledgments

The work of A.D. is partially supported by the Professional Staff Congress/City University of New York (PSC/CUNY). The work of J.R. is partially supported by the Natural Sciences and Engineering Research Council (NSERC)

References

[1] Bliem, T., "Weight multiplicities for $\mathfrak{so}(5,\mathbb{C})$," *Proceedings of the 2008 International Conference on Information Theory and Statistical Learning* (CSREA, Irvine, CA, 2008), pp. 80-86; e-print arXiv:0902.1744v1.

[2] Casati, P., Minniti, S. and Salari, V., "Indecomposable representations of the Diamond Lie algebra," J. Math. Phys. 51, 033515 (2010).

[3] Douglas, A. and Premat, A., "A class of nonunitary representations of the Euclidean algebra $\mathfrak{e}(2)$," Comm. Algebra, 35(5):1433-1448 (2007).

[4] Douglas, A. and Repka, J., "Embeddings of the Euclidean algebra $\mathfrak{e}(3)$ into $\mathfrak{sl}(4,\mathbb{C})$ and restrictions of irreducible representations of $\mathfrak{sl}(4,\mathbb{C})$," J. Math. Phys. 52, 013504 (2011).

[5] GAP-Groups, Algorithms, and Programming, Version 4.2, 2000 (http://www-gap.dcs.st-and.ac.uk/ gap)

[6] Hall, B. C., *Lie Groups, Lie Algebras, and Representations* (Springer-Verlag, New York, 2003).

[7] Humphreys, J., *Introduction to Lie Algebras and Representation Theory* (Springer-Verlag, New York, 1972).

An interface between physics and number theory

Gérard H.E. Duchamp[1], Hoang Ngoc Minh[1,2], Allan I. Solomon[3,4], Silvia Goodenough[1]

[1] Université Paris 13 – LIPN - UMR 7030 CNRS, F-93430 Villetaneuse, France.
[2] Université Lille 2, 1 Place Déliot, 59024 Lille Cédex, France.
[3] Physics and Astronomy Department,The Open University, Milton Keynes MK7 6AA, UK
[4] Université Pierre et Marie Curie, LPTMC, UMR 7600 CNRS, 4 pl. Jussieu, 75252 Paris 5, France.

Abstract. We extend the Hopf algebra description of a simple quantum system given previously, to a more elaborate Hopf algebra, which is rich enough to encompass that related to a description of perturbative quantum field theory (pQFT). This provides a *mathematical* route from an algebraic description of non-relativistic, non-field theoretic quantum statistical mechanics to one of relativistic quantum field theory.

Such a description necessarily involves treating the algebra of polyzeta functions, extensions of the Riemann Zeta function, since these occur naturally in pQFT. This provides a link between physics, algebra and number theory. As a by-product of this approach, we are led to indicate *inter alia* a basis for concluding that the Euler gamma constant γ may be rational.

1. Introduction

In an introductory paper delivered at this Conference[1], *"From Quantum Mechanics to Quantum Field Theory: The Hopf route"*, Allan I. Solomon [1, 2] *et al.* start their exposition with the Bell numbers $B(n)$ which count the number of *set partitions* within a given set of n elements. It is shown there that these very elementary combinatorial ideas are in a sense generic within quantum statistical mechanics. Further, it turns out that this approach also leads to a Hopf algebraic description of the physical system. Since recent work on relativistic field theory (perturbative quantum field theory or pQFT) also leads to a Hopf algebra description, the question naturally arise as to whether the algebraic structure, implicit in essentially the non-relativistic commutation relations, may be extended to the more complicated, relativistic system. Instead of extending the physical picture, we have chosen to follow a mathematical route. Doing so, it turns out that the basic structure, as exemplified in *"The Hopf route"* for the special case of Bell number diagrams (par. 3) provides the building blocks of a non-commutative parametric Hopf algebra which specializes on the one hand to a Hopf algebra of noncommutative symmetric functions, and on the other hand to a proper Hopf sub-algebra of that inherent in perturbative quantum field theory (pQFT). It well known that analysis of pQFT reveals that polyzeta functions, extensions of the Riemann zeta function of number theory, play an important role. It is not surprising therefore that the algebraic analysis embarked upon here should reveal properties of the polyzetas. And indeed it is the case that our algebra admits, as a homomorphic

[1] Talk delivered at the 28th International Colloquium on Group Theoretical Methods in Physics, Northumbria University, Newcastle, July 2010.

image, the algebra of polyzeta functions of number theory. This foray into number theory leads us to consider Euler's gamma constant γ, and conclude, on the basis of Lemmas given in the text, that γ may in fact be a rational number.

2. One-parameter groups and the exponential Hadamard product

As often in physics, one has to consider one-parameter groups with infinitesimal generator H,

$$G(t) := e^{tH} \ . \tag{1}$$

When we have two infinitesimal generators which commute $[H_1, H_2] = 0$ (for example, with hamiltonians constructed from two distinct boson modes), one can obtain the group $G_{12}(t) := e^{tH_1H_2}$ with operation

$$G_{12}(t) := G_1(t\frac{\partial}{\partial x})[G_2(x)]|_{x=0}) \ . \tag{2}$$

where $G_i(t) := e^{tH_i}$, $i = 1, 2$. Up to constant terms, this operation is a *Hadamard product* (i.e. the componentwise product, which we denote below by $\mathcal{H}$) but performed on the exponential generating series [3]. Specifically with

$$F(z) = \sum_{n\geq 0} a_n \frac{z^n}{n!}, \ G(z) = \sum_{n\geq 0} b_n \frac{z^n}{n!},$$
$$\mathcal{H}(F, G) := \sum_{n\geq 0} a_n b_n \frac{z^n}{n!} \ . \tag{3}$$

This formula is known as the *product formula* [4]. Since one-parameter groups are exponentials, we can try to find a universal formula for (3). To this end we set

$$F(z) = \exp\left(\sum_{n=1}^{\infty} L_n \frac{z^n}{n!}\right), \ G(z) = \exp\left(\sum_{n=1}^{\infty} V_n \frac{z^n}{n!}\right) . \tag{4}$$

and remark that F, G can simultaneously be expressed in terms of Bell[2] polynomials [5] B_n as

$$\exp\left(\sum_{n=1}^{\infty} X_n \frac{z^n}{n!}\right) = \sum_{n=0}^{\infty} \frac{z^n}{n!} B_n(X_1, X_2, \cdots X_n) \ . \tag{5}$$

The coefficients of the polynomials B_n are positive integers which can be interpreted as the number of *unordered set partitions* which are collections of mutually disjoint non-empty subsets (blocks) of a (finite) set F whose union is F. For example, the *set partitions* of $F = \{1, 2, 3, 4\}$ in two blocks of size two are

$$\{\{1, 2\}, \{3, 4\}\} \ \{\{1, 3\}, \{2, 4\}\} \ \{\{1, 4\}, \{2, 3\}\} \ . \tag{6}$$

The type of a set partition $P = \{A_1, A_2, \cdots A_k\}$ of F is the sequence of numbers

$$Type(P) := (\#\{l \ s.t. \ |A_l| = k\})_{k\geq 1} \ .$$

For example, the type of the set partition $P = \{\{2, 4, 6, 7\}, \{1, 5\}, \{3, 8\}\}$ is $(0, 2, 0, 1, 0, 0, 0, \cdots)$. We have the formula

$$B_n(X_1, X_2, \cdots X_n) = \sum_{P\in UP_n} \mathbb{X}^{Type(P)} \tag{7}$$

[2] Named after the mathematician and science fiction author Eric Temple Bell (1883, Scotland, to 1960, US).

(where UP_n is the set of unordered partitions of $[1 \cdots n]$) which can be proved (following the argument of Faà di Bruno) using the exponential formula [6]. Now from (4), (5), (7), one concludes that

$$\mathcal{H}(F,G) = \sum_{n\geq 0} \frac{z^n}{n!} \sum_{P_1,P_2\in UP_n} \mathbb{L}^{Type(P_1)}\mathbb{V}^{Type(P_2)} \ . \tag{8}$$

This formula can be rewritten as a diagrammatic expansion using the incidence matrix of the pairs of set partitions P_1, P_2 involved in (8) with respect to intersection numbers, i.e.

$$\text{Matrix of } (P_1,P_2) = \Big(\#B_1 \cap B_2\Big)_{B_1\in P_1,\ B_2\in P_2} \ . \tag{9}$$

These matrices, with unordered rows and columns, can be represented by a diagram with the following rule. Draw two rows of spots (white spots on the upper row and black on the lower row) label each black spot by a block of P_1 and each white spot by a block of P_2 and join two spots (one of each kind) by the number of common elements in the corresponding blocks[3]. the result is the diagram of the following figure.

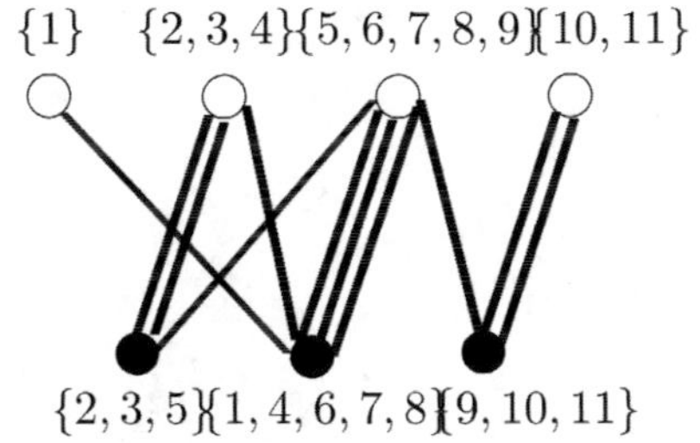

Fig 1. — *Diagram from P_1, P_2 (set partitions of $[1\cdots 11]$).*
$P_1 = \{\{2,3,5\},\{1,4,6,7,8\},\{9,10,11\}\}$ *and* $P_2 = \{\{1\},\{2,3,4\},\{5,6,7,8,9\},\{10,11\}\}$ *(respectively black spots for P_1 and white spots for P_2).*

The incidence matrix corresponding to the diagram (as drawn) or these partitions is $\begin{pmatrix} 0 & 2 & 1 & 0 \\ 1 & 1 & 3 & 0 \\ 0 & 0 & 1 & 2 \end{pmatrix}$ *but, due to the fact that the defining partitions are unordered, one can permute the spots (black and white) among themselves and so the lines and columns of this matrix can be permuted; the diagram could be represented by the matrix* $\begin{pmatrix} 0 & 0 & 1 & 2 \\ 0 & 2 & 1 & 0 \\ 1 & 0 & 3 & 1 \end{pmatrix}$ *as well.*

The set of such diagrams will be denoted by **diag**. This is precisely the set of bipartite graphs with multiple edges and no isolated vertex (and no order between the spots) and integer multiplicities. It is straightforward to verify that concatenation is associative with unit (the empty graph) within the set **diag**. We therefore denote the result of concatenation by $[d_1|d_2]_D$ (arbitrarily putting d_2 on the right of d_1 because the law is commutative). The monoid $(\mathbf{diag}, [-|-]_D, 1_{\mathbf{diag}})$ is a *free commutative monoid.* Its generators are the connected (non-empty) diagrams ($\mathbf{diag}_c$). The algebra of **diag** is the set of formal (finitely supported) sums

$$\{\sum_{d\in\mathbf{diag}} \alpha(d)d\}_{\alpha\in\mathbb{C}^{(\mathbf{diag})}} \tag{10}$$

The freeness of **diag** (as a commutative monoid) implies that this algebra is isomorphic to the algebra of polynomials $\mathbb{C}[\mathbf{diag}_c]$. We shall denote it differently (i. e. **DIAG**) in the sequel (see

[3] Compare this diagrammatic approach with the white and black spots of the preceding presentation[1].

[3] for details) as we wish to endow it with the structure of a Hopf algebra.
Indeed, the diagrams allow us to rewrite formula (8) as

$$\mathcal{H}(F,G) = \sum_{n\geq 0} \frac{z^n}{n!} \sum_{\substack{d\in diag \\ |d|=n}} mult(d)\mathbb{L}^{\alpha(d)}\mathbb{V}^{\beta(d)} \tag{11}$$

where $\alpha(d)$ (resp. $\beta(d)$) is the "white spot type" (resp. the "black spot type") i.e. the multi-index $(\alpha_i)_{i\in\mathbb{N}^+}$ (resp. $(\beta_i)_{i\in\mathbb{N}^+}$) such that α_i (resp. β_i) is the number of white spots (resp. black spots) of degree i (i lines connected to the spot) and $mult(d)$ is the number of pairs of unordered partitions of $[1\cdots|d|]$ (here $|d| = |\alpha(d)| = |\beta(d)|$ is the number of lines of d) with associated diagram d.

Now one may naturally ask
Q1) "Is there a (graphically) natural multiplicative structure on **diag** *such that the arrow*

$$d \mapsto mon(d) = \mathbb{L}^{\alpha(d)}\mathbb{V}^{\beta(d)} \tag{12}$$

is a morphism ?"

The answer is "yes" concerning the algebraic structure of **LDIAG** as it is easily seen that

$$\mathrm{mon}([d_1|d_2]_D) = \mathrm{mon}(d_1)\mathrm{mon}(d_2)\ . \tag{13}$$

But the algebra of polynomials is also endowed with the structure of a Hopf algebra (decomposition property [1, 2]). So it is natural to ask whether the coalgebra structure of the polynomials could be lifted to **DIAG** and, further, if this allows enriching the structure of **DIAG** to a Hopf algebra. Again, the answer is "yes" and the appropriate comultiplication is denoted by Δ_{BS} ("black spot" coproduct, see [3] for details). Computing $\Delta_{BS}(d)$ is immediate; one simply divides the black spots into two subsets, then obtain the tensors and sum all the results. More formally; noting the set of black spots of the diagram d by $BS(d)$ and denoting by $d[I]$ the sub-diagram whose black spots are in I and the white connected to I, one has

$$\Delta_{BS}(d) = \sum_{I+J=BS(d)} d[I]\otimes d[J] \tag{14}$$

For example, one labels by $\{a,b,c\}$ the three black spots of the diagram d of **Fig 1**. In this case, the copoduct reads

$$\begin{aligned}\Delta_{BS}(d) = d\otimes[\,]+[\,]\otimes d+\\ d[\{a\}]\otimes d[\{b,c\}]+d[\{b\}]\otimes d[\{a,c\}]+\\ d[\{c\}]\otimes d[\{a,b\}]++d[\{a,b\}]\otimes d[\{c\}]+\\ d[\{a,c\}]\otimes d[\{b\}]+d[\{b,c\}]\otimes d[\{a\}]\ .\end{aligned}$$

2.1. Labeling the nodes and the noncommutative analogue **LDIAG**

If one needs to label the black spots in order to compute the comultiplication, it is obvious that the precise form of labelling is ultimately irrelevant. It is more important to label the nodes endowing their sets with a linear order (one for the black spots between themselves and one for the white spots). This leads to **LDIAG**, the (non-deformed) noncommutative analogue of **DIAG** [7].
The solution (of this non-deformed problem [7]) is simple and consists in labeling the black (resp. white) spots from left to right and from "1" to p (resp. q); thus one obtains the *labelled*

diagrams. Again concatenation is associative and endows the set of the labelled diagrams (here denoted by **ldiag**) with the structure of a monoid (the unit is the empty diagram characterized by $p = q = 0$).
Again, and with the same formula (14) (but not applied to the same objects), the algebra of this monoid, **LDIAG** and the counit which consists in taking the coefficient of the empty diagram, is a Hopf algebra [4].

2.2. The deformed case.
The preceding coding is particularly well adapted to the deformation we want to construct here. The philosophy of the deformed product is expressed by the descriptive formula

$$[d_1|d_2]_{L(q_c,q_s)} = \sum_{CS(d_1,d_2)} q_c^{w1(CS)} q_s^{w2(CS)} CS([d_1|d_2]_{L(q_c,q_s)}) \tag{15}$$

where

- $q_c, q_s \in \mathbb{C}$ or q_c, q_s formal. These and other cases may be unified by considering the set of coefficients as belonging to a ring K
- the exponent of $w1(CS)$ is the number of crossings of "what crosses" times its weight
- the exponent of $w2(CS)$ is the product of the weights of "what is overlapped"
- $CS(d_1, d_2)$ are the diagrams obtained from $[d_1|d_2]_L$ by the process of crossing and superposing the black spots of d_2 on to those of d_1, the order and distinguishability of the black spots of d_1 (i.e. d_2) being preserved.

What is striking is that this law is associative. This fact is by no means trivial and three proofs have been given (one of them can be found in [3]). It can be shown that this new algebra (associative with unit denoted $\mathbf{LDIAG}(q_c, q_s)$) can be endowed with two comultiplications, Δ_0 and Δ_1, such that the result is a Hopf algebra which we denote by $\mathbf{LDIAG}(q_c, q_s, q_t)$.
It can be shown that this deformed Hopf algebra has two interesting specializations

$$\mathbf{LDIAG}(0,0,0) \simeq \mathbf{LDIAG};$$
$$\mathbf{LDIAG}(1,1,1) \simeq \mathbf{MQSym}\ ;$$

$\mathbf{LDIAG}(0,0,0)$ (no crossing and no superposition) is the undeformed case whereas $\mathbf{LDIAG}(1,1,1)$ is **MQSym**, the Hopf algebra of Noncommutative Matrix Quasisymmetric functions [8].

2.3. The arrow $\mathbf{LDIAG}(1,1) \to$ *polyzetas.*
We now explain how a refinement of the "black spot type" $\beta(d)$ of formula (11) evaluates the statistics $(\beta(d))$ by "number of nodes with outgoing degree k". We take advantage of the labelling and count these degrees node by node. Thus, for a labelled diagram d, $I(d,k)$ is simply the outgoing degree of the node with label k. It turns out, and is straightforward to show, that

$$I(d) := [I(d,1), I(d,2) \cdots I(d,p)]$$

(d is a diagram with p black spots) is a composition of the integer $|d|$ (number of edges of d).
Now, we define two structures which will be ubiquitous in the next section: *the shuffle and stuffle algebras*. Let X be any alphabet (i.e. set of variables). The shuffle algebra is defined

[4] At this stage, it is only a graded bialgebra, but, as it is graded and of finite dimension, a general theorem states that an antipode exists; (see [3] formula (19) for an explicit formula of this antimorphism).

on $\mathbb{C}\langle X\rangle := \mathbb{C}[X^*]$ by its values on the monomials (here the strings or words) by the following recursion

$$\begin{aligned} 1_{X^*} \sqcup\!\sqcup\, w &= w \sqcup\!\sqcup\, 1_{X^*} = w \text{ and} \\ xu \sqcup\!\sqcup\, yv &= x(u \sqcup\!\sqcup\, yv) + y(xu \sqcup\!\sqcup\, v) \ . \end{aligned} \tag{16}$$

Likewise, let $Y = y_{i\,i\geq 1}$ be a countable alphabet indexed by integers ≥ 1. The stuffle algebra [9] is defined on $\mathbb{C}\langle Y\rangle := \mathbb{C}[Y^*]$ by a recursion on the words as

$$\begin{aligned} 1_{Y^*} \uplus w &= w \uplus 1_{Y^*} = w \text{ and} \\ y_i u \uplus y_j v &= y_i(u \uplus y_j v) + y_j(y_i u \uplus v) + y_{i+j}(u \uplus v) \ . \end{aligned} \tag{17}$$

The shuffle (resp. stuffle) endow $\mathbb{C}\langle X\rangle := \mathbb{C}[X^*]$ (resp. $\mathbb{C}\langle Y\rangle := \mathbb{C}[Y^*]$) with the structure of an AAU[5]. We define mappings

$$\begin{aligned} \phi_{X2LDIAG} &: \mathbf{LDIAG}(1,0) \to \mathbb{C}\langle X\rangle \\ \phi_{Y2LDIAG} &: \mathbf{LDIAG}(1,1) \to \mathbb{C}\langle Y\rangle \end{aligned} \tag{18}$$

(here $X = \{x_i\}_{i\geq 1}$) by

$$\phi_{X2LDIAG}(d) = x_{I(d,1)} y_{I(d,2)} \cdots x_{I(d,p)}$$
$$\phi_{Y2LDIAG}(d) = y_{I(d,1)} y_{I(d,2)} \cdots y_{I(d,p)})$$

and have the following result.

Proposition 1 *The mappings*

$$\phi_{X2LDIAG} \ : \ (\mathbf{LDIAG}(1,0), [\ .|\ .]_{LD(1,0)} \to (\mathbb{C}\langle X\rangle, \uplus\,)$$
$$\phi_{Y2LDIAG} \ : \ (\mathbf{LDIAG}(1,1), [\ .|\ .]_{LD(1,1)} \to (\mathbb{C}\langle Y\rangle, \uplus\,)$$

are epimorphisms of AAU's.

This provides the opening to the next section.

3. Knizhnik-Zamolodchikov differential system

In 1986, in order to study the linear representation of the braid group B_n [10] coming from the monodromy of the Knizhnik-Zamolodchikov differential equations over $\mathbb{C}_*^n = \{(z_1,\ldots,z_n) \in \mathbb{C}^n | z_i \neq z_j \text{ for } i \neq j\}$:

$$dF(z) = \Omega_n F(z), \tag{19}$$

where

$$\Omega_n = \frac{1}{2\mathrm{i}\pi} \sum_{1\leq i<j\leq n} t_{i,j} \frac{d(z_i - z_j)}{z_i - z_j}, \tag{20}$$

Drinfel'd introduced a class of formal power series Φ on noncommutative variables over the finite alphabet $X = \{x_0, x_1\}$. Such a power series Φ is called an *associator*.

Since the system (19) is completely integrable then

$$d\Omega_n - \Omega_n \wedge \Omega_n = 0. \tag{21}$$

This is equivalent to the following braid relations

$$[t_{i,j}, t_{i,k} + t_{j,k}] = 0, \quad \text{for} \quad i \neq j \neq k, \tag{22}$$
$$[t_{i,j}, t_{k,l}] = 0, \quad \text{for} \quad i \neq j \neq k \neq l. \tag{23}$$

[5] Associative Algebra with Unit: Here they are, moreover, Hopf algebras [9] but we will not use this here.

- For $n = 2$, $\mathcal{T}_2 = \{t_{1,2}\}$, one has

$$\Omega_2(z) = \frac{t_{1,2}}{2\mathrm{i}\pi} \frac{d(z_1 - z_2)}{z_1 - z_2} \tag{24}$$

and a solution of (19) is given by

$$F(z_1, z_2) = (z_1 - z_2)^{\frac{t_{1,2}}{2\mathrm{i}\pi}}. \tag{25}$$

- For $n = 3$, $\mathcal{T}_3 = \{t_{1,2}, t_{1,3}, t_{2,3}\}$, there are two relations :

$$[t_{1,3}, t_{1,2} + t_{2,3}] = 0 \quad \text{and} \quad [t_{2,3}, t_{1,2} + t_{1,3}] = 0. \tag{26}$$

One has

$$\Omega_3(z) = \frac{1}{2\mathrm{i}\pi} \left[t_{1,2} \frac{d(z_1 - z_2)}{z_1 - z_2} + t_{1,3} \frac{d(z_1 - z_3)}{z_1 - z_3} + t_{2,3} \frac{d(z_2 - z_3)}{z_2 - z_3} \right] \tag{27}$$

and a solution of (19) is given by

$$F(z_1, z_2, z_3) = G\left(\frac{z_1 - z_2}{z_1 - z_3}\right) (z_1 - z_3)^{\frac{t_{1,2}+t_{1,3}+t_{2,3}}{2\mathrm{i}\pi}} \tag{28}$$

where G satisfies the following Fuchs differential equation with three regular singularities at $0, 1$ and ∞ :

$$dG(z) = [x_0\omega_0(z) + x_1\omega_1(z)]G(z), \tag{29}$$

with

$$\omega_0(z) := \frac{dz}{z} \quad \text{and} \quad \omega_1(z) := \frac{dz}{1-z}, \tag{30}$$

$$x_0 := \frac{t_{1,2}}{2\mathrm{i}\pi} \quad \text{and} \quad x_1 := -\frac{t_{2,3}}{2\mathrm{i}\pi}. \tag{31}$$

In the sequel, we set $X = \{x_0, x_1\}$; X^* denotes the set of words defined over X. The empty word is denoted by ε.

Proposition 2 ([11]) *If $G(z)$ and $H(z)$ are exponential solutions of (29) then there exists a Lie series $C \in \mathcal{L}ie_{\mathbb{C}}\langle\langle X \rangle\rangle$ such that $G(z) = H(z)\exp(C)$.*

Proof – Since $H(z)H(z)^{-1} = 1$ then by differentiating, we have $d[H(z)]H(z)^{-1} = -H(z)d[H(z)^{-1}]$. Therefore if $H(z)$ is solution of the Drinfel'd equation then

$$\begin{aligned}
d[H(z)^{-1}] &= -H(z)^{-1}[dH(z)]H(z)^{-1} \\
&= -H(z)^{-1}[x_0\omega_0(z) + x_1\omega_1(z)], \\
d[H(z)^{-1}G(z)] &= H(z)^{-1}[dG(z)] + [dH(z)^{-1}]G(z) \\
&= H(z)^{-1}[x_0\omega_0(z) + x_1\omega_1(z)]G(z) \\
&- H(z)^{-1}[x_0\omega_0(z) + x_1\omega_1(z)]G(z).
\end{aligned}$$

By simplification, we deduce that $H(z)^{-1}G(z)$ is a constant formal power series. Since the inverse and the product of group-like elements is group-like then we get the expected result. □

4. Iterated integral and Chen generating series

The iterated integral associated with $w = x_{i_1} \cdots x_{i_k} \in X^*$, over ω_0 and ω_1 and along the path $z_0 \rightsquigarrow z$, is defined by the following multiple integral

$$\int_{z_0}^{z} \cdots \int_{z_0}^{z_{k-1}} \omega_{i_1}(t_1) \ldots \omega_{i_k}(t_k), \tag{32}$$

where $t_1 \cdots t_{r-1}$ is a subdivision of the path $z_0 \rightsquigarrow z$. In an abbreviated notation, we denote this integral by $\alpha_{z_0}^{z}(w)$ and $\alpha_{z_0}^{z}(\varepsilon) = 1$.

Example 1

$$\begin{aligned}
\alpha_0^z(x_0x_1) &= \int_0^z \int_0^s \omega_0(s)\omega_1(t) \\
&= \int_0^z \int_0^s \frac{ds}{s}\frac{dt}{1-t} \\
&= \int_0^z \frac{ds}{s} \int_0^s dt \sum_{k\geq 0} t^k \\
&= \sum_{k\geq 1} \int_0^z ds \frac{s^{k-1}}{k} \\
&= \sum_{k\geq 1} \frac{z^k}{k^2}.
\end{aligned}$$

The last sum is nothing other than the Taylor expansion of the dilogarithm $\mathrm{Li}_2(z)$.

Example 2 *In the same way the classical polylogarithm of order* $n \geq 1$ *is the iterated integral associated with* $x_0^{n-1}x_1$, *over* ω_0 *and* ω_1 *and along the path* $0 \rightsquigarrow z$:

$$\mathrm{Li}_n(z) = \sum_{k\geq 1} \frac{z^k}{k^n} = \alpha_0^z(x_0^{n-1}x_1).$$

Generalizing to multi-indices $(n_1, \ldots, n_r)$, *one has :*

$$\mathrm{Li}_{n_1,\ldots,n_r}(z) = \sum_{k_1>\ldots>k_r>0} \frac{z^{k_1}}{k_1^{n_1}\ldots k_r^{n_r}} = \alpha_0^z(x_0^{n_1-1}x_1 \ldots x_0^{n_r-1}x_1).$$

This provides an analytic prolongation of $\mathrm{Li}_{n_1,\ldots,n_r}$ *over the Riemann surface of* $\mathbb{C} \setminus \{0,1\}$.

The Chen generating series along the path $z_0 \rightsquigarrow z$ associated with ω_0, ω_1 is the following power series

$$S_{z_0\rightsquigarrow z} = \sum_{w\in X^*} \alpha_{z_0}^z(w)\, w \tag{33}$$

and it is a solution of the differential equation (29) with the initial condition

$$S_{z_0\rightsquigarrow z}(z_0) = 1. \tag{34}$$

Any Chen generating series $S_{z_0\rightsquigarrow z}$ is group-like [12] and depends only on the homotopy class of $z_0 \rightsquigarrow z$ [13]. The product of two Chen generating series $S_{z_1\rightsquigarrow z_2}$ and $S_{z_0\rightsquigarrow z_1}$ is the Chen generating series

$$S_{z_0\rightsquigarrow z_2} = S_{z_1\rightsquigarrow z_2} S_{z_0\rightsquigarrow z_1}. \tag{35}$$

5. Polylogarithm, harmonic sum and polyzetas

To the polylogarithm $\mathrm{Li}_{n_1,\ldots,n_r}(z)$ we associate the following ordinary generating series

$$\mathrm{P}_{n_1,\ldots,n_r}(z) = \frac{\mathrm{Li}_{n_1,\ldots,n_r}(z)}{1-z} = \sum_{N\geq 0} \mathrm{H}_{n_1,\ldots,n_r}(N)\, z^N, \tag{36}$$

where

$$\mathrm{H}_{n_1,\ldots,n_r}(N) = \sum_{N\geq k_1>\ldots>k_r>0} \frac{1}{k_1^{n_1}\ldots k_r^{n_r}} \tag{37}$$

For $n_1 > 1$, the limit of $\mathrm{Li}_{n_1,\ldots,n_r}(z)$ and of $\mathrm{H}_{n_1,\ldots,n_r}(N)$, for $z \to 1$ and $N \to \infty$ exist and, by Abel's Theorem, are equal :

$$\lim_{z\to 1} \mathrm{Li}_{n_1,\ldots,n_r}(z) = \lim_{N\to\infty} \mathrm{H}_{n_1,\ldots,n_r}(N) = \zeta(n_1,\ldots,n_r), \tag{38}$$

where $\zeta(n_1,\ldots,n_r)$ is the convergent polyzeta

$$\zeta(n_1,\ldots,n_r) = \sum_{k_1>\ldots>k_r>0} \frac{1}{k_1^{n_1}\ldots k_r^{n_r}}. \tag{39}$$

Definition 1 *Let $\mathcal{Z}$ be the $\mathbb{Q}$-algebra generated by convergent polyzetas and let $\mathcal{Z}'$ be the $\mathbb{Q}[\gamma]$-algebra*[6] *generated by $\mathcal{Z}$.*

To any multi-index $\mathbf{n} = (n_1,\ldots,n_r)$ corresponds the word $v = y_{n_1}\ldots y_{n_r}$ over the infinite alphabet $Y = \{y_k\}_{k\geq 1}$. The word v itself corresponds to the word ending by the letter x_1, $u = x_0^{n_1-1}x_1\ldots x_0^{n_r-1}x_1 \in X^*x_1$. Then it is usual to index the polylogarithms, harmonic sums and polyzetas by words (over X or Y) :

$$\mathrm{Li}_{\mathbf{n}} = \mathrm{Li}_v(z) = \mathrm{Li}_u, \tag{40}$$

$$\mathrm{P}_{\mathbf{n}} = \mathrm{P}_v(z) = \mathrm{P}_u, \tag{41}$$

$$\mathrm{H}_{\mathbf{n}} = \mathrm{H}_v(N) = \mathrm{H}_u, \tag{42}$$

$$\zeta(\mathbf{n}) = \zeta(v) = \zeta(u). \tag{43}$$

Let us extend, over X^*, the definition of $\{\mathrm{Li}_w\}_{w\in X^*x_1}$ and $\{\mathrm{P}_w\}_{w\in X^*x_1}$ by putting

$$\forall k\geq 0,\quad \mathrm{Li}_{x_0^k}(z) = \frac{\log^k(z)}{k},\quad \mathrm{P}_{x_0^k}(z) = \frac{\mathrm{Li}_{x_0^k}(z)}{1-z}. \tag{44}$$

We get the following structures :

Theorem 1 ([15]) $(\mathbb{C}\{\mathrm{Li}_w\}_{w\in X^*},.) \cong (\mathbb{C}\langle X\rangle, ш)$.

Theorem 2 ([16]) $(\mathbb{C}\{\mathrm{P}_w\}_{w\in Y^*},\odot) \cong (\mathbb{C}\langle Y\rangle, ⧢)$.

Extended to $\mathcal{C} = \mathbb{C}[z, z^{-1}, (1-z)^{-1}]$, we also get as a consequence

[6] Here, γ stands for the Euler constant

$$\gamma = .577215664901532860606512090082402431042159335939923598805767234884867 7\ldots$$

- The polylogarithms $\{\mathrm{Li}_w\}_{w\in X^*}$ (resp. $\{\mathrm{P}_w\}_{w\in Y^*}$) are $\mathcal{C}$-linearly independent Then the harmonic sums $\{\mathrm{H}_w\}_{w\in Y^*}$ are linearly independent.
- The polylogarithms $\{\mathrm{Li}_l\}_{l\in\mathcal{L}ynX}$ (resp. $\{\mathrm{P}_l\}_{l\in\mathcal{L}ynY}$) are $\mathcal{C}$-algebraically independent Then the harmonic sums $\{\mathrm{H}_l\}_{l\in\mathcal{L}ynY}$ are algebraically independent.
- The polyzetas $\{\zeta(l)\}_{l\in\mathcal{L}ynX\setminus\{x_0,x_1\}(\text{resp. }\mathcal{L}ynY\setminus\{y_1\})}$, are generators of $\mathcal{Z}$.

Definition 2 *We put*

$$\mathrm{L}(z) := \sum_{w\in X^*} \mathrm{Li}_w(z)\, w \quad \textit{and} \quad \mathrm{H}(N) := \sum_{w\in Y^*} \mathrm{H}_w(N)\, w.$$

The noncommuting generating series of the polylogarithms is a solution of the differential equation (29) with the following boundary condition

$$\mathrm{L}(z) \widetilde{_{z\to 0}} \exp(x_0 \log z). \tag{45}$$

It follows that L is group-like and then H is also group-like, *i.e.*

$$\Delta_{\sqcup\!\sqcup}(\mathrm{L}) = \mathrm{L}\otimes\mathrm{L} \quad \text{and} \quad \Delta_{\uplus}(\mathrm{H}) = \mathrm{H}\otimes\mathrm{H}, \tag{46}$$

and we have

Theorem 3

$$\mathrm{L}(z) = e^{x_1 \log\frac{1}{1-z}}\mathrm{L}_{\mathrm{reg}}(z)e^{x_0\log z} \quad \textit{and} \quad \mathrm{H}(N) = e^{\mathrm{H}_1(N)\, y_1}\mathrm{H}_{\mathrm{reg}}(N),$$

where, denoting by $\{\hat{l}\}_{\mathcal{L}ynX}$ (resp. $\{\hat{l}\}_{\mathcal{L}ynY}$) the dual basis of $\mathcal{L}ynX$ (resp. $\mathcal{L}ynY$),

$$\mathrm{L}_{\mathrm{reg}}(z) := \prod_{l\in\mathcal{L}ynX, l\neq x_0,x_1}^{\searrow} e^{\mathrm{Li}_l(z)\,\hat{l}} \quad \textit{and} \quad \mathrm{H}_{\mathrm{reg}}(N) := \prod_{l\in\mathcal{L}ynY, l\neq y_1}^{\nearrow} e^{\mathrm{H}_l(N)\,\hat{l}}.$$

Corollary 1 *Let $z_0 \rightsquigarrow z$ be a differentiable path on $\mathbb{C}-\{0,1\}$ such that* L *admits an analytic continuation along this path. We have $S_{z_0\rightsquigarrow z} = \mathrm{L}(z)\mathrm{L}(z_0)^{-1}$.*

PROOF – By Theorem 3, $\mathrm{L}(z)$ is group-like. Hence $\mathrm{L}(z_0)$ is also group-like as is $\mathrm{L}(z_0)^{-1}$. Since the power series $\mathrm{L}(z)\mathrm{L}(z_0)^{-1}$ and $S_{z_0\rightsquigarrow z}$ satisfy (29) taking the same value at z_0 then we get the expected result. □

Definition 3 *We put*

$$Z_{\sqcup\!\sqcup} := \mathrm{L}_{\mathrm{reg}}(1) \quad \textit{and} \quad Z_{\uplus} := \mathrm{H}_{\mathrm{reg}}(\infty).$$

These two noncommuting generating series $Z_{\sqcup\!\sqcup}$ and $Z_{\uplus}$ induce the two following regularization morphisms respectively :

Theorem 4 ([17]) *Let $\zeta_{\uplus} : (\mathbb{C}\langle\langle Y\rangle\rangle, \uplus) \to (\mathbb{R}, .)$ be the morphism satisfying*

- *for $u, v \in Y^*, \zeta_{\uplus}(u \uplus v) = \zeta_{\uplus}(u)\zeta_{\uplus}(v)$,*
- *for all convergent words $w \in Y^* - y_1Y^*, \zeta_{\uplus}(w) = \zeta(w)$,*
- *$\zeta_{\uplus}(y_1) = 0$.*

Then

$$\sum_{w\in X^*} \zeta_{\uplus}(w)\, w = Z_{\uplus}.$$

Corollary 2 ([17]) *For any $w \in X^*, \zeta_{⊎}(w)$ belongs to the algebra $\mathcal{Z}$.*

Theorem 5 ([17]) *Let $\zeta_{\sqcup\!\sqcup} : (\mathbb{C}\langle\langle X\rangle\rangle, \sqcup\!\sqcup) \to (\mathbb{R}, .)$ be the morphism verifying*

- *for $u, v \in X^*, \zeta_{\sqcup\!\sqcup}(u \sqcup\!\sqcup v) = \zeta_{\sqcup\!\sqcup}(u)\zeta_{\sqcup\!\sqcup}(v)$,*
- *for all convergent word $w \in x_0X^*x_1, \zeta_{\sqcup\!\sqcup}(w) = \zeta(w)$,*
- *$\zeta_{\sqcup\!\sqcup}(x_0) = \zeta_{\sqcup\!\sqcup}(x_1) = 0$.*

Then

$$\sum_{w \in X^*} \zeta_{\sqcup\!\sqcup}(w)\, w = Z_{\sqcup\!\sqcup}.$$

Corollary 3 ([17]) *For any $w \in Y^*, \zeta_{\sqcup\!\sqcup}(w)$ belongs to the algebra $\mathcal{Z}$.*

6. Group of associators theorem

Drinfel'd proved that Eqn.(29) admits two particular solutions on the domain $\mathbb{C}\backslash]-\infty, 0] \cup [1, +\infty[$,

$$G_0(z) \widetilde{_{z \rightsquigarrow 0}} \exp[x_0 \log(z)] \text{ and } G_1(z) \widetilde{_{z \rightsquigarrow 1}} \exp[-x_1 \log(1-z)]. \tag{47}$$

He also proved that there exists Φ_{KZ} such that

$$G_0(z) = G_1(z)\Phi_{KZ}. \tag{48}$$

Lê and Murakami expressed the coefficients of the Drinfel'd associator Φ_{KZ} in terms of convergent polyzetas [18].

Let ρ be the monoid morphism verifying

$$\rho(x_0) = -x_1 \quad \text{and} \quad \rho(x_1) = -x_0. \tag{49}$$

We also have [19]

$$\mathrm{L}(z) = \rho[\mathrm{L}(1-z)]Z_{\sqcup\!\sqcup} = e^{x_0 \log z}\rho[\mathrm{L}_{\mathrm{reg}}(1-z)]e^{-x_1 \log(1-z)}Z_{\sqcup\!\sqcup}. \tag{50}$$

Thus,

$$\mathrm{L}(z) \widetilde{_{z \to 1}} \exp(-x_1 \log(1-z))\, Z_{\sqcup\!\sqcup}. \tag{51}$$

It follows from Eqn.(45) and Eqn. (51), with reference [17], that

$$\Phi_{KZ} = Z_{\sqcup\!\sqcup}\ , \tag{52}$$

and it is group-like, *i.e.*

$$\Delta_{\sqcup\!\sqcup}(\Phi_{KZ}) = \Phi_{KZ} \otimes \Phi_{KZ}, \tag{53}$$

and it can be graded in the adjoint basis [11] as follows

$$\Phi_{KZ} = \sum_{\substack{k \geq 0 \\ l_1, \cdots, l_k \geq 0}} \zeta_{\sqcup\!\sqcup}(x_1x_0^{l_1} \circ \cdots \circ x_1x_0^{l_k}) \prod_{i=0}^{k} \mathrm{ad}_{x_0}^{l_i} x_1, \tag{54}$$

where, for any $l \in \mathbb{N}$ and $P \in \mathbb{C}\langle X \rangle$, $\circ$ is defined by

$$x_1 x_0^l \circ P = x_1(x_0^l \sqcup\!\sqcup P), \tag{55}$$

and $\mathrm{ad}_{x_0}^l x_1$ is the iterated Lie bracket

$$\mathrm{ad}_{x_0}^0 x_1 = x_1 \quad \text{and} \quad \mathrm{ad}_{x_0}^l x_1 = [x_0, \mathrm{ad}_{x_0}^{l-1} x_1]. \tag{56}$$

Using the following expansion [20]

$$\mathrm{ad}_{x_0}^n x_1 = \sum_{i=0}^{n} \binom{i}{n} x_0^{n-i} x_1 x_0^i, \tag{57}$$

one then gets, via the regularization process of Theorem 5, the expression for the Drinfel'd associator Φ_{KZ} given by Lê and Murakami [18].

Finally, the asymptotic behaviour of L on (51) leads to

Proposition 3 ([15])

$$S_{\varepsilon \rightsquigarrow 1-\varepsilon} \widetilde{_{\varepsilon \to 0^+}} e^{-x_1 \log \varepsilon} Z_{\sqcup\!\sqcup} e^{-x_0 \log \varepsilon}.$$

In other words, $Z_{\sqcup\!\sqcup}$ is the regularized Chen generating series $S_{\varepsilon \rightsquigarrow 1-\varepsilon}$ of differential forms ω_0 and ω_1 : $Z_{\sqcup\!\sqcup}$ is the noncommutative generating series of the finite parts of the Chen generating series $e^{x_1 \log \varepsilon} S_{\varepsilon \rightsquigarrow 1-\varepsilon} e^{x_0 \log \varepsilon}$, the concatenation of $e^{x_0 \log \varepsilon}$ and then $S_{\varepsilon \rightsquigarrow 1-\varepsilon}$ and finally, $S_{\varepsilon \rightsquigarrow 1-\varepsilon}$.

Let $\{b_{n,k}(t_1, \ldots, t_{n-k+1})\}_{\substack{n \geq k \geq 1 \\ n \in \mathbb{N}_+}}$ be the Bell polynomials given by the following exponential generating series[7]

$$\sum_{n,k=0}^{\infty} b_{n,k}(t_1, \ldots, t_{n-k+1}) \frac{v^n u^k}{n!} = \exp\left(u \sum_{l=0}^{\infty} t_l \frac{v^l}{l!} \right). \tag{58}$$

We specify the variables $\{t_l\}_{l \geq 0}$

$$t_1 = \gamma, \quad \text{for } l > 1, \quad t_l = (-1)^{l-1}(l-1)!\zeta(l), \tag{59}$$

then let $B(x_1)$ be the following power series

$$B(x_1) = 1 + \sum_{n \geq 1} \sum_{k=1}^{n} b_{n,k}(\gamma, -\zeta(2), 2\zeta(3), \ldots) \frac{(-x_1)^n}{n!} \tag{60}$$

and

$$B'(x_1) = e^{-\gamma\, x_1} B(x_1). \tag{61}$$

We get

[7] Compare this expression with the less general Eq.(5). Here $b(n,k)$ is a refinement by the number of blocks (*i.e.* k) of Eq.(5). Thus one has

$$B_n(X_1, X_2, \cdots X_n) = \sum_{k=0}^{n} b_{n,k}(X_1, \ldots, X_{n-k+1}) \; .$$

Journal of Physics: Conference Series **284** (2011) 012023 doi:10.1088/1742-6596/284/1/012023

Proposition 4 *Let Ψ_{KZ} be an element of the quasi-shuffle algebra, $(\mathbb{C}\langle\langle Y\rangle\rangle, \uplus)$, such that*

$$\Pi_X\Psi_{KZ} = B(x_1)\Phi_{KZ},$$

where $\Pi_X\Psi_{KZ}$ is the projection of Ψ_{KZ} over X. Then Ψ_{KZ} is group-like, i.e.

$$\Delta_{\uplus}(\Psi_{KZ}) = \Psi_{KZ} \otimes \Psi_{KZ}, \tag{62}$$

and satisfies

- $\langle \Psi_{KZ} \mid \epsilon\rangle = 1$,
- $\langle \Psi_{KZ} \mid y_1\rangle = \gamma$,
- *for any* $r_1 > 1$, $\langle \Psi_{KZ} \mid y_{r_1} \dots y_{r_k}\rangle = \zeta(r_1, \dots, r_k)$,
- *for any* $u, v \in Y^*$, $\langle \Psi_{KZ}|u \uplus v\rangle = \langle \Psi_{KZ}|u\rangle\langle \Psi_{KZ}|v\rangle$.

Therefore we obtain,

Proposition 5 *The noncommutative generating series Ψ_{KZ} can be factorized by Lyndon words as follows*

$$\Psi_{KZ} = e^{\gamma\, y_1} Z_{\uplus}.$$

More generally, we have

Theorem 6 *Let A be a commutative $\mathbb{Q}$-algebra. For any $\Phi \in A\langle\langle X\rangle\rangle$ and $\Psi \in A\langle\langle Y\rangle\rangle$ such that $\Pi_X\Psi = B(x_1)\Phi$, there exists a unique $C \in \mathcal{L}ie_A\langle\langle X\rangle\rangle$ with coefficients in A such that*

$$\Phi = \Phi_{KZ}e^C \quad \text{and} \quad \Psi = B(y_1)\Pi_Y(\Phi_{KZ}e^C),$$

where $\Pi_Y(\Phi_{KZ}e^C)$ is the projection of $\Phi_{KZ}e^C$ over Y.

Proof – Let $C \in \mathcal{L}ie_A\langle\langle X\rangle\rangle$. Then, by Proposition 2, $\mathrm{L}' = \mathrm{L}e^C$ is solution of (29). Let H' be the noncommutative generating series of the Taylor coefficients, belonging to the harmonic algebra, of $\{(1-z)^{-1}\langle \mathrm{L}' \mid w\rangle\}_{w\in Y^*}$. Then $\mathrm{H}'(N)$ is also group-like.

By the asymptotic expansion on (51), we get

$$\mathrm{L}'(z) \widetilde{_{\varepsilon\to 1}} e^{-x_1\log(1-z)}Z_{\sqcup\!\sqcup}e^C.$$

We then put $\Phi := Z_{\sqcup\!\sqcup}e^C$ and deduce that

$$\frac{\mathrm{L}'(z)}{1-z} \widetilde{_{z\to 1}} \mathrm{Mono}(z)\Phi \quad \text{and} \quad \mathrm{H}'(N) \widetilde{_{N\to\infty}} \mathrm{Const}(N)\Pi_Y\Phi,$$

where [21]

$$\mathrm{Mono}(z) = e^{-(x_1+1)\log(1-z)} = \sum_{k\geq 0} \mathrm{P}_{y_1^k}(z)\; y_1^k$$

$$\mathrm{Const} = \sum_{k\geq 0} \mathrm{H}_{y_1^k}\; y_1^k = \exp\left[-\sum_{k\geq 1} \mathrm{H}_{y_k}\frac{(-y_1)^k}{k}\right].$$

Let κ_w be the constant part of $\mathrm{H}'_w(N)$. Then,

$$\sum_{w\in Y^*} \kappa_w\; w = B(y_1)\Pi_Y\Phi.$$

We now put $\Psi := B(y_1)\Pi_Y(Z_{\sqcup\!\sqcup} e^C)$ (and also $\Psi' := B'(y_1)\Pi_Y(Z_{\sqcup\!\sqcup} e^C)$. $\square$

Thus, any associator Φ and its image Ψ, can be determined from the Drinfel'd associator Φ_{KZ} and its image Ψ_{KZ} respectively, by the action of the group of constant Lie exponential series over X. This group is nothing other than the differential Galois group of the differential equation (29) and contains in particular [11] the monodromy group of (29) given by [15]

$$\mathcal{M}_0 = e^{2i\pi x_0} \quad \text{and} \quad \mathcal{M}_1 = \Phi_{KZ}^{-1} e^{-2i\pi x_1} \Phi_{KZ}. \tag{63}$$

By Proposition 3, we already saw that $Z_{\sqcup\!\sqcup}$ is the concatenation of the Chen generating series [13] $e^{x_0 \log \varepsilon}$ and then $S_{\varepsilon \rightsquigarrow 1-\varepsilon}$ and finally, $e^{x_1 \log \varepsilon}$:

$$Z_{\sqcup\!\sqcup} \underset{\varepsilon \to 0^+}{\widetilde{\quad}} e^{x_1 \log \varepsilon} S_{\varepsilon \rightsquigarrow 1-\varepsilon}\, e^{x_0 \log \varepsilon}. \tag{64}$$

From (63), the action of the monodromy group gives

$$e^{x_1\, 2k_1 i\pi} Z_{\sqcup\!\sqcup} e^{x_0\, 2k_0 i\pi} \underset{\varepsilon \to 0^+}{\widetilde{\quad}} e^{x_1(\log \varepsilon + 2k_1 i\pi)} S_{\varepsilon \rightsquigarrow 1-\varepsilon} e^{x_0(\log \varepsilon + 2k_0 i\pi)} \tag{65}$$

as being the concatenation of the Chen generating series [13] $e^{x_0(\log \varepsilon + 2k_0 i\pi)}$ (along a circular path turning k_0 times around 0), then the Chen generating series $S_{\varepsilon \rightsquigarrow 1-\varepsilon}$ and finally, the Chen generating series $e^{x_1(\log \varepsilon + 2k_1 i\pi)}$ (along a circular path turning k_1 times around 1). More generally, the action of the Galois differential group of polylogarithms shows that, for any Lie series C, the associator $\Phi = Z_{\sqcup\!\sqcup} e^C$ is the concatenation of some Chen generating series e^C and $e^{x_0 \log \varepsilon}$; then the Chen generating series $S_{\varepsilon \rightsquigarrow 1-\varepsilon}$ and finally, $e^{x_1 \log \varepsilon}$:

$$\Phi \underset{\varepsilon \to 0^+}{\widetilde{\quad}} e^{x_1 \log \varepsilon} S_{\varepsilon \rightsquigarrow 1-\varepsilon} e^{x_0 \log \varepsilon}\, e^C. \tag{66}$$

As in the proof of Theorem 6, by construction the associator Φ is then the noncommutative generating series of the finite parts of the coefficients of the Chen generating series $S_{z_0 \rightsquigarrow 1-z_0} e^C$, for $z_0 = \varepsilon \to 0^+$.

Lemma 1 *Let $\Phi \in dm(A) = \{\Phi_{KZ} e^C \mid C \in \mathcal{L}ie_A\langle\langle X \rangle\rangle$ and $\langle e^C \mid \epsilon \rangle = 1, \langle e^C \mid x_0 \rangle = \langle e^C \mid x_1 \rangle = 0\}$. Let $\Pi_Y \Phi$ be the projection of Φ over Y. One has*

$$\Psi = B(y_1)\Pi_Y \Phi \iff \Psi' = B'(y_1)\Pi_Y \Phi.$$

In particular,

$$\Psi_{KZ} = B(y_1)\Pi_Y \Phi_{KZ} \iff \Psi'_{KZ} = B'(y_1)\Pi_Y \Phi_{KZ}.$$

7. The route to properties of Euler's γ constant

We have already mentioned that as a remarkable by-product of our extension of the algebraic considerations related to simple quantum mechanics, we are led to algebras involving polyzeta functions. This excursion is not surprising, as it is already well known that evaluations in quantum field theory involve these functions [14]. In this context, an important related number is the Euler constant γ. We are thus led to to consider properties of γ. In fact, in this note we arrive at a proposition concerning the rationality of γ. The basis of this result depends on the following lemma[8]

Lemma 2 *Let $\Phi \in dm(A)$ and $\Psi' = B'(y_1)\Pi_Y \Phi$. The local coordinates (of the second kind) of Φ (resp. Ψ'), in the Lyndon-PBW basis, are polynomials of generators $\{\zeta(l)\}_{l \in \mathcal{L}ynX} - \{x_0, x_1\}$ (resp. $\{\zeta(l)\}_{l \in \mathcal{L}ynY} - \{y_1\}$) of $\mathcal{Z}$. While C describes $\mathcal{L}ie_A\langle\langle X \rangle\rangle$, these local coordinates describe $A[\{\zeta(l)\}_{l \in \mathcal{L}ynX} - \{x_0, x_1\}]$ (resp. $A[\{\zeta(l)\}_{l \in \mathcal{L}ynY} - \{y_1\}]$).*

[8] We warn the reader that Lemma 2 is not universally accepted.

Journal of Physics: Conference Series **284** (2011) 012023 doi:10.1088/1742-6596/284/1/012023

Lemma 3 *For any $\Phi \in dm(A)$, by identifying the local coordinates (of the second kind) on two members of the identities $\Psi = B(y_1)\Pi_Y\Phi$, or equivalently on $\Psi' = B'(y_1)\Pi_Y\Phi$, we get polynomial relations with coefficients in A of the convergent polyzetas.*

Therefore,

Theorem 7 *While Φ describes $dm(A)$, the identities $\Psi = B(y_1)\Pi_Y\Phi$ describe the ideal of polynomial relations of coefficients in A of the convergent polyzetas.*

Moreover, if the Euler constant γ does not belong to A then these relations are algebraically independent of γ.

Corollary 4 *If $\gamma \notin A$ then $\gamma \notin \bar{A}$.*

With $A = \mathbb{Q}$, it follows immediately that

Corollary 5 *γ is not an algebraic irrational number.*

Corollary 6 *γ is a rational number.*

PROOF – Since γ satisfies $t^2 - \gamma^2 = 0$ then γ is algebraic over $A = \mathbb{Q}(\gamma^2)$. If γ is transcendental then $\gamma \notin A = \mathbb{Q}(\gamma^2)$. Using Corollary 4, with $A = \mathbb{Q}(\gamma^2)$, γ is not algebraic over $A = \mathbb{Q}(\gamma^2)$. This contradicts the previous assertion. Thus, by Corollary 5, it is established that γ is rational over $\mathbb{Q}$. □

8. Structure of polyzetas

Let $\Phi \in dm(A)$ and let $\Psi = B'(y_1)\Pi_Y\Phi$. We introduce two algebra morphisms

$$\begin{array}{rcl} \phi : (A\langle X\rangle, ш) & \longrightarrow & A, \\ u & \longmapsto & \langle \Phi \mid u\rangle, \end{array} \tag{67}$$

and

$$\begin{array}{rcl} \psi : (A\langle Y\rangle, \text{ш}_{-}) & \longrightarrow & A, \\ v & \longmapsto & \langle \Psi \mid v\rangle, \end{array} \tag{68}$$

verifying respectively

$$\phi(\epsilon) = 1 \quad \text{and} \quad \phi(x_0) = \phi(x_1) = 0, \tag{69}$$

$$\psi(\epsilon) = 1 \quad \text{and} \quad \psi(y_1) = 0. \tag{70}$$

Hence,

$$\Phi = \sum_{u \in X^*} \phi(u)\, u = \prod_{\substack{l \in \mathcal{L}ynX \\ l \neq x_0, x_1}}^{\searrow} e^{\phi(l)\, \hat{l}}, \tag{71}$$

$$\Psi = \sum_{v \in Y^*} \psi(u)\, u = \prod_{\substack{l \in \mathcal{L}ynY \\ l \neq y_1}}^{\nearrow} e^{\psi(l)\, \hat{l}}. \tag{72}$$

Using these factorizations of the monoids by Lyndon words, we get

Journal of Physics: Conference Series **284** (2011) 012023 doi:10.1088/1742-6596/284/1/012023

Theorem 8 *For any $\Phi \in dm(A)$, let $\Psi = B'(y_1)\Pi_Y\Phi$. Then*

$$\prod_{\substack{l\in\mathcal{L}ynX\\ l\neq x_0,x_1}}^{\searrow} e^{\phi(l)\,\hat{l}} = e^{-\sum_{k\geq 2}\zeta(k)\frac{(-x_1)^k}{k}}\,\Pi_X \prod_{\substack{l\in\mathcal{L}ynY\\ l\neq y_1}}^{\nearrow} e^{\psi(l)\,\hat{l}}.$$

In particular, if $\Phi = Z_{\sqcup\!\sqcup}$ and $\Psi = Z_{⧢}$ then

$$\prod_{\substack{l\in\mathcal{L}ynX\\ l\neq x_0,x_1}}^{\searrow} e^{\zeta(l)\,\hat{l}} = e^{-\sum_{k\geq 2}\zeta(k)\frac{(-x_1)^k}{k}}\,\Pi_X \prod_{\substack{l\in\mathcal{L}ynY\\ l\neq y_1}}^{\nearrow} e^{\zeta(l)\,\hat{l}}.$$

Since

$$\forall l \in \mathcal{L}ynY \iff \Pi_X l \in \mathcal{L}ynX \setminus \{x_0\} \tag{73}$$

then identifying the local coordinates, we get polynomial relations among the generators which are algebraically independent of γ.

Corollary 7 *For $\ell \in \mathcal{L}ynY - \{y_1\}$ (resp. $\mathcal{L}ynX - \{x_0, x_1\}$), let $P_\ell \in \mathcal{L}ie_{\mathbb{Q}}\langle X\rangle$ (resp. $\mathcal{L}ie_{\mathbb{Q}}\langle Y\rangle$) be the decomposition of the polynomial $\Pi_X\hat{\ell} \in \mathbb{Q}\langle X\rangle$ (resp. $\Pi_Y\hat{\ell} \in \mathbb{Q}\langle Y\rangle$) in the Lyndon-PBW basis $\{\hat{l}\}_{l\in\mathcal{L}ynX}$ (resp. $\{\hat{l}\}_{l\in\mathcal{L}ynY}$) and let $\check{P}_\ell \in \mathbb{Q}[\mathcal{L}ynX - \{x_0, x_1\}]$ (resp. $\mathbb{Q}[\mathcal{L}ynY - \{y_1\}]$) be its dual. Then one obtains*

$$\Pi_X\ell - \check{P}_\ell \in \ker\phi \quad (\textit{resp.} \quad \Pi_Y\ell - \check{P}_\ell \in \ker\psi).$$

In particular, for $\phi = \zeta$ (resp. $\psi = \zeta$) then one also obtains

$$\Pi_X\ell - \check{P}_\ell \in \ker\zeta \quad (\textit{resp.} \quad \Pi_Y\ell - \check{P}_\ell \in \ker\zeta).$$

Moreover, for any $\ell \in \mathcal{L}ynY - \{y_1\}$ (resp. $\mathcal{L}ynX - \{x_0, x_1\}$), the polynomial $\Pi_Y\ell - \check{P}_\ell \in \mathbb{Q}\langle Y\rangle$ (resp. $\mathbb{Q}\langle X\rangle$) is homogenous of degree equal to $|\ \ell\ | > 1$.

Theorem 9 (Structure of polyzetas) *The $\mathbb{Q}$-algebra generated by convergent polyzetas is isomorphic to the* graded *algebra* $(\mathbb{Q} \oplus (Y - y_1)\mathbb{Q}\langle Y\rangle / \ker\zeta, ⧢)$.

PROOF – Since $\ker\zeta$ is an ideal generated by the homogenous polynomials then the quotient $\mathbb{Q} \oplus (Y - y_1)\mathbb{Q}\langle Y\rangle / \ker\zeta$ is graded. □

Corollary 8 *The $\mathbb{Q}$-algebra of polyzetas is freely generated by* irreducible *polyzetas.*

PROOF – For any $\lambda \in \mathcal{L}ynY$, if $\lambda = \check{P}_\lambda$ then one obtains the conclusion otherwise $\Pi_X\lambda - \check{P}_\lambda \in \ker\zeta$. Since $\check{P}_\lambda \in \mathbb{Q}[\mathcal{L}ynX]$ then $\check{P}_\lambda$ is polynomial in Lyndon words of degree less than or equal to $|\lambda|$. Since each Lyndon word does appear in this decomposition of $\check{P}_\lambda$, after applying Π_Y, we may repeat the same process until we obtain irreducible polyzetas. □

[1] Solomon A I, Duchamp G H E, Blasiak P, Horzela A and Penson K A 2010 From Quantum Mechanics to Quantum Field Theory: The Hopf route, These G28 proceedings.

[2] Blasiak P, Duchamp G H E, Solomon A I , Horzela A and Penson K A 2010 Combinatorial Algebra for second-quantized Quantum Theory *Adv. Theor. Math. Phys* **14**
arXiv : http://arxiv.org/abs/1001.4964

[3] Duchamp G H E, Blasiak P, Horzela A, Penson K A and Solomon A I 2010 A three-parameter Hopf deformation of the algebra of Feynman-like diagrams *J. Russian Laser Research* **31** 162

[4] Bender C M, Brody D C and Meister B K 1999 Quantum field theory of partitions *J. Math. Phys.* **40** 3239–3245

[5] Comtet L 1974 Advanced Combinatorics: The Art of Finite and Infinite Expansions *Reidel Publishing Company, Dordrecht-Holland/Boston-U.S.A.*

[6] Poinsot L, Duchamp G H E, Goodenough S and Penson K A 2010 Statistics on Graphs, Exponential Formula and Combinatorial Physics *J. Nonlinear Systems and Applications* (to appear) arXiv : 0910.0695v2 [cs.DM]

[7] Duchamp G H E, Blasiak P, Horzela A, Penson K A and Solomon A I 2006 Feynman graphs and related Hopf algebras *J.Phy.: Conf. Series* **30**107–118 // arXiv : cs.SC/0510041

[8] Duchamp G H E, Hivert F and Thibon J Y 2002 Non commutative functions VI: Free quasi-symmetric functions and related algebras *Int. J. Alg. Comp.* **12** 671–717

[9] Hoffman M E 2000 Quasi-shuffle products *J. Algebraic Combin.***11** 49–68

[10] Drinfel'd V 1986 Quantum groups *Proc. Int. Cong. Math., Berkeley*

[11] Minh H N 2003 Differential Galois groups and non commutative generating series of polylogarithms *Proc. Automata, Combinatorics and Geometry, 7th World Multi-conference on Systemics, Cybernetics and Informatics, Florida*

[12] Ree R 1958 Lie elements and an algebra associated with shuffles *Ann. Math* **68** 210–220.

[13] Chen K T 1977 Iterated path integrals *Bull. Amer. Math. Soc.* **83** 831–879

[14] Kreimer D 2000 Knots and Feynman Diagrams *Cambridge Lecture Notes in Physics CUP*

[15] Minh H N, Petitot M and Van der Hoeven J 1998 Polylogarithms and Shuffle Algebra *Proc. FPSAC'98, Toronto*

[16] Minh H N 2003 Finite polyzetas, Poly-Bernoulli numbers, identities of polyzetas and noncommutative rational power series *Proc.* 4th *International Conference on Words* 232–250

[17] Minh H N, Jacob G , Oussous N E and Petitot M 2001 De l'algèbre des ζ de Riemann multivariées à l'algèbre des ζ de Hurwitz multivariées *J. électronique du Séminaire Lotharingien de Combinatoire B44e*

[18] Lê T Q T and Murakami J 1996 Kontsevich's integral for Kauffman polynomial *Nagoya Math Jour.* **142** 39–65

[19] Minh H N, Petitot M and Van der Hoeven J 1999 L'algèbre des polylogarithmes par les séries génératrices *Proceedings of FPSAC'99*

[20] Bourbaki N 1974 *Algebra I* Chapters 1–3 Paris: Hermann.

[21] Minh H N 2007 Algebraic Combinatoric aspects of asymptotic analysis of nonlinear dynamical systems with singular inputs *Acta Academiae Aboensis, Ser. B* **67** 117–126

Journal of Physics: Conference Series **284** (2011) 012024 doi:10.1088/1742-6596/284/1/012024

The Construction of Field Operators

Valeriy V. Dvoeglazov

Universidad de Zacatecas, Apartado Postal 636, Suc. UAZ, Zacatecas 98062, Zac., México

E-mail: valeri@fisica.uaz.edu.mx

Abstract. We draw attention to some tune problems in constructions of the quantum-field operators for spins 1/2 and 1. They are related to the existence of negative-energy solutions and acausal solutions of relativistic wave equations. Particular attention is paid to neutral particles, chiral theories, and to the method of the Lorentz boosts.

1. Introduction.

The Dirac equation can be deduced either by the Dirac method [1] (the Hamiltonian should be linear in $\partial/\partial x^i$, and be compatible with $E^2 - \mathbf{p}^2c^2 = m^2c^4$); or by the Sakurai method [2] (based on the equation $(E - \sigma\cdot\mathbf{p})(E + \sigma\cdot\mathbf{p})\phi = m^2\phi$); or by the Ryder one [3] (the relation between 2-spinors at rest is $\phi_R(\mathbf{0}) = \pm\phi_L(\mathbf{0})$). The γ^μ obey the Clifford algebra anticommutation relations. Usually, everybody uses the following definition of the field operator [4]:

$$\Psi(x) = \frac{1}{(2\pi)^3}\sum_\sigma \int \frac{d^3\mathbf{p}}{2E_p}[u_\sigma(\mathbf{p})a_\sigma(\mathbf{p})e^{-ip\cdot x} + v_\sigma(\mathbf{p})b_\sigma^\dagger(\mathbf{p})]e^{+ip\cdot x}]\,, \tag{1}$$

as given *ab initio*. I studied in the previous works [13, 14, 17]: $\sigma \to h$ (the helicity basis); the modified Sakurai derivation (the additional $m_2\gamma^5$ term in the Dirac equation); the derivation of the Barut equation [18] from the first principles, namely based on the generalized Ryder relation, $(\phi_L^h(\mathbf{0}) = \hat{A}\phi_L^{-h\,*}(\mathbf{0}) + \hat{B}\phi_L^{h\,*}(\mathbf{0}))$. In fact, we have the second mass state (μ-meson) from that equation:

$$[i\gamma^\mu\partial_\mu - \alpha\partial_\mu\partial^\mu/m - \beta]\psi = 0\,; \tag{2}$$

the self/anti-self charge-conjugate Majorana 4-spinors [7, 8] in the momentum representation.

We begin with the method of constructions of the field operators given in [5]. The Wigner rules [6] of the Lorentz transformations for the $(0, S)$ left- $\phi_L(\mathbf{p})$ and the $(S, 0)$ right- $\phi_R(\mathbf{p})$ spinors are used. We can recover the Dirac equation in the matrix form:

$$\begin{pmatrix} \mp m\,1 & p_0 + \sigma\cdot\mathbf{p} \\ p_0 - \sigma\cdot\mathbf{p} & \mp m\,1 \end{pmatrix}\psi(p^\mu) = 0\,, \tag{3}$$

or $(\gamma\cdot p - m)u(\mathbf{p}) = 0$ and $(\gamma\cdot p + m)v(\mathbf{p}) = 0$. The solutions of the Dirac equation can be denoted as $u(\mathbf{p}) = column(\phi_R(\mathbf{p})\quad \phi_L(\mathbf{p}))$ and $v(\mathbf{p}) = \gamma^5 u(\mathbf{p})$, ref. [3, p.53] with $E = \sqrt{\mathbf{p}^2 + m^2} > 0$, $p_0 = \pm E$, $p^\pm = E \pm p_z$, $p_{r,l} = p_x \pm ip_y$. They are the parity eigenstates with the eigenvalues of ± 1. They also describe eigenstates of the charge operator. Thus, in this Section we have used the basis for charged particles in the $(S,0)\oplus(0,S)$ representation (in general)

$$u_{+\sigma}(\mathbf{0}) = N(\sigma)\begin{pmatrix}1\\0\\ \cdot\\ \cdot\\ \cdot\\0\end{pmatrix}, u_{\sigma-1}(\mathbf{0}) = N(\sigma)\begin{pmatrix}0\\1\\ \cdot\\ \cdot\\ \cdot\\0\end{pmatrix}, \ldots v_{-\sigma}(\mathbf{0}) = N(\sigma)\begin{pmatrix}0\\0\\ \cdot\\ \cdot\\ \cdot\\1\end{pmatrix} \tag{4}$$

Sometimes, the normalization factor is convenient to choose $N(\sigma) = m^\sigma$ in order the rest spinors to vanish in the massless limit. However, other constructs are possible in the $(1/2,0)\oplus(0,1/2)$ representation.

2. Majorana Spinors in the Momentum Representation.

During the 20th century various authors introduced *self/anti-self* charge-conjugate 4-spinors (including in the momentum representation), see [7, 8, 9, 10]. Later, they [11, 13, 14, 12] *etc* studied these spinors, they found corresponding dynamical equations, gauge transformations and other specific features of them. The definitions are:

$$C = e^{i\theta}\begin{pmatrix}0&0&0&-i\\0&0&i&0\\0&i&0&0\\-i&0&0&0\end{pmatrix}\mathcal{K} = -e^{i\theta}\gamma^2\mathcal{K} \tag{5}$$

is the anti-linear operator of charge conjugation. $\mathcal{K}$ is the complex conjugation operator. We define the *self/anti-self* charge-conjugate 4-spinors in the momentum space[1]

$$C\lambda^{S,A}(\mathbf{p}) = \pm\lambda^{S,A}(\mathbf{p})\,, C\rho^{S,A}(\mathbf{p}) = \pm\rho^{S,A}(\mathbf{p})\,. \tag{7}$$

The rest λ and ρ spinors are:

$$\lambda^S_\uparrow(\mathbf{0}) = \sqrt{\frac{m}{2}}\begin{pmatrix}0\\i\\1\\0\end{pmatrix}, \lambda^S_\downarrow(\mathbf{0}) = \sqrt{\frac{m}{2}}\begin{pmatrix}-i\\0\\0\\1\end{pmatrix}, \lambda^A_\uparrow(\mathbf{0}) = \sqrt{\frac{m}{2}}\begin{pmatrix}0\\-i\\1\\0\end{pmatrix}, \lambda^A_\downarrow(\mathbf{0}) = \sqrt{\frac{m}{2}}\begin{pmatrix}i\\0\\0\\1\end{pmatrix}, \tag{8}$$

$$\rho^S_\uparrow(\mathbf{0}) = \sqrt{\frac{m}{2}}\begin{pmatrix}1\\0\\0\\-i\end{pmatrix}, \rho^S_\downarrow(\mathbf{0}) = \sqrt{\frac{m}{2}}\begin{pmatrix}0\\1\\i\\0\end{pmatrix}, \rho^A_\uparrow(\mathbf{0}) = \sqrt{\frac{m}{2}}\begin{pmatrix}1\\0\\0\\i\end{pmatrix}, \rho^A_\downarrow(\mathbf{0}) = \sqrt{\frac{m}{2}}\begin{pmatrix}0\\1\\-i\\0\end{pmatrix}. \tag{9}$$

[1] Such definitions of 4-spinors differ, of course, from the original Majorana definition in x-representation:

$$\nu(x) = \frac{1}{\sqrt{2}}(\Psi_D(x) + \Psi^c_D(x))\,, \tag{6}$$

$C\nu(x) = \nu(x)$ that represents the positive real $C-$ parity field operator. However, the momentum-space Majorana-like spinors open various possibilities for description of neutral particles (with experimental consequences, see [12]). For instance, "for imaginary C parities, the neutrino mass can drop out from the single β decay trace and reappear in $0\nu\beta\beta$, a curious and in principle experimentally testable signature for a non-trivial impact of Majorana framework in experiments with polarized sources."

It is easy to derive the explicite forms of the 4-spinors of the second kind $\lambda^{S,A}_{\uparrow\downarrow}(\mathbf{p})$ and $\rho^{S,A}_{\uparrow\downarrow}(\mathbf{p})$ in this basis. Here they are:

$$\lambda^S_\uparrow(\mathbf{p}) = \frac{1}{2\sqrt{E+m}}\begin{pmatrix} ip_l \\ i(p^- + m) \\ p^- + m \\ -p_r \end{pmatrix}, \lambda^S_\downarrow(\mathbf{p}) = \frac{1}{2\sqrt{E+m}}\begin{pmatrix} -i(p^+ + m) \\ -ip_r \\ -p_l \\ (p^+ + m) \end{pmatrix}, \quad (10)$$

$$\lambda^A_\uparrow(\mathbf{p}) = \frac{1}{2\sqrt{E+m}}\begin{pmatrix} -ip_l \\ -i(p^- + m) \\ (p^- + m) \\ -p_r \end{pmatrix}, \lambda^A_\downarrow(\mathbf{p}) = \frac{1}{2\sqrt{E+m}}\begin{pmatrix} i(p^+ + m) \\ ip_r \\ -p_l \\ (p^+ + m) \end{pmatrix}, \quad (11)$$

$$\rho^S_\uparrow(\mathbf{p}) = \frac{1}{2\sqrt{E+m}}\begin{pmatrix} p^+ + m \\ p_r \\ ip_l \\ -i(p^+ + m) \end{pmatrix}, \rho^S_\downarrow(\mathbf{p}) = \frac{1}{2\sqrt{E+m}}\begin{pmatrix} p_l \\ (p^- + m) \\ i(p^- + m) \\ -ip_r \end{pmatrix}, \quad (12)$$

$$\rho^A_\uparrow(\mathbf{p}) = \frac{1}{2\sqrt{E+m}}\begin{pmatrix} p^+ + m \\ p_r \\ -ip_l \\ i(p^+ + m) \end{pmatrix}, \rho^A_\downarrow(\mathbf{p}) = \frac{1}{2\sqrt{E+m}}\begin{pmatrix} p_l \\ (p^- + m) \\ -i(p^- + m) \\ ip_r \end{pmatrix}. \quad (13)$$

As we showed, the λ and ρ 4-spinors are NOT the eigenspinors of the helicity. Moreover, the λ and ρ are NOT the eigenspinors of the parity (in this representation $P = \begin{pmatrix} 0 & 1 \\ 1 & 0 \end{pmatrix} R$), as opposed to the Dirac case. The indices $\uparrow\downarrow$ should be referred to the chiral helicity quantum number introduced in the 60s, $\eta = -\gamma^5 h$. While $Pu_\sigma(\mathbf{p}) = +u_\sigma(\mathbf{p})\,, Pv_\sigma(\mathbf{p}) = -v_\sigma(\mathbf{p})$, we have $P\lambda^{S,A}(\mathbf{p}) = \rho^{A,S}(\mathbf{p})\,, P\rho^{S,A}(\mathbf{p}) = \lambda^{A,S}(\mathbf{p})$ for the Majorana-like momentum-space 4-spinors on the first quantization level.

The dynamical coordinate-space equations are:

$$i\gamma^\mu\partial_\mu\lambda^S(x) - m\rho^A(x) = 0\,, \; i\gamma^\mu\partial_\mu\rho^A(x) - m\lambda^S(x) = 0\,, \quad (14)$$

$$i\gamma^\mu\partial_\mu\lambda^A(x) + m\rho^S(x) = 0\,, \; i\gamma^\mu\partial_\mu\rho^S(x) + m\lambda^A(x) = 0\,. \quad (15)$$

These are NOT the Dirac equation. However, they can be written in the 8-component form as follows:

$$[i\Gamma^\mu\partial_\mu - m]\,\Psi_{(+)}(x) = 0\,, \; [i\Gamma^\mu\partial_\mu + m]\,\Psi_{(-)}(x) = 0\,. \quad (16)$$

One can also re-write the equations into the two-component form. Similar formulations have been presented by M. Markov [15], and A. Barut and G. Ziino [9]. The group-theoretical basis for such doubling has been given in the papers by Gelfand, Tsetlin and Sokolik [16].

The Lagrangian is

$$\mathcal{L} = \frac{i}{2}\Big[\bar\lambda^S\gamma^\mu\partial_\mu\lambda^S - (\partial_\mu\bar\lambda^S)\gamma^\mu\lambda^S + \bar\rho^A\gamma^\mu\partial_\mu\rho^A - (\partial_\mu\bar\rho^A)\gamma^\mu\rho^A +$$
$$\bar\lambda^A\gamma^\mu\partial_\mu\lambda^A - (\partial_\mu\bar\lambda^A)\gamma^\mu\lambda^A + \bar\rho^S\gamma^\mu\partial_\mu\rho^S - (\partial_\mu\bar\rho^S)\gamma^\mu\rho^S -$$
$$-m(\bar\lambda^S\rho^A + \bar\lambda^S\rho^A - \bar\lambda^S\rho^A - \bar\lambda^S\rho^A)\Big]\,. \quad (17)$$

The connection with the Dirac spinors has been found. For instance,

$$\begin{pmatrix} \lambda^S_\uparrow(\mathbf{p}) \\ \lambda^S_\downarrow(\mathbf{p}) \\ \lambda^A_\uparrow(\mathbf{p}) \\ \lambda^A_\downarrow(\mathbf{p}) \end{pmatrix} = \frac{1}{2}\begin{pmatrix} 1 & i & -1 & i \\ -i & 1 & -i & -1 \\ 1 & -i & -1 & -i \\ i & 1 & i & -1 \end{pmatrix}\begin{pmatrix} u_{+1/2}(\mathbf{p}) \\ u_{-1/2}(\mathbf{p}) \\ v_{+1/2}(\mathbf{p}) \\ v_{-1/2}(\mathbf{p}) \end{pmatrix}. \quad (18)$$

See also ref. [16, 9]. A possibility was noted for generalizations of the concept of the Fock space, which lead to the "doubling" Fock space [16, 9].

It was shown [13] that the covariant derivative (and, hence, the interaction) can be introduced in this construct in the following way: $\partial_\mu \to \nabla_\mu = \partial_\mu - ig\L^5 A_\mu$, where $\L^5 = \mathrm{diag}(\gamma^5 \quad -\gamma^5)$, the 8×8 matrix. With respect to the transformations

$$\lambda'(x) \to (\cos\alpha - i\gamma^5 \sin\alpha)\lambda(x)\,, \; \overline{\lambda}'(x) \to \overline{\lambda}(x)(\cos\alpha - i\gamma^5 \sin\alpha)\,, \tag{19}$$

the spinors retain their properties to be self/anti-self charge conjugate spinors and the proposed Lagrangian [13, p.1472] remains to be invariant. This tells us that while self/anti-self charge conjugate states have zero eigenvalues of the ordinary (scalar) charge operator but they can possess the axial charge. [2]

Next, because the transformations

$$\lambda'_S(\mathbf{p}) = \begin{pmatrix} \Xi & 0 \\ 0 & \Xi \end{pmatrix} \lambda_S(\mathbf{p}) \equiv \lambda^*_A(\mathbf{p}), \tag{21}$$

$$\lambda''_S(\mathbf{p}) = \begin{pmatrix} i\Xi & 0 \\ 0 & -i\Xi \end{pmatrix} \lambda_S(\mathbf{p}) \equiv -i\lambda^*_S(\mathbf{p}), \tag{22}$$

$$\lambda'''_S(\mathbf{p}) = \begin{pmatrix} 0 & i\Xi \\ i\Xi & 0 \end{pmatrix} \lambda_S(\mathbf{p}) \equiv i\gamma^0\lambda^*_A(\mathbf{p}), \tag{23}$$

$$\lambda^{IV}_S(\mathbf{p}) = \begin{pmatrix} 0 & \Xi \\ -\Xi & 0 \end{pmatrix} \lambda_S(\mathbf{p}) \equiv \gamma^0\lambda^*_S(\mathbf{p}) \tag{24}$$

with the 2×2 matrix Ξ defined as (ϕ is the azimuthal angle related with $\mathbf{p}$)

$$\Xi = \begin{pmatrix} e^{i\phi} & 0 \\ 0 & e^{-i\phi} \end{pmatrix} \quad , \quad \Xi\Lambda_{R,L}(\mathbf{p} \leftarrow \mathbf{0})\Xi^{-1} = \Lambda^*_{R,L}(\mathbf{p} \leftarrow \mathbf{0})\,, \tag{25}$$

and because the corresponding transformations for λ^A do *not* change the properties of bispinors to be in the self/anti-self charge conjugate spaces, the Majorana-like field operator ($b^\dagger \equiv a^\dagger$) admits additional phase (and, in general, normalization) transformations:

$$\nu^{ML\,\prime}(x^\mu) = [c_0 + i(\tau \cdot \mathbf{c})]\, \nu^{ML\,\dagger}(x^\mu)\,, \tag{26}$$

where c_α are arbitrary parameters. The τ matrices are defined over the field of 2×2 matrices and the Hermitian conjugation operation is assumed to act on the c- numbers as the complex conjugation. One can parametrize $c_0 = \cos\phi$ and $\mathbf{c} = \mathbf{n}\sin\phi$ and, thus, define the $SU(2)$ group of phase transformations. One can select the Lagrangian which is composed from the both field operators (with λ spinors and ρ spinors) and which remains to be invariant with respect to this kind of transformations. The conclusion is: it is permitted a non-Abelian construct which is based on the spinors of the Lorentz group only. This is not surprising because both the $SU(2)$ group and $U(1)$ group are the sub-groups of the extended Poincaré group (cf. [3]).

[2] In fact, from this consideration one can also recover the Feynman-Gell-Mann equation (and its charge-conjugate equation). It is re-written in the two-component form

$$\begin{cases} \left[\pi^-_\mu \pi^{\mu\,-} - m^2 - \frac{g}{2}\sigma^{\mu\nu}F_{\mu\nu}\right]\chi(x) = 0\,, \\ \left[\pi^+_\mu \pi^{\mu\,+} - m^2 + \frac{g}{2}\widetilde{\sigma}^{\mu\nu}F_{\mu\nu}\right]\phi(x) = 0\,, \end{cases} \tag{20}$$

where already one has $\pi^\pm_\mu = i\partial_\mu \pm gA_\mu$, $\sigma^{0i} = -\widetilde{\sigma}^{0i} = i\sigma^i$, $\sigma^{ij} = \widetilde{\sigma}^{ij} = \epsilon_{ijk}\sigma^k$ and $\nu^{DL}(x) = \mathrm{column}(\chi \quad \phi)$.

The Dirac-like and Majorana-like field operators can be built from both $\lambda^{S,A}(\mathbf{p})$ and $\rho^{S,A}(\mathbf{p})$, or their combinations. The anticommutation relations are the following ones (due to the *bi-orthonormality*):

$$[a_{\eta'}(\mathbf{p}'), a^\dagger_\eta(\mathbf{p})]_\pm = (2\pi)^3 2E_p \delta(\mathbf{p}-\mathbf{p}')\delta_{\eta,-\eta'}\,, \tag{27}$$

and

$$[b_{\eta'}(\mathbf{p}'), b^\dagger_\eta(\mathbf{p})]_\pm = (2\pi)^3 2E_p \delta(\mathbf{p}-\mathbf{p}')\delta_{\eta,-\eta'} \tag{28}$$

Other (anti)commutators are equal to zero: ($[a_{\eta'}(\mathbf{p}'), b^\dagger_\eta(\mathbf{p})] = 0$).

In the Fock space operations of the charge conjugation and space inversions can be defined through unitary operators such that:

$$U^c_{[1/2]}\Psi(x^\mu)(U^c_{[1/2]})^{-1} = \mathcal{C}_{[1/2]}\Psi^\dagger_{[1/2]}(x^\mu)\,,\; U^s_{[1/2]}\Psi(x^\mu)(U^s_{[1/2]})^{-1} = \gamma^0\Psi(x'^{\mu}), \tag{29}$$

the time reversal operation, through *an antiunitary* operator

$$\left[V^T_{[1/2]}\Psi(x^\mu)(V^T_{[1/2]})^{-1}\right]^\dagger = S(T)\Psi^\dagger(x''^\mu) \quad , \tag{30}$$

with $x'^{\mu} \equiv (x^0, -\mathbf{x})$ and $x''^{\mu} = (-x^0, \mathbf{x})$. We further assume the vacuum state to be assigned an even P- and C-eigenvalue and, then, proceed as in ref. [4]. As a result we have the following properties of creation (annihilation) operators in the Fock space:

$$\begin{aligned} U^s_{[1/2]}a_\uparrow(\mathbf{p})(U^s_{[1/2]})^{-1} &= -ia_\downarrow(-\mathbf{p})\,,\; U^s_{[1/2]}a_\downarrow(\mathbf{p})(U^s_{[1/2]})^{-1} = +ia_\uparrow(-\mathbf{p})\,, && (31)\\ U^s_{[1/2]}b^\dagger_\uparrow(\mathbf{p})(U^s_{[1/2]})^{-1} &= +ib^\dagger_\downarrow(-\mathbf{p})\; U^s_{[1/2]}b^\dagger_\downarrow(\mathbf{p})(U^s_{[1/2]})^{-1} = -ib_\uparrow(-\mathbf{p})\,, && (32)\end{aligned}$$

what signifies that the states created by the operators $a^\dagger(\mathbf{p})$ and $b^\dagger(\mathbf{p})$ have very different properties with respect to the space inversion operation, comparing with Dirac states (the case also regarded in [9]):

$$\begin{aligned} U^s_{[1/2]}|\mathbf{p},\uparrow>^+ &= +i|-\mathbf{p},\downarrow>^+, U^s_{[1/2]}|\mathbf{p},\uparrow>^- = +i|-\mathbf{p},\downarrow>^-\,, && (33)\\ U^s_{[1/2]}|\mathbf{p},\downarrow>^+ &= -i|-\mathbf{p},\uparrow>^+, U^s_{[1/2]}|\mathbf{p},\downarrow>^- = -i|-\mathbf{p},\uparrow>^-\,. && (34)\end{aligned}$$

For the charge conjugation operation in the Fock space we have two physically different possibilities. The first one, *e.g.*,

$$\begin{aligned} U^c_{[1/2]}a_\uparrow(\mathbf{p})(U^c_{[1/2]})^{-1} &= +b_\uparrow(\mathbf{p})\,,\; U^c_{[1/2]}a_\downarrow(\mathbf{p})(U^c_{[1/2]})^{-1} = +b_\downarrow(\mathbf{p})\,, && (35)\\ U^c_{[1/2]}b^\dagger_\uparrow(\mathbf{p})(U^c_{[1/2]})^{-1} &= -a^\dagger_\uparrow(\mathbf{p})\,,\; U^c_{[1/2]}b^\dagger_\downarrow(\mathbf{p})(U^c_{[1/2]})^{-1} = -a^\dagger_\downarrow(\mathbf{p})\,. && (36)\end{aligned}$$

The action of this operator on the physical states are

$$\begin{aligned} U^c_{[1/2]}|\mathbf{p},\uparrow>^+ &= +|\mathbf{p},\uparrow>^-,\; U^c_{[1/2]}|\mathbf{p},\downarrow>^+ = +|\mathbf{p},\downarrow>^-, && (37)\\ U^c_{[1/2]}|\mathbf{p},\uparrow>^- &= -|\mathbf{p},\uparrow>^+,\; U^c_{[1/2]}|\mathbf{p},\downarrow>^- = -|\mathbf{p},\downarrow>^+\,. && (38)\end{aligned}$$

The second one acts in the following manner:

$$\begin{aligned} \widetilde{U}^c_{[1/2]}a_\uparrow(\mathbf{p})(\widetilde{U}^c_{[1/2]})^{-1} = -b_\downarrow(\mathbf{p})\,,\; \widetilde{U}^c_{[1/2]}a_\downarrow(\mathbf{p})(\widetilde{U}^c_{[1/2]})^{-1} = -b_\uparrow(\mathbf{p}), && (39)\\ \widetilde{U}^c_{[1/2]}b^\dagger_\uparrow(\mathbf{p})(\widetilde{U}^c_{[1/2]})^{-1} = +a^\dagger_\downarrow(\mathbf{p})\,,\; \widetilde{U}^c_{[1/2]}b^\dagger_\downarrow(\mathbf{p})(\widetilde{U}^c_{[1/2]})^{-1} = +a^\dagger_\uparrow(\mathbf{p})\,, && (40)\end{aligned}$$

and, therefore,

$$\widetilde{U}^c_{[1/2]}|\mathbf{p},\uparrow>^+ = -|\mathbf{p},\downarrow>^-,\ \widetilde{U}^c_{[1/2]}|\mathbf{p},\downarrow>^+ = -|\mathbf{p},\uparrow>^-, \quad (41)$$

$$\widetilde{U}^c_{[1/2]}|\mathbf{p},\uparrow>^- = +|\mathbf{p},\downarrow>^+,\ \widetilde{U}^c_{[1/2]}|\mathbf{p},\downarrow>^- = +|\mathbf{p},\uparrow>^+. \quad (42)$$

It is possible a situation when the operators of the space inversion and charge conjugation commute each other in the Fock space [19]. Finally, the time reversal *anti-unitary* operator in the Fock space should be defined in such a way that the formalism to be compatible with the CPT theorem. We obtain for the $\Psi(x^\mu)$ field:

$$V^T a^\dagger_\uparrow(\mathbf{p})(V^T)^{-1} = a^\dagger_\downarrow(-\mathbf{p}),\ V^T a^\dagger_\downarrow(\mathbf{p})(V^T)^{-1} = -a^\dagger_\uparrow(-\mathbf{p}), \quad (43)$$

$$V^T b_\uparrow(\mathbf{p})(V^T)^{-1} = b_\downarrow(-\mathbf{p}),\ V^T b_\downarrow(\mathbf{p})(V^T)^{-1} = -b_\uparrow(-\mathbf{p}). \quad (44)$$

Thus, this construct has very different properties with respect to C, P and T comparing with the Dirac construct.

The dependence of the physical results on the choice of the basis is a bit strange thing. Somewhat similar things have been presented in [17] when compared the Dirac-like constructs in the parity basis and the helicity basis. It was shown that the helicity eigenstates of the operator $(\sigma\cdot\mathbf{n})\otimes I$) are NOT the parity eigenstates, and vice versa, in the helicity basis (cf. with [20], while they obey the same Dirac equation. This is *not* surprising because the P operator does *not* commute with the h operator. The bases are connected by the unitary transformation. The both sets of 4-spinors form the complete system in a mathematical sense.

3. The Spin 1.

3.1. Maxwell Equations as Quantum Equations.

In refs. [21, 22] the Maxwell-like equations have been derived from the D'Alembert equation. Here they are:

$$\nabla\times\mathbf{E} = -\frac{1}{c}\frac{\partial\mathbf{B}}{\partial t} + \nabla Im\chi\,,\ \nabla\cdot\mathbf{E} = -\frac{1}{c}\frac{\partial}{\partial t}Re\chi\,, \quad (45)$$

$$\nabla\times\mathbf{B} = \frac{1}{c}\frac{\partial\mathbf{E}}{\partial t} + \nabla Re\chi\,,\ \nabla\cdot\mathbf{B} = \frac{1}{c}\frac{\partial}{\partial t}Im\chi\,. \quad (46)$$

Of course, similar equations can be obtained in the massive case $m\neq 0$, i.e., within the Proca-like theory. Similar to the spin-1/2 case we have in the spin-1 case:

$$(EI^{(3)} - \mathbf{S}\cdot\mathbf{p})(EI^{(3)} + \mathbf{S}\cdot\mathbf{p})\Psi^{(3)} - \mathbf{p}(\mathbf{p}\cdot\Psi^{(3)}) = m^2\Psi^{(3)}\,. \quad (47)$$

These lead to (45-46), when $m = 0$ provided that the $\Psi^{(3)}$ is chosen as a superposition of a vector (the electric field) and an axial vector (the magnetic field). When $\chi = 0$ we recover the common-used Maxwell equations.

Otherwise, we can start with $(c = \hbar = 1)$[3]

$$\frac{\partial\mathbf{E}}{\partial t} = curl\mathbf{B}\,,\quad \frac{\partial\mathbf{B}}{\partial t} = -curl\mathbf{E}\,. \quad (48)$$

[3] The question of both explicite and implicite dependences of the fields on the time (and, hence, the "whole-partial derivative") has been studied in [23, 24].

Since the spin-1 matrices can be presented in the form: $(\mathbf{S}^i)^{jk} = -i\epsilon^{ijk}$, we have

$$\frac{\partial(\mathbf{E}+i\mathbf{B})^i}{\partial t} + (\mathbf{S}\cdot\nabla)^{ik}(\mathbf{E}+i\mathbf{B})^k = 0\,, \tag{49}$$

$$\frac{\partial(\mathbf{E}-i\mathbf{B})^i}{\partial t} - (\mathbf{S}\cdot\nabla)^{ik}(\mathbf{E}-i\mathbf{B})^k = 0\,. \tag{50}$$

Finally, on using that $\hat{\mathbf{p}} = -i\hbar\nabla$ we have

$$i\frac{\partial\phi^i}{\partial t} = (\mathbf{S}\cdot\hat{\mathbf{p}})^{ik}\phi^k\,, \quad i\frac{\partial\xi^i}{\partial t} = -(\mathbf{S}\cdot\hat{\mathbf{p}})^{ik}\xi^k\,, \tag{51}$$

where $\phi = \mathbf{E}+i\mathbf{B}$ and $\xi = \mathbf{E}-i\mathbf{B}$. We can show that these equations can also be considered as the massless limit of the Weinberg $2(2S+1)$ quantum-field equation for spin-1.

Meanwhile, we can calculate the determinants of the above equations, $Det[E \mp (\mathbf{S}\cdot\mathbf{p})] = 0$, and we can find that we have both the causal $E = \pm|\mathbf{p}|$ and acausal $E = 0$ solutions. These results will be useful in analyzing the spin-1 quantum-field theory below.

3.2. The Weinberg $2(2S+1)$ *Theory for Spin-1.*

It is based on the relativity postulates [6]. We shall be mainly interested in the free-field theory for the moment. In such a way he proceeds to form the $2(2S+1)$-component object $\Psi = column(\Phi_\sigma \quad \Xi_\sigma)$ transforming according to the Wigner rules. For a given representation the matrices $\mathbf{S}$ can be constructed. Weinberg gives the high-spin equation:

$$[\gamma^{\mu_1\mu_2\ldots\mu_{2s}}\partial_{\mu_1}\partial_{\mu_2}\ldots\partial_{\mu_{2s}} + m^{2s}]\Psi(x) = 0\,, \tag{52}$$

with the Barut-Muzinich-Williams covariantly-defined matrices [26, 27]. For the spin-1 they are:

$$\gamma_{44} = \begin{pmatrix} 0 & 1 \\ 1 & 0 \end{pmatrix}, \quad \gamma_{i4} = \gamma_{4i} = \begin{pmatrix} 0 & iS_i \\ -iS_i & 0 \end{pmatrix}, \tag{53}$$

$$\gamma^{ij} = \begin{pmatrix} 0 & \delta_{ij} - S_iS_j - S_jS_i \\ \delta_{ij} - S_iS_j - S_jS_i & 0 \end{pmatrix}. \tag{54}$$

Later Sankaranarayanan and Good considered another version of this theory [27] (see also [29]). For the $S = 1$ case they introduced the Weaver-Hammer-Good sign operator, ref. [28], $m^2 \to m^2\,(i\partial/\partial t)/E$, which led to the different parity properties of an antiparticle with respect to a *boson* particle. Next, Tucker and Hammer *et al* [30] introduced another higher-spin equations. In the spin-1 case it is:

$$[\gamma_{\mu\nu}\partial_\mu\partial_\nu + \partial_\mu\partial_\mu - 2m^2]\Psi^{(s=1)} = 0 \tag{55}$$

(Euclidean metric is now used). In fact, they added the Klein-Gordon equation to the Weinberg equation. One can add the Klein-Gordon equation with arbitrary multiple factor to the Weinberg equation. The new equation is written ($p_\mu = -i\partial/\partial x^\mu$):

$$[\gamma_{\alpha\beta}p_\alpha p_\beta + Ap_\alpha p_\alpha + Bm^2]\Psi = 0\,. \tag{56}$$

It has solutions with relativistic dispersion relations $E^2 - \mathbf{p}^2 = m^2$, $(c = \hbar = 1)$ provided that $\frac{B}{A+1} = 1$, or $\frac{B}{A-1} = 1$. This can be proven by considering the algebraic equation $Det[\gamma_{\alpha\beta}p_\alpha p_\beta + Ap_\alpha p_\alpha + Bm^2] = 0$. It is of the 12th order in p_μ. Solving it with respect to energy one obtains the above conditions. **Unlike the Maxwell equations there are NO any $E = 0$ solutions.** The solutions in the momentum representation have been explicitly presented in [29] in the standard representation of $\gamma_{\mu\nu}$ matrices. If the 6-component $v(\mathbf{p})$ are defined in such way, we inevitably would get the additional energy-sign operator, refs. [28, 27], $\epsilon = i\partial_t/E = \pm 1$ in the dynamical equation, and the different parities of the corresponding boson and antiboson, $\hat{P}u_\sigma(\mathbf{p}) = +u_\sigma(\mathbf{p})$ and $\hat{P}v_\sigma(\mathbf{p}) = -v_\sigma(\mathbf{p})$.

Journal of Physics: Conference Series **284** (2011) 012024 doi:10.1088/1742-6596/284/1/012024

4. The Construction of Quantum Field Operators.

The method for constructions of field operators has been given in [5]:[4]

$$\phi(x) = \frac{1}{(2\pi)^{3/2}} \int dk e^{ikx} \tilde{\phi}(k) \,. \tag{57}$$

From the Klein-Gordon equation we know $(k^2 - m^2)\tilde{\phi}(k) = 0$. Finally, we come to

$$\begin{aligned} \phi(x) &= \frac{1}{(2\pi)^{3/2}} \int dk\, e^{ikx} \delta(k^2 - m^2)(\theta(k_0) + \theta(-k_0))\phi(k) = \\ &= \frac{1}{(2\pi)^{3/2}} \int dk \left[e^{ikx} \delta(k^2 - m^2)\phi^+(k) + e^{-ikx} \delta(k^2 - m^2)\phi^-(k) \right], \end{aligned} \tag{58}$$

where

$$\phi^+(k) = \theta(k_0)\phi(k)\,, \text{and } \phi^-(k) = \theta(k_0)\phi(-k)\,. \tag{59}$$

In the spinor case (the $(1/2, 0) \oplus (0, 1/2)$ representation space) we have more components. We have $(\hat{k} + m)\psi(k)|_{k^2=m^2} = 0$. However, we use again

$$\begin{aligned} \psi(x) &= \frac{1}{(2\pi)^{3/2}} \int dk\, e^{ikx} \delta(k^2 - m^2)(\theta(k_0) + \theta(-k_0))\psi(k) = \\ & \frac{1}{(2\pi)^3} \int \frac{d^3\mathbf{k}}{2E} \left[e^{ikx}\theta(k_0)\psi(k) + e^{-ikx}\theta(k_0)\psi(-k) \right], \end{aligned} \tag{60}$$

where $k_0 = E = \sqrt{\mathbf{k}^2 + m^2}$ is positive in this case, $(\hat{k} + m)\psi^+(\mathbf{k}) = 0\,,\quad (-\hat{k} + m)\psi^-(\mathbf{k}) = 0$.

Please note that the momentum-space Dirac equations have solutions $k_0 = \pm\sqrt{\mathbf{k}^2 + m^2}$, both for $u-$ and $v-$ spinors. This can be checked by calculating the determinants. Usually, one chooses $k_0 = E$ in the $u-$ and in the $v-$. This is because on the classical level (better to say, on the first quantization level) the negative-energy $u-$ can be transformed in the positive-energy $v-$, and vice versa. This is not precisely so, if we go to the secondary quantization level. The introduction of creation/annihilation noncommutating operators gives us more possibilities in constructing generalized theory even on the basis of the Dirac equation.

Various-type field operators are possible in the $(1/2, 1/2)$ representation. During the calculations below we have to present $1 = \theta(k_0) + \theta(-k_0)$ (as previously) in order to get positive- and negative-frequency parts.

$$A_\mu(x) = \frac{1}{(2\pi)^3} \int d^4k\, \delta(k^2 - m^2) e^{+ik\cdot x} A_\mu(k) = \tag{61}$$

$$\begin{aligned} &= \frac{1}{(2\pi)^3} \int \frac{d^4k}{2E} [\delta(k_0 - E_k) + \delta(k_0 + E_k)][\theta(k_0) + \theta(-k_0)] e^{+ik\cdot x} A_\mu(k) \\ &= \frac{1}{(2\pi)^3} \int \frac{d^3\mathbf{k}}{2E_k} \theta(k_0)[A_\mu(k)e^{+ik\cdot x} + A_\mu(-k)e^{-ik\cdot x}] = \\ &= \frac{1}{(2\pi)^3} \sum_\lambda \int \frac{d^3\mathbf{k}}{2E_k} [\epsilon_\mu(k,\lambda)a_\lambda(k)e^{+ik\cdot x} + \epsilon_\mu(-k,\lambda)a_\lambda(-k)e^{-ik\cdot x}]. \end{aligned} \tag{62}$$

[4] In this book a bit different notation for positive- (negative-) energy solutions has been used comparing with the general accepted one [4].

In general, due to theorems for integrals and for distributions the presentation $1 = \theta(k_0) + \theta(-k_0)$ is possible because we use this in the integrand. However, remember, that we have the $k_0 = E = 0$ solution of the Maxwell equations. Moreover, it has the experimental confirmation (for instance, the stationary magnetic field $curl\mathbf{B} = 0$). Meanwhile the θ function is NOT defined in $k_0 = 0$. Do we not loose this solution in the above construction of a quantum field operator? Of course, the same arguments can be applied in the construction of the quantum field operator for $F_{\mu\nu}$.

Next, we should transform the second part to $\epsilon^*_\mu(k,\lambda)b^\dagger_\lambda(k)$ as usual. In such a way we obtain the charge-conjugate states. In the Dirac case we should assume the following relation in the field operator:

$$\sum_\lambda v_\lambda(k)b^\dagger_\lambda(k) = \sum_\lambda u_\lambda(-k)a_\lambda(-k)\,. \tag{63}$$

Hence, $\Lambda_{\mu\lambda} = -im(\sigma\cdot\mathbf{n})_{\mu\lambda}$ and $b^\dagger_\mu(k) = i(\sigma\cdot\mathbf{n})_{\mu\lambda}a_\lambda(-k)$. In the $(1,0)\oplus(0,1)$ representation we have somewhat different situation. Namely, $a_\mu(k) = [1-2(\mathbf{S}\cdot\mathbf{n})^2]_{\mu\lambda}a_\lambda(-k)$. This signifies that in order to construct the Sankaranarayanan-Good field operator (which was used by Ahluwalia, Johnson and Goldman [29], it satisfies $[\gamma_{\mu\nu}\partial_\mu\partial_\nu - \frac{(i\partial/\partial t)}{E}m^2]\Psi = 0$, we need additional postulates. We can set for the 4-vector field operator:

$$\sum_\lambda \epsilon_\mu(-k,\lambda)a_\lambda(-k) = \sum_\lambda \epsilon^*_\mu(k,\lambda)b^\dagger_\lambda(k)\,. \tag{64}$$

However, in the $(\frac{1}{2},\frac{1}{2})$ representation we can also expand (apart the equation (64)) in the different way:

$$\sum_\lambda \epsilon_\mu(-k,\lambda)a_\lambda(-k) = \sum_\lambda \epsilon_\mu(k,\lambda)a_\lambda(k)\,. \tag{65}$$

From the first definition we obtain

$$b^\dagger_\sigma(k) = \frac{E_k^2}{m^2}\begin{pmatrix} 1+\frac{\mathbf{k}^2}{E_k^2} & \sqrt{2}\frac{k_r}{E_k} & -\sqrt{2}\frac{k_l}{E_k} & -\frac{2k_3}{E_k}\\ -\sqrt{2}\frac{k_r}{E_k} & -\frac{k_r^2}{\mathbf{k}^2} & -\frac{m^2k_3^2}{E_k^2\mathbf{k}^2}+\frac{k_rk_l}{E_k^2} & \frac{\sqrt{2}k_3k_r}{\mathbf{k}^2}\\ \sqrt{2}\frac{k_l}{E_k} & -\frac{m^2k_3^2}{E_k^2\mathbf{k}^2}+\frac{k_rk_l}{E_k^2} & -\frac{k_l^2}{\mathbf{k}^2} & -\frac{\sqrt{2}k_3k_l}{\mathbf{k}^2}\\ \frac{2k_3}{E_k} & \frac{\sqrt{2}k_3k_r}{\mathbf{k}^2} & -\frac{\sqrt{2}k_3k_l}{\mathbf{k}^2} & \frac{m^2}{E_k^2}-\frac{2k_3}{\mathbf{k}^2}\end{pmatrix}\begin{pmatrix} a_{00}(-k)\\ a_{11}(-k)\\ a_{1-1}(-k)\\ a_{10}(-k)\end{pmatrix}.$$

From the second definition $\Lambda^2_{\sigma\lambda} = \mp\sum_{\nu\mu}\epsilon^*_\nu(k,\sigma)[\gamma_{44}]_{\nu\mu}\epsilon_\mu(-k,\lambda)$ we have:

$$a_\sigma(k) = \begin{pmatrix} -1 & 0 & 0 & 0\\ 0 & \frac{k_3^2}{\mathbf{k}^2} & \frac{k_l^2}{\mathbf{k}^2} & \frac{\sqrt{2}k_3k_l}{\mathbf{k}^2}\\ 0 & \frac{k_r^2}{\mathbf{k}^2} & \frac{k_3^2}{\mathbf{k}^2} & -\frac{\sqrt{2}k_3k_r}{\mathbf{k}^2}\\ 0 & \frac{\sqrt{2}k_3k_r}{\mathbf{k}^2} & -\frac{\sqrt{2}k_3k_l}{\mathbf{k}^2} & 1-\frac{2k_3^2}{\mathbf{k}^2}\end{pmatrix}\begin{pmatrix} a_{00}(-k)\\ a_{11}(-k)\\ a_{1-1}(-k)\\ a_{10}(-k)\end{pmatrix}. \tag{66}$$

It is the strange case: the field operator will only destroy particles (like in the $(1,0)\oplus(0,1)$ case). Possibly, we should think about modifications of the Fock space in this case, or introduce several field operators for the $(\frac{1}{2},\frac{1}{2})$ representation. **However, other way is possible: to construct the left- and right- parts of the $(1,0)\oplus(0,1)$ field operator separately each other. In this case the commutation relations may be more complicated.**

Finally, going back to the rest $(S,0)\oplus(0,S)$ objects. **Bogoliubov *et al.* constructs them introducing the products with δ functions like $\delta(k_0 - m)$. Then, he makes the boost of the "spinors" only, and changes by hand the above δ to $\delta(k^2 - m^2)$ (where we**

already have $k_0 = E = \sqrt{\mathbf{k}^2 + m^2}$). Is it possible to make the boost of the δ functions consistently in such a way?

The conclusion is: we still have few questions unsolved in the bases of the quantum field theory, which open a room for generalized theories.

Acknowledgments

This paper is based on the invited talks given at the VIII International Workshop "Applied Category Theory. Graph-Operad-Logic", San Blas, Nayarit, México, January 9-16, 2010, at the 6th International Conference on the Dark Side of the Universe (DSU2010), Leon, Gto, México, June 1-6, 2010, and at the XXVIII ICGTMP, Newcastle, UK, July 25-30, 2010. I am grateful to participants of recent conferences for discussions.

References

[1] P. A. M. Dirac, Proc. Roy. Soc. A**114** (1927) 243; ibid. A**117** (1928) 610.
[2] J. J. Sakurai, *Advanced Quantum Mechanics.* (Addison-Wesley, 1967).
[3] L. H. Ryder, *Quantum Field Theory.* (Cambridge University Press, Cambridge, 1985).
[4] C. Itzykson and J.-B. Zuber, *Quantum Field Theory.* (McGraw-Hill Book Co., 1980), p. 156.
[5] N. N. Bogoliubov and D. V. Shirkov, *Introduction to the Theory of Quantized Fields.* 3rd Edition. (Nauka, Moscow, 1973), p.28, p, 55.
[6] E. P. Wigner, Ann. Math. **40** (1939) 149; S. Weinberg, Phys. Rev. **133** (1964) B1318.
[7] E. Majorana, Nuovo Cimento **14** (1937) 171.
[8] S. M. Bilenky and B. M. Pontekorvo, Phys. Repts **42** (1978) 224.
[9] A. Barut and G. Ziino, Mod. Phys. Lett. A**8** (1993) 1099; G. Ziino, Int. J. Mod. Phys. A**11** (1996) 2081.
[10] D. V. Ahluwalia, Int. J. Mod. Phys. A**11** (1996) 1855.
[11] P. Lounesto, *Clifford Algebras and Spinors.* (Cambridge University Press, 2002), Ch. 11 and 12; R. da Rocha and W. Rodrigues, Jr., *Where are Elko Spinor Fields in Lounesto Spinor Field Classification?* Preprint math-ph/0506075.
[12] M. Kirchbach, C. Compean and L. Noriega, *Beta Decay with Momentum-Space Majorana Spinors.* Eur. Phys. J. A**22** (2004) 149.
[13] V. V. Dvoeglazov, Int. J. Theor. Phys. **34** (1995) 2467; Nuovo Cim. **108**A (1995) 1467, Hadronic J. **20** (1997) 435; Acta Phys. Polon. B**29** (1998) 619.
[14] V. V. Dvoeglazov, Mod. Phys. Lett. A**12** (1997) 2741.
[15] M. Markov, ZhETF **7** (1937), 579, 603; Nucl. Phys. **55** (1964) 130.
[16] I. M. Gelfand and M. L. Tsetlin, ZhETF **31** (1956) 1107; G. A. Sokolik, ZhETF **33** (1957) 1515.
[17] V. V. Dvoeglazov, Int. J. Theor. Phys. **43** (2004) 1287.
[18] A. O. Barut, Phys. Lett. **73**B (1978) 310; Phys. Rev. Lett. **42** (1979) 1251.
[19] B. Nigam and L. L. Foldy, Phys. Rev. **102** (1956) 1410.
[20] V. B. Berestetskii, E. M. Lifshitz and L. P. Pitaevskii, *Quantum Electrodynamics.The IV Volume of the Landau Course of Theoretical Physics.* 2nd Edition (Butterworth-Heinemann, 1982).
[21] A. Gersten, Found. Phys. Lett. **12** (1999) 291.
[22] V. V. Dvoeglazov, J. Phys. A**33** (2000) 5011; Rev. Mex. Fis. Supl. **49**-S1 (2003) 99.
[23] K. R. Brownstein, Am. J. Phys. **67** (1999) 639.
[24] V. V. Dvoeglazov, Proc. the ICSSUR 2003 Conference, Puebla, México, June 2003 (Rinton Press, 2003), pp. 125-130, math-ph/0308017.
[25] V. V. Dvoeglazov, Fizika B**6** (1997) 111.
[26] A. Barut, I. Muzinich and D. N. Williams, Phys. Rev. **130** (1963) 442.
[27] A. Sankaranarayanan and Good, jr, Nuovo Cimento **36** (1965) 1303.
[28] D. L.Weaver, C. L. Hammer and R. H. Good, jr., Phys. Rev. B**135** (1964) 241.
[29] D. V. Ahluwalia, M. B. Johnson and T. Goldman, Phys. Lett. B**316** (1993) 102.
[30] R. H. Tucker and C. L. Hammer, Phys. Rev. D**3** (1971) 2448.

"Lost chains" in algebraic models

How the complete classification of subalgebras can lead to new unexpected physics

L. Fortunato[1] and W.A. de Graaf[2]

[1] European Centre for Theoretical Studies in Nuclear Physics and Related Areas, Strada delle Tabarelle 286, I-38123 Villazzano (TN), Italy

[2] Dip. Matematica, Università di Trento, via Sommarive 24 ,I-38123 Povo (Trento), Italy

E-mail: fortunat@pd.infn.it

Abstract. The algebraic structure of some of the simplest algebraic models $u(2), u(3)$ and $u(4)$, widely used in several branches of physics either as toy models or as working instruments, are reanalyzed under a new perspective that releases the requirement that chains should terminate or pass through the angular momentum algebra. Unitary algebras are non-semisimple, therefore we first apply the Levi-Malcev decomposition. Then we use the theory of weighted Dynkin diagrams to identify conjugacy classes of $A_1 \sim su(2) \sim so(3)$ subalgebras: a complete classification of new angular momentum non conserving (AMNC) dynamical symmetries follows that we substantiate with examples.

1. Introduction

Unitary algebras arising in bosonic models of quantum mechanics play a fundamental role in several branches of physics such as nuclear, atomic, molecular and particle physics[1, 2, 3, 4, 5]. They are not semisimple, therefore the very first step is to use the well-known Levi-Malcev decomposition to extract the special unitary subgroup, that instead possesses this property [6]. To this algebra we apply the theory of weighted Dynkin diagrams [7, 8, 9] to classify the different conjugacy classes of A_1-type subalgebra (that is Cartan's notation for the isomorphic algebras $sl(2) \sim su(2) \sim so(3)$) in terms of weighted Dynkin diagrams (WDD). Within any Lie algebra one can identify $sl(2)$-triples, that is a set of three elements that formally satisfy the commutation relations of $sl(2)$. There are infinitely many such triples, but there is only a finite number of distinct nilpotent orbits. Nilpotent orbits are in one-to-one correspondence with certain distinguished semisimple orbits through the Jacobson-Morozov theorem, a theorem of Kostant and another of Mal'cev [7]. These orbits in turn can be labeled by their WDD. This yields a complete classification of A_1-type subalgebras that brings in the possibility of additional chains that do not conserve the (usual) angular momentum, but rather have some sort of angular momentum of their own. A somewhat unexpected discovery of spinor operators, occurring in the tensor analysis of the whole algebra with respect to this new angular momentum, has generated a useful discussion at the conference, that helped us in understanding it more deeply. All the algebras we deal with have also been implemented in the GAP4 programming language [10], that allows symbolic manipulations of algebraic objects with the help of a package [11] that has been especially designed to provide relevant computational algorithms. We have summarized in Fig.1 the total number of subalgebras and the number of distinct conjugacy classes that are found for each A_n with $n = 1, \cdots, 7$.

Journal of Physics: Conference Series **284** (2011) 012025 doi:10.1088/1742-6596/284/1/012025

Type	# classes of subalgebras (A_n, B_n, C_n, D_n, etc.)	# nilpotent orbits # classes of A_1 subalg . # W.D.D .
A_1	1	1
A_2	2	2
A_3 - Vibron	8	4
A_4	14	6
A_5 - IBM	36	10
A_6	60	14
A_7	131	21

Figure 1. Number of classes of subalgebras and of nilpotent orbits, obtained using SLA (in GAP) from the recent classification of de Graaf [9]. Two examples that are used in algebraic models of molecular and nuclear physics are highlighted. See also next figure.

The dynamics of several physical system can be described in second quantization as being based on boson creation and annihilation operators $b_\alpha, b^\dagger_\alpha$ that are treated as elementary building blocks. The index specifies the number and any additional quantum number that can be useful to classify them. In particular they might carry an intrinsic angular momentum ℓ that has $2\ell+1$ components. They satisfy bosonic commutation rules of the form:

$$[b_\alpha, b^\dagger_{\alpha'}] = \delta_{\alpha,\alpha'} \tag{1}$$

$$[b_\alpha, b_{\alpha'}] = [b^\dagger_\alpha, b^\dagger_{\alpha'}] = 0 \tag{2}$$

The unitary algebra $u(n)$ is obtained by taking all the n^2 bilinear operators of the form $b^\dagger_\alpha b_{\alpha'}$.

Typical examples and very successful models are, for instance, the Interacting Boson Model for nuclei and the Vibron model for molecules, both of which have been investigated in great detail, mainly by F.Iachello and collaborators. A common feature of the majority of these models is that, once one has identified the relevant physical ingredients, that can be associated with boson creation and annihilation operators, one has a (larger) unitary algebra. Then one searches for all chains of subalgebras that end with the $so(3)$ angular momentum algebra, to ensure that the states, that are labeled by a suitable number of eigenvalues of Casimir operators of the chain, conserve angular momentum. This is certainly a very natural way to proceed, but the algebra is, mathematically speaking, richer than that. This richness can be appreciated in Fig. 2, where the A_3 and A_5 algebras (red points) and their subalgebras are plotted in a circular graph. The green points are A_1 subalgebras pertaining to different conjugacy classes. The number of chains of subalgebras that do not pass through the angular momentum algebra is much higher: although their existence has been known since the early days of algebraic models, they have not been much investigated in the past in connection with physical problems. The graph that corresponds to the IBM, albeit of huge importance in physics, is quite complicated, therefore we will restrict our discussion to some low rank cases, among which A_3 that corresponds to the vibron model. Our aim is to try and understand if and how the additional chains can be put in correspondence with the basic operators that define the algebra. In addition we will investigate their tensorial properties and suggest applications.

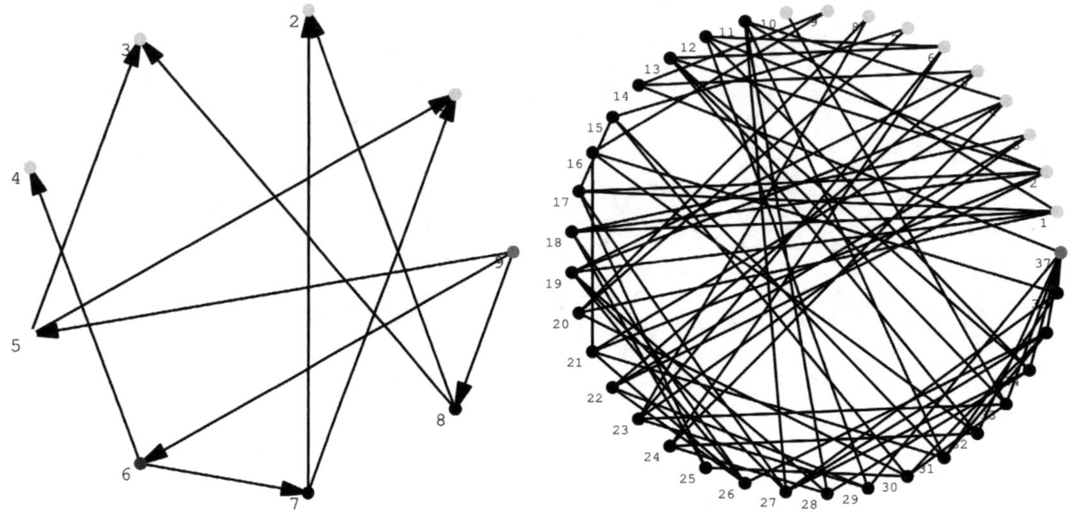

Figure 2. Circular graphs corresponding to the inclusions of subalgebras in A_3 (9 vertices, 11 edges) and A_5 (37 vertices, 79 edges) symbolized with red points. Green points are A_1 subalgebras. Black, yellow and blue in the left part indicate A_1A_1, A_2 and B_2 subalgebras respectively. Labels in the left part correspond to the ones in Fig. 5.

2. $u(2)$

The two dimensional harmonic oscillator problem arises in simple models of quantum mechanics [4, 12] and it is connected to the $u(2)$ Lie algebra. Two species of scalar bosons, called s and t can be invoked, that obey standard bosonic commutation rules of the form $[s, s^\dagger] = [t, t^\dagger] = 1$ and $[s, t] = [s, t^\dagger] = [s^\dagger, t] = [s^\dagger, t^\dagger] = 0$. The Lie algebra $u(2)$ is built with bilinear operators made up of these operators and it amounts to four elements:

$$\mathfrak{g}: \quad g_1 = s^\dagger s, \quad g_2 = s^\dagger t, \quad g_3 = t^\dagger s, \quad g_4 = t^\dagger t\,. \tag{3}$$

$\mathfrak{g}$ is not semisimple. Thanks to the Levi-Malcev decomposition it may be rewritten as the direct sum of a 1-dimensional subalgebra, the radical or maximal solvable ideal, and a 3-dimensional semisimple Levi subalgebra, as follows $u(2) \simeq \quad \mathfrak{r} \equiv u(1) \quad \oplus \quad \mathfrak{s} \equiv su(2)$, with $\mathfrak{r} = \langle g_1 + g_4 \rangle$ and $\mathfrak{s} = \langle g_2, g_3, g_1 - g_4 \rangle$. Notice that $\mathfrak{s}$ has a basis consisting of elements of the form given in Eq. (1.20) of Ref. [4], as the standard generators of $su(2)$ angular momentum algebra, namely $\hat{J}_x = (s^\dagger s - t^\dagger t)/2, \hat{J}_y = (t^\dagger s + s^\dagger t)/2, \hat{J}_z = i(t^\dagger s - s^\dagger t)/2$. Clearly the basis element of the radical is the total number of bosons operator, $\hat{N} = s^\dagger s + t^\dagger t$, that is known to commute with all the operators, thereby forming the center of the Lie algebra. Although in the present case it is almost trivial, the semisimple part is amenable to treatment with the theory of weighted Dynkin diagrams [7, 8]. It corresponds to A_1 in Cartan notation and admits only one nilpotent orbit, labeled by the weighted Dynkin diagram [2] (notice that WDD's are indicated in square brackets, not to be confused with references). Therefore there exists only one conjugacy class.

Two dynamical symmetries are present (see for example Ref. [2, 4, 12]), namely:

$$I) \quad \begin{array}{ccc} u(2) & \supset & u(1) \\ | & & | \\ {[N]} & & n_t \end{array} \qquad\qquad II) \quad \begin{array}{ccccc} u(2) & \supset & su(2) & \supset & so(2) \\ | & & | & & | \\ {[N]} & & j & & \mu \end{array}$$

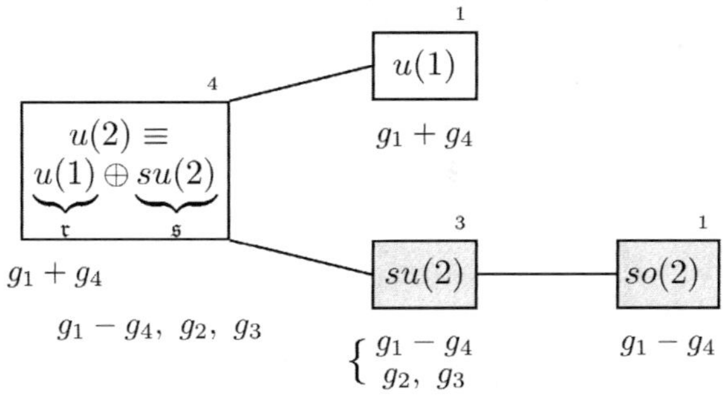

Figure 3. Classification of the $u(2)$ subalgebras. Semisimple algebras are indicated in yellow. Elements, according to the definitions in the text, are indicated below each frame and the order (dimension) is given as a small number in the upper-right corner.

Each one of them provides a basis that can be chosen for diagonalization. These two chains have a fundamental difference: one goes through the radical, the other through the semisimple Levi subalgebra. The eigenvalues of two invariants are needed to label the basis states and while they are fixed in the first chain $|\,[N]n_t\rangle$, there is more freedom in the second as they can be chosen either as $|\,[N]j\rangle$ or $|\,j\mu\rangle$. The second form exhibits in a natural way the splitting of eigenstates with different third component of angular momentum and it has prevailed in the specialized literature. The two subalgebras $u(1)$ and $so(2)$ are isomorphic, but when one is dealing with a complete classification of all subalgebras, some care must be taken. Although it is clear that the linear map that changes the sign of g_4 and leaves the rest unchanged, is actually swapping $u(1)$ and $so(2)$, *it does not send the whole reduction scheme of Fig. (3) into itself.* It can be easily proven that there is no Lie algebra homomorphism $\mathfrak{g} \to \mathfrak{g}$ mapping $u(1)$ into $so(2)$.

3. $u(3)$

The algebra $u(3)$ is built upon bilinear combinations of three boson operators. These can be chosen either i) as three scalar bosons, that might be physically interpreted as appropriate combinations of the three Cartesian coordinates and momenta in a three dimensional harmonic oscillator, or ii) as the three components of a vector boson or finally iii) as a scalar boson plus two, so-called, circular bosons [3, 13]. Each construction has practical applications, especially to the study of molecular spectra. We study here the construction ii) made up in terms of a p ($\ell = 1$) boson, because it is necessary for the construction of the Vibron model [2, 4, 5]. The operators $p^\dagger_\mu$ and $\tilde{p}_\mu$ with $\mu = -1, 0, 1$ transform as the $\ell = 1$ representation of the rotation group. The nine bilinear operators built from tensor couplings close into the $u(3)$ Lie algebra. One can apply the same arguments as in the preceding section, namely the Levi-Malcev decomposition and the classification of A_1 subalgebras in terms of weighted Dynkin diagrams, to show that the lattice of subalgebras of $u(3)$ takes the form displayed in Fig. (4). The elements of the algebra can be written as:

$$\begin{array}{lllll} g_1 = [p^\dagger \times \tilde{p}]^{(0)}_0 & g_2 = [p^\dagger \times \tilde{p}]^{(1)}_{-1} & g_3 = [p^\dagger \times \tilde{p}]^{(1)}_0 & g_4 = [p^\dagger \times \tilde{p}]^{(1)}_1 & \\ g_5 = [p^\dagger \times \tilde{p}]^{(2)}_{-2} & g_6 = [p^\dagger \times \tilde{p}]^{(2)}_{-1} & g_7 = [p^\dagger \times \tilde{p}]^{(2)}_0 & g_8 = [p^\dagger \times \tilde{p}]^{(2)}_1 & g_9 = [p^\dagger \times \tilde{p}]^{(2)}_2 \end{array} \tag{4}$$

where one can identify a scalar $\hat{N} = \sqrt{3}g_1$, a vector $\hat{L}_\kappa = \sqrt{2}[p^\dagger \times \tilde{p}]^{(1)}_\kappa$ (3 components) and a quadrupole tensor $\hat{Q}_\kappa = [p^\dagger \times \tilde{p}]^{(2)}_\kappa$ (rank 2, 5 components) to be further discussed below [3]. The Levi-Malcev decomposition is clear from the division into white and yellow blocks in Fig. (4) and the scalar operator g_1 is actually always responsible for the radical part of the classification. From Figs. (3) and (4) one can see that the whole scheme can be divided in

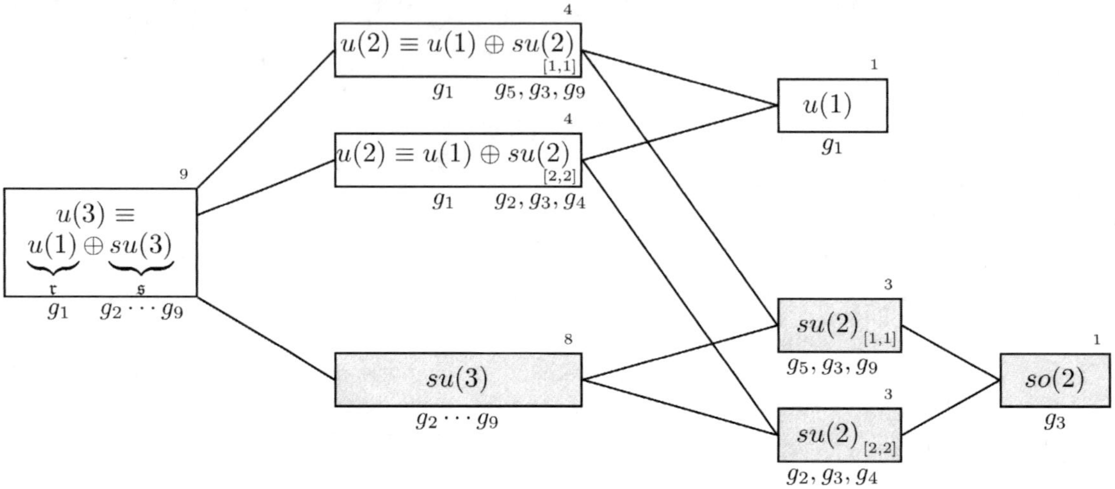

Figure 4. Classification of the $u(3)$ subalgebras. Semisimple algebras are indicated in yellow. Elements, according to the definitions in the text, are indicated below each frame and the order (dimension) is given as a small number above the upper-right corner. Weighted Dynkin diagrams are indicated in the lower-right corner: in the case of non-semisimple algebras they are referred only to the semisimple part. A total of 6 chains can be identified.

two parallel sheets, a *semisimple sheet*, containing all the semisimple Lie algebras, and a *non-semisimple sheet*, containing an exact copy of the structure of the lower one, where each algebra is multiplied by the radical. Of course $u(n)$ corresponds to $su(n)$ and so on, each corresponding pair is also connected by an inclusion relation from top to bottom with the exception of one dimensional subalgebras. The semisimple part starts at $su(3)$, called A_2 in Cartan's notation, that has two different types of A_1 subalgebras labeled by different weighted Dynkin diagrams, $[1,1]$ and $[2,2]$ respectively. Each of them is an entire conjugacy class of triplets of operators with the same WDD of which we choose just one representative (usually the simplest available or the one that has already been incorporated into an established model). The second one is the usual algebra of angular momentum, whose components are the components of the rank 1 tensor in Eq. (4). Indeed the three operators $\hat{L}_\kappa = \sqrt{2}[p^\dagger \times \tilde{p}]^{(1)}_\kappa$ satisfy the angular momentum algebra. This algebra and all the chains passing through it have been described by Iachello and the physical reasoning underlying this rightful choice is that the quantum mechanical description of our system and the basis states associated to the chain must conserve angular momentum.

It turns out, however, and this fact was unknown or mostly unnoticed till now, that the A_1 algebra labeled by $[1,1]$ is made up of three objects that are not components of a vector, but that, nevertheless, have commutation relations that formally identify them as an angular momentum algebra

$$[\hat{W}_+, \hat{W}_-] = 2W_0 \qquad [\hat{W}_0, \hat{W}_\pm] = \pm\hat{W}_\pm \tag{5}$$

where the operators are defined as $\{\hat{W}_- = g_5, W_0 = g_3/\sqrt{2}, W_+ = g_9\}$. In usual terminology they are the $\hat{Q}_{\pm 2}$ components of the quadrupole tensor and the $\hat{L}_0$ component of the angular momentum. Historically only two other theoretical works have introduced something of this sort: Chen and Arima [14] discuss the origin of cylindrical bosons within the Interacting Boson Model, where they introduce the Δ spin that is built upon the highest and lowest components of the quadrupole tensor plus the zero component of the angular momentum, the difference being that their operators are made up of s and d bosons, while ours are made up of p bosons [1]; Elliott discusses, in the fundamental Ref. [15], a similar algebraic structure in the context of

[1] I'd like to thank P.van Isacker for bringing Ref.[14] to my attention.

the collective motion in the nuclear shell model.

Now the crucial point is that once we have an angular momentum algebra, say $\hat{J}$, (to be replaced either by $\hat{L}$ or by $\hat{W}$) we can define spherical tensors with respect to *that* algebra by means of the two formulas [4]:

$$[J_z, T^k_q] \;=\; q\, T^k_q \qquad [J_\pm, T^k_q] \;=\; \sqrt{(k \pm q + 1)(k \mp q)}\, T^k_{q\pm 1}\,. \tag{6}$$

We will use the name of J-tensors to refer to tensors with respect to a particular J-set. The whole $u(3)$ algebra is made up by an L-scalar, an L-vector and and L-tensor of rank 2, as outlined above after Eq. (4), with respect to the L-set. In this L-set, both $p^\dagger$ and $\tilde{p}$ transform as L-vectors. The analysis of the elements of $u(3)$ with respect to the W-set reveals instead that, together with the obvious W-vector given by $\hat{W}$ itself, we have two W-scalars and two W-spinors: this might come as a surprise since we are not expecting spin 1/2 objects to arise in a purely bosonic model. The proof is obtained through the following steps: g_1 and g_7 commute with all components of $\hat{W}$: they are therefore scalars; the objects $sp_1 = \{g_6, g_4\}$ and $sp_2 = \{-g_8, -g_2\}$, with components in the $\{-,+\}$ order, have the following commutation relation with $\hat{W}_0$:

$$[\hat{W}_0, sp_i^\pm] = \pm\frac{1}{2} sp_i^\pm \tag{7}$$

with $i = 1, 2$ and therefore they are the $-1/2$ and $+1/2$ components of a spinor. Finally the commutator with $\hat{W}_+$ and $\hat{W}_-$ terminate after two steps:

$$\begin{aligned} [\hat{W}_+, sp_i^-] &= sp_i^+ \qquad [\hat{W}_+, sp_i^+] = 0 \\ [\hat{W}_-, sp_i^+] &= sp_i^- \qquad [\hat{W}_-, sp_i^-] = 0 \end{aligned} \tag{8}$$

with $i = 1, 2$. The coefficients are compatible with $k = 1/2$ in Eq. (6), therefore we have found two spin$-1/2$ spinors. We propose the name of Chen-Arima spinors for spinors arising in bosonic models, because, to our knowledge, these authors have found the first case. Notice that one can always form other scalars from tensor coupling of Chen-Arima spinors obtaining:

$$[sp_1 \times sp_1]_0^{(W=0)} \;=\; -[sp_2 \times sp_2]_0^{(W=0)} \;=\; -\frac{\sqrt{3}}{2} g_7 \tag{9}$$

that is a W-scalar consistent with the observation made above.

It is interesting to note that the vectors $p^\dagger$ and $\tilde{p}$ don't satisfy anymore good tensorial properties with respect to the W-set (in other words they do not transform as vectors). Their role is replaced by other two objects $V = \{\sqrt{2}p^\dagger_{-1}p^\dagger_{-1}, 2p^\dagger_{-1}p^\dagger_1, \sqrt{2}p^\dagger_1 p^\dagger_1\}$ and $U = \{\sqrt{2}\tilde{p}_{-1}\tilde{p}_{-1}, 2\tilde{p}_{-1}\tilde{p}_1, \sqrt{2}\tilde{p}_1\tilde{p}_1\}$ (with components ordered as $-1, 0, 1$) that are found to be good W-vectors by following the same argument outlined above. Their action creates or annihilates two bosons at the same time and this is consistent with the definitions of the components of the W-spin that are based on the highest and lowest ($\mu = \pm 2$) components of the quadrupole tensor.

4. $u(4)$

This is the case of the Vibron model [2, 3, 4, 5], that is built upon s and p bosons. Together with the nine generators in Eq. (4) one needs another seven generators containing s to close the algebra, namely:

$$\begin{aligned} &g_{10} = [s^\dagger \times \tilde{p}]^{(1)}_{-1} \quad g_{11} = [s^\dagger \times \tilde{p}]^{(1)}_0 \quad g_{12} = [s^\dagger \times \tilde{p}]^{(1)}_1 \\ &g_{13} = [p^\dagger \times \tilde{s}]^{(1)}_{-1} \quad g_{14} = [p^\dagger \times \tilde{s}]^{(1)}_0 \quad g_{15} = [p^\dagger \times \tilde{s}]^{(1)}_1 \quad g_{16} = [s^\dagger \times \tilde{s}]^{(0)}_0 \end{aligned} \tag{10}$$

Journal of Physics: Conference Series **284** (2011) 012025 doi:10.1088/1742-6596/284/1/012025

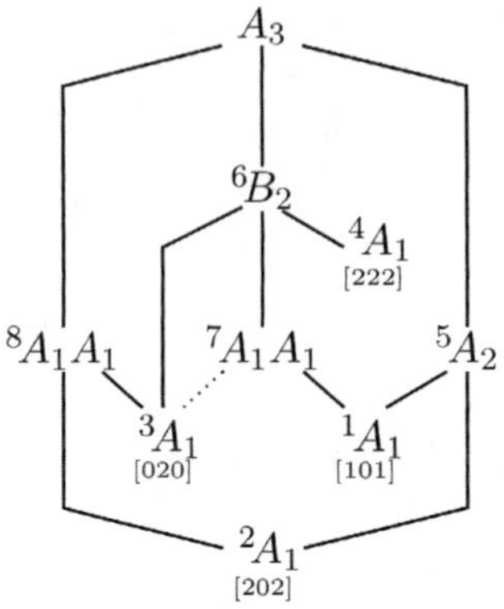

Figure 5. Classification of Lie subalgebras of A_3. Subalgebras carry an arbitrary upper left index. Three-dimensional subalgebras of type A_1 represent entire conjugacy classes. The corresponding WDD is given in square brackets. In theory all inclusions are possible, but in practice, in our particular realization, the dotted one is not possible simultaneously with all the others, although the direct inclusion $^6B_2 \supset {}^3A_1$ is still valid.

The radical of the $u(4)$ algebra is given by the total number of boson operator, $\hat{N} = \sqrt{3}g_1 + g_{16}$. The semisimple sheet consists (see Fig.5) of the seven chains that originate from A_3 and end up in one of the four possible A_1, the conjugacy classes of which are labeled by the WDD: [101], [202], [020] and [222]. In the Vibron model only two such chains, the ones passing through the standard angular momentum subalgebra [202], have been studied so far.

The standard well-known chains of the Vibron model are the most external paths of Fig. 5 connecting A_3 and $A_1[202]$ or, for the sake of clarity:

$$su(4) \sim A_3 \left\{ \begin{array}{ll} su(3) \sim & {}^5A_2 \\ & \\ so(4) \sim & {}^8A_1A_1 \end{array} \right\} so(3) \sim {}^2A_1 \tag{11}$$

that correspond to the nonrigid and rigid rotovibrator limits of the Vibron model. In the complete classification four classes of A_1 subalgebras are present and one has therefore four different "angular momenta" that can be used as J-sets to define spherical tensors. In Table 4 the usual tensor analysis of the algebra of the Vibron model with respect to the $^2A_1$ algebra with WDD [202] is given for reference. The W-angular momentum described above for $u(3)$ forms instead the $^1A_1$ algebra with WDD [101]. With respect to this W-set, the whole algebra $su(4)$ amounts to the W-vector $\hat{W}$, to four W-scalars and four W-spinors (cfr. Table 2). The scalars built with tensor couplings of $\hat{s}p_3$ with itself and $\hat{s}p_4$ with itself are identically zero, therefore one can define two new W-spinors $\hat{s}p_3' = \{-g_2 - g_{10} - g_{13}, g_8 - g_{12} + g_{15}\}$ and $\hat{s}p_4' = \{g_6 + g_{10} - g_{13}, g_4 + g_{12} + g_{15}\}$ as linear combinations. Their couplings give good W-scalars and they are also needed as elements of B_2.

We summarize in Table 2 the tensor analysis of A_3 with respect to $^1A_1[101]$ and we give in Tables 3 and 4 the tensor analysis of the whole algebra based on $^3A_1[020]$ and $^4A_1[222]$ respectively. A_3 is made up of 5 vectors with respect in the former case (among which the defining vector Y), while it is made up of a vector (the defining vector $T^{(1)}$), a quadrupole and an octupole tensor in the latter case.

One can write AMNC hamiltonians with the dynamical symmetry based for example on the

Table 1. Definitions and tensorial character of the operators forming the semisimple A_3 algebra with respect to the (standard) set forming the ${}^2A_1$ algebra with [202] WDD. The third column features alternative definitions for algebra elements that correspond to the second column with the following ordering $\mu = -k, .., 0, .., k$, where k is the rank.

operator	def^{rank}	alt.def.	components
$\hat{n}'$	$[s^\dagger \times \tilde{s}]^0_0 - \sqrt{\frac{1}{3}}[p^\dagger \times \tilde{p}]^0_0$	g_1	1
$\hat{L}_\mu/\sqrt{2}$	$[p^\dagger \times \tilde{p}]^1_\mu$	g_2, g_3, g_4	3
$\hat{Q}_\mu$	$[p^\dagger \times \tilde{p}]^2_\mu$	$g_5, \cdots, g_9$	5
$\hat{D}_\mu$	$i[p^\dagger \times \tilde{s} + s^\dagger \times \tilde{p}]^1_\mu$	g_{10}, g_{11}, g_{12}	3
$\hat{D}'_\mu$	$[p^\dagger \times \tilde{s} - s^\dagger \times \tilde{p}]^1_\mu$	g_{13}, g_{14}, g_{15}	3

Table 2. Definitions and tensorial character of the operators forming the A_3 algebra with respect to the non-standard set forming the ${}^1A_1$ algebra with [101] WDD (that can be obtained from the vector in the first line).

operator	definition	components
$\hat{s}_1, \cdots, \hat{s}_4$	g_1, g_7, g_{11}, g_{14}	1 (each)
$\hat{W}_\mu$	$-g_5/\sqrt{2}, g_3/\sqrt{2}, g_9/\sqrt{2}$	3
$\hat{sp}_1, \cdots, \hat{sp}_4$	$\{g_2, g_8\}, \{g_6, g_4\}, \{g_{10}, g_{12}\}, \{g_{13}, g_{15}\}$	2 (each)

Table 3. Definitions and tensorial character of the operators forming the A_3 algebra with respect to the non-standard set forming the ${}^3A_1$ algebra with [020] WDD (that can be obtained from the vector in the first line).

operator	definition	components
$\hat{Y}_\mu$	$i(g_2+g_6)/\sqrt{2}+g_{10}, (g_1+\sqrt{2}g_3+\sqrt{2/3}g_7+2ig_{11})/2, i\sqrt{2}g_4+g_{12}+g_{15}$	3
$\hat{B}_\mu$	$2\sqrt{2}g_9, \sqrt{2}i(g_4-g_8)-2g_{12}, 2i\sqrt{2}g_{11}$	3
$\hat{M}_\mu$	$-g_1-\sqrt{2/3}g_7+i(g_{14}-g_{11}), ig_6+(g_{13}-g_{10})/\sqrt{2}, g_5$	3
$\hat{R}_\mu$	$-i(g_4+g_8)/\sqrt{2}+g_{12}, -g_3/2-\sqrt{3/8}g_7-ig_{11}, i(g_2+g_6)/\sqrt{2}$	3
$\hat{S}_\mu$	$-i(g_4/g_8)/\sqrt{2}+g_{15}, -g_1/\sqrt{2}-g_3/2+\sqrt{1/12}g_7-ig_{11}, -g_{10}$	3

chain $A_3 \supset B_2 \supset {}^7A_1A_1 \supset {}^1A_1$ or

$$\begin{array}{ccccccc} su(4)\sim so(6) & \supset & so(5) & \supset & so(4) & \supset & so(3) \\ | & & | & & | & & | \\ N & & t & & u & & w \end{array} \tag{12}$$

where the labels in the last row are connected with the eigenvalues of the quadratic Casimir operators. The resulting energy formula for symmetric representations $E = \alpha N(N+4) + \beta t(t+3) + \gamma u(u+2) + \delta w(w+1)$ with $t = 0, \cdots, N$, $u = 0, \cdots, t$ and $w = 0, \cdots, u$, most probably is not a good choice for diatomic molecular spectra, but one can anyway use the basis states $| N, t, u, w\rangle$, that is actually a Gelfand-Tsetlin pattern for orthogonal algebras, to diagonalize hamiltonians based on the full spectrum generating algebra, with the *proviso* that, while N is the total boson number as in the Vibron model, the labels v, u and w *do not conserve angular momentum, but rather conserve* w. Following the established path, one would have not used

Table 4. Definitions and tensorial character of the operators forming the A_3 algebra with respect to the non-standard set forming the ${}^4A_1$ algebra with [222] WDD (that can be obtained from the vector in the first line).

oper.	definition	comp.
$\hat{T}^{(1)}_\mu$	$(g_2 - 7g_4 - 7g_6 - g_8 + 3g_{13})/2, -\frac{3}{2}g_1 - \sqrt{2}g_3 - 2g_5, g_2 + g_{10} + g_{12}$	3
$\hat{T}^{(2)}_\mu$	$-6\sqrt{2}(2g_3 + \sqrt{2}g_5 - \sqrt{2}g_9 + g_{14}), -3(g_2 + g_4 + g_6 - g_8 + g_{13}),$ $\sqrt{3/2}g_1 - \sqrt{3}g_3 - \sqrt{6}g_5 + g_7, (g_6 - g_2)/2 + g_{10} + g_{12}, (g_5 + \sqrt{2}g_{11})/4$	5
$\hat{T}^{(3)}_\mu$	$\frac{-72}{\sqrt{30}}(g_{13} - g_{15}), \frac{-12}{\sqrt{10}}(2g_2 + \sqrt{2}g_5 - \sqrt{2}g_9 - g_{14}), \frac{-6}{5\sqrt{2}}(3g_2 - g_4 - g_6 - 3g_8 - g_{13}),$ $(-\sqrt{6}g_1 + 2(\sqrt{3}g_3 + \sqrt{6}g_5 + 5g_7))/10, (g_2 + 5g_6 - 4g_{10} - 4g_{12})/(10\sqrt{2}),$ $(g_5/\sqrt{2} - g_{11})/\sqrt{40}, -g_{10}/\sqrt{120}$	7

this basis for the Vibron model, due to the difficulty in giving a precise physical meaning to the labels. One advantage is the ease of writing the branching rules due to the absence of missing labels. Several other AMNC dynamical symmetries, one for every possible path in Fig. 5 can be invoked with the same spirit: some of them would provide *at least* new and simpler diagonalization schemes and possibly applications to quantum many-body systems.

5. Conclusions

We have shown that the use of i) Levi-Malcev decomposition and ii) theory of weighted Dynkin diagrams allow a thorough classification of algebraic models arising in bosonic models of quantum mechanics providing i) a way of separating out semisimple from non-semisimple subalgebras and ii) a neat classification of all possible conjugacy classes of three dimensional subalgebras (A_1) respectively. Well-known algebraic models, either used as conventional or pedagogical toy models or actually applied to real systems, usually adopt subalgebra chains that end up in the standard angular momentum algebra. We have shown that one can "fill the gaps", i.e. write basis elements for the whole reduction scheme. In particular the elements of any additional A_1, having different WDD, can be used to define new angular momenta operators, with respect to which one can redefine tensors and give to the whole algebra elements a different tensorial character. With respect to one of these new angular momenta, it is found that other operators behave like spin-1/2 objects, a fact that was most surprising at first. This was hinted at by the old works of Elliott [15] and Chen-Arima [14] and our paper provides a more complete collocation for their findings. In particular Chen and Arima have found spinors arising within the Interacting Boson Model of the nucleus that is completely bosonic (built upon s and d bosons). We argue that they have found the $A_1[11011]$ subalgebra of $su(6)$, the analogous of our $A_1[101]$ subalgebra of $su(4)$. Another aspect worth mentioning again is that, although hamiltonian displaying dynamical symmetries based on AMNC chains might be unphysical, the basis states associated to them might provide an alternative (maybe easier) basis for diagonalization of complex hamiltonians.

Let us repeat once again that there is nothing wrong with the way algebraic models have been formulated: the chains that pass through the angular momentum algebra are certainly the most natural and the most useful. Since the mathematical structure is nevertheless larger, we simply wanted to understand if and how the other chains can be interpreted in terms of the basic operators that define the model, if this has some physical significance and if those alternative chains can be used either to model realistic hamiltonians (and this seems very improbable) or to furnish alternative diagonalization schemes (and this is clearly possible). We believe that there might be other insightful discoveries or advantages awaiting in the still little explored angular momentum non conserving chains of algebraic models.

6. Acknowledgments

L.F. thanks F.Iachello and A.Leviatan for the useful criticisms that have arisen during the conference. We would also like to express my gratitude to P.van Isacker, M.Carvajal and F.Pérez-Bernal for the suggestions that have arisen during the workshop "Quantum Phase Transitions in Molecular and Nuclear Structure 2010" held in Huelva (Spain).

References

[1] Iachello F and Arima A 1987 *The Interacting Boson Model* (Cambridge University Press)
[2] Iachello F and Levine R D 1995 *Algebraic Theory of Molecules* (Oxford University Press)
[3] Iachello F 2006 *Lie Algebras and Applications Lect. Notes Phys.* **11** (Springer, Berlin Heidelberg)
[4] Frank A and van Isacker P 2005 *Symmetry Methods in molecules and nuclei* (SyG Editores, Mexico)
[5] Oss S 1996 Advances in chemical physics **93** 455-649
[6] Barut A O and Raczka R 1977 *Theory of Group Representations and Applications* (PWN-Polish Scientific Publishers, Warsava)
[7] Collingwood D H and McGovern W M 1993 *Nilpotent Orbits in Semisimple Lie Algebras* (Van Nostrand Reinhold, New York)
[8] de Graaf W A 2008 LMS J. Comput. Math. **11** 280-297
[9] de Graaf W A 2010 *Preprint arXiv:1004.1972(2010) to appear in J.Algebra*
[10] The GAP Group, GAP – Groups, Algorithms, and Programming, Version 4.4.12; 2008. `http://www.gap-system.org`
[11] de Graaf W A *SLA - computing with Simple Lie Algebras. A GAP package*, `http://science.unitn.it/~degraaf/sla.html`
[12] Fortunato L and Sartori L 2010 *Commun. Theor. Phys.* **54** 589-593
[13] Iachello F and Oss S 1996 *J. Chem. Phys.* **104** (18) 6956-6963
[14] Chen H T and Arima A 1983 *Phys. Rev. Lett.* **51** 447-450
[15] Elliott J P 1958 *Proc. Roy. Soc. (London)* **245** 562-581

Journal of Physics: Conference Series **284** (2011) 012026 doi:10.1088/1742-6596/284/1/012026

Phase transition in generalized inhomogeneous 'cubic' systems

B R Gadjiev
International University for Nature, Society and Man,
19 Universitetskaya street, Dubna, 141980, Russia

E-mail: gadjiev@uni-dubna.ru

Abstract. In this paper we investigate peculiarities of phase transition in inhomogeneous systems. We consider a case of 'cubic' systems with anisotropy invariants in which the distribution of defects generates a small-world property. We define a fractional equation of motion for the order parameter for the systems with a small world property. Linearization of the equation of motion for the order parameter made it possible to define a non-linear dispersion law. A renormalization group analysis of phase transitions in the generalized inhomogeneous "cubic" systems is presented. We discuss the dependence of critical behavior on the nonextensivity parameter of the system.

1. Introduction

During the theoretical description of a spontaneous symmetry breaking in ideal crystals in terms of Landau inverse problem it is supposed that the symmetry groups of high-symmetry H_0 and low-symmetry H_1 phases are known [1, 2]. The group H_1 is a subgroup of the H_0, i.e. $H_1 \subset H_0$. If the groups H_0 and H_1 are known it is possible to determine the finite matrix group Ξ, which describes the transformation properties of the major order parameter η. In this case, in accordance with the symmetry arguments, the expansion of the free energy functional of a crystal depends on the integrity basis of the Ξ matrix group.

It is well known that the major order parameter is a certain linear combination of the system elements' coordinates. Thus, it could be expected that the equation of the motion of the order parameter is rather complex in the crystals with variable composition. Even if small deviations of the order parameter from its equilibrium value are taken into account, the equation of the motion of the order parameter could include time derivatives of an arbitrarily high order [3, 4].

It is normal for real crystals to have defects and impurities. That is why the issue of a weak disorder influence on critical behaviour of a system appears to be quite natural. The distribution of defects in crystals may be different. For example, a homogeneous random distribution of the defects in a structure may completely destroy large-scale fluctuations. As a result, either singularities of thermodynamic functions disappear or a shift of the critical temperature occurs [5]. However, the conditions of the crystal growth may lead to a inhomogeneous distribution of defects. In this case a new critical regime characterized by new universal critical indices may arise due to the impurities.

The analysis of the weak disorder effect on the system critical behaviour in the frame of the method of renormalization group and ε - expansion is given in [5]. It was shown that the critical indices of the system do not change (i.e. the universality class remains unchanged) [5]. In the same paper by using a replica symmetry broken method it was shown that in the case of a weak quenched disorder the universality class of the system may change.

In the present paper we study the effect of the weak disorder on the critical behaviour of the system with small-world properties [6] due to the distribution of defects. We study a fractional generalization of the dynamic equation for the major order parameter and obtain a nonlinear dispersion law. First we define a distribution function of the order parameter in the system with a small-world property. Then we introduce a fractional generalization for the equation of motion for the order parameter in such systems and define a non-linear dispersion law. Application of the renormalization group method allows us to show that a new critical regime may appear in the systems with a small-world property.

2. Structure with a small-world property

Usually in the continuous description of the defect influence on the phase transitions in a structure in terms of Ginzburg-Landau Hamiltonian we suppose that the presence of impurities may lead either to the space fluctuations of effective local temperature of the transition or to the space fluctuations inside a crystal field. For defining a distribution function for the order parameter we introduce the notation $f(\eta)=-\frac{\delta F[\eta]}{\delta\eta}$. Then the equation of motion for the order parameter η, with a fluctuation of the local temperature and local field may be given in the form

$$\frac{\partial\eta}{\partial t}=f(\eta)+u(\eta)\xi(t)+\varsigma(t), \tag{1}$$

where $\eta(t)$ is a stochastic variable, $u(\eta)$ is an arbitrary function and $\xi(t)$ and $\varsigma(t)$ are uncorrelated noises with a mean value equal to zero

$$\langle\xi(t)\xi(t')\rangle=2M\delta(t-t'), \tag{2}$$

$$\langle\varsigma(t)\varsigma(t')\rangle=2A\delta(t-t'), \tag{3}$$

where $M>0$ and $A>0$ are amplitudes of the multiplicative and additive noises, respectively [7].

Using Kramers-Moyal's expansion it is easy to show that Fokker-Plank equation for the probability density $p(\eta,t)$, connected to the equation (1) in Stratonovich calculus has the form

$$\frac{\partial p(\eta,t)}{\partial t}=-\frac{\partial(f(\eta)p(\eta,t))}{\partial\eta}+M\frac{\partial}{\partial\eta}\left(u(\eta)\frac{\partial}{\partial\eta}(u(\eta)p(\eta,t))\right)+A\frac{\partial^2 p(\eta,t)}{\partial\eta^2} \tag{4}$$

Suppose that $f(\eta)=-\tau u(\eta)\frac{\partial u(\eta)}{\partial\eta}$. In this case it is easy to show that the stationary solution of this equation takes the form

$$p_{st}(\eta)=p_0A^{-\frac{\tau+M}{2M}}\left(1+\frac{M}{A}u^2(\eta)\right)^{-\frac{\tau+M}{2M}}, \tag{5}$$

and, consequently,

$$p_{st}(\eta)=\frac{1}{Z}\left(1-(1-q)u^2(\eta)\right)^{\frac{1}{1-q}}, \tag{6}$$

where $q=\frac{\tau+3M}{\tau+M}$ and $\frac{M}{A}=q-1$.

It is clear that $u^2(\eta) = -\frac{2}{\tau}F[\eta]$. Thus, the multiplicative noise effect leads to the nonextensivity system [8].

Note that by definition $q = \frac{3M+\tau}{M+\tau}$, and in the case of absence of the multiplicative noise $M=0$ we obtain $q=1$. In this case the stationary distribution $p_{st}(\eta)$ is Boltzmann-Gibbs distribution and consequently the system nonextensivity is induced by multiplicative noises. Besides, if $q \neq 1$ for large $u^2(\eta)$ we obtain a power law distribution. If $q \neq 1$ and $u^2(\eta)$ are bounded, we have a density of distribution of q - type which is intermediate between a power and an exponential distribution. It should be stressed that the presence of a multiplicative noise means that some positions in the process of crystal growth are more preferable for the location of impurities. The saturation effect of the chemical bonds leads to the fact that actually we are dealing with a density distribution of q - type.

It is necessary to emphasize, that in [9] in the frame of a principle maximum entropy it is shown that the distribution (6) describes the structures with a small-world property with a satisfactory accuracy.

3. Derivation equation of motion for the order parameter in compounds with a small-world property

A free energy functional F is invariant concerning the structure symmetry group H_0 and consequently depends on the integrity basis of the irreducible representation Ξ and the corresponding invariant combinations of spatial and time derivatives of the order parameter.

Taking into account the presence of defects in a structure we define a free energy functional as

$$F[\eta] = F_0[\eta] + F_I[\eta], \tag{7}$$

where

$$F_0[\eta] = \int_R drdt \int_R dr'dt' \left\{ \frac{1}{2}\frac{\partial}{\partial t}\eta(r,t)\cdot g_0(r,t,r',t')\cdot\frac{\partial}{\partial t'}\eta(r',t') + \frac{1}{2}\frac{\partial}{\partial r}\eta(r,t)\cdot g_1(r,t,r',t')\cdot\frac{\partial}{\partial r'}\eta(r',t') \right\} \tag{8}$$

and

$$F_I[\eta] = -\int_R drdt \int_R dr'dt' V(\eta(r,t), u(r'.t')) \tag{9}$$

For the sake of simplicity, we consider a one-dimensional case. Here r is a coordinate and t is time and functions $g_0(r,t;r',t')$ and $g_1(r,t;r',t')$ describe the effect of small-world properties on the critical property of the system. The integration is carried out over a region R in the two dimensional space R^2 to which (r,t) belongs. Generalization of the obtained in this chapter results for high dimensional space is evident. The equation of motion for the field $\eta(r,t)$ is derived by using Gateaux derivative of the functional $F[\eta(r,t)]$, which is defined as

$$\delta F[\eta, h] = \left[\frac{d}{d\varepsilon}F[\eta+\varepsilon, h]\right]_{\varepsilon=0} = \lim_{\varepsilon \to 0}\frac{F[\eta+\varepsilon, h] - F[\eta]}{\varepsilon}, \tag{10}$$

where $h(r) = \delta\eta(r)$ is a smooth integrable function. The functional $F[\eta(r,t)]$ for $\eta + \varepsilon h(r,t)$ is

$$
\begin{aligned}
F[\eta+\varepsilon h] &= \int dr \int dt \int dr' \int dt' \left[\frac{1}{2}\frac{\partial}{\partial r}(\eta(r,t)+\varepsilon h(r,t))\cdot g_1(r,t,r',t')\cdot\frac{\partial}{\partial r'}(\eta(r',t')+\varepsilon h(r',t'))\right]+ \\
&\int dr \int dt dt' \int dr' \left[\frac{1}{2}\frac{\partial}{\partial t}(\eta(r,t)+\varepsilon h(r,t))\cdot g_0(r,t,r',t')\cdot\frac{\partial}{\partial t'}(\eta(r',t')+\varepsilon h(r',t'))\right] \\
&\int dr \int dt dt' \int dr V(\eta(r,t)+\varepsilon h(r,t),\eta(r',t')+\varepsilon h(r',t')).
\end{aligned}
\tag{11}
$$

This expression up to the order ε has the form

$$
\begin{aligned}
F[\eta+\varepsilon h] &= \int dt \int dr \int dt' \int dr' \left[\frac{1}{2}\frac{\partial}{\partial t}\eta(r,t)\cdot g_0(r,t,r',t')\cdot\frac{\partial}{\partial t'}\eta(r',t')\right]+ \\
&\int dt \int dr \int dt' \int dr' \left[\frac{1}{2}\frac{\partial}{\partial r}\eta(r,t)\cdot g_1(r,t,r',t')\cdot\frac{\partial}{\partial r'}\eta(r',t')\right]+ \\
&\int dt \int dr \int dt' \int dr' [-V(\eta(r,t),\eta(r',t'))]+\varepsilon\int dt \int dr \int dt' \int dr' \left[\frac{1}{2}\frac{\partial}{\partial t}\eta(r,t)\cdot g_0(r,t,r',t')\cdot\frac{\partial}{\partial t'}h(r',t')\right]+ \\
&\varepsilon\int dt \int dr \int dt' \int dr' \left[\frac{1}{2}\frac{\partial}{\partial t}h(r,t)\cdot g_0(r,t,r',t')\cdot\frac{\partial}{\partial t'}\eta(r',t')+\frac{1}{2}\frac{\partial}{\partial t}h(r,t)\cdot g_0(r,t,r',t')\cdot\frac{\partial}{\partial t'}\eta(r',t')\right]+ \\
&\varepsilon\int dt \int dr \int dt' \int dr' \left\{\frac{1}{2}\frac{\partial}{\partial r}\eta(r,t)\cdot g_1(r,t,r',t')\cdot\frac{\partial}{\partial r'}h(r',t')+\frac{1}{2}\frac{\partial}{\partial r}h(r,t)\cdot g_1(r,t,r',t')\cdot\frac{\partial}{\partial r'}\eta(r',t')\right\}+ \\
&\varepsilon\int dt \int dr \int dt' \int dr' \left[-\frac{\partial V(\eta(r,t),\eta(r',t'))}{\partial\eta(r,t)}h(r,t)-\frac{\partial V(\eta(r,t),\eta(r',t'))}{\partial\eta(r',t')}h(r',t')\right].
\end{aligned}
\tag{12}
$$

It is convenient to introduce the functions:

$$K_0(r,t,r',t')=\frac{1}{2}\big(g_0(r,t,r',t')+g_0(r',t',r,t)\big),$$

$$K_1(r,t,r',t')=\frac{1}{2}\big(g_1(r,t,r',t')+g_1(r',t',r,t)\big).$$

The dynamical equation for the order parameter follows from the stationary principle $\delta F[\eta,h]=0$, for any $h(r,t)$. For the symmetric function $U(\eta(r,t),\eta(r',t'))=U(r,t)\delta(r-r')\delta(t-t')$ the equation of motion for the order parameter has the form

$$\int dt' \int dr' \frac{\partial}{\partial t}K_0(r,t,r',t')\cdot\frac{\partial}{\partial t'}\eta(r',t')+\int dt' \int dr' \frac{\partial}{\partial r}K_1(r,t,r',t')\cdot\frac{\partial}{\partial r'}\eta(r',t')+\frac{\partial U(\eta(r,t))}{\partial\eta(r,t)}=0. \tag{13}$$

We suppose that $K_0(r,t,r',t')=\delta(r-r')\tilde{K}_0(t,t')$, $K_1(r,t,r',t')=\delta(t-t')\tilde{K}_1(r,r')$.
Introducing the notations

$$Z_t=\int dt' \frac{\partial}{\partial t}\tilde{K}_0(t,t')\cdot\frac{\partial}{\partial t'}\eta(r,t')=Z_t(r,t), \tag{14}$$

$$Z_r=\int dr' \frac{\partial}{\partial r}\tilde{K}_1(r,r')\cdot\frac{\partial}{\partial r'}\eta(r',t)=Z_r(r,t), \tag{15}$$

we can write the equation of motion for the order parameter $\eta(r,t)$ as

$$Z_t+Z_r+\frac{\partial U(\eta(r,t))}{\partial\eta(r,t)}=0. \tag{16}$$

It is suitable to represent the equation of motion for the order parameter in the form

$$\int_0^t G(t-t')\frac{\partial\eta(r,t')}{\partial t'}dt' = -\int_0^r dr' V(r-r')\frac{\partial^2\eta(r-r')}{\partial r'^2}dr' - \alpha\eta(r',t) - \beta\eta^3(r',t) = 0 . \qquad (17)$$

If a structure possesses a small-world property it is possible to assume that the functions $G(t-t')$ and $V(r-r')$ are Tsallis distributions and then for the large values of arguments it is possible to write

$$G(t-t') = \frac{1}{\Gamma(1-\nu)}\frac{1}{(t-t')^{\nu}} \qquad (18)$$

and

$$V(r-r') = \frac{1}{\cos\left(\frac{\pi\varepsilon}{2}\right)\Gamma(2-\varepsilon)}\frac{1}{|r-r'|^{\varepsilon-1}}, \qquad (19)$$

where $0<\nu<1$ and $1<\varepsilon<2$. After substituting (18) and (19) into the equation (17) we obtain

$$\frac{\partial^{\nu}\eta(r,t)}{\partial t^{\nu}} == -\kappa\frac{\partial^{\varepsilon}\eta(r,t)}{\partial r^{\varepsilon}} - \alpha\eta(r,t) - \beta\eta^3(r,t), \qquad (20)$$

where for $\frac{\partial^{\varepsilon}}{\partial r^{\varepsilon}}$ we used Caputo definition of the fractional derivative [9]. Using Fourier transformation we obtain the nonlinear dispersion law

$$(i\omega)^{\nu} = \left(k^2\right)^{\frac{\varepsilon}{2}} + \alpha . \qquad (21)$$

Hence, if a structure has a small-world property then the system exhibits a nonlinear dispersion law.

4. Renormalization group theory of the phase transition $P4_2/mnm \Rightarrow P2_1/c$

Consider the phase transition from high-symmetry phase with a space group symmetry $H_0 \equiv P4_2/mnm$ to low symmetric phase with a space group symmetry $H_1 \equiv P2_1/c$. The Hamiltonian of Ginzburg – Landau H depends on the integrity basis of invariants of the matrix group Ξ (the elements of Ξ are given in [10]). The thermodynamic theory of Landau type describing this phase transition is constructed in [10]. The corresponding irreducible representation Ξ of the space group $H_0 \equiv P4_2/mnm$ of high-symmetry phase has the fourth order invariants in the following form

$$\psi_1 = \eta_1^4 + \eta_2^4 + \eta_2^4 + \eta_4^4,\ \psi_2 = \eta_1^2\eta_2^2 + \eta_3^2\eta_4^2,$$

$$\psi_3 = \eta_1^2\eta_3^2 + \eta_2^2\eta_4^2,\ \psi_4 = \eta_1^2\eta_4^2 + \eta_2^2\eta_3^2 . \qquad (22)$$

The renormalization group calculations start with the classical Hamiltonian H, which is a functional of the “spins” η_i, where i = 1,2,3,4.

After using Fourier transformation the effective Hamiltonian of the system is represented by the expression

$$
\begin{aligned}
H = & -\frac{1}{2}\sum_{i=1}^{4}\int\left[r+q^{\gamma}\right]\eta_i(\vec{q})\mu_i(-\vec{q})d^d q - \\
& u_1\int[\eta_1(\vec{q}_1)\eta_1(\vec{q}_2)_1(\vec{q}_3)\eta_1(\vec{q}_4)+\eta_2(\vec{q}_1)\eta_2(\vec{q}_2)\eta_2(\vec{q}_3)\eta_2(\vec{q}_4)+\eta_3(\vec{q}_1)\eta_3(\vec{q}_2)\eta_3(\vec{q}_3)\mu_3(\vec{q}_4)+ \\
& \eta_4(\vec{q}_1)\eta_4(\vec{q}_2)\eta_4(\vec{q}_3)\eta_4(\vec{q}_4)]\delta(\vec{q}_1+\vec{q}_2+\vec{q}_3+\vec{q}_4)d^d\vec{q}_1 d^d\vec{q}_2 d^d\vec{q}_3 d^d\vec{q}_4 - \\
& u_2\int[\eta_1(\vec{q}_1)\eta_1(\vec{q}_2)\eta_2(\vec{q}_3)\eta_2(\vec{q}_4)+\eta_3(\vec{q}_1)\eta_3(\vec{q}_2)\eta_4(\vec{q}_3)\eta_4(\vec{q}_4)] \\
& \delta(\vec{q}_1+\vec{q}_2+\vec{q}_3+\vec{q}_4)d^d\vec{q}_1 d^d\vec{q}_2 d^d\vec{q}_3 d^d\vec{q}_4 - \\
& u_3\int[\eta_1(\vec{q}_1)\eta_1(\vec{q}_2)\eta_3(\vec{q}_3)\eta_3(\vec{q}_4)+\eta_2(\vec{q}_1)\eta_2(\vec{q}_2)\eta_4(\vec{q}_3)\eta_4(\vec{q}_4)] \\
& \delta(\vec{q}_1+\vec{q}_2+\vec{q}_3+\vec{q}_4)d^d\vec{q}_1 d^d\vec{q}_2 d^d\vec{q}_3 d^d\vec{q}_4 - \\
& u_4\int[\eta_1(\vec{q}_1)\eta_1(\vec{q}_2)\eta_4(\vec{q}_3)\eta_4(\vec{q}_4)+\eta_2(\vec{q}_1)\eta_2(\vec{q}_2)\eta_3(\vec{q}_3)\eta_3(\vec{q}_4)] \\
& \delta(\vec{q}_1+\vec{q}_2+\vec{q}_3+\vec{q}_4)d^d\vec{q}_1 d^d\vec{q}_2 d^d\vec{q}_3 d^d\vec{q}_4 .
\end{aligned}
\tag{23}
$$

where we are used an arbitrariness in the choice of invariants and here $\omega(q)=\gamma+q^{\gamma}$ is taken into account.

The effective Hamiltonian contains a number of anisotropic invariants and that is a distinctive peculiarity of 'cubic' systems .Therefore the derivation of the renormalization group equations is much more complicated. Application of the symmetry properties of the renormalization group transformation greatly simplifies a derivation of renormalization group equations. The relationship between the properties of transformations of the renormalization group and the symmetry of the Hamiltonian is investigated in [12]. A detailed study of this problem and a specific algorithm necessary for calculations are described in [13]. In [13] the $(m+1)$-dimensional parameter space is introduced:

$$
\Pi=\left\{u;u\in R^{m+1},\ \lim_{I_0\to+\infty}\sum_{j=0}^{m}u_j I_j=+\infty\right\},
\tag{24}
$$

where R is a real line and we denote by u a point $(u_0,u_1,\ldots,u_m)$. The definition (24) provides a normalization condition of the distribution function $\exp[H[\eta]]$.

In the space Π a nonlinear renormalization group transformation is defined as

$$
R: u\to\bar{u}=Ru,
\tag{25}
$$

Since the transformation R forms a semigroup, we can also write

$$
R_s R_t = R_{st}\text{, for all } s \text{ and } t \text{ from } [1,\infty).
\tag{26}
$$

In the fixed-point the equation (25) takes the form

$$
Ru^{*}=u^{*}.
\tag{27}
$$

In [13] it was shown that every diffeomorphism T in Π can generate the transformation R^T equivalent to the transformation R

$$
R^T=T^{-1}RT.
\tag{28}
$$

From the equation (28) follows that the symmetry properties of the renormalization group transformation R are defined by the operations that commute with T. These T - elements form a group which is denoted by G_T:

$$G_T = \{T : R^T = R\}. \tag{29}$$

Knowledge of the fourth-order invariants contained in the Hamiltonian, allows to determine the group G_T. Indeed, consider the transformation

$$\eta_j \Rightarrow V\eta_i, \tag{30}$$

where V is the orthogonal matrix $l \times l : V \in O(l)$, where $O(l)$ is a group of rotations in the l dimensional space. In general under the action of V other forms of the fourth order invariants are generated. The requirement of absence of these additional invariants leads to the equations for u. In the most general form $V(\theta_1,...) \in O(l)$, where $(\theta_1,...)$ are parameters. In the general case

$$V(\theta_1,...)\Pi \subseteq \Pi \tag{31}$$

defines a system of equations for the parameters $(\theta_1,...)$Therefore the solution of the equation (31) defines the group

$$G_g = \{V; V \in O(l), V \subseteq \Pi\}. \tag{32}$$

Further, it is easy to prove [13], that G_T is a linear representation of the group G_g, which commutes with the R transformation. Thus, we have

$$RTu = TRu \text{ for any } T \in G_T. \tag{33}$$

Consequently, Ru must be a covariant with respect to G_T. To introduce an equation of the renormalization group we can use the following theorem:

Let Γ_j be an irreducible representation (other than identity) of a point group, and the functions $\psi_\mu^{j\alpha}(\vec{r})$ ($\mu = 1,2,...,l_j$ and α is fixed) are the basis functions of Γ_j. Then for the arbitrary function $\psi_\mu^j(\vec{r})$ we can write the expansion

$$\psi_\mu^j(\vec{r}) = \sum_{\alpha=1}^{l_j} F^{j\alpha}(\vec{r})\psi_\mu^{j\alpha}(\vec{r}), \tag{34}$$

where $\vartheta(\nu)$ is an integrity basis of the group G_T, $c_\nu(u)$ is a linearly independent covariant function of G_T, f_ν are arbitrary functions[13]. Note that if G_T is a point group, the integrity basis forms a finite set. Thus, Ru is defined by the finite number of constants which can be determined from Feynman diagram expansion.

Define a group which commutes with the renormalization group operator. Let $(\Phi_1, \Phi_2, \Phi_3, \Phi_4)$ are transformed in accordance with a vector representation of $O(4)$. Then the requirement for the invariants $\{\Psi_1, \Psi_2, \Psi_3, \Psi_4\}$ to be expressed linearly via each other leads to a system of equations for the parameters of the $O(4)$ group the solution of which makes it possible to define the group G_T

The G_T group consists of the following elements

$$e = \begin{bmatrix} 1 & 0 & 0 & 0 \\ 0 & 1 & 0 & 0 \\ 0 & 0 & 1 & 0 \\ 0 & 0 & 0 & 1 \end{bmatrix}, \quad f_1 = \begin{bmatrix} 1 & 0 & 0 & 0 \\ 0 & 0 & 1 & 0 \\ 0 & 1 & 0 & 0 \\ 0 & 0 & 0 & 1 \end{bmatrix}, \quad f_2 = \begin{bmatrix} 1 & 0 & 0 & 0 \\ 0 & 0 & 0 & 1 \\ 0 & 0 & 1 & 0 \\ 0 & 1 & 0 & 0 \end{bmatrix},$$

$$f_3 = \begin{bmatrix} 1 & 0 & 0 & 0 \\ 0 & 1 & 0 & 0 \\ 0 & 0 & 0 & 1 \\ 0 & 0 & 1 & 0 \end{bmatrix},\ f_4 = \begin{bmatrix} 1 & 0 & 0 & 0 \\ 0 & 0 & 1 & 0 \\ 0 & 0 & 0 & 1 \\ 0 & 1 & 0 & 0 \end{bmatrix},\ f_5 = \begin{bmatrix} 1 & 0 & 0 & 0 \\ 0 & 0 & 0 & 1 \\ 0 & 1 & 0 & 0 \\ 0 & 0 & 1 & 0 \end{bmatrix}. \tag{35}$$

Thus, the finite matrix group $G_T = \{e, f_1, f_2, f_3, f_4, f_5\}$ is a group which commutes with a transformation of the renormalization group.

According to the symmetry arguments the equations of the renormalization group in the order $O(\varepsilon^2)$ presented in the form:

$$\begin{aligned}
\overline{u}_1 &= b^{-3d}\xi^4\left[u_1 + \alpha_2 u_1^2 + \alpha_4\left(u_2^2 + u_3^2 + u_4^2\right)\right],\\
\overline{u}_2 &= b^{-3d}\xi^4[u_2 + \beta u_3 u_4 + \gamma u_2^2 + \delta u_1 u_2],\\
\overline{u}_3 &= b^{-3d}\xi^4[u_3 + \beta u_2 u_4 + \gamma u_3^2 + \delta u_1 u_3],\\
\overline{u}_4 &= b^{-3d}\xi^4[u_4 + \beta u_2 u_3 + \gamma u_4^2 + \delta u_1 u_4].
\end{aligned} \tag{36}$$

For definition of this equation parameters it is necessary to calculate the integral

$$C(r) = K_d \int_{\Lambda/b}^{\Lambda} \frac{q^{d-1}dq}{\left(q^{2-\gamma}\right)^2} = K_3 \ln b(1 + 2\gamma \ln\Lambda - \gamma \ln b). \tag{37}$$

Note that the integrals are calculated in the three dimensional case $d = 3$. We define the parameter ξ from the condition that after transformation the coefficient q^γ in $v(q) = r + q^\gamma$ is equal to unity. It gives $\xi^2 b^{-d} b^{-\gamma} = 1$ and therefore $\xi^2 b^{-d} = b^\gamma$. In this case we obtain $b^{-3d}\xi^4 = b^{-d+2\gamma}$. Assume that $\gamma = \frac{3}{2} + \frac{\varepsilon}{2}$. Then we come to the result $b^{-3d}\xi^4 = b^\varepsilon$. Assuming that the parameter b^{-1} is small and leaving only the terms proportional to $\ln b$, we obtain

$$\begin{aligned}
\overline{u}_1 &= b^\varepsilon\left[u_1 - \frac{1}{2}\left(k_1 u_1^2 + k_2\left(u_2^2 + u_3^2 + u_4^2\right)\right)K_d \ln b\right],\\
\overline{u}_2 &= b^\varepsilon\left[u_2 - \frac{1}{2}\left(l u_2^2 + m u_3 u_4 + n u_1 u_2\right)K_d \ln b\right],\\
\overline{u}_3 &= b^\varepsilon\left[u_3 - \frac{1}{2}\left(l u_3^2 + m u_2 u_4 + n u_1 u_3\right)K_d \ln b\right],\\
\overline{u}_4 &= b^\varepsilon\left[u_4 - \frac{1}{2}\left(l u_4^2 + m u_2 u_4 + n u_1 u_4\right)K_d \ln b\right],
\end{aligned} \tag{38}$$

where $k_1 = 72, k_2 = 2, l = 16, m = 4, n = 48$. After the expansion $b^\varepsilon = 1 + \varepsilon \ln b$ and introducing the notation $u_i = \frac{\varepsilon}{K_d} x_i$ in the fixed point $\overline{x}_i = x_i = x_i^*$ we obtain:

$$
\begin{aligned}
x_1 &= 36x_1^2 + x_2^2 + x_3^2 + x_4^2, \\
x_2 &= 8x_2^2 + 2x_3x_4 + 24x_1x_2, \\
x_3 &= 8x_3^2 + 2x_2x_4 + 24x_1x_3, \\
x_4 &= 8x_1^2 + 2x_2x_3 + 24x_1x_4.
\end{aligned}
\tag{39}
$$

Note that we are interested in the real solutions of these equations only. Complex solutions are not considered because the initial Hamiltonian, being real, under the action of renormalization group transformations approaches the fixed points and remains real. The real solutions for the system renormalization group equations at the fixed points are given in table 1.

Table 1. Fixed points of the renormalization group equations

x_1	x_2	x_3	x_4
0	0	0	0
1/36	0	0	0
1/40	1/20	0	0
1/72	1/12	0	0
1/40	0	1/20	0
1/40	0	0	1/20
1/72	0	1/12	0
1/72	0	0	1/12

After linearization of the renormalization group equations we obtain

$$\Delta\bar{u}_i = M_{ij}\Delta u_j, \tag{40}$$

where

$$
M_{ij} = \begin{bmatrix}
1+\varepsilon\ln b - k_1u_1^*K_d\ln b; & -k_2K_d\ln bu_2^*; & -k_2K_d\ln bu_3^*; & -k_2K_d\ln bu_4^* \\
-\frac{n}{2}K_d\ln bu_2^*; & 1+\varepsilon\ln b - lu_2^*K_d\ln bu_1^*; & -\frac{m}{2}K_d\ln bu_4^*; & -\frac{m}{2}K_d\ln bu_3^* \\
-\frac{n}{2}u_3^*K_d\ln b; & -\frac{m}{2}K_d\ln bu_4^*; & 1+\varepsilon\ln b - lu_3^*K_d\ln bu_1^*; & -\frac{m}{2}K_d\ln bu_2^* \\
-\frac{n}{2}u_4^*K_d\ln b; & -\frac{m}{2}K_d\ln bu_3^*; & -\frac{m}{2}K_d\ln bu_2^*; & 1+\varepsilon\ln b - lu_4^*K_d\ln b - \frac{n}{2}u_1^*K_d\ln b
\end{bmatrix}
\tag{41}
$$

Now, it is necessary to calculate the eigenvalues of the matrix M_{ij} at the fixed points presented in table 1.

I. The fixed point $x_1 = x_2 = x_3 = x_4 = 0$. The eigenvalues of the matrix M_{ij} are equal to

$$\lambda_1 = \lambda_2 = \lambda_3 = \lambda_4 = b^{\varepsilon}.$$

II. The fixed point $x_1 = 1/36,\ x_2 = x_3 = x_4 = 0$.The eigenvalues of the matrix M_{ij} have the form:

$$\lambda_1 = b^{-\varepsilon};\ \lambda_2 = \lambda_3 = \lambda_4 = b^{\varepsilon/3}.$$

III. The fixed point $x_1 = 1/40,\ x_2 = 1/20,\ x_3 = x_4 = 0$. The eigenvalues of the matrix M_{ij} have the form:

$$\lambda_1 = b^{-0.8\varepsilon};\ \lambda_2 = b^{-0.4\varepsilon};\ \lambda_3 = \lambda_4 = b^{0.6\varepsilon}.$$

IV. The fixed point $x_1 = 1/72,\ x_2 = 1/12,\ x_3 = x_4 = 0$. The eigenvalues of the matrix M_{ij} are defined as:

$$\lambda_1 = b^0;\ \lambda_2 = b^{-(2/3)\varepsilon};\ \lambda_3 = \lambda_4 = b^{(2/3)\varepsilon}.$$

Thus the stability of the fixed points is determined by a sign of ε. If ε is positive $\varepsilon > 0$, all the fixed points are unstable whereas if ε is negative $\varepsilon < 0$, the fixed point $x_1 = x_2 = x_3 = x_4 = 0$ is stable, whereas the other fixed points are unstable.

5. Conclusion

In the framework of the standard method of the renormalization group and ε-expansion the analysis of the phase transitions with the effective Hamiltonian (25) is considered in [14, 15]. It is shown, that all the fixed points of the system are unstable. Consequently the phase transition is a first order transition in spite of the fact that the analysis in the framework of Landau theory leads to a phase transition of the second order.

If the presence of defects in such structures generates a small-world property the system spectrum has the form $\omega(q) = r + q^\gamma$. In this case if the parameter $\varepsilon = 2\gamma - 3 > 0$ then the fixed points of the system are unstable and we have a transition of the first order

However if $\varepsilon = 2\gamma - 3 < 0$ the fixed point $x_1 = x_2 = x_3 = x_4 = 0$ of the system is stable, therefore the phase transition in this case is a second order one. Note that in the case $\varepsilon = 2\gamma - 3 > 0$ the other fixed points of the system are unstable, so the corresponding transitions are of the first order. Thus a small-world property of systems taken into account leads to a new class of universality.

6. References

[1] Ascher E and Kobayashi J 1977 *J. Phys. C: Solid State Phys.* **10** 1349
[2] Ascher E 1977 *J. Phys. C: Solid State Phys.* **10** 1365
[3] Levanyuk A P and Sigov A S 1988 *Defects and Structural Phase Transitions.* (N. Y.: Gordon and Breach Sci. Publ.)
[4] Gadjiev B R 2002 *Proc. of the XXIII Int. Colloquium on Group Theoretical Methods in Physics* vol 2 (Dubna, JINR) p 527
[5] Dotsenko Vik S 1995 *Usp. Fiz. Nauk.* **165** 481
[6] Gitterman M and Halpern V 2006 *Phasetransitions: A Brief Account with Modern Applications* (World Scientific)
[7] Anteneodo C and Tsallis C 2003 *Journal of Mathematical Physics* **44** (11) 5194
[8] Tsallis C 1999 *Braz. J. Phys.* **29** 1
[9] Hilfer R 2008 *Threefold introduction to fractional derivative in Anomalous transport: Foundations and Applications* ed R Klagers et al (Wiley-VCN, Weinheim)
[10] Gufan Yu 1983 *Structural phase transition* (Moscow, Nauka)
[11] Aharony A 1976 *Dependence of universal critical behavior on symmetry and range of interaction* In *Phase Transition and Critical Phenomena* C Domb and S Green eds vol. 6 (Academic Press, NY) 357
[12] Korzhenevskii A L 1976 *Zh. Eksp. Teor. Fiz.* **71** 1434
[13] Jaric M V 1978 *Phys. Rev. B* **18** 2237
[14] Aksenov V L and Gadjiev B R 1998 The sequence of magnetic phase transition in mixed crystals $U(Pd_{1-x}Fe_x)_2Ge_2$ *Preprint JINR* P17-98-256 (Dubna)
[15] Gadjiev B R 2010 *Journal of Physics: Conference Series* **200** (1) 012044

GROUP 28: Physical and Mathematical Aspects of Symmetry IOP Publishing
Journal of Physics: Conference Series **284** (2011) 012027 doi:10.1088/1742-6596/284/1/012027

Pisot–Fibonacci q-coherent states

Jean Pierre Gazeau[1] **and Mariano A del Olmo**[2]

[1] Laboratoire APC, Université Paris Diderot *Paris 7*, 10 rue A. Domon et L. Duquet, 75205 Paris Cedex 13, France,

[2] Departamento de Física Teórica, Universidad de Valladolid, E-47005, Valladolid, Spain.

E-mail: gazeau@apc.univ-paris7.fr, olmo@fta.uva.es

Abstract. A family of q-coherent states is constructed allowing us to obtain a new quantized version of the harmonic oscillator. These q-states are normalized and form an overcomplete set resolving the unity with respect to the appropriate Jackson measure if $0 < q < 1$. We only consider here those values of q such that q^{-1} is a Pisot number. In this case the q-deformed integers ($[n]_q$) form Fibonacci-like sequences of integers. We study the main physical characteristics of the corresponding quantum oscillator: localization in the configuration and in the phase spaces, probability distributions and related statistical features and semi-classical phase space trajectories whose periodicity is related with the fact that q is an algebraic number.

1. Introduction

Starting from a separable (complex) Hilbert space, $\mathcal{H}$, with orthonormal basis $\{|e_n\rangle\}_{n=0}^{\infty}$, we can construct a family of "generalized" coherent states (CS) belonging to $\mathcal{H}$ by

$$|v_z\rangle := \sum_{n=0}^{\infty} \frac{1}{\sqrt{\mathcal{N}(|z|^2)}} \frac{z^n}{\sqrt{x_n!}} |e_n\rangle , \qquad \forall z \in \mathcal{D}_{\sqrt{R}} = \{z \in \mathbb{C} ,\, |z| < \sqrt{R}\}, \tag{1}$$

where R is the convergence radius of the "generalized" exponential

$$\mathcal{N}(t) := \sum_{n=0}^{+\infty} \frac{t^n}{x_n!} , \tag{2}$$

provided that the (Stieltjes) moment problem has a solution for the sequence of factorials $(x_n!)_{n\in\mathbb{N}}$ defined by $x_n! := x_1 \times x_2 \times \ldots \times x_n$ (and $x_0! := 1$), being $(x_n)_{n\in\mathbb{N}}$ an infinite and strictly increasing sequence of nonnegative real numbers such that $x_0 = 0$. From a physical point of view these x_n could be obtained through some experimental device, for instance, they could constitute the quantum energy spectrum of a given system, but they could be some other kind of observed data.

These vectors $|v_z\rangle$ enjoy properties similar to those obeyed by standard coherent states [1,2]:

(1) $\|v_z\| = 1$.

(2) The map $z \in \mathcal{D}_{\sqrt{R}} \mapsto |v_z\rangle$ is weakly continuous.

(3) The map $n \in \mathbb{N} \mapsto |\langle e_n|v_z\rangle|^2 = \dfrac{|z|^{2n}}{\mathcal{N}(|z|^2)\, x_n!}$ is a Poisson-like distribution in $|z|^2$ with average number of events $|z|^2$ [3].

(4) The map $z \in \mathcal{D}_{\sqrt{R}} \mapsto |\langle e_n|v_z\rangle|^2 = \dfrac{|z|^{2n}}{\mathcal{N}(|z|^2)\, x_n!}$ is a (gamma-like) probability distribution [3] (with respect to $|z|^2$) with x_n as a shape parameter and with respect to the measure $\nu(d^2z)$ on the open disk $\mathcal{D}_{\sqrt{R}}$ given by [4]

$$\nu(d^2z) := \mathcal{N}(|z|^2)\, w(|z|^2)\frac{d^2z}{\pi}\,, \qquad d^2z = d\,\mathrm{Re}z\, d\,\mathrm{Im}z. \tag{3}$$

(5) The family of vectors (1) resolves the unity, i.e. $\int_{\mathcal{D}_{\sqrt{R}}} \nu(d^2z)\, |v_z\rangle\langle v_z| = I_d$.

This last property is fundamental for the quantization of $\mathcal{D}_{\sqrt{R}}$ since it allows one to define:

(i) a normalized positive operator-valued measure on $\mathcal{D}_{\sqrt{R}}$ equipped with the measure $\nu(d^2z)$ (3) and its $\sigma-$algebra $\mathcal{F}$ of Borel sets

$$\Delta \in \mathcal{F} \mapsto \int_\Delta \nu(d^2z)\, |v_z\rangle\langle v_z| \ \in \mathcal{L}(\mathcal{H})^+\,,$$

where $\mathcal{L}(\mathcal{H})^+$ is the cone of positive bounded operators on $\mathcal{H}$;

(ii) a Berezin-Klauder-Toeplitz (or "anti-Wick") quantization (*coherent state quantization*) of the disk $\mathcal{D}_{\sqrt{R}}$ [5–7], i.e. to the function $f(z,\bar{z})$ in $\mathcal{D}_{\sqrt{R}}$ corresponds the operator A_f in $\mathcal{H}$ defined by

$$f \mapsto A_f := \int_{\mathcal{D}_{\sqrt{R}}} \nu(d^2z)\, f(z,\bar{z})\, |v_z\rangle\langle v_z| = \sum_{n,n'=0}^{\infty} (A_f)_{nn'}\, |e_n\rangle\langle e_{n'}| \tag{4}$$

with matrix elements

$$(A_f)_{nn'} = \frac{1}{\sqrt{x_n! x_{n'}!}} \int_{\mathcal{D}_{\sqrt{R}}} \frac{d^2z}{\pi}\, w(|z|^2)\, f(z,\bar{z})\, z^n \bar{z}^{n'}\,.$$

For instance, let us consider some interesting functions from the physical point of view:

1) $f(z,\bar{z}) = z$ and $f(z,\bar{z}) = \bar{z}$ give

$$A_z = a\,, \qquad a\,|e_n\rangle = \sqrt{x_n}|e_{n-1}\rangle\,, \qquad a\,|e_0\rangle = 0 \qquad \text{(lowering operator)},$$

$$A_{\bar{z}} = a^\dagger\,, \qquad a^\dagger\,|e_n\rangle = \sqrt{x_{n+1}}|e_{n+1}\rangle \qquad \text{(raising operator)}\,.$$

The operators a and $a^\dagger$ verify the non-canonical commutation rule

$$[a, a^\dagger] = x_{N+1} - x_N,$$

where the "deformed" number-operator x_N is

$$x_N = a^\dagger a \tag{5}$$

having spectrum $\{x_n\}_{n\in\mathbb{N}}$ and eigenvectors $\{|e_n\rangle\}_{n\in\mathbb{N}}$, i.e. $x_N\,|e_n\rangle = x_n|e_n\rangle$.

2) The function $f(z,\bar{z}) = |z|^2$ can be considered as the classical Hamiltonian of the harmonic oscillator when we have a particle moving on the line and, hence, the complex plane is the phase space of it, i.e. $z = \frac{1}{\sqrt{2}}(q+ip)$ and $|z|^2 = \frac{1}{2}(p^2+q^2)$. Now we get from (4) that

$$A_{\bar{z}z} = A_{z\bar{z}} = A_z\, A_{\bar{z}} = a\, a^\dagger = x_{N+1}\,. \tag{6}$$

Therefore, the spectrum of the quantized version of $|z|^2$ is the sequence $(x_n)_{n\geq 1}$. Note that $A_{\bar{z}}A_z = x_N$.

The operators Q and P, defined in terms of a and $a^\dagger$

$$Q = \frac{1}{\sqrt{2}}(a + a^\dagger)\,, \qquad P = \frac{1}{i\sqrt{2}}(a - a^\dagger)\,, \tag{7}$$

allow us to study the localization properties in $\mathbb{C}$, from the point of view of the sequence $\{x_n\}_{n\in\mathbb{N}}$ by considering the behaviour of the lower symbols $\check{Q}(z) := \langle v_z|Q|v_z\rangle$ and $\check{P}(z) := \langle v_z|P|v_z\rangle$, and the noncommutativity of the complex plane by $\langle v_z|[Q,P]|v_z\rangle$. The dispersions verify

$$(\Delta_{v_z}Q)^2 = (\Delta_{v_z}P)^2 = \frac{1}{2}\,\langle v_z|(x_{N+1} - x_N)|v_z\rangle = \frac{1}{2}|\langle v_z|[Q,P]|v_z\rangle|, \qquad \forall\,|v_z\rangle. \tag{8}$$

Notice that when $x_n = n$ and $R = \infty$ we recover the canonical quantization of $\mathcal{D}_\infty = \mathbb{C}$ as the classical phase space for the motion of a particle on the line, equipped with the usual Lebesgue measure $\nu(d^2z) = d^2z/\pi$. The quantized version of the classical harmonic oscillator Hamiltonian $H = \frac{1}{2}(p^2 + q^2)$ is the number operator $N + 1$ with spectrum $1, 2, 3, \ldots$.

A different statistical interpretation [4] of the phase space can be given by using the measure

$$\nu(d^2z) = w(|z|^2)\,\mathcal{N}(|z|^2)\,\mathrm{H}(|z|^2 - R)\,d^2z/\pi,$$

where $\mathrm{H}(\cdot)$ is the Heaviside function. In this case, classical states are described by the distribution $z \mapsto w(|z-z_0|^2)\mathcal{N}(|z-z_0|^2)\,\mathrm{H}(|z|^2-R)$ instead of Dirac distributions $z \mapsto \delta_{z_0}(z)$ (i.e. by points z_0) . Hence, from the "generalized" CS quantization we can consider the associated sequence $(x_n)_{n\in\mathbb{N}}$ as the energy spectrum according to $A_{\frac{1}{2}(p^2+q^2)} = A_{z\bar{z}} = a\,a^\dagger = x_{N+1}$ (6) .

On the other hand, in different versions of q-harmonic oscillators there appear q-integer numbers ($[n]_q$) like

$$\frac{1 - q^{\pm n}}{1 - q^{\pm}}, \qquad \frac{q^n - q^{-n}}{q - q^{-1}}, \qquad n \in \mathbb{N},\;\; 0 < |q| < 1,$$

as forming the energy spectrum, also q-coherent states have been constructed for these q-oscillators [9–13]. These q-integer numbers, $[n]_q$, are real numbers, deformations of nonnegative integers, that in the limit $q \to 1$ go to the original number n. In general, these sequences of q-numbers are non-integer numbers but we are going to search for those values of q such that their associated q-numbers are integer for all n. These integer sequences $\{[n]_q\}_{n\in\mathbb{N}}$ will be used to construct "generalized" CS and q-harmonic oscillators with spectrum made of positive integer numbers $[n]_q$.

The paper is organized as follows. In Section 2 we obtain integer sequences of q-deformed integers, which are Fibonacci-like sequences and are characterized by $0 < |q| < 1$ such that $1/|q|$ is a quadratic Pisot (more fairly Pisot-Vijayaraghavan) unit. Section 3 is devoted to the resolution of the moment problem, i.e, to finding a (discrete) measure such that $[n]_q! = \int_0^\infty t^n w_q(t)dt$. This point is crucial in order to have a family of vectors (1) resolving the unity. In section 4 we present some computations and graphics for different q-sequences showing the main physical properties of these q-CS. Finally some remarks end the paper.

2. q-Pisot–Fibonacci numbers

Let us start from the following pq-deformation [14] of natural numbers:

$$[[n]]_{qp} := \frac{q^n - p^n}{q - p}, \qquad n \in \mathbb{N}, \quad p, q \in \mathbb{R}.$$

Our task is to find particular values of p and q such that $u_n \equiv [[n]]_{qp} \in \mathbb{N}$. Obviously this is true for $u_0 = 0$ and $u_1 = 1$. From $u_2 = q + p \in \mathbb{N}$ and $u_3 = q^2 + qp + p^2 \in \mathbb{N}$, we see that q and p are quadratic integers, i.e. both are roots of the quadratic equation

$$X^2 - sX + r = 0\,, \tag{9}$$

with $s = u_2 = p + q \in \mathbb{N}$ and $r = pq = u_2^2 - u_3 \in \mathbb{Z}$. Imposing that $q \in \mathbb{R}^*$ and $p \in \mathbb{R}^*$ we get that $0 \le s^2 - 4r = 4u_3 - 3u_2^2 = u_3 - 3r$. So, from $u_3 \in \mathbb{N}$ and $r \in \mathbb{Z}$ we see that q and p are real if $r < 0$ and $u_3 \ge 3r$ if $r > 0$. The choice of $r, s \in \mathbb{Z}^*$ characterizes completely the sequence $\{u_n\}_{n\in\mathbb{N}}$ via a three-term recurrence

$$u_{n+1} - su_n + ru_{n-1} = 0\,, \quad u_0 = 0\,, \quad u_1 = 1\,.$$

Such sequences of numbers generalize the Fibonacci sequence corresponding to $s = 1$ and $r = -1$. In this case $p = \frac{1+\sqrt{5}}{2} \equiv \tau$ (the golden mean) and $q = \frac{1-\sqrt{5}}{2} = -1/\tau$.

We will only consider the cases $r = \pm 1$ and $s > 0$ ($s < 0$ corresponds to a change of sign for both roots):

1. $r = -1$, then $s \ge 1$ and the roots (q, p) of eq. (9) are such that $-1 < q < 0$ and $1 < p = -1/q$.

2. $r = 1$, then $s \ge 3$ and the roots of eq. (9) verify $0 < q < 1$ and $p = 1/q > 1$. The case $s = 2$ has been excluded since $p = q = 1$.

The algebraic integer $p > 1$ is a *quadratic* (the degree of the equation (9)) *Pisot-Vijayaraghavan* (since $p > 1$ and $|q| < 1$) *unit* (because in eq. (9) $r = \pm 1$) [8].

The antisymmetric or fermionic q-deformation of natural numbers appear in the case 1:

$${}^{[f]}[n]_q = \frac{q^n - (-1)^n q^{-n}}{q + q^{-1}} = (-1)^n\, {}^{[f]}[n]_{q^{-1}}\,.$$

The case 2 corresponds to the symmetric or bosonic q-deformation of natural numbers that we will use hereafter:

$${}^{[s]}[n]_q = \frac{q^n - q^{-n}}{q - q^{-1}} = {}^{[s]}[n]_{q^{-1}}\,. \tag{10}$$

3. Moment measure for the symmetric deformation of integers

In order to solve the moment problem for the sequence $\left(x_n \equiv {}^{[s]}[n]_q\right)_{n\in\mathbb{N}}$ given by (10) it is necessary to find a probability distribution $t \in [0, R) \mapsto w_q(t)$ such that

$$x_n! = {}^{[s]}[n]_q! = \int_0^R t^n\, w_q(t)\, dt\,.$$

In this case the associated exponential (2) defines an analytic entire function $\mathfrak{e}_q(z)$ in $\mathbb{C}$ for any $q > 0$ since $R = \lim_{n\to\infty} x_n = \infty\,, \ \ \forall\, q > 0$.

A useful tool is the "auxiliary" exponential

$$\mathfrak{E}_q(t) := \sum_{n=0}^{\infty} q^{\frac{n(n+1)}{2}} \frac{t^n}{x_n!}$$

whose radius of convergence is ∞ for $0 < q \le 1$ and $1/(q - q^{-1})$ for $q > 1$. It is connected with the two standard q-exponentials [15]

$$e_q^x := \sum_{n=0}^{\infty} \frac{x^n}{[n]_q!}\,, \qquad E_q^x := \sum_{n=0}^{\infty} q^{n(n-1)/2} \frac{x^n}{[n]_q!}\,, \qquad [n]_q := \frac{1 - q^n}{1 - q}\,,$$

by

$$\mathfrak{E}_q(t) = e^{qx}_{q^{-2}} = E^{qx}_{q^2} \,. \tag{11}$$

From the relation $e_q^{-x} E_q^x = 1$ we get that $[\mathfrak{E}_q(t)]^{-1} = \mathfrak{E}_{q^{-1}}(-t)$. The auxiliary exponential $\mathfrak{E}_q(t)$ is involved in the q-integral representation of the q-gamma function $\Gamma_q(x)$ defined by [15]

$$\Gamma_q(x) = \frac{(1-q)_q^{x-1}}{(1-q)^{x-1}} \,, \qquad x > 0 \,,$$

where

$$(1+a)_q^\infty = \prod_{j=0}^{\infty}(1+q^j a) \,, \qquad (1+a)_q^t = \frac{(1+a)_q^\infty}{(1+q^t a)_q^\infty} \,, \quad t \in \mathbb{C} \,.$$

The q-gamma function obeys

$$\Gamma_q(1) = 1 \,, \quad \Gamma_q(x+1) = [x]_q \, \Gamma_q(x) \,, \quad \Gamma_{q^2}(x+1) = q^{t-1\,[s]}[x]_q \, \Gamma_{q^2}(x) \quad \forall x > 0 \,.$$

A first integral representation of $\Gamma_q(x)$ is

$$\Gamma_q(x) = \int_0^{\frac{1}{1-q}} t^{x-1} E_q^{-qt} \, d_q t \,.$$

The q–integral [16, 17]

$$\int_0^a f(t) \, d_q t = (1-q) \sum_{j=0}^{\infty} a \, q^j \, f(a \, q^j)$$

becomes an ordinary integral with a discrete measure

$$\int_0^a f(t) \, d_q t = \int_0^\infty \rho_q(t,a) f(t) \, dt \,, \qquad \rho_q(t,a) = (1-q) \sum_{j=0}^{\infty} a q^j \, \delta(t - a q^j) \,.$$

Hence

$$\Gamma_q(x) = \int_0^\infty \rho_q(t, 1/(1-q)) \, t^{x-1} E_q^{-qt} \, dt \,. \tag{12}$$

Taking nonnegative integer values $x = n$ for $\Gamma_{q^2}(x+1)$ we get that

$$\Gamma_{q^2}(n+1) = q^{\frac{n(n-1)}{2}\,[s]}[n]_q! = q^{-n} \, q^{\frac{n(n+1)}{2}} \, x_n! \,. \tag{13}$$

From (13), (12) and (11) we get the solution for the moment problem with a positive measure for the sequence $\left(q^{\frac{n(n+1)}{2}} x_n!\right)_{n \in \mathbb{N}}$

$$q^{\frac{n(n+1)}{2}} x_n! = \int_0^\infty \varpi_q(t) \, t^n \, \mathfrak{E}_q(-t) \, dt \,, \qquad \varpi_q(t) = \sum_{j=0}^{\infty} q^{2j} \, \delta\left(t - \frac{q^{2j}}{q^{-1} - q}\right) . \tag{14}$$

In order to solve the moment problem for $(x_n!)_{n \in \mathbb{N}}$ it is necessary to solve it for $\left(q^{-\frac{n(n+1)}{2}}\right)_{n \in \mathbb{N}}$.
From [18, 19] the solution to the moment problem for $\left(q^{-\frac{n(n+1)}{2}}\right)_{n \in \mathbb{N}}$, with $0 < q < 1$, is given by

$$q^{-\frac{n(n+1)}{2}} = \int_0^\infty t^n g_q(t) \, dt \,, \qquad g_q(t) = \frac{1}{\sqrt{2\pi \, |\ln q|}} \exp\left[-\frac{(\ln(\sqrt{q} t)^2}{2|\ln q|} \right] . \tag{15}$$

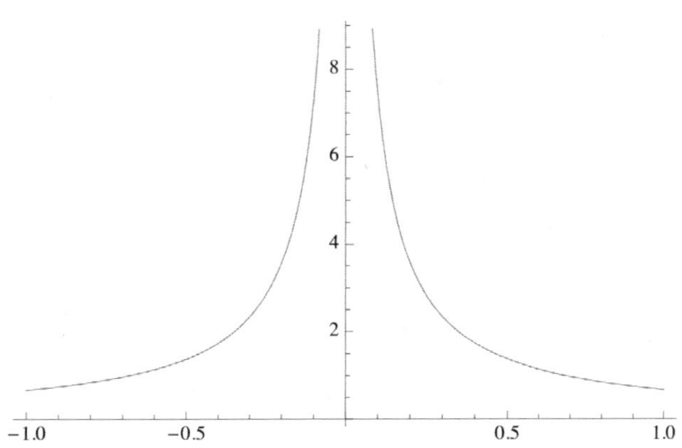

Figure 1. Representation of the function $g_q(t)$ for $q = (3 - \sqrt{5})/2$.

In Fig. 1 we display the function $g_q(z)$ when $q = (3 - \sqrt{5})/2$ which is solution of eq. (9) when $s = 3$ and $r = 1$.

Finally, the application of a composition formula for moments [18] to the product of (14) and (15) yields the solution to the moment problem for the sequence $(x_n!)$:

$$x_n! = \int_0^\infty t^n w_q(t)\, dt\,, \qquad w_q(t) = \int_0^\infty g_q(t/u)\, \varpi_q(u)\, \frac{du}{u}\,.$$

where $w_q(t)$ is a the positive measure density (for a detailed proof see Ref. [20]).

4. Pisot–Fibonacci q-CS quantization of the complex plane

After the moment problem for the sequence $\left(x_n \equiv {}^{[s]}[n]_q\right)_{n\in\mathbb{N}}$ is solved, we are able to construct a family of q-CS associated to these Pisot–Fibonacci symmetric sequences (section 2) according to formula (1). We now proceed with numerical explorations by choosing the lowest cases $s = 3, 4$ and 5 in eq. (9) with $r = 1$, that correspond to $q = (3-\sqrt{5})/2$ ($= 1/\tau^2$ with τ the golden mean), $q = 2 - \sqrt{3}$ and $q = (5 - \sqrt{21})/2$, respectively. The corresponding sequences of q-numbers ${}^{[s]}[n]_q$ are generalizations of the Fibonacci sequence: in particular for $q = (3 - \sqrt{5})/2$ we obtain Fibonacci numbers occupying the odd place in the Fibonacci series.

In the following we study some physical properties of these Pisot–Fibonacci q-CS for different values of q and we compare with the standard ($q = 1$) CS.

The explicit expression of these Pisot–Fibonacci q-coherent states (1) is

$$|v_z\rangle_q = \sum_{n=0}^{\infty} \frac{1}{\sqrt{\mathcal{N}_q(|z|^2)}} \frac{z^n}{\sqrt{{}^{[s]}[n]_q!}}\, |e_n\rangle\,.$$

It is easy to see that the Pisot–Fibonacci q-CS $|v_z\rangle_q$ goes to the standard CS $|v_z\rangle$ in the limit $q \to 1$ [5, 21, 22]

$$|v_z\rangle = \sum_{n=0}^{\infty} \frac{1}{\sqrt{\exp(|z|^2)}} \frac{z^n}{\sqrt{n!}}\, |e_n\rangle\,,$$

since ${}^{[s]}[n]_q \to n$ and $\mathcal{N}_q(|z|^2) \to \exp\left(|z|^2\right)$ when q goes to 1. A graphical way to see this limit is to consider the ratio-function $d_q(n) = {}^{[s]}[n]_q!/n!$ and display it versus n for different values of q [13] (see Fig. 2a and also Fig. 2b).

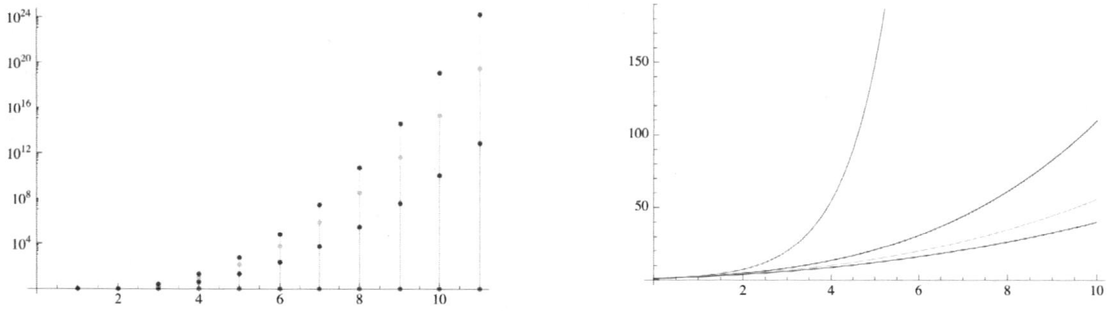

(a) $d_q(n)$ (b) $\mathcal{N}_q(t)$

Figure 2. Representation of $d_q(n)$ and $\mathcal{N}_q(t)$ with $t \geq 0$ for $q = 1$ (red), $(3-\sqrt{5})/2$ (blue), $2-\sqrt{3}$ (green), $(5-\sqrt{21})/2$ (purple).

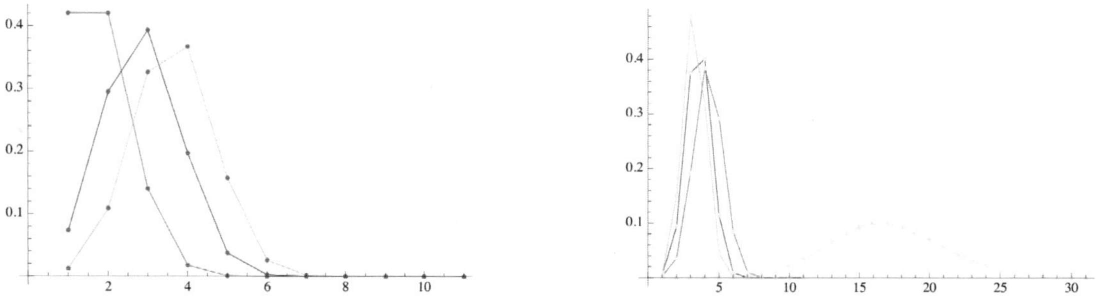

(a) $|z| = 4$ (red), 5 (blue), 6 (green) and $q = (3-\sqrt{5})/2$.

(b) $|z| = 4$, $q = 1$ (yellow), $(3-\sqrt{5})/2$ (red), $2-\sqrt{3}$ (blue), $(5-\sqrt{21})/2$ (green).

Figure 3. Representations of Poisson-like distributions $\rho_q(n, |z|)$ versus n for different values of the parameters $|z|$ and q.

Let $|e_n\rangle$ be the state of n bosons. The probability of finding n bosons in the q-CS $|v_z\rangle_q$ is given by the q-Poisson distribution

$$\rho_q(n, |z|) = \frac{|z|^{2n}}{\mathcal{N}(|z|^2)\, x_n!}.$$

Its limit when $q \to 1$ is the standard Poisson distribution. The q-Poisson distributions are left displaced with respect to the standard Poisson distribution, hence $\rho_q(n, |z|)$ is a sub-Poissonian distribution (see Fig. 3a and Fig. 3b).

Expression (8) shows us that these generalized-CS are intelligent states for the operators Q and P (7). In particular, for our Pisot–Fibonacci q-CS the variances of Q and P (8) are

$$(\Delta_{v_z} Q)^2 = (\Delta_{v_z} P)^2 = \frac{1}{2}\left| \frac{1}{\mathcal{N}(|z|^2)} + (s-1)|z|^2 - |z|^4 \left\langle \frac{1}{x_{n+2}} \right\rangle \right| \geq 1/2\,, \qquad \forall z \in \mathbb{C}. \qquad (16)$$

According to expression (16) only for $z = 0$, i.e. for the vacuum state $|v_0\rangle = |e_0\rangle$, $(\Delta_{v_0} Q)^2 = (\Delta_{v_0} P)^2 = 1/2$. So, the states $|v_z\rangle_q$ are not squeezed coherent states for any q and z.

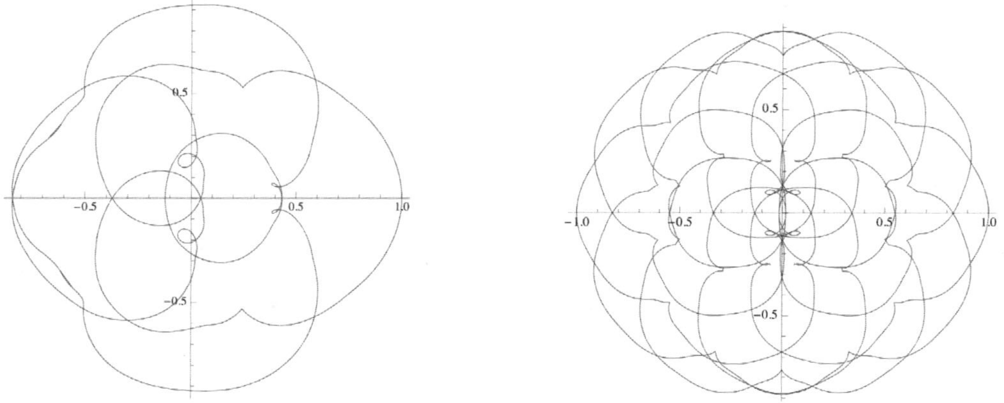

(a) $q = (3 - \sqrt{5})/2$. (b) $q = 2 - \sqrt{3}$.

Figure 4. Plots of $\Im \check{z}\,(t)$ versus $\Re \check{z}\,(t)$ for different values of q and $0 \leq t \leq 8\pi$.

Another point of interest is the behavior of the trajectories in phase space. So, we consider the time-evolution of $a = A_z$, i.e. $A_z(t) = e^{-i\hat{H}t}\, A_z\, e^{i\hat{H}t}$, which is given by the mean value in terms of the q-coherent states $|v_z\rangle_q$

$$\check{z}\,(t) = {}_q\langle v_z|\, e^{-i\hat{H}t}\, A_z\, e^{i\hat{H}t}\, |v_z\rangle_q = \frac{z}{\mathcal{N}_\lambda(|z|^2)} \sum_{n=0}^{+\infty} \frac{|z|^{2n}}{x_n!} \exp\left(-i\left(x_{n+2} - x_{n+1}\right)t\right) .$$

In Fig. 4 we plot $\Im \check{z}\,(t)$ versus $\Re \check{z}\,(t)$ for different values of q and $0 \leq t \leq 8\pi$. The phase space trajectories are periodic with period 2π because the q-numbers involved in these CS are integers.

5. Conclusions

In this paper we have introduced a q-dependent family of coherent states and we have explored some properties of these coherent states as well as the quantum harmonic oscillator obtained through the corresponding coherent state quantization. We have restricted our study to the case in which q^{-1} is a quadratic unit Pisot number, since then the spectrum of the quantum Hamiltonian is made of the q-deformed integers ${}^{[s]}[n]_q = (q^n - q^{-n})/(q - q^{-1})$ which are still integers and form sequences of the Fibonacci type. We have put into evidence interesting quantum features issued from these particular algebraic cases, concerning particularly the localization in the configuration space and in the phase space, probability distributions and related statistical features, time-evolution and semi-classical phase space trajectories. The periodicity of the latter nicely reflects the algebraic Pisot nature of the deformation parameter q. By contrast we present the semi-classical phase space trajectories for irrational values of q (Figs. 5). Obviously, the trajectories are not periodic.

Acknowledgments

This work was partially supported by the Ministerio de Educación y Ciencia of Spain (Project FIS2009-03959) and by the Junta de Castilla y León.

6. References

[1] Klauder J R and Skagerstam B S 1985 *Coherent States - Applications in Physics and Mathematical Physics* (Singapore: World Scientific)

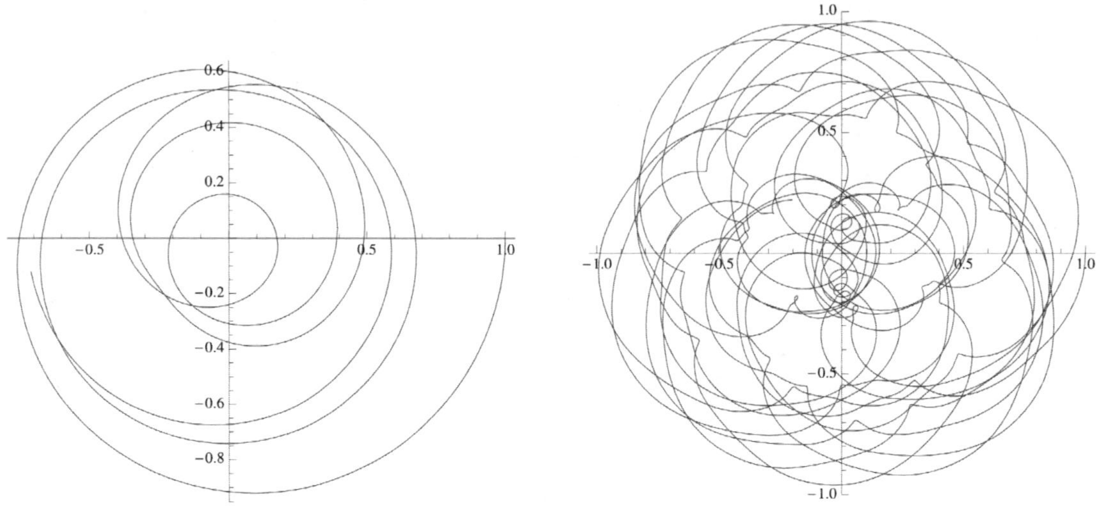

(a) $q = \sqrt{2}$. (b) $q = e$.

Figure 5. Plots of $\Im\tilde{z}(t)$ versus $\Re\tilde{z}(t)$ for irrational values of q and $0 \leq t \leq 8\pi$.

[2] Gazeau J P 2009 *Coherent States In Quantum Physics* (Berlin: Wiley-VCH)
[3] Ali S T, Balkova L, Curado E M F, Gazeau J P, Rego-Monteiro M A, Rodrigues L M C S and Sekimoto K, 2009 it J. Math. Phys. **50** 043517- 1-28;
Curado E M F, Gazeau J P and Rodrigues L M C S 2010 *Phys. Scr.* **82** 038108-1-9
[4] Ali S T, Gazeau J P and Heller B 2008 *J. Phys. A: Math. Theor.* **41** 365302
[5] Klauder J R 1963 *J. Math. Phys.* **4**, 1055 and 1058;
1995 *Ann. of Phys.* **237** 147
[6] Berezin F A 1975 *Commun. Math.Phys.* **40** 153
[7] Chakraborty B, Gazeau J P and Youssef A 2008 Coherent state quantization of angle, time, and more irregular functions and distributions (*Preprint* arXiv: 0805.1847v1)
[8] Bertin M J, Decomps-Guilloux A, Grandet-Hugot M, Pathiaux-Delefosse M, Schreiber J P 1992 *Pisot and Salem Numbers* (Birkhäuser);
Boyd D W 1978 *Math. Comp.* **32** 1244
[9] Arik M and Coon D D 1976 *J. Math. Phys.* **17** 524
[10] Jannussis A, Brodimas G, Sourlas D and Zisis V 1981 *Lett. Nuovo Cimento* **30** 123
[11] Biedenharn L C 1989 *J. Phys. A: Math. Gen.* **22** L873
[12] Macfarlane A J 1989 *J. Phys. A: Math. Gen.* **22** 4581
[13] Quesne C 2002 *J. Phys. A: Math. Gen.* **35** 9213
[14] Chakrabarti R and Jagannathan R 1991 *J. Phys. A: Math. Gen.* **24** L711
[15] De Sole A and Kac V 2005 *Rend. Mat. Acc. Lincei* **9** 11 (*Preprint* arXiv: math.QA/0302032)
[16] Thomae J 1869 *J. reine angew. Math.* **70** 258
[17] Jackson F H 1910 *Quart. J. Pure and Appli. Math.* **41** 193
[18] Bergeron H 2010 Nonlinear coherent states and their generalization: a resolvent-like definition (*Preprint* unpublished)
[19] Gazeau J P, Baldiotti M C and Gitman D M 2009 *Phys. Lett. A* **373** 1916
[20] del Olmo M A and Gazeau J P 2010 Pisot q-Coherent states quantization of the harmonic oscillator (Preprint)
[21] Glauber R J 1963 *Phys. Rev.* **130** 2529;
1963 *Phys. Rev.* **131** 2766
[22] Sudarshan E C G 1963 *Phys. Rev. Lett.* **10** 277.

Symplectic Dynamical Symmetries in Algebraic Models of Nuclear Structure

A.I. Georgieva[1,2], H.G. Ganev[1], M.I. Ivanov[1], V.P. Garistov[1] and J.P. Draayer[2]

[1] Institute for Nuclear Research and Nuclear Energy,
Bulgarian Academy of Sciences, Sofia, Bulgaria;
[2] Department of Physics and Astronomy,
Louisiana State University, Baton Rouge, LA70802

E-mail: anageorg@inrne.bas.bg

Abstract. Based on a generalized reduction scheme for boson representations of symplectic algebras of the type $Sp(4k, R)$, we consider the symplectic extension of a boson realization of compact unitary algebras for the $k = 1$, $k = 3$ and $k = 6$ cases, which have relevance in nuclear structure theory. First we review an application of the $k = 1$ case for the creation of a $Sp(4, R)$ classification scheme, which is used for obtaining a generalized phenomenological description of important nuclear characteristics in terms of the classification quantum numbers for large sets of nuclei. Then for the $k = 3$ and $k = 6$ cases we outline some of the new possibilities that appear in the symplectic extensions of the Interacting Vector Boson Model (IVBM) and the Interacting Boson Model (IBM-2), respectively. The examples presented are used to describe the collective modes of the nuclear spectra in individual nuclei as well as in sequences of nuclei.

1. Introduction

In terms of group theory – the language of symmetries, the concept of the dynamical symmetry of a nuclear system is represented by chains of group-subgroups that reduce the initial symmetry to the $SO(3)$ symmetry of the angular momentum. Each of these chains provide for a labeling of the basis states in terms of the representations of their constituents and for defining the respective Hamiltonians in terms of their Casimir invariants. This approach is commonly employed in two types of applications. The first is related to the classification of the many-body systems under consideration [1], and the second to the development of algebraic models for the description of the different collective modes in the individual nuclei or in sets of nuclei with specific properties. An important advantage of this approach is that exact analytic solutions of the associated eigenvalue problems can be given, which correspond to different limiting cases of physical phenomena.

Symplectic algebras have been applied extensively in the theory of nuclear structure. They are used generally to describe systems with changing number of particles or excitation quanta and in this way provide for larger representation spaces and richer subalgebraic structures. Based on a general reduction scheme of the boson representations of symplectic algebras of the type $Sp(4k, R)$ [2], we review the two types of applications in nuclear structure physics, starting with the $k = 1$ case where we present a classification scheme of the nuclei in the major nuclear shells, which leads to a generalized description of some of their properties.

Published under licence by IOP Publishing Ltd

Next we consider the $k = 3$ case illustrated by the Interacting Vector Boson Model (IVBM) [3] based on the embedding $Sp(12, R) \supset U(6)$, where the symplectic symmetry allows for a change in the number of bosons used to build basis states, which in turn leads to mixing of the collective modes in their description. Additional chains of subgroups that appear as a result of the symplectic generalization of the group of dynamical symmetry are also considered. This approach reveals new features of the collective spectra of the individual heavy even-even nuclei.

A new and different application of this general approach is presented by the $k = 6$ case, where the symplectic extension $Sp(24, R) \supset U(12)$ of the dynamical symmetry group of the Interacting Boson Model-2 (IBM-2) [4] is considered. In this case a change in the number of bosons is associated with a change of the number of valence protons and neutrons used to describe a nuclear system. New reduction chains that arise can be used for a classification of the even-even nuclei in the major nuclear shells, which results in a consistent group-theoretical investigation of the development of collectivity in their spectras and respectvelly in nuclear shapes and deformations, allowing these to be treated in a unified way.

2. Reductions of the boson representations of the $Sp(4k, R)$

The boson representation of the $Sp(4k, R)$ algebra is realized in terms of bilinear combinations of creation $a^{\dagger}_{\alpha i}$ and annihilation $a_{\alpha i}$ operators with two indexes, $\alpha = 1, 2;\ i = 1, 2, ..., k$, that satisfy Bose commutation relations: $[a_{\alpha i}, a^{\dagger}_{\beta j}] = \delta_{\alpha\beta}\delta_{ij}$ (all other commutators are zero).Specifically, by appending the operators

$$a^{\dagger}_{\alpha i}a^{\dagger}_{\beta j},\quad a_{\alpha i}a_{\beta j},\quad \alpha, \beta = 1, 2;\ i, j = 1, ..., k \tag{1}$$

to the Weyl generators $a^{\dagger}_{\alpha i}a_{\beta j}$ of the $U(2k)$ group, the boson representation of the $Sp(4k, R)$ algebra is obtained [5]. It is a reducible one that decomposes into two irreducible representations. One of them acts in the space $\mathcal{H}_+$, spanned over the vectors for which the number of bosons n is even, and the other acts in $\mathcal{H}_-$ defined by the condition n – odd so that $\mathcal{H} = \mathcal{H}_+ + \mathcal{H}_-$. By construction, each of the subspaces $\mathcal{H}_+$ and $\mathcal{H}_-$ spans a reducible representation of $U(2k)$ which decomposes into a direct sum of eigensubspaces of the first Casimir invariant $\mathcal{N} = \sum a^{\dagger}_{\alpha i}a_{\alpha i}$ of $U(2k)$. In this way the totally symmetric irreducible unitary representation (IUR) of $U(2k)$, denoted by $[n]_{2k}$ is realized.

Therefore, the group $U(2k)$ appears as a maximal compact subgroup, which further contains the direct product $U(2) \otimes U(k)$ of the two mutually complementary subgroups generated by the operators $F_{\alpha\beta} = \sum a^{\dagger}_{\alpha i}a_{\beta i}$ and $A_{ij} = \sum a^{\dagger}_{\alpha i}a_{\alpha j}$, respectively. The operator $\mathcal{N} = F_{\alpha\alpha} = A_{ii}$ is the first-order Casimir operator for the groups $U(2)$ as well as $U(k)$. Next, the IURs of the groups $SU(2)$, $SU(k)$, and $SU(2) \otimes SU(k)$ at $n =$ fixed, can be labeled by the eigenvalues $F(F + 1)$ of the operator $\mathbf{F^2}$, where $F = n/2, n/2 - 1, ..., 0\ or\ 1/2$ for n even or odd, respectively. Thus when n is fixed and T is fixed, $2F + 1$ equivalent representations of the group $SU(k)$ arise. Each of them is labeled by the eigenvalues of the operator F_0: $-F, -F + 1, \ldots, F$.

On the other hand, the so called ladder representation of the noncompact group $U(k, k)$ acts in the space of the boson representation of the $Sp(4k, R)$ algebra. There exists a connection between this ladder representation and the boson representation of $U(2k)$, which is realized through the third generator F_0 of the multiplier $SU(2)$ of the already mentioned direct product. This operator is also the first Casimir operator of the group $U(k, k)$. Different aspects of this relationship will be revealed in more details in the applications. As shown in [2], both reduction chains

$$Sp(4k, R) \supset U(2k) \supset U(2) \otimes U(k) \supset SU(k) \tag{2}$$

$$Sp(4k, R) \supset U(k, k) \supset U(k) \otimes U(k) \supset SU(k) \tag{3}$$

are equally convenient for the description of the representations of the final group $SU(k)$.

3. The $k = 1$ case and the $Sp(4, R)$ classification scheme with application

To illustrate the $k = 1$ case, we apply the above algebraic construction to a classification of the many-body nuclear systems. Similar methods are used for the classification of elementary particles. In this case we realize the generators in terms of two types of one-dimensional (scalar) creation $(\pi^\dagger, \nu^\dagger)$ and annihilation (π, ν) operators. The reduction of the boson representation of the classification group $Sp(4, R)$ to its compact $u(2)$ and non-compact $u(1,1)$ subalgebras [6],

$$sp(4,R) \begin{array}{ccc} N_t \nearrow & u(2) & \searrow F_0 \\ & & \\ F_0 \searrow & u(1,1) & \nearrow N_t \end{array} u_\pi(1) \oplus u_\nu(1), \qquad (4)$$

is the mathematical underpinning of the scheme. As illustrated by (4), the reduction is realized by means of the operator that counts the total number of particles, $N_t = (N_\pi + N_\nu)$ $(N_\pi = \pi^\dagger \pi, N_\nu = \nu^\dagger \nu\ )$ which is the first order invariant of $u(2)$, and the operator of the third projection of the F-spin, $F_0 = \frac{1}{2}(N_\pi - N_\nu)$, which does not differ essentially from the first order Casimir of $u(1,1)$. N_t reduces the space $\mathcal{H}$, in which the boson representation of $sp(4, R)$ acts, into a direct sum of a totally symmetric irreducible unitary representations /IUR/ of $su(2)$, labelled by $N_t = 0, 2, 4, ...$ (even $\mathcal{H}_+$) or $N_t = 1, 2, 3, ...$ (odd $\mathcal{H}_-$). The operator F_0 reduces the space $\mathcal{H}$ to the ladder series of $u(1,1)$, defined by its fixed eigenvalues. The same operator F_0 reduces each $u(2)$ representation (fixed value of N_t) to the representations of $u_\pi(1) \oplus u_\nu(1)$ labeled by N_π and N_ν, respectively. The same is obtained by reducing the $u(1,1)$ ladders with the operator N_t.

The relation of the algebraic operators used in the classification scheme to the nuclear characteristics in the valence shell, is quite natural when $N_\pi = \frac{1}{2}(N_p - Z^{(1)})$ and $N_\nu = \frac{1}{2}(N_n - N^{(1)})$ are counted as the numbers of proton and neutron valence pairs of the nucleus from a given shell, in which $Z^{(1)}$ and $N^{(1)}$ are the numbers of protons and neutrons of the double magic nucleus at the beginning of the shell. Then N_t and F_0 are exactly the operators reducing the $sp(4, R)$ spaces, and their interpretation corresponds to the one of the Interacting Boson Model - 2 (IBM-2) [4], as the total number of valence bosons and the third projection

Table 1. Nuclei from the $(28, 28|50, 50)_+$ shell mapped on the $\mathcal{H}_-$ (N_t-odd) subspace of $Sp(4, R)$. Each nuclei is defined by the total number of valence bosons N or boson holes $\bar{N}_t$, which label the rows from the left and from the right side respectively and the third projections F_0 $(\bar{F_0})$ of the $F-$ spin, which label the columns on the top and bottom, respectively.

		F_0			
N_t	0	-1	-2	-3	$\bar{N}_t$
0	^{56}Ni				22
2	^{60}Zn	^{60}Ni			20
4		^{64}Zn	^{64}Ni		18
6		^{68}Ge	^{68}Zn		16
8		^{72}Se	^{72}Ge	^{72}Zn	14
10		^{76}Kr	^{76}Se	^{76}Ge	12
12		^{80}Sr	^{80}Kr	^{80}Se	10
14		^{84}Zr	^{84}Sr	^{84}Kr	8
16		^{88}Mo	^{88}Zr	^{88}Sr	6
18			^{92}Mo		4
N_t	0	1	2	3	$\bar{N}_t$
		$\bar{F_0}$			

of the F-spin. This is illustrated on Table 1 for the even-even nuclei from the major shell with $28 < Z = N < 50$, labeled as $(28, 28|50, 50)_-$.

In the application of this classification scheme presented below [7], a component of the nuclear mass, called the "semiempirical microscopic mass" (SEM, where $SEM = M_{\rm exp} - M_{macr}$) is investigated empirically for all the nuclei from the shell $(28, 28|50, 50)$. The SEM values are defined as the differences between the experimental masses $M_{\rm exp}$ [8] and the calculated [9] spherical macroscopic mass M_{macr}. The structure independent macroscopic part M_{macr} is evaluated explicitly for spherical shapes through the dependence on the mass number A, neutron, N, or proton, Z, numbers of the nucleus considered. Obviously the residual microscopic part SEM of the nuclear masses contains the structure and deformation dependence, which is best studied in the above mentioned systematics related to the valence nuclear shells. We start with an empirical investigation of the behavior of the SEMs as functions of F_0 for extended multiplets containing all of the even as well as odd A nuclei from the major shell $(28, 28|50, 50)$. The plots of SEM versus F_0 (see Fig. 1) show a clearly expressed parabolic behavior in the isotopic chains. The parabolas form a dome with rather evenly spread points.

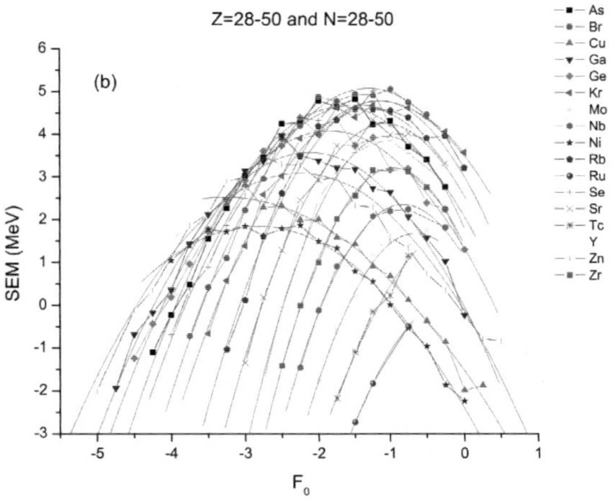

Figure 1. Experimental values of SEMs (shapes) in the isotopic chains (1) of the $(28, 28|50, 50)$ shell plotted versus $F_0^{l\{i\}}$ and their parabolic approximations (full lines) with function (5).

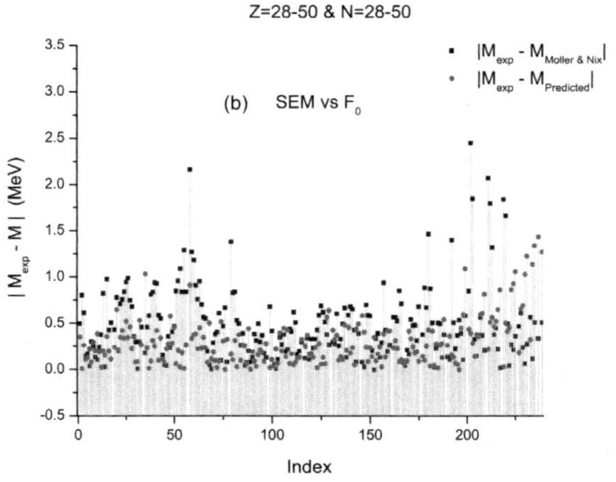

Figure 2. Comparison of the differences $|M_{Exp} - M_{Moller\&Nix}|$ between the experimental and predicted by Moller&Nix [9] masses with our results $|M_{Exp} - M_{Predicted}|$ with the parabolic $F^0_{l\{i\}}$ approximation.

Next we obtain a general functional dependence of the SEM_l on $F_0^{l\{i\}}$ (Fig. 1) in the isotopic chains (enumerated by l) of the considered shell under consideration. Obviously the most convenient fitting function is a parabolic one:

$$SEM(F_0^{l\{i\}}) = C_0^l + C_1^l F_0^{l\{i\}} + C_2^l (F_0^{l\{i\}})^2, l = 0, 1, 2, ..., 16. \quad (5)$$

The values of the parameters C_0^l, C_1^l and C_2^l were determined by means of a fitting procedure for each of the isotopic chains with their respective uncertainties. The latter were further approximated with second order polynomials in the number of valence protons $N_\pi^l = Z_l - 28$, defining the isotopic chains:

$$C_i^l = D_0^i + D_1^i N_\pi^l + D_2^i N_\pi^{l\,2}, i = 0, 1, 2. \quad (6)$$

Table 2. Values of the coefficients $D_k^i, k = 0, 1, 2$;$i = 0, 1, 2$ with their uncertainties ΔD_k^i.

$i \backslash D_k^i$	D_0^i	ΔD_0^i	D_1^i	ΔD_1^i	D_2^i	ΔD_2^i	R^2
0	-2.593	0.070	1.331	0.040	-0.079	0.004	0.975
1	-3.000	0.084	0.055	0.046	-0.006	0.005	0.896
2	-0.529	0.021	-0.048	0.011	-0.004	0.001	0.866

The isotopic chains with $N_\pi^l > 11$ were not included in the second step of the fitting procedure. In this way, the 9 coefficients $D_k^i, k = 0, 1, 2;\ i = 0, 1, 2$ (6) given in Table 2 with their respective uncertainties ΔD_k^i, were determined.
As a result, the general expression (5) for the SEMs of 188 nuclei from the $(28, 28|50, 50)$ shell, depending on 9 phenomenological parameters defined in (6) and with values given in Table 2 was obtained.

To test the accuracy and predictive power of the generalized formula (5), we compare the theoretical values of the nucleonic masses – $M_{Predicted} = SEM(F_0^{l\{i\}}) + M_{macro}$ to the experimental M_{Exp} ones [8]. In Fig. 2, the differences $|M_{Exp} - M_{Predicted}|$ are plotted and compared with the differences $|M_{Exp} - M_{Moller\&Nix}|$, where $M_{Moller\&Nix}$ are the masses obtained by the rather elaborate evaluation of Moller and Nix [9]. The results obtained from the parabolic dependence of SEMs as a function of F_0 (Fig. 1) are closer to the experimental ones than those of Moller and Nix for about 75% of the nuclei considered.

In summary, all the nuclei in the $A = 80$ region with $N = Z = 28 - 50$ can be described quite well by one phenomenological formula with 9 parameters. And in turn, the formula can be used to predict unknown masses of the nuclei from this region.

4. The $k = 3$ case and the symplectic extension of the Interacting Vector Boson Model

The $k = 3$ case of general reductions of boson representations of the $Sp(4k, R)$ (2), (3) algebra will be illustrated through the example of the Interacting Vector Boson Model (IVBM)[3]. The main assumption of the model is that the nuclear dynamics can be described by means of two types of vector "quasiparticles", which are also characterized by another quantum number – a "T - spin" (an analogue of the $F-$spin). The non-compact symplectic group $Sp(12, R)$ appears as the group of dynamical symmetry for the problem of two interacting vector bosons. The symplectic extension of the $U(6)$ algebra, which was the dynamical symmetry of the number preserving version of the model [10] allows the change in the number of phonons, needed to build the collective states, that results in larger model spaces, which can accommodate the more complex structural effects observed in the contemporary experiment.

Applications of the symplectic extension of the IVBM, are obtained by exploiting new chains in the reduction of $Sp(12, R)$ to the angular momentum subgroup $SO(3)$. The corresponding exactly solvable limiting cases are applied to achieve a description of complex nuclear collective spectra of even-even nuclei in the rare earth and actinide regions up to states of very high angular momentum. This generalized reduction scheme links the three dynamical symmetry "pathways" from $Sp(12, R)$ to $SO(3)$ and their respective physical interpretations. In mathematical terms, the relations in Fig. 3 are based on the appearance of the physically important $U(2)$ [11] group of the "T-spin"as the maximal compact subgroup of $Sp(4, R)$ [12] (4), as well as its non-compact $Sp(2, R)$ (or $SU(1, 1)$) counterpart [13].

The first reduction that we exploit is one that extends the rotational limit of the number preserving version of the model; namely, $Sp(12, R) \supset U(6) \supset U(2) \otimes U(3)$ [11]. Another limit of the symplectic IVBM, $Sp(12, R) \supset Sp(2, R) \otimes SO(6)$, contains in a natural way the 6-dimensional Davidson potential [13]. A common feature of these applications is the possibility of mixing with

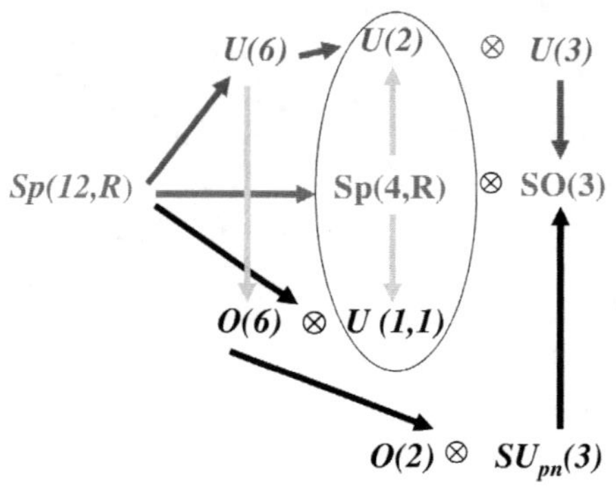

Figure 3. Generalized reduction scheme of the symplectic extension of IVBM.

varying strengths the two main collective modes – vibrational and rotational, which results in an accurate description even of nuclei at the critical points of phase/shape transitions in the framework of these exactly solvable cases, which is illustrated in Fig. 4. It has been established [3] that the two reduction schemes, which describe the developments of collective bands in various types of nuclear spectra, one through $U(6)$ [11] and one through $SO(6)$, [13] although using different realizations of the basis states and the Hamiltonians, because of their connection through the content of the $Sp(4,R)$ (4), yield very similar applications (see Figs. 4 and 5) for a description of the ground bands and the excited positive and negative parity bands. This reveals the important role of the structure of the band-heads, in particular the number of

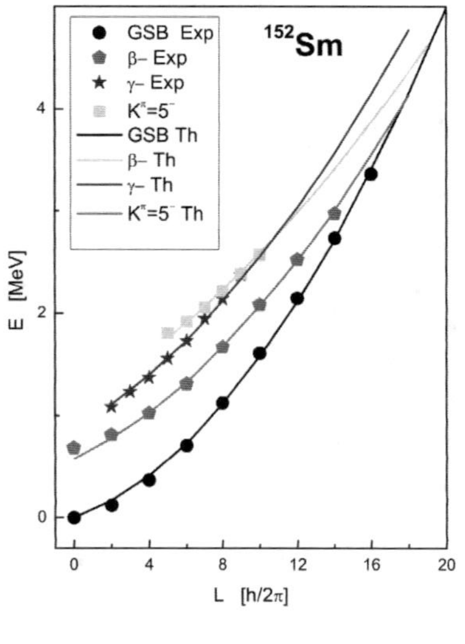

Figure 4. Comparison of the $U(6)$ limit theoretical and experimental energies of the states from several collective positive and negative parity bands of the ^{152}Sm nucleus at the critical point $X(5)$ symmetry.

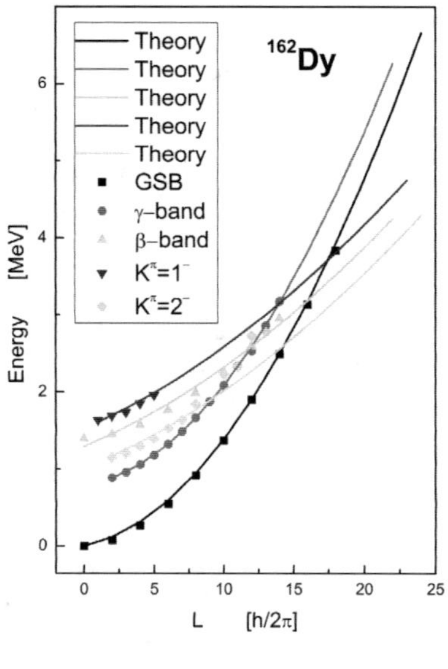

Figure 5. Comparison of theoretical energies calculated in the $Sp(2,R) \otimes SO(6)$ limit of the IVBM with experimental energies for ground and excited bands of the ^{162}Dy.

Journal of Physics: Conference Series **284** (2011) 012028 doi:10.1088/1742-6596/284/1/012028

bosons from which they are built, for the development of the excited bands, a feature that is due to the consideration of the symplectic extension of the model. The structure of band-head configurations, whose importance is established in the first two limits, is examined in the third reduction, $Sp(12, R) \supset Sp(4, R) \otimes SO(3)$ [12]. In this respect the energy distribution of states with fixed angular momenta, and in particular for low-lying ones with $L = 0_i^+, 2_i^+, 4_i^+$, most of which are band-head configurations, obtained in the connecting $Sp(4, R)$ [12], provide a tool to obtain their corresponding N_{L_i}-values upon which the excited bands can be build. From a physical point of view this structure also enables one to distinguish typical collective vibrational and rotational spectra. By means of this vertical structure (the second column in (3)) the dynamical symmetries describing the ground and excited bands are connected with the dynamical symmetry describing the sets of states with fixed angular momentum.

Because of the inclusion relations between the subgroups in the first and second rows of (3) and the relation between the second order Casimir operators of the mutually complementary groups, $SU(3)$ and $SU(2)$, we use the same Hamiltonian and basis as in the $U(6)$ limit [11], but in this case the Hamiltonian eigenvalues give the energies of the states with a fixed value of L with respect to N. In the former the dependence of the energies of the collective states on the number of phonons (vector bosons) N is parabolic. All the rest of the quantum numbers defining the states T, T_0 and L are expressed in terms N by means of the reduction procedure given in [12]. This result confirms the conclusions of the empirical investigation of the states with fixed angular momentum [14], that their energies are well described by the simple phenomenological formula $E_L(n) = an - bn^2$, where $a > 0$ and $b > 0$ are fitting parameters and $n = N_{L_i}/4$ is an integer number corresponding to each of the states with given L_i. The examples presented in Fig. 6 show that the procedure is accurate and appropriate for obtaining the values of the N_i from which the excited 0^+ band-heads can be obtained. As a demonstration of the predictive capability of this approach a second 0^+ excited state with energy 0.6813 MeV was experimentally

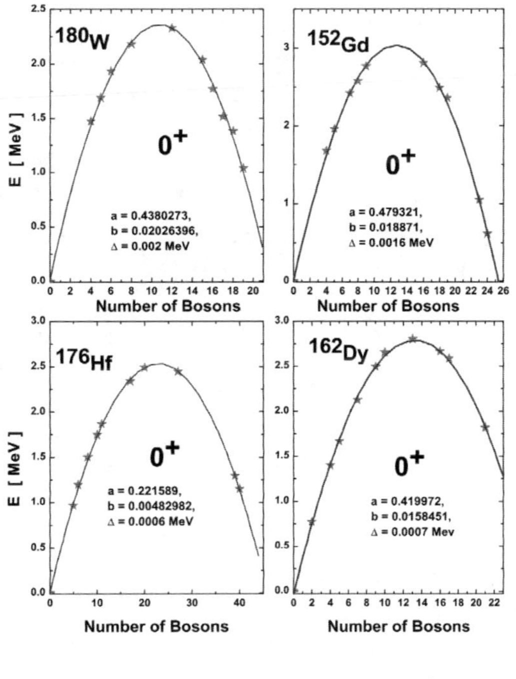

Figure 6. Parabolic distribution of the excited $K^\pi = 0^+$ states in ^{180}W, ^{152}Gd, ^{176}Hf and ^{162}Dy [16].

confirmed in the ^{160}Dy nucleus [15].

While a major appeal of this approach is the simplicity of its application in that it only requires the determination of a relatively small number of model parameters that can be achieved through fits to experimental data, its real power lies in its ability to predict the energetics of excited bands. This predictive capacity is related to the symplectic extension, which allows it to be used to identify the boson number of other band-head configurations and to predict the energies of states that belong to these bands. In order to obtain even greater predicting power of the symplectic extension of the IVBM, we need a further systematic investigation of the behavior of the model parameters as functions of the specific nuclear characteristics [17], as well as of the energy distributions of the band-head states in sequences of nuclei, which is a future aim. So, the symplectic extension of the IVBM permits a fuller classification of states than its unitary version and can be applied to rather diverse nuclear spectra. Another important characteristic of the model it is ability to yield a correct description of experimental energies when interactions between the proton and neutron subsystems are prominent, while still retaining the exact analytic nature of solutions in each of the applications considered.

5. The $k = 6$ case and the generalized symplectic dynamical symmetry of the Interacting Boson Model – 2

The building blocks of the IBM model [18] are the boson creation and annihilation operators $s^\dagger, s$ and $d^\dagger_m, d_m$ ($m = 0, \pm1, \pm2$), representing pairs of valence nucleons coupled to $l = 0$ and $l = 2$ respectively. The algebra that describes these systems is the $U(6)$ algebra. The version of the model, denoted as IBM-2, that considers the proton and neutron nuclear subsystems has dynamical symmetry represented by the direct product of the two algebras $U_\pi(6) \otimes U_\nu(6)$, which is obviously contained in $U(12)$ [19]. Based on the simple classification properties of the $Sp(4, R)$ group, outlined in Section 3, and the interpretation of the reduction operators in terms of the IBM-2 [4], it is straight forward to realize the symplectic extension of the group of dynamical symmetry of the model $U(12) \supset Sp(24, R)$ [1]. In this extension of the model, and in analogy with the generalized reduction scheme of the IVBM, (Fig. 3) we can explore the reduction scheme of the symplectic extension of IBM-2, given in Fig. 7. The formal algebraic

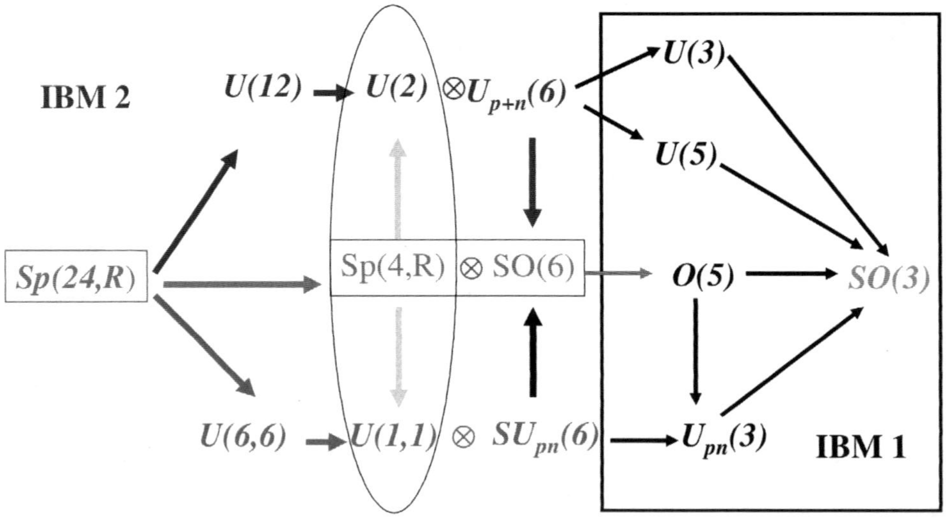

Figure 7. Generalized reduction scheme of the symplectic extension of IBM-2.

aspects of this construction are in a close analogy with the particle-hole version of the IBM and details on it can be found in [20].

From the $u(6)$ algebra down to the algebra of the angular momentum one can proceed with the reductions defining the three limiting cases of IBM-1 [18] through the $U(5)$, $U(3)$ and $O(6)$, which have the anharmonic vibrator, the axial rotor and the γ-unstable rotor as geometrical analogs. Actually the $O(6)$ multiplier of the algebra of $Sp(4,R)$ on the second row of (7) is exactly the one that will start the respective limit of the model. It was shown in [21] that using an infinite dimensional algebraic technique based on the relationship of the quantum numbers of representations and the second order Casimir invariants of the bases $U(5) \supset SO(5)$ and $SO(6) \supset SO(5)$ with the $SU^d(1,1) \supset U(1)$ and the $SU^{sd}(1,1) \supset U(1)$ respectively, exact analytic solutions can be obtained for the $O(6) \leftrightarrow U(5)$ transitional cases.

In this reduction scheme (Fig. 7) we have, as for the $k = 3$ case of the IVBM, the vertical structure in the reduction of $Sp(4,R)$ (4), but in this case related to the classification of all the even-even nuclei from a given major nuclear shell. This follows from the physical interpretation of the reduction operators as the operators of the total number of valence bosons (proton and neutron pairs) – $N = (N_\pi + N_\nu)$, the valence isospin – $F = \frac{N}{2}, \ldots, |\frac{N_\pi - N_\nu}{2}|$ and its third projection $F_0 = \frac{1}{2}(N_\pi - N_\nu)$. This construction involving a classification group in a larger dynamical group allows us to treat in a unified way the properties of sequences of nuclei. Furthermore, a way that is similar to the considerations in [21], it allows us to obtain analytic solutions for the energy spectrum and the transition operators of sequences of nuclei defined by the classification quantum numbers.

In this regard we are motivated by the empirical investigation of the experimental energies of the ground state bands of all even-even nuclei with $A > 20$, [6], based on the $Sp(4,R)$-classification scheme [1]. This reveals a smooth and periodic behavior of the energies as classified in fixed F_0 multiplets for changing values of N_t, which yields a generalized phenomenological description by means of the expression

$$E_L = \beta L\,(L+\Omega) = \beta L\,(L+1) + \beta\, L(\Omega - 1), \tag{7}$$

where the inertial parameter β is evaluated as a function of the classification quantum numbers and a parameter Ω in the geometrical part of the interaction is introduced to account for the mixing of rotational and vibrational degrees of freedom. In a more consistent application of the reduction scheme (Fig. 7), the interactions can be prescribed by the respective dynamical symmetry, as suggested in [6].

6. Conclusions

Based on general reduction schemes for boson representations of symplectic algebras of the type $Sp(4k,R)$ [2], we first considered applications of the simplest $k = 1$ case. Specifically, we showed that the $Sp(4,R)$ algebra is very convenient for classifying many-body nuclear systems within a major nuclear shell when we use an interpretation of the reduction operators in terms of IBM-2 bosons ([19]), representing pairs of valence protons and neutrons, the constituent particles of any nuclei. This interpretation was advanced for a generalized description [6], [1] of nuclear properties based on the quantum labels of the associated classification scheme. The scheme was illustrated through the example of a unified phenomenological formula with only 9 parameters for reproducing SEMs within the major $28 < Z = N < 50$ shell as functions of the third projection F_0 of the $F-$spin. It was also shown to have predictive capability for yet-to-be observed structures like excited band-head configurations and states within these bands [7].

Next we considered the $k = 3$ case by introducing the symplectic extension of the dynamical symmetry $U(6) \supset Sp(12,R)$ of the IVBM, that is constructed by means of two types of vector bosons and used it to describe the collective modes and their interactions of heavy even-even nuclear systems [3]. This extension yields a rich subgroup structure of $Sp(12,R)$ within which some new non-compact subgroup structures appear. The reduction through the direct product $Sp(4, R) \otimes SO(3)$ [12] is of particular importance as it is not only used to describe sequences of states with fixed angular momentum, but also to elucidate the connection of this chain with the other dynamical symmetries of the model through its $SU(2)$ and $SU(1,1)$ subgroups that are also substantial parts of the two other dynamical symmetries considered. In this way it plays the role of a generalized group classification scheme, one that orders (distributes) the collective excitations in spectra of individual nuclei in terms of the collective (boson) structure of their band-head configurations.

And finally we explored a generalized reduction scheme for the symplectic extension ($k = 6$) of the proton-neutron version of the IBM-2 [19]. A point of interest in this case is that the $k = 1$ results reappear with the same interpretation of the $Sp(4, R)$ structure and reductions as discussed in [1]. This limit of the theory provides us with an opportunity to describe, within the framework of the $Sp(24, R)$ dynamical symmetry, the development of collectivity across an entire nuclear shell, allowing for an investigation of transitions between the different limiting cases, while at the same time retaining the strategic advantage of dynamical symmetries for obtaining exact analytic solutions. This approach can also provide for an algebraic evaluation of critical point features, such as phase/shape transitions, in terms of the nuclear characteristics employed in their classification.

In summary, we explored the richer possibilities that the symplectic extensions of unitary algebras provide, and made use of their classification properties in order to achieve generalized descriptions of the nuclear collective behavior.

Acknowledgments This work was supported by the Bulgarian National Foundation (DID-02/16), the U.S. National Science Foundation (PHY-0500291 & OCI-0904874), and the Southeastern Universities Research Association (http:/www.sura.org).

References

[1] Georgieva A, Ivanov M, Raychev P and Roussev R 1989 *Int.J. Theor. Phys.* **28** 769
[2] Georgieva A, Ivanov M, Raychev P and Roussev R 1985 *Int.J. Theor. Phys.***25** 1181
[3] Georgieva A I, Ganev H G, Draayer J P and Garistov V P 2009 *Physics of Particles and Nuclei* **40**, No. 4, 461 - 501.
[4] Arima A, Otsuca T, Iachello F and Talmi I 1977 *Phys.Lett* B **66** 205 ; Elliott J P 1985 *Rep. Prog. Phys* **48** 171
[5] Asherova R M, Fursa D M, Georgieva A I and Smirnov Yu F 1993 *J. Phys.G: Nucl.Part. Phys.* **19** 1887
[6] Drenska S, Georgieva A, Gueorguiev V, Roussev R and Raychev P 1995 *Phys. Rev.* C **52** 1853 - 63
[7] Teymurazyan A, Aprahamian A and Georgieva A 2004 *BgNS Transactions* **9** 165-172.
[8] Wapstra A H , Audy G and Hoekstra R, Midstream (1986) *Atomic mass evaluation* available from the National Nuclear Data Center, Brookhaven National Laboratory
[9] Möller P and Nix J R, 1980 Los Alamos National Laboratory, Report LA-UR-90-1996 ; 1981 *At. Nucl. Data tables* **26** 165
[10] Georgieva A, Raychev P, Roussev R, 1982 *J. Phys. G: Nucl. Phys.* **8**, 1377
[11] Ganev H, Garistov V and Georgieva A 2004 *Phys. Rev.* C **69** 014305
[12] Ganev H, Garistov V P, Georgieva A I and Draayer J P 2004 *Phys. Rev.* C **70**, 054317
[13] Ganev H, Georgieva A I, and Draayer J P 2005 *Phys Rev.* C **72** 014314
[14] Garistov V, *Proceedings of the XXII International Workshop on Nuclear Theory*, 2003 ed. V. Nikolaev, (Heron Press Science Series, Sofia) 305
[15] Garistov V P, et al. 2010 *J. Phys.: Conf. Series* **205**, 012031
[16] Meyer D A et al. 2006 *Phys. Rev.* C **74** 044309
[17] Rowe D J, 2004 *Phys. Rev. Lett.* **93** 122502
[18] Iachello F and Arima A 1978 *The Interacting Boson Model* (Cambridge University Press, Cambridge)
[19] Frank A and Van Isacker P 1985 *Phys. Rev.* C **32** 1770; De Coster C and Heyde K 1989 *Int. J. Mod. Phys.* A **4** 3665
[20] De Coster C, Heyde K, Decroix B, Van Isacker P, Jolie J, Lehmann H, Wood J L 1986 *Nucl. Phys.* A **600**, 251-271
[21] Feng Pan and Draayer J P 1998 *Nucl.Phys.* A **636** 156 - 168

Journal of Physics: Conference Series **284** (2011) 012029 doi:10.1088/1742-6596/284/1/012029

Generalised dihedral subgroups of $\mathrm{SO}(3,\mathbb{Q})$

S Glied

Fakultät für Mathematik, Universität Bielefeld, Universitätsstr. 25, 33501 Bielefeld, Germany

E-mail: sglied@math.uni-bielefeld.de

Abstract. Subgroups of the special orthogonal group $\mathrm{SO}(3,\mathbb{R})$ that are generated by two rotations of finite order about axes that are also separated by an angle of finite order are sometimes called generalised dihedral groups. They appear as orientation groups in the theory of tilings of Euclidean 3-space. As our main result, we classify the generalised dihedral subgroups of $\mathrm{SO}(3,\mathbb{Q})$ as the finite subgroups of $\mathrm{SO}(3,\mathbb{Q})$ except the alternating group A_4.

Introduction

We consider those subgroups of $\mathrm{SO}(3,\mathbb{R})$ that are generated by two rotations of finite order about rotation axes that are themselves separated by an angle of finite order, i.e. an angle that is a rational multiple of π. These groups are called *generalised dihedral groups* in [9]. They appear as orientation groups in the theory of tilings of Euclidean 3-space. For example, the congruent triangular prisms in the "quaquaversal tiling" constructed in [3] appear in an infinite number of orientations. There exists a generalised dihedral group such that the orientations of any two prisms in that tiling are related by an element of this group.

The generalised dihedral subgroups of $\mathrm{SO}(3,\mathbb{R})$ are mostly free products or amalgamated free products of cyclic or dihedral groups and thus generally infinite, the only exceptions being the rotation symmetry groups of certain polyhedra. We use Cayley's parametrisation of $\mathrm{SO}(3,\mathbb{R})$ to represent rotations by quaternions and combine group theoretic results on amalgamated free products of groups with previous results on generalised dihedral subgroups of $\mathrm{SO}(3,\mathbb{R})$ to obtain our main result. It classifies the generalised dihedral subgroups of $\mathrm{SO}(3,\mathbb{Q})$ as the finite subgroups of $\mathrm{SO}(3,\mathbb{Q})$ except the alternating group A_4.

$\mathrm{SO}(3,\mathbb{Q})$ is the group of coincidence rotations of the cubic lattice $\mathbb{Z}^3$ and is thus of basic interest in crystallography for example when studying grain boundaries. The situation here is more complex than for the group $\mathrm{SO}(2,\mathbb{Q})$ of coincidence rotations of the square lattice. This is because $\mathrm{SO}(3,\mathbb{Q})$ is non-Abelian. Generally, the 2-dimensional case is, in contrast to the 3-dimensional one, rather well-understood [1, 6, 2]. Studying the structure of $\mathrm{SO}(3,\mathbb{Q})$ leads to the task of determining its subgroups. The classification of finite subgroups of $\mathrm{SO}(3,\mathbb{Q})$ is well known and this text focuses on a special class of 2-generator subgroups.

This article summarises results of [4] with an emphasis on presenting the main ideas rather than providing all the details. For those, the reader is referred to the original source.

Journal of Physics: Conference Series **284** (2011) 012029 doi:10.1088/1742-6596/284/1/012029

1. Amalgamated free products of groups

Let G_1 and G_2 be groups. If $G = G_1 \star G_2$ is the free product of G_1 and G_2, then every element $g \in G$ can be uniquely written in the form

$$g = g_1 g_2 \cdots g_\ell \quad (\ell \geqslant 0), \tag{1}$$

where $e \neq g_i \in G_{\lambda_i}$ and $\lambda_i \neq \lambda_{i+1}$ for $1 \leqslant i \leqslant \ell - 1$. Here, $\ell = 0$ is understood as $g = e$, the identity element of G. This is called the *normal form of* g. Note that the product of two elements of G is given by juxtaposition followed by the reduction to normal form; cf. [7, Ch. 6.2] for details.

In turn, the free product of groups is characterised by the existence of a normal form. Namely, if G is a group generated by subgroups G_λ, $\lambda \in \Lambda$, and every element has a unique expression of the form (1), then G is the free product of the G_λ's (cf. [7, Thm. 6.2.4]).

Given the free product $G_1 \star G_2$ of two groups G_1 and G_2, there are natural monomorphisms $\iota_i \colon G_i \longrightarrow G_1 \star G_2$ for $i \in \{1, 2\}$. Now let H be a group that is isomorphic to subgroups of G_1 and G_2 via monomorphisms $\varphi_i \colon H \longrightarrow G_i$, ($i \in \{1, 2\}$). Furthermore let N be the normal closure in $G_1 \star G_2$ of the set

$$\{\, \varphi_1(h)\, (\varphi_2(h))^{-1} \mid h \in H \,\}.$$

Then the factor group

$$G = (G_1 \star G_2)\, /N$$

is called the *free product of* G_1 *and* G_2 *with amalgamated subgroup* H *(with respect to* φ_1 *and* φ_2*).* Note that G depends on the choice of the monomorphisms φ_1 and φ_2. Nevertheless, we denote G by $G_1 \star_H G_2$.

Since $\varphi_1(h) \equiv \varphi_2(h) \bmod N$ for all $h \in H$, the subgroups $(\varphi_1(H)N)/N$ and $(\varphi_2(H)N)/N$ are identified in $G_1 \star_H G_2$.

Theorem 1. *[8, Thm. 11.58]* $G_1 \star_H G_2$ *satisfies the following universal property. Let* $\psi_i \colon G_i \longrightarrow T$ $(i \in \{1, 2\})$ *be homomorphisms into some group* T *with* $\psi_1|_H = \psi_2|_H$. *Then there exists a unique homomorphism* $\psi \colon G_1 \star_H G_2 \longrightarrow T$ *that makes the diagram below commutative.*

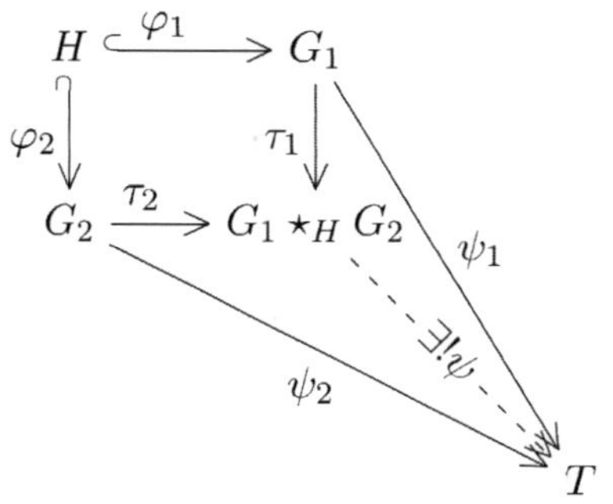

Here, $\tau_i := \mathrm{pr} \circ \iota_i$, *where* pr *is the canonical projection from* $G_1 \star G_2$ *onto* $G_1 \star_H G_2$.

In other words, $G_1 \star_H G_2$ *together with* τ_1 *and* τ_2 *is the pushout of* $\varphi_1 \colon H \longrightarrow G_1$ *and* $\varphi_2 \colon H \longrightarrow G_2$ *in the category of groups.* □

Corollary 1. *If* G' *is any other group with group homomorphisms* $\tau_i' \colon G_i \longrightarrow G'$ $(i \in \{1, 2\})$ *satisfying the universal property of Theorem 1, then* $G' \simeq G_1 \star_H G_2$. *Thus, the universal property characterises free products with amalgamated subgroup.* □

Taking a closer look at the involved homomorphisms in the diagram above, one finds that the homomorphisms $\tau_i \colon G_i \longrightarrow G_1 \star_H G_2$ are injective. Thus, one can identify G_i with its image $\tau_i(G_i)$, making G_i a subgroup of $G_1 \star_H G_2$. Moreover, $G_1 \star_H G_2$ is generated by $\tau_1(G_1)$ and $\tau_2(G_2)$, and their intersection is $\tau_1(H) = \tau_2(H)$, the latter being isomorphic to H; cf. [8, Thm. 10.67] for details. Therefore H can as well be viewed as a subgroup of $G_1 \star_H G_2$.

Theorem 2. *[8, Thm. 11.58] If G_1 has a presentation $\langle Y_1 \mid R_1 \rangle$ and G_2 has a presentation $\langle Y_2 \mid R_2 \rangle$ such that $Y_1 \cap Y_2 = \varnothing$, then the free product of G_1 and G_2 with amalgamated subgroup H has a presentation*

$$\big\langle\, Y_1 \cup Y_2 \;\big|\; R_1 \cup R_2 \cup \{\, \varphi_1(h)\, (\varphi_2(h))^{-1} \mid h \in H \,\}\, \big\rangle.$$

□

Example 1. *Let $C_4 = \langle x \mid x^4 \rangle$ and $C_6 = \langle y \mid y^6 \rangle$ be cyclic groups of order 4 and 6 respectively. Since x^2 and y^3 both have order 2, C_4 and C_6 have a common subgroup C_2. More precisely, let $C_2 = \langle z \mid z^2 \rangle$. The homomorphisms $\varphi_1 \colon C_2 \longrightarrow C_4$ defined by $\varphi_1(z) = x^2$ and $\varphi_2 \colon C_2 \longrightarrow C_6$ defined by $\varphi_2(z) = y^3$ are injective and the subgroups $\varphi_1(C_2)$ of C_4 and $\varphi_2(C_2)$ of C_6 are isomorphic. By Theorem 2, the free product of C_4 and C_6 with amalgamated subgroup C_2 has a presentation*

$$\begin{aligned} C_4 \star_{C_2} C_6 &= \big\langle x, y \,\big|\, x^4, y^6, \{\, \varphi_1(h)\, (\varphi(h))^{-1} \mid h \in H \,\}\big\rangle \\ &= \big\langle x, y \,\big|\, x^4, y^6, \varphi_1(z)\, (\varphi_2(z))^{-1} \,\big\rangle \\ &= \big\langle x, y \,\big|\, x^4, y^6, x^2 y^{-3} \big\rangle \\ &= \big\langle x, y \,\big|\, x^4, x^2 y^{-3} \big\rangle. \end{aligned}$$

Thus x^2 and y^3 are identified in $C_4 \star_{C_2} C_6$.

Hence it is easily possible to achieve a presentation of the free product $G_1 \star_H G_2$ with amalgamated subgroup H from presentations of G_1 and G_2. Given a presentation, it may be very difficult though (if not impossible) to determine the order of the group being presented; cf. [8, Ch. 12]. Without taking a presentation into account, there is a useful theorem to determine whether a free product with amalgamated subgroup has infinite order.

Theorem 3. *[7, Thm. 6.4.3] If both G_1 and G_2 are different from H, then $G_1 \star_H G_2$ has an element of infinite order. Moreover, any element of finite order is conjugate to an element of G_1 or G_2.*

Thus, the group $C_4 \star_{C_2} C_6$ of Example 1 is infinite. In the next lemma, we consider the case where one of the G_i's coincides with the amalgamated subgroup H.

Lemma 1. *One has $H \star_H G_2 \simeq G_2$.*

Proof. This follows easily using Corollary 1. □

2. Generalised dihedral groups

We review some results of [9], where all generalised dihedral groups are classified. To be more concrete, we adopt the notation of [9] and denote by ℓ the line in the x-y-plane of $\mathbb{R}^3$ through the origin that makes an angle of $2\pi n/m$ with the x-axis (where $n, m \in \mathbb{N}$ with $\gcd(n, m) = 1$ and $m > 2$). For $p \in \mathbb{N}$ let $R_x^{2\pi/p}$ be the rotation by $2\pi/p$ about the x-axis and define $R_\ell^{2\pi/q}$ for $q \in \mathbb{N}$ accordingly, see Figure 1 for an illustration.

Define $G_{n/m}(p, q)$ to be the subgroup of $\mathrm{SO}(3, \mathbb{R})$ generated by $R_x^{2\pi/p}$ and $R_\ell^{2\pi/q}$. From three simple facts on rotations by multiples of $\pi/2$ all relations between the two generators $R_x^{2\pi/p}$ and $R_\ell^{2\pi/q}$ are deduced, and thereby a group presentation for $G_{n/m}(p, q)$ is obtained. These simple facts are the following.

$$R_x^{\pi} R_y^{\theta} R_x^{\pi} = R_y^{-\theta}, \qquad \text{where } 0 \leqslant \theta < 2\pi \tag{2}$$

$$R_y^{\pi/2} R_x^{\pi/2} R_y^{\pi/2} = R_x^{\pi/2} R_y^{\pi/2} R_x^{\pi/2} \tag{3}$$

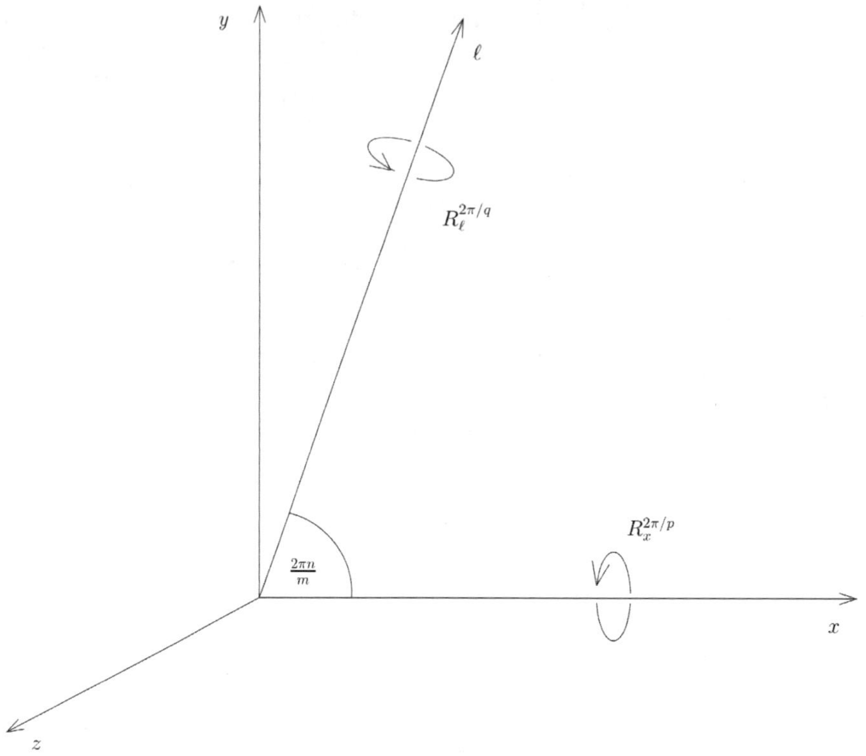

Figure 1. The generators of $G_{n/m}(p,q)$.

$$R_\ell^\pi R_x^\pi = R_z^{4\pi n/m} \tag{4}$$

Many finite groups are generalised dihedral groups. For example, all finite cyclic groups C_p, $p \in \mathbb{N}$, can be viewed as generalised dihedral groups by adding to $R_x^{2\pi/p}$ a second redundant generator, say $R_y^{2\pi}$. One obviously has $C_p \simeq G_{1/4}(p,1)$. The dihedral group D_p of order $2p$, $p \in \mathbb{N}$, has the group presentation $D_p = \langle \alpha, \beta \mid \alpha^p = e, \beta^2 = e, (\alpha\beta)^2 = e \rangle$. We use the simple fact (2) from Euclidean geometry and set $\theta = 2\pi/p$. This implies that $R_x^\pi =: \alpha$ and $R_y^\theta =: \beta$ generate D_m and hence $D_m \simeq G_{1/4}(p,2)$. Furthermore, one has $S_4 \simeq G_{1/4}(4,4)$. One can show that S_4 has the group presentation $S_4 = \langle u, v \mid u^4, v^4, (u^2v)^2, (uv^2)^2, (uv)^3 \rangle$. Here $(u^2v)^2$ and $(uv^2)^2$, respectively, are consequences of (2), whereas the last relation $(uv)^3$ is a consequence of (3).

But not all finite subgroups of $\mathrm{SO}(3,\mathbb{R})$ are generalised dihedral groups.

Lemma 2. *The rotation symmetry group of the tetrahedron A_4 and the rotation symmetry group of the icosahedron A_5 can each be generated by two rotations of finite order, but only with axes separated by an angle of infinite order. Thus, neither A_4 nor A_5 is a generalised dihedral group.*

Proof. We show that any two rotations of A_4 whose rotation axes are separated by an angle of finite order do not generate A_4. Consider the angles between the rotation axes of elements of the tetrahedral group A_4. The only angles of finite order that appear are the ones between two axes of 2-fold symmetry passing from the midpoint of an edge of the tetrahedron to the midpoint of the opposite edge. Two such axes form an orthogonal angle. The corresponding two rotations by π generate a group isomorphic to $G_{1/4}(2,2) \simeq D_2$. Thus A_4 is not a generalised dihedral group. One proceeds similarly for the group A_5. □

Remark 4. In view of our main objective we can already state that all finite subgroups of $\mathrm{SO}(3,\mathbb{Q})$ except the alternating group A_4 are generalised dihedral subgroups of $\mathrm{SO}(3,\mathbb{Q})$. This leaves the task to classify the infinite generalised dihedral subgroups of $\mathrm{SO}(3,\mathbb{Q})$, if there are any.

The classification of the generalised dihedral groups of $\mathrm{SO}(3,\mathbb{R})$ obtained in [9] can be summarised as follows. Mostly, these groups are free products, or amalgamated free products of cyclic or dihedral groups. In most cases, $G_{n/m}(p,q)$ is thus infinite (cf. Section 1), the only exceptions being the groups S_4, C_p and D_p, where $p \in \mathbb{N}$. Together with Lemma 2 and the classification of finite subgroups of $\mathrm{SO}(3,\mathbb{R})$, this implies the following.

Lemma 3. *The only finite subgroups of* $\mathrm{SO}(3,\mathbb{R})$ *are the alternating groups* A_4 *and* A_5 *and the finite generalised dihedral groups.* □

The generalised dihedral groups play a role in the theory of tilings of Euclidean 3-space. An example called "quaquaversal tilings" was constructed in [3], see [10] for an image. A quaquaversal tiling of $\mathbb{R}^3$ consists of congruent copies of a single triangular prism that appears in an infinite number of orientations. This set of orientations forms a dense subgroup of $\mathrm{SO}(3,\mathbb{R})$ that, in fact, is the generalised dihedral group $G_{1/4}(6,4) \simeq D_6 \star_{D_2} D_4$ [11]. (By Lemma 3, this group is infinite.) The quaquaversal tiling can be viewed as a 3-dimensional extension of the pinwheel tiling of the plane. Another example is a 3-dimensional "dite and kart tiling" constructed in [11], which is based on a version of the 2-dimensional kite and dart tiling and has $G_{1/4}(10,4)$ as orientation group.

3. Cayley's parametrisation

We recall some results on quaternions which can be found in [5, 1]. The Hamiltonian quaternion algebra is

$$\mathbb{H}(K) = \mathbb{R}e + \mathbb{R}i + \mathbb{R}j + \mathbb{R}k \tag{5}$$

where the defining relations for the generating elements e, i, j, k are given by

$$i^2 = j^2 = k^2 = -e \quad \text{and} \quad ij = -ji = k. \tag{6}$$

$\mathbb{H}(\mathbb{R})$ is a skew field with unit element e and its elements are called *quaternions*. For a quaternion q it is customary to write $q = (x_0, x_1, x_2, x_3)$ instead of $q = x_0e + x_1i + x_2j + x_3k$. Assign to a nonzero quaternion $\rho = (\kappa, \lambda, \mu, \nu)$ the following matrix $R(\rho)$ in $\mathrm{SO}(3,\mathbb{R})$:

$$R(\rho) = \frac{1}{|\rho|^2}\begin{pmatrix} \kappa^2+\lambda^2-\mu^2-\nu^2 & -2\kappa\nu+2\lambda\mu & 2\kappa\mu+2\lambda\nu \\ 2\kappa\nu+2\lambda\mu & \kappa^2-\lambda^2+\mu^2-\nu^2 & -2\kappa\lambda+2\mu\nu \\ -2\kappa\mu+2\lambda\nu & 2\kappa\lambda+2\mu\nu & \kappa^2-\lambda^2-\mu^2+\nu^2 \end{pmatrix}, \tag{7}$$

where $|\rho|^2 = \kappa^2+\lambda^2+\mu^2+\nu^2$. The next result goes back to Euler in 1770.

Theorem 5. *[5, Ch. 3.6] Every rotation matrix* $M \in \mathrm{SO}(3,\mathbb{R})$ *is of the form* $M = R(\rho)$ *for some nonzero* $\rho \in \mathbb{H}(\mathbb{R})$*, where* $R(\rho)$ *is defined as in (7). The mapping* $\rho \mapsto R(\rho)$ *is called* Cayley's parametrisation of $\mathrm{SO}(3,\mathbb{R})$. □

Thus, in order to obtain all rotation matrices of $\mathrm{SO}(3,\mathbb{R})$, it suffices to consider all nonzero quaternions $\rho = (\kappa, \lambda, \mu, \nu)$ such that $\kappa, \lambda, \mu, \nu \in \mathbb{R}$ and apply the mapping given by (7). But like this, any matrix in $\mathrm{SO}(3,\mathbb{R})$ is obtained multiple times since $R(t\rho) = R(\rho)$ for all nonzero $t \in \mathbb{R}$. This immediately suggests the question whether it is possible to further restrict the entries of ρ and still obtain all elements of $\mathrm{SO}(3,\mathbb{R})$. Since our aim is to classify the generalised dihedral subgroups of $\mathrm{SO}(3,\mathbb{Q})$, we shall restrict the entries of ρ even further. With this in mind, we call a quaternion $\rho = (\kappa, \lambda, \mu, \nu)$ *primitive*, if $\kappa, \lambda, \mu, \nu \in \mathbb{Z}$ and $\gcd(\kappa, \lambda, \mu, \nu) = 1$.

Theorem 6. *[4, Thm. 2.35]* $\mathrm{SO}(3,\mathbb{Q})$ *is parametrised by the set of primitive quaternions. Moreover,* $R(\rho) = R(\rho')$ *holds for two primitive quaternions if and only if* $\rho' = -\rho$. □

We list some properties of Cayley's parametrisation that will prove very useful in handling generalised dihedral subgroups $G_{n/m}(p,q)$ contained in $\mathrm{SO}(3,\mathbb{Q})$.

- The identity matrix E_3 is only parametrised by the nonzero quaternions of the form $(\kappa, 0, 0, 0) \in \mathbb{H}(\mathbb{R})$.
- For any other nonzero quaternion $\rho = (\kappa, \lambda, \mu, \nu)$ the vector $(\lambda, \mu, \nu)^t$ gives the rotation axis of $R(\rho)$. Consequently, if $\rho' = (\kappa', \lambda', \mu', \nu')$ is another nonzero quaternion such that $R(\rho)$ and $R(\rho')$ are "the two generators" of $G_{n/m}(p,q)$, the angle $2\pi n/m$ between those two generators satisfies

$$\cos\left(\frac{2\pi n}{m}\right) = \frac{\langle(\lambda,\mu,\nu)^t, (\lambda',\mu',\nu')^t\rangle}{\sqrt{\lambda^2+\mu^2+\nu^2}\cdot\sqrt{\lambda'^2+\mu'^2+\nu'^2}}$$

- The rotation angle ϕ of $R(\rho)$ is given by

$$\cos(\phi) = \frac{\kappa^2-\lambda^2-\mu^2-\nu^2}{\kappa^2+\lambda^2+\mu^2+\nu^2} \in \mathbb{Q}, \tag{8}$$

  due to $\mathrm{tr}(R(\rho)) = 1 + 2\cos(\phi)$.

This enables us to "translate" the information given on a generalised dihedral group $G_{n/m}(p,q)$ (i.e. the order of the generators and the enclosed angle between their rotation axes) into equations of $\kappa, \lambda, \mu, \nu$ over $\mathbb{Z}$.

Lemma 4. *Let* R *be an element of* $\mathrm{SO}(3,\mathbb{Q})$ *of finite order. Then* R *is a rotation of order* $1, 2, 3, 4$ *or* 6.

Proof. One easily sees that there are rotations of order 1 and 2 in $\mathrm{SO}(3,\mathbb{Q})$. We use Cayley's parametrisation of $\mathrm{SO}(3,\mathbb{Q})$ with primitive quaternions. Let $\rho = (\kappa,\lambda,\mu,\nu) \in \mathbb{H}(\mathbb{Q})$ be a primitive quaternion such that $R(\rho) = R$. Since R has finite order, its rotation angle $\phi \in [0, 2\pi)$ can be expressed as $\phi = 2\pi n/m$, where $\gcd(n,m) = 1$. Equation (8) implies

$$\cos(\phi) = \frac{\kappa^2-\lambda^2-\mu^2-\nu^2}{\kappa^2+\lambda^2+\mu^2+\nu^2} \in \mathbb{Q}.$$

Denote by φ Euler's totient function. One has $\mathbb{Q} = \mathbb{Q}(\cos(2\pi n/m)) = \mathbb{Q}(2\cos(2\pi/m))$, the maximal real subfield of $\mathbb{Q}(\zeta_m)$. If $m \geqslant 3$, its field degree over $\mathbb{Q}$ is $\varphi(m)/2$. Therefore, one has $\varphi(m)/2 = 1$ yielding $m \in \{3,4,6\}$. □

4. Generalised dihedral subgroups of $\mathrm{SO}(3,\mathbb{Q})$

We have already encountered some examples of generalised dihedral subgroups of $\mathrm{SO}(3,\mathbb{Q})$. The cyclic group C_k, where $k \in \{1,2,3,4,6\}$, is a subgroup of $\mathrm{SO}(3,\mathbb{Q})$ generated by a rotation by $2\pi/k$. It can be viewed as a generalised dihedral group by adding to this generator a second redundant one, given by the rotation by 2π about an axis orthogonal to the rotation axis of the first generator. The dihedral groups D_r with $r \in \{1,2,3,4,6\}$ are also generalised dihedral subgroups of $\mathrm{SO}(3,\mathbb{Q})$. They are generated by a rotation of order r and one of order 2 whose rotation axes are separated by an orthogonal angle, cf. Section 2. Furthermore, it was shown that the rotation symmetry group of the cube, S_4, is the generalised dihedral group $G_{1/4}(4,4)$.

It is generated by the two rotations by $\pi/2$ about the x-axis and y-axis, respectively. In matrix form, with respect to the canonical basis, these two generators are

$$\begin{pmatrix} 1 & 0 & 0 \\ 0 & 0 & -1 \\ 0 & 1 & 0 \end{pmatrix} \quad \text{and} \quad \begin{pmatrix} 0 & 0 & 1 \\ 0 & 1 & 0 \\ -1 & 0 & 0 \end{pmatrix}.$$

Therefore, S_4 is a generalised dihedral subgroup of $\mathrm{SO}(3,\mathbb{Q})$. All these examples are finite groups. Since A_4 is not a generalised dihedral group, this leaves no finite groups to consider.

What about other generalised dihedral groups, which are in most cases free products or amalgamated free products of cyclic or dihedral groups? It turns out that, in the rational case, no infinite generalised dihedral groups occur.

Theorem 7. *Let G be a nontrivial generalised dihedral subgroup of $\mathrm{SO}(3,\mathbb{Q})$, i.e., let G be a nontrivial subgroup of $\mathrm{SO}(3,\mathbb{Q})$ that is generated by two rotations of finite order about axes that are separated by an angle of finite order. Then G is one of the following groups:*

the cyclic group C_k with $k \in \{2,3,4,6\}$,
the dihedral group D_ℓ of order 2ℓ with $\ell \in \{2,3,4,6\}$,
or the symmetric group S_4.

In particular, all generalised dihedral subgroups of $\mathrm{SO}(3,\mathbb{Q})$ are finite.

The proof splits into numerous cases depending on the order of the two generators of the generalised dihedral group considered. For a complete proof, the reader is referred to [4, Thm. 2.42]. Here, we shall merely work through a single case that will nonetheless give an idea how the previous results on amalgamated free products, Cayley's parametrisation of $\mathrm{SO}(3,\mathbb{Q})$ and the classification of generalised dihedral groups of $\mathrm{SO}(3,\mathbb{R})$ are combined.

Prior to that, let us formulate one consequence of the main theorem with regard to the classification of finite subgroups of $\mathrm{SO}(3,\mathbb{Q})$.

Corollary 2. *The finite subgroups of $\mathrm{SO}(3,\mathbb{Q})$ are exactly the generalised dihedral subgroups of $\mathrm{SO}(3,\mathbb{Q})$ and the alternating group A_4.* □

4.1. Proof of Theorem 7

We need some further preliminaries. Consider the following equation.

$$\kappa^2 = \lambda^2 + \mu^2 + \nu^2 \tag{I}$$

Lemma 5. *If $\rho = (\kappa,\lambda,\mu,\nu)$ is a primitive quaternion, then the following statements hold.*

(i) $R(\rho)$ is a rotation of order 4 if and only if ρ satisfies Equation (I).

(ii) If Equation (I) holds, then κ is odd. Moreover, exactly one of λ,μ,ν is odd as well.

Proof. (i) If $R(\rho)$ has order 4, then straightforward calculations for its rotation angle ϕ show that $\phi = \pi/2$ or $\phi = 3\pi/2$. In both cases $\cos(\phi) = 0$ and, again by (8), Equation (I) follows. If conversely ρ fulfils (I), let $\varphi \in [0,2\pi)$ be the rotation angle of $R(\rho)$. We deduce from Equation (I) that $k^2 - \lambda^2 - \mu^2 - \nu^2 = 0$ and one therefore has $\cos(\varphi) = 0$ by (8). Thus $\varphi = \pi/2$ or $\varphi = 3\pi/2$ which implies that $R(\rho)$ has order 4.

(ii) The squares of integers are congruent to 0 or to 1 mod 4. If $\kappa^2 \equiv 0 \bmod 4$, then $\lambda^2+\mu^2+\nu^2 \equiv 0 \bmod 4$ and hence, $\lambda^2 \equiv \mu^2 \equiv \nu^2 \equiv 0 \bmod 4$. This implies that all λ,μ,ν are even. But κ is even as well, contradicting the fact that ρ is primitive. Thus $\kappa^2 \equiv 1 \bmod 4$, which shows that κ is odd. Then one furthermore has $1 \equiv \lambda^2+\mu^2+\nu^2 \bmod 4$. This implies that exactly one of λ^2,μ^2,ν^2 is congruent to 1 mod 4, and therefore exactly one of λ,μ,ν is odd.

□

Now let G be a nontrivial generalised dihedral subgroup of $\mathrm{SO}(3,\mathbb{Q})$ that is generated by a rotation of order p and a rotation of order q about rotation axes that enclose an angle of $2\pi n/m$. To prove the main theorem, it suffices to show that G is finite. Due to Lemma 4, only rotations of order $1, 2, 3, 4$ and 6 are possible. In other words, we only have to check for $p, q \in \{1, 2, 3, 4, 6\}$. Here, we merely consider the case $p = q = 4$.

We employ Cayley's parametrisation of $\mathrm{SO}(3,\mathbb{Q})$ with primitive quaternions. Let $\rho = (\kappa, \lambda, \mu, \nu)$ and $\rho' = (\kappa', \lambda', \mu', \nu')$ be two primitive quaternions such that $R(\rho)$ and $R(\rho')$ generate G and such that $R(\rho)$ and $R(\rho')$ are both rotations of finite order p. Let their rotation axes be separated by an angle of $2\pi n/m$ with $n, m \in \mathbb{N}$, $n < m$, $\gcd(n, m) = 1$ and $m > 2$.

$R(\rho)$ and $R(\rho')$ are both rotations of order 4. We translate the information given on $G_{n/m}(4, 4)$ into equations over the integers. By Lemma 5(i), one has

$$\kappa^2 = \lambda^2 + \mu^2 + \nu^2 \tag{I}$$

$$(\kappa')^2 = (\lambda')^2 + (\mu')^2 + (\nu')^2 \tag{I'}$$

Furthermore, one has

$$\cos\left(\frac{2\pi n}{m}\right) = \frac{\lambda\lambda' + \mu\mu' + \nu\nu'}{\sqrt{\lambda^2 + \mu^2 + \nu^2} \cdot \sqrt{(\lambda')^2 + (\mu')^2 + (\nu')^2}}. \tag{II}$$

$G_{n/m}(4, 4)$ is a subgroup of $\mathrm{SO}(3,\mathbb{Q})$ if and only if the three equations above have an integer solution $\{\kappa, \lambda, \mu, \nu, \kappa', \lambda', \mu', \nu'\}$ that is subject to the conditions $\gcd(\kappa, \lambda, \mu, \nu) = 1$ and $\gcd(\kappa', \lambda', \mu', \nu') = 1$.

We confine ourselves to the case where m is odd. But first consider the equations at hand. By Lemma 5(ii), Equations (I) and (I′) imply that κ and κ' are odd. Furthermore, inserting these two equations into Equation (II) yields

$$\cos\left(\frac{2\pi n}{m}\right) = \frac{\lambda\lambda' + \mu\mu' + \nu\nu'}{\kappa\kappa'}. \tag{9}$$

If m is odd, the classification of generalised dihedral groups of $\mathrm{SO}(3,\mathbb{R})$ [9] shows that

$$G_{n/m}(4, 4) \simeq C_4 \star_{C_2} D_m,$$

where $C_4 \star_{C_2} D_m$ denotes the free product of D_m and C_4 with amalgamated subgroup C_2. This group contains D_m as a subgroup as seen in Section 1. Assuming that $G_{n/m}(4, 4)$ is a subgroup of $\mathrm{SO}(3,\mathbb{Q})$, one has

$$D_m \subset G_{n/m}(4, 4) \subset \mathrm{SO}(3,\mathbb{Q}).$$

Now Corollary 4 yields $m \in \{1, 2, 3, 4, 6\}$ as a necessary condition. But m is odd and $m > 2$, hence $m = 3$. Using $\cos(2\pi n/3) = -1/2$ in Equation (9) yields

$$-\frac{1}{2} = \frac{\lambda\lambda' + \mu\mu' + \nu\nu'}{\kappa\kappa'},$$

and thus

$$-\kappa\kappa' = 2(\lambda\lambda' + \mu\mu' + \nu\nu').$$

The left hand side of this equation is odd, because κ and κ' are both odd. But the right hand side is even. This shows that the system of equations consisting of (I), (I′) and (II) does not have a solution over the integers if m is odd. Hence $G_{n/m}(4, 4)$ is not a subgroup of $\mathrm{SO}(3,\mathbb{Q})$ for m odd.

5. Generalised dihedral subgroups of $\mathrm{SO}(3,\mathbb{Q}(\tau))$

Classifications of all generalised dihedral groups of $\mathrm{SO}(3,\mathbb{R})$ and of the generalised dihedral subgroups of $\mathrm{SO}(3,\mathbb{Q})$ have been accomplished. Other real algebraic number fields K are of interest as well. For example, $K=\mathbb{Q}(\tau)$, where $\tau=\frac{1+\sqrt{5}}{2}$ is the golden ratio, is an important object of study in quasicrystallography when dealing with modules that display icosahedral symmetry. It is possible to parametrise also the group $\mathrm{SO}(3,\mathbb{Q}(\tau))$ via a certain type of quaternions. Not surprisingly, the finite generalised dihedral subgroups of $\mathrm{SO}(3,\mathbb{Q}(\tau))$ are precisely the cyclic groups C_k with $k\in\{1,2,3,4,5,6,10\}$, the dihedral groups D_ℓ of order 2ℓ with $\ell\in\{2,3,4,5,6,10\}$ and the symmetric group S_4; cf. [4, Thm. 2.48].

But in contrast to the rational case, $\mathrm{SO}(3,\mathbb{Q}(\tau))$ contains infinite generalised dihedral groups. For example, the group of orientations of the quaquaversal tiling $G_{1/4}(6,4)\simeq D_6 \star_{D_2} D_4$ is infinite by Lemma 3. It is generated by $R(\rho)$ and $R(\rho')$, where $\rho=(2\tau,-1,\tau,\tau+1)$ and $\rho'=(3\tau,\tau+1,0,1)$ [4, Rem. 2.49]. One easily checks that $R(\rho)$ and $R(\rho')$ are elements of $\mathrm{SO}(3,\mathbb{Q}(\tau))$.

However, the complete classification of all generalised dihedral subgroups of $\mathrm{SO}(3,\mathbb{Q}(\tau))$ remains an open problem so far.

Acknowledgments

This work was supported by the German Research Council (DFG), within the CRC 701.

References

[1] Baake M 1997 *The Mathematics of Long-Range Aperiodic Order* ed R V Moody, NATO-ASI Series C 489 (Dordrecht: Kluwer) pp 9–44; Revised version: `arXiv:math/0605222 [math.MG]`
[2] Baake M, Scharlau R and Zeiner P Similar sublattices of planar lattices *Can. J. Math.* in press (*Preprint* `arXiv:0908.2558v1 [math.MG]`)
[3] Conway J H and Radin C 1998 Quaquaversal tilings and rotations *Invent. math.* **132** 179–88
[4] Glied S 2010 *Coincidence and similarity isometries of modules in Euclidean space* `http://bieson.ub.uni-bielefeld.de/volltexte/2010/1749/`
[5] Koecher M and Remmert R 1991 *Numbers*, ed H-D Ebbinghaus et al (New York: Springer) ch 3 pp 189–220
[6] Pleasants P A B, Baake M and Roth J 1996 Planar coincidences for n-fold symmetry *J. Math. Phys.* **37** 1029–58 (*Revised version* `arXiv:math/0511147v1 [math.MG]`)
[7] Robinson D J S 1996 *A Course in the Theory of Groups* 2nd ed (New York: Springer)
[8] Rotman J J 1999 *An Introduction to the Theory of Groups* 4th ed, corrected 2nd printing (New York: Springer)
[9] Radin C and Sadun L 1999 On 2-generator subgroups of SO(3) *Trans. Amer. Math. Soc.* **351** 4469–80
[10] Radin C and Sadun L 1997 Rotating rationally `arXiv:math/9706203v1 [math.MG]`
[11] Radin C and Sadun L 1998 Subgroups of SO(3) associated with tilings *J. Algebra* **202** 611–33

Negative Even Grade mKdV Hierarchy and its Soliton Solutions

J.F. Gomes, G.R. de Melo, G. Starvaggi França and A.H. Zimerman
Instituto de Física Teórica IFT-Unesp
São Paulo - Brasil

E-mail: jfg@ift.unesp.br, gmelo@ift.unesp.br, guisf@ift.unesp.br, zimerman@ift.unesp.br

Abstract. In this paper we discuss the algebraic construction of the mKdV hierarchy in terms of an affine Lie algebra $\hat{sl}(2)$. An interesting novelty araises from the negative even grade sector of the affine algebra leading to nonlinear integro-differential equations admiting non-trivial vacuum configuration. These solitons solutions are constructed systematically from generalization of the dressing method based on non zero vacua. The sub-hierarchies admiting such class of solutions are classified.

1. Introduction

A subclass of non linear integrable models underlined by an affine $\hat{sl}(2)$ Lie algebra is well known to be connected to the mKdV equation. These are in fact higher flows obtained from the zero curvature representation [1] and by general algebraic arguments these flows are restricted to be related to certain positive odd grade generators. Another subclass of nonlinear integrable models containing for instance, the sinh-Gordon model may be formulated, also within the zero curvature representation, but now associated to negative odd generators. The relation between the mKdV and sine-Gordon models was observed some time ago [2], [3] in terms of conservation laws and by more algebraic arguments in [4]. In this paper we review an universal algebraic approach for constructing these models an show that there is a third consistent subclass of models associated to *negative even* grade generators. These later are shown to admit non trivial vacuum configuration.

In this notes we start by reviewing, in section 2, the algebraic construction of the mkdV hierarchy and show how the *positive and negative odd* sub-hierarchis naturally arises from the zero curvature representation. We then discuss the compatibility of the *negative even* sub-hierarchy and develop explicitly the first nontrivial equation. This is an integro-differential nonlinear equation proposed in [5] using recursion operator techniques and later developed further within the algebraic approach in [6]. Contrary to the known cases of positive and negative odd sub-hierarchies, it is observed that a constant (non-zero) configuration is a solution of such equation (non-trivial vacuum solution). This fact introduces extra terms within the zero curvature representation of the vacuum which in turn, induces deformation in the dressing method and in the vertex operators for constructing soliton solutions. In section 3 we discuss in detail the generalization required within the dressing formalism to incorporate this class of solutions and construct explicitly the one solitons solution. Finally in section 4 we discuss and classify the sub-hierarchies admiting non-trivial vacuum solutions.

2. The mKdV Hierarchy

In this section we review the basic ingredients for constructing the mkdV hierarchy. Let us start with an affine algebra $\mathcal{G} = \hat{sl}(2)$ with generatos $\{h^{(m)} = \lambda^m h, E_{\pm\alpha}^{(m)} = \lambda^m E_{\pm\alpha}\}$ satisfying

$$[h^{(m)}, E_{\pm\alpha}^{(n)}] = \pm 2E_{\pm\alpha}^{(n)}, \qquad [E_{\alpha}^{(m)} E_{-\alpha}^{(n)}] = h^{(m+n)} \tag{1}$$

A second important ingredient is a grading operator $Q = 2\lambda\frac{d}{d\lambda} + \frac{1}{2}h$ which decomposes the affine algebra $\mathcal{G}$ into graded subspaces $\mathcal{G} = \oplus \mathcal{G}_a$ such that $[Q, \mathcal{G}_a] = a\mathcal{G}_a$, where,

$$\begin{aligned} \mathcal{G}_{2m} &= \{h^{(m)} = \lambda^m h\}, \\ \mathcal{G}_{2m+1} &= \{\lambda^m (E_\alpha + \lambda E_{-\alpha}),\ \lambda^m (E_\alpha - \lambda E_{-\alpha})\} \end{aligned} \tag{2}$$

for $m = 0, \pm 1, \pm 2, \cdots$ and $[\mathcal{G}_a, \mathcal{G}_b] \subset \mathcal{G}_{a+b}$. The integrable hierarchy is then specified by a choice of a semi-simple element $E = E^{(1)}$, where

$$E^{(2n+1)} = \lambda^n (E_\alpha + \lambda E_{-\alpha}) \tag{3}$$

which decomposes $\mathcal{G} = \mathcal{K} + \mathcal{M}$ where $\mathcal{K} = \{x, /[x, E] = 0\}$ is the Kernel of E and $\mathcal{M}$ is its complement. From (2) it follows that

$$\mathcal{K} = \mathcal{K}_{2n+1} = \{\lambda^n (E_\alpha + \lambda E_{-\alpha})\} \tag{4}$$

has grade $2n + 1$. We assume that E is *semi-simple* in the sense that this second decomposition is such that

$$[\mathcal{K}, \mathcal{K}] \subset \mathcal{K}, \qquad [\mathcal{K}, \mathcal{M}] \subset \mathcal{M}, \qquad [\mathcal{M}, \mathcal{M}] \subset \mathcal{K}.$$

We now define the Lax operator

$$L = \partial + E + A_0, \qquad A_0 \in \mathcal{M} \in \mathcal{G}_0 \tag{5}$$

where $A_0 = vh^{(0)}$ and $v = v(x,t)$ is the physical field.

2.1. Zero Curvature for the Positive Hierarchy

Let us consider the zero curvature representation for the positive hierarchy

$$[\partial_x + E^{(1)} + A_0, \partial_{t_N} + D^{(N)} + D^{(N-1)} + \cdots + D^{(0)}] = 0, \tag{6}$$

where $D^{(j)} \in \mathcal{G}_j$. Eqn. (6) can be decomposed according to the graded structure as

$$\begin{aligned} [E, D^{(N)}] &= 0 \\ [E, D^{(N-1)}] + \partial_x D^{(N)} &= 0 \\ \vdots &= \vdots \\ [A_0, D^{(0)}] + \partial_x D^{(0)} - \partial_{t_N} A_0 &= 0 \rightarrow \text{Eq. of motion} \end{aligned} \tag{7}$$

and the unknown $D^{(j)}$'s can be solved starting from the highest to the lowest grade projections. Notice that, in particular the highest grade equation, namely $[E, D^{(N)}] = 0$ implies $D^{(N)} \in \mathcal{K}$

and since from (4), $\mathcal{K} \in \mathcal{G}_{2n+1}$, $N = 2n+1$. The component within the $\mathcal{M}$ of the zero grade projection leads to the evolutions equations according to time $t = t_{2n+1}$. Examples are,

$$N=3 \qquad 4v_{t_3} = v_{3x} - 6v^2 v_x, \qquad mKdV$$

$$N=5 \qquad 16v_{t_5} = v_{5x} - 10v^2 v_{3x} - 40 v v_x v_{2x} - 10 v_x^3 + 30 v^4 v_x,$$

$$\begin{aligned} N=7 \qquad 64v_{t_7} &= v_{7x} - 182 v_x v_{2x}^2 - 126 v_x^2 v_{3x} - 140 v v_{2x} v_{3x} \\ &- 84 v v_x v_{4x} - 14 v^2 v_{5x} + 420 v^2 v_{3x} + 560 v^3 v_x v_{2x} \\ &+ 70 v^4 v_{3x} - 140 v^6 v_x \end{aligned}$$

$$\cdots etc$$

2.2. Negative mKdV Hierarchy

For the negative mKdV hierarchy let us propose the following form for the zero curvature representation

$$[\partial_x + E^{(1)} + A_0, \partial_{t_{-N}} + D^{(-N)} + D^{(-N+1)} + \cdots + D^{(-1)}] = 0. \tag{8}$$

Differently from the positive hierarchy case, the lowest grade projection now yields,

$$\partial_x D^{(-N)} + [A_0, D^{(-N)}] = 0$$

a nonlocal equation for $D^{(-N)}$. Having solved for $D^{(-N)}$, the second lowest projection of grade $-N+1$, leads to

$$\partial_x D^{(-N+1)} + [A_0, D^{(-N+1)}] + [E^{(1)}, D^{(-N)}] = 0$$

which determines $D^{(-N+1)}$. The proccess follows recursively until we reach the zero grade projection

$$\partial_{t_{-N}} A_0 + [E^{(1)}, D^{(-1)}] = 0 \tag{9}$$

which yields the evolution equation for field A_0 according to time $t = t_{-N}$ Notice that in this case there is no condition upon N.

The simplest example is to take $N = 1$ when the zero curvature decomposes into

$$\begin{aligned} \partial_x D^{(-1)} + [A_0, D^{(-1)}] &= 0, \\ \partial_{t_{-1}} A_0 - [E^{(1)}, D^{(-1)}] &= 0. \end{aligned} \tag{10}$$

In order to solve the first equation, we define the zero grade group element $B = \exp(\mathcal{G}_0)$ and define

$$D^{(-1)} = B^{-1} E^{(-1)} B, \qquad A_0 = B^{-1} \partial_x B, \tag{11}$$

Under such parametrization the second eqn. (11) becomes the well known Leznov-Saveliev equation,

$$\partial_{t_{-1}} \left(B^{-1} \partial_x B \right) = [E^{(1)}, B^{-1} E^{(-1)} B] \tag{12}$$

which for $\hat{sl}(2)$ with principal gradation $Q = 2\lambda\frac{d}{d\lambda} + \frac{1}{2}h$, yields the sinh-Gordon equation (*relativistic*)

$$\partial_{t_{-1}}\partial_x\phi = e^{2\phi} - e^{-2\phi}, \qquad B = e^{\phi h}. \tag{13}$$

where $t_{-1} = z, x = \bar{z}, A_0 = vh = \partial_x\phi h$. An important fact to point out is that $\phi = 0$ corresponds to the vacuun solution of eqn. (13).

The next simple example is for $N = 2$ when the zero curvature decomposes as (see [6])

$$\begin{aligned} \partial_x D^{(-2)} + [A_0, D^{(-2)}] &= 0, \\ \partial_x D^{(-1)} + [A_0, D^{(-1)}] + [E^{(1)}, D^{(-2)}] &= 0, \\ \partial_{t_{-2}} A_0 - [E^{(1)}, D^{(-1)}] &= 0. \end{aligned} \tag{14}$$

Propose solution of the form

$$\begin{aligned} D^{(-2)} &= c_{-2}\lambda^{-1}h, \\ D^{(-1)} &= a_{-1}\left(\lambda^{-1}E_\alpha + E_{-\alpha}\right) + b_{-1}\left(\lambda^{-1}E_\alpha - E_{-\alpha}\right). \end{aligned}$$

we find $c_{-2} = const$ and

$$\begin{aligned} a_{-1} + b_{-1} &= 2c_{-2}\exp(-2d^{-1}v)d^{-1}\left(\exp(2d^{-1}v)\right), \\ a_{-1} - b_{-1} &= -2c_{-2}\exp(2d^{-1}v)d^{-1}\left(\exp(-2d^{-1}v)\right), \end{aligned}$$

Equation of motion is

$$\partial_{t_{-2}}v + 2c_{-2}e^{-2d^{-1}v}d^{-1}\left(e^{2d^{-1}v}\right) + 2c_{-2}e^{2d^{-1}v}d^{-1}\left(e^{-2d^{-1}v}\right) = 0. \tag{15}$$

where $d^{-1}f = \int^x f(x')dx'$. Oposite to the previous example $v = 0$ is not solution of (15), however, it is interesting to notice that $v = v_0 =$ const. $\neq 0$ is a solution.

3. Dressing Method

In order to employ the dressing method to construct systematically soliton solutions, we need to consider the full Kac-Moody algebra with central terms to ensure the existence of highest weight states, i.e.,

$$\begin{aligned} [h^{(m)}, h^{(n)}] &= 2m\delta_{m+n,0}\hat{c}, \\ [h^{(m)}, E_{\pm\alpha}^{(n)}] &= \pm 2E_{\pm\alpha}^{(m+n)}, \\ [E_\alpha^{(m)}, E_{-\alpha}^{(n)}] &= h^{(m+n)} + m\delta_{m+n,0}\hat{c} \end{aligned}$$

together with the derivation operator $\hat{d}$ such that,

$$[\hat{d}, T_a^{(n)}] = nT_a^{(n)}, \qquad T_a^{(n)} = \{h^{(n)}, \ E_{\pm\alpha}^{(n)}\}.$$

The grading operator now reads $Q = 2\hat{d} + 1/2h^{(0)}$.

The zero curvature condition implies pure gauge connections,

$$A_x = T^{-1}\partial_x T = E + A_0$$

and

$$A_{t_n} = T^{-1}\partial_{t_n}T = D^{(n)} + \cdots + D^{(0)}$$

or

$$A_{t_{-n}} = T^{-1}\partial_{t_{-n}}T = D^{(-n)} + \cdots + D^{(-1)},$$

according to positive or negative hierarchies respectively.

3.1. Dressing for zero vacua solution

Suppose there exists a vacuum solution $v = 0$ satisfying

$$A_{x,vac} = E^{(1)} - t_k \delta_{k+1,0} \hat{c}, \qquad A_{t_k,vac} = E^{(k)},$$

The solution for $A_{x,vac} = T_0^{-1} \partial_x T_0$ and $A_{t_k,vac} = T_0^{-1} \partial_{t_k} T_0$ is therefore given by

$$T_0 = \exp(x E^{(1)}) \exp(t_k E^{(k)}).$$

The dressing method is based on the assumption of the existence of two gauge transformations, generated by $\Theta_\pm$, mapping the vacuum into non trivial configuration, i.e.

$$\begin{aligned} A_x &= (\Theta_\pm)^{-1} A_{x,vac} \Theta_\pm + (\Theta_\pm)^{-1} \partial_x \Theta_\pm, \\ A_{t_k} &= (\Theta_\pm)^{-1} A_{t_k,vac} \Theta_\pm + (\Theta_\pm)^{-1} \partial_{t_k} \Theta_\pm. \end{aligned}$$

where $\Theta_\pm$ are group elements of the form

$$\Theta_-^{-1} = e^{p(-1)} e^{p(-2)} \ldots, \qquad \Theta_+^{-1} = e^{q(0)} e^{q(1)} e^{q(2)} \ldots,$$

$p^{(-i)}$ and $q^{(i)}$ are linear combinations of grade $(-i)$ and (i) generators, respectively.

As a consequence we relate

$$\Theta_- \Theta_+^{-1} = T_0^{-1} g T_0$$

where g is an arbitrary constant group element. For instance, from

$$A_x = E^{(1)} + B^{-1} \partial_x B + \partial_x \nu \hat{c} - t_k \delta_{k+1,0} \hat{c} = (\Theta_+)^{-1} (\partial_x + E) \Theta_+,$$

we find,

$$e^{q(0)} = B^{-1} e^{-\nu \hat{c}}$$

and henceforth

$$\ldots e^{-p(-2)} e^{-p(-1)} B^{-1} e^{-\nu \hat{c}} e^{q(1)} e^{q(2)} \ldots = T_0^{-1} g T_0$$

hence,

$$< \lambda' | B^{-1} | \lambda > e^{-\nu} = < \lambda' | T_0^{-1} g T_0 | \lambda >$$

where $|\lambda >$ and $< \lambda'|$ are annihilated by $\mathcal{G}_>$ and $\mathcal{G}_<$, respectively. Explicit space time dependence for the field in $\mathcal{G}_0$, is given by choosing specific matrix elements

$$\begin{aligned} e^{-\nu} &= < \lambda_0 | T_0^{-1} g T_0 | \lambda_0 > \equiv \tau_0, \\ e^{-\phi-\nu} &= < \lambda_1 | T_0^{-1} g T_0 | \lambda_1 > \equiv \tau_1. \end{aligned}$$

i.e.,

$$e^{\phi} = ln \left(\tau_0 / \tau_1 \right)$$

Suppose we now write the constant group element g as

$$g = \exp\{F(\gamma)\},$$

where γ is a complex parameter and we choose $F(\gamma)$ to be an eigenstate of $E^{(k)}$, i.e.

$$[E^{(k)}, F(\gamma)] = f^{(k)}(\gamma) F(\gamma)$$

where $f^{(k)}$ are specific functions of γ. It therefore follows that

$$T_0^{-1} g T_0 = \exp\{\rho(\gamma) F(\gamma)\}$$

where

$$\rho(\gamma) = \exp\{-t_k f^{(k)}(\gamma) - x f^{(1)}(\gamma)\}.$$

For more general cases in which

$$g = \exp\{F_1(\gamma_1)\} \exp\{F_2(\gamma_2)\} \ldots \exp\{F_N(\gamma_N)\}$$

with

$$[E^{(k)}, F_i(\gamma_i)] = f_i^{(k)}(\gamma_i) F_i(\gamma_i)$$

we find

$$T_0^{-1} g T_0 = \exp\{\rho_1(\gamma_1) F_1(\gamma_1)\} \exp\{\rho_2(\gamma_2) F_2(\gamma_2)\} \ldots \exp\{\rho_N(\gamma_N) F_N(\gamma_N)\}$$

where

$$\rho_i(\gamma_i) = \exp\{-t_k f_i^{(k)}(\gamma_i) - x f_i^{(1)}(\gamma_i)\}.$$

The specific eigenstate, in this case of $\hat{sl}(2)$, is given by

$$F(\gamma) = \sum_{n=-\infty}^{\infty} \left(h^{(n)} - \frac{1}{2}\delta_{n,0}\hat{c}\right) \gamma^{-2n} + \left(E_{\alpha}^{(n)} - E_{-\alpha}^{(n+1)}\right) \gamma^{-2n-1}$$

whose eigenvalues are obtained from

$$[E^{(k)}, F(\gamma)] = -2\gamma^k F(\gamma).$$

An important property of Vertex operators $F(\gamma)$ is nilpotency, i.e., It can be shown that

$$F(\gamma)^2 = 0$$

which implies truncation

$$\begin{aligned} T_0^{-1} g T_0 &= \exp\{\rho_1 F_1\} \exp\{\rho_2 F_2\} \ldots \exp\{\rho_N F_N\} \\ &= (1 + \rho_1 F_1)(1 + \rho_2 F_2) \cdots (1 + \rho_N F_N) \end{aligned}$$

for $\rho_i \equiv \rho_i(\gamma_i)$, $\quad F_i \equiv F_i(\gamma_i)$.

Examples soliton solution

$$\begin{aligned} \phi_{1-sol} &= ln\left(\frac{1 - a_1\rho_1}{1 + a_1\rho_1}\right) \\ \phi_{2-sol} &= ln\left(\frac{1 - a_1\rho_1 - a_2\rho_2 + a_1a_2\rho_1\rho_2}{1 + a_1\rho_1 + a_2\rho_2 + a_1a_2a_{12}\rho_1\rho_2}\right) \\ \vdots &= \vdots \end{aligned} \tag{16}$$

$a_{12} = \left(\frac{\gamma_1 - \gamma_2}{\gamma_1 + \gamma_2}\right)$ and

$$\rho_i(\gamma_i) = \exp\{2t_k\gamma_i^k + 2x\gamma_i\}. \tag{17}$$

3.2. Dressing Solution for non-trivial vacuun configuration

Propose the simplest non trivial vacuum configuration

$$A_{x,vac} = \left(E_{\alpha}^{(0)} + E_{-\alpha}^{(1)}\right) + v_0 h^{(0)} - \frac{1}{v_0} t_{-2m}\delta_{m-1,0}\hat{c},$$
$$A_{t_{-2m},vac} = \frac{1}{v_0}\left(E_{\alpha}^{(-m)} + E_{-\alpha}^{(1-m)}\right) + h^{(-m)}$$

with $v_0 = const. \neq 0$. It is straightforward to verify the zero curvature equation

$$[\partial_x + A_{x,vac}, \partial_{t_{-2m}} + A_{t_{-2m},vac}] = 0.$$

This *nontrivial vacuum* leads to the following deformation, i.e., $A_{x,vac} = T_0^{-1}\partial_x T_0$ and $A_{t_k,vac} = T_0^{-1}\partial_{t_k} T_0$,

$$T_0 = \exp\left\{x\left(E_{\alpha}^{(0)} + E_{-\alpha}^{(1)} + v_0 h^{(0)}\right)\right\}\cdot$$
$$\exp\left\{\frac{t_{-2m}}{v_0}\left(E_{\alpha}^{(-m)} + E_{-\alpha}^{(1-m)} + v_0 h^{(-1)}\right)\right\}.$$

which leads to

$$e^{q(0)} = B^{-1} e^{x v_0 h^{(0)}} e^{-\nu\hat{c}}.$$

The solution is then given by

$$e^{-\nu} = < \lambda_0 | T_0^{-1} g T_0 | \lambda_0 > \equiv \tau_0,$$
$$e^{-\phi + x v_0 - \nu} = < \lambda_1 | T_0^{-1} g T_0 | \lambda_1 > \equiv \tau_1$$

and hence,

$$v = v_0 - \partial_x \ln\left(\frac{\tau_0}{\tau_1}\right), \qquad v = \partial_x \phi. \tag{18}$$

In order to construct explicit soliton solutions we need a simultaneous eigenstate of

$$b_1 \equiv E_{\alpha}^{(0)} + E_{-\alpha}^{(1)} + v_0 h^{(0)},$$
$$b_{-2m} \equiv \left(E_{\alpha}^{(-m)} + E_{-\alpha}^{(-m+1)} + v_0 h^{(-m)}\right)$$

Let

$$F(\gamma, v_0) = \sum_{n=-\infty}^{\infty}\left(\gamma^2 - v_0^2\right)^{-n}[h^{(n)} + \frac{v_0 - \gamma}{2\gamma}\delta_{n,0}\hat{c} + E_{\alpha}^{(n)}(\gamma + v_0)^{-1} - E_{-\alpha}^{(n+1)}(\gamma - v_0)^{-1}].$$

be our deformed vertex operator. A direct calculation shows that

$$[b_1, F(\gamma, v_0)] = -2\gamma F(\gamma, v_0),$$
$$[b_{-2m}, F(\gamma, v_0)] = -2\gamma\left(\gamma^2 - v_0^2\right)^{-m} F(\gamma, v_0).$$

Therefore we find,

$$\rho(\gamma, v_0) = \exp\left\{2\gamma x + \frac{2\gamma t_{-2m}}{v_0\left(\gamma^2 - v_0^2\right)^m}\right\}. \tag{19}$$

if we take,

$$g = \exp\{F(\gamma, v_0)\}$$

We find the soliton solution from (18) with

$$\tau_0 = 1 + C_0\rho(\gamma, v_0), \qquad \tau_1 = 1 + C_1\rho(\gamma, v_0)$$

4. Models with non-trivial vacua

In this section we consider non-trivial vacum configuration, i.e., $v = v_0 \neq= 0$ and discuss which models within the hierarchy satisfy the zero curvature representation. Let us consider the following cases

- **Negative Even Hierarchy**

  The zero curvature representation in the vacuum is

$$[\partial_x + E + v_0 h^{(0)}, \partial_{t_{-2m}} + D_{vac}^{(-2m)} + D_{vac}^{(-2m+1)} + \cdots + D_{vac}^{(-2)} + D_{vac}^{(-1)}] = 0$$

with solution given as

$$D_{vac}^{(-2m)} = \lambda^{-m} h^{(0)} \qquad D_{vac}^{(-2m+1)} = \frac{\lambda^{-m} E}{v_0},$$
$$\vdots \qquad \vdots$$
$$D_{vac}^{(-2)} = \lambda^{-1} h^{(0)} \qquad D_{vac}^{(-1)} = \frac{\lambda^{-1} E}{v_0},$$

Since the r.h.s. of zero curvature equation requires $2m$ terms which combine together to form

$$D_{vac}^{(-2i)} + D_{vac}^{(-2i+1)} = \lambda^{-i}(h^{(0)} + \frac{E}{v_0})$$

such that

$$[\partial_x + E + v_0 h^{(0)}, \partial_{t_{-2m}} + \lambda^{-m}(h^{(0)} + \frac{E}{v_0}) + \cdots + \lambda^{-1}(h^{(0)} + \frac{E}{v_0})] = 0$$

all models within the negative even sub-hierarchy admit non-trivial vacuum configuration.

- **Positive Odd Hierarchy**

  This case contains the mKdV equation, etc and the zero curvature representation in the vacuum is

$$[\partial_x + E + v_0 h^{(0)}, \partial_{t_{2m+1}} + D_{vac}^{(2m+1)} + D_{vac}^{(2m)} + \cdots + D_{vac}^{(1)} + D_{vac}^{(0)}] = 0$$

with solution

$$D_{vac}^{(2m+1)} = \frac{\lambda^m E}{v_0} \lambda^m h^{(0)} \qquad D_{vac}^{(2m)} = \lambda^m h^{(0)},$$
$$\vdots \qquad \vdots$$
$$D_{vac}^{(1)} = \frac{\lambda E}{v_0} \qquad D_{vac}^{(0)} = h^{(0)},$$

The r.h.s. of zero curvature equation requires $2m$ terms which combine together to form

$$D_{vac}^{(2i+1)} + D_{vac}^{(2i)} = \lambda^{i}(h^{(0)} + \frac{E}{v_0})$$

such that

$$[\partial_x + E + v_0 h^{(0)}, \partial_{t_{2m+1}} + \lambda^{m}(h^{(0)} + \frac{E}{v_0}) + \cdots + \lambda^{0}(h^{(0)} + \frac{E}{v_0})] = 0$$

and therofore all models within the positive odd sub-hierarchy admit solutions with non-trivial vacuum configuration.

- **Negative Odd Hierarchy**
  This is a subcalss of models which contains the sinh-Gordon and the zero curvature representation is

$$
\begin{aligned}
[\partial_x + E + v_0 h^{(0)} \quad , \quad & \partial_{t_{-2m-1}} + D_{vac}^{(-2m-1)} \\
+ \quad & D_{vac}^{(-2m)} + D_{vac}^{(-2m+1)} + \cdots + D_{vac}^{(-2)} + D_{vac}^{(-1)}] = 0
\end{aligned}
$$

  Notice the r.h.s. of zero curvature equation has $2m+1$-terms that they cannot combine into terms proportional to $E + v_0 h^{(0)}$ and henceforth the negative odd hierarchy does not allow constant vacuum solutions.

5. Conclusions and Outlook

The general algebraic structure of the mKdV hierarchy has lead to the construction of sub-hierarchies and to the classification of its soliton solutions with trivial and non trivial vacuum. It is interesting to point out that within each sector, the soliton solutions for diferent members (equations) of the hierarchy are all related by keeping the general form (16) or (18), varying only from the time evolution parameter in (17) or (19).

More complicated examples may be constructed following the same line of reasoning by considering higher rank affine Lie algebras and generalized mixed gradations (see for instance [7]). On the other hand, it would be interesting to develop the dressing method to construct periodic solutions for the mkdV equation. These are known to be written in terms of ratios of Jacobi Theta functions and under certain specific limit, yield the soliton solutions constructed in (16).

References

[1] H. Aratyn, J.F. Gomes, E. Nissimov, S. Pacheva and A.H. Zimerman, " Symmetry flows, conservation laws and dressing approach to the integrable models.", NATO Advanced Research Workshop on Integrable Hierarchies and Modern Physical Theories, Chicago, Illinois, (2000), p., 243-275 e-Print: nlin/0012042
[2] A. Chodos, Phys. Rev. D **21**,(1980), 2818.
[3] C.A. Tracy and H. Widom, Comumm. Math. Phys. **179**(1976),1
[4] H. Aratyn, L.A. Ferreira, J.F. Gomes and A.H. Zimerman, J. of Physics **A33**(2000),L331, e-Print: nlin/0007002
[5] Z. Qiao and W. Strampp, Physica**A313**(2002), 365
[6] J.F. Gomes, G. Starvaggi França, G.R. de Melo and A.H. Zimerman, J. of Physics **A42**(2009),445204, arXiv:0906.5579 [hep-th]
[7] J.F. Gomes, G.M. Sotkov, A.H. Zimerman, Nucl.Phys. **B714**,(2005),179-216, hep-th/0405182

Journal of Physics: Conference Series **284** (2011) 012031 doi:10.1088/1742-6596/284/1/012031

On the surfaces associated with $\mathbb{C}P^{N-1}$ models

P. P. Goldstein[1] **and A. M. Grundland**[2]

[1] Theoretical Physics Department,
The Andrzej Soltan Institute for Nuclear Studies,
Hoza 69, 00-681 Warsaw, Poland

E-mail: Piotr.Goldstein@fuw.edu.pl

[2] Centre de Recherches Mathématiques, Université de Montréal,
C. P. 6128, Succ. Centre-ville, Montréal, (QC) H3C 3J7, Canada
Université du Québec, Trois-Rivières CP500 (QC) G9A 5H7, Canada

E-mail: grundlan@crm.umontreal.ca

Abstract. We study certain new properties of 2D surfaces associated with the $\mathbb{C}P^{N-1}$ models and the wave functions of the corresponding linear spectral problem. We show that $\mathfrak{su}(N)$-valued immersion functions expressed in terms of rank-1 orthogonal projectors are linearly dependent, but they span an $(N-1)$-dimensional subspace of the Lie algebra $\mathfrak{su}(N)$. Their minimal polynomials are cubic, except for the holomorphic and antiholomorphic solutions, for which they reduce to quadratic trinomials. We also derive the counterparts of these relations for the wave functions of the linear spectral problems. In particular, we provide a relation between the wave functions, which results from the partition of unity into the projectors. Finally, we show that the angle between any two position vectors of the immersion functions, corresponding to the same values of the independent variables, does not depend on those variables.

AMS classification scheme numbers: 53A07, 53B50, 53C43, 81T45

1. Introduction

Over the past decades there has been a significant progress in the study of immersion of 2D surfaces in multidimensional Euclidean spaces obtained from $\mathbb{C}P^{N-1}$ models. The most fruitful approach to this subject has been achieved through the description of these surfaces in terms of the homogeneous variables f_k [2, 3, 13, 19] and orthogonal projectors P_k [18, 7, 6, 4, 8]. Using this language, we have established the recurrence relations for the projectors satisfying the $\mathbb{C}P^{N-1}$ model equations, for the wave functions of their spectral problems and consequently the immersion functions of 2D surfaces in the Lie algebra $\mathfrak{su}(N)$. In this paper we add certain new properties, concerning both the immersion functions and the wave functions, in order to enhance the algebraic and geometric characterization of the studied surfaces. The surfaces are defined by a contour integral [6]

$$X_k(\xi,\bar{\xi}) = i\int_\gamma \left(-[\partial P_k, P_k]d\xi + [\bar{\partial} P_k, P_k]d\bar{\xi}\right), \quad k=0,...,N-1, \tag{1.1}$$

which is independent of the path of integration $\gamma \in \mathbb{C}$ according to the dynamics of the orthogonal rank-1 projectors P_k. The projectors P_k, $0 \leq k \leq N-1$, are successive solutions [19] of the Euler-

Lagrange (E-L) equations in the form of a conservation law [18]

$$\partial\,[\bar{\partial}P, P] + \bar{\partial}\,[\partial P, P] = \mathbf{0}, \tag{1.2}$$

corresponding to the action integral

$$\int d\xi d\bar{\xi}\,\mathcal{L} = \mathrm{tr}(\partial P \cdot \bar{\partial} P) \tag{1.3}$$

with the constraint

$$P^2 = P. \tag{1.4}$$

Equation (1.2) ensures that the integrand of (1.1) is an exact differential. This mapping of an area on a Riemann sphere S^2 into a set of $\mathfrak{su}(N)$ matrices: $\Omega \ni (\xi, \bar{\xi}) \mapsto X_k(\xi, \bar{\xi}) \in \mathfrak{su}(N) \simeq \mathbb{R}^{N^2-1}$ is a generalised Weierstrass formula for immersion (GWFI) of 2D surfaces in the Euclidean space $\mathbb{R}^{N^2-1}$ [9, 10, 12]. The target spaces of the projectors P_k are one-dimensional vector functions $f(\xi, \bar{\xi}) \in \mathbb{C}^N$, constituting an orthogonal basis in $\mathbb{C}^N$ [2, 19]

$$P_k = [1/(f_k^\dagger \cdot f_k)] f_k \otimes f_k^\dagger, \tag{1.5}$$

All the projectors may be obtained from the first projector P_0, whose target space is an arbitrary holomorphic vector function $f_0(\xi)$, by the recurrence formulae derived in our previous work [4]. The projectors are orthogonal to each other and they constitute a partition of unity [2, 19]

$$P_k P_l = \delta_{kl} P_k \quad \text{(no summation)}, \tag{1.6}$$

$$\sum_{k=0}^{N-1} P_k = \mathbb{I}, \tag{1.7}$$

where $\mathbb{I}$ is the N-dimensional unit matrix.

For the surfaces corresponding to the projectors P_k, the integration may be performed explicitly, with the result [6]

$$X_k = -i\left(P_k + 2\sum_{j=0}^{k-1} P_j\right) + ic_k\mathbb{I}, \quad c_k = \frac{1+2k}{N}, \quad 0 \le k \le N-1. \tag{1.8}$$

In this paper we use the inverse formulae for the projectors in terms of the surfaces X_k, obtained in our previous work [4]

$$P_k = X_k{}^2 - 2i\,(c_k - 1)\,X_k - c_k\,(c_k - 2)\,\mathbb{I}. \tag{1.9}$$

to turn the projective property, the partition of unity and the orthogonal property into corresponding properties of the surfaces. Namely, we obtain the minimal polynomials, dimensionality of the spanned subspace of $\mathbb{R}^{N^2-1}$, and the angle between the position vectors of the surfaces, respectively.

In a similar manner we obtain the inverse formulae for the projectors in terms of the wave functions and the spectral parameter of the spectral problems. These relations allow us to derive the corresponding relations for the wave functions. In particular, we determine the minimal polynomial and a linear-dependence equation for those functions.

2. Projectors and soliton surfaces

In order to obtain the relations between surfaces X_k, it is convenient to express them in terms of the orthogonal rank-1 projectors P_k. Making use of the expression (1.8) and the partition of unity in terms of the projectors (1.7), it can be shown that the algebraic condition

$$\sum_{k=0}^{N-1}(-1)^k X_k = \mathbf{0} \tag{2.1}$$

holds. This means that the $\mathfrak{su}(N)$-valued immersion functions X_j are linearly dependent.
The equation (2.1) follows directly from the GWFI (1.8) in terms of projectors P_k and the decomposition of unity (1.7). Indeed, subtracting (1.8) for neighbouring k, we obtain

$$X_k - X_{k-1} = -i(P_k + P_{k-1}) + \frac{2i}{N}\mathbb{I}. \tag{2.2}$$

By adding the equations (2.2) for every second k, we obtain an equation containing the sum of all the projectors P_k, which is the unit matrix, according to (1.7). The final result proves to be exactly eq. (2.1).

Note that we can obtain the projectors P_k from the surfaces X_k not only as quadratic functions of the surfaces (1.9), but also as linear combinations of the surfaces $X_0, \ldots, X_k$ [4]

$$P_k = i\sum_{j=1}^{k}(-1)^{k-j}\left(X_j - X_{j-1}\right) + (-1)^k iX_0 + \frac{1}{N}\mathbb{I}. \tag{2.3}$$

Thus we may regain all the projectors $P_0, \ldots, P_{N-1}$ from the appropriate linear combinations of the surfaces $X_0, \ldots, X_{N-1}$ and the unit matrix. This means that these surfaces span an $(N-1)$ dimensional subspace of the $\mathfrak{su}(N)$ Lie algebra, as the projectors are linearly independent.

The projective property $P_k^2 = P_k$ imposes an algebraic constraint on the surfaces X_k. To find the lowest order constraint on X_k, we compare $P_k \cdot X_k$ obtained from (1.9) multiplied by X_k with $P_k \cdot X_k$ obtained from (1.8) multiplied by P_k. This yields a cubic matrix equation

$$(X_k - ic_k\mathbb{I})[X_k - i(c_k-1)\mathbb{I}][X_k - i(c_k-2)\mathbb{I}] = \mathbf{0}, \quad 0 < k < N-1. \tag{2.4}$$

For holomorphic ($k=0$) and antiholomorphic ($k = N-1$) solutions of the $\mathbb{C}P^{N-1}$ equation (1.2) the minimal polynomial for the matrix-valued functions X_k is quadratic. Namely, for the surfaces corresponding to the holomorphic solutions we have

$$(X_0 - ic_0\mathbb{I})[X_0 - i(c_0-1)\mathbb{I}] = \mathbf{0}, \quad k=0 \tag{2.5}$$

and, using $c_0 + c_{N-1} = 2$, we get

$$(X_{N-1} + ic_0\mathbb{I})[X_{N-1} + i(c_0-1)\mathbb{I}] = \mathbf{0}, \quad k = N-1 \tag{2.6}$$

for the antiholomorphic ones. Although equation (2.6) is apparently the Hermitian conjugate of the equation (2.5), their solutions do not have to be the Hermitian conjugates of each other.

Condition (2.4), as well as two other conditions (2.5) and (2.6), have simple interpretation if we diagonalize them (which is always possible, as the matrices are anti-Hermitian). It follows that the following numbers are eigenvalues of X_k $k = 1, \ldots, N-2$

$$ic_k, \quad i(c_k-1) \quad \text{and} \quad i(c_k-2), \tag{2.7}$$

while only the first two are eigenvalues for $k=0$ and only the last two for $k = N-1$.

The three values listed in (2.7) are the only eigenvalues of X_k. More precisely

Journal of Physics: Conference Series **284** (2011) 012031 doi:10.1088/1742-6596/284/1/012031

- The non-degenerate eigenvalue $i\,(c_k-1)$ occurs at every X_k, $k=0,...,N-1$; the corresponding eigenvector is f_k (the latter follows directly from (1.8) and from the fact that f_l, $l=0,...,N-1$ are eigenvectors of the projectors P_k with the eigenvalue δ_{kl}).
- The k-fold degenerate eigenvalue $i\,(c_k-2)$ occurs at every X_k, except for $k=0$; the k corresponding eigenvectors are $f_0,...,f_{k-1}$.
- The $(N-1-k)$-fold degenerate eigenvalue ic_k occurs at every X_k, except for $k=N-1$; the corresponding eigenvectors are $f_{k+1},...,f_{N-1}$.

Equation (2.4), together with (2.5) and (2.6), constitute the lowest degree constraints on the immersion functions X_k of the surfaces (direct substitution of (1.9) into the projective property would yield a 4th degree one). Although equation (2.4) is obvious when we look at the source of X_k (1.8), it is nevertheless a nontrivial constraint imposed on the surfaces. Since all the eigenvalues are independent of the coordinates $(\xi,\bar{\xi})$, the whole kinematics of a moving frame may only be due to variation of the diagonalizing (unitary) matrix.

Let us now present certain geometrical aspects of surfaces immersed in the $\mathfrak{su}(N)$ Lie algebras. Once we have the immersion functions of the surfaces, we can describe their metric and curvature properties.

(i) Let g_k be the metric tensor corresponding to the surface X_k. Its components will be marked with indices outside the parentheses to distinguish them from the index of the surface. Then the diagonal elements of the metric tensor are zero. This property directly follows from vanishing of $\mathrm{tr}(\partial P_k \partial P_k)$ and its Hermitian conjugate, proven in [4]

$$\begin{aligned}(g_k)_{11} &= (\partial X_k,\partial X_k) = \frac{1}{2}\mathrm{tr}([\partial P_k,P_k]\cdot[\partial P_k,P_k]) = -\frac{1}{2}\mathrm{tr}(\partial P_k\cdot\partial P_k)=0,\\ (g_k)_{22} &= (\bar\partial X_k,\bar\partial X_k) = \frac{1}{2}\mathrm{tr}([\bar\partial P_k,P_k]\cdot[\bar\partial P_k,P_k]) = -\frac{1}{2}\mathrm{tr}(\bar\partial P_k\cdot\bar\partial P_k)=0,\end{aligned} \tag{2.8}$$

where the inner product (A,B) of the $\mathfrak{su}(N)$ matrices is defined by [5]

$$(A,B) = -\frac{1}{2}\mathrm{tr}(A\cdot B). \tag{2.9}$$

(ii) The nonzero off-diagonal element $(g_k)_{12}=(g_k)_{21}$ is equal to

$$(g_k)_{12} = -\frac{1}{2}\mathrm{tr}(\partial X_k\cdot\bar\partial X_k) = -\frac{1}{2}\mathrm{tr}([\partial P_k,P_k]\cdot[\bar\partial P_k,P_k]) = \frac{1}{2}\mathrm{tr}(\partial P_k\cdot\bar\partial P_k). \tag{2.10}$$

Thus the first fundamental form reduces to

$$I_k = \mathrm{tr}(\partial P_k\cdot\bar\partial P_k)\,d\xi d\bar\xi. \tag{2.11}$$

The second fundamental form

$$II_k = (\partial^2 X_k - (\Gamma_k)^1_{11}\partial X_k)d\xi^2 + 2\partial\bar\partial X_k d\xi d\bar\xi + (\bar\partial^2 X_k - (\Gamma_k)^2_{22}\bar\partial X_k)d\bar\xi^2, \tag{2.12}$$

is easy to find when we determine the Christoffel symbols $(\Gamma_k)^1_{11}$ and $(\Gamma_k)^2_{22}$. These are the only nonzero components of Γ_k. From equation (2.10), we get

$$(\Gamma_k)^1_{11} = \partial\ln(g_k)_{12}, \quad (\Gamma_k)^2_{22} = \bar\partial\ln(g_k)_{12}. \tag{2.13}$$

Using (1.1) and the E-L equations (1.2) together with (2.13), we can write (2.12) as

$$\begin{aligned} II_k = &-\mathrm{tr}(\partial P_k \cdot \bar{\partial} P_k)\, \partial \frac{[\partial P_k, P_k]}{\mathrm{tr}(\partial P_k \cdot \bar{\partial} P_k)} d\xi^2 + 2i\, [\bar{\partial} P_k, \partial P_k] d\xi d\bar{\xi} \\ &+ \mathrm{tr}(\partial P_k \cdot \bar{\partial} P_k)\, \bar{\partial} \frac{[\bar{\partial} P_k, P_k]}{\mathrm{tr}(\partial P_k \cdot \bar{\partial} P_k)} d\bar{\xi}^2. \end{aligned} \tag{2.14}$$

Implementation of the above result for the metric of the surfaces induced by Veronese solutions of the E-L equations (1.2) is presented in detail in [4].

Also the following 2nd order differential conditions hold:

$$(\partial\bar{\partial} X_k, \partial X_k) = 0, \qquad (\partial\bar{\partial} X_k, \bar{\partial} X_k) = 0, \tag{2.15}$$

$$(\partial\bar{\partial} X_k, \partial^2 X_k) = 0, \qquad (\partial\bar{\partial} X_k, \bar{\partial}^2 X_k) = 0. \tag{2.16}$$

Equations (2.15) follow from direct differentiation of (2.8). Hence, the mixed derivatives of the matrices X_k coincide and are normal to the surfaces [5]. The second order differential constraints (2.16) are calculated straightforwardly from the definition

$$\begin{aligned} (\partial\bar{\partial} X_k, \partial^2 X_k) &= -\frac{1}{2}\mathrm{tr}([\bar{\partial} P_k, \partial P_k] \cdot [\partial^2 P_k, P_k]) = -\frac{1}{2}\big[\mathrm{tr}(\bar{\partial} P_k \partial P_k \partial^2 P_k P_k) \\ &- \mathrm{tr}(\partial P_k \bar{\partial} P_k \partial^2 P_k P_k) - \mathrm{tr}(\bar{\partial} P_k \partial P_k P_k \partial^2 P_k) + \mathrm{tr}(\partial P_k \bar{\partial} P_k P_k \partial^2 P_k)\big] = 0, \end{aligned} \tag{2.17}$$

since the conditions

$$P_k \bar{\partial} P_k \partial P_k = \bar{\partial} P_k \partial P_k P_k, \qquad P_k \partial P_k \bar{\partial} P_k = \partial P_k \bar{\partial} P_k P_k, \tag{2.18}$$

hold. Similarly, the second relation in (2.16) holds for its respective Hermitian conjugates. Note that equations (2.15) and (2.16) are gauge-invariant since they are expressed in terms of the projectors P_k.

We now show that the surfaces X_k, X_l do not have common points for $k \neq l$, with the exception of X_0 and X_1 in the $\mathbb{C}P^1$ model, where simply X_0 coincides with X_1.

Indeed, let $l > k$ be two different indices of the surfaces. Subtracting (1.8) from the analogous expression for X_l, we obtain

$$-i\left[P_l - P_k + 2\sum_{j=k}^{l-1} P_j - \frac{2(l-k)}{N}\mathbb{I}\right] = \mathbf{0} \tag{2.19}$$

Multiplying eq. (2.19) by P_k, we obtain

$$P_k\left[1 - \frac{2(l-k)}{N}\right] = \mathbf{0}. \tag{2.20}$$

On the other hand, when we multiply both hand sides of (2.19) by P_{l-1}, we get

$$P_{l-1}\left[1 - \frac{l-k}{N}\right] = \mathbf{0} \quad \text{for } k < l-1, \text{ and} \tag{2.21}$$

$$P_k\left(1 - \frac{2}{N}\right) = \mathbf{0} \quad \text{for } k = l-1. \tag{2.22}$$

All of the equations (2.20), (2.21) and (2.22) may only be satisfied when $N = 2,\ l = 1,\ k = 0$. In that case (2.1) yields immediately $X_1 = X_0$.

We now show that the immersion functions X_k, X_m make a constant angle in the sense of the Euclidean inner product (2.9), i.e. the angle Φ_{km} between the immersion functions X_k and X_m does not depend on the particular choice of projector P_0, nor on the coordinates ξ, $\bar{\xi}$. Namely, the angle Φ_{km} between two different functions X_k, X_m, $k < m$, is given by the equation

$$\cos \Phi_{km} = \frac{c_k(2 - c_m)}{\{[c_k(2 - c_k) - 1/N]\,[c_m(2 - c_m) - 1/N]\}^{1/2}} \tag{2.23}$$

The formula (2.23) may be obtained in a straightforward way either by calculating the appropriate scalar product from the GWFI for X_k, X_m (1.8) (bearing in mind that the projectors P_0, P_1, ..., P_k are mutually orthogonal), or by direct operations on the eigenvalues of the immersion functions. In either way we obtain for $m > k$

$$(X_k, X_m) = -(1/2)\,\mathrm{tr}(X_k \cdot X_m) = (N/2)\,c_k(2 - c_m), \tag{2.24}$$

while

$$(X_k, X_k) = -(1/2)\,\mathrm{tr}(X_k \cdot X_k) = (1/2)\,[N\,c_k(2 - c_k) - 1]. \tag{2.25}$$

It may easily be proven that (2.23) always yields $\cos \Phi_{km} \in (0, 1)$ unless $N = 2$ (and obviously $k = 0,\ m = 1$), for which the surfaces coincide (and the cosine is obviously equal to 1). Equation (2.23) is symmetric with respect to a transformation $k \longleftrightarrow N - 1 - m$, which may be seen e.g. in the table of $\cos \Phi_{km}$ for the $\mathbb{C}P^3$ model

$k \backslash m$	1	2	3
0	$5/\sqrt{33}$	$\sqrt{3/11}$	$1/3$
1		$9/11$	$\sqrt{3/11}$
2			$5/\sqrt{33}$

We can regard the immersion functions X_k as position vectors, whose ends draw the two-dimensional surfaces in a $N^2 - 1$-dimensional $\mathfrak{su}(N)$ algebra. The above result means that the position vectors make a constant angle with each other, independent of the variables ξ, $\bar{\xi}$. Moreover, the angle is the same for all choices of the P_0 solutions of the E-L equations (1.2) within a particular $\mathbb{C}P^{N-1}$ model.

3. Projectors and the spectral problem

The spectral problem is closely related to the immersion functions of the surfaces. The relation between the wave functions and the immersion functions is given by the Sym-Tafel (ST) formula [14, 15, 16, 17]. The wave functions are also related to the immersion functions by their asymptotic properties. No wonder that the results of the previous section have their counterparts in corresponding relations between the wave functions.

Similarly to the surfaces, the wave functions of the spectral problem can also be expressed in terms of the projectors. The spectral problem found by Zakharov and Mikhailov [18] reads

$$\partial \Phi_k = \frac{2}{1+\lambda}[\partial P_k, P_k]\Phi_k, \qquad \bar{\partial} \Phi_k = \frac{2}{1-\lambda}[\bar{\partial} P_k, P_k]\Phi_k, \qquad k = 0, 1, \ldots, N-1, \tag{3.1}$$

where $\lambda \in \mathbb{C}$ is the spectral parameter. An explicit solution of the linear spectral problem (3.1) for which Φ_k tends to $\mathbb{I}$ as $\lambda \to \infty$ is given by [1]

$$\Phi_k = \mathbb{I} + \frac{4\lambda}{(1-\lambda)^2} \sum_{j=0}^{k-1} P_j - \frac{2}{1-\lambda} P_k, \quad {\Phi_k}^{-1} = \mathbb{I} - \frac{4\lambda}{(1+\lambda)^2} \sum_{j=0}^{k-1} P_j - \frac{2}{1+\lambda} P_k, \tag{3.2}$$

where λ is purely imaginary and thus $\Phi_k \in SU(N)$. This in turn yields the projectors P_k in terms of the wave functions [4]

$$P_k = (1/4)\left[2(1+\lambda^2)\mathbb{I} - (1-\lambda)^2\Phi_k - (1+\lambda)^2\Phi_k^{-1}\right]. \tag{3.3}$$

The projective property of P_k may be represented in terms of Φ_k as a factorisable 4th degree expression with one double (squared) factor

$$P_k^2 - P_k = (1/16)\Phi_k^{-2}(\mathbb{I} - \Phi_k)\left[(1+\lambda)^2 - (1-\lambda)^2\Phi_k\right]\left[(1+\lambda) - (1-\lambda)\Phi_k\right]^2 = \mathbf{0}. \tag{3.4}$$

Hence, the minimal polynomials of the matrices Φ_k are cubic and they satisfy the equation, resembling the corresponding equation for the surfaces X_k (2.4), but explicitly depending on the spectral parameter

$$(\mathbb{I} - \Phi_k)\left[(1+\lambda)\mathbb{I} - (1-\lambda)\Phi_k\right]\left[(1+\lambda)^2\mathbb{I} - (1-\lambda)^2\Phi_k\right] = \mathbf{0} \tag{3.5}$$

for $1 < k < N-1$. Similarly to the surfaces (2.5), (2.6), quadratic matrix equations are sufficient for $k = 0$ and $k = N-1$: the equation with only the first two factors of (3.5) is satisfied by Φ_0 and that with only the last two factors of (3.5) by Φ_{N-1}.

The immersion functions X_k may be expressed in terms of the wave functions Φ_k in two ways: either by the ST formula [14, 15, 16, 17]

$$X_k^{ST} = -\frac{i}{2}(1-\lambda^2)\Phi_k^{-1}\partial_\lambda\Phi_k \tag{3.6}$$

or as a limit [4]

$$X_k = i\lim_{\lambda\to\infty}\left[\frac{\lambda}{2}\Phi_k + \left(c_k - \frac{\lambda}{2}\right)\mathbb{I}\right]. \tag{3.7}$$

Using equation (3.7), one can check by explicit calculation that the cubic minimal polynomial (3.5) for the wave function Φ_k coincides with the cubic polynomial (2.4) for the immersion function X_k in the limit $\lambda \to \infty$.

We now show that the partition of unity (1.7) for the projectors P_k imposes constraints on the wave functions Φ_k, given by the relation

$$\sum_{j=0}^{N-1}\Phi_j\left(\frac{1-\lambda}{1+\lambda}\right)^j = \frac{1+\lambda}{2\lambda}\left[1 - \left(\frac{1-\lambda}{1+\lambda}\right)^{N-2}\right]\mathbb{I}. \tag{3.8}$$

Indeed, using the wave functions Φ_k, which can be expressed in terms of the projectors P_k through the formula (3.2), for the indices k and $k-1$, we obtain

$$\sum_{k=0}^{N-1}\sum_{j=0}^{k}(\Phi_{j-1} - \Phi_j)\left(\frac{1+\lambda}{1-\lambda}\right)^{k-j} = \frac{2}{1-\lambda}\mathbb{I}, \tag{3.9}$$

or equivalently

$$\sum_{j=0}^{N-1}(\Phi_{j-1} - \Phi_j)\left(\frac{1+\lambda}{1-\lambda}\right)^{N-j} - \sum_{j=0}^{N-1}(\Phi_{j-1} - \Phi_j) = \frac{4\lambda}{1-\lambda^2}\mathbb{I}, \tag{3.10}$$

or, factoring out the coefficients with respect to Φ_{j-1} and Φ_j, we get

$$\sum_{j=0}^{N-1}\Phi_{j-1}\left(\frac{1+\lambda}{1-\lambda}\right)^{N-j} - \sum_{j=0}^{N-1}\Phi_j\left(\frac{1+\lambda}{1-\lambda}\right)^{N-j} - \Phi_{N-1} = \left(\frac{1+\lambda}{1-\lambda}\right)^2\mathbb{I}. \tag{3.11}$$

Finally, this expression can be written in the form

$$\sum_{j=0}^{N-1} \Phi_j \left(\frac{1-\lambda}{1+\lambda}\right)^j = \frac{1+\lambda}{2\lambda}\left[1-\left(\frac{1-\lambda}{1+\lambda}\right)^{N-2}\right]\mathbb{I}. \tag{3.12}$$

Equation (3.12), transformed into an appropriate equation for the variable $(\lambda/2)\Phi_j - (c_j - \lambda/2)\mathbb{I}$, turns into equation (3.7) in the limit $\lambda \to \infty$.
Note that in view of the linear dependence of the immersion functions X_k, i.e. equation (2.1), the ST formula (3.6) leads to the following differential constraint on the wave functions Φ_k

$$\sum_{j=0}^{N-1}(-1)^j X_j^{ST} = -\frac{i}{2}(1-\lambda^2)\partial_\lambda \ln \prod_{j=0}^{N-1} \Phi_j^{(-1)^j} = \mathbf{0}. \tag{3.13}$$

This implies that the expression $\prod_{2l<N}\Phi_{2l}\prod_{2l+1<N}\Phi_{2l+1}^{-1}$ is independent of the spectral parameter λ but it may depend on the variables ξ and $\bar{\xi} \in \mathbb{C}$.

4. Concluding remarks

In our work we develop the approach proposed in [4], which relies on construction of the consecutive surfaces and immersion functions in terms of projectors. The inverse formulae found in that work allowed for deriving additional properties of the surfaces immersed in the $\mathfrak{su}(N)$ Lie algebra and functions immersed in the $SU(N)$ Lie group. In particular

- We have shown that the number of linearly independent surfaces, associated with the $\mathbb{C}P^{N-1}$ models is $N-1$.
- The angles between the position vectors of any two surfaces are constant (independent of ξ, $\bar{\xi}$ and independent of the choice of the holomorphic solution). The surfaces do not intersect with each other for $\mathbb{C}P^{N-1}$, $N \geq 2$; the only two surfaces of $\mathbb{C}P^1$ coincide.
- All the surfaces associated with the $\mathbb{C}P^{N-1}$ models satisfy a 3rd-degree matrix equation, which reduces to a 2nd-degree equation for the holomorphic and antiholomorphic solutions of the E-L equations (1.2).
- The corresponding relations hold for the wave functions of the spectral problem. Moreover the asymptotic properties of those functions while the spectral parameter tends to infinity connect the relations for the wave functions with those for the immersion functions.

The proposed approach opens a field of further research for other sigma models.

Acknowledgments

A.M.G.'s work was supported by a research grant from NSERC of Canada. This project was completed during A.M.G.'s visit to the École Normale Superieure de Cachan, and he would like to thank the CMLA for their kind invitation and hospitality.

References

[1] Din A M, Horvath Z and Zakrzewski W J 1984 *The Riemann-Hilbert problem and finite action* $\mathbb{C}P^{N-1}$ *solutions* Nucl. Phys. **B 233**, 269
[2] Din A M and Zakrzewski W J 1980 General classical solutions of the $\mathbb{C}P^{N-1}$ model *Nucl. Phys.* B **174** 397–403.
[3] Eells J and Wood J C 1983 Harmonic maps from surfaces to complex projective spaces *Adv. in Math.* **49** 217–263
[4] Goldstein P and Grundland A M Invariant recurrence relations for $\mathbb{C}P^{N-1}$ models 2010 *J. Phys. A: Math. Gen.* **43** 265206

[5] Grundland A M, Strasburger A and Zakrzewski W J 2005 Surfaces immersed in $\mathfrak{su}(N+1)$ Lie algebras obtained from the $\mathbb{C}P^N$ sigma models, *J. Phys. A: Math. Gen.* **39** 9187–9213.

[6] Grundland A M and Yurdusen I 2009 On analytic descriptions of two-dimensional surfaces associated with the CP^{N-1} sigma models *J. Phys. A: Math. Gen.* **42** 172001 (5pp).

[7] Guest M A 1997 *Harmonic Maps, Loop Groups and Integrable Systems* (Cambridge: Cambridge University Press)

[8] Hussin V, Yurducsen I and Zakrzewski W J 2010, Canonical surfaces associated with projectors in Grassmannian sigma models, J. Math. Phys. 51, 103509–103523.

[9] Konopelchenko B 1996 Induced surfaces and their integrable dynamics *Stud. Appl. Math.* **96** 9–51.

[10] Konopelchenko B and Taimanov I 1996 Constant mean curvature surfaces via an integrable dynamical system *J. Phys. A: Math. Gen.* **29** 1261–1265

[11] Manton N and Sutcliffe P 2004 *Topological Solitons* (Cambridge: Cambridge University Press).

[12] Nomizu K and Sasaki T 1994 *Affine differential geometry* (Cambridge: Cambridge University Press).

[13] Sasaki R 1983 *General class of solutions of the complex Grassmannian and* $\mathbb{C}P^{N-1}$ *sigma models* Phys. Lett. B 130 69–72

[14] Sym A 1982 *Soliton surfaces* Lett. Nuovo Cimento **33** 394–400

[15] Sym A 1983 *Soliton surfaces II: Geometric unification of solvable nonlinearities* Lett. Nuovo Cimento **36** 307–312

[16] Sym A 1985 *Soliton surfaces and their applications (Soliton geometry from spectral problems)*, Geometric Aspects of the Einstein Equations and Integrable systems (Lecture Notes in Physics vol 239) ed R Martini (Berlin: Springer), 154–231

[17] Tafel J 1995 *Surfaces in* R^3 *with prescribed curvature*, J. Geom. Phys. **17** 381–390

[18] Zakharov V E and Mikhailov A V 1979 Relativistically invariant two-dimensional models of field theory which are integrable by means of the inverse scattering problem method Sov. Phys.–JETP **47** 1017.

[19] Zakrzewski W J 1989 *Low Dimensional Sigma Models* (Bristol: Adam Hilger 1989), Chapter 3 (46–74).

Journal of Physics: Conference Series **284** (2011) 012032 doi:10.1088/1742-6596/284/1/012032

Some comments on pinwheel tilings and their diffraction

Uwe Grimm[1] **and Xinghua Deng**[2]

[1] Department of Mathematics and Statistics, The Open University, Walton Hall, Milton Keynes MK7 6AA, UK

[2] University of Waterloo, 200 University Avenue West, Waterloo, Ontario, Canada N2L 3G1

E-mail: u.g.grimm@open.ac.uk

Abstract. The pinwheel tiling is the paradigm for a substitution tiling with circular symmetry, in the sense that the corresponding autocorrelation is circularly symmetric. As a consequence, its diffraction measure is also circularly symmetric, so the pinwheel diffraction consists of sharp rings and, possibly, a continuous component with circular symmetry. We consider some combinatorial properties of the tiles and their orientations, and a numerical approach to the diffraction of weighted pinwheel point sets.

1. Introduction

Diffraction is an essential tool for the determination of the atomic structure of solids. The discovery of aperiodically ordered materials in the early 1980s [19, 10] added a new dimension of complexity to the inverse problem of structure determination [3, 4], and led to a renewed interest in mathematical diffraction theory, which focuses on kinematic diffraction [5, 6]. It also posed the question of what type of ordered structures are possible, and how one would be able to detect them. Regular model sets, which are structures derived from a (periodic) lattice in a higher-dimensional space, show pure point diffraction; see [11] and references therein. However, there are many interesting systems that are not model sets, such as tilings with circular symmetry; we refer to [7] for general background on the theory of aperiodic order.

Pinwheel tilings of the plane, first introduced by Conway and Radin [14, 15, 16, 17], are arguably the simplest examples of substitution tilings with circular symmetry. They are built entirely of triangular tiles of edge lengths 1, 2 and $\sqrt{5}$. The corresponding right triangles occur in both possible orientations, or chiralities, so in fact a pinwheel tiling comprises two tile shapes which are mirror images of each other.

A pinwheel tiling of the plane can be constructed as the fixed point of the inflation rule of Figure 1. It consists of an inflation with linear scaling factor $\sqrt{5}$ and a rotation by an angle $\varphi = -\arctan(\frac{1}{2}) \approx -26.565^\circ$, followed by a dissection into five triangles of the original size and shape, two of the same and three of the opposite chirality. The rotation ensures that the original triangle reappears in the inflated patch, and the sequence of iterates therefore converges to a space-filling tiling of the plane. At the same time, because φ is not a rational multiple of π, it produces an irrational rotation, which means that in each inflation step one obtains triangles in a new direction. In the limit, the set of available directions becomes dense on the unit circle,

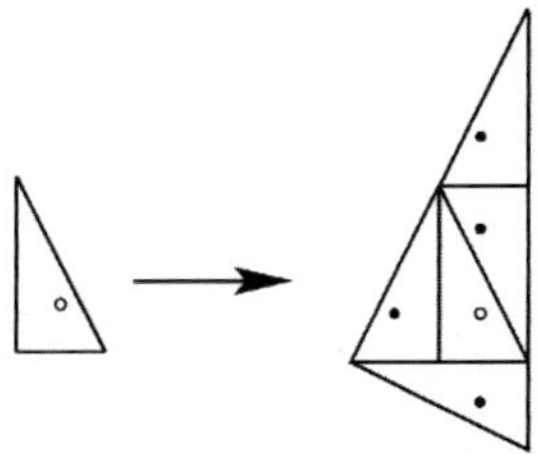

Figure 1. Inflation rule of the pinwheel tiling. Control points are denoted by dots, with the open dot specifying the origin. The inflation rule is compatible with reflection, in the sense that the triangle of different chirality is replaced by the mirror image of the patch shown here.

and this produces the circular symmetry of the structure; a proof of the circular symmetry of the autocorrelation can be found in [16, 12]; see also [9] for generalisations.

2. Pinwheel control point sets

Also shown in Figure 1 are control points, which form a point set Λ that is mutually locally derivable (MLD) with the pinwheel tiling, in the sense of [8]. The control point of a triangle is located at position $(\frac{1}{2}, \frac{1}{2})$ with respect to the right angle corner of the triangle. In other words, the triangle with control point $(0,0)$ and its short edge along the horizontal axis has vertices $(-\frac{1}{2}, -\frac{1}{2})$, $(\frac{1}{2}, -\frac{1}{2})$ and $(-\frac{1}{2}, \frac{3}{2})$, or $(-\frac{1}{2}, -\frac{1}{2})$, $(\frac{1}{2}, -\frac{1}{2})$ and $(\frac{1}{2}, \frac{3}{2})$, depending on its chirality. Given a control point $x \in \mathbb{R}^2 \simeq \mathbb{C}$, a triangle can be specified by a triple (x, ω, χ), where $x \in \mathbb{C}$ denotes the corresponding control point, ω denotes the angle of the short edge with respect to the horizontal axis and $\chi \in \{\pm 1\}$ specifies the chirality of the triangle.

The pinwheel point set Λ forms a substitutive point set with substitution rule

$$\sigma(x,\omega,\chi) = \begin{cases} (\sqrt{5}R_\varphi x, & \omega + \varphi - \chi\varphi, & \chi) \\ (\sqrt{5}R_\varphi x + R_{(\omega+\varphi-\chi\varphi+\frac{\chi\pi}{2})}, & \omega + \varphi - \chi\varphi + \pi, & \chi) \\ (\sqrt{5}R_\varphi x + 2R_{(\omega+\varphi-\chi\varphi+\frac{\chi\pi}{2})}, & \omega + \varphi - \chi\varphi + \pi, & -\chi) \\ (\sqrt{5}R_\varphi x + R_{(\omega+\varphi-\chi\varphi+\pi)}, & \omega + \varphi - \chi\varphi + \pi, & -\chi) \\ (\sqrt{5}R_\varphi x + R_{(\omega+\varphi-\chi\varphi-\frac{\chi\pi}{2})}, & \omega + \varphi - \chi\varphi - \frac{\chi\pi}{2}, & -\chi), \end{cases} \tag{1}$$

where $R_\theta = e^{i\theta}$ for $\theta \in \mathbb{R}$, so multiplication by R_θ corresponds to a rotation in the complex plane with the angle θ. A patch of the pinwheel tiling, with control points coloured according to the two chiralities of the triangular tiles, is shown in Figure 2.

The inflation rule σ gives rise to the substitution matrix

$$\begin{aligned} M &= \begin{pmatrix} e^{it0} + e^{it\pi} & 2e^{it(2\varphi-\pi)} + e^{it(2\varphi+\frac{\pi}{2})} \\ 2e^{it\pi} + e^{it(-\frac{\pi}{2})} & e^{it(2\varphi)} + e^{it(2\varphi-\pi)} \end{pmatrix} \\ &= \begin{pmatrix} 1 + y^2 & 2xy^{-1} + xy \\ 2y^2 + 1/y^{-1} & x + xy^{-2} \end{pmatrix} \end{aligned} \tag{2}$$

where we set $x = e^{2i\varphi t}$ and $y = e^{i\pi t/2}$. This matrix was used in [12] to derive the circular symmetry of the autocorrelation. If we only wish to distinguish orientations modulo 90 degrees, we can choose $t = 4$, so $y = 1$, and the matrix simplifies to

$$M = \begin{pmatrix} 2 & 3x \\ 3 & 2x \end{pmatrix}. \tag{3}$$

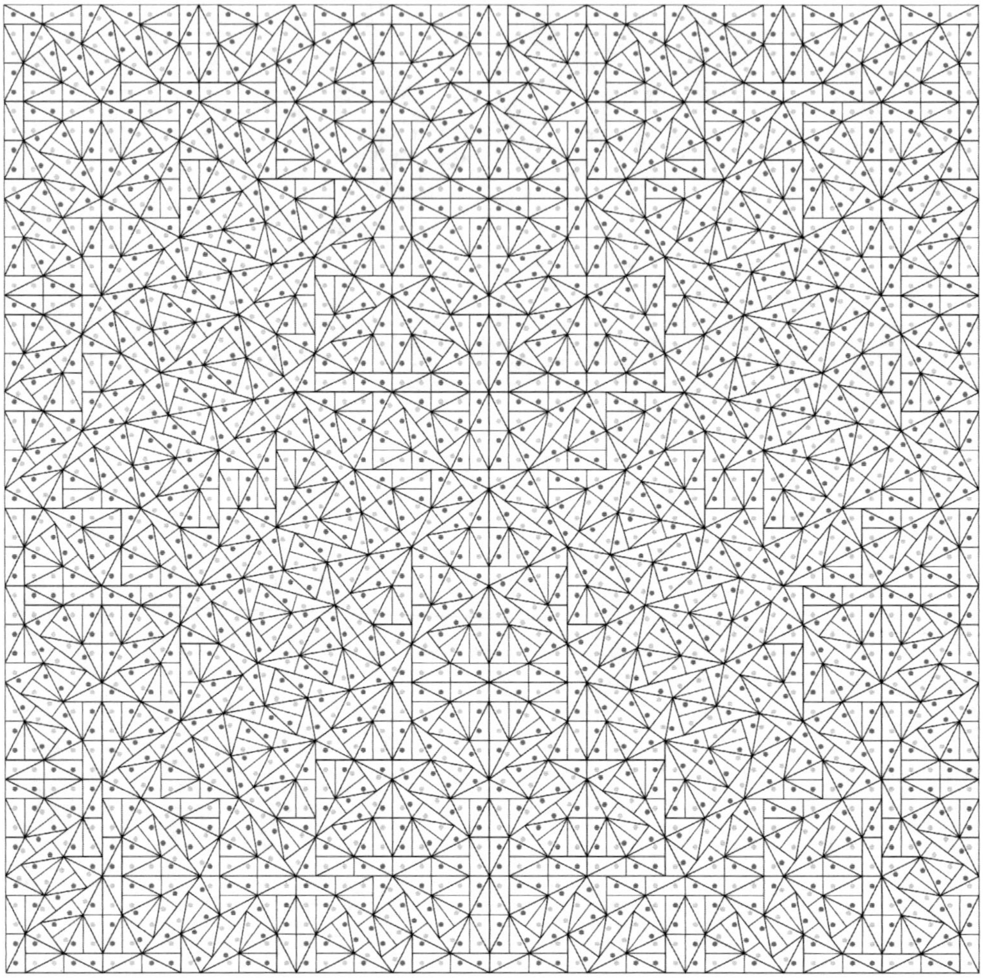

Figure 2. A patch of the pinwheel tiling. The control point in Λ are coloured according to the two chiralities of the triangles.

This matrix can be used to derive information about the subsets of control points that belong to the same direction, in the sense that they belong to the same power of x.

3. Hierarchies of subsets of different orientation

Let us consider what happens when we start from an initial patch containing one triangle of each chirality with horizontal orientation, and perform n inflation steps. Denoting by $f_n^{\pm}(k)$ the number of triangles of each chirality at step n with orientation factor x^k, where $0 \leq k \leq n$, we find

$$\begin{pmatrix} \sum_{k=0}^{n} f_n^{+}(k)\, x^k \\ \sum_{k=0}^{n} f_n^{-}(k)\, x^k \end{pmatrix} = M^n \begin{pmatrix} 1 \\ 1 \end{pmatrix}. \qquad (4)$$

Journal of Physics: Conference Series **284** (2011) 012032 doi:10.1088/1742-6596/284/1/012032

Multiplication by M gives the following recursion for the coefficients

$$f_{n+1}^{+}(k) \;=\; 2\,f_n^{+}(k) + 3\,f_n^{-}(k-1)\,, \tag{5a}$$
$$f_{n+1}^{-}(k) \;=\; 2\,f_n^{+}(k-1) + 3\,f_n^{-}(k)\,, \tag{5b}$$

for $0 \le k \le n+1$, where the appropriate initial conditions are $f_0^{\pm}(0) = 1$ and $f_n^{\pm}(k) = 0$ for all $k < 0$ and $k > n$.

Clearly, these numbers satisfy the relation $f_n^{+}(k) = f_n^{-}(n-k)$ for all $0 \le k \le n$, as can be seen from the relation

$$\begin{pmatrix} 0 & 1 \\ 1 & 0 \end{pmatrix} \begin{pmatrix} 2 & 3x \\ 3 & 2x \end{pmatrix} \begin{pmatrix} 0 & 1 \\ 1 & 0 \end{pmatrix} \;=\; \begin{pmatrix} 2x & 3 \\ 3x & 2 \end{pmatrix} \;=\; x \begin{pmatrix} 2 & 3x^{-1} \\ 3 & 2x^{-1} \end{pmatrix}. \tag{6}$$

Hence the total number of triangles $f_n(k) = f_n^{+}(k) + f_n^{-}(k)$ obeys the symmetry $f_n(k) = f_n(n-k)$ for all $0 \le k \le n$.

These observations have some interesting consequences. For instance, it can been shown that for even n

$$f_n^{+}(0) < f_n^{-}(0) < f_n^{+}(1) < f_n^{-}(1) < \ldots < f_n^{+}(\tfrac{n}{2}-1) < f_n^{-}(\tfrac{n}{2}-1) < f_n^{+}(\tfrac{n}{2}) = f_n^{-}(\tfrac{n}{2})$$

and for odd n

$$f_n^{+}(0) < f_n^{-}(0) < f_n^{+}(1) < f_n^{-}(1) < \ldots < f_n^{+}(\tfrac{n-1}{2}-1) < f_n^{-}(\tfrac{n-1}{2}-1) < f_n^{+}(\tfrac{n-1}{2}) = f_n^{-}(\tfrac{n+1}{2})\,.$$

It follows that $f_n(k) > f_n(k-1)$ for $1 \le k \le [\frac{n}{2}] - 1$, and the symmetry property $f_n(k) = f_n(n-k)$ then implies a corresponding inequality for large k.

4. Diffraction of pinwheel point sets

While it has been shown that the autocorrelation of the pinwheel point set is circularly symmetric [12], it is not known whether it is concentrated on sharp rings only, or whether there is also a continuous component. Direct numerical computations are of no help here, because any finite patch only contains a small number of independent directions, and is clearly not a good approximation of the limit structure. However, as shown in [1, 2], the knowledge that the diffraction is circularly symmetric can be exploited to derive an approximation of the diffraction based on a radial version of the Poisson summation formula; see [1] for details. This results in an expansion in terms of Bessel functions. As shown in [1, 2], this results in a radial distribution of the diffraction intensity that shows clear maxima for distances corresponding to those of a square lattice, so the radially discrete part resembles that of a powder diffraction of a square lattice. However, as the pinwheel structure contains a hierarchical sequence of scaled square lattices, the intensity distribution is not the same as for the square lattice, but deviates in a characteristic fashion. In addition, these numerical data provide an indication that there is also a continuous component.

Here, we consider the slightly generalised case of diffraction from weighted pinwheel sets. Circular symmetry also holds for the autocorrelation of a weighted pinwheel control point set, with weights depending on the two chiralities of the triangular tiles. This follows from [9, Thm. 6.1]: the triangles can be regarded as two prototiles, a left-handed and a right-handed one, and each individual prototile then shows circular symmetry, too.

Denoting the sets of pinwheel control points for the two chiralities by $\Lambda_{\pm}$, with $\Lambda_{+} \cup \Lambda_{-} = \Lambda$, we thus consider the weighted Dirac comb

$$\omega \;=\; \alpha_{+}\,\delta_{\Lambda_{+}} + \alpha_{-}\,\delta_{\Lambda_{-}} \;:=\; \sum_{x \in \Lambda_{+}} \alpha_{+}\,\delta_x + \sum_{x \in \Lambda_{-}} \alpha_{-}\,\delta_x\,, \tag{7}$$

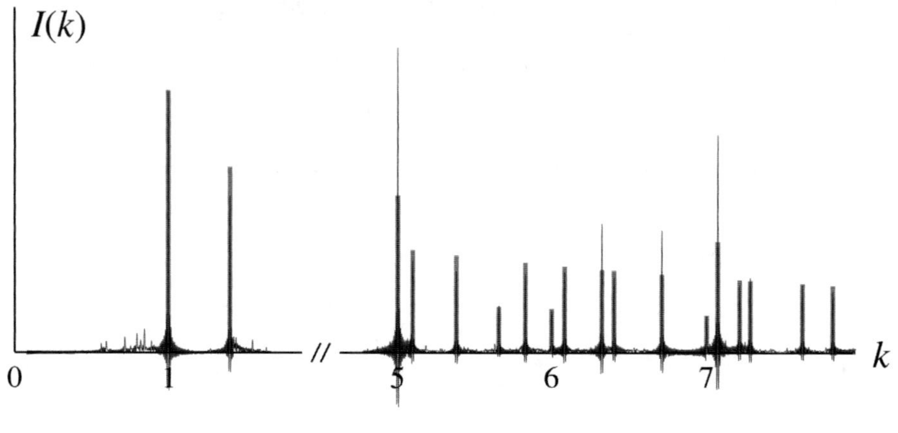

Figure 3. Part of the approximated radial intensity distribution for the pinwheel diffraction ($\alpha_+ = \alpha_- = 1$), compared with the exact diffraction for a square lattice powder (red bars).

where $\alpha_\pm$ are arbitrary, in general complex, weights. In this paper, we restrict ourselves to values $\alpha_\pm \in \{-1, 0, 1\}$, with the case $\alpha_+ = \alpha_- = 1$ corresponding to the pinwheel diffraction considered in [1, 2].

The approximations to the radial density distribution shown below are obtained by computing approximate autocorrelation coefficients $\eta(r)$ for distances r, obtained from finite approximants of the pinwheel tiling, up to a suitable maximum distance. The approximation to the autocorrelation then has the form

$$\gamma_\omega = \sum_{r \in \mathcal{D}} \eta(r)\, \mu_r \,, \tag{8}$$

where μ_r denotes the uniform measure on the unit circle, and

$$\mathcal{D} \subset \left\{ \sqrt{\frac{p^2+q^2}{5^\ell}} \;\middle|\; p, q, \ell \in \mathbb{N} \right\} \tag{9}$$

denotes the distance set for the pinwheel tiling. Here, the assumption of circular symmetry enters — the approximate autocorrelation coefficients obtained from a finite patch are used to define a circularly symmetric autocorrelation measure. From Equation (8), the approximate radial distribution of the diffraction intensity $I(k)$ is obtained as the weighted sum of the Fourier transform of the uniform measure μ_r. As the latter is given by $\widehat{\mu}_r(k) = J_0(2\pi|k|r)$, this results in finite sums of Bessel functions J_0 which show strong oscillations; compare [1, 2] for details. We note in passing that some values for frequencies of distances are known exactly, and they can in principle be computed from the inflation rule; see [13].

4.1. Systematics of peak intensities

Figure 3 shows the numerical approximation for the diffraction of δ_Λ, which is the pinwheel point sets with weights $\alpha_+ = \alpha_- = 1$. The corresponding result for a square lattice powder is also shown, with relative normalisation chosen to make the first peak match. Clearly, as observed in [1, 2], the main peaks in the intensity are well reproduced, but there are systematic deviations in the peak intensities. The latter can qualitatively be explained as arising from distances in $\mathcal{D}$ of Equation (9) with denominators 5^ℓ with $\ell > 0$.

To get a better picture of the peak intensities, we consider separately the integrated intensities on rings at values of k according to the highest power of 5 that divides k^2 (which will be 0, 1 or 2 in the examples below). Figure 4 shows the result for the case where k is not divisible by 5. On the left, you can clearly distinguish three groups of data, indicated by different colours, and for each group of data the decay of intensities with increasing k is well approximated by a function of the type c/k, where c is a constant.

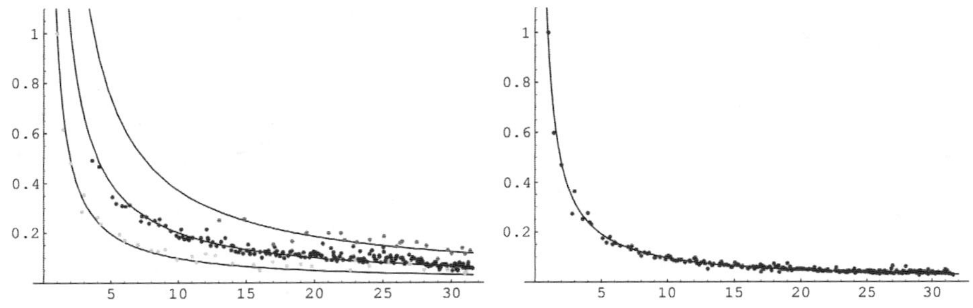

Figure 4. Intensity ratios $I(k)/I(1)$ of peaks for values of k such that k^2 is not divisible by 5, as a function of k. On the left, different colours refer to different number of prime factors equal to 1 modulo 4 (either 0 (green), 1 (blue) or 2 (red)). Note that equal factors are counted here; for instance, the lowest k in the second group is $k = \sqrt{13}$, and in the third group it is $k = \sqrt{169} = 13$. On the right, these data have been collapsed as described in the text. Lines are least square fits to functions of the form c/k with constant c.

The three groups of data are distinguished by the number of prime factors of k^2 that are equal to 1 modulo 4. In fact, dividing the intensity ratios by 2^s, where s is the sum of powers of all prime factors of this type, collapses the data set onto a single curve, as shown in the right part of Figure 4. This observation is in line with the similarity to the square lattice powder diffraction seen in [1, 2].

The corresponding collapsed data for peaks at values of k which are multiples of 5 and 5^2 are shown on the left and right of Figure 5, respectively. Again, the observed decay of the intensities is inversely proportional to k. This indicates that our qualitative understanding of the peak intensities is correct, though it is not clear whether this also holds quantitatively for the diffraction of an infinite pinwheel tiling point set.

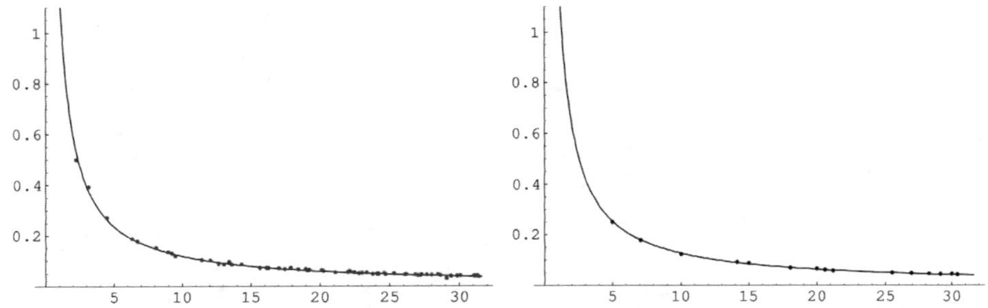

Figure 5. Same as right part of Figure 4, for values of k such that k^2 is divisible by 5 but not 5^2 (left) and such that k^2 is divisible by 5^2 but not by 5^3 (right).

4.2. *Diffraction of weighted pinwheel point sets*

In order to get a better indication of the possible continuous contribution of the pinwheel diffraction, it is advantageous to consider weighted pinwheel point sets. As discussed above, the resulting autocorrelation is still circularly symmetric, and hence the same approach as outlined above can be applied to obtain approximations of the radial intensity distribution. Figure 6 shows the result for the case $\omega = \delta_{\Lambda_+}$ ($\alpha_+ = 1$ and $\alpha_- = 0$), where only the half of the pinwheel point set is considered. The three spectra correspond to an increasing number of terms in the approximation, and increasing accuracy of the estimated correlation coefficients from larger patches of the pinwheel tiling constructed by successive inflation. The result very closely resembles that for the complete pinwheel point set, and may well be identical apart from the overall intensity normalisation due to the different density of scatterers.

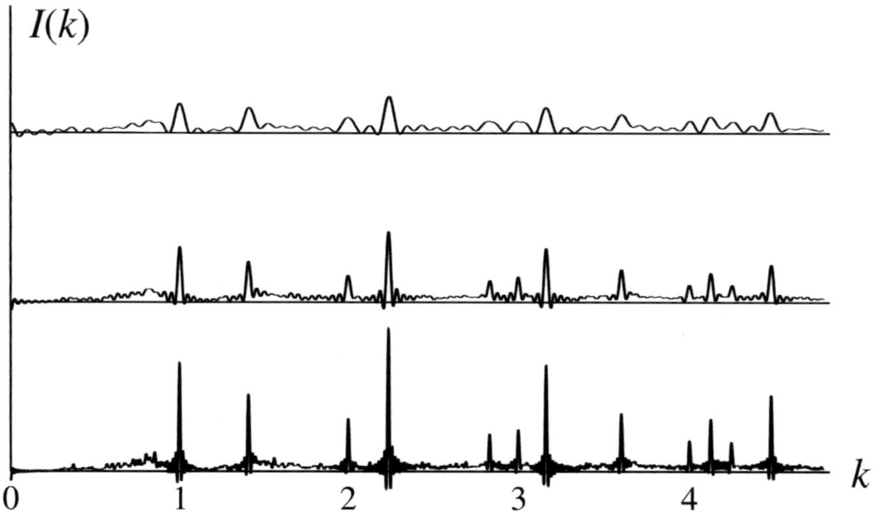

Figure 6. Three approximations of the radial intensity distribution of the diffraction of the Dirac comb δ_{Λ_+}.

More interesting is the case of $\alpha_+ = 1$ and $\alpha_- = -1$. Since this is a balanced case, the average scattering strength is zero, and there is no peak at $k = 0$. Therefore, a continuous background may be easier to spot in this case. Figure 7 shows three increasingly accurate approximations of the radial diffraction intensity for this case.

There is not only an indication of the presence of a continuous background, similar to what was found in [2], but also a number of new features appear to develop (in particular for small values of k), which may correspond to additional peaks for the infinite system. However, the approximation may not yet be good enough to draw firm conclusions.

5. Summary

The diffraction of substitution tilings with continuous rotational symmetry still remains to be completely understood. In this paper, we presented some observations on combinatorial properties of the pinwheel tiling and on its diffraction measure, using the approximation introduced in [1, 2]. From the numerical data, we corroborate the conclusions of [1, 2] on the similarities between the pinwheel diffraction and a square lattice powder. Considering the balanced case of zero average scattering strength provides an indication that there might not

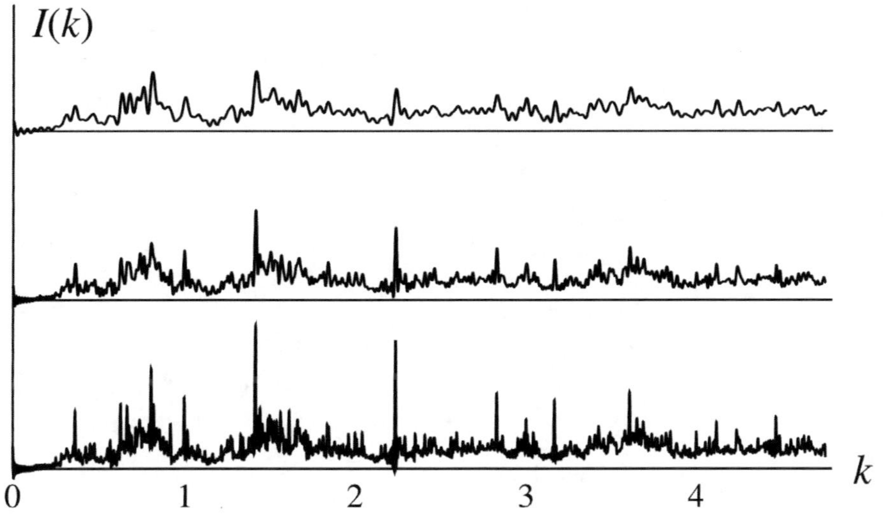

Figure 7. Three approximations of the radial intensity distribution of the diffraction of the Dirac comb ω for the balanced case with $\alpha_+ = 1$ and $\alpha_- = -1$.

only be a continuous contribution as conjectured in [2], but that there might also be additional sharp rings with finite intensity. This observation warrants further investigation. It would also be interesting to compare this result with the diffraction of other tilings, such as the generalised pinwheel tilings introduced by Sadun [18] or the tipi tilings in [9].

Acknowledgments

The authors thank Michael Baake, Dirk Frettlöh and Haïja Moustafa for useful discussions. XD is grateful to the Department of Mathematics and Statistics at the Open University for the hospitality during an extended visit in 2009, where some of this work was completed.

References

[1] Baake M, Frettlöh D and Grimm U 2007 A radial analogue of Poisson's summation formula with applications to powder diffraction and pinwheel patterns *J. Geom. Phys.* **57** 1331–1343
[2] Baake M, Frettlöh D and Grimm U 2007 Pinwheel patterns and powder diffraction *Philos. Mag.* **87** 2831–2838
[3] Baake M and Grimm U 2009 Kinematic diffraction is insufficient to distinguish order from disorder *Phys. Rev. B* **79** 020203(R) and **80** 029903(E)
[4] Baake M and Grimm U 2010 Surprises in aperiodic diffraction *J. Physics: Conf. Ser.* **226** 012023
[5] Baake M and Grimm U 2011 Mathematical diffraction theory of deterministic and stochastic structures: an informal summary *RIMS proceedings* (in press)
[6] Baake M and Grimm U Kinematic diffraction from a mathematical viewpoint (in preparation)
[7] Baake M and Grimm U *Theory of Aperiodic Order: A Mathematical Invitation* (Cambridge: CUP, in preparation)
[8] Baake M, Schlottmann M and Jarvis PD 1991 Quasiperiodic tilings with tenfold symmetry and equivalence with respect to local derivability *J. Phys. A: Math. Gen.* **24** 4637–4654
[9] Frettlöh D 2008 Substitution tilings with statistical circular symmetry *Eur. J. Combin.* **29** 1881–1893
[10] Ishimasa T, Nissen H-U and Fukano Y 1985 New ordered state between crystalline and amorphous in Ni-Cr particles *Phys. Rev. Lett.* **55** 511–513
[11] Lenz D 2008 Aperiodic order and pure point diffraction *Philos. Mag.* **88** 2059–2071
[12] Moody R V, Postnikoff D and Strungaru N 2006 Circular symmetry of pinwheel diffraction *Ann. Henri Poincaré* **7** 711–730

[13] Moustafa H 2009 PV cohomology of pinwheel tilings, their integer group of coinvariants and gap-labelling *Preprint* arXiv:0906.2107
[14] Radin C 1994 The pinwheel tilings of the plane *Ann. Math.* **139** 661–702
[15] Radin C 1995 Space tilings and substitutions *Geometriae Dedicata* **55** 257–264
[16] Radin C 1997 Aperiodic tilings, ergodic theory and rotations *The Mathematics of Long-Range Aperiodic Order* (NATO ASI C 489) ed RV Moody (Dordrecht: Kluwer) pp 499–519
[17] Radin C 1999 *Miles of Tiles* (Providence: AMS)
[18] Sadun L 1998 Some generalizations of the pinwheel tiling *Discrete Comput. Geom.* **20** 79–100
[19] Shechtman D, Blech I, Gratias D and Cahn J W 1984 Metallic phase with long-range orientational order and no translational symmetry *Phys. Rev. Lett.* **53** 1951–1953

Journal of Physics: Conference Series **284** (2011) 012033 doi:10.1088/1742-6596/284/1/012033

Generation of vector boson masses in Electroweak Model without dynamical symmetry breaking

Nikolay A. Gromov

Department of Mathematics, Komi Science Center UrD RAS, Kommunisticheskaya st., 24, Syktyvkar, 167982, Russia

E-mail: gromov@dm.komisc.ru

Abstract. The vector boson masses are generated by transformation of the *free* (without any potential term) Lagrangian of the Electroweak Model from Cartesian coordinates to a coordinates on the sphere S_3, which is defined by the gauge invariant quadratic form $\phi^\dagger\phi = \rho^2$ in the matter field space Φ_2. This transformation corresponds to transition from linear representation of the gauge group in the space Φ_2 to its nonlinear representation in the space of functions on S_3. Such modified Electroweak Model keep all experimentally verified fields of the standard Electroweak Model and does not include massive scalar field (Higgs boson), if the sphere radius does not depend on the space-time coordinates $\rho = R = const$. The concept of generation masses for vector bosons in Electroweak Model by transformation to radial coordinates is further developed in context of nonlinearly realized gauge groups, as well as in context of nonlinear sigma models. The limiting case of the modified Electroweak Model which corresponds to the contracted gauge group is discussed.

PACS 12.15–y.

1. Introduction

The Standard Electroweak Model based on gauge group $SU(2) \times U(1)$ gives a good description of electroweak processes. The massive vector bosons predicted by the model was experimentally observed and have the masses $m_W = 80GeV$ for charged W-boson and $m_Z = 91GeV$ for neutral Z-boson. One of the unsolved problems is the origin of the vector bosons masses. In the standard formulation these masses are arisen as a result of spontaneous symmetry breaking vie Higgs mechanism which include three steps: 1) the potential of the self-acting scalar field ϕ of the special form $V(\phi) = \mu^2\bar{\phi}\phi + \lambda(\bar{\phi}\phi)^2$ is introduced *by hand* in the Lagrangian; 2) its minimal values are considered for imaginary mass $\mu^2 < 0$ and are interpreted as degenerate vacuum; 3) one of the gauge equivalent vacuum is fixed and then all fields are regarded in the neighborhood of this vacuum.

Sufficiently artificial Higgs mechanism with its imaginary bare mass is a naive relativistic analog of the phenomenological description of superconductivity [1]. Therefore there are a serious doubt whether electroweak symmetry is broken by such a Higgs mechanism, or by something else. The existence of the Higgs boson is not yet experimentally verified. One expect that the future experiments on LHC will given definite answer on the question: does Higgs boson really exist or not. The emergence of large number Higgsless models [2]–[6] was stimulated by difficulties with Higgs boson. These models are mainly based on extra dimensions of different

Journal of Physics: Conference Series **284** (2011) 012033 doi:10.1088/1742-6596/284/1/012033

types or larger gauge groups. A finite electroweak model without a Higgs particle which is used a regularized quantum field theory [7],[8] was developed in [6].

The simple mechanism for generation of the vector boson masses in Electroweak Model was recently suggested [9]. It is based on the fact that the quadratic form $\phi^\dagger\phi = R^2$ in the matter fields space Φ_2 is invariant with respect to gauge transformations. This form define the 3-dimensional sphere S_3 of the radius R in the space Φ_2 which is C_2 or R_4 if real components are counted. The vector boson masses are automatically generated by the *free* (without any potential term) Lagrangian on the sphere S_3 and are expressed through the sphere radius R, so there is no need in special mechanism of spontaneous symmetry breaking. Higgs boson field does not presented in the model. The free Lagrangian on the sphere S_3 can be obtained as well from standard Electroweak Lagrangian by transformation of the Cartesian coordinates in Φ_2 to a some coordinates on the sphere S_3. This transformation corresponds to the transition from linear to nonlinear representation of the gauge group in the space of functions on S_3. The fermion Lagrangian of the Standard Electroweak Model are modified by replacing of the fields ϕ with the restricted on the quadratic form fields in such a way that its second order terms provide the electron mass and neutrino remain massless.

One of the important ingredient of the Electroweak Model is the simple group $SU(2)$. More then fifty years it is well known in physics the notion of group contraction [10], i.e. limit operation, which transforms, for example, a simple or semisimple group to a non-semisimple one. From general point of view for better understanding of a physical system it is useful to investigate its properties for limiting values of their physical parameters. In particular, for a gauge model one of the similar limiting case corresponds to a model with contracted gauge group. The gauge theories for non-semisimple groups which Lie algebras admit invariant non-degenerate metrics was considered in [11],[12]. In the present paper the limiting case of the modified Electroweak Model which corresponds to the contracted gauge group $SU(2; j) \times U(1)$ is regarded. We find transformation properties of gauge and matter fields under contractions. After that we obtain the Lagrangian of the contracted model from the noncontracted one by the substitution of the transformed fields. When contraction parameter tends to zero $j \to 0$ or takes nilpotent value $j = \iota$ the field space is fibered in such a way that electromagnetic, Z-boson and electron fields are in the base whereas charged W-bosons and neutrino fields are in the fiber. Within framework of the limit model the base fields can be interpreted as an external ones with respect to the fiber fields in the sense that the fiber fields do not effect on the base fields. The field interactions are simplified under contraction.

2. Modified Electroweak Model

The bosonic part of the Electroweak Model is given by the sum [13]

$$L_B = L_A + L_\phi. \tag{1}$$

where

$$L_A = \frac{1}{8g^2}\mathrm{Tr}(F_{\mu\nu})^2 - \frac{1}{4}(B_{\mu\nu})^2 = -\frac{1}{4}[(F^1_{\mu\nu})^2 + (F^2_{\mu\nu})^2 + (F^3_{\mu\nu})^2] - \frac{1}{4}(B_{\mu\nu})^2 \tag{2}$$

is the gauge fields Lagrangian for $SU(2) \times U(1)$ group. The gauge fields are

$$A_\mu(x) = -ig\sum_{k=1}^{3} T_k A^k_\mu(x) = -i\frac{g}{2}\begin{pmatrix} A^3_\mu & A^1_\mu - iA^2_\mu \\ A^1_\mu + iA^2_\mu & -A^3_\mu \end{pmatrix},$$

$$B_\mu(x) = -ig'YB_\mu(x) = -i\frac{g'}{2}\begin{pmatrix} B_\mu & 0 \\ 0 & B_\mu \end{pmatrix}, \tag{3}$$

where g, g' are coupling constants, $T_k = \frac{1}{2}\tau_k$, with τ_k being Pauli matrices, take their values in Lie algebras $su(2)$, $u(1)$ respectively. The stress tensors look as follows

$$F_{\mu\nu}(x) = \mathcal{F}_{\mu\nu}(x) + [A_\mu(x), A_\nu(x)], \quad \mathcal{F}_{\mu\nu} = \partial_\mu A_\mu - \partial_\nu A_\mu, \quad B_{\mu\nu} = \partial_\mu B_\nu - \partial_\nu B_\mu. \tag{4}$$

The second term in (1)

$$L_\phi = \frac{1}{2}(D_\mu\phi)^\dagger D_\mu\phi \tag{5}$$

is the *free* matter fields Lagrangian. Here D_μ are the covariant derivatives

$$D_\mu\phi = \partial_\mu\phi - ig\left(\sum_{k=1}^{3} T_k A_\mu^k\right)\phi - ig'YB_\mu\phi \tag{6}$$

of the fields ϕ from the space C_2 of fundamental representation of $SU(2)$.

For $\Omega \in SU(2)$, $e^{i\omega} \in U(1)$ the gauge transformations of the fields are as follows

$$\phi^\Omega = \Omega\phi, \quad \phi^\omega = e^{i\omega}\phi,$$

$$A_\mu^\Omega(x) = \Omega A_\mu(x)\Omega^{-1} - \partial_\mu\Omega \cdot \Omega^{-1}, \quad A_\mu^\omega(x) = A_\mu(x),$$

$$B_\mu^\Omega = B_\mu, \quad B_\mu^\omega = B_\mu - 2\partial_\mu\omega \tag{7}$$

and Lagrangian (1) is invariant under $SU(2) \times U(1)$ gauge group.

The Lagrangian $L_B = L_A + L_\phi$ describe massless fields. In a standard approach to generate mass terms for vector bosons the Higgs mechanism is used. We have proposed [9] new mechanism for generation masses of vector bosons in Electroweak Model which is based on the topological idea of the restriction of the free bosonic Lagrangian $L_B = L_A + L_\phi$ from the whole noncompact space R_4 to the compact sphere S_3. The complex two dimensional matter fields space C_2 can be regarded as four dimensional real Euclidean space R_4 if real components are counted. The quadratic form

$$\phi^\dagger\phi = \phi_1^*\phi_1 + \phi_2^*\phi_2 = R^2 \tag{8}$$

is invariant with respect to $SU(2)\times U(1)$ group and define the three dimensional sphere S_3 in C_2. Therefore the restriction of the Electroweak Model on the sphere S_3 will be the gauge model with the same gauge symmetry. Similar restriction (8) was appeared in a unified conformal model for fundamental interactions [14] as a consequence of the gauge fixing freedom connected with the local conformal symmetry group.

Let us introduce the real fields r, $\bar{\psi} = (\psi_1, \psi_2, \psi_3)$ by the equations

$$\phi_1 = r(\psi_2 + i\psi_1), \quad \phi_2 = r(1 + i\psi_3), \tag{9}$$

then quadratic form (8) is written as

$$r^2(1 + \bar{\psi}^2) = R^2, \tag{10}$$

where $\bar{\psi}^2 = \psi_1^2 + \psi_2^2 + \psi_3^2$, and define the sphere S_3 of the radius $R > 0$ in the space R_4. Three independent real fields $\bar{\psi}$ form intrinsic Beltrami coordinate system on S_3 with the following metric tensor

$$g_{kk}(\bar{\psi}) = \frac{1 + \bar{\psi}^2 - \psi_k^2}{(1 + \bar{\psi}^2)^2}, \quad g_{kl}(\bar{\psi}) = \frac{-\psi_k\psi_l}{(1 + \bar{\psi}^2)^2}, \quad k, l = 1, 2, 3. \tag{11}$$

Journal of Physics: Conference Series **284** (2011) 012033 doi:10.1088/1742-6596/284/1/012033

Let us define the *free* matter fields Lagrangian L_ϕ with the help of the metric tensor of the spherical space S_3 in the form

$$L_\psi = \frac{R^2}{2}\sum_{k,l=1}^{3} g_{kl} D_\mu\psi_k D_\mu\psi_l = \frac{R^2\left[(1+\bar\psi^2)(D_\mu\bar\psi)^2 - (\bar\psi, D_\mu\bar\psi)^2\right]}{2(1+\bar\psi^2)^2}. \tag{12}$$

The covariant derivatives are obtained with the help of the standard expressions (6), using now the nonlinear representations of generators for the algebras $su(2)$, $u(1)$ in the space S_3 [9]

$$T_1\bar\psi = \frac{i}{2}\begin{pmatrix} -(1+\psi_1^2) \\ \psi_3 - \psi_1\psi_2 \\ -(\psi_2+\psi_1\psi_3) \end{pmatrix}, \quad T_2\bar\psi = \frac{i}{2}\begin{pmatrix} -(\psi_3+\psi_1\psi_2) \\ -(1+\psi_2^2) \\ \psi_1 - \psi_2\psi_3 \end{pmatrix},$$

$$T_3\bar\psi = \frac{i}{2}\begin{pmatrix} -\psi_2+\psi_1\psi_3 \\ \psi_1+\psi_2\psi_3 \\ 1+\psi_3^2 \end{pmatrix}, \quad Y\bar\psi = \frac{i}{2}\begin{pmatrix} -(\psi_2+\psi_1\psi_3) \\ \psi_1 - \psi_2\psi_3 \\ -(1+\psi_3^2) \end{pmatrix} \tag{13}$$

and are as follows:

$$D_\mu\psi_1 = \partial_\mu\psi_1 + \frac{g}{2}\left[-(1+\psi_1^2)A_\mu^1 - (\psi_3+\psi_1\psi_2)A_\mu^2 - (\psi_2-\psi_1\psi_3)A_\mu^3\right] - \frac{g'}{2}(\psi_2+\psi_1\psi_3)B_\mu,$$

$$D_\mu\psi_2 = \partial_\mu\psi_2 + \frac{g}{2}\left[(\psi_3-\psi_1\psi_2)A_\mu^1 - (1+\psi_2^2)A_\mu^2 + (\psi_1+\psi_2\psi_3)A_\mu^3\right] + \frac{g'}{2}(\psi_1-\psi_2\psi_3)B_\mu,$$

$$D_\mu\psi_3 = \partial_\mu\psi_3 + \frac{g}{2}\left[-(\psi_2+\psi_1\psi_3)A_\mu^1 + (\psi_1-\psi_2\psi_3)A_\mu^2 + (1+\psi_3^2)A_\mu^3\right] - \frac{g'}{2}(1+\psi_3^2)B_\mu. \tag{14}$$

Let us stress that Lagrangian (12) can be obtain from the Lagrangian (5) by the transformation from the Cartesian coordinates in Euclidean space R_4 to the Beltrami coordinates on the sphere S_3, when sphere radius does not depend on space-time variables. The gauge fields Lagrangian (2) does not depend on the fields ϕ and therefore remains unchanged. So the full Lagrangian L_B is given by the sum of (2) and (12).

The particle content of the model is described by the second order part of the full Lagrangian. The ground state of L_B correspond to zero fields. For small fields in the neighborhood of ground state, the second order part of L_ψ is written as

$$L_\psi^{(2)} = \frac{R^2}{2}\sum_{k=1}^{3}\left[(D_\mu\psi_k)^{(1)}\right]^2, \tag{15}$$

where linear terms $(D_\mu\psi_k)^{(1)}$ in covariant derivatives have the form

$$(D_\mu\psi_1)^{(1)} = -\frac{g}{2}\left(A_\mu^1 - \frac{2}{g}\partial_\mu\psi_1\right) = -\frac{g}{2}\hat A_\mu^1,$$

$$(D_\mu\psi_2)^{(1)} = -\frac{g}{2}\left(A_\mu^2 - \frac{2}{g}\partial_\mu\psi_2\right) = -\frac{g}{2}\hat A_\mu^2,$$

$$(D_\mu\psi_3)^{(1)} = \partial_\mu\psi_3 + \frac{1}{2}(gA_\mu^3 - g'B_\mu) = \frac{1}{2}\sqrt{g^2+g'^2}Z_\mu.$$

For the new fields

$$W_\mu^\pm = \frac{1}{\sqrt2}\left(\hat A_\mu^1 \mp i\hat A_\mu^2\right), \quad (W_\mu^-)^* = W_\mu^+$$

$$Z_\mu = \frac{gA_\mu^3 - g'B_\mu + 2\partial_\mu\psi_3}{\sqrt{g^2+g'^2}}, \quad A_\mu = \frac{g'A_\mu^3 + gB_\mu}{\sqrt{g^2+g'^2}} \tag{16}$$

the quadratic part of the full Lagrangian

$$L_B^{(2)} = L_A^{(2)} + L_\psi^{(2)} =$$

$$= -\frac{1}{2}\mathcal{W}_{\mu\nu}^+\mathcal{W}_{\mu\nu}^- + m_W^2 W_\mu^+ W_\mu^- - \frac{1}{4}\mathcal{F}_{\mu\nu}\mathcal{F}_{\mu\nu} - \frac{1}{4}\mathcal{Z}_{\mu\nu}\mathcal{Z}_{\mu\nu} + \frac{m_Z^2}{2}Z_\mu Z_\mu, \tag{17}$$

where $\mathcal{X}_{\mu\nu} = \partial_\mu X_\nu - \partial_\nu X_\mu$ for $X_\mu = W_\mu^\pm, Z_\mu$, describes massive vector fields $W_\mu^\pm$ with identical mass $m_W = \frac{1}{2}gR$ (W-bosons), massless vector field A_μ (photon) and massive vector field Z_μ, $m_Z = \frac{R}{2}\sqrt{g^2+g'^2}$ (Z-boson). In other words it describes all the experimentally verified parts of the Electroweak Model, but does not include the scalar Higgs field. The particle masses are identical to those of the Standard Model and are expressed through the free parameter R of the model, which is now interpreted as the curvature radius of the spherical matter field space S_3.

The fermion Lagrangian of the standard Electroweek Model is taken in the form [13]

$$L_F = L_l^\dagger i\tilde{\tau}_\mu D_\mu L_l + e_r^\dagger i\tau_\mu D_\mu e_r - h_e[e_r^\dagger(\phi^\dagger L_l) + (L_l^\dagger\phi)e_r], \tag{18}$$

where $L_l = \begin{pmatrix} e_l \\ \nu_{e,l} \end{pmatrix}$ is the $SU(2)$-doublet, e_r the $SU(2)$-singlet, h_e is constant and e_r, e_l, ν_e are two component Lorentz spinor. Here τ_μ are Pauli matrices, $\tau_0 = \tilde{\tau}_0 = \mathbf{1}$, $\tilde{\tau}_k = -\tau_k$. The covariant derivatives $D_\mu L_l$ are given by (6) with L_l instead of ϕ and $D_\mu e_r = (\partial_\mu + ig'B_\mu)e_r$. The convolution on the inner indices of $SU(2)$-doublet is denoted by $(\phi^\dagger L_l)$.

The matter field ϕ appears in Lagrangian (18) only in mass terms. After expression ϕ through $\bar{\psi}$ the mass terms are rewritten in the form

$$h_e[e_r^\dagger(\phi^\dagger L_l) + (L_l^\dagger\phi)e_r] = \frac{h_e R}{\sqrt{1+\bar{\psi}^2}}\Big\{e_r^\dagger e_l + e_l^\dagger e_r +$$

$$+ i\psi_3\left(e_l^\dagger e_r - e_r^\dagger e_l\right) + i\left[\psi_1\left(\nu_{e,l}^\dagger e_r - e_r^\dagger \nu_{e,l}\right) + i\psi_2\left(\nu_{e,l}^\dagger e_r + e_r^\dagger \nu_{e,l}\right)\right]\Big\}. \tag{19}$$

Its second order terms $h_e R\left(e_r^\dagger e_l^- + e_l^{-\dagger} e_r\right)$ provide the electron mass $m_e = h_e R$, and neutrino remain massless.

The topological idea with compact matter field space S_3 instead of noncompact one R_4 was used and developed in several papers [15]–[20]. The transformation of the free Lagrangian from Cartesian to radial coordinates $R_+ \times S_3$ in R_4 was regarded in [15], where the sphere S_3 was parametrized by elements of $SU(2)$ groups. When sphere radius depend on space-time coordinates the real positive massless scalar field is presented in the model.

We are interested in parametrization of S_3 with $R = const$ by elements of $SU(2)$ groups, so write ϕ as

$$\phi = \begin{pmatrix} \phi_1 \\ \phi_2 \end{pmatrix} = R\begin{pmatrix} \chi_1 \\ \chi_2 \end{pmatrix} = R\begin{pmatrix} \chi_1 & -\chi_2^* \\ \chi_2 & \chi_1^* \end{pmatrix}\begin{pmatrix} 1 \\ 0 \end{pmatrix} \equiv Rh\varphi_0, \tag{20}$$

then from (8) it follows that

$$\chi^\dagger\chi = \chi_1^*\chi_1 + \chi_2^*\chi_2 = 1 \tag{21}$$

and the matrix h is unimodular $\det h = 1$ and unitary $h^\dagger h = 1$. So the vector $\chi = h\varphi_0 \in S_3$ or the matrix $h \in SU(2)$ defines the coordinates on the sphere (21). The matrix h transforms under the gauge transformations as

$$h^\Omega = \Omega h, \quad h^\omega = \begin{pmatrix} e^{i\omega}\chi_1 & -e^{-i\omega}\chi_2^* \\ e^{i\omega}\chi_2 & e^{-i\omega}\chi_1^* \end{pmatrix} = he^{i\omega\tau_3}, \tag{22}$$

where $\tau_3 = \mathrm{diag}(1, -1)$.

The matter fields Lagrangian (5) takes the form

$$L_\phi = \frac{1}{2}\frac{R^2}{4}\left(\sqrt{g^2+g'^2}\right)^2 (Z_\mu)^2 + \frac{R^2 g^2}{4} W_\mu^+ W_\mu^-, \tag{23}$$

where the new fields are introduced

$$W_\mu^\pm = \frac{1}{\sqrt{2}}\left(W_\mu^1 \mp i W_\mu^2\right), \quad Z_\mu = \frac{gW_\mu^3 + g'B_\mu}{\sqrt{g^2+g'^2}}, \quad A_\mu = \frac{g'W_\mu^3 - gB_\mu}{\sqrt{g^2+g'^2}},$$

$$W_\mu(x) = h^\dagger A_\mu(x) h + h^\dagger \partial_\mu h = -i\frac{g}{2}\sum_{k=1}^{3} W_\mu^k \tau_k. \tag{24}$$

These fields are invariant under $SU(2)$ transformations: $X^\Omega = X, \quad X = W_\mu^\pm, Z_\mu, A_\mu$ and are transforms under those of $U(1)$ as

$$\left(W_\mu^\pm\right)^\omega = e^{\mp 2i\omega} W_\mu^\pm, \quad Z_\mu^\omega = Z_\mu, \quad A_\mu^\omega = A_\mu + \frac{2}{e}\partial_\mu \omega, \quad e = \frac{gg'}{\sqrt{g^2+g'^2}}. \tag{25}$$

The stress tensors of the fields $A_\mu(x)$ and $W_\mu(x)$ (24) are connected by

$$W_{\mu\nu}(x) = h^\dagger F_{\mu\nu}(x) h \tag{26}$$

therefore $\mathrm{tr}(F_{\mu\nu}(x))^2 = \mathrm{tr}(W_{\mu\nu}(x))^2$and the gauge fields Lagrangian (2) can be written as

$$L_A = -\frac{1}{4}\left[(W_{\mu\nu}^1)^2 + (W_{\mu\nu}^2)^2\right] - \frac{1}{4}\left[(W_{\mu\nu}^3)^2 + (B_{\mu\nu})^2\right]. \tag{27}$$

Then the bosonic Lagrangian (1) assumes the form

$$L_B = -\frac{1}{2}\mathcal{W}_{\mu\nu}^+ \mathcal{W}_{\mu\nu}^- + m_W^2 W_\mu^+ W_\mu^- - \frac{1}{4}(\mathcal{F}_{\mu\nu})^2 - \frac{1}{4}(\mathcal{Z}_{\mu\nu})^2 + \frac{1}{2}m_Z^2 (Z_\mu)^2 +$$

$$+\frac{igP - g^2 S - (g\mathcal{Z}_{\mu\nu} + g'\mathcal{F}_{\mu\nu})H_{\mu\nu}}{2\sqrt{g^2+g'^2}} - \frac{1}{4}(H_{\mu\nu})^2, \tag{28}$$

where terms with third and fourth powers in fields are as follows

$$P = \mathcal{W}_{\mu\nu}^+\left[W_\mu^-(gZ_\nu + g'A_\nu) - W_\nu^-(gZ_\mu + g'A_\mu)\right] -$$

$$-\mathcal{W}_{\mu\nu}^-\left[W_\mu^+(gZ_\nu + g'A_\nu) - W_\nu^+(gZ_\mu + g'A_\mu)\right],$$

$$S = \left[W_\mu^+(gZ_\nu + g'A_\nu) - W_\nu^+(gZ_\mu + g'A_\mu)\right]\left[W_\mu^-(gZ_\nu + g'A_\nu) - W_\nu^-(gZ_\mu + g'A_\mu)\right],$$

$$H_{\mu\nu} = -ig\left(W_\mu^+ W_\nu^- - W_\mu^- W_\nu^+\right). \tag{29}$$

The second order terms of (28) coinside with thouse of (17), i.e. the particle contents of the model are identical. This demonstrate that the generation vector bosons masses do not depend on the choice of the specific coordinate system on S_3, but has topological origin. The parametrization of the sphere S_3 by the elements of $SU(2)$ group leads to more simple expressions for higher order terms. The Lagrangian (28) depend only on $SU(2)$-invariant fields, so the local $SU(2)$ symmetry is factored out of it. The concept of generation masses for vector bosons in Electroweak Model by transformation to radial coordinates is further developed in [16] and [17].

The Electroweak Model based on the nonlinearly realized $SU(2) \times U(1)$ gauge groups was suggested in [18],[19]. The $SU(2)$ matrix Ω and the matter field Φ are taken in the form

$$\Omega = \frac{1}{v}\left(\phi_0 + i\sum_{k=1}^{3}\tau_k\phi_k\right), \quad \Phi = \left(\begin{array}{c} i\phi_1 + \phi_2 \\ \phi_0 - i\phi_3 \end{array}\right), \tag{30}$$

where $\phi_0 = r, \phi_k = r\psi_k, k = 1,2,3, v = R$ in our notations (8)–(10). The nonlinearity of the representation comes from the constraint

$$\Omega^\dagger\Omega = 1 \Rightarrow \phi_0^2 + \bar{\phi}^2 = v^2, \tag{31}$$

which is the same as (11). The quantization of this model was consistently defined in the perturbative loop-wise expansion and satisfies Physical Unitarity.

The nonlinear partial-trace σ-model on G/H, which provides mass terms to the intermediate vector bosons associated with the quotient G/H and remain those of H massless, was developed in [20]. An infinite-dimensional symmetry, with non-trivial Noether invariants, which ensures quantum integrability of the model in a non-canonical quantization scheme was found. For $G = SU(2) \times U(1)$ and $H = U_{em}(1)$ this model gives Higgsless alternative to the Standard Model with a partial trace on a quotient manifold $G/H = SU(2) \cong S_3$.

3. Contraction of special unitary group $SU(2;j)$

Let us regard complex space $\Phi_2(j)$ with hermitian form

$$\phi^\dagger(j)\phi(j) = |\phi_1|^2 + j^2|\phi_2|^2, \tag{32}$$

where $\phi(j) = \left(\begin{array}{c} \phi_1 \\ j\phi_2 \end{array}\right) \in \Phi_2(j)$, $\phi^\dagger(j) = (\phi_1^*, j\phi_2^*)$ and star * denotes complex conjugation. The contraction parameter takes two values: $j = 1, \iota$, where $\iota \neq 0$ is nilpotent unit $\iota^2 = 0$. For $j = \iota$ the space $\Phi_2(\iota)$ is a fibered space with projection $pr : \{\phi_1, \phi_2\} \to \{\phi_1\}$, which has 1D base $\{\phi_1\}$ and 1D fiber $\{\phi_2\}$.

The group $SU(2;j)$ is defined as a transformation group of $\Phi_2(j)$, which keep invariant this hermitian form, i.e.

$$\phi'(j) = \left(\begin{array}{c} \phi_1' \\ j\phi_2' \end{array}\right) = \left(\begin{array}{cc} \alpha & j\beta \\ -j\beta^* & \alpha^* \end{array}\right)\left(\begin{array}{c} \phi_1 \\ j\phi_2 \end{array}\right) = u(j)\phi(j),$$

$$\det u(j) = |\alpha|^2 + j^2|\beta|^2 = 1, \quad u(j)u^\dagger(j) = 1. \tag{33}$$

Generators of the Lie algebra $su(2;j)$ in the fundamental representation are easily obtained

$$T_1(j) = j\frac{1}{2}\tau_1, \quad T_2(j) = j\frac{1}{2}\tau_2, \quad T_3(j) = \frac{1}{2}\tau_3,$$

and are subject of commutation relations

$$[T_1, T_2] = -j^2 T_3, \quad [T_3, T_1] = -T_2, \quad [T_2, T_3] = -T_1.$$

They form the general element of $su(2;j)$

$$T(j) = \sum_{k=1}^{3} a_k T_k(j) = \frac{i}{2}\left(\begin{array}{cc} a_3 & j(a_1 - ia_2) \\ j(a_1 + ia_2) & -a_3 \end{array}\right) = -T^\dagger(j),$$

which can be obtained from that of $su(2)$ by substitution

$$a_1 \to ja_1, \quad a_2 \to ja_2, \quad a_3 \to a_3. \tag{34}$$

For $j = \iota$, it follows from (33) that $\det u(\iota) = |\alpha|^2 = 1$, i.e. $\alpha = e^{i\varphi}$, therefore

$$u(\iota) = \begin{pmatrix} e^{i\varphi} & \iota\beta \\ -\iota\beta^* & e^{-i\varphi} \end{pmatrix} \in SU(2;\iota), \quad \beta \in C.$$

The simple group $SU(2)$ is contracted to the non-semisimple group $SU(2;\iota)$, which is isomorphic to the Euclid group $E(2)$.

The actions of the unitary group $U(1)$ with generator $Y = \frac{i}{2}\mathbf{1}$ and the electromagnetic subgroup $U(1)_{em}$ with generator $Q = Y + T_3$ in the fibered space $\Phi_2(\iota)$ are given by the same matrices

$$u(\beta) = e^{\beta Y} = \begin{pmatrix} e^{i\frac{\beta}{2}} & 0 \\ 0 & e^{i\frac{\beta}{2}} \end{pmatrix}, \quad u_{em}(\gamma) = e^{\gamma Q} = \begin{pmatrix} e^{i\gamma} & 0 \\ 0 & 1 \end{pmatrix},$$

as on the space Φ_2.

4. Electroweak Model for contracted gauge group $SU(2;j) \times U(1)$

The space $\Phi_2(j)$ can be obtained from Φ_2 by substitution $\phi_2 \to j\phi_2$, which induces the substitution (34) for Lie algebra $su(2)$ parameters. As far as the gauge fields take their values in Lie algebra, we can substitute gauge fields

$$W_\mu^\pm \to jW_\mu^\pm, \quad Z_\mu \to Z_\mu, \quad A_\mu \to A_\mu. \tag{35}$$

After that the bosonic Lagrangian (28) assumes the form

$$L_B(j) = -\frac{1}{4}(\mathcal{F}_{\mu\nu})^2 - \frac{1}{4}(\mathcal{Z}_{\mu\nu})^2 + \frac{1}{2}m_Z^2\,(Z_\mu)^2 +$$

$$+j^2\left\{-\frac{1}{2}\mathcal{W}_{\mu\nu}^+\mathcal{W}_{\mu\nu}^- + m_W^2 W_\mu^+ W_\mu^- + \frac{igP - g^2S - (g\mathcal{Z}_{\mu\nu} + g'\mathcal{A}_{\mu\nu})H_{\mu\nu}}{2\sqrt{g^2+g'^2}}\right\} - j^4\frac{1}{4}(H_{\mu\nu})^2 =$$

$$= L_b + j^2 L_f + j^4 L_h. \tag{36}$$

When contraction parameter takes nilpotent value $j = \iota$ or tends to zero $j^2 \to 0$, then the contribution L_f of W-bosons fields to the Lagrangian (36) will be small in comparison with the contribution L_b of Z-boson and electromagnetic fields. The term L_h being fourth order in j can be neglected. In other words the limit Lagrangian L_b includes Z-boson and electromagnetic fields and charded W-bosons fields does not effect on these fields. The part L_f form a new Lagrangian for W-bosons fields and their interactions with other fields. The appearance of two Lagrangians L_b and L_f for the limit model is in correspondence with two hermitian forms of fibered space $\Phi_2(\iota)$, which are invariant under the action of contracted gauge group $SU(2;\iota)$. Electromagnetic and Z-boson fields can be regarded as external ones with respect to the W-bosons fields.

In mathematical language the field space $\left\{A_\mu, Z_\mu, W_\mu^\pm\right\}$ is fibered after contraction $j = \iota$ to the base $\{A_\mu, Z_\mu\}$ and the fiber $\left\{W_\mu^\pm\right\}$. (In order to avoid terminological misunderstanding let us stress that we have in view locally trivial fibering, which is defined by the projection $pr: \left\{A_\mu, Z_\mu, W_\mu^\pm\right\} \to \{A_\mu, Z_\mu\}$ in the field space. This fibering is understood in the context of semi-Riemannian geometry [22] and has nothing to do with the principal fiber bundle.) Then L_b

in (36) presents Lagrangian in the base and L_f is Lagrangian in the fiber. In general, properties of a fiber are depend on a points of a base and not the contrary. In this sense fields in the base can be interpreted as external ones with respect to fields in the fiber.

Let us note that field interactions in contracted model are more simple as compared with the standard Electroweak Model due to nullification of some terms. For example, the last term $j^4 L_h$ in (36) disappears as having fourth order in $j \to 0$.

When the gauge group $SU(2)$ is contracted to $SU(2;j)$ and the matter field is fibered to $\phi(j)$ the same take place with doublet L_l in (18), namely, the first component e_l does not changed, but the second component is multiplied by contraction parameter: $\nu_{e,l} \to j\nu_{e,l}$. Then the mass terms (19) are rewritten in the form

$$h_e[e_r^\dagger(\phi^\dagger(j)L_l(j)) + (L_l^\dagger(j)\phi(j))e_r] = \frac{h_e R}{\sqrt{1+\bar{\psi}^2(j)}}\left\{e_r^\dagger e_l + e_l^\dagger e_r + \right.$$

$$\left. + i\psi_3\left(e_l^\dagger e_r - e_r^\dagger e_l\right) + ij^2\left[\psi_1\left(\nu_{e,l}^\dagger e_r - e_r^\dagger \nu_{e,l}\right) + i\psi_2\left(\nu_{e,l}^\dagger e_r + e_r^\dagger \nu_{e,l}\right)\right]\right\}, \quad (37)$$

where the $SU(2)$-singlet e_r does not transforms under contraction. For nilpotent value of the contraction parameter $j = \iota$ the fermion Lagrangian (18) is also split on electron part in the base and neutrino part in the fiber, i.e. in the limit model electron field is external one relative to neutrino field. This fact may probably explain why neutrino interaction with the matter is so weak. The mass terms (37) for $j = \iota$ are

$$h_e[e_r^\dagger(\phi^\dagger(\iota)L_l(\iota)) + (L_l^\dagger(\iota)\phi(\iota))e_r] = \frac{h_e R}{\sqrt{1+\psi_3^2}}\left[e_r^\dagger e_l^- + e_l^{-\dagger} e_r + i\psi_3\left(e_l^{-\dagger} e_r - e_r^\dagger e_l^-\right)\right]. \quad (38)$$

Its second order terms $h_e R\left(e_r^\dagger e_l^- + e_l^{-\dagger} e_r\right)$ provide the electron mass $m_e = h_e R$ and neutrino remain massless in exactly the same way as for the Standard Model.

5. Conclusions

The topological mechanism for generation of vector boson masses in the Electroweak Model is discussed. Vector boson masses are automatically generated by transformation of the free Lagrangian from the noncompact R_4 matter fields space to the compact sphere S_3. This model describes all experimentally observed fields and does not include the (up to now unobserved) scalar Higgs field. W- and Z-boson masses are expressed through the parameter R by the same formulas as in the standard case. The free parameter R of the model is now interpreted as the curvature radius of the spherical matter fields space. The development of this topological idea in different aspects, such as transformation to radial coordinates or nonlinearly realized gauge group or nonlinear partial-trace sigma-model, is briefly reviewed.

The limiting case of the modified Electroweak Model which corresponds to the contracted gauge group $SU(2;\iota) \times U(1)$ is presented. The masses of particles involved in the Electroweak Model remain the same under contraction, but field interactions are simplified. Electromagnetic and Z-boson fields can be regarded as external fields with respect to the W-boson fields, i.e. Z-boson and electromagnetic fields can act upon W-boson fields, but not on the contrary. In much the same way in the fermion sector electron field is external one relative to neutrino field.

Acknowledgments

This work is supported by the program "'Fundamental problems of nonlinear dynamics"' of Russian Academy of Sciences.

Journal of Physics: Conference Series **284** (2011) 012033 doi:10.1088/1742-6596/284/1/012033

References

[1] Shirkov D V 2009 *Usp. Fiz. Nauk* **179** No. 6 581.
[2] Casalbuoni R 2005 Deconstructed Higgsless models *Preprint* hep-ph/0509165
[3] Chivukala R S, Simmons E H, He H-J, Kurachi M and Tanabashi M 2004 The structure of corrections to electroweak interactions in Higgsless models *Preprint* hep-ph/0406073
[4] Slavnov A A 2006 Renormalizable electroweak model without fundamental scalar meson *Preprint* hep-th/0601125
[5] Cacciapaglia G, Csáki C, Marandella G and Terning J 2006 A new custodian for a realistic Higgsless model *Preprint* hep-ph/0607156
[6] Moffat J W and Toth V T A finite electroweak model without a Higgs particle *Preprint* arXiv:0812.1991 [hep-th]
[7] Efimov G V 1966 *Sov. J. Nucl. Phys.* **2** 138
[8] Efimov G V 1967 *Sov. J. Nucl. Phys.* **4** 309
[9] Gromov N A 2008 Higgsless Electroweak Model due to the Spherical Geometry *J. Phys.: Conf. Series* **128** 012005 (arXiv:0705.4575 [hep-th])
[10] Inonu E and Wigner E P 1953 *Proc. Nat. Acad. Sci. USA* **39** 510
[11] Nappi C R and Witten E A WZW model based on non-semisimple group, *Preprint* hep-th/9310112
[12] Tsetytlin A A On gauge theories for non-semisimple groups, *Preprint* hep-th/9505129
[13] Rubakov V A 1999 *Classical Gauge Fields* (Moscow, Editorial URSS) (in Russian)
[14] Pawlowski M and Raczka R 1994 A unified conformal model for fundamental interactions without dynamical Higgs field *Found. of Phys.* **24** 1305 (*Preprint* hep-th/9407137)
[15] Faddeev L D An alternative interpretation of the Weinberg-Salam model *Preprint* arXiv:0811.3311 [hep-th]
[16] Masson T and Wallet J - C A remark on the spontaneous symmetry breaking mechanism in the Standard Model *Preprint* arXiv:1001.1176 [hep-th]
[17] Ilderton A, Lavelle M and McMullan D J 2010 Symmetry breaking, conformal geometry and gauge invariance *J. Phys. A: Math. Theor.* **43**, 312002 (*Preprint* arXiv:1002.1170 [hep-th])
[18] Bettinelli D, Ferrari R and Quadri A 2009 The $SU(2) \otimes U(1)$ Electroweak model based on the nonlinearly realized gauge group *Int. J. Mod. Phys. A* **24**, 2639 (*Preprint* arXiv:0807.3882 [hep-th])
[19] Bettinelli D, Ferrari R and Quadri A 2010 The $SU(2) \otimes U(1)$ Electroweak model based on the nonlinearly realized gauge group. II *Acta Phys. Pol. B* **41** 597 (*Preprint* arXiv:0809.1994 [hep-th])
[20] Aldaya V, Calixto M and Lopez-Ruiz F F A quantizable model of massive gauge vector bosons without Higgs *Preprint* arXiv:0905.3688v1 [hep-th]
[21] Pimenov R I 1965 *Litov. Mat. Sb.* **5**, 457 (in Russian)
[22] Gromov N A 2009 The R.I. Pimenov unified gravitation and electromagnetism field theory as semi-Riemannian geometry *Phys. At. Nucl.* **72** 794 (*Preprint* arXiv:08100349 [gr-qc]).

Journal of Physics: Conference Series **284** (2011) 012034 doi:10.1088/1742-6596/284/1/012034

Covariant quantization of the Maxwell field in de Sitter space from $SO_0(2,4)$-invariance

E Huguet[1], S Faci[1], J Queva[2] and J. Renaud[3]

1 - Université Paris Diderot-Paris 7, APC-Astroparticule et Cosmologie (UMR-CNRS 7164), Batiment Condorcet, 10 rue Alice Domon et Léonie Duquet, F-75205 Paris Cedex 13, France.
2 Institüt fr Theoretische Physik, Universität Göttingen, Friedrich-Hund-Platz 1, 37077 Gtötingen, Germany.
3 - Université Paris-Est, APC-Astroparticule et Cosmologie (UMR-CNRS 7164), Batiment Condorcet, 10 rue Alice Domon et Léonie Duquet, F-75205 Paris Cedex 13, France.

E-mail: huguet@apc.univ-paris7.fr

Abstract. We present a $SO_0(2,4)$-invariant quantization of the free electromagnetic field in de Sitter space. Precisely, we quantize the Maxwell ("massless spin one") de Sitter field in a conformally invariant gauge. This result is obtained thanks to a canonical quantization scheme of the Gupta-Bleuler type and to a geometrical formalism in which the Minkowski, de Sitter and anti-de Sitter spaces are realized as intersections of the five dimensional null cone of $\mathbb{R}^6$ and a moving hyperplane. We obtain a new and simple de Sitter invariant two-point function.

1. Introduction and methodology

The aim of the work presented in [1] (to which the Reader is referred for details) is to provide a quantization of the free Maxwell field in de Sitter space in which the $SO_0(2,4)$-invariance is preserved. To this end we quantize the one-form field A^H_μ which fulfills the de Sitter Maxwell equations together with a conformal gauge condition:

$$\left\{ \begin{array}{c} \Box_H A^H_\mu - \nabla_\mu \nabla A^H + 3H^2 A^H_\mu = 0 \\ (\Box_H + 2H^2)\nabla A^H = 0, \end{array} \right. \tag{1}$$

where $12H^2 = R$, R being the Ricci scalar and $\Box_H$ the usual Laplace-Beltrami operator. As a result, we obtain the following de Sitter-covariant two-point function

$$D^H_{\mu\nu'}(p,p') = \frac{H^2}{8\pi^2}\left(\frac{1}{\mathcal{Z}-1} g_{\mu\nu'} - n_\mu n_{\nu'}\right), \tag{2}$$

where $g_{\mu\nu'}(p,p')$ is the parallel propagator, $\mathcal{Z}$, n_μ, $n_{\nu'}$ being related to the geodesic distance between the two points p and p' of the de Sitter space (see Appendix C).

The problem posed by such a quantization is twofold. First, we have to set up a geometrical framework in which the $SO_0(2,4)$-invariance is easily implemented. Then, we have to deal, in that context, with the known problem of quantizing an Abelian gauge field, that is to say to keep manifest the invariance in the quantized field.

Journal of Physics: Conference Series **284** (2011) 012034 doi:10.1088/1742-6596/284/1/012034

To answer the first part of the problem we use the formalism already described in [2, 3], which are summarized in Sec. 2. Essentially, we consider the null cone in $\mathbb{R}^6$ which is the natural geometric object related to the $\mathrm{SO}_0(2,4)$-invariance. Then, we realize the de Sitter and Minkowski spaces[1] as the intersections of the light cone and a moving hyperplane whose position is parameterized by the curvature H.

The second side of the problem is well known in the case of the usual covariant canonical quantization of the electromagnetic field in Lorenz gauge in Minkowski space. In short, it is not possible to build a quantum field operator which satisfies at the same time both the Maxwell equations and the Lorenz condition. A solution to this problem is to quantize a field which, in place of the Maxwell equations, satisfies another set of equations whose set of solutions contains as an invariant subset the solutions of the Maxwell equations in the Lorenz gauge. The quantum states which correspond to this subset are then specified by a condition which corresponds at the quantum level to the gauge condition . In that case the additional solutions correspond to the longitudinal and scalar photons and the quantum condition allows to select the physical transverse photons. This method is essentially the Gupta-Bleuler quantization [4, 5]. On a technical side, that method is implemented by adding a gauge-fixing term in the Lagrangian which leads to the equation mentioned above whose set of solutions is larger than that of the initial problem. In our conformal framework, that scheme is implemented in a different way which is an extention of the method used in [6]. There is no gauge-fixing term in the Lagrangian, instead, the Maxwell equations and the covariant gauge condition are obtained from a larger set of equations which in addition to the Maxwell field contains two more auxiliary fields. They encode both the gauge condition and transversality constraints. This is described in Sec. 3 and 4.

For convenience, our notations and conventions are collected in Appendix A, explicit formulas are given for reference in Appendix B and the definitions of quantities related to the geodesic distance which appears in the Wightman two-point function are given in Appendix C.

2. The geometrical framework in brief

Since our goal is to preserve the $\mathrm{SO}_0(2,4)$-invariance the natural geometric object, invariant under the action of that group, is the five dimensional null cone $\mathcal{C}$ of $\mathbb{R}^6$

$$\mathcal{C} = \left\{ y \in \mathbb{R}^6 : (y^0)^2 - \boldsymbol{y}^2 - (y^4)^2 + (y^5)^2 = 0 \right\}. \tag{3}$$

The de Sitter and Minkowski spaces are obtained as the intersection X_H of the five dimensional cone $\mathcal{C}$ of and a moving hyperplane P_H whose equation reads $f_H(y) = 2$ where

$$f_H(y) = (1 + H^2)y^5 + (1 - H^2)y^4, \tag{4}$$

where H is a non-negative real number which identifies with the curvature of X_H.

As already shown in [2], the manifold X_H is invariant under a subgroup of $\mathrm{SO}(2,4)$ isomorphic to the de Sitter group $\mathrm{SO}_0(1,4)$ for $H > 0$ and to the Poincaré group for $H = 0$. As a consequence, the manifold X_H is identified with the de Sitter space for $H > 0$ and Minkowski space for $H = 0$. This realization allows to make a continuous transition between spaces by a continuous change of the value of the parameter H. Note that, as a consequence of this property, the zero-curvature limit of all local objects depending on H in the de Sitter space (for instance, group generators, modes, two-point functions,...) is obtained by setting $H = 0$ in the concerned expression.

[1] The anti-de Sitter space is also obtained by this construction. In this work we are interested in quantization and thus we only consider globally hyperbolic spaces. At a classical level our results apply to the anti-de Sitter space as well. The expressions relevant for this space can be obtained directly from the substitution $H^2 \to -H^2$.

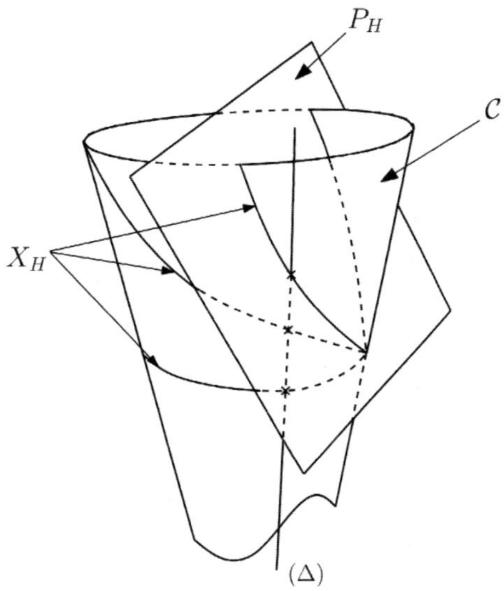

Figure 1. The main construction: to each point of $X_H = P_H \cap \mathcal{C}$ corresponds a unique half-line Δ, that is a point of $\mathcal{C}'$.

In this construction each point of a manifold X_H is associated to a half-line issued from the origin of $\mathbb{R}^6$ (see Fig.1), or, in other words, is associated to a point of the set $\mathcal{C}'$ of half-lines, which is the cone up to the dilations. The most important consequence of this fact is that each manifold X_H can be realized as a subset of $\mathcal{C}'$. Thus, all the manifolds X_H (i.e., for any value of H) are realized on the same underlying set $\mathcal{C}'$.

From the point of view of group theory note that thanks to the linearity of the action of $\mathrm{SO}_0(2,4)$, there is a natural action of this group on $\mathcal{C}'$. Another consequence of the relation between the points of X_H and the half-lines is that the functions on X_H and the homogeneous functions on the cone $\mathcal{C}$ are in one-to-one correspondence. Precisely, if y is a point of the cone and y^H the intersection of X_H with the half-line linking y to the origin of $\mathbb{R}^6$, then one can verify using the equations defining $\mathcal{C}$ and P_H that

$$y^H = 2y/f_H(y). \tag{5}$$

Thus, to any $\widetilde{F}$ homogeneous function of degree r defined on $\mathcal{C}$ corresponds a function F^H on X_H defined through

$$F^H(x) = \left(\frac{2}{f_H(y)}\right)^r \widetilde{F}(y), \tag{6}$$

where $y \in \mathcal{C}$ and $x = y^H$ is the corresponding point of X_H.

Finally, the metric $g^H_{\mu\nu}$ on X_H is inherited from $\tilde{\eta}_{\alpha\beta}$ on $\mathbb{R}^6$. For $H = 0$ this metric is the Minkowskian metric, while for $H > 0$ it is a de Sitterian metric hereafter denoted $g_{\mu\nu}$. It can be verified, for instance through a straightforward calculation in some coordinate system, that these metrics are related through a (local) Weyl rescaling. In the system $\{x^\mu\}$ (see (9) below) one has

$$g_{\mu\nu} = \left(K^H\right)^2 \eta_{\mu\nu}, K^H = \frac{1}{1 - \frac{H^2}{4}\eta_{\mu\nu}x^\mu x^\nu} \tag{7}$$

K^H being the Weyl factor and $\eta_{\mu\nu}$ the usual Minkowski metric.

Journal of Physics: Conference Series **284** (2011) 012034 doi:10.1088/1742-6596/284/1/012034

3. The Maxwell field in a conformal gauge

We use the Dirac's six-cone formalism [7, 8] as a starting point: we consider the $SO_0(2,4)$-invariant homogeneous wave equation

$$\tilde{\eta}^{\gamma\beta}\tilde{\partial}_\gamma\tilde{\partial}_\beta\tilde{a}_\alpha = 0, \tag{8}$$

where $\tilde{a} = \tilde{a}_\alpha dy^\alpha$ is a one-form field in $\mathbb{R}^6$. As shown in [7, 8], the one-forms $\tilde{a}$ must be homogeneous of degree -1 in order to consider the field and the equation on the cone $\mathcal{C}$.

In addition, we make use of auxiliary fields (see [6] for instance). Basically, they implement two constraints: they set the $SO_0(2,4)$-invariant gauge condition and ensure that the physical field A^H_μ is an intrinsic tensor of X_H. These fields are defined through the expansion of the one-form field $\tilde{a}$ on a convenient system of coordinates. For practical calculations we choose the system $\{x^I\}$, $I \in \{c, \mu, +\}$ defined through

$$\left\{\begin{array}{lcl} x^c & = & \dfrac{y_\alpha y^\alpha}{(y^4+y^5)^2} \\ x^\mu & = & 2\dfrac{y^\mu}{y^4+y^5} \\ x^+_H & = & f_H(y) = (1-H^2)y^4 + (1+H^2)y^5. \end{array}\right. \tag{9}$$

In this system, the restriction to the cone $\mathcal{C}$ is expressed by the constraint $x^c = 0$ and the restriction to the manifold X_H by the additional constraint $x^+_H = 2$. The auxiliary fields are then defined through

$$\tilde{a}_\alpha dy^\alpha = \widetilde{A}_c dx^c + \widetilde{A}_\mu dx^\mu + \frac{\widetilde{A}_+}{x^+_H} dx^+_H. \tag{10}$$

The $\widetilde{a}$ and the $\widetilde{A}$ being homogeneous, we can define the corresponding fields a^H_α and A^H on X_H, thanks to the relation (6).

As a consequence of (10) the fields a^H and A^H are related through linear relations. We will denotes hereafter these relations by $A^H = La^H$. They are given for reference in Appendix B.

The dynamical equations for the fields A^H_I lead, when the constraint $A^H_+ = 0$ is applied, to the Maxwell equations and to a conformal gauge condition. This is obtained as follow. We first push the equation (8) onto X_H using the homogeneity of the fields a with (6). One obtains,

$$\left(\Box^s_H + 2H^2\right) a^H_\alpha = 0, \tag{11}$$

where $\Box^s_H$ means that each component a^H_α must be considered as a scalar. In a second step, we transform the above equations, using $a^H = L^{-1}A^H$, in a system of partial differential equations for the A^H. This system is just an intermediate calculation which can be recasted (see [1]) under the form

$$\left\{\begin{array}{l} \Box_H A^H_\mu - \nabla_\mu \nabla A^H + 3H^2 A^H_\mu = -\dfrac{1}{2}\nabla_\mu \left(\Box_H + 2H^2\right) A^H_+ \\ (\nabla - W)A^H + (K^H)^{-2} A^H_c = \dfrac{1}{2}\left(\Box_H + 2H^2\right) A^H_+ \\ (\nabla - W)\nabla A^H_c = -K^H H^2 \left(\Box_H + 2H^2\right) A^H_+, \end{array}\right. \tag{12}$$

where ∇A^H is the divergence of A^H, and W is the one-form $W := \mathrm{d}\ln (K^H)^2$ of components $W_\mu = \nabla_\mu \ln (K^H)^2$.

Now, it is apparent that for $A^H_+ = 0$ the first line of (12) are precisely the Maxwell equations. Then, eliminating A^H_c from the remaining two equations and using the relation $\nabla_\mu (K^H)^2 = (K^H)^2 W_\mu$, one obtains

$$\left\{\begin{array}{l} \Box_H A^H_\mu - \nabla_\mu \nabla A^H + 3H^2 A^H_\mu = 0 \\ \nabla^\mu \left(\nabla_\mu + W_\mu\right)\left(\nabla^\nu - W^\nu\right) A^H_\nu = 0, \end{array}\right. \tag{13}$$

Note that, in the above system each equation is invariant under both the $\mathrm{SO}_0(2,4)$ transformations and the Weyl rescalling. One can show (see [1]) that this system is equivalent to

$$\begin{cases} \Box_H A^H_\mu - \nabla_\mu \nabla A^H + 3H^2 A^H_\mu = 0 \\ (\Box_H + 2H^2)\nabla A^H = 0. \end{cases} \tag{14}$$

In the second line we recognize the Eastwood-Singer gauge [9] specialized to our constant curvature space X_H. The gauge condition of the second line is invariant under both the $\mathrm{SO}_0(2,4)$ group and Weyl transformations only on the space of solutions of the Maxwell equations.

Two comments are in order before closing this section. First, it is apparent that the condition $A^H_+ = 0$ characterize the subset of solutions of the system (12) which verify both Maxwell's equations and the conformal gauge condition. This condition is invariant under the $\mathrm{SO}_0(2,4)$ transformation since it can be shown that $\widetilde{A}_+ = y^\alpha \tilde{a}_\alpha$ (see [1] again). Finally, the system (12) which is the the system we will quantize is in fact the equation (11) in which a^H appears as a conformal scalar field.

4. Quantization

Let us consider a tensor field F satisfying some linear equations: $\mathcal{E}F = 0$ on X_H. When F is not an abelian gauge field, one can obtain a covariant and causal quantized field $\widehat{F}$ satisfying the equations $\mathcal{E}\widehat{F} = 0$ by applying the following procedure:

(i) Select a Hilbert space $\mathcal{K}$ of solutions of the classical equation equipped with a scalar product $\langle\ ,\ \rangle$ and carrying a unitary representation of the symmetry group.

(ii) Obtain a causal reproducing kernel $\mathcal{W}$ for $\mathcal{K}$, the Wightman two-point function. More precisely, $\mathcal{W}$ is a bitensor such that, for each $x \in X_H$, $\mathcal{W}(x,\cdot) : x' \to \mathcal{W}(x,x')$ is, up to a smearing function on the variable x, an element of $\mathcal{K}$ satisfying

$$\langle \mathcal{W}(x,\cdot), \psi\rangle = \psi(x), \tag{15}$$

for any $\psi \in \mathcal{K}$, and such that $\mathcal{W}(x,x') = \mathcal{W}(x',x)$ as soon as x and x' are causally separated.

(iii) Define the quantum field $\widehat{F}$ through

$$\widehat{F}(x) = a\left(\mathcal{W}(x,\cdot)\right) + a^\dagger\left(\mathcal{W}(x,\cdot)\right), \tag{16}$$

where a and $a^\dagger$ are the usual creator and annihilator of the Fock space built onto $\mathcal{K}$.

In practice, one determines a family of modes $\{\phi_k\}$ solutions of $\mathcal{E}F = 0$ and finds a scalar product such that $\{\phi_k\}$ is an orthonormal Hilbert basis for $\mathcal{K}$. Then, the two-point function reads

$$\mathcal{W}(x,x') = \sum_k \phi^*_k(x) \otimes \phi_k(x'). \tag{17}$$

From this expression, using (16) and the anti-linearity and linearity of a and $a^\dagger$ respectively, one obtains the familiar expression of the quantum field:

$$\widehat{F}(x) = \sum_k \phi_k(x) b_k + \phi^*_k(x) b^\dagger_k, \tag{18}$$

where $b_k := a(\phi_k)$ and $b^\dagger_k := a^\dagger(\phi_k)$ are the annihilators and creators of the modes ϕ_k. The Hilbert space of quantum states $|\psi\rangle$, is then built as usual through the action of the $b^\dagger_k$ on the vacuum state of the theory.

As already mentioned in the introduction, in gauge context, due to the presence of pure gauge solutions, such a two-point reproducing kernel does not exist. The reason for that is the following: the pure gauge solutions (the gradients of complex functions in the case of the Maxwell equations) are known to be orthogonal to any modes including themselves, so replacing ψ by a pure gauge modes in (15) should make vanishing the left hand side and not the right hand side. In the Gupta-Bleuler scheme, one overcomes this problem by considering an enlarged space $\mathcal{H} \supset \mathcal{K}$ containing some elements not orthogonal to the pure gauges. This space, which is generally a Krein space, that is a space equipped with an indefinite inner product [10, 11], is defined through another equations $\mathcal{E}'F = 0$ also invariant under the group. The elements of $\mathcal{K}$, called in this context the physical solutions, satisfy, in addition to the new field equation, a constraint $\mathcal{G}F = 0$. This classical condition, which allows to characterize the classical physical solutions, translates into a quantum condition which allows in turn to characterize the subspace of physical states. For instance, in the usual Gupta-Bleuler quantization of the Maxwell field in Minkowski space: the equation defining the enlarged space $\mathcal{H}$ is $\Box A_\mu = 0$, the space $\mathcal{H}$ contains modes with negative norm (those corresponding to the scalar photons); the subspace of physical solutions (corresponding to the transverses photons) is characterized by the constraint $\nabla A = 0$, the Lorenz gauge condition, it's quantum counterpart reads $\partial A^{(+)} \mid \psi \rangle = 0$, where $\widehat{A}^{(+)}$ is the annihilator part of $\widehat{A}$ and $\mid \psi \rangle$ a physical state. The new quantum field $\widehat{F}$ is covariant and causal. It is obtained through

$$\widehat{F}(x) = \sum_k \zeta_k \left(\phi_k(x) b_k + \phi_k^*(x) b_k^\dagger \right), \tag{19}$$

where $\zeta_k = \pm 1$. One can show (see [1] again) that it satisfy:

$$\langle \psi_1 | \mathcal{E}\widehat{F} | \psi_2 \rangle = 0, \tag{20}$$

as soon as $|\psi_1\rangle, |\psi_2\rangle$ are physical states.

We now apply this quantization process to the Maxwell de Sitter field in conformal gauge (14). As for the non conformal case, it can be shown that the pure gauge solutions ($A^H_\mu = \nabla_\mu f$, with $(\Box_H + 2H^2)\Box_H f = 0$, f being an arbitrary function) are orthogonal to all the solutions including themselves. As a consequence, the space of solutions of (14) is degenerate and the canonical quantization process fails. Following the Gupta-Bleuler method, we consider the system (12), instead of (14). Now, as we have already pointed out at the end of Sec. 3, the system (12) is nothing but the equation (11) that is, basically, the equation for the conformal scalar field. It follows that the practical quantization scheme described above can be realized directly from the knowledge of the conformal scalar field (see [2] for details):

- The scalar product on the space of solutions of $(\Box^s_H + 2H^2)a^H = 0$ is defined through

$$\langle a^H, b^H \rangle := -\tilde{\eta}^{\alpha\beta} \langle a^H_\alpha, b^H_\beta \rangle_s, \tag{21}$$

where $\langle , \rangle_s$ is (with a slightly different notation from that used in [2]) the Klein-Gordon scalar product on the space of solutions of the conformal scalar equation on X_H.

- The modes on X_H reads

$$a^H_{LM(\gamma)}(x) = \epsilon_{(\gamma)} \Phi^H_{LM}(x), \tag{22}$$

where the one-forms $\epsilon_{(\gamma)}$ are defined trough $\epsilon_{(\gamma)\delta} = -\tilde{\eta}_{\gamma\delta}$ and $\Phi^H_{LM}(x)$ are the modes solutions of the scalar equation $(\Box_H + 2H^2)\Phi^H = 0$ (see [2] for details). These solutions are normalized with respect to (21) precisely

$$\langle a^H_{LM(\gamma)}, a^H_{L'M'(\delta)} \rangle = -\tilde{\eta}_{\gamma\delta} \delta_{LL'} \delta_{MM'}. \tag{23}$$

- The Wightman two-point function is obtained through the formula

$$\mathcal{W}^H = \sum_{LM\gamma} \zeta_\gamma \epsilon_{(\gamma)} \left(\Phi^H_{LM}\right)^* \otimes \epsilon_{(\gamma)} \Phi^H_{LM}, \tag{24}$$

with $\zeta_\gamma = -\tilde{\eta}_{\gamma\gamma}$, applied to the above modes. A straightforward calculation using the results of [2] gives

$$\mathcal{W}^H_{\alpha\beta}(x,x') = -\tilde{\eta}_{\alpha\beta} D^+_H(x,x'), \tag{25}$$

where $D^+_H(x,x')$ is the usual two-point function for the conformal scalar field in de Sitter space (see Appendix B).

- The quantum field $\widehat{a}^H$ reads

$$\widehat{a}^H(x) = \sum_{LM\gamma} a^H_{LM(\gamma)}(x) b_{LM(\gamma)} + a^{H\,*}_{LM(\gamma)}(x) b^\dagger_{LM(\gamma)}, \tag{26}$$

$b_{LM(\gamma)}$ and $b^\dagger_{LM(\gamma)}$ being the annihilators and creators of the mode $a^H_{LM(\gamma)}$.

- The physical states are created from physical solutions:

$$|\, a^H\,\rangle \text{ is a physical state iff } \ |\, a^H\,\rangle = a^\dagger(a^H)\,|\,0\,\rangle_H, \text{ and } A^H_+[a^H] := L^{-1} a^H_+ = 0. \tag{27}$$

The quantum field $\widehat{A}^H$ and the corresponding two-point function are obtained from the above relations thanks to the linear relations $A^H = La^H$. In particular, The Maxwell de Sitter two-point function defined as usual through

$$D^H_{\mu\nu'}(x,x') = {}_H\langle 0 \mid \widehat{A}^H_\mu(x) \widehat{A}^H_{\nu'}(x') \mid 0\rangle_H, \tag{28}$$

is found to be (see [1] for details)

$$D^H_{\mu\nu'}(x,x') = \frac{H^2}{8\pi^2}\left(\frac{1}{\mathcal{Z}_\varepsilon - 1} g_{\mu\nu'} - n_\mu n_{\nu'}\right), \tag{29}$$

where $\mathcal{Z}_\varepsilon := \mathcal{Z} - i\varepsilon(x^0 - x'^0)$, ε being a small positive real number, and the geometrical objects $\mathcal{Z}$, $g_{\mu\nu'}$, n_μ and n'_ν are explicitly defined in Appendix C.

Note finally that this two-point function possesses no other singular point than $\mathcal{Z} = 1$ and has clearly the Hadamard behavior (thus our vacuum is the Euclidean one).

5. Acknowledgements

E. H. wishes to thank the organizing committee for its efficiency and availability.

Appendix A. Conventions and notations

Here are the conventions:

$$\begin{aligned} \alpha,\beta,\gamma,\delta,\ldots &= 0,\ldots,5,\\ \mu,\nu,\rho,\sigma,\kappa\ldots &= 0,\ldots,3,\\ i,j,k,l,\ldots &= 1,\ldots,3. \end{aligned}$$

The indices and superscripts I, J stand for the set $\{c,\mu,+\}$, for instance $\{A_I\} = \{A_c, A_\mu, A_+\}$. he space $\mathbb{R}^6$, is provided with the natural orthogonal coordinates $\{y^\alpha\}$ and the metric $\tilde{\eta}_{\alpha\beta} = \mathrm{diag}(+,-,-,-,-,+)$. whose coefficients are denoted $\tilde{\eta}_{\alpha\beta}$:

$$\tilde{\eta}_{55} = \tilde{\eta}_{00} = 1 = -\tilde{\eta}_{ii} = -\tilde{\eta}_{44}. \tag{A.1}$$

For convenience we set $\eta_{\mu\nu} := \tilde{\eta}_{\mu\nu}$. Partial derivatives with respect to the variables $\{y^\alpha\}$ of $\mathbb{R}^6$ are denoted by $\tilde{\partial}_\alpha$.

Various spaces and maps are used throughout this paper. Except otherwise stated, quantities related to $\mathbb{R}^6$ and its null cone $\mathcal{C}$ are labeled with a tilde, those defined on X_H (see Sec. 2 hereafter) are denoted with a super or subscript H except when H takes the null value (Minkowski space) in which case the super or subscript 0 is omitted. The quantum operator associated with a classical quantity Q is denoted with a hat: $\widehat{Q}$.

Journal of Physics: Conference Series **284** (2011) 012034 doi:10.1088/1742-6596/284/1/012034

Appendix B. Explicit formulas

The linear relations between the A^H and the a^H reads

$$\left\{\begin{array}{lcl} A^H_c &=& (K^H)^2\left\{a^H_5\left(1-H^2\right)-a^H_4\left(1+H^2\right)-H^2a^H_\mu x^\mu\right\} \\ A^H_\mu &=& \dfrac{(K^H)^2}{2}\left\{\left(a^H_4\left(1+H^2\right)-a^H_5\left(1-H^2\right)\right)\eta_{\mu\nu}x^\nu+H^2a^H_\sigma x^\sigma\eta_{\mu\nu}x^\nu+\dfrac{2}{K^H}a^H_\mu\right\} \\ A^H_+ &=& K^H\left\{a^H_5\left(1-\dfrac{1}{4}\eta_{\mu\nu}x^\mu x^\nu\right)+a^H_4\left(1+\dfrac{1}{4}\eta_{\mu\nu}x^\mu x^\nu\right)+a^H_\mu x^\mu\right\}, \end{array}\right. \tag{B.1}$$

from which the matrix L can be read. These relations are obtained by first expressing the basis $\{dy\}$ in the left hand side of (10) in terms of the basis $\{dx\}$ and identifying both sides and then applying the correspondence (6).

The above system can be inverted in

$$\left\{\begin{array}{lcl} a^H_5 &=& \dfrac{1}{2K^H}\left\{A^H_c\left(1-\dfrac{1}{4}\eta_{\mu\nu}x^\mu x^\nu\right)-A^H_\sigma x^\sigma+A^H_+K^H\left(1+H^2\right)\right\} \\ a^H_4 &=& \dfrac{1}{2K^H}\left\{A^H_c\left(-1-\dfrac{1}{4}\eta_{\mu\nu}x^\mu x^\nu\right)-A^H_\sigma x^\sigma+A^H_+K^H\left(1-H^2\right)\right\} \\ a^H_\mu &=& \dfrac{1}{2K^H}\left\{A^H_c\eta_{\mu\nu}x^\nu+2A^H_\mu\right\}. \end{array}\right. \tag{B.2}$$

The Wightman two-point function for the conformal scalar field in the de Sitter space reads

$$D^+_H(x,x')=-\frac{H^2}{8\pi^2}\frac{1}{\left(\mathcal{Z}(x,x')-1-i\varepsilon\mathrm{sgn}(t-t')\right)}, \tag{B.3}$$

in which ε is a small positive real number and $\mathcal{Z}(x,x')$ is defined in Appendix C.

Appendix C. Intrinsic quantities for bitensors in de Sitter space

Following Allen and Jacobson [12] (where the reader is referred for proofs and details), any maximally symmetric bitensor (that is, invariant under the isometry group of a maximally symmetric manifold, here the de Sitter space) can be decomposed in a unique way as sum of products of fundamental objects. They are: the metric at points p and p' of the manifold and three quantities related to the length $\mu(p,p')$, of the geodesic from p to p' (μ being imaginary when the geodesic is spacelike), namely:

$n_\mu(p,p')=\nabla_\mu\mu(p,p')$ is the unit tangent vector to the geodesic at the point p,

$n_{\nu'}(p,p')=\nabla_{\nu'}\mu(p,p')$ is the unit tangent vector to the geodesic at the point p',

$g_{\mu\nu'}(p,p')=\frac{1}{C}\nabla_\mu n_{\nu'}(p,p')-n_\mu(p,p')n_{\nu'}(p,p')$ is the parallel propagator along the geodesic,

where we use the usual convention that a primed (resp. not primed) index refers to a primed (resp. not primed) point. The factor C will be given in what follows.

In order to define the standard variable $\mathcal{Z}$, let us introduce the five-dimensional "ambient" Minkowski space with metric $\bar\eta=\mathrm{diag}(+,-,-,-,-)$. We will use small roman letters $a,b,c,\ldots$ to denote indices running from 0 to 4. The de Sitter space can be viewed as the sub-manifold defined by the equation

$$\bar\eta_{ab}X^aX^b=-H^{-2}, \tag{C.1}$$

where $\{X^a\}$ denotes ambient space cartesian coordinates. A point p on the de Sitter space is associated to the vector $X(p)$ of coordinates $X^a(p)$. The ambient coordinates are related to the coordinates $\{x^\mu\}$ through

$$\begin{array}{ll} X^\mu &= K^Hx^\mu, \\ X^4 &= \dfrac{1}{H}\left(2K^H-1\right). \end{array} \tag{C.2}$$

The function $\mathcal{Z} = \mathcal{Z}(p, p')$ is then defined through

$$\mathcal{Z} := -H^2 \bar{\eta}_{ab} X^a X'^b, \tag{C.3}$$

where $X = X(p)$ and $X' = X(p')$. The geodesic distance $\mu(p, p')$ is related to $\mathcal{Z}$ by

$$\mathcal{Z} = \cosh(H\mu), \mathcal{Z} \geqslant -1. \tag{C.4}$$

The case $\mathcal{Z} < -1$ corresponds to the situation where p' is lying in the interior of the light cone of the antipodal of p and, in this case, there is no geodesic connecting p and p'. Nevertheless, $\mathcal{Z}$ is always defined and one can define $\mu(p, p')$ through an analytic continuation (see [12] again). As a function of $\mathcal{Z}$ the factor C reads

$$C = \frac{-H}{\sqrt{(\mathcal{Z}^2 - 1)}}. \tag{C.5}$$

References

[1] S. Faci, E. Huguet, J. Queva, and J. Renaud, Phys. Rev. D **80**, 124005 (2009).
[2] E. Huguet, J. Queva, and J. Renaud, Phys. Rev. D **73**, 084025 (2006).
[3] E. Huguet, J. Queva, and J. Renaud, Phys. Rev. D **77**, 044025 (2008).
[4] S.N. Gupta, Proc. Phys. Soc. London A **63**, 681 (1950).
[5] K. Bleuler, Helv. Phys. Acta **23**, 567 (1950).
[6] F. Bayen *et al.*, Phys. Rev. D **32**, 2673 (1985).
[7] P.A.M. Dirac, Ann. Math. **37**, 429 (1936).
[8] G. Mack and A. Salam, Ann. Phys. (N.Y.) **53**, 174 (1969).
[9] M. Eastwood and M. Singer, Phys. Lett. **107A**, 73 (1985).
[10] J. Bognar, *Indefinite inner product spaces*, Springer-Verlag, Berlin (1974).
[11] T. Garidi, E. Huguet, and J. Renaud, J. Phys. A, **38**, 245 (2005).
[12] B. Allen, T. Jacobson, Comm. Math. Phys., **103**, 669 (1986).

Journal of Physics: Conference Series **284** (2011) 012035 doi:10.1088/1742-6596/284/1/012035

A convenient criterion under which $\mathbb{Z}_2$-graded operators are Hamiltonian

Véronique Hussin[1] **and Arthemy V Kiselev**[2,3]

[1] Département de Mathématiques et de Statistique, Université de Montréal, C.P. 6128, succ. Centre-ville, Montréal, Québec H3C 3J7, Canada.

[2] Mathematical Institute, University of Utrecht, P.O.Box 80.010, 3508 TA Utrecht, The Netherlands.

[3] J. Bernoulli Institute for Mathematics and Computer Science, University of Groningen, P.O.Box 407, 9700 AK Gtoningen, The Netherlands.

E-mail: `hussin@dms.umontreal.ca, A.V.Kiselev@rug.nl`

Abstract. We formulate a simple and convenient criterion under which skew-adjoint $\mathbb{Z}_2$-graded total differential operators are Hamiltonian, provided that their images are closed under commutation in the Lie algebras of evolutionary vector fields on the infinite jet spaces for vector bundles over smooth manifolds.

In this short note we consider Hamiltonian differential operators that induce the Poisson brackets on the spaces of Hamiltonian functionals on the infinite jet spaces for $\mathbb{Z}_2$-graded vector bundles over smooth (super-)manifolds. In other words, we study the structures that are related to the bundles in which the fibres are split, as vector spaces, in the even and the odd components so that, in particular, the even components of the local sections commute with everything whereas the odd components of the sections anti-commute between themselves. In addition, the base of the bundles can be a supermanifold itself, whence the super-derivatives emerge; we always assume that the operators at hand are polynomial in the (super-)derivatives.

We extend a very simple criterion ([1] and [2, p. 130]), under which linear differential operators are Hamiltonian, to the $\mathbb{Z}_2$-graded setup. This will be helpful, in particular, in the study of supersymmetric integrable systems. Obviously, the same criterion allows us to check the compatibility [3] of two given $\mathbb{Z}_2$-graded Hamiltonian operators A_1, A_2 by verifying that the linear combinations $A_1 + \lambda A_2$ remain Hamiltonian at all λ. The tool which we elaborate is very practical and efficient: indeed, its "hardest" component amounts to the calculation of the commutator of two evolutionary vector fields (c.f. [4]). It is important that the procedure is purely algorithmic and is applicable immediately without any further adaptations (handled, e.g., by the software [5]). We recall that other methods for checking whether a given operator is Hamiltonian are available from the literature: e.g. one can re-derive the algorithmic verification procedure from [6] in the $\mathbb{Z}_2$-graded setup; that concept is based on the use of variational polyvectors which are already endowed with their own grading. We finally recall that the book [7] contains another step-by-step verification procedure but (especially in the $\mathbb{Z}_2$-graded case) in practice it is much more involved.

This note is structured as follows. We first extend the criterion of [1] to bosonic super-fields and super-operators, see (5). Theorem 1 in section 2 is our main result that covers the general

Journal of Physics: Conference Series **284** (2011) 012035 doi:10.1088/1742-6596/284/1/012035

setup of $\mathbb{Z}_2$-graded fields. Its proof, which is given here in full detail, is considerably simplified with respect to the one in [1].

All notions and constructions from the geometry of differential equations are standard ([7] and [2, 8]). We follow the notation of [1] which agrees with that of [2] but initially covered only the non-graded case. In the sequel, everything is real and C^∞-smooth.

1. Bosonic super-fields and Hamiltonian super-operators

Let $B^n \ni x = (x^1, \ldots, x^n)$ be an n-dimensional orientable manifold and let $\pi\colon E^{m+n} \xrightarrow[F^m]{} B^n$ be a vector bundle over B^n with m-dimensional fibres $F^m \ni u = (u^1, \ldots, u^m)$. By $J^\infty(\pi)$ we denote the infinite jet space over π. We denote by u_σ, $|\sigma| \geq 0$, its fibre coordinates. We also denote by $\mathfrak{g}$ the Lie algebra of evolutionary vector fields ∂_φ on $J^\infty(\pi)$ and by Ω the linear space of variational covectors, which contains the variational derivatives $\delta\mathcal{H}/\delta u$ of the Hamiltonian functionals $\mathcal{H} \in \bar{H}$ and which is dual to $\mathfrak{g}$ with respect to the coupling $\langle\, ,\, \rangle$ that takes values in $\bar{H}$.

Let $A\colon \Omega \to \mathfrak{g}$ be a total differential operator the image of which is closed with respect to the commutation in $\mathfrak{g}$,

$$[\operatorname{im} A, \operatorname{im} A] \subseteq \operatorname{im} A. \tag{1}$$

This is indeed so for $\mathbb{Z}_2$-graded Hamiltonian operators; the criterion in Theorem 1, see below, makes it clear that condition (1) is not superfluous for their definition. Further examples of *non-Hamiltonian* differential operators, the images of which in the Lie algebras of evolutionary vector fields are subject to the collective commutation closure but the domains of which are different from Ω, are studied in [1, 9] and [10].

The operator A transfers the Lie algebra structure $[\, ,\,]\big|_{\operatorname{im} A}$ to the skew-symmetric bracket $[\, ,\,]_A$ in the quotient $\operatorname{dom} A/\ker A$,

$$\big[A(\boldsymbol{p}), A(\boldsymbol{q})\big] = A\big([\boldsymbol{p}, \boldsymbol{q}]_A\big), \qquad \boldsymbol{p}, \boldsymbol{q} \in \Omega. \tag{2}$$

By the Leibnitz rule, two sets of summands appear in the bracket $\big[\partial_{A(\boldsymbol{p})}, \partial_{A(\boldsymbol{q})}\big] = \partial_{[A(\boldsymbol{p}),A(\boldsymbol{q})]}$ of evolutionary vector fields $\partial_{A(\boldsymbol{p})}$ and $\partial_{A(\boldsymbol{q})}$:

$$\big[A(\boldsymbol{p}), A(\boldsymbol{q})\big] = A\big(\partial_{A(\boldsymbol{p})}(\boldsymbol{q}) - \partial_{A(\boldsymbol{q})}(\boldsymbol{p})\big) + \big(\partial_{A(\boldsymbol{p})}(A)(\boldsymbol{q}) - \partial_{A(\boldsymbol{q})}(A)(\boldsymbol{p})\big). \tag{3}$$

In the first term we have used the permutability of evolutionary derivations, which are of the form $\partial_\varphi = \varphi\,\frac{\partial}{\partial u} + \frac{\mathrm{d}}{\mathrm{d}x}(\varphi)\,\frac{\partial}{\partial u_x} + \cdots$, and total derivatives. The second term hits the image of A by construction. Consequently, the Lie algebra structure $[\, ,\,]_A$ on the domain of A equals

$$[\boldsymbol{p}, \boldsymbol{q}]_A = \partial_{A(\boldsymbol{p})}(\boldsymbol{q}) - \partial_{A(\boldsymbol{q})}(\boldsymbol{p}) + \{\{\boldsymbol{p}, \boldsymbol{q}\}\}_A. \tag{4}$$

Example 1. The second Hamiltonian operator for the Korteweg–de Vries equation is $A = -\frac{1}{2}\frac{\mathrm{d}^3}{\mathrm{d}x^3} + u\,\frac{\mathrm{d}}{\mathrm{d}x} + \frac{\mathrm{d}}{\mathrm{d}x} \circ u$, where $\frac{\mathrm{d}}{\mathrm{d}x} = \frac{\partial}{\partial x} + u_x\,\frac{\partial}{\partial u} + \cdots$. The image of A is closed under commutation, and the Lie algebra structure $[\, ,\,]_A$ on its domain is related by the homomorphisms $\delta/\delta u$ and A to the Lie algebra $\big(\bar{H}, \{\, ,\, \}_A\big)$ of Hamiltonians, endowed with the Poisson bracket, and to the Lie algebra $\big(\mathfrak{g}, [\, ,\,]\big)$ of evolutionary vector fields, respectively (see [11]). It can easily be checked [12] that, for the above operator, the bracket $[\, ,\,]_A$ on the domain Ω of A is $[p, q]_A = \partial_{A(p)}(q) - \partial_{A(q)}(p) + \frac{\mathrm{d}}{\mathrm{d}x}(p) \cdot q - p \cdot \frac{\mathrm{d}}{\mathrm{d}x}(q)$, here $p, q \in \Omega$.

The bracket $\{\{\, ,\, \}\}_A$ for Hamiltonian operators A can be obtained explicitly from the Jacobi identity $[\![\boldsymbol{A}, \boldsymbol{A}]\!] = 0$ for the Lie algebra $\big(\bar{H}, \{\, ,\, \}_A\big)$ of the Hamiltonian functionals endowed by A with the Poisson bracket $\{\, ,\, \}_A$; here $\boldsymbol{A}$ is the representation of A by the variational Poisson bi-vector and $[\![\, ,\,]\!]$ is the variational Schouten bracket, see [1, 2, 4]. We now write the result of such a derivation in local coordinates but in a properly ordered way which is slightly different from Eq. (5) in [1]: For a Hamiltonian operator $A = \|A^{ij}_{\boldsymbol{\tau}} \cdot \frac{\mathrm{d}^{|\tau|}}{\mathrm{d}x^{\boldsymbol{\tau}}}\|$, the k-th ($1 \leq k \leq m$) component of $\{\!\{\,,\,\}\!\}_A$ equals

$$\{\!\{\boldsymbol{p},\boldsymbol{q}\}\!\}^k_A = \sum_{|\boldsymbol{\sigma}|,|\boldsymbol{\tau}|\geq 0} \sum_{i,j=1}^{m} \Bigl(\frac{\mathrm{d}^{|\sigma|}}{\mathrm{d}x^{\boldsymbol{\sigma}}}\Bigr)^{\dagger} \Bigl[q_i \cdot \frac{\partial A^{ij}_{\boldsymbol{\tau}}}{\partial u^k_{\boldsymbol{\sigma}}} \cdot \frac{\mathrm{d}^{|\tau|}}{\mathrm{d}x^{\boldsymbol{\tau}}}(p_j)\Bigr], \tag{5}$$

where † denotes the adjoint. The benefit of this notation is that formula (5) covers the super-setup of bosonic super-fields and parity-preserving Hamiltonian operators that endow the spaces of bosonic functionals with Poisson brackets. Here the multi-indices $\boldsymbol{\sigma}$ and $\boldsymbol{\tau}$ can run through the super-derivations as well, and the partial derivatives $\partial/\partial u^k_{\boldsymbol{\sigma}}$ in (5) act according to the graded Leibniz rule.

Example 2. Let $\boldsymbol{u} = u_0(x,t)\cdot\mathbf{1} + \theta_1\cdot u_1(x,t) + \theta_2\cdot u_2(x,t) + \theta_1\theta_2\cdot u_{12}(x,t)$ be a scalar bosonic super-field, that is, a mapping of $\mathbb{R}^2 \ni (x,t)$ to the four-dimensional Grassmann algebra generated over $\mathbb{R}$ by θ_1 and θ_2 satisfying $\theta_i\theta_j = -\theta_j\theta_i$. By definition, put $\mathcal{D}_i = \partial/\partial\theta_i + \theta_i\cdot\mathrm{d}/\mathrm{d}x$, here $1 \leq i,j \leq 2$ and it is readily seen that $\mathcal{D}_i\mathcal{D}_j + \mathcal{D}_j\mathcal{D}_i = 2\delta_{ij}\cdot\mathrm{d}/\mathrm{d}x$.

Consider the super-operator $\boldsymbol{A}_2$ that comes from the N=2 classical super-conformal algebra [13] and yields the second Hamiltonian structure for the triplet of integrable N=2 supersymmetric Korteweg–de Vries equations ([14], see also [15])

$$\boldsymbol{A}_2 = \mathcal{D}_1\mathcal{D}_2\tfrac{\mathrm{d}}{\mathrm{d}x} + 2\boldsymbol{u}\tfrac{\mathrm{d}}{\mathrm{d}x} - \mathcal{D}_1(\boldsymbol{u})\mathcal{D}_1 - \mathcal{D}_2(\boldsymbol{u})\mathcal{D}_2 + 2\boldsymbol{u}_x. \tag{6}$$

Let the bosonic super-sections $\boldsymbol{p},\boldsymbol{q} \in \Omega$ be two arguments of $\boldsymbol{A}_2$. Then formula (5) yields their skew-symmetric bracket

$$\{\!\{\boldsymbol{p},\boldsymbol{q}\}\!\}_{\boldsymbol{A}_2} = 2\bigl(\tfrac{\mathrm{d}}{\mathrm{d}x}\boldsymbol{p}\cdot\boldsymbol{q} - \boldsymbol{p}\cdot\tfrac{\mathrm{d}}{\mathrm{d}x}\boldsymbol{q}\bigr) - \mathcal{D}_1(\boldsymbol{p})\cdot\mathcal{D}_1(\boldsymbol{q}) - \mathcal{D}_2(\boldsymbol{p})\cdot\mathcal{D}_2(\boldsymbol{q}), \tag{7}$$

and the validity of (4) confirms that the super-operator $\boldsymbol{A}_2$ is indeed Hamiltonian.

2. $\mathbb{Z}_2$-graded fields and the Hamiltonianity criterion

The purely bosonic setup of [1, 2] and the N=2 supersymmetry invariance in Example 2 are particular cases in the general $\mathbb{Z}_2$-graded framework of $(m_0 \mid m_1)$-dimensional fibre bundles π and parity-preserving Hamiltonian operators $A\colon \Omega \to \mathfrak{g}$ for bosonic Hamiltonian functionals.

Let $\langle\,,\,\rangle$ denote the standard coupling $\Omega \times \mathfrak{g} \to \bar{H}$ and define $\langle\,|\,\rangle$ by setting $\langle \boldsymbol{b} \mid \boldsymbol{d}\rangle := \langle \boldsymbol{d}, \boldsymbol{b}\rangle$. Namely, if $\boldsymbol{b} = (\boldsymbol{b}^0, \boldsymbol{b}^1)$ and $\boldsymbol{d} = (\boldsymbol{d}^0, \boldsymbol{d}^1)$ are decomposed to even and odd-graded components, then $\langle \boldsymbol{b}, \boldsymbol{d}\rangle = \boldsymbol{b}^0\cdot\boldsymbol{d}^0 + \boldsymbol{b}^1\cdot\boldsymbol{d}^1$ and $\langle \boldsymbol{b} \mid \boldsymbol{d}\rangle = \boldsymbol{b}^0\cdot\boldsymbol{d}^0 - \boldsymbol{b}^1\cdot\boldsymbol{d}^1$. The definition of adjoint graded operators implies $\langle \boldsymbol{b}, A(\boldsymbol{d})\rangle = \langle \boldsymbol{d}, A^{\dagger}(\boldsymbol{b})\rangle = \langle A^{\dagger}(\boldsymbol{b}) \mid \boldsymbol{d}\rangle$.

Theorem 1. *A $\mathbb{Z}_2$-graded parity-preserving skew-adjoint total differential operator $A\colon \Omega \to \mathfrak{g}$ is Hamiltonian if and only if its image is closed under commutation and, for all $\mathbf{p},\boldsymbol{q},\mathbf{r} \in \Omega$, the bracket $\{\!\{\,,\,\}\!\}_A$ in (4) satisfies the equality*

$$\bigl\langle A(\{\!\{\boldsymbol{p},\boldsymbol{q}\}\!\}_A) \mid \boldsymbol{r}\bigr\rangle = :\bigl\langle \boldsymbol{p}, \partial_{\underleftarrow{A}(\underrightarrow{\boldsymbol{r}})}(A)(\boldsymbol{q})\bigr\rangle:, \tag{8}$$

where the normal order : : suggests that all derivations are thrown off $A(\boldsymbol{r})$ by the graded Green formula and the arrows indicate that first $A(\boldsymbol{r})$ is moved to the right of $\boldsymbol{q}$, and then the operator A is pushed to the left of $\boldsymbol{p}$ by Green's formula again (this is explained in the proof below). The arising argument of the skew-adjoint operator A is the bracket $\{\!\{\boldsymbol{p},\boldsymbol{q}\}\!\}_A$.

Proof. Let us expand each of the three terms of the Jacobi identity,

$$\sum_{\circlearrowright} \partial_{A(\boldsymbol{p})}(\langle \boldsymbol{q}, A(\boldsymbol{r})\rangle) = 0,$$

by using the Leibniz rule. We obtain

$$\sum_{\circlearrowright}\Big[\langle\partial_{A(\boldsymbol{p})}(\boldsymbol{q}),A(\boldsymbol{r})\rangle+\langle\boldsymbol{q},\partial_{A(\boldsymbol{p})}(A)(\boldsymbol{r})\rangle+\langle\boldsymbol{q},A(\partial_{A(\boldsymbol{p})}(\boldsymbol{r}))\rangle\Big]=0. \tag{9}$$

Consider the third term in (9) and, by the substitution principle [7], suppose that $\boldsymbol{r}$ is the variational derivative of a Hamiltonian functional, whence the linearization $\ell_{\boldsymbol{r}}$ is self-adjoint in the graded sense. Consequently,

$$\begin{aligned}\langle\boldsymbol{q},A(\partial_{A(\boldsymbol{p})}(\boldsymbol{r}))\rangle&=-\langle A(\boldsymbol{q})\mid\partial_{A(\boldsymbol{p})}(\boldsymbol{r})\rangle=-\langle A(\boldsymbol{q})\mid\ell_{\boldsymbol{r}}(A(\boldsymbol{p}))\rangle=-\langle A(\boldsymbol{p})\mid\ell_{\boldsymbol{r}}^{\dagger}(A(\boldsymbol{q}))\rangle\\&=-\langle A(\boldsymbol{p})\mid\ell_{\boldsymbol{r}}(A(\boldsymbol{q}))\rangle=-\langle\ell_{\boldsymbol{r}}(A(\boldsymbol{q})),A(\boldsymbol{p})\rangle=-\langle\partial_{A(\boldsymbol{q})}(\boldsymbol{r}),A(\boldsymbol{p})\rangle.\end{aligned}$$

Substituting this back in (9) and taking the sum over the cyclic permutations, we cancel 3×2 terms, except for

$$\langle\boldsymbol{q},\partial_{A(\boldsymbol{p})}(A)(\boldsymbol{r})\rangle+\langle\boldsymbol{r},\partial_{A(\boldsymbol{q})}(A)(\boldsymbol{p})\rangle+\langle\boldsymbol{p},\partial_{A(\boldsymbol{r})}(A)(\boldsymbol{q})\rangle=0. \tag{10}$$

Now we consider separately the first and second summands in (10), paying due attention to the order of graded objects and the directions the derivations act in. First, applying the even vector field $\partial_{A(\boldsymbol{p})}$ to the equality $\langle\boldsymbol{q},A(\boldsymbol{r})\rangle=\langle A^{\dagger}(\boldsymbol{q})\mid\boldsymbol{r}\rangle$ and using $A^{\dagger}=-A$, we conclude that

$$\langle\boldsymbol{q},\partial_{A(\boldsymbol{p})}(A)(\boldsymbol{r})\rangle=-\langle\partial_{A(\boldsymbol{p})}(A)(\boldsymbol{q})\mid\boldsymbol{r}\rangle.$$

Likewise, the second summand in (10) gives

$$\langle\boldsymbol{r},\partial_{A(\boldsymbol{q})}(A)(\boldsymbol{p})\rangle=\langle\partial_{A(\boldsymbol{q})}(A)(\boldsymbol{p})\mid\boldsymbol{r}\rangle.$$

Hence from (10) we obtain

$$\langle\partial_{A(\boldsymbol{p})}(A)(\boldsymbol{q})\mid\boldsymbol{r}\rangle-\langle\partial_{A(\boldsymbol{q})}(A)(\boldsymbol{p})\mid\boldsymbol{r}\rangle=\langle\boldsymbol{p},\partial_{A(\boldsymbol{r})}(A)(\boldsymbol{q})\rangle.$$

Integrating the right-hand side by parts, we move the skew-adjoint operator A off $\boldsymbol{r}$ and obtain the bracket $\{\![\boldsymbol{p},\boldsymbol{q}]\!\}_A$ as its argument.

We have shown that if the bracket induced on the domain of a given graded skew-adjoint operator A with involutive image, see (4), coincides with the bracket $\{\![\,,\,]\!\}_A$ emerging from (8), then A is indeed Hamiltonian, and *vice versa*. This concludes the proof.

Example 3. Writing the super-operator (6) in components (see Appendix A below), now with $p_i=\delta\mathcal{H}/\delta u_i$, whence p_0 and p_{12} are even and p_1, p_2 are odd, we obtain the (4×4)-matrix operator [13]

$$A_2=\begin{pmatrix}-\frac{\mathrm{d}}{\mathrm{d}x} & -u_2 & u_1 & 2u_0\frac{\mathrm{d}}{\mathrm{d}x}+2u_{0;x}\\ -u_2 & \left(\frac{\mathrm{d}}{\mathrm{d}x}\right)^2+u_{12} & -2u_0\frac{\mathrm{d}}{\mathrm{d}x}-u_{0;x} & 3u_1\frac{\mathrm{d}}{\mathrm{d}x}+2u_{1;x}\\ u_1 & 2u_0\frac{\mathrm{d}}{\mathrm{d}x}+u_{0;x} & \left(\frac{\mathrm{d}}{\mathrm{d}x}\right)^2+u_{12} & 3u_2\frac{\mathrm{d}}{\mathrm{d}x}+2u_{2;x}\\ 2u_0\frac{\mathrm{d}}{\mathrm{d}x} & -3u_1\frac{\mathrm{d}}{\mathrm{d}x}-u_{1;x} & -3u_2\frac{\mathrm{d}}{\mathrm{d}x}-u_{2;x} & \left(\frac{\mathrm{d}}{\mathrm{d}x}\right)^3+4u_{12}\frac{\mathrm{d}}{\mathrm{d}x}+2u_{12;x}\end{pmatrix}. \tag{11}$$

The application of Theorem 1 is particularly transparent since the coefficients of (11) are linear functions. The right-hand side of (8) yields the four components of the skew-symmetric bracket $\{\![\boldsymbol{p},\boldsymbol{q}]\!\}_{A_2}$,

$$\begin{aligned}\{\![\boldsymbol{p},\boldsymbol{q}]\!\}_{A_2}^{0}&=2(p_{0;x}q_{12}-p_{12}q_{0;x})-(p_{1;x}q_2+p_2q_{1;x})+(p_{2;x}q_1+p_1q_{2;x}),\\ \{\![\boldsymbol{p},\boldsymbol{q}]\!\}_{A_2}^{1}&=2(p_{1;x}q_{12}-p_{12}q_{1;x})+(p_0q_2-p_2q_0)+(p_{12;x}q_1-p_1q_{12;x}),\\ \{\![\boldsymbol{p},\boldsymbol{q}]\!\}_{A_2}^{2}&=2(p_{2;x}q_{12}-p_{12}q_{2;x})+(p_1q_0-p_0q_1)+(p_{12;x}q_2-p_2q_{12;x}),\\ \{\![\boldsymbol{p},\boldsymbol{q}]\!\}_{A_2}^{12}&=2(p_{12;x}q_{12}-p_{12}q_{12;x})-p_1q_1-p_2q_2.\end{aligned}$$

This is the component expansion of (7); see [15] for further results on the geometry of the N=2 supersymmetric a=4-Korteweg–de Vries equation.

Conclusion

Theorem 1 provides an exact and exhaustive answer on the question whether a given skew-adjoint $\mathbb{Z}_2$-graded differential operator with involutive image in $\mathfrak{g}$ is Hamiltonian:

- take sections $\boldsymbol{p}, \boldsymbol{q} \in \Omega$ and calculate the commutator $\big[A(\boldsymbol{p}), A(\boldsymbol{q})\big]$, omitting the standard terms $\partial_{A(\boldsymbol{p})}(\boldsymbol{q}) - \partial_{A(\boldsymbol{q})}(\boldsymbol{p})$;
- calculate the m-tuple $A\big(\{\{\boldsymbol{p}, \boldsymbol{q}\}\}_A\big)$ by using formula (8).

If the two expressions coincide, the operator A is Hamiltonian.

Appendix A. Field–superfield correlation for variational derivatives

Given a Hamiltonian functional $\boldsymbol{\mathcal{H}} = \int \boldsymbol{h}[\boldsymbol{u}]\,\mathrm{d}\boldsymbol{\theta}\mathrm{d}x$, $\mathrm{d}\boldsymbol{\theta} = \mathrm{d}\theta_1 \cdot \ldots \cdot \mathrm{d}\theta_N$ whose density $\boldsymbol{h}$ is a differential superfunction in $(m_0 \mid m_1)$ super-fields $\boldsymbol{u}^\alpha$ of 2^N components each, what is the correlation between the components p_I^α of the variational derivatives

$$\frac{\delta \boldsymbol{\mathcal{H}}}{\delta \boldsymbol{u}^\alpha} = p_\varnothing^\alpha \cdot \mathbf{1} + \ldots + \theta_1 \cdots \theta_N \cdot p_{(1,\ldots,N)}^\alpha = \sum_{|I|=0}^{N} \boldsymbol{\theta}_I \cdot p_I^\alpha$$

with respect to the super-fields

$$\boldsymbol{u}^\alpha = u_\varnothing^\alpha \cdot \mathbf{1} + \ldots + \theta_1 \cdots \theta_N \cdot u_{(1,\ldots,N)}^\alpha = \sum_{|J|=0}^{N} \boldsymbol{\theta}_J \cdot u_J^\alpha \tag{A.1}$$

and, on the other hand, the variational derivatives $\psi_J^\alpha = \delta\boldsymbol{\mathcal{H}}/\delta u_J^\alpha$ of the functional $\boldsymbol{\mathcal{H}}$ with respect to the $(m_0 + m_1) \cdot 2^N$ components u_J^α of the super-fields (here $1 \le \alpha \le m_0 + m_1$)? We note that the answer to this question (for which it suffices to consider only one super-field, hence we shall omit the superscripts α) also encodes the Hamiltonian super-operators in their matrix component form (e.g., see (11)).

Proposition. Let $\boldsymbol{u}$ be an $N \ge 1$ super-field (A.1) and $\boldsymbol{\mathcal{H}} = \int \boldsymbol{h}[\boldsymbol{u}]\,\mathrm{d}\boldsymbol{\theta}\mathrm{d}x$ be a Hamiltonian super-functional. For all multi-indexes J of length $|J|$ such that $0 \le |J| \le N$, denote by I the multiindex of length $|I| = N - |J|$ such that their disjoint union is $I \sqcup J = \{1, \ldots, N\}$. Then the sought correlation between p_I and ψ_J is

$$\psi_J = (-1)^{(|\boldsymbol{\mathcal{H}}|-|\boldsymbol{u}|-|I|)\cdot|J|} \cdot (-1)^{\overline{I,J}} \cdot p_I, \tag{A.2}$$

where $|\boldsymbol{\mathcal{H}}|$ is the parity of the Hamiltonian, $|\boldsymbol{u}|$ is the parity of the super-field, and the ordered concatenation of the multi-indexes $\overline{I,J}$ is a permutation of $1, \ldots, N$.

Example 4. Suppose N=2 as in Examples 2 and 3. Then the correlation between the component expansion $\boldsymbol{p} = p_0 \cdot \mathbf{1} + \theta_1 p_1 + \theta_2 p_2 + \theta_1\theta_2 p_{12}$ of the variational derivative $\delta\boldsymbol{\mathcal{H}}/\delta\boldsymbol{u}$ and the variations $\psi_I = \delta\boldsymbol{\mathcal{H}}/\delta u_I$ with respect to the components u_I, $I \in \{0, 1, 2, 12\}$, of the super-field $\boldsymbol{u} = u_0 \cdot \mathbf{1} + \theta_1 u_1 + \theta_2 u_2 + \theta_1\theta_2 u_{12}$ is given by the formula[1]

$$\psi_0 = p_{12}, \qquad \psi_1 = p_2, \qquad \psi_2 = -p_1, \qquad \text{and} \quad \psi_{12} = p_0. \tag{A.2$'$}$$

We thus recover the (4×4)-matrix operator (11) by writing in components both $\boldsymbol{u}$ and the argument $\boldsymbol{p}$ of the Hamiltonian super-operator (6), by inserting the correlation (??$'$) for the

[1] Although formula (??$'$) is valid for all $\boldsymbol{\mathcal{H}}$, it is particularly transparent for $\boldsymbol{h} = \frac{1}{2}\boldsymbol{u}^2$ such that the identity $\frac{1}{2}\int \boldsymbol{u}^2\,\mathrm{d}\boldsymbol{\theta}\mathrm{d}x = \int (u_0 u_{12} - u_1 u_2)\,\mathrm{d}x$ yields $\boldsymbol{p} = \boldsymbol{u}$ and $\vec{\psi} = {}^{\mathrm{t}}(u_{12}, u_2, -u_1, u_0)$, the superscript ${}^{\mathrm{t}}$ denoting the transposition.

components of $\boldsymbol{p}$, and then reordering the columns of the matrix operator A_2 so that its argument $\vec{\psi}$ acquires the standard form ${}^{\mathrm{t}}(\psi_0, \psi_1, \psi_2, \psi_{12})$. To let the notation of Example 3 match Theorem 1, we re-denote by p_i the variational derivatives $\delta\boldsymbol{\mathcal{H}}/\delta u_i$.

Proof of Proposition. Consider the Hamiltonian $\boldsymbol{\mathcal{H}} = \int \boldsymbol{h}\,\mathrm{d}\boldsymbol{\theta}\mathrm{d}x$. Varying the super-field $\boldsymbol{u}$ by $\delta\boldsymbol{u}$, we throw all the derivatives off $\delta\boldsymbol{u}$ using multiple integration by parts in x, which yields $\int \boldsymbol{p} \cdot \delta\boldsymbol{u}\,\mathrm{d}\boldsymbol{\theta}$. Next, let us insert the expansions $\boldsymbol{p} = \sum_I \theta_{i_1} \cdots \theta_{i_{|I|}} \cdot p_I$ and $\delta\boldsymbol{u} = \sum_J \theta_{j_1} \cdots \theta_{j_{|J|}} \cdot u_J$ in this super-integral. By its definition, only the coefficient of $\theta_1 \cdots \theta_N$ in the product $\boldsymbol{p} \cdot \delta\boldsymbol{u}$ contributes to the integral's value, hence only the complementary multi-indexes $I \sqcup J = \{1, \ldots, N\}$ count. Pushing p_I through $\boldsymbol{\theta}^J$, we accumulate the sign $(-1)^{|p_I| \cdot |J|}$, where $|p_I| = |\boldsymbol{\mathcal{H}}| - |\boldsymbol{u}| - |I|$. Finally, reordering the product $\boldsymbol{\theta}_I \cdot \boldsymbol{\theta}_J$ to $\theta_1 \cdots \theta_N$, we obtain the sign of the permutation $\overline{I, J}$.

Acknowledgments
The authors thank J. W. van de Leur for helpful discussions and constructive criticisms. This research is partially supported by NSERC (for V. H.) and NWO grants B61–609 and VENI 639.031.623 (for A. K.). A part of this research was done while A. K. was visiting at CRM (Montréal); the financial support and hospitality of this institution are gratefully acknowledged.

References

[1] Kiselev A V and van de Leur J W 2009 *J. Phys.* A: *Math. Theor.* **42** 404011 (*Preprint* nlin.SI/0903.1214)
[2] Krasil'shchik I and Verbovetsky A 1998 *Homological methods in equations of mathematical physics* (Opava: Open Education and Sciences)
[3] Magri F 1978 *J. Math. Phys.* **19** 1156–62.
[4] Krasil'shchik J and Verbovetsky A 2011 Geometry of jet spaces and integrable systems *J. Geom. Phys.* 54pp accepted (*Preprint* math.DG/1002.0077)
[5] Kiselev A V and Wolf T 2007 *Comput. Phys. Commun.* **177** 315–28 (*Preprint* nlin.SI/0609065)
[6] Kersten P, Krasil'shchik I and Verbovetsky A 2004 *J. Phys.* A: *Math. Gen.* **37** 5003–19
[7] Olver P J 1993 *Applications of Lie groups to differential equations* Grad. Texts in Math. vol 107 (2nd ed) (NY: Springer–Verlag)
[8] Dubrovin B A 1996 Lect. Notes in Math. vol 1620 *Integrable systems and quantum groups* (Montecatini Terme 1993) (Berlin: Springer) pp 120–348
[9] Kiselev A V and van de Leur J W 2010 *Theor. Math. Phys.* **162** 149–62 (*Preprint* nlin.SI/0902.3624)
[10] Kiselev A V and van de Leur J W 2011 Variational Lie algebroids and homological evolutionary vector fields *Theor. Math. Phys.* **168** 18pp accepted (*Preprint* math.DG/1006.4227)
[11] Dorfman I Ya 1993 *Dirac structures* (J. Whiley & Sons)
[12] Zhiber A V and Sokolov V V 2001 *Russ. Math. Surveys* **56** 61–101
[13] Chaichian M and Kulish P P 1987 *Phys. Lett. B* **183** 169–74
[14] Labelle P and Mathieu P 1991 *J. Math. Phys.* **32** 923–7
[15] Hussin V, Kiselev A V, Krutov A O and Wolf T 2010 *J. Math. Phys.* **51** 083507 (*Preprint* nlin.SI/0911.2681)

Journal of Physics: Conference Series **284** (2011) 012036 doi:10.1088/1742-6596/284/1/012036

Quantization in terms of symplectic groups: The harmonic oscillator as a generic example

H A Kastrup

DESY Theory Group, Notkestr. 85, 22603 Hamburg, Germany

E-mail: hans.kastrup@desy.de

Abstract. The conventional quantization of the harmonic oscillator in terms of operators Q and P can be implemented with the help of irreducible unitary representations of the Heisenberg-Weyl group which acts transitively and effectively on the simply connected classical phase space $\mathcal{S}_{q,p} \cong \mathbb{R}^2$. In the description of the harmonic oscillator in terms of angle and action variables φ and I the associated phase space $\mathcal{S}_{\varphi,I}$ corresponds to the multiply connected punctured plane $\mathbb{R}^2 - \{0\}$, on which the 3-dimensional symplectic group $Sp(2,\mathbb{R})$ acts transitively, leaving the origin invariant. As this group contains the compact subgroup $U(1)$ it has infinitely many covering groups. In the here relevant irreducible unitary representations (positive discrete series) the self-adjoint generator K_0 of $U(1)$ represents the classical action variable I. It has the possible spectra $n+k$, $n=0,1,\ldots;\, k>0$, where k depends on the covering group. This implies different possible spectra for the action variable Hamiltonian $\hbar\omega K_0$ of the harmonic oscillator. On the other hand, expressing the operators Q and P (non-linearly) in terms of the three generators K_0 etc. of $Sp(2,\mathbb{R})$ leads to the usual framework. Possible physical (experimental) implications and generalizations to higher dimensions are discussed briefly.

1. Introduction: Angle and action variable for the HO

It may appear as a provocation or even a joke to present a contribution on the old-fashioned harmonic oscillator (HO in the following) to an established international conference devoted to current and relevant research in physics! But see for yourself! The basics for the present paper are described in my long article [1] which contains a wealth of references to the work of others which will not be quoted here again. The present contribution will sketch the main idea and emphasize some new important physical aspects.
The canonical Eqs. of motion

$$\dot{p} = -b\,q\,,\; p = M\,\dot{q}\,,\; b > 0\,, \tag{1}$$

for the HO can be simplified by the canonical transformation

$$q(\varphi, I) = \sqrt{\frac{2\,I}{M\,\omega}}\,\cos\varphi\,,\quad p(\varphi, I) = -\sqrt{2\,M\,\omega\,I}\,\sin\varphi\,,\quad \omega = \sqrt{b/M}\,, \tag{2}$$

which is *locally* symplectic:

$$dq \wedge dp = d\varphi \wedge dI\,,\;\text{ or }\; \frac{\partial(q,p)}{\partial(\varphi, I)} = 1\,. \tag{3}$$

It is singular for $(q,p) = (0,0)$, $I = 0$ (equilibrium or critical point) corresponding to the branch point $I = 0$ of $\sqrt{I}$.

The Hamiltonian is now given by

$$H(q,p) = \frac{1}{2M}\,p^2 + \frac{1}{2}\,M\,\omega^2\,q^2 = H(\varphi, I) = \omega\, I\,, \tag{4}$$

implying the (φ, I)-Eqs. of motion

$$\dot{\varphi} = \frac{\partial H}{\partial I} = \omega\,, \quad \dot{I} = -\frac{\partial H}{\partial \varphi} = 0\,. \tag{5}$$

They have the solutions

$$\varphi(t) = \omega\, t + \varphi_0\,, \quad I = \text{ const.} > 0\,. \tag{6}$$

Here the angle φ is essentially the time variable t. Notice that the time t does not stop after a period $\omega\,\hat{t} = 2\pi + \varphi_0$!

We now have two *globally* different phase spaces for the HO:

$$\mathcal{S}_{q,p} \;=\; \{(q,p) \in \mathbb{R}^2\} \cong \mathbb{R}^2 \tag{7}$$

$$\text{and}\;\; \mathcal{S}_{\varphi,I} \;=\; \{(\varphi, I), \varphi \in \mathbb{R} \bmod 2\pi\,,\, I > 0\} \cong S^1 \times \mathbb{R}^+ \cong \mathbb{R}^2 - \{0\}\,. \tag{8}$$

The phase space $\mathcal{S}_{\varphi,I}$ is topologically a simple cone with the tip deleted:

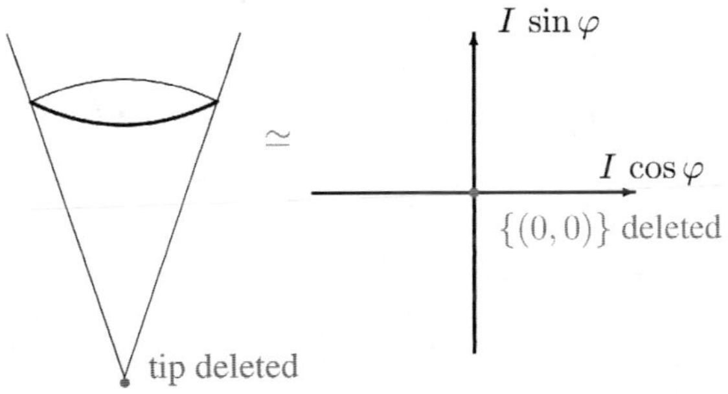

It may also be interpreted as the orbifold

$$\mathcal{S}_{\varphi,I} = \mathcal{S}_{q,p}/Z_2\,. \tag{9}$$

It is shown in Ref. [1] (the reasons will briefly be indicated in the next Sec.) that the proper global coordinates on $\mathcal{S}_{\varphi,I}$ are the following ones:

$$h_0(\varphi, I) = I > 0\,, \;\; h_1(\varphi, I) = I\cos\varphi\,, \;\; h_2(\varphi, I) = -I\sin\varphi\,, \tag{10}$$

which obey the Pythagorean relation

$$\vec{h}^2 \equiv h_1^2 + h_2^2 = h_0^2\,. \tag{11}$$

The invariant measure on $\mathcal{S}_{\varphi,I}$ is

$$d\mu(\varphi, I) = 2\theta(h_0)\,\delta(h_0^2 - \vec{h}^2)\,dh_0 dh_1 dh_2 = d\varphi dI = dq dp\,. \tag{12}$$

The $h_j(\varphi, I)$ obey the Poisson Lie algebra

$$\{h_0, h_1\}_{\varphi,I} = -h_2\,,\ \{h_0, h_2\}_{\varphi,I} = h_1\,,\ \{h_1, h_2\}_{\varphi,I} = h_0\,, \tag{13}$$

where

$$\{h_j, h_k\}_{\varphi,I} \equiv \partial_\varphi h_j\, \partial_I h_k - \partial_I h_j\, \partial_\varphi h_k. \tag{14}$$

Thus, the h_j generate the isomorphic *simple* Lie algebras

$$\mathfrak{so}(1,2) = \mathfrak{sp}(2,\mathbb{R}) = \mathfrak{sl}(2,\mathbb{R}) = \mathfrak{su}(1,1)\,. \tag{15}$$

This is an analogue of the *nilpotent* Poisson Lie algebra of the Heisenberg-Weyl group:

$$\{q,p\} = 1\,,\ \{q,1\} = 0\,,\ \{p,1\} = 0\,. \tag{16}$$

On $\mathcal{S}_{\varphi,I}$ the basic "observables" for the construction of Hamilton functions are now the $h_j, j = 0,1,2$. Simple examples are:

(i) Time-dependent frequency ω :

$$H = \omega(t)\, I\,,\ \dot{\omega}(t) \neq 0\,, \tag{17}$$

with the Eqs. of motion

$$\dot{\varphi} = \partial_I H = \omega(t)\,,\ \dot{I} = -\partial_\varphi H = 0\,, \tag{18}$$

so that

$$\varphi(t) = \int_{t_0}^{t} d\tau \omega(\tau) + \varphi_0\,,\ I(t) = I_0 = \text{ const. }. \tag{19}$$

The energy

$$E(t) = \omega(t)\, I_0 \tag{20}$$

is *not* conserved! Conserved is the action variable $I = I_0$.

(ii) If the Hamilton function

$$H(q,p) = \frac{1}{2M}p^2 + V(q) = E = \text{ const.} \tag{21}$$

has bounded periodic motions on $\mathcal{S}_{q,p}$ then the associated action variable is defined as

$$I(E) = \oint_{C(E)} dq\, p(q;E)\,,\ p(q;E) = \pm\sqrt{2M(E - V(q))}\,, \tag{22}$$

where $C(E)$ is the closed contour in phase space, determined by the relation $H(q,p) = E$. Here $I(E)$ is the area of the phase space region with boundary $C(E)$.
E.g. for the Morse potential

$$V_{\text{Mo}}(q) = V_0 \tanh(aq) \tag{23}$$

one gets

$$H(I) = \omega_0\, I (1 - \frac{\omega_0\, I}{4V_0})\,,\ \omega_0 = a\sqrt{2V_0/M}\,. \tag{24}$$

(iii) Example where $H = H(\varphi, I)$ depends only on some h_j:

$$H = \omega(h_0 + g\, h_1) = \omega I\,(1 + g\, \cos\varphi)\,,\ |g| < 1\,. \tag{25}$$

Here

$$H(\varphi, I) = E = \text{ const. }. \tag{26}$$

Solutions of the associated Eqs. of motion are:

$$\tan[\varphi(\tilde{t})/2] = \sqrt{\frac{1+g}{1-g}}\tan[(\sqrt{1-g^2})\,(\tilde{t}/2)]\,, \tag{27}$$

$$I(\tilde{t}) = I_0\, [1 + g\, \cos\varphi(\tilde{t})\,]^{-1};\ I_0 = E/\omega\,,\ \tilde{t} = \omega\, t\,,\ \cos\varphi = \frac{1 - \tan^2(\varphi/2)}{1 + \tan^2(\varphi/2)}\,. \tag{28}$$

2. Brief justification of the global coordinates $\mathbf{h_j, j = 1, 2, 3}$ on $\mathcal{S}_{\varphi,\mathbf{I}}$

I sketch briefly the justification for the choice (10) of the global basic coordinates h_j on the phase space $\mathcal{S}_{\varphi,I}$. The main tool is "group theoretical quantization" as developed by Isham, Sternberg and others. Details and references can be found in Ref. [1]. One has to look for an appropriate symplectic and transitive transformation group on the phase space in question. In our case the 3-parametric symplectic group

$$G \equiv Sp(2,\mathbb{R}) \tag{29}$$

(being isomorphic to the groups $SL(2,\mathbb{R}) \cong SU(1,1)$) is defined by

$$g = \begin{pmatrix} a_{11} & a_{12} \\ a_{21} & a_{22} \end{pmatrix}, \; a_{jk} \in \mathbb{R}, \; \det g = 1, \tag{30}$$

$$g^T \cdot \begin{pmatrix} 0 & 1 \\ -1 & 0 \end{pmatrix} \cdot g = \begin{pmatrix} 0 & 1 \\ -1 & 0 \end{pmatrix}. \tag{31}$$

It acts on the real plane $\mathbb{R}^2$ in an obvious way by matrix multiplication of $(\tilde{q},\tilde{p})^T \in \mathbb{R}^2$. The action of G leaves the origin $(0,0)$ invariant and therefore is the proper "canonical" group of the punctured plane

$$\mathcal{S}_{\tilde{q},\tilde{p};0} \equiv \mathcal{S}_{\tilde{q},\tilde{p}} - \{(\tilde{q},\tilde{p}) = (0,0)\} \cong \mathbb{R}^2 - \{(0,0)\}, \tag{32}$$

with g acting symplectically:

$$d[g(\tilde{q})] \wedge d[g(\tilde{p})] = d\tilde{q} \wedge d\tilde{p}. \tag{33}$$

N.b.: quantities with "tilde" are made dimensionless by using the intrinsic length $\lambda_0 = \sqrt{\hbar/(\omega M)}$ and $\hbar$: $\tilde{q} = q/\lambda_0$; $\tilde{p} = \lambda_0\, p/\hbar$.

In order to derive the coordinates (10) one needs the first three of the 1-dimensional subgroups

$$R: \quad r = \begin{pmatrix} \cos(\theta/2) & \sin(\theta/2) \\ -\sin(\theta/2) & \cos(\theta/2) \end{pmatrix}, \; \theta \in (-2\pi, +2\pi], \tag{34}$$

$$A: \quad a = \begin{pmatrix} e^{-\tau/2} & 0 \\ 0 & e^{\tau/2} \end{pmatrix}, \; \tau \in \mathbb{R}, \tag{35}$$

$$B: \quad b = \begin{pmatrix} \cosh(s/2) & \sinh(s/2) \\ \sinh(s/2) & \cosh(s/2) \end{pmatrix}, \; s \in \mathbb{R}, \tag{36}$$

$$N: \quad n = \begin{pmatrix} 1 & \xi \\ 0 & 1 \end{pmatrix}, \; \xi \in \mathbb{R}. \tag{37}$$

Given a smooth function $f(x)$ each subgroup $\{g(u), g(u=0) = 1\}$ induces a *global* Hamiltonian vector field

$$[Af](x) = \lim_{u\to 0} \frac{1}{u}[f(g(-u)\cdot x) - f(x)], \tag{38}$$

where A is a vector field of the type

$$A = -[\partial_{\tilde{p}} j(x)\, \partial_{\tilde{q}} - \partial_{\tilde{q}} j(x)\, \partial_{\tilde{p}}], \; x = (\tilde{q},\tilde{p})^T. \tag{39}$$

For the above subgroups R, A and B the corresponding Hamiltonian functions $j(x)$ are

$$R: \; j_0(x) = \frac{1}{4}(\tilde{q}^2 + \tilde{p}^2) = \frac{I}{2\hbar}, \tag{40}$$

$$A: \; j_2(x) = -\frac{1}{2}\tilde{q}\,\tilde{p} = \frac{I}{2\hbar}\sin 2\varphi, \tag{41}$$

$$B: \; j_1(x) = \frac{1}{4}(-\tilde{q}^2 + \tilde{p}^2) = -\frac{I}{2\hbar}\cos 2\varphi, \tag{42}$$

Journal of Physics: Conference Series **284** (2011) 012036 doi:10.1088/1742-6596/284/1/012036

where the relations (2) have been used. Thus, by a very simple symplectic rescaling $I/2 \to I$, $2\varphi \to \varphi$, $j_1 \to -j_1$, $j_2 \to -j_2$ one obtains the basic coordinates (10). In other words: the global coordinate functions $h_j(\varphi, I)$ on $\mathcal{S}_{\varphi,I}$ are determined by the action of the "canonical" symplectic group $Sp(2,\mathbb{R})$ on $\mathcal{S}_{q,p;0}$!

Compare this to the analogous global coordinate functions q and p on $\mathcal{S}_{q,p}$, where the canonical group consists of the translations

$$q \to q + a,\;\; p \to p;\;\; q \to q,\;\; p \to p - b,\;\; a, b \in \mathbb{R}\,, \tag{43}$$

which generate on $\mathcal{S}_{q,p}$ the vector fields

$$A_q = -\partial_q,\;\; A_p = \partial_p, \tag{44}$$

with the associated global Hamiltonian functions

$$j_q(x) = p,\;\;\; j_p(x) = q\,. \tag{45}$$

Let me add some structural relations on the classical level: The coordinates $h_j, j = 0, 1, 2$, transform as *vectors* with respect to the group $SO^{\uparrow}(1,2) = Sp(2,\mathbb{R})/Z_2$, whereas the coordinates q, p transform as vectors with respect to the group $Sp(2,\mathbb{R})$ which is a twofold covering of $SO^{\uparrow}(1,2)$.
The phase spaces $\mathcal{S}_{\varphi,I}$ and $\mathcal{S}_{q,p}$ may be represented as homogeneous spaces as follows:

$$\mathcal{S}_{\varphi,I} = SO^{\uparrow}(1,2)/N\,,\;\; \mathcal{S}_{q,p} = Sp(2,\mathbb{R})/N\,, \tag{46}$$

which again yield the orbifold

$$\mathcal{S}_{\varphi,I} = \mathcal{S}_{q,p}/Z_2\,. \tag{47}$$

3. Quantum mechanics for angle and action variables of the HO

The quantization of the global "coordinates" h_j from Eq. (10) is implemented by promoting them to self-adjoint operators,

$$h_j \to K_j = \hbar\,\tilde{K}_j \tag{48}$$

which obey the associated Lie algebra (13):

$$[\tilde{K}_0,\, \tilde{K}_1] = i\,\tilde{K}_2\,,\;\; [\tilde{K}_0,\, \tilde{K}_2] = -i\,\tilde{K}_1\,,\;\; [\tilde{K}_1,\, \tilde{K}_2] = -i\,\tilde{K}_0\,. \tag{49}$$

The self-adjoint generators $\tilde{K}_j$ may be obtained from unitary irreducible representations of the corresponding groups $SO^{\uparrow}(1,2), Sp(2,\mathbb{R})$ $[= SL(2,\mathbb{R}), SU(1,1)]$ or one of their infinitely many covering groups.

As $\tilde{K}_0$ is the generator of a maximal compact abelian subgroup, its eigenstates may used as a Hilbert space basis:

$$\tilde{K}_0|k,\, n\rangle = (n+k)\,|k,\, n\rangle\,, \tag{50}$$

where k is some real number ("Bargmann index"). With

$$\tilde{K}_{\pm} = \tilde{K}_1 \pm i\,\tilde{K}_2 \tag{51}$$

it follows from the relations (49) that

$$\tilde{K}_+|k,\, n\rangle = [(2k+n)(n+1)]^{1/2}\,|k,\, n+1\rangle\,,\;\; \tilde{K}_-|k,\, n\rangle = [(2k+n-1)n]^{1/2}\,|k,\, n-1\rangle\,. \tag{52}$$

If there exists a $|k, 0\rangle$ such that

$$\tilde{K}_0|k,0\rangle = k\,|k,0\rangle\,,\;\;\; \tilde{K}_-|k,0\rangle = 0\,, \tag{53}$$

then

$$\tilde{K}_0 \,|k,n\rangle = (n+k)\,|k,n\rangle\,,\; n=0,1,\ldots;\;\; k>0\,. \tag{54}$$

This is the so-called "positive discrete series" $D_k^{(+)}$ among the different types of possible irreducible unitary representations of $Sp(2,\mathbb{R})$ (see Ref. [1] for details). The Bargmann index k characterizes an irred. repr. $D_k^{(+)}$.

The Group $SO^{\uparrow}(1,2)$ has *infinitely many covering groups* because its compact subgroup $SO(2) \cong S^1$ is not simply connected! Let us denote the m-fold covering by

$$SO^{\uparrow}_{[m]}(1,2)\,,\; m=1,2,\ldots\,. \tag{55}$$

Its irreducible unitary representations $D_k^{(+)}$ have the indices

$$k=\frac{\mu}{m}\,,\; \mu\in\mathbb{N}=\{1,2,\ldots\}\,. \tag{56}$$

This shows that $k_{min} = 1/m$ can be arbitrarily small > 0 if m is large enough!
The 2-fold coverings

$$Sp(2,\mathbb{R}) = SL(2,\mathbb{R}) \cong SU(1,1) \tag{57}$$

have $k = 1/2$. The related (q,p)-Hamiltonian

$$H(q,p) \to H_{osc}(Q,P) = \frac{1}{2\,M}\,P^2 + \frac{1}{2}\,M\,\omega^2\,Q^2 \tag{58}$$

$$= -\frac{\hbar^2}{2\,M}\,\frac{d^2}{d\,q^2} + \frac{1}{2}\,M\,\omega^2\,q^2 \tag{59}$$

has eigenvalues

$$E_n = \hbar\,\omega\,(n+\frac{1}{2})\,,\;\; n=0,1,\ldots\,. \tag{60}$$

However, the (φ,I)-Hamiltonian

$$H(\varphi,I) \to H_{osc}(\vec{K}) = \omega\,K_0\,,\;\; \vec{K} = \hbar\,(\tilde{K}_0,\tilde{K}_1,\tilde{K}_2) \tag{61}$$

can have spectra

$$E_{k,n}(\varphi,I) = \hbar\,\omega\,(n+k)\,,\; n=0,1,\ldots\,;\;\; k\in\mathbb{R}^+\,! \tag{62}$$

How to reconcile the two results if $k \neq 1/2$?
Interesting enough there exists an operator version of the mapping $(\varphi,I)\to(q,p)$ from Eq. (2), namely

$$Q = Q(\vec{K}) = \frac{\lambda_0}{\sqrt{2}}\,(A^\dagger + A)(\vec{K})\,,\;\; P = P(\vec{K}) = \frac{i\,\hbar}{\sqrt{2}\,\lambda_0}\,(A^\dagger - A)(\vec{K})\,, \tag{63}$$

$$\lambda_0 = \sqrt{\hbar/(M\,\omega)}\,, \tag{64}$$

where the operators

$$A(\vec{K}) = (\tilde{K}_0 + k)^{-1/2}\tilde{K}_-\,,\;\; A^\dagger(\vec{K}) = \tilde{K}_+(\tilde{K}_0+k)^{-1/2} \tag{65}$$

act as usual:

$$A^\dagger\,|k,\,n\rangle = \sqrt{n+1}\,|k,\,n+1\rangle\,,\;\; A\,|k,\,n\rangle = \sqrt{n}\,|k,\,n-1\rangle\,. \tag{66}$$

This means

$$[A,\,A^\dagger] = \mathbf{1}\;\;\forall\, D_k^{(+)}\,, \tag{67}$$

Journal of Physics: Conference Series **284** (2011) 012036 doi:10.1088/1742-6596/284/1/012036

independent of the value of k!!
So again

$$H_{osc}(Q,P)|k,n\rangle = \hbar\,\omega\,(n+1/2)\,|k,n\rangle\,, \tag{68}$$

where now

$$H_{osc}(Q,P) = \hbar\omega\,(A^\dagger A + 1/2)\,. \tag{69}$$

Thus, the (φ, I)-variable QM of the HO is more subtle than its conventional (q,p)-variable one, due to the topologically non-trivial classical phase space $\mathcal{S}_{\varphi,I}$! As the transformation (65) "erases" the k-dependence, the usual Stone-von Neumann uniqueness theorem for self-adjoint Q and P still holds!
Eqs. (65) are the operator version of the classical relations

$$q(\varphi,I) = \sqrt{\frac{2}{M\,\omega}}\,\frac{h_1(\varphi,I)}{\sqrt{h_0(\varphi,I)}}\,,\quad p(\varphi,I) = \sqrt{2\,M\,\omega}\,\frac{h_2(\varphi,I)}{\sqrt{h_0(\varphi,I)}}\,. \tag{70}$$

The most important physical question is, of course, whether one can detect the k-dependence experimentally. More about that below.

3.1. Simple consequences

The Casimir operator of a representation $D_k^{(+)}$ is

$$\mathfrak{C} = \tilde{K}_1^2 + \tilde{K}_2^2 - \tilde{K}_0^2 = k(1-k)\,. \tag{71}$$

This means that the "classical Pythagoras" (11) is violated quantum mechanically for $k \neq 1$, e.g. for the HO with $k = 1/2$!
The unitary time evolution is given by

$$U(\tilde{t}) = e^{-i\,\tilde{H}\,\tilde{t}}\,,\ \ \tilde{H} = \tilde{K}_0 = N + k\,,\ \ \omega\,t = \tilde{t} = \theta\,, \tag{72}$$

i.e. time is an angle variable here!
The unitary operator (72) implies the usual Heisenberg Eqs. of motion:

$$U(-\tilde{t})\,\tilde{K}_1\,U(\tilde{t}) = \cos\tilde{t}\,\tilde{K}_1 + \sin\tilde{t}\,\tilde{K}_2\,,\quad U(-\tilde{t})\,\tilde{K}_2\,U(\tilde{t}) = -\sin\tilde{t}\,\tilde{K}_1 + \cos\tilde{t}\,\tilde{K}_2\,; \tag{73}$$

$$U(-\tilde{t})\,\tilde{Q}\,U(\tilde{t}) = \cos\tilde{t}\,\tilde{Q} + \sin\tilde{t}\,\tilde{P}\,,\quad U(-\tilde{t})\,\tilde{P}\,U(\tilde{t}) = -\sin\tilde{t}\,\tilde{Q} + \cos\tilde{t}\,\tilde{P}\,. \tag{74}$$

3.1.1. Covering groups

For $\tilde{t} = 2\pi$ the operator (72) becomes

$$U(\tilde{t} = 2\pi) = e^{-2\pi i k}\mathbf{1}\,. \tag{75}$$

If

$$k = n/m\,,\ n,m \in \mathbb{N}\,, \tag{76}$$

this implies for $SO^{\uparrow}_{[m]}(1,2)$:

$$U(\tilde{t} = m\,2\pi) = \mathbf{1}\,. \tag{77}$$

The ground state $|k,0\rangle$ has the time evolution

$$U(\tilde{t})\,|k,0\rangle = e^{-i\,k\,\tilde{t}}\,|k,0\rangle\,, \tag{78}$$

with the associated time period

$$T_{2\pi,\,k} = \frac{2\pi}{\omega_k}\,,\ \ \omega_k \equiv k\,\omega\,, \tag{79}$$

which can become arbitrarily large for $k = 1/m$, $m \to \infty$.

Journal of Physics: Conference Series **284** (2011) 012036 doi:10.1088/1742-6596/284/1/012036

3.1.2. Some matrix elements It follows from the relations (51)-(53) that

$$\langle k,n|\tilde{K}_j|k,n\rangle = 0\,,\; j=1,2, \tag{80}$$

$$(\Delta\tilde{K}_j)^2_{k,n} = \frac{1}{2}(n^2+2nk+k)\,,\; j=1,2, \tag{81}$$

so that

$$(\Delta\tilde{K}_1)_{k,n}\,(\Delta\tilde{K}_2)_{k,n} = \frac{1}{2}(n^2+2kn+k)\,, \tag{82}$$

$$(\Delta\tilde{K}_1)_{k,n=0}\,(\Delta\tilde{K}_2)_{k,n=0} = \frac{k}{2}\,. \tag{83}$$

The last relation shows that $k \to 0$ is a kind of classical limit in the angle-action framework!
The composite $\tilde{Q}$ and $\tilde{P}$ have the usual k-independent properties

$$\langle k,n|\tilde{Q}|k,n\rangle = 0\,,\;\; \langle k,n|\tilde{P}|k,n\rangle = 0\,, \tag{84}$$

$$(\Delta\tilde{Q})^2_{k,n} = (\Delta\tilde{P})^2_{k,n} = n+1/2\,, \tag{85}$$

$$(\Delta\tilde{Q})_{k,n}\,(\Delta\tilde{P})_{k,n} = n+1/2\,. \tag{86}$$

4. Possible applications

For applications of the angle-action framework it is important to use the variables h_j and K_j as primary observables, e.g. as building blocks for the construction of Hamiltonians! An additional experimental problem is how to avoid the dominance of the q- and p-degrees of freedom!
For $k = 1/2$ a possible Hilbert space for the HO is the Hardy space characterized by

$$(f_2,f_1)_+ = \int_0^{2\pi}\frac{d\vartheta}{2\pi}\,f_2^*(\vartheta)f_1(\vartheta)\,,\;\; \text{basis}: e_n(\vartheta) = e^{i\,n\,\vartheta}\,,\; n=0,1,\ldots \tag{87}$$

Now

$$\tilde{K}_0 = \frac{1}{i}\partial_\vartheta + 1/2\,. \tag{88}$$

For $k \neq 1/2$ one can use the associated Hilbert space with the scalar product (see Ref. [1])

$$(f_2,f_1)_{+,k} = (f_2,A_kf_1)_+\,,\;\; (A_k\,e_n)(\vartheta) = \frac{n!}{(2k)_n}\,e_n(\vartheta)\,. \tag{89}$$

It has the orthonormal basis

$$\hat{e}_{k,n}(\vartheta) = \sqrt{\frac{(2k)_n}{n!}}\,e_n(\vartheta)\,,\;\; (2k)_n = \frac{\Gamma(2k+n)}{\Gamma(2k)}\,. \tag{90}$$

Now one has

$$\tilde{K}_0 = \frac{1}{i}\partial_\vartheta + k\,, \tag{91}$$

$$\tilde{K}_+ = e^{i\vartheta}(\frac{1}{i}\partial_\vartheta + 2k)\,,\;\; \tilde{K}_- = e^{-i\vartheta}\,\frac{1}{i}\partial_\vartheta\,. \tag{92}$$

Examples:

Journal of Physics: Conference Series **284** (2011) 012036 doi:10.1088/1742-6596/284/1/012036

(i) For experimental tests a HO with a time-dependent frequency $\omega(t)$ may be of considerable interest (see examples (iii)). The associated Schrödinger equation ($\hbar = 1$)

$$H = \omega(t)\tilde{K}_0\,,\; i\partial_t\psi(t,\vartheta) = H\psi(t,\vartheta) \tag{93}$$

may easily be solved by a separation of variables:

$$\psi(t,\vartheta) = \sigma(t)\, f(\vartheta)\,,\; \tilde{K}_0 f(\vartheta) = I_0\, f(\vartheta)\,, \tag{94}$$

where

$$\sigma(t) = \sigma(t_0)\, e^{-iI_0\int_{t_0}^{t} d\tau\,\omega(\tau)}\,. \tag{95}$$

If $f(\vartheta)$ is an eigenfunction (90), then

$$I_0 = n + k\,. \tag{96}$$

Here I_0 is constant, not the energy $E = \omega(t)\, I_0$!

(ii) The Hamiltonian

$$H = \omega(\tilde{K}_0 + g\tilde{K}_1)\,,\; |g| < 1\,, \tag{97}$$

is the quantized version of the classical one (25). It has the eigenvalues

$$E_n = \omega(n+k)\sqrt{1-g^2} \tag{98}$$

and the eigenfunctions

$$f_{k,n}(\vartheta) = C_k\,(1 + g\,\cos\vartheta)^{-k}\, e^{i\,\{(n+k)2\arctan[\sqrt{\frac{1-g}{1+g}}\tan(\vartheta/2)]-k\vartheta\}}\,,\; C_k = \text{const.}. \tag{99}$$

First-order perturbation theory yields

$$\langle k,m|H|k,n\rangle = \omega[\langle k,m|\tilde{K}_0|k,n\rangle + g\langle k,m|\tilde{K}_1|k,n\rangle]\,; \tag{100}$$

so that

$$\langle k,m|H|k,m\rangle = \omega(n+k); \tag{101}$$

notice that the exact eigenvalues (98) are of second order in g! Furthermore

$$\langle k,m|H|k,m-1\rangle = \frac{g\omega}{2}[m(2k+m-1)]^{1/2}\,,\;\; \langle k,1|H|k,0\rangle = g\,\omega\sqrt{k/2}\,. \tag{102}$$

(iii) Possible experiments should make use of the essential property that the action I or K_0 is the basic observable, not the Hamiltonian $\omega\, K_0$! This is an old idea which was already exploited in the framework of the old quantum mechanics (e.g. by Mulliken in 1924) even before the "correct" modern QM was found in 1925. The main idea is as follows: The frequency $\omega = \sqrt{b/M}$ of a given HO type may be changed, either by modifying the mass M or (and) the oscillator strength b! Consider two different oscillator frequencies ω_1 and ω_2 and the transitions from some "exterior" levels as indicated in the following figure:

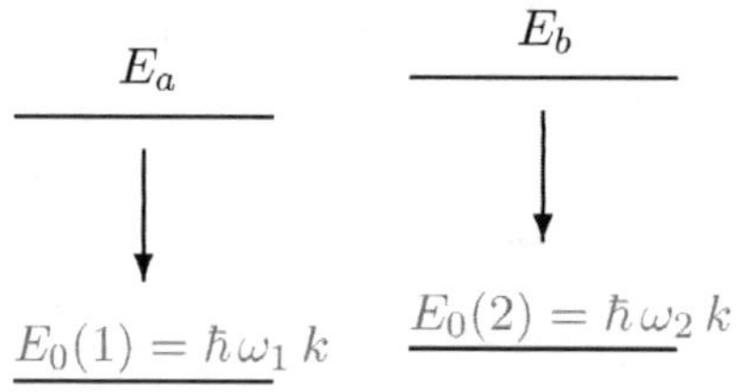

The transitions are associated with the frequencies

$$\omega_{a,1} = [E_a - E_0(1)]/\hbar\,, \quad \omega_{b,2} = [E_b - E_0(2)]/\hbar\,, \tag{103}$$

which lead to the k-dependent differences

$$\omega_{a,1} - \omega_{b,2} = (E_a - E_b)/\hbar - k\,(\omega_1 - \omega_2)\,, \tag{104}$$

which - in principle - can serve to determine k!
Examples:

- The relation (104) has been used by Mulliken and others in order to determine the vibrational levels (infrared bands) of diatomic molecules with isotopic atoms which lead to different (reduced) masses $M_i\,, i = 1, 2$, for the oscillator: the isotopes were $B^{10}O$ and $B^{11}O$; $AgCl^{35}$ and $AgCl^{37}$.
  The conclusion then was that $k \approx 1/2$. Nowadays it should be possible to make corresponding experiments with much higher precision!
- Presently one can build sophisticated 1-dimensional harmonic traps for ultra-cold ions, atoms and BE-condensates for which the trap-frequency $\omega(t)$ can be tuned! Determining the ratio $E(t)/\omega(t)$ could - in principle - yield information about k.

5. Problems and generalizations

The previous discussion raises a large number of questions and suggests possible generalizations. Only a few will be mentioned here:

- Most important above all are ideas for appropriate experiments!
- What about fermions?
- How are essential features of the usual quantum field theories affected: locality; Casimir effect; etc. etc.?
- Generalizations: The symplectic group $Sp(2n, \mathbb{R})$ of a $2n$-dimensional symplectic space $\mathcal{S}^{2n}$ has dimension dimension of $2n^2 + n$ and rank n.
  Its maximal compact subgroup is $U(n) = SU(n) \times U(1)$, where $SU(n)$ is simply connected; the rank of $U(n)$ is n; thus, rank $[Sp(2n, \mathbb{R})]$ = rank $[U(n)]$; according to a famous theorem by Harish-Chandra this implies the existence of a positive discrete series among the irreducible unitary representations of $Sp(2n, \mathbb{R})$.

Conclusion: HO has an interesting "hidden side" which appears worthwhile to be investigated further!

6. References

[1] Kastrup H A 2007 A new look at the quantum mechanics of the harmonic oscillator *Ann. Physik (Leipzig)* **16** 437-580; [arXiv:quant-ph/0612032]

Journal of Physics: Conference Series **284** (2011) 012037 doi:10.1088/1742-6596/284/1/012037

Boundary conditions for the wave functions and Gamow vectors

Piotr Kielanowski

Institute of Mathematics, University of Białystok, Poland

and

Departamento de Física, Centro de Investigación y de Estudios Avanzados del IPN Mexico City, Mexico

Abstract

We introduce a new kind of boundary conditions for the wave functions, which admit solutions in the form of Gamow vectors for a wide class of potentials. We discuss in detail an example of an unstable quantum oscillator.

1 Introduction

The main equation of the non-relativistic quantum mechanics is the Schrödinger equation[1]

$$H\psi = -\frac{\hbar^2}{2m}\Delta\psi + V(x)\psi = E\psi. \tag{1}$$

Eq. (1) is not sufficient for the construction of the physically meaningful solutions. For bound states one uses the Hilbert space condition

$$\int |\psi(x)|^2 dx < \infty, \tag{2}$$

which leads to the real eigenvalues of the Hamiltonian. Gamow solutions on the other hand are obtained if one assumes that the asymptotic behavior of the wave function $\psi(x)$ is a purely outgoing wave. Such a boundary condition

[1]From now on I will use the units, where $\hbar = 1$.

leads to complex eigenvalues [1] of the Hamiltonian and the corresponding discrete solutions are interpreted as wave functions of unstable states.

The other kind of the continuity conditions for the wave function $\psi(x)$ is related to the form of the potential $V(x)$. The wave function and its first derivative are assumed to be continuous at finite jumps of the potential. The only notable exception is when the potential becomes infinite and at such a point the wave function is assumed to vanish and its first derivative is not assumed to be continuous.

In this paper we consider the potential wall in the form of the Dirac delta function [2, 3]

$$\Omega\delta(x) \tag{3}$$

in the presence of an infinite oscillator wall. We then present explicit solutions for such a system with the boundary condition in the form of the outgoing wave.This leads to the Gamow wave functions with complex eigenvalues of the Hamiltonian. From the explicit solution we can study the behavior of the wave function $\psi(x)$ at the point $x = 0$, where the Dirac delta potential wall is present. We find that at this point the wave function is continuous and the first derivative has a finite jump. We generalize these results to an arbitrary form of the potential.

The plan of the paper is the following. In Section 2 we present the case of the quantum unstable oscillator. In Section 3 we discuss the wave function of the quantum unstable oscillator and analyze its properties. In Section 4 we generalize the results obtained in In Section 3 and discuss the new boundary conditions for the wave function and draw conclusions from our results.

2 Unstable quantum oscillator – general discussion

Harmonic oscillator is one of the best known physical systems. It is characterized by the equidistant energy levels. We consider here the system which is the superposition of two potentials: on the left hand side the motion is bounded by the "oscillator wall"

$$V_o = \begin{cases} \frac{m\omega^2}{2}x^2 & \text{for } x \leq 0 \\ 0 & \text{for } x > 0 \end{cases}. \tag{4}$$

At the origin we insert the semi-transparent potential wall in the form of the Dirac delta function

$$V_\delta(x) = \Omega\delta(x). \tag{5}$$

The total Hamiltonian is thus equal

$$H = \frac{p^2}{2m} + V_\text{o}(x) + V_\delta(x). \tag{6}$$

It turns out that the system described by the Hamiltonian (6) is exactly soluble [4] and one can find the energy levels and eigenfunctions. We seek the solution by the method of the Green's function [5, 3] for the Hamiltonian (6). For the determination of $G(x,x';E)$ we must impose boundary conditions and we choose

$$G(x,x';E) \to \begin{cases} \text{is bounded} & \text{for } x \to -\infty \\ \sim \mathrm{e}^{ikx} & \text{for } x \to +\infty \end{cases}. \tag{7}$$

It turns out that one can treat the admixture of the potential $V_\delta(x)$ in the Hamiltonian (6) as a perturbation. It means that first we find the Green's function $G_0(x,x';E)$ of the potential H_0

$$H_0 = \frac{p^2}{2m} + V_\text{o}(x) \tag{8}$$

and then the full Green's function $G(x,x';E)$ of (6) is determined from the formula

$$G(x,x';E) = G_0(x,x';E) + \frac{G_0(x,0;E)G_0(0,x';E)}{\frac{1}{\Omega} - G_0(0,0;E)}. \tag{9}$$

The function $G(x,x';E)$ given in Eq. (9) is an exact result, i.e., we have the explicit expression for the Green's function of the Hamiltonian (6).

From Eq. (6) one can also find the eigenvalues E_n of the Hamiltonian from the poles of the Green's function in the complex plane of energy. In our case it means that we must find zeros of the denominator in Eq. (9)

$$\frac{1}{\Omega} - G_0(0,0;E) = 0. \tag{10}$$

The eigenfunctions $\psi_n(x)$ for the n-th pole are obtained from the residue of the pole of the Green's function at $E = E_n$

$$G(x,x';E) \sim \frac{\psi_n(x)\psi_n(x')}{E - E_n}. \tag{11}$$

In our case we find from Eq. (9) that the eigenfunction $\psi_n(x)$ is equal

$$\psi_n(x) = \frac{G_0(x, 0; E)}{N_n}, \tag{12}$$

where E_n is the position of the n-th pole and the normalization factor N_n is equal

$$N_n^2 = -\frac{\partial}{\partial E} G_0(0, 0; E). \tag{13}$$

3 Energy eigenfunctions of the quantum unstable oscillator

We will now apply the formalism sketched in Section 2 to the Hamiltonian (6). The Green's function $G_0(x, x'; E)$ is obtained from the solutions $\psi^{\pm\infty}(x)$ of the Schrödinger equation for the Hamiltonian H_0 in Eq. (8)

$$\left(\frac{p^2}{2m} + V_o(x)\right) \psi^{\pm\infty}(x) = E\psi^{\pm\infty}(x), \tag{14}$$

which fulfill the boundary conditions

$$\begin{array}{ll} \psi^{-\infty}(x) & \text{is bounded for } x \to -\infty \\ \psi^{+\infty}(x) & \sim \mathrm{e}^{ikx} \text{ for } x \to +\infty \end{array}. \tag{15}$$

Such boundary conditions are sufficient for the determination of the function $\psi^{+\infty}(x)$ uniquely and of the function $\psi^{-\infty}(x)$ up to a multiplicative constant[2]. The functions $\psi^{\pm\infty}(x)$ are equal

$$\psi^{-\infty}(x) = \begin{cases} \mathrm{e}^{-\frac{1}{2}\xi^2} \left({}_1F_1\left(\frac{1-2\epsilon}{2}, \frac{1}{2}; \xi^2\right) - 2|\xi| \frac{\Gamma\left(\frac{3-2\epsilon}{4}\right)}{\Gamma\left(\frac{1-2\epsilon}{4}\right)} {}_1F_1\left(\frac{3-2\epsilon}{2}, \frac{3}{2}; \xi^2\right) \right) & \text{for } x \leq 0 \\ \frac{1}{2} \left(1 - \frac{2i}{k} \sqrt{\frac{m\omega}{\hbar}} \frac{\Gamma\left(\frac{3-2\epsilon}{4}\right)}{\Gamma\left(\frac{1-2\epsilon}{4}\right)} \right) \mathrm{e}^{ikx} + \text{c.c. for } x \geq 0 \end{cases} \tag{16a}$$

and

$$\psi^{+\infty}(x) = \begin{cases} \mathrm{e}^{-\frac{1}{2}\xi^2} \left({}_1F_1\left(\frac{1-2\epsilon}{2}, \frac{1}{2}; \xi^2\right) - ik|\xi| \sqrt{\frac{\hbar}{m\omega}}\, {}_1F_1\left(\frac{3-2\epsilon}{2}, \frac{3}{2}; \xi^2\right) \right) & \text{for } x \leq 0 \\ \mathrm{e}^{ikx} \text{ for } x \geq 0 \end{cases}, \tag{16b}$$

[2]This constant is determined from the condition $\psi^{-\infty}(0) = 1$

where ${}_1F_1(\alpha, \beta; z)$ is the confluent hypergeometric function, $\Gamma(z)$ is the Euler's gamma function and ξ, ϵ and k are equal

$$\xi = \sqrt{\frac{m\omega}{\hbar}}, \quad \epsilon = \frac{E}{\hbar\omega}, \quad k = \frac{\sqrt{2mE}}{\hbar} = \sqrt{\frac{m\omega}{\hbar}}\sqrt{2\epsilon}. \tag{17}$$

The Green's function $G_0(x, x'; E)$ is obtained from the functions $\psi^{\pm\infty}(x)$:

$$G_0(x, x'; E) = C_0 \begin{cases} \psi^{-\infty}(x)\psi^{+\infty}(x') & \text{for } x \leq x' \\ \psi^{+\infty}(x)\psi^{-\infty}(x') & \text{for } x \leq x' \end{cases} \tag{18}$$

and the normalization constant C_0 is determined from the Wronskian of two solutions $\psi^{\pm\infty}(x)$ and is equal

$$C_0 = i\frac{2m}{\hbar^2}\frac{1}{k + 2i\sqrt{\frac{m\omega}{\hbar}}\frac{\Gamma\left(\frac{3-2\epsilon}{4}\right)}{\Gamma\left(\frac{1-2\epsilon}{4}\right)}}. \tag{19}$$

From (12) and (13) we see that the n-th eigenfunction of the Hamiltonian (6) with the purely outgoing wave boundary conditions at infinity is equal:

$$\psi_n(x) = \frac{G_0(x, 0; E_n)}{N_n} = \frac{C_0}{N_n}\begin{cases} \psi^{-\infty}(x)\psi^{+\infty}(0) & \text{for } x \leq 0 \\ \psi^{+\infty}(x)\psi^{-\infty}(0) & \text{for } x \geq 0 \end{cases}. \tag{20}$$

The point $x = 0$ is where the semi-transparent Dirac delta potential is inserted and now we will study the properties of the wave function at this point. The function $\psi_n(x)$ is continuous at the point $x = 0$ since

$$\lim_{\varepsilon\to 0}(\psi_n(\varepsilon) - \psi_n(-\varepsilon)) = \lim_{\varepsilon\to 0}(\psi^{+\infty}(\varepsilon)\psi^{-\infty}(0) - \psi^{+\infty}(0)\psi^{-\infty}(-\varepsilon)) = 0. \tag{21}$$

The derivative of the function $\psi_n'(x)$ is not continuous since differentiating the function $\psi_n(x)$ from the left one obtains

$$\lim_{\varepsilon\to 0}\psi_n'(\varepsilon) = \lim_{\varepsilon\to 0}\frac{C_0}{N_n}\psi^{+\infty\prime}(\varepsilon)\psi^{-\infty}(0) = \frac{C_0}{N_n}\psi^{+\infty\prime}(0)\psi^{-\infty}(0) \tag{22a}$$

and for the derivative from the right we get

$$\lim_{\varepsilon\to 0}\psi_n'(-\varepsilon) = \lim_{\varepsilon\to 0}\frac{C_0}{N_n}\psi^{+\infty}(0)\psi^{-\infty\prime}(-\varepsilon) = \frac{C_0}{N_n}\psi^{+\infty}(0)\psi^{-\infty\prime}(0). \tag{22b}$$

The discontinuity of the derivative of the wave function $\psi_n(x)$ at the point $x = 0$ is thus equal

$$\lim_{\varepsilon\to 0}(\psi'_n(\varepsilon) - \psi'_n(-\varepsilon)) = \frac{C_0}{N_n}(\psi^{+\infty\prime}(0)\psi^{-\infty}(0) - \psi^{+\infty}(0)\psi^{-\infty\prime}(0))$$
$$= \frac{C_0}{N_n}W(\psi^{-\infty}(x), \psi^{+\infty}(x))\bigg|_{x=0}, \tag{23}$$

where $W(\psi^{-\infty}(x), \psi^{+\infty}(x))$ is the Wronskian of the wave functions $\psi^{-\infty}(x)$, $\psi^{+\infty}(x)$ and it is equal to $1/C_0$ from the condition on the Green's function. Thus finally we obtain

$$\lim_{\varepsilon\to 0}(\psi'_n(\varepsilon) - \psi'_n(-\varepsilon)) = \frac{1}{N_n}. \tag{24}$$

4 Generalization and discussion of the results

Eq. (24) contains our main result and it demonstrates that for the potential of the Dirac delta type (3) the wave function is continuous and its first derivative has a finite jump. This discontinuity can be identified with the presence of the semi-transparent wall and such a potential can be used in the construction of models of unstable system that decay in time.

Eq. (24) was obtained for a special case of of the potential in the shape of the Dirac delta function, but Eq. (24) can serve as the starting point for a generalization to an arbitrary potential to produce decaying systems. Let us suppose that we have a potential $V(x)$. We can make the walls of such a potential to be semi-transparent if we solve the corresponding Schrödinger equation with the following boundary conditions for the wave function $\psi(x)$:

1. The wave function $\psi(x)$ is continuous.
2. The first derivative of the wave function $\psi'(x)$ is discontinuous at the point x_0, where $V(x_0) = E$ and the jump of the derivative is given by the formula
$$\lim_{\varepsilon\to 0}(\psi'(x_0+\varepsilon) + \psi'(x_0-\varepsilon)) = g(E), \tag{25}$$
where $g(E)$ is a given function.
3. $\psi(x) \sim \mathrm{e}^{\pm ikx}$ for $x \to \pm\infty$.

The wave function obtained from these boundary conditions will describe an unstable system with complex energy eigenvalues.

References

[1] Such solutions do not belong to the Hilbert space, because they have an infinite norm. The mathematically precise formulation requires the use of the rigged Hilbert spaces (Gel'fand triplets), A. Bohm, Lett. Math. Phys, **3**, 455 (1978), A. Bohm, J. Math. Phys. **22**, 2813 (1981); for the description of unstable states one can also use the rigged Hilbert space with a different dense subspace of the Hilbert space. An introduction to such rigged Hilbert spaces is given in A. Bohm and M. Gadella, *Dirac Kets, Gamow Vectors and Gel'fand Triplets.* Lecture Notes in Physics. Springer-Verlag, Berlin, Heidelberg and New York, 1989. A recent review of the physical and mathematical properties of the rigged Hilbert spaces can be found in O. Civitarese and M. Gadella, Phys. Rep., **396**, 41 (2004).

[2] F.A. Berezin and L.D. Fadeev, Dokl. Akad. Nauk CCCP, **137**, 1011 (1961) ; S. Albeverio, F. Gesztesy, R. Høegh-Krohn and H. Holden, *Solvable Models of Quantum Mechanics*, AMS Chelsea Publishing, American Mathematical Society, Providence, Rhode Island, 1988, ch I.3; S. Albeverio and P. Kurasov, *Singular Perturbations of Differential Operators*, London Mathematical Society, Lecture Notes Series 271, Cambridge University Press, Cambridge, UK (2000).

[3] A discussion of the the influence of the $\delta-$type term in the potential on the properties of the bound states and resonances is contained in M.L. Glasser, W. Jaskólski, F. García Moliner and V.R. Velasco, Phys. Rev. B, **42**, 7630 (1990). These results were obtained by the analysis of the Green's function of the system.

[4] M.G. Espinosa and P. Kielanowski, J. Phys.: Conf. Ser. **128** 012037 (2008).

[5] A. Ludviksson, J. Phys. A, **20** 4733 (1987).

Covariant Transform

Vladimir V. Kisil
School of Mathematics, University of Leeds, Leeds LS2 9JT, UK
(On leave from the Odessa University)
E-mail: kisilv@maths.leeds.ac.uk

Dedicated to the memory of Cora Sadosky

Abstract. The paper develops theory of covariant transform, which is inspired by the wavelet construction. It was observed that many interesting types of *wavelets* (or *coherent states*) arise from group representations which *are not* square integrable or vacuum vectors which *are not* admissible. Covariant transform extends an applicability of the popular wavelets construction to classic examples like the Hardy space H_2, Banach spaces, covariant functional calculus and many others.

Keywords: Wavelets, coherent states, group representations, Hardy space, Littlewood–Paley operator, functional calculus, Berezin calculus, Radon transform, Möbius map, maximal function, affine group, special linear group, numerical range, characteristic function, functional model.

A general group-theoretical construction [1, 3, 6, 7, 13, 17, 21] of *wavelets* (or *coherent states*) starts from an irreducible square integrable representation—in the proper sense or modulo a subgroup. Then a mother wavelet is chosen to be admissible. This leads to a wavelet transform which is an isometry to L_2 space with respect to the Haar measure on the group or (quasi)invariant measure on a homogeneous space.

The importance of the above situation shall not be diminished, however an exclusive restriction to such a setup is not necessary, in fact. Here is a classical example from complex analysis: the Hardy space $H_2(\mathbb{T})$ on the unit circle and Bergman spaces $B_2^n(\mathbb{D})$ in the unit disk produce wavelets associated with representations ρ_1 and ρ_n of the group $SL_2(\mathbb{R})$ respectively [11]. While representations ρ_n are from square integrable discrete series, the mock discrete series representation ρ_1 is not square integrable [20, § VI.5; 23, § 8.4]. However it would be natural to treat the Hardy space in the same framework as Bergman ones. Some more examples will be presented below.

1. Covariant Transform

To make a sharp but still natural generalisation of wavelets we give the following definition.

Definition 1. [15] Let ρ be a representation of a group G in a space V and F be an operator from V to a space U. We define a *covariant transform* $\mathcal{W}$ from V to the space $L(G, U)$ of U-valued functions on G by the formula:

$$\mathcal{W} : v \mapsto \hat{v}(g) = F(\rho(g^{-1})v), \qquad v \in V,\ g \in G. \tag{1}$$

Operator F will be called *fiducial operator* in this context.

We borrow the name for operator F from fiducial vectors of Klauder and Skagerstam [17].

Remark 2. We do not require that fiducial operator F shall be linear. Sometimes the homogeneity, i.e. $F(tv) = tF(v)$ for $t > 0$, alone can be already sufficient, see Example 12.

Remark 3. Usefulness of the covariant transform is in the reverse proportion to the dimensionality of the space U. The covariant transform encodes properties of v in a function $\mathcal{W}v$ on G. For a low dimensional U this function can be ultimately investigated by means of harmonic analysis. Thus $\dim U = 1$ (scalar-valued functions) is the ideal case, however, it is unattainable sometimes, see Example 9 below. We may have to use a higher dimensions of U if the given group G is not rich enough.

As we will see below covariant transform is a close relative of wavelet transform. The name is chosen due to the following common property of both transformations.

Theorem 4. *The covariant transform* (1) *intertwines* ρ *and the left regular representation* Λ *on* $L(G,U)$*:*

$$\mathcal{W}\rho(g) = \Lambda(g)\mathcal{W}.$$

Here Λ *is defined as usual by:*

$$\Lambda(g) : f(h) \mapsto f(g^{-1}h). \tag{2}$$

Proof. We have a calculation similar to wavelet transform [13, Prop. 2.6]. Take $u = \rho(g)v$ and calculate its covariant transform:

$$\begin{aligned}
[\mathcal{W}(\rho(g)v)](h) &= [\mathcal{W}(\rho(g)v)](h) = F(\rho(h^{-1})\rho(g)v)\\
&= F(\rho((g^{-1}h)^{-1})v)\\
&= [\mathcal{W}v](g^{-1}h)\\
&= \Lambda(g)[\mathcal{W}v](h).
\end{aligned}$$

□

The next result follows immediately:

Corollary 5. *The image space* $\mathcal{W}(V)$ *is invariant under the left shifts on* G.

2. Examples of Covariant Transform

We start from the classical example of the group-theoretical wavelet transform:

Example 6. Let V be a Hilbert space with an inner product $\langle\cdot,\cdot\rangle$ and ρ be a unitary representation of a group G in the space V. Let $F : V \to \mathbb{C}$ be a functional $v \mapsto \langle v, v_0\rangle$ defined by a vector $v_0 \in V$. The vector v_0 is oftenly called the *mother wavelet* in areas related to signal processing or the *vacuum state* in quantum framework.

Then the transformation (1) is the well-known expression for a *wavelet transform* [1, (7.48)] (or *representation coefficients*):

$$\mathcal{W} : v \mapsto \hat{v}(g) = \left\langle \rho(g^{-1})v, v_0\right\rangle = \left\langle v, \rho(g)v_0\right\rangle, \qquad v \in V,\ g \in G. \tag{3}$$

The family of vectors $v_g = \rho(g)v_0$ is called *wavelets* or *coherent states*. In this case we obtain scalar valued functions on G, thus the fundamental rôle of this example is explained in Rem. 3.

This scheme is typically carried out for a square integrable representation ρ and v_0 being an admissible vector [1,3,6,7,21]. In this case the wavelet (covariant) transform is a map into the square integrable functions [5] with respect to the left Haar measure. The map becomes an isometry if v_0 is properly scaled.

However square integrable representations and admissible vectors does not cover all interesting cases.

Example 7. Let $G = \mathrm{Aff}$ be the "$ax + b$" (or *affine*) group [1, § 8.2]: the set of points (a, b), $a \in \mathbb{R}_+$, $b \in \mathbb{R}$ in the upper half-plane with the group law:

$$(a, b) * (a', b') = (aa', ab' + b) \tag{4}$$

and left invariant measure $a^{-2}\, da\, db$. Its isometric representation on $V = L_p(\mathbb{R})$ is given by the formula:

$$[\rho_p(g)\, f](x) = a^{\frac{1}{p}} f\,(ax + b)\,, \qquad \text{where } g^{-1} = (a, b). \tag{5}$$

We consider the operators $F_\pm : L_2(\mathbb{R}) \rightarrow \mathbb{C}$ defined by:

$$F_\pm(f) = \frac{1}{2\pi i} \int_{\mathbb{R}} \frac{f(t)\, dt}{x \mp \mathrm{i}}. \tag{6}$$

Then the covariant transform (1) is the Cauchy integral from $L_p(\mathbb{R})$ to the space of functions $\hat{f}(a, b)$ such that $a^{-\frac{1}{p}} \hat{f}(a, b)$ is in the Hardy space in the upper/lower half-plane $H_p(\mathbb{R}^2_\pm)$. Although the representation (5) is square integrable for $p = 2$, the function $\frac{1}{x \pm \mathrm{i}}$ used in (6) is not an admissible vacuum vector. Thus the complex analysis become decoupled from the traditional wavelet theory. As a result the application of wavelet theory shall relay on an extraneous mother wavelets [9].

Many important objects in complex analysis are generated by inadmissible mother wavelets like (6). For example, if $F : L_2(\mathbb{R}) \rightarrow \mathbb{C}$ is defined by $F : f \mapsto F_+ f + F_- f$ then the covariant transform (1) reduces to the *Poisson integral.* If $F : L_2(\mathbb{R}) \rightarrow \mathbb{C}^2$ is defined by $F : f \mapsto (F_+ f, F_- f)$ then the covariant transform (1) represents a function f on the real line as a jump:

$$f(z) = f_+(z) - f_-(z), \qquad f_\pm(z) \in H_p(\mathbb{R}^2_\pm) \tag{7}$$

between functions analytic in the upper and the lower half-planes. This makes a decomposition of $L_2(\mathbb{R})$ into irreducible components of the representation (5). Another interesting but non-admissible vector is the Gaussian e^{-x^2}.

Example 8. For the group $G = SL_2(\mathbb{R})$ [20] let us consider the unitary representation ρ on the space of square integrable function $L_2(\mathbb{R}^2_+)$ on the upper half-plane through the Möbius transformations:

$$\rho(g) : f(z) \mapsto \frac{1}{(cz + d)^2} f\left(\frac{az + b}{cz + d}\right), \qquad g^{-1} = \begin{pmatrix} a & b \\ c & d \end{pmatrix}. \tag{8}$$

This is a representation from the discrete series and $L_2(\mathbb{D})$ and irreducible invariant subspaces are parametrised by integers. Let F_k be the functional $L_2(\mathbb{R}^2_+) \rightarrow \mathbb{C}$ of pairing with the lowest/highest k-weight vector in the corresponding irreducible component $B_k(\mathbb{R}^2_\pm)$, $k \geq 2$ of the discrete series [20, Ch. VI]. Then we can build an operator F from various F_k similarly to the previous Example. In particular, the jump representation (7) on the real line generalises to the representation of a square integrable function f on the upper half-plane as a sum

$$f(z) = \sum_k a_k f_k(z), \qquad f_k \in B_n(\mathbb{R}^2_\pm)$$

for prescribed coefficients a_k and analytic functions f_k in question from different irreducible subspaces.

Covariant transform is also meaningful for principal and complementary series of representations of the group $SL_2(\mathbb{R})$, which are not square integrable [11].

Example 9. A straightforward generalisation of Ex. 6 is obtained if V is a Banach space and $F : V \to \mathbb{C}$ is an element of V^*. Then the covariant transform coincides with the construction of wavelets in Banach spaces [13].

Example 10. The next stage of generalisation is achieved if V is a Banach space and $F : V \to \mathbb{C}^n$ is a linear operator. Then the corresponding covariant transform is a map $\mathcal{W} : V \to L(G, \mathbb{C}^n)$. This is closely related to M.G. Krein's works on *directing functionals* [18], see also *multiresolution wavelet analysis* [2], Clifford-valued Bargmann spaces [4] and [1, Thm. 7.3.1].

Example 11. Let F be a projector $L_p(\mathbb{R}) \to L_p(\mathbb{R})$ defined by the relation $(Ff)\hat{}(\lambda) = \chi(\lambda)\hat{f}(\lambda)$, where the hat denotes the Fourier transform and $\chi(\lambda)$ is the characteristic function of the set $[-2,-1] \cup [1,2]$. Then the covariant transform $L_p(\mathbb{R}) \to C(\mathrm{Aff}, L_p(\mathbb{R}))$ generated by the representation (5) of the affine group from F contains all information provided by the *Littlewood–Paley operator* [8, § 5.1.1].

Example 12. A step in a different direction is a consideration of non-linear operators. Take again the "$ax + b$" group and its representation (5). We define F to be a homogeneous but non-linear functional $V \to \mathbb{R}_+$:

$$F(f) = \frac{1}{2}\int_{-1}^{1} |f(x)|\, dx.$$

The covariant transform (1) becomes:

$$[\mathcal{W}_p f](a,b) = F(\rho_p(a,b)f) = \frac{1}{2}\int_{-1}^{1} \left| a^{\frac{1}{p}} f\,(ax+b)\right| dx = a^{\frac{1}{p}} \frac{1}{2a}\int_{b-a}^{b+a} |f\,(x)|\, dx. \qquad (9)$$

Obviously $M_f(b) = \max_a [\mathcal{W}_\infty f](a,b)$ coincides with the Hardy *maximal function*, which contains important information on the original function f. From the Cor. 5 we deduce that the operator $M : f \mapsto M_f$ intertwines ρ_p with itself $\rho_p M = M\rho_p$.

Of course, the full covariant transform (9) is even more detailed than M. For example, $\|f\| = \max_b [\mathcal{W}_\infty f](\frac{1}{2}, b)$ is the shift invariant norm [10].

Example 13. Let $V = L_c(\mathbb{R}^2)$ be the space of compactly supported bounded functions on the plane. We take F be the linear operator $V \to \mathbb{C}$ of integration over the real line:

$$F : f(x,y) \mapsto F(f) = \int_{\mathbb{R}} f(x,0)\, dx.$$

Let G be the group of Euclidean motions of the plane represented by ρ on V by a change of variables. Then the wavelet transform $F(\rho(g)f)$ is the *Radon transform*.

Example 14. Let a representation ρ of a group G act on a space X. Then there is an associated representation ρ_B of G on a space $V = B(X,Y)$ of linear operators $X \to Y$ defined by the identity:

$$(\rho_B(g)A)x = A(\rho(g^{-1})x), \qquad x \in X,\ g \in G,\ A \in B(X,Y). \qquad (10)$$

Following the Remark 3 we take F to be a functional $V \to \mathbb{C}$, for example F can be defined from a pair $x \in X$, $l \in Y^*$ by the expression $F : A \mapsto \langle Ax, l\rangle$. Then the covariant transform is:

$$\mathcal{W} : A \mapsto \hat{A}(g) = F(\rho_B(g)A).$$

This is an example of *covariant calculus* [13, 14].

Example 15. A modification of the previous construction is obtained if we have two groups G_1 and G_2 represented by ρ_1 and ρ_2 on X and Y^* respectively. Then we have a covariant transform $B(X,Y) \to L(G_1 \times G_2, \mathbb{C})$ defined by the formula:

$$\mathcal{W} : A \mapsto \hat{A}(g_1, g_2) = \langle A\rho_1(g_1)x, \rho_2(g_2)l \rangle .$$

This generalises *Berezin functional calculi* [13].

Example 16. Let us restrict the previous example to the case when $X = Y$ is a Hilbert space, $\rho_1 = \rho_2 = \rho$ and $x = l$ with $\|x\| = 1$. Than the range of the covariant transform:

$$\mathcal{W} : A \mapsto \hat{A}(g) = \langle A\rho(g)x, \rho(g)x \rangle$$

is a subset of the *numerical range* of the operator A. As a function on a group $\hat{A}(g)$ provides a better description of A than the set of its values—numerical range.

Example 17. The group $SU(1,1) \simeq SL_2(\mathbb{R})$ consists of 2×2 matrices of the form $\begin{pmatrix} \alpha & \beta \\ \bar{\beta} & \bar{\alpha} \end{pmatrix}$ with the unit determinant [20, § IX.1]. Let T be an operator with the spectral radius less than 1. Then the associated Möbius transformation

$$g : T \mapsto g \cdot T = \frac{\alpha T + \beta I}{\bar{\beta} T + \bar{\alpha} I}, \qquad \text{where} \quad g = \begin{pmatrix} \alpha & \beta \\ \bar{\beta} & \bar{\alpha} \end{pmatrix} \in SL_2(\mathbb{R}), \tag{11}$$

produces a well-defined operator with the spectral radius less than 1 as well. Thus we have a representation of $SU(1,1)$. A choice of an operator F will define the corresponding covariant transform. In this way we obtain generalisations of *Riesz–Dunford functional calculus* [14].

Example 18. Consider again the action (11) of the Moebius transformations on operators from the previous Example. Let us introduce the defect operators $D_T = (I - T^*T)^{1/2}$ and $D_{T^*} = (I - TT^*)^{1/2}$. For the case $F = D_{T^*}$ the covariant transform is, cf. [22, § VI.1, (1.2)]:

$$[\mathcal{W}T](g) = F(g \cdot T) = -e^{\mathrm{i}\phi}\, \Theta_T(z)\, D_T, \qquad \text{for } g = \begin{pmatrix} e^{\mathrm{i}\phi/2} & 0 \\ 0 & e^{-\mathrm{i}\phi/2} \end{pmatrix} \begin{pmatrix} 1 & -z \\ -\bar{z} & 1 \end{pmatrix},$$

where the *characteristic function* $\Theta_T(z)$ [22, § VI.1, (1.1)] is:

$$\Theta_T(z) = -T + D_{T^*}\, (I - zT^*)^{-1}\, z\, D_T.$$

Thus we approached the *functional model* of operators from the covariant transform. In accordance with Remark 3 the model is most fruitful for the case of operator $F = D_{T^*}$ being one-dimensional.

3. Induced Covariant Transform

The choice of fiducial operator F can significantly influence the behaviour of the covariant transform. Let G be a group and H be its closed subgroup with the corresponding homogeneous space $X = G/H$. Let ρ be a representation of G by operators on a space V, we denote by ρ_H the restriction of ρ to the subgroup H.

Definition 19. Let χ be a representation of the subgroup H in a space U and $F : V \to U$ be an intertwining operator between χ and the representation ρ_H:

$$F(\rho(h)v) = F(v)\chi(h), \qquad \text{for all } h \in H,\ v \in V.$$

Then the covariant transform (1) generated by F is called the *induced covariant transform.*

The following is the main motivating example.

Example 20. Consider the traditional wavelet transform as outlined in Ex. 6. Chose a vacuum vector v_0 to be a joint eigenvector for all operators $\rho(h)$, $h \in H$, that is $\rho(h)v_0 = \chi(h)v_0$, where $\chi(h)$ is a complex number depending of h. Then χ is obviously a character of H.

The image of wavelet transform (3) with such a mother wavelet will have a property:

$$\hat{v}(gh) = \langle v, \rho(gh)v_0 \rangle = \langle v, \rho(g)\chi(h)v_0 \rangle = \chi(h)\hat{v}(g).$$

Thus the wavelet transform is uniquely defined by cosets on the homogeneous space G/H. In this case we can speak about the *reduced wavelet transform* [12]. A representation ρ_0 is *square integrable* mod H if the induced wavelet transform $[\mathcal{W}f_0](w)$ of the vacuum vector $f_0(x)$ is square integrable on X.

The image of induced covariant transform have the similar property:

$$\hat{v}(gh) = F(\rho((gh)^{-1})v) = F(\rho(h^{-1})\rho(g^{-1})v) = F(\rho(g^{-1})v)\chi(h^{-1}). \tag{12}$$

Thus it is enough to know the value of the covariant transform only at a single element in every coset G/H in order to reconstruct it for the entire group G by the representation χ. Since coherent states (wavelets) are now parametrised by points homogeneous space G/H they are referred sometimes as coherent states which are not connected to a group [16], however this is true only in a very narrow sense as explained above.

Example 21. To make it more specific we can consider the representation of $SL_2(\mathbb{R})$ defined on $L_2(\mathbb{R})$ by the formula, cf. (8):

$$\rho(g) : f(z) \mapsto \frac{1}{(cx+d)} f\left(\frac{ax+b}{cx+d}\right), \qquad g^{-1} = \begin{pmatrix} a & b \\ c & d \end{pmatrix}.$$

Let $K \subset SL_2(\mathbb{R})$ be the compact subgroup of matrices $h_t = \begin{pmatrix} \cos t & \sin t \\ -\sin t & \cos t \end{pmatrix}$. Then for the fiducial operator $F_\pm$ (6) we have $F_\pm \circ \rho(h_t) = e^{\mp it} F_\pm$. Thus we can consider the covariant transform only for points in $SL_2(\mathbb{R})/K$, however this set can be naturally identified with the $ax+b$ group. Thus we do not obtain any advantage of extending the group in Ex. 7 from $ax+b$ to $SL_2(\mathbb{R})$ if we will be still using the fiducial operator $F_\pm$ (6).

Functions on the group G, which have the property $\hat{v}(gh) = \hat{v}(g)\chi(h)$ (12), provide a space for the representation of G induced by the representation χ of the subgroup H. This explains the choice of the name for induced covariant transform.

Remark 22. Induced covariant transform uses the fiducial operator F which passes through the action of the subgroup H. This reduces information which we obtained from this transform in some cases.

There is also a simple connection between a covariant transform and right shifts:

Proposition 23. *Let G be a Lie group and ρ be a representation of G in a space V. Let $[\mathcal{W}f](g) = F(\rho(g^{-1})f)$ be a covariant transform defined by the fiducial operator $F : V \to U$. Then the right shift $[\mathcal{W}f](gg')$ by g' is the covariant transform $[\mathcal{W}'f](g) = F'(\rho(g^{-1})f)]$ defined by the fiducial operator $F' = F \circ \rho(g^{-1})$.*

In other words the covariant transform intertwines right shifts with the associated action ρ_B (10) on fiducial operators.

Although the above result is obvious, its infinitesimal version has interesting consequences.

Corollary 24. *Let G be a Lie group with a Lie algebra $\mathfrak{g}$ and ρ be a smooth representation of G. We denote by $d\rho_B$ the derived representation of the associated representation ρ_B (10) on fiducial operators.*

Let a fiducial operator F be a null-solution, i.e. $AF = 0$, for the operator $A = \sum_J a_j d\rho_B^{X_j}$, where $X_j \in \mathfrak{g}$ and a_j are constants. Then the wavelet transform $[\mathcal{W}f](g) = F(\rho(g^{-1})f)$ for any f satisfies:

$$DF(g) = 0, \qquad \text{where} \quad D = \sum_j \bar{a}_j \mathfrak{L}^{X_j}.$$

Here $\mathfrak{L}^{X_j}$ are the left invariant fields (Lie derivatives) on G corresponding to X_j.

Example 25. Consider the representation ρ (5) of the $ax+b$ group with the $p=1$. Let A and N be the basis of the corresponding Lie algebra generating one-parameter subgroups $(e^t, 0)$ and $(0, t)$. Then the derived representations are:

$$[d\rho^A f](x) = f(x) + xf'(x), \qquad [d\rho^N f](x) = f'(x).$$

The corresponding left invariant vector fields on $ax+b$ group are:

$$\mathfrak{L}^A = a\partial_a, \qquad \mathfrak{L}^N = a\partial_b.$$

The mother wavelet $\frac{1}{x+\mathrm{i}}$ is a null solution of the operator $d\rho^A + \mathrm{i}d\rho^N = I + (x+\mathrm{i})\frac{d}{dx}$. Therefore the covariant transform with the fiducial operator F_+ (6) will consist with the null solutions to the operator $\mathfrak{L}^A - \mathrm{i}\mathfrak{L}^N = -\mathrm{i}a(\partial_b + \mathrm{i}\partial_a)$, that is in the essence the Cauchy-Riemann operator in the upper half-plane.

There is a statement which extends the previous Corollary from differential operators to integro-differential ones. We will formulate it for the wavelets setting.

Corollary 26. *Let G be a Lie group with a Lie algebra $\mathfrak{g}$ and ρ be a unitary representation of G, which can be extended to a vector space V of functions or distributions on G.*

Let a mother wavelet $w \in V'$ satisfy the equation

$$\int_G a(g)\, \rho(g) w \, dg = 0,$$

for a fixed distribution $a(g) \in V$. Then any wavelet transform $F(g) = \mathcal{W}f(g) = \langle f, \rho(g) w_0 \rangle$ obeys the condition:

$$DF = 0, \qquad \text{where} \quad D = \int_G \bar{a}(g)\, R(g)\, dg,$$

with R being the right regular representation of G.

Clearly the Corollary 24 is a particular case of Corollary 26.

4. Inverse Covariant Transform

An object invariant under the left action Λ (2) is called *left invariant*. For example, let L and L' be two left invariant spaces of functions on G. We say that a pairing $\langle\cdot,\cdot\rangle : L \times L' \to \mathbb{C}$ is *left invariant* if

$$\langle \Lambda(g)f, \Lambda(g)f' \rangle = \langle f, f' \rangle, \qquad \text{for all} \quad f \in L,\ f' \in L'. \tag{13}$$

Remark 27. (i) We do not require the pairing to be linear in general.

(ii) If the pairing is invariant on space $L \times L'$ it is not necessarily invariant (or even defined) on the whole $C(G) \times C(G)$.

(iii) In a more general setting we shall study an invariant pairing on a homogeneous spaces instead of the group. However due to length constraints we cannot consider it here beyond the Example 30.

(iv) An invariant pairing on G can be obtained from an invariant functional l by the formula $\langle f_1, f_2\rangle = l(f_1\bar{f}_2)$.

For a representation ρ of G in V and $v_0 \in V$ we fix a function $w(g) = \rho(g)v_0$. We assume that the pairing can be extended in its second component to this V-valued functions, say, in the weak sense.

Definition 28. Let $\langle\cdot,\cdot\rangle$ be a left invariant pairing on $L \times L'$ as above, let ρ be a representation of G in a space V, we define the function $w(g) = \rho(g)v_0$ for $v_0 \in V$. The *inverse covariant transform* $\mathcal{M}$ is a map $L \to V$ defined by the pairing:

$$\mathcal{M} : f \mapsto \langle f, w\rangle\,, \qquad \text{where } f \in L. \tag{14}$$

Example 29. Let G be a group with a unitary square integrable representation ρ. An invariant pairing of two square integrable functions is obviously done by the integration over the Haar measure:

$$\langle f_1, f_2\rangle = \int_G f_1(g)\bar{f}_2(g)\,dg.$$

For an admissible vector v_0 [5], [1, Chap. 8] the inverse covariant transform is known in this setup as a *reconstruction formula.*

Example 30. Let ρ be a square integrable representation of G modulo a subgroup $H \subset G$ and let $X = G/H$ be the corresponding homogeneous space with a quasi-invariant measure dx. Then integration over dx with an appropriate weight produces an invariant pairing. The inverse covariant transform is a more general version [1, (7.52)] of the *reconstruction formula* mentioned in the previous example.

Let ρ be not a square integrable representation (even modulo a subgroup) or let v_0 be inadmissible vector of a square integrable representation ρ. An invariant pairing in this case is not associated with an integration over any non singular invariant measure on G. In this case we have a *Hardy pairing.* The following example explains the name.

Example 31. Let G be the "$ax+b$" group and its representation ρ (5) from Ex. 7. An invariant pairing on G, which is not generated by the Haar measure $a^{-2}da\,db$, is:

$$\langle f_1, f_2\rangle = \lim_{a\to 0} \int_{-\infty}^{\infty} f_1(a,b)\,\bar{f}_2(a,b)\,db. \tag{15}$$

For this pairing we can consider functions $\frac{1}{2\pi i(x+i)}$ or e^{-x^2}, which are not admissible vectors in the sense of square integrable representations. Then the inverse covariant transform provides an *integral resolutions* of the identity.

Similar pairings can be defined for other semi-direct products of two groups. We can also extend a Hardy pairing to a group, which has a subgroup with such a pairing.

Example 32. Let G be the group $SL_2(\mathbb{R})$ from the Ex. 8. Then the "$ax+b$" group is a subgroup of $SL_2(\mathbb{R})$, moreover we can parametrise $SL_2(\mathbb{R})$ by triples (a, b, θ), $\theta \in (-\pi, \pi]$ with the respective Haar measure [20, III.1(3)]. Then the Hardy pairing

$$\langle f_1, f_2\rangle = \lim_{a\to 0} \int_{-\infty}^{\infty} f_1(a,b,\theta)\,\bar{f}_2(a,b,\theta)\,db\,d\theta. \tag{16}$$

is invariant on $SL_2(\mathbb{R})$ as well. The corresponding inverse covariant transform provides even a finer resolution of the identity which is invariant under conformal mappings of the Lobachevsky half-plane.

A further study of covariant transform shall be continued elsewhere.

Acknowledgement. Author is grateful to the anonymous referee for many helpful suggestions.

References

[1] Syed Twareque Ali, Jean-Pierre Antoine, and Jean-Pierre Gazeau, *Coherent states, wavelets and their generalizations*, Graduate Texts in Contemporary Physics, Springer-Verlag, New York, 2000. MR2002m:81092

[2] Ola Bratteli and Palle E. T. Jorgensen, *Isometries, shifts, Cuntz algebras and multiresolution wavelet analysis of scale N*, Integral Equations Operator Theory **28** (1997), no. 4, 382–443. E-print: `arXiv:funct-an/9612003`.

[3] Jens Gerlach Christensen and Gestur Ólafsson, *Examples of coorbit spaces for dual pairs*, Acta Appl. Math. **107** (2009), no. 1-3, 25–48. MR2520008

[4] Jan Cnops and Vladimir V. Kisil, *Monogenic functions and representations of nilpotent Lie groups in quantum mechanics*, Math. Methods Appl. Sci. **22** (1999), no. 4, 353–373. E-print: `arXiv:math/9806150`. Zbl 1005.22003. MR1671449 (2000b:81044)

[5] M. Duflo and Calvin C. Moore, *On the regular representation of a nonunimodular locally compact group*, J. Functional Analysis **21** (1976), no. 2, 209–243. MR52#14145

[6] Feichtinger, Hans G. and Groechenig, K.H., *Banach spaces related to integrable group representations and their atomic decompositions, I*, J. Funct. Anal. **86** (1989), no. 2, 307–340. Zbl 691.46011.

[7] Hartmut Führ, *Abstract harmonic analysis of continuous wavelet transforms*, Lecture Notes in Mathematics, vol. 1863, Springer-Verlag, Berlin, 2005. MR2130226 (2006m:43003)

[8] Loukas Grafakos, *Classical Fourier analysis*, Second, Graduate Texts in Mathematics, vol. 249, Springer, New York, 2008. MR2445437

[9] Ondrej Hutník, *On Toeplitz-type operators related to wavelets*, Integral Equations Operator Theory **63** (2009), no. 1, 29–46. MR2480637

[10] Andreas Johansson, *Shift-invariant signal norms for fault detection and control*, Systems Control Lett. **57** (2008), no. 2, 105–111. MR2378755 (2009d:93035)

[11] Vladimir V. Kisil, *Analysis in $\mathbf{R}^{1,1}$ or the principal function theory*, Complex Variables Theory Appl. **40** (1999), no. 2, 93–118. E-print: `arXiv:funct-an/9712003`. MR1744876 (2000k:30078)

[12] ______, *Two approaches to non-commutative geometry*, Complex methods for partial differential equations (Ankara, 1998), 1999, pp. 215–244. E-print: `arXiv:funct-an/9703001`. MR1744440 (2001a:01002)

[13] ______, *Wavelets in Banach spaces*, Acta Appl. Math. **59** (1999), no. 1, 79–109. E-print: `arXiv:math/9807141`, On-line. MR1740458 (2001c:43013)

[14] ______, *Spectrum as the support of functional calculus*, Functional analysis and its applications, 2004, pp. 133–141. E-print: `arXiv:math.FA/0208249`. MR2098877

[15] ______, *Wavelets beyond admissibility*, Proceedings of the 10th ISAAC Congress, London 2009, 2010, pp. 6 pp. E-print: `arXiv:0911.4701`.

[16] John R. Klauder, *Coherent states for the hydrogen atom*, J. Phys. A **29** (1996), no. 12, L293–L298. MR1398598 (97f:81052)

[17] John R. Klauder and Bo-Sture Skagerstam (eds.), *Coherent states. applications in physics and mathematical physics.*, World Scientific Publishing Co., Singapur, 1985.

[18] M. G. Kreĭn, *On Hermitian operators with directed functionals*, Akad. Nauk Ukrain. RSR. Zbirnik Prac′ Inst. Mat. **1948** (1948), no. 10, 83–106. MR14:56c, reprinted in [19].

[19] ______, *Избранные Труды. II*, Akad. Nauk Ukrainy Inst. Mat., Kiev, 1997. MR96m:01030.

[20] Serge Lang, $SL_2(\mathbf{R})$, Graduate Texts in Mathematics, vol. 105, Springer-Verlag, New York, 1985. Reprint of the 1975 edition. MR803508 (86j:22018)

[21] A. Perelomov, *Generalized coherent states and their applications*, Texts and Monographs in Physics, Springer-Verlag, Berlin, 1986. MR87m:22035

[22] Béla Sz.-Nagy and Ciprian Foiaş, *Harmonic analysis of operators on Hilbert space*, North-Holland Publishing Company, Amsterdam, 1970.

[23] Michael E. Taylor, *Noncommutative harmonic analysis*, Mathematical Surveys and Monographs, vol. 22, American Mathematical Society, Providence, RI, 1986. MR88a:22021

Quasi regular polygons and their duals with Coxeter symmetries D_n represented by complex numbers

M Koca[1] and N **O Koca**[2]

[1, 2] Department of Physics, College of Science, Sultan Qaboos University, P.O. Box 36, Al- Khoud, 123 Muscat, Sultanate of Oman

E-mail: [1]kocam@squ.edu.om ; [2]nazife@squ.edu.om

Abstract. This paper deals with tiling of the plane by quasi regular polygons and their duals. The problem is motivated from the fact that the graphene, infinite number of carbon molecules forming a honeycomb lattice, may have states with two bond lengths and equal bond angles or one bond length and different bond angles. We prove that the Euclidean plane can be tiled with two tiles consisting of quasi regular hexagons with two different lengths (isogonal hexagons) and regular hexagons. The dual lattice is constructed with the isotoxal hexagons (equal edges but two different interior angles) and regular hexagons. We also give similar tilings of the plane with the quasi regular polygons along with the regular polygons possessing the Coxeter symmetries D_n, n=2,3,4,5. The group elements as well as the vertices of the polygons are represented by the complex numbers.

1. Introduction

The graphene [1], an infinite sheet of carbon atoms [2], tiled with regular hexagons has attracted much attention. In this paper we approach the problem from the mathematical point of view and deal with tiling of the plane by quasi regular polygons possessing dihedral symmetries. We use the rank-2 Coxeter diagrams to describe the symmetries of the polygons. The polygons are of two types: the isogonal polygons consisting of two alternating unequal edges with equal interior angles; the isotoxal polygons consisting of equal edges but with alternating unequal interior angles. The isogonal polygon with 2n sides is vertex transitive under the dihedral group D_n. Its dual polygon, the isotoxal polygon, is edge transitive under the same symmetry. The tiling of the plane with the isogonal and the isotoxal hexagons is an interesting problem by itself. The problem can be extended to the tiling of the plane by the isogonal polygons with the Coxeter symmetries D_n, n=4 and 5. We also give the dual tilings of the plane consisting of isotoxal polygons as well as corresponding regular polygons. The tiling of the plane with two isotoxal hexagons and one regular hexagon at the same vertex suggests that the graphene may have a state with equal bond lengths but with different bond angles. This is a slightly modified version of the honeycomb model of graphene. One can find some quasi regular 2D tiles in the reference [3]. In three dimensions a superspace-group approach has been formulated for the description of the composite crystals [4].

In section 2 we introduce rank-2 Coxeter diagrams [5] describing the root spaces as well as dual spaces by complex numbers. In section 3 we construct some even-sided polygons with alternating two edge lengths (isogonal polygons) having equal interior angles. We also construct their dual polygons (isotoxal polygons) with equal edge lengths but with alternating two interior angles. With this

Published under licence by IOP Publishing Ltd

approach we observe that in the case of $W(A_2) \approx D_3$ symmetry the Euclidean plane can be tiled by two isogonal hexagons having two different edge lengths and a regular hexagon all sharing the same vertex. The quasi regular polygons with dihedral symmetries $W(B_2) \approx D_4$ and $W(H_2) \approx D_5$ are also constructed and the tiling of the plane with the isogonal and isotoxal octagons and decagons are discussed. Section 4 is devoted to the concluding remarks.

2. Coxeter diagrams with complex numbers

All Coxeter diagrams are represented with one type of simple roots [6] contrary to the Dynkin diagrams representing the Lie algebra root systems having long and short roots. In 2D space we have an infinite number of Coxeter diagrams shown in figure 1. It is customary to use the notations $I_2(n)$ for the rank-2 Coxeter diagrams but we will continue using the notations $A_1 \oplus A_1$, A_2, B_2 and H_2 for $I_2(n)$, when n=2,3,4,5 respectively.

$$\bullet \overset{n}{} \bullet$$

Figure 1. Rank-2 Coxeter diagram $I_2(n)$.

Figure 1 implies that the angle between the simple roots α_1 and α_2 is $\frac{(n-1)\pi}{n}$. We choose the norm of the roots $\sqrt{2}$ to be consistent with the Dynkin diagrams when they coincide. The Cartan matrix defined by the scalar product $C_{ij} = (\alpha_i, \alpha_j)$ of the rank-2 diagrams and its inverse $(C^{-1})_{ij} = (\omega_i, \omega_j)$, $(i, j = 1, 2)$ will read

$$C = \begin{bmatrix} 2 & 2\cos\frac{(n-1)\pi}{n} \\ 2\cos\frac{(n-1)\pi}{n} & 2 \end{bmatrix},$$

$$C^{-1} = \frac{1}{4\sin^2\frac{(n-1)\pi}{n}} \begin{bmatrix} 2 & -2\cos\frac{(n-1)\pi}{n} \\ -2\cos\frac{(n-1)\pi}{n} & 2 \end{bmatrix}. \tag{1}$$

The fundamental weights ω_i are the basis vectors of the dual space defined by the relation $(\alpha_i, \omega_j) = \delta_{ij}$ [6] where δ_{ij} is the Kronecker delta. The simple roots and the fundamental weights are related to each other by the relations:

$$\omega_i = (C^{-1})_{ij}\alpha_j, \; \alpha_i = C_{ij}\omega_j. \tag{2}$$

Summation over the repeated index is implicit. Action of the reflection generator r_i on an arbitrary vector Λ is defined by the relation

$$r_i\Lambda = \Lambda - (\Lambda, \alpha_i)\alpha_i \quad (\text{no summation over } i). \tag{3}$$

They generate the dihedral group D_n of order $2n$ satisfying the relations $r_1^2 = r_2^2 = (r_1 r_2)^n = 1$. When Λ and α_i are represented by complex numbers, equation (3) can be written as

$$r_i\Lambda = -\frac{1}{2}\alpha_i\bar{\Lambda}\alpha_i = -\frac{1}{2}\alpha_i^{2}\bar{\Lambda}. \tag{4}$$

Here the roots are complex numbers for rank-2 Coxeter diagrams, a subset of quaternions, where the unit complex numbers $\alpha_1' = \alpha_1/\sqrt{2}$ and $\alpha_2' = \alpha_2/\sqrt{2}$ generate the cyclic group of order *n*; however, the generators in equation (4) generate the dihedral group of order 2*n*. Let $\beta = \exp(i\frac{\pi}{n})$ be the unit complex number. Then all integer powers $\{\beta^k, k = 1,2,...,2n\}$ constitute a scaled copy of the root system of rank-2 Coxeter diagrams. If Λ represents any complex number, the generators act on Λ as follows:

$$r_1\Lambda = -\bar{\Lambda}\ ; \qquad r_2\Lambda = -e^{i\frac{2(n-1)\pi}{n}}\bar{\Lambda}. \tag{5}$$

Let us follow the Lie algebraic technique to obtain the orbits of the Coxeter groups. If *g* is the Lie algebra of rank *l* then the highest weight

$$\Lambda = a_1\omega_1 + a_2\omega_2 + ... + a_l\omega_l \equiv (a_1, a_2, ..., a_l) \tag{6}$$

is represented by the *l* non-negative integers [7]. Applying the Coxeter- Weyl group $W(g)$ on the highest weight one can generate the $\Lambda - orbit \equiv O(\Lambda) = W(g)\Lambda$. In 2D space the orbits of the Coxeter group is either regular polygons or even-sided quasi regular polygons. In what follows, we will discuss the orbits of the Coxeter groups with $n = 2,3,4,5$. The fundamental weights in terms of complex numbers can be written as

$$\omega_1 = \frac{1}{\sqrt{2}}\left(1 - i\frac{\cos\frac{(n-1)\pi}{n}}{\sin\frac{(n-1)\pi}{n}}\right), \quad \omega_2 = \frac{i}{\sqrt{2}\sin\frac{(n-1)\pi}{n}}. \tag{7}$$

A general vertex $\Lambda = a_1\omega_1 + a_2\omega_2$ is then an arbitrary complex number which can be written as

$$\Lambda = \frac{1}{\sqrt{2}}\left[a_1 + i\frac{1}{\sin\frac{(n-1)\pi}{n}}\left(-a_1\cos\frac{(n-1)\pi}{n} + a_2\right)\right] \tag{8a}$$

and the generators act as follows,

$$r_1\Lambda = -\bar{\Lambda}, \quad r_2\Lambda = -e^{\frac{-2\pi}{n}i}\bar{\Lambda}, \quad r_1r_2\Lambda = e^{\frac{2\pi}{n}i}\Lambda, \quad r_2r_1\Lambda = e^{-\frac{2\pi}{n}i}\Lambda. \tag{8b}$$

It is clear from equation (8b) that $r_1^2 = r_2^2 = (r_1r_2)^n = (r_2r_1)^n = 1$.

The 2*n* vertices of the polygon can be determined as

$$(r_1r_2)^k\Lambda = e^{\frac{2\pi k}{n}i}\Lambda, \quad (r_1r_2)^k(r_1\Lambda) = -e^{\frac{2\pi k}{n}i}\bar{\Lambda}, \quad k = 1,2,...,n. \tag{9}$$

For $a_1 = a_2 = a$ the 2*n* vertices of a regular polygon of edge length $\sqrt{2}a$ are given in terms of complex numbers by a simple formula

$$\frac{a}{\sqrt{2}\cos(\frac{(n-1)\pi}{n})}e^{i\frac{\pi}{2n}(n-1+4k)}, \quad \frac{-a}{\sqrt{2}\cos(\frac{(n-1)\pi}{n})}e^{i\frac{\pi}{2n}(1-n+4k)}, \quad k = 1,2,...,n. \tag{10}$$

3. Construction of the quasi regular polygons and the tiling of the plane

3.1. $n=2$ with $D_2 = C_2 \times C_2$ symmetry

Two simple roots $\alpha_1 = \sqrt{2}$ and $\alpha_2 = \sqrt{2}i$, representing the Coxeter diagram $A_1 \oplus A_1$, are orthogonal to each other where the generators form the Klein's four-group $D_2 = C_2 \times C_2$. A general orbit is obtained by acting the group elements on the vector $\Lambda = \frac{1}{\sqrt{2}}(a_1 + ia_2)$. For $\Lambda = \omega_1 \equiv (10)$ and $\Lambda = \omega_2 \equiv (01)$, the orbits are two segments of straight lines perpendicular to each other. The orbit $O(\Lambda = \omega_1 + \omega_2) \equiv O(11)$ involves the vectors $\pm\omega_1 = \pm\frac{1}{\sqrt{2}}$, $\pm\omega_2 = \pm\frac{1}{\sqrt{2}}i$ which form a square. For all rank-2 Coxeter groups when the orbit is derived from the vector $\Lambda = a(\omega_1 + \omega_2) \equiv a(11)$, where a is an arbitrary real number, then the polygon has an additional symmetry. It is the symmetry of the Coxeter-Dynkin diagram which can be defined by the generator $\gamma : \alpha_1 \leftrightarrow \alpha_2$ leading to a larger symmetry $W(g):C_2$ where (:) denotes the semi-direct product of two groups. In the above case the group is $D_4 \approx (C_2 \times C_2):C_2 \approx C_4 : C_2'$ of order 8.

When we consider the most general case, namely, $\Lambda = a_1\omega_1 + a_2\omega_2$ the orbit represents a rectangle of sides $\sqrt{2}a_1$, $\sqrt{2}a_2$. The rectangle is an isogonal polygon with the D_2 symmetry. The dual of the rectangle is a rhombus (an isotoxal polygon) whose vertices can be determined by taking the mid point of one of the edge of the rectangle, say, the vector $a_1\omega_1$. We take the other vector $\lambda\omega_2$ bisecting the edge of length $\sqrt{2}a_1$ of the rectangle. To determine the dual of the rectangle the line joining these vectors $\lambda\omega_2 - a_1\omega_1$ must be orthogonal to the vector $\Lambda = a_1\omega_1 + a_2\omega_2$ which determines the scale factor $\lambda = \frac{a_1^2}{a_2}$ the vertices of which consist of two fundamental orbits $O(a_1 0) = \{\pm a_1\omega_1\}$ and $\lambda O(10) = \{\pm\lambda\omega_2\}$ representing a rhombus. The rectangle is vertex transitive under the Klein's group $C_2 \times C_2$ and its dual rhombus is edge transitive. For the values $a_1 = 1$ and $a_2 = 2$ the rectangle and its dual rhombus are given in figure 2.

(a) (b)

Figure 2. The rectangle (a) and its dual rhombus (b) possessing the symmetry $D_2 = C_2 \times C_2$.

3.2 $n=3$ with $W(A_2) \approx D_3 \approx S_3$ symmetry

The Coxeter group $W(A_2) \approx D_3 \approx S_3$ consists of six elements. Three elements r_1, r_2, $r_1r_2r_1 = r_2r_1r_2$ represent reflections with respect to the lines orthogonal to the roots α_1, α_2, $\alpha_1 + \alpha_2$ respectively and the rotational group elements 1, r_1r_2 *and* $(r_1r_2)^2$ represent the cyclic group C_3 which rotates the system by 120^0. We have two fundamental orbits $O(10) = \{\omega_1, \omega_2 - \omega_1, -\omega_2\}$ and $O(01) = \{\omega_2, -\omega_1, \omega_1 - \omega_2\}$. Each orbit represents an equilateral triangle, dual to each other which are transformed to each other by the Coxeter-Dynkin diagram symmetry.

The orbit $O(11) = O(\omega_1 + \omega_2)$ represents a regular hexagon which has a larger symmetry $D_3 : C_2 \approx D_6$ because of the diagram symmetry $\gamma : \alpha_1 \leftrightarrow \alpha_2$. Now we discuss the orbit obtained from the vector $\Lambda = a_1\omega_1 + a_2\omega_2$ where $a_1 \neq a_2 \neq 0$. The Coxeter group $W(A_2)$ generates the six elements of the orbit $O(\Lambda) = O(a_1, a_2)$ arranged in the counter clockwise order as follows:

$$O(a_1, a_2) = \{\Lambda_1,\ \Lambda_2,\ \Lambda_3,\ \Lambda_4,\ \Lambda_5,\ \Lambda_6\}$$

where

$$\begin{array}{l} \Lambda_1 = a_1\omega_1 + a_2\omega_2,\ \ \Lambda_2 = -a_1\omega_1 + (a_1 + a_2)\omega_2,\ \ \Lambda_3 = -(a_1 + a_2)\omega_1 + a_1\omega_2 \\ \Lambda_4 = -a_2\omega_1 - a_1\omega_2,\ \Lambda_5 = a_2\omega_1 - (a_1 + a_2)\omega_2,\ \Lambda_6 = (a_1 + a_2)\omega_1 - a_2\omega_2 \end{array}. \tag{11}$$

They can also be obtained as complex numbers from equation (10) by substituting n=3. These vectors represent the vertices of an isogonal hexagon with 120^0 interior angles and the alternating edge lengths $\sqrt{2}a_1$ and $\sqrt{2}a_2$.

The dual of the isogonal hexagon is an isotoxal hexagon in which the edges are equal, however, it has two different alternating interior angles α and β such that $\alpha + \beta = 240^0$. The vertices of the isotoxal hexagon lie on two fundamental orbits $O(a_1 0)$ and $\lambda O(01)$ .The scale factor λ is determined by the relation

$$(\lambda\omega_2 - a_1\omega_1).(a_1\omega_1 + a_2\omega_2){=}0 \quad \Rightarrow \lambda = \frac{a_1(2a_1 + a_2)}{a_1 + 2a_2}. \tag{12}$$

The vertices of the isotoxal hexagon can then be written in the counter clockwise order as follows:

$$\{B_1 = a_1\omega_1,\ B_2 = \lambda\omega_2,\ B_3 = a_1(-\omega_1 + \omega_2), B_4 = -\lambda\omega_1,\ B_5 = -a_1\omega_2,\ B_6 = \lambda(\omega_1 - \omega_2)\}. \tag{13}$$

Defining $\eta = \dfrac{\lambda}{a_1}$ one can check that the edge length of the isotoxal hexagon is given by

$$a_1[\frac{2}{3}(\eta^2 - \eta + 1)]^{\frac{1}{2}} \tag{14}$$

with two different interior angles given by

$$\alpha = \cos^{-1}\frac{2\eta^2 - 2\eta - 1}{2(\eta^2 - \eta + 1)},\quad \beta = \cos^{-1}\frac{-\eta^2 - 2\eta + 2}{2(\eta^2 - \eta + 1)}. \tag{15}$$

One can prove that for any value of $\eta, \alpha + \beta = 240^0$. The isogonal hexagon and its dual isotoxal hexagon are shown in figure 3 for the values $a_1 = 1$ and $a_2 = 2$.

(a) (b)

Figure 3. The isogonal hexagon (a) and its dual isotoxal hexagon (b) possessing the D_3 symmetry.

One can tile the plane with regular and isogonal hexagons provided two isogonal hexagons and one regular hexagon share the same vertex regardless of the values of a_1 and a_2. One type of tiling of the plane is shown in figure 4(a) with the values $a_1 = 1$ and $a_2 = 2$. Another tiling is shown in figure 4 (b) for the isogonal hexagon for $a_1 = 2$ and $a_2 = 1$. In the limit $a_1 \to 0$ and $a_2 = 1$ isogonal hexagon turns out to be an equilateral triangle and such tiling is depicted in figure 4(c). The other extreme limit is shown in figure 4(d) where $a_1 = 1$ and $a_2 \to 0$. The honeycomb lattice corresponds to the case where $a_1 = a_2$ which is shown in the figure 4(e). The honeycomb lattice representing the tiling of the plane with regular hexagons possesses translational invariance, that is to say, it is invariant under the affine Coxeter group $\hat{W}(A_2)$. We anticipate that a neutral state of the graphene consisting of infinite number of carbon atoms can be represented by a tiling similar to the one in figure 4(a) provided one of the parameters represents the double bonds (say a_1) and the other is representing the single bonds (a_2). Experimentally one expects $a_1 \prec a_2$ but nevertheless they are nearly equal to each other contrary to the isogonal hexagon in figure 4(a) which is an exaggerated version of this quasi crystal lattice. As we will discuss in the paper [8] the C_{60} molecule represents tiling of the sphere with isogonal hexagons and the pentagons for it has two bond lengths.

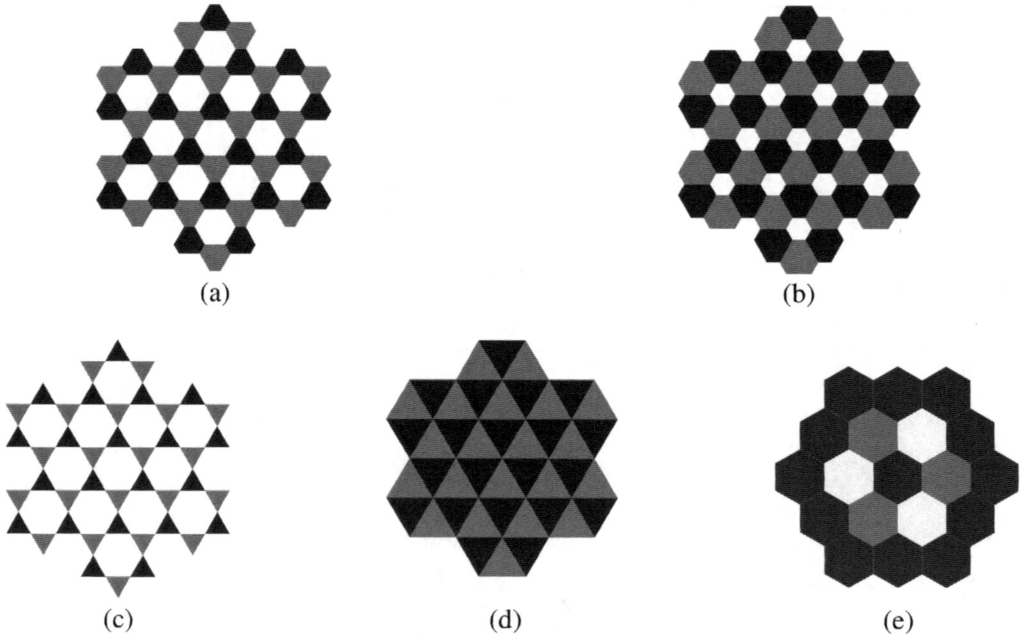

Figure 4. Tiling of the plane with isogonal hexagons-regular hexagons (a-b), regular hexagons and triangles (c), tiling with triangles (d) and the honeycomb lattice (e).

Tiling of the plane can also be made with the isotoxal hexagons by joining its two vertices with two alternating angles α and β so that one obtains an exterior angle of 120^0 at each vertex. This way we create a regular hexagon surrounded by six isotoxal hexagons as shown in figure 5. This tiling is the dual of the tiling in figure 4(a). In this tiling all hexagons have the same edge lengths. However at each vertex the three angles satisfy, as expected, the relation $\alpha + \beta + 120^0 = 360^0$. Here again when

$\eta \to 1$ we obtain the honeycomb lattice of figure 4(a).This tiling can be considered as a conducting graphene model allowing slight modifications of the bond angles.

Figure 5. Tiling of the plane with isotoxal and regular hexagons.

3.3 n= 4 with $W(B_2) \approx D_4$ symmetry

We can repeat similar arguments raised for the case n=3. Here also we have two fundamental orbits

$$O(10) = \{\omega_1, -\omega_1 + \sqrt{2}\omega_2, -\omega_1, \omega_1 - \sqrt{2}\omega_2\} \quad (16a)$$

$$O(01) = \{\omega_2, -\omega_2 + \sqrt{2}\omega_1, -\omega_2, \omega_2 - \sqrt{2}\omega_1\}. \quad (16b)$$

Each orbit represents a square. The orbit $O(\omega_1 + \omega_2) = O(11)$ is a regular octagon possessing the symmetry $D_4 : C_2 \approx D_8$ .One can tile the plane with regular octagon and the square as shown in figure 6.

Figure 6. Tiling of the plane by regular octagons and squares.

The general orbit $O(\Lambda) = O(a_1 a_2)$ with $a_1 \neq a_2 \neq 0$ is an isogonal octagon with interior angles 135^0. The vertices of an isogonal octagon can be generated from the complex number $\Lambda = a_1\omega_1 + a_2\omega_2 = \frac{1}{\sqrt{2}}[a_1 + (a_1 + \sqrt{2}a_2)i]$. The vertices of the isotoxal octagons are determined by finding the factor $\lambda = \frac{a_1(\sqrt{2}a_1 + a_2)}{(a_1 + \sqrt{2}a_2)}$. The vertices of isotoxal octagon is the union of the orbits $O(a_1 0)$ and $\lambda O(01)$. The alternating angles satisfy the relation $\alpha + \beta = 270^0$. The isogonal octagon with values $a_1 = 1$ and $a_2 = 2$ and its dual isotoxal octagon are depicted in figure 7.

(a) (b)
Figure 7. The isogonal octagon (a) and its dual isotoxal octagon (b).

The tiling of the plane with two isogonal octagons and one square at each vertex is shown in figure 8(a). The tiling of the plane with two isotoxal octagons and one square at one vertex is shown in figure 8(b).

(a) (b)
Figure 8. Aperiodic tiling of the plane by isogonal octagons (a) and isotoxal octagons (b) with squares.

3.4 n =5 with $W(H_2) \approx D_5$ symmetry

This case corresponds to the Coxeter group $W(H_2) \approx D_5$, the dihedral group of order 10. Here the Cartan matrix and its inverse corresponding to the Coxeter diagram H_2 can be written in terms of the golden ratio $\tau = \frac{1+\sqrt{5}}{2}$ and $\sigma = \frac{1-\sqrt{5}}{2}$ as

$$C = \begin{bmatrix} 2 & -\tau \\ -\tau & 2 \end{bmatrix} \text{ and } C^{-1} = \frac{1}{2+\sigma}\begin{bmatrix} 2 & \tau \\ \tau & 2 \end{bmatrix}. \tag{17}$$

The orbit of a general vector $\Lambda = a_1\omega_1 + a_2\omega_2$ can be obtained as the set of 10 vectors $O(a_1, a_2) = D_5(a_1, a_2)$. The fundamental orbits are the regular pentagons which are the duals of each other. The regular decagon is obtained by letting $a_1 = 1$ and $a_2 = 1$. The regular decagon is both vertex and edge transitive under the dihedral group $W(H_2) : C_2 \approx D_{10}$ because of the diagram symmetry.

The tiling of the plane with five-fold symmetry is introduced by Penrose [9] since the plane cannot be tiled only with regular pentagons. It has been recently shown that the Islamic tiling of the plane [10] with five-fold symmetry dates back to the medieval time. The Islamic architecture used five different tiles, decagon, pentagon, rhombus, nonregular hexagon and bow tie. Here we give in figure 9 one of those tilings of the plane with regular decagons and bow ties.

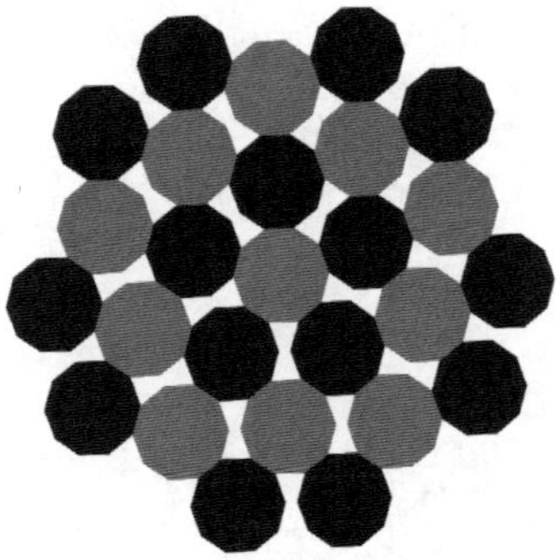

Figure 9. The tiling of the plane by regular decagons and bow ties.

One can also construct an isogonal decagon represented by alternating edge lengths $\sqrt{2}a_1$ and $\sqrt{2}a_2$ with $a_1 \neq a_2 \neq 0$. Its interior angles are all equal to 144^0. One such isogonal decagon is shown in figure 10(a) corresponding to the alternating edge lengths for $a_1 = 1$ and $a_2 = 2$. The dual of the isogonal decagon is the isotoxal decagon with equal edge lengths but with alternating angles $\alpha + \beta = 288^0$. The isotoxal decagon is shown in figure 10(b).

(a) (b)

Figure 10. The isogonal decagon (a) and its dual the isotoxal decagon (b).

One type of aperiodic tiling of the plane with isogonal decagons is shown in figure 11.

Figure 11. Tiling of the plane with isogonal decagons.

4. Conclusion

We have displayed a method to construct the quasi regular polygons and their duals using the rank-2 Coxeter diagrams. The isogonal polygons with *2n* sides and their duals isotoxal polygons possess the dihedral symmetry D_n. It is tempting to suggest that the tiling of the plane by the isogonal hexagon and regular hexagon may represent a state of graphene where double bond and single bonds can be represented by two edges of the isogonal hexagons. We have constructed a number of isogonal polygons with their duals possessing various dihedral symmetries.The corresponding tilings of the plane with the tiles chosen as isogonal and isotoxal polygons are studied.

References

[1] Novoselov K S, Geim A K, Morozov S V, Jiang D, Zhang Y, Dubonos SV, Gregorieva I V and Forsov AA 2004 *Science*, **306** 666; Geim A K and Novoselov K S 2007 *Nature Materials* **6** 183
[2] Wallace P R 1947 *Phys. Rev.* **71** 622
[3] Senechal M 1989 *Introduction to the Mathematics of Quasicrystals* ed Marko V. Jaric (Academic Press) p 1
[4] Janner A and Janssen T 1980 *Acta Cryst.* **A36** 399; ibid **A36** 408
[5] Coxeter H S M and Moser W O J 1965 *Generators and relations for discrete groups* (Springer Verlag) ; Coxeter H S M 1973 *Regular complex polytopes* (Cambridge: Cambridge University Press)
[6] Carter R W 1972 *Simple groups of lie type* (John Wiley & Sons Ltd.); Humphreys J E 1990 *Reflection groups and coxeter groups* (Cambridge University Press, Cambridge)
[7] Slansky R 1981 *Phys. Rep.***79**, 1
[8] Koca M, Al-Ajmi M and Shidani S 2010 Quasi regular polyhedra and their duals with Coxeter symmetries represented by quaternions II *arXiv:1006.2973*
[9] Penrose R 1974 *Bull.Inst. Math. Appl.* **10**, 266; Penrose R 1989 *Introduction to the Mathematics of Quasicrystals* vol 2 ed Marko V. Jaric (Academic Press) p 53
[10] Lu P J and Steinhardt PJ 2007 *Science* **315**, 1106

Journal of Physics: Conference Series **284** (2011) 012040 doi:10.1088/1742-6596/284/1/012040

4d-polytopes described by Coxeter diagrams and quaternions

Mehmet Koca

Sultan Qaboos University
College of Science, Physics Department
P.O Box 36, Al-Khoudh, 123 Muscat
Sultanate of Oman

E-mail: kocam@squ.edu.om

Abstract 4D-polytopes and their dual polytopes can be described as the orbits of the rank-4 Coxeter-Weyl groups. Their symmetries follow from the quaternionic descriptions of the rank-4 Coxeter-Dynkin diagrams. There exists a one to one correspondence between the finite subgroups of quaternions and the rank-4 Coxeter-Weyl groups.

1. Introduction

Discovery of the Platonic solids; tetrahedron, cube, octahedron, icosahedron and dodecahedron dates back to the people of Scotland lived 1000 years earlier than the ancient Greeks and the models curved on the stones are now kept in the Ashmolean Museum at Oxford [1]. Plato associated tetrahedron with fire, cube with earth, air with octahedron, and water with icosahedron. Archimedes discovered the semi-regular convex solids and several centuries later they were rediscovered by the renaissance mathematicians. By introducing prisms and anti-prisms as well as four regular non-convex polyhedra, Kepler completed the work in 1620. Nearly two centuries later, in 1865, Catalan constructed the dual solids of the Archimedean solids now known as Catalan solids [2]. Extensions of the platonic solids to 4D dimensions have been made in 1855 by L. Schlaffli [3] and their generalizations to higher dimensions in 1900 by T.Gosset [4]. Further important contributions are made by W. A. Wythoff [5] among many others and in particular by the contemporary mathematicians H.S.M. Coxeter [6] and J.H. Conway [7].

The 3D and 4D convex polytopes single out as compared to the polytopes in higher dimensions. The number of Platonic solids is five in 3D and there exist six regular polytopes in 4D contrary to the higher dimensional cases where there exist only three platonic polytopes which are the generalizations of tetrahedron, octahedron and cube to higher dimensions. The Platonic and Archimedean solids [8] as well as the Catalan solids [9] can be described with the rank-3 Coxeter groups $W(A_3)$, $W(B_3)$ and $W(H_3)$. The 4D polytopes are described by the rank-4 Coxeter groups $W(A_4)$, $W(B_4)$, $W(H_4)$ and the group $W(F_4)$.

This paper studies the regular and semi regular 4D polytopes as the orbits of the rank-4 Coxeter groups expressed in terms of quaternions. There is a one-to-one correspondence between the finite subgroups of quaternions and the symmetries of the regular and semi regular 4D polytopes. In Section 2 we introduce the finite subgroups of quaternions. Constructions of the rank-4 Coxeter groups in terms of quaternions are given in Section 3. The analysis of the 4D polytopes requires a study of the 3D polyhedra which will be

Published under licence by IOP Publishing Ltd

introduced in Section 4. Examples of the 4D polytopes and their dual polytopes will be given in Section 5. Concluding remarks are made in Section 6.

2. Finite subgroups of quaternions

In this section we introduce the quaternions and its relevance to the orthogonal transformations in 4 dimensions and give the list of finite subgroups of quaternions.

2.1. Quaternions and O (4) Transformations

Let $q = q_0 + q_i e_i$, (i=1,2,3) be a real unit quaternion with its conjugate defined by $\bar{q} = q_0 - q_i e_i$ and the norm $q\bar{q} = \bar{q}q = 1$. The imaginary units satisfy

$$e_i e_j = -\delta_{ij} + \varepsilon_{ijk} e_k, \; (i, j, k = 1, 2, 3). \tag{1}$$

Let p, q be unit quaternions and r represents an arbitrary quaternion. Then the transformations [10]

$$r \rightarrow prq : [p, q]; \quad r \rightarrow p\bar{r}q : [p, q]^* \tag{2}$$

define the orthogonal group $O(4)$ which preserves the norm $r\bar{r} = \bar{r}r$. The first term in (2) represents a proper rotation and the second includes also the reflection, generally called rotary reflection. In particular, the group element

$$r \rightarrow -p\bar{r}p : [p, -p]^* \tag{3}$$

represents the reflection with respect to the hyperplane orthogonal to the quaternion p.

2.2. *Finite Subgroups of Quaternions*

The finite subgroups of quaternions are well known and its classification can be found in the Coxeter's book on regular complex polytopes and in the du Val's book on quaternions [11]. They are given as follows.

(a) Cyclic group of order $2n$ is generated by $\langle p = \exp(e_1 \frac{\pi}{n}) \rangle$ and dicyclic group of order $4n$ is generated by the generators $\langle p = \exp(e_1 \frac{\pi}{n}), e_2 \rangle$.

(b) The binary tetrahedral group can be represented by the set of 24 unit quaternions:

$$T = \{ \pm 1, \pm e_1, \pm e_2, \pm e_3, \frac{1}{2}(\pm 1 \pm e_1 \pm e_2 \pm e_3) \}. \tag{4}$$

(c) The binary octahedral group consists of 48 unit quaternions. Let the set $T' = (V_1 \oplus V_2 \oplus V_3)$ represents the 24 unit quaternions where

$$V_1 = \{\tfrac{1}{\sqrt{2}}(\pm 1 \pm e_1), \tfrac{1}{\sqrt{2}}(\pm e_2 \pm e_3)\},\; V_2 = \{\tfrac{1}{\sqrt{2}}(\pm 1 \pm e_2), \tfrac{1}{\sqrt{2}}(\pm e_3 \pm e_1)\} \tag{5}$$

$$V_3 = \{\tfrac{1}{\sqrt{2}}(\pm 1 \pm e_3), \tfrac{1}{\sqrt{2}}(\pm e_1 \pm e_2)\}.$$

Then the union of the set $O = T \oplus T'$ represents the binary octahedral group.

(d) The binary icosahedral group $I = \langle b, c \rangle$ of order 120 can be generated by two unit quaternions $b = \frac{1}{2}(\tau + \sigma e_1 + e_2) \in 12(1)_+$ and $c = \frac{1}{2}(\tau - \sigma e_1 + e_2) \in 12(1)_+$. We display its elements in table 1 as

Table 1. Conjugacy classes of the binary icosahedral group I represented by quaternions

Order of the elements	Conjugacy classes denoted by the number of elements and order of elements
1	1
2	-1
10	$12_+ : \frac{1}{2}(\tau \pm e_1 \pm \sigma e_3), \frac{1}{2}(\tau \pm e_2 \pm \sigma e_1), \frac{1}{2}(\tau \pm e_3 \pm \sigma e_2)$
5	$12_- : \frac{1}{2}(-\tau \pm e_1 \pm \sigma e_3), \frac{1}{2}(-\tau \pm e_2 \pm \sigma e_1), \frac{1}{2}(-\tau \pm e_3 \pm \sigma e_2)$
10	$12'_+ : \frac{1}{2}(\sigma \pm e_1 \pm \tau e_2), \frac{1}{2}(\sigma \pm e_2 \pm \tau e_3), \frac{1}{2}(\sigma \pm e_3 \pm \tau e_1)$
5	$12'_- : \frac{1}{2}(-\sigma \pm e_1 \pm \tau e_2), \frac{1}{2}(-\sigma \pm e_2 \pm \tau e_3), \frac{1}{2}(-\sigma \pm e_3 \pm \tau e_1)$
6	$20_+ : \frac{1}{2}(1 \pm e_1 \pm e_2 \pm e_3), \frac{1}{2}(1 \pm \tau e_1 \pm \sigma e_2), \frac{1}{2}(1 \pm \tau e_2 \pm \sigma e_3), \frac{1}{2}(1 \pm \tau e_3 \pm \sigma e_1)$
3	$20_- : \frac{1}{2}(-1 \pm e_1 \pm e_2 \pm e_3), \frac{1}{2}(-1 \pm \tau e_1 \pm \sigma e_2), \frac{1}{2}(-1 \pm \tau e_2 \pm \sigma e_3), \frac{1}{2}(-1 \pm \tau e_3 \pm \sigma e_1)$
4	$30 : \pm e_1, \pm e_2, \pm e_3, \frac{1}{2}(\pm \sigma e_1 \pm \tau e_2 \pm e_3), \frac{1}{2}(\pm \sigma e_2 \pm \tau e_3 \pm e_1), \frac{1}{2}(\pm \sigma e_3 \pm \tau e_1 \pm e_2)$

In table 1 we used the golden ratio $\tau = \frac{1+\sqrt{5}}{2}$ and $\sigma = \frac{1-\sqrt{5}}{2}$ which satisfy

$$\tau + \sigma = 1, \quad \tau\sigma = -1, \quad \tau^2 = \tau + 1, \quad \sigma^2 = \sigma + 1. \tag{6}$$

3. Constructions of the rank-4 Coxeter-Weyl groups by finite subgroups of quaternions

Let $I_2(n)$ denotes the Coxeter diagram representing two vectors with the angle $\frac{2\pi}{n}$ between them. Then the Coxeter diagram $I_2(n) \oplus I_2(n)$ of rank-4 is represented by the figure 1.

Figure 1. The Coxeter diagram $I_2(n) \oplus I_2(n)$.

Let $p = \exp(e_1 \frac{\pi}{n})$ and $q = \exp(e_1 \frac{\pi}{n})e_2$ be two orthogonal unit quaternions. Then the following set of quaternions describes the root system of the diagram of figure1

$$I_2(n) \oplus I_2(n) = \{p^k, q^k\}; \; k = 1, 2, ..., 2n. \tag{7}$$

If $s,t \in (I_2(n) \oplus I_2(n))$ are arbitrary elements of the root system then the group $Aut(I_2(n) \oplus I_2(n)) = \{[s,t] \oplus [s,t]^*\}$ of order $4n \times 4n$ is represented by the elements of the dicyclic group.

The F_4 diagram shown in figure 2 with its quaternionic roots [12] leads to its automorphism group $Aut(F_4) = W(F_4) : \gamma = \{[O,O] \oplus [O,O]^*\}$ which is of the order 48×48=2304. Here γ is the Dynkin diagram symmetry generator.

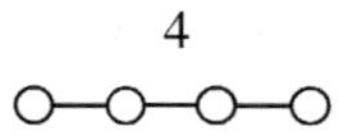

Figure 2. The Coxeter diagram of F_4.

The automorphism group of H_4 with its quaternionic simple roots [13] is given as $W(H_4) = \{[I,I] \oplus [I,I]^*\}$ with the order of 120×120=14,400. The rank-4 Coxeter-Weyl groups $W(B_4)$ of order 384 and $W(D_4)$ of order 192 are the subgroups of the group $W(F_4)$. On the other hand the Coxeter-Weyl group $W(A_4) \approx S_5$ of order 120 is a subgroup of the group $W(H_4)$ [13].

4. Rank-3 Coxeter-Weyl groups and polyhedra

These groups are discussed extensively in the references [8-9]. Quaternionic representations of these groups can be classified as follows:

(a) Icosahedral group: $W(H_3) = \{[I,\bar{I}] \oplus [I,\bar{I}]^*\} \approx A_5 \times C_2$

(b) Octahedral group: $W(B_3) \approx Aut(A_3) = \{[T,\bar{T}] \oplus [T',\bar{T}'] \oplus [T,\bar{T}]^* \oplus [T',\bar{T}']^*\}$

(c) Tetrahedral group: $W(A_3) = \{[T,\bar{T}] \oplus [T',\bar{T}']^*\}$

(d) Symmetry of prisms: $W(I_2(n) \oplus A_1) \approx D_n \times C_2$.

Below we display some polyhedra as the orbits of these Coxeter groups. Let α_i and ω_i denote respectively the set of simple roots and basis vectors of the dual space satisfying the relations

$$(\alpha_i, \omega_j) = \delta_{ij} ,\ (\alpha_i, \alpha_j) = C_{ij} ,\ (\omega_i, \omega_j) = (C^{-1})_{ij} . \tag{8}$$

Define an arbitrary vector in the dual space by $\Lambda = a_1\omega_1 + a_2\omega_2 + a_3\omega_3 \equiv (a_1 a_2 a_3)$ and the $\Lambda - orbit$ by $O(\Lambda) = W(G)\Lambda$.

The Platonic and Archimedean solids as well as Catalan solids can be described as the orbits of the rank-3 Coxeter groups. Some examples are given below.

4.1. The tetrahedral group $W(A_3) \approx S_4$

The orbits $O(100) = \frac{1}{2}(\pm e_1 \pm e_2 \pm e_3)$ with even number of (-) sign and $O(001) = \frac{1}{2}(\pm e_1 \pm e_2 \pm e_3)$ with odd number of (-) sign represent two tetrahedra dual to each other. The union of two tetrahedra constitutes a cube. Similarly the orbits $O(110)$ and $O(011)$ represent two truncated tetrahedra.

Journal of Physics: Conference Series **284** (2011) 012040 doi:10.1088/1742-6596/284/1/012040

4.2. The octahedral group $W(B_3) \approx S_4 \times C_2$

The orbit $O(100) = (\pm e_1, \pm e_2, \pm e_3)$ represents an octahedron and the orbit $O(001) = \frac{1}{2}(\pm e_1 \pm e_2 \pm e_3)$ represents a cube which is the dual of octahedron. All non-chiral Archimedean solids with octahedral symmetry can be constructed as the orbits of the octahedral group.

4.3. The icosahedral group $W(H_3) \approx A_5 \times C_2$

The platonic as well as Archimedean solids possessing the icosahedral symmetry can be constructed as the orbits of the Coxeter group $W(H_3)$. For example, two dual platonic solids, the icosahedron and dodecahedron, are represented by the orbits $O(100)$ and $O(001)$ respectively.

The Catalan solids, as we already stated, are the duals of the Archimedean solids [9]. Two examples which are well known in crystallography and viral structures are given below.

(a) Dual of the cuboctahedron possessing the octahedral symmetry is the rhombic dodecahedron. Its 14 vertices is the union of the orbits $(100) \oplus \frac{1}{\sqrt{2}} O(001)$ which is represented in figure 3.

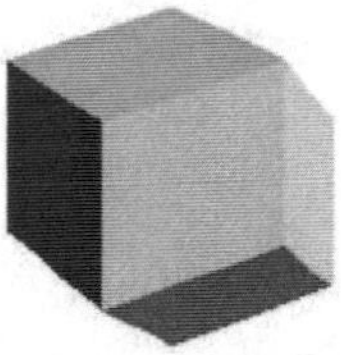

Figure 3. The rhombic dodecahedron.

(b) Dual of the icosidodecahedron is the rhombic triacontahedron**.** The 32 vertices is the union of the orbits $O(100) \oplus \sqrt{\frac{\tau+2}{3}} O(001)$ which is displayed in figure 4.

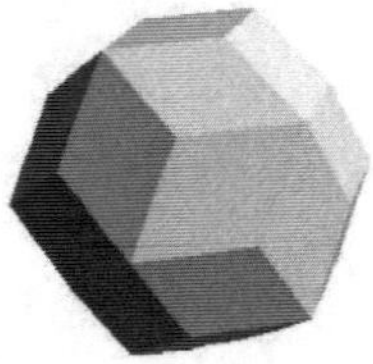

Figure 4. The rhombic triacontahedron.

5. Rank-4 Coxeter groups and 4D Polytopes

5.1. The Coxeter group $W(B_4)$ *of order 384*

This group is an extension of the octahedral group $W(B_3)$ to four dimensions. As it is expected the generalizations of the octahedron and the cube in terms of quaternions are straightforward. The hyper octahedron, also known as (16-cell), **is** given by the orbit $O(1000) = \{\pm 1, \pm e_1, \pm e_2, \pm e_3\}$**.** It consists of 16 tetrahedra like $\{1, e_1, e_2, e_3\}$. Similarly the hyper cube, known as 8-cell, is represented by the orbit $O(1000) = \frac{1}{2}(\pm 1 \pm e_1 \pm e_2 \pm e_3)$. It consists of 8 cubic cells. Hyper octahedron and hyper cube are dual to each other as expected.

5.2. *The Coxeter group $W(F_4)$ of order 1152*

It is a unique Coxeter group in the sense that it has no correspondence in any dimensions. It has a self dual polytope with 24 vertices with 24 cells made of octahedra. It is simply called 24-cell which is represented in terms of quaternions either by the orbit

$$O(1000) = T = \{\pm 1, \pm e_1, \pm e_2, \pm e_3, \frac{1}{2}(\pm 1 \pm e_1 \pm e_2 \pm e_3)\} \text{ or by } O(0001) = T' = (V_1 \oplus V_2 \oplus V_3).$$

A Dynkin diagram symmetry of F_4 transforms these two sets of vertices to each other. Thus the polytope 24-cell is said to be self dual.

5.3. The Coxeter group $W(H_4) \subset W(E_8)$

The Coxeter group $W(H_4)$ is a maximal subgroup of the group $W(E_8)$. Its dual platonic polytopes are called 120-cell and 600-cell represented by quaternions as follows:

120- cell: $O(1000) = \sum_{i,j=0}^{4} \oplus p^i T' p^j$ which has 600 vertices. Here $p \in I, p^5 = \pm 1$.

Each cell is a dodecahedron. At each vertex there are 4 dodecahedra. The 600-cell is represented by the set of quaternions I which also represents the binary icosahedral group:

600-cell: $O(0001) = I = \sum_{i=0}^{4} \oplus p^i T$.

Each cell is a tetrahedron. At each vertex there are 20 tetrahedra. When the sphere S^3 representing the 600-cell is sliced by parallel hyperplanes one obtains 3D polyhedra represented by the conjugacy classes of the binary icosahedral group I.

They consist of two points ± 1 corresponding to the poles of the sphere S^3, 4 icosahedra, 2 dodecahedra and one icosidodecahedron which can be readily seen from the table 1. The 120-cell and 600-cell are dual to each other. In addition to these two platonic solids we will study two more polytopes possessing the symmetry $W(H_4)$.

$720-cell = 120 + 600$: This is the orbit represented by $O(0100)$ consisting of 1200 vertices. Its cells are made of 120 icosidodecahedra and 600 tetrahedra. At any vertex there are 3 icosidodecahedra and 2 tetrahedra whose centers are represented respectively by the vectors $\omega_4, r_3 r_4 \omega_4, (r_3 r_4)^2 \omega_4$ up to a scale factor and by the vectors $\omega_1, r_1\omega_1$.

The dual polytope of the 720-cell consists of 720 vertices which is the union of the orbits [14] $\frac{3}{2\sqrt{2}} O(0001) \oplus O(1000)$. Each cell of the dual polytope is a dipyramid with a triangular base. The dual polytope is the projection of the Voronoi cell of E_8 [15] with 19,440 vertices into 4D space. Another semi regular polytope of the Coxeter group $W(H_4)$ has 720 vertices and 720 cells which we denote by $720' - cell = 120 + 600$. The $720' - cell$ is represented by the orbit $O(0010)$. It consists of the cells of 120 icosahedra and 600 octahedra. At each vertex there are 2 icosahedra and 5 octahedra. Their centers can be represented respectively by $\omega_4, r_4\omega_4$ up to a scale factor $\frac{\sqrt{2}}{2+\sigma}$ and $\omega_1, r_1 r_2 \omega_1, (r_1 r_2)^2 \omega_1, (r_1 r_2)^3 \omega_1, (r_1 r_2)^4 \omega_1$. These 7 vertices form a dipyramid with a pentagonal base. The 720 vertices of the dual polytope consist of the set $\frac{\sqrt{2}}{2+\sigma} O(0001) \oplus O(1000)$.

It is also interesting to note that the symmetries of the semi regular 4D polytopes, the grand antiprism and the snub 24-cell, are represented by two maximal subgroups of the Coxeter group $W(H_4)$. We display the maximal subgroups of the Coxeter group $W(H_4)$ in table 2 [13].

Table 2. Maximal subgroups of the Coxeter group $W(H_4)$.

Maximal subgroup	Order	Polytope
$Aut(A_2 \oplus A_2)$	144	
$Aut(H_2 \oplus H_2)$	400	Grand antiprism with 100 vertices
$W(H_3 \oplus A_1)$	240	
$Aut(A_4)$	240	
$W(D(4))/C_2 : S_3$	576	Snub 24-cell with 96 vertices

One of these semi regular polytope is the Grand Antiprism (GA) which is first determined by a computer calculation by Conway and Guy in 1965 [16] .This semi regular polytope has 100 vertices which can be obtained as a subset of the set of quaternions I (600-cell) under the decomposition of its maximal subgroup $Aut(H_2 \oplus H_2)$ [17]. Let us take the quaternions

$$b = \frac{1}{2}(\tau + \sigma e_1 + e_2) \in 12(1)_+ \text{ and } c = \frac{1}{2}(\tau - \sigma e_1 + e_2) \in 12(1)_+ .$$

Then the vertices of the **GA** are given by the set of quaternions

$$\text{GA}=\{b^m c b^n, b^m e_3 c b^n\}, \quad m,n=0,1,...,9 \tag{9}$$

which consist of 100 elements of the set of quaternions I from which the set of 20 quaternionic roots of the Coxeter diagram $H_2 \oplus H_2' = \{b^m, \ e_3 b^m\}$; $m=0,1,...,9$ are removed. The GA has 100 vertices, 320 cells (20 pentagonal antiprisms+300 tetrahedra), 500 edges and 720 faces (700 triangles+20 pentagons). Dual Polytope of the GA consists of 320 vertices which are given as the orbits of the group $Aut(H_2 \oplus H_2)$ as follows:

200 vertices like $J_1 = \sum_{i,j=0}^{4} \oplus b^i V_1 b^j$; 100 vertices like $J_3' = \sum_{i,j=0}^{4} \oplus b^i \frac{1}{\sqrt{2}}(\pm 1 \pm e_3) b^j$ and

20 vertices like $\{\frac{\tau}{\sqrt{2}} b^m, \ \frac{\tau}{\sqrt{2}} e_3 b^m\}$, $m = 0,1,...,9$.

The typical cell of the dual polytope has 14 vertices with faces of 4 pentagons, 4 kites and 2 isosceles trapezoids as shown in figure 5.

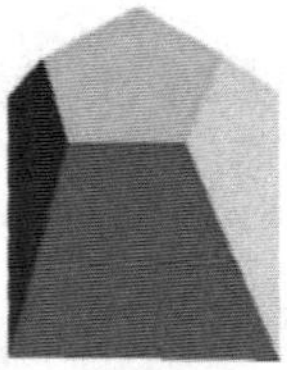

Figure 5. A typical cell of the dual polytope of the grand antiprism.

Journal of Physics: Conference Series **284** (2011) 012040 doi:10.1088/1742-6596/284/1/012040

Another interesting semi regular polytope is the snub 24-cell. The snub 24-cell [18] consists of 96 vertices $S = I - T = \sum_{i=1}^{4} p^i T$ with the symmetry group $\left(W(D_4)/C_2\right) : S_3 = \{[T,T] \oplus [T,T]^*\}$ of order 576. Its 144 cells consist of 24 icosahedra and 120 tetrahedra. The dual of the snub 24-cell has the vertices $T \oplus T' \oplus S'$ with $S' = \sum_{i=1}^{4} \oplus p^i \bar{p}^{\dagger i} T'$ where $p^\dagger$ is obtained from p by exchanging $\tau \leftrightarrow \sigma$. One cell of the dual polytope is a solid with 8 vertices as shown in figure 6.

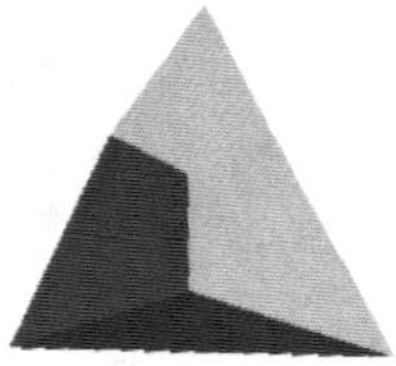

Figure 6. A typical cell of the dual polytope of the snub 24-cell.

5.4. The Coxeter group $W(A_4)$ of order 120

The automorphism group $Aut(A_4) \approx W(A_4):\gamma$ is an extension of the Coxeter group $W(A_4)$ by the Dynkin diagram symmetry γ. It is a maximal subgroup of the group $W(H_4)$ and has its own Platonic and Archimedean polytopes. We discuss the following examples. The 5-cell is an extension of tetrahedron to 4D and can be represented either by the orbit $O(1000)$ or $O(0001)$. It has 5 vertices consisting of 5 tetrahedra 4 of which are meeting at one vertex. Similarly the orbit $O(0100)$ represents the rectified 5-cell consisting of 5 tetrahedral and 5 octahedral cells. The rectified cell has 10 vertices. At each vertex there are 2 tetrahedra and 3 octahedra whose centers respectively consist of the vertices $\omega_1, r_1\omega_1$ and $\omega_4, r_3r_4\omega_4, (r_3r_4)^2\omega_4$ up to a scale factor. These vertices form a dipyramid which is a typical cell of the dual polytope [19]. The vertices of the dual polytope consist of the union of the orbits $O(0100) \oplus \frac{3}{2} O(0001)$. Another polytope with the symmetry group $Aut(A_4)$ is called runcinated 5-cell consisting of 20 vertices and 30 cells composed of 10 tetrahedra and 20 triangular prisms. It is represented by the orbit $O(1001)$. At each vertex there are 2 tetrahedra and 6 triangular prisms. The dual polytope has 30 vertices consisting of the union of the orbits $O(100) \oplus O(0100) \oplus O(0010) \oplus O(0001)$. The dual polytope has 20 cells each made of rhombohedra.

6. Concluding Remarks

We presented a few examples of the 4D polytopes along with their duals. Their cells are made of Platonic and Archimedean polyhedra. A systematic study of the 4D polytopes and their dual polytopes will follow [14, 19].

References

[1] Atiyah M and Sutcliffe P 2003 *Milan Journal of Math.* **71** 33; arXiv:math-ph/0303071 v1 31 March (2003).
[2] Catalan E 1865 *Mémoire sur la Théorie des Polyèdres.* J. l'École Polytechnique (Paris) **41** 1
[3] Schlaffli L 1855 *Journal de Mathematiques* 20 359
[4] Gosset T 1900 *Messenger of Mathematics* 29 43

[5] Wythoff W A 1907 *Koninklijke Akademie van Wetenshappen te Amsterdam, Proc. of the Section of Sciences* **9** 529
[6] Coxeter H S M 1973 *Regular Polytopes, Third Edition, Dover Publications, New York*
[7] Conway J H, Burgiel H and Goodman-Strauss C 2008 *The Symmetries of Things, A K Peters Ltd. Wellesley, Massachusetts* ISBN-13: 978-1-56881-220-5
[8] Koca M, Koç R and Al-Ajmi M 2007 *J. Math. Phys.* **48** 113514
[9] Koca M, Koca N O and Koç R 2010 *J. Math. Phys.* **51** 043501
[10] Conway J H and Smith D A 2003 On *Quaternions and Octonions A K Peters Ltd Natick*, Massachusetts; Koca M, Koç R, Al-Barwani M 2001 *J. Phys. A: Math. & Gen.* **34** 1120
[11] Coxeter H S M 1974 *Regular complex polytopes* (Cambridge Univ. Press); du Val P 1964 *Homographies, Quaternions and Rotations,* (Oxford University Press)
[12] Koca M, Koç R and Al-Barwani M 2003 *J. Math. Phys.* **44** 3123
[13] Koca M, Koç R, Al-Barwani M and Al-Farsi S 2006 *Linear Algebra Appl.* **412** 441
[14] Koca M, Koca N O and Al-Ajmi M $W(H_4)$ *Polytopes and Their Dual Polytopes* (under preparation)
[15] Elser V and Sloane N JA 1987 *J.Phys.A Math. Gen.* **20** 6161
[16] Conway J H and Guy M J T 1965 *Four-Dimensional Archimedean Polytopes*, *Proc. of the Colloquium on Convexity at Copenhagen* p 38-39, *Kobenhavns University Mathematics Institute, Copenhagen*, 1967
[17] Koca M, Koca N O and Al-Ajmi M 2009 *J. Phys.A: Math. Theor.* **42** 495201
[18] Koca M, Al-Ajmi M and Koca N O 2011 *Linear Algebra Appl.* **434** 977
[19] Koca N O and Koca M *4D Polytopes and Their Dual Polytopes of the Coxeter Group* $W(A_4)$ *Represented by Quaternions* (under preparations)

Journal of Physics: Conference Series **284** (2011) 012041 doi:10.1088/1742-6596/284/1/012041

Free field approach to diagonalization of boundary transfer matrix : recent advances

Takeo Kojima

Department of Mathematics and Physics, Graduate School of Science and Engineering, Yamagata University, Jonan 4-3-16, Yonezawa 992-8510, Japan

E-mail: kojima@yz.yamagata-u.ac.jp

Abstract. We diagonalize in nitely many commuting operators $T_B(z)$. We call these operators $T_B(z)$ the boundary transfer matrix associated with the quantum group and the elliptic quantum group. The boundary transfer matrix is related to the solvable model with a boundary. When we diagonalize the boundary transfer matrix, we can calculate the correlation functions for the solvable model with a boundary. We review the free eld approach to diagonalization of the boundary transfer matrix $T_B(z)$ associated with $U_q(A_2^{(2)})$ and $U_{q,p}(\widehat{sl_N})$. We construct the free eld realizations of the eigenvectors of the boundary transfer matrix $T_B(z)$. This paper includes new unpublished formula of the eigenvector for $U_q(A_2^{(2)})$. It is thought that this diagonalization method can be extended to more general quantum group $U_q(g)$ and elliptic quantum group $U_{q,p}(g)$.

1. Introduction

We study in nitely many commuting operators $T_B(z)$ that we call the boundary transfer matrix. The boundary transfer matrix is related to the solvable model with a boundary. There have been many developments in solvable models in the last 30 years. Various models were found to be exactly solvable and various methods were invented to solve these models. The Free eld approach is a powerful method to study exactly solvable models [1]. This paper is devoted to the free eld approach to diagonalization of the boundary transfer matrix $T_B(z)$. When we diagonalize the boundary transfer matrix, we can calculate the correlation functions for the solvable model with a boundary. The rst paper on this subject was devoted to the XXZ chain with a boundary [2], in which the boundary transfer matrix $T_B(z)$ acts on the highest representation of the quantum group $U_q(\widehat{sl_2})$. It is thought that this basic theory for the quantum group $U_q(\widehat{sl_2})$ can be extended to the quantum group $U_q(g)$ for arbitrary a ne Lie algebra g. It is thought that the theory on the quantum group $U_q(g)$ can be generalized to those on the elliptic quantum group $U_{q,p}(g)$. In this paper we summarize the generalization on this direction. This paper includes a review on free eld approach to the boundary transfer matrix [2, 3, 4, 5, 6, 7, 8] and new unpublished formula of the boundary state for the quantum group $U_q(A_2^{(2)})$. The plan of this paper is as follows. In section 2 we summarize the results for the quantum group $U_q(A_2^{(2)})$ [4] and give new unpublished formulae of the boundary state associated with nontrivial K-matrix $K_\pm(z)$. In section 3 we review the results for the elliptic quantum group $U_{q,p}(\widehat{sl_N})$, which gives a generalization of the papers [2, 3, 6, 7].

2. Quantum group $U_q(A_2^{(2)})$

In this section we diagonalize the boundary transfer matrix $T_B(z)$ for quantum group $U_q(A_2^{(2)})$.

2.1. Boundary transfer matrix $T_B(z)$

We x q and z such that $0 < |q| < 1$ and $|q^2| < |z| < |q \;\; ^2|$. Let us set the q-integers

$$[a]_q = \frac{q^a \quad q \;\; ^a}{q \quad q \;\; ^1}.$$

We use the abbreviation.

$$(z; p_1, p_2, \cdots, p_M)_\infty = \prod_{k_1, k_2, \cdots, k_M = 0}^{\infty} (1 \quad p_1^{k_1} p_2^{k_2} \cdots p_M^{k_M} z).$$

The R-matrix $R(\zeta)$ for the twisted quantum group $U_q(A_2^{(2)})$ is given by following [9].

$$R(z) = \frac{1}{\kappa(z)} \begin{pmatrix} 1 &&&&&&&& \\ & b(z) && c(z) &&&&& \\ && d(z) && e(z) && f(z) && \\ & zc(z) && b(z) &&&&& \\ && q^2 ze(z) && j(z) && e(z) && \\ &&&&& b(z) && c(z) & \\ && n(z) && q^2 ze(z) && d(z) && \\ &&&&& zc(z) && b(z) & \\ &&&&&&&& 1 \end{pmatrix}. \tag{1}$$

Here we have set

$$b(z) = \frac{q(z \quad 1)}{q^2 z \quad 1}, \quad c(z) = \frac{q^2 \quad 1}{q^2 z \quad 1}, \quad d(z) = \frac{q^2(z \quad 1)(qz+1)}{(q^2 z \quad 1)(q^3 z+1)}, \quad e(z) = \frac{q^{\frac{1}{2}}(z \quad 1)(q^2 \quad 1)}{(q^2 z \quad 1)(q^3 z+1)},$$
$$f(z) = \frac{(q^2 \quad 1)((q^3+q)z \quad (q \quad 1))}{(q^2 z \quad 1)(q^3 z+1)}, \quad n(z) = \frac{(q^2 \quad 1)((q^3 \quad q)z+(q^2+1))}{(q^2 z \quad 1)(q^3 z+1)},$$
$$j(z) = \frac{q^4 z^2 + (q^5 \quad q^4 \quad q^3+q^2+q \quad 1)z \quad q}{(q^2 z \quad 1)(q^3 z+1)},$$
$$\kappa(z) = z \frac{(q^6 z; q^6)_\infty (q^2/z; q^6)_\infty (\;\; q^5 z; q^6)_\infty (\;\; q^3/z; q^6)_\infty}{(q^6/z; q^6)_\infty (q^2 z; q^6)_\infty (\;\; q^5/z; q^6)_\infty (\;\; q^3 z; q^6)_\infty}.$$

Let $\{v_+, v_0, v \;\; \}$ denote the natural basis of $V = \mathbf{C}^3$. When viewed as an operator on $V \otimes V$, the matrix element of $R(z)$ are de ned by $R(z) v_{k_1} \otimes v_{k_2} = \sum_{j_1, j_2 = \pm, 0} v_{j_1} \otimes v_{j_2} R(z)_{j_1, j_2}^{k_1, k_2}$, where the ordering of the index is given by $(+,+), (+,0), (+, \;\;), (0,+), (0,0), (0, \;\;), (\;\; ,+), (\;\; ,0), (\;\; , \;\;)$. The R-matrix $R(z)$ satis es the Yang-Baxter equation.

$$R_{1,2}(z_1/z_2) R_{1,3}(z_1/z_3) R_{2,3}(z_2/z_3) = R_{2,3}(z_2/z_3) R_{1,3}(z_1/z_3) R_{1,2}(z_1/z_2). \tag{2}$$

The R-matrix $R(z)$ is characterized by the intertwiner as the quantum group $U_q(A_2^{(2)})$. We set the normalization function $\kappa(z)$ such that the minimal eigenvalue of the corner transfer matrix becomes 1 [20]. There exist three diagonal solutions of the K-matrix for the quantum group $U_q(A_2^{(2)})$. The K-matrix $K_\epsilon(z)$, $(\epsilon = \pm, 0)$ are given by following [12].

$$K_0(z) = \frac{\varphi_0(z)}{\varphi_0(z \;\; ^1)} \begin{pmatrix} 1 && \\ & 1 & \\ && 1 \end{pmatrix}, \quad K_\pm(z) = \frac{\varphi_\pm(z)}{\varphi_\pm(z \;\; ^1)} \begin{pmatrix} z^2 && \\ & \frac{\pm\sqrt{\quad 1} q^{\frac{3}{2}} + z}{\pm\sqrt{\quad 1} q^{\frac{3}{2}} + z \;\; ^1} & \\ && 1 \end{pmatrix}. \tag{3}$$

Journal of Physics: Conference Series **284** (2011) 012041 doi:10.1088/1742-6596/284/1/012041

where

$$\varphi_0(z)=\frac{(q^8z;q^{12})_\infty(\ q^9z^2;q^{12})_\infty}{(q^{12}z;q^{12})_\infty(\ q^5z^2;q^{12})_\infty},\varphi_\pm(z)=\frac{(\pm\sqrt{\ 1}q^{\frac{9}{2}}z;q^6)_\infty(\mp\sqrt{\ 1}q^{\frac{7}{2}}z;q^6)_\infty}{(\pm\sqrt{\ 1}q^{\frac{1}{2}}z;q^6)_\infty(\mp\sqrt{\ 1}q^{\frac{3}{2}}z;q^6)_\infty}\varphi_0(z). \tag{4}$$

The K-matrix $K(z)=K_\epsilon(z)$ satis es the boundary Yang-Baxter equation in $\mathrm{End}(V\otimes V)$ [10].

$$K_2(z_2)R_{2,1}(z_1z_2)K_1(z_1)R_{1,2}(z_1/z_2)=R_{2,1}(z_1/z_2)K_1(z_1)R_{1,2}(z_1z_2)K_2(z_2). \tag{5}$$

We set the normalization function $\varphi_\epsilon(z)$, $(\epsilon=\pm,0)$ such that the minimal eigenvalue of the boundary transfer matrix $T_B(z)$ becomes 1. The Izergin-Korepin model associated with the identity solution $\bar{K}_0(z)=id$ was studied in [4]. In this paper we give the free eld realization of the boundary state for nontrivial solutions $K_\pm(z)$. In the gapless regime $|q|=1$ the Izergin-Korepin model with the nontrivial solutions $K_\epsilon(z)$, $(\epsilon=\pm,0)$ was studied in [5].

Let us introduce the vertex operators $\Phi_\epsilon(z)$, $(\epsilon=\pm,0)$ [13, 15, 16] that satisfy the following commutation relation

$$\Phi_{\epsilon_2}(z_2)\Phi_{\epsilon_1}(z_1)=\sum_{\epsilon_1',\epsilon_2'=\pm,0}R_{\epsilon_1\epsilon_2}^{\epsilon_1'\epsilon_2'}(z_1/z_2)\Phi_{\epsilon_1'}(z_1)\Phi_{\epsilon_2'}(z_2). \tag{6}$$

The vertex operator $\Phi_\epsilon(z)$ and its dual $\Phi_\epsilon^*(z)=q^{\ \frac{\epsilon}{2}}\Phi_{\ \epsilon}(\ q^{\ 3}z)$, $(\epsilon=\pm,0)$ satisfy the inversion relation

$$g\Phi_{\epsilon_1}^*(z)\Phi_{\epsilon_2}(z)=\delta_{\epsilon_1,\epsilon_2}id, \tag{7}$$

where $g=\frac{1}{1+q}\frac{(q^2;q^6)_\infty(\ q^3;q^6)_\infty}{(q^6;q^6)_\infty(\ q^5;q^6)_\infty}$. Let us introduce the boundary transfer matrix $T_B(z)$ for $U_q(A_2^{(2)})$. The boundary transfer matrix $T_B(z)$ is given by

$$T_B(z)=g\sum_{\epsilon,\epsilon'=\pm,0}\Phi_\epsilon^*(z^{\ 1})K_\epsilon^{\epsilon'}(z)\Phi_{\epsilon'}(z). \tag{8}$$

The boundary transfer matrix $T_B(z)$ is related to the boundary Izergin-Korepin model [4, 5]. From the commutation relations of the vertex operators (6) and the boundary Yang-Baxter equation (5), the commutativity of the boundary transfer matrix $T_B(z)$ is ensured.

$$[T_B(z_1),T_B(z_2)]=0,\quad \text{for any } z_1,z_2. \tag{9}$$

We call the eigenvector $|B\rangle_\epsilon,(\epsilon=\pm,0)$ with eigenvalue 1 the boundary state.

$$T_B(z)|B\rangle_\epsilon=|B\rangle_\epsilon. \tag{10}$$

From the de nition of $\varphi_\epsilon(z)$, the boundary state $|B\rangle$ is the eigenvector with the minimal eigenvalue. In the following section we give the free eld realization of the boundary state $|B\rangle_\epsilon$ for the diagonal boundary K-matrix $K_\epsilon(z)$, $(\epsilon=\pm,0)$. The boundary state $|B\rangle_\epsilon$ for the identity K-matrix $K_0(z)$ was constructed in [4]. Let us introduce the type-II vertex operators $\Psi\ (\xi)$, $(\ =\pm,0)$ that satisfy the following commutation relation.

$$\Phi_\epsilon(z)\Psi\ (\xi)=\ (z/\xi)\Psi\ (\xi)\Phi_\epsilon(z),\quad (\epsilon,\ =\pm,0), \tag{11}$$

where

$$(z)=z^{\ 1}\frac{\Theta_{q^6}(q^5z)\Theta_{q^6}(\ q^4z)}{\Theta_{q^6}(q^5z^{\ 1})\Theta_{q^6}(\ q^4z^{\ 1})},\quad \Theta_p(z)=(p;p)_\infty(z;p)_\infty(pz^{\ 1};p)_\infty.$$

Multiplying the type-II vertex operators $\Psi\ (\xi)$ to the boundary state $|B\rangle_\epsilon$,

$$|\xi_1, \xi_2, \cdots, \xi_N\rangle_{\ 1,\ 2, \cdots,\ N} = \Psi_{\ 1}(\xi_1)\Psi_{\ 2}(\xi_2)\cdots\Psi_{\ N}(\xi_N)|B\rangle_\epsilon, \tag{12}$$

we have many eigenvectors.

$$T_B(z)|\xi_1, \xi_2, \cdots, \xi_N\rangle_{\ 1,\ 2, \cdots,\ N} = \prod_{s=1}^{N} \quad (z/\xi_s) \ (\quad 1/q^3 z\xi_s)|\xi_1, \xi_2, \cdots, \xi_N\rangle_{\ 1,\ 2, \cdots,\ N}.$$

The vectors $\{|\xi_1, \cdots, \xi_N\rangle_{\ 1, \cdots,\ N}\}$ are the basis of the space of the state of the boundary Izergin-Korepin model.

2.2. Free eld realization

In this section we give the free eld realization of the boundary state $|B\rangle_\epsilon$. Let us introduce bosons $a_m, (m \in \mathbf{Z}_{\neq 0})$ as following [13, 14, 15, 16].

$$[a_m, a_n] = \delta_{m+n}\frac{[m]_q}{m}([2m]_q \quad (\quad 1)^m [m]_q). \tag{13}$$

Let us set the zero-mode operators P, Q by

$$[a_m, P] = [a_m, Q] = 0, \ [P, Q] = 1. \tag{14}$$

Level-1 irreducible highest representation $V(\ _1)$ of $U_q(A_2^{(2)})$ is realized by

$$V(\ _1) = \mathbf{C}[a \ _1, a \ _2, \cdots] \oplus_{n\in\mathbf{Z}} e^{nQ}| \ _1\rangle, \quad | \ _1\rangle = e^{\frac{Q}{2}}|0\rangle.$$

The vacuum vector $|0\rangle$ is characterized by

$$a_m|0\rangle = 0, \ (m > 0), \quad P|0\rangle = 0.$$

Let us set the auxiliary operators $P(z), Q(z), R \ (w), S \ (w)$ by

$$P(z) = \sum_{m>0} \frac{a \ _m q^{\frac{9m}{2}} z^m}{[2m]_q \quad (\quad 1)^m [m]_q}, \qquad Q(z) = \sum_{m>0} \frac{a_m q^{\ \frac{7m}{2}} z^{\ \ m}}{[2m]_q \quad (\quad 1)^m [m]_q},$$

$$R \ (w) = \sum_{m>0} \frac{a \ _m}{[m]_q} q^{\frac{m}{2}} w^m, \qquad S \ (w) = \sum_{m>0} \frac{a_m}{[m]_q} q^{\frac{m}{2}} w^{\ \ m}.$$

Let us set $\epsilon(q) = ([2]_{q^{\frac{1}{2}}})^{\frac{1}{2}}$. Let us set the current,

$$X \ (w) = \epsilon(q) e^{R^-(w)} e^{S^-(w)} e^{\ \ Q} w^{\ \ P+\frac{1}{2}}.$$

The free eld realizations of the vertex operators $\Phi_\epsilon(z)$ [13, 15, 16] are given by

$$\Phi \ (z) = \frac{1}{\epsilon(q)} e^{P(z)} e^{Q(z)} e^{Q} (\quad zq^4)^{P+\frac{1}{2}},$$

$$\Phi_0(z) = \oint_{C_1} \frac{dw}{2 \ \sqrt{\ \ 1} w} \frac{(q^2 \quad 1)}{q^4 z(1 \quad qw/z)(1 \quad qz/w)} : \Phi \ (z) X \ (q^4 w) :,$$

$$\Phi_+(z) = \oint\oint_{C_2} \frac{dw_1}{2 \ \sqrt{\ \ 1} w_1} \frac{dw_2}{2 \ \sqrt{\ \ 1} w_2} \frac{q^{1/2}(1 \quad q^2)^2}{q^4 z^2 w_1 w_2}$$

$$\frac{(w_1 \quad w_2)^2 (q(w_1 + w_2) \quad (1 + q^2)z)}{(1 + qw_1/w_2)(1 + qw_2/w_1)(1 \quad qw_1/z)(1 \quad qz/w_1)(1 \quad qw_2/z)(1 \quad qz/w_2)}$$

$$: \Phi \ (z) X \ (q^4 w_1) X \ (q^4 w_2) : .$$

The integrand contour C_1 encircles $w = 0, qz$ but not $w = q \ ^1 z$. The integrand contour C_2 encircles $w_1 = 0, qz, qw_2$ and $w_2 = 0, qz, qw_1$ but not $w_1 = q \ ^1 z, q \ ^1 w_2$ and $w_2 = q \ ^1 z, q \ ^1 w_1$. The free eld realization of type-II vertex operators $\Psi \ (\xi)$ are given as similar way [13]. Now we have the free eld realization of the boundary transfer matrix $T_B(z)$, using those of the vertex operators. We construct the free eld realization of the boundary state $|B\rangle_\epsilon$, analyzing those of the boundary transfer matrix $T_B(z)$. The following is **main result** of this section. The free eld realization of the boundary states $|B\rangle_\epsilon$, $(\epsilon = \pm, 0)$ are given by

$$|B\rangle_\epsilon = e^{F_\epsilon} e^{\ \frac{Q}{2}} |0\rangle, \quad (\epsilon = \pm, 0). \tag{15}$$

Here we have set

$$\begin{aligned} F_\epsilon &= \frac{1}{2} \sum_{m>0} \frac{m q^{8m}}{[2m]_q \quad (\ 1)^m [m]_q} a^2 \ _m \\ &+ \sum_{m>0} \left\{ \ _m \left(\frac{(q^m \quad q^{\ m} \quad (\ 1)^m) q^{8m}}{[2m]_q \quad (\ 1)^m [m]_q} \right) \quad \frac{(\epsilon \sqrt{\ 1})^m q^{3m}}{[2m]_q \quad (\ 1)^m [m]_q} \right\} a \ _m . \end{aligned} \tag{16}$$

Here we have used $\ _m(x) = \begin{cases} x, & m : \text{even} \\ 0, & m : \text{odd} \end{cases}$. The boundary state $|B\rangle_0$ for the identity K-matrix $\bar{K}_0(z) = id$ was constructed in [4]. The realizations of $|B\rangle_\epsilon$ for the nontrivial K-matrix $K_\epsilon(z)$ are new. Multiplying the type-II vertex operators $\Psi \ (\xi)$ to the boundary state $|B\rangle_\epsilon$, we get the diagonalization of the boundary transfer matrix $T_B(z)$. It is thought that this method can be extended to the case of the a ne quantum group $U_q(g)$.

3. Elliptic quantum group $U_{q,p}(\widehat{sl_N})$

In this section we diagonalize the boundary transfer matrix $T_B(z)$ associated with the elliptic quantum group $U_{q,p}(\widehat{sl_N})$ [8]. It gives a generalization of the papers [2, 3, 6, 7].

3.1. Boundary transfer matrix

Let us set the integer $N = 2, 3, \cdots$. We assume that $0 < x < 1$ and $r \quad N + 2$ $(r \in \mathbf{Z})$. We set $z = x^{2u}, x = e^{\ \pi i / r}$. We set the elliptic theta function $[u]$ by

$$[u] = x^{\frac{u^2}{r} \ u} \Theta_{x^{2r}}(x^{2u}), \quad \Theta_q(z) \quad = \quad (q; q)_\infty (z; q)_\infty (q/z; q)_\infty .$$

Let $\epsilon \ (1 \leq \quad \leq N)$ be the orthonormal basis of $\mathbf{R}^N$ with the inner product $(\epsilon \ |\epsilon \ ) = \delta \ _, $. Let us set $\bar{\epsilon} \ = \epsilon \qquad \epsilon$ where $\epsilon = \frac{1}{N} \sum_{=1}^N \epsilon$. Note that $\sum_{=1}^N \bar{\epsilon} \ = 0$. Let $\quad (1 \leq \quad \leq N \quad 1)$ the simple root : $\quad = \bar{\epsilon} \qquad \bar{\epsilon}_{\ +1}$. Let $\omega \ (1 \leq \quad \leq N \quad 1)$ be the fundamental weights, which satisfy

$$(\ \ |\omega \ ) = \delta \ _, \ , \quad (1 \leq \ , \ \leq N \quad 1).$$

Explicitly we set $\omega \ = \sum_{\ =1} \bar{\epsilon}$. The type $A_{N \ 1}$ weight lattice is the linear span of $\bar{\epsilon}$ or ω .

$$P = \sum_{=1}^{N \ 1} \mathbf{Z} \bar{\epsilon} \ = \sum_{=1}^{N \ 1} \mathbf{Z} \omega \ .$$

For $a \in P$ we set a and $a \ _,$ by

$$a \ _, \ = a \qquad a \ , \quad a \ = (a + \rho | \bar{\epsilon} \ ), \quad (\ , \ \in P).$$

Here we set $\rho = \sum_{=1}^{N\ 1} \omega$. Let us set the restricted path $P^+_{r\ N}$ by

$$P^+_{r\ N} = \{a = \sum_{=1}^{N\ 1} c\ \omega\ \in P | c\ \in \mathbf{Z}, c\quad 0, \sum_{=1}^{N\ 1} c\ \leq r\quad N\}.$$

For $a \in P^+_{r\ N}$, condition $0 < a_{\ ,}\ < r$, $(1 \leq\ <\ \leq N\quad 1)$ holds.

We recall elliptic solutions of the Yang-Baxter equation of face type. An ordered pair $(b, a) \in P^2$ is called admissible if and only if there exists $\ (1 \leq\ \leq N)$ such that $b\quad a = \bar{\epsilon}$. An ordered set of four weights $(a, b, c, d) \in P^4$ is called an admissible con guration around a face if and only if the ordered pairs (b, a), (c, b), (d, a) and (c, d) are admissible. Let us set the Boltzmann weight functions $W\begin{pmatrix} c & d \\ b & a \end{pmatrix}\Big| u$ associated with admissible con guration $(a, b, c, d) \in P^4$ [11]. For $a \in P^+_{r\ N}$, we set

$$W\begin{pmatrix} a + 2\bar{\epsilon} & a + \bar{\epsilon} \\ a + \bar{\epsilon} & a \end{pmatrix}\Big| u = R(u), \tag{17}$$

$$W\begin{pmatrix} a + \bar{\epsilon}\ + \bar{\epsilon} & a + \bar{\epsilon} \\ a + \bar{\epsilon} & a \end{pmatrix}\Big| u = R(u)\frac{[u][a_{\ ,}\quad 1]}{[u\quad 1][a_{\ ,}\ ]}, \tag{18}$$

$$W\begin{pmatrix} a + \bar{\epsilon}\ + \bar{\epsilon} & a + \bar{\epsilon} \\ a + \bar{\epsilon} & a \end{pmatrix}\Big| u = R(u)\frac{[u\quad a_{\ ,}\ ][1]}{[u\quad 1][a_{\ ,}\ ]}. \tag{19}$$

The normalizing function $R(u)$ is given by

$$R(u) = z^{\frac{r-1}{r}\frac{N-1}{N}}\frac{\varphi(z\ \ 1)}{\varphi(z)}, \quad \varphi(z) = \frac{(x^2 z; x^{2r}, x^{2N})_\infty (x^{2r+2N\ \ 2} z; x^{2r}, x^{2N})_\infty}{(x^{2r} z; x^{2r}, x^{2N})_\infty (x^{2N} z; x^{2r}, x^{2N})_\infty}.$$

Because $0 < a_{\ ,}\ < r$ $(1 \leq\ <\ \leq N\quad 1)$ holds for $a \in P^+_{r\ N}$, the Boltzmann weight functions are well de ned. The Boltzmann weight functions satisfy the Yang-Baxter equation of the face type.

$$\sum_g W\begin{pmatrix} d & e \\ c & g \end{pmatrix}\Big| u_1\ W\begin{pmatrix} c & g \\ b & a \end{pmatrix}\Big| u_2\ W\begin{pmatrix} e & f \\ g & a \end{pmatrix}\Big| u_1\quad u_2$$

$$= \sum_g W\begin{pmatrix} g & f \\ b & a \end{pmatrix}\Big| u_1\ W\begin{pmatrix} d & e \\ g & f \end{pmatrix}\Big| u_2\ W\begin{pmatrix} d & g \\ c & b \end{pmatrix}\Big| u_1\quad u_2\ . \tag{20}$$

We set the normalization function $\varphi(z)$ such that the minimal eigenvalue of the corner transfer matrix becomes 1 [20]. An order set of three weights $(a, b, g) \in P^3$ is called an admissible con guration at a boundary if and only if the ordered pairs (g, a) and (g, b) are admissible. Let us set the boundary Boltzmann weight functions $K\left(\begin{matrix} & a \\ g & \\ & b \end{matrix}\Big| u\right)$ for admissible weights (a, b, g) as following [12].

$$K\left(\begin{matrix} & a \\ a + \bar{\epsilon} & \\ & b \end{matrix}\Big| u\right) = z^{\frac{r-1}{r}\frac{N-1}{N}\ \ \frac{2}{r}a_1}\frac{h(z)}{h(z\ \ 1)}\frac{[c\quad u][a_{1,}\ + c + u]}{[c + u][a_{1,}\ + c\quad u]}\delta_{a,b}. \tag{21}$$

In this paper, we consider the case of continuous parameter $0 < c < 1$. The normalization function $h(z)$ is given by following [8].

$$h(z) = \frac{(x^{2r+2N\ \ 2}/z^2; x^{2r}, x^{4N})_\infty (x^{2N+2}/z^2; x^{2r}, x^{4N})_\infty}{(x^{2r}/z^2; x^{2r}, x^{4N})_\infty (x^{4N}/z^2; x^{2r}, x^{4N})_\infty}$$

$$\frac{(x^{2N+2c}/z;x^{2r},x^{2N})_\infty(x^{2r\ \ 2c}/z;x^{2r},x^{2N})_\infty}{(x^{2N+2r\ \ 2c\ \ 2}/z;x^{2r},x^{2N})_\infty(x^{2c+2}/z;x^{2r},x^{2N})_\infty}$$
$$\prod_{j=2}^{N}\frac{(x^{2r+2N\ \ 2c\ \ 2a_{1,j}}/z;x^{2r},x^{2N})_\infty(x^{2c+2a_{1,j}}/z;x^{2r},x^{2N})_\infty}{(x^{2r+2N\ \ 2c\ \ 2a_{1,j}\ \ 2}/z;x^{2r},x^{2N})_\infty(x^{2c+2+2a_{1,j}}/z;x^{2r},x^{2N})_\infty}. \tag{22}$$

The boundary Boltzmann weight functions and the Boltzmann weight functions satisfy the Boundary Yang-Baxter equation.

$$\sum_{f,g} W\begin{array}{cc} c & f \\ b & a\end{array}\Big| u_1\ \ u_2\ \ W\begin{array}{cc} c & d \\ f & g\end{array}\Big| u_1+u_2\ \ K\left(\begin{array}{ccc} & g & \\ f & & \\ & a & \end{array}\Big| u_1\right)K\left(\begin{array}{ccc} & e & \\ d & & \\ & g & \end{array}\Big| u_2\right)$$
$$=\sum_{f,g} W\begin{array}{cc} c & d \\ f & e\end{array}\Big| u_1\ \ u_2\ \ W\begin{array}{cc} c & f \\ b & g\end{array}\Big| u_1+u_2\ \ K\left(\begin{array}{ccc} & e & \\ f & & \\ & g & \end{array}\Big| u_1\right)K\left(\begin{array}{ccc} & g & \\ b & & \\ & a & \end{array}\Big| u_2\right). \tag{23}$$

We set the normalization function $h(z)$ such that the minimal eigenvalue of the boundary transfer matrix $T_B(z)$ becomes 1.

The vertex operator $\Phi^{(b,a)}(z)$ and the dual vertex operator $\Phi\ ^{(a,b)}(z)$ associated with the elliptic quantum group $U_{q,p}(\widehat{sl_N})$, are the operators which satisfy the following commutation relations

$$\Phi^{(a,b)}(z_1)\Phi^{(b,c)}(z_2) = \sum_g W\begin{array}{cc} a & g \\ b & c\end{array}\Big| u_2\ \ u_1\ \ \Phi^{(a,g)}(z_2)\Phi^{(g,c)}(z_1), \tag{24}$$
$$\Phi\ ^{(a,b)}(z_1)\Phi\ ^{(b,c)}(z_2) = \sum_g W\begin{array}{cc} c & b \\ g & a\end{array}\Big| u_2\ \ u_1\ \ \Phi\ ^{(a,g)}(z_2)\Phi\ ^{(g,c)}(z_1), \tag{25}$$
$$\Phi^{(a,b)}(z_1)\Phi\ ^{(b,c)}(z_2) = \sum_g W\begin{array}{cc} g & c \\ a & b\end{array}\Big| u_1\ \ u_2\ \ \Phi\ ^{(a,g)}(z_2)\Phi^{(g,c)}(z_1), \tag{26}$$

and the inversion relation

$$\Phi^{(a,g)}(z)\Phi\ ^{(g,b)}(z) = \delta_{a,b}. \tag{27}$$

We de ne the boundary transfer matrix $T_B(z)$ for the elliptic quantum group $U_{q,p}(\widehat{sl_N})$.

$$T_B(z) = \sum_{=1}^{N}\Phi\ ^{(a,a+\epsilon\ )}(z\ ^1)K\left(\begin{array}{ccc} & a & \\ a+\bar{\epsilon} & & \\ & a & \end{array}\Big| u\right)\Phi^{(a+\epsilon\ ,a)}(z). \tag{28}$$

From the commutation relations of the vertex operators (24), (25), (26), and the boundary Yang-Baxter equation (23), the boundary $T_B(z)$ commute with each other.

$$[T_B(z_1), T_B(z_2)] = 0, \quad \text{for any } z_1, z_2. \tag{29}$$

We call the eigenvector $|B\rangle$ with the eigenvalue 1 the boundary state.

$$T_B(z)|B\rangle = |B\rangle. \tag{30}$$

Let us introduce the type-II vertex operators $\Psi\ ^{(b,a)}(z)$ by

$$\Phi^{(d,c)}(z_1)\Psi\ ^{(b,a)}(z_2) = \chi(z_2/z_1)\Psi\ ^{(b,a)}(z_2)\Phi^{(d,c)}(z_1), \tag{31}$$
$$\Phi\ ^{(c,d)}(z_1)\Psi\ ^{(b,a)}(z_2) = \chi(z_1/z_2)\Psi\ ^{(b,a)}(z_2)\Phi\ ^{(c,d)}(z_1). \tag{32}$$

where we have set

$$\chi(z) = z^{\ \frac{N-1}{N}} \frac{\Theta_{x^{2N}}(\ xz)}{\Theta_{x^{2N}}(\ xz^{\ 1})}.$$

We set the vectors $|\xi_1, \xi_2, \cdots, \xi_M\rangle_{\ 1,\ 2,\cdots,\ M}$ $(1 \leq \ _1,\ _2, \cdots,\ _M \leq N)$.

$$|\xi_1, \xi_2, \cdots, \xi_M\rangle_{\ 1,\ 2,\cdots,\ M} \tag{33}$$
$$= \Psi^{\ (b+\epsilon\ _1+\epsilon\ _2+\cdots+\epsilon\ _M, b+\epsilon\ _2+\cdots+\epsilon\ _M)}(\xi_1)\cdots\Psi^{\ (b+\epsilon\ _{M-1}+\epsilon\ _M, b+\epsilon\ _M)}(\xi_{M\ 1})\Psi^{\ (b+\epsilon\ _M, b)}(\xi_M)|B\rangle.$$

Now we have many eigenvectors of $T_B(z)$.

$$T_B(z)|\xi_1, \xi_2, \cdots, \xi_M\rangle_{\ 1,\ 2,\cdots,\ M} = \prod_{j=1}^{M} \chi(\xi_j/z)\chi(1/\xi_j z)\ |\xi_1, \xi_2, \cdots, \xi_M\rangle_{\ 1,\ 2,\cdots,\ M}.$$

The vectors $|\xi_1, \xi_2, \cdots, \xi_M\rangle_{\ 1,\ 2,\cdots,\ M}$ are the basis of the space of the state of the boundary $U_{q,p}(\widehat{sl_N})$ face model.

3.2. Free eld realization

In this section we give the free eld realizations of the boundary state $|B\rangle$. Let us introduce the bosons $\ ^i_m, (i = 1, 2, \cdots, N\ \ 1; m \in \mathbf{Z})$ as following [17].

$$[\ ^j_m,\ ^k_n] = \begin{cases} m\frac{[(r\ \ 1)m]_x}{[rm]_x}\frac{[(N\ \ 1)m]_x}{[Nm]_x}\delta_{m+n,0} & (j = k) \\ mx^{Nm\ sgn(j\ \ k)}\frac{[(r\ \ 1)m]_x}{[rm]_x}\frac{[m]_x}{[Nm]_x}\delta_{m+n,0} & (j \neq k). \end{cases} \tag{34}$$

Let us set $\ ^N_m$ by $\sum_{j=1}^{N} x^{\ 2jm}\ ^j_m = 0$. The above commutation relations are valid for all $1 \leq j, k \leq N$. We also introduce the zero-mode operators P_α, Q_α, $(\ \in P)$ by

$$[\sqrt{\ 1}P_\alpha, Q_\beta] = (\ |\ ),\ \ (\ ,\ \ \in P). \tag{35}$$

In what follows we deal with the bosonic Fock space $\mathcal{F}_{l,k}$, generated by $\ ^j_m (m > 0)$ over the vacuum vector $|l, k\rangle$, where $l = b + \rho$, $k = a + \rho$ for $a \in P_r^+\ _N, b \in P_r^+\ _{1\ N}$.

$$\mathcal{F}_{l,k} = \mathbf{C}[\{\ ^j_1,\ ^j_2, \cdots\}_{j=1,\cdots,N\ \ 1}]|l, k\rangle,\ \ |l, k\rangle = e^{\sqrt{\ 1}\sqrt{\frac{r}{r-1}}Q_l\ \ \sqrt{\ 1}\sqrt{\frac{r-1}{r}}Q_k}|0, 0\rangle.$$

where

$$\ ^j_m|l, k\rangle = 0,\ (m > 0),\ P_\alpha|l, k\rangle = \left(\ \left|\sqrt{\frac{r}{r\ \ 1}}l\ \ \sqrt{\frac{r\ \ 1}{r}}k\right)|l, k\rangle.\right.$$

The commutation relation of bosons $\ ^j_m$ is not symmetric. It is convenient to introduce new generators of bosons $\ ^j_m$ $(m \in \mathbf{Z}_{\neq 0}; 1 \leq j \leq N\ \ 1)$ by

$$\ ^j_m = x^{\ jm}(\ ^j_m\ \ \ ^{j+1}_m). \tag{36}$$

They satisfy the following commutation relations.

$$[\ ^j_m,\ ^k_n] = m\frac{[(r\ \ 1)m]_x}{[rm]_x}\frac{[A_{j,k}m]_x}{[m]_x}\delta_{m+n,0},$$

Journal of Physics: Conference Series **284** (2011) 012041 doi:10.1088/1742-6596/284/1/012041

where $A_{j,k}$ is a matrix element of the Cartan matrix of sl_N type. We give a free eld realization of the vertex operators $\Phi^{(b,a)}(z)$. Let us set the operators $P\ (z), Q\ (z), R^j\ (z), S^j\ (z)$, $(1 \le j \le N\quad 1)$ by

$$P\ (z) = \sum_{m>0} \frac{1}{m}\ {}^{1}{}_{m} z^{m},\ Q\ (z) = \sum_{m>0} \frac{1}{m}\ {}^{1}_{m} z^{\ m},$$

$$R^j\ (z) = \sum_{m>0} \frac{1}{m}\ {}^{j}{}_{m} z^{m},\ S^j\ (z) = \sum_{m>0} \frac{1}{m}\ {}_{m} z^{\ m}.$$

Let us set the basic operators $U(z)$,$F_{\alpha_j}(z)$, $(1 \le j \le N\quad 1)$ on the Fock space $\mathcal{F}_{l,k}$.

$$U(z) = z^{\frac{r-1}{2r}\frac{N-1}{N}} e^{\ \sqrt{\ 1}\sqrt{\frac{r-1}{r}}Q_{\bar{\epsilon}_1}} z^{\ \sqrt{\frac{r-1}{r}}P_{\bar{\epsilon}_1}} e^{P_-(z)} e^{Q_-(z)},$$

$$F_{\alpha_j}(z) = z^{\frac{r-1}{r}} e^{\sqrt{\ 1}\sqrt{\frac{r-1}{r}}Q_{\alpha_j}} z^{\sqrt{\frac{r-1}{r}}P_{\alpha_j}} e^{R^j_-(z)} e^{S^j_-(z)}.$$

In what follows we set $l = b + \rho, k = a + \rho,\ (a \in P^+_{r\ N}, b \in P^+_{r\ N\ 1})$ and $=$ $\sqrt{r(r\quad 1)}P_{\epsilon}\ ,\quad , =\quad$. Then acts on $\mathcal{F}_{l,k}$ as an integer ($\epsilon\quad \epsilon\ |rl\quad (r\quad 1)k$). We give the free eld realization of the vertex operators $\Phi^{(a+\epsilon\ ,a)}(z)$, $(1 \le\ \ \le N\quad 1)$ [17] by

$$\Phi^{(a+\epsilon_1,a)}(z_0^{\ 1}) = U(z_0),$$

$$\Phi^{(a+\epsilon\ ,a)}(z_0^{\ 1}) = \oint \cdots \oint \prod_{j=1}^{\ 1} \frac{dz_j}{2\ iz_j} U(z_0)F_{\alpha_1}(z_1)F_{\alpha_2}(z_2)\cdots F_{\alpha\ -1}(z\quad 1)$$

$$\prod_{j=1}^{\ 1} \frac{[u_j\quad u_{j\ 1} + \frac{1}{2}\qquad {}_{j,\ }]}{[u_j\quad u_{j\ 1}\quad \frac{1}{2}]}.$$

Here we set $z_j = x^{2u_j}$. We take the integration contour to be simple closed curve that encircles $z_j = 0, x^{1+2rs}z_{j\ 1}, (s \in \mathbf{N})$ but not $z_j = x^{\ 1\ \ 2rs}z_{j\ 1}, (s \in \mathbf{N})$ for $1 \le j \le\qquad 1$. The $\Phi^{(a+\epsilon\ ,a)}(z)$ is an operator such that $\Phi^{(a+\epsilon\ ,a)}(z) : \mathcal{F}_{l,k} \to \mathcal{F}_{l,k+\epsilon}$. The free eld realization of the dual vertex operator $\Phi\ ^{(a,b)}(z)$ and the type-II vertex operator $\Psi\ ^{(a,b)}(z)$ are given by similar way [17, 18]. Now we have the free eld realization of the boundary transfer matrix $T_B(z)$, using those of the vertex operators. We construct the free eld realization of the boundary state $|B\rangle$, analyzing those of the transfer matrix $T_B(z)$. The following is **main result** of this section. The free eld realization of the boundary state $|B\rangle$ is given as following [8].

$$|B\rangle = e^F|k,k\rangle. \tag{37}$$

Here we have set

$$F = \frac{1}{2}\sum_{m>0}\sum_{j=1}^{N\ 1}\sum_{k=1}^{N\ 1} \frac{1}{m}\frac{[rm]_x}{[(r\quad 1)m]_x} I_{j,k}(m)\ {}^{j}{}_{m}\ {}^{k}{}_{m} + \sum_{m>0}\sum_{j=1}^{N\ 1}\frac{1}{m}D_j(m)\ {}^{j}{}_{m}, \tag{38}$$

where

$$\begin{aligned} D_j(m) = &\ {}_{m}\left(\frac{[(N\quad j)m/2]_x[rm/2]^+_x x^{\frac{(3j-N-1)m}{2}}}{[(r\quad 1)m/2]_x}\right) \\ &+\frac{x^{(j\ \ 1)m}[(\ \ r+2\ {}_{1,j}+2c\quad j+2)m]_x}{[(r\quad 1)m]_x} \\ &+\frac{[m]_x x^{(r\ \ 2c+2j\ \ 2)m}}{[(r\quad 1)m]_x}\left(\sum_{k=j+1}^{N\ 1} x^{\ \ 2m\pi_{1,k}}\right) \\ &+\frac{x^{(2j\ \ N)m}[(r\quad 2\ {}_{1,N}\quad 2c+N\quad 1)m]_x}{[(r\quad 1)m]_x}, \end{aligned} \tag{39}$$

and

$$I_{j,k}(m) = \frac{[jm]_x[(N \quad k)m]_x}{[m]_x[Nm]_x} = I_{k,j}(m) \quad (1 \leq j \leq k \leq N \quad 1). \tag{40}$$

Here we have used

$$[a]_x^+ = x^a + x^{\ \ a}, \qquad {}_m(x) = \begin{array}{ll} x, & m: \text{ even}, \\ 0, & m: \text{ odd}. \end{array}$$

Multiplying the type-II vertex operators $\Psi^{\ (a,b)}(\xi)$ to the boundary state $|B\rangle$, we get the diagonalization of the boundary transfer matrix $T_B(z)$ on the space of state of the boundary $U_{q,p}(\widehat{sl_N})$ face model. It is thought that this method can be extended to the case of the elliptic quantum group $U_{q,p}(g)$ for a ne Lie algebra g [19].

Acknowledgments

This work is supported by the Grant-in-Aid for Scienti c Research **C** (21540228) from Japan Society for Promotion of Science.

References

[1] Jimbo M and Miwa T 1995 *Algebraic Analysis of Solvable Lattice Models* (American Mathematical Society)
[2] Jimbo M, Kedem R, Kojima T, Konno H and Miwa T 1995 XXZ chain with a boundary *Nucl.Phys.***B441** 437
[3] Furutsu H and Kojima T 2000 The $U_q(\widehat{sl_n})$ analogue of the XXZ chain with a boundary *J.Math.Phys.***41** 4413
[4] Yang W.-L. and Zhang Y.-Z. 2001 Izergin-Korepin model with a boundary *Nucl.Phys.***B596** 495
[5] Kojima T 2010 Boundary Izergin-Korepin model in gapless regime *Preprint* submitted for publication
[6] Miwa T and Weston R 1997 Boundary ABF models *Nucl.Phys.***B486** 517
[7] Kojima T 2010 Diagonalization of boundary transfer matrix for $U_{q,p}(\widehat{sl}(3,\mathbf{C}))$ ABF model *AIP Conf.Proc.* **1243** 241
[8] Kojima T 2010 Diagonalization of in nite transfer matrix of boundary $U_{q,p}(A_{N-1}^{(1)})$ face model *Preprint* [arXiv.1007.3795] accepted for publication in *J.Math.Phys.*
[9] Izergin A and Korepin V 1981 The inverse scattering method approach to the quantum Shabat-Mikhailov model *Commun.Math.Phys.* **79** 303
[10] Sklyanin E 1988 Boundary condition for integrable quantum system *J.Phys.***A21** 2375
[11] Jimbo M, Miwa T and Okado M 1988 Local State Probabilities of Solvable Lattice Models : An $A_{n-1}^{(1)}$ family *Nucl.Phys.***B300** 74
[12] Batchelor M, Fridkin V, Kuniba A and Zhou Y 1996 Solutions of the reflection for face and vertex models associated with $A_n^{(1)}$, $B_n^{(1)}$, $C_n^{(1)}$ and $A_n^{(2)}$ *Phys.Lett.***B376** 266
[13] Matsuno Y 1997 *Thesis (Kyoto University)*
[14] Jing N 1990 Twisted vertex representation of quantum affine algebras *Invent.Math.***102** 663
[15] Jing N and Misra K 1999 Vertex operators for twisted quantum affine algebras *Trans.Amer.Math.Soc.***351** 1663
[16] Hou B, Yang W.-L. and Zhang Y.-Z. 1999 The twisted quantum affine algebra $U_q(A_2^{(2)})$ and correlation functions for the Izergin-Korepin model, *Nucl.Phys.***B556** 485
[17] Asai Y, Jimbo M, Miwa T and Pugai Ya 1996 Bosonization of vertex operators for the $A_{n-1}^{(1)}$ face model *J.Phys.***A29** 501
[18] Furutsu H, Kojima T and Quano Y 2000 Type-II vertex operators for the $A_{n-1}^{(1)}$ face model *Int.J.Mod.Phys.***A15** 1533
[19] Jimbo M, Konno H, Odake S and Shiraishi J 1999 The elliptic algebra $U_{q,p}(\widehat{sl_2})$: Drinfeld currents and vertex operators *Commun.Math.Phys.***199** 605
[20] Baxter R 1982 *Exactly Solved Models in Statistical Mechanics* (Academic Press, London)

Constructing a parasupersymmetric Virasoro algebra

S Kuwata

Department of Frontier Sciences, Hiroshima City University, Asaminami-ku, Hiroshima 731-3194, Japan

E-mail: `kuwata@hiroshima-cu.ac.jp`

Abstract. We construct a para SUSY Virasoro algebra by generalizing the ordinary fermion in SUSY Virasoro algebra (Ramond or Neveu-Schwarz algebra) to the parafermion. First, we obtain a polynomial relation (PR) between di erent-mode parafermion f_i's by generalizing the corresponding single-mode PR to such that is invariant under the unitary transformation of f_i (Green's condition). Di erently from a usual context, where the Green's condition is imposed only on the de ning relation of f_i (degree three with respect to f_i and $f_i^\dagger$), we impose it on any degree of PR. For the case of order-two parafermion (the simplest case of para SUSY), we calculate a PR between the parasupercharge G_0, the bosonic hamiltonian L_0^B and parafermionic one L_0^F, although it is difficult to obtain a PR between G_0 and the total hamiltonian L_0 $(= L_0^B + L_0^F)$. Finally, we construct a para SUSY Virasoro algebra by generalizing L_0 to the L_n's such that form a Virasoro algebra.

1. Introduction

Parafermion (SUSY) conformal eld theory has been rst introduced as an exactly solvable model in statistical systems, where a two-dimensional $\mathbb{Z}_n$ symmetric model is developed [1] by generalizing an ordinary spin to fractional spins (parafermions). In a usual context of para SUSY eld theory, parafermions are formulated using the paraGrassmann variables [2, 3]. Along this line, para SUSY Virasoro algebra has been constructed using paraGrassmann variables [4]. However, the paraGrassmann algebra is not necessarily equivalent to the parafermion algebra in its original sense of the word. Recalling that in quantizing a relativistic quantum eld, the most general statistics is parastatistics [5], we should use the paraparticles themselves to have a physically realizable relativistic eld theory model.

To construct a Virasoro algebra using (para) particle creation and annihilation operators c_i's (Feigin-Fuks realization), it is necessary to obtain a polynomial relation between the c_i's of di erent modes, although it is not so trivial a matter to obtain the polynomial relation. To obtain it, we use the Green's condition [6], where the polynomial relation is required to be invariant under the unitary transformation $c_i \to c_i' = \sum_j U_{ij} c_j$. However, it is found that the Green's condition is incompatible with the paraboson-including Virasoro algebra. This is because the commutation relation between the multi-mode parabose elds cannot be obtained, under the Green's condition, from the corresponding single-mode commutation relation (Calogero-Vasiliev oscillator algebra [7]). Considering that the Green's condition guarantees the realization of a Virasoro algebra, we may state that it is di cult to construct a paraboson-including Virasoro algebra. In this sense, we keep bosons intact, while generalizing fermions to parafermions.

The aim of this article is to construct the para SUSY Virasoro algebra by generalizing the ordinary fermion in SUSY Virasoro algebra to the parafermion f_i. First, we obtain the multi-mode parafermi polynomial relation from the corresponding single-mode relation under the Green's condition. Di erently from a usual context, where the Green's condition is imposed only on the de ning relation of f_i (degree three with respect to f_i and $f_i^\dagger$), we apply the Green's condition to the polynomial relation of any degree. Then we calculate the polynomial relation between the parasupercharge G_0, bosonic hamiltonian L_0^B, and parafermionic one L_0^F. As opposed to the ordinary SUSY case, it seems to be impossible to obtain the polynomial relation between G_0 and the total hamiltonian L_0 $(= L_0^B + L_0^F)$. Finally, we obtain the para SUSY Virasoro algebra by generalizing L_0 to the L_n's (for $n \in \mathbb{Z}$) which form a Virasoro algebra.

2. Para SUSY Virasoro Algebra

In this section, we rst review the Green's condition, under which the parabose single-mode commutation relation cannot be generalized to a multi-mode one. Next, we obtain the multi-mode parafermi polynomial relation for the parafermionic order $p_f = 2$, followed by the calculation of the polynomial relation between G_0, L_0^B, and L_0^F. Finally, we generalize L_0 to the L_n's that form a Virasoro algebra, as in the ordinary SUSY case.

2.1. Green's condition

Denote by c_i and $c_i^\dagger$ the annihilation and creation operators, respectively. The Green's condition is given by [6]

$$\begin{cases} [c_i, [c_j^\dagger, c_k]_\pm] = 2\delta_{ij}c_k, \\ [c_i, [c_j, c_k]_\pm] = 0, \end{cases} \tag{1}$$

for parabosons (upper sign) and parafermion (lower sign). As was pointed out in [8], the relation of Eq. (1) can be derived from the requirement that the relation

$$[c_i, h_j] = \delta_{ij}c_i, \quad \text{where } h_j = \tfrac{1}{2}[c_j^\dagger, c_j]_\pm \tag{2}$$

is invariant under the unitary transformation of $c_i \to c_i' = \sum_j U_{ij}c_j$, where $\sum_j U_{ij}\overline{U}_{kj} = \sum_j \overline{U}_{ji}U_{jk} = \delta_{ik}$. This requirement seems reasonable in the sense that the total hamiltonian $\sum_i h_i$ is invariant under the unitary transformation of the c_i's.

First we deal with the parabosons for a parabosonic order p_b. For the single mode, the commutation relation can be given by the Calogero-Vasiliev oscillator algebra [7]. This algebra can be generalized to such that satis es the relation of Eq. (2) as

$$\begin{cases} [c_i, c_j] = 0, \\ [c_i, c_j^\dagger] = (1 + \ R_i)\delta_{ij}, \end{cases} \tag{3}$$

where $\ = p_b \quad 1$ and $R_i = e^{i\pi(h_i \ \frac{p_b}{2})} = \cos(i\ (h_i \ \frac{p_b}{2}))$. More explicitly, c_i can be written as [9, 10] $c_i = \frac{1}{\sqrt{2}}(x_i + \frac{d}{dx_i} \ \frac{\ }{2x_i}R_i)$, where $[R_i, x_i]_+ = [R_i, \frac{d}{dx_i}]_+ = 0$. However, it is found that Eq. (3) is not compatible with the rst relation of Eq. (1) unless $\ = 0$, due to

$$[c_i, R_j] = 0 \quad (i \neq j) \tag{4}$$

by Eq. (2). If Eq. (3) were compatible with the rst relation of Eq. (1), it would be required that $\ [c_i, R_j]_+ = 0$ (for all i, j), which, together with Eq. (4), leads to $\ c_i = 0$ (due to $R_j^2 = 1$). Thus, Eq. (3) is incompatible with the rst relation of Eq. (1) unless $\ = 0$. Considering that Eq. (1) is su cient for the realization of a Virasoro algebra, as in the ordinary bosonic case, we may state that it is di cult to realize a Virasoro algebra constructed from the paraboson (with a de nite value of p_b). In this sense, we concentrate on the ordinary bosonic case, where $p_b = 1$.

Next, we deal with the parafermion. As opposed to the paraboson, the polynomial relation between the c_i's and $c_i^\dagger$'s can be obtained from the corresponding single-mode polynomial relation by requiring that the polynomial relation is invariant under the unitary transformation of the c_i's (we call this requirement the generalized Green's condition, or the Green's condition for short). Some useful anti-commutators are introduced.

$$[A,B,C]_+ = ABC + CBA,$$
$$[A,B,C,D]_+ = [A,[B,C]_+,D]_+.$$

Notation 1 *For the parafermions,* $f_i = c_i$ *and* $f_{\ i} = c_i^\dagger$ *(for* $i > 0$*).*

In what follows, we concentrate on the case of $p_f = 2$, for simplicity.

Lemma 2 *For* $p_f = 2$*, the polynomial relations between the* f_i*'s are given under the generalized Green's condition by*

$$[f_i, f_j, f_k]_+ = 2(\delta_{i+j,0} f_k + (i \leftrightarrow k)), \tag{5}$$
$$[f_i, f_j, f_k, f_\ell]_+ = (\mathcal{F}_{ijk\ell} + (j \leftrightarrow k)) + (i \leftrightarrow \ell), \tag{6}$$

where $\mathcal{F}_{ijk\ell} = 2\delta_{i+j,0}[f_k, f_\ell]_+ \quad \delta_{i+\ell,0}[f_j, f_k]_+ \quad 4\delta_{i+j,0}\delta_{k+\ell,0} + 4\delta_{i+\ell,0}\delta_{j+k,0}$.

Proof. The corresponding single-mode relations are given by

$$\begin{cases} ff^\dagger f = 2f, \ [f,f,f^\dagger]_+ = 2f, \ f^3 = 0, \\ ff^\dagger f^\dagger f = 2[f,f^\dagger]_+ \quad 4, \end{cases}$$

where the other tetralinear relations such as $f^4 = 0$, $[f^3, f^\dagger]_+ = 0$, $f[f,f^\dagger]_+ f = 4f^2$, $[f,f,f^\dagger,f^\dagger]_+ = 8$ can be derived from the remaning relations. First we show Eq. (5). Under the generalized Green's condition, $[f_i, f_j, f_k]_+$ can be written as in the form $[f_i, f_j, f_k]_+ = a\,\delta_{i+k,0} f_j + b(\delta_{i+j,0} f_k + (i \leftrightarrow k))$, where $a, b \in \mathbb{R}$. The coe cients a and b can be obtained from the single-mode relations as $2b = 4$ and $a + b = 2$. Hence we get $a = 0, b = 2$. Eq. (6) can be obtained in a similar way. □

The relation of Eq. (5), which is found in many references [5, 11], is often derived from the assumption of the Green's ansatz [6], in which the f_i's for a parafermionic order p_f is decomposed as $f_i = \sum_{\alpha=1}^{p_f} f_i^{(\alpha)}$, where the(anti)commutation relations between the $f_i^{(\alpha)}$'s and $f_i^{(\alpha)\dagger}$'s are so chosen that f_i satis es the relation of Eq. (1). However, the representation of Eq. (1) under the Green's ansatz is not an irreducible representation. In the case of $p_f = 2$, for example, the vector space V associated with the representation under the Green's ansatz can be decomposed into $V_1 \oplus V_2$, where V_1 and V_2 are associated with the irreducible representation of f_i for $p_f = 2$ and that for $p_f = 0$, respectively. Actually, the relation of Eq. (5) is satis ed by the irreducible representation of f_i for $p_f = 0$, that is, $f_i = 0$. On the other hand, the relation of Eq. (6) cannot be obtained from the Green's ansatz, because the relation of $f_i = 0$ does not satisfy Eq. (6).

As will be found in the next subparagraph, the relation of Eq. (6) is indispensable to the para SUSY relation for $p_f = 2$.

2.2. Para SUSY

Notation 3 *For the (para)bosons, let the* b_k*'s be given by* $b_k = \sqrt{k} c_k$ *and* $b_{\ k} = \sqrt{k} c_k^\dagger$ *(for* $k > 0$*).*

Notation 4 *Let* G_r*,* L_n^B*, and* L_n^F *(*$r \in \mathbb{Z} + \kappa$*,* $n \in \mathbb{Z}$*) be given by* $G_r = \sum_{k \in \mathbb{Z}} b_k f_{r\ k}$*,* $L_n^B = \frac{1}{4}\sum_{k\in\mathbb{Z}}[b_{n\ k}, b_k]_+$*, and* $L_n^F = \frac{1}{4}\sum_{r\in\mathbb{Z}+\kappa}(r \quad \frac{n}{2})[f_{n\ r}, f_r]$*, where* $\kappa = 0$ *(Ramond type),* $\frac{1}{2}$ *(Neveu-Schwarz type).*

To obtain the polynomial relation between the G_r's, L_n^B's, and L_n^F's, we should specify the polynomial relations between the b_k's and f_r's. Considering the Green's condition of Eq. (1), one may point out that it seems natural to choose the trilinear relations between the b_k's and f_r's, as is found in Ref. [11]. However, we adopt a simpler one: the bilinear commutation relation

$$[b_k, f_r] = 0 \quad (\text{for all } k \in \mathbb{Z},\ r \in \mathbb{Z} + \kappa), \tag{7}$$

as in the ordinary SUSY.

In the rest of this subsection, we deal with the Ramond type ($\kappa = 0$) for somewhat simplicity. The Neveu-Schwarz type ($\kappa = \frac{1}{2}$) can be dealt with in quite an analogous way.

Notation 5 $\overline{G}_0 = \frac{1}{2}\sum_{k\in\mathbb{Z}} k f_{\ k} G_0 f_k$ *and* $N = \sum_{k\in\mathbb{Z}} k^2 f_{\ k} f_k$.

For $p_f = 2$ and $p_b = 1$, the anticommutation of G_0 with $\overline{G}_0$ (and with N) gives (see Appendix)

$$\begin{cases} [\overline{G}_0, G_0]_+ = \quad 5N + [[L_0^F, G_0], G_0] \quad 2(L_0^F)^2 + 12\,\zeta(\ 2) \\ [N, G_0]_+ = 2[[G_0, L_0^F], L_0^F] + 4\,\zeta(\ 2) G_0 \end{cases} \quad (\text{for } p_f = 2, p_b = 1), \tag{8}$$

where $\zeta(\ 2) = \frac{1}{2}\sum_{k\in\mathbb{Z}} k^2$. Although $\zeta(\ 2)$ can be chosen as vanishing through the analytic continuation of the zeta function, we keep it remaining to check that the following relation is correct in the single-mode limit.

Proposition 6 *For* $p_b = 2$ *and* $p_b = 1$, G_0, L_0^B, *and* L_0^F *satisfy the polynomial relation*

$$\varphi(G_0, L_0^B + L_0^F) = \Delta(G_0, L_0^F; \zeta(\ 2)),$$

where $\varphi(x, y) = x^5 \quad 4x^3 y$ *and* $\Delta(x, y; \ ) = \quad [x, xyx]_+ \quad 3[x, y^2]_+ + 5yxy + \quad x$.

Proof. Substituting $b_{\ i} b_{\ j} b_{\ k}$ with both sides of Eq. (5) and summing over $i, j, k \in \mathbb{Z}$, we obtain

$$G_0^3 \quad 4G_0(L_0^B + L_0^F) = \overline{G}_0 \quad [G_0, L_0^F]_+.$$

Applying $[\cdot, [\cdot, G_0]_+]_+$ to both sides of the above relation, and using Eq. (8), we obtain the relation to be proven, where use has been made of $[G_0, L_0^B + L_0^F] = 0$. □

Now we check the validity of Prop. 6 by showing that it has a correct relation in the single-mode limit, where the sum over k in G_0, L_0^B, and L_0^F is restricted to ± 1. Even in this limit, the polynomial relation remains the same form as the original relation, so that it follows from Prop. 6 that the corresponding single-mode relation satis es

$$\varphi(Q, H) = \Delta(Q, H_f; 1),$$

where we have replaced as $G_0 \to Q$, $(L_0^B + L_0^F) \to H$, $L_0^F \to H_f$, and $\zeta(\ 2) \to 1$. For $p_f = 2, p_b = 1$, it is found that $\Delta(Q, H_f; 1) = 0$ from $[Q, QH_fQ]_+ = \quad 2Q$, $[Q, H_f^2]_+ = Q$, and $H_f Q H_f = 0$. As a consequence, we get $\varphi(Q, H) = 0$, which is equivalent to

$$Q(Q^2 \quad 4H) = 0,$$

where the hermiticity of Q and H has been used. The relation of $Q(Q^2 \quad 4H) = 0$ coincides with a well-known relation found in the $p_f = 2$ para SUSY quantum mechanics [12]. Thus, the relation of Prop. 6 turns out to be correct in the single-mode limit.

2.3. Conformality

Notation 7 *Let L_n ($n \in \mathbb{Z}$) be given by*

$$L_n = (L_n^B \quad \langle L_n^B \rangle) + (L_n^F \quad \langle L_n^F \rangle + \quad \delta_{n,0}),$$

where $\langle \ldots \rangle$ denotes the vacuum expectation value, and $\; = \frac{1}{16} p_f$ *(for $\kappa = 0$) and* 0 *(for $\kappa = \frac{1}{2}$).*

In the ordinary SUSY, $L_n^S \quad \langle L_n^S \rangle$ (for $S = B, F$) can be replaced by the creation-annihilation normal ordering $\overset{\circ}{\circ} L_n^S \overset{\circ}{\circ}$. For the para SUSY case, however, the di erence between L_n^S and $\overset{\circ}{\circ} L_n^S \overset{\circ}{\circ}$ does not necessarily amount to a constant, so we use instead the vacuum expectation value to avoid the ambiguity caused from $\infty \quad \infty$ in the calculation of $[L_m, L_n]$.

Under the Green's condition of Eq. (1), together with the commutativity of Eq. (7), it is found that the L_n's ($n \in \mathbb{Z}$) and G_r's ($r \in \mathbb{Z} + \kappa$) satisfy the following commutation relations:

$$\begin{cases} [L_m, L_n] = (m \quad n) L_{m+n} + \frac{c}{12} m(m^2 \quad 1) \delta_{m+n,0}, \\ [G_r, L_n] = (r \quad \frac{n}{2}) G_{r+n} \end{cases} \tag{9}$$

where $c = p_b + \frac{1}{2} p_f$. In deriving Eq. (9), we have used

$$\begin{cases} [b_m, L_n] = m\, b_{m+n}, \\ [f_r, L_n] = (r + \frac{n}{2}) f_{r+n}, \end{cases} \qquad \begin{cases} \langle [b_{n \quad k}, b_k]_+ \rangle = p_b\, |k|\, \delta_{n,0}, \\ \langle [f_{n \quad r}, f_r] \rangle = \quad p_f \operatorname{sgn}(r)\, \delta_{n,0}, \end{cases}$$

where the second set of the relations is given by $\forall i, j > 0$, $c_i c_j^\dagger |0\rangle = p\, \delta_{i,j} |0\rangle$ with $c_i |0\rangle = 0$ (where $p = p_b, p_f$ for $c_i = b_i, f_i$, respectively). Rearranging the terms in L_n, and using the formula $\sum_{k=1}^{\infty} (k \quad \kappa) = \frac{1}{24} \quad \frac{1}{8} (2\kappa \quad 1)^2$ (for all $\kappa \in \mathbb{C}$), we can formally rewrite L_n as

$$\begin{aligned} L_n &= L_n^B + L_n^F + \tfrac{c}{24} \delta_{n,0} \\ &= \sum_{S=B,F} \left(L_n^S + \tfrac{c^S}{24} \delta_{n,0} \right) \qquad c^B = p_b,\ c^F = \tfrac{1}{2} p_f \quad , \end{aligned}$$

which indicates that

$$[L_m, L_n^S] = (m \quad n) L_{m+n}^S + \frac{c^S}{12} m^3\, \delta_{m+n,0} \qquad (S = B, F). \tag{10}$$

Now we are in a position to obtain the polynomial relation between the G_r's, L_n^B's, and L_n^F's. For the Ramond type ($\kappa = 0$), it is su cient to obtain the polynomial relation between G_0, L_0^B, and L_0^F. The repeated application of the commutator with the L_n's to the polynomial relation $^R(G_0, L_0^B, L_0^F) = 0$, yields from Eq. (10) a result that we want. For $p_f = 2$ and $p_b = 1$, the result is given by

$$_{r_1 \ldots r_5} + (\mathrm{perm})_{r_1 \ldots r_5} = 0 \qquad (\text{for } p_f = 2, p_b = 1), \tag{11}$$

where $(\mathrm{perm})_{r_1 \ldots r_5}$ represents the permutation with respect to $r_1, \ldots, r_5$, and

$$\begin{aligned} {}_{r_1 \ldots r_5} &= G_{r_1} G_{r_2} G_{r_3} G_{r_4} G_{r_5} \quad 4 G_{r_1} G_{r_2} G_{r_3} (\tilde{L}_{r_4, r_5}^B + \tilde{L}_{r_4, r_5}^F) \\ &+ [G_{r_1}, G_{r_2} \tilde{L}_{r_4, r_5}^F G_{r_3}]_+ + 3 [G_{r_1}, \tilde{L}_{r_2, r_3}^F \tilde{L}_{r_4, r_5}^F]_+ \quad 5 \tilde{L}_{r_2, r_3}^F G_{r_1} \tilde{L}_{r_4, r_5}^F, \end{aligned} \tag{12}$$

with $\tilde{L}_{r,s}^S := L_{r+s}^S + \frac{c^S}{6} r^2\, \delta_{r+s,0}$ ($S = B, F$).

For the Neveu-Schwarz type ($\kappa = \frac{1}{2}$), on the other hand, the starting polynomial relation would be of the form $\ ^{NS}(G_{\frac{1}{2}}, L_1^B, L_1^F) = 0$. However, this is not su cient to obtain the polynomial relation between the G_r's, L_n^B's, and L_n^F's by the repeated application of the commutator with the L_m's. The polynomial relation of the form $\ ^{NS}_{(n)}(G_{\frac{1}{2}+n}, L_{1+2n}^B, L_{1+2n}^F) = 0$ (for $n \in \mathbb{Z}$) is furthermore necessary. Fortunately, the functional form of $\ ^{NS}_{(n)}(x, y, z)$, if chosen as x-monic, is independent of n, and is the same as $\ ^{R}(x, y, z)$, as is observed in the ordinary SUSY case. Eventually, the relation of Eq. (11) with (12) holds for all $r_i \in \mathbb{Z} + \kappa$ ($i = 1, \ldots, 5$).

It should be noticed that, in obtaining $\ _{r_1 \ldots r_5}$, we have only to replace the G_κ's and $L_{2\kappa}^S$'s in $\ (G_\kappa, L_{2\kappa}^B, L_{2\kappa}^F)$ ($=$ $\ ^R$ for $\kappa = 0$ and $=$ $\ ^{NS}$ for $\kappa = \frac{1}{2}$) by

$$\begin{cases} (G_\kappa)^n & \to G_{s_1} \ldots G_{s_n}, \\ (L_{2\kappa}^S)^m & \to \tilde{L}_{t_1,t_2}^S \ldots \tilde{L}_{t_{2m-1},t_{2m}}^S \qquad (S = B, F), \end{cases} \tag{13}$$

where the set of the subscripts in each term of $\ _{r_1 \ldots r_5}$, $\{s_1, \ldots, s_n, \ldots; t_1, \ldots, t_{2m}, \ldots\}$, is so chosen that it is equal to the set of the subscripts of $\ _{r_1 \ldots r_5}$ itself, $\{r_1, \ldots, r_5\}$. This substitution is valid for an arbitrary $\ (x, y, z)$, provided that satis es the homogeneity of the form

$$\forall \ \in \mathbb{C}, \quad (\ x, \ ^2 y, \ ^2 z) = \ ^{\mathfrak{n}} \ (x, y, z), \tag{14}$$

where $\mathfrak{n} = \deg_x \ (x, y, z)$ for an x-monic $\ (x, y, z)$. The relation of Eq. (14) guarantees that all the numbers of the subscripts in each term of $\ _{r_1 \ldots r_{\mathfrak{n}}}$ are the same (as $\mathfrak{n}$).

At the end of this subsection, we verify the validity of the substitution of Eq. (13) for an arbitrary $\ (x, y, z)$ such that Eq. (14) is satis ed. Let $\psi_{r_1 \ldots r_{\mathfrak{n}}}$ be de ned by

$$\psi_{r_1 \ldots r_{\mathfrak{n}}} := \ _{r_1 \ldots r_{\mathfrak{n}}} + (\mathrm{perm})_{r_1 \ldots r_{\mathfrak{n}}}, \tag{15}$$

where $\ _{r_1 \ldots r_{\mathfrak{n}}}$ is given by $\ ^R(G_0, L_0^B, L_0^F)$ or $\ ^{NS}(G_{\frac{1}{2}}, L_1^B, L_1^F)$ under the substitution of Eq. (13). In particular, $\ (G_\kappa, L_{2\kappa}^B, L_{2\kappa}^F) = 0$ corresponds to

$$\psi_{\kappa \ldots \kappa} = 0 \qquad (\kappa = 0, \tfrac{1}{2}). \tag{16}$$

We verify the validity of the substitution of Eq. (13) by showing that $\psi_{r_1 \ldots r_{\mathfrak{n}}} = 0$ for all $r_i \in \mathbb{Z} + \kappa$ ($i = 1, \ldots \mathfrak{n}$). To show it, we rst calculate the commutation relation of $[L_m, \psi_{r_1 \ldots r_{\mathfrak{n}}}]$. For this purpose, it may be convenient to align the parameters s_i's and t_i's as

$$(s_1, \ldots, s_n, \ldots; t_1, \ldots, t_{2m}, \ldots) = (r_1, \ldots, r_{\mathfrak{n}}).$$

In this case, the term including $\tilde{L}_{t,t'}^S$ in $\ _{r_1 \ldots r_{\mathfrak{n}}}$ is symmetric under ($r_{\mathfrak{n}\ 1} \leftrightarrow r_{\mathfrak{n}}$), ($r_{\mathfrak{n}\ 3} \leftrightarrow r_{\mathfrak{n}\ 2}$), and so on, due to $\tilde{L}_{t,t'}^S = \tilde{L}_{t',t}^S$. To make the other terms in $\ _{r_1 \ldots r_{\mathfrak{n}}}$ preserve the same symmetry, we rewrite $\psi_{r_1 \ldots r_{\mathfrak{n}}}$ by symmetrizing $\ _{r_1 \ldots r_{\mathfrak{n}}}$ as

$$\psi_{r_1 \ldots r_{\mathfrak{n}}} = \ '_{r_1 \ldots r_{\mathfrak{n}}} + (\mathrm{perm})'_{r_1 \ldots r_{\mathfrak{n}}}, \qquad \ '_{r_1 \ldots r_{\mathfrak{n}}} = \begin{cases} \ _{\{r_1 r_2\}\{r_3 r_4\} \ldots \{r_{\mathfrak{n}-1} r_{\mathfrak{n}}\}} & (\mathfrak{n} \in 2\mathbb{N}), \\ \ _{r_1 \{r_2 r_3\} \ldots \{r_{\mathfrak{n}-1} r_{\mathfrak{n}}\}} & (\mathfrak{n} \in 2\mathbb{N} + 1), \end{cases}$$

where $(\mathrm{perm})'_{r_1 \ldots r_{\mathfrak{n}}}$ represents the ($r_1 \ldots r_{\mathfrak{n}}$)-permutation where the cases of ($r_{\mathfrak{n}\ 1} \leftrightarrow r_{\mathfrak{n}}$), ($r_{\mathfrak{n}\ 3} \leftrightarrow r_{\mathfrak{n}\ 2}$), ... are excluded; and $\ _{\{r_1 r_2\}} = \ _{r_1 r_2} + \ _{r_2 r_1}$ and the like. Hence, it is found that $[L_m, \psi_{r_1 \dots r_n}]$ can be written as a linear combination of the $\psi_{r_1 \dots r_n}$'s:

$$
\begin{aligned}
[L_m, \psi_{r_1\dots r_n}] &= [L_m, \;\; '_{r_1\dots r_n} + (\text{perm})'_{r_1\dots r_n}\;] \\
&= [L_m, \;\; '_{r_1\dots r_n}] + (\text{perm})'_{r_1\dots r_n} \\
&= \sum_{i=1}^{n} \left(\frac{m}{2} \quad r_i\right) \;\; '_{r_1\dots r_{i-1}, r_i+m, r_{i+1}\dots r_n} + (\text{perm})'_{r_1\dots r_n} \\
&= \sum_{i=1}^{n} \left(\frac{m}{2} \quad r_i\right) \left(\;\; '_{r_1\dots r_{i-1}, r_i+m, r_{i+1}\dots r_n} + (\text{perm})'_{r_1\dots r_{i-1}, r_i+m, r_{i+1}\dots r_n}\right) \\
&= \sum_{i=1}^{n} \left(\frac{m}{2} \quad r_i\right) \psi_{r_1\dots r_{i-1}, r_i+m, r_{i+1}\dots r_n},
\end{aligned}
\tag{17}
$$

where in the third line, we have used the identity

$$
\frac{1}{6}\left(\frac{m}{2} \quad r\right)(m+r)^2 + \frac{1}{6}\left(\frac{m}{2} \quad s\right)(m+s)^2 = \frac{1}{12}m^3 \qquad (\text{for } m+r+s=0).
$$

The point of the de nition of Eq. (15) is that under the map $[L_m, \cdot]$, $\psi_{r_1 \dots r_n}$ transforms in a same way as $G_{r_1} \dots G_{r_n} + (\text{perm})_{r_1 \dots r_n}$.

Then the repeated application of the commutator with the L_m's to Eq. (16) (for the Neveu-Schwarz type, together with $\psi_{\kappa_n \dots \kappa_n} = 0$ ($\kappa_n = n + \frac{1}{2}$)) yields from Eq. (17) the relation

$$
\psi_{r_1 \dots r_n} = 0 \qquad (\text{for } r_i \in \mathbb{Z} + \kappa).
$$

This ends the veri cation of the validity of Eq. (13) by showing the vanishing of $\psi_{r_1 \dots r_n}$.

3. Summary

We have constructed the para SUSY Virasoro algebra under Eq. (7) and the Green's condition, where the polynomial relations between the c_i's are invariant under the unitary transformation of the c_i's. However, we neglect the paraboson-including Virasoro algebra, because the commutation relation between the parabosonic operators b_i's cannot be obtained from the corresponding single-mode commutation relation (Calogero-Vasiliev oscillator algebra, which has a de nite value of p_b) under the Green's condition. Even if we assume that the Green's condition is applied only to the de ning relation of the b_i's, that is, Eq. (1), it is found that the paraboson-including para SUSY Virasoro algebra is not favorable by the following reason.

Consider, for example, the parabose trilinear relation found in many literatures [5, 11] as

$$
[b_i, b_j, b_k] = 2(i\,\delta_{i+j,0} b_k \quad (i \leftrightarrow k)),
\tag{18}
$$

where $[A, B, C] := ABC \quad CBA$. Using Eq. (18), which indeed satis es the relation of Eq. (1), and Eq. (5), we obtain a para SUSY relation of $G_0^3 = 4G_0(L_0^B + L_0^F)$, leading to another para SUSY Virasoro algebra. However, the b_i's in Eq. (18) does not have a de nite parabosonic order p_b, because the relation of Eq. (18) is di erent from the trilinear commutation relation constructed from the c_i's and $c_i^\dagger$'s in Eq. (3), where the c_i's and $c_i^\dagger$'s have a de nite value of p_b. Recalling that the parabosonic central charge c^B is given by p_b, we nd that the Virasoro algebra constructed from Eq. (18) has not a well-de ned central charge. If we assume that a Virasoro algebra should have a well-de ned central charge, it follows that the relation of Eq. (18) is not favorable in constructing a Virasoro algebra, although Eq. (18) in itself may be available as a parabosonic relation.

For $p_f = 2$ (with $p_b = 1$), we calculate the polynomial relation between the G_r's, L_n^B's, and L_n^F's. The key point of the calculation is that we have not only Eq. (5) but also Eq. (6) available; without Eq. (6), we would not obtain the rst relation of Eq. (8); without Eq. (8), it would be di cult to obtain the polynomial relation between G_0, L_0^B, and L_0^F. As contrasted with the ordinary SUSY case, it is di cult to write the polynomial relation in terms of the G_r's and $(L_n^B + L_n^F)$'s only. For $p_f = 3$, it is still more di cult to obtain the polynomial relation between the G_r's, L_n^B's, and L_n^F's, due to the lack of the (symmetric) trilinear relation corresponding Eq. (5). Bearing these in mind, we nd it natural to conjecture that there is a polynomial relation between the G_r's and $(L_n^B + L_n^F)$'s if and only if $p_f = 1$, the ordinary SUSY case.

Acknowledgments
The author is indebted to K. Terada for invaluable discussions and suggestions.

Appendix. Derivation of Eq. (8)

To begin with, we show the rst relation of Eq. (8). From $[b_i, b_j] = i\delta_{i+j,0}$, $G_0 f_k G_0$ can be rewritten as

$$\begin{aligned} G_0 f_k G_0 &= \frac{1}{2}\sum_{\ell,m\in\mathbb{Z}} b \ {}_\ell b \ {}_m [f_\ell, f_k, f_m]_+ + \frac{1}{2}\sum_{\ell\in\mathbb{Z}} \ell f \ {}_\ell f_k f_\ell \\ &= [b_k, G_0]_+ + \frac{1}{2}\sum_{\ell\in\mathbb{Z}} \ell f \ {}_\ell f_k f_\ell, \end{aligned}$$

where in the second equality, use has been made of Eq. (5). Then it is found that

$$\begin{aligned} 2[\overline{G}_0, G_0]_+ &= \sum_{k\in\mathbb{Z}} kf \ {}_k \left([b_k, G_0]_+ + \frac{1}{2}\sum_{\ell\in\mathbb{Z}} f \ {}_\ell f_k f_\ell \right) + (\text{h.c.}) \\ &= \ \ 2[L_0^F, G_0]G_0 \quad N + \frac{1}{2}(\ \ 8N \quad 4(L_0^F)^2 + 24\,\zeta(\ \ 2)) \quad + (\text{h.c.}) \\ &= 2[[F_0^F, G_0], G_0] \quad 10N \quad 4(L_0^F)^2 + 24\,\zeta(\ \ 2), \end{aligned}$$

where we have used $[b_k, G_0]_+ = 2b_k G_0 \quad kf_k$ by $[b_i, b_j] = i\delta_{i+j,0}$, $\sum_{k\in\mathbb{Z}} kf \ {}_k b_k = [L_0^F, G_0]$ by $[f_r, L_0^F] = rf_r$ (for $r \in \mathbb{Z} + \kappa$), and $\sum_{k,\ell\in\mathbb{Z}} k\ell f \ {}_k f \ {}_\ell f_k f_\ell = \ \ 8N \quad 4(L_0^F)^2 + 24\,\zeta(\ \ 2)$ by Eq. (6) with $i = \ \ k, j = \ \ \ell$. This ends the proof of the rst relation of Eq. (8).

The second relation of Eq. (8) can be derived from Eq. (5).

References

[1] Zamolodchikov A B and Fateev V A 1985 *Sov. Phys. JETP* **62** 215
[2] Beckers J and Debergh N 1993 *Int. J. Mod. Phys.* A **8** 5041
[3] Nikitin A G and Galkin A V 2001 *Nucl. Phys.* B **102&103** 322
[4] Durand S, Floreanini R and Vinet M 1989 *Phys. Lett.* B **233** 158
[5] Ohnuki Y and Kamefuchi S 1982 *Quantum Field Theory and Parastatistics* (Tokyo/Berlin: Univ. of Tokyo/Springer)
[6] Green H S 1953 *Phys. Rev.* **90** 270
[7] Vasiliev M A 1991 *Int. J. Mod. Phys.* A **6** 1115
[8] Bialynicki-Birula I 1963 *Nucl. Phys.* **49** 605
[9] Wigner E 1950 *Phys. Rev.* **77** 711
[10] Yang L M 1951 *Phys. Rev.* **84** 788
[11] Greenberg O W and Messiah A M L 1965 *Phys. Rev.* **138B** 1155
[12] Ruvakov V A and Spiridonov V P 1988 *Mod. Phys. Lett.* A **3** 1337

GROUP 28: Physical and Mathematical Aspects of Symmetry IOP Publishing
Journal of Physics: Conference Series **284** (2011) 012043 doi:10.1088/1742-6596/284/1/012043

Branching rules for Weyl group orbits involving the Lie algebra $U(1)$

M Larouche

Department of Mathematics and Statistics, Université de Montréal, 2920 chemin de la Tour, Montreal, Quebec, H3T 1J4, Canada

E-mail: `larouche@dms.umontreal.ca`

Abstract. The orbits of Weyl groups $W(L)$ of all simple Lie algebras L are reduced to the union of orbits of the Weyl groups of maximal reductive non-semisimple subalgebras L' of L, whenever such subalgebras are present. Matrices transforming points of the orbits of $W(L)$ into points of $W(L')$ orbits are listed for the infinite series of algebra-subalgebra pairs $A_n \supset A_{n-k-1} \times A_k \times U(1)$, $B_n \supset B_{n-1} \times U(1)$, $C_n \supset A_{n-1} \times U(1)$, $D_n \supset A_{n-1} \times U(1)$, $D_n \supset D_{n-1} \times U(1)$, and for $E_6 \supset D_5 \times U(1)$ and $E_7 \supset E_6 \times U(1)$. Examples of branching rules are shown for all cases.

1. Introduction

Finite groups generated by reflections of a finite-dimensional Euclidean space are commonly known as finite Coxeter groups [1, 2]. They are of two types, crystallographic and non-crystallographic. Finite crystallographic Coxeter groups are the Weyl groups of semisimple Lie groups or Lie algebras. They are the symmetry groups of lattices of $\mathbb{R}^n$. They fall in four families – A_n, B_n, C_n, and D_n – with five exceptions E_6, E_7, E_8, F_4, and G_2. The index n, 6, 7, 8, 4 or 2, is the rank of the corresponding Lie algebra and the dimension of the finite Euclidean space in which the reflections take place.

We consider the orbits of Weyl groups $W(L)$ of semisimple Lie algebras L. An orbit of $W(L)$ is a finite set of points of $\mathbb{R}^n$ obtained from the action of $W(L)$ on a single point of $\mathbb{R}^n$. The number of points in the orbit is at most the order of the Weyl group. Since the reflections take place in mirrors that all go through the origin, all the points of an orbit are equidistant from the origin. Geometrically, points of an orbit can be seen as vertices of a polytope. A method has been developed in [3] to describe the faces of the polytopes using recursive decoration of the Dynkin diagrams.

Weyl group orbits are closely related to weight systems of finite-dimensional irreducible representations of semisimple Lie algebras. Indeed a weight system consists of many orbits of the corresponding Weyl group, a specific orbit often appearing more than once. Which orbits a particular representation is comprised of is well known, and extensive tables of multiplicities of dominant weights can be found in [4].

Reduction of weight systems of representations of simple Lie algebras to weight systems of representations of their maximal reductive subalgebras has been addressed several times in the literature [5, 6, 7, 8]. In physics that problem is often referred to as computation of branching rules. We consider here a related problem for individual orbits of Weyl groups. In both problems transformation of orbit points (weights in the case of weight systems) is involved. However, considering the reduction of individual orbits rather than of entire weight systems offers some advantages. Firstly, while the number weights of a weight system grows without limits with the dimension of the representation, the number of points of an individual orbit is at most the order of the corresponding Weyl group. When dealing with large-scale computation for representations, one often needs to break down the problem into smaller ones for individual orbits. Secondly, while weights of a weight system must be on a weight lattice of the Lie algebra, points of an individual orbit can be anywhere in $\mathbb{R}^n$. That allows orbits to be as close to one another as desired, as it is discussed in [9].

The list of possible reductions of orbits of $W(L)$ is the result of a major classification carried out more than sixty years ago, when the maximal reductive subalgebras of simple Lie algebras were determined [10]. We will only consider maximal reductive subalgebras that are not semisimple. Those are exactly the subalgebras that contain the one-parameter Lie algebra $U(1)$. We will discuss all the cases when such a subalgebra is present :

$$\begin{array}{lll}
A_n \supset A_{n-1} \times U(1)\,, & n \geq 2\,, & \\
A_n \supset A_{n-k-1} \times A_k \times U(1)\,, & n \geq 3\,, & 1 \leq k \leq [\frac{n-1}{2}]\,, \\
B_n \supset B_{n-1} \times U(1)\,, & n \geq 3\,, & \\
C_n \supset A_{n-1} \times U(1)\,, & n \geq 2\,, & \\
D_n \supset A_{n-1} \times U(1)\,, & n \geq 4\,, & \\
D_n \supset D_{n-1} \times U(1)\,, & n \geq 5\,, & \\
E_6 \supset D_5 \times U(1)\,, & & \\
E_7 \supset E_6 \times U(1)\,. & &
\end{array}$$

The branching rules for $W(L) \supset W(L')$, where L' is a maximal reductive non-semisimple subalgebra of L, is a linear transformation between Euclidean spaces $\mathbb{R}^n \to \mathbb{R}^m$, where n and m are the ranks of L and L' respectively. The branching rules are unique, unlike transformations of individual orbit points, which depend on the relative choice of bases. We provide the linear transformation in the form of an $m \times n$ matrix, the 'projection matrix'. A suitable choice of bases allows one to obtain integer matrix elements in all the projection matrices listed here. Note that we use Dynkin notations and numberings for roots, weights and diagrams.

The method we use to compute the branching rules of orbits of $W(L)$ is an extension of the one used in [5, 6, 7, 8] for branching rules of representations of Lie algebras. Orbit-orbit branching rules have been discussed for one of the first times in the literature in [8]. They were then addressed in [11, 12, 13], where specific methods were developed for different algebra-subalgebra pairs. The main advantage of the projection matrix method is its uniformity, as it can be used for any algebra-subalgebra pair. The present paper can be viewed as a continuation of [14], where orbit-orbit branching rules are computed with the projection matrix method for orbits of $W(A_n)$.

2. The projection matrix method

Consider the pair $W(L) \supset W(L')$, where L' is a maximal reductive non-semisimple subalgebra of the simple Lie algebra L. The orbit reduction problem is solved when the $m \times n$ matrix P is found with the property that points of any orbit of the Weyl group $W(L)$ are projected by P into points of the corresponding orbits of $W(L')$. Computation of the branching rule for a specific orbit of $W(L)$ amounts to applying P to the points of the orbit, and to sorting out the projected points into a sum (union) of orbits of $W(L')$.

The projection matrix P is calculated from one known branching rule. The classification of maximal reductive subalgebras of simple Lie algebras [10] provides the information to find that branching rule. The projection matrix is then obtained using the weight systems of the representations, by requiring that weights of L be transformed by P to weights of L'. Since any ordering of the weights is admissible, the projection matrix is not unique. We choose the natural lexicographical ordering of the weights. The projection matrix obtained can then be used to project points of any orbit of $W(L)$ into points of orbits of $W(L')$. An example of the construction of a projection matrix is presented in [14] for the case $W(A_3) \supset W(C_2)$.

To compute the branching rule for a specific orbit of $W(L)$, one has to list all the points of that orbit and to multiply them by the projection matrix. A standard method to calculate points of an orbit of any finite Coxeter group is given in [9], where the points are given in the corresponding basis of fundamental weights, the so-called ω-basis. All the orbits appearing here will be given in that ω-basis, linked to the basis of simple roots by the Cartan matrix of the group. Since every orbit contains precisely one point with nonnegative coordinates in the ω-basis, we can identify the orbit by that point, called the dominant point of the orbit. Hence when referring to an orbit, one does not have to list all the points it contains.

The Weyl group of the one-parameter Lie algebra $U(1)$ is trivial, consisting of the identity element only. Its irreducible representations are all 1-dimensional, hence its orbits consist of one element. They are labeled by integers, which can also take negative values. The symbol (k) may stand for either the orbit $\{k, -k\}$ of $W(A_1)$, or for the $W(U(1))$ orbit of one point $\{k\}$. Distinction should be made from the context. For example, the orbit $(p)(q)$, where $p \in \mathbb{Z}^{>0}$, $q \in \mathbb{Z}$, of $W(A_1 \times U(1))$, has two elements, $\{(p)(q), (-p)(q)\}$. Since we are working with orbits of the Weyl group of $U(1)$ and the compactness of the Lie group is of no interest to us here, we can allow the orbits of $W(U(1))$ to take real values.

Reductions in this paper are all of the form $W(L) \supset W(L'') \times U(1)$, where the ranks of L and L'' are n and $n-1$ respectively. Hence reductions will all be to orbits of the same dimension. Geometrically, the orbit points are not displaced in this case; rather, they are relabeled by the coordinates given in the standard basis of the subgroup.

To alleviate notation, we will simply write L instead of $W(L)$ to refer to the Weyl group of the Lie algebra L, and λ instead of W_λ to refer to the orbit of the dominant point λ of the Weyl group W.

Many examples of projection matrices and branching rules can be found in [14] for the lowest dimensional cases for orbits of the Weyl group of A_n. We give here an example for orbits of the Weyl group of B_3.

Example. *Consider the case $B_3 \supset B_2 \times U(1)$. The projection matrix P for that case is constructed using the known branching rule $(1,0,0) \supset (1,0)(0) + (0,0)(2) + (0,0)(-2)$, and we obtain the 3×3 matrix*

$$P = \begin{pmatrix} 1 & 0 & 0 \\ 0 & 2 & 1 \\ 0 & 0 & 1 \end{pmatrix}. \tag{1}$$

Let us illustrate the actual computation of branching rules for the B_3 orbit $(0,1,0)$ containing 12 points. We write the coordinates of the points (given in the ω-basis of B_3) as column vectors:

$$\begin{gathered}\begin{pmatrix} 0 \\ 1 \\ 0 \end{pmatrix}, \begin{pmatrix} 1 \\ -1 \\ 2 \end{pmatrix}, \begin{pmatrix} -1 \\ 0 \\ 2 \end{pmatrix}, \begin{pmatrix} 1 \\ 1 \\ -2 \end{pmatrix}, \begin{pmatrix} -1 \\ 2 \\ -2 \end{pmatrix}, \begin{pmatrix} 2 \\ -1 \\ 0 \end{pmatrix}, \\ \begin{pmatrix} 1 \\ -2 \\ 2 \end{pmatrix}, \begin{pmatrix} -2 \\ 1 \\ 0 \end{pmatrix}, \begin{pmatrix} 1 \\ 0 \\ -2 \end{pmatrix}, \begin{pmatrix} -1 \\ -1 \\ 2 \end{pmatrix}, \begin{pmatrix} -1 \\ 1 \\ -2 \end{pmatrix}, \begin{pmatrix} 0 \\ -1 \\ 0 \end{pmatrix}.\end{gathered} \tag{2}$$

Multiplying each of the points of (2) by the matrix (1), one gets the points of the $B_2 \times U(1)$ orbits written as column vectors. Rewriting them in the horizontal form and keeping in mind that the first two coordinates belong to B_2 orbits, the third one belonging to orbits of $U(1)$, we have the set of projected points. It remains to distribute the points into individual orbits. Practically it *suffices to select just the dominant ones (no negative coordinates, except for the orbits of $U(1)$) because they represent the orbits that are present. The result is given by (3), for $b = 1$.*

We give the branching rules for orbits of B_3 of size 6, 12, 8 and 48 respectively:

$$\begin{aligned}
(a,0,0) &\supset (a,0)(0) + (0,0)(2a) + (0,0)(-2a)\,,\\
(0,b,0) &\supset (0,2b)(0) + (b,0)(2b) + (b,0)(-2b)\,,\\
(0,0,c) &\supset (0,c)(c) + (0,c)(-c)\,,\\
(a,b,c) &\supset (a,2b{+}c)(c) + (a{+}b,c)(2b{+}c)\\
&\quad + (a,2b{+}c)(-c) + (b,c)(2a{+}2b{+}c)\\
&\quad + (a{+}b,c)(-2b{-}c) + (b,c)(-2a{-}2b{-}c)\,,
\end{aligned} \tag{3}$$

where $a, b, c \in \mathbb{R}^{>0}$.

3. Infinite series of classical cases

For each case involving a classical Lie algebra and the Lie algebra $U(1)$, we give the projection matrix for a general rank n and list branching rules for 3 different orbits of $\mathbb{R}^n$. The orbit of the highest weight of the adjoint representation of the classical Lie algebra is included as one of the special cases.

3.1. $A_n \supset A_{n-1} \times U(1), \quad n \geq 2$

The projection matrix is

$$\left(\begin{array}{ccccccc|c}
 & & & I_{n-1} & & & & \mathbf{0}\\
\hline
1 & 2 & 3 & \dots & n{-}2 & n{-}1 & & n
\end{array}\right).$$

Note that, here and everywhere below, I_k denotes the $k \times k$ identity matrix and $\mathbf{0}$ represents the zero matrix of appropriate dimensions.

We give branching rules for this case for orbits of A_n of order $n{+}1$, $(n^2{+}n)/2$ and $n^2{+}n$ respectively:

$$\begin{aligned}
(a,0,0,\dots,0) &\supset (a,0,0,\dots,0)(a) + (0,\dots,0)(-na)\,,\\
(0,b,0,\dots,0) &\supset (0,b,0,\dots,0)(2b) + (b,0,0,\dots,0)(-(n{-}1)b)\,,\\
(c,0,\dots,0,d) &\supset (c,0,0,\dots,0)(c{+}nd) + (c,0,0,\dots,0,d)(c{-}d)\\
&\quad + (0,0,\dots,0,d)(-nc{-}d)\,.
\end{aligned}$$

Note that, here and everywhere below, $a, b, c, d \in \mathbb{R}^{>0}$.

3.2. $A_n \supset A_{n-k-1} \times A_k \times U(1), \quad 1 \leq k \leq [\frac{n-1}{2}]$

The previous case can be viewed as a special case of the present one where $k = 0$. Here we are considering the cases with general rank $n \geq 3$ and $1 \leq\ k \leq\ [\frac{n-1}{2}]$.

The projection matrix is

$$\left(\begin{array}{c|c|c} \multicolumn{2}{c|}{I_{n-k-1}} & \mathbf{0} \\ \hline \multicolumn{2}{c|}{\mathbf{0}} & I_k \\ \hline \underbrace{k{+}1\ \ 2(k{+}1)\ \cdots\ (n{-}k{-}1)(k{+}1)}_{n-k-1} & \underbrace{(n{-}k)(k{+}1)}_{1} & \underbrace{k(n{-}k)\ \cdots\ 2(n{-}k)\ \ n{-}k}_{k} \end{array}\right)\begin{array}{l}\}\,n-k-1\\ \}\,k\\ \}\,1\end{array}$$

We give branching rules for this case for orbits of A_n of order $n{+}1$, $(n^2{+}n)/2$ and $n^2{+}n$ respectively:

$$\begin{aligned}
(a,0,0,\ldots,0) &\supset (a,0,\ldots,0)(0,\ldots,0)((k{+}1)a) + (0,\ldots,0)(a,0,\ldots,0)((k{-}n)a)\,,\\
(0,b,0,\ldots,0) &\supset (0,b,0,\ldots,0)(0,\ldots,0)(2(k{+}1)b) + (b,0,\ldots,0)(b,0,\ldots,0)((2k{+}1{-}n)b)\\
&\quad + (0,\ldots,0)(0,b,0,\ldots,0)(2(k{-}n)b)\,,\\
(c,0,\ldots,0,d) &\supset (c,0,\ldots,0)(0,\ldots,0,d)((k{+}1)c{+}(n{-}k)d)\\
&\quad + (c,0,\ldots,0,d)(0,\ldots,0)((k{+}1)(c{-}d))\\
&\quad + (0,\ldots,0)(c,0,\ldots,0,d)((n{-}k)(d{-}c))\\
&\quad + (0,\ldots,0,d)(c,0,\ldots,0)((-k{-}1)d{+}(k{-}n)c)\,.
\end{aligned}$$

3.3. $B_n \supset B_{n-1} \times U(1)$, $\quad n \geq 3$
The projection matrix is

$$\left(\begin{array}{c|c} I_{n-2} & \mathbf{0} \\ \hline \mathbf{0} & \begin{smallmatrix} 2 & 1 \\ \cdot & 1 \end{smallmatrix} \end{array}\right).$$

Note that, here and everywhere below, a dot in a matrix denotes a zero matrix element.
We give branching rules for this case for orbits of B_n of order $2n$, $2n(n{-}1)$ and 2^n respectively:

$$\begin{aligned}
(a,0,0,\ldots,0) &\supset (a,0,\ldots,0)(0) + (0,\ldots,0)(2a) + (0,\ldots,0)(-2a)\,,\\
(0,b,0,\ldots,0) &\supset (0,b,0,\ldots,0)(0) + (b,0,\ldots,0)(2b) + (b,0,\ldots,0)(-2b)\,,\\
(0,0,\ldots,0,c) &\supset (0,\ldots,0,c)(c) + (0,\ldots,0,c)(-c)\,.
\end{aligned}$$

Note that in the case of B_2, the orbit $(0,b,0,\ldots,0)$ is $(0,2b)$.

3.4. $C_{2n} \supset A_{2n-1} \times U(1)$, $\quad n \geq 1$
The projection matrix is

$$\begin{pmatrix}
1 & 1 & \cdot & \cdot & \cdot & \cdot & \cdots & \cdot & \cdot & \cdot & \cdot & \cdot & \cdot \\
\cdot & \cdot & 1 & 1 & \cdot & \cdot & & \cdot & \cdot & \cdot & \cdot & \cdot & \cdot \\
 & & & & & & \vdots & & & & & & \\
\cdot & \cdot & \cdot & \cdot & \cdot & \cdot & & \cdot & \cdot & 1 & 1 & \cdot & \cdot \\
\cdot & \cdot & \cdot & \cdot & \cdot & \cdot & \cdots & \cdot & \cdot & \cdot & \cdot & 1 & 2 \\
\cdot & \cdot & \cdot & \cdot & \cdot & \cdot & & \cdot & \cdot & \cdot & 1 & 1 & \cdot \\
\cdot & \cdot & \cdot & \cdot & \cdot & \cdot & & \cdot & 1 & 1 & \cdot & \cdot & \cdot \\
 & & & & & & \vdots & & & & & & \\
\cdot & 1 & 1 & \cdot & \cdot & \cdot & \cdots & \cdot & \cdot & \cdot & \cdot & \cdot & \cdot \\
1 & \cdot & 1 & \cdot & 1 & \cdot & & 1 & \cdot & 1 & \cdot & 1 & \cdot
\end{pmatrix}.$$

We give branching rules for this case for orbits of C_{2n} of order $4n$, $4n(2n-1)$ and 2^{2n} respectively:

$$\begin{aligned}
(a,0,0,\dots,0) &\supset (a,0,\dots,0)(a) + (0,\dots,0,a)(-a)\,,\\
(0,b,0,\dots,0) &\supset (b,0,\dots,0,b)(0) + (0,b,0,\dots,0)(2b) + (0,\dots,0,b,0)(-2b)\,,\\
(0,0,\dots,0,c) &\supset (\underbrace{0,\dots,0}_{n-1},2c,\underbrace{0,\dots,0}_{n-1})(0) + (\underbrace{0,\dots,0}_{n},2c,\underbrace{0,\dots,0}_{n-2})(2c)\\
&\quad + (\underbrace{0,\dots,0}_{n-2},2c,\underbrace{0,\dots,0}_{n})(-2c) + (\underbrace{0,\dots,0}_{n+1},2c,\underbrace{0,\dots,0}_{n-3})(4c)\\
&\quad + (\underbrace{0,\dots,0}_{n-3},2c,\underbrace{0,\dots,0}_{n+1})(-4c) + \dots + (0,\dots,0,2c)((2n-2)c)\\
&\quad + (2c,0,\dots,0)(-(2n-2)c) + (0,\dots,0)(2nc) + (0,\dots,0)(-2nc)\,.
\end{aligned}$$

3.5. $C_{2n+1} \supset A_{2n} \times U(1), \quad n \geq 1$

The projection matrix is

$$\left(\begin{array}{cccccccccccccc}
1 & 1 & \cdot & \cdot & \cdot & \cdot & \cdots & \cdot & \cdot & \cdot & \cdot & \cdot & \cdot & \cdot\\
\cdot & \cdot & 1 & 1 & \cdot & \cdot & & \cdot & \cdot & \cdot & \cdot & \cdot & \cdot & \cdot\\
 & & & & & & \vdots & & & & & & & \\
\cdot & \cdot & \cdot & \cdot & \cdot & \cdot & & \cdot & \cdot & \cdot & \cdot & 1 & 1 & \cdot\\
\cdot & \cdot & \cdot & \cdot & \cdot & \cdot & \cdots & \cdot & \cdot & \cdot & \cdot & \cdot & 1 & 2\\
\cdot & \cdot & \cdot & \cdot & \cdot & \cdot & & \cdot & \cdot & \cdot & 1 & 1 & \cdot & \cdot\\
\cdot & \cdot & \cdot & \cdot & \cdot & \cdot & & \cdot & 1 & 1 & \cdot & \cdot & \cdot & \cdot\\
 & & & & & & \vdots & & & & & & & \\
\cdot & 1 & 1 & \cdot & \cdot & \cdot & \cdots & \cdot & \cdot & \cdot & \cdot & \cdot & \cdot & \cdot\\
1 & \cdot & 1 & \cdot & 1 & \cdot & & 1 & \cdot & 1 & \cdot & 1 & \cdot & 1
\end{array}\right).$$

We give branching rules for this case for orbits of C_{2n+1} of order $2(2n+1)$, $4n(2n+1)$ and 2^{2n+1} respectively:

$$\begin{aligned}
(a,0,0,\dots,0) &\supset (a,0,\dots,0)(a) + (0,\dots,0,a)(-a)\,,\\
(0,b,0,\dots,0) &\supset (b,0,\dots,0,b)(0) + (0,b,0,\dots,0)(2b) + (0,\dots,0,b,0)(-2b)\,,\\
(0,0,\dots,0,c) &\supset (\underbrace{0,\dots,0}_{n},2c,\underbrace{0,\dots,0}_{n-1})(c) + (\underbrace{0,\dots,0}_{n-1},2c,\underbrace{0,\dots,0}_{n})(-c)\\
&\quad + (\underbrace{0,\dots,0}_{n+1},2c,\underbrace{0,\dots,0}_{n-2})(3c) + (\underbrace{0,\dots,0}_{n-2},2c,\underbrace{0,\dots,0}_{n+1})(-3c)\\
&\quad + \dots + (0,\dots,0,2c)((2n-1)c) + (2c,0,\dots,0)(-(2n-1)c)\\
&\quad + (0,\dots,0)((2n+1)c) + (0,\dots,0)(-(2n+1)c)\,.
\end{aligned}$$

3.6. $D_{2n} \supset A_{2n-1} \times U(1), \quad n \geq 2$

The projection matrix is

$$\left(\begin{array}{ccccccccccccc}
1 & 1 & \cdot & \cdot & \cdot & \cdot & \cdots & \cdot & \cdot & \cdot & \cdot & \cdot & \cdot\\
\cdot & \cdot & 1 & 1 & \cdot & \cdot & & \cdot & \cdot & \cdot & \cdot & \cdot & \cdot\\
 & & & & & & \vdots & & & & & & \\
\cdot & \cdot & \cdot & \cdot & \cdot & \cdot & & \cdot & \cdot & 1 & 1 & \cdot & \cdot\\
\cdot & \cdot & \cdot & \cdot & \cdot & \cdot & \cdots & \cdot & \cdot & \cdot & \cdot & \cdot & 1\\
\cdot & \cdot & \cdot & \cdot & \cdot & \cdot & & \cdot & \cdot & \cdot & 1 & 1 & \cdot\\
\cdot & \cdot & \cdot & \cdot & \cdot & \cdot & & \cdot & 1 & 1 & \cdot & \cdot & \cdot\\
 & & & & & & \vdots & & & & & & \\
\cdot & 1 & 1 & \cdot & \cdot & \cdot & \cdots & \cdot & \cdot & \cdot & \cdot & \cdot & \cdot\\
1 & \cdot & 1 & \cdot & 1 & \cdot & & 1 & \cdot & 1 & \cdot & 1 & \cdot
\end{array}\right).$$

Journal of Physics: Conference Series **284** (2011) 012043 doi:10.1088/1742-6596/284/1/012043

We give branching rules for this case for orbits of D_{2n} of order $4n$, $4n(2n{-}1)$ and 2^{2n-1} respectively:

$$(a,0,0,\dots,0)\supset(a,0,\dots,0)(a)+(0,\dots,0,a)(-a)\,,$$
$$(0,b,0,\dots,0)\supset(b,0,\dots,0,b)(0)+(0,b,0,\dots,0)(2b)+(0,\dots,0,b,0)(-2b)\,,$$
$$(0,0,\dots,0,c)\supset\begin{cases}(\underbrace{0,\dots,0}_{n-1},c,\underbrace{0,\dots,0}_{n-1})(0)+(\underbrace{0,\dots,0}_{n+1},c,\underbrace{0,\dots,0}_{n-3})(2c)\\ +(\underbrace{0,\dots,0}_{n-3},c,\underbrace{0,\dots,0}_{n+1})(-2c)+(\underbrace{0,\dots,0}_{n+3},c,\underbrace{0,\dots,0}_{n-5})(4c)\\ +(\underbrace{0,\dots,0}_{n-5},c,\underbrace{0,\dots,0}_{n+3})(-4c)+\ldots+(0,\dots,0,c,0)((n{-}2)c)\\ +(0,c,0,\dots,0)(-(n{-}2)c)+(0,\dots,0)(nc)+(0,\dots,0)(-nc) & n\text{ even}\\ \\ (\underbrace{0,\dots,0}_{n-1},c,\underbrace{0,\dots,0}_{n-1})(0)+(\underbrace{0,\dots,0}_{n+1},c,\underbrace{0,\dots,0}_{n-3})(2c)\\ +(\underbrace{0,\dots,0}_{n-3},c,\underbrace{0,\dots,0}_{n+1})(-2c)+(\underbrace{0,\dots,0}_{n+3},c,\underbrace{0,\dots,0}_{n-5})(4c)\\ +(\underbrace{0,\dots,0}_{n-5},c,\underbrace{0,\dots,0}_{n+3})(-4c)+\ldots+(0,\dots,0,c)((n{-}1)c)\\ +(c,0,\dots,0)(-(n{-}1)c) & n\text{ odd}\end{cases}$$

3.7. $D_{2n+1}\supset A_{2n}\times U(1),\quad n\geq 2$

The projection matrix is

$$\left(\begin{array}{cccccccccccccc}1&1&\cdot&\cdot&\cdot&\cdot&\cdots&\cdot&\cdot&\cdot&\cdot&\cdot&\cdot&\cdot\\ \cdot&\cdot&1&1&\cdot&\cdot&&\cdot&\cdot&\cdot&\cdot&\cdot&\cdot&\cdot\\ &&&&&&\vdots&&&&&&&\\ \cdot&\cdot&\cdot&\cdot&\cdot&\cdot&&\cdot&\cdot&\cdot&\cdot&1&1&\cdot\\ \cdot&\cdot&\cdot&\cdot&\cdot&\cdot&\cdots&\cdot&\cdot&\cdot&\cdot&\cdot&\cdot&1\\ \cdot&\cdot&\cdot&\cdot&\cdot&\cdot&&\cdot&\cdot&\cdot&1&1&\cdot&\cdot\\ \cdot&\cdot&\cdot&\cdot&\cdot&\cdot&&\cdot&1&1&\cdot&\cdot&\cdot&\cdot\\ &&&&&&\vdots&&&&&&&\\ \cdot&1&1&\cdot&\cdot&\cdot&\cdots&\cdot&\cdot&\cdot&\cdot&\cdot&\cdot&\cdot\\ 2&\cdot&2&\cdot&2&\cdot&&2&\cdot&2&\cdot&2&-1&1\end{array}\right).$$

We give branching rules for this case for orbits of D_{2n+1} of order $2(2n{+}1)$, $4n(2n{+}1)$ and 2^{2n} respectively:

$$(a,0,0,\dots,0)\supset(a,0,\dots,0)(2a)+(0,\dots,0,a)(-2a)\,,$$
$$(0,b,0,\dots,0)\supset(b,0,\dots,0,b)(0)+(0,b,0,\dots,0)(4b)+(0,\dots,0,b,0)(-4b)\,,$$
$$(0,0,\dots,0,c)\supset\begin{cases}(\underbrace{0,\dots,0}_{n},c,\underbrace{0,\dots,0}_{n-1})(c)+(\underbrace{0,\dots,0}_{n-2},c,\underbrace{0,\dots,0}_{n+1})(-3c)\\ +(\underbrace{0,\dots,0}_{n+2},c,\underbrace{0,\dots,0}_{n-3})(5c)+(\underbrace{0,\dots,0}_{n-4},c,\underbrace{0,\dots,0}_{n+3})(-7c)\\ +\ldots+(c,0,\dots,0)(-(2n{-}1)c)+(0,\dots,0)((2n{+}1)c) & n\text{ even}\\ \\ (\underbrace{0,\dots,0}_{n},c,\underbrace{0,\dots,0}_{n-1})(c)+(\underbrace{0,\dots,0}_{n-2},c,\underbrace{0,\dots,0}_{n+1})(-3c)\\ +(\underbrace{0,\dots,0}_{n+2},c,\underbrace{0,\dots,0}_{n-3})(5c)+(\underbrace{0,\dots,0}_{n-4},c,\underbrace{0,\dots,0}_{n+3})(-7c)\\ +\ldots+(0,\dots,0,c)((2n{-}1)c)+(0,\dots,0)(-(2n{+}1)c) & n\text{ odd}\end{cases}$$

Journal of Physics: Conference Series **284** (2011) 012043 doi:10.1088/1742-6596/284/1/012043

3.8. $D_n \supset D_{n-1} \times U(1), \quad n \geq 5$
The projection matrix is

$$\left(\begin{array}{c|c} I_{n-3} & \mathbf{0} \\ \hline \mathbf{0} & \begin{smallmatrix} 1 & \cdot & \cdot \\ 1 & 1 & 1 \\ \cdot & 1 & -1 \end{smallmatrix} \end{array}\right).$$

We give branching rules for this case for orbits of D_n of order $2n$, $2n(n-1)$ and 2^{n-1} respectively:

$$\begin{aligned}
(a,0,0,\dots,0) &\supset (a,0,\dots,0)(0) + (0,\dots,0)(2a) + (0,\dots,0)(-2a)\,,\\
(0,b,0,\dots,0) &\supset (0,b,0,\dots,0)(0) + (b,0,\dots,0)(2b) + (b,0,\dots,0)(-2b)\,,\\
(0,0,\dots,0,c) &\supset (0,\dots,0,c,0)(c) + (0,\dots,0,c)(-c)\,.
\end{aligned}$$

4. Exceptional cases

For each exceptional case involving the Lie algebra $U(1)$, we give the projection matrix and list branching rules for different orbits of $\mathbb{R}^n$. The orbit of the highest weight of the adjoint representation of the exceptional Lie algebra is included as one of the special cases.

4.1. $E_6 \supset D_5 \times U(1)$
The projection matrix is

$$\begin{pmatrix} \cdot & 1 & 1 & 1 & \cdot & \cdot \\ \cdot & \cdot & \cdot & \cdot & \cdot & 1 \\ \cdot & \cdot & 1 & \cdot & \cdot & \cdot \\ \cdot & \cdot & \cdot & 1 & 1 & \cdot \\ 1 & 1 & \cdot & \cdot & \cdot & \cdot \\ 1 & -1 & \cdot & 1 & -1 & \cdot \end{pmatrix}.$$

We give branching rules for this case for orbits of E_6 of order 27, 216, 720 and 72 respectively:

$$\begin{aligned}
(a,0,0,0,0,0) &\supset (0,0,0,0,a)(a) + (a,0,0,0,0)(-2a) + (0,0,0,0,0)(4a)\,,\\
(0,b,0,0,0,0) &\supset (b,0,0,0,b)(-b) + (0,0,b,0,0)(2b) + (0,b,0,0,0)(-4b)\\
&\quad + (0,0,0,0,b)(5b)\,,\\
(0,0,c,0,0,0) &\supset (c,0,c,0,0)(0) + (0,c,0,c,0)(3c) + (0,c,0,0,c)(-3c)\\
&\quad + (0,0,c,0,0)(-6c) + (0,0,c,0,0)(6c)\,,\\
(0,0,0,0,0,d) &\supset (0,d,0,0,0)(0) + (0,0,0,d,0)(3d) + (0,0,0,0,d)(-3d)\,.
\end{aligned}$$

Many more examples can be found in [8].

4.2. $E_7 \supset E_6 \times U(1)$
The projection matrix is

$$\begin{pmatrix} \cdot & \cdot & \cdot & \cdot & 1 & 1 & \cdot \\ \cdot & \cdot & 1 & 1 & \cdot & \cdot & \cdot \\ \cdot & 1 & \cdot & \cdot & \cdot & \cdot & \cdot \\ \cdot & \cdot & 1 & \cdot & \cdot & \cdot & 1 \\ \cdot & \cdot & \cdot & 1 & 1 & \cdot & \cdot \\ 1 & \cdot & \cdot & \cdot & \cdot & \cdot & \cdot \\ \cdot & \cdot & \cdot & 1 & \cdot & 1 & 1 \end{pmatrix}.$$

We give branching rules for this case for orbits of E_7 of order 126, 2016, 4032, 756, 56 and

576 respectively:

$$\begin{aligned}
(a,0,0,0,0,0,0) &\supset (0,0,0,0,0,a)(0) + (0,0,0,0,a,0)(2a) + (a,0,0,0,0,0)(-2a)\,,\\
(0,b,0,0,0,0,0) &\supset (0,0,b,0,0,0)(0) + (0,0,0,0,b,b)(2b) + (b,0,0,0,0,b)(-2b)\\
&\quad + (0,0,0,b,0,0)(4b) + (0,b,0,0,0,0)(-4b)\,,\\
(0,0,0,c,0,0,0) &\supset (0,c,0,0,c,0)(c) + (c,0,0,c,0,0)(-c) + (0,0,c,0,0,0)(3c)\\
&\quad + (0,0,c,0,0,0)(-3c) + (0,c,0,0,0,0)(5c) + (0,0,0,c,0,0)(-5c)\,,\\
(0,0,0,0,d,0,0) &\supset (d,0,0,0,d,0)(0) + (0,d,0,0,0,0)(2d) + (0,0,0,d,0,0)(-2d)\\
&\quad + (d,0,0,0,0,0)(4d) + (0,0,0,0,d,0)(-4d)\,,\\
(0,0,0,0,0,e,0) &\supset (e,0,0,0,0,0)(e) + (0,0,0,0,e,0)(-e) + (0,0,0,0,0,0)(3e)\\
&\quad + (0,0,0,0,0,0)(-3e)\,,\\
(0,0,0,0,0,0,f) &\supset (0,0,0,f,0,0)(f) + (0,f,0,0,0,0)(-f) + (0,0,0,0,0,f)(3f)\\
&\quad + (0,0,0,0,0,f)(-3f)\,,
\end{aligned}$$

where e and f are also in $\mathbb{R}^{>0}$. Many more examples can be found in [8].

5. Concluding remarks

- The pairs $W(L) \supset W(L')$ in this paper involve a maximal subalgebra L' in L. A chain of maximal subalgebras linking L and any of its reductive non-maximal subalgebras L'' can be found. Corresponding projection matrices combine, by the common matrix multiplication, into the projection matrix for $W(L) \supset W(L'')$.
- Projection matrices of $W(L) \supset W(L')$ when the ranks of L and L' are the same are square matrices with determinant different from zero. Hence they can be inverted and used in the opposite direction. The inverse matrix transforms an orbit of $W(L')$ into the linear combination of orbits of $W(L)$, where $L' \subset L$. The linear combination has integer coefficients of both signs in general. We know of no interpretation of such 'branching rules' in applied literature, although they have their place in the Grothendieck rings of representations.
- Curious and completely unexplored relations between pairs of maximal subalgebras, say L' and L'', of the same Lie algebra L can be found by combining the projection matrices $P(L \supset L')$ and $P(L \supset L'')$ as

  $$P(L' \to L'') = P(L \supset L'')P^{-1}(L \supset L').$$

  Here L' must be of the same rank as L for $P(L \supset L')$ to be invertible. We write $L' \to L''$ instead of $L' \supset L''$ here because L'' is obviously not a subalgebra of L'.
- The index of a semisimple subalgebra in a simple Lie algebra is an invariant of all branching rules for a fixed algebra-subalgebra pair. It was introduced in [15], see Equation (2.26). It is also an invariant for any pair $W(L) \supset W(L')$.

Acknowledgments

This work was supported by the Alexander Graham Bell Doctoral Scholarship of the Natural Sciences and Engineering Research Council of Canada. I am grateful for the opportunity to speak at the XXVIII International Colloquium on Group-Theoretical Methods in Physics. I wish to thank Prof. Jiri Patera for his precious help and support.

References

[1] Humphreys J E 1990 *Reflection groups and Coxeter groups* (*Cambridge Studies in Advanced Mathematics* vol 29) (Cambridge: Cambridge University Press)

[2] Kane R 2001 *Reflection groups and invariant theory* CMS Books in Mathematics/Ouvrages de Mathématiques de la SMC, 5 (New York: Springer-Verlag)

[3] Champagne B, Kjiri M, Patera J and Sharp R 1995 *Can. J. Phys.* **73** 566–584

[4] Bremner M R, Moody R V and Patera J 1985 *Tables of dominant weight multiplicities for representations of simple Lie algebras* (*Monographs and Textbooks in Pure and Applied Mathematics* vol 90) (New York: Marcel Dekker Inc.)

[5] Patera J and Sankoff D 1973 *Tables of branching rules for representations of simple Lie algebras* (Les Presses de l'Université de Montréal, Montreal, Que.)

[6] McKay W, Patera J and Sankoff D 1977 *Computers in nonassociative rings and algebras* (New York: Academic Press) pp 235–277

[7] McKay W G and Patera J 1981 *Tables of dimensions, indices, and branching rules for representations of simple Lie algebras* (*Lecture Notes in Pure and Applied Mathematics* vol 69) (New York: Marcel Dekker Inc.)

[8] McKay W G, Patera J and Rand D W 1990 *Tables of representations of simple Lie algebras. Vol. I* (Montreal, QC: Université de Montréal Centre de Recherches Mathématiques)

[9] Háková L, Larouche M and Patera J 2008 *J. Phys. A* **41** 495202, 21

[10] Borel A and De Siebenthal J 1949 *Comment. Math. Helv.* **23** 200–221

[11] Gingras F, Patera J and Sharp R T 1992 *J. Math. Phys.* **33** 1618–1626

[12] Thoma M and Sharp R T 1996 *J. Math. Phys.* **37** 4750–4757

[13] Thoma M and Sharp R T 1996 *J. Math. Phys.* **37** 6570–6581

[14] Larouche M, Nesterenko M and Patera J 2009 *J. Phys. A* **42** 485203, 15

[15] Dynkin E B 1952 *Mat. Sbornik N.S.* **30(72)** 349–462 (3 plates)

Journal of Physics: Conference Series **284** (2011) 012044 doi:10.1088/1742-6596/284/1/012044

Exact dynamical and partial symmetries

A Leviatan

Racah Institute of Physics, The Hebrew University, Jerusalem 91904, Israel

E-mail: ami@phys.huji.ac.il

Abstract. We discuss a hierarchy of broken symmetries with special emphasis on partial dynamical symmetries (PDS). The latter correspond to a situation in which a non-invariant Hamiltonian accommodates a subset of solvable eigenstates with good symmetry, while other eigenstates are mixed. We present an algorithm for constructing Hamiltonians with this property and demonstrate the relevance of the PDS notion to nuclear spectroscopy, to quantum phase transitions and to mixed systems with coexisting regularity and chaos.

1. Introduction

Symmetries play an important role in dynamical systems. They provide quantum numbers for the classification of states, determine spectral degeneracies and selection rules, and facilitate the calculation of matrix elements. An exact symmetry occurs when the Hamiltonian of the system commutes with all the generators (g_i) of the symmetry-group G, $[\hat{H}\,,\,g_i] = 0$. In this case, all states have good symmetry and are labeled by the irreducible representations (irreps) of G. The Hamiltonian admits a block structure so that inequivalent irreps do not mix and all eigenstates in the same irrep are degenerate. In a dynamical symmetry the Hamiltonian commutes with the Casimir operator of G, $[\hat{H}\,,\,\hat{C}_G] = 0$, the block structure of $\hat{H}$ is retained, the states preserve the good symmetry but, in general, are no longer degenerate. When the symmetry is completely broken then $[\hat{H}\,,\,g_i] \neq 0$, and none of the states have good symmetry. In-between these limiting cases there may exist intermediate symmetry structures, called partial (dynamical) symmetries, for which the symmetry is neither exact nor completely broken. This novel concept of symmetry and its implications for dynamical systems are the focus of the present contribution.

Models based on spectrum generating algebras form a convenient framework to examine different types of symmetries and have been used extensively in diverse areas of physics [1–3]. In such models the Hamiltonian is expanded in elements of a Lie algebra, (G_0), called the spectrum generating algebra. A dynamical symmetry occurs if the Hamiltonian can be written in terms of the Casimir operators of a chain of nested algebras, $G_0 \supset G_1 \supset \ldots \supset G_n$. The following properties are then observed. (i) All states are solvable and analytic expressions are available for energies and other observables. (ii) All states are classified by quantum numbers, $|\alpha_0, \alpha_1, \ldots, \alpha_n\rangle$, which are the labels of the irreps of the algebras in the chain. (iii) The structure of wave functions is completely dictated by symmetry and is independent of the Hamiltonian's parameters.

The merits of a dynamical symmetry are self-evident. However, in most applications to realistic systems, the predictions of an exact dynamical symmetry are rarely fulfilled and one is compelled to break it. More often one finds that the assumed symmetry is not obeyed uniformly, *i.e.*, is fulfilled by only some states but not by others. The required symmetry-breaking is

exact symmetry **degeneracy**

all states with good symmetry

$$[\hat{H}\,,\,g_i] = 0 \qquad\qquad g_i \in G$$

dynamical symmetry **splitting no mixing**

all states solvable, with good symmetry

$$G_0 \supset G_1 \supset \ldots \supset G_n$$
$$|\alpha_0, \alpha_1, \ldots, \alpha_n\rangle$$
$$[\hat{H}\,,\,\hat{C}_{G_k}] = 0 \qquad\qquad \hat{H} = \sum_k a_k \hat{C}_{G_k}$$

partial symmetry

part of the **degenerate** states (solvable) with good symmetry

$$[\hat{H}\,,\,g_i] \;\neq\; 0$$
$$[\hat{H}\,,\,g_i]|\Psi\rangle \;=\; 0 \qquad g_i \in G$$

partial dynamical symmetry

part of **non-degenerate** states (solvable) with good symmetry

$$[\hat{H}\,,\,\hat{C}_G] \;\neq\; 0$$
$$[\hat{H}\,,\,\hat{C}_G]|\Psi\rangle \;=\; 0$$

broken symmetry

none of the states solvable, without good symmetry

$$[\hat{H}\,,\,g_i] \neq 0 \qquad\qquad g_i \in G$$

Figure 1. Hierarchy of symmetries.

achieved by including in the Hamiltonian terms associated with different sub-algebra chains of the parent spectrum generating algebra. In general, under such circumstances, solvability is lost, there are no remaining non-trivial conserved quantum numbers and all eigenstates are expected to be mixed. A partial dynamical symmetry (PDS) corresponds to a particular symmetry breaking for which some (but not all) of the virtues of a dynamical symmetry are retained. The essential idea is to relax the stringent conditions of *complete* solvability so that the properties (i)–(iii) are only partially satisfied [4].

The notion of partial dynamical symmetry generalizes the concepts of exact and dynamical symmetries. In making the transition from an exact to a dynamical symmetry, states which are degenerate in the former scheme are split but not mixed in the latter, and the block structure of the Hamiltonian is retained. Proceeding further to partial symmetry, some blocks or selected states in a block remain pure, while other states mix and lose the symmetry character. A partial dynamical symmetry lifts the remaining degeneracies, but preserves the symmetry-purity of the

Journal of Physics: Conference Series **284** (2011) 012044 doi:10.1088/1742-6596/284/1/012044

selected states. The hierarchy of broken symmetries is depicted in Fig. 1.

The existence of Hamiltonians with partial symmetry or partial dynamical symmetry is by no means obvious. An Hamiltonian with the above property is not invariant under the group G nor does it commute with the Casimir invariants of G, so that various irreps are in general mixed in its eigenstates. However, it posses a subset of solvable states, denoted by $|\Psi\rangle$ in Fig. 1, which respect the symmetry. The commutator $[\,\hat{H}\,,\,g_i\,]$ or $[\,\hat{H}\,,\,\hat{C}_G\,]$ vanishes only when it acts on these 'special' states with good G-symmetry.

2. Partial dynamical symmetries

When a partial dynamical symmetry (PDS) occurs, the defining properties of a dynamical symmetry (DS), namely, solvability, good quantum numbers, and symmetry-dictated structure are fulfilled exactly, but by only a subset of states. An algorithm for constructing Hamiltonians with PDS has been developed in [5] and further elaborated in [6]. The analysis starts from the chain of nested algebras

$$\begin{array}{ccccccc} G_{\rm dyn} & \supset & G & \supset & \cdots & \supset & G_{\rm sym} \\ \downarrow & & \downarrow & & & & \downarrow \\ [h] & & \langle\Sigma\rangle & & & & \Lambda \end{array} \tag{1}$$

where, below each algebra, its associated labels of irreps are given. Eq. (1) implies that $G_{\rm dyn}$ is the dynamical (spectrum generating) algebra of the system such that operators of all physical observables can be written in terms of its generators; a single irrep of $G_{\rm dyn}$ contains all states of relevance in the problem. In contrast, $G_{\rm sym}$ is the symmetry algebra and a single of its irreps contains states that are degenerate in energy. Assuming, for simplicity, that particle number is conserved, then all states, and hence the representation $[h]$, can then be assigned a definite particle number N. For N identical particles the representation $[h]$ of the dynamical algebra $G_{\rm dyn}$ is either symmetric $[N]$ (bosons) or antisymmetric $[1^N]$ (fermions) and will be denoted, in both cases, as $[h_N]$. The occurrence of a DS of the type (1) signifies that the Hamiltonian is written in terms of the Casimir operators of the algebras in the chain, $\hat{H}_{DS} = \sum_G a_G\,\hat{C}_G$, and the eigenstates can be labeled as $|[h_N]\langle\Sigma\rangle\ldots\Lambda\rangle$; additional labels (indicated by $\ldots$) are suppressed in the following. The eigenvalues of the Casimir operators in these basis states determine the eigenenergies $E_{DS}([h_N]\langle\Sigma\rangle\Lambda)$ of $\hat{H}_{DS}$. Likewise, operators can be classified according to their tensor character under (1) as $\hat{T}_{[h_n]\langle\sigma\rangle\lambda}$.

Of specific interest in the construction of a PDS associated with the reduction (1), are the n-particle annihilation operators $\hat{T}$ which satisfy the property

$$\hat{T}_{[h_n]\langle\sigma\rangle\lambda}|[h_N]\langle\Sigma_0\rangle\Lambda\rangle = 0\ , \tag{2}$$

for all possible values of Λ contained in a given irrep $\langle\Sigma_0\rangle$ of G. Equivalently, this condition can be phrased in terms of the action on a lowest weight (LW) state of the G-irrep $\langle\Sigma_0\rangle$, $\hat{T}_{[h_n]\langle\sigma\rangle\lambda}|LW;\,[h_N]\langle\Sigma_0\rangle\rangle = 0$, from which states of good Λ can be obtained by projection. Any n-body, number-conserving normal-ordered interaction written in terms of these annihilation operators and their Hermitian conjugates (which transform as the corresponding conjugate irreps), $\hat{H} = \sum_{\alpha,\beta} A_{\alpha\beta}\,\hat{T}^{\dagger}_{\alpha}\hat{T}_{\beta}$, has a partial G-symmetry. This comes about since for arbitrary coefficients, $A_{\alpha\beta}$, $\hat{H}$ is not a G-scalar, hence most of its eigenstates will be a mixture of irreps of G, yet relation (2) ensures that a subset of its eigenstates $|[h_N]\langle\Sigma_0\rangle\Lambda\rangle$, are solvable and have good quantum numbers under the chain (1). An Hamiltonian with partial dynamical symmetry is obtained by adding to $\hat{H}$ the dynamical symmetry Hamiltonian, $\hat{H}_{PDS} = \hat{H}_{DS} + \hat{H}$, still preserving the solvability of states with $\langle\Sigma\rangle = \langle\Sigma_0\rangle$.

If the operators $\hat{T}_{[h_n]\langle\sigma\rangle\lambda}$ span the entire irrep $\langle\sigma\rangle$ of G, then the annihilation condition (2) is satisfied for all Λ-states in $\langle\Sigma_0\rangle$, if none of the G irreps $\langle\Sigma\rangle$ contained in the $G_{\rm dyn}$ irrep $[h_{N-n}]$

belongs to the G Kronecker product $\langle\sigma\rangle \times \langle\Sigma_0\rangle$. So the problem of finding interactions that preserve solvability for part of the states (1) is reduced to carrying out a Kronecker product. If relation (2) holds only for some states Λ in the irrep $\langle\Sigma_0\rangle$ and/or some components λ of the tensor $\hat{T}_{[h_n]\langle\sigma\rangle\lambda}$, then the Kronecker product rule does not apply. However, the PDS Hamiltonian is still of the indicated normal-ordered form, but now the solvable states span only part of the corresponding G-irrep. The arguments for choosing the special irrep $\langle\Sigma\rangle = \langle\Sigma_0\rangle$ in Eq. (2), which contains the solvable states, are based on physical grounds. A frequently encountered choice is the irrep which contains the ground state of the system.

In what follows we illustrate the above procedure and demonstrate the relevance of the PDS notion to dynamical systems. For that purpose, we employ the interacting boson model (IBM) [2], widely used in the description of low-lying collective states in nuclei in terms of N interacting monopole (s) and quadrupole (d) bosons representing valence nucleon pairs. The dynamical algebra is $G_{\text{dyn}} = \text{U}(6)$ and the symmetry algebra is $G_{\text{sym}} = \text{O}(3)$. The Hamiltonian commutes with the total number operator of s- and d- bosons, $\hat{N}$, which is the linear Casimir of U(6). Three DS limits occur in the model with leading subalgebras U(5), SU(3), and O(6), corresponding to typical collective spectra observed in nuclei, vibrational, rotational, and γ-unstable, respectively. A geometric visualization of the model is obtained by an energy surface

$$E_N(\beta,\gamma) \;=\; \langle\beta,\gamma;N|\hat{H}|\beta,\gamma;N\rangle \;, \tag{3}$$

defined by the expectation value of the Hamiltonian in the coherent (intrinsic) state [7, 8]

$$|\beta,\gamma;N\rangle \;=\; (N!)^{-1/2}(b_c^{\dagger})^N\,|0\,\rangle \;, \tag{4a}$$

$$b_c^{\dagger} \;=\; (1+\beta^2)^{-1/2}[\beta\cos\gamma d_0^{\dagger} + \beta\sin\gamma(d_2^{\dagger}+d_{-2}^{\dagger})/\sqrt{2} + s^{\dagger}] \;. \tag{4b}$$

Here (β,γ) are quadrupole shape parameters whose values, (β_0,γ_0), at the global minimum of $E_N(\beta,\gamma)$ define the equilibrium shape for a given Hamiltonian. For a Hamiltonian with one- and two-body interactions, the shape can be spherical ($\beta = 0$) or deformed ($\beta > 0$) with $\gamma = 0$ (prolate), $\gamma = \pi/3$ (oblate), or γ-independent. The equilibrium deformations associated with the DS limits are $\beta_0 = 0$ for U(5), $(\beta_0 = \sqrt{2}, \gamma_0 = 0)$ for SU(3) and $(\beta_0 = 1, \gamma_0\,\text{arbitrary})$ for O(6).

3. PDS and nuclear spectroscopy

The SU(3) DS chain of the IBM and related quantum numbers are given by [2]

$$\begin{array}{ccccc} \text{U}(6) & \supset & \text{SU}(3) & \supset & \text{O}(3) \\ \downarrow & & \downarrow & & \downarrow \\ [N] & & (\lambda,\mu) & K & L \end{array} \;, \tag{5}$$

where K is a multiplicity label needed for complete classification in the $SU(3) \supset O(3)$ reduction. The spectrum of the SU(3) DS Hamiltonian resembles that of an axially-deformed rotovibrator and the corresponding eigenstates are arranged in SU(3) multiplets. The label K corresponds geometrically to the projection of the angular momentum on the symmetry axis. In a given SU(3) irrep (λ,μ), each K-value is associated with a rotational band and states with the same angular momentum L, in different K-bands, are degenerate. The lowest SU(3) irrep is $(2N,0)$, which describes the ground band $g(K = 0)$ of a prolate deformed nucleus. The first excited SU(3) irrep $(2N-4,2)$ contains degenerate $\beta(K = 0)$ and $\gamma(K = 2)$ bands. This β-γ degeneracy is a characteristic feature of the SU(3) limit which, however, is not commonly observed. In most deformed nuclei the β band lies above the γ band. In the IBM framework, with at most two-body interactions, one is therefore compelled to break SU(3) in order to conform with the experimental data.

The construction of Hamiltonians with SU(3)-PDS is based on identification of n-boson operators which annihilate all states in a given SU(3) irrep (λ,μ), chosen here to be the ground band irrep $(2N,0)$. For that purpose, we consider the following two-boson SU(3) tensors, $B^{\dagger}_{[n](\lambda,\mu)K;Lm}$, with $n=2$, $(\lambda,\mu)=(0,2)$ and angular momentum $L=0,\,2$

$$B^{\dagger}_{[2](0,2)0;00} \;\propto\; P^{\dagger}_{0} = d^{\dagger}\cdot d^{\dagger} - 2(s^{\dagger})^2 \;, \tag{6a}$$

$$B^{\dagger}_{[2](0,2)0;2m} \;\propto\; P^{\dagger}_{2m} = 2d^{\dagger}_{m}s^{\dagger} + \sqrt{7}\,(d^{\dagger}\,d^{\dagger})^{(2)}_{m} \;. \tag{6b}$$

The corresponding Hermitian conjugate boson-pair annihilation operators, P_0 and P_{2m}, transform as $(2,0)$ under SU(3), and satisfy

$$\begin{aligned} P_0\,|[N](2N,0)K=0,L\rangle &= 0\;,\\ P_{2m}\,|[N](2N,0)K=0,L\rangle &= 0\;. \end{aligned} \tag{7}$$

The indicated L-states span the entire SU(3) irrep $(\lambda,\mu)=(2N,0)$. They can be obtained by angular momentum projection from the coherent state, $|\beta=\sqrt{2},\gamma=0;N\rangle$, of Eq. (4), which is the lowest-weight state of this irrep and serves as an intrinsic state for the SU(3) ground band. The relations in Eq. (7) follow from the fact that the action of the operators P_{Lm} leads to a state with $N-2$ bosons in the U(6) irrep $[N-2]$, which does not contain the SU(3) irreps obtained from the product $(2,0)\times(2N,0)=(2N+2,0)\oplus(2N,1)\oplus(2N-2,2)$. In addition, P_0 satisfies

$$P_0\,|[N](2N-4k,2k)K=2k,LM\rangle \;=\; 0\;, \tag{8}$$

where for $k>0$ the indicated L-states span only part of the SU(3) irreps $(\lambda,\mu)=(2N-4k,2k)$ and form the rotational members of excited $\gamma^k(K=2k)$ bands.

Following the general algorithm, a two-body Hamiltonian with partial SU(3) symmetry can now be constructed as [5,9]

$$\hat{H}(h_0,h_2) \;=\; h_0\,P^{\dagger}_{0}P_0 + h_2\,P^{\dagger}_{2}\cdot\tilde{P}_2\;, \tag{9}$$

where $\tilde{P}_{2m}=(-)^{m}P_{2,-m}$ and the dot denotes a scalar product. For $h_2=h_0$, the Hamiltonian is an SU(3) scalar, related to the quadratic Casimir operator of SU(3), for $h_0=-5h_2$ it transforms as a $(2,2)$ SU(3) tensor component and has the form

$$\hat{H}(h_0,h_2) \;=\; h_2\left[-\hat{C}_{\mathrm{SU(3)}} + 2\hat{N}(2\hat{N}+3)\right] + (h_0-h_2)\,P^{\dagger}_{0}P_0\;. \tag{10}$$

The first term in Eq. (10) belongs to the SU(3) DS Hamiltonian. The $P^{\dagger}_{0}P_0$ term is not diagonal in the SU(3) chain, however, Eqs. (7)-(8) ensure that $\hat{H}(h_0,h_2)$ retains selected solvable states with good SU(3) symmetry. Specifically, the solvable states are members of the ground $g(K=0)$ and $\gamma^k(K=2k)$ bands with the following characteristics

$$\begin{array}{lll} |N,(2N,0)K=0,L\rangle & E=0 & L=0,2,4,\ldots,2N\\ |N,(2N-4k,2k)K=2k,L\rangle & E=h_2\,6k\,(2N-2k+1) & L=K,K+1,\ldots,(2N-2k)\;. \end{array} \tag{11}$$

The remaining eigenstates of $\hat{H}(h_0,h_2)$ do not preserve the SU(3) symmetry and therefore get mixed. This situation corresponds precisely to that of partial SU(3) symmetry. An Hamiltonian $\hat{H}(h_0,h_2)$ which is not an SU(3) scalar has a subset of *solvable* eigenstates which continue to have good SU(3) symmetry. One can add the Casimir operator of O(3), $\hat{C}_{\mathrm{O(3)}}$, to $\hat{H}(h_0,h_2)$ and by doing so, convert the partial SU(3) symmetry into partial dynamical SU(3) symmetry. The

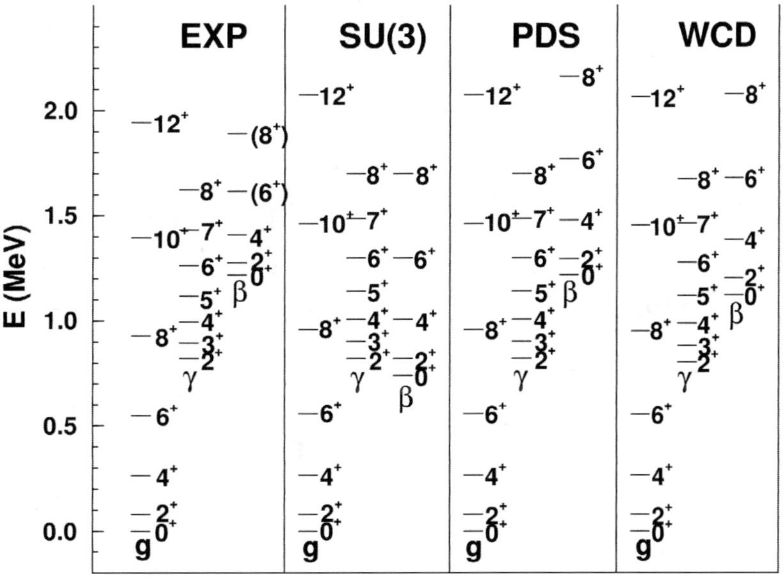

Figure 2. Spectra of ^{168}Er ($N = 16$). Experimental energies (EXP) are compared with IBM calculations in an exact SU(3) dynamical symmetry [SU(3)], in a broken SU(3) symmetry (WCD) and in a partial dynamical SU(3) symmetry (PDS). The latter employs the Hamiltonian of Eq. (9), $\hat{H}(h_0, h_2) + C\,\hat{C}_{O(3)}$, with $h_0 = 8$, $h_2 = 4$, $C = 13$ keV.

additional rotational term contributes just an $L(L + 1)$ splitting but does not affect the wave functions.

The empirical spectrum of ^{168}Er is shown in Fig. 2 and compared with SU(3)-DS, SU(3)-PDS and broken SU(3) calculations [9]. The SU(3)-PDS spectrum shows an improvement over the schematic, exact SU(3) dynamical symmetry description, since the β-γ degeneracy is lifted. The quality of the calculated PDS spectrum is similar to that obtained in the broken-SU(3) calculation, however, in the former the ground $g(K = 0_1)$ and $\gamma(K = 2_1)$ bands remain solvable with good SU(3) symmetry, $(\lambda, \mu) = (2N, 0)$ and $(2N - 4, 2)$ respectively. At the same time, the excited $K = 0_2^+$ band involves about 13% SU(3) admixtures into the dominant $(2N - 4, 2)$ irrep. Since the wave functions of the solvable states (11) are known, one can obtain *analytic* expressions for matrix elements of observables between them. In particular, the calculated B(E2) ratios for $\gamma \to g$ transitions lead to parameter-free predictions in excellent agreement with experiment [9], thus confirming the relevance of SU(3)-PDS to the spectroscopy of ^{168}Er.

4. PDS and quantum phase transitions

Quantum phase transitions (QPT) occur at zero temperature as a function of a coupling constant in the Hamiltonian. Such ground-state energy phase transitions are a pervasive phenomenon observed in many branches of physics, and are realized empirically in nuclei as transitions between different shapes. QPTs occur as a result of a competition between terms in the Hamiltonian with different symmetry character, which lead to considerable mixing in the eigenfunctions, especially at the critical-point where the structure changes most rapidly. An interesting question to address is whether there are any symmetries (or traces of) still present at the critical points of QPT. As shown below, unexpectedly, partial dynamical symmetries can survive at the critical point in spite of the strong mixing [10].

A convenient framework to study symmetry aspects of QPTs in nuclei is the IBM [2], whose

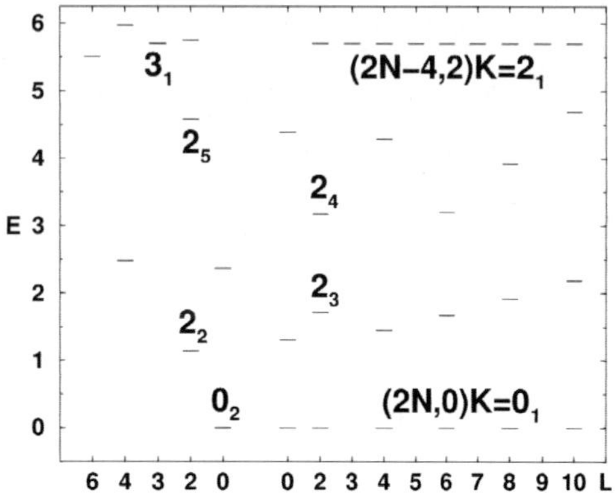

Figure 3. Spectrum of $\hat{H}_{cri}$, Eq. (13), with $h_2 = 0.05$ and $N = 10$. $L(K = 0_1)$ and $L(K = 2_1)$ are the solvable SU(3) states of Eq. (11) with $k = 0$ and $k = 1$, respectively. $L = 0_2, 3_1$ are the solvable U(5) states of Eq. (14).

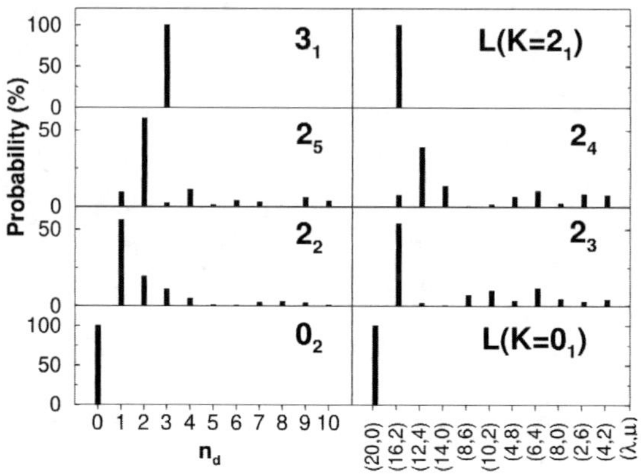

Figure 4. U(5) (n_d) and SU(3) $[(\lambda, \mu)]$ decomposition for selected spherical and deformed states in Fig. 3.

dynamical symmetries correspond to possible phases of the system. The relevant Hamiltonian in such a study involves terms with from different DS chains. The nature of the phase transition is governed by the topology of the corresponding surface (3), which serves as a Landau's potential with the equilibrium deformations as order parameters. The surface at the critical-point of a first-order transition is required to have two-degenerate minima, corresponding to the two coexisting phases. For example, the following first-order critical surface

$$E_N(\beta, \gamma = 0) \;=\; 2h_2 N(N-1)(1+\beta^2)^{-2}\beta^2 \left(\beta - \sqrt{2}\right)^2 , \tag{12}$$

has degenerate spherical and deformed minima at $\beta = 0$ and $(\beta = \sqrt{2}, \gamma = 0)$, corresponding to spherical and axially-deformed shapes. A barrier of height $h = 2h_2 N(N-1)(1-\sqrt{3}/2)$ separates the two minima. Such a surface can be obtained from Eq. (3) with the following Hamiltonian

$$\hat{H}_{cri} \;=\; h_2\, P_2^\dagger \cdot \tilde{P}_2 , \tag{13}$$

which is qualified to be a critical Hamiltonian of a first-order transition between these shapes. $\hat{H}_{cri}$ is recognized to be a special case of the Hamiltonian of Eq. (9), shown to have SU(3)-PDS. As such, it has a subset of solvable eigenstates, Eq. (11), which are members of deformed ground $g(K = 0)$ and $\gamma^k(K = 2k)$ bands with good SU(3) symmetry, $(\lambda, \mu) = (2N - 4k, 2k)$. In addition, H_{cri} has also the following solvable spherical eigenstates with good U(5) symmetry

$$|N, n_d = \tau = L = 0\rangle \qquad E = 0 , \tag{14a}$$
$$|N, n_d = \tau = L = 3\rangle \qquad E = 6h_2(2N - 1) . \tag{14b}$$

These are selected basis states, $|[N], n_d, \tau, L\rangle$, of the U(5)-DS chain $U(6) \supset U(5) \supset O(5) \supset O(3)$. H_{cri} (13) is not invariant under $U(5)$, nor does it have a U(5)-DS, yet it has solvable states with good U(5) symmetry. By definition, it posses a U(5)-PDS. The spherical $L = 0$ state, Eq. (14a), is exactly degenerate with the SU(3) ground band, Eq. (11) with $k = 0$, and the spherical $L = 3$ state, Eq. (14b), is degenerate with the SU(3) γ-band, Eq. (11) with $k = 1$. The remaining levels of $\hat{H}_{cri}$, shown in Fig. 3 are calculated numerically. Their wave functions are spread over many U(5) and SU(3) irreps, as is evident from the n_d and (λ, μ) decomposition shown in Fig. 4. This

Journal of Physics: Conference Series **284** (2011) 012044 doi:10.1088/1742-6596/284/1/012044

situation, where some states are solvable with good U(5) symmetry, some are solvable with good SU(3) symmetry and all other states are mixed with respect to both U(5) and SU(3), defines a U(5) PDS coexisting with a SU(3) PDS. The presence in the spectrum of both spherical states (dominated by a single n_d component) and deformed states arranged in bands (with a broad n_d distribution) signals a first-order transition.

The above results demonstrate the relevance of the PDS notion to critical-points of QPT, with phases characterized by Lie-algebraic symmetries. In the example considered, first-order critical Hamiltonians exhibit distinct subsets of solvable states with good symmetries, giving rise to a coexistence of different PDS. The ingredients of an algebraic description of QPT is a spectrum generating algebra and an associated geometric space, formulated in terms of coherent (intrinsic) states. The same ingredients are used in the construction of Hamiltonians with PDS. These, in accord with the present discussion, can be used as tools to explore the role of partial symmetries in governing the critical behaviour of dynamical systems undergoing QPT.

5. PDS and mixed regular and chaotic dynamics

Partial dynamical symmetries can play a role not only for discrete spectroscopy but also for analyzing statistical aspects of nonintegrable systems. Hamiltonians with a dynamical symmetry are always completely integrable. The Casimir invariants of the algebras in the chain provide a set of constants of the motion in involution. The classical motion is purely regular. A symmetry-breaking is connected to nonintegrability and may give rise to chaotic motion. Hamiltonians with PDS are not completely integrable, hence can exhibit stochastic behavior, nor are they completely chaotic, since some eigenstates preserve the symmetry exactly. Consequently, such Hamiltonians are optimally suitable to the study of mixed systems with coexisting regularity and chaos.

The dynamics of a generic classical Hamiltonian system is mixed; KAM islands of regular motion and chaotic regions coexist in phase space. In the associated quantum system, if no separation between regular and irregular states is done, the statistical properties of the spectrum are usually intermediate between the Poisson and the Gaussian orthogonal ensemble (GOE) statistics. In a PDS, the symmetry of the subset of solvable states is exact, yet does not arise from invariance properties of the Hamiltonian. If the fraction of solvable states remains finite in the classical limit, one might expect that a corresponding fraction of the phase space would consist of KAM tori and exhibit regular motion. It turns out that PDS has an even greater effect on the dynamics. It is strongly correlated with suppression (*i.e.*, reduction) of chaos even though the fraction of solvable states approaches zero in the classical limit [11, 12].

We consider the following IBM Hamiltonian

$$\hat{H}(\beta_0) \;=\; h_0\, P_0^\dagger(\beta_0)P_0(\beta_0) + h_2\, P_2^\dagger(\beta_0)\cdot\tilde{P}_2(\beta_0)\;, \tag{15}$$

where $P_0^\dagger(\beta_0) = d^\dagger\cdot d^\dagger - \beta_0^2(s^\dagger)^2$ and $P_{2m}^\dagger(\beta_0) = \beta_0\sqrt{2}\,d_m^\dagger s^\dagger + \sqrt{7}\,(d^\dagger\, d^\dagger)^{(2)}_m$. For $\beta_0 = \sqrt{2}$, $\hat{H}(\beta_0 = \sqrt{2})$ reduces to the Hamiltonian of Eq. (9), which has SU(3)-PDS with a subset of solvable states listed in Eq. (11). At a given spin per boson $l = L/N$, and to leading order in $1/N$, the fraction f of solvable states decreases like $1/N^2$ with boson number. However, at a given boson number N, this fraction increases with l, a feature which is valid also for finite N [11]. The classical Hamiltonian is obtained from (15) by replacing operators by c-numbers, $s^\dagger, d_\mu^\dagger \to \alpha_s^*, \alpha_\mu^*$ and taking $N\to\infty$, with $1/N$ playing the role of $\hbar$.

To study the effect of the SU(3) PDS on the dynamics, we fix the ratio h_2/h_0 at a value far from the exact SU(3) symmetry (for which $h_0/h_2 = 1$). We then change β_0 in the range $1 \le \beta_0 \le 2$. Classically, we determine the fraction σ of chaotic volume and the average largest Lyapunov exponent $\bar{\lambda}$. To analyze the quantum Hamiltonian, we study spectral and transition intensity distributions. The nearest neighbors level spacing distribution is fitted

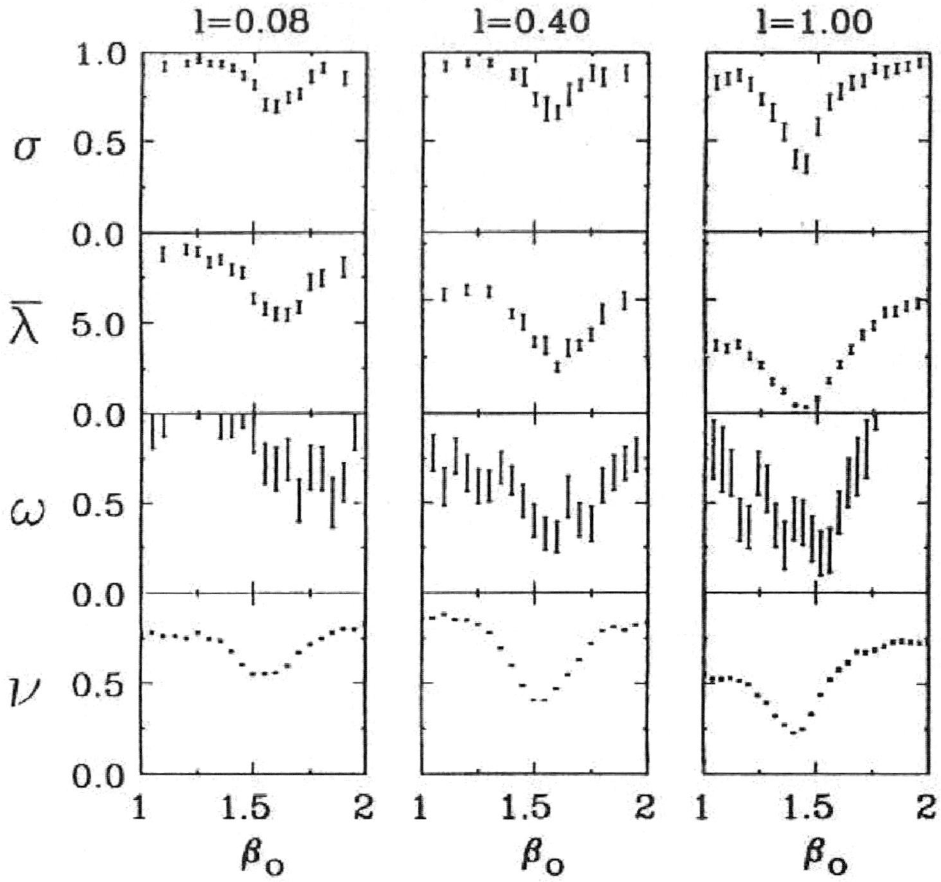

Figure 5. Classical $(\sigma, \bar{\lambda})$ and quantal (ω, ν) measures of chaos versus β_0 for the Hamiltonian (15) with $h_2/h_0 = 7.5$. Shown are three cases with classical spins $l = 0.08$, 0.4, and 1. The quantal calculations (ω, ν) are done for $N = 25$ bosons and spins $L = 2, 10$, and 25, respectively. Notice that with increasing spin the minimum gets deeper and closer to $\beta_0 = \sqrt{2}$. The suppression of chaos near $\beta_0 = \sqrt{2}$ is seen both for finite N through the measures ω, ν and in the classical limit $N \to \infty$ through the measures σ, $\bar{\lambda}$.

by a Brody distribution, $P_\omega(S) = AS^\omega \exp(-\alpha S^{1+\omega})$, where A and α are determined by the conditions that $P_\omega(S)$ is normalized to 1 and $\langle S \rangle = 1$. For the Poisson statistics $\omega = 0$ and for GOE $\omega = 1$, corresponding to integrable and fully chaotic classical motion, respectively. The intensity distribution of the E2 operator is fitted by a χ^2 distribution in ν degrees of freedom, $P_\nu(y) = [(\nu/2\langle y \rangle)^{\nu/2}/\Gamma(\nu/2)]y^{\nu/2-1}\exp(-\nu y/2\langle y \rangle)$. For the GOE, $\nu = 1$ and ν decreases as the dynamics become more regular.

Fig. 5 shows the two classical measures σ, $\bar{\lambda}$ and the two quantum measures ω, ν for the Hamiltonian (15) as a function of β_0. The parameters of the Hamiltonian are taken to be $h_2/h_0 = 7.5$ and the number of bosons is $N = 25$. Shown are three classical spins $l = 0.08$, 0.4 and 1, which correspond in the quantum case to $L = 2$, 10 and 25. All measures show a pronounced minimum which gets deeper and closer to $\beta_0 = \sqrt{2}$ [where the partial SU(3) symmetry occurs] as the classical spin increases. This behaviour is correlated with the fraction of solvable states (at a constant N) being larger at higher l. We remark that the classical measures show a clear enhancement of the regular motion near $\beta_0 = \sqrt{2}$ even though the fraction of solvable states vanishes as $1/N^2$ in the classical limit $N \to \infty$. That the observed suppression of chaos is related to the SU(3) PDS, is confirmed by the fact that the SU(3) entropy averaged over all eigenstates of $\hat{H}(\beta_0)$ displays a minimum which is well correlated with the minimum in Fig. 5. The existence of an SU(3) PDS seems to have an effect of increasing the SU(3) symmetry of all states, not just those with an exact SU(3) symmetry.

The following physical picture emerges from the analysis of low-dimensional systems [11, 12]. At the quantum level, PDS by definition implies the existence of a "special" subset of states, which observe the symmetry. The PDS affects the purity of other states in the system; in particular, neighboring states, accessible by perturbation theory, possess approximately good symmetry. Analogously, at the classical level, the region of phase space near the "special" torus also has toroidal structure. As a consequence of having PDS, a finite region of phase space is regular and a finite fraction of states is approximately "special". This clarifies the observed suppression of chaos.

6. Concluding remarks

Underlying the PDS notion, is the recognition that a non-invariant Hamiltonian can have selected eigenstates with good symmetry and good quantum numbers. In such a case, the symmetry in question is preserved in some states but is broken in the Hamiltonian (an opposite situation to that encountered in a spontaneously-broken symmetry). PDSs appear to be a common feature in algebraic descriptions of dynamical systems. They are not restricted to a specific model but can be applied to any quantal systems of interacting particles, bosons and fermions [4, 13, 14].

In PDS of type I described above, only part of the eigenspectrum is analytically solvable and retains all the dynamical symmetry (DS) quantum numbers. Additional types of PDS are possible. In PDS of type II, the entire eigenspectrum retains some of the DS quantum numbers [15]. PDS of type III has a hybrid character, in the sense that some (solvable) eigenstates keep some of the quantum numbers [16]. General algorithms for selecting and constructing Hamiltonians with PDSs of various types are available. The advantage of using interactions with a PDS is that they can be introduced, in a controlled manner, without destroying results previously obtained with a DS for a segment of the spectrum. These virtues generate an efficient tool which greatly enhance the scope of applications of algebraic modeling of quantum many-body systems.

Acknowledgments

This work is supported by the Israel Science Foundation and the BSF.

[1] Bohm A, Néeman Y and Barut A O eds 1988 *Dynamical Groups and Spectrum Generating Algebras* (Singapore: World Scientific)
[2] Iachello F and Arima A 1987 *The Interacting Boson Model* (Cambridge: Cambridge Univ. Press)
[3] Iachello F and Levine R D 1994 *Algebraic Theory of Molecules* (Oxford: Oxford Univ. Press)
[4] Leviatan A 2010 *Prog. Part. Nucl. Phys.*, in press
[5] Alhassid Y and Leviatan A 1992 *J. Phys. A* **25** L1265
[6] García-Ramos J E, Leviatan A and Van Isacker P 2009 *Phys. Rev. Lett.* **102** 112502
[7] Ginocchio J N and Kirson M W 1980 *Phys. Rev. Lett.* **44** 1744
[8] Dieperink A E L, Scholten O and Iachello F 1980 Phys. Rev. Lett. **44** 1747
[9] Leviatan A 1996 *Phys. Rev. Lett.* **77** 818
[10] Leviatan A 2007 *Phys. Rev. Lett.* **98** 242502
[11] Whelan N , Alhassid Y and Leviatan A 1993 *Phys. Rev. Lett.* **71** 2208
[12] Leviatan A and Whelan N D 1996 *Phys. Rev. Lett.* **77** 5202
[13] Escher J and Leviatan A 2000 *Phys. Rev. Lett.* **84** 1866
[14] Van Isacker P and Heinze S 2008 *Phys. Rev. Lett.* **100** 052501
[15] Van Isacker P 1999 *Phys. Rev. Lett.* **83** 4269
[16] Leviatan A and Van Isacker P 2002 *Phys. Rev. Lett.* **89** 222501

Noninertial symmetry group with invariant Minkowski line element consistent with Heisenberg quantum commutation relations

Stephen G. Low
www.Stephen-Low.net
E-mail: `Stephen.Low@alumni.utexas.net`

Abstract.
The maximal symmetry of a quantum system with Heisenberg commutation relations is given by the projective representations of the automorphism group of the Weyl-Heisenberg algebra. The automorphism group is the central extension of the inhomogeneous symplectic group with a conformal scaling that acts on extended phase space. We determine the subgroup that also leaves invariant a degenerate Minkowski orthogonal line element. This defines noninertial relativistic symmetry transformations that have the expected classical limit as $c \rightarrow \infty$.

1. Introduction

The inhomogeneous Lorentz group defines the relation between inertial states. Clocks locally at rest relative to an inertial state are related to the clocks of other inertial states through the Minkowski proper time line element. Quantum states are rays in a Hilbert space and therefore inertial quantum states are related through the projective representation of the inhomogeneous Lorentz group. Projective representations are equivalence classes of unitary representations of the central extension of the group. The central extension of the inhomogeneous Lorentz group is its cover, the Poincaré group, as this group does not admit an algebraic extension. [1],[2]

The equivalence principle of general relativity enables the noninertial frames of a particle accelerating under gravity to be understood as locally inertial states on a curved manifold. Particles under gravity follow geodesics that are locally inertial trajectories and neighboring locally inertial frames are related by the connection. The clock locally at inertial rest is related to the local clocks of other neighboring locally inertial states in the gravitating system through the Riemannian proper time line element.

Neither general relativity nor special relativity addresses the issue of noninertial states that are not due to gravity, but rather one of the other forces. A special case is a region in which gravity is negligible and the underlying manifold may be considered to be flat. Consider for example an electron in a region that gravity is negligible that encounters an electromagnetic field and therefore has a noninertial trajectory. How are the clocks of such noninertial states related?

We hypothesize that the noninertial relativistic symmetry group relating these states is the most general group consistent with the requirements that

1) the Heisenberg uncertainty principle holds in the noninertial as well as inertial states

2) the proper time given by the Minkowski line element that is invariant for noninertial as well as inertial states

Consider a quantum system in which the position, momentum, energy and time degrees of freedom are represented by the Hermitian representation of the algebra of the Weyl-Heisenberg group $\mathcal{H}(n+1)$ where the number of spacial dimensions is $n = 3$. The requirement that the algebra transforms into itself under the action of the relativistic symmetry group means that the symmetry group is a subgroup of the automorphism group of the Weyl-Heisenberg algebra. [3],[4],[5] This automorphism group is

$$\mathcal{A}ut_{\mathcal{H}} \simeq \mathbb{Z}_2 \otimes_s \mathcal{D} \otimes_s \overline{\mathcal{S}p}(2n+2) \otimes_s \mathcal{H}(n+1). \tag{1}$$

$\mathbb{Z}_2$ is the 2 element discrete group, $\mathcal{D}$ is the abelian group isomorphic to the reals under multiplication, $\mathcal{S}p(2n+2)$ is the symplectic group with the over-bar denoting the universal cover and $\mathcal{H}(n+1)$ is the Weyl-Heisenberg group. At this point there is no concept of a relativistic symmetry.

The Minkowski line element is $d\tau^2 = dt^2 - \frac{1}{c^2}dq^2$. This is an invariant for states that are inertially related and the second assertion is that this continues to be true for general noninertial states.

We will show that the homogeneous relativistic symmetry group that is a subgroup of the automorphism group of the Weyl-Heisenberg group that leaves the Minkowski line element invariant is

$$\mathcal{O}a(1,n) \simeq \mathcal{O}(1,n) \otimes_s \mathcal{A}(m), \tag{2}$$

where $m = (n+1)(n+2)/2$ and $\mathcal{A}(m)$ is the abelian group isomorphic to $\mathbb{R}^m$ under addition. The additional generators of the abelian group behave as a power-force stress tensor that is the proper time derivative of the energy-momentum stress tensor. We show that this leads to expected relativistic results in transforming to noninertial states.

This relativistic theory must lead to expected classical results in the limit $c \to \infty$ where the Minkowski line element reduces to the invariant Newtonian time line element dt^2. We have previously studied the most general group that leaves invariant the Newtonian time line element dt^2 that is a subgroup of the automorphisms of the Weyl-Heisenberg group. This results in a group that leads directly to Hamilton's equations and, with the additional requirement of orthonormal position frames, describes the Hamilton symmetry group for noninertial transformations in a classical context. [4]

2. Consistency between a relativistic symmetry group and quantum mechanics

States in quantum mechanics are represented by rays Ψ in a Hilbert space $\mathbf{H}$ that are the equivalence class of states $|\psi\rangle$ in the Hilbert space that are related by a phase

$$\Psi \simeq \left\{ e^{i\omega} |\psi\rangle \,|\omega \in \mathbb{R} \right\}, \qquad |\psi\rangle \in H. \tag{3}$$

A relativistic symmetry group $g \in \mathcal{G}$ acts on the states through a projective representation π, $\tilde{\Psi} = \pi(g)\Psi$, with the property that it also has a phase,

$$\pi(\tilde{g} \cdot g) = e^{i\omega(\tilde{g},g)}\pi(\tilde{g})\pi(g), \omega(\tilde{g},g) \in \mathbb{R}. \tag{4}$$

Projective representations are equivalence classes of the unitary representations ϱ of the central extension $\check{\mathcal{G}}$ (denoted by the inverted hat) of the group $\mathcal{G}$ that act on the states, [6],[7]

$$|\tilde{\psi}\rangle = \varrho(g) \left|\psi\right\rangle, \quad g \in \check{\mathcal{G}}, \qquad |\psi\rangle \in \mathbf{H}^{\varrho}. \tag{5}$$

The Hilbert space is determined by the unitary representation ϱ and so we label it as $\mathbf{H}^{\varrho}$. Observables corresponding to a the relativistic symmetry group $\mathcal{G}$ are represented by the Hermitian representations ϱ' of elements Z of the algebra of the centrally extended group$\check{\mathcal{G}}$, $\hat{Z} = \varrho'(Z)$. The action of the projective representation of the group element $g \in \check{\mathcal{G}}$ on these observables is

$$\hat{\tilde{Z}}\,|\tilde{\psi}\rangle = \varrho(g)\hat{Z}\,|\psi\rangle = \varrho(g)\hat{Z}\varrho(g)^{-1}\varrho(g)\,|\psi\rangle = \varrho(g)\hat{Z}\varrho(g)^{-1}\,|\tilde{\psi}\rangle \tag{6}$$

and so $\hat{\tilde{Z}} = \varrho'(\tilde{Z}) = \varrho'(gZg^{-1})$. Therefore, if the representation ϱ is faithful, we have $\tilde{Z} = gZg^{-1}$.

Position, momentum, energy and time observables are the Hermitian representation of the algebra of the Weyl-Heisenberg group $\mathcal{H}(n+1)$ with a general element given by $Z = z^{\alpha} Z_{\alpha}$, $\alpha = 1, ..2n{+}2$ where $\{z^{\alpha}\} \in \mathbb{P} \simeq \mathbb{R}^{2n+2}$ and Z_{α} are a dimensionless basis for the Weyl-Heisenberg algebra that satisfy the commutation relations

$$[Z_{\alpha}, Z_{\beta}] = \zeta_{\alpha,\beta} I, \tag{7}$$

where $\zeta_{\alpha,\beta}$ are the components of a symplectic metric. The Hermitian representation of the algebra satisfies $[\hat{Z}_{\alpha}, \hat{Z}_{\beta}] = i\zeta_{\alpha,\beta}\hat{I}$, where $\hat{Z}_{\alpha} = \varrho'(Z_{\alpha})$ and $\hat{I} = \varrho'(I)$ is the unit operator on the Hilbert space. The position-time and momentum-energy Hermitian operators are $\{\hat{Z}_{\alpha}\} = \{\hat{X}_a, \hat{P}_a\}$ with $a = 0, ..n$. Note that there is no relativistic symmetry yet and hence no concept of a mass shell.

The basic physical assumption is that the Heisenberg commutation relations are satisfied by any state related by a relativistic symmetry group $\mathcal{G}$. That is, position, momentum, energy and time observables satisfying the Heisenberg quantum commutation relations will also satisfy the Heisenberg quantum commutation relations for any states related by the projective representations of the symmetry group (6). This implies that if $\{X_a, P_a, I\}$ are a basis of the Weyl-Heisenberg algebra, then $\{\tilde{X}_a, \tilde{P}_a, \tilde{I}\}$ are also a basis of the Weyl-Heisenberg algebra where

$$\tilde{X}_a = gX_a g^{-1}, \tilde{P}_a = gP_a g^{-1}, \tilde{I} = gIg^{-1} = I. \tag{8}$$

and $g \in \check{\mathcal{G}}$ and ϱ is a faithful representation. The maximal group for which this property is true is the automorphism group of the Weyl-Heisenberg group. This results in basic consistency condition that the central extension $\check{\mathcal{G}}$ of the relativistic symmetry group $\mathcal{G}$ must be a subgroup of the automorphism group of the Weyl-Heisenberg group and algebra, $\check{\mathcal{G}} \subseteq \mathcal{A}ut_{\mathcal{H}(n+1)}$.

The automorphism group of the Weyl-Heisenberg group is [3],[4],[5]

$$\mathcal{A}ut_{\mathcal{H}(n+1)} = \overline{\mathcal{DS}p}(2n+2) \otimes_s \mathcal{H}(n+1). \tag{9}$$

The Heisenberg group itself are the inner automorphisms. The outer automorphisms are the cover of the homogeneous group

$$\mathcal{DS}p(2n+2) \simeq \mathbb{Z}_2 \otimes \mathcal{D} \otimes \mathcal{S}p(2n+2). \tag{10}$$

The matrix realization of this group and the group properties are given in [4],[5]. The inhomogeneous group includes translations on extended phase space $\mathbb{P}$,

$$\begin{aligned} \mathcal{IDS}p(2n+2) &= \mathcal{DS}p(2n+2) \otimes_s \mathcal{A}(2n+2) \\ &= \mathbb{Z}_2 \otimes_s \mathcal{D} \otimes_s \mathcal{S}p(2n+2) \otimes_s \mathcal{A}(2n+2) \end{aligned} . \tag{11}$$

The automorphism group is the central extension of this inhomogeneous group, $\mathcal{A}ut_{\mathcal{H}(n+1)} \simeq \mathcal{I}\check{\mathcal{D}}\mathcal{S}p(2n+2)$.

Projective representations of the classical inhomogeneous group are equivalence classes of the unitary representations of the automorphism group that is its central extension. The unitary representation determines the Hilbert space of quantum states. These representations are the largest symmetry of a quantum system that preserves the Weyl-Heisenberg commutation relations.

3. Homogeneous relativistic symmetry group

We determine in this section the homogeneous relativistic symmetry group for noninertial frames that satisfies two conditions

1) It leaves invariant the Minkowski proper time line element.

2) It is a subgroup of the automorphism group of the Weyl-Heisenberg group.

The group $\mathcal{O}a(1,n)$ is dependent on the scale c. We show a homomorphism parameterized by c satisfies the conditions to define a Inönü-Wigner contraction. This contraction results in the Hamilton group that we have previously shown is the relativistic symmetry group for noninertial states in the nonrelativistic ($c \to \infty$) context.

3.1. The group $\mathcal{O}a(1,n)$ and its algebra

The postulates of special relativity requires the invariance of the Minkowski proper time line element

$$d\tau^2 = \eta_{a,b} dx^a dx^b \tag{12}$$

with $a, b.. = 0, ..n$ and η is the diagonal matrix $\eta = [\eta_{a,b}] = \text{diag}\{-1, 1, ...1\}$ with units where $c = 1$.

Consider the $2n + 2$ dimensional time, position, energy, momentum extended phase space $\mathbb{P} \simeq \mathbb{R}^{2n+2}$ with coordinates $\{z^\alpha\} = \{x^a, p^a\}$ where $\alpha, \beta = 1, ...2n + 2$, $a, b = 0, 1..n$. The Minkowski metric is a degenerate line element on the cotangent space $T_z^*\mathbb{P}$

$$d\tau^2 = \tilde{\eta}_{\alpha,\beta} dz^\alpha dz^\beta \tag{13}$$

where $\tilde{\eta}_{\alpha,\beta}$ are the components of the $(2n + 2) \times (2n + 2)$ dimensional matrix $\tilde{\eta}$

$$\tilde{\eta} = [\tilde{\eta}_{\alpha,\beta}] = \begin{pmatrix} [\eta_{a,b}] & 0 \\ 0 & 0 \end{pmatrix}.$$

The group $\mathcal{GL}(2n + 2, \mathbb{R})$ of nonsingular $(2n + 2) \times (2n + 2)$ matrices acts naturally on the cotangent space $T_z^*\mathbb{P}$ with basis $\{dz^\alpha|_z\}$. Elements Γ of the subgroup $\mathcal{OGL}a(1,n) \subset \mathcal{GL}(2n+2, \mathbb{R})$ that is defined as the subgroup that leaves invariant the degenerate line element (13) satisfies $\Gamma^{\rm t}\tilde{\eta}\Gamma = \tilde{\eta}$. Γ may be written in terms of $(n + 1) \times (n + 1)$ submatrices as $\Gamma = \begin{pmatrix} \Lambda & B \\ \Xi & A \end{pmatrix}$ and therefore

$$\begin{pmatrix} \eta & 0 \\ 0 & 0 \end{pmatrix} = \begin{pmatrix} \Lambda^{\rm t} & \Xi^{\rm t} \\ B^{\rm t} & A^{\rm t} \end{pmatrix} \begin{pmatrix} \eta & 0 \\ 0 & 0 \end{pmatrix} \begin{pmatrix} \Lambda & B \\ \Xi & A \end{pmatrix} = \begin{pmatrix} \Lambda^{\rm t}\eta\Lambda & \Lambda^{\rm t}\eta B \\ B^{\rm t}\eta\Lambda & B^{\rm t}\eta B \end{pmatrix}. \tag{14}$$

It follows immediately that $B = 0$ and $\Lambda \in \mathcal{O}(1,n)$. As $\det\Gamma = \det\Lambda \det A$, $\det A \neq 0$ and $A \in \mathcal{GL}(n + 1, \mathbb{R})$. Therefore,

$$\mathcal{OGL}a(1,n) \simeq (\mathcal{O}(1,n) \otimes \mathcal{GL}(n + 1, \mathbb{R})) \otimes_s \mathcal{A}((n + 1)^2) \tag{15}$$

that has elements Γ and Γ^{-1} of the form

$$\Gamma = \begin{pmatrix} \Lambda & 0 \\ \Xi & A \end{pmatrix}, \quad \Gamma^{-1} = \begin{pmatrix} \Lambda^{-1} & 0 \\ -A^{-1}\Xi\Lambda^{-1} & A^{-1} \end{pmatrix}, \quad \begin{matrix} \Lambda \in \mathcal{O}(1,n) \\ A \in \mathcal{GL}(n+1, \mathbb{R}) \end{matrix} \quad \Xi \in \mathcal{A}((n + 1)^2) \tag{16}$$

It can be verified that this has the semidirect product structure as claimed following exactly the same steps as given explicitly below for the $\mathcal{O}a(1,n)$ group. The homogeneous relativistic symmetry group, that is called $\mathcal{O}a(1,n)$, is the homogenous subgroup of the automorphism group that leaves invariant the degenerate line element,

$$\mathcal{O}a(1,n) = \mathcal{DS}p(2n + 2) \cap \mathcal{OGL}a(1,n). \tag{17}$$

The elements of the $\mathcal{DS}p(2n+2)$ group are of the form $\Delta\Sigma$ where $\Delta \in \mathbb{Z}_2 \otimes \mathcal{D}$ and $\Sigma \in \mathcal{S}p(2n+2)$. The symplectic matrices satisfy the condition $\Sigma^{\mathrm{t}}\zeta\Sigma = \zeta$ and so $\Sigma^{-1} = -\zeta\Sigma^{\mathrm{t}}\zeta$ with $\zeta = \begin{pmatrix} 0 & \eta \\ -\eta & 0 \end{pmatrix}$. This may also be written in terms of $(n+1)\times(n+1)$ submatrices $\Sigma_{\mu,\nu}$ $\mu,\nu = 1,2$ with the matrix and inverse having the form

$$\Sigma = \begin{pmatrix} \Sigma_{1,1} & \Sigma_{1,2} \\ \Sigma_{2,1} & \Sigma_{2,2} \end{pmatrix}, \quad \Sigma^{-1} = \begin{pmatrix} \eta\Sigma^{\mathrm{t}}_{2,2}\eta & -\eta\Sigma^{\mathrm{t}}_{1,2}\eta \\ -\eta\Sigma^{\mathrm{t}}_{2,1}\eta & \eta\Sigma^{\mathrm{t}}_{1,1}\eta \end{pmatrix}. \tag{18}$$

If Γ in (16) is a subgroup of the outer automorphism group, we have $\Sigma = \Delta^{-1}\Gamma$ or expanding,

$$\begin{pmatrix} \Sigma_{1,1} & \Sigma_{1,2} \\ \Sigma_{2,1} & \Sigma_{2,2} \end{pmatrix} = \Delta^{-1}\begin{pmatrix} \Lambda & 0 \\ \Xi & A \end{pmatrix}. \tag{19}$$

$\Gamma^{-1} = \Delta^{-1}\Sigma^{-1}$ that we can compute from (18) and equate to the inverse calculated in (16)

$$\Gamma^{-1} = \begin{pmatrix} \Lambda^{-1} & 0 \\ -\Lambda^{-1}\Xi A^{-1} & A^{-1} \end{pmatrix} = \Delta^{-2}\begin{pmatrix} \eta A^{\mathrm{t}}\eta & 0 \\ -\eta\Xi^{\mathrm{t}}\eta & \eta\Lambda^{\mathrm{t}}\eta \end{pmatrix} \tag{20}$$

from which it follows that $\Lambda^{-1} = \Delta^{-2}\eta A^{\mathrm{t}}\eta$ and $A^{-1} = \Delta^{-2}\eta\Lambda^{\mathrm{t}}\eta$. This has a solution if and only if $\Delta = \pm 1_n \in \mathbb{Z}_2 \subset \mathcal{D}$ and $A^{\mathrm{t}} = \eta\Lambda^{-1}\eta$. 1_n is the $n\times n$ unit matrix. Noting that $\Lambda^{-1} = \eta\Lambda^{\mathrm{t}}\eta$ this gives $A^{\mathrm{t}} = \Lambda^{\mathrm{t}}$ and therefore $A = \Lambda$. Finally,

$$\Xi^{\mathrm{t}} = \eta\Lambda^{-1}\Xi\Lambda^{-1}\eta = \Lambda^{\mathrm{t}}\eta\Xi\eta\Lambda^{\mathrm{t}}. \tag{21}$$

Thus elements of $\mathcal{O}a(1,n)$ have the form

$$\Gamma(\Lambda,\Xi) = \begin{pmatrix} \Lambda & 0 \\ \Xi & \Lambda \end{pmatrix}, \quad \Lambda \in \mathcal{O}(1,n), \quad \Xi \in \mathcal{A}(m) \ . \tag{22}$$

with $m = (n+1)(n+2)/2$. The group multiplication and inverse of $\mathcal{O}a(1,n)$ are

$$\begin{aligned} \Gamma(\Lambda,\Xi) &= \Gamma(\Lambda',\Xi')\Gamma(\Lambda'',\Xi'') \\ &= \Gamma(\Lambda'\Lambda'',\Xi'\Lambda''+\Lambda'\Xi''), \end{aligned} \qquad \Gamma(\Lambda,\Xi)^{-1} = \Gamma(\Lambda^{-1},-\Lambda^{-1}\Xi\Lambda^{-1}). \tag{23}$$

The Lorentz group is the subgroup $\Gamma(\Lambda,0)$. The matrix components of the Lorentz matrices may be given as the usual expressions in regular and hyperbolic trigonometry terms of the rotation angles and boost angles.

The elements $\Gamma(1_n,\Xi)$ define an abelian normal subgroup with group multiplication, inverse and automorphisms given by

$$\Gamma(1_n,\Xi')\Gamma(1_n,\Xi'') = \Gamma(1_n,\Xi'+\Xi''), \qquad \Gamma(1_n,\Xi)^{-1} = \Gamma(1_n,-\Xi) \tag{24}$$

$$\Gamma(\Lambda',\Xi')\Gamma(1_n,\Xi)\Gamma(\Lambda',\Xi')^{-1} = \Gamma(1_n,\Lambda'\Xi\Lambda'^{-1}) \ . \tag{25}$$

The intersection $\Gamma(\Lambda,0)\cap\Gamma(1_n,\Xi) = 1_n$ and $\Gamma(\Lambda,\Xi) = \Gamma(1_n,\Xi)\Gamma(\Lambda,0)$. Therefore it is the semidirect product

$$\mathcal{O}a(1,n) \simeq \mathcal{O}(1,n) \otimes_s \mathcal{A}((n+1)\,(n+2)\,/2). \tag{26}$$

It can be shown that it does not admit an algebraic central extension and therefore the central extension of this group is simply its cover.

A general element of the algebra of $\mathcal{O}a(1,n)$ is $Z = \lambda^{a,b}L_{a,b} + \xi^{a,b}M_{a,b}$. Note that as $\xi^{a,b} = \xi^{b,a}$, that $M_{a,b} = M_{b,a}$. The Lie algebra relations may be directly computed to be

$$\begin{array}{l} [L_{a,b}, L_{c,d}] = -L_{b,d}\eta_{a,c} + L_{b,c}\eta_{a,d} + L_{a,d}\eta_{b,c} - L_{a,c}\eta_{b,d}, \\ [L_{a,b}, M_{c,d}] = -M_{b,d}\eta_{a,c} - M_{b,c}\eta_{a,d} + M_{a,d}\eta_{b,c} + M_{a,c}\eta_{b,d}, \quad [M_{a,b}, M_{c,d}] = 0. \end{array} \tag{27}$$

The $M_{a,b}$ abelian generators transform as a symmetric $(0,2)$ tensor under the Lorentz generators $L_{a,b}$.

Returning to the group, the transformation equations are $d\tilde{z} = \Gamma dz$. Using the definition of Γ in (26) results in

$$d\tilde{x} = \Gamma dx, \qquad d\tilde{p} = \Gamma dp + \Xi dx, \tag{28}$$

that in component form are (with units where $c = 1$)

$$d\tilde{x}^a = \lambda^a_b dx^b, \quad d\tilde{p}^a = \lambda^a_b dp^b + \xi^a_b dx^b. \tag{29}$$

Then, the proper time line element is invariant as required,

$$d\tau^2 = \eta_{a,b} d\tilde{x}^a d\tilde{x}^b = \eta_{a,b}\lambda^a_c dx^c \lambda^b_d dx^d = \eta_{a,b} dx^a dx^b. \tag{30}$$

The λ^a_c are the components of the Lorentz transformation that, as usual, depend on the relative rotation angle and hyperbolic boost angle. The mass μ satisfies

$$\begin{array}{rl} c^2 d\tilde{\mu}^2 & = \eta_{a,b} d\tilde{p}^a d\tilde{p}^b = \eta_{a,b}(\lambda^a_c dp^c + \xi^a_c dx^c)\left(\lambda^c_d dp^d + \xi^b_d dx^d\right) \\ & = c^2 d\mu^2 + \eta_{a,b}\xi^a_c\xi^b_d dx^c dx^d + 2\eta_{a,b}\xi^a_c\lambda^b_d dx^c dp^d. \end{array} \tag{31}$$

From basic dimensional analysis, the ξ^a_c have the dimensions of force or power (in units with $c = 1$, these are the same). It is a symmetric tensor satisfying $\xi^a_b = \eta^{a,c}\eta_{b,d}\xi^d_c$ that transforms as an (1,1) tensor under the Lorentz transformation

$$\tilde{\xi}^a_b = \lambda^a_c \lambda^d_b \xi^c_d. \tag{32}$$

These are the properties of a power-force stress tensor that is the proper time derivative of the energy-momentum stress tensor.

The rate of change of the mass squared with respect to the proper time is given by

$$\frac{d\tilde{\mu}^2}{d\tau^2} = \frac{d\mu^2}{d\tau^2} + \frac{1}{c^2}\eta_{a,b}\xi^a_c V^c(\xi^b_d V^d + 2\ \lambda^b_d F^d) \tag{33}$$

where $V^a = \frac{dx^a}{d\tau}$ is the *four* velocity and $F^a = \frac{dp^a}{d\tau}$ is the *four* force for the case $n = 3$.

3.2. Three notation

Further insight into the physical meaning of the group may be obtained by converting to $n+1$ notation that for $n = 3$ is the familiar *three* notation $\{x^a\} = \{t, \frac{1}{c}q^i\}$, $\{p^a\} = \{\frac{1}{c}e, p^i\}$, $i, j = 1, ..n$. The Lorentz matrix $\Lambda(\alpha, \beta)$ parameterized by rotation angles $\alpha^{i,j} = -\alpha^{j,i}$ and hyperbolic boost rotations β^i that have the usual form. As usual, we identify velocity as $v^i = c\frac{\beta^i}{|\beta|}\tanh(\beta)$ and define $\gamma(\beta) = \cosh(\beta) = \lambda^0_0$ or equivalently $\gamma(v) = (1 - (\frac{v}{c})^2)^{-\frac{1}{2}}$.

The velocity *four* vectors are given as usual by $\{V^0, V^i\} = \{\gamma, \gamma v^i\} = \gamma\{1, \frac{dx^i}{dt}\}$ where $\gamma = \frac{dt}{d\tau}$. The *four* force likewise is $\{F^0, F^i\} = \{\gamma r, \gamma f^i\}$ where $f^i = \frac{dp^i}{dt}$ and $r = \frac{de}{dt}$ and f^i has the dimensions of force and r has the dimensions of power. The power-force-stress components are

$$\Xi = \begin{pmatrix} \xi^0_0 & \xi^0_i \\ \xi^j_0 & \xi^j_i \end{pmatrix} = \gamma \begin{pmatrix} \frac{1}{c}r & -f_i \\ f^j & \frac{1}{c}m^j_i \end{pmatrix}. \tag{34}$$

Therefore, the transformation equations for the position, time, momentum, energy basis is

$$\begin{array}{ll} d\tilde{t} = \gamma dt + \frac{1}{c}\lambda_i^0 \ dq^i, & d\tilde{p}^i = \lambda_j^i dp^j + \lambda_i^0 \ \ dt, \\ d\tilde{q}^i = \lambda_j^i dq^j + c\lambda_0^i \ \ dt, & d\tilde{e} = \gamma de + c\lambda_i^0 \ \ dp^i - \gamma f_i dq^i + c\gamma r dt \ \ . \end{array} \tag{35}$$

For $n = 1$ these are simply

$$\begin{array}{ll} d\tilde{t} = \gamma(v) \left(\ dt + \frac{1}{c^2} v dq \right), & d\tilde{p} = \gamma(v) \left(dp^j + \frac{1}{c^2} v de + f dt + \frac{1}{c^2} \ \ m \ \ dq \right), \\ d\tilde{q} = \gamma(v) \left(dq + v dt \right), & d\tilde{e} = \gamma(v) \left(de + v dp - f dq + r dt \right). \end{array} \tag{36}$$

and the corresponding group parameter transformations using (23) are

$$\gamma(v) \begin{pmatrix} 1 & v \\ v & 1 \end{pmatrix} = \gamma(v'')\gamma(v') \begin{pmatrix} 1 & v' \\ v' & 1 \end{pmatrix} \begin{pmatrix} 1 & v'' \\ v'' & 1 \end{pmatrix} \tag{37}$$

$$\gamma(v) \begin{pmatrix} r & f \\ f & m \end{pmatrix} = \gamma(v'')\gamma(v') \left(\begin{pmatrix} r' & f' \\ f' & m' \end{pmatrix} \begin{pmatrix} 1 & v'' \\ v'' & 1 \end{pmatrix} + \begin{pmatrix} 1 & v' \\ v' & 1 \end{pmatrix} \begin{pmatrix} r'' & f'' \\ f'' & m'' \end{pmatrix} \right) \tag{38}$$

and so

$$\begin{array}{l} v = \left(v'' + v' \right) / \left(1 + \frac{v'v''}{c^2} \right), \quad f = \left(f'' + f' + \frac{1}{c^2} \left(r'v'' - v'r'' \right) \right) / \left(1 + \frac{v'v''}{c^2} \right), \\ r = \left(r'' + r' - f'v'' + v'f'' \right) / \left(1 + \frac{v'v''}{c^2} \right), m = \left(m'' + m' + f'v'' - v'f'' \right) / \left(1 + \frac{v'v''}{c^2} \right). \end{array} \tag{39}$$

Consider next the algebra. We have the infinitesimal parameter correspondence

$$\lambda^{0,i} = \frac{1}{c}\beta^i, \lambda^{j,i} = \alpha^{i,j}, \xi^{0,0} = \frac{1}{c}r, \xi^{j,0} = f^j, \xi^{j,i} = \frac{1}{c}m^{j,i} \tag{40}$$

where $\alpha^{i,j} = -\alpha^{i,j}$ and $m^{i,j} = m^{j,i}$ with the corresponding generators

$$L_{0,j} = cK_j, L_{i,j} = J_{i,j}, M_{0,0} = cR, M_{i,0} = N_i, M_{i,j} = cM^\circ_{i,j}. \tag{41}$$

A general element of the algebra is $Z = \alpha^{i,j} J_{i,j} + \beta^i K_i + f^i N_i + rR + m^{i,j} M^\circ_{i,j}$. The nonzero commutators of the Lie algebra (27) written in terms of these generators.

$$\begin{array}{ll} [J_{i,j}, J_{k,l}] = -J_{j,l}\delta_{i,k} + J_{j,k}\delta_{i,l} + J_{i,l}\delta_{j,k} - J_{i,k}\delta_{j,l}, & [K_i, K_k] = \frac{1}{c^2} J_{i,k}, \\ [J_{i,j}, K_k] = -K_j\delta_{i,k} + K_i\delta_{j,k}, & [K_i, N_k] = -M^\circ_{i,k} - R\delta_{i,k}, \\ [J_{i,j}, N_k] = -N_j\delta_{i,k} + N_i\delta_{j,k}, & [K_i, R] = -\frac{2}{c^2} N_i, \\ \left[J_{i,j}, M^\circ_{k,l}\right] = M^\circ_{j,l}\delta_{i,k} - M^\circ_{j,k}\delta_{i,l} + M^\circ_{i,l}\delta_{j,k} + M^\circ_{i,k}\delta_{j,l} & \\ \left[K_i, M^\circ_{k,l}\right] = \frac{-1}{c^2} \left(N_l\delta_{i,k} + N_k\delta_{i,l} \right), & \end{array} \tag{42}$$

3.3. Contraction in the limit $c \to \infty$

The scaling with c given in (42) satisfies the conditions for an Inönü-Wigner [8] contraction $c \to \infty$ for which the nonzero contracted commutators are

$$\begin{array}{ll} [J_{i,j}, J_{k,l}] = -J_{j,l}\delta_{i,k} + J_{j,k}\delta_{i,l} + J_{i,l}\delta_{j,k} - J_{i,k}\delta_{j,l}, & [K_i, N_k] = -M^\circ_{i,k} - R\delta_{i,k}, \\ [J_{i,j}, K_k] = -K_j\delta_{i,k} + K_i\delta_{j,k}, & [J_{i,j}, N_k] = -N_j\delta_{i,k} + N_i\delta_{j,k}, \\ \left[J_{i,j}, M^\circ_{k,l}\right] = -M^\circ_{j,l}\delta_{i,k} - M^\circ_{j,k}\delta_{i,l} + M^\circ_{i,l}\delta_{j,k} + M^\circ_{i,k}\delta_{j,l}. & \end{array} \tag{43}$$

The subgroup spanned by $\{J_{i,j}, K_i, N_i, R\}$ is the algebra of the Hamilton group $\mathcal{H}a(n)$. The full algebra of the group defined by

$$\mathcal{O}(n) \otimes_s \mathcal{A}(n(n+1)/2) \otimes_s \mathcal{H}(n) = \hat{\mathcal{H}a}(n) \otimes_s \mathcal{A}(n(n+1)/2) \tag{44}$$

where $\hat{\mathcal{H}a}(n)$ is the extended Hamilton group

$$\hat{\mathcal{H}a}(n) = \mathbb{Z}_2 \otimes_s \mathcal{H}a(n), \quad \mathcal{H}a(n) = \mathcal{SO}(n) \otimes_s \mathcal{H}(n) \tag{45}$$

where $\{J_{i,j}\}$ are the generators of $\mathcal{O}(n)$, $\{M_{i,j}\}$ are the generators of $\mathcal{A}(n(n+1)/2)$ and $\{K_i, N_i, R\}$ are the generators of the Weyl-Heisenberg group $\mathcal{H}(n)$.

The basis transformation equations (35) contract to the expected transformation equations in the limit [5]

$$\begin{array}{ll} d\tilde{t} = \ \ dt, & d\tilde{p}^i = \lambda(\alpha)^i_j dp^j + f^i dt, \\ d\tilde{q}^i = \lambda(\alpha)^i_j dq^j + v^i \ \ dt, & d\tilde{e} = de + v_i dp^i - f_i dq^i + r dt. \end{array} \tag{46}$$

where the $\lambda(\alpha)^i_j$ are now the components of a rotation matrix, $\mathcal{O}(n)$.

4. Summary

We started by noting that neither special relativity nor general relativity address the problem of how clocks of noninertial states due to forces other than gravity are related.

The hypothesis that the Minkowski proper time line element is invariant in these states that are not necessarily inertial and requiring that the Heisenberg commutation relations hold in all noninertial states results in the noninertial relativistic symmetry group $\mathcal{O}a(1,n)$. This group gives the expected transformations to noninertial states in terms of a power-force stress tensor that is the proper time derivative of the energy-momentum stress tensor.

The $\mathcal{O}a(1,n)$ group is also the $b \to \infty$ of the $\mathcal{U}(1,n)$ group of reciprocal relativity described in [9]. This gives an understanding of the behavior of reciprocal relativity in the small interaction limit (that is, small forces relative to b) that is analogous to the manner in which the Euclidean group that is the homogeneous group of the Galilei group gives the small velocity limit, relative to c, of the Lorentz group.

Spacetime is an invariant subspace under the actions of the $\mathcal{O}a(1,n)$ group and therefore is observer independent or *absolute*. In this limit, there is an apparent global inertial frame that all observers agree on. Forces appear to be relative to this frame rather than being strictly relative to particle states. Forces and the power-force-stress energy tensor are simply additive and unbounded. Velocities are bounded by c and strictly relative to particle states.

In the $c \to \infty$ limit yields the classical *nonrelativistic* Hamilton theory that describes particles undergoing general noninertial motion. In this case, there is an apparent global inertial rest frame that all observers agree on. Forces and velocities appear to be relative to this frame rather than being strictly relative to particle states. Forces and velocities are simply additive and unbounded.

[1] Wigner, E. (1939). On the unitary representations of the inhomogeneous Lorentz group. *Annals of Mathematics*, **40**, 149–204.
[2] Weinberg, S. (1995). *The Quantum Theory of Fields, Volume 1.* Cambridge: Cambridge.
[3] Folland, G. B. (1989). *Harmonic Analysis on Phase Space.* Princeton: Princeton University Press.
[4] Low, S. G. (2007). Relativity group for noninertial frames in Hamilton's mechanics. *J. Math. Phys.*, **48**, 102901.
[5] Low, S. G. (2008). Hamilton relativity group for noninertial states in quantum mechanics. *J. Phys. A: Math Theor.*, **41**, 304034.
[6] Bargmann, V. (1954). On Unitary Ray Representations of Continuous Groups. *Annal. Math.*, **59**, 1–46.
[7] Mackey, G. W. (1958). Unitary Representations of Group Extensions. I. *Acta Math.*, **99**, 265–311.
[8] Gilmore, R. (1974). *Lie Groups, Lie Algebras and Some of their Applications.* New York: John Wiley.
[9] Low, S. G. (2007). Reciprocal relativity of noninertial frames: quantum mechanics. *J. Phys A: Math. Theor.*, **40**, 3999–4016.

Journal of Physics: Conference Series **284** (2011) 012046 doi:10.1088/1742-6596/284/1/012046

Dark Matter: A Relativistic Inertial Effect in Einstein Canonical Gravity?

Luca Lusanna
Sezione INFN di Firenze, Via Sansone 1, 50019 Sesto Fiorentino (FI), Italy
E-mail: lusanna@fi.infn.it

Abstract: After consideration of non-inertial frames in special relativity with emphasis on the problem of clock synchronization (i.e. of how to define 3-space), an overview is given of Einstein canonical gravity in the York canonical basis and of its Hamiltonian Post-Minkowskian (PM) linearization in 3-orthogonal gauges. It is shown that the York time (the trace of the extrinsic curvature of 3-spaces) is the inertial gauge variable describing the general relativistic remnant of the clock synchronization gauge freedom. The dark matter implied by the rotation curves of galaxies can be explained with a choice of the York time implying a PM extension of the Newtonian International Celestial Reference System (ICRS).

I. INTRODUCTION

We can speak of predictability in physics when there is a well-posed Cauchy problem for the partial differential equations involved in the problem under investigation. A pre-requisite for the Cauchy problem is a sound definition of an instantaneous 3-space where the Cauchy data are given. While in the Galilei space-time of Newtonian physics this is not a problem due to the absolute nature of time and of Euclidean 3-space, no intrinsic notion of 3-space exists in special and general relativistic theories. In special relativity (SR), where there is the absolute Minkowski space-time, only the conformal structure (the light-cone) is intrinsically given. The standard way out from the problem of 3-space is to choose the Euclidean 3-space of an inertial frame centered on an inertial observer and then use the kinematical Poincare' group to connect different inertial frames. However, this is not possible in general relativity (GR), where there is no absolute notion since also space-time becomes dynamical (with the metric structure satisfying Einstein's equations). The equivalence principle implies the absence of global inertial frames: in the restricted class of globally hyperbolic, asymptotically Minkowskian at spatial infinity, space-times the best we can have are global non-inertial frames connected by 4-diffeomorphisms (the gauge group of GR). As a consequence, also in SR we have to face the problem of reformulating physics in non-inertial frames centered on accelerated observers [1] as a first step before facing GR.

II. NON-INERTIAL FRAMES AND RADAR 4-COORDINATES IN SPECIAL AND GENERAL RELATIVITY

In Minkowski space-time the Euclidean 3-spaces of the inertial frames centered on an inertial observer A are identified by means of Einstein convention for the synchronization of clocks: the inertial observer A sends a ray of light at x_i^o towards the (in general accelerated) observer B; the ray is reflected towards A at a point P of B world-line and then reabsorbed by A at x^o_f; by convention P is synchronous with the mid-point between emission and absorption on A's world-line, i.e. $x^o_P = x^o_i + \frac{1}{2}(x^o_f - x^o_i) = \frac{1}{2}(x^o_i + x^o_f)$. Therefore the description of non-inertial frames must replace Einstein convention with a more general one allowing non-Euclidean 3-spaces. Since in GR the gauge freedom is the arbitrariness in the choice of the 4-coordinates, in the non-inertial frames of SR a similar arbitrariness is expected. However, the experimental description of matter in both theories is based on *metrological conventions*: a) an atomic clock as a standard of time; b) the velocity of light in place of a standard of length; c) a conventional reference frame centered on a given observer as a standard of space-time (think to the Global Positioning System, GPS!). The description of satellites around the Earth is done by means of NASA coordinates [2] either in International Terrestrial Reference System (ITRS2003) [3] (frame fixed on the Earth surface) or in Geocentric Celestial Reference System (GCRS) (the geocentric frame, centered on the Earth center, of the conventions of the International Astronomy Union (IAU2000) [4]). The description of planets and spacecrafts uses the Barycentric Celestial Reference System (BCRS) of IAU2000 [4] centered on the barycenter of the Solar System (a quasi-inertial frame in a nearly Minkowski space-time due to gravity). Both GCRS and BCRS are connected to Post-Newtonian solutions of Einstein equations in special harmonic gauges. Instead in astronomy the positions of stars and galaxies are determined from the data (luminosity, light spectrum, angles) on the sky as living in a 4-dimensional nearly-Galilei space-time with the International Celestial Reference System (ICRS) [5] considered as a "quasi-inertial frame" (all galactic dynamics is Newtonian gravity), in accord with the assumed validity of the cosmological and Copernican principles. Namely one assumes a homogeneous and isotropic cosmological Friedmann-Robertson - Walker solution of Einstein equations (the standard ΛCDM cosmological model [6]). In it the constant intrinsic 3-curvature of instantaneous 3-spaces is nearly zero as implied by the Cosmic Microwave Background (CMB) data [6], so that Euclidean 3-spaces (and Newtonian gravity) can be used. However, to reconcile all the data with this 4-dimensional reconstruction one must postulate the existence of dark matter and dark energy as the dominant components of the classical universe after the recombination 3-surface!

As a consequence, we need a metrology-oriented description of non-inertial frames already in SR [1, 7]. This can be done with the *3+1 point of view* and the use of observer-dependent Lorentz scalar radar 4-coordinates. Let us give the world-line $x^\mu(\tau)$ of an arbitrary time-like observer carrying a standard atomic clock: τ is an arbitrary monotonically increasing function of the proper time of this clock. Then we give an admissible 3+1 splitting of Minkowski space-time, namely a nice foliation with space-like instantaneous 3-spaces Σ_τ: it is the mathematical idealization of a protocol for clock synchronization (all the clocks in the points of Σ_τ sign the same time of the atomic clock of the observer). On each 3-space Σ_τ we choose curvilinear 3-coordinates σ^r having the observer as origin. These are the radar 4-coordinates $\sigma^A = (\tau; \sigma^r)$. If $x^\mu \mapsto \sigma^A(x)$ is the coordinate transformation from the Cartesian 4-coordinates x^μ of a reference inertial observer to radar coordinates, its inverse $\sigma^A \mapsto x^\mu = z^\mu(\tau, \sigma^r)$ defines the *embedding* functions $z^\mu(\tau, \sigma^r)$ describing the 3-spaces Σ_τ as embedded 3-manifold into Minkowski space-time. The induced 4-metric on Σ_τ is the following functional of the embedding ${}^4g_{AB}(\tau, \sigma^r) = [z^\mu_A\, \eta_{\mu\nu}\, z^\nu_B](\tau, \sigma^r)$, where $z^\mu_A = \partial\, z^\mu / \partial\, \sigma^A$ and ${}^4\eta_{\mu\nu} = \epsilon\,(+ - --)$ is the flat metric ($\epsilon = \pm 1$ according to either the particle physics $\epsilon = 1$ or the general relativity $\epsilon = -1$ convention). While the 4-vectors $z^\mu_r(\tau, \sigma^u)$ are tangent to Σ_τ, so that the unit normal $l^\mu(\tau, \sigma^u)$ is proportional to $\epsilon^\mu{}_{\alpha\beta\gamma}\, [z^\alpha_1\, z^\beta_2\, z^\gamma_3](\tau, \sigma^u)$, we have $z^\mu_\tau(\tau, \sigma^r) = [N\, l^\mu + N^r\, z^\mu_r](\tau, \sigma^r)$ ($N(\tau, \sigma^r) = \epsilon\, [z^\mu_\tau\, l_\mu](\tau, \sigma^r)$ and $N_r(\tau, \sigma^r) = -\epsilon\, g_{\tau r}(\tau, \sigma^r)$

Journal of Physics: Conference Series **284** (2011) 012046 doi:10.1088/1742-6596/284/1/012046

are the lapse and shift functions). The foliation is nice and admissible if it satisfies the conditions:
1) $N(\tau,\sigma^r) > 0$ in every point of Σ_τ (the 3-spaces never intersect, avoiding the coordinate singularity of Fermi coordinates);
2) $\epsilon\, {}^4g_{\tau\tau}(\tau,\sigma^r) > 0$, so to avoid the coordinate singularity of the rotating disk, and with the positive-definite 3-metric ${}^3g_{rs}(\tau,\sigma^u) = -\epsilon\, {}^4g_{rs}(\tau,\sigma^u)$ having three positive eigenvalues (these are the Møller conditions [1, 8] forbidding global *rigid* rotations);
3) all the 3-spaces Σ_τ must tend to the same space-like hyper-plane at spatial infinity (so that there are always asymptotic inertial observers to be identified with the fixed stars).

In the 3+1 point of view the 4-metric ${}^4g_{AB}(\tau,\vec{\sigma})$ on Σ_τ has the components $\epsilon\, {}^4g_{\tau\tau} = N^2 - N_r\, N^r$, $-\epsilon\, {}^4g_{\tau r} = N_r = {}^3g_{rs}\, N^s$, ${}^3g_{rs} = -\epsilon\, {}^4g_{rs} = \sum_{a=1}^3 {}^3e_{(a)r}\, {}^3e_{(a)s} = \tilde{\phi}^{2/3} \sum_{a=1}^3 e^{2\sum_{\bar{b}=1}^2 \gamma_{\bar{b}a} R_{\bar{b}}}\, V_{ra}(\theta^i)\, V_{sa}(\theta^i))$, where ${}^3e_{(a)r}(\tau,\sigma^u)$ are cotriads on Σ_τ, $\tilde{\phi}^2(\tau,\sigma^r) = det\, {}^3g_{rs}(\tau,\sigma^r)$ is the 3-volume element on Σ_τ, $\lambda_a(\tau,\sigma^r) = [\tilde{\phi}^{1/3}\, e^{\sum_{\bar{b}=1}^2 \gamma_{\bar{b}a} R_{\bar{b}}}](\tau,\sigma^r)$ are the positive eigenvalues of the 3-metric ($\gamma_{\bar{a}a}$ are suitable numerical constants) and $V(\theta^i(\tau,\sigma^r))$ are diagonalizing rotation matrices depending on three Euler angles. The components ${}^4g_{AB}$ or the quantities N, N_r, $\tilde{\phi}$, $R_{\bar{a}}$, θ^i, play the role of the *inertial potentials* generating the relativistic apparent forces in the non-inertial frame. It can be shown [1, 8] that the Newtonian inertial potentials are hidden in the functions N, N_r and θ^i. The extrinsic curvature ${}^3K_{rs}(\tau,\sigma^u) = [\frac{1}{2N}\,(N_{r|s} + N_{s|r} - \partial_\tau\, {}^3g_{rs})](\tau,\sigma^u)$, describing the *shape* of the instantaneous 3-spaces of the non-inertial frame as embedded 3-manifolds of Minkowski space-time, is a secondary inertial potential functional of the independent inertial potentials ${}^4g_{AB}$.

Instead in GR the 4-metric is described by ten dynamical fields ${}^4g_{\mu\nu}(x)$: it is not only a (pre)potential for the gravitational field but also determines the chrono-geometrical structure of space-time through the line element $ds^2 = {}^4g_{\mu\nu}\, dx^\mu\, dx^\nu$ (it teaches relativistic causality to the other fields) [1]. To get its Hamiltonian description in the quoted restricted class of space-times the same 3+1 point of view and the radar 4-coordinates employed in SR can be used: *this allows to separate the inertial (gauge) degrees of freedom of the gravitational field (playing the role of inertial potentials) from the dynamical tidal ones.* But now the admissible embeddings $x^\mu = z^\mu(\tau,\sigma^r)$ are not dynamical variables: instead their gradients $z^\mu_A(\tau,\sigma^r)$ give the transition coefficient from radar to world 4-coordinates, ${}^4g_{AB}(\tau,\sigma^r) = [z^\mu_A\, z^\nu_B](\tau,\sigma^r)\, {}^4g_{\mu\nu}(z(\tau,\sigma^r))$. As shown in Ref.[10], the dynamical nature of space-time implies that each solution of Einstein's equations dynamically selects a preferred 3+1 splitting of the space-time (modulo coordinate transformations) and preferred 3-spaces, whose extrinsic curvature is a mixture of dynamical (tidal) pieces and inertial gauge variables playing the role of inertial potentials.

III. ISOLATED SYSTEMS IN SPECIAL RELATIVISTIC NON-INERTIAL FRAMES AND THE REST-FRAME INSTANT FORM OF DYNAMICS

The description of isolated systems (particles, strings, fields, fluids) admitting a Lagrangian formulation in the non-inertial frames of SR is done by means of *parametrized*

---

[1] The ACES mission of ESA [9] will give the first precision measurement of the gravitational redshift of the geoid, namely of the $1/c^2$ deformation of Minkowski light-cone caused by the geo-potential (a non-pertubative effect in QFT on a fixed background).

Minkowski theories [1, 7]. The matter variables are replaced with new ones knowing the 3-spaces Σ_τ. For instance a Klein-Gordon field $\tilde{\phi}(x)$ will be replaced with $\phi(\tau, \sigma^r) = \tilde{\phi}(z(\tau, \sigma^r))$; the same for every other field. Instead for a relativistic particle with world-line $x^\mu(\tau)$ we must make a choice of its energy sign: then it will be described by 3-coordinates $\eta^r(\tau)$ defined by the intersection of the world-line with Σ_τ: $x^\mu(\tau) = z^\mu(\tau, \eta^r(\tau))$. Differently from all the previous approaches to relativistic mechanics, the dynamical configuration variables are the 3-coordinates $\eta^r_i(\tau)$ and not the world-lines $x^\mu_i(\tau)$ (to rebuild them in an arbitrary frame we need the embedding defining that frame!). Then the matter Lagrangian is coupled to an external gravitational field and the external 4-metric is replaced with the 4-metric $g_{AB}(\tau, \sigma^r)$ of an admissible 3+1 splitting of Minkowski space-time. With this procedure we get a Lagrangian depending on the given matter and on the embedding $z^\mu(\tau, \sigma^r)$, which is invariant under *frame-preserving diffeomorphisms*. As a consequence, there are four first-class constraints (an analogue of the super-Hamiltonian and super-momentum constraints of canonical gravity) implying that the embeddings $z^\mu(\tau, \sigma^r)$ are *gauge variables*, so that all the admissible non-inertial or inertial frames are gauge equivalent, namely physics does *not* depend on the clock synchronization convention and on the choice of the 3-coordinates σ^r: only the appearances of phenomena change by changing the notion of instantaneous 3-space. Even if the gauge group is formed by the frame-preserving diffeomorphisms, the matter energy-momentum tensor allows the determination of the ten conserved Poincare' generators P^μ and $J^{\mu\nu}$ (assumed finite) of every configuration of the system. If we restrict ourselves to inertial frames, we can define the *inertial rest-frame instant form of dynamics for isolated systems* by choosing the 3+1 splitting corresponding to the intrinsic inertial rest frame of the isolated system centered on an inertial observer: the instantaneous 3-spaces, named Wigner 3-space due to the fact that the 3-vectors inside them are Wigner spin-1 3-vectors [7], are orthogonal to the conserved 4-momentum P^μ of the configuration. In Ref.[1] there is the extension to admissible *non-inertial rest frames*, where P^μ is orthogonal to the asymptotic space-like hyper-planes to which the instantaneous 3-spaces tend at spatial infinity. This non-inertial family of 3+1 splittings is the only one admitted by the asymptotically Minkowskian space-times covered by canonical gravity formulation discussed below.

The framework of the inertial rest frame allowed the solution of the following old open problems:
A) The explicit form of the Lorentz boosts for some interacting systems [11, 12].
B) The classification of the relativistic collective variables, replacing the Newtonian center of mass, that can be built in terms of the Poincare' generators and their non measurability due to the non-local character of such generators (they know the whole 3-space Σ_τ) [1, 13].
C) The description of every isolated system as a decoupled (non-measurable) canonical non-covariant (Newton-Wigner) center of mass carrying a pole-dipole structure: the invariant mass and the rest spin of the system expressed in terms of Wigner-covariant relative variables inside the Wigner 3-spaces [1, 13–15].
D) The formulation of classical relativistic atomic physics [11, 14] (the electro-magnetic field in the radiation gauge plus charged scalar particles with Grassmann-valued electric charges to regularize the self-energies) and the identification of the Darwin potential at the classical level by evading Haag's theorem.
E) A new formulation of *relativistic quantum mechanics* [15] englobing all the known results about relativistic bound states and a first formulation of *relativistic entanglement* taking into account the *non-locality and spatial non-separability coming from the Poincare' group.*

Journal of Physics: Conference Series **284** (2011) 012046 doi:10.1088/1742-6596/284/1/012046

IV. HAMILTONIAN FORMULATION OF EINSTEIN'S GENERAL RELATIVITY IN ASYMPTOTICALLY MINKOWSKIAN SPACETIMES

Let us now consider Einstein's general relativity where space-time is dynamical. Since all the properties of the standard model of elementary particles are connected with properties of the representations of the Poincare' group in inertial frames of Minkowski space-time, we shall restrict ourselves to globally hyperbolic, asymptotically Minkowskian at spatial infinity, topologically trivial space-times, for which a well defined Hamiltonian formulation of gravity is possible if we replace the Hilbert action with the ADM one. The 4-metric tends in a suitable way to the flat Minkowski 4-metric ${}^4\eta_{\mu\nu}$ at spatial infinity: having an *asymptotic* Minkowskian background we can avoid to split the 4-metric in the bulk in a background plus perturbations in the weak field limit. In these space-times we can use admissible 3+1 splittings and observer-dependent radar 4-coordinates. Since tetrad gravity is more natural for the coupling of gravity to the fermions, the 4-metric is decomposed in terms of cotetrads, ${}^4g_{AB} = E^{(\alpha)}_A \, {}^4\eta_{(\alpha)(\beta)} \, E^{(\beta)}_B$, and the ADM action, now a functional of the 16 fields $E^{(\alpha)}_A(\tau, \sigma^r)$, is taken as the action for ADM tetrad gravity. In tetrad gravity the diffeonorphism group is enlarged with the O(3,1) gauge group of the Newman-Penrose approach (the extra gauge freedom acting on the tetrads in the tangent space of each point of space-time and reducing from 16 to 10 the number of independent fields like in metric gravity). This leads to an interpretation of gravity based on a congruence of time-like observers endowed with orthonormal tetrads: in each point of space-time the time-like axis is the unit 4-velocity of the observer, while the spatial axes are a (gauge) convention for observer's gyroscopes. This framework was developed in the works in Refs.[16].

In these space-times we assume direction-independent boundary conditions on the 4-metric and its conjugate momenta able to kill *super-translations* [17], so that the SPI group of asymptotic symmetries [18] is reduced to the ADM Poincare' group with the generators P^A_{ADM}, J^{AB}_{ADM} given as boundary conditions. It turns out that the admissible 3+1 splittings are the *non-inertial rest frames* (with the 3-spaces asymptotically orthogonal to the ADM 4-momentum; $P^r_{ADM} \approx 0$ is the rest-frame condition) of the 3-universe with a mass and a rest spin fixed by the boundary conditions [2]. In absence of matter Christodoulou - Klainermann space-times [19] are compatible with this description. With this kind of formalism we can get a deparametrization of general relativity: if we switch off the Newton constant and we choose the flat Minkowski 4-metric in Cartesian coordinates as solution of Einstein's equations, we get the description of the matter present in the 3-universe in the non-inertial rest frames of Minkowski space-time with the weak ADM Poincare' group collapsing in the Poincare' group of particle physics. As shown in Refs.[16], with the previous boundary conditions the DeWitt surface term at spatial infinity in the Dirac Hamiltonian turns out to be the *strong* ADM energy (a flux through a 2-surface at spatial infinity), which is equal to the *weak* ADM energy (expressed as a volume integral over the 3-space) plus constraints. Therefore in this family of space-times there is *not a frozen picture*, like in the family of spatially compact without boundary space-times considered in loop quantum gravity, where the Dirac Hamiltonian is a combination of constraints.

---

[2] There are asymptotic inertial observers identified by the fixed stars of star catalogues. If ϵ^μ_A are a set of asymptotic flat tetrads, the simplest embedding adapted to the 3+1 splitting of space-time is $x^\mu = z^\mu(\tau, \sigma^r) = x^\mu(\tau) + \epsilon^\mu_r \, \sigma^r = x^\mu_o + \epsilon^\mu_A \, \sigma^A$ and we have ${}^4g_{AB}(\tau, \sigma^r) = \epsilon^\mu_A \, \epsilon^\nu_B \, {}^4g_{\mu\nu}(x)$.

In this framework the configuration variables are cotetrads, which are connected to cotetrads adapted to the 3+1 splitting of space-time (so that the adapted time-like tetrad is the unit normal to the 3-space Σ_τ) by standard Wigner boosts for time-like vectors of parameters $\varphi_{(a)}(\tau,\sigma^r)$, $a=1,2,3$: $E^{(\alpha)}_A = L^{(\alpha)}{}_{(\beta)}(\varphi_{(a)})\,\overset{o\,(\beta)}{E}_A$. The adapted cotetrads have the following expression in terms of cotriads ${}^3e_{(a)r}$ on Σ_τ and of the lapse $N=1+n$ and shift $n_{(a)} = N^r\,{}^3e_{(a)r}$ functions: $\overset{o\,(o)}{E}_\tau = 1+n$, $\overset{o\,(o)}{E}_r = 0$, $\overset{o\,(a)}{E}_\tau = n_{(a)}$, $\overset{o\,(a)}{E}_r = {}^3e_{(a)r}$. The 4-metric becomes ${}^4g_{\tau\tau} = \epsilon\,[(1+n)^2 - \sum_a n^2_{(a)}]$, ${}^4g_{\tau r} = -\epsilon\sum_a n_{(a)}\,{}^3e_{(a)r}$, ${}^4g_{rs} = -\epsilon\,{}^3g_{rs} = -\epsilon\sum_a {}^3e_{(a)r}\,{}^3e_{(a)s}$. The 16 configurational variables in the ADM action are $\varphi_{(a)}$, $1+n$, $n_{(a)}$, ${}^3e_{(a)r}$. There are ten primary constraints (the vanishing of the 7 momenta of boosts, lapse and shift variables plus three constraints describing the rotation on the flat indices (a) of the cotriads) and four secondary ones (the super-Hamiltonian and super-momentum constraints): all of them are first class in the phase space spanned by 16+16 fields. This implies that there are 14 gauge variables describing *inertial effects* and 2 canonical pairs of physical degrees of freedom describing the *tidal effects* of the gravitational field (namely gravitational waves in the weak field limit). In this canonical basis only the momenta ${}^3\pi^r_{(a)}$ conjugated to the cotriads are not vanishing.

V. THE YORK CANONICAL BASIS OF ADM CANONICAL TETRAD GRAVITY

Then in Ref.[18] we have found a canonical transformation to a canonical basis adapted to ten of the first class constraints. It implements the York map of Ref.[20] and diagonalizes the York-Lichnerowicz approach [21]. Its final form is ($\alpha_{(a)}(\tau,\sigma^r)$ are angles; ${}^3e_{(a)r} = R_{(a)(b)}(\alpha_{(c)})\,{}^3\bar{e}_{(b)r} = R_{(a)(b)}(\alpha_{(c)})\,V_{rb}(\theta^i)\,\tilde{\phi}^{1/3}\,e^{\sum_{\bar{a}}^{1,2}\gamma_{\bar{a}a}R_{\bar{a}}}$, ${}^4g_{\tau\tau} = \epsilon\,[(1+n)^2 - \sum_a \bar{n}^2_{(a)}]$, ${}^4g_{\tau r} = -\epsilon\,\bar{n}_{(a)}\,{}^3\bar{e}_{(a)r}$, ${}^4g_{rs} = -\epsilon\,{}^3g_{rs} = -\epsilon\,\tilde{\phi}^{2/3}\sum_a V_{ra}(\theta^i)\,V_{sa}(\theta^i)\,e^{2\sum_{\bar{a}}^{1,2}\gamma_{\bar{a}a}R_{\bar{a}}}$)

$\varphi_{(a)}$ $\quad\alpha_{(a)}$	n $\quad\bar{n}_{(a)}$	θ^r	$\tilde{\phi}$	$R_{\bar{a}}$
$\pi_{\varphi_{(a)}} \approx 0$ $\;\pi^{(\alpha)}_{(a)} \approx 0$	$\pi_n \approx 0$ $\;\pi_{\bar{n}_{(a)}} \approx 0$	$\pi^{(\theta)}_r$	$\pi_{\tilde{\phi}} = \frac{c^3}{12\pi G}\,{}^3K$	$\Pi_{\bar{a}}$

In this York canonical basis the *inertial effects* are described by the arbitrary gauge variables $\alpha_{(a)}$, $\varphi_{(a)}$, $1+n$, $\bar{n}_{(a)}$, θ^i, ${}^3K$, while the *tidal effects*, i.e. the physical degrees of freedom of the gravitational field, by the two canonical pairs $R_{\bar{a}}$, $\Pi_{\bar{a}}$, $\bar{a}=1,2$. The momenta $\pi^{(\theta)}_r$ and the 3-volume element $\tilde{\phi} = \sqrt{det\,{}^3g_{rs}}$ have to be found as solutions of the super-momentum and super-hamiltonian (i.e. the Lichmerowicz equation) constraints, respectively. The gauge variables $\alpha_{(a)}$, $\varphi_{(a)}$ parametrize the extra O(3,1) gauge freedom of the tetrads (the gauge freedom for each observer to choose three gyroscopes as spatial axes and to choose the law for their transport along the world-line). The gauge angles θ^i (i.e. the director cosines of the tangents to the three coordinate lines in each point of Σ_τ) describe the freedom in the choice of the 3-coordinates σ^r on each 3-space: their fixation implies the determination of the shift gauge variables $\bar{n}_{(a)}$, namely the appearances of gravito-magnetism in the chosen 3-coordinate system.

The final basic gauge variable is a momentum, namely the trace ${}^3K(\tau,\sigma^r)$ of the extrinsic curvature (also named the *York time*) of the non-Euclidean 3-space Σ_τ. The Lorentz signature of space-time implies that ^3K is a momentum variable: it is a time coordinate, while θ^i are spatial coordinates. Differently from SR ^3K is *an independent inertial gauge variable describing the remnant in GR of the freedom in clock synchronization*! The other components of the extrinsic curvature are dynamically determined. This gauge variable has no Newtonian counterpart (the Euclidean 3-space is absolute), because its fixation determines the final shape of the non-Euclidean 3-space. Moreover this gauge variable gives rise to a negative kinetic term in the weak ADM energy $\hat{E}_{ADM}$, vanishing only in the gauges $^3K(\tau, \vec{\sigma}) = 0$ [18].

In the York canonical basis the Hamilton equations generated by the Dirac Hamiltonian $H_D = \hat{E}_{ADM} + (constraints)$ are divided in four groups: A) the contracted Bianchi identities, namely the evolution equations for $\tilde{\phi}$ and $\pi_i^{(\theta)}$ (they say that given a solution of the constraints on a Cauchy surface, it remains a solution also at later times); B) the evolution equation for the four basic gauge variables θ^i and ^3K: these equations determine the lapse and the shift functions once the basic gauge variables are fixed; C) the evolution equations for the tidal variables $R_{\bar{a}}$, $\Pi_{\bar{a}}$; D) the Hamilton equations for matter, when present. Once a gauge is completely fixed, the Hamilton equations become deterministic. Given a solution of the super-momentum and super-Hamiltonian constraints and the Cauchy data for the tidal variables on an initial 3-space, we can find a solution of Einstein's equations in radar 4-coordinates adapted to a time-like observer.

VI. THE POST-MINKOWSKIAN LINEARIZATION OF ADM CANONICAL GRAVITY WITH MATTER IN THE NON-HARMONIC 3-ORTHOGONAL SCHWINGER TIME GAUGES

In the first paper of Ref.[22], we studied the coupling of N charged scalar particles plus the electro-magnetic field to ADM tetrad gravity in this class of asymptotically Minkowskian space-times without super-translations. To regularize the self-energies both the electric charge and the sign of the energy of the particles are Grassmann-valued. The introduction of the non-covariant radiation gauge allows to reformulate the theory in terms of transverse electro-magnetic fields and to extract the generalization of the action-at-a-distance Coulomb interaction among the particles in the Riemannian instantaneous 3-spaces of global non-inertial frames. After the reformulation of the whole system in the York canonical basis, we give the restriction of the Hamilton equations and of the constraints to the family of *non-harmonic 3-orthogonal* Schwinger time gauges, in which the instantaneous Riemannian 3-spaces have a non-fixed trace ^3K of the extrinsic curvature but a diagonal 3-metric. This family of gauges is determined by the gauge fixings $\theta^i(\tau, \sigma^r) \approx 0$ and $^3K(\tau, \sigma^r) \approx (arbitrary\, numerical\, function)$.

In the second paper of Ref.[22] it was shown that in this family of non-harmonic 3-orthogonal Schwinger gauges it is possible to define a consistent *linearization* of ADM canonical tetrad gravity plus matter in the weak field approximation, to obtain a formulation of Hamiltonian Post-Minkowskian gravity with non-flat Riemannian 3-spaces and asymptotic Minkowski background. This means that the 4-metric tends to the asymptotic Minkowski metric at spatial infinity, $^4g_{AB} \rightarrow {}^4\eta_{AB}$. The decomposition $^4g_{AB} = {}^4\eta_{AB} + {}^4h_{(1)AB}$, with a first order perturbation $^4h_{(1)AB}$ vanishing at spatial infinity, is only used for calculations, but has no intrinsic meaning. Moreover, due to the presence of a ultra-violet cutoff for matter, we can avoid to make Post-Newtonian expansions, namely we get fully relativistic expressions. We have found solutions for the first order quantities $\pi^{(\theta)}_{(1)r}$, $\tilde{\phi} = 1 + 6\,\phi_{(1)}$, $1 + n_{(1)}$, $\bar{n}_{(1)(a)}$ (the action-at-a-distance part of the gravitational interaction). Then we can show that the tidal variables $R_{\bar{a}}$ satisfy a wave equation $\Box\, R_{\bar{a}} = (known\, functional\, of\, matter)$ with the D'Alambertian associated to the asymptotic Minkowski 4-metric. Therefore, by using a no-incoming radiation condition based on the asymptotic Minkowski light-cone, we get a description of gravitational waves in these non-harmonic gauges, which can be connected to generalized TT(transverse traceless) gauges, as retarded functions of the matter. These gravitational waves do not propagate in inertial frames of the background (like it happens in the standard harmonic gauge description), but in non-Euclidean instantaneous 3-spaces differing from Euclidean 3-spaces at the first order (their intrinsic 3-curvature is determined by the gravitational waves) and dynamically determined by the linearized solution of Einstein equations. These 3-spaces have a first order extrinsic curvature (with ${}^3K_{(1)}(\tau, \sigma^r)$ describing the clock synchronization convention) and a first order modification of Minkowski light-cone. We can write explicitly the linearized Hamilton equations for the particles and for the electro-magnetic field: among the forces there are both the inertial potentials and the gravitational waves.

VII. THE POST-NEWTONIAN LIMIT OF THE PARTICLE EQUATIONS OF MOTION AND THE EMERGENCE OF DARK MATTER AS A RELATIVISTIC INERTIAL EFFECT

In the third paper of Ref.[22] we disregarded electro-magnetism and we studied the non-relativistic limit of the particle equations. We found that the particle 3-coordinates $\eta^r_i(\tau = ct) = \tilde{\eta}^r_i(t)$ satisfy the equation $m\,\frac{d^2\tilde{\eta}^r_i(t)}{dt^2} = \sum_{j\neq i} F^r_{Newton}(\vec{\tilde{\eta}}_i(t) - \vec{\tilde{\eta}}_j(t)) + \frac{1}{c}\,\frac{d\tilde{\eta}^r_i(t)}{dt}\,\Big(\frac{1}{\triangle}\,c^2\,\partial^2_\tau\,{}^3K_{(1)}(\tau = ct, \vec{\sigma})\Big)|_{\vec{\sigma}=\vec{\tilde{\eta}}_i(t)}$, where $\vec{F}_{Newton}$ is the Newton gravitational force. Therefore the (arbitrary in these gauges) double rate of change in time of the trace of the extrinsic curvature creates a post-Newtonian damping (or anti-damping since the sign of ${}^3K_{(1)}$ is not fixed) effect on the motion of particles. This is a inertial effect (hidden in the lapse function) not existing in Newton theory where the Euclidean 3-space is absolute. In the 2-body case we get that for Keplerian circular orbits of radius r the modulus of the relative 3-velocity can be written in the form $\sqrt{\frac{G\,(m+\triangle\,m(r))}{r}}$ with $\triangle\,m(r)$ function only of ${}^3K_{(1)}$. Now the rotation curves of galaxies (see Ref.[23] for a review) imply that this quantity goes to constant for large r (instead of vanishing): as a consequence $\triangle\,m(r)$ is interpreted as a *dark matter halo* around the galaxy. With our approach this dark matter would be a *relativistic inertial effect* consequence of the non-Euclidean nature of 3-space. This option would differ: 1) from the non-relativistic MOND approach [24] (where one modifies Newton equations); 2) from modified gravity theories like the $f(R)$ ones (see for instance Refs.[25]; here one gets a modification of the Newton potential); 3) from postulating the existence of WIMP particles [26].

At the experimental level *the description of baryon matter is intrinsically coordinate-dependent*. That is, it depends on the conventions used by physicists, engineers and astronomers to model of space-time. As a consequence of the dependence on coordinates of the description of matter, our proposal for solving the gauge problem in our Hamiltonian framework with non-Euclidean 3-spaces is to choose a gauge (i.e. a 4-coordinate system) in non-modified Einstein gravity which is in agreement with the observational conventions in astronomy. Since ICRS [5] has diagonal 3-metric, our 3-orthogonal gauges are a good choice. We are left with the inertial gauge variable ${}^3\mathcal{K}_{(1)} = \frac{1}{\triangle}\, {}^3K_{(1)}$ not existing in Newtonian gravity. As already said the suggestion is to try to fix ${}^3\mathcal{K}_{(1)}$ in such a way to eliminate dark matter as much as possible, by reinterpreting it as a relativistic inertial effect induced by the shift from Euclidean 3-spaces to non-Euclidean ones (independently from cosmological assumptions). As a consequence, ICRS should be reformulated not as a *quasi-inertial* reference frame in Galilei space-time, but as a reference frame in a PM space-time with ${}^3K$ (i.e. the clock synchronization convention) deduced from the data connected to dark matter. Then automatically BCRS would be its quasi-Minkowskian approximation for the Solar System. This point of view could also be useful for the ESA GAIA mission (cartography of the Milky Way) [27].

VIII. CONCLUSIONS

The Hamiltonian formulation of Einstein theory done in a form which takes into account the problem of 3-space (i.e. of clock synchronization) and the coordinate-dependent description of matter (i.e. metrology) opens a new scenario for dark matter. Besides looking for other experimental signatures of the York time, we also have the possibility to explore its role in the back-reaction approach [28] to dark energy, according to which dark energy is a byproduct of the non-linearities of general relativity when one considers spatial mean values on large scales to get a cosmological description of the universe taking into account the inhomogeneity of the observed universe. In the York canonical basis all the relevant quantities are 3-scalars and it is possible to study the mean value of nearly all the Hamilton equations.

---

[1] D.Alba and L.Lusanna, Int.J.Geom.Methods in Physics **7**, 33 (2010) (0908.0213) and **7**, 185 (2010).

[2] T.D.Moyer, *Formulation for Observed and Computed Values of Deep Space Network Data Types for Navigation* (John Wiley, New York, 2003).

[3] *IERS Conventions (2003)*, eds. D.D.McCarthy and G.Petit, IERS TN 32 (2004), Verlag des BKG. G.H.Kaplan, *The IAU Resolutions on Astronomical Reference Systems, Time Scales and Earth Rotation Models*, U.S.Naval Observatory circular No. 179 (2005) (astro-ph/0602086).

[4] M.Soffel, S.A.Klioner, G.Petit, P.Wolf, S.M.Kopeikin, P.Bretagnon, V.A.Brumberg, N.Capitaine, T.Damour, T.Fukushima, B.Guinot, T.Huang, L.Lindegren, C.Ma, K.Nordtvedt, J.Ries, P.K.Seidelmann, D.Vokroulicky', C.Will and Ch.Xu, *The IAU 2000 Resolutions for Astrometry, Celestial Mechanics and Metrology in the Relativistic Framework: Explanatory Supplement* Astron.J., **126**, pp.2687-2706, (2003) (astro-ph/0303376).

[5] J.Kovalevski, I.I.Mueller and B.Kolaczek, *Reference Frames in Astronomy and Geophysics* (Kluwer, Dordrecht, 1989). O.J.Sovers and J.L.Fanselow, Rev.Mod.Phs. **70**, 1393 (1998).

[6] M.Bartelmann, *The Dark Universe*, (0906.5036).

[7] L. Lusanna, Int. J. Mod. Phys. **A12**, 645 (1997); *The Chrono-Geometrical Structure of Special and General Relativity: A Re-Visitation of Canonical Geometrodynamics*, lectures at 42nd Karpacz Winter School of Theoretical Physics: Current Mathematical Topics in Gravitation

and Cosmology, Ladek, Poland, 6-11 Feb 2006, Int.J.Geom.Methods in Mod.Phys. **4**, 79 (2007) (gr-qc/0604120).

[8] D.Alba and L.Lusanna, Int.J.Mod.Phys. **D16**, 1149 (2007) (gr-qc/0501090).

[9] L.Cacciapuoti and C.Salomon, *ACES: Mission Concept and Scientific Objective*, 28/03/2007, ESA document, Estec (ACES_Science_v1_printout.doc). L.Blanchet, C.Salomon, P.Teyssandier and P.Wolf, Astron.Astrophys. **370**, 320 (2000).

[10] L.Lusanna and M.Pauri, Gen.Rel.Grav. **38**, 187 and 229 (2006) (gr-qc/0403081 and 0407007); *Dynamical Emergence of Instantaneous 3-Spaces in a Class of Models of General Relativity*, in *Relativity and the Dimensionality of the World*, ed. V.Petkov (Springer Series Fundamental Theories of Physics, Berlin, 2007) (gr-qc/0611045).

[11] H.Crater and L.Lusanna, Ann.Phys.(N.Y.) **289**, 87 (2001)(hep-th/0001046). D.Alba, H.Crater and L.Lusanna, Int.J.Mod.Phys. **A16**, 3365 (2001) (hep-th/0103109); J.Phys. **A40**, 9585 (2007) (gr-qc/0610200).

[12] D.Alba, H.W.Crater and L.Lusanna, *The Rest-Frame Instant Form and Dirac Observables for the Open Nambu String*, arXiv 1005.3653; *Massless Particles plus Matter in the Rest-Frame Instant Form of Dynamics*, J.Phys. **A43**, 405203 (2010) (arXiv 1005.5521).

[13] D.Alba, L.Lusanna and M.Pauri, *New Directions in Non-Relativistic and Relativistic Rotational and Multipole Kinematics for N-Body and Continuous Systems* (2005), in *Atomic and Molecular Clusters: New Research*, ed.Y.L.Ping (Nova Science, New York, 2006) (hep-th/0505005); J.Math.Phys. **43**, 1677-1727 (2002) (hep-th/0102087); J. Math.Phys. **46**, 062505, 1-36 (2004) (hep-th/0402181).

[14] D.Alba, H.W.Crater and L.Lusanna, Canad.J.Phys. **88**, 379 (2010) (arXiv: 0806.2383) and **88**, 425 (2010) (arXiv: 0811.0715).

[15] D.Alba, H.W.Crater and L.Lusanna, *Relativistic Quantum Mechanics and Relativistic Entanglement in the Rest-Frame Instant Form of Dynamics*, (arXiv 0907.1816).

[16] L.Lusanna, Gen.Rel.Grav. **33**, 1579 (2001)(gr-qc/0101048). L.Lusanna and S.Russo, Gen.Rel.Grav. **34**, 189 (2002)(gr-qc/0102074). R.DePietri, L.Lusanna, L.Martucci and S.Russo, Gen.Rel.Grav. **34**, 877 (2002) (gr-qc/0105084).

[17] R.Beig and Ó Murchadha, Ann.Phys.(N.Y.) **174**, 463 (1987).

[18] D.Alba and L.Lusanna, Gen.Rel.Grav. **39**, 2149 (2007) (gr-qc/0604086, v2).

[19] D.Christodoulou and S.Klainerman, *The Global Nonlinear Stability of the Minkowski Space* (Princeton, Princeton, 1993).

[20] J.Isenberg and J.E.Marsden, J.Geom.Phys. **1**, 85 (1984).

[21] I.Ciufolini and J.A.Wheeler, *Gravitation and Inertia* (Princeton Univ.Press, Princeton, 1995).

[22] D.Alba and L.Lusanna, *The Einstein-Maxwell-Particle System in the York Canonical Basis of ADM Tetrad Gravity: I) The Equations of Motion in Arbitrary Schwinger Time Gauges.*, 2009 (arXiv 0907.4087); *II) The Weak Field Approximation in the 3-Orthogonal Gauges and Hamiltonian Post-Minkowskian Gravity: the N-Body Problem and Gravitational Waves with Asymptotic Background.*, (1003.5143); *III) The Post-Minkowskian N-Body Problem, its Post-Newtonian Limit in Non-Harmonic 3-Orthogonal Gauges and Dark Matter as an Inertial Effect.* (1009.1794).

[23] E.Battaner and E.Florido, Fund.Cosmic Phys. **21**, 1 (2000). D.G.Banhatti, Current Science **94**, 986 (2008). Astron.Astrophys. **385**, 816 (2002) (astro-ph/0201276).

[24] M.Milgrom, *New Physics at Low Accelerations (MOND): An Alternative to Dark Matter*, (0912.2678).

[25] S.Capozziello, V.F.Cardone and A.Troisi, Mon.Not.R.Astron.Soc. **375**, 1423 (2007) (astro-

ph/0603522).

[26] J.L.Feng, *Dark Matter Candidates from Particle Physics and Methods of Detection* , (arXiv 1003.0904).

[27] S.Jordan, Astron.Nachr. **329**, 875 (2008) (DOI 10.1002/asna.200811065).

[28] R.J.van den Hoogen, *Averaging Spacetime: Where do we go from here ?*, (1003.4020). T.Buchert, Gen.Rel.Grav. **40**, 467 (2008) (0707.2153).

Journal of Physics: Conference Series **284** (2011) 012047 doi:10.1088/1742-6596/284/1/012047

An infinite family of superintegrable systems from higher order ladder operators and supersymmetry

Ian Marquette

Department of Mathematics, University of York, Heslington, York, UK. YO10 5DD

E-mail: im553@york.ac.uk

Abstract. We discuss how we can obtain new quantum superintegrable Hamiltonians allowing the separation of variables in Cartesian coordinates with higher order integrals of motion from ladder operators. We also discuss how higher order supersymmetric quantum mechanics can be used to obtain systems with higher order ladder operators and their polynomial Heisenberg algebra. We present a new family of superintegrable systems involving the fifth Painlevé transcendent which possess fourth order ladder operators constructed from second order supersymmetric quantum mechanics. We present the polynomial algebra of this family of superintegrable systems.

1. Introduction

The most well known superintegrable systems are the Kepler-Coulomb system and the harmonic oscillator [1]. Superintegrable systems with second order integrals of motion in two-dimensional [2] and three dimensional Euclidean spaces [3,4] were classified. However, a systematic search for superintegrable systems with third order integrals of motion in classical and quantum mechanics is more recent and was started in Ref.5. Classical and quantum systems with a second and a third order integrals of motion allowing separation of variables in Cartesian coordinates were obtained by S.Gravel [6] and five of the 14 quantum systems were written in terms of the first, second and fourth Painlevé transcendents [7]. These quantum systems were studied [8-10] from the point of view of supersymmetric quantum mechanics and their polynomial algebras. For a review of two-dimensional superintegrable systems we refer the reader to the following paper [11]. In recent years, many articles were devoted to integrable and superintegrable systems with higher order integrals of motion [12-31].

A relation between first and second order ladder operators and integrals of motion of superintegrable systems were acknowledged by many authors [1-4,10,15,16,32-34]. A relation between first order supercharges [35] and integrals of motion were also discussed [36-39]. Recently, we pointed out [8-10,12,13] how superintegrability with higher order integrals, higher order ladder operators and higher order supersymmetric quantum mechanics [40-44] are related. We discussed [10] how from known one-dimensional Hamiltonians that possess ladder operators we can form multi-dimensional superintegrable systems, their integrals and their polynomial algebra. We constructed a new infinite family of superintegrable systems involving the fourth Painlevé transcendent from third order ladder operators. We presented a classical analog of this construction and obtained an infinite family of classical superintegrable using third order ladder operators [12]. The factorization of Hamiltonians in supersymmetric quantum mechanics is not necessarily unique and can be used as a tool for obtaining from known systems new superintegrable systems with higher order integrals of motion [13].

The purpose of this paper is to discuss how from systems allowing two intertwining relations with second order supercharges [44], which are related to fourth order ladder operators, we can generate an infinite family of superintegrable systems involving the fifth Painlevé transcendent. In Section 2, we present a method to construct superintegrable systems introduced in Ref.10. In Section 3, we discuss the relation between supersymmetric quantum mechanics and ladder operators. In Section 4, we recall results of Ref.44 and using these results we present explicitly the system with fourth order ladder operators, its ladder operators and zero modes in terms of the fifth Painlevé transcendent. In Section 5, using results of Section 2 and Section 4, we obtain a new family of superintegrable systems and we present their integrals of motion and the polynomial algebra generated by these integrals.

2. Superintegrability and ladder operators

Let us consider the following two-dimensional Hamiltonian allowing separation of variables in Cartesian coordinates:

$$H = H_1 + H_2 = \frac{P_1^2}{2} + \frac{P_2^2}{2} + V_1(x_1) + V_2(x_2). \tag{1}$$

We suppose the existence of polynomial ladder operators (ω_i are constants and Q_i polynomials of order k_i) :

$$[H_i, A_i^\dagger] = \omega_i A_i^\dagger, \quad [H_i, A_i] = -\omega_i A_i, \tag{2}$$

$$A_i A_i^\dagger = Q_i(H_i + \omega_i), \quad A_i^\dagger A_i = Q_i(H_i), \quad i = 1, 2. \tag{3}$$

The polynomial Heisenberg algebra generated by $\{H_i, A_i, A_i^\dagger\}$ provides information on the spectrum of H_i. The annihilation operator can allow at most k_i zero modes (i.e. a state such $A_i\psi = 0$). In each axis we can have at most k_i infinite ladders by acting iteratively with the creation operator. The creation operator can also allow zero modes and we can have finite ladders.

From these operators, we construct the following integrals of motion (of order 2, $k_1n_1 + k_2n_2$ and $k_1n_1 + k_2n_2$) with $n_1\omega_1 = n_2\omega_2 = \omega$, $n_1, n_2 \in \mathbb{N}^*$:

$$K = \frac{1}{2\omega}(H_1 - H_2), \quad I_- = A_1^{n_1} A_2^{\dagger n_2}, \quad I_+ = A_1^{\dagger n_1} A_2^{n_2}. \tag{4}$$

The systems is thus superintegrable. We can also consider the integrals $I_1 = I_- - I_+$ and $I_2 = I_- + I_+$. These results can be used to study known superintegrable systems and also construct new ones. Moreover, this construction can be extended for N-dimensional quantum systems and also in context of classical systems [12]. These integrals of motion generate a polynomial algebra [8-12,45-48] of order $(k_1n_1 + k_2n_2 - 1)$

$$[K, I_\pm] = \pm I_\pm, \quad [I_-, I_+] = F_{n_1,n_2}(K + 1, H) - F_{n_1,n_2}(K, H), \tag{5}$$

$$F_{n_1,n_2}(K, H) = \prod_{i=1}^{n_1} Q_1\left(\frac{H}{2} + n_1\omega_1 K - (n_1 - i)\omega_1\right) \prod_{j=1}^{n_2} Q_2\left(\frac{H}{2} - n_2\omega_2 K + j\omega_2\right). \tag{6}$$

The equations (5) and (6) form a deformed $u(2)$ algebra and the finite dimensional representation modules can be found using a realization as a generalized deformed oscillator algebra [49]. The operators $b^\dagger = I_+$, $b = I_-$, $N = K - u$ (u is a constant) and $\Phi(N, u, H) = F_{n_1,n_2}(K, H)$ satisfy a deformed oscillator algebra :

$$[N, b^t] = b^t, \quad [N, b] = -b, \quad b^t b = \Phi(N, u, H), \quad bb^t = \Phi(N + 1, u, H) \quad . \tag{7}$$

The function $\Phi(N)$ is called the "structure function". Following [7-10,47,48] we have an energy dependent Fock space of dimension p+1 if

$$\Phi(0,u,E)=0,\quad \Phi(p+1,u,E)=0,\quad \Phi(n,u,E)>0,\qquad n=1,2,...,p. \tag{8}$$

The Fock space is defined by

$$H|E,n>=E|E,n>,\quad N|E,n>=n|E,n>\quad b|E,0>=0, \tag{9}$$

$$b^t|n>=\sqrt{\Phi(n+1,u,E)}|E,n+1>,\quad b|n>=\sqrt{\Phi(n,u,E)}|E,n-1>. \tag{10}$$

The relations given by equation (8) can be used to obtain the finite dimensional unitary representations and the corresponding degenerate energy spectrum of superintegrable systems [8-11,47,48].

2.1. Systems with ladder operators of first, second and third order

Infinite families of superintegrable systems in two-dimensional Euclidean space from first [1], second [15] and third [10] order ladder operators were discussed in previous articles (with $n_1\omega_1=n_2\omega_2=\omega$, n_1,$n_2\in\mathbb{N}^*$):

$$V(x_1,x_2)=\frac{1}{2}(\omega_1^2x_1^2+\omega_2^2x_2^2), \tag{11}$$

$$V(x_1,x_2)=\frac{1}{2}(\omega_1^2x_1^2+\omega_2^2x_2^2)+\frac{l_1}{x_1^2}+\frac{l_2}{x_2^2}, \tag{12}$$

$$V(x_1,x_2)=\frac{\omega_1^2}{2}(x_1+\sqrt{\frac{\hbar}{\omega_1}}P_{4(1)}(\sqrt{\frac{\omega_1}{\hbar}}x_1),\alpha_1,\beta_1)^2+\frac{\epsilon_1\hbar\omega_1}{2}P'_{4(1)}(\sqrt{\frac{\omega_1}{\hbar}}x_1,\alpha_1,\beta_1)+\frac{\hbar\omega_1}{3}(\epsilon_1-\alpha_1) \tag{13}$$

$$+\frac{\omega_2^2}{2}(x_2+\sqrt{\frac{\hbar}{\omega_2}}P_{4(2)}(\sqrt{\frac{\omega_2}{\hbar}}x_2,\alpha_2,\beta_2))^2+\frac{\epsilon_2\hbar\omega_2}{2}P'_{4(2)}(\sqrt{\frac{\omega_2}{\hbar}}x_2,\alpha_2,\beta_2)+\frac{\hbar\omega_2}{3}(\epsilon_2-\alpha_2)$$

$$P''_{4(i)}(z_i)=\frac{P'^2_{4(i)}(z_i)}{2P_{4(i)}(z_i)}+\frac{3}{2}P^3_{4(i)}(z_i)+4zP^2_{4(i)}(z_i)+2(z_i^2-\alpha_i)P_{4(i)}(z_i)+\frac{\beta_i}{P_{4(i)}(z_i)}, \tag{14}$$

$$P_{4(i)}(z_i)=P_{4(i)}(z_i,\alpha_i,\beta_i),\quad z_i=\sqrt{\frac{\omega_i}{\hbar}}x_i.$$

The integrals of motion, the polynomial algebras , the finite-dimensional unitary repsentations and the corresponding degenerate energy spectrum were given respectively for superintegrable systems with first order ladder operators in Ref. 33 and systems with second and third order ladder operators in Ref. 10.

3. Supersymmetry and ladder operators

Let us recall some results of first and second order supersymmetric quantum mechanics (also known as intertwining or factorization method) [35]

Let us consider the following intertwining relation

$$H_aL^\dagger=L^\dagger H_b, \tag{15}$$

with a first order intertwining operator (also called supercharge) and two Hamiltonians of the form

$$H_i=-\frac{\hbar^2}{2}\frac{d^2}{dx^2}+V_i(x),\quad L^\dagger=\frac{1}{\sqrt{2}}(-\hbar\frac{d}{dx}+\alpha(x)),\quad i=a,b. \tag{16}$$

These relations lead to

$$V_b = V_a - \hbar\alpha', \quad 2\alpha V_b - \hbar^2\alpha'' = 2\alpha V_a - 2\hbar V_a'. \tag{17}$$

We obtain by putting the first relation in the second and integrating the result

$$\hbar\alpha' + \alpha^2 = 2(V_a(x) - \epsilon). \tag{18}$$

These equations are equivalent to the factorization method

$$H_a = L^\dagger L + \epsilon, \quad H_b = LL^\dagger + \epsilon. \tag{19}$$

If one of these systems is exactly solvable and allows ladder operators, we can obtained the wavefunctions (up to a possible zero mode state), the ladder operators and the energy spectrum of its superpartners using the supercharges. The operators $A_a^\dagger$ and $A_b^\dagger$ (A_a and A_b) respectively the raising (lowering) operators of the Hamiltonians H_a and H_b are related by $A_a^\dagger = L^\dagger A_b^\dagger L$ ($A_a = L^\dagger A_b L$). Supersymmetric quantum mechanics can thus be used to generate systems with higher order ladder operators.

Let us now consider two Hamiltonians [40-44]

$$L^\dagger = \frac{1}{2}(\hbar^2\frac{d^2}{dx^2} - \hbar g(x)\frac{d}{dx} + h(x)). \tag{20}$$

We obtain the following three equations

$$V_b = V_a - \hbar g(x)', \tag{21}$$

$$\hbar^2\frac{g(x)''}{2} - \hbar h(x)' - g(x)V_b = 2\hbar V_a' - g(x)V_a, \tag{22}$$

$$2h(x)V_b - \hbar^2 h(x)'' = 2\hbar^2 V_0'' - 2\hbar g(x)V_a' + 2h(x)V_a. \tag{23}$$

From the equations (21), (22) and (23) we can obtain

$$V_{a,b}(x) = \hbar^2\frac{g(x)''}{4g(x)} - \hbar^2\frac{g(x)'^2}{8g(x)^2} \pm \hbar\frac{g(x)'}{2} + \frac{g(x)^2}{8} + \frac{d}{2\hbar} + \frac{c}{2g(x)^2\hbar}, \tag{24}$$

$$h(x) = -\hbar^2\frac{g(x)''}{2g(x)} + \hbar^2\frac{g(x)'^2}{4g(x)^2} - \hbar\frac{g(x)'}{2} + \frac{g(x)^2}{4} - \frac{c}{g(x)^2\hbar}. \tag{25}$$

Moreover, we have

$$LL^\dagger = ((H_a - \frac{d}{2\hbar})^2 - \frac{c}{4\hbar}), \quad L^\dagger L = ((H_b - \frac{d}{2\hbar})^2 + \frac{c}{4\hbar}). \tag{26}$$

The ladder operators of the Hamiltonians H_a and H_b are thus related by the second order supercharges.

A first and second order supersymmetry can also be combined [41] to obtain systems with third order ladder operators related to superintegrable systems obtained by S.Gravel [6,8-10]. If we consider a first $H_a L_1^\dagger = L_1^\dagger(H_b + \hbar\omega)$ and a second order intertwining relations $H_a L_2^\dagger = L_2^\dagger H_b$, we can construct the following third order supercharges $A^\dagger = L_1^\dagger L_2$ and $A = L_2^\dagger L_1$. They satisfy respectively the following relations : $[H_a, A^\dagger] = \hbar\omega A^\dagger$ and $[H_a, A] = -\hbar\omega A$. These operators can thus be interpreted as raising and lowering operators. These potentials can be written in terms of the fourth Painlevé transcendent. The equation (13) consists in a sum of two such one-dimensional systems.

4. Systems with fourth order ladder operators

Let us recall results from Ref.44 and consider two Hamiltonians

$$H_i = -\frac{\hbar^2}{2}\frac{d^2}{dx^2} + V_i(x), \quad i = a, b, \tag{27}$$

intertwined by two second order supercharges L_1 and L_2

$$L_1^\dagger = \frac{1}{2}(\hbar^2\frac{d^2}{dx^2} - \hbar g_1(x)\frac{d}{dx} + h_1(x)), \quad L_2^\dagger = \frac{1}{2}(\hbar^2\frac{d^2}{dx^2} - \hbar g_2(x)\frac{d}{dx} + h_2(x)). \tag{28}$$

The interwining equations have the following form

$$H_b L_2^\dagger = L_2^\dagger H_a, \tag{29}$$

$$H_a L_1^\dagger = L_1^\dagger(H_b + \hbar\omega). \tag{30}$$

We construct from these supercharges the following fourth order ladder operators of the Hamiltonian H_a and their polynomial Heisenberg algebra

$$A^\dagger = L_1^\dagger L_2^\dagger, \quad A = L_2 L_1, \tag{31}$$

$$[H_a, A^\dagger] = \hbar\omega A^\dagger, \quad [H_a, A] = -\hbar\omega A, \tag{32}$$

$$A^\dagger A = Q(H_a) = \prod_j^4 (H_a - \epsilon_j), \quad AA^\dagger = Q(H_a + \hbar\omega), \tag{33}$$

with $\epsilon_{1,2} = \hbar\omega + \frac{d_2}{2\hbar} \pm \sqrt{\frac{c_2}{4\hbar}}$, $\epsilon_{3,4} = \frac{d_1}{2\hbar} \pm \sqrt{\frac{c_1}{4\hbar}}$. The operators ($A^\dagger = L_2^\dagger L_1^\dagger$ and $A = L_1 L_2$) are ladder operators of the Hamiltonian H_2.

The intertwining relations (29) and (30) give six equations. Using the following changes of variables $z = \sqrt{\frac{\omega}{\hbar}}x$, $g_i = \sqrt{\hbar\omega}\tilde{g_i}$, $h_i = \omega\hbar\tilde{h_i}$ (with i=1,2), $V_k = \hbar\omega\tilde{V_k}$ (with k=a,b) and $\epsilon_j' = \frac{\epsilon_j}{\hbar\omega}$ (with $j = 1, 2, 3, 4$) we obtain

$$\frac{\tilde{g_1}''}{2\tilde{g_1}} - (\frac{\tilde{g_1}'}{2\tilde{g_1}})^2 \mp \tilde{g_1}' + \frac{\tilde{g_1}^2}{4} + \frac{(\epsilon_3' - \epsilon_4')^2}{\tilde{g_1}^2} + \epsilon_3' + \epsilon_4' = 2\tilde{V}_{a,b}, \tag{34}$$

$$\frac{\tilde{g_2}''}{2\tilde{g_2}} - (\frac{\tilde{g_2}'}{2\tilde{g_2}})^2 \pm \tilde{g_2}' + \frac{\tilde{g_2}^2}{4} + \frac{(\epsilon_1' - \epsilon_2')^2}{\tilde{g_2}^2} + \epsilon_1' + \epsilon_2' - 1 \mp 1 = 2\tilde{V}_{a,b}, \tag{35}$$

$$-\frac{\tilde{g_1}''}{2\tilde{g_1}} + (\frac{\tilde{g_1}'}{2\tilde{g_1}})^2 - \frac{\tilde{g_1}'}{2} + \frac{\tilde{g_1}^2}{4} - \frac{(\epsilon_3' - \epsilon_4')^2}{\tilde{g_1}^2} = h_1, \tag{36}$$

$$-\frac{\tilde{g_2}''}{2\tilde{g_2}} + (\frac{\tilde{g_2}'}{2\tilde{g_2}})^2 - \frac{\tilde{g_2}'}{2} + \frac{\tilde{g_2}^2}{4} - \frac{(\epsilon_1' - \epsilon_2')^2}{\tilde{g_2}^2} = h_2, \tag{37}$$

$$\tilde{V_b} = \tilde{V_a} + \tilde{g_1}' - 1 = \tilde{V_a} - \tilde{g_2}'. \tag{38}$$

From the equations (34), (35) and (38), we can obtain the following nonlinear second order differential equation [44]

$$\tilde{g_2}'' = \frac{(-2\tilde{g_2} + z)}{2\tilde{g_2}(-\tilde{g_2} + z)}(\tilde{g_2}')^2 + \frac{\tilde{g_2}}{z(-\tilde{g_2} + z)}\tilde{g_2}' + +[2z\tilde{g_2}(-\tilde{g_2} + z)]^{-1}[-2z\tilde{g_2}^5 + (5z^2 + 8\epsilon + 4)\tilde{g_2}^4 \tag{39}$$
$$-4z(z^2 + 4\epsilon + 2)\tilde{g_2}^3 + [z^4 + 4(2\epsilon + 1)z^2 + 4(\Delta_2^2 - \Delta_1^2) - 1]\tilde{g_2}^2 - 4\Delta_1^2 z(-2\tilde{g_2} + z)]$$

with ($\epsilon = \frac{1}{2}(\epsilon_3' + \epsilon_4') - \frac{1}{2}(\epsilon_1' + \epsilon_2')$, $\Delta_1 = \epsilon_1' - \epsilon_2'$, $\Delta_2 = \epsilon_3' - \epsilon_4'$).

The equation (39) can be transformed into the fifth Painlevé transcendent by considering ($\tilde{g_2} = \frac{-z}{W-1}$ and $y = z^2$)

$$\frac{d^2W}{dy^2} = (\frac{1}{2W} + \frac{1}{W-1})(\frac{dW}{dy})^2 - \frac{1}{y}\frac{dW}{dy} + \frac{(W-1)^2}{y^2}(\frac{aW^2 + b}{W}) + \frac{cW}{y} + \frac{dW(W+1)}{W-1}, \tag{40}$$

with the following parameters $a = \frac{\Delta_1^2}{2}$, $b = -\frac{\Delta_2^2}{2}$, $c = -\epsilon - \frac{1}{2}$, and $d = -\frac{1}{8}$.

Journal of Physics: Conference Series **284** (2011) 012047 doi:10.1088/1742-6596/284/1/012047

4.1. Systems in terms of the fifth Painlevé transcendent

Using these results, we present the Hamiltonians H_a and H_b, the supercharges L_1, L_2, the ladder operators L and $L^\dagger$ and their zero modes in terms of the fifth Painlevé transcendent.

The potential $V(x) = V_a(x) - \frac{1}{2}(\epsilon_1 + \epsilon_2)$ is given by the following expression

$$V(x) = \frac{\omega^2}{8}(1 + \frac{4(P_5 + P_5')^2 - P_5^2}{(P_5 - 1)^2 P_5})x^2 + \frac{\hbar^2}{x^2}(a - b - \frac{1}{8} + \frac{b - aP_5^2}{P_5}) - \hbar\omega(1 + \frac{(1 + 2cP_5)}{2(P_5 - 1)}), \quad (41)$$

with $(P_5 = P_5(\frac{\omega}{\hbar}x^2, a, b, c, -\frac{1}{8}))$.

The functions h_1,h_2,g_1 and g_2 in terms of the fifth Painlevé transcendent are given by

$$g_1 = \frac{\omega x}{P_5 - 1} + \sqrt{\frac{\omega}{\hbar}}x, \quad h_1 = \hbar\omega(\frac{2(4c - P_5)}{4(-1 + P_5)}) + \hbar^2(\frac{-8b + P_5(1 - 8a + 8b + 8aP_5)}{4x^2 P_5}) \quad (42)$$

$$+\omega^2(\frac{x^2(P_5(-1 + (-1 + P_5)P_5) + 4P_5P_5' - 4P_5'^2)}{4(-1 + P_5)^2 P_5}),$$

$$g_2 = -\frac{\omega x}{P_5 - 1}, \quad h_2 = \hbar\omega(\frac{2(1 + 4cP_5)}{4(-1 + P_5)}) + \hbar^2(\frac{-8b + P_5(1 - 8a + 8b + 8aP_5)}{4x^2 P_5}) \quad (43)$$

$$+\omega^2(\frac{-x^2(P_5^2 + P_5^3 + 4P_5'^2 + P_5(-1 + 4P_5'))}{4(-1 + P_5)^2 P_5}).$$

There are four possible zero modes for the annihilation operator

$$\psi_{1,2}^{(i)} \propto (\pm\sqrt{\frac{a}{2}}(-1 + P_5) - (\frac{1}{2} + c) + \frac{\frac{\omega}{\hbar}x^2(P_5 + 2P_5')}{4(-1 + P_5)})e^{\int \frac{-1 + \frac{\omega}{\hbar}x^2 \mp \sqrt{8a}(-1 + P_5)^2 + P_5 - 2\frac{\omega}{\hbar}x^2 P_5'}{2x(-1 + P_5)}dx}, \quad (44)$$

$$\psi_{3,4} \propto e^{\int (\frac{\pm\sqrt{-8b}(-1 + P_5)^2 - \frac{\omega}{\hbar}x^2(P_5^2 - 2P_5')}{2x(-1 + P_5)P_5} + \frac{1}{2x})dx}, \quad (45)$$

with the corresponding energy levels

$$E_{\psi_{1,2}} = \pm\hbar\omega\sqrt{\frac{a}{2}}, \quad E_{\psi_{3,4}} = \hbar\omega(-c - \frac{1}{2} \mp \sqrt{\frac{-b}{2}}), \quad (46)$$

and also four possible zero modes for the creation operator

$$\phi_{1,2} \propto e^{\int \frac{-1 - \frac{\omega}{\hbar}x^2 \pm \sqrt{8a}(-1 + P_5)^2 + P_5 - 2\frac{\omega}{\hbar}x_i^2 P_5'}{2x(-1 + P_5)}dx}, \quad (47)$$

$$\phi_{3,4} \propto (\mp\sqrt{\frac{b}{2}}(-1 + P_5) + (\frac{1}{2} + c) - \frac{\frac{\omega}{\hbar}x^2(P_5 + 2P_5')}{4(-1 + P_5)})e^{\int (\frac{((\mp\sqrt{-8b})(-1 + P_5)^2 + \frac{\omega}{\hbar}x^2(P_5^2 - 2P_5'))}{2xP_5(-1 + P_5)} + \frac{1}{2x})dx}, \quad (48)$$

with the following energy levels

$$E_{\phi_{1,2}} = \pm\hbar\omega(\sqrt{\frac{a}{2}} - 1), \quad E_{\phi_{3,4}} = \hbar\omega(-c - \frac{3}{2} \mp \sqrt{\frac{-b}{2}}). \quad (49)$$

Thus, the system can have one, two, three or four infinite sequences of levels and finite sequence of levels (singlet, doublet or triplet states) depending of the parameters (a,b,c and ω). A more detailed study of these cases should be performed and we let it to a future article. The fifth Painlevé transcendent also allows rational solutions [50] depending of the parameters.

5. Superintegrable systems

Using results from Section 2, we can construct an infinite family of two-dimensional superintegrable systems (with $n_1\omega_1 = n_2\omega_2 = \lambda$, n_1,$n_2 \in \mathbb{N}^*$)

$$H = \sum_{i}^{2} \frac{P_{x_i}^2}{2} + \frac{\omega_i^2}{8}(1+\frac{4(P_{5(i)} + P'_{5(i)})^2 - P_{5(i)}^2}{(P_{5(i)} - 1)^2 P_{5(i)}})x_i^2 + \frac{\hbar^2}{x_i^2}(a_i - b_i - \frac{1}{8} + \frac{b_i - a_i P_{5(i)}^2}{P_{5(i)}}) - \hbar\omega_i(1+\frac{(1 + 2c_i P_{5(i)})}{2(P_{5(i)} - 1)}), \tag{50}$$

with ($P_{5(i)} = P_{5(i)}(\frac{\omega_i}{\hbar}x_i^2, a_i, b_i, c_i, -\frac{1}{8})$ i=1,2). The integrals of motion are given by (using (28), (31), (42), (43))

$$K = \frac{1}{2\lambda}(H_1 - H_2), \quad I_- = A_1^{n_1} A_2^{\dagger n_2}, \quad I_+ = A_1^{\dagger n_1} A_2^{n_2}. \tag{51}$$

These integrals of motion are of order 2 and $4(k_1 + k_2)$. The structure function of the polynomial algebra (of order $4(k_1 + k_2) - 1$) is given by

$$\Phi(N, u, H) = \prod_{i=1}^{n_1} Q_1(\frac{H}{2} + n_1\omega_1(N + u) - (n_1 - i)\omega_1) \prod_{j=1}^{n_2} Q_2(\frac{H}{2} - n_2\omega_2(N + u) + j\omega_2), \tag{52}$$

$$Q_i(H_i) = 16 \prod_{k=1}^{4} (H_i - E_{\psi_k}^{(i)}), \quad E_{\psi_{1,2}}^{(i)} = \mp\hbar\omega_i\sqrt{\frac{a_i}{2}}, \quad E_{\psi_{3,4}}^{(i)} = \hbar\omega_i(-c_i - \frac{1}{2} \mp \sqrt{\frac{-b_i}{2}}). \tag{53}$$

6. Conclusion

We presented a new family of superintegrable systems with higher integrals of motion from results obtained in context of supersymmetric quantum mechanics [44] using a method discussed Ref.10. The integrals of motion were constructed from the fourth order ladder operators that are product of second order supercharges. We presented also the polynomial algebra generated by these integrals of motion. The classical case of systems allowing fourth order ladder operators would differ of the quantum case. We let to a future article to obtain and study such systems and the corresponding classical superintegrable system.

The relation between superintegrability, ladder operators and supersymmetric quantum mechanics can be used to study known superintegrable systems or obtain new superintegrable systems with higher order integrals of motion [10,13,39] and need to be more explored.

The general equations for third order interwining relations for stationary Schrödinger operators [51] and second order intertwining relations for non-stationary Schrödinger operators [52] were obtained and may be used to construct superintegrable systems.

Moreover, systems written in terms of the Painlevé transcendents appear in many contexts in quantum mechanics [6-11,13,14,52-58] and recently a superintegrable system involving the sixth Painlevé transcendent was obtained [14]. The results obtained in this paper also extend the number of known superintegrable systems involving the Painlevé transcendents.

Acknowledgments

The research of I.M. was supported by a postdoctoral research fellowship from FQRNT of Quebec.

References

[1] Jauch J M and Hill E L 1940 *Phys.Rev.* **57** 641
[2] Winternitz P, Smorodinsky Ya A, Uhlir M and Fris I 1967 *Sov. J.Nucl.Phys.* **4** 444
[3] Evans N W 1990 *Phys.Rev.* A **41** 5666
[4] Evans N W 1991 *J.Math.Phys.* **32** 3369
[5] Gravel S and Winternitz P 2002 *J.Math.Phys.* **43** (12) 5902
[6] Gravel S 2004 *J.Math.Phys.* **45** (3) 1003

[7] Ince E L 1956 *Ordinary Differential Equations* (New York: Dover)
[8] Marquette I 2009 *J.Math.Phys.* **50** 012101
[9] Marquette I 2009 *J.Math.Phys.* **50** 095202
[10] Marquette I 2010 *J.Phys.A: Math. Gen.* **43** 135203
[11] Marquette I and Winternitz P 2008 *J. Phys. A: Math. Theor.* **41** 304031
[12] Marquette I 2010 *J.Math.Phys.* **51** 092903
[13] Marquette I 2009 *J.Math.Phys.* **50** 122102
[14] Tremblay F and Winternitz P 2010 *J. Phys. A: Math. Theor.* **43** 175206
[15] Verrier P E and Evans N W 2008 *J.Math.Phys.* **49** 092902
[16] Rodriguez m A, Tempesta P and Winternitz P 2008 *Phys.Rev.* E **78** 046608
[17] Kalnins E G, Miller Jr W and Post S 2010 *J.Phys.A: Math. theor.* **43** 035202
[18] Tremblay F, Turbiner V A and Winternitz P 2010 *J.Phys.A: Math. Theor.* **42** 242001
[19] Post S and Winternitz P 2010 *J. Phys. A: Math. Theor.* **43** 222001
[20] Quesne C 2010 *J.Phys. A:Math. Theor.* **43** 082001
[21] Ballesteros A and Herranz F J 2009 *J.Phys. A: Math.Theor.* **42** 245203
[22] Kalnins E G, Kress J M and Miller Jr W 2010 *J.Phys.A: Math. Theor.* **43** 265205
[23] Kalnins E G, Kress J M and Miller Jr W 2010 *J.Phys.A: Math. Theor.* **43** 092001
[24] Ioffe M V and Nishnianidze D N 2007 *Phys.Rev.* A **76** 052114
[25] Ioffe M V, Negro J, Nieto L M and Nishnianidze D N 2006 *J.Phys.* A **39** 9297
[26] Cannata F, Ioffe M V and Nishnianidze D N 2005 *Phys.Lett.* A **340** 31
[27] Ioffe m V and Valinevich P A 2005 *J.Phys.* A **38** 2497
[28] Correa F, Jakubsky V, Plyushchay M S 2009 *Annals Phys.* **324** 1078
[29] Alvarez P D, Gomis J, Kamimura K and Plyushchay M S 2008 *Phys.Lett.* B **659** 906
[30] Correa F, Nieto L M and Plyushchay M S 2007 *Phys.Lett.* B **644** 94
[31] Correa F, Jakubsky V and Plyushchay M S 2008 *J.Phys.* A **41** 485303
[32] Boyer C P and Miller Jr. W 1974 *J.Math.Phys.* **15** 9
[33] Bonatsos D and Daskaloyannis C 1999 *Prog.Part.Nucl.Phys.* **43** 537
[34] Ranada M F, Rodriguez M A and Santander M 2010 *J.Math.Phys.* **51** 042901
[35] Junker G 1995 *Supersymmetric Methods in Quantum and Statistical Physics* (New York: Springer)
[36] Mota R D, Garcia J and Granados V D 2001 *J.Phys. A: Math.Gen.* **34** 2041
[37] Lyman J M and Aravind P K 1993 *J.Phys. A: Math.Gen.* **26** 3307
[38] Calzada J A, Kuru S, Negro J and Del Olmo M A 2008 *J.Phys. A: Math. Theor.* **41** 255201
[39] Demircioglu B, Kuru S, Onder M and Vercin A 2002 *J.Math.Phys.* **43** 5
[40] Andrianov A, Ioffe M and Spiridonov V P 1003 *Phys.Lett.* A **174** 273; Andrianov A, Cannata F, Dedonder J P and Ioffe M 1995 *Int.Mod.Phys.* A **10** 2683
[41] Andrianov A, Cannata F, Ioffe M and Nishnianidze D 2000 *Phys.Lett.* A **266** 341 (*Preprint* quant-ph/9902057)
[42] Plyushchay M 2000 *Int.J.Mod.Phys.* A **15** 3679
[43] Fernandez D J and Hussin V 1999 *J.Phys.A: Math.Gen.* **32** 3603
[44] Carballo J M, Fernandez D J, Negro J and Nieto L M 2004 *J.Phys.A : Math.Gen.* **37** 10349
[45] Granovskii Ya I, Zhedanov A S and Lutzenko I M 1991 *J. Phys.* A **24** 3887
[46] Letourneau P and Vinet L 1995 *Ann.Phys.* **243** 144
[47] Bonatsos D, Daskaloyannis C and Kokkotas K 1993 *Phys.Rev.* A **48** R3407
[48] Daskaloyannis C 2001 *J.Math.Phys.* **42** 1100
[49] Daskaloyannis C 1991 *J.Phys.A: Math.Gen.* **24** L789
[50] Kitaev A V, Law C K and McLeod J B 1994 *Differential Integral Equations* **7** 967
[51] Ioffe M V and Mishnianidze D N 2004 *PhysLett.* A **327** 425
[52] Cannata F, Ioffe M, Junker G and Nishnianidze D 1999 *J.Phys.A.Math.Gen.* **32** 3583
[53] Veselov A P and Shabat A 1993 *Funkt.Analiz.Prilozh* **27** (2) 1
[54] Fushchych W I and Nikitin A G 1997 *J.Math.Phys.* **38** 11 5944
[55] Cannata F, Ioffe M, Junker G and Nishnianidze D 1999 *J.Phys.A.Math.Gen.* **32** 3583
[56] Willox R and Hietarinta J 2003 *J.Phys.A: Math.Gen.* **36** 10615
[57] Yurov A V and Yurov V A 2005 Phys.Rev. D **72** 026003
[58] Carlisle J E, Johnson C V and Pennington J S 2008 *J.Phys.A: Math.Theor.* **41** 085401

Journal of Physics: Conference Series **284** (2011) 012048 doi:10.1088/1742-6596/284/1/012048

The Flavour Symmetry: S_3

A. Mondragón

Instituto de Física, Universidad Nacional Autónoma de México, Apdo. Postal 20-364, México 01000 D.F., México

E-mail: mondra@fisica.unam.mx

Abstract. By introducing in the theory three Higgs fields that are $SU(2)$ doublets and a flavour permutational symmetry, S_3, we extend the concepts of flavour and generations to the Higgs sector and formulate a Minimal S_3–Invariant Extension of the Standard Model. The mass matrices of the neutrinos and charged leptons are re-parameterized in terms of their eigenvalues, then the neutrino mixing matrix, V_{PMNS}, is computed and exact, explicit analytical expressions for the neutrino mixing angles as functions of the masses of neutrinos and charged leptons are obtained in excellent agreement with the latest experimental data. We also compute the branching ratios of some selected flavour-changing neutral current (FCNC) processes, as well as the contribution of the exchange of neutral flavour-changing scalars to the anomaly of the magnetic moment of the muon, as functions of the masses of charged leptons and the neutral Higgs bosons. We find that the $S_3 \times Z_2$ flavour symmetry and the strong mass hierarchy of the charged leptons strongly suppress the FCNC processes in the leptonic sector, well below the present experimental bounds by many orders of magnitude. The contribution of FCNC's to the anomaly of the muon's magnetic moment is small, but non-negligible.

1. Introduction

The observation of flavour oscillations of solar, atmospheric, reactor, and accelerator neutrinos established that they have non-vanishing masses and mix among themselves, much like the quarks do [1, 2, 3, 4, 5, 6, 7, 8, 9, 10, 11, 12, 13, 14, 15, 16, 17, 18, 19, 20]. This discovery brought out very forcibly the need of extending the Standard Model (SM) in order to accomodate in the theory the new data on neutrino physics in a consistent way that would allow a unified and systematic treatment of the observed hierarchy of masses and mixings of all fermions. At the same time, the number of free parameters in the extended form of the SM had to be drastically reduced in order to give predictive power to the theory. These two seemingly contradictory demands are met by a flavour symmetry under which the families transfom in a non-trivial fashion.

In the Minimal S_3–Invariant Extension of the Standard Model [21, 22, 23, 24, 25, 26, 27], the concept of flavour and generations is extended to the Higgs sector in such a way that all the matter fields –Higgs, quarks, and lepton fields, including the right-handed neutrino fields– have three species and transform under the flavour symmetry group as the three dimensional representation $1 \oplus 2$ of the permutational group S_3. A model with more than one Higgs $SU(2)$ doublet has tree level flavour changing neutral currents whose exchange may give rise to lepton flavour violating processes and may also contribute to the anomalous magnetic moment of the muon. An effective test of the phenomenological success of the model is obtained by verifying that all flavour changing neutral current processes and the magnetic anomaly of the muon, computed in the S_3–Invariant extended form of the Standard Model, agree with the experimental values.

2. The Minimal S_3-invariant Extension of the Standard Model

In the Standard Model analogous fermions in different generations have identical couplings to all gauge bosons of the strong, weak and electromagnetic interactions. Prior to the introduction of the Higgs boson and mass terms, the Lagrangian is chiral and invariant with respect to permutations of the left and right fermionic fields.

The six possible permutations of three objects (f_1, f_2, f_3) are elements of the permutational group S_3. This is the discrete, non-Abelian group with the smallest number of elements. The three-dimensional real representation is not an irreducible representation of S_3. It can be decomposed into the direct sum of a doublet f_D and a singlet f_s, where

$$\begin{aligned} f_s &= \tfrac{1}{\sqrt{3}}(f_1 + f_2 + f_3), \\ f_D^T &= \left(\tfrac{1}{\sqrt{2}}(f_1 - f_2), \tfrac{1}{\sqrt{6}}(f_1 + f_2 - 2f_3)\right). \end{aligned} \tag{1}$$

The direct product of two doublets $\mathbf{p_D}^T = (p_{D1}, p_{D2})$ and $\mathbf{q_D}^T = (q_{D1}, q_{D2})$ may be decomposed into the direct sum of two singlets $\mathbf{r_s}$ and $\mathbf{r_{s'}}$, and one doublet $\mathbf{r_D}^T$ where

$$\mathbf{r_s} = p_{D1}q_{D1} + p_{D2}q_{D2}, \quad \mathbf{r_{s'}} = p_{D1}q_{D2} - p_{D2}q_{D1}, \tag{2}$$

$$\mathbf{r_D}^T = (r_{D1}, r_{D2}) = (p_{D1}q_{D2} + p_{D2}q_{D1}, p_{D1}q_{D1} - p_{D2}q_{D2}). \tag{3}$$

The antisymmetric singlet $\mathbf{r_{s'}}$ is not invariant under S_3.

Since the Standard Model has only one Higgs $SU(2)_L$ doublet, which can only be an S_3 singlet, it can only give mass to the quark or charged lepton in the S_3 singlet representation, one in each family, without breaking the S_3 symmetry.

Hence, in order to impose S_3 as a fundamental symmetry, unbroken at the Fermi scale, we are led to extend the Higgs sector of the theory. The quark, lepton and Higgs fields are

$$\begin{aligned} Q^T &= (u_L, d_L)\ ,\ u_R\ ,\ d_R\ , \\ L^T &= (\nu_L, e_L)\ ,\ e_R\ ,\ \nu_R \ \text{ and } \ H, \end{aligned} \tag{4}$$

in an obvious notation. All of these fields have three species, and we assume that each one forms a reducible representation $\mathbf{1}_S \oplus \mathbf{2}$. The doublets carry capital indices I and J, which run from 1 to 2, and the singlets are denoted by Q_3, u_{3R}, d_{3R}, L_3, e_{3R}, ν_{3R} and H_S. Note that the subscript 3 denotes the singlet representation and not the third generation. The most general renormalizable Yukawa interactions of this model are given by

$$\mathcal{L}_Y = \mathcal{L}_{Y_D} + \mathcal{L}_{Y_U} + \mathcal{L}_{Y_E} + \mathcal{L}_{Y_\nu}, \tag{5}$$

where

$$\begin{aligned} \mathcal{L}_{Y_E} \;=\; & -Y_1^e \overline{L}_I H_S e_{IR} - Y_3^e \overline{L}_3 H_S e_{3R} \\ & -Y_2^e [\, \overline{L}_I \kappa_{IJ} H_1 e_{JR} + \overline{L}_I \eta_{IJ} H_2 e_{JR}\,] \\ & -Y_4^e \overline{L}_3 H_I e_{IR} - Y_5^e \overline{L}_I H_I e_{3R} + \text{ h.c.}, \end{aligned} \tag{6}$$

$$\begin{aligned} \mathcal{L}_{Y_\nu} \;=\; & -Y_1^\nu \overline{L}_I (i\sigma_2) H_S^* \nu_{IR} - Y_3^\nu \overline{L}_3 (i\sigma_2) H_S^* \nu_{3R} \\ & -Y_2^\nu [\, \overline{L}_I \kappa_{IJ} (i\sigma_2) H_1^* \nu_{JR} + \overline{L}_I \eta_{IJ} (i\sigma_2) H_2^* \nu_{JR}\,] \\ & -Y_4^\nu \overline{L}_3 (i\sigma_2) H_I^* \nu_{IR} - Y_5^\nu \overline{L}_I (i\sigma_2) H_I^* \nu_{3R} + \text{ h.c.}, \end{aligned} \tag{7}$$

Table 1. Z_2 assignment in the leptonic sector.

$-$	$+$
H_S, ν_{3R}	H_I, L_3, L_I, e_{3R}, e_{IR}, ν_{IR}

and

$$\kappa = \begin{pmatrix} 0 & 1 \\ 1 & 0 \end{pmatrix} \quad \text{and} \quad \eta = \begin{pmatrix} 1 & 0 \\ 0 & -1 \end{pmatrix}. \tag{8}$$

$\mathcal{L}_{Y_D}$ and $\mathcal{L}_{Y_U}$ have similar expressions to $\mathcal{L}_{Y_E}$ and $\mathcal{L}_{Y_\nu}$ respectively.

Furthermore, we add to the Lagrangian the Majorana mass terms for the right-handed neutrinos

$$\mathcal{L}_M = -M_1 \nu_{IR}^T C \nu_{IR} - M_3 \nu_{3R}^T C \nu_{3R}. \tag{9}$$

The extended Higgs sector has three $SU(2)$ doublets, in a reducible representation $\mathbf{1}_S \oplus \mathbf{2}$ of the flavour group S_3. The Higgs potential, invariant under S_3, has an additional reflection symmetry $R: \ H_s \rightarrow \ -H_s$. and an accidental permutational symmetry S_2': $H_1 \leftrightarrow H_2$. Hence, $\langle H_1 \rangle = \langle H_2 \rangle$. Then the Yukawa interactions yield mass matrices for all fermions in the theory, of the general form [21]

$$\mathbf{M} = \begin{pmatrix} \mu_1 + \mu_2 & \mu_2 & \mu_5 \\ \mu_2 & \mu_1 - \mu_2 & \mu_5 \\ \mu_4 & \mu_4 & \mu_3 \end{pmatrix}. \tag{10}$$

The Majorana masses for the left handed neutrinos ν_L are generated by the see-saw mechanism. The corresponding mass matrix is given by

$$\mathbf{M}_\nu = \mathbf{M}_{\nu_\mathbf{D}} \tilde{\mathbf{M}}^{-1} (\mathbf{M}_{\nu_\mathbf{D}})^T, \tag{11}$$

where $\tilde{\mathbf{M}} = \mathrm{diag}(M_1, M_1, M_3)$.

In principle, all entries in the mass matrices can be complex since there is no restriction coming from the flavour symmetry S_3. The mass matrices are diagonalized by bi-unitary transformations as

$$\begin{aligned} U^\dagger_{d(u,e)L} \mathbf{M}_{d(u,e)} U_{d(u,e)R} &= \mathrm{diag}(m_{d(u,e)}, m_{s(c,\mu)}, m_{b(t,\tau)}), \\ U_\nu^T \mathbf{M}_\nu U_\nu &= \mathrm{diag}(m_{\nu_1}, m_{\nu_2}, m_{\nu_3}). \end{aligned} \tag{12}$$

The entries in the diagonal matrices may be complex, so the physical masses are their absolute values.

The mixing matrices are, by definition,

$$V_{CKM} = U^\dagger_{uL} U_{dL}, \quad V_{PMNS} = U^\dagger_{eL} U_\nu K. \tag{13}$$

where K is the diagonal matrix of the Majorana phase factors.

3. The mass matrices in the leptonic sector and Z_2 symmetry

A further reduction of the number of parameters in the leptonic sector may be achieved by means of an Abelian Z_2 symmetry. A possible set of charge assignments of Z_2, compatible with the experimental data on masses and mixings in the leptonic sector is given in Table 1. These Z_2 assignments forbid the following Yukawa couplings Y_1^e, Y_3^e, Y_1^ν and Y_5^ν. Therefore, the corresponding entries in the mass matrices vanish, *i.e.*, $\mu_1^e = \mu_3^e = 0$ and $\mu_1^\nu = \mu_5^\nu = 0$.

3.1. The mass matrix of the charged leptons

The remaining three parameters in the mass matrix of the charged leptons $|\tilde{\mu}_2|$, $|\tilde{\mu}_4|$ and $|\tilde{\mu}_5|$ may readily be expressed in terms of the charged lepton masses [23]. The resulting expression for M_e, written to order $(m_\mu m_e/m_\tau^2)^2$ and $x^4 = (m_e/m_\mu)^4$ is

$$M_e \approx m_\tau \begin{pmatrix} \frac{1}{\sqrt{2}}\frac{\tilde{m}_\mu}{\sqrt{1+x^2}} & \frac{1}{\sqrt{2}}\frac{\tilde{m}_\mu}{\sqrt{1+x^2}} & \frac{1}{\sqrt{2}}\sqrt{\frac{1+x^2-\tilde{m}_\mu^2}{1+x^2}} \\ \frac{1}{\sqrt{2}}\frac{\tilde{m}_\mu}{\sqrt{1+x^2}} & -\frac{1}{\sqrt{2}}\frac{\tilde{m}_\mu}{\sqrt{1+x^2}} & \frac{1}{\sqrt{2}}\sqrt{\frac{1+x^2-\tilde{m}_\mu^2}{1+x^2}} \\ \frac{\tilde{m}_e(1+x^2)}{\sqrt{1+x^2-\tilde{m}_\mu^2}}e^{i\delta_e} & \frac{\tilde{m}_e(1+x^2)}{\sqrt{1+x^2-\tilde{m}_\mu^2}}e^{i\delta_e} & 0 \end{pmatrix}. \tag{14}$$

This approximation is numerically exact up to order 10^{-9} in units of the τ mass. Notice that this matrix has no free parameters other than the Dirac phase δ_e.

The unitary matrix U_{eL} that diagonalizes $M_e M_e^\dagger$ and enters in the definition of the neutrino mixing matrix V_{PMNS} may be written in the polar form as $U_{eL} = P_{eL}\mathbf{O_{eL}}$ [24], where P_{eL} is a diagonal matrix of phases and the orthogonal matrix $\mathbf{O}_{eL}$ can be written as M_e, as follows

$$\mathbf{O}_{eL} \approx \begin{pmatrix} \frac{1}{\sqrt{2}}x\frac{(1+2\tilde{m}_\mu^2+4x^2+\tilde{m}_\mu^4+2\tilde{m}_e^2)}{\sqrt{1+\tilde{m}_\mu^2+5x^2-\tilde{m}_\mu^4-\tilde{m}_\mu^6+\tilde{m}_e^2+12x^4}} & -\frac{1}{\sqrt{2}}\frac{(1-2\tilde{m}_\mu^2+\tilde{m}_\mu^4-2\tilde{m}_e^2)}{\sqrt{1-4\tilde{m}_\mu^2+x^2+6\tilde{m}_\mu^4-4\tilde{m}_\mu^6-5\tilde{m}_e^2}} & \frac{1}{\sqrt{2}} \\ -\frac{1}{\sqrt{2}}x\frac{(1+4x^2-\tilde{m}_\mu^4-2\tilde{m}_e^2)}{\sqrt{1+\tilde{m}_\mu^2+5x^2-\tilde{m}_\mu^4-\tilde{m}_\mu^6+\tilde{m}_e^2+12x^4}} & \frac{1}{\sqrt{2}}\frac{(1-2\tilde{m}_\mu^2+\tilde{m}_\mu^4)}{\sqrt{1-4\tilde{m}_\mu^2+x^2+6\tilde{m}_\mu^4-4\tilde{m}_\mu^6-5\tilde{m}_e^2}} & \frac{1}{\sqrt{2}} \\ -\frac{\sqrt{1+2x^2-\tilde{m}_\mu^2-\tilde{m}_e^2}(1+\tilde{m}_\mu^2+x^2-2\tilde{m}_e^2)}{\sqrt{1+\tilde{m}_\mu^2+5x^2-\tilde{m}_\mu^4-\tilde{m}_\mu^6+\tilde{m}_e^2+12x^4}} & -x\frac{(1+x^2-\tilde{m}_\mu^2-2\tilde{m}_e^2)\sqrt{1+2x^2-\tilde{m}_\mu^2-\tilde{m}_e^2}}{\sqrt{1-4\tilde{m}_\mu^2+x^2+6\tilde{m}_\mu^4-4\tilde{m}_\mu^6-5\tilde{m}_e^2}} & \frac{\sqrt{1+x^2}\tilde{m}_e\tilde{m}_\mu}{\sqrt{1+x^2-\tilde{m}_\mu^2}} \end{pmatrix}, \tag{15}$$

where, as before, $\tilde{m}_\mu = m_\mu/m_\tau$, $\tilde{m}_e = m_e/m_\tau$ and $x = m_e/m_\mu$.

3.2. The mass matrix of the neutrinos

According to the Z_2 selection rule, the mass matrix of the Dirac neutrinos takes the form

$$\mathbf{M}_{\nu_\mathbf{D}} = \begin{pmatrix} \mu_2^\nu & \mu_2^\nu & 0 \\ \mu_2^\nu & -\mu_2^\nu & 0 \\ \mu_4^\nu & \mu_4^\nu & \mu_3^\nu \end{pmatrix}. \tag{16}$$

Then, the mass matrix for the left-handed Majorana neutrinos, $\mathbf{M}_\nu$, obtained from the see-saw mechanism, $\mathbf{M}_\nu = \mathbf{M}_{\nu_\mathbf{D}}\tilde{\mathbf{M}}^{-1}(\mathbf{M}_{\nu_\mathbf{D}})^T$, is

$$\mathbf{M}_\nu = \begin{pmatrix} 2(\rho_2^\nu)^2 & 0 & 2\rho_2^\nu\rho_4^\nu \\ 0 & 2(\rho_2^\nu)^2 & 0 \\ 2\rho_2^\nu\rho_4^\nu & 0 & 2(\rho_4^\nu)^2+(\rho_3^\nu)^2 \end{pmatrix}, \tag{17}$$

where $\rho_2^\nu = (\mu_2^\nu)/M_1^{1/2}$, $\rho_4^\nu = (\mu_4^\nu)/M_1^{1/2}$ and $\rho_3^\nu = (\mu_3^\nu)/M_3^{1/2}$; M_1 and M_3 are the masses of the right handed neutrinos appearing in (9).

The non-Hermitian, complex, symmetric neutrino mass matrix M_ν may be brought to a diagonal form by a unitary transformation, as

$$U_\nu^T M_\nu U_\nu = \mathrm{diag}\left(|m_{\nu_1}|e^{i\phi_1}, |m_{\nu_2}|e^{i\phi_2}, |m_{\nu_3}|e^{i\phi_\nu}\right), \tag{18}$$

where U_ν is the matrix that diagonalizes the matrix $M_\nu^\dagger M_\nu$.

As in the case of the charged leptons, the matrices M_ν and U_ν can be reparametrized in terms of the complex neutrino masses. Then [23, 24]

$$M_\nu = \begin{pmatrix} m_{\nu 3} & 0 & \sqrt{(m_{\nu 3}-m_{\nu 1})(m_{\nu 2}-m_{\nu 3})}e^{-i\delta_\nu} \\ 0 & m_{\nu 3} & 0 \\ \sqrt{(m_{\nu 3}-m_{\nu 1})(m_{\nu 2}-m_{\nu 3})}e^{-i\delta_\nu} & 0 & (m_{\nu 1}+m_{\nu 2}-m_{\nu 3})e^{-2i\delta_\nu} \end{pmatrix} \quad (19)$$

and

$$U_\nu = \begin{pmatrix} 1 & 0 & 0 \\ 0 & 1 & 0 \\ 0 & 0 & e^{i\delta_\nu} \end{pmatrix} \begin{pmatrix} \cos\eta & \sin\eta & 0 \\ 0 & 0 & 1 \\ -\sin\eta & \cos\eta & 0 \end{pmatrix}, \quad (20)$$

where

$$\sin^2\eta = \frac{m_{\nu 3}-m_{\nu 1}}{m_{\nu 2}-m_{\nu 1}}, \quad \cos^2\eta = \frac{m_{\nu 2}-m_{\nu 3}}{m_{\nu 2}-m_{\nu 1}}. \quad (21)$$

The unitarity of U_ν constrains $\sin\eta$ to be real and thus $|\sin\eta| \leq 1$, this condition fixes the phases ϕ_1 and ϕ_2 as

$$|m_{\nu 1}|\sin\phi_1 = |m_{\nu 2}|\sin\phi_2 = |m_{\nu 3}|\sin\phi_\nu. \quad (22)$$

The only free parameters in the matrices M_ν and U_ν, are the phase ϕ_ν, implicit in $m_{\nu 1}$, $m_{\nu 2}$ and $m_{\nu 3}$, and the Dirac phase δ_ν.

3.3. The neutrino mixing matrix

The neutrino mixing matrix V_{PMNS}, is the product $U_{eL}^\dagger U_\nu K$, where K is the diagonal matrix of the Majorana phase factors, defined by

$$diag(m_{\nu 1}, m_{\nu 2}, m_{\nu 3}) = K^\dagger diag(|m_{\nu 1}|, |m_{\nu 2}|, |m_{\nu 3}|)K^\dagger. \quad (23)$$

Except for an overall phase factor $e^{i\phi_1}$, which can be ignored, K is

$$K = diag(1, e^{i\alpha}, e^{i\beta}), \quad (24)$$

where $\alpha = 1/2(\phi_1 - \phi_2)$ and $\beta = 1/2(\phi_1 - \phi_\nu)$ are the Majorana phases.

Therefore, the theoretical mixing matrix V_{PMNS}^{th}, is given by

$$V_{PMNS}^{th} = \begin{pmatrix} O_{11}\cos\eta + O_{31}\sin\eta e^{i\delta} & O_{11}\sin\eta - O_{31}\cos\eta e^{i\delta} & -O_{21} \\ -O_{12}\cos\eta + O_{32}\sin\eta e^{i\delta} & -O_{12}\sin\eta - O_{32}\cos\eta e^{i\delta} & O_{22} \\ O_{13}\cos\eta - O_{33}\sin\eta e^{i\delta} & O_{13}\sin\eta + O_{33}\cos\eta e^{i\delta} & O_{23} \end{pmatrix} \times K, \quad (25)$$

where $\cos\eta$ and $\sin\eta$ are given eq. (21) O_{ij} are given in (15), and $\delta = \delta_\nu - \delta_e$.

To find how our results are related to the neutrino mixing angles we make use of the equality of the absolute values of the elements of V_{PMNS}^{th} and V_{PMNS}^{PDG} [28], that is

$$|V_{PMNS}^{th}| = |V_{PMNS}^{PDG}|. \quad (26)$$

This relation allows us to derive expressions for the mixing angles in terms of the charged lepton and neutrino masses.

The magnitudes of the reactor and atmospheric mixing angles, θ_{13} and θ_{23}, are determined by the masses of the charged leptons only. Keeping only terms of order (m_e^2/m_μ^2) and $(m_\mu/m_\tau)^4$, we get

$$\sin\theta_{13} \approx \frac{1}{\sqrt{2}} x \frac{(1+4x^2-\tilde{m}_\mu^4)}{\sqrt{1+\tilde{m}_\mu^2+5x^2-\tilde{m}_\mu^4}}, \quad \sin\theta_{23} \approx \frac{1}{\sqrt{2}} \frac{1+\frac{1}{4}x^2-2\tilde{m}_\mu^2+\tilde{m}_\mu^4}{\sqrt{1-4\tilde{m}_\mu^2+x^2+6\tilde{m}_\mu^4}}. \quad (27)$$

Journal of Physics: Conference Series **284** (2011) 012048 doi:10.1088/1742-6596/284/1/012048

The magnitude of the solar angle depends on charged lepton and neutrino masses, as well as, on the Dirac and Majorana phases,

$$|\tan\theta_{12}|^2 = \frac{m_{\nu_2}-m_{\nu_3}}{m_{\nu_3}-m_{\nu_1}}\left(\frac{1-2\frac{O_{11}}{O_{31}}\cos\delta\sqrt{\frac{m_{\nu_3}-m_{\nu_1}}{m_{\nu_2}-m_{\nu_3}}}+\left(\frac{O_{11}}{O_{31}}\right)^2\frac{m_{\nu_3}-m_{\nu_1}}{m_{\nu_2}-m_{\nu_3}}}{1+2\frac{O_{11}}{O_{31}}\cos\delta\sqrt{\frac{m_{\nu_2}-m_{\nu_3}}{m_{\nu_3}-m_{\nu_1}}}+\left(\frac{O_{11}}{O_{31}}\right)^2\frac{m_{\nu_2}-m_{\nu_3}}{m_{\nu_3}-m_{\nu_1}}}\right). \tag{28}$$

The dependence of $\tan\theta_{12}$ on the Dirac phase δ, see (28), is very weak, since $O_{31}\sim 1$ but $O_{11}\sim 1/\sqrt{2}(m_e/m_\mu)$. Hence, we may neglect it when comparing (28) with the data on neutrino mixings.

The dependence of $\tan\theta_{12}$ on the phase ϕ_ν and the physical masses of the neutrinos enters through the ratio of the neutrino mass differences, it can be made explicit with the help of the unitarity constraint on U_ν, eq. (22),

$$\frac{m_{\nu_2}-m_{\nu_3}}{m_{\nu_3}-m_{\nu_1}} = \frac{(|m_{\nu_2}|^2-|m_{\nu_3}|^2\sin^2\phi_\nu)^{1/2}-|m_{\nu_3}||\cos\phi_\nu|}{(|m_{\nu_1}|^2-|m_{\nu_3}|^2\sin^2\phi_\nu)^{1/2}+|m_{\nu_3}||\cos\phi_\nu|}. \tag{29}$$

4. Neutrino mass spectrum

In the present model, the numerical values of $\sin\theta_{13}$ and $\sin^2\theta_{23}$ are determined by the masses of the charged leptons only, in very good agreement with the experimental values [12, 13, 29, 30, 31, 32],

$$(\sin\theta_{13})^{th} = 3.4\times 10^{-3}, \qquad -0.179 \le (\sin\theta_{13})^{exp} \le 0.11, \tag{30}$$

and

$$(\sin^2\theta_{23})^{th} = 0.5, \quad (\sin^2\theta_{23})^{exp} = 0.5^{+0.06}_{-0.05}. \tag{31}$$

In this model, the experimental restriction $|\Delta m_{12}^2| < |\Delta m_{13}^2|$ implies an inverted neutrino mass spectrum, $|m_{\nu_3}| < |m_{\nu_1}| < |m_{\nu_2}|$ [21].

As can be seen from eqs. (28) and (29), the solar mixing angle is sensitive to the neutrino mass differences and the phase ϕ_ν, but is only very weakly sensitive to the charged lepton masses. If we neglect the small terms proportional to O_{11} and O_{11}^2 in (28), we get

$$\tan^2\theta_{12} = \frac{(\Delta m_{12}^2+\Delta m_{13}^2+|m_{\nu_3}|^2\cos^2\phi_\nu)^{1/2}-|m_{\nu_3}||\cos\phi_\nu|}{(\Delta m_{13}^2+|m_{\nu_3}|^2\cos^2\phi_\nu)^{1/2}+|m_{\nu_3}||\cos\phi_\nu|}. \tag{32}$$

From this expression, we may readily derive expressions for the neutrino masses in terms of $\tan\theta_{12}$ and ϕ_ν and the differences of the squared masses of the neutrinos masses

$$|m_{\nu_3}| = \frac{\sqrt{\Delta m_{13}^2}}{2\cos\phi_\nu\tan\theta_{12}}\frac{1-\tan^4\theta_{12}+r^2}{\sqrt{1+\tan^2\theta_{12}}\sqrt{1+\tan^2\theta_{12}+r^2}}, \tag{33}$$

where $r^2 = \Delta m_{12}^2/\Delta m_{13}^2 \approx 3\times 10^{-2}$.

The other two masses, $|m_{\nu_1}|$ and $|m_{\nu_2}|$ are immediately obtained from the knowledge of $|m_{\nu_3}|$ and Δm_{12}^2 and Δm_{13}^2.

The cosmological upper bound on the sum of neutrino masses sets a lower bound for $\cos\phi_\nu$ [18]

$$\sum|m_\nu| \le 0.17\ eV \quad\longrightarrow\quad \cos\phi_\nu \ge 0.55\ . \tag{34}$$

Since, for small values of ϕ_ν, the neutrino masses change very slowly with $\cos\phi_\nu$, in the absence of any other experimental information, we set $\phi_\nu = 0$ in our formulas. Hence, we find

$$|m_{\nu_2}| \approx 0.056eV \quad |m_{\nu_1}| \approx 0.055eV \quad |m_{\nu_3}| \approx 0.022eV, \tag{35}$$

where we used the values $\Delta m_{13}^2 = 2.6\times 10^{-3}eV^2$, $\Delta m_{21}^2 = 7.9\times 10^{-5}eV^2$ and $\tan\theta_{12} = 0.667$, taken from [14].

5. V^{th}_{PMNS} and the tri-bimaximal form

Once the numerical values of the neutrino masses are determined, we may readily verify that the theoretical mixing matrix, V^{th}_{PMNS}, is very close to the tri-bimaximal form of the mixing matrix [33],

$$V^{th}_{PMNS} = \begin{pmatrix} \sqrt{\frac{2}{3}} & \sqrt{\frac{1}{3}} & 0 \\ -\sqrt{\frac{1}{6}} & \sqrt{\frac{1}{3}} & -\sqrt{\frac{1}{2}} \\ -\sqrt{\frac{1}{6}} & \sqrt{\frac{1}{3}} & \sqrt{\frac{1}{2}} \end{pmatrix} + \delta V^{tri}_{PMNS}, \tag{36}$$

where $\delta V^{tri}_{PMNS} = V^{th}_{PMNS} - V^{tri}_{PMNS}$. From eq. (25), the correction term to the tri-bimaximal form of the mixing matrix comes out as

$$\delta V^{tri}_{PMNS} \approx \begin{pmatrix} 1.94\times10^{-2} & -2.84\times10^{-2} & -3.4\times10^{-3} \\ 2.21\times10^{-2} & 1.5\times10^{-2} & -8.2\times10^{-6} \\ 1.8\times10^{-2} & 1.24\times10^{-2} & 3.1\times10^{-10} \end{pmatrix}. \tag{37}$$

A more complete discussion of the deviation from the tri-bimaximal form in the framework of the minimal S_3-invariant extension of the SM can be found in [26].

6. Flavour Changing Neutral Currents (FCNC)

Models with more than one Higgs $SU(2)$ doublet have tree level flavour changing neutral currents. In the Minimal S_3-invariant Extension of the Standard Model considered here, there is one Higgs $SU(2)$ doublet per generation coupling to all fermions. The flavour changing Yukawa couplings may be written in a flavour labelled, symmetry adapted weak basis as

$$\begin{aligned} \mathcal{L}^{\rm FCNC}_{Y} &= \left(\overline{E}_{aL} Y^{ES}_{ab} E_{bR} + \overline{U}_{aL} Y^{US}_{ab} U_{bR} + \overline{D}_{aL} Y^{DS}_{ab} D_{bR}\right) H^0_S \\ &+ \left(\overline{E}_{aL} Y^{E1}_{ab} E_{bR} + \overline{U}_{aL} Y^{U1}_{ab} U_{bR} + \overline{D}_{aL} Y^{D1}_{ab} D_{bR}\right) H^0_1 + \\ &\left(\overline{E}_{aL} Y^{E2}_{ab} E_{bR} + \overline{U}_{aL} Y^{U2}_{ab} U_{bR} + \overline{D}_{aL} Y^{D2}_{ab} D_{bR}\right) H^0_2 + \text{h.c.} \end{aligned} \tag{38}$$

The Yukawa couplings of immediate physical interest in the computation of the flavour changing neutral currents are those defined in the mass basis, according to $\tilde{Y}^{EI}_m = U^{\dagger}_{eL} Y^{EI}_w U_{eR}$, where U_{eL} and U_{eR} are the matrices that diagonalize the charged lepton mass matrix defined in eqs. (12). We obtain [24]

$$\tilde{Y}^{E1}_m \approx \frac{m_\tau}{v_1} \begin{pmatrix} 2\tilde{m}_e & -\frac{1}{2}\tilde{m}_e & \frac{1}{2}x \\ -\tilde{m}_\mu & \frac{1}{2}\tilde{m}_\mu & -\frac{1}{2} \\ \frac{1}{2}\tilde{m}_\mu x^2 & -\frac{1}{2}\tilde{m}_\mu & \frac{1}{2} \end{pmatrix}_m, \tag{39}$$

and

$$\tilde{Y}^{E2}_m \approx \frac{m_\tau}{v_2} \begin{pmatrix} -\tilde{m}_e & \frac{1}{2}\tilde{m}_e & -\frac{1}{2}x \\ \tilde{m}_\mu & \frac{1}{2}\tilde{m}_\mu & \frac{1}{2} \\ -\frac{1}{2}\tilde{m}_\mu x^2 & \frac{1}{2}\tilde{m}_\mu & \frac{1}{2} \end{pmatrix}_m, \tag{40}$$

where $\tilde{m}_\mu = 5.94\times10^{-2}$, $\tilde{m}_e = 2.876\times10^{-4}$ and $x = m_e/m_\mu = 4.84\times10^{-3}$. All the non-diagonal elements are responsible for tree-level FCNC processes. If the S_2' symmetry in the Higgs sector is preserved [34], $\langle H^0_1\rangle = \langle H^0_2\rangle = v$.

Table 2. Leptonic FCNC processes, calculated with $M_{H_{1,2}} \sim 120$ GeV.

FCNC processes	Theoretical BR	Experimental upper bound BR	References
$\tau \to 3\mu$	8.43×10^{-14}	2×10^{-7}	B. Aubert *et. al.* [37]
$\tau \to \mu e^+ e^-$	3.15×10^{-17}	2.7×10^{-7}	B. Aubert *et. al.* [37]
$\tau \to \mu\gamma$	9.24×10^{-15}	6.8×10^{-8}	B. Aubert *et. al.* [38]
$\tau \to e\gamma$	5.22×10^{-16}	1.1×10^{-11}	B. Aubert *et. al.* [39]
$\mu \to 3e$	2.53×10^{-16}	1×10^{-12}	U. Bellgardt *et al.* [36]
$\mu \to e\gamma$	2.42×10^{-20}	1.2×10^{-11}	M. L. Brooks *et al.* [40]

The amplitude of the flavour violating process $\mu \to 3e$, is proportional to $\tilde{Y}^E_{\mu e}\tilde{Y}^E_{ee}$ [35]. Then, the leptonic branching ratio,

$$Br(\mu \to 3e) = \frac{\Gamma(\mu \to 3e)}{\Gamma(\mu \to e\nu\bar{\nu})} \tag{41}$$

and

$$\Gamma(\mu \to 3e) \approx \frac{m_\mu^5}{3 \times 2^{10}\pi^3} \frac{\left(Y^{1,2}_{\mu e} Y^{1,2}_{ee}\right)^2}{M^4_{H_{1,2}}}, \tag{42}$$

which is the dominant term, and the well known expression for $\Gamma(\mu \to e\nu\bar{\nu})$ [28], give

$$Br(\mu \to 3e) \approx 2(2 + \tan^2\beta)^2 \left(\frac{m_e m_\mu}{m_\tau^2}\right)^2 \left(\frac{m_\tau}{M_H}\right)^4, \tag{43}$$

taking for $M_H \approx 120$ GeV and $\tan\beta = 1$ we obtain $Br(\mu \to 3e) = 2.53 \times 10^{-16}$, well below the experimental upper bound for this process, which is 1×10^{-12} [36].

Similar computations give the numerical estimates of the branching ratios for some others flavour violating processes in the leptonic sector. These results, and the corresponding experimental upper bounds are shown in Table 2. In all cases considered, the theoretical estimations made in the framework of the minimal S_3-invariant extension of the SM are well below the experimental upper bounds [24].

7. Muon anomalous magnetic moment

In the minimal S_3-invariant extension of the Standard Model we are considering here, we have three Higgs $SU(2)$ doublets, one in the singlet and the other two in the doublet representations of the S_3 flavour group. The Z_2 symmetry decouples the charged leptons from the Higgs boson in the S_3 singlet representation. Therefore, in the leading order of perturbation theory there are two neutral scalars and two neutral pseudoscalars whose exchange will contribute to the anomalous magnetic moment of the muon. Since the heavier generations have larger flavour-changing couplings, the largest contribution comes from the heaviest charged leptons coupled to the lightest of the neutral Higgs bosons.

A straightforward computation gives

$$\delta a_\mu^{(H)} = \frac{Y_{\mu\tau} Y_{\tau\mu}}{16\pi^2} \frac{m_\mu m_\tau}{M_H^2} \left(log\left(\frac{M_H^2}{m_\tau^2}\right) - \frac{3}{2}\right). \tag{44}$$

With the help of (39) and (40) we may write $\delta a_\mu^{(H)}$ as

$$\delta a_\mu^{(H)} = \frac{m_\tau^2}{(246\ GeV)^2} \frac{(2 + \tan^2\beta)}{32\pi^2} \frac{m_\mu^2}{M_H^2} \left(log\left(\frac{M_H^2}{m_\tau^2}\right) - \frac{3}{2}\right), \tag{45}$$

Taking again $M_H = 120\ GeV$ and the upper bound for $\tan\beta = 14$ gives an estimate of the largest possible contribution of the FCNC to the anomaly of the muon's magnetic moment $\delta a_\mu^{(H)} \approx 1.7 \times 10^{-10}$. This number has to be compared with the difference between the experimental value and the Standard Model prediction for the anomaly of the muon's magnetic moment [41]

$$\Delta a_\mu = a_\mu^{exp} - a_\mu^{SM} = (28.7 \pm 9.1) \times 10^{-10}, \tag{46}$$

which means

$$\frac{\delta a_\mu^{(H)}}{\Delta a_\mu} \approx 0.06. \tag{47}$$

Hence, the contribution of the flavour changing neutral currents to the anomaly of the magnetic moment of the muon is smaller than or of the order of 6% of the discrepancy between the experimental value and the Standard Model prediction.

8. Conclusions

By introducing three Higgs fields that are $SU(2)_L$ doublets in the theory, we extended the concept of flavour and generations to the Higgs sector and formulated a minimal S_3-invariant extension of the Standard Model [21]. A well defined structure of the Yukawa couplings is obtained, which permits the calculations of mass and mixings matrices for quark and leptons in a unified way. A further reduction of redundant parameters is achieved in the leptonic sector by introducing a Z_2 symmetry. The flavour symmetry group $S_3 \times Z_2$ relates the nuetrino mass spectrum and mixings. This allowed us to compute the neutrino mixing matrix, V_{PMNS}, explicitly in terms of the masses of the charged leptons and neutrinos and two phases δ and ϕ_ν [23, 24]. In this model, the tri-bimaximal mixing structure of V_{PMNS} and the magnitudes of the three mixing angles are determined by the interplay of the flavour $S_3 \times Z_2$, the see-saw mechanism and the lepton mass hierarchy. We also found that V_{PMNS} has three CP-violating phases, namely, one Dirac phase $\delta = \delta_\nu - \delta_e$ and two Majorana phases, α and β, which are functions of the neutrino masses and the phase ϕ_ν, which is independent of the Dirac phase. The numerical values of the reactor, θ_{13}, and the atmospheric, θ_{23}, mixing angles are determined by the masses of the charged leptons only, in very good agreement with the latest analysis of the experimental data on neutrino oscillations and mixings [30, 32, 31]. The solar mixing angle, θ_{12}, is almost insensitive to the values of the masses, but its experimental value allowed us to fix the scale and origin of the neutrino mass spectrum, which leads to the prediction of an inverted hierarchy, with the values $|m_{\nu_2}| = 0.056eV$, $|m_{\nu_1}| = 0.055eV$ and $|m_{\nu_3}| = 0.022eV$. We also obtained explicit expressions for the matrices of the Yukawa couplings of the lepton sector parametrized in terms of the charged lepton masses and the VEV's of the neutral Higgs bosons in the S_3-doublet representation. These Yukawa matrices are closely related to the fermion mass matrices and have a structure of small and very small entries reflecting the observed charged lepton mass hierarchy. With the help of the Yukawa matrices, we computed the branching ratios of a number of FCNC processes and found that the branching ratios of all FCNC processes considered here are strongly suppressed by powers of the small mass ratios m_e/m_τ and m_μ/m_τ, and by the ratio $\left(m_\tau/M_{H_{1,2}}\right)^4$, where $M_{H_{1,2}}$ is the mass of the neutral Higgs bosons in the S_3-doublet. Taking for $M_{H_{1,2}}$ a very conservative value ($M_{H_{1,2}} \approx 120\ GeV$), we found that the numerical values of the branching ratios of the FCNC in the leptonic sector are well below the corresponding experimental upper bounds by many orders of magnitude. It has already been argued that small FCNC processes mediating non-standard quark-neutrino interactions could be important in the theoretical description of the gravitational core collapse and shock generation in the explosion stage of a supernova [42, 43, 44]. Finally, the contribution of the flavour changing neutral currents to the anomalous magnetic moment of the muon is small but non-negligible, and it is compatible with the best, state of the art measurements and theoretical computations.

9. Acknowledgements

We thank Prof. Jens Erler and Dr. Genaro Toledo-Sánchez for helpful discussions about $g-2$. This work was partially supported by CONACYT México under contracts No 51554-F and 82291 and by DGAPA-UNAM under contract PAPIIT-IN112709.

References

[1] Jung C K, McGrew C, Kajita T and Mann T 2001 *Ann. Rev. Nucl. Part. Sci.* **51** 451
[2] Mohapatra R N and Smirnov A Y 2006 *Ann. Rev. Nucl. Part. Sci.* **56** 569 [hep-ph/0603118]
[3] GNO, Altmann M *'et al'.* 2005 *Phys. Lett.* B **616** 174 [hep-ex/0504037]
[4] Super-Kamiokande, Smy M B *'et al'.* 2004 *Phys. Rev.* D **69** 011104 [hep-ex/0309011]
[5] SNO, Ahmad Q R *'et al'.* 2002 *Phys. Rev. Lett.* **89** 011302 [nucl-ex/0204009]
[6] SNO, Ahmad Q R *'et al'.* 2002 *Phys. Rev. Lett.* **89** 011301 [nucl-ex/0204008]
[7] SNO, Aharmim B *'et al'.* 2005 *Phys. Rev.* C **72** 055502 [nucl-ex/0502021]
[8] Super-Kamiokande, Fukuda S *'et al'.* 2002 *Phys. Lett.* B **539** 179 [hep-ex/0205075]
[9] Super-Kamiokande, Ashie Y *'et al'.* 2005 *Phys. Rev.* D **71** 112005 [hep-ex/0501064]
[10] Bemporad C, Gratta G and Vogel P 2002 *Rev. Mod. Phys.* **74** 297 [hep-ph/0107277]
[11] KamLAND, Araki T *'et al'.* 2005 *Phys. Rev. Lett.* **94** 081801 [hep-ex/0406035]
[12] Maltoni M, Schwetz T, Tortola M A and Valle J W F 2004 *New J. Phys.* **6** 122 [hep-ph/0405172]
[13] Schwetz T 2005 *Acta Phys. Polon.* **B36** 3203 [hep-ph/0510331]
[14] Gonzalez-Garcia M C and Maltoni M 2008 *Phys. Rept.* **460** 1 [0704.1800]
[15] CHOOZ, Apollonio M *'et al'.* 2003 *Eur. Phys. J.* C **27** 331 [hep-ex/0301017]
[16] Eitel K 2005 *Nucl. Phys. Proc. Suppl.* **143** 197
[17] Elliott S R and Engel J 2004 *J. Phys.* G **30** R183 [hep-ph/0405078]
[18] Seljak U, Slosar A and McDonald P 2006 *JCAP* **0610** 014 [astro-ph/0604335]
[19] Elgaroy O and Lahav O 2005 *New J. Phys.* **7** 61 [hep-ph/0412075]
[20] Lesgourgues J and Pastor S 2006 *Phys. Rept.* **429** 307 [astro-ph/0603494]
[21] Kubo J, Mondragon A, Mondragon M and Rodriguez-Jauregui E 2003 *Prog. Theor. Phys* **109** 795 [hep-ph/0302196]
[22] Kubo J *'et al'.* 2005 *J. Phys. Conf. Series:* **18** 380
[23] Felix O, Mondragon A, Mondragon M and Peinado E 2007 *AIP Conf. Proc.* **917** 383 [hep-ph/0610061]
[24] Mondragon A, Mondragon M and Peinado P 2007 *Phys. Rev.* D **76** 076003 [0706.0354]
[25] Mondragon A, Mondragon M and Peinado E 2008 *J. Phys.* A **41** 304035 [0712.1799]
[26] Mondragon A, Mondragon M and Peinado E 2008 *AIP Conf. Proc.* **1026** 164 [0712.2488]
[27] Mondragon A, Mondragon M and Peinado E 2008 *Rev. Mex. Fis.* S **54** **3** 81 [0805.3507]
[28] Particle Data Group, Yao W M *'et al".* 2006 *J. Phys.* G **33** 1
[29] Fogli G L, Lisi E, Marrone A and Palazzo A 2006 *Prog. Part. Nucl. Phys.* **57** 742 [hep-ph/0506083]
[30] Fogli G L, Lisi E, Marrone A, Palazzo A and Rotunno A M 2009 SNO, KamLAND and neutrino oscillations: theta(13), *talk presented at NEUTEL 2009, XIII International Workshop on Neutrino Telescopes* [0905.3549]
[31] Ge H L, Giunti C and Liu Q Y 2009 *Phys. Rev.* D **80** 053009 [0810.5443]
[32] Roa J E, Latimer D C and Ernst D J 2009 *Phys. Rev. Lett.* **103** 061804 [0904.3930]
[33] Harrison P F, Perkins D H and Scott W G 2002 *Phys. Lett.* B **530** 167 [hep-ph/0202074]
[34] Pakvasa S and Sugawara H 1978 *Phys. Lett.* B **73** 61
[35] Sher M and Yuan Y 1991 *Phys. Rev.* D **44** 1461
[36] SINDRUM, Bellgardt U *'et al'.* 1988 *Nucl. Phys.* B **299** 1
[37] BABAR, Aubert B *'et al'.* 2004 *Phys. Rev. Lett.* **92** 121801 [hep-ex/0312027]
[38] BABAR, Aubert B *'et al'.* 2005 *Phys. Rev. Lett.* **95** 041802 [hep-ex/0502032]
[39] BABAR, Aubert B *'et al'.* 2006 *Phys. Rev. Lett.* **96** 041801 [hep-ex/0508012]
[40] MEGA, Brooks M L *'et al'.* 1999 *Phys. Rev. Lett.* **83** 1521 [hep-ex/9905013]
[41] Jegerlehner F 2007 *Acta Phys. Polon.* B **38** 3021 [hep-ph/0703125]
[42] Raffelt G G 2007 Supernova neutrino observations: What can we learn? *Proceedings of the 22nd International Conference on Neutrino Physics and Astrophysics (Neutrino 2006)* [astro-ph/0701677]
[43] Amanik P S, Fuller G M and Grinstein B 2005 *Astropart. Phys.* **24** 160 [hep-ph/0407130]
[44] Esteban-Pretel A, Tomas R and Valle J W F 2007 *Phys. Rev.* D **D76** 053001 [0704.0032]

Journal of Physics: Conference Series **284** (2011) 012049 doi:10.1088/1742-6596/284/1/012049

Novel results from an algebraic approach to molecular bending dynamics

Francisco Pérez-Bernal[1], Osiris Álvarez-Bajo[1], José M Arias[2], Miguel Carvajal[1], José E García-Ramos[1], Danielle Larese[3], and Pedro Pérez-Fernández[2]

[1] Departamento de Física Aplicada, Fac. de CC. Experimentales, Universidad de Huelva, Huelva 21071 (SPAIN)

[2] Departamento de Física Atómica, Molecular y Nuclear, Facultad de Física, Universidad de Sevilla, Apdo. 1065, Sevilla E-41080 (SPAIN)

[3] Department of Chemistry, Yale University, P.O. Box 208107, New Haven, CT 06520-8107 (USA)

E-mail: `francisco.perez@dfaie.uhu.es`

Abstract. We present a brief review of research topics of current interest that depend on an algebraic approach to molecular bending dynamics. This approach is based on a $u(3)$ spectrum generating algebra. In particular, we briefly present results on three topics: the calculation of finite-size analytical corrections to mean field results, the application of the model to the large-amplitude vibrational bending mode of the NCNCS molecule, and the analysis of the influence of quadratic Casimir operators on excited state quantum phase transitions.

1. Introduction

The modeling of n-dimensional systems using a $u(n+1)$ Lie algebra as the system's spectrum generating algebra (SGA) has proved successful in an ample set of physical situations [1]. This approach has caused a major impact in nuclear structure studies, where the interacting boson model (IBM), based on a $u(6)$ SGA and formulated by Iachello and Arima in the seventies, has become a standard theoretical tool [2].

Following the trail of the IBM, the Vibron Model was born in the eighties, treating molecules as bosonic systems where bosons represent vibrational and rotational excitations (vibrons) [3]. The SGA in this case is a $u(4)$ Lie algebra, due to the three-dimensional nature of the problem.

The Vibron Model, though useful, proved quite cumbersome when applied to molecular species composed by more than four atoms. In the cases that do not require a simultaneous treatment of molecular rotations and vibrations, a possible alternative approach is the one-dimensional (1D) limit of the vibron model, based on a $u(2)$ SGA [4]. In this limit rotational degrees of freedom are neglected and each local vibrational degree of freedom has an associated $u(2)$ Lie algebra. Therefore this model considers anharmonicity from the outset, due to an isomorphism between the $u(2)$ Lie algebra and the 1D Morse potential [5]. In addition, this approach naturally embeds discrete symmetry considerations [6].

The bending vibrational dynamics of linear and quasi-linear molecules imply the interplay of vibrational and rotational degrees of freedom. In this case an appropriate model is the two-dimensional (2D) limit of the vibron model, with a $u(3)$ Lie algebra as the system's SGA [7].

Journal of Physics: Conference Series **284** (2011) 012049 doi:10.1088/1742-6596/284/1/012049

The 2D limit of the vibron model has successfully modeled spectral term energies and intensities of various molecular species, from rigidly-linear to rigidly-bent molecules, including quasi-linear, and non-rigid intermediate situations [8, 9, 10]. A thorough description of both quantal and classical aspects for this model can be found in Ref. [11].

In recent years the study and characterization of quantum phase transitions (QPTs) in algebraic models has attracted much attention, a line of research that can be traced back to the seminal work of Gilmore [12]. These transitions, also called shape phase transitions and ground state transitions, are zero-temperature phase transitions between different geometric limits of the system's ground state upon variation of one or several control parameters in the Hamiltonian. This research line has been recently bolstered up with the discovery of precursors of QPTs in nuclear systems. Reference [13] is a excellent and recent review on this topic.

More recently, the study of excited state quantum phase transitions (ESQPTs), transitions that take place upon the variation of the system's excitation energy instead of the Hamiltonian parameters, has been proposed [14]. In this case the transition happens for some particular excited states of the system. This is especially relevant in the field of molecular spectroscopy, where novel spectroscopic techniques allow the access to highly-excited states. In fact, quantum monodromy effects, whose appearance in several molecular systems has been experimentally confirmed [15, 16], can be regarded as an ESQPT from the algebraic model perspective [11].

We finish this section with a brief outline of the present work. Section 2 is devoted to a short presentation of the bosonic algebraic approach to 2D systems and its classical limit. Analytical corrections to the classical limit of the model, going beyond its mean field limit are presented in section 3. An application to a realistic case can be found in section 4, where we fit vibrational bending energies of a large amplitude bending mode of NCNCS. Section 5 contains some interesting results concerning the influence of quadratic Casimir operators on the model ESQPT. In section 6 we finish by giving some concluding remarks.

2. Algebraic approach to 2D systems

In this section we present, in an abridged form, the 2D limit of the vibron model and its classical limit.

2.1. Basic results

The building bricks for the $u(3)$ Lie algebra used for modeling 2D systems are three bosons of two different types. One of them is a scalar boson, $\sigma^\dagger$, and the other two are Cartesian bosons, $\{\tau_x^\dagger, \tau_y^\dagger\}$. The operators follow the usual bosonic commutation relations: $[\sigma, \sigma^\dagger] = 1$ and $\left[\tau_i, \tau_j^\dagger\right] = \delta_{i,j}$ with $i, j = x, y$. All other commutators are zero.

For the sake of convenience we introduce circular bosons [11]

$$\tau_\pm^\dagger = \mp \frac{\tau_x^\dagger \pm i\tau_y^\dagger}{\sqrt{2}} \quad , \quad \tau_\pm = \mp \frac{\tau_x \mp i\tau_y}{\sqrt{2}} \; . \tag{1}$$

The nine $u(3)$ generators are built with the possible bilinear products of $\sigma^\dagger$ and $\tau_x^\dagger$, $\tau_y^\dagger$ creation and annihilation operators. They are conventionally expressed as [7]:

$$\begin{array}{ll}
\hat{n} = \tau_+^\dagger \tau_+ + \tau_-^\dagger \tau_- & , \quad \hat{n}_s = \sigma^\dagger \sigma \\
\hat{l} = \tau_+^\dagger \tau_+ - \tau_-^\dagger \tau_- & \\
\hat{D}_+ = \sqrt{2}(\tau_+^\dagger \sigma - \sigma^\dagger \tau_-) & , \quad \hat{D}_- = \sqrt{2}(-\tau_-^\dagger \sigma + \sigma^\dagger \tau_+) \\
\hat{Q}_+ = \sqrt{2}\,\tau_+^\dagger \tau_- & , \quad \hat{Q}_- = \sqrt{2}\,\tau_-^\dagger \tau_+ \\
\hat{R}_+ = \sqrt{2}(\tau_+^\dagger \sigma + \sigma^\dagger \tau_-) & , \quad \hat{R}_- = \sqrt{2}(\tau_-^\dagger \sigma + \sigma^\dagger \tau_+) \; .
\end{array} \tag{2}$$

The operator $\hat{l}$ is the 2D angular momentum, as one can easily see once it is expressed in terms of Cartesian boson operators.

There are two possible dynamical symmetries that conserve 2D angular momentum, starting in the $u(3)$ SGA and ending in the $so(2)$ symmetry algebra.

$$u(3) \supset u(2) \supset so(2) \quad \text{Chain I} , \tag{3}$$
$$u(3) \supset so(3) \supset so(2) \quad \text{Chain II} . \tag{4}$$

Numerical calculations in the present article have been carried out using the basis associated with dynamical symmetry I. A detailed discussion of both dynamical symmetries can be found in [11].

Another ingredient of the algebraic approach is the Casimir (or invariant) operators associated to each subalgebra chain [5]. The first and second order Casimir operators of the subalgebras in chains (3) and (4) are

$$\begin{array}{l} \hat{C}_1[u(2)] = \hat{n} \quad , \quad \hat{C}_2[u(2)] = \hat{n}(\hat{n}+1) \ , \\ \hat{C}_2[so(3)] = \hat{W}^2 = (\hat{D}_+\hat{D}_- + \hat{D}_-\hat{D}_+)/2 + \hat{l}^2 \quad , \\ \hat{C}_1[so(2)] = \hat{l} \quad , \quad \hat{C}_2[so(2)] = \hat{l}^2 \ . \end{array} \tag{5}$$

The $so(3)$ Casimir operator $\hat{W}^2$ can be replaced by the pairing operator $\hat{P} = N(N+1) - \hat{W}^2$, where the total number operator, $\hat{N} = \hat{n}_s + \hat{n}$, has been replaced by its value, N, due to the fact that we consider systems with a fixed number of bosons.

The most general rotation and parity-invariant, one and two-body Hamiltonian can be written as a linear combination of the four possible first and second order Casimir operators (5)

$$\hat{H} = E_0 + \epsilon\,\hat{C}_1[u(2)] + \alpha\,\hat{C}_2[u(2)] + \beta\,\hat{C}_2[so(2)] + A\,\hat{C}_2[so(3)] \ . \tag{6}$$

Matrix elements in the two possible bases for the Casimir operators (5) can be found in Ref. [11].

2.2. Mean field limit of the model

The mean field limit of the model, also called the thermodynamical limit, the large N limit, or the classical limit of the model, can be found following an algorithm established by Gilmore [17] known as the coherent (or intrinsic) state approach.

In making use of the coherent state approach, we use projective coherent states that define an intrinsic ground state

$$|[N]; r, \theta\rangle = \frac{1}{\sqrt{N!}} \left(b_c^\dagger\right)^N |0\rangle \ , \tag{7}$$

where r and θ are polar coordinates associated to Cartesian coordinates x and y, and $b_c^\dagger$ is the boson condensate

$$b_c^\dagger = \frac{1}{\sqrt{1+r^2}} \left(\sigma^\dagger + x\tau_x^\dagger + y\tau_y^\dagger\right) \ . \tag{8}$$

The coherent state (7) is the number projected generalized coherent state of $u(3)$ [2]. The coherent state approach provides an energy functional that is an approximation for the ground state energy of a given Hamiltonian. This approximation becomes exact for $N \to \infty$.

The existence of a second order QPT between the two possible dynamical symmetries can be easily proven using the aforementioned coherent state approach [11]. In order to do so, it is convenient to define a simplified model Hamiltonian, with one operator for each chain and properly normalized

$$\hat{\mathcal{H}} = \varepsilon \left[(1-\xi)\hat{n} + \frac{\xi}{N-1}\hat{P} \right] \ . \tag{9}$$

Journal of Physics: Conference Series **284** (2011) 012049 doi:10.1088/1742-6596/284/1/012049

The parameter ε is an energy scale and ξ is a control parameter that takes values between zero ($u(2)$ dynamical symmetry) and one ($so(3)$ dynamical symmetry).

The energy per boson for the essential Hamiltonian (9) is [9, 11]

$$\mathcal{E}_\xi(r) \equiv \frac{1}{N}\frac{\langle[N];r,\theta|\hat{\mathcal{H}}|[N];r,\theta\rangle}{\langle[N];r,\theta|[N];r,\theta\rangle} = \epsilon\left[(1-\xi)\frac{r^2}{1+r^2} + \xi\left(\frac{1-r^2}{1+r^2}\right)^2\right] . \tag{10}$$

The extremization of the energy functional (10) reveals that there are two different geometric limits: symmetric (or linear) and deformed (or bent). In the symmetric case the energy minimum is at $r_e = 0$, while in the second case $r_e = \sqrt{(5\xi-1)/(3\xi+1)} \neq 0$. The symmetric phase takes place for control parameter values $\xi \leq \xi_c = 0.2$, and the deformed phase for $\xi > \xi_c = 0.2$. When evaluated at $r = r_e$, the energy functional is

$$\mathcal{E}_\xi(r_e) = \left\{ \begin{array}{cc} \xi & 0 \leq \xi \leq \xi_c \\ \frac{-9\xi^2+10\xi-1}{16\xi} & \xi_c < \xi \leq 1 \end{array} \right. . \tag{11}$$

By evaluating the derivatives of $\mathcal{E}_\xi(r_e)$ with respect to ξ, one finds that, according to Ehrenfest's classification scheme, the $u(2) - so(3)$ phase transition is of second order [11].

Another observable of interest is the expectation value of the $u(2)$ number operator in the system ground state

$$\langle[N];r,\theta|\hat{n}|[N];r,\theta\rangle = N\frac{r_e^2}{1+r_e^2} = \left\{ \begin{array}{cc} 0 & 0 \leq \xi \leq \xi_c \\ \frac{5\xi-1}{8\xi} & \xi_c < \xi \leq 1 . \end{array} \right. \tag{12}$$

This operator can act as a classical order parameter for the transition between the symmetric and deformed phases [11].

3. Beyond mean field analytical corrections

We evaluate analytical corrections to the mean field limit of the algebraic approach to go beyond the results presented in the previous section. The mean field or classical limit is only exact in the large N limit, and the calculated corrections allow us to calculate the N^0 corrections to the ground state energy (11) and the number of τ bosons (12).

In order to compute these corrections we perform a Holstein-Primakoff expansion and a shift transformation, followed by a Bogoliubov transformation, following Ref. [18]. This reference is a general work, aimed at bosonic two-level Hamiltonians with SGA $u(2L+2)$ with $L = 1, 2, \ldots$. In the symmetric region Dusuel and collaborators go a step further than us, making use of the continuous unitary transformations approach (CUTS) [19].

The Holstein-Primakoff expansion implies the definition of a new pair of Cartesian bosons, $b_i^\dagger(b_i)$, $i = x, y$, with the usual bosonic commutation relations, $[b_i, b_j^\dagger] = \delta_{ij}$. The inclusion of this boson removes the dependence on the scalar $\sigma^\dagger$ boson of the Hamiltonian

$$\begin{aligned} \tau_i^\dagger\tau_j &= b_i^\dagger b_j , \\ \tau_i^\dagger\sigma &= \sqrt{N}\, b_i^\dagger\sqrt{1-\hat{n}_b/N} = \left(\sigma^\dagger\tau_i\right)^\dagger , \\ \hat{n}_\sigma &= \sigma^\dagger\sigma = N - \hat{n}_b , \end{aligned} \tag{13}$$

where $i, j = x, y$, and $\hat{n}_b = b_x^\dagger b_x + b_y^\dagger b_y$. A third set of bosons, $c_i^\dagger(c_i)$, with $i = x, y$ is defined via a shift transformation

$$b_i^\dagger = \sqrt{N}\lambda\delta_{ix} + c_i^\dagger \; ; \quad i = x, y . \tag{14}$$

The parameter λ, known as the shift parameter, is zero in the spherical phase and nonzero in the deformed phase.

Once the model Hamiltonian (9) is transformed, we obtain an expansion in powers of N, with terms

$$\mathcal{H} = \hat{H}_1 + \hat{H}_{1/2} + \hat{H}_0 + \mathcal{O}(1/\sqrt{N}) , \tag{15}$$

where $\hat{H}_i$ incorporates the terms with a N^i dependence.

The first term is

$$\hat{H}_1 = N\left[\xi + (1-5\xi)\lambda^2 - 4\xi\lambda^4\right] . \tag{16}$$

Setting $\lambda = r/\sqrt{1+r^2}$ this term is nothing more than the mean field result (10). The next order in the expansion of the model Hamiltonian, $\hat{H}_{1/2}$, is zero once the equilibrium values of r (or λ) are substituted. This is due to the relation $\hat{H}_{1/2} = \frac{2}{\sqrt{N}}\frac{d\hat{H}_1}{d\lambda}$. Therefore, the first finite-size correction to the mean field limit is the $\hat{H}_0$ term, that is quadratic in the c boson operators

$$\begin{aligned} H_0 &= \lambda^2(4\lambda^2-1)\xi + (1-3\xi+14\lambda^2\xi)c_x^\dagger c_x + (1-3\xi+4\lambda^2\xi)c_y^\dagger c_y \\ &+ (5\lambda^2-1)\xi\left(c_x^\dagger c_x^\dagger + c_x c_x\right) + (2\lambda^2-1)\xi\left(c_y^\dagger c_y^\dagger + c_y c_y\right) . \end{aligned} \tag{17}$$

The Hamiltonian (17) can be diagonalized with a Bogoliubov transformation

$$\begin{aligned} c_i^\dagger &= u_i a_i^\dagger + v_i a_i , \\ c_i &= u_i a_i + v_i a_i^\dagger , \end{aligned} \tag{18}$$

that should be independently performed in the symmetric and deformed phases [18, 20]. The final result provides the correction to the ground state energy per particle in the symmetric and deformed regions

$$\mathcal{E}_0^{sym} = \xi + N^{-1}\left[3\xi - 1 + \Xi^{sym}(\xi)^{1/2}\right], \tag{19}$$

$$\mathcal{E}_0^{def} = \frac{-9\xi^2+10\xi-1}{16\xi} + N^{-1}\left[\frac{1-6\xi-27\xi^2+8\xi\Xi^{def}(\xi)^{1/2}}{16\xi}\right], \tag{20}$$

where $\Xi^{sym} = 5(\xi_c - \xi)(1-\xi)$ and $\Xi^{def}(\xi) = 5(\xi-\xi_c)(1+3\xi)$. In the deformed phase the y coordinate contribution is a spurious Goldstone boson, associated with a ground state rotation.

Compared to the mean field limit, the deduced finite-size corrections greatly improve the agreement with numerical results. This can be clearly seen in Fig. 1 where the mean field and beyond mean field (BMF) corrections for the ground state energy are compared to a numerical calculation for $N = 20$.

The finite-size correction can be calculated for other observables. We include results for the expectation value in the ground state of $\hat{n}$ (5). In this case we can make use of the Hellman-Feynman theorem, defining a new control parameter x such that $\xi = 1/(1+x)$, we obtain for the symmetric and deformed phases

$$\frac{\langle\hat{n}\rangle}{N} = \frac{d}{dx}\left[(1+x)\mathcal{E}_0^{sym}(x)\right] = \frac{1}{N}\frac{1-3\xi-\Xi^{sym}(\xi)^{1/2}}{\Xi^{sym}(\xi)^{1/2}} + \mathcal{O}(N^{-2}) , \tag{21}$$

$$\frac{\langle\hat{n}\rangle}{N} = \frac{d}{dx}\left[(1+x)\mathcal{E}_0^{def}(x)\right] = \frac{5\xi-1}{8\xi} + \frac{1}{N}\frac{4\xi(\xi-1)+(1-3\xi)\Xi^{def}(\xi)^{1/2}}{8\xi\Xi^{def}(\xi)^{1/2}} + \mathcal{O}(N^{-2}) . \tag{22}$$

We compare this result, the mean field limit and the numerical calculation for $N = 20$ in Fig. 2. As in the previous case, the obtained correction notably improves the mean field result, in particular in the vicinity of the critical control parameter.

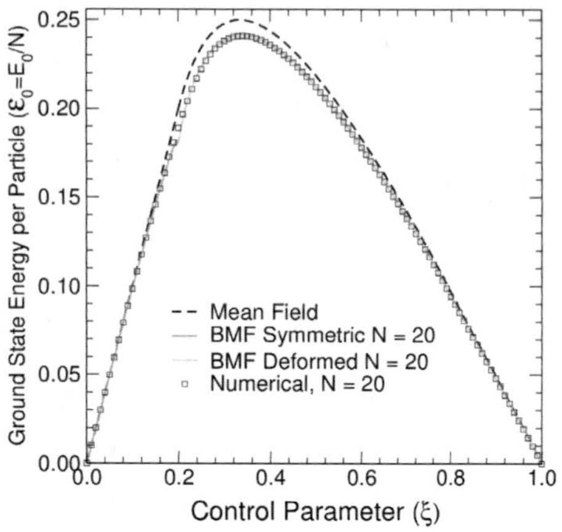

Figure 1. (Color online) Ground state energy per particle ($\mathcal{E}_0$) in arbitrary units as a function of the control parameter ξ.

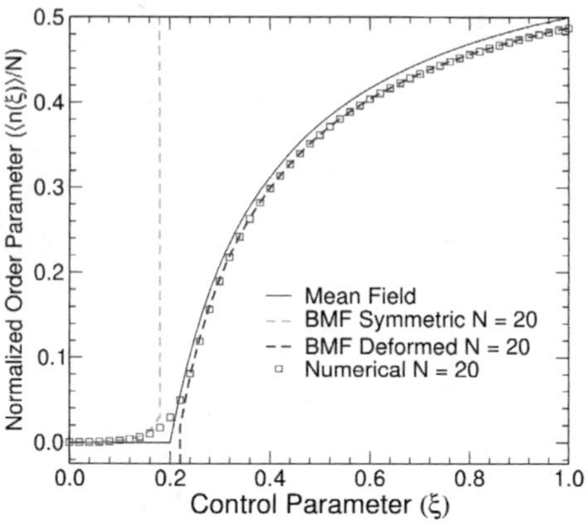

Figure 2. (Color online) Expectation value of the number operator $\hat{n}$ in the ground state of Hamiltonian (9) as a function of the control parameter ξ.

The results presented are encompassed in an open line of research [21]. We are currently working on computing the BMF correction for other observables, on analytically deriving finite-size scaling exponents for the $u(2) - so(3)$ second order phase transition, and on establishing a connection between Ref. [18] results and ours.

4. Application to the ν_7 bending mode of cyanogen isothiocyanate

The bending dynamics of several molecular species [7, 9, 10, 22, 23, 24, 25] have been modeled using the two-dimensional limit of the vibron model. We are currently interested in the analysis of the spectra of non-rigid molecules, as these probably provide examples of ESQPT. As the excitation energy increases, the non-rigid molecule's excited states have to overcome a potential hump in the origin, and change from a bent-like to a linear-like character. This fact was known long ago [26], but nowadays such systems have focused an increasing degree of attention due to quantum monodromy effects and its implications [16].

We have performed a fitting to the bending energy spectrum of cyanogen isothiocyanate (NCNCS), a non-rigid molecule whose large amplitude bending spectrum displays monodromy effects and is experimentally accessible [27]. We fit Hamiltonian (6) to the NCNCS term values obtained with the General Semirigid Bender Hamiltonian model in Ref. [27]. The final *rms* is 2.2 cm^{-1}, in a fit to 70 term energies. The optimized parameters are given in Table 1. The calculations were performed using the FORTRAN code `triat_u3` [28].

Table 1. Optimized parameters of the one- and two-body Hamiltonian (6) and their associated uncertainties obtained in the fit to ν_7 term values of NCNCS [27]. All parameters, excepting N, are in units of cm^{-1}. The fit root mean square deviation is $rms = 2.2$ cm^{-1}.

N	ϵ	α	β	A
70	203.6(19)	-2.58(3)	1.496(10)	0.813(6)

The obtained results are promising, especially if the simplicity of the model is considered.

The results in Ref. [27] and our fit are compared in a quantum monodromy plot in Fig. 3 where the good agreement between both approaches is remarkable. We reproduce the change in slope in the origin that characterizes the critical monodromy point.

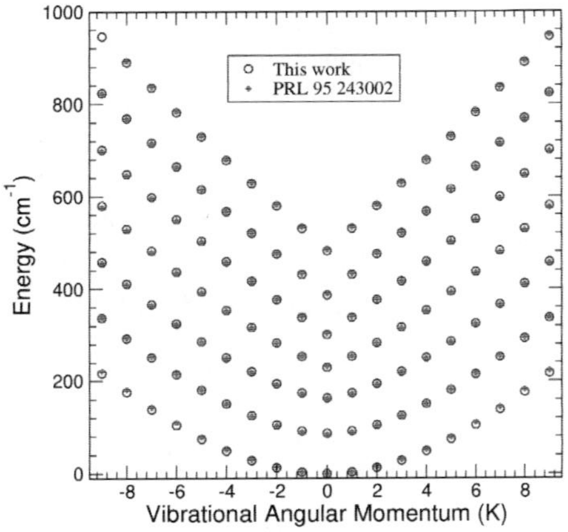

Figure 3. (Color online) Quantum monodromy plot of calculated ν_7 mode bending vibrational levels of NCNCS. The vibrational angular momentum l is labelled K following the usual spectroscopic notation.

We are currently working in the fit of the vibrational bending spectrum of water, where quantum monodromy effects have been experimentally reported [15].

5. Influence of the quadratic $u(2)$ Casimir operator in ESQPT

As shown in Table 1, the second order Casimir operator of the $u(2)$ subalgebra plays an important role when fitting experimental data. This also happens in other cases [9, 10] and has suggested us to explore the role of this anharmonic term in the model, in particular its effects on ground state QPT and ESQPT [29].

This can be accomplished with a new model Hamiltonian, conveniently scaled, that incorporates a second control parameter

$$\hat{\mathcal{H}} = \varepsilon\left[(1-\xi)\hat{n} + \frac{\alpha}{N-1}\hat{n}(\hat{n}+1) + \frac{\xi}{N-1}\hat{P}\right]. \tag{23}$$

The coherent state approach, when applied to this case, gives as a result that the ground state transition is mostly unperturbed by the addition of the new term. The energy functional $\mathcal{E}_0(\xi,\alpha)$ in the mean field limit becomes

$$\mathcal{E}_0(\xi,\alpha;r) = \varepsilon\frac{\xi + (1-3\xi)r^2 + (1+\alpha)r^4}{(1+r^2)^2}. \tag{24}$$

The minimization of the energy functional (24) provides the equilibrium values $r_e = 0\,,\ \sqrt{\frac{5\xi-1}{3\xi+2\alpha+1}}$ that implies a lower bound $\alpha > -(1+3\xi)/2$ if $\xi > 0.2$. The critical value of the control parameter is, once more, $\xi_c = 0.2$. For $\xi \le \xi_c$ the energy functional minimum lies at the origin, while for values of $\xi > \xi_c$ a new minimum appears and the minimum at the origin becomes a maximum. When evaluated for $r = r_e$ and $\varepsilon = 1$, the energy functional is

$$\mathcal{E}_0(\xi,\alpha;r_e) = \begin{cases} \xi & 0 \le \xi \le \xi_c \\ \frac{-9\xi^2+10\xi+4\alpha\xi-1}{16\xi+4\alpha} & \xi_c < \xi \le 1 \end{cases}, \tag{25}$$

which has a discontinuous second order derivative in $\xi = \xi_c$.

Although the second order ground state transition is basically unaffected by the new control parameter, there are conspicuous effects in the system's ESQPT for negative α values. This is specially noticeable once the excitation energy diagram for the Hamiltonian (23) is plotted for different α values. Instead of one separatrix between the configuration with linear and bent character, like in the $\alpha = 0$ case [11], there are two separatrices. In Figs. 4 and 5 we plot the excitation energy diagram of zero angular momentum ($l = 0$) states for and $\alpha = 0$ and -0.4, respectively. The number of bosons in both cases is $N = 100$ and the energy is plotted as a function of the control parameter ξ. The color code in the curves is assigned in accordance to the magnitude of the maximum squared-component of the eigenvector in the two bases of choice. If the eigenvector is close to dynamical symmetry (3), the maximal component will be close to one in the $u(2)$ basis and the point will be plotted in red color. If the maximal component is in the $so(3)$ basis the mapped color is blue. This way of representing the data clearly marks the states that correspond to the separatrix in the monodromy plot, in the high level density region that separates the linear- and bent-like states. The appearance of two separatrices in Fig. 5 is markedly clear.

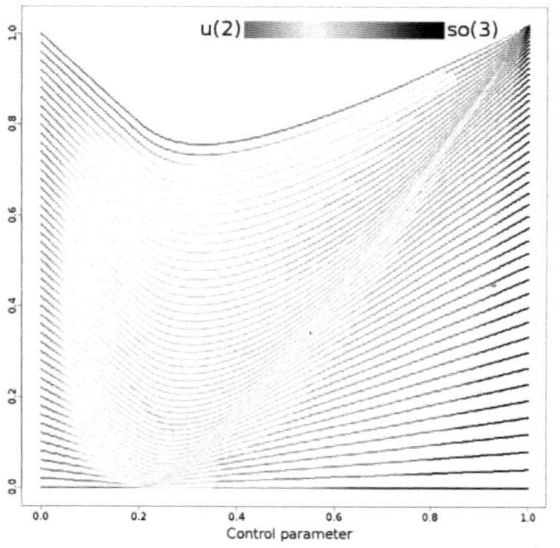

Figure 4. (Color online) Excitation energy diagram in arbitrary units ($\varepsilon = 1$) for zero angular momentum eigenstates of Hamiltonian (23) as a function of the control parameter ξ for $\alpha = 0$. The number of bosons is $N = 100$ and energies are normalized by N. The colorcode indicates the state proximity to the $u(2)$ or $so(3)$ dynamical symmetries (see text).

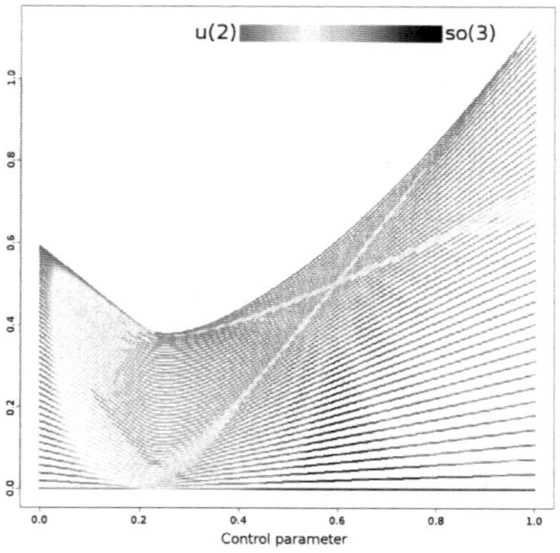

Figure 5. (Color online) Excitation energy diagram in arbitrary units ($\varepsilon = 1$) for zero angular momentum eigenstates of Hamiltonian (23) as a function of the control parameter ξ for $\alpha = -0.4$. The number of bosons is $N = 100$ and energies are normalized by N. The colorcode indicates the state proximity to the $u(2)$ or $so(3)$ dynamical symmetries (see text).

The position of the two separatrices in Fig. 5 depends on the values of the energy functional (25) maximum at the origin and its asymptotic value [29]. The value at the origin is only a function of ξ, while the asymptotic value depends only on α. In particular, there is a critical value of α, $\alpha_c = \xi - 1$, where the two separatrices cross and it is associated with equal values of the energy functional at zero and for large r values.

In the case of ESQPT there is not a clear definition of an order parameter [14]. The expectation value of the number operator $\hat{n}$ in the different eigenstates provides a possible approximation. This observable changes abruptly in the critical excitation energy but it does not go to zero in any of the phases [14]. We plot in Fig. 6 the expected value of $\hat{n}$ for the eigenstates of Hamiltonian (9) as a function of the normalized state excitation energy. We have fixed the number of bosons ($N = 800$) and the control parameter ξ ($\xi = 0.6$). For $\alpha = 0$ (red curve in Fig. 6) we recover the results published in [14], where the critical energy is marked by the warped minimum in the plotted function. For $0 > \alpha > \alpha_c$, in addition to the displacement to larger energies of the initial minimum, a maximum appears for larger excitation energies (orange curve in Fig. 6). As α tends to α_c both extremes approach. We found the surprising result that for $\alpha = \alpha_c$ the expectation value of $\hat{n}$ becomes constant for the full energy range except a sudden variation around the critical energy that corresponds to the crossing point of the two separatrices (green curve in Fig. 6).

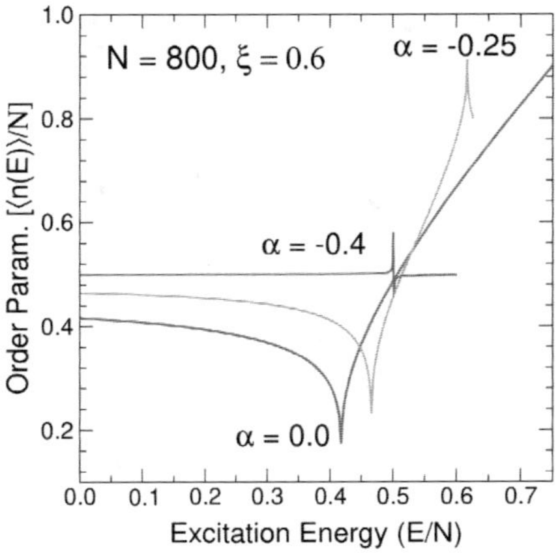

Figure 6. (Color online) Expectation value of the number operator $\hat{n}$ in the eigenstates of model Hamiltonian (23) for $N = 800$ and $\xi = 0.6$ as a function of the excitation energy of the state normalized by N for three different values of α.

6. Concluding remarks

From our point of view, the two dimensional limit of the vibron model is a simple algebraic model that still encompasses enough complexity to be applied to interesting real physical systems.

The calculation of finite-size corrections to the mean field limit greatly deepens the knowledge of the model. These corrections are a fundamental help to characterize the QPTs between the different geometric limits of a model and to analytically extract the scaling exponents for the model, as it will be shown for the model under study in a forthcoming publication [21].

We have applied the model to the bending dynamics of a non-rigid molecule. Despite the simplicity of the approach, it is capable of coping with complicated situations, as it has been shown in the case of NCNCS, where quantum monodromy effects appear. We are especially interested in the description with this model of water's vibrational bending levels.

Being the simplest bosonic two-level model with a non-trivial angular momentum, this limit of the vibron model is a very convenient aid in the study of ESQPTs and their implications. The experimental access to bending excitation energies above the monodromy critical energy makes this feature especially attractive. The obtainment of experimental data for such states is not possible in other fields, as nuclear spectroscopy, which have been the traditional test ground for ground state transitions.

Acknowledgments

This work was supported in part by the Spanish Junta de Andalucía under projects P07-FQM-02962, P07-FQM-03014, and P07-FQM-02894 and by the Spanish MICINN and the European regional development fund (FEDER) under projects FIS2008-04189 and CPAN-Ingenio (CSD2007-00042). PPF acknowledge the Spanish MEC for a FPU grant. The authors thank Jorge Dukelsky, Francesco Iachello, Pieter van Isacker, and Amiram Leviatan for valuable comments and suggestions.

References

[1] Iachello F 1993 *Rev. Mod. Phys.* **65** 569
[2] Iachello F and Arima A 1987 *The Interacting Boson Model* (Cambridge University Press, Cambridge)
[3] Iachello F and Levine R D 1995 *Algebraic Theory of Molecules* (Oxford University Press, Oxford)
[4] van Roosmalen O S, Benjamin I and Levine R D 1984 *J. Chem. Phys.* **81** 5986
[5] Frank A and van Isacker P 1994 *Algebraic Methods in Molecular and Nuclear Structure Physics* (John Wiley and Sons, New York)
[6] Frank A, Lemus R, Bijker R, Pérez-Bernal F and Arias J M 1996 *Ann. Phys.* **252** 211
[7] Iachello F and Oss S 1996 *J. Chem. Phys.* **104** 6956
[8] Ishikawa H, Toyosaki H, Mikami N, Pérez-Bernal F, Vaccaro P H and Iachello F 2002 *Chem. Phys. Lett.* **365** 57–68
[9] Iachello F, Pérez-Bernal F and Vaccaro P H 2003 *Chem. Phys. Lett.* **375** 309–320
[10] Pérez-Bernal F, Santos L F, Vaccaro P H and Iachello F 2005 *Chem. Phys. Lett.* **414** 398–404
[11] Pérez-Bernal F and Iachello F 2008 *Phys. Rev. A* **77** 032115
[12] Gilmore R and Feng D 1978 *Nucl. Phys. A* **25** 189
[13] Cejnar P, Jolie J and Casten R F 2010 *Rev. Mod. Phys.* **82** 2155–2212
[14] Caprio M A, Cejnar P and Iachello F 2008 *Ann. Phys.* **323** 1106–1135
[15] Zobov N F *et al.* 2005 *Chem. Phys. Lett.* **414** 193–197
[16] Winnewisser M *et al.* 2006 *J. Mol. Struct.* **798** 1–26
[17] Gilmore R 1979 *J. Math. Phys.* **20** 891
[18] Dusuel S, Vidal J, Arias J M, Dukelsky J and García-Ramos J E 2005 *Phys. Rev. C* **72** 064332
[19] Dusuel S and Vidal J 2004 *Phys. Rev. Lett.* **93** 237204
[20] Arias J M, Dukelsky J, García-Ramos J E and Vidal J 2007 *Phys. Rev. C* **75** 014301
[21] Pérez-Fernández P, Arias J M, García-Ramos J E and Pérez-Bernal F 2010 Finite-size corrections in the bosonic algebraic approach to two-dimensional systems, submitted to *Phys. Rev. A.*
[22] Champion J M, Abbouti Temsamani M and Oss S 1999 *Chem. Phys. Lett.* **308** 274–282
[23] Abbouti Temsamani M, Champion J M and Oss S 1999 *J. Chem. Phys.* **110** 2893–2902
[24] Sánchez-Castellanos M, Lemus R, Carvajal M and Pérez-Bernal F 2008 *J. Mol. Spectrosc.* **253** 1–15
[25] Iachello F and Pérez-Bernal F 2009 *J. Phys. Chem. A* **113** 13273–13286
[26] Dixon R N 1964 *Trans. Far. Soc.* **60** 1363
[27] Winnewisser B P *et al.* 2005 *Phys. Rev. Lett.* **95** 243002
[28] Pérez-Bernal F 2010 triat_u3_min v. 2.2.0 `http://www.uhu.es/gem/clinux/descargas.php`
[29] Pérez-Bernal F and Álvarez-Bajo O 2010 *Phys. Rev. A* **81** 050101(R)

Journal of Physics: Conference Series **284** (2011) 012050 doi:10.1088/1742-6596/284/1/012050

Modular geometry of the symplectic group attached to a q-level system and to multiple q-dit mixtures

Michel Planat

Institut FEMTO-ST, CNRS, 32 Avenue de l'Observatoire,
F-25044 Besançon, France.

E-mail: `michel.planat@femto-st.fr`

Abstract. We study the commutation relations within the Pauli groups built on all decompositions of a given Hilbert space dimension q, containing a square, into its factors. Illustrative low dimensional examples are the quartit ($q = 4$) and two-qubit ($q = 2^2$) systems, the octit ($q = 8$), qubit/quartit ($q = 2\times 4$) and three-qubit ($q = 2^3$) systems, and so on. In the single qudit case, e.g. $q = 4, 8, 12, \ldots$, one defines a bijection between the $\sigma(q)$ maximal commuting sets [with $\sigma[q)$ the sum of divisors of q] of Pauli observables and the maximal submodules of the modular ring $\mathbb{Z}_q^2$, that arrange into the projective line $P_1(\mathbb{Z}_q)$ and a independent set of size $\sigma(q) - \psi(q)$ [with $\psi(q)$ the Dedekind psi function]. In the multiple qudit case, e.g. $q = 2^2, 2^3, 3^2, \ldots$, the Pauli graphs rely on symplectic polar spaces such as the generalized quadrangles $GQ(2,2)$ (if $q = 2^2$) and $GQ(3,3)$ (if $q = 3^2$). More precisely, in dimension p^n (p a prime) of the Hilbert space, the observables of the Pauli group (modulo the center) are seen as the elements of the $2n$-dimensional vector space over the field $\mathbb{F}_p$. In this space, one makes use of the commutator to define a symplectic polar space $W_{2n-1}(p)$ of cardinality $\sigma(p^{2n-1})$, that encodes the maximal commuting sets of the Pauli group by its totally isotropic subspaces. Building blocks of $W_{2n-1}(p)$ are punctured polar spaces (i.e. a observable and all maximum cliques passing to it are removed) of size given by the Dedekind psi function $\psi(p^{2n-1})$. For multiple qudit mixtures (e.g. qubit/quartit, qubit/octit and so on), one finds multiple copies of polar spaces, ponctured polar spaces, hypercube geometries and other intricate structures. Such structures play a role in the science of quantum information.

1. Introduction

The q-level quantum systems (also denoted q-dits, or qudits), and tensor products of them, possibly with a different number of levels in each factor, are basic constituents of quantum information processing. Multiple qubits, that are tensor products of two-qubit systems are routinely employed in quantum algorithms, but multiple copies of q-dits (with $q > 2$) may turn to be more interesting in terms of self error-correction, and in relation to multipartite communication, as on the quantum Internet. The most general system would be a mixture of multiple qudits corresponding to the factors of a integer factorization of the Hilbert space dimension as $q = \prod_i q_i^{p_i}$. Let us point out that, for a given dimension q, there exists several such factorizations, leading to distinct quantum systems. In the lowest dimensional case involving a square, one has either $q = 4$ or $q = 2^2$, corresponding to the single quartit and two-qubit systems, respectively. It may be convenient to use a 4-level system (like the states of a nuclear spin $\frac{3}{2}$) to physically implement the two-qubit CNOT gate [1], and in some respect both systems display similar symmetries (like in the Bloch sphere representation) [2], but in general they have distinctive features (like in the Pauli group of observables and in the structure of the maximal commuting sets).

In this paper, we focus on the commutation relations of observables attached to a selected decomposition of the Hilbert space dimension q. The observables in a factor are defined from the action on a vector $|s\rangle$ of the q_i-dimensional Hilbert space of the q_i-dit Pauli group generated by two unitary X (shift) and clock Z operators via $X|s\rangle = |s+1\rangle$ and $Z|s\rangle = \omega^s|s\rangle$, with ω a primitive q_i-th root of unity. Then the observables in dimension q are obtained by taking tensor products over the q_i-dimensional observable of each factor. A Pauli graph is constructed by taking the observables as vertices and a edge joining two commuting observables. Maximal sets of mutually commuting observables, i.e. maximum cliques of the Pauli graph, are used to define a point/line incidence geometry with observables as points and maximum cliques as lines.

In recent papers, multiple qubits [3, 4], single qudits [5, 6, 7] and a few examples of qudit mixtures [9] were already explored. Further work was published to clarify this earlier work dealing with symplectic polar spaces of multiple qudits [10, 11, 12] and, in what concerns multiple qubits, its link to units in Clifford algebras [13] and to Lie algebras [14]. Prior to the advent of quantum information science, the incidence properties of the q-dimensional geometry and the relations to Clifford algebras were published in [15, 16]. The link of mutual unbiasedness to the general theory of angular momentum is explored in [17], and its link to Feymann's path integral may be found in [18].

In this paper, we focus on quantum systems of Pauli observables defined over the Hilbert space of dimension q containing a square. In the single qudit case, studied in Sec. 2, the maximal mutually commuting sets of observables in the Hilbert space of dimension q are mapped bijectively to the maximal submodules over the ring $\mathbb{Z}_q$ [5, 6]. If q contains a square, there are $\psi(q) = q\prod_{p|q}(1+\frac{1}{p})$ points on the projective line $P_1(\mathbb{Z}_q)$ (in the Dedekind finction $\psi(q)$, the product is taken over all primes p dividing q) and the remaining $\sigma(q) - \psi(q) \neq 0$ independent points (with $\sigma(q)$ the sum of divisors function) is playing the role of a reference frame and possess their own modular substructure. The number theoretical properties of the modular ring $\mathbb{Z}_q$ are used to count the cardinality of the symplectic group $\mathrm{Sp}(2,\mathbb{Z}_q)$ [7, 19, 20]. In Sec. 3, we remind the established results concerning the point/line geometries attached to multiple qudit systems in dimension p^n, that symplectic polar spaces $W_{2n-1}(p)$ of order p and rank n govern the commutation structure of the observables. Here, the number theoretical functions $\sigma(p^{2n-1})$ and $\psi(p^{2n-1})$ are found to count the number of observables in the symplectic polar space and in the *punctured* polar space, respectively. In Sec. 4, we study composite systems when at least one of the factors q_i of the Hilbert space dimension is a square. It is shown, that the non-modularity leads to a natural splitting of the Pauli graph/geometry into several copies of basic structures such as polar spaces, punctured polar spaces and related hyperdimensional structures.

A few properties of the structures we have checked are in table 1. Details are given in the subsequent sections.

Most calculations are performed on Magma [21]. High dimensional computations have been made possible thanks to the supercomputer facilities of the Mésocentre de calcul at University of Franche-Comté.

2. Pauli graph/geometry of a single qudit

A single qudit is defined by a Weyl pair (X, Z) of *shift* and *clock* cyclic operators satisfying

$$ZX - \omega XZ = 0, \tag{1}$$

Journal of Physics: Conference Series **284** (2011) 012050 doi:10.1088/1742-6596/284/1/012050

q	name [Ref.]	# cliques	geometry	spectrum	aut. group
4	quartit [6, 2]	$6+1$	$P_1(\mathbb{Z}_4)^\dagger$	$\{4^1, 0^{3+1}, -2^2\}$	$G_{48} = \mathbb{Z}_2 \times S_4$
2^2	2-qubit [3, 4]	15	$GQ(2,2)$	$\{6^1, 1^9, -3^5\}$	S_6
8	octit [6, 2]	$12+3$	$P_1(\mathbb{Z}_8)^\dagger$	$\{8^1, 0^{9+3}, -4^2\}$	$\mathbb{Z}_2^6 \rtimes (\mathbb{Z}_3^3 \rtimes G_{48})$
2.2^2	qubit/quartit [9]	$36+3$	$3 \times GQ(2,2)'$	$\{5^1, 1^6, -1^2, -3^3\}^3$	$G_{48}^3 \rtimes S_3$
2^3	3-qubit [3, 4]	135	$W_5(2)$	$\{30^1, 3^{35}, -5^{27}\}$	$\mathrm{Sp}(6,2)$
9	9-dit [6]	$12+1$	$P_1(\mathbb{Z}_9)^\dagger$	$\{9^1, 0^{8+1}, -3^3\}$	$G_{648} \rtimes G_{48}$
3^2	2-qutrit [3]	40	$GQ(3,3)$	$\{25^1, 5^{24}, -1^{40}, -7^{15}\}$	$\mathbb{Z}_2^{40}.W(E_6)$
12	12-dit [6]	$24+4$	$P_1(\mathbb{Z}_{12})^\dagger$	$\{12^1, 2^6, 0^{12+4}, -4^3, -6^2\}$	$\mathbb{Z}_2^{12} \rtimes G_{144}$
3.4	qutrit/quartit	$24+4$	as above	as above	as above
$2^2.3$	2-qubit/qutrit	60	$4 \times GQ(2,2)$	$\{6^1, 1^9, -3^5\}^4$	$S_6^4 \rtimes S_4$
16	16-dit [6]	$24+7$	$P_1(\mathbb{Z}_{16})^\dagger$	$\{16^1, 0^{21+7}, -8^2\}$	$A_8^3 \rtimes G_{48}$
2.8	qubit/octit	$72+15$	$6 \times GQ(2,2)'$	$\{5^1, 1^6, -1^2, -3^3\}^6$	$G_{48}^6 \rtimes S_6$
4.4	2-quartit	$120+31$	15-cube	$\{-3^1, 3^1, -1^3, 1^3\}^{15}$	$G_{48}^{15} \rtimes S_{15}$
$2^2.4$	2-qubit/quartit	$360+15$	$3 \times W_5(2)'$	$\{13^1, 5^{25}, 3^9, -1^{70}, -5^5, -7^{10}\}^3$	$(\mathbb{Z}_2^5 \rtimes S_6)^3 \rtimes S_3$
2^4	4-qubit [3, 4]	2295	$W_7(2)$	$\{126^1, 7^{135}, -9^{119}\}$	$\mathrm{Sp}(8,2)$
18	18-dit [6]	$36+3$	$P_1(\mathbb{Z}_{18})^\dagger$	$\{18^1, 3^6, 0^{24+3}, -6^3, -9^2\}$	$\mathbb{Z}_3^{12} \rtimes (\mathbb{Z}_2^{12} \rtimes G_{144})$
2.9	qubit/9-dit	$36+3$	as above	as above	as above
2.3^2	2-qutrit/qubit [9]	120	$3 \times GQ(3,3)$	$\{12^1, 2^{24}, -4^{15}\}^3$	$W'(E_6)^3.G_{48}$
24	24-dit [6]	$48+12$	$P_1(\mathbb{Z}_{24})^\dagger$	$\{24^1, 4^6, 0^{36+12}, -8^3, -12^2\}$	$G_{2^{24}3^{12}} \rtimes (\mathbb{Z}_2^{12} \rtimes G_{144})$
2.3.4	2-dit/3-dit/4-dit	$144+12$	see Sec. 4	see Sec. 4	
$2^3.3$	3-qubit/qutrit	540	$4 \times W_5(2)$	$\{56^1, 14^{15}, 2^{35}, -4^{84}\}^4$	$\mathrm{Sp}(6,2)^4.S_4$

Table 1. The main properties of the studied Pauli graphs. The first and second column gives the selected decomposition of q and the name of the corresponding Pauli system, respectively. Third column represents the number of maximal sets of mutually commuting observables of size $q-1$ (i.e. the number of maximum cliques in the corresponding Pauli graph) and how it splits into two numbers of geometrical significance explained in the paper. The fourth column provides a geometry that may be identified. The fifth column provides the spectrum of the Pauli graph, that of its dual geometry or that of an important subgraph, depending on context (see the corresponding section for details). The automorphism group of the selected geometry is given in the last column. The notation S_n, A_n and D_n is for the symmetric, alternating and dihedral group, respectively. Symbols $\times$, $\rtimes$ and . are for the direct, semidirect and not semidirect products of groups, respectively.
The notation $W_{2n-1}(p)$ is for the symplectic polar space of order p and rank n [3, 4]. The polar space $W_3(2)$ is the generalized (self-dual) quadrangle of order two GQ(2,2), also called the doily. The notation $W_{2n-1}(p)'$ means the polar space $W_{2n-1}(p)$ minus a perp-set (i.e. a point and the maximum cliques passing through it). Whenever multiple polar spaces are featured in the table, it means that we are dealing with the mutual incidence of cliques at multiple points (see Sec. 4 for details).
† The incidence geometry is associated to the maximum cliques of the Pauli graph and the spectrum is that of all cliques (see Sec. 2 for details).

where $\omega = \exp \frac{2i\pi}{q}$ is a primitive q-th root of unity and 0 is the null q-dimensional matrix. In the standard computational basis $\{|s\rangle, s \in \mathbb{Z}_q\}$, the explicit form of the pair is as follows

$$X = \begin{pmatrix} 0 & 0 & \dots & 0 & 1 \\ 1 & 0 & \dots & 0 & 0 \\ . & . & \dots & . & . \\ . & . & \dots & . & . \\ 0 & 0 & \dots & 1 & 0 \end{pmatrix}, \quad Z = \mathrm{diag}(1, \omega, \omega^2, \dots, \omega^{q-1}). \tag{2}$$

The Weyl pair generates the single qudit Pauli group $\mathcal{P}_q = \langle X, Z\rangle$, of order q^3, where each element may be written in a unique way as $\omega^a X^b Z^c$, with $a, b, c \in \mathbb{Z}_q$.

It will be shown in this section that the study of commutation relations in a arbitrary single qudit system may be based on the study of symplectic modules over the modular ring $\mathbb{Z}_q^2$, and conversely that the elegant number theoretical relations underlying the *isotropic* lines of $\mathbb{Z}_q^2$ have their counterpart in the maximal commuting sets of a qudit system. Our results may be found in various disguises in several publications where the proofs are given [6, 7, 19, 20].

Let us start with the Weyl pair property (1) and write the group theoretical commutator as $[X, Z] = XZX^{-1}Z^{-1} = \omega^{-1}I_q$ (where I_q is the q-dimensional identity matrix), so that one gets the expression

$$\left[\omega^a X^b Z^c, \omega^{a'} X^{b'} Z^{c'}\right] = \omega^{cb'-c'b} I_q, \tag{3}$$

meaning that two elements of $\mathcal{P}_q$ commute if only if the determinant $\Delta = \det\begin{pmatrix} b' & b \\ c' & c \end{pmatrix}$ vanishes. Two vectors such that their symplectic inner product $[(b', c').(b, c)] = \Delta = b'c - bc'$ vanishes are called perpendicular. Thus, from (3), one can transfer the study of commutation relations within the group $\mathcal{P}_q$ to the study of perpendicularity of vectors in the ring $\mathbb{Z}_q^2$ [6].

From (3), one gets the important result that the set $\mathcal{P}'_q$ of commutators (also called the derived subgroup) and the center $Z(\mathcal{P}_q)$ of the Pauli group $\mathcal{P}_q$ are identical, and one is led to the isomorphism

$$(\mathcal{P}_q/Z(\mathcal{P}_q), \times) \cong (\mathbb{Z}_q^2, +), \tag{4}$$

i.e. multiplication of observables taken in the central quotient $\mathcal{P}_q/Z(\mathcal{P}_q)$ transfers to the algebra of vectors in the $\mathbb{Z}_q$ module $\mathbb{Z}_q^2$ endowed with the symplectic inner product ".".

Isotropic lines of the lattice $\mathbb{Z}_q^2$

Let us now define a *isotropic line* as a set of q points on the lattice $\mathbb{Z}_q^2$ such that the symplectic product of any two of them is $0(\text{mod } q)$. From (4), to such an isotropic line corresponds a maximal commuting set in $\mathcal{P}_q/Z(\mathcal{P}_q)$.

Taking the prime power decomposition of the Hilbert space dimension as $q = \prod_i p_i^{s_i}$, it is shown in (18) of [7] that the number of isotropic lines of the lattice $\mathbb{Z}_q^2$ reads

$$\eta(q) = \prod_i \frac{p_i^{s_i+1} - 1}{p_i - 1} \equiv \sigma(q), \tag{5}$$

where $\sigma(q)$ denotes the sum of divisor function [1]

It may be checked from table 1 (colum 3), that the number of maximum cliques in the Pauli graph of $\mathcal{P}_q$ [i.e. the number of maximal commuting set in $\mathcal{P}_q/Z(\mathcal{P}_q)$] in the considered single qudit decompositions $q = 4, 8, 9, 12, 16$ and 18 are $\sigma(4) = 1+2+4 = 7$, $\sigma(8) = 1+2+4+8 = 15$, $\sigma(9) = 13$, $\sigma(12) = 27$, $\sigma(16) = 31$ and $\sigma(18) = 39$, respectively.

Another important quantity is the number $\eta(q; x)$ of isotropic lines through a given point $x = (b, c)$ of the lattice. Denoting by $t_i = v_{p_i}(x)$ the p_i-valuation [2] of x, it is shown in (36) of [7] that one obtains

$$\eta(q; x) = \prod_i \frac{p_i^{t_i+1} - 1}{p_i - 1} \equiv \sigma(\tilde{q}(x)), \tag{6}$$

[1] The identification of $\eta(q)$ to $\sigma(q)$ is not provided in [7]. However, it is easy to see that the factors in (5) are $\frac{p_i^{s+1}-1}{p_i-1} = 1 + p_i + p_i^2 + \cdots + p_i^s = \sigma(p_i^s)$ and, since $\sigma(q)$ is multiplicative, (5) immediately follows. Similarly, the identification of $\eta(q; x)$ to $\sigma(\tilde{q}(x))$ given in (6) is easy to establish.

[2] The p-adic valuation $v_p(x)$ of a integer number x is the highest exponent t suct that the power of prime p^t divides x.

where $\tilde{q}(x) = \prod_i p_i^{t_i} \leq q$ is a *local* dimension defined at the selected point x.

The projective line $P_1(\mathbb{Z}_q)$ and the symplectic group $Sp(2, \mathbb{Z}_q)$
As shown in [7], a isotropic line of $\mathbb{Z}_q^2$ corresponds to a *Lagrangian submodule*, i.e. a maximal module such that the perpendicular module $M^\perp = M$. Let us now specialize to Lagrangian submodules that are *free cyclic submodules*

$$\mathbb{Z}_q(b, c) = \{(ub, uc) | u \in \mathbb{Z}_q\}, \tag{7}$$

for which the application $u \to (ub, uc)$ is injective. Not all Lagrangian submodules are free cyclic submodules. A point $x = (b, c)$ such that $\mathbb{Z}_q(b, c)$ is free is called an *admissible point*, and the set of admissible points is called the projective line

$$\mathbb{P}_1(\mathbb{Z}_q) = \{\mathbb{Z}_q(b, c) | (b, c) \text{ is admissible}\}. \tag{8}$$

Following theorem 5 in [6], the number of points of the projective line is

$$|\mathbb{P}_1(\mathbb{Z}_q)| = \prod_i (p_i^{s_i} + p_i^{s_i - 1}) \equiv \psi(q), \tag{9}$$

where $\psi(q) = q \prod_{p|q}(1 + \frac{1}{p})$ and the product is taken over all primes p dividing q [3]. Note that one has $\psi(q) \leq \sigma(q)$, where the equality holds if q is square-free integer.

In the considered single qudit decompositions $q = 4, 8, 9, 12, 16$ and 18, that contain a square, one gets $\psi(4) = 4(1 + \frac{1}{2}) = 6$, $\psi(8) = 8(1 + \frac{1}{2}) = 12$, $\psi(9) = 12$, $\psi(12) = 24$, $\psi(16) = 24$, $\psi(18) = 36$, as it it is also shown in table 1 (column 3).

Then, still using theorem 5 in [6], the number of points of the projective line containing a selected vector $x = (b, c)$ of the lattice reads as

$$|\mathbb{P}_1(\mathbb{Z}_q; x)| = \psi(\tilde{q}(x)), \tag{10}$$

where $\tilde{q}(x)$ is the local dimension introduced in (6).

As for the projective line $\mathbb{P}_1(\mathbb{Z}_q)$, the symplectic group $\mathrm{Sp}(2, \mathbb{Z}_q)$ contains interesting number theoretical features.

We defined an admissible vector (b, c) as one leading to a point of the projective line $\mathbb{P}_1(\mathbb{Z}_q)$. If $q = p^s$, there are $p^{2s} - p^{2(s-1)}$ admissible vectors and, for arbitrary dimensions $q = \prod_i p_i^{s_i}$, the number of admissible vectors is

$$q^2 \prod_i (1 - \frac{1}{p^2}) = \phi(q)\psi(q) = J_2(q), \tag{11}$$

where $\phi(q) = q \prod_i (1 - \frac{1}{p_i})$ is the Euler totient function and $J_2(q)$ is known as the Jordan totient function.

Following the same line of reasoning than (6) and (10), one may also define a finer structure of admissibility from the number $J_2(\tilde{q}(x))$, with $\tilde{q}(x) = \prod_i p_i^{t_i} \leq q$ is the local dimension. If $q = p^s$, one has $\tilde{q}(x) = q$ and the structure is simpler than in the composite case such as $q = 12$ and $q = 18$.

The symplectic group $\mathrm{Sp}(2, \mathbb{Z}_q)$ is built from all matrices $\begin{pmatrix} b' & b \\ c' & c \end{pmatrix}$ such that (b, c) is an admissible vector and the symplectic inner product, i.e. the determinant $\Delta = b'c - bc' = 1$. The cardinality of such a group is is $|\mathrm{Sp}(2, \mathbb{Z}_q)| = qJ_2(q)$ [19, 20].

[3] As for the relation (5), the identification of $|\mathbb{P}_1(\mathbb{Z}_q)|$ to the Dedekind psi function $\psi(q)$ is not provided in [6]. The proof is easy to establish since $\psi(q)$ is a multiplicative function

The Pauli graph of a qudit
In the previous subsections, we investigated the bijection between sets of operators of the Pauli group $\mathcal{P}_q$ and vectors defined over the modular ring $\mathbb{Z}_q$. More precisely, from (4), elements of the central quotient of the Pauli group $\mathcal{P}_q/Z(\mathcal{P}_q)$ were mapped to vectors of the lattice $\mathbb{Z}_q^2$ and, from (5) the $\sigma(q)$ isotropic lines of $\mathbb{Z}_q^2$ were mapped to its maximal commuting sets.

One can see these bijections in a clearer way by defining the Pauli graph $\mathcal{G}_q$ of the qudit system. The Pauli graph $\mathcal{G}_q$ is constructed by taking the observables as vertices and a edge joining two commuting observables. A maximal set of mutually commuting observables corresponds to a maximum clique of $\mathcal{G}_q$, and one further defines a point/line incidence geometry with observables as points and maximum cliques as lines. One characterizes this geometry by creating a dual graph $\mathcal{G}_q^\star$ such that the vertices are the cliques and a edge joins two non-intersecting cliques. The connected component of $\mathcal{G}_q^\star$ corresponds to the graph of the projective line $\mathbb{P}_1(\mathbb{Z}_q)$ (as defined in previous papers [3]-[12]).

In the subsequent sections, we shall also introduce the graph $\mathcal{G}_q^{(k)}$, in which the vertices are the maximum cliques of the Pauli graph $\mathcal{G}_q$ and a edge joins two maximum cliques intersecting at k points.

The quartit system For the four-level system, there are $4^2 - 1$ observables/vertices in the Pauli graph $\mathcal{G}_4$. The $\sigma(4) = 7$ maximum cliques

$$cl := \{(X^2, Z^2, Z^2X^2), (X, X^2, X^3), (X^2, Z^2X, Z^2X^3), (Z, Z^2, Z^3),\\ (ZX, Z^2X^2, Z^3X^3), (ZX^2, Z^2, Z^3X^2), (ZX^3, Z^2X^2, Z^3X)\} \quad (12)$$

are mapped to the following isotropic lines of $\mathbb{Z}_4^2$

$$il := \{\{(0,2),(2,0),(2,2)\}, \{(0,1),(0,2),(0,3)\}, \{(0,2),(2,1),(2,3)\},\\ \{(1,0),(2,0),(3,0)\}, \{(1,1),(2,2),(3,3)\}, \{(1,2),(2,0),(3,2)\},\\ \{(1,3),(2,2),(3,1)\}\}. \quad (13)$$

From the latter list, one easily observes that non-admissible vectors belong to the first line $\{(0,2),(2,0),(2,2)\}$, that corresponds to the maximum clique (X^2, Z^2, Z^2X^2). The remaing vectors in $\mathbb{Z}_4^2$ generate free cyclic submodules of the form (7).

The sequence of degrees in $\mathcal{G}_q^\star$ is obtained as $(1, 0, 0, 0, 6)$, meaning that the first clique given in (12) (of degree 0) intersects all the remaing ones, and that cliques number 2 to 7 in (12) (of degrees 4) form the projective line $\mathbb{P}_1(\mathbb{Z}_4)$. Indeed, one has $|\mathbb{P}_1(\mathbb{Z}_4)| = \psi(4) = 6$. There are $J_2(4) = \phi(4)\psi(4) = 12$ admissible points.

The graph $\mathcal{G}_4^\star$ is strongly regular, with spectrum $\{4^1, 0^{3+1}, -2^2\}$ (in the notations of [3]); the notation 0^{3+1} in the spectrum means that 0^3 belongs to the projective line subgraph and there exists an extra 0 eigenvalue in the spectrum of $\mathcal{G}_4^\star$. The automorphism group of $\mathbb{P}_1(\mathbb{Z}_4)$ is found to be the direct product $G_{48} = \mathbb{Z}_2 \times S_4$ (where S_4 is the four-letter symmetric group).

The 12-dit system The main results for all qudit systems with $4 \le q \le 18$, such that q contains a square, are given in table 1. We take the composite dimension $q = 2^2 \times 3$ as our second illustration. There are $12^2 - 1 = 143$ observables in the Pauli graph $\mathcal{G}_{12}$. There are $\sigma(12) = 28$ maximum cliques in $\mathcal{G}_{12}$, as expected. The sequence of degrees in the dual graph $\mathcal{G}_{12}^\star$ is found as $(4, 0, \ldots, 24)$, i.e. there are four cliques of degree 0 and the remaining $\psi(12) = 24$ ones have degree 12 (as also seen from the spectrum given in Table 1).

Owing to the composite character of the dimension, the structure of $\mathcal{G}_{12}^\star$ is more complex than in the quartit case, see Fig. 1 of [6] for a picture. All four independent cliques intersect at the three vectors $(0,6), (6,0), (6,6)$, corresponding to the three observables X^6, Z^6, X^6Z^6. The remaining 24 cliques intersect at 0, 1, 2, 3 or 5 points. The automorphism group of $\mathbb{P}_1(\mathbb{Z}_{12})$ is found to be $\mathbb{Z}_2^{12} \rtimes G_{144}$, with $G_{144} = A_4 \rtimes D_6$.

Remarkably, the automorphism groups of $\mathbb{P}_1(\mathbb{Z}_{18})$ and $\mathbb{P}_1(\mathbb{Z}_{24})$ encompass that of $\mathbb{P}_1(\mathbb{Z}_{12})$, as shown in Table 1.

3. Pauli graph/geometry for multiple qudits

In this section, we specialize on multiple qudits $q = p^n$, when the qudit is a p-dit (with p a prime number). The multiple qudit Pauli group $\mathcal{P}_q$ is generated from the n-fold tensor product of Pauli operators X and Z [defined in (2) with $\omega = \exp(\frac{2i\pi}{p})$]. One has $|\mathcal{P}_q| = p^{2n+1}$ and the derived group $\mathcal{P}'_q$ equals the center $Z(\mathcal{P}_q)$ so that $|\mathcal{P}'_q| = p$.

Following [4, 10], the observables of $\mathcal{P}_q/Z(\mathcal{P}_q)$ are seen as the elements of the $2n$-dimensional vector space $V(2n,p)$ defined over the field $\mathbb{F}_p$, and one makes use of the commutator

$$[.,.]: \; V(2n,p) \times V(2n,p) \to \mathcal{P}'_q \tag{14}$$

to induce a non-singular alternating bilinear form on $V(2n,p)$, and simultaneously a symplectic form on the projective space $PG(2n-1,p)$ over $\mathbb{F}_p$.

Doing this, the $|V(2n,q)| = p^{2n}$ observables of $\mathcal{P}_q/Z(\mathcal{P}_q)$ are mapped to the points of the symplectic polar space $W_{2n-1}(p)$ of cardinality [4]

$$|W_{2n-1}(p)| = \frac{p^{2n}-1}{p-1} \equiv \sigma(p^{2n-1}), \tag{15}$$

and two elements of $[\mathcal{P}_q/Z(\mathcal{P}_q), \times]$ commute iff the corresponding points of the polar space $W_{2n-1}(p)$ are collinear.

A subspace of $V(2n,p)$ is called totally isotropic if the symplectic form vanishes identically on it. The polar space $W_{2n-1}(p)$ can be regarded as the space of totally isotropic subspaces of the $(2n-1)$-dimensional projective space $PG(2n-1,p)$. Such totally isotropic subspaces, also called generators G, have dimension $p^n - 1$ and their number is

$$|\Sigma(W_{2n-1}(p))| = \prod_{i=1}^{n}(1+p^i). \tag{16}$$

Let us call a spread S of a vector space a set of generators partitioning its points. The size of a spread of $V(2n,p)$ is $|S| = p^n + 1$ and one has $|V(2n,p)| - 1 = |S| \times |G| = (p^n+1) \times (p^n - 1) = p^{2n} - 1$, as expected.

Going back to the Pauli observables, a generator G corresponds to a maximal commuting set and a spread S corresponds to a maximum (and complete) set of disjoint maximal commuting sets. Two generators in a spread are mutually disjoint and the corresponding maximal commuting sets are mutually unbiased [3, 23].

Let us define the punctured polar space $W_{2n-1}(p)'$ as the polar space $W_{2n-1}(p)$ minus a perp-set (i.e. a point u and all the totally isotropic spaces passing though it) [5]. Then, one gets

$$|W_{2n-1}(p)'| = \sigma(p^{2n-1}) - \sigma(p^{2n-3}) = \psi(p^{2n-1}), \tag{17}$$

where $\sigma(p^{2n-3})$ is the size of a perp-set and $\psi(q)$ is the Dedekind psi function.

[4] The proof of this statement is given in [10]. The identification of $|W_{2n-1}(p)|$ to $\sigma(p^{2n-1})$ is new in this context. It is reminiscent of (5) and has still unoticed consequences about the structure of the polar space, as explained in the sequel of the paper. For q-level systems (single qudits), $\sigma(q)$ and $\psi(q)$ refer to the number of isotropic lines and the number of points of the projective line, respectively (as in (5) and (9)). For multiple qudits, one has $q = p^{2n-1}$ and $\sigma(q)$ and $\psi(q)$ refer to the number of points of the symplectic polar space $W_{2n-1}(p)$ and of *punctured polar space* $W_{2n-1}(p)'$, respectively (as in (15) and (17)).

[5] In the graph context the symbol ' means a puncture in the graph. It is not the same symbol as in the derived subgroup G' of the group G.

Journal of Physics: Conference Series **284** (2011) 012050 doi:10.1088/1742-6596/284/1/012050

The Pauli graph of a multiple qudit

The symmetries carried by multiple qudit systems may also be studied with Pauli graphs. We define the Pauli graph $\mathcal{G}_{p^n}$ of a multiple p^n-dit, as we did for the single qudit case, by taking the observables as vertices and a edge joining two commuting observables. A dual graph $\mathcal{G}^\star_{p^n}$ is such that the vertices are the maximum cliques and a edge joins two non-interesting cliques. One denotes $\mathcal{G}'^\star_{p^n}$ the corresponding graph attached to the punctured polar space. Finally, one denotes $\mathcal{G}^{(k)}_{p^n}$ the graph whose vertices are the maximum cliques of the Pauli graph $\mathcal{G}_{p^n}$ and whose edges join two maximum cliques intersecting at k points.

Actual calculations have been performed for two- and three-qubits, and for two- and three-qutrits. Main results are in table 2 (see details in the corresponding subsections). Denoting c the ratio between the cardinalities of $\mathrm{aut}(\mathcal{G}^*_{p^n})$ and $\mathrm{aut}(\mathcal{G}'^*_{p^n})$, one observes that c identifies to the size $\sigma(p^{2n-1})$ of the polar space $W_{2n-1}(p)$, except for the case of the 3 -qubit system where c is twice the number of cliques of the Pauli graph $\mathcal{G}_{2^3}$. Thus, the space $W'_{2n-1}(p)$ may be seen as a building block of Pauli systems. One may remind that $W_{2n-1}(p)$ *contracts* to $W'_{2n-1}(p)$, as the size $\sigma(p^{2n-1})$ to the size $\psi(p^{2n-1})$, that the ratio of cardinalities of their automorphism groups is the number c, and anticipate on the structural role of $W'_{2n-1}(p)$ in qudit mixtures, shown in Table 1 and Sec. 4. For more details, see [22].

$q = p^n$	name	$\mathrm{aut}(\mathcal{G}_{p^n})$	$\mathrm{aut}(\mathcal{G}^*_{p^n})$	$\mathrm{aut}(\mathcal{G}'^*_{p^n})$	$c = \frac{\lvert\mathrm{aut}(\mathcal{G}^*_{p^n})\rvert}{\lvert\mathrm{aut}(\mathcal{G}'^*_{p^n})\rvert}$
2	qubit	S_3	S_3	S_2	$3 \equiv \sigma(2)$
2^2	2-qubit	S_6	S_6	$G_{48} = \mathbb{Z}_2 \times S_4$	$15 \equiv \sigma(2^3)$
2^3	3-qubit	$Sp(6,2)$	$O^+(8,2)$	$\mathbb{Z}_2^6 \rtimes A_8$	$2 \times 135 \neq 63 = \sigma(2^5)$
3	qutrit	$\mathbb{Z}_2^3 \rtimes G_{48}$	S_4	S_3	$4 \equiv \sigma(3)$
3^2	2-qutrit	$\mathbb{Z}_2^{40}.W(E_6)$	$W(E_6)$	$G_{648} \rtimes \mathbb{Z}_2$	$40 \equiv \sigma(3^3)$
3^3	3-qutrit	$\mathbb{Z}_2^{364}.G$	G	$(E_{243} \rtimes \mathbb{Z}_2).W(E_6)$	$364 \equiv \sigma(3^5)$

Table 2. Comparison of the automorphism group of the dual Pauli graph $\mathcal{G}^*_{p^n}$ and that of its *building block* $\mathcal{G}'^*_{p^n}$, defined by removing a perp-set in the symplectic polar space. The ratio of sizes of both groups turns out to be the number of observables $\sigma(p^{2n-1})$ of the space, except for the case of the 3-qubit system where it is twice the number 135 of cliques in the Pauli graph $\mathcal{G}_{2^3}$.

4. Pauli graph/geometry of multiple qudit mixtures

As before, $\mathcal{G}_q$ is the Pauli graph whose vertices are the observables and whose edges join two commuting observables. A dual graph of the Pauli graph is $\mathcal{G}^*_q$ whose vertices are the maximum cliques and whose edges join two non-intersecting cliques. In this section, we also introduces $\mathcal{G}^{(k)}_q$, the graph whose vertices are the maximum cliques and whose edges join two maximum cliques intersecting at k points.

First of all, as shown in Sec. 6 of [5], a qudit mixture in composite dimension $q = p_1 \times p_2 \times \cdots \times p_r$ (p_i a prime number), identifies to a single q-dit. Since the ring $\mathbb{Z}_q$ is isomorphic to the direct product $\mathbb{Z}_{p_1} \times \mathbb{Z}_{p_2} \times \cdots \mathbb{Z}_{p_r}$ the commutation relations arrange as the $\sigma(q) \equiv \psi(q)$ isotropic lines of the lattice $\mathbb{Z}_q^2$, that reproduce the projective line $\mathbb{P}_1(\mathbb{Z}_q) = \mathbb{P}_1(\mathbb{Z}_{p_1}) \times \mathbb{P}_1(\mathbb{Z}_{p_2}) \times \cdots \times \mathbb{P}_1(\mathbb{Z}_{p_r})$.

The sextit system

The simplest non-trivial case is in dimension $q = 6 = 2 \times 3$. The projective line may be pictured by the dual Pauli graph $\mathcal{G}^\star_6$ of spectrum $\{6^1, 1^6, -2^3, -3^2\}$. It represents the complement of a

Journal of Physics: Conference Series **284** (2011) 012050 doi:10.1088/1742-6596/284/1/012050

3×4 grid, or in graph theoretical language the complement $\hat{L}(K_{3,4})$ of the line graph over the complete bipartite graph $K_{3,4}$ (see Fig. 1 of [8]). One finds 24 maximum cliques of size 3 in $\mathcal{G}_6^\star$ corresponding to the same number of non-complete sets of mutually unbiased bases. The symmetry of this new configuration is the semi-direct product $G_{144} = A_4 \rtimes D_6$ of two groups of order twelve, namely the four-letter alternating group A_4 and the dihedral group D_6. Until now, it is not known whether sets of mutually unbiased bases of size larger than three can be built [23, 24].

A summary of the main results for mixtures where at least one factor in the prime number decomposition of q contains a square is in Table 1. See [22] for details.

The qubit/quartit and qubit/octit systems

Let us illustrate the lowest case of a mixture where a factor is not a prime: the qubit/quartit system living in the Hilbert space dimension $q = 2\times 4$. As shown in Table 1, the maximum cliques studied from the dual Pauli graph $\mathcal{G}_{2\times 4}^\star$ split into two parts, that are a set of 3 independent (non-intersecting) cliques and a connected component of 36 cliques. The single qudit in dimension $q = 18$ has a similar splitting since $\sigma(18) = 39$ and $\psi(18) = 36$.

Let us denote $\mathcal{G}_{2\times 4}^{\star(c)}$ the connected subgraph of the dual Pauli graph. Its spectrum is again that $\{16^1, 0^{27}, -8^4, 4^4\}$ of a regular graph. Maximum cliques of the Pauli graph intersect each other at 0, 1 or 3 points and there is a subgeometry of the qubit/quartit system found by taking the incidence graph of maximum cliques interesting at 3 points. The spectrum of this latter graph $\mathcal{G}_{2\times 4}^{(7)}$ is $\{5^1, 1^6, -1^2, -3^3\}^3$ corresponding to three copies of the punctured Pauli graph associated to $GQ(2,2)'$.

Similarly, one considers the qubit/octit system living in the Hilbert space dimension $q = 2\times 8$. As shown in Table 1, the maximum cliques studied from the Pauli graph $\mathcal{G}_{2\times 8}^\star$ split into two parts, that are a set of 15 independent (non-intersecting) cliques and a connected component of 72 cliques. Here, there exists no single qudit with such a splitting. Let us denote $\mathcal{G}_{2\times 8}^{\star(c)}$ the connected subgraph of the dual Pauli graph. Its spectrum is $\{32^1, 8^4, 0^{63}, -16^4\}$. Maximum cliques of the Pauli graph intersect each other at 0, 1, 3 or 7 points and there is a subgeometry of the qubit/octit system found by taking the incidence graph $\mathcal{G}_{2\times 8}^{(7)}$ of maximum cliques interesting at 7 points. The spectrum of this latter graph corresponds to six copies of of $GQ(2,2)'$. The automorphism group of this latter configuration is found to be the semidirect product of groups $G_{48}^6 \rtimes S_6$.

5. Conclusion

It has been shown for the first time that number theoretical functions $\sigma(q)$ and $\psi(q)$ enter into the structure of commutation relations of Pauli graphs and geometries. For single q-dits (in section 2), $\sigma(q)$ and $\psi(q)$ refer to the number of maximal commuting sets and the cardinality of the projective line $\mathbb{P}_1(\mathbb{Z}_q)$, respectively. For multiple qudits, with dimension p^n, p a prime number, (in section 3) the parameter $q = p^{2n-1}$ enters in the function $\sigma(q)$ to count the size of the symplectic polar space $W_{2n-1}(p)$ (that carries the multiple qudit system), and enters in the function $\psi(q)$ to count the size of the basic constituent: the punctured polar space $W'_{2n-1}(p)$. For multiple qudit mixtures, spaces $W_{2n-1}(p)$ and $W'_{2n-1}(p)$ are also found to arise as constituents of the commutation structure.

The structural role of symplectic groups $Sp(2n, p)$ has been found, as expected. Other important symmetry groups are $G_{48} = \mathbb{Z}_2 \times S_4$, $G_{144} = A_4 \times D_6$ and $W(E_6)$. The group G_{48} is first of all the automorphism group of the single qudit Pauli group $\mathcal{P}_1$ and is important in understanding the CPT symmetry [25]. In this paper, it arises as the symmetry group of the quartit, of the punctured generalized quadrangle $GQ(2,2)'$ and as a normal subgroup of many systems of qudits (as shown in Table 1). The torus group G_{144} occurs in the symmetries of the

6-dit, 12-dit, 18-dit and 24-dit systems. The Weyl group $W(E_6)$ happens to be central in the symmetries of three-qubit and multiple qutrit systems. The understanding of symmetries in the Hilbert space is important for the applications in quantum information processing.

Acknowledgements
The author thanks Fabio Anselmi for his feedback on this topic and for his critical reading of the manuscript.

Bibliography

[1] Hirayama Y, Miranovwicz A, Ota T, Yusa G, Muraki K, Ozdemir S K and Imoto N 2006 Nanometer-scale nuclear-spin device for quantum information processing. *J. Phys.: Condens. Matter* **18** S885.
[2] Planat M 2010 On the geometry and invariants of qubits, quartits and octits *Preprint 1005.1997* (quant-ph); to appear in Int. J. Geom. Meth. Mod. Phys. **8**, March 2011.
[3] Planat M and Saniga M 2008 On the Pauli graphs of N-qudits *Quant. Inf. Comp.* **8** 127-146.
[4] Saniga M and Planat M 2007 Multiple qubits as symplectic polar spaces of order two *Adv. Stud. Theor. Phys.* **1** 1-4.
[5] Havlicek H ans Saniga M 2007 Projective line of a specific qudit *J. Phys. A: Math. Theor.* **40** F943-F952;
[6] Havlicek H and Saniga M 2008 Projective line of an arbitrary single qudit *J. Phys. A: Math. Theor.* **41** 015302-14.
[7] Albouy O 2009 The isotropic lines of $\mathbb{Z}_d^2$ *J. Phys. A: Math. Theor.* **42** 072001.
[8] Planat M, Baboin A C and Saniga M 2008 Multi-line geometry of qubit-qutrit and higher order Pauli operators *Int. J. Theor. Phys.* **47** 1127-35.
[9] Planat M and Baboin A C 2007 Qudits of composite dimension, mutually unbiased bases and projective ring geometry *J. Phys. A: Math. Theor.* **40** F1-F8.
[10] Thas K 2009 The geometry of generalized Pauli operators of N-qudit Hilbert space, and an application to MUBs *Europh. Lett. EPL* **86** 6005-7.
[11] Havlicek H, Odehnal B and Saniga M 2009 Factor-group-generated polar spaces and (multi-)qudits *SIGMA* **5** 096, 15 pages.
[12] Havlicek H, Odehnal B and Saniga M 2010 Möbius pair of simplices and commuting Pauli operators Math. Panonica **21** 1-14.
[13] Sengupta A R 2009 Finite geometries with qubit operators *Infin. Dimens. Anal., Quant. Probab. and Rela. Top.* **12** 359-366.
[14] Rau A R P 2009 Mapping two-qubit operators onto projective geometries *Phys. Rev. A* **79** 042323-28.
[15] Shaw R 1989 Finite geometries and Clifford algebras *J. Math. Phys.* **30** 1971-1984.
[16] Shaw R 1990 Finite geometries and Clifford algebras II *J. Math. Phys.* **31** 1315-1324.
[17] Kibler M 2009 An angular momentum approach to quadratic Fourier transform, Hadamard matrices, Gauss sums, mutually unbiased bases, the unitary group and the Pauli group *J. Phys. A: Math. Theor.* **42** 353001.
[18] Tolar J and Chadzitaskos G 2009 Feyman's path integral and mutually unbiased bases *J. Phys. A: Math. Theor.* **42** 245306.1-11.
[19] Vourdas A and Banderier C 2010 Symplectic transformations and quantum tomography in finite quantum systems *J. Phys. A: Math. Theor.* **43** 042001-10.
[20] Novotný P and Hrivnàk 2005 On orbits of the ring Z_n^M under action of the group $SL(m, Z_n)$ *Acta Polytechnica* **45** 39-43.
[21] Bosma W, Cannon J and Playoust C 1997 The Magma algebra system. I. The user language *J. Symb. Comp.* **24** 235-265.
[22] Planat M 2010 Pauli graphs when the Hilbert space dimension contains a square: why the Dedekind psi function? Preprint 1009.3958 [math.ph].
[23] Planat M, Rosu H and Perrine S 2006 A survey of finite algebraic grometric structures underlying mutually unbiased quantum measurements. *Found. Phys.* **36** 1662.
[24] Brierley S and Weigert S 2009 Constructing mutually unbiased bases in dimension six *Phys. Rev. A* **79** 052316.
[25] Planat M 2010 Three-qubit entangled embeddings of CPT and Dirac groups within E_8 Weyl group *Int. J. Theor. Phys.* **49** 1044-54.
[26] Cameron P J 1991 Projective and polar spaces, http://www.maths.qmul.ac.uk/~pjc/pps/.
[27] Levay P, Saniga M, Vrna P and Pracna P 2009 Black hole entropy and finite geometry *Phys. Rev. D* **79** 084036.1-12.

Journal of Physics: Conference Series **284** (2011) 012051 doi:10.1088/1742-6596/284/1/012051

Optical waveguide Hamiltonians leading to step-2 difference equations

Juvenal Rueda-Paz[1] **and Kurt Bernardo Wolf**[1]

[1] Instituto de Ciencias Físicas, Universidad Nacional Autónoma de México, Av. Universidad s/n, Cuernavaca, Morelos 62251, Mexico

E-mail: bwolf@fis.unam.mx

Abstract.

We examine the evolution of an N-point signal produced and sensed at finite arrays of points transverse to a planar waveguide, within the framework of the finite quantization of geometric optics. In contradistinction to the common mechanical Hamiltonians (kinetic plus potential energy terms), the classical waveguide Hamiltonian is the square root of a difference of squares of the refractive index profile minus the optical momentum. The finitely quantized model requires the solution of the square eigenvalue and eigenfunction problem, which leads to a step-two difference equation that contains two solutions and two signs of energy. We find the proper linear combinations to fit the Kravchuk functions of the finite oscillator model.

1. Introduction

Most studies in symmetry, supersymmetry, and separation of variables, pertain Hamiltonian systems of mechanical nature, *i.e.*, whose Hamiltonian functions separate kinetic from potential energy terms,

$$H^{\rm mec}(\vec{x},\vec{p}) = |\vec{p}|^2/2M + V(\vec{x}), \tag{1}$$

where M is the mass. Schrödinger quantization of this classical Hamiltonian provides the well-known Schrödinger second-order differential equation, whose eigenvalues are the energies and the stationary states are the eigenvectors of such systems. Waveguides with finite sensor arrays provide Hamiltonian functions which are of a different structure, and whose quantization is proposed to be distinct from the Schrödinger one, where now the geometric optical system is associated to a model where the observables of position, momentum and 'energy' have a discrete, finite set of eigenvalues. This we call 'so(3)-' or 'finite quantization.' This is taylored for parallel analysis of finite signals and the processing of pixellated images.

We set up the Hamilton equations of geometric optics starting from the two similar right triangles in Figure 1. Let light rays be abstracted as lines in 3-space $\vec{x}(s)$, $s \in \Re$; and let their tangent vectors be $\vec{p}(s)$. In [1, Chap. 1] we prove succintly that, postulating the change in ray direction to be in linear response to the gradient of the *refractive index* $n(\vec{x})$, as $\mathrm{d}\vec{p} = \vec{\nabla} n(\vec{x})\,\mathrm{d}s$, leads to the restriction $|\vec{p}| = n(\vec{x})$. Then, writing

$$\vec{p} = \begin{pmatrix} \mathbf{p} \\ p_z \end{pmatrix}, \qquad \vec{x} = \begin{pmatrix} \mathbf{x} \\ z \end{pmatrix}, \tag{2}$$

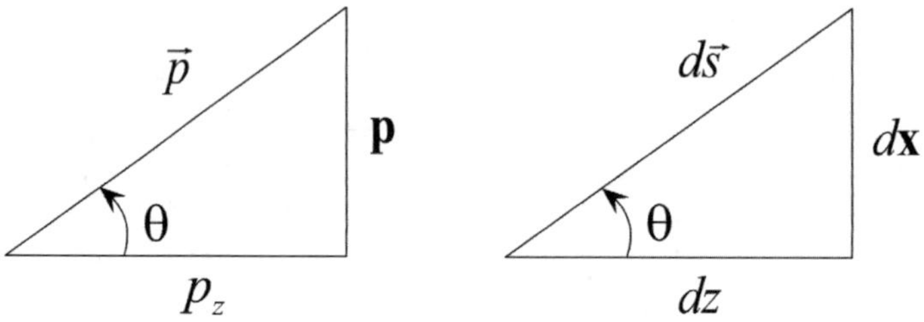

Figure 1. *Left*: The ray direction vector $\vec{p}$ decomposed into its component $\mathbf{p}$ along the screen $z = 0$, and p_z normal to it; the length of $\vec{p}$ is the refractive index $|\vec{p}| = n(\vec{x})$. *Right*: the differential length $\mathrm{d}\vec{s}$ along the ray, decomposed into the differential $\mathrm{d}\mathbf{x}$ on the screen, and $\mathrm{d}z$ normal to it; $\mathrm{d}z$ is the differential evolution parameter. The two triangles are similar. The angle θ opens between the optical z-axis and the ray.

the similarity of the triangles in Figure 1 implies that

$$\mathrm{d}\mathbf{x}/\mathrm{d}z = \mathbf{p}/p_z, \quad \mathrm{d}\mathbf{p}/\mathrm{d}z = \frac{\mathrm{d}\mathbf{p}/\mathrm{d}s}{\mathrm{d}z/\mathrm{d}s} = n\,\nabla n/p_z, \quad \mathrm{d}p_z^2/\mathrm{d}z = \partial n^2/\partial z. \tag{3}$$

These equations fit into the Hamiltonian form

$$\frac{\mathrm{d}\mathbf{x}}{\mathrm{d}z} = \frac{\partial h}{\partial \mathbf{p}}, \quad \frac{\mathrm{d}\mathbf{p}}{\mathrm{d}z} = -\frac{\partial h}{\partial \mathbf{x}}, \quad \frac{\mathrm{d}p_z}{\mathrm{d}z} = -\frac{\partial h}{\partial z}, \tag{4}$$

with the Hamiltonian function

$$h(\mathbf{x}, \mathbf{p}, z) := -\sigma\sqrt{n(\mathbf{x}, z)^2 - |\mathbf{p}|^2} = -p_z = -n(\vec{x})\cos\theta, \tag{5}$$

where $\sigma = +$ applies to *forward* rays (in the $+z$ direction), $\sigma = -$ to *backward* rays (in the $-z$ direction), and θ is the angle at the *screen* $z = 0$ between the ray and the *optical* axis $+z$. (The sign assignment can be seen easily for free propagation in a medium $n =$ constant.) The square-root form of the Hamiltonian (5) is characteristic of optics, and we shall particularly devote our attention to the case when n is independent of z —namely, when the optical medium is a *waveguide*, so its refractive index is $n(\mathbf{x})$.

We point to the resemblance of (5) with the total relativistic energy of a freely moving mass point,

$$E = c\sqrt{M^2c^2 + |\mathbf{p}|^2} = Mc^2 + \frac{1}{2M}|\mathbf{p}|^2 - \frac{1}{8M^3c^2}|\mathbf{p}|^4 + \cdots. \tag{6}$$

The difference between both is of course the sign of the momentum term $|\vec{p}|^2$ in the square root: in (6) the evolution parameter is time t, and the metric of $(\mathbf{x}, t)$ and $(\mathbf{p}, E)$ is Lorentzian, while in (5) the metric of $(\mathbf{x}, z)$ and $(\mathbf{p}, p_z)$ is Euclidean. The Schrödinger quantization of the square of (6) leads to the Klein-Gordon equation as a relativistic quantum mechanical model; it is the Dirac quantization however, which posits a *linear* matrix equation that is most productive to describe Nature. Similarly, while the wavization of the square of (5) produces the Helmholtz wave equation, we feel that its matrix square-root form can yield interesting results provided with a proper interpretation as a model of finite optics.

In Section 2 we introduce the specific family of waveguides to be treated here, whose refractive index profile is elliptic, arriving at their symmetry expressed in algebraic terms through Poisson brackets. Then in Section 3 we quantize the classical system to a finite system given by a Hamiltonian matrix defined by the square root of a self-adjoint matrix. The signs of the eigenvalues and eigenvector problem that stems from a step-two difference equation are analyzed and resolved in Section 4. The concluding Section 5 refers to results on the evolution of finite Kravchuk coherent states in such waveguides, and the possible use of aberration expansions in modeling nonlinear optical media.

2. Waveguide Hamiltonians

In a two-dimensional (planar) geometric optical system $\vec{x} = (x, z)^\top$ where the index of refraction is $n(x, z)$, the x-coordinate measures the screen $z = 0$, and its z-coordinate is the independent evolution parameter along the optical axis of the device. In this paper we shall consider media that are invariant under z-translations, *i.e.* waveguides, where $n \equiv n(x)$; and moreover, for reasons of parametric proximity, we propose waveguides whose refractive index profile be *elliptic*, namely

$$n^{\nu,\mu}(x) := +\sqrt{\nu^2 - \mu^2 x^2}, \quad \nu \geq 1,\ \mu \geq 0,\ |x| \leq \nu/\mu. \tag{7}$$

At the center of the guide $n(0) = \nu$, while $1/\mu$ estimates the width of the guide. Although physically $n \geq 1$ this is not necessary for the analysis, so we consider the guide to extend between $\pm\nu/\mu$.

The Hamiltonian (5) of the waveguide medium (7) is

$$h^{\nu,\mu}(x, p) = -\sigma\sqrt{\nu^2 - (p^2 + \mu^2 x^2)} \tag{8}$$

$$= -\sigma\nu + \sigma\frac{p^2 + \mu^2 x^2}{2\nu} + \sigma\frac{(p^2 + \mu^2 x^2)^2}{8\nu^3} + \cdots, \tag{9}$$

where we have expanded the square root into a series of powers of the oscillator Hamiltonian $k_\mu := \frac{1}{2}(p^2 + \mu^2 x^2)$. In the waveguide (8), the evolution of phase space along z will be that of an incompressible fluid, flowing along the ellipses $p^2 + \mu^2 x^2 = \text{constant} < \nu^2$. The trajectory of lowest energy $h = -\nu$ is a ray along the $+z$ axis —at the center of phase space, and that of highest energy ν is along the $-z$ axis. Two particlar cases of waveguide with $\nu = 1$ will serve as reference:

$$\text{circular profile} \quad \mu = 1, \quad k_1 = \tfrac{1}{2}(p^2 + x^2), \tag{10}$$

$$\text{free particle} \quad \mu = 0, \quad k_0 = \tfrac{1}{2}p^2. \tag{11}$$

Written with Poisson brackets, the Hamilton equations are

$$\{k_\mu, x\} = -p, \quad \{k_\mu, p\} = \mu^2 x, \qquad \{x, p\} = 1. \tag{12}$$

The four quantities $\{k_\mu, x, p, 1\}$ form, for $\mu = 1$, the oscillator Lie algebra osc, while for $\mu = 0$ they form the nilpotent algebra free.

3. Finite $\mathsf{so}(3)$-quantization

Finite quantization involves a *deformation* (or 'pre-contraction') of the Lie algebra that contains the classical system [2, 3, 4]. Poisson brackets between the classical phase space observables of position and momentum (x, p) are replaced by i times the commutator Lie brackets between operators $(\mathcal{X}, \mathcal{P})$, closing with a *pseudo*-Hamiltonian $\mathcal{K}_\mu$ —for which we assume no specific form— that obey the same geometry and dynamics. Corresponding to (12), the two Hamilton equations are:

$$[\mathcal{K}_\mu, \mathcal{X}] = -\mathrm{i}\,\mathcal{P}, \qquad [\mathcal{K}_\mu, \mathcal{P}] = \mathrm{i}\,\mu^2\,\mathcal{X}. \tag{13}$$

The deformation is contained in the third —*non-standard*— commutator

$$[\mathcal{X}, \mathcal{P}] = -\mathrm{i}\,\mathcal{K}_\mu. \tag{14}$$

For $\mu = 1$, the three generators $\{\mathcal{X}, \mathcal{P}, \mathcal{K}_\mu\}$ close into the 'rotation' Lie algebra $\mathsf{so}(3)$, which we shall use as reference for several considerations below. For $\mu = 0$ these generators close into the Euclidean algebra $\mathsf{iso}(2)$.

Among the realizations of $\mathsf{so}(3)$ we choose the self-adjoint irreducible representation matrices in the eigenspaces of the Casimir operator [5, 6],

$$\mathcal{C} := \mathcal{X}^2 + \mathcal{P}^2 + \mathcal{K}^2 = j(j+1)\, 1, \tag{15}$$

whose eigenvalues (numbered by j) determine the vector space to be of dimension $N = 2j+1$. Thus we represent the classical quantities by the self-adjoint matrices that belong to the representation j of $\mathsf{so}(3)$: $x \sim \mathbf{X} = \|X_{m,m'}\|$ will be diagonal, $p \sim \mathbf{P} = \|P_{m,m'}\|$ skew-symmetric, and $h + \text{const.} \sim \mathbf{K} = \|K_{m,m'}\|$ symmetric, with elements [6]

$$\begin{aligned} X_{m,m'} &= m\,\delta_{m,m'}, \qquad m, m' \in \{-j, -j+1, \ldots, j\}, && (16)\\ P_{m,m'} &= -\mathrm{i}\tfrac{1}{2}\sqrt{(j-m)(j+m+1)}\,\delta_{m+1,m'} \\ &\quad + \mathrm{i}\tfrac{1}{2}\sqrt{(j+m)(j-m+1)}\,\delta_{m-1,m'}, && (17)\\ K_{m,m'} &= \tfrac{1}{2}\sqrt{(j-m)(j+m+1)}\,\delta_{m+1,m'} \\ &\quad + \tfrac{1}{2}\sqrt{(j+m)(j-m+1)}\,\delta_{m-1,m'}, && (18) \end{aligned}$$

and their sum of squares (15) is $\mathbf{C} := \mathbf{X}^2 + \mathbf{P}^2 + \mathbf{K}^2 = j(j+1)\,\mathbf{1}$. The spectrum of the position operator $\mathbf{X}$ is thus $x_m \equiv m \in \{-j, -j+1, \ldots, j\}$; the eigenvalues $\{\varpi_r\}_{r=-j}^{j}$ of the momentum $\mathbf{P}$, and $\{\kappa_n\}_{n=-j}^{j}$ of the *pseudo-energy* operator $\mathbf{K}$, take values in the same range. Since we consider important to have sensor arrays that include one point $x_0 = 0$ at the center, we consider only integer j's, to the exclusion of the $\mathsf{su}(2)$ spin representations.

The elliptic-profile waveguide Hamiltonian $h^{\nu,\mu}(x,p)$ in (8) will be thus finitely-quantized to an $N \times N$ hermitian matrix which, taking (15) into account, provides the equation for its eigenvalues and eigenvectors:

$$\begin{aligned} \mathbf{H}^{\nu,\mu} &:= -\sqrt{[(\nu^2-1)\mathbf{C} - (\mu^2-1)\mathbf{X}^2] + \mathbf{K}^2}, && (19)\\ \mathbf{H}^{\nu,\mu}\,\Psi_\eta^{\nu,\mu}(m) &= \eta^{\nu,\mu}\,\Psi_\eta^{\nu,\mu}(m), \quad m \in \{-j, -j+1, \ldots, j\}, && (20) \end{aligned}$$

where the matrix between the square brackets in (19) is diagonal. We shall call $\eta^{\nu,\mu}$ the *energy* of the state $\Psi_\eta^{\nu,\mu}(m)$, corresponding to the value of the classical Hamiltonian (8). In the special case $\nu = 1 = \mu$ we have $\mathbf{H}^{1,1} = -\sqrt{\mathbf{K}^2}$, and we refer to this as a *Kravchuk* guide, since then the eigenvalues are $\eta_n^{1,1} = \pm\kappa_n \in \{-j, -j+1, \ldots, j\}$, and the eigenfunctions $\Psi_\eta^{1,1}(m)$ are the Kravchuk functions studied in Refs. [4, 7, 8, 9], which are finite deformations of the harmonic oscillator (Hermite-Gauss) wavefunctions.

When $\mu = 0$, the free Hamiltonian (19) has eigenvalues $\eta_r^{\nu,0} = \sqrt{[\nu^2 j(j+1) - \varpi_r^2]}$, with $\varpi_r = r$ having the same range as κ_n. We restrict the waveguide parameters to $0 \le \mu \le \nu \ge 1$ to keep the eigenvalues of the radicand positive. Since the square root of a matrix is not uniquely defined, we propose to use its square and solve (20) through

$$(\mathbf{H}^{\nu,\mu})^2\,\Psi_\eta(m) = \lambda^{\nu,\mu}\,\Psi_\eta(m), \quad \lambda^{\nu,\mu} := (\eta^{\nu,\mu})^2, \tag{21}$$

knowing that the eigenvectors of $\mathbf{H}$ and of its square are the same. We thus diagonalize (21) by its eigenvector matrix,

$$\begin{aligned} \mathbf{\Psi}^\dagger\,(\mathbf{H}^{\nu,\mu})^2\mathbf{\Psi} &= \mathbf{\Lambda}^{\nu,\mu}, \qquad \mathbf{\Psi} = \|\Psi_{\eta_n}(m)\| \\ \mathbf{\Lambda}^{\nu,\mu} &= \operatorname{diag}(\lambda_{\max}^{\nu,\mu}, \lambda_{\max-1}^{\nu,\mu}, \ldots \lambda_{\min+1}^{\nu,\mu}, \lambda_{\min}^{\nu,\mu}), \end{aligned} \tag{22}$$

where the $N = 2j+1$ eigenvalues $\{\lambda_n^{\nu,\mu}\}$ are ordered from maximum to minimum, observing that away from $\mu = 0, 1$, they are non-degenerate, although j pairs are very close in various ranges of μ. In Figure 2 (left) we show the computed eigenvalues $\{\lambda_n^{\nu,\mu}\}$ for $\nu = 1$ and $\nu = \sqrt{2}$, in guides with $\mu = \nu$, and for free $\mu = 0$.

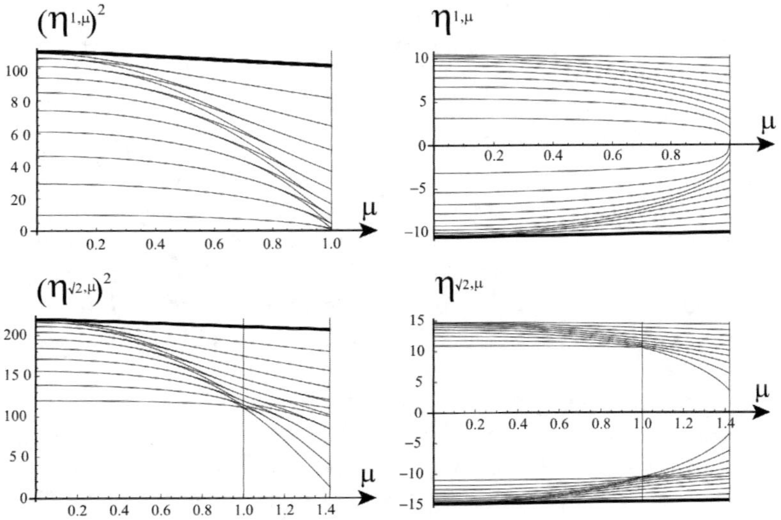

Figure 2. *Left column*: The squared eigenvalues $\{\eta^2\}$ of $(\mathbf{H}^{\nu,\mu})^2$. *Right column*: eigenvalues $\{\eta\}$ of the root $\mathbf{H}^{\nu,\mu}$ with the sign asignments detailed in the text. *Top row*: The guide $\nu = 1$. *Bottom row*: A guide with $\nu = \sqrt{2}$. Heavy lines mark the ground state eigenvalues η_0^2 and η_0. Throughout, $j = 10$ ($N = 21$ points) and $0 \le \mu \le \nu$.

4. Eigenvalue signs and eigenvectors

We now proceed to analyze the sign of the square roots $\sqrt{(\eta^{\nu,\mu})^2}$. Notice that the ground state of a waveguide has the lowest energy eigenvalue $\eta_{\rm gr}^{\nu,\mu} < 0$ of $\mathbf{H}^{\nu,\mu}$, and this appears as the highest $\lambda_{\rm max}^{\nu,\mu} = (\eta_{\rm gr}^{\nu,\mu})^2 > 0$ of $(\mathbf{H}^{\nu,\mu})^2$. In the Kravchuk case this eigenvalue is $\eta_{-j}^{1,1} = -j$, whose square j^2 is degenerate with that of the highest energy that the waveguide can carry, $\eta_j^{1,1} = j$. In the other extreme case $\mu = 0$, the ground state has zero transverse momentum $\varpi_0 = 0$, so $\eta_{\rm gr}^{\nu,0} = -\nu\sqrt{j(j+1)}$; this state also has the highest eigenvalue $\lambda_{\rm max}^{\nu,\mu} = (\eta_{\rm gr}^{\nu,\mu})^2 > 0$ under $(\mathbf{H}^{\nu,\mu})^2$, and is nondegenerate. Since eigenvalues do not cross, we conclude that for all $0 \le \mu < 1$ the *highest* eigenvalue of $(\mathbf{H}^{\nu,\mu})^2$ corresponds to the ground state. We thus take the square root of this eigenvalue with a negative sign: $\eta_{\rm gr} := -\sqrt{\lambda_{\rm max}}$. On the other hand, the state of highest energy that the waveguide can carry, $\eta_{\rm top}^{\nu,\mu}$, is positive; and this corresponds with the *next-highest* square eigenvalue, so we should opt for the positive sign: $\eta_{\rm top}^{\nu,\mu} := +\sqrt{\lambda_{\rm max-1}^{\nu,\mu}}$. Again, since eigenvalues do not cross, $\lambda_{\rm max-1}^{\nu,\mu} < \lambda_{\rm max}^{\nu,\mu}$. Continuing in this way with the $2j+1$ eigenvalues $\{\lambda_n^{\nu,\mu}\}$ of $(\mathbf{H}^{\nu,\mu})^2$ from top to bottom, we conclude that $\eta_{{\rm gr}+n}^{\nu,\mu} = -\sqrt{\lambda_{{\rm max}-2n}^{\nu,\mu}}$ and $\eta_{{\rm top}-n}^{\nu,\mu} = +\sqrt{\lambda_{{\rm max}-2n-1}^{\nu,\mu}}$, for $n \in \{0, 1, \ldots, j\}$. In Figure 2 (right) we show the η's with this alternating sign selection.

To compute the eigenvectors $\Psi_\eta \equiv \Psi_\eta^{\nu,\mu}$ of $(\mathbf{H}^{\nu,\mu})^2$ in (21) we must solve a *difference* equation [10], which is of *step-two*:

$$\begin{aligned}
&\tfrac{1}{4}\sqrt{(j-m-1)(j-m)(j+m+1)(j+m+2)}\,\Psi_\eta(m+2)\\
&+ [j(j+1)(\nu^2-\tfrac{1}{2}) - m^2(\mu^2-\tfrac{1}{2}) - \eta^2]\,\Psi_\eta(m)\\
&+ \tfrac{1}{4}\sqrt{(j+m-1)(j+m)(j-m+1)(j-m+2)}\,\Psi_\eta(m-2) = 0,
\end{aligned} \tag{23}$$

where we notice that both $\Psi_\eta(m)$ and $\Psi_{-\eta}(m)$ are solutions to the same equation, as well as any linear combination of the two. Replacing $m \leftrightarrow -m$ in (23) returns the same equation for $\Psi_{\pm\eta}(-m)$, so the solutions will have definite parity. Equation (23) stands in effect for a *pair* of

Journal of Physics: Conference Series **284** (2011) 012051 doi:10.1088/1742-6596/284/1/012051

difference equations. Assume first that j is odd; then, the values of one solution, $\Psi^{\rm e}_\eta(m)$, can be determined at all the even points $x_m \equiv m$ by setting $\Psi^{\rm e}_\eta(\pm[j+1]) = 0$, using the recurrence (23) to determine its values $\Psi^{\rm e}_\eta(m)$ for all other even $|m| \le j-1$, while its values at all odd m's are zero. A second solution, $\Psi^{\rm e}_\eta(m)$, nonzero at all odd points m stems from setting $\Psi^{\rm o}_\eta(\pm[j+2]) = 0$ and then using the same (23) to determine its values at all odd points $|m| \le j$, and zero at the even m's. (When j is even, a corresponding separation into two independent solutions is made.)

Since both $\Psi^{\rm e}_\eta(m)$ and $\Psi^{\rm o}_\eta(m)$ have zeros at alternating points, they appear as 'porcupine' signals, such as shown in Figure 3 (*top row*). A question thus arises on the relation between these two orthogonal solutions to (23) and the two solutions $\Psi_{\pm\eta}(m)$ mentioned earlier. What we expect is to reproduce ground states with a Gaussian shape —in particular the ground Kravchuk function in a Kravchuk guide [2]; and for $\mu = 0$ a state resembling a free wave. In all cases, definite parity is guaranteed by (23).

For general ν, μ it does not seem possible to find a known discrete special function that will solve (23) 'exactly,' so we must resort to numerical computation to plot results. In Figure 3 (*rows 2–4*) the normalized sum and difference of the first three lowest energy ($\eta < 0$) states (24), and the three highest ($\eta > 0$) states (25), in the range $0 \le \mu \le \nu = 1$. The resulting states

$$\Psi_\eta(m) = \frac{1}{\sqrt{2}}[\Psi^{\rm e}_\eta(m) + \Psi^{\rm o}_\eta(m)], \tag{24}$$

$$\Psi_{-\eta}(m) = \frac{1}{\sqrt{2}}[\Psi^{\rm e}_\eta(m) - \Psi^{\rm o}_\eta(m)], \tag{25}$$

have recognizable shapes and definite energy eigenvalues.

The $\mu \to 1$ limit to the 'Kravchuk' spectrum $\{-j, -j+1, \ldots, j\}$ thus defines the 'proper' square root of the diagonal matrix $\mathbf{\Lambda}^{\nu,\mu}$ in (22),

$$\widetilde{\mathbf{H}}^{\nu,\mu} := \mathrm{diag}\,(\eta_{\rm gr},\, \eta_{\rm gr+1},\, \ldots,\, \eta_{\rm top-1},\, \eta_{\rm top}). \tag{26}$$

The Hamiltonian matrix in (19) is finally obtained through de-diagonalizing (26) by means of the transpose of the eigenvector matrix,

$$\mathbf{H}^{\nu,\mu} = \mathbf{\Psi}\, \widetilde{\mathbf{H}}^{\nu,\mu}\, \mathbf{\Psi}^\dagger. \tag{27}$$

5. Concluding remarks

The z-evolution along the waveguide is generated by the proper Hamiltonian matrix (27), exponentiated a 1-parameter group of unitary $N \times N$ matrices

$$\mathbf{U}^{\nu,\mu}(z) := \exp(\mathrm{i}z\,\mathbf{H}^{\nu,\mu}) \in \mathsf{U}(N), \tag{28}$$

for $z \in \Re$. This is a line within the N^2-dimensional compact manifold of $\mathsf{U}(N)$ that will in general not close. In Ref. [10] we computed the z-evolution of the finite Kravchuk coherent states in a Kravchuk guide, and plotted the result using the $\mathsf{so}(3)$ Wigner function [11] to ascertain that one reproduces the usual *forward* $z > 0$ circulation of phase space. The movement of *backward* rays is obtained simply by letting $z < 0$, which is equivalent to exchanging the signs of all energies in (26). Also, we have let the finite coherent states evolve in a non-Kravchuk guide with $\nu = \sqrt{2}$ and $\mu = \frac{1}{2}\sqrt{2}$ to observe the nonlinear aberration of the Wigner function that results from classical phase space points moving on concentric ellipses $p^2 + \frac{1}{2}x^2 = \text{constant} < 2$ at different velocities. Presently we are investigating the behavior of the coherent states in waveguides whose refractive index is different from the elliptic one in (7), such as a rectangular index profile with zeros at the endpoints, and an intermediate family of 'smooth' profiles.

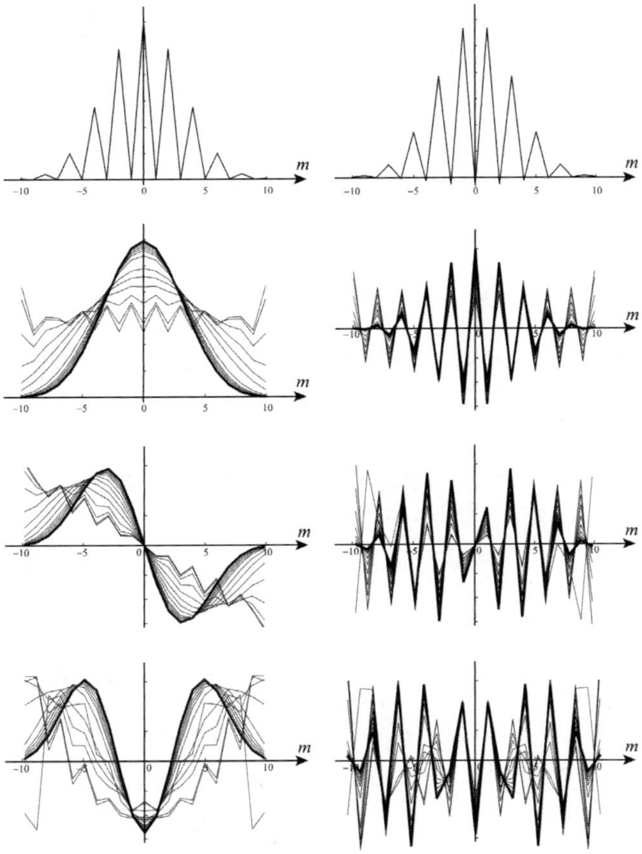

Figure 3. Solutions $\Psi_\eta(m)$ to the step-two difference equation (23), for $N = 21$ points ($|m| \leq j = 10$); lines connect their values at the integer m's. *Top row*: The ground 'porcupine' states, $\Psi^{\rm e}_{\rm gr}(m)$ and $\Psi^{\rm o}_{\rm gr}(m)$, in a 'Kravchuk' guide $\nu = 1 = \mu$, corresponding to the largest eigenvalue η^2 in Fig. 2, and yielding both the ground and top energy states. *Second row*: The sum and difference (24)–(25) of the previous porcupines produces the ground, and top states of the Kravchuk guide —indicated by a heavy line; with light lines we show the ground and top states of guides with $\mu = 0.95, 0.9, \ldots, 0.8, 0.7, \ldots, 0.0$. *Third row*: The same as the previous row, but for the next-to-largest eigenvalue η^2 of Fig. 2, corresponding to the next-to-lowest ground and next-to-top states. *Fourth row*: Similarly, for the following lowest and highest states.

In this paper our interest was to examine optical Hamiltonian operators that are defined by the square root of a positive, self-adjoint and otherwise well known operator, which is a problem seldom addressed in mechanics. The resulting step-two difference equation separates solutions into two 'porcupine' components that seem to be inseparable from the sign ambiguity of the square root operation. Since several group-theoretic developments exist which (rightly or wrongly) include the square root of a positive operator —for purposes of normalization, usually—, this operation may shown to have some hidden difficulties.

As suggested by the aberration expansion (9) for the geometric optical Hamiltonian, we also considered the perturbative description of waveguide evolution as a sum of powers of harmonic oscillator Hamiltonians. The main difficulty in using this expansion for finite waveguide models is that the matrix series does not converge: in the Kravchuk $\mathsf{so}(3)$ case the maximum element of $\mathbf{H}^{\rm osc} := \mathbf{P}^2 + \mathbf{X}^2 = \mathbf{C} - \mathbf{K}^2$ is $j(j+1) - j^2 = j$, and hence a bound for the maximum of $(\mathbf{H}^{\rm osc})^n$ is $j^n(2j+1)^{n-1}$; meanwhile, the coefficient of this term in the $\sqrt{(1-x)}$ series is $(2n-3)!!/(2n)!!$. Yet, the first two powers can be used to model paraxial and nonlinear Kerr media [12, 13, 14], as done in Ref. [15]. The purpose of our exploration of discrete and finite systems based on representations of compact groups such as $\mathsf{so}(3)$ and $\mathsf{so}(4)$, and also of discrete infinite systems [4], is to understand finite signal analysis and pixellated image processing in terms of the symmetries behind their geometric optical realizations.

Acknowledgments
We acknowledge the support of the *Óptica Matemática* projects, DGAPA-UNAM IN-105008 and SEP-CONACYT 79899, and we thank Guillermo Krötzsch (ICF-UNAM) for his careful help with the graphics.

References

[1] K.B. Wolf, *Geometric Optics on Phase Space* (Springer-Verlag, Heidelberg, 2004).
[2] N. M. Atakishiyev, G. S. Pogosyan and K. B. Wolf, "Finite models of the oscillator," *Phys. Part. Nuclei* **36**, 521–555 (2005).
[3] C.A. Muñoz, J. Rueda-Paz and K.B. Wolf, "Discrete repulsive oscillator wavefunctions," *J. Phys. A* **42**, art. 485210 (12 p.) (2009).
[4] K.B. Wolf, "Discrete systems and signals on phase space," *Appl. Math. & Information Science* **4**, 141–181 (2010).
[5] R. Gilmore, *Lie Groups, Lie Algebras, and Some of their Applications* (Wiley-Interscience, New York, 1978).
[6] L.C. Biedenharn and J.D. Louck, *Angular Momentum in Quantum Physics*, Encyclopedia of Mathematics and Its Applications, Vol. 8 (Addison-Wesley Publ. Co., Reading, Mass., 1981).
[7] M. Krawtchouk, "Sur une géneralization des polinômes d'Hermite," *C.R. Acad. Sci. Paris* **189**, 620-622 (1929).
[8] N. M. Atakishiyev and K. B. Wolf, "Fractional Fourier-Kravchuk transform," *J. Opt. Soc. Am. A* **14**, 1467–1477 (1997).
[9] N.M. Atakishiyev, G.S. Pogosyan and K.B. Wolf, "Contraction of the finite one-dimensional oscillator," *Int. J. Mod. Phys. A* **18**, 317–327 (2003).
[10] J. Rueda-Paz and K.B. Wolf , "Finite signals in planar waveguides" *J. Opt. Soc. Am. A* **submitted**, ().
[11] N. M. Atakishiyev, S. M. Chumakov and K. B. Wolf, "Wigner distribution function for finite systems," *J. Math. Phys.* **39**, 6247–6261 (1998).
[12] R. Tanaś, In: *Coherence and Quantum Optics V*, L. Mandel and E. Wolf Eds. (New York: Plenum, 1984), p. 645.
[13] G.J. Milburn, "Quantum and classical Liouville dynamics of the anharmonic oscillator," *Phys. Rev. A* **33**, 674–685 (1986).
[14] G.J. Milburn and C.A. Holmes, "Quantum coherence and classical chaos in a pulsed parametric oscillator with Kerr nonlinearity," *Phys. Rev. A* **44**, 4704–4711 (1991).
[15] S.M. Chumakov, A. Frank and K.B. Wolf, "Finite Kerr medium: Macroscopic quantum superposition states and Wigner functions on the sphere", *Phys. Rev. A* **60**, 1817–1823 (1999).

Journal of Physics: Conference Series **284** (2011) 012052 doi:10.1088/1742-6596/284/1/012052

Finsler metrics and CPT

Sarben Sarkar

King's College London, Department of Physics, Strand, London WC2R 2LS, UK.

E-mail: sarben.sarkar@kcl.ac.uk

Abstract. The role of Finsler-like metrics in situations where Lorentz symmetry breaking and also CPT violation are discussed. Various physical instances of such metrics both in quantum gravity and analogue systems are discussed. Both differences and similarities between the cases will be emphasised. In particular the medium of D-particles that arise in string theory will be examined. In this case the breaking of Lorentz invariance, at the level of quantum fluctuations, together with concomitant CPT in certain situations will be analysed. In particular it will be shown correlations for neutral meson pairs will be modified and a new contribution to baryogenesis will appear.

1. Introduction

Paul Finsler in 1918 wrote a thesis, "ber Kurven und Flchen in Allgemeiner Rmen", which is a generalisation of Riemannian manifolds. It remained a topic in the province of pure mathematics until quite recently. However, within physics, it has been realised more recently that a framework more general than pseudo-Riemannian geometry is required if causal relations, which are more complicated that special relativity, are involved. It is interesting that this realisation has sprung up in somewhat diverse contexts. We shall be primarily interested in the particular context known as space-time D foam, where it appears in a more generalised way. In general, the mathematical structure that is unveiled in physical applications, is different from the original Finsler theory in that the associated norm is not positive definite; hence, following physics tradition, it can be dubbed pseudo-Finsler. However, in the foamy context, a stochastic aspect to the norm is introduced.

At each point x in a (standard) Finsler manifold M [1] there is a norm on the tangent space TM_x which has not been induced by an inner product. Rather the norm itself induces an inner product. However this inner product is not parametrised by points in M but by directions in TM. Specifically a Finsler norm F on TM is a smooth mapping on $TM \setminus \{0\} (\equiv \cup \{T_x M \setminus \{0\} : x \in M\})$. Now $F|_{T_xM}: TM \rightarrow [0,\infty)$is such that $F|_{T_xM} (\equiv f_x)$ is homogeneous of degree 1 and, for all y in $TM \setminus \{0\}$, a form $g_y : T_xM \times T_xM \rightarrow R$ can be defined by

$$g_y : (u,v) \rightarrow \frac{1}{2} \frac{\partial^2 \left\{ \left[f_x \left(y + pu + qv \right) \right]^2 \right\}}{\partial p \partial q} |_{p=q=0}.$$

The form g_y is bilinear and required to be positive defnite. A Riemannian manifold (M,G) ,G being the metric, trivially can be recast as a Finsler manifold with

$$F(x,y) = \sqrt{G_x(y,y)}.$$

Here x is a point in M.

1.1. *The stochastic Finsler metric*

String theory currently attracts much attention in that quantum states of gravitons are part of its spectrum together with other states which are required for the standard model of particle physics. The discovery of new solitonic structures in superstring theory [2] has dramatically changed the understanding of target space structure. These new non-perturbative objects are known as D-branes and their inclusion leads to a scattering picture of space-time fluctuations. Typically open strings interact with D-particles and satisfy Dirichlet boundary conditions when attached to them. Closed strings may be cut by D-particles. D-particles are allowed in certain string theories such as bosonic, type IIA and type I and here we will here consider them to be present in string theories of phenomenological interest. Furthermore even when elementary D particles cannot exist consistently there can be effective D-particles formed by the compactification of higher dimensional D branes. Moreover D particles are non-perturbative constructions since their masses are inversely proportional to the the string coupling g_s. The study of D-brane dynamics has been made possible by Polchinski's realization that such solitonic string backgrounds can be described in a conformally invariant way in terms of world sheets with boundaries. On these boundaries Dirichlet boundary conditions for the collective target-space coordinates of the soliton are imposed. Heuristically, when low energy matter given by a closed (or open) string propagating in a $(D+1)$-dimensional space-time collides with a very massive D-particle embedded in this space-time, the D-particle recoils as a result. Since there are no rigid bodies in general relativity the recoil fluctuations of the brane and their effectively stochastic back-reaction on space-time cannot be neglected.Based on these considerations, a model for a supersymmetric space-time foam has been suggested in [3] . The model is based on parallel brane worlds (with three spatial large dimensions), moving in a bulk space time which contains a "gas" of D-particles. The number of parallel branes used is dictated by the requirements of target-space supersymmetry in the limit of zero-velocity branes. One of these branes represents allegedly our Observable Universe. As the brane moves in the bulk space, D-particles cross the brane in a random way. From the point of view of an observer in the brane the crossing D-particles will appear as flashing on and off space-time defects, that is microscopic space-time fluctuations. This will give the four-dimensional brane world a "D-foamy" structure.

Interactions in string theory, at the present moment, are not treated systematically as a second quantised formalism is lacking. An important consistency requirement of first quantised string theory is conformal invariance, which determines the space-time dimension and/or structure. On the brane there are closed and open strings propagating. Each time these strings "meet" a D-particle, there is a possibility of being attached to it. The entangled state causes a back reaction onto the space-time, which can be calculated perturbatively using logarithmic conformal field theory formalism [4]. Some details are reviewed in Appendix A.

Let $M_{QG}\left(=\frac{M_S}{g_s}\right)$ be the quantum gravity mass scale with $g_s$being the string coupling and M_S the string mass scale. Even at low energies E, decoherence effects of this foam can have magnitude $O\left(\left[\frac{E}{M_{QG}}\right]^n\right)$, where n is model-dependent with values. Usually M_{QG} is taken to be M_P, the Planck mass. Hence, if M_{QG} can be decreased by many orders of magnitude, then the decoherence effect is duly enhanced. This decoherence is due to the topologically non-trivial interactions of the D-particles with strings. We shall see that in some models, involving large extra dimensions, M_{QG} can reduce to TeV scale. Such models were devised to deal with the hierarchy of scales between the Planck and electro-weak scales. Indeed, in the absence of large extra dimensions, at the heuristic level within the context of microsocpic black holes, two quarks inside a proton can be absorbed by a virtual black hole with a re-emission into an antiquark and lepton. An estimate for the probability of this process can be given in terms of

Journal of Physics: Conference Series **284** (2011) 012052 doi:10.1088/1742-6596/284/1/012052

the quantum chromodynamics scale (Λ_{QCD}) and M_{QG}. Typically the proton size is Λ_{QCD}^{-1}, and so, the probability of two quarks to come within a distance l_P within the lifetime of the black hole, is $O\left(\left(\frac{\Lambda_{QCD}}{M_{QG}}\right)^4\right)$. The constraint on proton lifetime being 10^{33}yr or longer imples that $M_{QG} > 10^{16}$GeV[5]. Fast baryon decay can be inhibited by separating the quark and lepton in a large extra dimension, resulting in overlaps of wavefunctions being reduced and a smaller M_{QG}. A model[6] which can produce M_{QG} of TeV size has been proposed and involves a $4+n$ dimensional Universe with structure $M_4 \times K_n$, where M_4 is a $3+1$ dimensional domain wall (brane) and $K_n$is a flat n-dimensional manifold with extra dimensions of size R. In the model, all standard model interactions in particle physics would be confined on M_4, but not gravity. This leads to an M_{QG} satisfying:

$$M_{QG} = \frac{1}{2\pi R}\left(2\pi R M_P\right)^{\frac{2}{n+2}}.$$

For example if $n = 2$ and $R = 0.1$mm then $M_{QG} = 10$TeV.

Using the brane model [7] for space-time fluctuations one can obtain the following expression for the induced space-time distortion as a result of the D-particle recoil, in the weakly coupled string limit, which will be appropriately used in :

$$g_{ij} = \delta_{ij},\, g_{00} = -1,\, g_{0i} = \varepsilon\left(\varepsilon y_i + u_i t\right)\Theta_\varepsilon\left(t\right),\; i = 1,\ldots,D \tag{1}$$

where the suffix 0 denotes temporal (Liouville) components and

$$\Theta_\varepsilon\left(t\right) = \frac{1}{2\pi i}\int_{-\infty}^{\infty}\frac{dq}{q - i\varepsilon}e^{iqt}, \tag{2}$$

$$u_i = \left(k_1 - k_2\right)_i\,, \tag{3}$$

with $k_1\,(k_2)$ the momentum of the propagating closed-string state before (after) the recoil; y_i are the spatial collective coordinates of the D particle and ε^{-2} is identified with the target Minkowski time t for $t \gg 0$ after the collision [4] (see Appendix A). These relations have been calculated for non-relativistic branes where u_i is small and require the machinery of logarithmic conformal field theory. Now for large t, to leading order,

$$g_{0i} \simeq \overline{u}_i \equiv \frac{u_i}{\varepsilon} \propto \frac{\Delta p_i}{M_P} \tag{4}$$

where Δp_i is the momentum transfer during a collision and M_P is the Planck mass (actually, to be more precise $M_P = M_s/g_s$, where $g_s < 1$ is the (weak) string coupling, and M_s is a string mass scale); so g_{0i} is constant in space-time but depends on the energy content of the low energy particle.

1.2. Decoherence and non-unitary evolution

The recoilng D-particle background leads to a fluctuating background. In string theory for fixed backgrounds to be consistent with a classical space-time interpretation the central charge needs to have the critical value. Quantum scattering off the D-particle excites an open string state which attaches itself to the D-particle. Closed-to-open string amplitudes determine these excitation and emission processes. To allow in principle for fluctuating backgrounds it is necessary to deform the conformal points through vertex operatorsV_{g^I} associated with background fieldsg^I.The world-sheet action S_σ has a structure

$$S_\sigma = S^* + g^I\int_\Sigma V_I\left(\Xi\right)d^2\xi,$$

which is a (weakly) non-conformal deformation of S^*, the conformal action,V_I is a vertex (logarithmic conformal field theory) operator associated with the D-particle, Ξ are target space matter fields, Ξ is the usual holomorphic world sheet co-ordinate, and $\Sigma$is the world-sheet surface. In our case the relevant $g^I = u_i$ with $i = 1, 2, 3$, if the D-particle recoil is confined to the D3 brane on which matter strings propagate. Since we are interested in the dynamics of matter, we will consider the reduced density matrix of matter ρ_M on a fixed genus world-sheet.

The renormalization group for field theories in two dimensions has properties which will have an interesting reinterpretation for string theories [8, 9]. In theory space, i.e. in the infinite dimensional space of couplings $\left\{g^I\right\}$, for all renormalizable two-dimensional unitary quantum field theories, there exists a function $c(\{g^I\})$ which has the following properties:

- $c(\{g^I\})$ is non-negative and non-increasing on renormalization group flows towards an infrared fixed point
- the renormalization group fixed points are also critical points of $c(\{g^I\})$
- critical value of $c(\{g^I\})$ is the conformal anomaly

More precisely in terms of a renormalization group flow parameter (scale) t

$$\frac{d}{dt}c(\{g^I\}) = \beta^I G_{IJ}\beta^J \tag{5}$$

where the renormalization group β function is defined by

$$\beta^I = \frac{d}{dt}g^I(t)$$

and G_{IJ} is referred to as the Zamolodchikov metric. $G_{IJ}$is negative definite and is the matrix of second derivatives of the free energy. In the approach to perturbative string theory based on conformal field theory on the world sheet , it has long been thought that the time evolution of the string backgrounds and world sheet renormalization group flows are connected [10][11]. In non-critical strings conformal invariance is restored by gravitational dressing with a Liouville field φ (which can be viewed as a local world sheet scale) so that (for convenience now dropping the index I)

$$\int d^2z g V_g(\Xi) \to \int d^2z g e^{\alpha_g \varphi} V_g(\Xi). \tag{6}$$

However V_g has a scaling dimension α_g to $O(g)$ with $\alpha_g = -gC_{ggg}+\ldots$ and C_{ggg} is the expansion coefficient in the operator product expansion of V_g with itself. In a small g expansion

$$\int d^2z g V_g(\Xi) \to \int d^2z g V_g(\Xi) - \int d^2z g^2 C_{ggg}\varphi V_g(\Xi).$$

Scale invariance is restored by defining a renormalized coupling g_R

$$g_R = g - C_{ggg}\varphi g^2 \tag{7}$$

The local scale interpretation of φ is clearly consistent with the renormalisation group β function that we have defined earlier. On noting that $\gamma_{\alpha\beta} = e^{\varphi}\widehat{\gamma}_{\alpha\beta}$ where $\widehat{\gamma}_{\alpha\beta}$ is a fiducial metric, the integration over world sheet metrics $\gamma_{\alpha\beta}$ (in the Polyakov string action) implies an integration over φ . In this way φ becomes a dynamical variable with a kinetic term. For matter fields with central charge $c_m > 25$ the signature of this term is opposite to the kinetic terms for the fields Ξ and it has been suggested that in this case the zero mode of φ is a target time t. The requirement of renormalizability of the world sheet σ model implies that for the density matrix ρ_M of a string state propagating in a background $\{g_i\}$

$$\frac{d}{dt}\rho_M(g^I, p_I, t) = 0 \tag{8}$$

where p_i is the conjugate momentum to g^i within the framework of a dynamical system with hamiltonian H and action the Zamolodchikov c-function [12], i.e.

$$c[g] = \int dt(p_I \dot{g}^I - H). \tag{9}$$

From (\ref{density}) we deduce that

$$\frac{\partial\rho}{\partial t} + \dot{g}^I\frac{\partial\rho}{\partial g^I} + \dot{p}_I\frac{\partial\rho}{\partial p^I} = 0. \tag{10}$$

The piece $\dot{p}_I\frac{\partial\rho}{\partial p^I}$ in (10) can be written as $G_{IJ}\beta^J\frac{\partial\rho}{\partial p^I}$; using the canonical relationship of g^I and p^I this can be recast as $-iG_{IJ}\beta^J[\rho, g^J]$. As discussed in [12], such a term leads to a non-unitary evolution of ρ. This analysis cannot be regarded as being conclusive concerning the issue of non-unitarity, even within the framework of non-critical string theory, since there are issues relating to the time like signature of the Liouville field and the identification of the local renormalization group scale. However we will asssume that these caveats are not sufficient to invalidate the main consequence and we shall explore the implications for $CPT\,(\equiv\Theta)$ symmetry where where the operators C, P, and T denote charge conjugation, parity and time reversal respectively.

1.3. CPT

Currently the successful theories are based on local Lorentz invariant lagrangians, and it has been shown given the spin-statistics connection that CPT is a symmetry. The symmetry implies that the solution set of a theory is invariant under reversal of parity, time and interchange of particle and antiparticle. Consequently any violations of the consequences of CPT symmetry [13] would entail physics beyond the standard model of particle physics which is based on lagrangians. Typically the consequences of CPT that are considered are those of equal masses and lifetimes for particles and antiparticles. Recently it was noted that when the CPT operator is not well defined there are implications for the symmetry structure of the initial entangled state of two neutral mesons in meson factories such as DAΦNE, the Frascati ϕ factory. Indeed, if CPT can be defined as a quantum mechanical operator, then the decay of a (generic) meson with quantum numbers $J^{PC} = 1^{--}$ [14], leads to a pair state of neutral mesons $|i\rangle$ having the form of the entangled state

$$|i\rangle = \frac{1}{\sqrt{2}}\left(\left|\overline{M_0}\left(\vec{k}\right)\right\rangle\left|M_0\left(-\vec{k}\right)\right\rangle - \left|M_0\left(\vec{k}\right)\right\rangle\left|\overline{M_0}\left(-\vec{k}\right)\right\rangle\right). \tag{11}$$

This state has the Bose symmetry associated with particle-antiparticle indistinguishability $C\mathcal{P} = +$, where C is the charge conjugation and $\mathcal{P}$ is the permutation operation. If, however, CPT is not a good symmetry (i.e. ill-defined due to space-time foam), then M_0 and $\overline{M_0}$ may not be identified and the requirement of $C\mathcal{P}$ is relaxed˜[15]. Consequently, in a perturbative framework, the state of the meson pair can be parametrised to have the following form:

$$|i\rangle = \frac{1}{\sqrt{2}}\left(\left|\overline{M_0}\left(\vec{k}\right)\right\rangle\left|M_0\left(-\vec{k}\right)\right\rangle - \left|M_0\left(\vec{k}\right)\right\rangle\left|\overline{M_0}\left(-\vec{k}\right)\right\rangle\right) + \frac{\omega}{\sqrt{2}}\left|\Delta\left(\vec{k}\right)\right\rangle$$

where

$$\left|\Delta\left(\vec{k}\right)\right\rangle \equiv \left|\overline{M}\left(\vec{k}\right)\right\rangle\left|M_0\left(-\vec{k}\right)\right\rangle + \left|M_0\left(\vec{k}\right)\right\rangle\left|\overline{M}\left(-\vec{k}\right)\right\rangle$$

Journal of Physics: Conference Series **284** (2011) 012052 doi:10.1088/1742-6596/284/1/012052

and $\omega = |\omega| e^{i\Omega}$ is a complex CPT violating (CPTV) parameter. For definiteness in what follows we shall term this quantum-gravity effect in the initial state [15].

Decoherence or non-unitary evolution leads to a new method of breaking Θ. Unitary time evolution is an implicit assumption in the proof of the CPT theorem; Wald argued that non-unitary evolution would allow Θ not to be a symmetry.Within the context of scattering theory let $\mathcal{H}_{in}$ denote the Hilbert space of in states, and $\bar{\mathcal{H}}_{in}$ the dual space. We can define the analogous entities for the out states. Let S be a mapping from the set of in-states $\mathcal{G}_{in}$ to the set of out-states $\mathcal{G}_{out}$. In our framework $\mathcal{G}_{in}$ is isomorphic to $\mathcal{H} \bigotimes \bar{\mathcal{H}}_{in}$ since the states are represented as density matrices ρ^{A}_{B} where A is a vector index associated with $\mathcal{H}_{in}$ and B is a vector index associated with $\bar{\mathcal{H}}_{in}$. The indexed form of S is S^{aD}_{bC} where the lower case indices refer to $\mathcal{H}_{out}$ and $\bar{\mathcal{H}}_{out}$. If probability is conserved

$$tr\,(S\rho) = (\rho) \tag{12}$$

which in index notation can be written as $S^{aD}_{aC} = \delta^{D}_{C}$ where we have adopted the summation convention for repeated indices. Consider now operators Θ_{in} and Θ_{out} which implement the CPT transformation on $\mathcal{G}_{in}$ and $\mathcal{G}_{out}$ respectively i.e.

$$\Theta_{in} \;:\; \mathcal{G}_{in} \rightarrow \mathcal{G}_{out} \tag{13}$$
$$\Theta_{out} \;:\; \mathcal{G}_{out} \rightarrow \mathcal{G}_{in}. \tag{14}$$

In particular under a CPT transformation let $\rho_{in} \in \mathcal{G}_{in}$ be mapped into $\rho^{'}_{out}$ and $\rho_{out} \in \mathcal{G}_{out}$ be mapped into $\rho^{'}_{out}$. If the theory (including quantum gravity) is assumed to be invariant under CPT then

$$\rho_{out} = S\rho_{in}, \tag{15}$$
$$\rho^{'}_{out} = S\rho^{'}_{in} \tag{16}$$
$$\Theta_{in}\rho_{in} = \rho^{'}_{out}, \tag{17}$$
$$\Theta_{out}\rho_{out} = \rho^{'}_{in}, \tag{18}$$

and

$$\Theta_{in}\Theta_{out} = I,\; \Theta_{out}\Theta_{in} = I. \tag{19}$$

Since

$$\Theta_{out} = \Theta^{-1}_{in} \tag{20}$$

it is convenient to drop the suffix *in* in this last relation. Hence from (15), (16),(17)and (18)we can deduce that

$$\Theta\rho_{in} = \rho^{'}_{out}, \tag{21}$$
$$= S\rho^{'}_{in} \tag{22}$$
$$= S\Theta^{-1}\rho_{out} \tag{23}$$
$$= S\Theta^{-1}S\rho_{in},. \tag{24}$$

From (21) and (24) we can deduce the important result that S has an inverse given by $\Theta^{-1}S\Theta^{-1}$. Hence if Θexists then time reversed evolution is permitted. Consequently for non-unitary evolution Θ cannot be defined.

Another possible contribution of breakdown of CPT invariance is towards the generation of cosmological charge asymmetry (baryogenesis). This would provide an additional mechanism to that proposed by Sakhorov based on: non-conservation of baryon number; deviation from thermal equilibrium; and C and CP violation. This has, in the past, been addressed within thermal equilibrium by using a mass difference between particle and anti-particle, a possible consequence of CPT violation. In our case of D-foam the decoherence effects on particle and anti-particle may be different, a fact we exploit.

2. Effective model of D-foam

For an observer on the brane world the crossing D-particles will appear as twinkling space-time defects, i.e. microscopic space-time fluctuations. This will give the four-dimensional brane world a "D-foamy" structure. In phase space, for a D3-brane world, the function u_i, involving a momentum transfer, Δk_i, can be modelled by a local operator using the following parametrization [16]:

$$u_i = g_s \frac{\Delta k_i}{2M_s} = r_i k_i \ , \text{ no sum } i = 1, 2, 3 \ , \tag{25}$$

where the (dimensionful) variables $r_i, i = 1, 2, 3$, appearing above, are related to the fraction of momentum that is transferred at a collision with a D-particle in each spatial direction i .The target space-time metric state, which is close to being flat, can be represented schematically as a density matrix

$$\rho_{\text{grav}} = \int d^4\tilde{r}\ f(\tilde{r}_\mu)\,|g(\tilde{r}_\mu)\rangle\,\langle g(\tilde{r}_\mu)|\,. \tag{26}$$

The parameters $\tilde{r}_\mu$ $(\mu = 0, \ldots, 4)$ are stochastic with a gaussian distribution $f(\tilde{r}_\mu)$ characterised by the averages%

$$\langle \tilde{r}_\mu \rangle = 0,\ \langle \tilde{r}_\mu \tilde{r}_\nu \rangle = \Delta_\mu \delta_{\mu\nu}\,. \tag{27}$$

The fluctuations experienced by the two entangled neutral mesons will be assumed to be independent and $\Delta_\mu \sim O\left(\frac{E^2}{M_P^2}\right)$i.e. very small. As matter moves through the space-time foam, assuming ergodicity, the effect of time averaging is assumed to be equivalent to an ensemble average. As far as our present discussion is concerned we will consider a semi-classical picture for the metric and so $|g(\tilde{r}_\mu)\rangle$ in (26) will be a coherent state. In order to address two "flavours" the fluctuations of each component of the metric tensor $g^{\alpha\beta}$ will not be simply given by the simple recoil distortion (4), but instead will be taken to have a 2×2 ("flavour") structure:

$$\begin{aligned} g_{00} &= -1, \\ g_{01} &= g_{10} = \tilde{r}_0\sigma_0 + \tilde{r}_1\sigma_1 + \tilde{r}_2\sigma_2 && (28) \\ g_{jj} &= 1 && (29) \end{aligned}$$

where $\sigma_0 = 1_2$, (the identity matrix). The modelling of the metric can be made more elaborate, but the salient feature is the structure of g^{01}(where for simplicity we have taken the momentum to be in the 1-direction). For the case of the omega effect this restriction is acceptable since the K-mesons are produced collinearly in a ϕ-meson factory (in the centre of mass frame).A more general ansatz than (28) (which will reduce to this when momenta are collinear) has the form

$$g_{0j} = r_\mu \sigma_\mu \hat{k}_j$$

where $\hat{k}_j$ is the momentum operator of a particle moving in this background; this is the Finsler nature of the metric.We use the term flavour in a general sense. For K mesons flavour would

Journal of Physics: Conference Series **284** (2011) 012052 doi:10.1088/1742-6596/284/1/012052

mean $K_L$or K_S. An elementary D-particle will typically affect neutral particles because it does not carry charge and a charge string attched to it would not have anywhere for the flux to flow. Elementary D-particles appear in some string theories such as IIa but in other more phenomenologically relevant string theories they do not. In such cases, effective D-particle behaviour can occur from D3 branes wrapped around 3-cycles. Consequently such D-particles could interact with charged strings and hence contribute to baryogenesis. However the treatment of D-particles, arising from compactifications, is technically more involved and the orders of magnitude will depend on the details of the compactification process. We nonetheless will point out how the D-foam formally can give rise to particle-antiparticle asymmetry.

For the neutral Kaon system, the case of interest, $K_0 - \overline{K}_0$, is produced by a ϕ-meson at rest,i.e. $K_0 - \overline{K}_0$ in their C.M. frame. The CP eigenstates (on choosing a suitable phase convention for the states $|K_0\rangle$ and $\left|\overline{K_0}\right\rangle$) are, in standard notation, $|K_\pm\rangle$ with

$$|K_\pm\rangle = \frac{1}{\sqrt{2}}\left(|K_0\rangle \pm \left|\overline{K_0}\right\rangle\right).$$

The mass eigensates $|K_S\rangle$ and $|K_L\rangle$ are written in terms of $|K_\pm\rangle$ as

$$|K_L\rangle = \frac{1}{\sqrt{1+|\varepsilon_2|^2}}\left[|K_-\rangle + \varepsilon_2 |K_+\rangle\right]$$

and

$$|K_S\rangle = \frac{1}{\sqrt{1+|\varepsilon_1|^2}}\left[|K_+\rangle + \varepsilon_1 |K_+\rangle\right].$$

In terms of the mass eigenstates

$$|i\rangle = \mathcal{C}\left\{\begin{array}{l} \left(\left|K_L\left(\overrightarrow{k}\right)\right\rangle\left|K_S\left(-\overrightarrow{k}\right)\right\rangle - \left|K_S\left(\overrightarrow{k}\right)\right\rangle\left|K_L\left(-\overrightarrow{k}\right)\right\rangle\right) + \\ \omega\left(\left|K_S\left(\overrightarrow{k}\right)\right\rangle\left|K_S\left(-\overrightarrow{k}\right)\right\rangle - \left|K_L\left(\overrightarrow{k}\right)\right\rangle\left|K_L\left(-\overrightarrow{k}\right)\right\rangle\right) \end{array}\right\}$$

where $\mathcal{C} = \frac{\sqrt{(1+|\varepsilon_1|^2)(1+|\varepsilon_2|^2)}}{\sqrt{2}(1-\varepsilon_1\varepsilon_2)}$ [15]. In the notation of two level systems (on suppressing the $\overrightarrow{k}$ label) we write

$$|K_L\rangle = |\uparrow\rangle \tag{30}$$

$$|K_S\rangle = |\downarrow\rangle. \tag{31}$$

These will be our "flavours" and represent the two physical eigenstates, with masses $m_1 \equiv m_L$, $m_2 \equiv m_S$, with

$$\Delta m = m_L - m_S \sim 3.48 \times 10^{-15} \text{ GeV}. \tag{32}$$

The Klein-Gordon equation for a spinless neutral meson field $\Phi = \begin{pmatrix} \phi_1 \\ \phi_2 \end{pmatrix}$ with mass matrix $m = \frac{1}{2}(m_1 + m_2)\,1 + \frac{1}{2}(m_1 - m_2)\,\sigma_3$ in a gravitational background is

$$(g^{\alpha\beta} D_\alpha D_\beta - m^2)\Phi = 0 \tag{33}$$

where D_α is the covariant derivative. Since the Christoffel symbols vanish for a_i independent of space time the D_α coincide with ∂_α. Hence

$$\left(g^{00}\partial_0^2 + 2g^{01}\partial_0\partial_1 + g^{11}\partial_1^2\right)\Phi - m^2\Phi = 0. \tag{34}$$

Journal of Physics: Conference Series **284** (2011) 012052 doi:10.1088/1742-6596/284/1/012052

It is useful at this stage to rewrite the state $|i\rangle$ in terms of the mass eigenstates.

The unnormalised state $|i\rangle$ will then be an example of an initial state

$$|\psi\rangle = |k,\uparrow\rangle^{(1)}|-k,\downarrow\rangle^{(2)} - |k,\downarrow\rangle^{(1)}|-k,\uparrow\rangle^{(2)} + |\Delta\rangle \tag{35}$$

with

$$|\Delta\rangle = \xi\,|k,\uparrow\rangle^{(1)}|-k,\uparrow\rangle^{(2)} + \xi'\,|k,\downarrow\rangle^{(1)}|-k,\downarrow\rangle^{(2)}$$

where $\left|K_L\left(\overrightarrow{k}\right)\right\rangle = |k,\uparrow\rangle$ and we have taken $\overrightarrow{k}$ to have only a non-zero component k in the x-direction; superscripts label the two separated detectors of the collinear meson pair, ξ and ξ' are complex constants and we have left the state $|\psi\rangle$ unnormalised. The evolution of this state is governed by a hamiltonian $\widehat{H}$

$$\widehat{H} = g^{01}\left(g^{00}\right)^{-1}\widehat{k} - \left(g^{00}\right)^{-1}\sqrt{\left(g^{01}\right)^2 k^2 - g^{00}\left(g^{11}k^2 + m^2\right)} \tag{36}$$

which is the natural generalisation of the standard Klein-Gordon hamiltonian in a one particle situation. Moreover $\widehat{k}\,|\pm k,\uparrow\rangle = \pm k\,|k,\uparrow\rangle$ together with the corresponding relation for $\downarrow$. We next note that the Hamiltonian interaction terms

$$\widehat{H_I} = -\left(r_1\sigma_1 + r_2\sigma_2\right)\widehat{k} \tag{37}$$

are the leading order contribution in the small parameters r_μ in the Hamiltonian H (36), since the corresponding variances $\sqrt{\Delta_\mu}$ are small. The term (37), has been used in [15] as a perturbation in the framework of non-degenerate perturbation theory, in order to derive the "gravitationally-dressed" initial entangled meson states, immediately after the ϕ decay. The result is:

$$|k,\uparrow\rangle^{(1)}_{QG}|-k,\downarrow\rangle^{(2)}_{QG} - |k,\downarrow\rangle^{(1)}_{QG}|-k,\uparrow\rangle^{(2)}_{QG} = |\Sigma\rangle + |\widetilde{\Delta}\rangle \tag{38}$$

where

$$|\Sigma\rangle = |k,\uparrow\rangle^{(1)}|-k,\downarrow\rangle^{(2)} - |k,\downarrow\rangle^{(1)}|-k,\uparrow\rangle^{(2)},$$

$$|\widetilde{\Delta}\rangle = \begin{array}{c} |k,\downarrow\rangle^{(1)}|-k,\downarrow\rangle^{(2)}\left(\beta^{(1)} - \beta^{(2)}\right) + |k,\uparrow\rangle^{(1)}|-k,\uparrow\rangle^{(2)}\left(\alpha^{(2)} - \alpha^{(1)}\right) \\ +\beta^{(1)}\alpha^{(2)}\,|k,\downarrow\rangle^{(1)}|-k,\uparrow\rangle^{(2)} - \alpha^{(1)}\beta^{(2)}\,|k,\uparrow\rangle^{(1)}|-k,\downarrow\rangle^{(2)} \end{array}$$

and

$$\alpha^{(i)} = \frac{{}^{(i)}\left\langle\uparrow,k^{(i)}\right|\widehat{H_I}\left|k^{(i)},\downarrow\right\rangle^{(i)}}{E_2 - E_1}, \quad \beta^{(i)} = \frac{{}^{(i)}\left\langle\downarrow,k^{(i)}\right|\widehat{H_I}\left|k^{(i)},\uparrow\right\rangle^{(i)}}{E_1 - E_2}, \quad i = 1,2 \tag{39}$$

where the index (i) runs over meson species ("flavours") ($1 \to K_L$, $2 \to K_S$). The reader should notice that the terms proportional to $\left(\alpha^{(2)} - \alpha^{(1)}\right)$ and $\left(\beta^{(1)} - \beta^{(2)}\right)$ in (\ref{entangl}) generate ω-like effects. We concentrate here for brevity and concreteness in the strangeness conserving case of the ω-effect in the initial decay of the ϕ meson [15], which corresponds to $r_i \propto \delta_{i2}$. We should mention, however, that in general quantum gravity does not have to conserve this quantum number, and in fact strangeness-violating ω-like terms are generated in this problem through time evolution [15].

We next remark that, on averaging the density matrix over the random variables r_i, which are treated as independent variables between the two meson particles of the initial state , we observe that only terms of order $|\omega|^2$ will survive, with the order of $|\omega|^2$ being

$$|\omega|^2 = \widetilde{\Delta}_{(1),(2)}\left(\mathcal{O}\left(\frac{1}{(E_1-E_2)}(\langle\downarrow,k|H_I|k,\uparrow\rangle)^2\right)\right),$$
$$= \widetilde{\Delta}_{(1),(2)}\left(\mathcal{O}\left(\frac{\Delta_2 k^2}{(E_1-E_2)^2}\right)\right) \sim \widetilde{\Delta}_{(1),(2)}\left(\frac{\Delta_2 k^2}{(m_1-m_2)^2}\right) \tag{40}$$

for the physically interesting case of non-relativistic Kaons in ϕ factories, in which the momenta are of order of the rest energies. The notation $\widetilde{\Delta}_{(1),(2)}(\ldots)$ above indicates that one considers the differences of the variances Δ_2 between the two mesons 1 - 2, in that order.

The variances in our model of D-foam, which are due to quantum fluctuations of the recoil velocity variables about the zero average (dictated by the imposed requirement on Lorentz invariance of the string vacuum) lead for the square of the amplitude of the (complex) ω-parameter:

$$|\omega|^2 \sim g_0^2\frac{(m_1^2-m_2^2)}{M_s^2}\frac{k^2}{(m_1-m_2)^2} = \frac{m_1+m_2}{m_1-m_2}\frac{k^2}{(M_s^2/g_0^2)}, \tag{41}$$

where $M_P \equiv M_s/g_0$ represents the (average) quantum gravity scale, which may be taken to be the four-dimensional Planck scale. In general, M_s/g_0 is the (average) D-particle mass, as already mentioned. In the modern version of string theory, M_s is arbitrary and can be as low as a few TeV, but in order to have phenomenologically correct string models with large extra dimensions one also has to have in such cases very weak string couplings g_0, such that even in such cases of low M_s, the D-particle mass M_s/g_0 is always close to the Planck scale 10^{19} GeV. But of course one has to keep an open mind about ways out of this pattern, especially in view of the string landscape.

The result (41), implies, for neutral Kaons in a ϕ factory, for which $\Delta m = m_L - m_S \sim 3.48\times 10^{-15}$GeV is valid, a value of: $|\omega| \sim 10^{-5}$, which in the sensitive η^{+-} bi-pion decay channel, is enhanced by three orders of magnitude, as a result of the fact that the $|\omega|$ effect always appears in the corresponding observables [15] in the form $|\omega|/|\eta^{+-}|$, and the CP-violating parameter $|\eta^{+-}| \sim 10^{-3}$. Unfortunately, this value is still some two orders of magnitude away from current bounds of the ω-effect of the KLOE collaboration at DAΦNE, but is within the projected sensitivity of the proposed upgrades.

A similar calculation can be done for the particle-antiparticle asymmetry except that we will use the more general metric resulting in

$$H = E_r\left(\overrightarrow{k}\right)1_2 + r_\mu\sigma_\mu\left(\overrightarrow{k}\right)^2 \tag{42}$$

where

$$E_r\left(\overrightarrow{k}\right) = \sqrt{m^2+\sum_j k_j^2 + (r_\mu)^2\left(\sum_j k_j^2\right)^2 - \sum r_j^2 k_j^4}$$

This leads to gravitational dressing of the particle and antiparticle states so that the masses get shifted so as to induce a mass difference; in thermodyanmic equilibrium the canonical number distribution (Bose-Einstein or Fermi-Direc) would give a difference between the particle and anti-particle number. Explcitly, if the eigenvalues of H are denoted by $a_1\left(\overrightarrow{k},r_\mu\right)$ and $a_2\left(\overrightarrow{k},r_\mu\right)$ then to lowest order we can show that $a_2\left(\overrightarrow{k},r_\mu\right) - a_1\left(\overrightarrow{k},r_\mu\right) = 2\left|\overrightarrow{k}\right||\overrightarrow{r}|$ and clearly there will then be a difference between the grand canonical number distribution functions

$$n_{a_j}\left(\overrightarrow{k},r_\mu,\xi\right) = \frac{1}{\exp\left(\beta\left[a_j\left(\overrightarrow{k},r_\mu\right)-\mu\right]\right)+\xi},\quad j=1,2$$

where $\xi = \pm 1$ for different j.

Conclusions

In our D-foam model, string matter on a brane world interacts with D-particles in the bulk. Recoil of the heavy D particles owing to interactions with the stringy matter produces a gravitational distortion which has a back-reaction on matter. There is information loss at the recoil which leads to non-unitary evolution for the matter. From the general considerations of Wald, CPT symmetry may then be violated. This new mechanism of violation is explored for the omega effect as well as matter-anti-matter asymmetry. For the omega effect there is a chance that the upgrade to the the DAΦNE detector will allow a stringent test of the D-foam prediction. Particle-antiparticle asymmetry has more theoretical uncertainities in D-foam predictions and further investigations are needed using D-brane instantons for example.

Acknowledgements

My work is partially supported by the European Union through the Marie Curie Research and Training Network *UniverseNet* (MRTN-2006-035863). The work presented in this paper has arisen due to discussion as well as collaboration with N.E. Mavromatos.

Appendix A: Calculation of back reaction in D-particle foam

D-particle foam contributions to master equation for Liouville-decoherence

The material in this Appendix is a review based on [4], where we refer the reader for further details. Let us consider a *D*-particle, located at $y^i(t=0) \equiv y_i$ of the spatial coordinates of a $(d+1)$-dimensional space time (which could be a D3-brane world), which at a time $t=0$ experiences an impulse, as a result of scattering with a matter string state . In a σ-model framework, the trajectory of the *D*-particle $y^i(t)$, $i = 1, 2, \ldots d$, a spatial index, is described by inserting the following vertex operator in the σ-model of a free string:

$$V = \int_{\partial\Sigma} g_{ij} y^j(t) \partial_n X^i \tag{43}$$

where g_{ij} denotes the spatial components of the metric, $\partial\Sigma$ denotes the world-sheet boundary, ∂_n is a normal world-sheet derivative, X^i are σ-model fields obeying Dirichlet boundary conditions on the world sheet, and t is a σ-model field obeying Neumann boundary conditions on the world sheet, whose zero mode is the target time. The space-time prior to Liouville dressing is assumed Euclidean for formal reasons (convergence of the corresponding σ-model path integral). We note, however, that the final Liouville-dressed target space-time acquires Minkowski signature as a result of the time-like signature of the Liouville mode [17].

In the non-relativistic approximation, appropriate for a heavy D-particle defect of mass M_s/g_s, with M_s the string scale, and g_s the string coupling, assumed weak ($g_s \ll 1$), the path $y^i(t)$ corresponding to the impulse is given by:

$$y_i(t) = (\varepsilon y_i + u_i t)\,\Theta_\varepsilon(t) \tag{44}$$

$$u_i = (k_1 - k_2)_i \;, \tag{45}$$

with k_1 (k_2) the momentum of the propagating string state before (after) the recoil y_i are the spatial collective coordinates of the D particle, and the regularized Heaviside functional operator $\Theta_\varepsilon(t)$ is given by (2) in the text [4]:

$$\Theta_\varepsilon(t) = \frac{1}{2\pi i}\int_{-\infty}^{\infty} \frac{dq}{q - i\varepsilon} e^{iqt}, \tag{46}$$

Eq. (44) contains actually a *pair* of deformations corresponding to the σ-model couplings y_i and u_i. These deformations are relevant in a world-sheet renormalization-group sense, having anomalous scaling dimension $-\frac{\varepsilon^2}{2}$, i.e. to leading order in a coupling constant expansion their renormalization-group β-functions read:

$$\beta^{y^i} = -\frac{\varepsilon^2}{2} y^i \,, \qquad \beta^{u^i} = -\frac{\varepsilon^2}{2} u^i \,. \tag{47}$$

The deformations form a logarithmic conformal algebra (superconformal algebra in the case of superstrings) which *closes* if and only if one identifies [4] the regulating parameter ε^{-2} with the world-sheet renormalization-group scale $\ln|L/a|^2$ ($L(a)$ is the Infrared (Ultraviolet) world-sheet scale):

$$\varepsilon^{-2} = \eta \ln|L/a|^2 \tag{48}$$

where η denotes the signature of time t of the target-space manifold of the σ-model (prior to Liouville dressing). For Euclidean manifolds, assumed here for path-integral convergence, $\eta = +1$.

Upon the identification (48) the re-scaled couplings $\overline{y}_i \equiv \frac{y_i}{\varepsilon}$ and $\overline{u}_i \equiv \frac{u_i}{\varepsilon}$ are *marginal*, that is independent of the scale ε. It is these marginal couplings that are connected to target-space quantities of physical significance, such as the space-time back reaction of recoil. In the limit $\varepsilon \to 0$ (the long-time limit) the dominant contributions come from the u_i recoil deformation in (44).

[1] see, for instance: D. Bao, S. S. Chern and Z. Shen, *An introduction to Finsler Geometry* (Springer-Verlag (NY, 2000)); For a recent review, with applications to cosmology, see S. Vacaru, "Principles of Einstein-Finsler Gravity and Cosmology," arXiv: 1012. 4148 [physics.gen-ph].

[2] J. Polchinski,"Dirichlet branes and Ramond-Ramond charges", Phys. Rev. Lett. 75 4724 (1995).

[3] J. R. Ellis, N. E. Mavromatos and M.Westmuckett, "A supersymmetric D-brane model of space-time foam," Phys. Rev. D 70, 044036 (2004) [arXiv:gr-qc/0405066]; "Potentials between D-branes in a supersymmetric model of space-time foam", Phys.Rev. D 71, 106006 (2005) [arXiv:gr-qc/0501060].

[4] I. I. Kogan, N. E. Mavromatos and J. F. Wheater, "D-brane recoil and logarithmic operators," Physics Letters B 387, 483 (1996) [arXiv:hep-th/9606102]; For a review focusing on D-brane recoil, including supermembranes, see: N. E. Mavromatos, "Logarithmic conformal field theories and strings in changing backgrounds," arXiv:hep-th/0407026, in Shifman, M. (ed.) et al.: From fields to strings, I. Kogan memorial Volume 2}, 1257-1364. (World Sci. 2005), and references therein.

[5] L. A. Anchordoqui, "Spacetime foam at a TeV," J. Phys.Conf. Ser. 60 (2007) 191 [arXiv:hep-ph/0610025].

[6] N. Arkani-Hamed, S. Dimopoulos and G. R. Dvali, "The hierarchy problem and new dimensions at a millimeter," Phys. Lett. B 429, 263 (1998) [arXiv:hep-ph/9803315].

[7] J. R. Ellis, N. E. Mavromatos and D. V. Nanopoulos, Gen. Rel. Grav. **32**, 127 (2000) [arXiv:gr-qc/9904068]; Phys. Rev. D **61**, 027503 (2000) [arXiv:gr-qc/9906029]; Phys. Rev. D **62**, 084019 (2000) [arXiv:gr-qc/0006004]. J. R. Ellis, K. Farakos, N. E. Mavromatos, V. A. Mitsou and D. V. Nanopoulos, Astrophys. J. **535**, 139 (2000) [arXiv:astro-ph/9907340]; J. R. Ellis, N. E. Mavromatos, D. V. Nanopoulos and G. Volkov, Gen. Rel. Grav. **32**, 1777 (2000) [arXiv:gr-qc/9911055]. J. R. Ellis, N. E. Mavromatos and M. Westmuckett, Phys. Rev. D **70**, 044036 (2004) [arXiv:gr-qc/0405066].

[8] G. M. Shore,"A local renormalization group equation, diffeomorphisms and conformal invariance in sigma-models", Nuclear Physics B **286** 349 (1987)

[9] A. B. Zamolodchikov, "'Irreversibility' Of The Flux Of The Renormalization Group In A 2-D Field Theory," JETP Lett. 43, 730 (1986) [Pisma Zh. Eksp. Teor. Fiz.43, 565 (1986)].

[10] M. Gutperle, M. Headrick, S. Minwalla and M. Schomerus, "Space-time energy decreases under world-sheet renormalization group flow", Journal of High Energy Physics, 1 073 (2003)

[11] J. Ellis, N. E. Mavromatos and D. V. Nanopoulos, "A non-critical string approach to black holes, time and quantum dynamics",in "From supersymmetry to the origin of space-time" Ed A. Zichichi, World Scientific (1995), arXiv:hep-th/9403133

[12] N. E. Mavromatos, "Decoherence and CPT Violation in a Stringy Model of Space-Time Foam," Found.Phys. 40, 917 (2010) [arXiv:0906.2712 [hep-th]].

[13] R. F. Streater and A. S. Wightman, PCT, Spin and Statisitics, and All That, Benjamin, New York, (1964); O. W. Greenberg, "Why is CPT fundamental?," Found. Phys. 36, 1535 (2006) [arXiv:hep-ph/0309309].

[14] H. J. Lipkin, "CP violation and coherent decays of kaon pairs," Phys. Rev. 176 1715 (1968)

[15] J. Bernabeu, N. E. Mavromatos and S. Sarkar, "Decoherence induced CPT violation and entangled neutral mesons," Physical\ Review D 74, 045014 (2006) [arXiv:hep-th/0606137].

[16] N.E.~Mavromatos and Sarben Sarkar, "Liouville decoherence in a model of flavour oscillations in the presence of dark energy," Physical Review D 72, 065016 (2005) [arXiv:hep-th/0506242].
[17] E. Gravanis and N. E. Mavromatos, "Vacuum energy and cosmological supersymmetry breaking in brane worlds," Physics Letters B 547, 117 (2002) [arXiv:hep-th/0205298].

Journal of Physics: Conference Series **284** (2011) 012053 doi:10.1088/1742-6596/284/1/012053

Axially-symmetric sphaleron solutions of the Skyrme model

Ya Shnir[1] and D H Tchrakian[2]

[1] Department of Mathematical Sciences, Durham University, UK

[2] School of Theoretical Physics – DIAS, 10 Burlington Road, Dublin 4, Ireland

E-mail: `shnir@maths.tcd.ie`

Abstract. Static axially symmetric sphaleron-type solutions describing chains of interpolating Skyrmion–anti-Skyrmions have been constructed numerically. The configurations are characterized by two integers n and m, where $\pm n$ are the winding numbers of the constituent Skyrmion and anti-Skyrmion and the second integer m defines type of the solution, it has zero topological charge for even m and for odd values of m the Skyrmion–anti-Skyrmion chain has topological charge n. For the vanishing mass term we confirm the existence of such chain solutions for winding number $|n| \geq 2$. The similarity with monopole–anti-monopole pairs is highlighted.

1. Introduction

The purpose of this note based on our recent work [1] is to consider axially symmetric chain lump–anti-lump solutions in the usual Skyrme model [2] on $\mathbb{R}^3$.

It was agrued over the last decade that the multi-monopole (MM) solutions of the Yang-Mills-Higgs model and the multi-Skyrmion (MS) solutions of the Skyrme model have many features in common [3]. The question of constructing a Skyrme–anti-Skyrme pair (SAS) follows naturally from the corresponding construction of a monopole–anti-monopole (MAP) pair [4, 5] and monopole–anti-monopole chains [6], and like the former, it would describe a sphaleron like configuration. Likewise unsurprisingly, the resulting solutions are not given in closed form but are constructed numerically. The construction of an axially symmetric Skyrme–anti-Skyrme pair was carried out recently by Krusch and Sutcliffe [8].

In our paper [1] we extended the analysis presented in ref. [8] by inclusion of the pion mass term in the Lagrangian of the model. Besides the Skyrme–anti-Skyrme pair solution, we constructed new axially symmetric saddlepoint solutions which represent chains of m single Skyrmions and anti-Skyrmions, each carrying charge n in alternating order. For an equal number of Skyrmions and anti-Skyrmions, the chains reside in the topologically trivial sector. When the number of Skyrmions exceeds the number of anti-Skyrmions by one, the chains reside in the sector with topological charge n. The aim here is to reveal qualitative and quantitative similarities of chains of Skyrmions and anti-Skyrmions, with those of chains of monopoles and antimonopoles. Unlike for the $m = 2$ chain in [8] however, we cannot expect that the axially symmetric chain configurations with more numerous ($m \geq 2$) constituents we construct are of the minimal energy. Such sphaleron of the lowest energy may well exhibit discrete symmetries.

The boundary value problem we consider applies to a two dimensional nonlinear partial differential equation system and the numerical technique employed is exactly that applied in the corresponding monopole problem [5, 6]. The two dimensional system of equations is the result of the imposition of *axial symmetry* to this problem. We recover of course the MS's carrying topological charge n, which were found a long time ago for $n = 2$ in [10, 12], and for up to $n = 5$ in [11].

It should be pointed out that certain qualitative features expected here are more similar to instanton solutions of the Yang–Mills system, rather than the corresponding features of the monopole solutions of the Yang-Mills–Higgs system, inspite of the fact that the latter are described on $\mathbb{R}^3$ like the Skyrmion, while the former are described on $\mathbb{R}^4$. There is a fundamental qualitative difference between 'instanton' and 'monopole' type solutions, namely that the gauge connection in the former is asymptotically *pure gauge* while in the latter case it is *one half pure gauge*. It turns out that the asymptotic properties of the solitons of sigma models are 'instanton' like, irrespective of dimensions, in that the *composite connection* is asymptotically *pure gauge*. Thus we expect from the outset, that the Skyrme–anti-Skyrme lumps should be more akin to the instanton–anti-instanton [13], rather than the monopole–anti-monopole. The property we have in mind is that there exists a monopole–anti-monopole consisting of a charge n and a charge $-n$ pair with $n = 1$, and, with winding number $n > 1$ [14]. It turns out that the corresponding instanton–anti-instanton exists only for values $n \geq 2$. This was found in [13] only as a result of a numerical construction, which is in agreement with the same nonexistence result for $n = 1$ in the analytic proof for the existence of such nonselfdual lumps given in [15, 16]. Thus we would not expect here, to find a Skyrme–anti-Skyrme lump consisting of a charge 1 and charge -1 pair. We do however find such a pair when the pion-mass potential in introduced. We will return to elaborate on this feature of the solutions that we find, in our conclusions. In section 2 we impose axial symmetry and calculate the resulting reduced two dimensional energy density functional whose second order equations will be integrated, as well as the topological (baryonic) charges of these configurations. Our numerical results are presented in section 3, and our conclusions in section 4.

2. The model and imposition of symmetry

In terms of the order parameter multiplet $\phi^a = (\phi^\alpha, \phi^3, \phi^4)$, $\alpha = 1, 2$, of the nonlinear $O(4)$ sigma model field subject to $|\phi^a|^2 = 1$, the rescaled static Hamiltonian of the Skyrme [2] model is expressed as

$$\mathcal{H}_{\rm stat} = |\partial_i \phi^a|^2 + \frac{1}{4}|\partial_{[i}\phi^a \partial_{j]}\phi^b|^2 + \mu^2(1 - \phi^3) \tag{1}$$

$i = x, y, z$, with the notation $[ij]$ implying antisymmetrisation. Here $\mu^2(1 - \phi^3)$ is the pion mass term.

The energy density (1) is bounded from below by the topological charge density

$$\varrho_0 = \frac{1}{24\pi^2}\, \varepsilon_{ijk}\varepsilon^{abcd}\partial_i\phi^a\partial_j\phi^b\partial_k\phi^c\phi^d \tag{2}$$

whose integral is the integer winding number n, namely the baryon number. It is well known that this lower bound cannot be saturated and hence we are concerned only with the second order Euler-Lagrange equations.

Next, we state the axially symmetric Ansatz parametrised by two functions $f = f(\rho, z)$ and $g = g(\rho, z)$, $\rho^2 = x^2 + y^2$, and in terms of the two component unit vector n^α imposing axial symmetry on the $O(4)$ sigma model field $\phi^a = (\phi^\alpha, \phi^3, \phi^4)$

$$n^\alpha = \begin{pmatrix} \cos n\varphi \\ \sin n\varphi \end{pmatrix} \tag{3}$$

with the integer n counting the winding of the azimutal angle φ, being the baryon number.

This Ansatz is expressed compactly as

$$\phi^\alpha = \sin f \sin g\, n^\alpha \equiv a(\rho, z)\, n^\alpha;\ \phi^3 = \sin f \cos g \equiv b(\rho, z);\ \phi^4 = \cos f \equiv c(\rho, z) \tag{4}$$

with the functions $f = f(\rho, z)$ and $g = g(\rho, z)$ dependent on the radial coordinate $\rho = \sqrt{|x_\alpha|^2}$ of the $\mathbb{R}^2$ subspace, with the index $i = 1, 2$, and $z = x_3$. $\hat{x}^\alpha = x^\alpha/\rho$ is the unit radius vector in the $\mathbb{R}^2$ subspace and n^α is the unit vector with vortex number n. The Ansatz (4) has the same general structure as the parametrization used in [8]. Furthermore, one readily verifies that the parametrisation (4) is consistent, i.e., the complete set of the field equations which follows from the variation of the original action of the Skyrme model, is compatible with two equations which follow from variation of the reduced action on the Ansatz (4). In discussing the asymptotics we employ polar coordinates replacing $r = \sqrt{\rho^2 + z^2}$, $\theta = \arctan \frac{z}{\rho}$. It is also convenient to use the trigonometric parametrization of the Skyrme model in terms of the functions f, g to represent the energy functional of the model and the topological charge (baryon number) density [17], although it is not appropriate from the point of view of numerical calculations because of the numerical errors which originate from the disagreement between the boundary conditions on the angular type function $g(r, \theta)$ on the ρ-axis and the boundary points $r = 0, \infty$, respectively[1].

Indeed, the reduced two dimensional energy density functional resulting from the imposition of axial symmetry stated in the Ansatz (4), is given by [17],

$$\begin{aligned} E &= \frac{1}{r^2}\left((r\partial_r f)^2 + (\partial_\theta f)^2 + \left[(r\partial_r g)^2 + (\partial_\theta g)^2 + \frac{n^2 \sin^2 g}{\sin^2 \theta}\right]\sin^2 f\right) \\ &+ \frac{\sin^2 f}{r^4}\left((r\partial_r f\partial_\theta g - r\partial_r g\partial_\theta f)^2 + \frac{n^2 \sin^2 g}{\sin^2 \theta}\left[(r\partial_r f)^2 + (r\sin f\partial_r g)^2\right.\right. \\ &+ \left.\left.(\partial_\theta f)^2 + (\sin f\partial_\theta g)^2\right]\right) + \mu^2(1 - cos f) \end{aligned} \tag{5}$$

or, equivalently

$$\begin{aligned} E &= \left[1 + \left(\frac{na}{\rho}\right)^2\right]\left[(a_\rho^2 + b_\rho^2 + c_\rho^2) + (a_z^2 + b_z^2 + c_z^2)\right] \\ &+ \left(\frac{na}{\rho}\right)^2 + \left((a_{[\rho}\, b_{z]})^2 + (b_{[\rho}\, c_{z]})^2 + (c_{[\rho}\, a_{z]})^2\right) + \mu^2(1 - c) \end{aligned} \tag{6}$$

where $a_{[\rho}\, b_{z]} \equiv \partial_\rho a \partial_z b - \partial_\rho b \partial_z a = a_\rho b_z - b_\rho a_z$. The latter truncated functional represents some modification of the $O(3)$ sigma model on the half-plane.

[1] For the MS solution it results in the appearance of step function like dependence at these points: $g(0, \theta) = g(\infty, \theta) = \pi\ \Theta(\theta - \pi/2)$, Θ being the step function.

We shall be choosing our boundary conditions such that the resulting multi-Skyrmion, or, the Skyrmion–anti-Skyrmion chain solution has the appropriate baryon charge. In the case of the multi-Skyrmions this is the (topological) winding number n appearing in (3) and (4). In the case of Skyrmion-antiSkyrmion chains with an odd number of lumps the baryon number is again n, and it vanishes for Skyrmion–anti-Skyrmion chains with an even number of lumps. The topological charge is labeled with two distinct integers, the winding number n and a second integer m which specifies the asymptotic value of the function $g(r,\theta)$ in (4) when $r \to \infty$, namely

$$\lim_{r\to\infty} g(r,\theta) = m\,\theta\,. \tag{7}$$

The formula for the baryon number that we use is

$$B = \int \varrho_0 d^3x = -\frac{1}{2}\,n\,\left[\cos m\theta\right]_{\theta=0}^{\theta=\pi} = \frac{1}{2}\,n\,\left[1-(-1)^m\right]. \tag{8}$$

the configurations being classified according to the values of two integer numbers, n and m. The consideration above indicates that the case $m = 1$ corresponds to the (multi)skyrmions of topological charge n, while $m = 2$ yields a configuration with zero net topological charge consisting of two lumps; a sphaleron like axially symmetric static solution of the Skyrme model, consisting of a charge n Skyrmion and charge $-n$ anti-Skyrmion. More general, for odd values of m the winding number n coincides with the topological charge of the Skyrme field $B = n$ whereas even values of m correspond to the deformations of the topologically trivial sector. In the following we shall see that the value of the integer m defines the number of the constituents of the configuration which can be identified with individual charge n Skyrmions and charge $-n$ antiSkyrmions placed along the axis of symmetry in alternating order.

3. Numerical results

The Euler-Lagrange equations arising from the variations of (6) have been integrated by imposing the boundary conditions, which respect finite mass-energy and finite energy density conditions as well as regularity and symmetry requirements. Also the sigma-model constraint is imposed.

The numerical calculations are performed employing the package FIDISOL/CADSOL, based on the Newton-Raphson iterative procedure [18]. We solve the system of three coupled nonlinear partial differential equations numerically, on a non-equidistant grid in x and θ, employing the compact radial coordinate $x = r/(1+r) \in [0:1]$. Typical grids used have sizes 75×60.

It will be correct to impose boundary conditions on the fields a, b, c as [8]

$$a|_{r=\infty} = 0\,, \quad b|_{r=\infty} = 0\,, \quad c|_{r=\infty} = 1 \tag{9}$$

at infinity, and for the multi-Skyrmions and Skyrmion–anti-Skyrmion chains with odd number of constituents we require

$$a|_{r=0} = 0\,, \quad b|_{r=0} = 0 \quad c|_{r=0} = -1, \tag{10}$$

at the origin. For the Skyrmion–anti-Skyrmion pair and the chains with even number of constituents the Neumann boundary conditions must be imposed there on the fields b, c, i.e.,

$$a|_{r=0} = 0\,, \quad \partial_r b|_{r=0} = 0 \quad \partial_r c|_{r=0} = 0, \tag{11}$$

The boundary conditions along the z-axis for odd and even values of the integer number m appearing in (8) are

$$\begin{aligned} a|_{\theta=0} &= 0, \qquad \partial_\theta b|_{\theta=0} = 0, \quad \partial_\theta c|_{\theta=0} = 0; \\ a|_{\theta=\pi} &= 0, \qquad \partial_\theta b|_{\theta=\pi} = 0, \quad \partial_\theta c|_{\theta=\pi} = 0 \end{aligned} \tag{12}$$

To produce a configuration with correct topology and boundary condition which can be used as an input into the system of Euler-Lagrange equations, we implemented following algorithm. The trigonometric parametrisation of the triplet a, b, c given by the Ansatz (4) is used with linear dependence of the profile function $f(x,\theta) = \pi(1-x)$, $x \in [0,1]$ on the compact radial variable x and linear dependence of the second angular function $g(x,\theta) = m\theta$, $\theta \in [0,\pi]$ on the polar angle θ. Note we do not impose additional symmetry restrictions on the triplet of fields a, b, c in the input configuration, however the resulting numerical solutions reveal such a discrete symmetry with respect to the reflection $z \to -z$. In addition to the solution of the boundary problem performed with the package FIDISOL/CADSOL, the gradient flow equations are solved to check the numerical results although in the latter case convergence is somewhat slower.

As a first step we reproduced the well known results for charge n multi-Skyrmions [10, 11, 12] and for the charge 2 Skyrmion and charge -2 anti-Skyrmion pair [8]. Evidently, this algorithm can be implemented to construct Skyrmion–anti-Skyrmion chains, the configurations similar to the monopole-antimonopole chains constructed in [4, 5, 6, 14]. For example, the value of the integer number $m = 3$ corresponds to the charge $-n$ anti-Skyrmion located at the origin and two charge n Skyrmions located on the z-axis symmetrically with respect to the xy-plane, the system with $m = 4$, which resides in the topologically trivial sector, corresponds to the chain of 2 SAS pairs alternating on the symmetry axis, etc.

Indeed, for the values of the winding number $m \geq 3$ in the input configuration we produce the Skyrmion-anti-Skyrmion chain solutions with winding numbers $n \geq 2$. The relative error is estimated to be lower than 10^{-3}. Another check of the correctness of our results was performed by verifying that the virial relation [19], namely the identity that ensues from the Derrick scaling requirenment, is satisfied. This was done both for multi-Skyrmions (MS's) of different topological charges and for Skyrmion-antiSkyrmion chain configurations.

Although the initial configuration satisfies the boundary condition (7), we do not impose it as a boundary condition in our numerical calculations, nevertheless the solutions asymptotically tend to satisfy (7).

The numerical algorithm we impemented to solve the boundary problem allows us to evaluate the dipole moment of the solutions. Indeed, the leading term in the asymptotic expansion of the axially symmetric multi-Skyrmion solution is a dipole [9], i.e., $\phi^3 \sim \frac{d\cos\theta}{r^2} + O(r^{-3})$, so in terms of the compact radial coordinate x we can extract the value of the dipole moment from the first and second derivatives of the field ϕ^3 at the boundary $x = 1$: $d = \partial_x \phi^3(\theta, x)\Big|_{x=1,\theta=0} + \frac{1}{2}\partial^2_{xx}\phi^3(\theta, x)\Big|_{x=1,\theta=0}$ Fig. 1 displays the energy density surfaces for the charge 2 axially symmetric Skyrmion ($n = 2, m = 1$), charge 2 Skyrmion and charge -2 anti-Skyrmion pair ($n = 2, m = 2$) and for the Skyrmion–anti-Skyrmion chains where Skyrmions and anti-Skyrmions alternate along the symmetry axis ($n = 2, m = 3, 4, 5, 6$). Positions of the constituents can be identified as points in space where the field $\phi^4 \equiv c(\rho_0, z_0)$ is equal to -1 [8]. Indeed, Fig 2 demonstrates that these points almost coincide with the maxima of the energy density distribution. The higher energy Skyrmion–anti-Skyrmion chains are formed

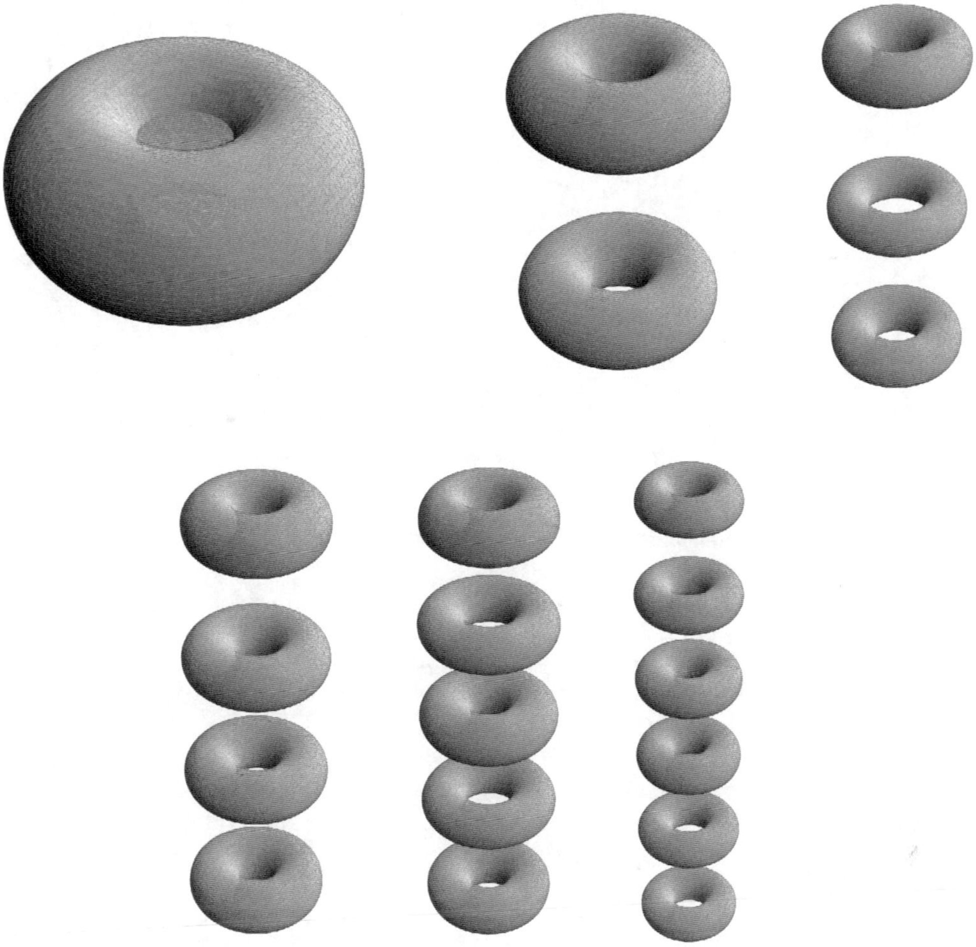

Figure 1. The 3d energy isosurfaces of the charge 2 axially symmetric Skyrmion and different Skyrmion–anti-Skyrmion chains for $n = 2$, $m = 1, \ldots 6$ are shown for $\mu = 0$. (in different scales, cf. Table 1).

from m constituents of charge $\pm n$. Numerical values of masses of these configurations, the corresponding interaction energy ΔE, the loci z_i of the field $c(0, z_i)$ and the values of the dipole moment d for the solutions with θ winding number $1 \leq m \leq 6$ are given in [1].

A qualitative similarity between these Skyrmion–antiSkyrmion chains and the monopole–anti-monopole chain solutions [6] is that in both cases there is a picture of an effective interaction between the constitutents which allows the sphaleron-type solution to exist, although the nature of the interaction is different. In the former case it is an effective electromagnetic interaction between the constituents [6, 7] while in the latter case there is a dipole-dipole interaction between the Skyrmions [9]. We observe that the distances between

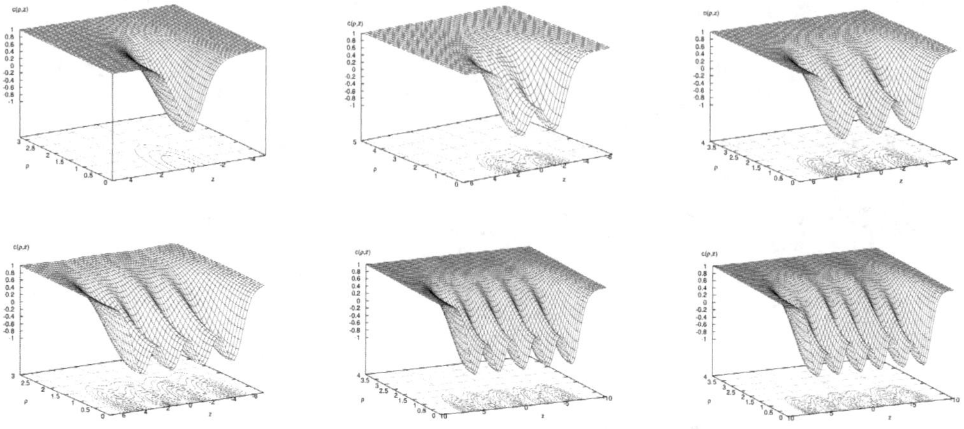

Figure 2. The field function $c(\rho, z)$ of the $m = 1$ charge 2 Skyrmion and Skyrmion-anti-Skyrmion chains for $n = 2$, $m = 1, \ldots 6$ are shown for $\mu = 0$ as functions of the coordinates ρ, z.

positions of the constituents do not vary much within a chain. One can try to model the Skyrmion–antiSkyrmion chains in the framework of the effective dipole-dipole interaction of the Skyrmions as was conjectured in [8].

Indeed the numerical results suggest the dipole moment of the Skyrmion-antiSkyrmion pair associated with the asymptotic behaviour of the field b, is vanishing. We observe similar results for chains with even number of constituents. For odd values of m there is only one component of the dipole moment of the configuration, directed along the symmetry axis. For these chains it almost coincides with the dipole moment of the single charge n Skyrmion. Considering multiSkyrmions of charge n, we can verify the conjecture of [8] concerning the additivity of the dipole moments. Indeed for the axially symmetric Skyrmions with topological charge $n = 1, \ldots, 4$ the dipole moment behaves approximately as $d = nd_1$ where $d_1 = 2.16$ is the dipole moment of the spherically symmetric charge one Skyrmion, although the results of the numerical calculations indicate it is slightly higher than this additive estimate for $n \geq 4$. On the other hand it is known that the axially symmetric configurations of charge $n \geq 3$ are not the minimal energy states of the system [3].

Note that both types of the chain solutions with even and odd values of m possess the symmetry

$$(a(\rho, z), b(\rho, z), c(\rho, z)) \to (-a(\rho, z), -b(\rho, z), c(\rho, z))$$

which corresponds to the inversion of the asymptotic pion dipole fields.

For $n = 2, 3, 4$ these m-Skyrmion chains possess m points on the z-axis where the field $\phi^4(0, z_0) = c(0, z_0) = -1$ and a soliton is placed. Due to reflection symmetry, each such a point $-z_0$ on the negative z-axis corresponds to a point z_0 on the positive z-axis. For even values of m the triplet of the fields a, b, c has reflection symmetry with respect to the xy plane

$$(a(\rho, z), b(\rho, z), c(\rho, z)) \to (-a(\rho, -z), b(\rho, -z), c(\rho, -z))$$

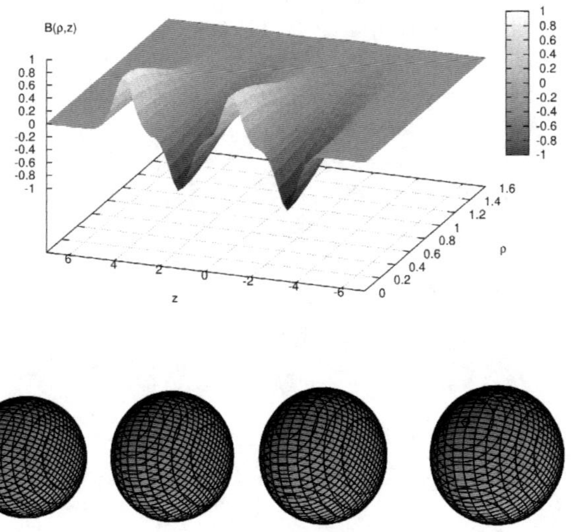

Figure 3. The topological charge density (top) $B(\rho, z)$ and the energy density isosurface of the $m = 4$ charge 1 Skyrmion and Skyrmion-anti-Skyrmion chain are shown for $\mu = 0.1$.

while for odd values of of m the reflection symmetry is

$$(a(\rho, z), b(\rho, z), c(\rho, z)) \to (a(\rho, -z), -b(\rho, -z), c(\rho, -z))$$

Note that these symmetries are not imposed by the boundary conditions on the fields but arise when an initial configuration relaxes into a solution. For even m the field $c(0, 0)$ is far from the value -1 at the origin, although its value there decreases with increasing of n.

Inclusion of the pion mass term in the Lagrangian (1) makes it possible for $n = 1$ Skyrmion-anti-Skyrmion chains to exist. In Fig. 3 we present the distribution of the topological charge density and the isosurface of the energy density for such a chain with $m = 4$ at $\mu = 0.1$. Note that such solutions do not exist either in Yang-Mills theory, nor in Skyrme theory in the absence of a pion mass potential. In these cases only topological charge pairs of $n \geq 2$ and $n \leq -2$ are found.

An interesting analogy with the multi-monopoles (MM) and the monopole-antimonopole (MA) chain solutions is that, the component of the Skyrme field ϕ^4 shows a clear relation to the corresponding behaviour of the magnitude $|\Phi|$ of the Higgs field of MM's and MAP pair, respectively [6], as seen in the plots in Fig 2. Indeed, these systems both are parametrised by two integer numbers, one of those being associated with the topological charge of the constituents and corresponds to the n-fold rotation about the symmetry axis, and the other winding number, m, appears in the boundary conditions imposed on the field configuration. More specifically, in order to construct the MA chains we impose asymptotic boundary conditions on the YMH system which correspond to the m-fold rotation of the fields as the polar angle θ varies from 0 to π [6]. Similarly, we impose the boundary condition

$g(x,\pi) = m\pi$ on the angular function in the Ansatz (4) to construct the Skyrmion-anti-Skyrmion chains. Furthermore, both in the YMH model and in the Skyrme model the structure of the configurations changes as the winding number n increases beyond some critical value at which the positions of the constituents are no longer associated with some set of isolated points on the symmetry axis but form circles around it. Note that the similarity between the monopole-antimonopole chains and chain solutions in the Skyrme model becomes even more transparent if we consider axially symmetric YM caloron solutions where monopoles and antimonopoles are constituents of the caloron [20]. Then there is a holonomy operator $\mathrm{Tr}\,\mathcal{P}(\vec{r}) = \cos||A_0(\vec{r})||$ whose loci $\mathcal{P}(\vec{r}_0) = -1$ are associated with the maxima of the action density [21]. On the other hand, in 1989 M. Atiyah and N. Manton observed that the holonomy of the Yang-Mills field yields good approximations to static Skyrmions [22], so our results provide another interesting link between the non-self dual sphaleron solutions of the Yang-Mills-Higgs model and the Skyrme model.

4. Conclusions

We have constructed new static axially symmetric Skyrmion–anti-Skyrmion chain solutions numerically. This was done by solving the two dimensional nonlinear partial differential equations as a boundary value problem, using the same formalism and numerical techniques as were used in the construction of the monopole–antimonopole chains [5, 6] and non–self-dual instantons [13]. In implementing the requisite boundary conditions for this task, two integers (m, n) are employed. Both these examples are solved as 2 dimensional problems, the former arising from the imposition of axial symmetry in $\mathbb{R}^3$ and the latter of bi-azimuthal symmetry in $\mathbb{R}^4$. The integer n labels the topological charge of each constituent lump or anti-lump. In the monopole case [5, 6] n is the winding of the azimuthal angle in $\mathbb{R}^3$, while in the instanton case [13] it is the winding of the two azimuthal angles in $\mathbb{R}^4$, taken to be equal. In both cases, it is the topological charge descendening from the second Chern-Pontryagin charge subject to the respective symmetries. The integer m enters the asymptotic value of the (form factor) function which maps on to the remaining angular coordinate, namely the first (polar) angle different from the azimuthal angle(s). m is not a topological charge. The total topological charge of any such finite action/energy configuration vanishes when m is *even*, and equals n when m is *odd*.

In the present work, we have restricted to $m = 2, \ldots, 6$. Clearly, for $m = 1$ we simply recover the axially symmetric MS of charge n, which we have done as a warmup. We have examined the cases with values of n, starting from $n = 1$, through to $n = 6$.

Our numerical investigations indicate fairly clearly that in the usual Skyrme model with no pion-mass term, there exist no zero baryon number solutions when each of the constituents carries baryon number $|n| < 2$. However, inclusion of the pion-mass term results in the existance of chains with $n = 1$. This is perhaps not surprising in the background of the known close analogy between solutions to Yang-Mills and sigma models respectively, *e.g.*, the usual YM model and the Skyrme model with no pion mass term. This analogy is demonstrated most simply by considering the unit charge solutions (BPST instantons) of the YM system parametrised by the radial function $w(r)$ on the one hand, and the unit charge solutions (hedgehog) of the Skyrme system parametrised by the radial chiral function $f(r)$ on the other. This corresponce is $w = \cos f$, and it manifests itself already in the one dimensional reduced actions of the two systems, but only in the absence of the pion mass potential in the (Skyrme) sigma model. This is clear since the one dimensional reduced action does not feature a pion mass like term $(1 - \cos f) \equiv (1 - w)$, in the presence of which the analogy between YM instantons and Skyrmions disappears.

The $n = 2, \mu = 0$ example was the one examined most intensively, since it is the first non marginal case where we could reliably verify the existence of the Skyrmion-antiSkyrmion chains. However, unlike in the two previous examples in [5, 6] and in [13], the binding energy of the constituents of the Skyrmion chain turned out to be quite weak, so the $n = 2$ chains, especially for the odd numbers of the consitutuents, are very unstable w.r.t. perturbations.

The nonexistence of the Skyrmion-antiSkyrmion chains with constituents carrying baryon numbers $\pm n$, for values of $n \leq n_{\rm min}$ is not surprising. In the corresponding analytic proof [15, 16] of existence for non–self-dual instantons of zero Pontyagin charge, the existence of the case where the constituents carried Pontyagin charge $n = |1|$ was not established, indicating its nonexistence. In that case $|n_{\rm min}| = 2$ which coincides with previous result of [8].

Acknowledgements
We would like to acknowledge numerous valuable discussions with Eugen Radu and Paul Sutcliffe. This work was carried out in the framework of Science Foundation Ireland (SFI) Research Frontiers Programme (RFP) project RFP07/FPHY330. Ya.S. is very grateful to the Organizers of the GROUP28 Colloquium for invitation and hospitality in Newcastle.

[1] Shnir Ya and Tchrakian D H 2010 *J. Phys.* A **43**, 025401
[2] Skyrme T H R 1961 *Proc. Roy. Soc. Lond.* A **260** 127
[3] Houghton C J, Manton N S and Sutcliffe P M 1998 *Nucl. Phys.* B **510** 507
[4] Nahm W, unpublished;
Rüber B 1985 *Eine axialsymmetrische magnetische Dipollsung der Yang-Mills-Higgs-Gleichungen* (Bonn)
[5] Kleihaus B and Kunz J 2000 *Phys. Rev.* D **61** 025003
[6] Kleihaus B, Kunz J and Shnir Ya 2003 *Phys. Lett.* B **570** 237;
Kleihaus B, Kunz J and Shnir Ya 2003 *Phys. Rev.* D **68** 101701;
Kleihaus B, Kunz J and Shnir Ya 2004 *Phys. Rev.* D **70** 065010.
[7] Shnir Ya 2005 *Phys. Rev.* D **72**, 055016
[8] Krusch S and Sutcliffe P 2004 *J. Phys.* A **37** 9037
[9] Manton N S 1994 *Acta Phys. Polon.* B **25**, 1757
[10] Weigel H, Schwesinger B and Holzwarth G 1986 *Phys. Lett.* B **168** 321.
[11] Kopeliovich V and Stern B 1987 *JETP Lett.* **45** 203
[12] Verbaarschot J 1987 *Phys. Lett.* B **195** 235
[13] Radu E and Tchrakian D H 2006 *Phys. Lett.* B **636** 201
[14] Paturyan V and Tchrakian D H 2004 *J. Math. Phys.* **45** 302
[15] Sadun L and Segert J 1992 *Commun. Math. Phys.* **145** 363
[16] Bor G 1992 *Commun. Math. Phys.* **145** 393
[17] Sawado N, Shiiki N, Maeda K I and Torii T 2004 *Gen. Rel. Grav.* **36** 1361
[18] Schönauer W and WeißR 1989 *J. Comput. Appl. Math.* **27** 279;
Schönauer W and Schnepf E 1987 *ACM Trans. on Math. Soft.* **13** 333
[19] Jaffe A and Taubes C H 1980 *Monopoles and Vortices* (Birkhäuser, Zurich)
[20] Shnir Ya 2007 *Europhys. Lett.* **77** 21001
[21] Shnir Ya and Ilgenfritz E M 2008 *PoS CONFINEMENT8* 038
[22] Atiyah M F and Manton N S 1989 *Phys. Lett.* B **222**, 438

Unified theory for external and internal attributes and symmetries of fundamental fermions

Ikuo S. Sogami

Maskawa Institute for Science and Culture, Kyoto Sangyo University, Kita-Ku, Kyoto 603-8555, JAPAN

E-mail: sogami@cc.kyoto-su.ac.jp

Abstract. An unorthodox unified theory is developed to describe external and internal attributes and symmetries of fundamental fermions, *quarks and leptons*. Basic ingredients of the theory are an algebra which consists of all the triple-direct-products of Dirac γ-matrices and a triple-spinor-field, called a *triplet field*, defined on the algebra. The algebra possesses three commutative subalgebras which describe, respectively, the external spacetime symmetry, the family structure and the internal color symmetry of quarks and leptons. The triplet field includes threefold (fourfold) repetitional modes of spin $\frac{1}{2}$ component fields with SU(3) (SU(4)) color symmetry. It is possible to qualify the Yukawa interaction and to make a new interpretation of its coupling constants naturally in an intrinsic mechanism of the triplet field formalism. The Dirac mass matrices with quasi-democratic structure are derived as an illustration.

1. Introduction

In the orthodox way of unification, such as the standard model and the grand unified theories, fundamental fermions belonging individually to the irreducible representations of the external Lorentz group are organized into definite multiplets of the internal symmetry group. Here, we propose an unorthodox scheme of unification where a family of fundamental fermions is described by a master spinor field obeying to a generalized Dirac equation with coefficients belonging to an algebra, called a *triplet algebra*, which consists of all the triple-direct-products of the Dirac matrices. In accordance with the Coleman-Mandula theorem [1, 2] insisting that "spacetime and internal symmetries cannot be combined in any but a trivial way", we decompose the triplet algebra into mutually-commutative subalgebras. The subalgebra related with internal property of fundamental fermions, *quarks and leptons*, is made up of two subalgebras which can be related, respectively, to repetitional family structure and color symmetry. The master spinor field, named a *triplet field*, is proved to include threefold (fourfold) repetitional modes of spin $\frac{1}{2}$ component fields with SU(3) (SU(4)) color symmetry.

We construct explicitly the triplet algebra and its subalgebra for the external spacetime symmetry in §2 and define the triplet field forming a multiple representation of the Lorentz group in §3. Characteristic of the subalgebra for internal properties of fundamental fermions is closely examined in §4 and §5. First, chirality operators acting on the triplet field are decomposed further into the sums of the projection operators which can be interpreted to discriminate repetitional degrees of freedom for family structure. Second, we find the algebra for color symmetry in the set of the elements which are commutative both with the chirality operators and all elements of the algebra for external symmetry. In the present scheme, the triplet field has no degrees of freedom for the symmetry of electroweak interaction. In §6, we reformulate the standard model by introducing the set of chiral triplet fields for the left-handed doublet and the right-handed singlets of the Weinberg-Salam symmetry. It turns out possible to qualify the Yukawa interaction and make a new interpretation on the Yukawa coupling constants naturally in an intrinsic mechanism of the triplet field formalism. As an explicit illustration, the Dirac mass matrices with quasi-democratic structure are derived. Possible applications of the present scheme to the systems of fundamental fermions with different attributes and symmetries are discussed in §7.

2. Triplet algebra and subalgebra for external Lorentz symmetry

The Dirac's γ-matrices γ_μ ($\mu = 0,\ 1,\ 2,\ 3$) satisfying the Clifford anticommutation relations $\gamma_\mu\gamma_\nu + \gamma_\nu\gamma_\mu = 2\eta_{\mu\nu}1$ with $(\eta_{\mu\nu}) = \text{diagonal}\,(1,-1,-1,-1)$ are introduced solely as the mathematical quantities which carry no direct physical meaning. Hermite conjugate of γ_μ is defined by $\gamma_\mu^\dagger = \gamma_0\gamma_\mu\gamma_0$. The γ-matrices generate the 16 dimensional manifold, named a *Dirac algebra*, as

$$A_\gamma = \langle\gamma_\mu\rangle = \{1,\ \gamma_\mu,\ \sigma_{\mu\nu},\ \gamma_5\gamma_\mu,\ \gamma_5\} \tag{1}$$

where $\sigma_{\mu\nu} = \frac{i}{2}(\gamma_\mu\gamma_\nu - \gamma_\nu\gamma_\mu)$ and $\gamma_5 = i\gamma_0\gamma_1\gamma_2\gamma_3 = \gamma^5$.

Let us call the triple-direct-product of the bases 1, γ_μ, $\sigma_{\mu\nu}$, $\gamma_5\gamma_\mu$ and γ_5 a *primitive triplet* and the triple-direct-product of arbitrary elements of A_γ a *triplet.* Then, we define the *triplet algebra* A_T as the 16^3 dimensional manifold spanned by all the linear combinations of triplets. In other words, the triplet algebra A_T is generated in terms of the 12 primitive triplets $\gamma_\mu\otimes1\otimes1$, $1\otimes\gamma_\mu\otimes1$ and $1\otimes1\otimes\gamma_\mu$ as follows:

$$A_T = \langle\gamma_\mu\otimes1\otimes1,\ 1\otimes\gamma_\mu\otimes1,\ 1\otimes1\otimes\gamma_\mu\rangle. \tag{2}$$

The primitive triplets $A_1\otimes B_1\otimes C_1$ and $A_2\otimes B_2\otimes C_2$ obey the multiplication rule

$$(A_1\otimes B_1\otimes C_1)(A_2\otimes B_2\otimes C_2) = A_1A_2\otimes B_1B_2\otimes C_1C_2 \tag{3}$$

and the triplets satisfy the distribution law

$$\left(\sum_i a_iA_i\right)\otimes\left(\sum_j b_jB_j\right)\otimes\left(\sum_k c_kC_k\right) = \sum_{ijk} a_ib_jc_kA_i\otimes B_j\otimes C_k \tag{4}$$

where $A's$, $B's$ and $C's$ are the bases of A_γ in (1), and $a's$, $b's$ and $c's$ are complex numbers. The transpose (Hermite conjugate) of the primitive triplet is defined by the triple-direct-product of its transposed (Hermite conjugate) components of A_γ as

$$(A\otimes B\otimes C)^T = A^T\otimes B^T\otimes C^T,\quad (A\otimes B\otimes C)^\dagger = A^\dagger\otimes B^\dagger\otimes C^\dagger \tag{5}$$

and the trace of the triplet is set to be

$$\mathrm{Tr}\,(A\otimes B\otimes C) = \mathrm{Tr}\,(A)\,\mathrm{Tr}\,(B)\,\mathrm{Tr}\,(C)\,. \tag{6}$$

The triplet algebra A_T is large enough to form a variety of realizations of the Lorentz group representation. To obtain such kind of multiple spinor representation that describes exclusively the spin $\frac{1}{2}$ states, we note that the primitive triplets

$$\Gamma_\mu = \gamma_\mu\otimes\gamma_\mu\otimes\gamma_\mu \qquad (\mu = 0,\ 1,\ 2,\ 3) \tag{7}$$

satisfy the Clifford relations [3]

$$\Gamma_\mu\Gamma_\nu + \Gamma_\nu\Gamma_\mu = 2\eta_{\mu\nu} I \tag{8}$$

where $I = 1 \otimes 1 \otimes 1$. Hermite conjugate of Γ_μ is defined by

$$\Gamma_\mu^\dagger = \Gamma_0\Gamma_\mu\Gamma_0 \tag{9}$$

and the traces are calculated to be

$$\mathrm{Tr}\,\Gamma_\mu = 0, \quad \mathrm{Tr}\,(\Gamma_\mu\Gamma_\nu) = 64\eta_{\mu\nu}. \tag{10}$$

With these new elements, we are now able to construct a subalgebra of A_T being isomorphic to the Dirac algebra A_γ as follows:

$$A_\Gamma = \ \langle\Gamma_\mu\rangle = \ \{1,\, \Gamma_\mu,\, \Sigma_{\mu\nu},\, \Gamma_5\Gamma_\mu,\, \Gamma_5\} \tag{11}$$

where

$$\Sigma_{\mu\nu} = \frac{i}{2}(\Gamma_\mu\Gamma_\nu - \Gamma_\nu\Gamma_\mu) = -\sigma_{\mu\nu} \otimes \sigma_{\mu\nu} \otimes \sigma_{\mu\nu} \tag{12}$$

and

$$\Gamma_5 = i\Gamma_0\Gamma_1\Gamma_2\Gamma_3 = \Gamma^5 = -\gamma_5 \otimes \gamma_5 \otimes \gamma_5. \tag{13}$$

The set of elements which are commutative with the subalgebra A_Γ, *i.e.*,

$$C_\Gamma = C(A_\Gamma : A_T) = \{\, X \in A_T : [\, X,\ \Gamma_\mu\,] = 0\,\} \tag{14}$$

forms a subalgebra of A_T, called a *centralizer* of A_Γ in A_T. The primitive triplet in C_Γ is proved to have even numbers of components γ_μ-matrices for all μ [3]. Therefore, the composition of C_Γ is determined to be

$$C_\Gamma = \langle\ 1 \otimes \gamma_\mu \otimes \gamma_\mu,\ \gamma_\mu \otimes 1 \otimes \gamma_\mu\ \rangle. \tag{15}$$

The triplet algebra has a non-intersecting decomposition $A_\Gamma \cup C_\Gamma = A_T$.

Note that all of the antisymmetric operators $\Sigma_{\mu\nu}$ form a subalgebra

$$A_\Sigma = \left\{\Sigma_{\mu\nu}\right\} \subset A_\Gamma \subset A_T. \tag{16}$$

We identify this subalgebra A_Σ with the *external algebra* which generates the Lorentz group in our spacetime. It is readily proved that the operators

$$M_{\mu\nu} = \frac{1}{2}\Sigma_{\mu\nu} \tag{17}$$

are subject to the commutation relations of the Lie algebra for the orthogonal group O(1, 3) as

$$[M_{\kappa\lambda},\ M_{\mu\nu}] = -i\eta_{\kappa\mu}M_{\lambda\nu} + i\eta_{\kappa\nu}M_{\lambda\mu} - i\eta_{\lambda\nu}M_{\kappa\mu} + i\eta_{\lambda\mu}M_{\kappa\nu} \tag{18}$$

and $M_{\mu\nu}$ and Γ_λ satisfy the relations

$$[M_{\mu\nu},\ \Gamma_\lambda] = i\eta_{\lambda\nu}\Gamma_\mu - i\eta_{\lambda\mu}\Gamma_\nu. \tag{19}$$

Accordingly, it is possible to postulate that the operators $M_{\mu\nu}$ generate the Lorentz transformations and that the subscripts of operators consisting of Γ_μ refer directly to the superscripts of the spacetime coordinates $\{\, x^\mu\, \}$ of the 4 dimensional world where we exist as observers. Therefore, it is natural to define the raising and lowering operations of index of Γ_μ by the metric $\eta_{\mu\nu}$ as follows:

$$\Gamma^\mu = \eta^{\mu\nu}\Gamma_\nu, \quad \Gamma_\mu = \eta_{\mu\nu}\Gamma^\nu. \tag{20}$$

As is evident from the construction so far, the component γ-matrices carry no direct physical meaning and regarded as playing only the role of *alphabets*. It is the elements of the triplet algebra A_T that play the role of *codons* carrying the direct physical meanings.

Journal of Physics: Conference Series **284** (2011) 012054 doi:10.1088/1742-6596/284/1/012054

3. Multiple spinor representation of the Lorentz group

At this stage, we are allowed to introduce the triplet field, $\Psi(x)$, on the spacetime point x^μ, which spans a 4^3 dimensional vector space for multiple spinor representations of the Lorentz group. The triplet field is proved to possess 4^2 component fields with spin $\frac{1}{2}$ carrying different internal attributes. For the triplet field $\Psi(x)$ and its adjoint field

$$\overline{\Psi}(x) = \Psi^\dagger(x)\Gamma_0, \tag{21}$$

the scalar product is defined as

$$\overline{\Psi}(x)\Psi(x) = \sum_{abc} \overline{\Psi}_{abc}(x)\Psi_{abc}(x). \tag{22}$$

Under the proper Lorentz transformation $x'^\mu = \Omega^\mu{}_\nu x^\nu$ where $\Omega_{\lambda\mu}\Omega^\lambda{}_\nu = \eta_{\mu\nu}$ and $\det\Omega = 1$, the triplet field and its adjoint are presumed to transform as

$$\Psi'(x') = S(\Omega)\Psi(x), \quad \overline{\Psi}'(x') = \overline{\Psi}(x)S^{-1}(\Omega) \tag{23}$$

where the transformation matrix $S(\Omega)$ belongs to the external algebra A_Σ. For the bilinear form $\overline{\Psi}(x)\Gamma_\mu\Psi(x)$ to behave as the vector under the Lorentz transformation, $S(\Omega)$ must satisfy the conditions

$$S^{-1}(\Omega)\Gamma_\mu S(\Omega) = \Omega_\mu{}^\nu\Gamma_\nu, \quad S^{-1}(\Omega) = \Gamma_0 S^\dagger(\Omega)\Gamma_0 \tag{24}$$

which determine its explicit form to be

$$S(\Omega) = \exp\left(-\frac{i}{2}M_{\mu\nu}\omega^{\mu\nu}\right) \tag{25}$$

for the spacetime rotations with angles $\omega^{\mu\nu}$ in the μ-ν planes. For the discrete spacetime transformations such as the space inversion, the time reversal and the charge conjugation, the present scheme retains exactly the same structure as the ordinary Dirac theory.

The internal attributes of fundamental fermions should be fixed independently of the inertial frame of reference in which observations are made. This means that the generators specifying their internal attributes must be commutative with the elements of the external algebra A_Σ for the Lorentz transformations. Accordingly, we are led naturally to postulate that the algebra for internal symmetry should be related to the centralizer of A_Σ defined by

$$C_\Sigma = C(A_\Sigma : A_T) = \{X \in A_T : [X, \Sigma_{\mu\nu}] = 0\}. \tag{26}$$

The primitive triplet of the centralizer C_Σ is proved to consist of either an even number or an odd number of component γ_μ-matrix for all μ [3]. Consequently, the composition of the centralizer C_Σ is determined as follows:

$$C_\Sigma = \langle\, 1\otimes\gamma_\mu\otimes\gamma_\mu,\ \gamma_\mu\otimes 1\otimes\gamma_\mu,\ \gamma_5\otimes\gamma_5\otimes\gamma_5\,\rangle. \tag{27}$$

It is now recognized that the triplet algebra has another non-intersecting decomposition $A_\Sigma \cup C_\Sigma = A_T$. We have to examine detailed structures of C_Σ, which is identified with *internal algebra*, and to inquire what sorts of internal attributes of the fundamental fermions can be inscribed on the triplet field $\Psi(x)$ forming the multiple spinor representation of the Lorentz group. The subalgebras of the triplet algebra A_T defined so far, A_Γ, A_Σ, C_Γ and C_Σ, obey the inclusion relations

$$A_\Sigma \subset A_\Gamma \subset A_T : \ C_\Gamma \subset C_\Sigma \subset A_T. \tag{28}$$

Connectedly, the element Γ_5 belongs to the subalgebras A_Γ and C_Σ as follows:

$$\Gamma_5 \notin A_\Sigma,\ \Gamma_5 \in A_\Gamma :\ \Gamma_5 \notin C_\Gamma,\ \Gamma_5 \in C_\Sigma. \tag{29}$$

Journal of Physics: Conference Series **284** (2011) 012054 doi:10.1088/1742-6596/284/1/012054

4. Subdivision of chirality operators and repetitional modes of family structure

Equation (29) shows that the chirality operators

$$L=\frac{1}{2}(I-\Gamma_5),\quad R=\frac{1}{2}(I+\Gamma_5) \tag{30}$$

belong to both of the subalgebras A_Γ and C_Σ. It is straightforward to verify that these operators have the following subdivisions in the internal algebra C_Σ as

$$L=\ell\otimes r\otimes r+r\otimes \ell\otimes r+r\otimes r\otimes \ell+\ell\otimes\ell\otimes\ell \tag{31}$$

and

$$R=r\otimes \ell\otimes \ell+\ell\otimes r\otimes \ell+\ell\otimes\ell\otimes r+r\otimes r\otimes r \tag{32}$$

where

$$\ell=\frac{1}{2}(1-\gamma_5),\quad r=\frac{1}{2}(1+\gamma_5). \tag{33}$$

Note that this subdivision cannot be realized in the algebra A_Γ.

To give independent physical roles and meanings to those subdivided terms in (31) and (32), let us introduce the following operators in C_Σ as

$$\begin{cases} \Pi_{1L}=\ell\otimes r\otimes r, & \Pi_{1R}=r\otimes\ell\otimes\ell, \\ \Pi_{2L}=r\otimes\ell\otimes r, & \Pi_{2R}=\ell\otimes r\otimes \ell, \\ \Pi_{3L}=r\otimes r\otimes \ell, & \Pi_{3R}=\ell\otimes\ell\otimes r, \\ \Pi_{4L}=\ell\otimes\ell\otimes\ell, & \Pi_{4R}=r\otimes r\otimes r, \end{cases} \tag{34}$$

which are chiral projection operators obeying the relations

$$\Pi_{ih}\Pi_{jh'}=\delta_{ij}\delta_{hh'}\Pi_{ih},\quad \sum_{ih}\Pi_{ih}=I \tag{35}$$

for $i,\ j=1,\ 2,\ 3,\ 4$ and $h,\ h'=L,\ R$. Next, we define the non-chiral projection operators

$$\Pi_i=\Pi_{iL}+\Pi_{iR} \tag{36}$$

satisfying the relations

$$\Pi_i\Pi_j=\delta_{ij}\Pi_i,\quad \sum_i\Pi_i=I. \tag{37}$$

All of these operators form the subalgebra $\{\Pi_{ih}\}=\langle\ \gamma_5\otimes 1\otimes 1,\ 1\otimes\gamma_5\otimes 1,\ 1\otimes 1\otimes\gamma_5\ \rangle$ of the internal algebra C_Σ.

Here, we make interpretation that the projected modes of triplet field given by

$$\Psi_{iL}(x)=L\Pi_i\Psi(x)=\Pi_{iL}\Psi(x),\quad \Psi_{iR}(x)=R\Pi_i\Psi(x)=\Pi_{iR}\Psi(x) \tag{38}$$

and

$$\Psi_i(x)=\Pi_i\Psi(x)=\Psi_{iL}(x)+\Psi_{iR}(x) \tag{39}$$

represent the repetitional family structures of quarks and leptons. To apply this scheme to the three generations of quarks and leptons observed in the low energy regime, we impose the constraint condition, a posteriori,

$$\Pi_4\Psi(x)=0 \tag{40}$$

on the triplet field to suppress the fourth mode. Henceforth, we describe this three generation model in the present paper. Note, however, that it is possible also to describe the model including the fourth mode whose existence could be confirmed in sufficiently high energy regime.

Journal of Physics: Conference Series **284** (2011) 012054 doi:10.1088/1742-6596/284/1/012054

5. Ordinary and extended color symmetries

In the previous section, one aspect of the internal algebra C_Σ was clarified by examining its subalgebra $\{\Pi_{ih}\}$. Consequently, it is necessary and reasonable to investigate the structure of the centralizer $C(\{\Pi_{ih}\} : C_\Sigma)$ of $\{\Pi_{ih}\}$ in C_Σ. For its purpose, let us notice that the Dirac algebra A_γ has the three elements

$$\rho_1 = i\gamma_2\gamma_3, \ \rho_2 = i\gamma_3\gamma_1, \ \rho_3 = i\gamma_1\gamma_2 \tag{41}$$

which satisfy the multiplication rules of the Pauli algebra, *i.e.*, $\rho_a\rho_b = \delta_{ab}I + i\epsilon_{abc}\rho_c$. Taking the triple-direct-products of these elements, we are able to construct the bases for a 16 dimensional submanifold included exclusively in the centralizer $C(\{\Pi_{ih}\} : C_\Sigma)$ as follows:

$$\begin{cases} \lambda_1 = \frac{1}{2}(\rho_1 \otimes \rho_1 \otimes 1 + \rho_2 \otimes \rho_2 \otimes 1), & \lambda_2 = \frac{1}{2}(\rho_1 \otimes \rho_2 \otimes \rho_3 - \rho_2 \otimes \rho_1 \otimes \rho_3), \\ \lambda_3 = \frac{1}{2}(1 \otimes \rho_3 \otimes \rho_3 - \rho_3 \otimes 1 \otimes \rho_3), & \\ \lambda_4 = \frac{1}{2}(\rho_1 \otimes 1 \otimes \rho_1 + \rho_2 \otimes 1 \otimes \rho_2), & \lambda_5 = \frac{1}{2}(\rho_1 \otimes \rho_3 \otimes \rho_2 - \rho_2 \otimes \rho_3 \otimes \rho_1), \\ \lambda_6 = \frac{1}{2}(1 \otimes \rho_1 \otimes \rho_1 + 1 \otimes \rho_2 \otimes \rho_2), & \lambda_7 = \frac{1}{2}(\rho_3 \otimes \rho_1 \otimes \rho_2 - \rho_3 \otimes \rho_2 \otimes \rho_1), \\ \lambda_8 = \frac{1}{2\sqrt{3}}(1 \otimes \rho_3 \otimes \rho_3 + \rho_3 \otimes 1 \otimes \rho_3 - 2\rho_3 \otimes \rho_3 \otimes 1), & \\ \lambda_9 = \frac{1}{2}(1 \otimes \rho_1 \otimes \rho_1 - 1 \otimes \rho_2 \otimes \rho_2), & \lambda_{10} = -\frac{1}{2}(\rho_3 \otimes \rho_1 \otimes \rho_2 + \rho_3 \otimes \rho_2 \otimes \rho_1), \\ \lambda_{11} = \frac{1}{2}(\rho_1 \otimes 1 \otimes \rho_1 - \rho_2 \otimes 1 \otimes \rho_2), & \lambda_{12} = -\frac{1}{2}(\rho_1 \otimes \rho_3 \otimes \rho_2 + \rho_2 \otimes \rho_3 \otimes \rho_1), \\ \lambda_{13} = \frac{1}{2}(\rho_1 \otimes \rho_1 \otimes 1 - \rho_2 \otimes \rho_2 \otimes 1), & \lambda_{14} = \frac{1}{2}(\rho_1 \otimes \rho_2 \otimes \rho_3 + \rho_2 \otimes \rho_1 \otimes \rho_3), \\ \lambda_{15} = -\frac{1}{\sqrt{6}}(1 \otimes \rho_3 \otimes \rho_3 + \rho_3 \otimes 1 \otimes \rho_3 + \rho_3 \otimes \rho_3 \otimes 1). & \end{cases} \tag{42}$$

The 15 operators λ_j $(j = 1, 2, \cdots, 15)$ satisfy the commutation relations

$$[\lambda_j, \lambda_k] = 2f_{jkl}\lambda_l \tag{43}$$

and the anticommutation relations

$$\{\lambda_j, \lambda_k\} = \delta_{jk}I + 2d_{jkl}\lambda_l \tag{44}$$

of the Lie algebra A_3, where f_{jkl} and d_{jkl} are the symmetric and antisymmetric constants characterizing the algebra. The operators λ_j are self-adjoint and have the traces

$$\mathrm{Tr}\lambda_j = 0, \quad \mathrm{Tr}\lambda_j\lambda_k = 32\delta_{jk}. \tag{45}$$

We postulate, here, that the Lie group generated by the algebra

$$A_{c4} = \{I, \lambda_1, \lambda_2, \cdots, \lambda_{15}\} \tag{46}$$

describes an extended color $\mathrm{SU}_c(4)$ symmetry. Evidently, this is one realization of the Pati-Salam symmetry in which the leptons are interpreted as the 4-th *lilac* color mode.

The projection operators to the quark and lepton sectors are constructed, respectively, by

$$\Lambda^{(q)} = \frac{1}{4}(3I - \otimes\rho_3 \otimes \rho_3 - \rho_3 \otimes 1 \otimes \rho_3 - \rho_3 \otimes \rho_3 \otimes 1) \tag{47}$$

and

$$\Lambda^{(\ell)} = \frac{1}{4}(I + \otimes\rho_3 \otimes \rho_3 + \rho_3 \otimes 1 \otimes \rho_3 + \rho_3 \otimes \rho_3 \otimes 1) \equiv \Lambda_\ell \tag{48}$$

which are subject to the relations

$$\Lambda^{(a)}\Lambda^{(b)} = \delta_{ab}\Lambda^{(a)} \tag{49}$$

for $a, b = q, \ell$. Accordingly, the operator for *baryon number minus lepton number* takes the form

$$B - L = \frac{1}{3}\Lambda^{(q)} - \Lambda^{(\ell)} = -\frac{1}{3}(1 \otimes \rho_3 \otimes \rho_3 + \rho_3 \otimes 1 \otimes \rho_3 + \rho_3 \otimes \rho_3 \otimes 1) = \sqrt{\frac{2}{3}}\lambda_{15}. \tag{50}$$

It is straightforward to verify that the operator $\Lambda^{(q)}$ has the subdivision as

$$\Lambda^{(q)} = \Lambda_r + \Lambda_y + \Lambda_g \tag{51}$$

where

$$\begin{cases} \Lambda_r = \frac{1}{4}\left(I + \otimes\rho_3 \otimes \rho_3 - \rho_3 \otimes 1 \otimes \rho_3 - \rho_3 \otimes \rho_3 \otimes 1\right), \\ \Lambda_y = \frac{1}{4}\left(I - \otimes\rho_3 \otimes \rho_3 + \rho_3 \otimes 1 \otimes \rho_3 - \rho_3 \otimes \rho_3 \otimes 1\right), \\ \Lambda_g = \frac{1}{4}\left(I - \otimes\rho_3 \otimes \rho_3 - \rho_3 \otimes 1 \otimes \rho_3 + \rho_3 \otimes \rho_3 \otimes 1\right) \end{cases} \tag{52}$$

are the projection operators into tricolor quark modes. The operators Λ_a satisfy the relations

$$\Lambda_a\Lambda_b = \delta_{ab}\Lambda_a, \quad \sum_a \Lambda_a = I \tag{53}$$

for $a,\ b = r,\ y,\ g,\ \ell$.

The extended color symmetry $\mathrm{SU}_c(4)$ is presumed to be broken spontaneously at an appropriate energy scale down to the symmetry of semi-simple group $\mathrm{SU}_c(3)\otimes\mathrm{U}_{B-L}(1)$. To see how $\mathrm{SU}_c(3)$ is embedded in $\mathrm{SU}_c(4)$, it is necessary to note the following property of the constant d_{jkl} of SU(4) group as $d_{jkl} = \frac{1}{\sqrt{6}}\delta_{k15}$ for $j \leq 8$, and the relation between λ_{15} and $\Lambda^{(q)}$ in (50). For the eight operators λ_j $(j = 1,\ 2,\ \cdots, 8)$, we obtain the commutation relations in (43) with the structure constants f_{jkl} of the SU(3) group and anticommutation relations

$$\{\lambda_j,\ \lambda_k\} = \frac{4}{3}\delta_{jk}\Lambda^{(q)} + 2d_{jkl}\lambda_l. \tag{54}$$

Note that the projection operator $\Lambda^{(q)}$ appears in the right hand side of this relation. Without it, this relation runs into contradiction, when applied to the operator $\Lambda^{(\ell)}$ from the right. The identities

$$\Lambda^{(a)}\lambda_j = \delta_{aq}\lambda_j \tag{55}$$

which hold for $a = q, \ell$ and $j = 1, 2, \cdots, 8$ imply that $\lambda_j(\leq 8)$ are simultaneous eigenvectors of $\Lambda^{(q)}$ and $\Lambda^{(\ell)}$ with respective eigenvalues 1 and 0, and that the operators $\lambda_j(\leq 8)$ annihilate the leptonic mode. These results prove that the 9 dimensional submanifold of the extended color algebra A_{c4} defined by

$$A_{c3} = \{I, \lambda_1, \lambda_2, \cdots, \lambda_8\} \tag{56}$$

generates the color $\mathrm{SU}_c(3)$ group. The $\mathrm{U}_{B-L}(1)$ group is generated by the operator $B - L$ given in (50).

From the triplet spinor field $\Psi(x)$, the quark and lepton fields are projected out by

$$\Psi^{(a)}(x) = \Lambda^{(a)}\Psi(x), \tag{57}$$

respectively, for $a = q,\ \ell$, and four color modes of $\mathrm{SU}_c(4)$ symmetry are given by

$$\Psi_a(x) = \Lambda_a\Psi(x) \tag{58}$$

for $a = r,\ y,\ g,\ \ell$.

6. Standard model in the triplet field formalism

To separate the component fields of internal modes of the triplet field, it is convenient to introduce the braket symbols for the projection operators Λ_a and Π_{ih} by

$$\Lambda_a =| a\rangle\langle a |, \quad \Pi_{ih} =| i\, h\rangle\langle i\, h | \,. \tag{59}$$

Then, the decomposition of the bilinear scalar and vector forms of the triplet fields can be achieved as follows:

$$\bar{\Psi}(x)\Psi(x) = \sum_a \sum_{ih} \bar{\Psi}(x)\Lambda_a\Pi_{ih}\Psi(x) = \sum_a \sum_{ih} \bar{\Psi}_{ai\bar{h}}(x)\Psi_{aih}(x) \tag{60}$$

and

$$\bar{\Psi}(x)\Gamma_\mu\Psi(x) = \sum_a \sum_{ih} \bar{\Psi}(x)\Gamma_\mu\Lambda_a\Pi_{ih}\Psi(x) = \sum_a \sum_{ih} \bar{\Psi}_{aih}(x)\Gamma_\mu\Psi_{aih}(x) \tag{61}$$

where $\bar{h}$ implies that $\bar{L} = R$ and $\bar{R} = L$. The chiral component fields $\Psi_{aih}(x)$ and $\bar{\Psi}_{aih}(x)$ are defined by

$$\Lambda_a\Pi_{ih}\Psi(x) = |aih\rangle\langle aih|\Psi(x)\rangle = |aih\rangle\Psi_{aih}(x) \tag{62}$$

and

$$\bar{\Psi}(x)\Lambda_a\Pi_{ih} = \langle\bar{\Psi}(x)|aih\rangle\langle aih| = \bar{\Psi}_{ai\bar{h}}(x)\langle aih|. \tag{63}$$

We have extracted maximally 4^2 component spinor fields with spin $\frac{1}{2}$, $\Psi_{ai} = \sum_h \Psi_{aih}$, from the original triplet spinor field $\Psi(x)$.

In the present formalism, the triplet field has no degrees of freedom for the Weinberg-Salam symmetry $\mathrm{SU}_L(2)\otimes\mathrm{U}_Y(1)$. Therefore, to describe the actual system of quarks and leptons in this algebraic scheme, it is required to prepare a set of triplet fields to make up for the deficit degrees of freedom for electroweak interaction.

Here we incorporate the Weinberg-Salam symmetry into our formalism by introducing the set of chiral triplet fields so that the left-handed triplets constitute the doublet

$$\Psi_L(x) = \begin{pmatrix} \Psi^{(\uparrow)}(x) \\ \Psi^{(\downarrow)}(x) \end{pmatrix}_L \tag{64}$$

and the right-handed triplets form the singlets

$$\Psi_R^{(\uparrow)}(x), \quad \Psi_R^{(\downarrow)}(x) \tag{65}$$

of the electroweak isospin, where $(\uparrow)$ and $(\downarrow)$ signify the up and down states. The quark and lepton parts of the electroweak doublet $\Psi_L(x)$ have the component modes expressed schematically as

$$\Psi_L^{(q)}(x) = \Lambda^{(q)}\Psi_L(x) = \begin{pmatrix} \Lambda^{(q)}\Lambda_a\Pi_{iL}\Psi^{(\uparrow)} \\ \Lambda^{(q)}\Lambda_a\Pi_{iL}\Psi^{(\downarrow)} \end{pmatrix}_L = \begin{pmatrix} u_a & c_a & t_a \\ d_a & s_a & b_a \end{pmatrix}_L \tag{66}$$

and

$$\Psi_L^{(\ell)}(x) = \Lambda^{(\ell)}\Psi_L(x) = \begin{pmatrix} \Lambda^{(\ell)}\Pi_{iL}\Psi^{(\uparrow)} \\ \Lambda^{(\ell)}\Pi_{iL}\Psi^{(\downarrow)} \end{pmatrix}_L = \begin{pmatrix} \nu_e & \nu_\mu & \nu_\mu \\ e & \mu & \tau \end{pmatrix}_L . \tag{67}$$

Similarly, the electroweak singlets $\Psi_R^{(\uparrow)}(x)$ and $\Psi_R^{(\downarrow)}(x)$ possess the quark and lepton parts with contents of the component modes as follows:

$$\Psi_R^{(u)}(x) = \Lambda^{(q)}\Psi_R^{(\uparrow)}(x) = \left(\Lambda^{(q)}\Lambda_a\Pi_{iR}\Psi_R^{(\uparrow)}\right) = \left(\, u_a \quad c_a \quad t_a \,\right)_R , \tag{68}$$

Journal of Physics: Conference Series **284** (2011) 012054 doi:10.1088/1742-6596/284/1/012054

$$\Psi_R^{(d)}(x) = \Lambda^{(q)}\Psi_R^{(\downarrow)}(x) = \left(\Lambda^{(q)}\Lambda_a\Pi_{iR}\Psi_R^{(\downarrow)}\right) = \left(\begin{array}{ccc} d_a & s_a & b_a \end{array}\right)_R, \tag{69}$$

$$\Psi_R^{(\nu)}(x) = \Lambda^{(\ell)}\Psi_R^{(\uparrow)}(x) = \left(\Lambda^{(\ell)}\Pi_{iR}\Psi_R^{(\uparrow)}\right) = \left(\begin{array}{ccc} \nu_e & \nu_\mu & \nu_\tau \end{array}\right)_R, \tag{70}$$

and

$$\Psi_R^{(e)}(x) = \Lambda^{(\ell)}\Psi_R^{(\downarrow)}(x) = \left(\Lambda^{(\ell)}\Pi_{iR}\Psi_R^{(\downarrow)}\right) = \left(\begin{array}{ccc} e & \mu & \tau \end{array}\right)_R. \tag{71}$$

The chiral triplets in (64) and (65) enable us to express the Lagrangian density of the kinetic part including the gauge interactions of quarks and leptons in the generic form

$$\mathcal{L}_{kg} = \bar{\Psi}_L(x)\Gamma^\mu\mathcal{D}_\mu\Psi_L(x) + \bar{\Psi}_R^{(\uparrow)}(x)\Gamma^\mu\mathcal{D}_\mu\Psi_R^{(\uparrow)}(x) + \bar{\Psi}_R^{(\downarrow)}(x)\Gamma^\mu\mathcal{D}_\mu\Psi_R^{(\downarrow)}(x) \tag{72}$$

with the covariant derivatives $\mathcal{D}_\mu$. In the low energy regime where the standard model can be applied, the covariant derivatives act on the triplet fields as follows:

$$\mathcal{D}_\mu\Psi_L(x) = \left(\partial_\mu - gA_\mu^a(x)\frac{1}{2}\tau_a - g'B_\mu(x)\frac{1}{2}(B-L) - g_cA_{c\mu}^a(x)\frac{1}{2}\lambda_a\right)\Psi_L(x), \tag{73}$$

$$\mathcal{D}_\mu\Psi_R^{(\uparrow)}(x) = \left(\partial_\mu - g'B_\mu(x)\frac{1}{2}(B-L+1) - g_cA_{c\mu}^a(x)\frac{1}{2}\lambda_a\right)\Psi_R^{(\uparrow)}(x) \tag{74}$$

and

$$\mathcal{D}_\mu\Psi_R^{(\downarrow)}(x) = \left(\partial_\mu - g'B_\mu(x)\frac{1}{2}(B-L-1) - g_cA_{c\mu}^a(x)\frac{1}{2}\lambda_a\right)\Psi_R^{(\downarrow)}(x) \tag{75}$$

where $A_\mu^a(x)$ and $B_\mu(x)$ are the gauge fields which interact, respectively, to the electroweak isospin $\frac{1}{2}\tau_a$ of $\mathrm{SU}_L(2)$ and the hypercharge of $\mathrm{U}_Y(1)$, and $A_{c\mu}^a(x)$ is the gauge field of color symmetry. This density $\mathcal{L}_{kg}$ is readily proved to reproduce the gauge interactions of the standard model with right-handed neutrino species.

In the present scheme, the chiral triplet fields include the repetitional degrees of freedom for family structure. This feature enables us to formulate the Yukawa interactions in an intrinsic mechanism without bringing in so many unknown parameters from outside. In terms of chiral triplet fields, the Lagrangian density of the Yukawa interactions can be expressed to be

$$\begin{aligned}\mathcal{L}_Y &= g_u\bar{\Psi}_L^{(q)}(x)\mathcal{Y}_u\tilde{\phi}(x)\Psi_R^{(u)}(x) + g_d\bar{\Psi}_L^{(q)}(x)\mathcal{Y}_d\phi(x)\Psi_R^{(d)}(x) \\ &+ g_\nu\bar{\Psi}_L^{(\ell)}(x)\mathcal{Y}_\nu\tilde{\phi}(x)\Psi_R^{(\nu)}(x) + g_e\bar{\Psi}_L^{(\ell)}(x)\mathcal{Y}_e\phi(x)\Psi_R^{(e)}(x) + \mathrm{h.c}\end{aligned} \tag{76}$$

where

$$\phi(x) = \begin{pmatrix} \phi^+ \\ \phi^0 \end{pmatrix}, \quad \tilde{\phi}(x) = \begin{pmatrix} \phi^{0*} \\ -\phi^- \end{pmatrix} \tag{77}$$

are the Higgs field and its conjugate of the standard model, and the kernel operator $\mathcal{Y}_a$ ($a = u, d, \nu, e$) works to induce mixings among repetitional modes. To be commutative with the generators of color symmetry, the operator must be composed of triple-direct-products of the Dirac matrices γ_0 and γ_5. To frame its generic construction, we impose the conditions that it is self-adjoint and it brings forth the quasi-democratic mass matrices in the low energy regime. As a natural candidate for $\mathcal{Y}_a$, we adopt

$$\begin{aligned}\mathcal{Y}_a &= \tfrac{1}{2}\left\{I + y_{a1}\left(1\otimes\gamma_0e^{i\theta_{a2}\gamma_5}\otimes\gamma_0e^{i\theta_{a3}\gamma_5}\right)\right. \\ &\left. + y_{a2}\left(\gamma_0e^{i\theta_{a1}\gamma_5}\otimes 1\otimes\gamma_0e^{i\theta_{a3}\gamma_5}\right) + y_{a3}\left(\gamma_0e^{i\theta_{a1}\gamma_5}\otimes\gamma_0e^{i\theta_{a2}\gamma_5}\otimes 1\right)\right\}\end{aligned} \tag{78}$$

with real parameters y_{ai} and θ_{ai} $(a = u,\ d,\ \nu,\ e;\ i = 1,\ 2,\ 3)$.

The quark part of the Lagrangian density in (76) is readily decomposed into the Yukawa interactions among component quark fields as

$$\sum_a g_a \bar{\Psi}_L^{(q)} \mathcal{Y}_a \phi \Psi_R^{(a)} = \sum_a g_a \sum_{ij} \bar{\Psi}_L^{(q)} \Pi_{iR} \mathcal{Y}_a \Pi_{jR} \phi \Psi_R^{(a)} = \sum_{aij} \bar{\Psi}_{iL}^{(q)} \left[g_a \langle iR \mid \mathcal{Y}_a \mid jR \rangle \right] \phi \Psi_{jR}^{(a)}. \quad (79)$$

This result and the similar decomposition of the lepton part lead us to the interpretation that the quantities

$$g_a \langle iR | \mathcal{Y}_a | jR \rangle \quad (80)$$

are the Yukawa coupling constants of the a sector $(a = u,\ d,\ \nu,\ e)$.

In the low energy regime where the Higgs field takes the vacuum expectation value $\langle \phi \rangle$ and, consequently, the Weinberg-Salam symmetry is broken, the Yukawa interaction with the kernel in (78) results in the Dirac mass matrices

$$\mathcal{M}_a = \left(g_a \langle \phi \rangle \left(\langle iR \mid \mathcal{Y}_a \mid jR \rangle \right) \right) = m_a \begin{pmatrix} 1 & \alpha_{a3} e^{i\vartheta_{a3}} & \alpha_{a2} e^{-i\vartheta_{a2}} \\ \alpha_{a3} e^{-i\vartheta_{a3}} & 1 & \alpha_{a1} e^{i\vartheta_{a1}} \\ \alpha_{a2} e^{i\vartheta_{a2}} & \alpha_{a1} e^{-i\vartheta_{a1}} & 1 \end{pmatrix} \quad (81)$$

where $m_a = \frac{1}{2} g_a \langle \phi \rangle$, $\alpha_{ai} = y_{ai} |\langle \ell | \gamma_0 | r \rangle|^2$ and $\vartheta_{a1} = \theta_{a2} - \theta_{a3}$ (cyclic). This type of the Hermitian quasi-democratic matrices are recognized to explain the hierarchical structures of mass spectra and the weak mixing matrix for the quark sector [4, 5].

7. Discussion

In this way, we have formulated an algebraic theory for unified description of external and internal symmetries of quarks and leptons. The basic ingredients of the theory are the *triplet algebra* consisting of all the triple-direct-products of the Dirac matrices and the *triplet field* forming the multiple spinor representation of the Lorentz group. The triplet field has the 4^3 component fields which afford the freedom for the repetitional family structure as well as the external spacetime symmetry and the internal color symmetry.

By postulating the triplet fields of the electroweak doublet and singlets to exist, the standard model with three generations is successfully reproduced so that the Yukawa interactions are formulated, without bringing in many unknown coupling constants from outside, and the Dirac mass matrices with quasi-democratic structure are naturally derived.

The present formalism is flexible enough to be applicable to various field theories suitable for descriptions for different energy scales. Removing of the constraint in (40) enables us formulate the four generation theory of quarks and leptons and introduction of the right-handed electroweak doublet in place of the singlets in (65) allows us to obtain the gauge theory of left-right symmetry. Further, we can bring in the gauge symmetry for repetitional family modes. Note that the three Dirac matrices $\rho'_1 = \gamma_0$, $\rho'_2 = i\gamma_0\gamma_5$ and $\rho'_3 = \gamma_5$ generate the Pauli algebra. Taking the triple-direct-products of these elements just as in (42), we can construct the 16 dimensional submanifold which generates the SU(4) group for family symmetry. Therefore, it is possible to formulate the gauge field theory with the symmetry $\mathrm{SU}_c(4) \otimes \mathrm{SU}_L(2) \otimes \mathrm{SU}_R(2) \otimes \mathrm{SU}_{\mathrm{family}}(4)$ in the triplet field formalism.

Acknowledgments
This work was supported by the JSPS Grant-in-Aid for Scientific Research (S) No. 22224003.

[1] Coleman S and Mandula J 1967 *Phys. Rev.* **159** 1251
[2] Weinberg S 2000 *The Quantum Field Theory of Fields* vol III(Cambridge University Press) chapter 24
[3] Sogami I S 1987 *Prog. Theor. Phys.* **78** 487
[4] Branco G C, Silva-Marcos J I and Rebelo M N 1990 *Phys. Lett.* **B237** 446
[5] Sogami I S, Nishida K, Tanaka H and Shinohara T 1998 *Prog. Theor. Phys.* **99** 281

Journal of Physics: Conference Series **284** (2011) 012055 doi:10.1088/1742-6596/284/1/012055

From Quantum Mechanics to Quantum Field Theory: The Hopf route

A. I. Solomon[1 2], **G. H. E. Duchamp**[3], **P. Blasiak**[4], **A. Horzela**[4] **and K.A. Penson**[2]

[1] Physics and Astronomy Department,The Open University, Milton Keynes MK7 6AA, UK
[2] Lab.de Phys.Théor. de la Matière Condensée, University of Paris VI, France
[3] Institut Galilee, LIPN, CNRS UMR 7030 99 Av. J.-B. Clement, F-93430 Villetaneuse, France
[4] H. Niewodniczański Institute of Nuclear Physics, Polish Academy of Sciences, Division of Theoretical Physics, ul. Eliasza-Radzikowskiego 152, PL 31-342 Kraków, Poland

E-mail: a.i.solomon@open.ac.uk, gduchamp2@free.fr, pawel.blasiak@ifj.edu.pl, andrzej.horzela@ifj.edu.pl, penson@lptl.jussieu.fr

Abstract. We show that the combinatorial numbers known as *Bell numbers* are generic in quantum physics. This is because they arise in the procedure known as *Normal ordering* of bosons, a procedure which is involved in the evaluation of quantum functions such as the canonical partition function of quantum statistical physics, *inter alia*. In fact, we shall show that an evaluation of the non-interacting partition function for a single boson system is identical to integrating the *exponential generating function* of the Bell numbers, which is a device for encapsulating a combinatorial sequence in a single function.

We then introduce a remarkable equality, the Dobinski relation, and use it to indicate why renormalisation is necessary in even the simplest of perturbation expansions for a partition function.

Finally we introduce a global algebraic description of this simple model, giving a Hopf algebra, which provides a starting point for extensions to more complex physical systems.

1. Introduction

The motivation for this work is as follows: Recent analysis of perturbative quantum field theory (pQFT) has shown that at the heart of this theory there lies a Hopf algebra description. Since pQFT is a subtle and as yet not completely solved area, it is of pedagogical value to examine simpler systems to illustrate how these algebraic constructs may arise. In this note we examine the simple theory given by the quantum partition function Z of a non-relativistic boson system, a completely solvable model (at least in the free boson case). On the basis of a combinatorial methodology, we show that certain combinatorial numbers, known as Bell numbers, play a central role in such an analysis. The graphical approach to the description of these combinatorial numbers leads to a sort of non-relativistic analogue of Feynman diagrams. En route, we also remark how that even in this simple case, uncritical expansion methods may lead to divergences, which can be cured by regularization; this imitates the renormalization methods of pQFT. This theory does not depend on space or time; so it is a non-field theory. However, the description of the Quantum Partition Function that we employ uses the coherent state parameter z as

Journal of Physics: Conference Series **284** (2011) 012055 doi:10.1088/1742-6596/284/1/012055

integration variable; and this plays the role of integration over space and time in the relativistic case.

The core of the note then deals with how an elementary Hopf algebra arises naturally from this approach. Although the resulting Hopf algebra, which we call **BELL** (since it arises directly from a graphical description of the combinatorial Bell numbers), is rather too uncomplicated, (technically it is commutative and co-commutative), it serves as a starting point for an extension to more complicated algebraic systems. Indeed, in later work we show that extensions and deformations of this algebra can emulate that of pQFT, thus supplying an alternate, algebraic, route from the physical path which takes us from the non-relativistic arena to that of relativistic quantum field theory.

2. Bell and Stirling Numbers; Exponential Generating Function

The *Bell number* $B(n)$ simply gives the number of partitions of $\{1, 2, \ldots, n\}$. Thus for the set $\{1, 2, 3\}$, we have

$$\{1,2,3\}, \quad \{1,2\}\{3\} \quad \{1,3\}\{2\} \quad \{3,2\}\{1\} \quad \{1\}\{2\}\{3\}$$

so that $B(3) = 5$. Equivalently, $B(n)$ gives the number of ways in which n distinct objects may be distributed among n identical containers, some of which may remain empty. The first few values of $B(n)$ are

$$B(n) = 1,\ 1,\ 2,\ 5,\ 15,\ 52,\ 203,\ \ldots, \qquad n = 0,\ 1,\ 2,\ \ldots .$$

A compact way of defining a combinatorial sequence is by means of a *generating function*. For the Bell numbers, the *exponential generating function* is given by

$$\sum_{n=0}^{\infty} B(n) \frac{x^n}{n!} = \exp(e^x - 1). \tag{1}$$

We shall give a proof of Eq.(1) after we introduce a graphical description for the Bell numbers in Section 3.2. A slight generalization of the exponential generating function for the Bell numbers is given by that defining the *Bell polynomials* $B_n(y)$:

$$\exp\left(y(e^x - 1)\right) = \sum_{n=0}^{\infty} B_n(y) \frac{x^n}{n!} \tag{2}$$

With this notation $B(n) \equiv B_n(1)$.

Stirling numbers (of the second kind) $S(n,k)$ are related to the Bell numbers. These are defined as the number of ways of putting n different objects into k identical containers, leaving none empty. From this definition we have $B(n) = \sum_{k=1}^{n} S(n,k)$.

Bell and Stirling numbers are fundamental in quantum theory. This is because they arise naturally in the *normal ordering problem* of bosonic operators. By the normal ordering operation $\mathcal{N} f(a, a^\dagger)$ we mean reorder the boson operators in $f(a, a^\dagger)$ so that all annihilation operators are on the right. For canonical boson creation and annihilation operators $\{a^\dagger, a\}$ satisfying $[a, a^\dagger] = 1$, the Stirling numbers of the second kind $S(n,k)$ intervene through [2, 3]

$$\mathcal{N}(a^\dagger a)^n = \sum_{k=1}^{n} S(n,k)(a^\dagger)^k a^k \,. \tag{3}$$

The corresponding Bell numbers $B(n) = \sum_{k=1}^{n} S(n,k)$ are simply the expectation values

$$B(n) = \langle z|(a^\dagger a)^n|z\rangle_{z=1} \tag{4}$$

taken in the coherent state defined by

$$a|z\rangle = z|z\rangle \tag{5}$$

for $z = 1$.

Journal of Physics: Conference Series **284** (2011) 012055 doi:10.1088/1742-6596/284/1/012055

3. Graphs

3.1. Bell number diagrams

We now give a graphical representation of the Bell numbers, based on work of Brody, Bender and Meister [4, 5], which we have extended in [6]. Consider labelled lines which emanate from a white dot, the origin, and finish on a black dot, the vertex. We shall allow only one line from each white dot but impose no limit on the number of lines ending on a black dot. Clearly this simulates the definition of $S(n, k)$ and $B(n)$, with the white dots playing the role of the distinguishable objects, whence the lines are labelled, and the black dots that of the indistinguishable containers. The identification of the graphs for 1, 2 and 3 lines is given in Figure 1 below. We have concentrated

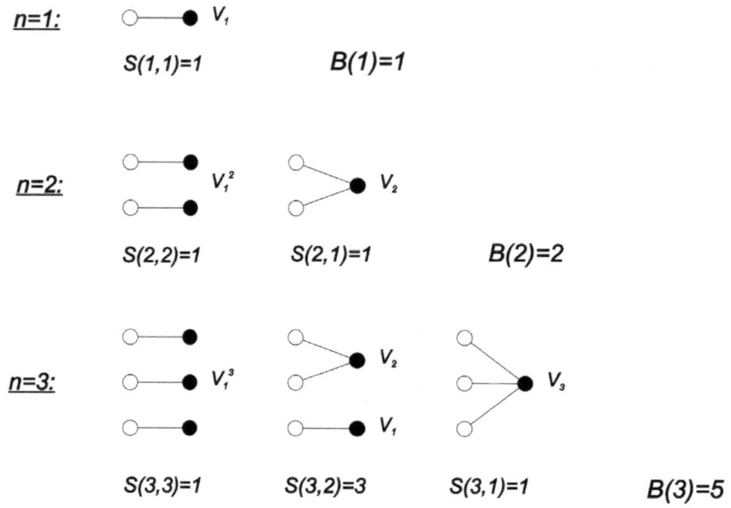

Figure 1. Graphs for $B(n)$, $n = 1, 2, 3$.

on the Bell number sequence and its associated graphs since, as we shall show, there is a sense in which this sequence of graphs is *generic* in the evaluation of the Quantum Partition Function. By this we mean the following: The exponential generating function for the Bell polynomials is identical to the (integrand of) the Canonical Quantum Partition Function of a *non-interacting* single-mode boson model. Then, when interactions are present these may be incorporated by the use of suitable strengths associated with the vertices of the graphs. That is, we can represent the perturbation expansion of an interacting model by the same sequence of graphs as in Figure 1, with suitable vertex multipliers (denoted by the V terms in the same figure).

3.2. The Connected Graph theorem

We introduce this important theorem[7] in order to obtain the exponential generating function of the Bell numbers. The theorem states:

Theorem 1 *Connected Graph Theorem: If* $\mathcal{C}(x) = \sum_{n=1}^{\infty} c(n)x^n/n!$ *is the exponential generating function of a set of CONNECTED labelled graphs then* $\mathcal{A}(x) = \exp \mathcal{C}(x)$ *is the exponential generating function of ALL the graphs.*

In Figure 2 we see that there is only *one* connected graph for each n, so

$$\mathcal{C}(x) = x/1! + x^2/2! + x^3/3! + \ldots = e^x - 1$$

whence

$$\mathcal{A}(x) = \exp(e^x - 1)$$

gives the exponential generating function for the Bell numbers.

Journal of Physics: Conference Series **284** (2011) 012055 doi:10.1088/1742-6596/284/1/012055

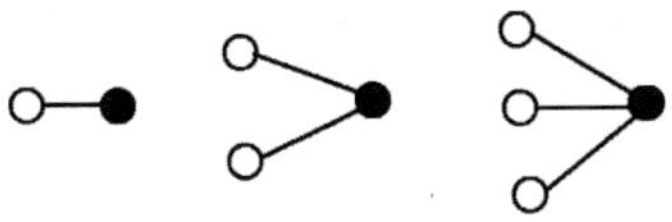

Figure 2. Connected Graphs for $B(n)$, $n = 1, 2, 3$.

4. Quantum Partition Function

We now show that the Canonical Partition Function of Quantum Statistical mechanics is simply related to the exponential generating function of the Bell polynomials.

4.1. Free boson gas and Bell polynomials

The canonical partition function Z associated with a quantum mechanical hamiltonian H is given by

$$Z = \mathrm{Tr}\exp(-\beta H) \tag{6}$$

where β is the usual inverse temperature ($1/k_{Boltzmann}T$). We start with the elementary case of the hamiltonian for the single–mode free boson gas $H = \epsilon a^\dagger a$ (ignoring an additive constant), $\epsilon > 0$. The usual computation of the partition function, exploiting the completeness property $\sum_{n=0}^{\infty} |n\rangle\langle n| = I$, is immediate:

$$\begin{aligned} Z &= \mathrm{Tr}\exp(-\beta\epsilon a^\dagger a) \\ &= \sum_{n=0}^{\infty} \langle n|e^{-\beta\epsilon\hat{n}}|n\rangle \\ &= \sum_{n=0}^{\infty} e^{-\beta\epsilon n} \\ &= \left(1 - e^{-\beta\epsilon}\right)^{-1}. \end{aligned} \tag{7}$$

However, we may use *any* complete set to perform the trace. We choose coherent states as defined in Eq.(5) above, which are explicitly given by

$$|z\rangle = e^{-|z|^2/2}\sum_n (z^n/n!){a^\dagger}^n|0\rangle. \tag{8}$$

For these states the completeness or *resolution of unity* property is

$$\frac{1}{\pi}\int d^2z|z\rangle\langle z| = I \equiv \int d\mu(z)|z\rangle\langle z|. \tag{9}$$

The appropriate trace calculation is

$$Z = \frac{1}{\pi}\int d^2z\langle z|\exp(-\beta\epsilon a^\dagger a)|z\rangle = \tag{10}$$

$$= \frac{1}{\pi}\int d^2z\langle z| :\exp(a^\dagger a(e^{-\beta\epsilon} - 1)): |z\rangle\,, \tag{11}$$

where we have used the following well-known relation [8, 9] for the *forgetful* normal ordering operator $:f(a, a^\dagger):$ which means *normally order the creation and annihilation operators in f* forgetting[1] *the commutation relation* $[a, a^\dagger] = 1$:

$$\mathcal{N}exp(xa^\dagger a) =: \exp(a^\dagger a(e^x - 1)):. \tag{12}$$

[1] This procedure generally alters the value of the operator to which it is applied.

Journal of Physics: Conference Series **284** (2011) 012055 doi:10.1088/1742-6596/284/1/012055

We therefore obtain, integrating over the angle variable θ and the radial variable $r = |z|$,

$$Z = \frac{1}{\pi} \int_0^{2\pi} d\theta \int_0^{\infty} r\, dr \exp\left(r^2(e^{-\beta\epsilon} - 1)\right), \tag{13}$$

which gives us $Z = (1 - e^{-\beta\epsilon})^{-1}$ as before.

We rewrite the above equation to show the connection with our previously–defined combinatorial numbers. Writing $y = r^2$ and $x = -\beta\epsilon$, Eq.(13) becomes

$$Z = \int_0^{\infty} dy \exp(y(e^x - 1)). \tag{14}$$

This is an integral over the classical exponential generating function for the Bell polynomials as given in Eq.(2). This leads to the combinatorial form for the partition function

$$Z = \int_0^{\infty} dy \sum_{n=0}^{\infty} B_n(y) \frac{x^n}{n!}. \tag{15}$$

4.2. General partition functions

We now apply this graphical approach to the general partition function in second quantized form, with the usual definition for the partition function Eq.(6). In general the hamiltonian is given by $H = \epsilon w(a, a^\dagger)$, where ϵ is the energy scale, and w is a string (= sum of products of positive powers) of boson creation and annihilation operators. The partition function integrand F for which we seek to give a graphical expansion, is

$$Z(x) = \int F(x, z)\, d\mu(z), \tag{16}$$

where

$$\begin{aligned} F(x,z) &= \langle z| \exp(xw) |z\rangle = \qquad\qquad (x = -\beta\epsilon) \\ &= \sum_{n=0}^{\infty} \langle z| w^n |z\rangle \frac{x^n}{n!} \\ &= \sum_{n=0}^{\infty} W_n(z) \frac{x^n}{n!} \\ &= \exp\Big(\sum_{n=1}^{\infty} V_n(z) \frac{x^n}{n!} \Big), \end{aligned} \tag{17}$$

with obvious definitions of W_n and V_n. The sequences $\{W_n\}$ and $\{V_n\}$ may each be recursively obtained from the other [10]. This relates the sequence of multipliers $\{V_n\}$ of Figure 1 to the hamiltonian of Eq.(6). The lower limit 1 in the V_n summation is a consequence of the normalization of the coherent state $|z\rangle$.

Although Eq.(15) is remarkably simple in form, it is often by no means a straightforward matter to evaluate the analogous integral for other than the free boson system considered here. Further, as we show below, we may not interchange the integration and the summation, as each individual y integral diverges. At this point the analogue of this model to perturbative quantum field theory becomes more transparent. For example, to calculate the partition Function $Z(x)$, we must integrate over the Partition Function Integrand $F(x, z)$. Here, the integration over the coherent state parameter z plays the role of the space-time integration of Quantum Field

Theory. As we have just remarked, with reference to Eq.(15) for the non-interacting case, and as in pQFT (perturbative quantum field theory), term-by-term integration, corresponding to fallaciously interchanging the summation and integration actions, results in infinities at each term, as we now verify for the non-interacting case. Thus even here an elementary form of renormalization is necessary, if we resort to term-wise integration.

Considerations such as these have led many authors, as noted in reference [1], to consider a global, algebraic approach to pQFT; and this we shall do in Section 6.

But first we show how divergences arise.

5. The Dobiński Relation and Divergence

5.1. Divergence

The celebrated Dobiński relation for the Bell numbers is given by (see, for example, [11]):

$$B(n) = \frac{1}{e}\sum_{k=0}^{\infty}\frac{k^n}{k!}, \qquad n = 1, 2, \ldots \tag{18}$$

and for the Bell polynomials by

$$B_n(y) \;=\; e^{-y}\sum_{k=0}^{\infty}\frac{k^n}{k!}y^k. \tag{19}$$

These relations have been extended in [12].

Using the Dobiński relation Eq.(19) we can rewrite the expression for the partition function for the free single-mode boson Eq.(15) as

$$Z \;=\; \int_0^{\infty} dy \sum_{n=0}^{\infty} B_n(y)\,\frac{x^n}{n!} \tag{20}$$

$$\;=\; \int_0^{\infty}\sum_{n=0}^{\infty}\frac{x^n}{n!}\{e^{-y}\sum_{k=0}^{\infty}\frac{k^n}{k!}y^k\}dy \tag{21}$$

We may *attempt* to write this as a perturbation expansion (in $x \equiv -\beta\epsilon$) by interchanging the integration and summation, thus:

$$Z \;=\; \sum_{n=0}^{\infty}\frac{x^n}{n!}\sum_{k=0}^{\infty}\frac{k^n}{k!}\int_0^{\infty}\{e^{-y}y^k\}dy \tag{22}$$

$$\;=\; \sum_{n=0}^{\infty}\frac{x^n}{n!}\sum_{k=0}^{\infty}\frac{k^n}{k!}\Gamma(k+1) \tag{23}$$

$$\;=\; \sum_{n=0}^{\infty}\frac{x^n}{n!}\sum_{k=0}^{\infty}k^n. \tag{24}$$

Therefore the resulting "perturbation expansion" is of the form

$$Z = \sum_{0}^{\infty} b_n x^n \qquad b_n = \frac{1}{n!}\sum_{k=0}^{\infty}k^n = \infty$$

with each term in the expansion being infinite.

Thus even in this simple case of a free boson expansion, a perturbation expansion obtained from the illegal interchange of integration and summation is divergent in every term.

5.2. Regularization

In this section we illustrate how the method of regularization may be applied to alleviate the problem of divergences. Writing the partition function Eq.(14) as

$$Z = \int_0^\infty e^{-\alpha y}\, dy \qquad (\alpha = 1 - e^{-\beta\epsilon}) \tag{25}$$

immediate evaluation gives $Z = 1/\alpha$ as in Eq.(7).

However, if we expand

$$Z = \int_0^\infty \sum_{n=0}^\infty (-\alpha y)^n/n!\, dy \tag{26}$$

and exchange the order of integration, we obtain

$$\text{(FALSE)} \qquad Z = \sum_{n=0}^\infty (-\alpha)^n/n! \int_0^\infty y^n\, dy \tag{27}$$

each term of which diverges. The *regularization* method is to write

$$Z = \lim_{M\to\infty} \int_0^M \sum_{n=0}^\infty (-\alpha y)^n/n!\, dy \tag{28}$$

and, since the integration is over a finite range, we may interchange integration and summation, to obtain

$$\begin{aligned} Z &= \lim_{M\to\infty} \sum_{n=0}^\infty (-\alpha)^n/n! \int_0^M y^n\, dy & (29)\\ &= \lim_{M\to\infty} \sum_{n=0}^\infty \{(-\alpha)^n/n!\}\{M^{(n+1)}/(n+1)\} & (30)\\ &= \lim_{M\to\infty} \sum_{n=0}^\infty \{(-\alpha M)^{(n+1)}/(n+1)!\}(-1/\alpha) & (31)\\ &= \lim_{M\to\infty} \{(e^{-M\alpha} - 1)(-1/\alpha)\} & (32)\\ &= 1/\alpha & (33) \end{aligned}$$

as before.

Note that y bounded by a finite M implies the eminently reasonable assumption of a maximum energy for the coherent state as defined in Eq.(5).

6. Hopf Algebra structure

So far we have discussed the analytic and combinatorial structure of our simple model. Modern approaches to perturbative quantum field theory introduce a global algebraic structure. We describe the analogue of this *Hopf Algebra structure* in the context of our model, which we refer to as BELL below. We first introduce a basic Hopf Algebra, which we call POLY, generated by a single parameter x. This is essentially a standard one-variable polynomial algebra on which we impose the Hopf operation of co-product, co-unit and antipode, as a useful pedagogical device for describing their properties. The Hopf structure associated with our model described above is simply a multi-variable extension of **POLY**.

Journal of Physics: Conference Series **284** (2011) 012055 doi:10.1088/1742-6596/284/1/012055

6.1. The Hopf Algebra POLY

POLY consists of polynomials in x (say, over the real field $\mathcal{R}$, for example).The standard algebra structure of addition and associative multiplication is obtained in the usual way, by polynomial addition and multiplication. The additional Hopf operations are:

(i) The coproduct $\Delta : \mathbf{POLY} \longrightarrow \mathbf{POLY} \times \mathbf{POLY}$ is defined by

$$\begin{aligned} \Delta(e) &= e \times e \quad (\text{unit } e) \\ \Delta(x) &= x \times e + e \times x \quad (\text{generator } x) \\ \Delta(AB) &= \Delta(A)\Delta(B) \quad \text{otherwise} \end{aligned}$$

so that Δ is an algebra homomorphism.

(ii) The co-unit ϵ satisfies $\epsilon(e) = 1$ otherwise $\epsilon(A) = 0$.

(iii) The antipode $\mathcal{S} : \mathbf{POLY} \longrightarrow \mathbf{POLY}$ satisfies $\mathcal{S}(e) = e$; on the generator x, $\mathcal{S}(x) = -x$. It is an *anti-homomorphism*, i.e. $\mathcal{S}(AB) = \mathcal{S}(B)\mathcal{S}(A)$.

It may be shown that the foregoing structure **POLY** satisfies the axioms of a commutative, co-commutative Hopf algebra.

6.2. BELL

We now briefly describe the Hopf algebra **BELL** which is appropriate for the diagram structure introduced in this note, defined by the diagrams of Figure 1.

(i) Each distinct diagram is an individual basis element of **BELL**; thus the dimension is infinite. (Visualise each diagram in a "box".) The sum of two diagrams is simply the two boxes containing the diagrams. Scalar multiples are formal; for example, they may be provided by the V coefficients.

(ii) The identity element e is the empty diagram (an empty box).

(iii) Multiplication is the juxtaposition of two diagrams within the same "box". **BELL** is generated by the *connected* diagrams; this is a consequence of the Connected Graph Theorem (as discussed above in 3.2). Since we have not here specified an order for the juxtaposition, multiplication is commutative.

(iv) The coproduct $\Delta : \mathbf{BELL} \longrightarrow \mathbf{BELL} \times \mathbf{BELL}$ is defined by

$$\begin{aligned} \Delta(e) &= e \times e \quad (\text{unit } e) \\ \Delta(y) &= y \times e + e \times y \quad (\text{generator } y) \\ \Delta(AB) &= \Delta(A)\Delta(B) \quad \text{otherwise} \end{aligned}$$

so that Δ is an algebra homomorphism.

(v) The co-unit ϵ satisfies $\epsilon(e) = 1$ otherwise $\epsilon(A) = 0$.

(vi) The antipode $\mathcal{S} : \mathbf{BELL} \longrightarrow \mathbf{BELL}$ satisfies $\mathcal{S}(e) = e$; on a generator x, $\mathcal{S}(y) = -y$. It is an *anti-homomorphism*, i.e. $\mathcal{S}(AB) = \mathcal{S}(B)\mathcal{S}(A)$.

6.3. BELL *as an extension of* POLY

It may be seen that **BELL** is a multivariable version of **POLY**. To show this, we code the diagrams by letters. We use an infinite alphabet $Y = \{y_1, y_2, y_3 \cdots\} = \{y_k\}_{k \geq 1}$ and code each connected diagram with one black spot and k white spots with the letter y_k. An unconnected diagram will be coded by the product of its letters.

In this way, the diagrams of Figure 1 are coded as follows:

- first line : y_1
- second line : y_1^2 and y_2
- third line : y_1^3 and $y_1 y_2$ and y_3 .

Thus one sees that each diagram of weight n with k connected components is coded bijectively by a monomial of weight n (the weight of a monomial $y_{i_1} y_{i_2} \cdots y_{i_r}$ is just the sum of the indices $\sum_{j=1}^{r} i_j$) and k letters. The algebra **BELL** is coded by commutative polynomials in the infinite alphabet Y; that is, the coding is an isomorphism **BELL** $\to \mathcal{R}[Y]$. As an aside, one may note that the basis elements of **BELL** are sometimes referred to as *forests*. As above, it may be shown that the foregoing structure **BELL** satisfies the axioms of a commutative, co-commutative Hopf algebra.

7. Discussion

In this note we have shown the value of the combinatorial method, which leads *inter alia* through the graphical approach, to a Hopf algebra description of a simple non-relativistic free boson system. We named the resulting (commutative, co-commutative) Hopf algebra **BELL**. To extend this simple description to more realistic models, it is convenient to introduce a symmetrized form of **BELL**; i.e. where the black and white spots appear symmetrically[13]. Thus we have introduced the Hopf algebra **DIAG** where a typical series of graphs is given in Figure 3 below. Note that since the black and white spots play symmetrical roles, in the

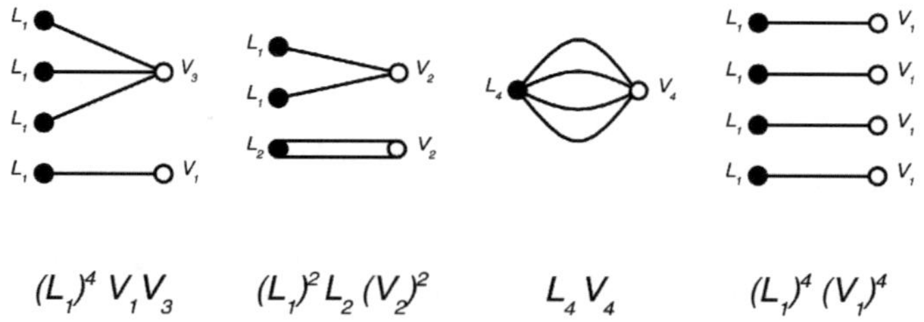

Figure 3. Some Graphs of DIAG

figure we permit ourselves to exchange their order from that used in, for example, Figure 1. We may also apply "weights" L_m to the black spots as well. Although the resulting Hopf algebra **DIAG** is still commutative/co-commutative, it may in turn be extended to a non-commutative Hopf algebra **LDIAG**. A three-parameter *deformation* of this algebra **LDIAG**(q_c, q_s, q_t) can be endowed with a non-commutative product and non-cocommutative coproduct and is therefore sufficiently general to emulate those algebras which describe pQFT. This extension is described in detail in [14], as well in a subsequent article in this conference [15]. We finally note that a combinatorial and Hopf algebra description of Quantum Field Theory, introducing however the concept of a Hopf superalgebra which goes beyond the remit of the current approach, has also been proposed by Brouder *et al* [16].

Acknowlegements
The authors wish to acknowledge support from the Agence Nationale de la Recherche (Paris, France) under Program No. ANR-08-BLAN-0243-2 and from PAN/CNRS Project PICS No.4339(2008-2010) as well as the Polish Ministry of Science and Higher Education Grant No.202 10732/2832.

[1] Kreimer D 2000 Knots and Feynman Diagrams, *Cambridge Lecture Notes in Physics*, CUP
[2] Katriel J 1974 *Lett. Nuovo Cimento* **10** 565
[3] Katriel J 2000 *Phys. Lett. A.* **273** 159
[4] Bender C M, Brody D C and Meister B K 1999 Quantum field theory of partitions *Journal of Mathematical Physics* **40** 3239
[5] Bender C M, Brody D C and Meister B K 2000 Combinatorics and field theory, *Twistor Newsletter* **45** 36
[6] Blasiak P, Penson K A, Solomon A I, Horzela A and Duchamp G H E 2005 *J.Math.Phys.* **46**, 052110 and arXiv:quant-ph/0405103
[7] Ford G W and Uhlenbeck G E 1956 *Proc.Nat.Acad.***42**122
[8] Klauder J R and Sudarshan E C G 1968 Fundamentals of Quantum Optics *Benjamin, New York*
[9] Louisell W H 1990 *Quantum Statistical Properties of Radiation* (J. Wiley, New York)
[10] Pourahmadi M 1984 *Amer. Math. Monthly* **91**, 303
[11] Wilf H S 1994 Generatingfunctionology *New York, Academic Press*
[12] Penson K A, Blasiak P, Duchamp G H E, Horzela A and Solomon A I 2004 *Hierarchical Dobiński-type relations via substitution and the moment problem. J.Phys.A:Math.Gen.* **37** 3475
[13] Duchamp G H E, Blasiak P, Horzela A, Penson K A and Solomon A I 2006 Feynman graphs and related Hopf algebras *J. Phys.: Conf. Ser.* **30** 107
[14] Duchamp G H E, Blasiak P, Horzela A, Penson K A and Solomon A I 2010 A three-parameter Hopf deformation of the algebra of Feynman-like diagrams *J. Russian Laser Research* **31** 162
[15] Minh H N, Duchamp G H E, Solomon A I and Goodenough S 2010 An interface between physics and number theory *These proceedings*
[16] Brouder C, Fauser B, Frabetti A and Oeckl R 2004 *J. Phys. A: Math. Gen.* **37** 5895

Journal of Physics: Conference Series **284** (2011) 012056 doi:10.1088/1742-6596/284/1/012056

Group-theoretical approach to interaction renormalization for *ab initio* solutions of the nuclear problem

Kristina D. Sviratcheva[1], **Jerry P. Draayer**[1], **Tomáš Dytrych**[1]

[1] Department of Physics and Astronomy, Louisiana State University, Baton Rouge, LA 70803, USA

Abstract. For many problems in nuclear, high-energy and solid-state physics a crucial step toward a solution involves a diagonalization or simplification of the Hamiltonian of a many-body quantum-mechanical system. Problems of interest include renormalization of the Hamiltonian to accommodate excluded basis spaces, decoupling of a model space region of special interest from irrelevant configurations, or near-diagonalization to invoke perturbative treatments. In such applications, the Similarity Renormalization Group (SRG) approach (or *flow equations*) can provide solutions where more conventional methods fail. We show that when SRG employs a near symmetry of a system, utilizing the Casimir invariant operators, it evolves the Hamiltonian of the system toward a (block-)diagonal form that can be realized in terms of a finite number of integrity basis operators. In particular, for atomic nuclei, an SRG flow equation that utilizes the SU(3) symmetry, which often dominates the nuclear dynamics, is found to naturally evolve a nucleon-nucleon (NN) interaction toward a block-diagonal form. A remarkable consequence is that a renormalized SU(3)-preserving many-body Hamiltonian expressed in terms of only the SU(3) integrity basis realizes energy spectra identical to the ones of the original realistic NN interaction, which reflects the low-energy symmetries as well as symmetry-breaking patterns of quantum chromodynamics. Such a scheme is especially suitable for *ab initio* SU(3) symmetry-adapted shell model calculations, which can further allow for perturbative treatments of the nuclear interaction. Moreover, the use of symmetries dramatically reduces the size of the problem, as the flow equation can be rewritten in terms of SU(3) reduced matrix elements of the NN interaction. This is crucial for maintaining the unitarity of the SRG transformations that generate many-body interactions.

1. Introduction

Analogous to the Jacobi numerical method for diagonalization in a stepwise fashion, the Similarity Renormalization Group (SRG) or *flow equations* approach has been designed as a nonperturbative renormalization/diagonalization method by Glazek and Wilson [1, 2] and independently by Wegner [3]. Recent studies of interest include quantum nonequilibrium systems [4], the sine-Gordon strong-coupling model [5], the SU(2) Lipkin model [6], the Hubbard model together with a focus on the electron-phonon problem [7, 8], and effective field theory for, e.g., quantum electro- and chromo-dynamics [9, 10].

In nuclear physics, modern realistic internucleon interactions tied to quantum chromodynamics (QCD) considerations currently limit *ab initio* nuclear structure calculations [11, 12, 13, 14] to slow convergence with increasing model spaces, restricted many-body correlations, or small nucleon numbers. While innovative *ab initio* approaches (e.g., [15, 16, 17]) attempt to address these limitations, the use of techniques for renormalization (e.g., Lee-Suzuki [18, 19], SRG [20], and UCOM [21]) of two-nucleon (NN or $2N$), and perhaps $3N$ and $4N$, interactions prove essential.

In this study we show that using symmetries that dominate the many-body dynamics yields controllable SRG-generated higher-order coupling terms (many-body forces), which are essential to retaining the unitarity of the SRG transformations and key to the method's applicability. A novel perspective, based on the spectral distribution theory [22], provides further insight into the SRG approach. Specifically, for an arbitrary basis of dimension N_d the traceless (many-body) Hamiltonian matrix representation can be mapped onto a vector in a multi-dimensional linear vector space with coordinates specified by the independent matrix elements of H. The σ_H vector "length" (specifying the interaction "strength") is related to the Hilbert-Schmidt norm, $\sigma_H^2 = \langle (H - \langle H \rangle)^\dagger (H - \langle H \rangle) \rangle$ with $\langle \ldots \rangle \equiv \frac{1}{N_d}\mathrm{Tr}(\ldots)$, while the spatial orientation of two operators, H and H', is given by their correlation coefficient (specifying the similarity between the two interactions), $\zeta_{H,H'} = \frac{\langle (H-\langle H \rangle)^\dagger (H' - \langle H' \rangle) \rangle}{\sigma_H \sigma'_H} = \cos\theta$, with θ being the angle between H and H'. Hence, the positive square root of σ_H^2 is a natural measure of the size of the H operator and realizes the width (or spread) of the H eigenvalue distribution, namely, the smaller the width (the weaker the interaction), the more compressed the energy spectrum of H. In the framework of spectral distribution theory, we present first estimates for the strength of the SRG-induced many-body forces for a general case [23]. Furthermore, a realization of the SRG-evolved Hamiltonian in terms of a finite set of integrity basis (IB) operators with usually known analytical matrix elements and reduced dimensions that capitalize on the use of symmetry make many-body SRG calculations feasible. Such a scheme applied to realistic NN interactions is especially suitable for *ab initio* symmetry-adapted shell models as the interaction becomes prediagonalized in the basis while allowing for further perturbative treatments.

2. Theoretical Framework

The SRG concept performs a continuous sequence of $U(s)$ infinitesimal unitary transformations of a H Hamiltonian that is generated by the antihermitian η operator, $\eta^\dagger = -\eta$,

$$H_s = U(s) H U(s)^\dagger \text{ with } U(s) = e^{s\eta}, \tag{2.1}$$

which preserves the trace and strength of the Hamiltonian, $\langle H_s \rangle = \langle H \rangle$ and $\sigma_{H_s} = \sigma_H$. The differential SRG procedure is based on the following class of equations,

$$\frac{d}{ds} H_s = [\eta, H_s]. \tag{2.2}$$

As shown below, the generator η is chosen in a way to drive the Hamiltonian toward a specific – usually, (block-)diagonal – form. In the framework of spectral distribution theory, for any general hermitian operator, C_s, which varies with s, Eq. (2.2) yields,

$$\begin{aligned} \frac{d}{ds}\langle (C_s - \langle C_s \rangle)(H_s - \langle H_s \rangle) \rangle &= \left\langle \frac{d(C_s - \langle C_s \rangle)}{ds} H_s \right\rangle + \langle C_s [\eta, H_s] \rangle \\ \sigma_H (\dot{\sigma}_{C_s} \zeta_{H_s,C_s} + \sigma_{C_s} \dot{\zeta}_{H_s,C_s}) &= \left\langle \frac{d(C_s - \langle C_s \rangle)}{ds} H_s \right\rangle + \langle [H_s, C_s] \eta \rangle , \end{aligned} \tag{2.3}$$

where we use that the trace of a commutator is zero ($\mathrm{Tr}([\eta, H_s]) = 0$). We consider two choices for C_s, (i) C_s does not vary with s (fixed), and (ii) C_s is the diagonal part of H_s, D_s, as originally proposed by Wegner [3].

(i) $C_s = C$ does not vary with s (fixed). Eq. (2.3) becomes,

$$\sigma_H \sigma_C \dot{\zeta}_{H_s,C} = \left\langle [C, H_s]^\dagger \eta \right\rangle . \tag{2.4}$$

(ii) $C_s = D_s$ is the diagonal part of H_s. As $H_s = D_s + O_s$, where O_s is the traceless non-diagonal part of H_s, the relations $\langle D_s O_s \rangle = 0$ and $\sigma_H^2 = \sigma_{D_s}^2 + \sigma_{O_s}^2$ hold. In this case, Eq (2.3) becomes,

$$\begin{aligned}
\sigma_H \sigma_{D_s} \dot{\zeta}_{H_s,D_s} &= -\sigma_{D_s} \dot{\sigma}_{D_s} + \left\langle \frac{d(D_s - \langle D_s \rangle)}{ds} H_s \right\rangle + \langle [H_s, D_s] \eta \rangle \\
&= \left\langle \dot{D}_s O_s \right\rangle + \langle [H_s, D_s] \eta \rangle \\
\sigma_H \sigma_{D_s} \dot{\zeta}_{H_s,D_s} &= \left\langle [D_s, H_s]^\dagger \eta \right\rangle ,
\end{aligned} \tag{2.5}$$

where we use that $\sigma_{D_s} = \sigma_H \zeta_{H_s,D_s}$, $\sigma_{D_s} \dot{\sigma}_{D_s} = \left\langle \frac{d(D_s - \langle D_s \rangle)}{ds} (D_s - \langle D_s \rangle) \right\rangle = \left\langle \frac{d(D_s - \langle D_s \rangle)}{ds} D_s \right\rangle$ and $\left\langle \frac{d(D_s - \langle D_s \rangle)}{ds} O_s \right\rangle = \left\langle \dot{D}_s O_s \right\rangle = 0$, as $\sum_{ij} \dot{H}_{s,ij} \delta_{ij} O_{s,ji} = 0$ $(O_{s,ii} = 0)$.

Clearly, for both cases, $C_s = C$ (2.4) and $C_s = D_s$ (2.5), by choosing,

$$\eta = [C_s, H_s] \tag{2.6}$$

the rate of change of ζ_{H_s,C_s} is always nonnegative ($\langle \eta^\dagger \eta \rangle \geq 0$),

$$\frac{d}{ds} \zeta_{H_s,C_s} = \frac{\sigma_{\eta_s}^2}{\sigma_{H_s} \sigma_{C_s}} \geq 0, \tag{2.7}$$

where the η_s strength is, $\sigma_{\eta_s} = \sqrt{\langle [H_s, C_s][C_s, H_s] \rangle} = \sqrt{2 \langle C_s^2 H_s^2 \rangle - 2 \langle C_s H_s C_s H_s \rangle}$. This dependence (2.7) is significant as it reveals that the ζ_{H_s,C_s} correlation between H_s and C_s that is bounded from above monotonically increases until $\sigma_{\eta_s}^2 = 0$, that is until H_s commutes with C_s ($[C_s, H_s] = 0$) when ζ_{H_s,C_s} is maximum. Hence, the ζ_{H_s,C_s} evolution (2.7) with decreasing "decoupling" energy parameter $\lambda_d = 1/\sqrt{s}$ manifests the increasing similarity between the evolved H_s and the C_s, and equally, traces the dynamics of the decoupling between states of low-C_s eigenvalues from those with high-C_s eigenvalues. Therefore, if C_s is chosen to be diagonal in the representation of the initial H, H_s is driven toward a (block-)diagonal form in this representation. In the case of $C_s = D_s$, the H_s is driven toward its diagonal part and eventually coincides with D_s when $\zeta_{H_s,D_s} = 1$. We choose C to be the second-order Casimir invariant, C_2^{su3}, diagonal in a SU(3) shell model basis. Other choices for C include the diagonal part of H_s, D_s [3], as well as, for nuclear structure physics, the $T_{\rm rel}$ relative kinetic energy [20] and the free $H_{\rm HO}$ harmonic oscillator (HO) Hamiltonian.

Furthermore, in the case when C is a linear combination of the Casimir invariants of the symmetry under consideration [e.g., of the C_2^{su3} and the third-order C_3^{su3} for the rank two SU(3) group], the symmetry-based SRG approach offers an improvement over the use of the diagonal D_s. Specifically, for $\sigma_{\eta_s} \approx 0$, H_s becomes a superposition of operators that respect the symmetry and therefore commutes with C. These operators lie in a "plane" (*symmetry plane*) spanned by a countable number of integrity basis operators [24]. This remarkable result points to the fact that H_s can then be expressed as a finite polynomial in the generators and scalars of the symmetry. For nuclear dynamics, this implies that a many-body H_s, that realizes an equivalent energy spectrum of a realistic NN interaction, H (or H_0) $= H_{NN}$, can be chosen to be SU(3)-symmetric and expressed in terms of SU(3) generators and scalars, while reflecting the low-energy symmetry patterns of QCD inherent to the NN realistic interaction.

In short, the spectral distribution theory brings forward a geometrical interpretation of the SRG approach. Clearly, η generates infinitesimal rotations of H_s in a perpendicular multi-dimensional "plane", which contains H_s and C together with any many-body operator X that commutes with C or H_s. This follows from the fact that $\zeta_{C_s,\eta} = 0$, $\zeta_{H_s,\eta} = 0$, and $\zeta_{X,\eta} = 0$, since $\mathrm{Tr}(H_s\eta) = \mathrm{Tr}(C[H_s, H_s]) = 0$ and $\mathrm{Tr}(X\eta) = \mathrm{Tr}([X, C]H_s) = 0$.

The discussions that follow refer to a fixed C not varying with s (such as C_2^{su3}, C_3^{su3}, T_{rel}, and H_{HO}) and may not apply to the diagonal D_s.

Dimensionless SRG evolution – In the framework of spectral distribution theory, a dimensionless SRG evolution is readily available. In particular, we introduce dimensionless unit operators, $H'_s = \frac{H_s}{\sigma_H}$ and $C' = \frac{C}{\sigma_C}$ ($\sigma_{H'_s} = 1$ and $\sigma_{C'} = 1$), and hence Eq. (2.2) becomes,

$$\frac{d}{ds'} H'_s = [[C', H'_s], H'_s], \tag{2.8}$$

with dimensionless $\eta'_s = [C', H'_s]$ and $s' = s\sigma_C\sigma_H$, together with $\lambda_d = \frac{1}{\sqrt{s}} = \sqrt{\frac{\sigma_C\sigma_H}{s'}}$ in units of energy. The correlation coefficient rate of change (2.7) is thus given by,

$$\frac{d}{ds'}\zeta_{H'_s,C'} = \sigma^2_{\eta'_s}. \tag{2.9}$$

Traceless operators – Without loss of generality (*w.l.o.g.*), we can assume traceless H_s and C matrices, that is $\langle H_s\rangle$=0 and $\langle C\rangle$=0. This is because the SRG flow (2.2) is independent of these traces as a result of the unitarity of transformations ($\langle H_s\rangle$=$\langle H\rangle$ and hence, $d\langle H_s\rangle/ds = 0$) together with the fact that traces commute.

In what follows, we will use – *w.l.o.g.* as shown above – a dimensionless flow and traceless operators.

3. Hierarchy of higher-order flow-generated terms

For an infinitesimal transformation, dependent upon a small parameter ϵ, of an operator K that is generated by a hermitian operator G and evolving continuously from the identity, $U = 1 - i\epsilon G$, the transformed $K' = UKU^\dagger = K - i\epsilon[G, K]$. Clearly, with $\eta = -iG$, one obtains the SRG rotations (2.1).

For finite unitary transformations (as used in the numerical flow evolutions), the interaction is rotated at a finite $\delta s = h$ dimensionless step,

$$H_{n+1} = e^{h\eta_n} H_n e^{-h\eta_n} = \sum_{k=0}^{\infty} \frac{h^k}{k!} H_{n(k)} \tag{3.1}$$

with $H_{n(k)} \equiv \underbrace{[\eta_n, [\eta_n, \ldots [\eta_n}_{k}, H_n] \ldots]]$ or equivalently, $H_{n(k)} = [\eta_n, H_{n(k-1)}]$ with $H_{n(0)} = H_n$. Here, the k-th term involves k commutators yielding a many-body interaction of a strength $\sigma_{H_{n(k)}}$, which can be shown to be at the most of order of $(S_\eta)^{2k}$ for a S_η upper bound of σ_η (Appendix A).

Clearly, with a 1-body C and a 2-body H_0 initial Hamiltonian, the first rotation renders $H_{0(1)}$ 3-body, $H_{0(2)}$ 4-body and *etc.* $H_{0(1)}$ is of particle rank 3, which includes 2-body interactions that are present in a flow evolved in a two-particle system ($A = 2$). However, such a flow does not take into account the pure 2-body part of the 3-body terms[1] in $H_{0(1)}$. The lowest particle

[1] A definite ν particle rank interaction (pure ν-body interaction), e.g., for a scalar distribution, is a U(N) irreducible tensor $\mathcal{H}^k(\nu)$ of rank $\nu = 0, 1, \ldots, k$, such that for a k-body interaction, $H^k = \oplus\mathcal{H}^k(\nu)$ and N is the single-particle basis space dimension.

Journal of Physics: Conference Series **284** (2011) 012056 doi:10.1088/1742-6596/284/1/012056

ranks (LPR) of the induced interaction that are not present in H_0 enter in $hH_{0(1)}$, while higher particle ranks, such as the next-to-LPR (NLPR) in $h^2H_{0(2)}$ and N^{2}LPR in $h^3H_{0(3)}$, are brought forward in the next terms. In general, for a p_1-b C and a p_2-b H_0, the maximum particle rank of H_n at $\mathcal{O}(h^K)$ is $K^n(p_1+p_2-2)+(2-p_1)$.

It is sufficient to choose $hS_\eta^2 < 1$ to observe a hierarchy of the many-body terms within a rotation and to evolve (2.2),

$$H_{n+1(I)} \equiv H_{n+1} - H_n = hH_{n(1)} + \frac{h^2}{2!}H_{n(2)} + \frac{h^3}{3!}H_{n(3)} + \mathcal{O}(h^4). \tag{3.2}$$

As $H_{n(k+1)} = [\eta_n, H_{n(k)}]$, every two $H_{n(k+1)}$ and $H_{n(k)}$ are orthogonal (that is, $\langle H_{n(k+1)}H_{n(k)}\rangle = 0$) and in addition, $\langle H_{n(k+2)}H_{n(k)}\rangle = -\sigma^2_{H_{n(k+1)}}$. While $H_{n(2)}$ does not project onto $H_{n(1)}$, $H_{n(3)}$ contains $H_{n(1)}$-like contributions, $H_{n(3)} = w_nH_{n(1)} + \tilde{H}_{n(3)}$ with $w_n = \frac{\langle H_{n(3)}H_{n(1)}\rangle}{\sigma^2_{H_{n(1)}}} = -\frac{\sigma^2_{H_{n(2)}}}{\sigma^2_{H_{n(1)}}}$, where $\tilde{H}_{n(3)}$ is the $H_{n(3)}$ portion orthogonal to $H_{n(1)}$ (clearly, it is also orthogonal to $H_{n(2)}$). Hence,

$$H_{n+1(I)} = (1+\frac{h^2}{3!}w_n)hH_{n(1)} + \frac{h^2}{2!}H_{n(2)} + \frac{h^3}{3!}\tilde{H}_{n(3)} + \mathcal{O}(h^4), \tag{3.3}$$

and the induced interaction strength at the $(n+1)^{\rm st}$ rotation is given by $\sigma^2_{H_{n+1(I)}} = \left\langle H^\dagger_{n+1(I)}H_{n+1(I)}\right\rangle$,

$$\begin{aligned}\sigma^2_{H_{n+1(I)}} &= (1+\frac{h^2}{3!}w_n)^2h^2\sigma^2_{H_{n(1)}} + \frac{h^4}{4}\sigma^2_{H_{n(2)}} + \mathcal{O}(h^6),\\ &= (1+\frac{h^2}{3}w_n)h^2\sigma^2_{H_{n(1)}} + \frac{h^4}{4}\sigma^2_{H_{n(2)}} + \mathcal{O}(h^6).\end{aligned} \tag{3.4}$$

The first term in (3.4) is of the $H_{n(1)}$ particle rank, while the second term introduces higher-order couplings. The strengths of the interactions in (3.3) are related as, $\sigma^2_{H_{n+1(I)}} = h^2\sigma^2_{H_{n(1)}} - \frac{h^4}{12}\sigma^2_{H_{n(2)}} + \mathcal{O}(h^6)$.

To estimate the LPR contribution to $hH_{n(1)}$, the latter can be expressed as [using (3.1)],

$$\begin{aligned}hH_{n(1)} &= hH_{n-1(1)} + h^2V_{n-1(2)} + h^3V_{n-1(3)} + \mathcal{O}(h^4), \text{with}\\ V_{n-1(2)} &\equiv H_{n-1(2)} + [[C, H_{n-1(1)}], H_{n-1}]\\ V_{n-1(3)} &\equiv \frac{1}{2}H_{n-1(3)} + [[C, H_{n-1(1)}], H_{n-1(1)}] + \frac{1}{2}[[C, H_{n-1(2)}], H_{n-1}]\end{aligned} \tag{3.5}$$

and thus directly related, in first order, to the LPR in $hH_{0(1)}$. Similarly to the orthogonalization we perform above, one can identify the $H_{n(1)}$-like contributions of $V_{n(2)}$ and $V_{n(3)}$,

$$\begin{aligned}V_{n(2)} &= q_nH_{n(1)} + \tilde{V}_{n(2)}, \text{with } q_n = \frac{\langle H_{n(1)}V_{n(2)}\rangle}{\sigma^2_{H_{n(1)}}}\\ V_{n(3)} &= r_nH_{n(1)} + \tilde{V}_{n(3)}, \text{with } r_n = \frac{\langle H_{n(1)}V_{n(3)}\rangle}{\sigma^2_{H_{n(1)}}},\end{aligned} \tag{3.6}$$

where $\tilde{V}_{n(2)}$ [$\tilde{V}_{n(3)}$] is the $V_{n(2)}$ [$V_{n(3)}$] portion orthogonal to $H_{n(1)}$, and thus obtain,

$$\begin{aligned}hH_{n(1)} = \ & p_nhH_{n-1(1)} + h^2\tilde{V}_{n-1(2)} + h^3\tilde{V}_{n-1(3)} + \mathcal{O}(h^4), \text{ with}\\ & p_n \equiv 1 + hq_{n-1} + h^2r_{n-1}.\end{aligned} \tag{3.7}$$

Note that p_n gives the projection of $H_{n(1)}$ onto $H_{n-1(1)}$, $p_n = \left\langle H_{n-1(1)} H_{n(1)} \right\rangle / \sigma^2_{H_{n-1(1)}}$. It follows that,

$$\begin{aligned} h^2\sigma^2_{H_{n(1)}} &= p_n^2 h^2 \sigma^2_{H_{n-1(1)}} + h^4\sigma^2_{\tilde{V}_{n-1(2)}} + \mathcal{O}(h^5) \\ &= ((1+hq_{n-1})^2 + 2h^2 r_{n-1}) h^2\sigma^2_{H_{n-1(1)}} + h^4\sigma^2_{\tilde{V}_{n-1(2)}} + \mathcal{O}(h^5), \end{aligned} \tag{3.8}$$

where, to $\mathcal{O}(h^3)$, $p_n^2 = (1+hq_{n-1})^2 + 2h^2 r_{n-1}$ with

$$q_n = \frac{d}{ds}\ln(\sigma_{H_{s(1)}})\bigg|_n \quad \text{and } r_n \approx -\frac{\sigma^2_{H_{n(2)}}}{\sigma^2_{H_{n(1)}}}. \tag{3.9}$$

This follows from, $q_n\sigma^2_{H_{n(1)}} = \langle H_{n(1)} V_{n(2)}\rangle = \langle H_{n(1)}[[C, H_{n(1)}], H_n]\rangle = \left\langle H_{n(1)}[\frac{d\eta_s}{ds}\Big|_n, H_n]\right\rangle = \langle H_{n(1)}\ \frac{d}{ds}[\eta_s, H_s]\big|_n\rangle = \left\langle H_{n(1)}\ \frac{d}{ds}H_{s(1)}\big|_n\right\rangle = \frac{1}{2}\frac{d}{ds}\ \sigma^2_{H_{s(1)}}\Big|_n$, where we use that $\frac{dH_s}{ds}\big|_n = H_{n(1)}$, $\frac{d\eta_s}{ds} = [C, \frac{dH_s}{ds}]$, and $\langle H_{n(1)}[\eta_n, \frac{d}{ds}H_s\big|_n]\rangle = 0$, together with an approximate r_n value, for which typically $\langle H_{n(1)}V_{n(3)}\rangle \approx \langle H_{n(1)}H_{n(3)}\rangle$. Note that as $h \to 0$ (infinitesimal rotations), the h^4 terms in (3.8) vanish and the approximation in r_n becomes irrelevant.

Therefore, the main contribution to $H_{n(1)}$ is due to the LPR interactions that emerge in $H_{0(1)}$,

$$\begin{aligned} h^2\sigma^2_{H_{n(1)}} &= h^2\sigma^2_{H_{0(1)}}\prod_{k=1}^{n} p_k^2 + h^4(\sum_{k=2}^{n}\sigma^2_{\tilde{V}_{k-2(2)}}\prod_{l=k}^{n} p_l^2 + \sigma^2_{\tilde{V}_{n-1(2)}}) + \mathcal{O}(nh^5), \\ &= h^2\sigma^2_{H_{0(1)}}\prod_{k=1}^{n} p_k^2 + nh^4\sigma^2_{\tilde{V}_{0(2)}} + \mathcal{O}(nh^5), \end{aligned} \tag{3.10}$$

where we use that, $p_l^2 = 1 + \mathcal{O}(h)$ in the h^4-term, together with the fact that, to $\mathcal{O}(h^6)$, $h^4\sigma^2_{\tilde{V}_{n(2)}} \sim h^4\sigma^2_{\tilde{V}_{n-1(2)}}$ and hence, related to the NLPR of $\tilde{V}_{0(2)}$. The induced interaction strength (3.4) at the (n+1)st rotation is then,

$$\begin{aligned} \sigma^2_{H_{n+1(I)}} &= \underbrace{h^2\sigma^2_{H_{0(1)}}(1+\frac{h^2}{3}w_n)\prod_{k=1}^{n}p_k^2}_{\equiv\sigma^2_{n,\text{LPR}}} \\ &+ \underbrace{nh^4\sigma^2_{\tilde{V}_{0(2)}} + \frac{h^4}{4}\sigma^2_{H_{0(2)}}}_{\equiv\sigma^2_{n,\text{NLPR}}} + \mathcal{O}(nh^5), \end{aligned} \tag{3.11}$$

with a major contribution ($\sim h^2$) of the LPR induced interactions ($\sigma^2_{n,\text{LPR}}$) and a much smaller contribution ($\sim h^4$) of the NLPR induced interactions ($\sigma^2_{n,\text{LPR}}$).

To estimate the total strength of the (N)LPR interaction, $\sigma_{\text{(N)LPR}}$, induced during the SRG flow evolved to $\lambda_{d,\text{min}}$ (in units of energy), the induced interactions at each rotation are summed as aligned 'vectors' (that lie on $H_{0(1)}$ regardless of the sign of their projection), which is an upper bound, $|\mathbf{a}+\mathbf{b}| \leq |\mathbf{a}| + |\mathbf{b}|$, while the sum of the perpendicular vectors, projections on $H_{0(1)}$ and projections on $\tilde{V}_{0(2)}$, give the total induced interaction strength, σ_I,

$$\sigma_{\text{(N)LPR}} = \sum_{n=0}^{n_{\max}-1}\sigma_{n,\text{(N)LPR}} \tag{3.12}$$

$$\sigma_I = \sqrt{\sigma^2_{\text{LPR}} + \sigma^2_{\text{NLPR}}}, \tag{3.13}$$

Journal of Physics: Conference Series **284** (2011) 012056 doi:10.1088/1742-6596/284/1/012056

where $n_{\max}h = \sigma_H\sigma_C/\lambda^2_{d,\min}$ for dimensionless h and the error term for $\sigma_{(\mathrm{N})\mathrm{LPR}}$ is of order of $n^2_{\max}h^3 \sim \mathcal{O}(h)$. Note that σ_{LPR} is of order of $n_{\max}h$, which is a constant for given $\lambda^2_{d,\min}$.

For infinitesimal rotations, $h \to 0$ and $n_{\max}h = \sigma_H\sigma_C/\lambda^2_{d,\min}$, $\sigma_{\mathrm{NLPR}} \to 0$. The only contribution to σ_I is due to the LPR interactions. The latter include the LPR component of the NLPR $V_{0(2)}$ (that enters through q_n), while contributions from NNLPR (r_n and w_n) vanish.

4. Flow dynamics and estimates for the strengths of the LPR and NLPR induced higher-order coupling terms

To obtain estimates for the total strength of the (N)LPR induced interaction, one needs to determine p_n [q_n and $r_n \approx w_n$ (3.9)], that is, $\sigma_{H_{s(1)}}$, $\sigma_{H_{s(2)}}$, and the rate of change of $\sigma_{H_{s(1)}}$. We employ $\sigma_{H_{s(k)}}$ typically $\sim \sigma^{2k}_{\eta_s}$ (analogous to their S^{2k}_η upper limits) and an approximate *upper* limit of σ_{η_s},

$$\sigma_{\eta_s} \approx \frac{2\sigma_{\eta_0}}{1+\zeta_{H_0,C}} \frac{e^{-\frac{\sigma^2_{\eta_0}\frac{s}{\sigma_{H_0}\sigma_C}}{1-\zeta^2_{H_0,C}}}}{1+\frac{1-\zeta_{H_0,C}}{1+\zeta_{H_0,C}}e^{-2\frac{\sigma^2_{\eta_0}\frac{s}{\sigma_{H_0}\sigma_C}}{1-\zeta^2_{H_0,C}}}}, \tag{4.1}$$

which can be deduced from (2.7) and the smallest decrease rate of σ_{η_s} (Appendix B),

$$\frac{d}{ds}\sigma^2_{\eta_s} \lesssim -\frac{2\sigma^4_{\eta_s}}{\sigma_H\sigma_C}\frac{\zeta_{H_s,C}}{1-\zeta^2_{H_s,C}}, \tag{4.2}$$

where *w.l.o.g.* we assume $\zeta_{H_0,C} \geq 0$. This approximation excludes local increases in σ_{η_s} and uses that typically for near symmetries (strongly correlated C and the evolved H_s), $\zeta_{H_{s(2)},C} \approx \zeta_{H_{s(2)},H_s}\zeta_{H_s,C}$.

In particular, for $\delta = 0.01$, $\tilde{\sigma}_{H_{0(1)}} \approx \tilde{\sigma}_{H_{0(2)}} \approx 1$ (with $h \equiv \delta/\sigma^2_{\eta_0}$ and $\tilde{\sigma}_{H_{0(k)}} \equiv \sigma_{H_{0(k)}}/\sigma^{2k}_{\eta_0}$) and a typical $\zeta_{H_0,C} \sim 0.2$, the total NLPR induced during an infinitely evolved flow turn out to be only 0.45% of the total induced interactions ($\frac{\sigma^2_{\mathrm{NLPR}}}{\sigma^2_{\mathrm{I}}}$). This percentage is proportional to δ and $\tilde{\sigma}_{H_{0(2)}}$, and hence for larger $\tilde{\sigma}_{H_{0(2)}}$ smaller δ steps are preferable. It is also possible to estimate the contribution of 3-b interactions to the induced σ_{LPR}. Namely, three-shell $T = 1$ calculations (s, p and sd) with random 1-b C and 2-b H_0 initial interaction with $\zeta_{H_0,C} = 0.2$ reveal that $\sigma^2_{H_{0(1)}}$ contains around 40% (60%) 3-b (2-b) interactions (incidentally, the pure 2-b part of the 3-b interaction is only $\sim 3\%$). This implies that for the example parameters (and with dimensionless $\sigma_{H(C)} = 1$), the induced 2-b LPR interaction strength is $\sigma_{\mathrm{LPR-2b}} \simeq 0.62$ (59.73% of the total induced σ^2_{I}), the induced 3-b LPR interaction strength is $\sigma_{\mathrm{LPR-3b}} \simeq 0.51$ (39.82% of σ^2_{I}), and $\sigma_{\mathrm{NLPR}} = 0.05$, which includes 4-body interactions. These strengths further decrease for larger $\zeta_{H_0,C}$.

Therefore, while it is clear that SRG evolving H_{NN} with a 1-b C in a $A = 2$ system neglects 3-b terms with a contribution $\lesssim 67\%$ as compared to the SRG-induced 2-b interactions and hence not suitable for $A \geq 3$ applications, SRG evolved in a $A = 3$ system will neglect, in addition to an even smaller contribution of higher particle rank interactions, net 4-b forces of significance of less than a percent (σ_{NLPR}) together with the pure 3-b *part* of the 4-b NLPR forces (note that the dominant 3-b NLPR forces are accounted for). The forces neglected appear to be insignificant in applications of such renormalized interactions to $A = 4$ [25], further rendering applications to $A > 4$ systems feasible. Note that the second in importance term ($hH_{0(1)}$) is 5-b if evolving H_{NN+3N} requiring at least $A = 5$ SRG calculations. This term is 4-b for evolving H_{NN} with the 2-b C^{su3}_2. Fortunately, in the SU(3) case, the use of symmetry renders 4-body terms manageable.

Journal of Physics: Conference Series **284** (2011) 012056 doi:10.1088/1742-6596/284/1/012056

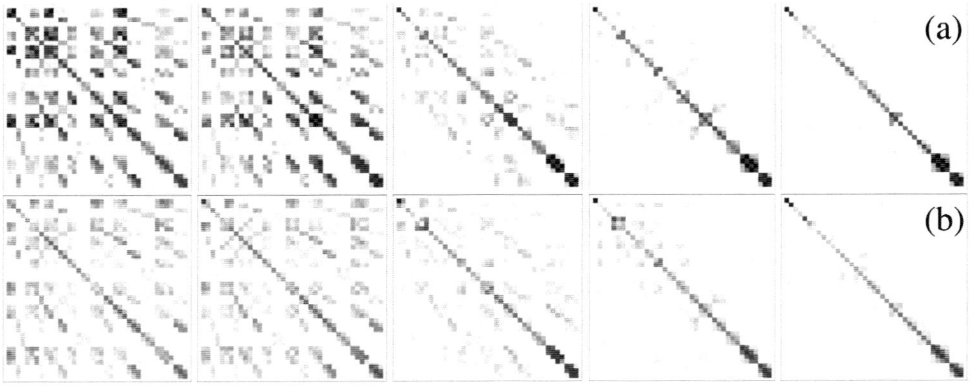

Figure 1. SRG evolution ($J = T = 0$, 8 shells) for $\lambda_d = \infty$, 31.6, 10, 3.2 and 1 MeV of $T_{\rm rel}+$ (a) AV18 and (b) N^{3}LO NN bare interactions. Nonzero matrix elements (filled squares) are shown for SU(3)-coupled two-nucleon basis states, ordered by increasing C_2^{su3} eigenvalues.

The C_2^{su3}-SRG is applied to a general 2-b interaction in a representation of a SU(3)-coupled HO basis (using conventional labels and $\Gamma = JT$), $|n_r n_u \omega\kappa(LS)\Gamma M_\Gamma\rangle = \frac{\{a^\dagger_{n_r}\times a^\dagger_{n_u}\}^{\omega\kappa(LS)\Gamma M_\Gamma}|0\rangle}{\sqrt{1+\delta_{n_r,n_u}}}$ with SU(3) quantum numbers, $\omega \equiv (\lambda\,\mu) = (n_r\,0)\times(n_u\,0)$, and κ multiplicity of L for a given ω. This basis is obtained via a unitary transformation from the JT-coupled HO basis, $|ru\Gamma M_\Gamma\rangle$ with $r(u) = \{n(l\,\frac{1}{2})jt = \frac{1}{2}\}$, and yields the (isoscalar) Hamiltonian matrix elements ($\chi = n_r n_u$ for a two-particle system), $V^\Gamma_{fi} = \langle \chi_f\omega_f\kappa_f(L_fS_f)\Gamma\|H\|\chi_i\omega_i\kappa_i(L_iS_i)\Gamma\rangle/\sqrt{2\Gamma+1}$. The C_2^{su3} (C_3^{su3} and $H_{\rm HO}$) is diagonal in this basis and can be used to drive H toward a block-diagonal form.

The two-particle C_2^{su3}-SRG method, performed separately for each Γ, is applied to various realistic NN interactions, such as CD-Bonn ($\hbar\Omega$=15 MeV) and N^{3}LO (11 MeV) [26, 27], as well as JISP16 (15 MeV) [28] and AV18 (18 MeV) [29], with the relative kinetic energy added prior to the SRG transformation. As an illustrative example, we consider $6\hbar\Omega$ for 'p-shell' nuclei (Fig. 1). Using the aforementioned estimates based only on C_2^{su3} and H_{NN} initial properties, it is expected that, e.g., for N^{3}LO evolved to $\lambda_d = 1$ MeV with $\delta \sim 0.001$, the net NLPR contribution to the total induced interaction is less than 0.8%. In Fig. 1, the order of states is chosen in increasing C_2^{su3} eigenvalues, that is, in increasing deformation. Typically, large deformations involve high-energy configurations, which correspond to large excitation numbers N. Hence, the SRG brings forward decoupling of higher-energy configurations (large $N\hbar\Omega$) from the low-energy ones.

5. Conclusions

We introduced symmetry considerations into the Similarity Renormalization Group (SRG) theory for nuclear physics applications. Specifically, invariants of SU(3), which is a symmetry that is known to dominate the dynamics, were used to evolve the SRG flow. First we gave quantitative estimates for the strength of the SRG-induced many-body interactions for a general flow, and then showed that an application of the SU(3)-SRG approach to various two-nucleon interactions decouples low-energy configurations that are especially relevant for the regime of ordinary bound systems from higher-energy ones, with manageable induced many-body forces.

Acknowledgments

Discussions with R.J. Furnstahl and C. Bahri are gratefully acknowledged. This work was supported by the U.S. National Science Foundation (PHY-0500291 & OCI-0904874) and the

Southeastern Universities Research Association, as well as benefited from the INT-07-3 and INT-09-01 programs at the National Institute for Nuclear Theory (University of Washington).

Appendix A. Estimate for the strength of the SRG generator and $H_{s(1)}$
To estimate the upper limit of the $\eta_s = [C, H_s]$ strength,

$$\sigma^2_{\eta_s} = -\langle \eta_s \eta_s \rangle = 2\left\langle C^2 H_s^2 \right\rangle - 2\langle CH_sCH_s\rangle \quad (A.1)$$

we use bounds of correlation coefficients,

$$\begin{aligned} -1 \leq \zeta_{H_sC,CH_s} &= \frac{\langle CH_sCH_s\rangle - \langle CH_s\rangle^2}{\sigma^2_{CH_s}} \leq 1, \\ -1 \leq \zeta_{C^2,H_s^2} &= \frac{\left\langle C^2H_s^2\right\rangle - \left\langle C^2\right\rangle\left\langle H_s^2\right\rangle}{\sigma_{C^2}\sigma_{H_s^2}} \leq 1 \end{aligned} \quad (A.2)$$

and

$$\sigma^2_{CH_s} = \left\langle C^2H_s^2\right\rangle - \langle CH_s\rangle^2 . \quad (A.3)$$

Hence, $\sigma^2_{\eta_s} \leq 4\left\langle C^2H_s^2\right\rangle \leq 4(\sigma_{C^2}\sigma_{H_s^2} + \left\langle C^2\right\rangle\left\langle H_s^2\right\rangle)$. During a flow with a s-independent C, the strength of η_s for traceless C and H_s is found to have a S_η upper bound,

$$\sigma_{\eta_s} \leq S_\eta = 2\sigma_C\sigma_H\sqrt{\sqrt{(\mathcal{M}_C^{(4)} - 1)(\mathcal{M}_H^{(4)} - 1)} + 1}, \quad (A.4)$$

where the k-th spectral moment of an operator K (C or H_s) scaled by σ_K is $\mathcal{M}_K^{(k)} = \frac{1}{\sigma_K^k}\left\langle (K - \langle K\rangle)^k\right\rangle$. If an operator follows a Gaussian distribution, $\mathcal{M}^{(4)} = 3$, $\mathcal{M}^{(6)} = 15$, and $\mathcal{M}^{(8)} = 105$. Clearly, the second moment is always $\mathcal{M}^{(2)} = 1$ and for the unitary transformation of H, $\mathcal{M}_{H_s}^{(k)} = \mathcal{M}_H^{(k)}$ and $\sigma_{H_s} = \sigma_H$. This is an important result, as spectral moments need to be calculated only for the initial H.

The moments for C_2^{su3} and H are estimated for the 8-shell JT-coupled scheme of various realistic interactions, such as CD-Bonn ($\hbar\Omega$=15 MeV), N^{3}LO (11 MeV) [26, 27] and JISP16 (15 MeV) [28], with the relative kinetic energy added. In addition, the C_2^{su3} analytical form is used to calculate moments in JT-coupled scheme up through $N_{\max} = 100$, which agrees with the 8-shell calculations for C_2^{su3}. Hence, estimates for H in larger spaces are expected to remain of the same order. The spectral moments calculated are typically less than,

$$\mathcal{M}^{(4)}_{C_2^{su3}} \sim 3,\ \mathcal{M}^{(6)}_{C_2^{su3}} \sim 15,\ \mathcal{M}^{(8)}_{C_2^{su3}} \sim 10^2,\ \mathcal{M}^{(4)}_H \sim 10, \text{ and } \mathcal{M}^{(6)}_H \sim 10^2. \quad (A.5)$$

These estimates are used to provide upper limits for various strengths.

For example, the maximum strength of the dimensionless η_s is found from (A.4) and (A.5) to be $S_\eta \sim 5$ for $C = C_2^{su3}$. Similarly, the $\sigma_{H_{s(1)}}$ strength of $H_{s(1)} = dH_s/ds$ is less than,

$$\sigma_{H_{s(1)}} \lesssim S_{H_{(1)}} = 4\sigma_C\sigma_H^2\left((\sqrt{\mathcal{M}_C^{(8)}\mathcal{M}_H^{(4)}} + \mathcal{M}_C^{(4)})\mathcal{M}_H^{(6)}\right)^{1/4}, \quad (A.6)$$

which gives the estimate for the upper bound $S_{H_{(1)}} \sim S_\eta^2$ for $C = C_2^{su3}$ and the realistic NN interactions under consideration (A.5).

Appendix B. Rate of change of σ_{η_s}

The rate of change of σ_{η_s} is given as,

$$\begin{aligned} \frac{1}{2}\frac{d}{ds}\sigma^2_{\eta_s} &= -\left\langle \eta_s \frac{d\eta_s}{ds}\right\rangle = -\left\langle C[\frac{dH_s}{ds}, \eta_s]\right\rangle = \left\langle H_{s(2)}C\right\rangle \\ &\approx \sigma_C \sigma_{H_{s(2)}} \zeta_{H_{s(2)},H_s} \zeta_{H_s,C} = -\frac{\sigma_C \sigma^2_{H_{s(1)}}}{\sigma_H}\zeta_{H_s,C}, \end{aligned} \tag{B.1}$$

where we use that $\zeta_{H_{s(2)},C} \approx \zeta_{H_{s(2)},H_s}\zeta_{H_s,C}$ and $\sigma_{H_{s(2)}}\zeta_{H_{s(2)},H_s} = \frac{1}{\sigma_H}\left\langle H_s H_{s(2)}\right\rangle = -\frac{\sigma^2_{H_{s(1)}}}{\sigma_H}$.

The smallest decrease of η_s is given by the lower bound of $\sigma_{H_{s(1)}}$, which can be derived from,

$$\sigma^2_{\eta_s} = -\left\langle \eta_s [C, H_s]\right\rangle = \left\langle CH_{s(1)}\right\rangle \le \sigma_C \sigma_{H_{s(1)}} \sqrt{1-\zeta^2_{H_s,C}}. \tag{B.2}$$

This follows from $\zeta_{C,H_{s(1)}} = \zeta_{C,H_{s(1)\|}}\zeta_{H_{s(1)\|},H_{s(1)}} \le \sin\theta_s$, where $H_{s(1)\|}$ is the projection of $H_{s(1)}$ in the (H_s, C) plane. The angle between $H_{s(1)\|}$ and C is $\frac{\pi}{2} - \theta_s$ ($H_{s(1)}$ is perpendicular to H_s and $\cos\theta_s = \zeta_{H_s,C}$). Hence, $-\sigma^2_{H_{s(1)}} \le -\frac{\sigma^4_{\eta_s}}{\sigma^2_C(1-\zeta^2_{H_s,C})}$ and,

$$-\frac{\sigma_C \sigma^2_{H_{s(1)}}}{\sigma_H}\zeta_{H_s,C} \le -\frac{\sigma^4_{\eta_s}}{\sigma_H \sigma_C}\frac{\zeta_{H_s,C}}{1-\zeta^2_{H_s,C}}, \tag{B.3}$$

which together with (B.1) yields (4.2).

References

[1] Glazek S and Wilson K 1993 *Phys. Rev. D* **48** 5863
[2] Glazek S and Wilson K 1994 *Phys. Rev. D* **49** 4214
[3] Wegner F 1994 *Ann. Phys. (Leipzig)* **3** 77
[4] Kehrein S 2005 *Phys. Rev. Lett.* **95** 056602
[5] Sabio J and Kehrein S 2010 *New J. Phys.* **12** 055008
[6] Scholtz F, Bartlett B and Geyer H 2003 *Phys. Rev. Lett.* **91** 080602
[7] Grote I, Koerding E and Wegner F 2002 *J. Low Temp. Phys.* **126** 1385
[8] Aprea G, Castro C D, Grilli M and Lorenzana J 2006 *Nucl. Phys. B* **744** 277
[9] Glazek S D 2007 *Phys. Rev. D* **75** 025005
[10] Gubankova E and Wegner F 1998 *Phys. Rev. D* **58** 025012
[11] Navrátil P, Vary J P and Barrett B R 2000 *Phys. Lett.* **84** 5728
[12] Navrátil P, Vary J P and Barrett B R 2000 *Phys. Rev. C* **62** 054311
[13] Wloch M, Dean D J, Gour J R, Hjorth-Jensen M, Kowalski K, Papenbrock T and Piecuch P 2005 *Phys. Rev. Lett.* **94** 212501
[14] Pieper S C, Varga K and Wiringa R B 2002 *Phys. Rev. C* **66** 044310
[15] Dytrych T, Sviratcheva K D, Bahri C, Draayer J P and Vary J P 2007 *Phys. Rev. Lett.* **98** 162503
[16] Navrátil P, Quaglioni S, Stetcu I and Barrett B R 2009 *J. Phys. G: Nucl. Part.* **36** 083101
[17] Roth R and Navrátil P 2007 *Phys. Rev. Lett.* **99** 092501
[18] Suzuki K and Lee S Y 1980 *Prog. Theor. Phys.* **64** 2091
[19] Navrátil P and Barrett B R 1998 *Phys. Rev. C* **57** 562
[20] Bogner S, Furnstahl R and Perry R 2007 *Phys. Rev. C* **75** 061001(R)
[21] Roth R, Reinhardt S and Hergert H 2008 *Phys. Rev. C* **77** 064003
[22] French J and Ratcliff K 1971 *Phys. Rev. C* **3** 94
[23] Sviratcheva K, Dytrych T and Draayer J 2010 *submitted*
[24] Draayer J and Rosensteel G 1985 *Nucl. Phys. A* **439** 61
[25] Jurgenson E D, Navrátil P and Furnstahl R J 2009 *Phys. Rev. Lett.* **103** 082501
[26] Machleidt R 2001 *Phys. Rev. C* **63** 024001
[27] Entem D R and Machleidt R 2003 *Phys. Rev. C* **68** 041001
[28] Shirokov A, Vary J, Mazur A and Weber T 2007 *Phys. Lett.* **B 644** 33
[29] Wiringa R B, Stoks V G J and Schiavilla R 1995 *Phys. Rev. C* **51** 38

Journal of Physics: Conference Series **284** (2011) 012057 doi:10.1088/1742-6596/284/1/012057

Invariant theory to study symmetry breakings in the adjoint representation of E_8

Vittorino Talamini

Dipartimento di Fisica, Università di Udine, via delle Scienze 208, 33100 Udine, Italy and INFN, Sezione di Trieste, Trieste, Italy

E-mail: talamini@uniud.it

Abstract. To study analytically the possible symmetry breakings that may take place in the adjoint representation of the compact Lie group E_8 one may take advantage of the knowledge of the basic polynomials of its Weyl group, that is of the finite reflection group E_8, and of the (polynomial) equations and inequalities defining its orbit space. This orbit space is stratified and each of its strata represents a possible symmetry breaking, both for the adjoint representation of the Lie group E_8 and for the defining representation of the finite reflection group E_8. The concrete determination of the equations and inequalities defining the strata is not yet done but it is possible in principle using rank and positive semi-definiteness conditions on a symmetric 8×8 matrix whose matrix elements are real polynomials in 8 variables. This matrix, called the $\widehat{P}$-matrix, has already been determined completely. In this article it will be reviewed how one may determine the equations and inequalities defining the strata of the orbit space and how these equations and inequalities may help to study spontaneous symmetry breakings. Here the focus is on E_8, because some of the results for E_8 are new, but all what it is here said for E_8 may be repeated for any other compact simple Lie group. The $\widehat{P}$-matrices of all irreducible reflection groups have in fact already been determined.

1. Basic invariant polynomials and orbit spaces

Orbit spaces may be useful to study with clarity and simplicity functions that are invariant under transformations of a compact symmetry group, in particular in problems concerning spontaneous symmetry breakings or structural phase transitions. The invariant functions are in fact constant on the orbits of the group, and may be studied by means of functions defined on the orbit space of the group action.

It may happen that different groups share the same orbit space structure. For example, this may happen for the adjoint representation of a compact connected Lie group G and the defining representation of the Weyl group W of G. When different groups share the same orbit space, the study of invariant functions in the orbit space requires exactly the same calculations and yields, in a certain sense, exactly the same results, for all these groups. Moreover, one may consider the simplest of all these groups to determine the algebraic and geometric structure of the orbit space.

Let's assume that the symmetry group G is a compact group (finite or continuous) acting linearly on a finite dimensional Euclidean space V. Without loss of generality, with respect to a orthonormal basis in V, one may identify V with $\mathbb{R}^n$, for a certain $n \in \mathbb{N}$, and G with a group of real orthogonal $n \times n$ matrices acting on $\mathbb{R}^n$ with the matrix multiplication. In addition, one

Journal of Physics: Conference Series **284** (2011) 012057 doi:10.1088/1742-6596/284/1/012057

may assume that this G-action be effective, that is, with no non-zero vector left invariant by all the transformations of G.

Let $f(x)$, $x \in \mathbb{R}^n$, be a real function of n real variables. f is said to be *G-invariant* if $f(gx) = f(x)$, $\forall g \in G$ and $x \in \mathbb{R}^n$ in the domain of f (and this implies that in the domain of a G-invariant function there are complete orbits of the G-action). The polynomial functions defined in $\mathbb{R}^n$ form an algebra, indicated with $\mathbb{R}[\mathbb{R}^n]$. G acts naturally on $\mathbb{R}[\mathbb{R}^n]$ by defining $(gp)(x) = p(g^{-1}x)$, $\forall p \in \mathbb{R}[\mathbb{R}^n]$ and $x \in \mathbb{R}^n$. $p \in \mathbb{R}[\mathbb{R}^n]$ is then G-invariant if $gp = p$ $\forall g \in G$. The G-invariant real polynomial functions on $\mathbb{R}^n$ form an algebra, indicated by $\mathbb{R}[\mathbb{R}^n]^G$. For the linearity of the G-action, if $p \in \mathbb{R}[\mathbb{R}^n]$ is homogeneous of degree d, then $gp \in \mathbb{R}[\mathbb{R}^n]$ is homogeneous of degree d.

By Hilbert's theorem on invariants, $\mathbb{R}[\mathbb{R}^n]^G$ is finitely generated, that is, there exist $p_1, \ldots, p_q \in \mathbb{R}[\mathbb{R}^n]^G$, such that $\forall p \in \mathbb{R}[\mathbb{R}^n]^G$, there exists a unique polynomial $\widehat{p}$ in q indeterminates such that $p(x) = \widehat{p}(p_1(x), \ldots, p_q(x))$, $\forall x \in \mathbb{R}^n$. The invariant polynomials $p_1, \ldots, p_q$, are usually called *basic invariants*, *basic (invariant) polynomials* or *basic generators*, and they can in all generality be chosen real and homogeneous. Especially in the physical literature, a set of basic polynomials is sometimes called an *integrity basis* of the G-action. I will usually write MIB for a *minimal integrity basis* $p_1, \ldots, p_q$. For the minimality condition, no proper subset of the basic polynomials in a MIB can itself be a MIB. The choice of a set $p_1, \ldots, p_q$ of basic polynomials is not unique, but the number q and the degrees $d_1, \ldots, d_q$ of the basic polynomials $p_1, \ldots, p_q$, do not depend on the MIB chosen and are uniquely determined by the group G. Any other MIB, formed by a set $p'_1, \ldots, p'_q$, of q homogeneous invariant polynomials of the same degrees $d_1, \ldots, d_q$, may differ from the MIB $p_1, \ldots, p_q$ either for a different choice of the (orthonormal) coordinate system in $\mathbb{R}^n$, or because the p'_a can be expressed as polynomials in the p_a, with the jacobian matrix $J(p) = ||\partial p'(p)/\partial p||$ not singular.

It is usual to order the basic polynomials according to their degrees, for example in such a way that $2 = d_1 \leq d_2 \leq \ldots \leq d_q$. One may also put $p_1(x) = (x, x) = \sum_{i=1}^n x_i^2$, where $(\cdot, \cdot)$ is the canonical scalar product in $\mathbb{R}^n$ (a natural choice for real orthogonal actions).

A MIB allows one to determine the algebraic equations and inequalities defining a semi-algebraic set $\mathcal{S} \subset \mathbb{R}^q$, that can be considered as a concrete representation of the abstract orbit space $\mathbb{R}^n/G$ of the G-action. Moreover, a MIB allows one to determine a one to one correspondence between the smooth G-invariant functions $f(x)$, $x \in \mathbb{R}^n$ and the smooth functions $\widehat{f}(p)$, $p \in \mathcal{S}$, in such a way that $f(x) = \widehat{f}(p_1(x), \ldots, p_q(x))$, $\forall x \in \mathbb{R}^n$ in the domain of f. Then, one may study the G-invariant functions on the set $\mathcal{S}$, and this may sometimes be convenient and/or enlightening.

The rest of this Section reports some known definitions and results, concerning orbit spaces of effective actions on $\mathbb{R}^n$ of a general compact group $G \subset O(n)$.

(i) The *orbit* of $x \in \mathbb{R}^n$ is the set $\Omega(x) = \{gx, \forall g \in G\} \subset \mathbb{R}^n$.

(ii) The *orbit space* of the G-action is the quotient space $\mathbb{R}^n/G$, in which each orbit is reduced to a single point.

(iii) The *isotropy subgroup* of $x \in \mathbb{R}^n$ is the subgroup $G_x = \{g \in G \mid gx = x\}$ of G.

(iv) The *orbit type* $[G_x]$ of an orbit $\Omega(x)$ is the conjugacy class of the isotropy subgroups of the points $x \in \Omega(x)$: $[G_x] = \{gG_xg^{-1}, \forall g \in G\}$. One has in fact $G_{gx} = gG_xg^{-1}$, $\forall g \in G$.

(v) The *stratum* of type $[G_x]$ is the set $\Sigma_{[G_x]} = \{\overline{x} \in \mathbb{R}^n \mid G_{\overline{x}} \in [G_x]\}$, containing all orbits of orbit type $[G_x]$. The set of strata form a partition of $\mathbb{R}^n$, called *stratification*. The number of different orbit types and strata is finite.

(vi) An orbit type is *smaller* (or *greater*) of another orbit type, $[H] < [K]$ (or $[K] > [H]$), if $H' \subset K'$ for some $H' \in [H]$ and $K' \in [K]$. Note that this is a partial ordering in the set of orbit types, given in fact two arbitrary orbit types, it may happen that neither one of the two orbit types is smaller or greater than the other one. There is a unique smallest orbit type $[G_p]$, the corresponding stratum $\Sigma_{[G_p]}$ is called the *principal stratum* and is dense in $\mathbb{R}^n$, all other strata are called *singular*. There is a unique greatest orbit type $[G]$, the corresponding stratum $\Sigma_{[G]}$, for all effective actions, is formed by just one point, the origin of $\mathbb{R}^n$: $\Sigma_{[G]} \equiv 0 \in \mathbb{R}^n$.

(vii) The basic polynomials *separate the orbits.* This means that, given two different orbits, at least one of the q basic polynomials takes a different value on the two orbits. As a consequence, the *orbit map* $\bar{p} : \mathbb{R}^n \to \mathbb{R}^q \mid x \to \bar{p}(x) = (p_1(x), \dots, p_q(x))$ maps $\mathbb{R}^n$ into a subset $\mathcal{S} \subset \mathbb{R}^q$ in such a way that there is a one to one correspondence between orbits in $\mathbb{R}^n$ and points in $\mathcal{S}$. The orbit map induces a diffeomorphism between the orbit space $\mathbb{R}^n/G$ and the set $\mathcal{S}$, and for this reason $\mathcal{S}$ can be identified with the orbit space. $\mathcal{S}$ is connected, even if G is not connected or is finite. If Σ is a stratum of $\mathbb{R}^n$ then $\bar{p}(\Sigma)$ is said to be a *stratum* of $\mathcal{S}$. If $[K] > [H]$, then $\bar{p}(\Sigma_{[K]})$ lies in the boundary of $\bar{p}(\Sigma_{[H]})$. The interior of $\mathcal{S}$ coincides with the principal stratum $\bar{p}(\Sigma_{[G_p]})$ and the boundary of $\mathcal{S}$ contains all singular strata. The origin of $\mathbb{R}^q$ is the image through the orbit map of the origin of $\mathbb{R}^n$, in fact $\bar{p}(\Sigma_{[G]}) = \bar{p}(0) = 0 \in \mathbb{R}^q$, and is the only stratum of $\mathcal{S}$ that lies in the boundary of all other strata of $\mathcal{S}$. The orbit map determines a natural grading of the coordinates $p_1, \dots, p_q$ of $\mathbb{R}^q$, such that $\deg(p_a) = \deg(p_a(x))$, $\forall a = 1, \dots, q$. Both the orbit map and $\mathcal{S}$ are defined through a MIB and depend on the MIB chosen.

(viii) Every polynomial (or C^∞) G-invariant function $f(x)$ is constant along the orbits of the G-action and may be expressed in a unique way as a polynomial (or a C^∞) function of the MIB: $f(x) = \widehat{f}(\bar{p}(x)) = \widehat{f}(p_1(x), \dots, p_q(x))$, $\forall x \in \mathbb{R}^n$ in the domain of f. The domain of $\widehat{f}(p)$ in $\mathbb{R}^q$ may be extended also outside $\mathcal{S}$, possibly in many ways, but to study $f(x)$ one is only interested in the restriction of $\widehat{f}(p)$ to $\mathcal{S}$. If $f(x)$ is a G-invariant homogeneous polynomial function, then $\widehat{f}(p)$ is a homogeneous polynomial function, and $\deg(\widehat{f}(p)) = \deg(f(x))$, taking into account the degrees of the graded variables $p_1, \dots, p_q$. The correspondence $\pi : \mathbb{R}[\mathbb{R}^n]^G \to \mathbb{R}[\mathbb{R}^q] : f \to \widehat{f}$ is then one to one and degree preserving.

(ix) For the linearity of the G-action, the points x and λx, $\forall\, \lambda \in \mathbb{R}$ and $x \in \mathbb{R}^n$, belong to the same stratum. Then each ball centered in the origin of $\mathbb{R}^q$ intersects all the strata of $\mathcal{S}$ and each hyperplane of $\mathbb{R}^q$ of equation $p_1 =$constant > 0 intersects all strata of $\mathcal{S}$ different than the origin. The intersection of $\mathcal{S}$ or of each connected component of its strata with each one of these spheres or hyperplanes is compact and connected. Roughly, $\mathcal{S}$ appears like a pyramid in $\mathbb{R}^q$ with vertex at the origin, curved faces, and infinite extension just towards the positive direction of the p_1 axis.

(x) For any stratum $\Sigma \subset \mathbb{R}^n$, and a general point $x \in \Sigma$, the dimension of the stratum $\bar{p}(\Sigma) \subset \mathcal{S}$ coincides with the rank of the jacobian matrix $j(x) = ||\partial p(x)/\partial x||$ and, consequently, with the rank of the matrix $P(x) = j(x)^T j(x)$, evaluated in x. The explicit expressions of the matrix elements of $j(x)$ and $P(x)$ are the following:

$$j_{ai}(x) = \frac{\partial p_a(x)}{\partial x_i}, \quad P_{ab}(x) = \sum_{i=1}^{n} \frac{\partial p_a(x)}{\partial x_i} \frac{\partial p_b(x)}{\partial x_i}, \quad a, b = 1, \dots, q, \quad i = 1, \dots, n.$$

$P(x)$ is a real symmetric positive semi-definite matrix whose matrix elements are homogeneous G-invariant polynomial functions of degrees $\deg(P_{ab}(x)) = d_a + d_b - 2$.

(xi) The matrix elements of $P(x)$ may be expressed as polynomials of the MIB: $P_{ab}(x) = \widehat{P}_{ab}(\bar{p}(x))$, $\forall\, a, b = 1, \dots, q$, $x \in \mathbb{R}^n$. One so obtains a real symmetric $q \times q$ matrix $\widehat{P}(p)$, that is positive semi-definite in $\mathcal{S} \subset \mathbb{R}^q$. The matrix $\widehat{P}(p)$ is called the *$\widehat{P}$-matrix* (associated to the given MIB) and it is a matrix function of the points $p \in \mathbb{R}^q$. The $\widehat{P}$-matrix clearly depends on the choice of the MIB. Other names that have been used in the literature for the $\widehat{P}$-matrix are *displacement matrix* and *discriminant matrix*.

(xii) If the basic polynomials are algebraically independent (this will be the only case considered in this article because it is the case of the adjoint representations of the compact simple Lie groups), the set $\mathcal{S}$ coincides with the subset of $\mathbb{R}^q$ where the $\widehat{P}$-matrix is positive semi-definite:

$$\mathcal{S} = \{p \in \mathbb{R}^q \mid \widehat{P}(p) \geq 0\}.$$

Given $x \in \Sigma \subset \mathbb{R}^n$ and $p = \bar{p}(x)$ its image in $\mathcal{S}$, one has $\dim(\bar{p}(\Sigma)) = \mathrm{rank}(\widehat{P}(p))$, $\forall\, p \in \bar{p}(\Sigma)$. The union $\mathcal{S}_k$ of all k-dimensional strata of $\mathcal{S}$ ($k \leq q$) is then the following set:

$$\mathcal{S}_k = \{p \in \mathbb{R}^q \mid \widehat{P}(p) \geq 0,\ \mathrm{rank}(\widehat{P}(p)) = k\}.$$

Clearly, $\mathcal{S}_q$ coincides with the principal stratum and $\mathcal{S}_0$ with the origin of $\mathbb{R}^q$, and there are strata of $\mathcal{S}$ of all dimensions k such that $0 \leq k \leq q$. To find practically the equations and inequalities defining $\mathcal{S}_k$, one may require that: 1) all the principal minors of $\widehat{P}(p)$ of order $k+1$ are 0; 2) at least one of those of order k is > 0; 3) all principal minors of order $\leq k$ are ≥ 0 (remind that the non negativity of just the leading minors of a real symmetric matrix is not a sufficient condition for its positive semi-definiteness). To determine $\mathcal{S}_k$ one has then to solve a system of algebraic equations and inequalities that may be quite big.

(xiii) The set $\mathcal{S}$ is a q-dimensional *semi-algebraic set* because it is defined just using polynomial equations and inequalities. As all semi-algebraic sets, $\mathcal{S}$ is stratified in *primary strata*, that is, it has a connected q-dimensional interior, bordered by connected $(q-1)$-dimensional faces, that are bordered by connected $(q-2)$-dimensional faces, and so on, down to its unique 0-dimensional vertex at the origin of $\mathbb{R}^q$. $\mathcal{S}_k$ is then the union of all the k-dimensional primary strata of $\mathcal{S}$. There is a one to one correspondence among the connected components of the strata determined by the orbit types of the group action and the primary strata of $\mathcal{S}$.

(xiv) A *MIB transformation* is a change of the MIB: $\{p_a(x)\} \to \{p'_a(x)\}$, such that $\forall a = 1,\ldots,q$, $p'_a(x) = \widehat{p}'_a(\bar{p}(x))$, with the $\widehat{p}'_a(p)$ homogeneous polynomial functions of the variables $p \in \mathbb{R}^q$ of degrees $d_a = \deg(p_a)$, and the jacobian matrix $J(p) = ||\partial\widehat{p}'(p)/\partial p\,||$ not singular. A MIB transformation always implies a change of the system of coordinates in $\mathbb{R}^q$: $p \to p'$, such that $\forall a = 1,\ldots,q$, $p'_a = \widehat{p}'_a(p)$. A MIB transformation usually implies a change of the $\widehat{P}$-matrix and a consequent change of the set $\mathcal{S} \subset \mathbb{R}^q$. A change of the MIB consequent just to a change of the orthonormal system of coordinates in $\mathbb{R}^n$ is not necessarily a MIB transformation and never implies a change of the $\widehat{P}$-matrix and of the set $\mathcal{S}$.
Let us write $\widehat{P}(p')$ and $\widehat{P}(p)$ for the $\widehat{P}$-matrices determined by the bases $\{p'_a(x)\}$ and $\{p_a(x)\}$ related by a MIB transformation. The relation between the $\widehat{P}$-matrices $\widehat{P}(p')$ and $\widehat{P}(p)$ is expressed by the following *$\widehat{P}$-matrix transformation formula*:

$$\widehat{P}(p') = J^T(p)\ \widehat{P}(p)\ J(p)\Big|_{p\to\widehat{p}(p')}.$$

This formula is very useful to determine the matrix $\widehat{P}(p')$ if one knows the matrix $\widehat{P}(p)$, corresponding to some other basis $\{p_a(x)\}$. In fact, using the $\widehat{P}$-matrix transformation formula, there is no need to calculate the matrix $P(x)$ corresponding to the basis $\{p'_a(x)\}$, a calculation that is often very long and tedious. The equations and inequalities defining the strata of $\mathcal{S}$ in the new variables p' may be obtained either through the coordinate transformation $p \to p'$ applied to the equations and inequalities defining the strata of $\mathcal{S}$ in the old variables p, or by solving the system of algebraic equations and inequalities expressing the rank and positive semi-definiteness conditions on the new $\widehat{P}$-matrix $\widehat{P}(p')$.

Details and proofs of most of the statements here recalled may be found in [2, 14, 1] and in the references therein.

2. The Chevalley isomorphism of the algebras of polynomial invariants

Let G be a compact connected simple Lie group and $\mathfrak{g}$ its Lie algebra. Let l be the rank of G and $\mathfrak{g}$. The Weyl group W of G is an irreducible finite reflection group of rank l. We are here interested only in the defining representation of W, so $W \subset O(l)$. It is known that W has a MIB with l algebraically independent elements of known degrees $2 = d_1 \leq \ldots \leq d_l$ [6, 5].

There is an isomorphism, found by Chevalley [4], between the algebra $\mathbb{R}[\mathfrak{g}]^{\mathrm{Ad}(G)}$ of the real polynomial functions on $\mathfrak{g}$, invariant for the adjoint representation of G in $\mathfrak{g}$, and the algebra $\mathbb{R}[\mathbb{R}^l]^W$ of the real polynomial functions on $\mathbb{R}^l$, invariant for the action of W in $\mathbb{R}^l$. Independently, Harish-Chandra, in [8], Part III, proved the existence of an isomorphism between $\mathbb{R}[\mathbb{R}^l]^W$ and the center $Z(\mathfrak{g})$ of the universal enveloping algebra $U(\mathfrak{g})$ of G (the elements of $Z(\mathfrak{g})$ are often called *Casimir elements*). Because of the Chevalley and Harish-Chandra isomorphisms, both $\mathbb{R}[\mathfrak{g}]^{\mathrm{Ad}(G)}$ and $Z(\mathfrak{g})$ are generated by l algebraically independent elements of degrees $d_1, \ldots, d_l$. In this Section I will review how Chevalley found the isomorphism between $\mathbb{R}[\mathfrak{g}]^{\mathrm{Ad}(G)}$ and $\mathbb{R}[\mathbb{R}^l]^W$.

The Lie algebra $\mathfrak{g}$ is a real vector space of dimension n. In $\mathfrak{g}$ one can consider a basis $\{T_i\}$, $i = 1, \ldots, n,\ T_i \in \mathfrak{g}$, such that any element $X \in \mathfrak{g}$ can be written in the following way:

$$X = \sum_{i=1}^{n} x_i T_i, \qquad x_i \in \mathbb{R}.$$

The basic elements $\{T_i,\}$, $i = 1, \ldots, n$, are often called the *generators* of $\mathfrak{g}$. A one to one correspondence between $\mathfrak{g}$ and $\mathbb{R}^n$ is obtained by identifying $X \in \mathfrak{g}$ with the vector $x = (x_1, \ldots, x_n) \in \mathbb{R}^n$.

The group elements $g \in G$ are obtained by the exponential law $g = \exp(X)$ from the elements $X \in \mathfrak{g}$. In what follows we think of $X \in \mathfrak{g}$ and $g \in G$ to belong to a given representation (of $\mathfrak{g}$ and of G). More precisely, we suppose that the elements of $\mathfrak{g}$ are represented by anti-hermitian matrices (so $X^\dagger = -X$) and that the group elements are represented by unitary matrices (so $g^\dagger = g^{-1}$). The order m of these matrices is the dimension of the representation.

In a representation of $\mathfrak{g}$ realized by anti-hermitian matrices the Killing form $K(T_i, T_j) = -\mathrm{Tr}(T_i^\dagger T_j)$ is a real symmetric $n \times n$ matrix that can be diagonalized and reduced to the identity matrix by choosing a proper basis $\{T_i\}$. In that case one has $K(T_i, T_j) = \delta_{ij}$ and the Killing form defines a scalar product in $\mathfrak{g}$ for which the basis $\{T_i\}$ is orthonormal.

The *adjoint representation* $\mathrm{Ad}(G)$ of the group G is given by the following linear action of G in the given representation space of $\mathfrak{g}$:

$$\mathrm{Ad}(g)\, X = gXg^{-1} = gXg^\dagger \qquad \forall g \in G,\ X \in \mathfrak{g},$$

where the g in the formula, except that one in $\mathrm{Ad}(g)$, means the $m \times m$ unitary matrix that represents the abstract element $g \in G$ in the given representation of G. Sometimes the group $\mathrm{Ad}(G)$, whose elements are $\mathrm{Ad}(g)$, is called the *adjoint group* of G.

Using the Killing form as the scalar product in $\mathfrak{g}$, this action (of G on $\mathfrak{g}$) doesn't change the norms of the vectors $X \in \mathfrak{g}$, in fact, by the cyclic property of the trace and the unitarity of $g \in G$, one has:

$$||\mathrm{Ad}(g)X||^2 = ||g\, X\, g^\dagger||^2 = -\mathrm{Tr}\left((g\, X\, g^\dagger)^\dagger\, g\, X\, g^\dagger\right) = -\mathrm{Tr}(X^\dagger\, X) = ||X||^2.$$

Hence, by a standard result of linear algebra, the linear action of the group $\mathrm{Ad}(G)$ in $\mathfrak{g}$ can be represented by a group Γ of real orthogonal $n \times n$ matrices acting in $\mathbb{R}^n$ with the matrix multiplication. In fact, if $Y = \mathrm{Ad}(g)X$, and $x, y \in \mathbb{R}^n$ represent $X, Y \in \mathfrak{g}$ with respect to the orthonormal basis $\{T_i\}$, and $y = O(g)x$, where $O(g) \in \Gamma$ is the real orthogonal $n \times n$ matrix that represents the linear transformation induced by $\mathrm{Ad}(g)$ in $\mathbb{R}^n$, one has:

$$||Y||^2 = ||y||^2 = (y, y) = (O(g)x, O(g)x) = (x, O(g)^T O(g)x) = (x, x) = ||x||^2 = ||X||^2.$$

Note that in the last formula there are two different, but equivalent, ways to evaluate the norms: $||X||^2 = K(X,X) = -\mathrm{Tr}(X^\dagger X)$, if $X \in \mathfrak{g}$, and $||x||^2 = (x,x) = \sum_{i=1}^n x_i^2$, if $x \in \mathbb{R}^n$.

In the following we will always consider the action of Γ in $\mathbb{R}^n$ instead of the action of $\mathrm{Ad}(G)$ in $\mathfrak{g}$. For what has just been said, the two descriptions are equivalent and the use of Γ allows to employ all the results reviewed in Section 1.

It is known that, if $l = \mathrm{rank}(\mathfrak{g})$, there exist an l-dimensional subalgebra $\mathfrak{h}$ of $\mathfrak{g}$, called a Cartan subalgebra of $\mathfrak{g}$, such that all the elements of $\mathfrak{h}$ commute among themselves. It is possible, and that will be here supposed, to choose the basic orthonormal elements $\{T_i\}$ in such a way that the first l of them generate $\mathfrak{h}$, so that $\mathfrak{g} = \mathfrak{h} \oplus \mathfrak{e}$, where $\mathfrak{h}$ and $\mathfrak{e}$ are subspaces of $\mathfrak{g}$ generated by $\{T_1, \ldots, T_l\}$ and by $\{T_{l+1}, \ldots, T_n\}$, respectively. $\mathfrak{h}$ may then be identified with the l-dimensional subspace of $\mathbb{R}^n$ of the first l coordinates: for $x = (x_1, \ldots, x_l, 0, \ldots, 0) \in \mathbb{R}^n$, we set $\xi_i = x_i$, $\forall i = 1, \ldots, l$, then $\xi = (\xi_1, \ldots, \xi_l) \in \mathbb{R}^l$ is in a one to one correspondence with the element $X = \sum_{i=1}^l \xi_i T_i \in \mathfrak{h}$. For this reason we will sometimes write $\mathfrak{h}$ and $\mathfrak{g}$ also to mean the vector spaces $\mathbb{R}^l$ and $\mathbb{R}^n$ of the coefficients ξ_i and x_i, respectively. The l one-parameter subgroups of G with elements of the form $\exp(\xi_i T_i)$, $i = 1, \ldots, l$, (in which no sum over i has to be understood), generate an l-dimensional Abelian subgroup H of G, the Cartan subgroup of G that has Lie algebra $\mathfrak{h}$.

For the compactness of G, every $g \in G$ is conjugated to an element $h \in H$. This means that all the orbits of the action of the group Γ in $\mathfrak{g}$ intersect $\mathfrak{h}$. As every $\xi \in \mathfrak{h}$ completely specifies the orbit $\Omega(\xi)$ of the action of Γ in $\mathfrak{g}$ that contains ξ, all the orbits of the action of Γ in $\mathfrak{g}$ can be specified by points $\xi \in \mathfrak{h}$.

Let $N(H) = \{g \in G \mid gHg^{-1} \subseteq H\}$ be the normalizer of H in G. If $X \in \mathfrak{h}$ and $\epsilon \in \mathbb{R}$, $h = \exp(\epsilon X) = \mathbb{1}_m + \epsilon X + O(\epsilon^2) \in H$, and $ghg^{-1} = \mathbb{1}_m + \epsilon(gXg^{-1}) + O(\epsilon^2)$. If $g \in N(H)$, $ghg^{-1} \in H$ and then $gXg^{-1} \in \mathfrak{h}$. This implies that the restriction of Γ to the elements $g \in N(H)$, is a group Γ_N of inner automorphisms of $\mathfrak{h}$, under which $\mathfrak{h}$ is invariant. The matrices of the linear group Γ_N are then block diagonal matrices, formed by the direct sum of an $l \times l$ orthogonal matrix, in correspondence to $\mathfrak{h}$, and an $(n-l) \times (n-l)$ orthogonal matrix, in correspondence to $\mathfrak{e}$. An invariant polynomial $p(x) \in \mathbb{R}[\mathfrak{g}]^\Gamma$ must also be an element of $\mathbb{R}[\mathfrak{g}]^{\Gamma_N}$, because Γ_N is a subgroup of Γ. Moreover, also the restriction $\tilde{p}(\xi)$ of $p(x)$ to $\mathfrak{h}$ is in $\mathbb{R}[\mathfrak{g}]^{\Gamma_N}$, because $\mathfrak{h}$ is Γ_N-invariant. Vice versa, a polynomial $\tilde{p}(\xi)$, $\xi \in \mathfrak{h}$, invariant for the action of Γ_N in $\mathfrak{g}$, defines a unique polynomial $p(x) \in \mathbb{R}[\mathfrak{g}]^\Gamma$, such that $\tilde{p}(\xi)$ is its restriction to $\mathfrak{h}$. In fact, $\forall x \in \mathfrak{g}$, $\exists g \in G \mid O(g)x \in \mathfrak{h}$, so that one may define $p(x) = \tilde{p}(O(g)x)$. Of course, it may exist a $g_1 \in G$, $g_1 \neq g$, such that $O(g_1)x \in \mathfrak{h}$, but then $\exists\, g' \in N(H) \mid g'g = g_1$, and both $O(g)x$ and $O(g_1)x$ must lie in the same orbit of Γ_N, so that $\tilde{p}(O(g)x) = \tilde{p}(O(g_1)x)$ and this means that $p(x)$ is well defined. The polynomial $p(x)$ so defined is Γ-invariant, in fact, the value of p in all the points of the orbit $\Omega(x)$ of the Γ-action is the same as in the points ξ of the intersection $\Omega(x) \cap \mathfrak{h}$. All this proves the existence of an isomorphism between the algebra $\mathbb{R}[\mathfrak{g}]^\Gamma$ and the restriction $\mathbb{R}[\mathfrak{g}]^{\Gamma_N}\big|_{\mathfrak{h}}$ of the algebra $\mathbb{R}[\mathfrak{g}]^{\Gamma_N}$ to $\mathfrak{h}$.

The image of the Abelian subgroup H of G in $\mathrm{Ad}(G)$ is represented by an Abelian subgroup Γ_H of Γ that acts trivially on $\mathfrak{h}$ (that is, the elements of Γ_H have the identity matrix on the first l-dimensional block). Then the quotient group Γ_N/Γ_H acts naturally on $\mathfrak{h}$. The *Weyl group* W of G is defined as the quotient group $N(H)/H$, a definition that does not depend on the Cartan subgroup $H \subset G$. W is in any case a finite group, even if G and H are not. The action of the group Γ_N/Γ_H in $\mathfrak{h}$ is then isomorphic to the action of the group W in $\mathfrak{h}$ and this also implies $\mathbb{R}[\mathfrak{g}]^{\Gamma_N}\big|_{\mathfrak{h}} \equiv \mathbb{R}[\mathfrak{h}]^W$.

All these facts prove that there is an isomorphism between $\mathbb{R}[\mathfrak{g}]^\Gamma$ and $\mathbb{R}[\mathfrak{h}]^W$ and that the restriction to $\mathfrak{h}$ of a MIB for $\mathbb{R}[\mathfrak{g}]^\Gamma$ gives a MIB for $\mathbb{R}[\mathfrak{h}]^W$ and vice versa, a MIB for $\mathbb{R}[\mathfrak{h}]^W$ uniquely specifies a MIB for $\mathbb{R}[\mathfrak{g}]^\Gamma$.

The relation between the action of Γ in $\mathbb{R}^n$ ($\simeq \mathfrak{g}$) and of the action of W in $\mathbb{R}^l$ ($\simeq \mathfrak{h}$), may be further described using a Cartan Weyl basis for the complexification $\mathfrak{g}_c = \mathfrak{g} + i\mathfrak{g}$ of $\mathfrak{g}$.

With a basis for $\mathfrak{h}$ one may determine the roots and a Cartan Weyl basis for $\mathfrak{g}_c$, in which it is usual to write the basic elements as $\{H_1, \dots, H_l, E_{\alpha_i}, E_{-\alpha_i}, \dots\}$, with $H_i \in \mathfrak{h}$, $\forall i = 1, \dots, l$, and $E_{\alpha_i}, E_{-\alpha_i} \in \mathfrak{e}_c$, where α_i is ranging in the set of the positive roots of $\mathfrak{h}$ in $\mathfrak{g}$. This basis may be expressed in terms of the basis $\{T_i\}$ in the following way: $H_i = T_i$, $\forall i = 1, \dots, l$, $E_{\alpha_i} = \frac{1}{\sqrt{2}}(T_{l+2i} - iT_{l+2i-1})$, $E_{-\alpha_i} = E^\dagger_{\alpha_i}$, $\forall i = 1, \dots, (n-l)/2$, if the $\{T_i\}$ are ordered properly.

Every root β is a non vanishing vector of $\mathbb{R}^l$ and defines a reflection S_β in $\mathbb{R}^l$,

$$S_\beta \xi = \xi - \frac{2(\xi, \beta)}{(\beta, \beta)}\beta, \qquad \forall \xi \in \mathbb{R}^l,$$

sending each point of $\xi \in \mathbb{R}^l$ to its symmetric with respect to the hyperplane through the origin of $\mathbb{R}^l$ orthogonal to β. The *Weyl group* W of G (and $\mathfrak{g}$) may be defined as the group generated by all the reflections S_β, with β a root. All the elements $w \in W$ may then be expressed as a product of a finite number of these reflections. W is then a group of real orthogonal $l \times l$ matrices, irreducible if G is simple. This definition of the Weyl group turns out to be equivalent to the former, given as the quotient group $N(H)/H$. All elements $w \in W$ send the set of the roots to itself and the set of the reflecting hyperplanes to itself. The action of w in $\mathbb{R}^l$ corresponds then to a permutation of the roots: $\alpha_i \to w\alpha_i$, and induces a permutation of the generators of $\mathfrak{e}_c$ according to the rule: $E_{\alpha_i} \to E_{w\alpha_i}$, for every root α_i.

3. Basic invariant polynomials and orbit spaces of the adjoint representations

In this Section I will review how one may determine the equations and inequalities defining the orbit spaces of the adjoint representations of the compact simple Lie groups. I will continue to use the notation introduced in the previous Section.

Elements of $\mathbb{R}[\mathbb{R}^n]^\Gamma$, that is Γ-invariant real polynomial functions defined on $\mathbb{R}^n$, are obtained by writing the explicit expressions of the Casimir elements in $Z(\mathfrak{g})$ [10] in terms of the n real coefficients x_i of the expansions of the elements $X \in \mathfrak{g}$ in terms of the generators T_i, $i = 1, \dots, n$. The explicit expressions of these *Casimir polynomials* are the following:

$$I^{(k)}(x) = \mathrm{Tr}(X^k), \qquad X = \sum_{i=1}^{n} x_i T_i, \qquad x_i \in \mathbb{R},\ k \in \mathbb{N}.$$

The $I^{(k)}(x)$ so defined are Γ-invariant polynomials, one in fact has:

$$I^{(k)}(O(g)x) = \mathrm{Tr}((gXg^\dagger)^k) = \mathrm{Tr}(gX^kg^\dagger) = \mathrm{Tr}(X^k) = I^{(k)}(x), \quad \forall g \in G,\ x \in \mathbb{R}^n.$$

For the results in the previous Section, the restriction of a Casimir polynomial $I^{(k)}(x)$ to the l dimensional subspace corresponding to the Cartan subalgebra $\mathfrak{h} \subset \mathfrak{g}$, gives an invariant polynomial $\tilde{I}^{(k)}(\xi)$ for the action of the Weyl group W in $\mathbb{R}^l$:

$$\tilde{I}^{(k)}(\xi) = \mathrm{Tr}(X^k), \qquad X = \sum_{i=1}^{l} \xi_i T_i, \qquad \xi_i \in \mathbb{R},\ k \in \mathbb{N}.$$

Vice versa, $\tilde{I}^{(k)}(\xi)$ uniquely extends and defines the Casimir polynomial $I^{(k)}(x)$.

For a certain set of degrees $d_1, \dots, d_l$, the l Casimir polynomials $I^{(d_a)}(x)$, $a = 1, \dots, l$, are algebraically independent and form a MIB of $\mathbb{R}[\mathfrak{g}]^\Gamma$. Their restrictions to $\mathfrak{h}$, $\tilde{I}^{(d_a)}(\xi)$, form then a MIB of l algebraically independent polynomials for $\mathbb{R}[\mathfrak{h}]^W$ with the same set of degrees.

An important result is that both the MIB $I^{(d_a)}(x)$, $a = 1, \dots, l$, of $\mathbb{R}[\mathbb{R}^n]^\Gamma$, and the MIB $\tilde{I}^{(d_a)}(\xi)$, $a = 1, \dots, l$, of $\mathbb{R}[\mathbb{R}^l]^W$, determine the same $\widehat{P}$-matrix. The proof is in [12], Section

III. For the results in Section 1, this implies that, both the actions of Γ in $\mathbb{R}^n$ and of W in $\mathbb{R}^l$ have the same orbit space, with the same geometric stratification. The strata of $\mathcal{S}$ correspond in the two cases of Γ or of W to different orbit types, that is to the conjugacy classes of the isotropy subgroups of the actions of Γ or of W. Obviously, to a MIB transformation for $\mathbb{R}[\mathbb{R}^n]^\Gamma$ (for $\mathbb{R}[\mathbb{R}^l]^W$) it corresponds a MIB transformation for $\mathbb{R}[\mathbb{R}^l]^W$ (for $\mathbb{R}[\mathbb{R}^n]^\Gamma$), given by the same transforming equations, if these are written in terms of the basic invariants. This MIB transformation implies a change of the $\widehat{P}$-matrix (according to the $\widehat{P}$-matrix-transformation formula) and a consequent change of the equations and inequalities defining the orbit space $\mathcal{S}$ and its strata. It happens then that if the MIB $\{\tilde{p}_a(\xi)\}$ of $\mathbb{R}[\mathbb{R}^l]^W$ is obtained by restricting to $\mathfrak{h}$ a MIB $\{p_a(x)\}$ of $\mathbb{R}[\mathbb{R}^n]^\Gamma$, then the MIB's $\{\tilde{p}_a(\xi)\}$ and $\{p_a(x)\}$ determine the same $\widehat{P}$-matrix and the same orbit space $\mathcal{S} \subset \mathbb{R}^l$ for the actions of W and Γ. In fact, there exists a unique MIB transformation that transforms the MIB $\{\tilde{I}^{(d_a)}(\xi)\}$ to the MIB $\{\tilde{p}_a(\xi)\}$ and the MIB $\{I^{(d_a)}(x)\}$ to the MIB $\{p_a(x)\}$, this last one differs from the first one just for the substitutions $\tilde{I}^{(d_a)} \to I^{(d_a)}$, and $\tilde{p}_a \to p_a$, $\forall a = 1, \ldots, l$.

The arguments described in this Section suggest that to determine the orbit space of the adjoint representation of a simple compact Lie group G, it is possible, and more convenient, to determine the orbit space of the Weyl group W of G. The main reasons of this convenience are the following: 1) W is finite while Γ is not; 2) W acts in $\mathbb{R}^l$, while Γ acts in $\mathbb{R}^n$, and $l < n$; 3) both W and Γ have the same $\widehat{P}$-matrix, the same orbit space $\mathcal{S}$ and the same equations and inequalities defining all strata of $\mathcal{S}$, if these equations and inequalities are expressed in terms of the MIB, and if the MIB of $\mathbb{R}[\mathbb{R}^l]^W$ is obtained by restricting to $\mathfrak{h}$ a MIB of $\mathbb{R}[\mathbb{R}^n]^\Gamma$.

If the rank l is large, given a general MIB $\{p_a(\xi)\}$ of $\mathbb{R}[\mathbb{R}^l]^W$, it may be very hard to determine the $\widehat{P}$-matrix directly from the MIB $\{p_a(\xi)\}$ (that is, through the determination of the matrix $P(\xi)$). It is often convenient to use a known $\widehat{P}$-matrix, corresponding, in general, to another MIB $\{p'_a(\xi)\}$ of $\mathbb{R}[\mathbb{R}^l]^W$, determine the MIB transformation relating the MIB $\{p'_a(\xi)\}$ to the MIB $\{p_a(\xi)\}$ and make use of the $\widehat{P}$-matrix transformation formula. For example, if one knows the $\widehat{P}$-matrix corresponding to the MIB $\{p'_a(\xi)\}$ of $\mathbb{R}[\mathbb{R}^l]^W$, and one wishes to determine the $\widehat{P}$-matrix corresponding to the MIB $\{I^{(d_a)}(x)\}$ of $\mathbb{R}[\mathbb{R}^n]^\Gamma$, it is convenient to determine the MIB transformation relating the MIB's $\{p'_a(\xi)\}$ and $\{\tilde{I}^{(d_a)}(\xi)\}$ and use the $\widehat{P}$-matrix transformation formula. The $\widehat{P}$-matrices of the finite reflection groups E_7 and E_8 have been calculated in [16]. The $\widehat{P}$-matrices of all the other irreducible reflection groups have been calculated in [7, 11, 17, 13].

4. Symmetry breakings in the orbit spaces

We continue here to use the notation of the previous Section and write Γ for the real orthogonal group acting on $\mathbb{R}^n$ that realizes the adjoint representation $\mathrm{Ad}(G)$, and W for the Weyl group of G, acting on $\mathbb{R}^l$ as a reflection group. Clearly, $n = \dim(\mathfrak{g})$ and $l = \mathrm{rank}(\mathfrak{g}) = \mathrm{rank}(W)$.

Let f be a function invariant for the group Γ (or for the group W). f may be expressed in a unique way in terms of a given MIB of Γ (or of W), so f uniquely defines a function $\widehat{f}$ of l variables such that $f(x) = \widehat{f}(p_1(x), \ldots, p_l(x))$, $\forall x$ in the domain of f. The function $\widehat{f}$ represents both a Γ-invariant function and a W-invariant function, the only difference is in considering the l variables in $\widehat{f}$ as the l basic polynomials of Γ or of W. When the MIB of $\mathbb{R}[\mathbb{R}^l]^W$ is the restriction to $\mathfrak{h}$ of a MIB of $\mathbb{R}[\mathbb{R}^n]^\Gamma$, the actions of W and Γ determine the same orbit space $\mathcal{S} \subset \mathbb{R}^l$, with the same stratification. Only this case will be considered in the following. In this case, when one studies the function $\widehat{f}$ in $\mathcal{S}$, there is no difference between the problem in which the symmetry group is Γ and the problem in which the symmetry group is W. In both cases one has in fact to study the function $\widehat{f}$ in the same semi-algebraic connected subset $\mathcal{S} \subset \mathbb{R}^l$ that represents the orbit space, of Γ and of W.

$\widehat{f}$ depends in general on the l variables corresponding to the l basic polynomials (of Γ or of W) and to a certain number of other parameters related to the physical problem one is dealing with. If f is an invariant potential function, and one wants to study spontaneous symmetry breakings or phase transitions, one is interested in the location of the minima of $\widehat{f}$ and in the orbit type of the stratum hosting the minimum point. Let the minimum point of $\widehat{f}$ be in a given point of $\mathcal{S}$, belonging to a given stratum that corresponds to a given orbit type. If for some reasons the parameters in $\widehat{f}$ change with continuity, the minimum point of $\widehat{f}$, in general, changes its position with continuity, and it may happen that it changes stratum, changing in this way the orbit type of the orbit of minimum points (and consequently the residual symmetry), realizing in this way what is called a spontaneous symmetry breaking or a second order structural phase transition.

This changing may take place only between neighboring strata of $\mathcal{S}$, because of the continuity of the changing of the location of the minimum point of a continuous function.

What one has to do to study the spontaneous symmetry breakings in the orbit space $\mathcal{S}$ is summarized in the following points.
1) Determine the equations and inequalities defining the various strata of $\mathcal{S}$. 2) Determine the orbit types (of the Γ- or of the W-action) corresponding to the various strata of $\mathcal{S}$. 3) Express the invariant potential function f in terms of the MIB, obtaining in this way a function $\widehat{f}$ of l variables. 4) Determine the constrained minima of $\widehat{f}$ in the various strata of $\mathcal{S}$, and the range of the parameters in $\widehat{f}$ for which the absolute minimum of $\widehat{f}$ falls into one or another stratum. Point 1) is solved using rank and positive semi-definiteness conditions of the $\widehat{P}$-matrix. Point 2) is solved by determining the isotropy subgroups in convenient points $\xi \in \mathbb{R}^l$ (or $x \in \mathbb{R}^n$), where $\mathbb{R}^l$ (or $\mathbb{R}^n$) is the space on which W (or Γ) acts, chosen in such a way to have images through the orbit map in each one of the primary strata of $\mathcal{S}$. It may be useful to remind that points lying in the boundary of $\mathcal{S}$ are images through the orbit map of orbits lying in the set of the reflecting hyperplanes of W. To each stratum of $\mathcal{S}$ it corresponds both an orbit type of W and an orbit type of Γ, depending on which group action one considers. The two sets of orbit types are clearly related. Point 4) is solved for example with the method of Lagrange multipliers to find constrained extrema on the various primary strata of $\mathcal{S}$.

The study of minimization of invariant functions here described can be done analytically, that is with full precision, because from the $\widehat{P}$-matrices one may determine all the equations and inequalities defining the various strata of the orbit space. The calculations are not straightforward but do not present conceptual difficulties. In practice, one has to work with systems of algebraic equations and inequalities that may be quite complicated to handle.

5. Results for the adjoint representation of E_8

During the last 30 years, the adjoint representation of the compact Lie group E_8 has been many times proposed in models of grand unification theories and of string theories and an exact computational method to employ in the study of spontaneous symmetry breakings in these models may be of a certain interest.

As before, let's write $W \subset O(8)$ for the finite reflection group E_8 and $\Gamma \subset O(248)$ for the linear group realizing the adjoint representation of the Lie group E_8.

The degrees of a set of basic polynomials $p_1, \ldots, p_8$, both for $\mathbb{R}[\mathbb{R}^8]^W$ and for $\mathbb{R}[\mathbb{R}^{248}]^\Gamma$, are known to be 2, 8, 12, 14, 18, 20, 24, 30. The center $Z(\mathfrak{e}_8)$ of the universal enveloping algebra $U(\mathfrak{e}_8)$ of E_8 is also generated by 8 algebraically independent Casimir elements of the same set of degrees. Besides the quadratic and the octic (of degree 8) Casimir element [3], to my knowledge, no other generator of $Z(\mathfrak{e}_8)$ have been explicitly calculated so far. Many authors described how to construct a set of basic polynomials for $\mathbb{R}[\mathbb{R}^8]^W$. A simple description was given for example by Mehta in [9], in terms of the linear functions defining the invariant hyperplanes of W. These basic polynomials were calculated explicitly in [16] and their expressions may be found in the supplementary material of that article (Ref. 30 of [16]). Their expressions are quite large, for example, $p_8(x)$ is a real homogeneous polynomial function of 8 variables of degree 30 that has

10592 terms when expanded. The explicit expression of the corresponding $\widehat{P}$-matrix is also reported in Ref. 30 of [16]. This $\widehat{P}$-matrix, using the $\widehat{P}$-matrix transformation formula, allows to determine the $\widehat{P}$-matrices corresponding to any other MIB of $\mathbb{R}[\mathbb{R}^8]^W$ (in [16], for example, this formula was used to calculate the $\widehat{P}$-matrix corresponding to a flat basis). The geometric structure of the orbit space $\mathcal{S}$, corresponding both to the group W and to the group Γ can then be determined explicitly, as described in Section 1 above, and in particular it is possible to determine explicitly the equations and inequalities of all the strata of $\mathcal{S}$. The calculations to do that are not yet done and are not trivial because the $\widehat{P}$-matrix is an 8×8 matrix with elements that are real homogeneous polynomials in 8 variables with degrees ranging from 2 to 58 (by considering the degrees of the graded variables $p_1, \ldots, p_8$). Anyway, these calculations are possible because one knows completely the explicit expression of the $\widehat{P}$-matrix.

The study of the minima of a Γ- (or W-) invariant potential function and of the related spontaneous symmetry breakings may be performed following the method outlined in Section 4, however, because of the high degrees of the equations that are involved, one encounters quite complicated systems of algebraic equations and inequalities.

Acknowledgments

The author wishes to thank prof. Pietro Corvaja for the numerous enlightening discussions on Lie group theory and for the reading of the manuscript. Many thanks also to the referee that helped to improve both the English and the format of the article.

References

[1] Abud M and Sartori G 1983 The geometry of spontaneous symmetry breaking *Ann. Phys., NY* **150** 307–72
[2] Bredon G 1972 *Introduction to Compact Transformation Groups* (New York: Academic Press)
[3] Cederwall M and Palmkvist J 2007 The octic E_8 invariant *J. Math. Phys.* **48** 073505-1-7
[4] Chevalley C 1952 The Betti numbers of the exceptional simple Lie groups *Proc. of the Int. Congress of Mathematicians, Cambridge, Mass., 1950, vol. 2* (Providence, R. I.: Amer. Math. Soc.) pp 21–4
[5] Chevalley C 1955 Invariants of finite groups generated by reflections *Amer. J. of Math.* **77** 778–82
[6] Coxeter H S M 1951 The product of the generators of a finite group generated by reflections *Duke Math. J.* **18** 765–82
[7] Givental' A B 1980 Displacement of invariants of groups that are generated by reflections and are connected with simple singularities of functions *Funct. Anal. Appl.* **14** 81–9
[8] Harish-Chandra 1951 On some applications of the universal enveloping algebra of a semisimple Lie group *Trans. Amer. Math. Soc.* **70** 28–96
[9] Mehta M L 1988 Basic sets of invariant polynomials for finite reflection groups *Comm. Algebra* **16** 1083–98
[10] Racah G 1950 Sulla caratterizzazione delle rappresentazioni irriducibili dei gruppi semisemplici di Lie *Atti Accad. Naz. Lincei. Rend. Cl. Sci. Fis. Mat. Nat.* **8** 108–12
[11] Saito K, Yano T and Sekiguchi J 1980 On a certain generator system of the ring of invariants of a finite reflection group *Comm. Algebra* **8** 373–408
[12] Sartori G and Talamini V 1998 Orbit spaces of compact coregular simple Lie groups with 2, 3 and 4 basic polynomial invariants: effective tools in the analysis of invariant potentials *J. Math. Phys.* **39** 2367–401
[13] Sartori G and Valente G 1996 Orbit spaces of reflection groups with 2, 3 and 4 basic polynomial invariants, *J. Phys. A: Math. Gen.* **29** 193–223
[14] Schwarz G W 1980 Lifting smooth homotopies of orbit spaces, *I.H.E.S. Publ. Math.* **51** 37–135
[15] Springer T A 1977 *Invariant Theory* (Lecture Notes in Mathematics vol 585) (Berlin: Springer-Verlag)
[16] Talamini V 2010 Flat bases of invariant polynomials and $\widehat{P}$-matrices of E_7 and E_8 *J. Math. Phys.* **51** 023520-1–20
[17] Yano T and and Sekiguchi J 1981 The microlocal structure of weighted homogeneous polynomials associated with Coxeter systems, II *Tokyo J. of Math.* **4** 1–34

New aspects of one-dimensional $\mathcal{N} = 4$-extended supersymmetry

Francesco Toppan

CBPF, Rua Dr. Xavier Sigaud 150, Urca, cep 22290-180, Rio de Janeiro (RJ), Brazil.

E-mail: toppan@cbpf.br

Abstract. In this talk I will present very recent results concerning the $\mathcal{N} = 4$-extended supersymmetry in one-dimension. The new features consist in the construction of the non-minimal linear multiplets realized by 8 bosonic and 8 fermionic components, the construction of non-linear realizations induced by the supersymmetrization of the second Hopf fibration and, finally, the construction of a new superalgebra with $\mathcal{N} = 4$ fermionic generators (two of them nilpotent) obtained by the dimensional reduction of the four-dimensional twisted super-Yang-Mills theory.

1. Introduction

In this talk I will summarize the main results, concerning the $\mathcal{N} = 4$ extended one-dimensional supersymmetry, obtained in the last year in three separate papers.[1, 2, 3]

These new results are part of a systematic investigation of the properties of the Extended one-dimensional supersymmetry, started approximately ten years ago, focused on investigating the properties of the supersymmetry representations and of their associated invariant actions. One of the aim of the investigation consists in providing constraints on the construction of higher-dimensional supersymmetric theories based on the information encoded in the rich one-dimensional structural data, which allows the use of powerful mathematical tools. This line of research has been worked out and discussed in several works, see. [4]–[17].

The three main results I will discuss in this talk concern:

i) the construction of the non-minimal linear representations of the $\mathcal{N} = 4$-extended supersymmetry realized on a multiplet of 8 bosonic and 8 fermionic fields. Since this multiplet can (minimally) carry at most $\mathcal{N} = 8$ supersymmetries, these results finds application in the context of partial supersymmetry breaking (we recall that a $D = 4$ $N = 2$ supersymmetric theory admits $N = 8$ supercharges, while a $D = 4$ $N = 1$ supersymmetric theory admits $\mathcal{N} = 4$ supercharges);

ii) the geometrical construction of non-linear realization of the $N = 4$ supersymmetry obtained by the supersymmetrization of the 2nd Hopf fibration;

iii) a non-trivial realization of a new superalgebra with 4 supercharges, obtained by dimensional reduction of the twisted super-Yang-Mills theory. The superalgebra under consideration, contrary to the dimensional reduction of the ordinary supersymmetry, possesses generators which are nilpotent (therefore behaving like Grassmann parameters or BRST charges).

For completeness in an appendix the main definitions and properties of the representations of the one-dimensional supersymmetry are reported.

2. $\mathcal{N}=4$ Non-minimal linear supermultiplets

The $1D$ $\mathcal{N}$-Extended Superalgebra, with $\mathcal{N}$ odd generators Q_I ($I=1,2,\ldots,\mathcal{N}$) and a single even generator H satisfying the (anti)-commutation relations

$$\begin{aligned} \{Q_I,Q_J\} &= \delta_{IJ}H, \\ [H,Q_I] &= 0, \end{aligned} \tag{1}$$

is the superalgebra underlying the Supersymmetric Quantum Mechanics [18]. In recent years the structure of its linear representations has been unveiled by a series of works (upon which the present investigation is based) [4]–[15].

The linear representations under considerations (supermultiplets) contain a finite, equal number of bosonic and fermionic fields depending on a single coordinate (the time). The operators Q_I and H act as differential operators. The linear representations are characterized by a series of properties which, for sake of consistency, are reviewed in the appendix.

The *minimal* linear representations (also called *irreducible supermultiplets*) are given by the minimal number n_{min} of bosonic (fermionic) fields for a given value of $\mathcal{N}$. The value n_{min} is given [4] by the formula

$$\begin{aligned} \mathcal{N} &= 8l+m, \\ n_{min} &= 2^{4l}G(m), \end{aligned} \tag{2}$$

where $l=0,1,2,\ldots$ and $m=1,2,3,4,5,6,7,8$.

$G(m)$ appearing in (2) is the Radon-Hurwitz function

m	1	2	3	4	5	6	7	8
$G(m)$	1	2	4	4	8	8	8	8

(3)

Non-minimal linear representations have been discussed in [6, 14, 15, 1]. The maximal *finite* number n_{max} of bosonic (fermionic) fields entering a non-minimal representation is given by [14, 15]

$$n_{max} = 2^{\mathcal{N}-1}. \tag{4}$$

An important subclass of non-minimal representations is given by the reducible but indecomposable supermultiplets (see [6]). For this subclass the associated graph is connected (there is a path connecting any two given vertices).

For $\mathcal{N}=4$ we have that $n_{min}=4$ and $n_{max}=8$. As a consequence, there are only two subclasses of non-minimal $\mathcal{N}=4$ representations. Besides the irreducible but indecomposable subclass, we have the subclass of fully reducible representations given by the direct sum of two minimal $\mathcal{N}=4$ representations (the associated graph is disconnected and given by two separate minimal $\mathcal{N}=4$ graphs. We present here the complete list of inequivalent $\mathcal{N}=4$ non-minimal supermultiplets associated to connected graphs (therefore providing reducible but indecomposable representations).

Up to equivalence, there exists a unique length-2 $\mathcal{N}=4$ root supermultiplet, based on a connected graph, of field content $(8,8)$.

The inequivalent non-minimal supermultiplets of length-3 are given by the table below. One should note that, for field content $(k,8,8-k)$ with $k=2,3,4,5,6$, inequivalent supermultiplets are discriminated by their respective connectivity symbol (see the appendix) and are named according to the given label. The table further reports the dually related supermultiplet (see the appendix). We have

field content:	*label:*	*connectivity symbol:*	*dual supermultiplet:*
$(1,8,7)$		4_4+4_3	$(7,8,1)$
$(2,8,6)$	A	$2_4+4_3+2_2$	$(6,8,2)_A$
	B	8_3	$(6,8,2)_B$
$(3,8,5)$	A	$1_4+3_3+3_2+1_1$	$(5,8,3)_A$
	B	4_3+4_2	$(5,8,3)_B$
$(4,8,4)$	A	$1_4+6_2+1_0$	self-dual
	B	4_3+4_1	self-dual
	C	$2_3+4_2+2_1$	self-dual
	D	8_2	self-dual
$(5,8,3)$	A	$1_3+3_2+3_1+1_0$	$(3,8,5)_A$
	B	4_2+4_1	$(3,8,5)_B$
$(6,8,2)$	A	$2_2+4_1+2_0$	$(2,8,6)_A$
	B	8_1	$(2,8,6)_B$
$(7,8,1)$		4_1+4_0	$(1,8,7)$

(5)

3. $\mathcal{N}=4$ Non-linear realizations from the $2nd$ Hopf fibration

The geometrical construction of the $\mathcal{N}=4$ non-linear realization acting on the multiplets of field-content $(3,4,1)$ and $(2,4,2)$ (see the appendix) is obtained by the supersymmetrization of the second Hopf fibration.[2]

We present at first the four (bosonic) Hopf maps (for $k=1,2,4,8$), associated to the four division algebras, that can be illustrated by the following commutative diagram

$$\begin{array}{ccc} \mathbf{R}^{2k} & \stackrel{p}{\longrightarrow} & \mathbf{R}^{k+1} \\ {\scriptstyle\rho}\downarrow & & \downarrow{\scriptstyle\rho'} \\ \mathbf{S}^{2k-1} & \stackrel{h}{\longrightarrow} & \mathbf{S}^{k} \end{array} \tag{6}$$

which connects four spaces (two Euclidean spaces and two spheres) which, for later convenience, can be identified as I, II, III, IV according to

$$\begin{array}{ccc} I & \stackrel{p}{\longrightarrow} & II \\ {\scriptstyle\rho}\downarrow & & \downarrow{\scriptstyle\rho'} \\ III & \stackrel{h}{\longrightarrow} & IV \end{array} \tag{7}$$

The four arrows correspond to the following maps:
- the bilinear map $p: I \to II$, sending coordinates $\vec{u} \in \mathbf{R}^{2k}$ into coordinates $\vec{x} \in \mathbf{R}^{k+1}$ according to

$$p: \quad \vec{u} \mapsto x_i = u^T \gamma_i u, \tag{8}$$

(γ_i are the Euclidean gamma matrices of $\mathbf{R}^{k+1}$);
- the restrictions ρ, ρ' on spheres, where $\rho: I \to III$ and $\rho': II \to IV$;
- the hopf map $h: II \to IV$, admitting $\mathbf{S}^{k-1}$ as a fiber (for $k=8$, $\mathbf{S}^7$ is a parallelizable manifold but not, properly speaking, a group-manifold due to the nonassociativity of the octonions; for $k=1$, $\mathbf{S}^0 \equiv \mathbf{Z}_2$).

For $k = 1, 2, 4, 8$ the map (8) preserves the norm, allowing to induce the map h from p:

$$u^T u = R \mapsto x^T x = r, \quad with \quad r = R^2. \tag{9}$$

By setting $k = 2^l$, the four Hopf maps h will be referred to (for $l = 0, 1, 2, 3$ respectively) as the $0^{th}, 1^{st}, 2^{nd}$ and 3rd Hopf map.

In the following we will give a detailed description of the supersymmetric extension of the first Hopf map ($k = 2$), corresponding to the diagram

$$\begin{array}{ccc} \mathbf{R}^4 & \xrightarrow{p} & \mathbf{R}^3 \\ {}^{\rho}\downarrow & & \downarrow^{\rho'} \\ \mathbf{S}^3 & \xrightarrow{h} & \mathbf{S}^2 \end{array} \tag{10}$$

In the supersymmetric extension $\mathbf{R}^4$ is replaced by the $\mathcal{N} = 4$ root supermultiplet $(4, 4)$ whose four bosonic (target) coordinates correspond to the coordinates of $\mathbf{R}^4$. The off-shell supermultiplets extending II, III and IV are induced by applying, respectively, the map p and the restriction ρ to $(4, 4)$, as well as the restriction ρ' on the induced supermultiplet generalizing III.

For our purposes it will be convenient to define the target coordinates of the supermultiplets extending III (IV) in terms of the stereographic projection of the I (II) target coordinates.

For $k = 2$ we can express the three Euclidean gamma matrices γ_i as

$$\gamma_1 = \tau_1 \otimes \mathbf{1}_2, \quad \gamma_2 = \tau_A \otimes \tau_A, \quad \gamma_3 = \tau_2 \otimes \mathbf{1}_2, \tag{11}$$

where

$$\tau_1 = \begin{pmatrix} 0 & 1 \\ 1 & 0 \end{pmatrix}, \quad \tau_2 = \begin{pmatrix} 1 & 0 \\ 0 & -1 \end{pmatrix}, \quad \tau_A = \begin{pmatrix} 0 & 1 \\ -1 & 0 \end{pmatrix}. \tag{12}$$

With this convention the bilinear map p is explicitly presented as

$$\begin{array}{rcl} x_1 & = & 2(u_1u_3 + u_2u_4), \\ x_2 & = & 2(u_1u_4 - u_2u_3), \\ x_3 & = & {u_1}^2 + {u_2}^2 - {u_3}^2 - {u_4}^2. \end{array} \tag{13}$$

It is invariant under the σ transformation ($\sigma^2 = -\mathbf{1}$), given by

$$\sigma: \quad u_1 \mapsto u_2, \quad u_2 \mapsto -u_1, \quad u_3 \mapsto u_4, \quad u_4 \mapsto -u_3. \tag{14}$$

Induced by the $\mathcal{N} = 4$ $(4, 4)$ root supermultiplet, three more (inequivalent) $\mathcal{N} = 4$ off-shell supermultiplets are obtained. They correspond to the supersymmetric extensions of II, III and IV. The supertransformations extending II are all linear. This supermultiplet has field content $(3, 4, 1)$ and will therefore be denoted as $(3, 4, 1)_{lin}$. The supermultiplets extending III and IV possess supertransformations which are differential polynomials in their component fields. Since their field content is, respectively, $(3, 4, 1)$ and $(2, 4, 2)$, the supermultiplets will be denoted as $(3, 4, 1)_{nl}$ and $(2, 4, 2)_{nl}$, respectively.

Schematically, we have

$$\begin{array}{ccc} (4,4) & \longrightarrow & (3,4,1)_{lin} \\ \downarrow & & \downarrow \\ (3,4,1)_{nl} & \longrightarrow & (2,4,2)_{nl} \end{array} \tag{15}$$

Journal of Physics: Conference Series **284** (2011) 012058 doi:10.1088/1742-6596/284/1/012058

Their component fields are parametrized according to

$$\begin{array}{ccc} (u_1, u_2, u_3, u_4; \psi_1, \psi_2, \psi_3, \psi_4) & \longrightarrow & (x_1, x_2, x_3; \mu_1, \mu_2, \mu_3, \mu_4; f) \\ \downarrow & & \downarrow \\ (w_1, w_2, w_3; \xi_1, \xi_2, \xi_3, \xi_4; g) & \longrightarrow & (z_1, z_2; \eta_1, \eta_2, \eta_3, \eta_4; h_1, h_2) \end{array} \quad (16)$$

The greek letters have been employed to denote the fermionic fields; $\vec{u}, \vec{x}, \vec{w}, \vec{z}$ denote the bosonic target coordinates of the respective supermultiplets, while $f, g, h_{1,2}$ denote the auxiliary fields.

We present at first the supersymmetry transformations of the four supermultiplets above (the presentation of the transformations explicitly connecting their component fields, namely the "arrows" in (16), will be given in the next Section). In the following tables the entries give the supertransformations of the component fields under the action of the Q_I supersymmetry operator.

The $\mathcal{N} = 4$ (linear) $(4,4)$ root supermultiplet can be explicitly presented as

	Q_1	Q_2	Q_3	Q_4
u_1	ψ_1	ψ_2	ψ_3	ψ_4
u_2	ψ_2	$-\psi_1$	ψ_4	$-\psi_3$
u_3	ψ_3	$-\psi_4$	$-\psi_1$	ψ_2
u_4	ψ_4	ψ_3	$-\psi_2$	$-\psi_1$
ψ_1	$\dot{u}_1$	$-\dot{u}_2$	$-\dot{u}_3$	$-\dot{u}_4$
ψ_2	$\dot{u}_2$	$\dot{u}_1$	$-\dot{u}_4$	$\dot{u}_3$
ψ_3	$\dot{u}_3$	$\dot{u}_4$	$\dot{u}_1$	$-\dot{u}_2$
ψ_4	$\dot{u}_4$	$-\dot{u}_3$	$\dot{u}_2$	$\dot{u}_1$

(17)

The $(3,4,1)_{lin}$ supermultiplet is given by

	Q_1	Q_2	Q_3	Q_4
x_1	μ_1	$-\mu_2$	$-\mu_3$	μ_4
x_2	μ_2	μ_1	$-\mu_4$	$-\mu_3$
x_3	μ_3	μ_4	μ_1	μ_2
μ_1	$\dot{x}_1$	$\dot{x}_2$	$\dot{x}_3$	$-f$
μ_2	$\dot{x}_2$	$-\dot{x}_1$	f	$\dot{x}_3$
μ_3	$\dot{x}_3$	$-f$	$-\dot{x}_1$	$-\dot{x}_2$
μ_4	f	$\dot{x}_3$	$-\dot{x}_2$	$\dot{x}_1$
f	$\dot{\mu}_4$	$-\dot{\mu}_3$	$\dot{\mu}_2$	$-\dot{\mu}_1$

(18)

The $(3,4,1)_{nl}$ supermultiplet is given by

	Q_1	Q_2	Q_3	Q_4
w_1	$\xi_1 + \frac{1}{R}(w_1\xi_4)$	$\xi_2 + \frac{1}{R}(w_1\xi_3)$	$\xi_2 - \frac{1}{R}(w_1\xi_2)$	$\xi_4 - \frac{1}{R}(w_1\xi_1)$
w_2	$\xi_2 + \frac{1}{R}(w_2\xi_4)$	$-\xi_1 + \frac{1}{R}(w_2\xi_3)$	$\xi_4 - \frac{1}{R}(w_2\xi_2)$	$-\xi_3 - \frac{1}{R}(w_2\xi_1)$
w_3	$\xi_3 + \frac{1}{R}(w_3\xi_4)$	$-\xi_4 + \frac{1}{R}(w_3\xi_3)$	$-\xi_1 - \frac{1}{R}(w_3\xi_2)$	$\xi_2 - \frac{1}{R}(w_3\xi_1)$
ξ_1	$\dot{w}_1 - \frac{1}{R}(w_1 g + \xi_1\xi_4)$	$-\dot{w}_2 + \frac{1}{R}(w_2 g - \xi_1\xi_3)$	$-\dot{w}_3 + \frac{1}{R}(w_3 g + \xi_1\xi_2)$	$-g$
ξ_2	$\dot{w}_2 - \frac{1}{R}(w_2 g + \xi_2\xi_4)$	$\dot{w}_1 - \frac{1}{R}(w_1 g + \xi_2\xi_3)$	$-g$	$\dot{w}_3 - \frac{1}{R}(w_3 g + \xi_1\xi_2)$
ξ_3	$\dot{w}_3 - \frac{1}{R}(w_3 g + \xi_3\xi_4)$	g	$\dot{w}_1 - \frac{1}{R}(w_1 g + \xi_2\xi_3)$	$-\dot{w}_2 + \frac{1}{R}(w_2 g - \xi_1\xi_3)$
ξ_4	g	$-\dot{w}_3 + \frac{1}{R}(w_3 g + \xi_3\xi_4)$	$\dot{w}_2 - \frac{1}{R}(w_2 g + \xi_2\xi_4)$	$\dot{w}_1 - \frac{1}{R}(w_1 g + \xi_1\xi_4)$
g	$\dot{\xi}_4$	$-\dot{\xi}_3$	$-\dot{\xi}_2$	$-\dot{\xi}_1$

(19)

Finally, the $(2,4,2)_{nl}$ supermultiplet is given by

	Q_1	Q_2	Q_3	Q_4
z_1	$\eta_1+\frac{1}{r}(z_1\eta_3)$	$-\eta_2+\frac{1}{r}(z_1\eta_4)$	$-\eta_3+\frac{1}{r}(z_1\eta_1)$	$\eta_4+\frac{1}{r}(z_1\eta_2)$
z_2	$\eta_2+\frac{1}{r}(z_2\eta_3)$	$\eta_1+\frac{1}{r}(z_2\eta_4)$	$-\eta_4+\frac{1}{r}(z_2\eta_1)$	$-\eta_3+\frac{1}{r}(z_2\eta_2)$
η_1	$\dot{z}_1-\frac{1}{r}(z_1h_1+\eta_1\eta_3)$	$\dot{z}_2-\frac{1}{r}(z_2h_1+\eta_1\eta_4)$	h_1	$-h_2-\frac{1}{r}(\eta_1\eta_2)$
η_2	$\dot{z}_2-\frac{1}{r}(z_2h_1+\eta_2\eta_3)$	$-\dot{z}_1+\frac{1}{r}(z_1h_1-\eta_2\eta_4)$	$h_2-\frac{1}{r}(\eta_1\eta_2)$	h_1
η_3	h_1	$-h_2-\frac{1}{r}(\eta_3\eta_4)$	$-\dot{z}_1+\frac{1}{r}(z_1h_1+\eta_1\eta_3)$	$-\dot{z}_2+\frac{1}{r}(z_2h_1+\eta_2\eta_3)$
η_4	$h_2-\frac{1}{r}(\eta_3\eta_4)$	h_1	$-\dot{z}_2+\frac{1}{r}(z_2h_1+\eta_1\eta_4)$	$\dot{z}_1-\frac{1}{r}(z_1h_1+\eta_2\eta_4)$
h_1	$\dot{\eta}_3$	$\dot{\eta}_4$	$\dot{\eta}_1$	$\dot{\eta}_2$
h_2	$\dot{\eta}_4-\frac{1}{r}(h_1\eta_4-h_2\eta_3)$	$-\dot{\eta}_3+\frac{1}{r}(h_1\eta_3+h_2\eta_4)$	$\dot{\eta}_2-\frac{1}{r}(h_1\eta_2+h_2\eta_1)$	$-\dot{\eta}_1+\frac{1}{r}(h_1\eta_1+h_2\eta_2)$

(20)

The constant parameters R (entering $(3,4,1)_{nl}$) and r (entering $(2,4,2)_{nl}$) can be reabsorbed (set equal to 1) through a suitable rescaling of the component fields. It is however convenient to present them explicitly to show that in the contraction limit (for $R, r \to \infty$) the linear supermultiplets $(3,4,1)_{lin}$ and, respectively, $(2,4,2)_{lin}$ are recovered.

4. New $1D$ superalgebras from dimensional reduction of twisted SYM

On a Euclidean 4-manifold with $SU(2)$ holonomy the $N=1$ superYang-Mills theory can be expressed in twisted form, both for the vector and the scalar multiplets $(3,4,1)$ and $(2,4,2)$. [19, 20, 21, 22, 23, 24]. In the twisted formulation the spinors are described as holomorphic and antiholomorphic forms, in such a way that both multiplets are decomposed as follows

$$\begin{array}{lcl} (3,4,1) & : & A_m, A_{\bar m}; \psi_m, \chi_{\bar m\bar n}, \chi; h. \\ (2,4,2) & : & \Phi, \bar\Phi; \psi_{\bar m}, \chi_{mn}, \bar\chi; B_{\bar m\bar n}, T_{mn}. \end{array} \qquad (21)$$

Here we use the complex space coordinates $z^m, z^{\bar m}$, where $m=1,2$ are $SU(2)$ indices and $\bar m = \bar 1, \bar 2$ are the complex conjugate ones. One has a complex structure J^i_j, with $J^2 = -\mathbf{1}$, $J_{m\bar n} = -J_{\bar n m}$, $J_{mn} = 0 = J_{\bar m\bar n} = 0$, which can be used as a metric, according to $X_m Y_{\bar m} \equiv X_m J^{m\bar n} Y_{\bar n}$.

The counting of the degrees of freedom is as follows. For the first multiplet, $(3,4,1)$, we have $4 = 2+2$ gauge components $A_m, A_{\bar m}$ with one gauge degree of freedom (leading to $4-1=3$ physical degrees of freedom), 4 fermions $(\psi_m, \chi_{\bar m\bar n}, \chi)$ and 1 bosonic auxiliary field (h). For the $(2,4,2)$ multiplet we have 2 propagating bosons $(\Phi, \bar\Phi)$, 4 fermions $(\psi_{\bar m}, \chi_{mn}, \bar\chi)$ and 2 bosonic auxiliary fields $(B_{\bar m\bar n}, T_{mn})$.

The link between analytically continued Majorana spinors $\lambda^\alpha, \lambda_{\dot\alpha}$ and their holomorphic-antiholomorphic decompositions as in Eq. (21) is given by the following formula

$$\begin{array}{rcl} \psi_m & = & \lambda^\alpha \sigma_{\mu\ \alpha 1} e^\mu_m, \\ \chi_{\bar m\bar n} & = & \bar\lambda_{\dot\alpha}\, \bar\sigma_{\mu\nu}{}^{\dot\alpha}{}_{\dot 2}\, e^\mu_{\bar m} e^\nu_{\bar n}, \\ \chi & = & \delta_{\dot 2}{}^{\dot\alpha} \bar\lambda_{\dot\alpha}. \end{array} \qquad (22)$$

The twist formula (22) can be considered as a mere (complex) change of variables. The Dirac Lagrangian is rexpressed as

$$\bar\lambda \gamma^\mu D_\mu \lambda = \chi_{\bar m\bar n} D_{[m}\psi_{n]} + \chi D_{\bar m}\psi_m. \qquad (23)$$

The four (ordinary) Poincaré supercharges are rexpressed, in the twisted supersymmetry formulation, as the operators $s, s_{\bar p}$ and s_{mn} which close, up gauge transformations of the r.h.s., the superalgebra

$$\begin{aligned} \{s, s_{\overline{m}}\} &= \partial_{\overline{m}} + \delta_{gau.}(A_{\bar{m}}), \\ \{s_{\overline{p}}, s_{mn}\} &= J_{\bar{p}[m}(\partial_{n]} + \delta_{gau.}(A_{n]})) \end{aligned} \tag{24}$$

(the remaining anticommutators are all vanishing).

The balance between bosonic and fermionic degrees of freedom can be reached through a gauge-fixing, which can be imposed by setting, e.g., $A_{\bar{2}} = 0$. On the other hand, further constraints are necessary in order to mantain the compatibility of the supersymmetry algebra with the gauge-fixing. The net result is that the dependence on the coordinates $m = 1, 2, \bar{m} = \bar{1}$ must be frozen, leaving a one-dimensional theory which depnds on the single variable $\bar{2}$. The resulting 4-generator dimensionally reduced superalgebra is expressed (for the twisted $(3, 4, 1)$ multiplet) by

	s	$s_{\bar{1}}$	$s_{\bar{2}}$	s_{12}
A_1	ψ_1	0	$-a\chi$	0
A_2	ψ_2	$a\chi$	0	0
$A_{\bar{1}}$	0	0	$-\chi_{\overline{12}}$	$a\psi_1$
ψ_1	0	0	$\dot{A}_1 + ah$	0
ψ_2	0	$-ah$	$\dot{A}_2$	0
$\chi_{\overline{12}}$	$-\dot{A}_{\bar{1}}$	0	0	$a(\dot{A}_1 + ah)$
χ	h	0	0	0
h	0	0	$\dot{\chi}$	0

(25)

The mixed transformations can be eliminated by field redefinitions

$$\begin{array}{ll} z_1 = A_1, & \xi_1 = \psi_1 - a\chi, \\ z_2 = A_{\bar{1}} - aA_2, & \xi_2 = \psi_1 + a\chi, \\ z_3 = A_{\bar{1}} + aA_2, & \xi_3 = \chi_{\overline{12}}, \\ g = \dot{A}_1 + 2ah, & \xi_4 = \chi_{\overline{12}} - a\psi_2. \end{array} \tag{26}$$

By introducing the basis of four operators

$$s_{\pm} = s \pm s_{\bar{2}}, \qquad N_{\pm} = \frac{1}{a}(s_{12} \pm s_{\bar{1}}), \tag{27}$$

one ends up with a non-trivial realization of a generalized supersymmetry which consists of ordinary supersymmetry operators and nilpotent operators. The four operators $s_{\pm}$ and $N_{\pm}$ are indeed mutually anticommuting and satisfy

$$s_{\pm}{}^2 = \pm\partial_t \quad , \quad N_{\pm}{}^2 = 0. \tag{28}$$

The $(3, 4, 1)$ twisted multiplet transforms as

	s_+	s_-	N_+	N_-
z_1	ξ_1	ξ_2	0	0
z_2	$-\xi_3$	ξ_4	ξ_1	ξ_2
z_3	$-\xi_4$	ξ_3	$-\xi_2$	ξ_1
ξ_1	$\dot{z}_1$	$-g$	0	0
ξ_2	g	$-\dot{z}_1$	0	0
ξ_3	$-\dot{z}_2$	$-\dot{z}_3$	$\dot{z}_1$	g
ξ_4	$-\dot{z}_3$	$-\dot{z}_2$	g	$\dot{z}_1$
g	$\dot{\xi}_2$	$\dot{\xi}_1$	0	0

(29)

This is non-trivial realization (a similar results holds for the twisted $(2,4,2)$ multiplet as well), of the $\mathcal{N}\equiv[n_+=1,n_-=1,n_0=2]$ generalized supersymmetry

$$\begin{aligned}\{Q_I,Q_J\} &= \eta_{IJ}H,\\ [H,Q_I] &= 0,\end{aligned} \tag{30}$$

where η_{IJ} is a diagonal metric with $n_\pm$ diagonal entries ±1 e n_0 diagonal entries 0.

The presence of a nilpotent operator implies that the twisted theory is a topological theory defined in terms of a BRST cohomology. On the other hand the transformations (22) allows us to relate the topological theory to the ordinary supersymmetric theory.

5. Conclusions

The new $\mathcal{N}=4$ one-dimensional supermultiplets which have been presented in this talk, namely the non-minimal linear supermultiplets, the non-linear supermultiplets induced by the second Hopf fibration and the twisted supermultiplets induced by the dimensional reduction of the twisted superYang-Mills theory, all admit invariant actions which define their associated $\mathcal{N}=4$ sigma-models. The actions are not reported here for lack of space.

The supermultiplets here presented can be regarded as building blocks to construct different realizations of supersymmetries (non-minimal, non-linear, twisted) for larger values of $\mathcal{N}>4$.

5.1. Acknowledgments

I am grateful to my collaborators L. Baulieu, L. Faria Carvalho, M. Gonzales, S. Khodaee and Z. Kuznetsova. The work was partially supported by Edital Universal CNPq, Proc. 472903/2008-0.

5.2. Appendix

For completeness we report the definitions, applied to the cases used in the text, of the properties characterizing the linear representations of the one-dimensional $\mathcal{N}$-Extended Superalgebra. In particular the notions of *mass-dimension*, *field content*, *dressing transformation*, *connectivity symbol*, *dual supermultiplet* and so on, as well as the association of linear supersymmetry transformations with graphs, will be reviewed following [4, 6, 5, 11, 12, 1]. The Reader can consult these papers for broader definitions and more detailed discussions.

Mass-dimension:

A grading, the mass-dimension d, can be assigned to any field entering a linear representation (the hamiltonian H, proportional to the time-derivative operator $\partial\equiv\frac{d}{dt}$, has a mass-dimension 1). Bosonic (fermionic) fields have integer (respectively, half-integer) mass-dimension.

Field content:

Each finite linear representation is characterized by its "field content", i.e. the set of integers $(n_1,n_2,\ldots,n_l)$ specifying the number n_i of fields of mass-dimension d_i ($d_i=d_1+\frac{i-1}{2}$, with d_1 an arbitrary constant) entering the representation. Physically, the n_l fields of highest dimension are the auxiliary fields which transform as a time-derivative under any supersymmetry generator. The maximal value l (corresponding to the maximal dimensionality d_l) is defined to be the *length* of the representation (a root representation has length $l=2$). Either $n_1,n_3,\ldots$ correspond to the bosonic fields (therefore $n_2,n_4,\ldots$ specify the fermionic fields) or viceversa.
In both cases the equality $n_1+n_3+\ldots=n_2+n_4+\ldots=n$ is guaranteed.

Dressing transformation:

Higher-length supermultiplets are obtained by applying a dressing transformation to the length-2 root supermultiplet. The root supermultiplet is specified by the $\mathcal{N}$ supersymmetry operators $\widehat{Q}_i$ ($i=1,\ldots,\mathcal{N}$), expressed in matrix form as

$$\widehat{Q}_j=\frac{1}{\sqrt{2}}\begin{pmatrix}0 & \gamma_j\\ -\gamma_j\cdot H & 0\end{pmatrix},\quad \widehat{Q}_{\mathcal{N}}=\frac{1}{\sqrt{2}}\begin{pmatrix}0 & \mathbf{1}_n\\ \mathbf{1}_n\cdot H & 0\end{pmatrix}, \tag{31}$$

where the γ_j matrices ($j = 1, \ldots, \mathcal{N} - 1$) satisfy the Euclidean Clifford algebra

$$\{\gamma_i, \gamma_j\} \quad = \quad -2\delta_{ij}\mathbf{1}_n. \tag{32}$$

The length-3 supermultiplets are specified by the $\mathcal{N}$ operators Q_i, given by the dressing transformation

$$Q_i \quad = \quad D\widehat{Q}_i D^{-1}, \tag{33}$$

where D is a diagonal dressing matrix such that

$$D \quad = \quad \left(\begin{array}{cc} \widetilde{D} & 0 \\ 0 & \mathbf{1}_n \end{array}\right), \tag{34}$$

with $\widetilde{D}$ an $n \times n$ diagonal matrix whose diagonal entries are either 1 or the derivative operator ∂.

Association with graphs:

The association between linear supersymmetry transformations and $\mathcal{N}$-colored oriented graphs goes as follows. The fields (bosonic and fermionic) entering a representation are expressed as vertices. They can be accommodated into an $X-Y$ plane. The Y coordinate can be chosen to correspond to the mass-dimension d of the fields. Conventionally, the lowest dimensional fields can be associated to vertices lying on the X axis. The higher dimensional fields have positive, integer or half-integer values of Y. A colored edge links two vertices which are connected by a supersymmetry transformation. Each one of the $\mathcal{N}$ Q_i supersymmetry generators is associated to a given color. The edges are oriented. The orientation reflects the sign (positive or negative) of the corresponding supersymmetry transformation connecting the two vertices. Instead of using arrows, alternatively, solid or dashed lines can be associated, respectively, to positive or negative signs. No colored line is drawn for supersymmetry transformations connecting a field with the time-derivative of a lower dimensional field. This is in particular true for the auxiliary fields (the fields of highest dimension in the representation) which are necessarily mapped, under supersymmetry transformations, in the time-derivative of lower-dimensional fields.

Each irreducible supersymmetry transformation can be presented (the identification is not unique) through an oriented $\mathcal{N}$-colored graph with $2n$ vertices. The graph is such that precisely $\mathcal{N}$ edges, one for each color, are linked to any given vertex which represents either a 0-mass dimension or a $\frac{1}{2}$-mass dimension field. An unoriented "color-blind" graph can be associated to the initial graph by disregarding the orientation of the edges and their colors (all edges are painted in black).

Connectivity symbol:

A characterization of length $l = 3$ color-blind, unoriented graphs can be expressed through the connectivity symbol ψ_g, defined as follows

$$\psi_g \quad = \quad (m_1)_{s_1} + (m_2)_{s_2} + \ldots + (m_Z)_{s_Z}. \tag{35}$$

The ψ_g symbol encodes the information on the partition of the n $\frac{1}{2}$-mass dimension fields (vertices) into the sets of m_z vertices ($z = 1, \ldots, Z$) with s_z edges connecting them to the $n - k$ 1-mass dimension auxiliary fields. We have

$$m_1 + m_2 + \ldots + m_Z \quad = \quad n, \tag{36}$$

while $s_z \neq s_{z'}$ for $z \neq z'$.

Dual supermultiplet:

A dual supermultiplet is obtained by mirror-reversing, upside-down, the graph associated to the original supermultiplet.

Journal of Physics: Conference Series **284** (2011) 012058 doi:10.1088/1742-6596/284/1/012058

References

[1] M. Gonzales, S. Khodaee and F. Toppan, arXiv:1006.4678[hep-th].
[2] L. Faria Carvalho, Z. Kuznetsova and F. Toppan, *Nucl. Phys.* **B 834** (2010), 237 (arXiv:0912.3279[hep-th]).
[3] L. Baulieu and F. Toppan, arXiv:1010.3004[hep-th].
[4] A. Pashnev and F. Toppan, *J. Math. Phys.* **42** (2001), 5257 (hep-th/0010135).
[5] M. Faux and S. J. Gates Jr., *Phys. Rev.* **D 71** (2005), 065002 (hep-th/0408004).
[6] Z. Kuznetsova, M. Rojas and F. Toppan, *JHEP* **0603** (2006), 098 (hep-th/0511274).
[7] C.F. Doran, M. G. Faux, S. J. Gates Jr., T. Hubsch, K. M. Iga and G. D. Landweber, math-ph/0603012.
[8] C.F. Doran, M. G. Faux, S. J. Gates Jr., T. Hubsch, K. M. Iga and G. D. Landweber, hep-th/0611060.
[9] F. Toppan, *POS* **IC2006**, 033 (hep-th/0610180).
[10] F. Toppan, in *Quantum, Super and Twistors*, Proc. 22*nd* Max Born Symp., Wrocław 2006. Eds. Kowalski-Glikman and Turko (2008), 143 (hep-th/0612276).
[11] Z. Kuznetsova and F. Toppan, *Mod. Phys. Lett.* **A 23** (2008), 37 (hep-th/0701225).
[12] Z. Kuznetsova and F. Toppan, *Int. J. Mod. Phys.* **A 23** (2008), 3947 (arXiv:0712.3176).
[13] F. Toppan, *Acta Polyt.* **48** (2008), 56.
[14] C.F. Doran, M. G. Faux, S. J. Gates Jr., T. Hubsch, K. M. Iga, G. D. Landweber and R. L. Miller, arXiv:08060050.
[15] C.F. Doran, M. G. Faux, S. J. Gates Jr., T. Hubsch, K. M. Iga and G. D. Landweber, arXiv:08060051.
[16] M. G. Faux, K. M. Iga and G. D. Landweber, arXiv:0907.3605.
[17] M. G. Faux and G. D. Landweber, *Phys. Lett.* **B 681** (2009), 161 (arXiv:0907.4543).
[18] E. Witten, *Nucl. Phys.* B 188 (1981), 513.
[19] A. Johansen, *Int. J. Mod. Phys.* A 10 (1995) 4325 (hep-th/9403017).
[20] E. Witten, *J. Math. Phys.* 35 (1994) 5101 (hep-th/9403195).
[21] A. D. Popov, *Phys. Lett.* B 473 (2000) 65 (hep-th/9909135); T. A. Ivanova and A. D. Popov, *J. Math. Phys.* 41 (2000) 2604 (hep-th/0002120).
[22] C. Hofman and J. S. Park, *Nucl. Phys.* B 600 (2001) 133 (hep-th/0010103); J. S. Park, *Commun. Math. Phys.* 163 (1994) 113 (hep-th/9304060); J. S. Park, *Nucl. Phys.* B 423 (1994) 559 (hep-th9305095).
[23] L. Baulieu, A. Tanzini, *J. Geom. Phys.* 56 (2006) 2379-2401 (hep-th/0412014).
[24] L. Baulieu, G. Bossard, *Phys. Lett.* B 632 (2005) 138 (hep-th/0507004).

Journal of Physics: Conference Series **284** (2011) 012059 doi:10.1088/1742-6596/284/1/012059

Quantum communication and state transfer in spin chains

Joris Van der Jeugt

Department of Applied Mathematics and Computer Science, Ghent University, Krijgslaan 281-S9, B-9000 Gent, Belgium

E-mail: `Joris.VanderJeugt@UGent.be`

Abstract. We investigate the time evolution of a single spin excitation state in certain linear spin chains, as a model for quantum communication. We consider first the simplest possible spin chain, where the spin chain data (the nearest neighbour interaction strengths and the magnetic field strengths) are constant throughout the chain. The time evolution of a single spin state is determined, and this time evolution is illustrated by means of an animation. Some years ago it was discovered that when the spin chain data are of a special form so-called perfect state transfer takes place. These special spin chain data can be linked to the Jacobi matrix entries of Krawtchouk polynomials or dual Hahn polynomials. We discuss here the case related to Krawtchouk polynomials, and illustrate the possibility of perfect state transfer by an animation showing the time evolution of the spin chain from an initial single spin state. Very recently, these ideas were extended to discrete orthogonal polynomials of q-hypergeometric type. Here, a remarkable result is a new analytic model where perfect state transfer is achieved: this is when the spin chain data are related to the Jacobi matrix of q-Krawtchouk polynomials. This case is discussed here, and again illustrated by means of an animation.

1. Introduction

The transfer of a quantum state (i.e. a qubit) from one site to another is a key requirement in quantum communication. This process is important for connecting distinct quantum processors or for the (short distance) mapping of quantum states to elements of a quantum register.

S. Bose [1, 2] introduced linear spin chains as a channel for such short distance quantum communication. An excellent review paper on the topic was published in 2007 [3]. Since the pioneering work of Bose, the study of transmission of data in such linear quantum registers has been the subject of many investigations [1, 4, 5, 6, 7].

The linear spin chains as introduced by Bose consist of a linear sequence of permanently coupled qubits, i.e. quantum two-state systems. These are realized as spin-1/2 systems, where $|0\rangle$ is the notation for spin down and $|1\rangle$ for spin up. The question is whether the chain can act as a quantum data-bus if the interactions in the chain are permanent (i.e. the couplings are always "on" and fixed in strength).

From the mathematical point of view, a linear spin chain is determined by its Hamiltonian. A system of $N+1$ interacting qubits (spin 1/2 particles) in a quantum register is usually modeled

Journal of Physics: Conference Series **284** (2011) 012059 doi:10.1088/1742-6596/284/1/012059

by an isotropic Hamiltonian of XY type:

$$\hat{H} = \frac{1}{2}\sum_{k=0}^{N-1} J_k(\sigma_k^x \cdot \sigma_{k+1}^x + \sigma_{k+1}^y \cdot \sigma_k^y) + \frac{1}{2}\sum_{k=0}^{N} h_k(\sigma_k^z + 1), \tag{1}$$

where J_k is the coupling strength between the kth and $(k+1)$th qubit, and h_k is the "Zeeman" energy of the kth qubit. For such Hamiltonians, it is common to use the Jordan-Wigner transformation, mapping the Pauli matrices to spinless lattice fermions [8, 9]. Then the Hamiltonian takes the form

$$\hat{H} = \sum_{k=0}^{N-1} J_k(a_k^\dagger a_{k+1} + a_{k+1}^\dagger a_k) + \sum_{k=0}^{N} h_k a_k^\dagger a_k. \tag{2}$$

We shall be dealing with the problem in this form; the constants J_k and h_k are referred to as the spin chain data.

Initially, the system is in its completely polarized ground state $|\mathbf{0}\rangle = |00\cdots 0\rangle = |0\rangle\otimes|0\rangle\otimes\cdots\otimes|0\rangle$, where $|0\rangle$ denotes the spin down state. Let $|k) = |00\cdots 010\cdots 0\rangle = a_k^\dagger|\mathbf{0}\rangle$ $(k = 0, 1, \ldots, N)$ denote a state in which there is a single fermion at the site k and all other sites are empty, i.e. $|k)$ describes the state in which the spin at the site k has been flipped to $|1\rangle$. These notions are illustrated in the following figures.

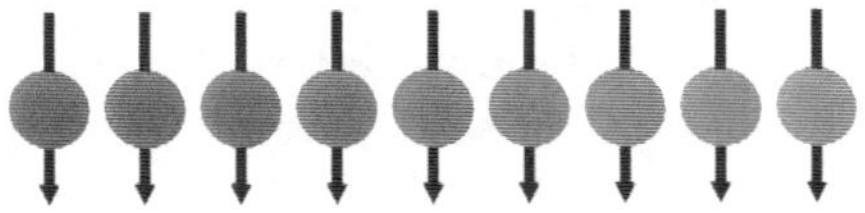

Figure 1. Illustration of the ground state $|\mathbf{0}\rangle$ of the linear spin chain, with all spins down.

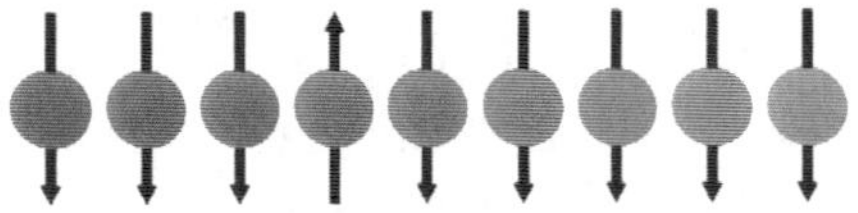

Figure 2. Illustration of the single fermion state $|k)$, with the spin at site k up and all others down (here $k = 3$).

The set of states $|k)$ $(k = 0, 1, \ldots, N)$ forms a basis for the single-fermion states of the system. In this single-fermion basis, the Hamiltonian $\hat{H}$ takes the matrix form

$$M = \begin{pmatrix} h_0 & J_0 & 0 & \cdots & 0 \\ J_0 & h_1 & J_1 & \cdots & 0 \\ 0 & J_1 & h_2 & \ddots & \\ \vdots & \vdots & \ddots & \ddots & J_{N-1} \\ 0 & 0 & & J_{N-1} & h_N \end{pmatrix}. \tag{3}$$

The dynamics (time evolution) of the system is completely determined by the eigenvalues ϵ_j and eigenvectors φ_j of this matrix M. Once the single-fermion eigenstates φ_j are determined, the n-fermion eigenstates of $\hat{H}$ $(n \leq N)$ can be constructed using the technique of Slater determinants [9, 10]. In order to determine the single-fermion states, note that the matrix M in (3) is real and symmetric, so the spectral theorem [11] implies that it can be written as $M = UDU^T$, where D is a diagonal matrix and U an orthogonal matrix:

$$D = \mathrm{diag}(\epsilon_0, \epsilon_1, \ldots, \epsilon_N), \tag{4}$$
$$UU^T = U^T U = I. \tag{5}$$

The entries of D are the single-fermion energy eigenvalues, and the columns of the matrix U are the (orthonormal) eigenvectors of M, i.e. the single-fermion eigenstates $\varphi_j = \sum_{k=0}^{N} U_{kj}\, |k)$ with $\hat{H}\varphi_j = M\varphi_j = \epsilon_j\, \varphi_j$.

The *dynamics* of the system under consideration is described by the unitary time evolution operator $\mathcal{U}(t) \equiv \exp(-it\hat{H})$. Assume that the "state sender" is located at site s of the spin chain, and the "state receiver" at site r (s and r are site labels, belonging to $\{0, 1, \ldots, N\}$). At time $t = 0$ the sender turns the system into the single spin state $|s)$. After a certain time t, the system evolves to the state $\mathcal{U}(t)|s)$ which may be expressed as a linear superposition of all the single spin states. So the transition amplitude of an excitation from site s to site r of the spin chain is given by the time-dependent correlation function

$$f_{r,s}(t) = (r|\mathcal{U}(t)|s). \tag{6}$$

Using the orthogonality of the states φ_j, one finds [12]:

$$f_{r,s}(t) = \langle \sum_{k=0}^{N} U_{rk}\varphi_k | \exp(-it\hat{H}) \sum_{j=0}^{N} U_{sj}\varphi_j \rangle = \sum_{j=0}^{N} U_{rj} U_{sj} e^{-it\epsilon_j}. \tag{7}$$

One says that there is *perfect state transfer* at time t from one end of the chain to the other end when $|f_{N,0}(t)| = 1$. The conditions for perfect state transfer can quite easily be described in terms of the "mirror symmetry" of the matrix M in (3), see [13, 7]. However, if our aim is to study analytical solutions, we should also require the conditions that the eigenvalues ϵ_j and the eigenvector components U_{kj} should be analytic (closed form) expressions, for arbitrary N (i.e. the value of N is not fixed, and appears as a parameter in the closed form expressions of ϵ_j and U_{kj}). We shall refer to such cases as "analytically solvable Hamiltonians $\hat{H}$".

The purpose of this contribution is to present some results and examples of spin chains allowing perfect state transfer in the context of orthogonal polynomials. These examples are taken from [12] and [14]. The main addition for the current contribution is that we present some animations illustrating the ideas of perfect state transfer in this context.

2. Spin chain with constant interaction

The simplest possible example of a Hamiltonian $\hat{H}$ that is analytically solvable has zero background magnetic field (i.e. all $h_k = 0$) and constant nearest-neighbour interaction in the chain (i.e. all $J_k = 1$). For an interaction matrix M with these simple spin chain data, there is indeed an analytically closed form expression for its eigenvalues and eigenvectors [1]. In particular,

$$\epsilon_j = 2\cos\left(\frac{(j+1)\pi}{N+2}\right), \qquad (j = 0, 1, \ldots, N) \tag{8}$$

and the eigenvectors φ_j are the columns of the (orthogonal) matrix U given by

$$U_{ij} = \sqrt{\frac{2}{N+2}} \sin\left(\frac{(i+1)(j+1)\pi}{N+2}\right) \qquad (i, j = 0, 1, \ldots, N). \tag{9}$$

Following (7), the correlation function becomes

$$f_{r,s}(t) = \sum_{j=0}^{N} \frac{2}{N+2} \sin\left(\frac{(r+1)(j+1)\pi}{N+2}\right) \sin\left(\frac{(s+1)(j+1)\pi}{N+2}\right) \exp\left(-2it\cos\left(\frac{(j+1)\pi}{N+2}\right)\right). \tag{10}$$

Let us consider this function in more detail for the case $s = 0$ (i.e. the sender is at the left end of the chain). For any time t, one can compute the function $f_{r,0}(t)$ or its absolute value $|f_{r,0}(t)|$, and this for all r-values ($r = 0, 1, \ldots, N$). If $|f_{r,0}(t)| = 0$, the rth spin of the system is spin down (depicted by an arrow down for site r). If $|f_{r,0}(t)| = 1$, the rth spin of the system is spin up (depicted by an arrow up). In general, when $0 < |f_{r,0}(t)| < 1$, the rth spin of the system is in a superposition state of the form

$$\cos(\alpha)|0\rangle + e^{i\phi}\sin(\alpha)|1\rangle.$$

Such a state can be depicted by an arrow in the direction of the angle α; following our convention of spin up and spin down, this means $\alpha = -\frac{\pi}{2} + \pi|f_{r,0}(t)|$.

We are now in a position to illustrate the time evolution of such a spin chain by means of an animation. Consider the spin chain with $N = 8$. In the following figure, we depict the evolution of the system for time t running from 0 to 3π. At time $t = 0$, the system is in the initial state $|0)$ with the leftmost spin in a pure state with spin up, and all the others spin down. Then, at any time t, we compute $f_{r,0}(t)$ for $r = 0, 1, \ldots, N$, and depict the corresponding arrows according to the above-mentioned convention. The result is shown in Figure 3, as an animation.

In this case, due to the nature of the correlation function $f_{r,0}(t)$ given by (10), there is clearly no perfect state transfer. In other words, at any time $t > 0$ all values $f_{r,0}(t)$ satisfy $0 < |f_{r,0}(t)| < 1$.

animation1

Figure 3. Time evolution of a linear spin chain with constant data. Click on the picture to see the animation running, from $t = 0$ up to $t = 3\pi$. This movie is also available as animation1.gif in the *Supplementary Data* accompanying this paper.

3. Spin chains allowing perfect state transfer

The previous example did not allow perfect state transfer, due to the simple choice of the spin chain data. It is known, however, that if the spin chain data J_k and h_k are chosen appropriately, perfect state transfer is possible. Such an example was given in [10, 4], with spin chain data

$$h_k = N/2, \qquad J_k = \frac{1}{2}\sqrt{(k+1)(N-k)}. \tag{11}$$

It turns out that in this case, the eigenvectors of M are related to Krawtchouk polynomial evaluations. For a second known example allowing perfect state transfer [10], dual Hahn polynomials play a role. In this context, we studied more generally systems where the spin chain data is related to the Jacobi matrix of a set of discrete orthogonal polynomials [12]. This general approach also leads to closed form expressions for the correlation function $f_{r,s}(t)$.

Let us present one example here, based upon the Krawtchouk polynomials. The Krawtchouk polynomial of degree n ($n = 0, 1, \ldots, N$) in the variable x, with parameter $0 < p < 1$ is given by

$$K_n(x) \equiv K_n(x; p, N) = {}_2F_1\left[\begin{matrix} -x, -n \\ -N \end{matrix} ; \frac{1}{p} \right]. \tag{12}$$

The function ${}_2F_1$ is the classical hypergeometric series [15, 16], and in this case it is a terminating series because of the appearance of the negative integer $-n$ as a numerator parameter. Krawtchouk polynomials satisfy a (discrete) orthogonality relation [17]:

$$\sum_{x=0}^{N} w(x) K_n(x) K_m(x) = d_n \delta_{mn}, \tag{13}$$

where $w(x)$ is the weight function in x and d_n is a function depending on n:

$$w(x) = \binom{N}{x} p^x (1-p)^{N-x} \qquad (x = 0, 1, \ldots, N); \qquad d_n = \frac{1}{\binom{N}{n}} \left(\frac{1-p}{p}\right)^n. \tag{14}$$

They also satisfy the following three-term recurrence relation:

$$-xK_n(x) = n(1-p)\, K_{n-1}(x) - \big[p(N-n) + n(1-p)\big]\, K_n(x) + p(N-n)\, K_{n+1}(x). \tag{15}$$

It is convenient to introduce orthonormal Krawtchouk functions by

$$\tilde{K}_n(x) \equiv \frac{\sqrt{w(x)} K_n(x)}{\sqrt{d_n}}, \tag{16}$$

satisfying

$$x\tilde{K}_n(x) = J_{n-1}\, \tilde{K}_{n-1}(x) + h_n\, \tilde{K}_n(x) + J_n\, \tilde{K}_{n+1}(x), \tag{17}$$

where

$$h_n = Np + (1-2p)n, \qquad J_n = -\sqrt{p(1-p)}\sqrt{(n+1)(N-n)}. \tag{18}$$

In other words, for the interaction matrix M_K with the above spin chain data, and a matrix U with $U_{jk} = \tilde{K}_j(k)$, one has $M_K U = UD$, where $D = \mathrm{diag}(0, 1, 2, \ldots, N)$ (due to the recurrence relation), and U is an orthogonal matrix (due to the orthogonality relations of the polynomials). So the eigenvectors of the Hamiltonian (in the single-fermion case) corresponding to the quantities (18) have components equal to normalized Krawtchouk polynomials, and the corresponding energy eigenvalues are $\epsilon_k = k$ $(k = 0, 1, \ldots, N)$. This implies that the correlation function (7), where we put $z = e^{-it}$, takes the following form:

$$\begin{aligned} f_{r,s}(t) &= \sum_{k=0}^{N} U_{rk} U_{sk} z^{\epsilon_k} = \sum_{k=0}^{N} \tilde{K}_r(k) \tilde{K}_s(k) z^k \\ &= \frac{1}{\sqrt{d_r d_s}} \sum_{k=0}^{N} w(k) K_r(k) K_s(k) z^k \qquad (z = e^{-it}). \end{aligned} \tag{19}$$

Now one can use summation formulas for hypergeometric series, leading to [12]

$$\begin{aligned} f_{r,s}(t) &= \sqrt{\binom{N}{r}\binom{N}{s}} (\sqrt{p(1-p)})^{r+s} (1-z)^{r+s} (1-p+pz)^{N-r-s} \\ &\times {}_2F_1 \left[\begin{matrix} -r, -s \\ -N \end{matrix} ; \frac{-z}{p(1-p)(1-z)^2} \right]. \end{aligned} \tag{20}$$

This is a closed form expression for the general correlation function. Note that it can also be obtained by means of a group theoretical computation [12]: the Jacobi matrix M_K can be identified with an element of the Lie algebra $su(2)$ in the $(N+1)$-dimensional representation. Then the calculation of

$$f_{r,s}(t) = (r|\exp(-itM_K)|s)$$

simply leads to the computation of a matrix element of an $SU(2)$ group element, and these are known as d-functions, related to Jacobi polynomials, i.e. expressed as ${}_2F_1$-functions [18, Chapter 4].

Let us now consider the possibility of perfect state transfer. From (20), one finds

$$f_{N,0}(t) = \left(\sqrt{p(1-p)}\right)^N (1-e^{-it})^N, \tag{21}$$

so in particular $f_{N,0}(\pi) = 1$ when $p = 1/2$. In other words, this model allows perfect state transfer when the parameter $p = 1/2$, and the transfer takes place at time $t = \pi$. This is illustrated as an animation in Figure 4.

animation2

Figure 4. Time evolution of a linear spin chain ($N = 8$) with data determined by the Jacobi matrix of Krawtchouk polynomials (18). Here $p = 1/2$. Click on the picture to see the animation running, from $t = 0$ up to $t = \pi$. Note that there is indeed perfect state transfer. (view as animation2.gif in *Supplementary Data*)

When the parameter p is different from $1/2$, one can see from (21) that $|f_{N,0}(t)|$ never reaches the value 1. Once again, let us illustrate this as an animation, see Figure 5. Note that at time $t = 2\pi$ the system is back in its initial state. This is because in general the function (19) is a periodic function with period 2π.

animation3

Figure 5. Time evolution of a linear spin chain ($N = 8$) with data determined by the Jacobi matrix of Krawtchouk polynomials (18). Now we have chosen $p = 1/3$. Click on the picture to see the animation running, from $t = 0$ up to $t = 2\pi$, and note that there is no perfect state transfer. (view as animation3.gif in *Supplementary Data*)

It should be mentioned that general conditions for perfect state transfer can be set up in terms of the interaction matrix M [4, 10, 7]. Following this, one can in principle use a numerical procedure known as the inverse eigenvalue problem, and design spin chains for perfect state

Journal of Physics: Conference Series **284** (2011) 012059 doi:10.1088/1742-6596/284/1/012059

transfer numerically [13]. This is, however, always for a chosen numerical value of N, whereas here we have worked with arbitrary N.

In [12] systems for which the interaction matrix coincides with the Jacobi matrix of a set of orthogonal polynomials were systematically investigated. Although no other solutions with perfect state transfer were found this way, the theoretical analysis gave rise to a number of interesting results: explicit formulas for transition amplitudes (or correlation functions), an explanation of why these two systems discovered in [10] are so special, and a group theoretical approach of the problem. For the case where the Jacobi matrix coincides with that of Krawtchouk polynomials (described above), some interesting limiting cases were also discussed in [12]. When the Jacobi matrix is that of Hahn polynomials (then two parameters α and β are involved), a compact formula for $f_{r,s}(t)$ was obtained, but no perfect state transfer is possible (although there were interesting limiting cases). When the Jacobi matrix is that of dual Hahn polynomials (with two parameters γ and δ), again a compact formula for $f_{r,s}(t)$ was obtained. This time, perfect state transfer is possible when $\gamma = \delta = \frac{2P+1}{2Q}$, $P, Q \in \mathbb{N}$, at time $t = Q\pi$. This corresponds to the second example of [10]. Note that in this case the single-fermion energy eigenvalues are of the form $\epsilon_k = k(k + \gamma + \delta + 1)$. Finally, when the Jacobi matrix is that of Racah polynomials (now four parameters α, β, γ and δ are present), a (more complicated) formula for $f_{r,s}(t)$ was obtained, but no new transfer results compared to earlier cases followed.

4. Spin chains with q-numbers in the spin chain data

The approach of spin chains with data originating from Jacobi polynomials of discrete orthogonal polynomials, as initiated in [12], turned out to be very illuminating. Apart from several known systems of discrete orthogonal polynomials of hypergeometric type in the Askey-scheme [17], there is also a list of discrete orthogonal polynomials of q-hypergeometric type. So it is a natural question to ask whether Jacobi matrices of these q-orthogonal polynomials could also function as interaction matrices for spin chains, and whether they would give rise to new solutions with perfect state transfer. This topic was treated in [14]. Among the main results there is indeed a new analytical solution for a spin chain with perfect state transfer. This new solution occurs in the context of q-Krawtchouk polynomials. In [14] all cases of the q-Askey-scheme were studied, and their analysis has shown that this is the only new case with perfect state transfer. For the complete analysis, we refer to [14]. Here, we shall just present the simple case related to q-Krawtchouk polynomials, that leads to a new model for perfect state transfer.

Let us first fix some notation. In the context of q-series, q is a positive real number with $q \neq 1$, and for us it can be considered as an extra parameter in the model. We use the common notation for q-numbers:

$$[n] \equiv [n]_q = \frac{1-q^n}{1-q} \qquad (n \in \mathbb{Z}) \tag{22}$$

and $[n] \to n$ in the limit $q \to 1$. For any complex number a and any nonnegative integer n, the q-shifted factorial is defined by

$$(a;q)_n = (1-a)(1-aq)\cdots(1-aq^{n-1}), \tag{23}$$

and the product is just 1 when $n = 0$. Sometimes, in the context when $0 < q < 1$, one also uses the infinite product

$$(a;q)_\infty = \prod_{k=0}^{\infty}(1-aq^k). \tag{24}$$

For products of q-shifted factorials, it is common to use the abbreviation

$$(a_1, a_2, \ldots, a_A; q)_n = (a_1;q)_n(a_2;q)_n \cdots (a_A;q)_n. \tag{25}$$

The q-hypergeometric series or *basic hypergeometric series* ${}_A\phi_B$ depends on A numerator parameters a_i, B denominator parameters b_i and a variable z and is defined as [19]:

$$ {}_A\phi_B\left[\begin{array}{c} a_1, a_2, \ldots, a_A \\ b_1, \ldots, b_B \end{array} ; q, z\right] = \sum_{n=0}^{\infty} \frac{(a_1, a_2, \ldots, a_A; q)_n}{(q, b_1, \ldots, b_B; q)_n} \left[(-1)^n q^{\binom{n}{2}}\right]^{1+B-A} z^n. \tag{26} $$

When the numerator contains a parameter of the form q^{-m}, with m a nonnegative integer, the series is terminating.

The q-Krawtchouk polynomials $\mathcal{K}_n(q^{-x}; p, N; q)$ are characterized by a positive integer parameter N and a positive real parameter p: $p > 0$. This polynomial of degree n in q^{-x} is defined as [17]

$$ \mathcal{K}_n\left(q^{-x}\right) \equiv \mathcal{K}_n(q^{-x}; p, N; q) = {}_3\phi_2\left[\begin{array}{c} q^{-n}, q^{-x}, -pq^n \\ q^{-N}, 0 \end{array} ; q, q\right], \qquad n = 0, 1, \ldots, N. \tag{27} $$

The q-Krawtchouk polynomials satisfy a discrete orthogonality relation, namely

$$ \sum_{x=0}^{N} w(x)\mathcal{K}_m(q^{-x})\mathcal{K}_n(q^{-x}) = d_n \delta_{mn}, \tag{28} $$

where the weight function is

$$ w(x) = \frac{(q^{-N}; q)_x}{(q; q)_x}(-p)^{-x}, \tag{29} $$

and the square norm is

$$ d_n = \frac{(q, -pq^{N+1}; q)_n}{(-p, q^{-N}; q)_n} \frac{(1+p)}{(1+pq^{2n})}(-pq; q)_N \, p^{-N} q^{-\binom{N+1}{2}}(-pq^{-N})^n q^{n^2}. \tag{30} $$

It is easy to see that $d_n > 0$ for $0 < q < 1$ and also for $q > 1$ (since $p > 0$). The polynomials $\mathcal{K}_n(q^{-x})$ also satisfy the following three term recurrence relation:

$$ -(1 - q^{-x})\mathcal{K}_n(q^{-x}) = A_n \mathcal{K}_{n+1}(q^{-x}) - (A_n + C_n)\mathcal{K}_n(q^{-x}) + C_n \mathcal{K}_{n-1}(q^{-x}), \tag{31} $$

with [17]

$$ A_n = \frac{(1 - q^{n-N})(1 + pq^n)}{(1 + pq^{2n})(1 + pq^{2n+1})}, \qquad C_n = -pq^{2n-N-1} \frac{(1 + pq^{n+N})(1 - q^n)}{(1 + pq^{2n-1})(1 + pq^{2n})}. \tag{32} $$

The corresponding orthonormal q-Krawtchouk functions

$$ \tilde{\mathcal{K}}_n(q^{-x}) \equiv \sqrt{\frac{w(x)}{d_n}} \mathcal{K}_n(q^{-x}) \tag{33} $$

satisfy the recurrence relation

$$ -[-x]\tilde{\mathcal{K}}_n(q^{-x}) = J_{n-1}\tilde{\mathcal{K}}_{n-1}(q^{-x}) + h_n \tilde{\mathcal{K}}_n(q^{-x}) + J_n \tilde{\mathcal{K}}_{n+1}(q^{-x}), \tag{34} $$

where

$$ J_n = \frac{A_n}{1-q}\sqrt{\frac{d_{n+1}}{d_n}}, \qquad h_n = -\frac{A_n + C_n}{1-q}. \tag{35} $$

As in the previous section, for the interaction matrix M_{qK} with the above spin chain data, and a matrix U with $U_{jk} = \tilde{\mathcal{K}}_j(q^{-k})$, one has $M_{qK}U = UD$, where $D = \mathrm{diag}(\epsilon_0, \epsilon_1, \epsilon_2, \ldots, \epsilon_N)$ with $\epsilon_k = -[-k]$ (due to the recurrence relation), and U is an orthogonal matrix (due to the orthogonality relations of the polynomials). In other words, the eigenvectors of the Hamiltonian (in the single-fermion case) corresponding to the quantities (35) have components equal to normalized q-Krawtchouk polynomials, and the corresponding energy eigenvalues are

$$\epsilon_k = -[-k] = -\frac{1-q^{-k}}{1-q} = q^{-1} + q^{-2} + \cdots + q^{-k}. \tag{36}$$

As before, the correlation function becomes

$$f_{r,s}(t) = \sum_{k=0}^{N} U_{rk}U_{sk}z^{\epsilon_k} = \frac{1}{\sqrt{d_r d_s}} \sum_{k=0}^{N} w(k)\mathcal{K}_r(q^{-k})\mathcal{K}_s(q^{-k})z^{-[-k]} \qquad (z = e^{-it}). \tag{37}$$

In order to evaluate this sum, consider first the simple case when sender and receiver are at different ends of the chain, namely when $s = 0$ and $r = N$. One finds:

$$\sum_{k=0}^{N} w(k)\mathcal{K}_N(q^{-k})\mathcal{K}_0(q^{-k})z^{-[-k]} = \sum_{k=0}^{N} \frac{(q^{-N};q)_k}{(q;q)_k} q^{Nk} z^{-[-k]}. \tag{38}$$

Because of the factor $z^{-[-k]}$, the sum in (38) is not of q-hypergeometric type, hence there is no hope that it can be simplified any further for arbitrary values of z (i.e. for arbitrary values of t). For certain specific values, however, simplification does take place. Since $z = e^{-it}$, one has

$$z^{-[-k]} = e^{-it(q^{-1}+q^{-2}+\cdots+q^{-k})}.$$

Assume now that the deformation parameter q is a rational number of the following form:

$$q^{-1} = \frac{P}{Q}, \quad \text{with } P \text{ and } Q \textit{ odd} \text{ positive integers (having no common factors).} \tag{39}$$

Then, for each value of the index k with $k \leq N$:

$$\begin{aligned} q^{-1} + q^{-2} + \cdots + q^{-k} &= \frac{1}{Q^N}(PQ^{N-1} + P^2Q^{N-2} + \cdots + P^kQ^{N-k}) \\ &= \frac{1}{Q^N} \times k \times (\text{an odd integer}). \end{aligned} \tag{40}$$

Suppose now that we consider the system at time

$$t = T \equiv Q^N\pi,$$

then

$$z^{-[-k]} = e^{-it(q^{-1}+q^{-2}+\cdots+q^{-k})} = e^{-i\pi(PQ^{N-1}+P^2Q^{N-2}+\cdots+P^kQ^{N-k})} = (-1)^k.$$

In this case, the expression (38) simplifies drastically, since on can apply the q-binomial theorem [19, (II.4)]:

$$\sum_{k=0}^{N} \frac{(q^{-N};q)_k}{(q;q)_k} q^{Nk}(-1)^k = (-1;q)_N = \prod_{j=0}^{N-1}(1+q^j). \tag{41}$$

Taking into account the expressions for d_0 and d_N from (30), one has

$$f_{N,0}(T) = \frac{(-1;q)_N}{\sqrt{d_0 d_N}} = (-1;q)_N \sqrt{\frac{p^N q^{N(N+1)/2}}{(-pq,-pq^N;q)_N}}.$$

It is easy to see that this expression takes its maximum value when $p = q^{-N}$, and in that case

$$f_{N,0}(T)\Big|_{p=q^{-N}} = \sqrt{\frac{q^{-N(N-1)/2}(-1;q)_N}{(-q^{-N+1};q)_N}} = 1,$$

allowing perfect state transfer. So the q-Krawtchouk polynomials with parameter $p = q^{-N}$ and q of the special form (39) yield a new model of a spin chain with perfect state transfer.

Once again, let us illustrate this as an animation, see Figure 6.

animation4

Figure 6. Time evolution of a linear spin chain ($N = 8$) with data determined by the Jacobi matrix of q-Krawtchouk polynomials. Here $p = q^{-N} = q^{-8}$, and $q = Q/P = 3$. Click on the picture to see the animation running, from $t = 0$ up to $t = T = Q^N\pi = 3^8\pi$. Note that there is perfect state transfer. (view as animation4.gif in *Supplementary Data*)

Acknowledgments
This research was supported by project P6/02 of the Interuniversity Attraction Poles Programme (Belgian State – Belgian Science Policy).

References
[1] Bose S 2003 *Phys. Rev. Lett.* **91** 207901
[2] Bose S, Jin B-Q and Korepin V E 2005 *Phys. Rev. A* **72** 022345
[3] Bose S 2007 *Contemp. Phys.* **48** 13
[4] Christandl M, Datta N, Ekert A and Landahl A J 2004 *Phys. Rev. Lett.* **92** 187902
[5] Christandl M, Datta N, Dorlas T C, Ekert A, Kay A and Landahl A J 2005 *Phys. Rev. A* **71** 032312
[6] Paternostro M, Palma G M, Kim M S and Falci G 2005 *Phys. Rev. A* **71** 042311
[7] Kay A 2009 A review of perfect state transfer and its application as a constructive tool (*Preprint* arXiv:0903.4274v2).
[8] Jordan P and Wigner E 1928 *Z. Phys.* **47** 631
[9] Lieb E and Wu F 1968 *Phys. Rev. Lett.* **20** 1445-1448
[10] Albanese C, Christandl M, Datta N and Ekert A 2004 *Phys. Rev. Lett.* **93** 230502
[11] Golub G H and Van Loan C F 1996 *Matrix Computations* (Baltimore: Johns Hopkins University Press)
[12] Chakrabarti R and Van der Jeugt J 2010 *J. Phys. A* **43** 085302
[13] Karbach P and Stolze J 2005 *Phys. Rev. A* **72** 030301
[14] Jafarov E I and Van der Jeugt J 2010 *J. Phys. A* **43** 405301
[15] Bailey W N 1964 *Generalized hypergeometric series* (Cambridge: Cambridge University Press)
[16] Slater L J 1966 *Generalized hypergeometric functions* (Cambridge: Cambridge University Press)
[17] Koekoek R, Lesky P A and Swarttouw R F 2010 *Hypergeometric orthogonal polynomials and their q-analogues* (Berlin: Springer)
[18] Edmonds A R 1960 *Angular momentum in quantum theory* (Princeton: Princeton University Press)
[19] Gasper G and Rahman M 2004 *Basic hypergeometric series* (Cambridge: Cambridge University Press)

Journal of Physics: Conference Series **284** (2011) 012060 doi:10.1088/1742-6596/284/1/012060

d-Orthogonal Charlier Polynomials and the Weyl Algebra

Luc Vinet[1] and Alexei Zhedanov[2]

[1] Centre de recherches mathématiques, Université de Montréal, C.P. 6128, succ. Centre-ville, Montréal, Qc, H3C 3J7, Canada

E-mail: luc.vinet@umontreal.ca

[2] Department of Electronic and Kinetic Properties, Donetsk Institute for Physics and Technology R.Luxemburg str. 72, Donetsk, 83114, Ukraine

E-mail: zhedanov@yahoo.com

Abstract. It is shown that d-orthogonal Charlier polynomials arise as matrix elements of non unitary automorphisms of the Weyl algebra. The structural formulas that these polynomials obey are derived from this algebraic setting.

1. Charlier polynomials

The (normalized) Charlier polynomials $p_n(k)$ are functions of a discrete variable k that satisfy the 3-term recurrence relation [1]

$$p_{n+1}(k) = (k - n - a)p_n(k) - na\, p_{n-1}(k) \tag{1}$$

and obey the orthogonality relation

$$\sum_{k=0}^{\infty} p_m(k)p_n(k)\mathrm{e}^{-a}\frac{a^k}{k!} = a^n n!\delta_{m,n}, \tag{2}$$

with a a real parameter.

2. Weyl algebra

The Weyl algebra is generated by 3 elements a, a^* and I that satisfy the commutation relations

$$[a, a^*] = 1, \quad [a, I] = [a^*, I] = 0. \tag{3}$$

It possesses a canonical irreducible representation defined as follows on basis states $|n\rangle$, $n = 0, 1, 2, \ldots$.

$$a|n\rangle = \sqrt{n}|n-1\rangle, \quad a^*|n\rangle = \sqrt{n+1}|n+1\rangle, \quad I|n\rangle = |n\rangle, \tag{4}$$

$$N = a^*a, \quad N|n\rangle = n|n\rangle, \quad a|0\rangle = 0. \tag{5}$$

3. Charlier polynomials and Weyl algebra

We recall how Charlier polynomials arise in the context of the Weyl algebra [2]. This algebra admits the following oscillator realization:

$$\begin{gathered} a = \frac{1}{\sqrt{2}}\Big(\frac{\mathrm{d}}{\mathrm{d}x} + x\Big), \quad a^* = \frac{1}{\sqrt{2}}\Big(-\frac{\mathrm{d}}{\mathrm{d}x} + x\Big), \\ N = -\frac{1}{2}\frac{\mathrm{d}^2}{\mathrm{d}x^2} + \frac{1}{2}x^2 - \frac{1}{2}. \end{gathered} \tag{6}$$

Now effect the shift $x \to x - \lambda$. It induces the following transformations:

$$\begin{gathered} \widetilde{a}(\lambda) = a - \frac{\lambda}{\sqrt{2}}, \quad \widetilde{a}^*(\lambda) = a^* - \frac{\lambda}{\sqrt{2}}, \\ \widetilde{N}(\lambda) = N - \lambda x + \frac{\lambda^2}{2}. \end{gathered} \tag{7}$$

Let $\widetilde{N}|\widetilde{n}\rangle = n|\widetilde{n}\rangle$ and set

$$|\,k\rangle = \sum_n \widetilde{\langle n|}\,k\rangle |\widetilde{n}\rangle. \tag{8}$$

From

$$N|\,k\rangle = \sum_n \widetilde{\langle n|}\,k\rangle \left[\widetilde{N} + \frac{\lambda}{\sqrt{2}}(\widetilde{a} + \widetilde{a}^*) + \frac{\lambda^2}{2}\right]|\widetilde{n}\rangle \tag{9}$$

it is found that the overlaps $\widetilde{\langle n|}\,k\rangle$ satisfy the 3-term recurrence relation

$$k\widetilde{\langle n|}\,k\rangle = \Big(n + \frac{\lambda^2}{2}\Big)\widetilde{\langle n|}\,k\rangle + \frac{\lambda}{\sqrt{2}}\sqrt{n+1}\,\widetilde{\langle n+1|}\,k\rangle + \frac{\lambda}{\sqrt{2}}\sqrt{n}\,\widetilde{\langle n-1|}\,k\rangle \tag{10}$$

and hence are given in terms of Charlier polynomials

$$\widetilde{\langle n|}\,k\rangle = \Big(\frac{\sqrt{2}}{\lambda}\Big)^n \frac{1}{\sqrt{n!}}\nu(k)p_n(k) \tag{11}$$

where $\nu(k)$ is a normalization factor.

Let us remark that translations obviously provide an automorphism of the Weyl algebra and that various properties of the Charlier polynomials can be derived from the above context (see [2]).

4. d-orthogonal polynomials

Monic d-orthogonal polynomials [3, 4] of degree n are defined through recurrence relations of order $\mathrm{d}+1$ that are of the form

$$P_{n+1}(x) = xP_n(x) - \sum_{s=0}^{\mathrm{d}} a_{n,n-s}P_{n-s}(x) \tag{12}$$

with initial conditions

$$P_k(x) = 0, \quad k < 0; \qquad P_0(x) = 1. \tag{13}$$

For $\mathrm{d} = 1$, one recovers the standard orthogonal polynomials. When $\mathrm{d} > 1$, they satisfy vector orthogonality relations. Namely there exist d linear functionals σ_k, $k = 0, 1, \ldots, \mathrm{d}-1$ such that

$$\begin{gathered} \langle \sigma_k, P_nP_m\rangle = 0, \quad m > \mathrm{d}n + k \\ \langle \sigma_k, P_nP_{\mathrm{d}n+k}\rangle \neq 0, \quad n \geq 0. \end{gathered} \tag{14}$$

These have applications in various areas, in particular in Padé approximation problems and in random matrix theory.

5. A nonlinear automorphism of the Weyl algebra

We shall show that such polynomials also arise as matrix elements of certain automorphisms of the Weyl algebra [5].

Take this automorphism to be

$$S = \mathrm{e}^{\beta a^*}\,\mathrm{e}^{Q(a)}, \quad S^{-1} = \mathrm{e}^{-Q(a)}\,\mathrm{e}^{-\beta a^*} \tag{15}$$

where Q is a polynomial of degree $N > 1$ and β a parameter.

We have

$$\begin{aligned} SaS^{-1} &= a - \beta, & S^{-1}aS &= a + \beta, \\ Sa^*S^{-1} &= a^* + Q'(a-\beta), & S^{-1}a^*S &= a^* - Q'(a). \end{aligned} \tag{16}$$

6. d-Charlier polynomials

We shall calculate the matrix element

$$\psi_{nk} = \langle k|S|n\rangle \tag{17}$$

by obtaining for it a recurrence relation. From

$$\begin{aligned} \langle k|a^*aS|n\rangle &= \langle k|SS^{-1}a^*aS|n\rangle \\ &= \langle k|S\big(a^* - Q'(a)\big)(a+\beta)|n\rangle \end{aligned} \tag{18}$$

and using the fact that $Q'(x)(x+\beta)$ is a polynomial of degree N if this is so for Q, we find

$$k\psi_{n,k} = (n - \xi_0)\psi_{n,k} + \beta\sqrt{n+1}\psi_{n+1,k} - \sum_{i=1}^{N}\xi_i\sqrt{n(n-1)\dots(N-i+1)}\psi_{n-i,k} \tag{19}$$

where the coefficients ξ_i are defined by the expansion

$$Q'(x)(x+\beta) = \sum_{i=1}^{N}\xi_i x^i. \tag{20}$$

Formula (19) gives $\psi_{n,k}$ recursively from $\psi_{0,k}$ and we can write

$$\psi_{n,k} = P_n(k)\psi_{0,k} \tag{21}$$

where $P_n(k)$ is a polynomial that satisfy

$$\beta\sqrt{n+1}P_{n+1}(k) = (k - n + \xi_0)P_n(k) + \sum_{i=1}^{N}\xi_i\sqrt{n(n-1)\dots(n-i+1)}P_{n-i}(k),$$
$$P_0(k) = 1, \quad P_n(k) = 0, \qquad n < 0 \tag{22}$$

that is a recurrence relation of order $N+1$.

The matrix element $\psi_{0,k}$ is straightforwardly found to be

$$\psi_{0,k} = \mathrm{e}^{Q(0)}\frac{\beta^k}{\sqrt{k!}}. \tag{23}$$

To make the polynomials monic, set

$$\widehat{P}_n(k) = \beta^n \sqrt{n!} P_n(k) \tag{24}$$

so as to find the defining $N+1$ recurrence relation

$$\widehat{P}_{n+1}(k) = (k - n + \xi_0)\widehat{P}_n(k) + \sum_{i=1}^{N} \beta^i \xi_i n(n-1)\ldots(n-i+1)\widehat{P}_{n-1}(k) \tag{25}$$

for the d-Charlier polynomials [6].

For $N = 1$, we recover the relation for the ordinary Charlier polynomials (1) up to the shift $k \to k + \beta\xi_1 - \xi_0$.

7. Properties

The characterization of the d-Charlier polynomials is easily performed using the algebraic tools now at our disposal.

• Difference equation

Start in this case from the "dual" recurrence relation in the variable k that the following equation entail

$$\begin{aligned} \langle k|Sa^* a|n\rangle &= \langle k|Sa^* a S^{-1} S|n\rangle \\ &= \langle k\big(a^* + Q'(a-\beta)\big)(a-\beta)S|n\rangle \end{aligned} \tag{26}$$

to find that

$$(n-k)\widehat{P}_n(k) + k\widehat{P}_n(k-1) = \sum_{i=0}^{N} \eta_i \beta^i \widehat{P}_n(k+1). \tag{27}$$

Here the parameters η_i are defined through the expansion

$$(x-\beta)Q'(x-\beta) = \sum_{i=1}^{N} \eta_i x^i. \tag{28}$$

Using the difference operators

$$\Delta f(k) = f(k+1) - f(x), \quad \nabla f(k) = f(k) - f(k-1) \tag{29}$$

and the identity

$$F(x+m) = \sum_{i=0}^{m} \binom{m}{i} \Delta^i F(x). \tag{30}$$

Eq. (27) becomes

$$n\widehat{P}_n(k) = k\nabla \widehat{P}_n(k) + \sum_{i=0}^{N} \eta_i \beta^i \sum_{\ell=0}^{i} \binom{i}{\ell} \Delta^\ell \widehat{P}_n(k). \tag{31}$$

Recalling (28), we have

$$\begin{aligned} yQ'(y) &= \sum_{i=0}^{N} \eta_i (y+\beta)^i = \sum_{i=0}^{N} \eta_i \sum_{\ell=0}^{i} \binom{i}{\ell} y^\ell \beta^{i-\ell} \\ &= \sum_{i=0}^{N} \eta_i \beta^i \sum_{\ell=0}^{i} \binom{i}{\ell} (y/\beta)^\ell \equiv \sum_{j=0}^{N} \mu_j (y/\beta)^j \end{aligned} \tag{32}$$

Journal of Physics: Conference Series **284** (2011) 012060 doi:10.1088/1742-6596/284/1/012060

thus defining new parameters μ_i.

As a result the d-Charlier polynomials $\widehat{P}_n$ are observed to satisfy a finite difference eigenvalue equation

$$H\widehat{P}_n = n\widehat{P}_n \tag{33}$$

with

$$H = k\nabla + \sum_{i=0}^{N} \mu_i \Delta^i. \tag{34}$$

• Ladder operators

The relation

$$\langle k|Sa|n\rangle = \langle k|(a-\beta)S|n\rangle \tag{35}$$

is easily seen to give

$$\Delta\widehat{P}_n(k) = n\widehat{P}_{n-1}(k) \tag{36}$$

thereby showing that the polynomials $\widehat{P}_n(k)$ satisfy the difference Appell property.

Similarly

$$\langle k|Sa^*|n\rangle = \langle k|\big(a^* + Q'(a-\beta)\big)S|n\rangle \tag{37}$$

yields the Rodrigues' formula

$$\widehat{P}_n(k) = R^n\{1\} \tag{38}$$

with

$$R = kT_- + \sum_{i=0}^{N-1} \nu_i \beta^{i+1} T_+^i \tag{39}$$

where $Q'(x-\beta) = \sum_{i=0}^{N-1} \nu_i x^i$ and $T_\pm F(k) = F(k \pm 1)$.

• Generating function

Consider the generating function

$$F(z;k) = \mathrm{e}^{-Q(0)} \beta^{-k} \sqrt{k!} \sum_{n=0}^{\infty} \frac{\psi_{n,k} z^n}{\sqrt{n!}}. \tag{40}$$

Recalling that

$$\psi_{n,k} = \mathrm{e}^{Q(0)} \frac{\beta^{k-n}}{\sqrt{n!k!}} \widehat{P}_n(k) \tag{41}$$

it yields

$$F(z,k) = \sum_{n=0}^{\infty} \frac{1}{n!} \left(\frac{z}{\beta}\right)^n \widehat{P}_n(k). \tag{42}$$

The oscillator coherent states enjoy the following properties:

$$\begin{aligned} |z\rangle &= \mathrm{e}^{za^*}|0\rangle, & \langle n|z\rangle &= \frac{z^n}{\sqrt{n!}} \\ a|z\rangle &= z|z\rangle, & F(a)|z\rangle &= F(z)|z\rangle. \end{aligned} \tag{43}$$

In terms of these coherent states, $F(z,k)$ can be written

$$F(z,k) = \mathrm{e}^{-Q(0)} \beta^{-k} \sqrt{k!} \langle k|S|z\rangle. \tag{44}$$

Journal of Physics: Conference Series **284** (2011) 012060 doi:10.1088/1742-6596/284/1/012060

The matrix element $\langle k|S|z\rangle$ is easily computed using (43):

$$\langle k|\,\mathrm{e}^{\beta a^*}\,\mathrm{e}^{Q(a)}|z\rangle = \mathrm{e}^{Q(z)}\langle k|z+\beta\rangle = \mathrm{e}^{\varphi(x)}\frac{(z+\beta)^k}{\sqrt{k!}} \tag{45}$$

to find the following generating function for the d-Charlier polynomials:

$$\mathrm{e}^{Q(z)-Q(0)}\left(1+\frac{z}{\beta}\right)^k = \sum_{n=0}^{\infty}\left(\frac{z}{\beta}\right)^n \frac{\widehat{P}_n(k)}{n!}. \tag{46}$$

• Characterization

It can be proved that the $N+1$ order recurrence relation, the matrix elements of S, the Appell difference property, the eigenvectors of H can all be derived from one another and represent equivalent characterizations of d-orthogonal Charlier polynomials.

8. Explicit expression in a special case

For the special case where the polynomial $Q(x)$ is a monomial, i.e.

$$Q(x) = \sigma x^N \tag{47}$$

$$F(z;k) = \left(1+\frac{z}{\beta}\right)^k \exp(\sigma z^N) = \sum_{n=0}^{\infty}\left(\frac{z}{\beta}\right)^n \frac{\widehat{P}_n(k)}{n!}. \tag{48}$$

Expanding F and performing some transformations give the explicit expression for $\widehat{P}_n(k)$:

$$\begin{gathered}\widehat{P}_n(k) = \frac{(-1)^q\beta^{Nj}\sigma^j(-k)_q}{q!j!}\,{}_{N+1}F_{N-1}\left(\begin{matrix}-j,\{a_m\}\\ \{b_m\}\end{matrix}\;;\frac{(-1)^{N+1}}{\sigma\beta^n}\right)\\ j=0,1,\ldots \qquad q=0,1,\ldots N-1\\ a_m=(q-k+m)/N, \quad b_m=(q+m+1)/N\\ m=0,\ldots,N-1, \quad (-k)_q=(-k)(-k+1)\ldots(-k+q+1).\end{gathered} \tag{49}$$

For $N=1$, the standard expression

$$\widehat{P}_n(k) = \frac{\sigma^n\beta^n}{n!}\,{}_2F_0\left(\begin{matrix}-n,-k\\ -\end{matrix}\;;\frac{1}{\sigma\beta}\right) \tag{50}$$

for the ordinary Charlier polynomial is recovered.

9. Dual functions and orthogonality

Much more can be obtained from this algebraic framework. Most interesting is probably what it allows to say about the orthogonality properties of the d-orthogonal polynomials that the matrix elements $\psi_{n,k} = \langle k|S|n\rangle$ entail.

Assume S invertible and define

$$\varphi_{n,k} = \langle n|S^{-1}|k\rangle. \tag{51}$$

It can be shown, proceeding as above, that the matrix elements are similarly expressible in terms of d-Charlier polynomials, albeit with different parameters. A generating function for $\varphi_{n,k}$ can also be obtained.

As a result, the identity

$$\sum_{k=0}^{\infty} \varphi_{m,k}\psi_{n,k} = \sum_{k=0}^{\infty} \langle m|S^{-1}|k\rangle \langle k|S|n\rangle = \delta_{m,n} \tag{52}$$

is tantamount to a biorthogonality relation for the d-Charlier polynomials.

Moreover, it is possible to relate this biorthogonality between $\varphi_{m,k}$ and $\psi_{n,k}$ to the vector orthogonality of the underlying d-orthogonal polynomials. In fact, from the biorthogonality relation and the knowledge of the matrix elements $\varphi_{m,k}$ and $\psi_{n,k}$, it is possible to construct explicit realizations of the linear functionals expressing the vector orthogonality. (See ref. [5]).

References

[1] Koekoek R and Swarttouw R F 1994 The Askey-scheme of hypergeometric orthogonal polynomials and its q-analogue Tech. Rep. 94-05 Delft University of Technology
[2] Floreanini R, LeTourneux J and Vinet L 1993 *Ann. Phys.* **226** 331
[3] van Iseghem J 1987 *J. Comput. Appl. Math.* **19** 141
[4] Maroni P 1989 *Ann. Fac. Sci. Toulouse Math.* **10** 105
[5] Vinet L and Zhedanov A 2009 *J. Math. Phys.* **50** 33511
[6] Cheikh Y B and Zaghouani A 2003 *J. Comput. Appl. Math.* **156** 253

Journal of Physics: Conference Series **284** (2011) 012061 doi:10.1088/1742-6596/284/1/012061

Engineering of quantum systems with variables in $GF(p^\ell)$

A. Vourdas

Department of Computing,

University of Bradford,

Bradford BD7 1DP, United Kingdom

A system comprised of ℓ subsystems, each of which is described with variables in $\mathbb{Z}_p$, is considered. Position and momentum states in this ℓ-partite system can be labelled with elements in $GF(p^\ell)$. It is shown that the whole quantum formalism can be expressed in terms of Galois arithmetic, if and only if the Hamiltonian belongs to a particular subset of the set of all Hamiltonians of this ℓ-partite system (we say that these Hamiltonians are compatible with $GF(p^\ell)$). Several examples of such Hamiltonians are presented.

I. INTRODUCTION

There is currently a lot of interest on finite quantum systems i.e., systems with positions and momenta in the ring $\mathbb{Z}_d$ of integers modulo d (reviews are presented in [1, 2]). When $d = p$ where p is a prime number, $\mathbb{Z}_p$ is a field and the system has stronger properties. (e.g., the number of mutually unbiased bases is equal to the maximum possible value $d + 1$ [3–9]).

In this general line of research we study systems with variables in a Galois field $GF(p^\ell)$ [10]. The aim is firstly to understand their properties, secondly to engineer such systems, and thirdly to use them for applications in the general area of information processing (both classsical and quantum).

With regard to quantum engineering of these systems, in recent work [11] we have considered a system comprised of ℓ subsystems, each of which is described with variables in $\mathbb{Z}_p$ (e.g., spins with $j = (p-1)/2$). Position and momentum states in this ℓ-partite system can be labelled with elements in $GF(p^\ell)$. We have shown that the whole formalism can be expressed in terms of Galois arithmetic, if and only if the Hamiltonian belongs to a particular subset of the set of all Hamiltonians of these systems. We say that these Hamiltonians are compatible with $GF(p^\ell)$.

$GF(p^\ell)$ is similar to $[\mathbb{Z}_p]^\ell$ with regard to addition, but there is a lot of extra structure in $GF(p^\ell)$ related to multiplication and also other operations like the trace. Consequently, position and momentum states in the ℓ-partite system which are labelled with elements in $[\mathbb{Z}_p]^\ell$, can also be labelled with elements in $GF(p^\ell)$. But only some of the Hamiltonians of the ℓ-partite system can accommodate this extra mathematical structure, i.e., they are compatible with $GF(p^\ell)$. Related problem (which is also a problem in its own right) is the study of transformations of various quantum mechanical quantities, as the basis in $GF(p^\ell)$ changes.

In this paper we review briefly and extend this work, with emphasis on the physical aspects and with a minimum of technical details in terms of Galois theory. We also give several examples of specific Hamiltonians which are compatible with $GF(p^\ell)$. They can be used for engineering these systems. In this sense the paper aims to bridge the gap between the mathematical aspects and the applied/engineering aspects of these systems.

In section II we present some aspects of Galois fields which are needed later. In particular we discuss transformations of the basis in $GF(p^\ell)$. In section III we present very briefly quantum systems with variables in $\mathbb{Z}_p$ in order to establish the notation. In section IV we discuss quantum systems with positions and momenta in $GF(p^\ell)$, and show that a change in the basis in $GF(p^\ell)$ is related to unitary transformations of the quantum states. In section V we give Hamiltonians which are compatible with $GF(p^\ell)$. The emphasis here is on examples rather than general theorems which have been presented in [11]. We conclude in section VI with a discussion of our results.

II. GALOIS FIELDS

We consider the Galois field $GF(p^\ell)$ (where p is a prime number and $p \neq 2$). Its elements can be written in terms of the 'indeterminate' ϵ as

$$\alpha = \sum_{\lambda=0}^{\ell-1} \alpha_\lambda \epsilon^\lambda; \qquad \alpha_\lambda \in \mathbb{Z}_p, \tag{1}$$

and they are defined modulo an irreducible polynomial

$$P(\epsilon) \equiv c_0 + c_1\epsilon + ... + c_{\ell-1}\epsilon^{\ell-1} + \epsilon^\ell; \qquad c_i \in \mathbb{Z}_p. \tag{2}$$

We use here the polynomial basis $\{1, \epsilon, ..., \epsilon^{\ell-1}\}$. We can go to another basis[12]

$$\mathcal{E}_\kappa = \sum_\lambda \epsilon^\lambda (V^{-1})_{\lambda\kappa}. \tag{3}$$

where V is a matrix in $GL(\ell, \mathbb{Z}_p)$ (the group of $\ell \times \ell$ invertible matrices with elements in $\mathbb{Z}_p$). Then

$$\alpha = \sum_{\lambda=0}^{\ell-1} A_\lambda \mathcal{E}_\lambda; \qquad A_\lambda = \sum_{\kappa=0}^{\ell-1} V_{\lambda\kappa}\alpha_\kappa. \tag{4}$$

Given a basis $\mathcal{E}_\kappa$ we define its dual basis $\overline{\mathcal{E}}_\kappa$ through the relation

$$\mathrm{Tr}(\overline{\mathcal{E}}_\kappa \mathcal{E}_\lambda) = \delta_{\kappa\lambda}, \tag{5}$$

Then

$$\alpha = \sum_{\lambda=0}^{\ell-1} \overline{A}_\lambda \overline{\mathcal{E}}_\lambda; \qquad \overline{A}_\lambda = \mathrm{Tr}[\alpha \mathcal{E}_\lambda] = \sum_{\kappa=0}^{\ell-1} \overline{\alpha}_\kappa (V^{-1})_{\kappa\lambda}. \tag{6}$$

We use the notation $\{E_\lambda\}$ for the basis which is dual to ϵ^λ (i.e., $\mathrm{Tr}(\epsilon^\kappa E_\lambda) = \delta_{\kappa\lambda}$).

For later use we define the symmetric matrices g_V and G_V:

$$(g_V)_{\kappa\lambda} = \mathrm{Tr}(\mathcal{E}_k \mathcal{E}_\lambda); \qquad G_V = g_V^{-1} = \mathrm{Tr}(\overline{\mathcal{E}}_k \overline{\mathcal{E}}_\lambda) \tag{7}$$

We use the index V in the notation, because these matrices are defined with respect to the basis $\mathcal{E}_k$, which is defined through the matrix V. We denote the matrices corresponding to the polynomial basis of Eq.(1), as g and G. Then

$$g_V = (V^{-1})^T g V^{-1}; \qquad G_V = g_V^{-1} = V G V^T \tag{8}$$

A. Trace

The trace of $\alpha \in GF(p^\ell)$ is defined as:

$$\mathrm{Tr}(\alpha) = \alpha + \alpha^p + ... + \alpha^{p^{\ell-1}}; \qquad \mathrm{Tr}(\alpha) \in \mathbb{Z}_p. \tag{9}$$

The trace does not depend on the basis.

We can express the trace of the product of $\alpha, \beta \in GF(p^\ell)$ in terms of their components as

$$\mathrm{Tr}(\alpha\beta) = \sum_{\lambda,\kappa} (g_V)_{\lambda\kappa} A_\lambda B_\kappa = \sum_{\lambda,\kappa} (G_V)_{\lambda\kappa} \overline{A}_\lambda \overline{B}_\kappa = \sum_\lambda A_\lambda \overline{B}_\lambda. \tag{10}$$

More generally we introduce the symmetric tensors

$$\left(g_V^{(N)}\right)_{\lambda_1...\lambda_N} \equiv \mathrm{Tr}\left[\mathcal{E}_{\lambda_1}...\mathcal{E}_{\lambda_N}\right]; \qquad \left(G_V^{(N)}\right)_{\lambda_1...\lambda_N} \equiv \mathrm{Tr}\left[\overline{\mathcal{E}}_{\lambda_1}...\overline{\mathcal{E}}_{\lambda_N}\right]; \qquad \lambda_i = 0, ..., \ell - 1 \tag{11}$$

and express the trace of a product of N elements of $GF(p^\ell)$, as

$$\mathrm{Tr}\left[\alpha^{(1)}...\alpha^{(N)}\right] = \sum \left(g_V^{(N)}\right)_{\lambda_1...\lambda_N} A_{\lambda_1}^{(1)}...A_{\lambda_N}^{(N)} = \sum \left(G_V^{(N)}\right)_{\lambda_1...\lambda_N} \overline{A}_{\lambda_1}^{(1)}....\overline{A}_{\lambda_N}^{(N)}. \tag{12}$$

As above, the index V in the notation indicates the basis used for these tensors (if there is no index it means that the polynomial basis of Eq.(1) has been used, i.e., $V = \mathbf{1}$).

We call $\mathfrak{G}(V)$ and $\overline{\mathfrak{G}}(V)$ the following sets of tensors

$$\mathfrak{G}(V) = \{g_V, g_V^{(3)}, g_V^{(4)}, ...\}; \qquad \overline{\mathfrak{G}}(V) = \{G_V, G_V^{(3)}, G_V^{(4)}, ...\}. \tag{13}$$

B. Characters

Additive characters in $GF(p^\ell)$ are given by

$$\chi(\alpha) = \omega[\mathrm{Tr}(\alpha)]; \tag{14}$$

where we use the notation

$$\omega(m) = \exp\left(i\frac{2\pi m}{p}\right); \qquad m \in \mathbb{Z}_p. \tag{15}$$

Using Eq.(10) we get

$$\chi(\alpha\beta) = \omega\left[\sum_{\lambda,\kappa} (g_V)_{\lambda\kappa} A_\lambda B_\kappa\right] = \omega\left[\sum_{\lambda,\kappa} (G_V)_{\lambda\kappa} \overline{A}_\lambda \overline{B}_\kappa\right] = \omega\left[\sum_\lambda A_\lambda \overline{B}_\lambda\right]. \tag{16}$$

More generally Eq.(12) leads to

$$\chi\left[\alpha^{(1)}...\alpha^{(N)}\right] = \omega\left[\sum\left(g_V^{(N)}\right)_{\lambda_1...\lambda_N} A_{\lambda_1}^{(1)}...A_{\lambda_N}^{(N)}\right] = \omega\left[\sum\left(G_V^{(N)}\right)_{\lambda_1...\lambda_N} \overline{A}_{\lambda_1}^{(1)}....\overline{A}_{\lambda_N}^{(N)}\right]. \tag{17}$$

We also define characters of diagonal matrices with elements in $GF(p^\ell)$. Let Θ be the $N \times N$ diagonal matrix

$$\Theta = \mathrm{diag}(\theta_{ii}). \tag{18}$$

The character of Θ is defined to be the $N \times N$ complex diagonal matrix

$$\chi(\Theta) = \mathrm{diag}[\chi(\theta_{ii})]. \tag{19}$$

The matrix Θ can also be written as

$$\Theta = \Theta_0\mathcal{E}_0 + ... + \Theta_{\ell-1}\mathcal{E}_{\ell-1}, \tag{20}$$

where Θ_λ are diagonal matrices with elements in $\mathbb{Z}_p$. The Galois trace of Θ is a diagonal matrix with elements in $\mathbb{Z}_p$, defined by

$$\mathrm{Tr_G}\Theta = \Theta + \Theta^p + ... + \Theta^{p^{\ell-1}}. \tag{21}$$

Then the character of Θ can be expressed as

$$\chi(\Theta) = \omega\left(\mathrm{Tr_G}\Theta\right). \tag{22}$$

III. QUANTUM SYSTEMS WITH VARIABLES IN $\mathbb{Z}_p$

We consider a quantum system with positions and momenta in $\mathbb{Z}_p$, described by a p-dimensional Hilbert space $\mathcal{H}$. Let $|\mathcal{X}; m\rangle$ be the orthonormal basis of position states and $|\mathcal{P}; m\rangle$ the orthonormal basis of momentum states. Here $m \in \mathbb{Z}_p$ and the $\mathcal{X}$,$\mathcal{P}$ are used to indicate position and momentum bases. They are related through the Fourier transform

$$|\mathcal{P}; m\rangle = \mathcal{F}|\mathcal{X}; m\rangle; \qquad \mathcal{F} = p^{-1/2}\sum_{m=0}^{p-1}\sum_{n=0}^{p-1}\omega(mn)|\mathcal{X}; m\rangle\langle\mathcal{X}; n| \tag{23}$$

Position and momentum operators are defined as

$$\mathcal{Q} = \sum_m m|\mathcal{X}; m\rangle\langle\mathcal{X}; m|; \qquad \mathcal{P} = \sum_m m|\mathcal{P}; m\rangle\langle\mathcal{P}; m|. \tag{24}$$

Displacement operators are defined as $\mathcal{D}(\alpha, \beta) = \mathcal{Z}^\alpha\mathcal{X}^\beta\omega(-2^{-1}\alpha\beta)$ where

$$\begin{aligned} \mathcal{Z} &= \omega(\mathcal{Q}); \qquad \mathcal{X} = \omega(-\mathcal{P}) \\ \mathcal{X}^p &= \mathcal{Z}^p = \mathbf{1}; \qquad \mathcal{X}^\beta\mathcal{Z}^\alpha = \mathcal{Z}^\alpha\mathcal{X}^\beta\omega(-\alpha\beta); \qquad \alpha, \beta \in \mathbb{Z}_p. \end{aligned} \tag{25}$$

The $\mathcal{D}(\alpha, \beta)\omega(\gamma)$ form a representation of the Heiseberg-Weyl group.

IV. QUANTUM SYSTEMS WITH VARIABLES IN $GF(p^\ell)$

We consider an ℓ-partite system comprised of ℓ component systems of the type described in section III. Its Hilbert space $H = \mathcal{H} \otimes ... \otimes \mathcal{H}$ is p^ℓ-dimensional. Using the polynomial basis in $GF(p^\ell)$, we label the position state $|\mathcal{X}; m_0\rangle \otimes ... \otimes |\mathcal{X}; m_{\ell-1}\rangle$ in H, as follows:

$$|X;m\rangle \equiv |\mathcal{X};m_0\rangle \otimes ... \otimes |\mathcal{X};m_{\ell-1}\rangle; \qquad m = \sum_{\kappa=0}^{\ell-1} m_\kappa \epsilon^\kappa; \qquad m_\kappa \in \mathbb{Z}_p. \tag{26}$$

The Fourier operator, is given by

$$\begin{aligned} F &= p^{-\ell/2} \sum_{m,n \in GF(p^\ell)} \chi(mn)|X;m\rangle\langle X;n| \\ &= p^{-\ell/2} \sum \omega \left[\sum g_{\kappa\lambda} m_\kappa n_\lambda\right] |\mathcal{X};m_0\rangle\langle\mathcal{X};n_0| \otimes ... \otimes |\mathcal{X};m_{\ell-1}\rangle\langle\mathcal{X};n_{\ell-1}| \end{aligned} \tag{27}$$

where $m = \sum m_\kappa \epsilon^\kappa$ and $n = \sum n_\kappa \epsilon^\kappa$.

The momentum states are given by

$$|P;m\rangle = F|X;m\rangle = p^{-\ell/2} \sum_n \chi(mn)|X;n\rangle \tag{28}$$

They can be expressed in terms of momentum states in the various component systems as

$$|P;m\rangle = |\mathcal{P};\overline{m}_0\rangle \otimes ... \otimes |\mathcal{P};\overline{m}_{\ell-1}\rangle. \tag{29}$$

The dual components $\overline{m}_i$ of m appear in the momentum states (see Eq.(16)).

The position and momentum operators are given by

$$Q = \sum_m m|X;m\rangle\langle X;m|; \qquad P = \sum_m m|P;m\rangle\langle P;m| = FQF^\dagger \tag{30}$$

For later use we define characters of powers of Q, P:

$$\chi(\alpha Q^r) = \sum_{m \in GF(p^\ell)} \chi(\alpha m^r)|X;m\rangle\langle X;m|; \qquad \chi(\beta P^r) = \sum_{m \in GF(p^\ell)} \chi(\beta m^r)|P;m\rangle\langle P;m| \tag{31}$$

where $\alpha, \beta \in GF(p^\ell)$. They are $p^\ell \times p^\ell$ complex matrices.

Displacement operators are given by:

$$Z^\alpha = \chi(\alpha Q); \qquad X^\beta = \chi(-\beta P), \tag{32}$$

We can prove that

$$X^\beta Z^\alpha = Z^\alpha X^\beta \chi(-\alpha\beta). \tag{33}$$

General displacement operators are given by:

$$D(\alpha,\beta) = Z^\alpha X^\beta \chi(-2^{-1}\alpha\beta). \tag{34}$$

The displacement operators can be expressed in terms of displacement operators acting on the various subsystems as:

$$D(\alpha,\beta) = \mathcal{D}(\overline{\alpha}_0,\beta_0)\otimes ... \otimes \mathcal{D}(\overline{\alpha}_{\ell-1},\beta_{\ell-1}); \qquad \alpha = \sum_\lambda \overline{\alpha}_\lambda E_\lambda; \qquad \beta = \sum_\lambda \beta_\lambda \epsilon_\lambda, \tag{35}$$

where $\overline{\alpha}_\lambda$ are the dual components of α, and β_λ the components of β.

In this section up to now, we used a labelling method with the polynomial basis of $GF(p^\ell)$. We can go to a labelling method with a different basis, with the unitary transformation

$$U_V = \sum_{n\in GF(p^\ell)} |X; m(n)\rangle\langle X; n|; \qquad m(n) = \sum \epsilon^\kappa \mathrm{Tr}(n\overline{\mathcal{E}}_\kappa) \tag{36}$$

For example

$$|X;V;n\rangle \equiv U_V|X;n\rangle = |X;m(n)\rangle = |\mathcal{X};\mathrm{Tr}(n\overline{\mathcal{E}}_0)\rangle \otimes ... \otimes |\mathcal{X};\mathrm{Tr}(n\overline{\mathcal{E}}_{\ell-1})\rangle \tag{37}$$

Since $\sum \mathcal{E}_\kappa \mathrm{Tr}(n\overline{\mathcal{E}}_\kappa) = n$, we label this state as $|X;V;n\rangle$. The V in the notation indicates that the basis $\{\mathcal{E}_\kappa\}$ (defined through V in Eq.(3)) has been used in the labelling method. We can also introduce the displacecement operators

$$D_V(\alpha,\beta) \equiv U_V D(\alpha,\beta)U_V^\dagger = D(\alpha',\beta'); \qquad \alpha' = \sum E_\kappa \mathrm{Tr}(\alpha\mathcal{E}_\kappa); \qquad \beta' = \sum \epsilon^\kappa \mathrm{Tr}(\beta\overline{\mathcal{E}}_\kappa). \tag{38}$$

The U_V transformations form a unitary representation of the $GL(\ell,\mathbb{Z}_p)$ group.

V. HAMILTONIANS

We consider a system comprised of ℓ subsystems, each of which is described with variables in $\mathbb{Z}_p$. We use the notation

$$\begin{aligned} \mathfrak{Q}_\lambda &= \mathbf{1}\otimes ... \otimes \mathbf{1} \otimes \mathcal{Q}^{(\lambda)} \otimes \mathbf{1} \otimes ... \otimes \mathbf{1} \\ \mathfrak{P}_\lambda &= \mathbf{1}\otimes ... \otimes \mathbf{1} \otimes \mathcal{P}^{(\lambda)} \otimes \mathbf{1} \otimes ... \otimes \mathbf{1}, \end{aligned} \tag{39}$$

It is convenient to express the Hamiltonian of this system as

$$h = h\left[f_n\left(\sigma^{(n)}\right), f'_n\left(\tau^{(n)}\right)\right] \tag{40}$$

where $f_n\left(\sigma^{(n)}\right)$, $f'_n\left(\tau^{(n)}\right)$ are the complex matrices

$$\begin{aligned} f_n\left(\sigma^{(n)}\right) &= \omega\left[\sum \sigma^{(n)}_{\lambda_1...\lambda_n} \mathfrak{Q}_{\lambda_1}...\mathfrak{Q}_{\lambda_n}\right] \\ f'_n\left(\tau^{(n)}\right) &= \omega\left[\sum \tau^{(n)}_{\lambda_1...\lambda_n} \mathfrak{P}_{\lambda_1}...\mathfrak{P}_{\lambda_n}\right] \end{aligned} \tag{41}$$

and the $\sigma^{(n)}, \tau^{(n)}$ are symmetric tensors which take values in $\mathbb{Z}_p$. The $\sum \sigma^{(n)}_{\lambda_1...\lambda_n} \mathfrak{Q}_{\lambda_1}...\mathfrak{Q}_{\lambda_n}$ and $\sum \tau^{(n)}_{\lambda_1...\lambda_n} \mathfrak{P}_{\lambda_1}...\mathfrak{P}_{\lambda_n}$ are matrices with elements in $\mathbb{Z}_p$ whilst the $f_n\left(\sigma^{(n)}\right)$ and $f'_n\left(\tau^{(n)}\right)$ are complex matrices. It is convenient to work with complex matrices and for this reason we have written the Hamiltonian as a function of the latter.

In [11] we have proved that the Hamiltonian of Eq.(40) is compatible with $GF(p^\ell)$ if and only if there exists $V \in GL(\ell, \mathbb{Z}_p)$ (which does not depend on n) such that $\sigma^{(n)} \in \mathfrak{G}(V)$ and $\tau^{(n)} \in \overline{\mathfrak{G}}(V)$, for all n that enter in the Hamiltonian. In this case the Hamiltonian of Eq.(40) can be written in terms of the characters in Eq.(31), as

$$h = h\left[\chi(\alpha_N Q_V^N), \chi(\beta_N P_V^N)\right] \tag{42}$$

Here the Q_V^N, P_V^N are matrices with elements in $GF(p^\ell)$ but they enter in the Hamiltonian through their characters which are finite complex matrices. Then the time evolution operator is $\exp(ith)$. Examples are

$$\begin{aligned} \exp(ith_1) &= \left[\chi(\alpha Q_V^2)\, \chi(\beta P_V^2)\right]^t \\ \exp(ith_2) &= \left[\chi(\alpha P_V^2)\, \chi(\beta Q_V^2)\right]^t = F \exp(ith_1) F^\dagger \\ \exp(ith_3) &= \left[\chi(\alpha P_V^2)\, \chi(\beta Q_V^2)\, \chi(\gamma Q_V^4)\right]^t \end{aligned} \tag{43}$$

The $\exp(ith_1)$ and $\exp(ith_2)$ contain quadratic terms of the position and momentum operators, while the $\exp(ith_3)$ also contains a quartic term of the position operator. There is no simple analytic relation between $\exp(ith_1)$ and $\exp(ith_2)$ (as in the harmonic oscillator case), but they are related through a Fourier transform, and therefore their eigenvalues are the same. We note that Eqs.(43), involve a real power of a complex matrix which is defined through logarithms and therefore is multivalued, but the principal value can be used.

A. Example with the $GF(9)$ in the polynomial basis

We consider the $GF(9)$ (i.e., $p = 3$ and $\ell = 2$) and its polynomial basis $1, \epsilon$. We choose the irreducible polynomial $P(\epsilon) = \epsilon^2 + \epsilon + 2$ and we calculate the matrices g, G:

$$g = \begin{pmatrix} -1 & -1 \\ -1 & 0 \end{pmatrix}; \quad G = \begin{pmatrix} 0 & -1 \\ -1 & 1 \end{pmatrix} \tag{44}$$

The dual basis in this case is $E_0 = -\epsilon$, $E_1 = -1 + \epsilon$ and an arbitrary element $m \in GF(9)$, can be written as

$$m = m_0 + \epsilon m_1 = E_0 \overline{m}_0 + E_1 \overline{m}_1; \quad \overline{m}_0 = -m_0 - m_1; \quad \overline{m}_1 = -m_0. \tag{45}$$

A quantum system with variables in $GF(9)$ can be constructed as a bipartite system where each component system is described with variables in $\mathbb{Z}_3$. We have seen in Eqs(26), (29), that position and momentum states in this bipartite system, can be labelled with elements of $GF(9)$. We now consider Hamiltonians describing this system. We first consider the polynomial basis and Hamiltonians with quadratic terms. Such Hamiltonians are compatible with $GF(9)$ if $h = h[f_2', f_2]$ where

$$\begin{aligned} f_2' &= \omega\left[\sum G_{\lambda\mu} \mathfrak{P}_\lambda \mathfrak{P}_\mu\right] = \omega\left[\mathbf{1} \otimes (\mathcal{P}^{(1)})^2 - 2\mathcal{P}^{(0)} \otimes \mathcal{P}^{(1)}\right] \\ f_2 &= \omega\left[\sum g_{\lambda\mu} \mathfrak{Q}_\lambda \mathfrak{Q}_\mu\right] = \omega\left[-(\mathcal{Q}^{(0)})^2 \otimes \mathbf{1} - 2\mathcal{Q}^{(0)} \otimes \mathcal{Q}^{(1)}\right] \end{aligned} \tag{46}$$

Indeed, let

$$|X;m\rangle = |\mathcal{X};m_0\rangle \otimes |\mathcal{X};m_1\rangle; \qquad |P;m\rangle = |\mathcal{P};\overline{m}_0\rangle \otimes |\mathcal{P};\overline{m}_1\rangle$$
$$Q = (\mathcal{Q}^{(0)} \times \mathbf{1}) + \epsilon(\mathbf{1} \times \mathcal{Q}^{(1)}); \qquad P = E_0(\mathcal{P}^{(0)} \times \mathbf{1}) + E_1(\mathbf{1} \times \mathcal{P}^{(1)}) \tag{47}$$

Then it is easily seen that

$$f_2' = \chi(P^2); \qquad f_2 = \chi(Q^2) \tag{48}$$

where the characters have been defined in Eq.(22). Therefore the bipartite system, is described entirely in terms of variables in $GF(9)$. We note that the two component systems are coupled (the terms $\mathcal{Q}^{(0)} \otimes \mathcal{Q}^{(1)}$ and $\mathcal{P}^{(0)} \otimes \mathcal{P}^{(1)}$ in Eq.(46)).

We next add quartic terms of the position in the Hamiltonian. Then

$$h = h[f_2', f_2, f_4]; \qquad f_4 = \omega\left[\sum g^{(4)}_{\lambda_1\lambda_2\lambda_3\lambda_4}\mathfrak{Q}_{\lambda_1}\mathfrak{Q}_{\lambda_2}\mathfrak{Q}_{\lambda_3}\mathfrak{Q}_{\lambda_4}\right]; \quad \lambda_i = 0, 1 \tag{49}$$

The sum in f_4 has 16 terms but the fact that $g^{(4)}$ is symmetric reduces the number of terms. We first calculate the

$$g^{(4)}_{0000} = \mathrm{Tr}(1) = -1; \quad g^{(4)}_{1000} = \mathrm{Tr}(\epsilon) = -1; \quad g^{(4)}_{1100} = \mathrm{Tr}(\epsilon^2) = 0$$
$$g^{(4)}_{1110} = \mathrm{Tr}(\epsilon^3) = -1; \quad g^{(4)}_{1111} = \mathrm{Tr}(\epsilon^4) = 1 \tag{50}$$

and then

$$f_4 = \omega\left[-(\mathcal{Q}^{(0)})^4 \otimes \mathbf{1} - 4(\mathcal{Q}^{(0)})^3 \otimes \mathcal{Q}^{(1)} - 4\mathcal{Q}^{(0)} \otimes (\mathcal{Q}^{(1)})^3 + \mathbf{1} \otimes (\mathcal{Q}^{(1)})^4\right] = \chi(Q^4) \tag{51}$$

Here the factors 4 are related to the symmetry (but since all coefficients are defined modulo 3, they can be replaced with 1). The term $(\mathcal{Q}^{(0)})^2 \otimes (\mathcal{Q}^{(1)})^2$ is absent because $g^{(4)}_{1100} = 0$.

B. Example with the $GF(9)$ in an arbitrary basis

Let $V \in GL(2, \mathbb{Z}_3)$ be the matrix

$$V = \begin{pmatrix} a & b \\ c & d \end{pmatrix}; \quad V^{-1} = \frac{1}{\Delta}\begin{pmatrix} d & -b \\ -c & a \end{pmatrix}; \quad \Delta = ad - bc \neq 0; \quad a, b, c, d \in \mathbb{Z}_3 \tag{52}$$

and $\mathcal{E}_\kappa$ the corresponding basis defined by Eq.(3). We can calculate the corresponding g_V and G_V using Eqs.(8), and then an arbitrary element $m \in GF(9)$, expressed in eq.(45) in the polynomial basis, can also be written in the $\mathcal{E}_\kappa$ basis and its dual, as

$$m = \mathcal{E}_0 M_0 + \mathcal{E}_1 M_1 = \overline{\mathcal{E}}_0 \overline{M}_0 + \overline{\mathcal{E}}_1 \overline{M}_1$$
$$M_0 = am_0 + bm_1; \qquad M_1 = cm_0 + dm_1$$
$$\overline{M}_0 = \frac{d\overline{m}_0 - c\overline{m}_1}{\Delta}; \quad \overline{M}_1 = \frac{-b\overline{m}_0 + a\overline{m}_1}{\Delta} \tag{53}$$

We also define the

$$
\begin{aligned}
&|X;m\rangle = |\mathcal{X};M_0\rangle \otimes |\mathcal{X};M_1\rangle; \qquad |P;m\rangle = |\mathcal{P};\overline{M}_0\rangle \otimes |\mathcal{P};\overline{M}_1\rangle \\
&Q_V = \mathcal{E}_0(\mathcal{Q}^{(0)} \times \mathbf{1}) + \mathcal{E}_1(\mathbf{1} \times \mathcal{Q}^{(1)}); \qquad P_V = \overline{\mathcal{E}}_0(\mathcal{P}^{(0)} \times \mathbf{1}) + \overline{\mathcal{E}}_1(\mathbf{1} \times \mathcal{P}^{(1)})
\end{aligned} \tag{54}
$$

Here Hamiltonians with quadratic terms, which are compatible with $GF(9)$ are

$$
\begin{aligned}
&h = h[f_2'(\kappa_1,\lambda_1,\mu_1), f_2(\kappa_2,\lambda_2,\mu_2)] \\
&f_2'(\kappa_1,\lambda_1,\mu_1) = \omega\left[\kappa_1(\mathcal{P}^{(0)})^2 \otimes \mathbf{1} + \lambda_1 \mathbf{1} \otimes (\mathcal{P}^{(1)})^2 + \mu_1 \mathcal{P}^{(0)} \otimes \mathcal{P}^{(1)}\right] = \chi(P_V^2) \\
&f_2(\kappa_2,\lambda_2,\mu_2) = \omega\left[\kappa_2(\mathcal{Q}^{(0)})^2 \otimes \mathbf{1} + \lambda_2 \mathbf{1} \otimes (\mathcal{Q}^{(1)})^2 + \mu_2 \mathcal{Q}^{(0)} \otimes \mathcal{Q}^{(1)}\right] = \chi(Q_V^2) \\
&\kappa_1 = b^2 - 2ab; \qquad \lambda_1 = d^2 - 2dc; \qquad \mu_1 = 2(bd - ad - bc) \\
&\kappa_2 = -\frac{\lambda_1}{\Delta^2}; \quad \lambda_2 = -\frac{\kappa_1}{\Delta^2}; \quad \mu_2 = \frac{\mu_1}{\Delta^2}; \qquad \kappa_i, \lambda_i, \mu_i, \Delta \in \mathbb{Z}_3
\end{aligned} \tag{55}
$$

Using various values of $a, b, c, d \in \mathbb{Z}_3$ we get a wide class of Hamiltonians with quadratic terms. These Hamiltonians are compatible with $GF(9)$ and they are a subset of the set of all Hamiltonians with quadratic terms, in the corresponding bipartite system. For example, a Hamiltonian $h[f_2'(\kappa_1,\lambda_1,\mu_1), f_2(\kappa_3,\lambda_3,\mu_3)]$ where the $\kappa_3, \lambda_3, \mu_3$ are different from the $\kappa_2, \lambda_2, \mu_2$, is not compatible with $GF(9)$.

In general the two component systems are coupled (the terms $\mu_2 \mathcal{Q}^{(0)} \otimes \mathcal{Q}^{(1)}$ and $\mu_1 \mathcal{P}^{(0)} \otimes \mathcal{P}^{(1)}$ in Eq.(55)). Below we consider a diagonal basis where the two component systems become decoupled (in the case of quadratic coupling).

C. Example with the $GF(9)$ in a diagonal basis

We consider the matrix of Eq.(52) with $a = 0$ and $d = c \neq 0$, $b \neq 0$. This gives the diagonal basis $\mathfrak{E}_\kappa$ and its dual, as

$$
\begin{aligned}
&m = \mathfrak{E}_0\mathfrak{M}_0 + \mathfrak{E}_1\mathfrak{M}_1 = \overline{\mathfrak{E}}_0\overline{\mathfrak{M}}_0 + \overline{\mathfrak{E}}_1\overline{\mathfrak{M}}_1 \\
&\mathfrak{M}_0 = bm_1; \qquad \mathfrak{M}_1 = c(m_0 + m_1) \\
&\overline{\mathfrak{M}}_0 = \frac{\overline{m}_1 - \overline{m}_0}{b}; \quad \overline{\mathfrak{M}}_1 = \frac{\overline{m}_0}{c}
\end{aligned} \tag{56}
$$

In this case $\mu_1 = \mu_2 = 0$ and if the Hamiltonian contains only quadratic terms, the two subsystems become decoupled:

$$
\begin{aligned}
&f_2' = \omega\left[b^2(\mathcal{P}^{(0)})^2 \otimes \mathbf{1} - c^2 \mathbf{1} \otimes (\mathcal{P}^{(1)})^2\right] \\
&f_2 = \omega\left[\frac{1}{b^2}(\mathcal{Q}^{(0)})^2 \otimes \mathbf{1} - \frac{1}{c^2}\mathbf{1} \otimes (\mathcal{Q}^{(1)})^2\right]
\end{aligned} \tag{57}
$$

If the Hamiltonian has higher order terms, then even in this basis there is coupling between the subsystems.

VI. DISCUSSION

We have studied how to engineer a finite quantum system which is described with variables in $GF(p^\ell)$. This system is an ℓ-partite system comprised of component systems each of which is described with variables in $\mathbb{Z}_p$ (e.g., spins with $j = (p-1)/2$). The position and momentum states in such a system can be labelled with elements in $GF(p^\ell)$, as in Eqs(26), (29). We also require that the Hamiltonian (and more generally the whole formalism) should be expressed in terms of Galois arithmetic as opposed to arithmetic in $[\mathbb{Z}_p]^\ell$. $GF(p^\ell)$ is similar to $[\mathbb{Z}_p]^\ell$ with regard to addition, but it has a lot of extra structure related to multiplication, the trace, etc. Consequently, only some of the Hamiltonians of the ℓ-partite system can be expressed within the Galois formalism and we called them Hamiltonians compatible with $GF(p^\ell)$. In [11] we gave general theorems about the compatibility of a Hamiltonian with $GF(p^\ell)$, and here we extended this work by giving several examples of such Hamiltonians, in a more physical and less technical language.

The next step in this line of research is to consider realistic physical systems (e.g., spins, superconducting Josephson devices, etc) and to study how we can adjust the parameters in order to get one of the desirable Hamiltonians. Another direction is to study applications of these devices in information processing. There are many applications of Galois fields in classical information processing (e.g., coding and cryptography in classical communications) and devices which operate with Galois arithmetic can be useful in this area. Applications to quantum information processing, could also be studied.

[1] A. Vourdas, Rep. Prog. Phys. 67, 267 (2004)
[2] M. Kibler, J. Phys. A42, 353001 (2009)
[3] W. Wootters, B.D. Fields, Ann. Phys. (NY), 191, 363 (1989)
K. Gibbons, M.J. Hoffman, W. Wootters, Phys. Rev. A70, 062101 (2004)
[4] S. Chaturvedi, Phys. Rev. A65, 044301 (2002)
[5] S. Bandyopadhyay, P.O. Boykin, V.Roychowdhury, F. Vatan, Algorithmica 34, 512 (2002)
[6] G. Bjork, A.B. Klimov, L.L. Sanchez-Soto, Prog. Optics 51, 469 (2008)
[7] M. Saniga, M. Planat, H. Rosu, J. Opt. B-Quantum Semiclass. Optics 6, L19 (2004)
[8] J. Tolar, G. Hadzitaskos, J. Phys. A42, 245306 (2009)
[9] T. Durt, B.G. Englert, I. Bengtsson, K. Zyczkowski, Int. J. Quantum Comp. 8, 535 (2010)
[10] A. Vourdas, J. Phys. A40, R285 (2007)
A. Vourdas, J. Fourier Anal. Appl. 14, 102 (2008)
[11] A. Vourdas, J. Math. Phys. 51, 052102 (2010)
[12] Z.X. Wan, 'Lectures on finite fields and Galois rings' (World Scientific, Singapore, 2003)

Journal of Physics: Conference Series **284** (2011) 012062 doi:10.1088/1742-6596/284/1/012062

A round trip from Caldirola to Bateman systems

J. Guerrero[1,2], F.F. López-Ruiz[2], V. Aldaya[2] and F. Cossío[2]

[1]Departamento de Matemática Aplicada, Campus de Espinardo, Universidad de Murcia, 30100 Murcia, Spain

[2]Instituto de Astrofísica de Andalucía, CSIC, Apartado Postal 3004, 18080 Granada, Spain

E-mail: juguerre@um.es

Abstract. For the quantum Caldirola-Kanai Hamiltonian, describing a quantum damped harmonic oscillator, a couple of constant of motion operators generating the Heisenberg algebra can be found. The inclusion in this algebra, in a unitary manner, of the standard time evolution generator $i\hbar\frac{\partial}{\partial t}$, which is not a constant of motion, requires a non-trivial extension of this basic algebra and the physical system itself, which now includes a new dual particle. This enlarged algebra, when exponentiated, leads to a group, named the Bateman group, which admits unitary representations with support in the Hilbert space of functions satisfying the Schrödinger equation associated with the quantum Bateman Hamiltonian, either as a second order differential operator as well as a first order one. The classical Bateman Hamiltonian describes a dual system of a damped (losing energy) particle and a dual (gaining energy) particle. The classical Bateman system has a solution submanifold containing the trajectories of the original system as a submanifold. When restricted to this submanifold, the Bateman dual classical Hamiltonian leads to the Caldirola-Kanai Hamiltonian for a single damped particle. This construction can also be done at the quantum level, and the Caldirola-Kanai Hamiltonian operator can be derived from the Bateman Hamiltonian operator when appropriate constraints are imposed.

1. Introduction

The interest in dissipative systems at the quantum level has remained constant since the early days of Quantum Mechanics. The difficulties in describing damping, which intuitively could be understood as a mesoscopic property, within the fundamental quantum framework, have motivated a huge amount of papers.

Applications of quantum dissipation abound. For example, in quantum optics, where the quantum theory of lasers and masers makes use of models including damping [1], or in the study of decoherence phenomena [2]. Some authors have modeled dissipation by means of the theory of open systems or the thermal bath approach, in which a damped system is considered to be a subsystem of a bigger one with infinite degrees of freedom [3, 2]. However, damped systems are interesting in themselves as fundamental ones. In particular, the quantum damped harmonic oscillator, frequently described by the Caldirola-Kanai equation [4, 5], has attracted much attention, as it could be considered one of the simplest and paradigmatic examples of dissipative system.

The description of the quantum damped harmonic oscillator by the Caldirola-Kanai model, which includes a time-dependent Hamiltonian, has been considered to have some flaws. For instance, uncertainty relations are not preserved under time evolution and could eventually be violated [6, 7]. Many considerations were made in this direction. For example, Dekker in [8] introduced complex variables and noise operators to tackle the problem, claiming that no dynamical description in terms of a Schrödinger wave function can be expected to exist. [9] proposed a non-linear Schrödinger-Langevin wave equation as the starting point in formulating the quantum theory.

Despite the problems for the Caldirola-Kanai model, many developments went ahead. Coherent states were calculated in [10] by finding creation and annihilation operators, built out of operators which commute with the Schrödinger equation. The corresponding number operator turns out to be an auxiliary, conserved operator, obviously different from the time-dependent Hamiltonian. This paper also defined the so-called loss energy states for the damped harmonic oscillator. The famous report by Dekker [11] provides a historical overview of some relevant results.

The analysis of damping from the symmetry point of view has proved to be especially fruitful. In a purely classical context, the symmetries of the equation of the damped harmonic oscillator with time-dependent parameters were found in [12]. Two comprehensive articles, [13, 14], are of special interest. In those papers the authors find, for the damped harmonic oscillator, finite-dimensional point symmetry groups for the corresponding Lagrangian (the un-extended Schrödinger group [15]) and the equations of motion ($SL(3,\mathbb{R})$) respectively, and an infinite contact one for the set of trajectories of the classical equation. They singularize a "non-conventional" Hamiltonian from those generators of the symmetry, recovering some results from [10]. Then, they conclude that the damped harmonic oscillator should not be claimed to be dissipative at all at the quantum level, as this true, "non-conventional" Hamiltonian is conserved, and should be related to an oscillator with variable frequency.

Many papers related to the Caldirola-Kanai model keep appearing, showing that the debate about fundamental quantum damping is far from being closed. We can mention [16], where the driven damped harmonic oscillator is analyzed, or the review [17]. Even the possible choices of classical Poisson structures and Hamiltonians, or generalizations to the non-commutative plane, have deserved attention as recently as in [18] and [19], respectively.

In fact, in [20] we provided a neat framework to study this model, based on a quantum generalization of the Arnold transformation [21]. The integrals of motion and symmetries were identified and exploited to calculate wave functions, basic operators and the exact time evolution operator.

Besides the Caldirola-Kanai model, the Bateman's dual system appears as an alternative description of dissipation in the damped harmonic oscillator. In his original paper [26], Bateman looked for a variational principle for equations of motion with a friction term linear in velocity, but he allowed the presence of extra equations. This trick effectively doubles the number of degrees of freedom, introducing a time-reversed version of the original damped harmonic oscillator, which acts as an energy reservoir and could be considered as an effective description of a thermal bath. The Hamiltonian that describes this system was rediscovered by Feschbach and Tikochinsky [27, 28, 11] and the corresponding quantum theory was immediately analyzed.

Some issues regarding the Bateman's system arose. The Hamiltonian presents a set of complex eigenvalues of the energy (see [30] and references therein), and the vacuum of the theory decays with time. This last feature was treated in [29], where Celeghini et al. suggested that the quantum theory of the dual system could find a more natural framework in quantum field theory. On the other hand, in [30] the generalized eigenvectors corresponding to the complex eigenvalues are interpreted as resonant states.

Bateman's dual system is still frequently discussed [31]. Many authors have considered this model as a *good* starting point for the formulation of the quantum theory of dissipation. One of the aims of this paper will be to show that the study of the symmetries of the Caldirola-Kanai model leads to the Bateman's dual system, thus to be considered as a *natural* starting point for the study of quantum dissipation.

Historically, Caldirola and Kanai derived their Hamiltonian from the Bateman one by means of time-dependent canonical transformations. In this work we are going to proceed in the opposite direction, deriving Bateman Hamiltonian from Caldirola-Kanai one by purely symmetry considerations.

Finally, we shall return back to Caldirola-Kanai by imposing constraints.

2. Caldirola-Kanai system

The Lagrangian for the Caldirola-Kanai system reads

$$L_{CK} = e^{\gamma t}(\frac{1}{2}m\dot{x}^2 - \frac{1}{2}m\omega^2 x^2)\,, \tag{1}$$

and from this the classical equation of motion are derived:

$$\ddot{x} + \gamma\dot{x} + \omega^2 x = 0\,. \tag{2}$$

Two independent solutions, from which all solution can be obtained by linear combinations with appropriate dimensional coefficients, are

$$u_1(t) = e^{-\frac{\gamma}{2}t}\sin\Omega t, \qquad u_2(t) = e^{-\frac{\gamma}{2}t}\cos\Omega t\,, \tag{3}$$

where

$$\Omega = \sqrt{\omega^2 - \frac{\gamma^2}{4}}\,. \tag{4}$$

The Caldirola-Kanai Hamiltonian is [4, 5]:

$$H_{CK}(t) = \frac{p^2}{2m}e^{-\gamma t} + \frac{1}{2}m\omega^2 x^2 e^{\gamma t}\,, \tag{5}$$

and from this the quantum Hamiltonian can be obtained via canonical quantization, providing the Schrödinger equation for the Caldirola-Kanai system:

$$\begin{aligned} i\hbar\frac{\partial}{\partial t}\phi(x,t) &= \hat{H}_{CK}(t)\,\phi(x,t) \\ &\equiv \left[-\frac{\hbar^2}{2m}e^{-\gamma t}\frac{\partial^2}{\partial x^2} + \frac{1}{2}m\omega^2 x^2 e^{\gamma t}\right]\phi(x,t)\,. \end{aligned} \tag{6}$$

The Caldirola-Kanai Hamiltonian has been widely used to describe quantum dissipative systems, although it has many detractors, arguing that there are serious inconsistencies like that uncertainty relations are violated when $t \to \infty$. However, this inconsistency seems to be associated with a confusion between canonical momentum and "physical" momentum (see, for instance, [32]).

Let $\mathcal{H}_0 = L^2(\mathbb{R})$. Then any solution of the Schrödinger equation can be written as

$$\phi(x,t) = \hat{U}(t)\phi(x)\,, \tag{7}$$

where $\phi(x) \in \mathcal{H}_0$ and $\hat{U}(t) \equiv \hat{U}(t,t_0)$ is the unitary time evolution operator associated with the Hamiltonian $\hat{H}_{CK}(t)$, and t_0 is arbitrary (for instance $t_0 = 0$). Denote by $\mathcal{H}_t$ the Hilbert space built up with solutions of the Schrödinger equation at time t. Then $\mathcal{H}_t$ is isomorphic to $\mathcal{H}_0$, and $\mathcal{H}_t = \hat{U}(t)\mathcal{H}_0$.

Note that $\hat{U}(t)$ does not constitute a one parameter unitary group of operators, since $\hat{H}_{CK}(t)$ does not commute with itself at different times, $[\hat{H}_{CK}(t_1), \hat{H}_{CK}(t_2)] \neq 0$. This makes the computation of $\hat{U}(t)$ more involved, and time-ordered exponential expansion or Magnus expansion must be used to compute it. See, however, [20] for an exact computation of $\hat{U}(t)$ as a product of exponential operators.

It should be stressed that

$$\mathcal{H}_t \neq \tilde{\mathcal{H}}_t \equiv \left\{ \phi(x,t) \,\text{such that} \int_{-\infty}^{+\infty} |\phi(x,t)|^2 dx < \infty \right\}, \tag{8}$$

where $\tilde{\mathcal{H}}_t$ is the standard Hilbert space in quantum mechanics, usually identified with $L^2(\mathbb{R})$. In fact $\mathcal{H}_t \subset \tilde{\mathcal{H}}_t$, and $\mathcal{H}_t$ can be seen as the quantum counterpart of the classical solution manifold in symplectic mechanics, made up with the constants of motion.

Symmetry (self-adjoint) operators, i.e. constants or integral of motion, preserve $\mathcal{H}_t$, while generic (self-adjoint) operators do not, only preserve $\tilde{\mathcal{H}}_t$. The basic quantum operators, preserving $\mathcal{H}_t$ at each t, are given by (see [20]):

$$\begin{aligned} \hat{p}(t) &= -i\hbar e^{-\frac{\gamma t}{2}}(\cos(\Omega t) + \frac{\gamma}{2\Omega}\sin(\Omega t))\frac{\partial}{\partial x} + m e^{\frac{\gamma t}{2}}\frac{\omega^2}{\Omega}\sin(\Omega t)\, x\,, \\ \hat{x}(t) &= e^{\frac{\gamma t}{2}}(\cos(\Omega t) - \frac{\gamma}{2\Omega}\sin(\Omega t))\, x + i\hbar\frac{e^{-\frac{\gamma t}{2}}}{m\Omega}\sin(\Omega t)\frac{\partial}{\partial x}\,. \end{aligned} \tag{9}$$

These operators match those found in [10] looking for integrals of motion commuting with the Schrödinger equation.

Note that

$$\begin{aligned} \hat{p}(t) &= \hat{U}(t)\hat{p}\hat{U}(t)^\dagger \\ \hat{x}(t) &= \hat{U}(t)\hat{x}\hat{U}(t)^\dagger\,, \end{aligned} \tag{10}$$

where $\hat{p} \equiv \hat{p}(0) = -i\hbar\frac{\partial}{\partial x}$ and $\hat{x} \equiv \hat{x}(0) = x$.

Neither $\hat{H}_{CK}(t)$ nor $i\hbar\frac{\partial}{\partial t}$ make sense as operators acting on $\mathcal{H}_t$, only $i\hbar\frac{\partial}{\partial t} - \hat{H}_{CK}(t) \equiv 0$ does, while $\hat{p}(t)$ and $\hat{x}(t)$ act properly on $\mathcal{H}_t$. Notwithstanding this, $\hat{H}_{CK}(t)$ and $i\hbar\frac{\partial}{\partial t}$ are well defined operators on $\tilde{\mathcal{H}}_t$, but they are not identical. The basic operators $\hat{x}(t)$, $\hat{p}(t)$, and the identity operator $\hat{I}$, close the Heisenberg algebra. There are also other operators acting on $\mathcal{H}_t$, which are quadratic in the basic operators, that close the Schrödinger (or Weyl-symplectic) algebra (see [20]).

3. Including time evolution as a symmetry

The generator of time evolution, $i\hbar\frac{\partial}{\partial t}$, nor $\hat{H}_{CK}(t)$ close under commutation with $\hat{x}(t)$ and $\hat{p}(t)$. We wonder if it is possible to incorporate them into the basic Lie algebra of operators, trying to close an enlarged Lie algebra acting on the (possibly enlarged) Hilbert space $\mathcal{H}_t$. The answer to this question is positive, but it requires a delicate analysis. The resulting enlarged algebra includes $\hat{X} \equiv \hat{x}(t), \hat{P} \equiv \hat{p}(t), \hat{H} \equiv i\hbar\frac{\partial}{\partial t}$ and four more generators (plus $\hat{I}$), denoted by $\hat{Q}, \hat{\Pi}, \hat{G}_1$ and $\hat{G}_2$.

The operators $\hat{Q}$ and $\hat{\Pi}$ (plus $\hat{I}$) expand a Heisenberg algebra, and $\hat{H}, \hat{G}_1$ and $\hat{G}_2$ expand a 2-D affine algebra (with $\hat{H}$ acting as dilations). However, in this realization $\hat{Q}$ and $\hat{\Pi}$ are not basic (this can be seen as an anomaly), and $\hat{H}$ and $\hat{G}_1$ are basic, resulting in time being a canonical variable. Clearly, this is not satisfactory, and an alternative description should be looked for.

A detailed study of the (projective) representations of the enlarged (7+1) dimensional Lie algebra shows that there are three relevant kinds of representations, describing systems with different degrees of freedom:

- A generic family with 3 degrees of freedom: $(\hat{X},\hat{P})$, $(\hat{Q},\hat{\Pi})$ and $(\hat{H},\hat{G}_1)$, time being a canonical variable.
- An anomalous family with 2 degrees of freedom: $(\hat{X},\hat{P})$ and $(\hat{H},\hat{G}_1)$, time being a canonical variable (the one already described).
- A family with 2 degrees of freedom: $(\hat{X},\hat{P})$ and $(\hat{Q},\hat{\Pi})$.

Clearly, the interesting case is the third one, since it contains two degrees of freedom and time is not a canonical variable. Its algebra is given by:

$$\begin{aligned}
\left[\hat{X},\hat{P}\right] &= i\hbar\hat{I} & \left[\hat{Q},\hat{\Pi}\right] &= i\hbar\hat{I} \\
\left[\hat{X},\hat{Q}\right] &= -\frac{i\hbar}{m}\hat{G}_2 & \left[\hat{X},\hat{\Pi}\right] &= i\hbar\hat{G}_1 \\
\left[\hat{Q},\hat{P}\right] &= i\hbar\hat{G}_1 - i\hbar\gamma\hat{G}_2 & \left[\hat{P},\hat{\Pi}\right] &= -i\hbar m\omega^2\hat{G}_2 \\
\left[\hat{H},\hat{X}\right] &= \frac{i\hbar}{m}\hat{\Pi} & \left[\hat{H},\hat{P}\right] &= i\hbar m\omega^2\hat{Q} \\
\left[\hat{H},\hat{Q}\right] &= \frac{i\hbar}{m}(\hat{P}+2\hat{\Pi}) & \left[\hat{H},\hat{\Pi}\right] &= i\hbar m\omega^2(\hat{X}-2\hat{Q}) \\
&\quad + i\hbar\gamma\hat{Q} & &\quad - i\hbar\gamma\hat{\Pi} \\
\left[\hat{H},\hat{G}_1\right] &= -i\hbar\gamma\hat{G}_1 + 2i\hbar\omega^2\hat{G}_2 & \left[\hat{H},\hat{G}_2\right] &= -2i\hbar\hat{G}_1 + i\hbar\gamma\hat{G}_2\,.
\end{aligned} \tag{11}$$

In this case the operators $\hat{G}_1$ and $\hat{G}_2$ are gauge, and therefore are represented trivially. The effective dimension of the algebra is $5+1$: $(\hat{X},\hat{P}),(\hat{Q},\hat{\Pi}),\hat{H}$ and $\hat{I}$.

$$\begin{aligned}
\left[\hat{X},\hat{P}\right] &= i\hbar\hat{I} & \left[\hat{Q},\hat{\Pi}\right] &= i\hbar\hat{I} \\
\left[\hat{X},\hat{Q}\right] &= 0 & \left[\hat{X},\hat{\Pi}\right] &= 0 \\
\left[\hat{Q},\hat{P}\right] &= 0 & \left[\hat{P},\hat{\Pi}\right] &= 0 \\
\left[\hat{H},\hat{X}\right] &= \frac{i\hbar}{m}\hat{\Pi} & \left[\hat{H},\hat{P}\right] &= i\hbar m\omega^2\hat{Q} \\
\left[\hat{H},\hat{Q}\right] &= \frac{i\hbar}{m}(\hat{P}+2\hat{\Pi}) & \left[\hat{H},\hat{\Pi}\right] &= i\hbar m\omega^2(\hat{X}-2\hat{Q}) \\
&\quad + i\hbar\gamma\hat{Q} & &\quad - i\hbar\gamma\hat{\Pi}\,.
\end{aligned} \tag{12}$$

Here $\hat{H}$ is not a basic operator, and can be written in terms of the basic ones in an irreducible representation:

$$\hat{H} = -\frac{1}{m}\hat{\Pi}\hat{P} - \frac{\gamma}{2}(\hat{Q}\hat{\Pi}+\hat{\Pi}\hat{Q}) - \frac{\hat{\Pi}^2}{m} + m\omega^2\hat{X}\hat{Q} - m\omega^2\hat{Q}^2\,. \tag{13}$$

The classical version of the Hamiltonian is:

$$H = -\frac{1}{m}\Pi P - \gamma Q\Pi - \frac{\Pi^2}{m} + m\omega^2 XQ - m\omega^2 Q^2\,. \tag{14}$$

4. Bateman system

This classical Hamiltonian can be transformed, using the linear, constant, canonical transformation:

$$\begin{aligned} X &= \frac{m\omega^2 y - (p_y + m\frac{\gamma}{2}x)i\Omega}{m\omega\sqrt{-\gamma i\Omega}} \\ P &= \frac{\omega(p_x - m\frac{\gamma}{2}y + mxi\Omega)}{\sqrt{-\gamma i\Omega}} \\ Q &= \frac{m\omega^2 y - (p_y - m\frac{\gamma}{2}x)i\Omega}{m\omega\sqrt{-\gamma i\Omega}} \\ \Pi &= -\frac{\omega(p_x + m\frac{\gamma}{2}y + mxi\Omega)}{\sqrt{-\gamma i\Omega}}\,, \end{aligned} \tag{15}$$

into the Bateman dual Hamiltonian

$$H_B = \frac{p_x p_y}{m} + \frac{\gamma}{2}(yp_y - xp_x) + m\Omega^2 xy\,, \tag{16}$$

that describes a damped particle (x, p_x) and its time reversal (y, p_y):

$$\ddot{x} + \gamma\dot{x} + \omega^2 x = 0\,, \qquad \ddot{y} - \gamma\dot{y} + \omega^2 y = 0\,. \tag{17}$$

The quantum Bateman Hamiltonian is:

$$\hat{H}_B = \frac{\hat{p}_x\hat{p}_y}{m} + \frac{\gamma}{2}(\hat{y}\hat{p}_y - \hat{x}\hat{p}_x) + m\Omega^2\hat{x}\hat{y}\,, \tag{18}$$

and the Schrödinger equation for the Bateman system is given by[1]:

$$i\hbar\frac{\partial\phi(x,y,t)}{\partial t} = \left[-\frac{\hbar^2}{m}\frac{\partial^2}{\partial x\partial y} - i\hbar\frac{\gamma}{2}(y\frac{\partial}{\partial y} - x\frac{\partial}{\partial x}) + m\Omega^2 xy\right]\phi(x,y,t)\,. \tag{19}$$

The Bateman system is conservative, so our objective of including time evolution among the symmetries has been accomplished. $\hat{H}_B$ closes a 5+1 dimensional algebra with $(\hat{x}, \hat{p}_x)$ and $(\hat{y}, \hat{p}_y)$:

$$\begin{aligned} &[\hat{x},\hat{p}_x] = i\hbar\hat{I} && [\hat{y},\hat{p}_y] = i\hbar\hat{I} \\ &[\hat{x},\hat{y}] = 0 && [\hat{x},\hat{p}_y] = 0 \\ &[\hat{y},\hat{p}_x] = 0 && [\hat{p}_x,\hat{p}_y] = 0 \\ &\left[\hat{H}_B,\hat{x}\right] = \frac{i\hbar}{m}(-\hat{p}_y + m\frac{\gamma}{2}x) && \left[\hat{H}_B,\hat{p}_x\right] = i\hbar(-\frac{\gamma}{2}\hat{p}_x + m\Omega^2\hat{y}) \\ &\left[\hat{H}_B,\hat{y}\right] = \frac{i\hbar}{m}(-\hat{p}_x - m\frac{\gamma}{2}y) && \left[\hat{H}_B,\hat{p}_y\right] = i\hbar(\frac{\gamma}{2}\hat{p}_y + m\Omega^2\hat{x})\,. \end{aligned} \tag{20}$$

However, it has been argued that the quantum Bateman system possesses inconsistencies, like complex eigenvalues and non-normalizable eigenstates. But Chruściński & Jurkowski [30] showed that $\hat{H}_B$ has real, continuous spectrum, and that the complex eigenvalues are associated with resonances, which in last instance are the responsible of dissipation.

[1] The Bateman system admits an equivalent description in terms of a real, first order, Schrödinger equation (see [35])

5. Turning back to Caldirola-Kanai

Historically, Bateman first derived H_B, and later Caldirola and Kanai obtained H_{CK} using time-dependent canonical transformations. Here we have gone the opposite way, started from H_{CK} and derived H_B closing a finite Lie algebra. Now we wonder if we can do the way back to the Caldirola-Kanai system. The answer, again, is positive, and can be achieved by using constraints. To know how to proceed, let us analyse first the classical case.

Classically, Bateman system and a pair of dual Caldirola-Kanai systems share the same second order equations of motion. If we impose them to share the first-order, Hamilton equations, the following constraint must be satisfied:

$$\begin{aligned} y &= \frac{\omega^2}{\Omega^2} e^{\gamma t} x + \frac{\gamma}{2m\Omega^2} p_x \\ p_y &= e^{\gamma t} p_x + m\frac{\gamma}{2} x \,. \end{aligned} \tag{21}$$

These constraints, although time dependent, preserve the equations of motion since they are equivalent to a relation among initials constants:

$$\begin{aligned} y_0 &= \frac{\omega^2}{\Omega^2} x_0 + \frac{\gamma}{2m\Omega^2} p_{x0} \\ p_{y0} &= p_{x0} + m\frac{\gamma}{2} x_0 \,. \end{aligned} \tag{22}$$

These constraints can be seen to be of second order type, besides being time-dependent, therefore care should be taken when imposing them: Dirac theory for constraints can be used or we can embed the constraints in a time-dependent canonical transformation before applying them.

But we are interested in the quantum derivation. Therefore we try to impose the operator constraints:

$$\begin{aligned} \hat{y} - \frac{\omega^2}{\Omega^2}\hat{x} - \frac{\gamma}{2m\Omega^2}\hat{p}_x &= 0 \\ \hat{p}_y - \hat{p}_x - m\frac{\gamma}{2}\hat{x} &= 0 \,, \end{aligned} \tag{23}$$

but only one of them can be imposed, since the operators at the lhs of the equations canonically commute: they are of second order type. At the quantum level, only one of them can be imposed, therefore we must select one of them. If we impose the constraint,

$$\hat{y} = \frac{\omega^2}{\Omega^2}\hat{x} + \frac{\gamma}{2m\Omega^2}\hat{p}_x \,, \tag{24}$$

the Hilbert space reduces to those functions verifying:

$$\phi(x,y,t) = e^{\frac{i e^{-\gamma t} m\Omega y \mathrm{Csc}^2(\Omega t)\left(\gamma\Omega y \mathrm{Cos}(2\Omega t) + 2\left(\omega^2 e^{\gamma t} x' - \Omega^2 y\right)\mathrm{Sin}(2\Omega t)\right)}{4\hbar\omega^2}} \psi(x',t) \,, \tag{25}$$

where $x' = x + \frac{\Omega^2}{2\omega^2} y e^{-\gamma t}\mu(t)$, and $\mu(t) = (2 - \frac{\gamma}{\Omega}\mathrm{Cot}(\Omega t))$. The Schrödinger equation for the Bateman system reduces to:

$$i\hbar\frac{\partial\psi(x',t)}{\partial t} = \left[-\frac{\Omega^2\hbar^2}{2m\omega^2} e^{-\gamma t}\mu(t)\frac{\partial^2}{\partial x'^2} - \frac{1}{2} i\hbar x'\Omega\mu(t)\frac{\partial}{\partial x'} + i\hbar\frac{\Omega^2}{\gamma}(\mu(t) - 2)\right]\psi(x',t) \,. \tag{26}$$

When constraints are imposed, not all the operators acting on the original Hilbert space preserve the constrained Hilbert space. The notion of "good" (usually denoted gauge-independent in constrained gauge theories) operators as those preserving the constrained Hilbert space naturally emerges.

In most of the cases "good" operators are characterized as those commuting with the constraints (see [40] for a detailed account of quantum constraints in a group-theoretical setting and a more general characterization of "good" operators). In this case they are:

$$\hat{p}_x + \frac{2m\omega^2}{\gamma}\hat{x} \qquad \hat{p}_y - \frac{2m\Omega^2}{\gamma}\hat{x}\,. \tag{27}$$

Note that $\hat{H}_B$ (nor $i\hbar\frac{\partial}{\partial t}$) is not among the "good" operators since it does not preserve the constrained Hilbert space. Therefore, time invariance is lost in the process of going from the Bateman system to the Caldirola-Kanai system due to the very nature of the constraints imposed.

Now let us perform the transformation

$$\psi(x',t) = e^{-i\frac{m\omega^2}{\hbar\Omega}x'^2 f(t)}g(t)\chi(\kappa,\tau)\,, \tag{28}$$

where

$$f(t) = -\frac{e^{\gamma t}}{4\Omega\mu(t)^2\tau'(t)}\left((-\gamma(2+\mathrm{Cos}(2\Omega t)) + 2\Omega\mathrm{Sin}(2\Omega t))\tau'(t) - \gamma\mu(t)\tau'(t)^2 + \mu(t)\tau''(t)\right) \tag{29}$$

$$g(t) = e^{-\frac{1}{4}\gamma\tau}\left(-\frac{\tau'(t)}{\Omega\mathrm{Sin}^2(\Omega t)\mu(t)}\right)^{1/4} \tag{30}$$

$$\kappa = x'e^{\frac{\gamma}{2}(t-\tau)}\frac{\omega}{\Omega}\sqrt{\frac{\tau'(t)}{\mu(t)}} \tag{31}$$

$$\tau(t) = \frac{1}{\Omega}\mathrm{ArcTan}\left[\frac{A\frac{\gamma^2}{\Omega^2}}{\mu(t)^2}\right]\,,\ A \in \mathbb{R} - \{0\}\,. \tag{32}$$

The Schrödinger equation finally transforms into:

$$i\hbar\frac{\partial}{\partial\tau}\chi(\kappa,\tau) = \left[-\frac{\hbar^2}{2m}e^{-\gamma\tau}\frac{\partial^2}{\partial\kappa^2} + \frac{1}{2}m\omega^2\kappa^2 e^{\gamma\tau}\right]\chi(\kappa,\tau)\,, \tag{33}$$

which is the Caldirola-Kanai equation in the variables (κ,τ). Even more, the two independent operators (27) preserving the constrained Hilbert space turn, under the previous transformation, to the basic operators for the Caldirola-Kanai system $\hat{x}(t)$ and $\hat{p}(t)$. Therefore, we have recovered completely the Caldirola-Kanai system from the Bateman system by imposing one constraint.

It should be stressed that $\tau'(0) = 0$, therefore the time transformation is singular at the origin and there are two disconnected regions, one with $t > 0$ and other with $t < 0$. It also turns out that $sign(\tau) = sign(A)$, therefore choosing appropriately the sign of A in each case we can map $t > 0$ to $\tau > 0$ and $t < 0$ to $\tau < 0$, respectively.

This kind of behavior coincides with the results of other authors (see [30]) where, starting with the Bateman system, they obtain two subspaces $\mathcal{S}^{\pm}$ for which the restriction of the one parameter group of unitary time-evolution operators $\hat{U}(t) = e^{-\frac{i}{\hbar}t\hat{H}_B}$ produces two semigroups of operators, for $t < 0$ and $t > 0$.

Therefore, starting from the quantum, conservative, Bateman system we have arrived to the quantum, time-dependent, Caldirola-Kanai system. All the process we have performed can be schematically showed as:

$$\begin{array}{ccc} \text{Bateman} & \overset{\text{Constraint}}{\Rightarrow} & \text{Caldirola-Kanai} \\ t \in \mathbb{R} & & t \in \mathbb{R}^+ \text{ or } t \in \mathbb{R}^- \\ \text{Conservative} & \underset{\text{Closing algebra}}{\Leftarrow} & \text{Dissipative} \end{array} \tag{34}$$

Acknowledgments
Work partially supported by the Fundación Séneca, Spanish MICINN and Junta de Andalucía under projects 08814/PI/08, FIS2008-06078-C03-01 and FQM219-FQM1951, respectively.

[1] Scully M.O., Zubairy M.S., *Quantum Optics* (Cambridge University Press, 1997)
[2] Breuer H-P., Petruccione F., *The Theory of Open Quantum System* (Oxford University Press 2002)
[3] Dittrich T. et al., *Quantum Transport and Dissipation* (Wiley-VCH 1998)
[4] Caldirola P., Nuovo Cimento (1941) 18, 393
[5] Kanai E. (1948) Prog. Theor. Phys. 3, 440
[6] Brittin W.E., Phys. Rev. **77**, 396 (1950)
[7] Razavy M., *Classical and Quantum Dissipative Systems* (Imperial College Press 2005)
[8] Dekker H., Phys. Rev. A **16**, 2126 (1977)
[9] Razavy M., Z. Physik B **26**, 201 (1977)
[10] Dodonov V.V., Man'ko V.I., Phys. Rev. A **20**, 550 (1979)
[11] Dekker H., Phys. Rep. **80**, 1 (1981)
[12] Martini R., Kersten P.H.M., J. Phys. A **16**, 455 (1983)
[13] Cerveró J.M., Villarroel J., J. Phys. A **17**, 1777 (1984)
[14] Cerveró J.M., Villarroel J., J. Phys. A **17**, 2963 (1984)
[15] Niederer U., Helv. Phys. Acta **45**, 802 (1972); **46**, 191 (1973); **47**, 167 (1974)
[16] Um C.I., Yeon K.H. and Kahng W.H., J. Phys. A **20** 611 (1987)
[17] Um C-I., Yeon K-H., George T.F., Phys. Rep. **362**, 63 (2002)
[18] Hone A.N.W., Senthilvelan M., J. Math. Phys. **50**, 102902 (2009)
[19] Streklas A., Physica A, **385**, 124 (2007)
[20] V. Aldaya, F. Cossío, J. Guerrero and F. F. López-Ruiz, *On a quantum version of the Arnold transformation*, arXiv:1010.5521v1[quant-ph]
[21] Arnold V. I., *Geometrical methods in the theoy of ordinary differential equations*, (Springer,1998)
[22] Lutzky M. J. Phys. A, **11**, 249 (1978)
[23] Takagi S., Prog. Theor. Phys. **84**, 1019 (1990)
[24] Huang M-C., Wu M-C., Chin. J. Phys. **36**, 566 (1998)
[25] Kanasugi H., Okada H., Prog. Theor. Phys. **93**, 949 (1995)
[26] Bateman H., Phys. Rev. **38**, 815 (1931)
[27] Tikochinsky Y., J. Math. Phys. **19**, 888 (1978); Feshbach H. and Tikochinsky Y., in: A Festschrift for I.I. Rabi, Trans. New York Ac. Sc. Ser. 2. Vol. 38 (1977) 44
[28] Morse P.M. and Feshbach H., *Methods of Theoretical Physics Vol. I* (McGraw-Hill, New York, 1953), p. 298.
[29] Celeghini E., Rasetti M., Vitielo G., Ann. Phys. **215**, 156 (1992)
[30] Chruściński D., Jurkowski J., Ann. Phys. **321**, 854 (2006)
[31] Blasone M., Jizba P., Ann. Phys. **312**, 312 (2004)
[32] Schuch D., Phys. Rev. A **55**, 935 (1997)
[33] Guerrero J., López-Ruiz F.F., Aldaya V. and Cossío F., *Discrete basis of localized quantum states for the free particle*, arXiv:1010.5525.
[34] Blanes S., Casas F., Oteo J.A., Ros J., Phys. Rep. **470**, 151 (2009)
[35] Aldaya V., Cossío F., Guerrero J. and López-Ruiz F.F., *A symmetry trip from Caldirola to Bateman damped systems*, to be published.
[36] Aldaya V., de Azcárraga J.A., J. Math. Phys. **23**, 1297 (1982)
[37] Aldaya V., Guerrero J. Rep. Math. Phys. **47**, 213 (2001)
[38] Aldaya V., Calixto M., Guerrero J., López-Ruiz F. F., J. Nonl. Math. Phys. **15**, 1 (2008)
[39] Aldaya V., Navarro-Salas J., Bisquert J., Loll R., J.Math. Phys. **33**, 3087 (1992)
[40] V. Aldaya, M. Calixto and J. Guerrero, Comm. Math. Phys. **178**, 399-424 (1996)

AUTHOR INDEX

AUTHOR INDEX